WILEY ENCYCLOPEDIA OF

TELECOMMUNICATIONS

VOLUME 2

WILEY ENCYCLOPEDIA OF TELECOMMUNICATIONS

WILEY ENCYCLOPEDIA OF

TELECOMMUNICATIONS

VOLUME 2

John G. Proakis
Editor

A John Wiley & Sons Publication

The *Wiley Encyclopedia of Telecommunications* is available online at
http://www.mrw.interscience.wiley.com/eot

Published by John Wiley & Sons, Inc., Hoboken, New Jersey.
Published simultaneously in Canada.

Library of Congress Cataloging in Publication Data:

Wiley encyclopedia of telecommunications / John G. Proakis, editor.
p. cm.
includes index.
ISBN 0-471-36972-1
1. Telecommunication — Encyclopedias. I. Title: Encyclopedia of telecommunications. II. Proakis, John G.
TK5102 .W55 2002
621.382′03 — dc21

2002014432

Printed in the United States of America

10 9 8 7 6 5 4 3 2 1

WILEY ENCYCLOPEDIA OF

TELECOMMUNICATIONS

VOLUME 2

D

DATA COMPRESSION

JOHN KIEFFER
University of Minnesota
Minneapolis, Minnesota

1. INTRODUCTION

A modern-day data communication system must be capable of transmitting data of all types, such as text, speech, audio, image or video data. The block diagram in Fig. 1 depicts a data communication system, consisting of *source, encoder, channel*, and *decoder*:

The source generates the data sequence that is to be transmitted through the data communication system. The encoder converts the data sequence into a binary codeword for transmission through the channel. The decoder generates a reconstructed data sequence that may or not be equal to the original data sequence. The encoder/decoder pair in Fig. 1 is the *code* of the data communication system. In Fig. 1, the source and channel are fixed; the choice of code is flexible, in order to accomplish the twin goals of *bandwidth efficiency* and *reliable transmission*, described as follows:

1. *Bandwidth efficiency* — the portion of the available channel bandwidth that is allocated in order to communicate the given data sequence should be economized.
2. *Reliable transmission* — the reconstructed data sequence should be equal or sufficiently close to the original data sequence.

Unfortunately, these are conflicting goals; less use of bandwidth makes for less reliable transmission, and conversely, more reliable transmission requires the use of more bandwidth. It is the job of the data communication system designer to select a code that will yield a good tradeoff between these two goals. Code design is typically done in one of the following two ways.

1. *Separated Code Design.* Two codes are designed, a *source code* and a *channel code*, and then the source code and the channel code are cascaded together. Figure 2 illustrates the procedure. The source code is the pair consisting of the source encoder and source decoder; the channel code is the (channel encoder, channel decoder) pair. The source code achieves the goal of bandwidth efficiency: The source encoder removes a large amount of redundancy from the data sequence that can be restored (or approximately restored) by the source decoder. The channel code achieves the goal of reliable transmission: The channel encoder inserts a small amount of redundancy in the channel input stream that will allow the channel decoder to correct the transmission errors in that stream that are caused by the channel.

2. *Combined Code Design.* One code (as in Fig. 1) is designed to accomplish the twin goals of bandwidth efficiency and reliable transmission. Clearly, combined code design is more general than separated code design. However, previously separated codes were preferred to combined codes in data communication system design. There were two good reason for this: (a) Claude Shannon showed that if the data sequence is sufficiently long, and if the probabilistic models for the source and the channel are sufficiently simple, then there is no loss in the bandwidth versus reliability tradeoff that is achievable using separated codes of arbitrary complexity instead of combined codes of arbitrary complexity; and (b) the code design problem is made easier by separating it into the two decoupled problems of source code design and channel code design. For the communication of short data sequences, or for the scenario in which the complexity of the code is to be constrained, there can be an advantage to using combined codes as opposed to separated codes; consequently, there much attention has focused on the combined code design problem since the mid-1980s. At the time of the writing of this article, however, results on combined code design are somewhat isolated and have not yet been combined into a nice theory. On the other hand, the two separate theories of source code design and channel code design are well developed. The purpose of the present article is to provide an introduction to source code design.

In source code design, one can assume that the communication channel introduces no errors, because the purpose of the channel code is to correct whatever channel errors occur. Thus, we may use Fig. 3 below, which contains no channel, as the conceptual model guiding source code design.

The system in Fig. 3 is called a *data compression system* — it consists of the source and the source code consisting of the (source encoder, source decoder) pair. The data sequence generated by the source is random and is denoted X^n; the notation X^n is a shorthand for the following random sequence of length n:

$$X^n = (X_1, X_2, \ldots, X_n) \tag{1}$$

The X_i values $(i = 1, 2, \ldots, n)$ are the individual *data samples* generated by the source. In Fig. 3, B^K is a

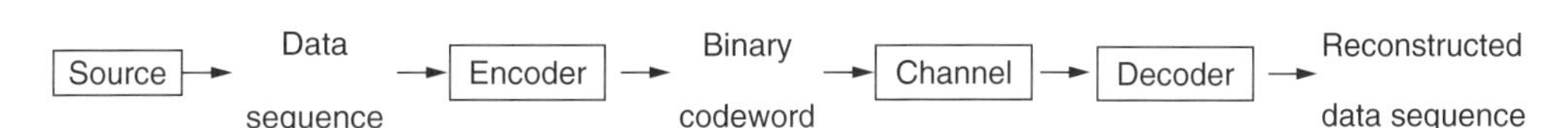

Figure 1. Block diagram of data communication system.

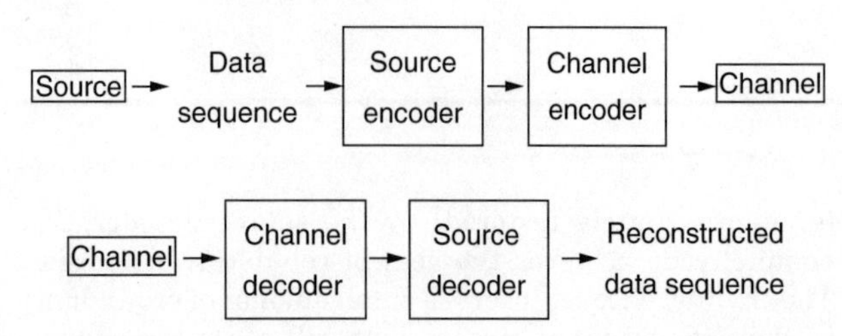

Figure 2. Data communication system with separated code.

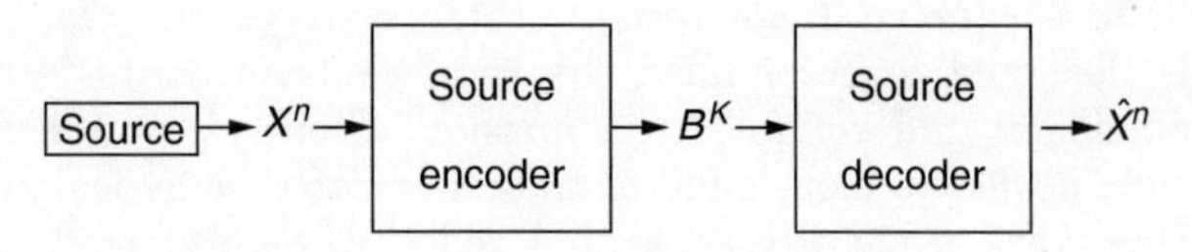

Figure 3. Block diagram of data compression system.

binary codeword generated by the source encoder as a deterministic function of X^n;B^K is random, and its distribution can be computed from the distribution of X^n. Our notational convention means that B^K is a shorthand for

$$B^K = (B_1, B_2, \ldots, B_K) \tag{2}$$

where each $B_i (i = 1, 2, \ldots, K)$, is a *code bit* belonging to the binary alphabet {0, 1}. The fact that K is capitalized means that K is random; that is, a *variable-length* codeword is used. The reconstructed data sequence generated by the source decoder in Fig. 3 is of length n and has been denoted $\hat{X}^n$. The sequence $\hat{X}^n$ is a deterministic function of B^K, and therefore also a deterministic function of X^n;$\hat{X}^n$ is random and its distribution can be computed from the distribution of X^n. Notationally

$$\hat{X}^n = (\hat{X}_1, \hat{X}_2, \ldots, \hat{X}_n) \tag{3}$$

where each $\hat{X}_i (i = 1, 2, \ldots, n)$ is an approximation to X_i.

1.1. Mathematical Description of Source

We need to give a formal mathematical model describing the probabilistic nature of the source in Fig. 3. Accordingly, a source is defined to be a triple $[n, A_n, P_n]$, where

- n is a positive integer that is the length of the random data sequence X^n generated by the source $[n, A_n, P_n]$.
- A_n is the set of all sequences of length n which are realizations of X^n. (The set A_n models the set of all possible deterministic sequences that could be processed by the data compression system driven by the source $[n, A_n, P_n]$.)
- P_n denotes the probability distribution of X^n; it is a probability distribution on A_n. We have

$$\Pr[X^n \in S_n] = P_n(S_n), S_n \subset A_n$$

$$\Pr[X^n \in A_n] = P_n(A_n) = 1$$

The *alphabet* of the source $[n, A_n, P_n]$ is the smallest set A such that $A_n \subset A^n$, where A^n denotes the set of all sequences of length n whose entries come from A. For a fixed positive integer n, a source $[n, A_n, P_n]$ shall be referred to as an n*th-order source*.

1.2. Memoryless Source

The most common type of source model is the *memoryless source*. In an nth-order memoryless source, the data samples $X_1, X_2, \ldots, X_n$ are taken to be independent, identically distributed random variables. Therefore, the joint probability density function $f(x_1, x_2, \ldots, x_n)$ of the memoryless source output X^n factors as

$$f(x_1, x_2, \ldots, x_n) = f_1(x_1)f_1(x_2)\cdots f_1(x_n)$$

where f_1 is a fixed probability density function.

1.3. Markov Source

The second most common type of source model is the *stationary Markov source*. For an nth-order source, the stationary Markov source assumption means that the joint probability density function $f(x_1, x_2, \ldots, x_n)$ of the source output X^n factors as

$$f(x_1, x_2, \ldots, x_n) = \frac{f_2(x_1, x_2)f_2(x_2, x_3)\cdots f_2(x_{n-1}, x_n)}{f_1(x_2)f_1(x_3)\cdots f_1(x_{n-1})}$$

where f_2 is a fixed 2D (two-dimensional) probability density function, and f_1 is a 1D probability density function related to f_2 by

$$f_1(x_1) = \int_{-\infty}^{\infty} f_2(x_1, x_2)\, dx_2 = \int_{-\infty}^{\infty} f_2(x_2, x_1)\, dx_2$$

1.4. Lossless and Lossy Compression Systems

Two types of data compression systems are treated in this article: *lossless compression systems* and *lossy compression systems*. In a lossless compression system, the set A_n of possible data sequence inputs to the system is finite, the encoder is a one-to-one mapping (this means that there is a one-to-one correspondence between data sequences and their binary codewords), and the decoder is the inverse of the encoder; thus, in a lossless compression system, the random data sequence X^n generated by the source and its reconstruction $\hat{X}^n$ at the decoder output are the same:

$$\Pr[X^n = \hat{X}^n] = 1$$

In a lossy compression system, two or more data sequences in A_n are assigned the same binary codeword, so that

$$\Pr[X^n \neq \hat{X}^n] > 0$$

Whether one designs a lossless or lossy compression system depends on the type of data that are to be transmitted in a data communication system. For example, for textual data, lossless compression is used because one typically wants perfect reconstruction of the transmitted text; on the other hand, for image data, lossy compression would be appropriate if the reconstructed image is required only to be perceptually equivalent to the original image.

This article is divided into two halves. In the first half, we deal with the design of lossless codes, namely, source codes for lossless compression systems. In the second half, design of lossy codes (source codes for lossy compression systems) is considered.

2. LOSSLESS COMPRESSION METHODOLOGIES

In this section, we shall be concerned with the problem of designing lossless codes. Figure 4 depicts a general lossless compression system. In Fig. 4, the pair consisting of source encoder and source decoder is called a *lossless code*. A lossless code is completely determined by its source encoder part, since the source decoder is the inverse mapping of the source encoder.

Let $[n, A_n, P_n]$ be a given source with finite alphabet. When a lossless code is used to compress the data generated by the source $[n, A_n, P_n]$, a lossless compression system S_n results as in Fig. 4. The effectiveness of the lossless code is then evaluated by means of the figure of merit

$$R(S_n) \triangleq n^{-1} \sum_{x^n \in A_n} P_n(x^n) K(x^n)$$

where $K(x^n)$ is the length of the codeword assigned by the lossless code to the data sequence $x^n \in A_n$. The figure of merit $R(S_n)$ is called *compression rate* and its units are "code bits per data sample." An efficient lossless code for compressing data generated by the source $[n, A_n, P_n]$ is a lossless code giving rise to a compression system S_n for which the compression rate $R(S_n)$ is minimized or nearly minimized. In this section, we put forth various types of lossless codes that are efficient in this sense.

In lossless code design, we make the customary assumption that the codewords assigned by a lossless code must satisfy the *prefix condition*, which means that no codeword is a prefix of any other codeword. If $K_1, K_2, \ldots, K_j$ are the lengths of the codewords assigned by a lossless code, then *Kraft's inequality*

$$2^{-K_1} + 2^{-K_2} + \cdots + 2^{-K_j} \leq 1 \tag{4}$$

must hold. Conversely, if positive integers $K_1, K_2, \ldots, K_j$ obey Kraft's inequality, then there exists a lossless code whose codewords have these lengths. In this case, one can build a rooted tree T with j leaves and at most two outgoing edges per internal vertex, such that $K_1, K_2, \ldots, K_j$ are the lengths of the root-to-leaf paths; the codewords are obtained by labeling the edges along these paths with 0s and 1s. The tree T can be found by applying the Huffman algorithm (covered in Section 2.2) to the set of probabilities $2^{-K_1}, 2^{-K_2}, \ldots, 2^{-K_j}$.

There are two methods for specifying a lossless code for the source $[n, A_n, P_n]$: (1) an encoding table can be given which lists the binary codeword to be assigned to each sequence in A_n (decoding is then accomplished by using the encoding table in reverse), or (2) encoding and decoding algorithms can be given that indicate how to compute the binary codeword for each sequence in $x^n \in A_n$ and how to compute x^n from its codeword. We will specify each lossless code discussed in this section using either method 1 or 2, depending on which method is more convenient. Method 1 is particularly convenient if the codeword lengths are known in advance, since, as pointed out earlier, a tree can be constructed that yields the codewords. Method 2 is more convenient if the data length n is large (which makes the storing of an encoding table impractical).

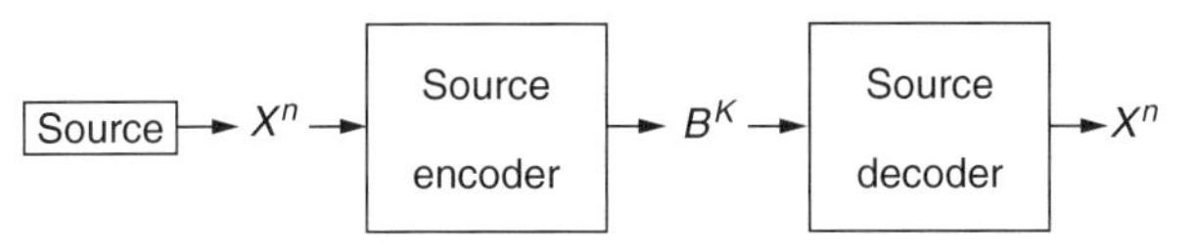

Figure 4. Lossless data compression system.

2.1. Entropy Bounds

It is helpful to understand the entropy upper and lower bounds on the performance of lossless codes. With these bounds, one can determine before designing a lossless code what kind of performance it is possible for such a code to achieve, as well as what kind of performance it is not possible to achieve.

The entropy of the source $[n, A_n, P_n]$ is defined by

$$H_n \triangleq \sum_{x^n \in A_n} (-\log_2 P_n(x^n)) P_n(x^n)$$

Suppose that the random data sequence generated by the source $[n, A_n, P_n]$ is compressed by an arbitrary lossless code and let S_n be the resulting lossless compression system (as in Fig. 4). Then, the compression rate is known to satisfy the relationship

$$R(S_n) \geq \frac{H_n}{n} \tag{5}$$

Conversely, it is known that there exists at least one lossless code for which

$$R(S_n) \leq \frac{H_n + 1}{n} \tag{6}$$

Assume that the data length n is large. We can combine the bounds (5) and (6) to assert that a lossless code is efficient if and only if the resulting compression rate satisfies

$$R(S_n) \approx \frac{H_n}{n}$$

The quantity H_n/n is called the *entropy rate* of the source $[n, A_n, P_n]$. Our conclusion is that a lossless code is efficient for compressing source data if and only if the resulting compression rate is approximately equal to the entropy rate of the source. For the memoryless source and the Markov source, this result can be sharpened, as the following discussion shows.

2.1.1. Efficient Lossless Codes for Memoryless Sources.

Assume that the given source $[n, A_n, P_n]$ is a memoryless source; let A be the finite source alphabet. There is a probability mass function $[p(a): a \in A]$ on A such that

$$P_n(x_1, x_2, \ldots, x_n) = \prod_{i=1}^{n} p(x_i), \ (x_1, x_2, \ldots, x_n) \in A_n \tag{7}$$

Let H_0 be the number

$$H_0 \triangleq \sum_{a \in A} (-\log_2 p(a)) p(a) \tag{8}$$

It is easy to show from (7) that the source entropy satisfies

$$H_n = nH_0$$

and therefore H_0 is the entropy rate of the source $[n, A_n, P_n]$. Assuming that the data length n is large, we conclude that a lossless code for compressing the data generated by the memoryless source $[n, A_n, P_n]$ is efficient if and only if the resulting compression rate is approximately equal to H_0.

2.1.2. Efficient Lossless Codes for Markov Sources. Now assume that the source $[n, A_n, P_n]$ is a stationary Markov source; let A be the finite source alphabet. There is a probability mass function $[p(a): a \in A]$ on A, and a nonnegative matrix $[\pi(a_1, a_2): a_1, a_2 \in A]$ whose rows each sum to one such that both of the following are true:

$$p(a_2) = \sum_{a_1 \in A} p(a_1)\pi(a_1, a_2), \quad a_2 \in A \tag{9}$$

$$P_n(x_1, x_2, \ldots, x_n) = p(x_1) \prod_{i=2}^{n} \pi(x_{i-1}, x_i), \quad (x_1, x_2, \ldots, x_n) \in A_n \tag{10}$$

Let H_0 be the number (8) and H_1 be the number

$$H_1 \triangleq \sum_{a_1 \in A} \sum_{a_2 \in A} (-\log_2 \pi(a_1, a_2)) p(a_1)\pi(a_1, a_2)$$

It can be shown from these last two equations that

$$H_n = H_0 + (n-1)H_1$$

Thus, for large n, the entropy rate H_n/n of the source is approximately equal to H_1. Assuming that the data length n is large, we conclude that a lossless code for compressing the data generated by the stationary Markov source $[n, A_n, P_n]$ is efficient if and only if the resulting compression rate is approximately equal to H_1.

In the rest of this section, we survey each of the following efficient classes of lossless codes:

- Huffman codes
- Enumerative codes
- Arithmetic codes
- Lempel–Ziv codes

2.2. Huffman Codes

Fix a source $[n, A_n, P_n]$ with A_n finite; let S_n denote a lossless compression system driven by this source (see Fig. 4). In 1948, Claude Shannon put forth the following monotonicity principle for code design for the system S_n: The length of the binary codeword assigned to each data sequence in A_n should be inversely related to the probability with which that sequence occurs. According to this principle, data sequences with low probability of occurrence are assigned long binary codewords, whereas data sequences with high probability of occurrence are assigned short binary codewords. In Shannon's 1948 paper [42], a code called the *Shannon–Fano code* was put forth that obeys the monotonicity principle; it assigns a codeword to data sequence $(x_1, x_2, \ldots, x_n) \in A_n$ of length

$$\lceil -\log_2 P_n(x_1, x_2, \ldots, x_n) \rceil$$

The compression rate $R(S_n)$ resulting from the use of the Shannon–Fano code in system S_n is easily seen to satisfy

$$\frac{H_n}{n} \le R(S_n) \le \frac{H_n + 1}{n} \tag{11}$$

However, the Shannon–Fano code does not yield the minimal compression rate. The problem of finding the code that yields the minimal compression rate was solved in 1952 by David Huffman [22], and this code has been named the "Huffman code" in his honor.

We discuss the simplest instance of Huffman code design, namely, design of the Huffman code for a first-order source $[1, A, P]$; such a code encodes the individual letters in the source alphabet A and will be called a *first-order Huffman code*. If the letters in A are $a_1, a_2, \ldots, a_j$ and K_i denotes the length of the binary codeword into which letter a_i is encoded, then the Huffman code is the code for which

$$P(a_1)K_1 + P(a_2)K_2 + \cdots + P(a_j)K_j$$

is minimized. The Huffman algorithm constructs the encoding table of the Huffman code. The Huffman algorithm operates recursively in the following way. First, the letters a_i, a_j with the two smallest probabilities $P(a_i), P(a_j)$ are removed from the alphabet A and replaced with a single "superletter" $a_i a_j$ of probability $P(a_i) + P(a_j)$. Then, the Huffman code for the reduced alphabet is extended to a Huffman code for the original alphabet by assigning codeword $w0$ to a_i and codeword $w1$ to a_j, where w is the codeword assigned to the superletter $a_i a_j$.

Table 1 gives an example of the encoding table for the Huffman code for a first-order source with alphabet $\{a_1, a_2, a_3, a_4\}$. It is easy to deduce that the code given by Table 1 yields minimum compression rate without using the Huffman algorithm. First, notice that the codeword lengths $K_1, K_2, \ldots, K_j$ assigned by a minimum compression rate code must satisfy the equation

$$2^{-K_1} + 2^{-K_2} + \cdots + 2^{-K_j} = 1 \tag{12}$$

Table 1. Example of a Huffman Code

Source Letter	Probability	Codeword
a_1	$\frac{1}{2}$	0
a_2	$\frac{1}{5}$	10
a_3	$\frac{3}{20}$	110
a_4	$\frac{3}{20}$	111

[Kraft's inequality (2.4) is satisfied; if the inequality were strict, at least one of the codewords could be shortened and the code would not yield minimum compression rate.] Since the source has only a four-letter alphabet, there are only two possible solutions to (12), and therefore only two possible sets of codeword lengths for a first-order code, namely, 1,2,3,3 and 2,2,2,2. The first choice yields compression rate (or, equivalently, expected codeword length) of

$$1\tfrac{1}{2} + 2\tfrac{1}{5} + 3\tfrac{3}{20} + 3\tfrac{3}{20} = 1.8$$

code bits per data sample, whereas the second choice yields the worse compression rate of 2 code bits per data sample. Hence, the code given by Table 1 must be the Huffman code. One could use the Huffman algorithm to find this code. Combining the two least-probable letters a_3 and a_4 into the superletter a_3a_4, the following table gives the Huffman encoder for the reduced alphabet:

Source Letter	Probability	Codeword
a_1	$\frac{1}{2}$	0
a_3a_4	$\frac{6}{20}$	11
a_2	$\frac{1}{5}$	10

(This is immediate because for a three-letter alphabet, there is only one possible choice for the set of codeword lengths, namely, 1,2,2.) Expanding the codeword 11 for a_3a_4 into the codewords 110 and 111, we obtain the Huffman code in Table 1.

2.3. Enumerative Codes

Enumerative coding, in its present state of development, is due to Thomas Cover [11]. Enumerative coding is used for a source $[n, A_n, P_n]$ in which the data sequences in A_n are equally likely; the best lossless code for such a source is one that assigns codewords of equal length. Here is Cover's approach to enumerative code design:

Step 1. Let N be the number of sequences in A_n, and let A be the source alphabet. Construct the rooted tree T with N leaves, such that the edges emanating from each internal vertex have distinct labels from A, and such that the N sequences in A_n are found by writing down the labels along the N root-to-leaf paths of T.

Step 2. Locate all paths in T that visit only unary vertices in between and that are not subpaths of other such paths. Collapse each of these paths to single edges, labeling each such single edge that results with the sequence of labels along the collapsed path. This yields a tree T^* with N leaves (the same as the leaves of T). Label each leaf of T^* with the sequence obtained by concatenating together the labels on the root-to-leaf path to that leaf; these leaf labels are just the sequences in A_n.

Step 3. Assign an integer weight to each vertex v of T^* as follows. If v has no siblings or is further to the left than its siblings, assign v a weight of zero. If v has siblings further to the left, assign v a weight equal to the number of leaves of T^* that are equal to or subordinate to the siblings of v that are to the left of v.

Step 4. To encode $x^n \in A_n$, follow the root-to-leaf path in T^* terminating in the leaf of T^* labeled by x^n. Let I be the sum of the weights of the vertices along this root-to-leaf path. The integer I is called the *index* of x^n, and satisfies $0 \le I \le N - 1$. Encode x^n into the binary codeword of length $\lceil \log_2 N \rceil$ obtained by finding the binary expansion of I and then padding that expansion (if necessary) to $\lceil \log_2 N \rceil$ bits by appending a prefix of zeros.

For example, suppose that the source $[n, A_n, P_n]$ satisfies

$$A_n = \{aaa, aba, abb, abc, baa, bba, caa, cba, cbb, cca\} \tag{13}$$

Then steps 1–3 yield the tree T^* in Fig. 5, in which every vertex is labeled with its weight from step 3. [The 10 leaves of this tree, from left to right, correspond to the 10 sequences in (13), from left to right.] The I values along the 10 paths are just the cumulative sums of the weights, which are seen to give $I = 0,1,2,3,4,5,6,7,8,9$. The codeword length is $\lceil \log_2 10 \rceil = 4$, and the respective codewords are

0000, 0001, 0010, 0011, 0100, 0101, 0110, 0111, 1000, 1001

Sometimes, the tree T^* has such a regular structure that one can obtain an explicit formula relating a data sequence and its index I, thereby dispensing with the need for the tree T^* altogether. A good example of this occurs with the source $[n, A_n, P]$ in which A_n is the set of all binary sequences of length n and having a number of ones equal to m (m is a fixed positive integer satisfying

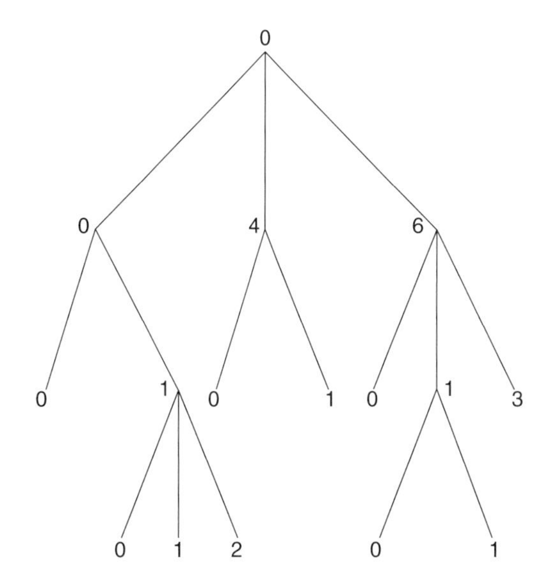

Figure 5. Weighted enumerative coding tree T^*.

$0 \le m \le n$). For a data sequence $(x_1, x_2, \ldots, x_n) \in A_n$, Cover showed that its index I is computable as

$$I = \sum_{j=1}^{n} x_j \binom{j-1}{x_1 + x_2 + \cdots + x_j} \quad (14)$$

There are $\binom{n}{m}$ sequences in A_n. Therefore, Eq. (14) provides a one-to-one correspondence between these sequences and the integers

$$0, 1, 2, 3, \ldots, \binom{n}{m} - 1$$

Having obtained I from $(x_1, x_2, \ldots, x_n)$ via this equation, $(x_1, x_2, \ldots, x_n)$ is encoded by expanding integer I in binary out to $\lceil \log_2 \binom{n}{m} \rceil$ bits. Conversely, $(x_1, x_2, \ldots, x_n)$ is decoded from its bit representation by first finding I, and then finding the unique expansion of I in (14) given by the right-hand side. To illustrate, suppose that $n = 8$ and $m = 4$. Then, A_n contains $\binom{8}{4} = 70$ sequences and I can be any integer between 0 and 69, inclusively. Suppose $I = 52$. To decode back into the data sequence in A_n that gave rise to the index $I = 52$, the decoder finds the unique integers $0 \le j_1 < j_2 < j_3 < j_4 < 8$ satisfying

$$52 = \binom{j_1}{1} + \binom{j_2}{2} + \binom{j_3}{3} + \binom{j_4}{4}$$

The solution is $j_1 = 1, j_2 = 4, j_3 = 5$, and $j_4 = 7$. The data sequence we are looking for must have ones in positions $j_1 + 1 = 2, j_2 + 1 = 5, j_3 + 1 = 6, j_4 + 1 = 8$; this is the sequence (0,1,0,0,1,1,0,1).

2.4. Arithmetic Codes

Arithmetic codes were invented by Peter Elias in unpublished work around 1960, but his schemes were not practical. In the 1970s, other people put Elias' ideas on a practical footing [33,36,37]. This section gives an introduction to arithmetic codes.

Arithmetic coding presents a whole new philosophy of coding. As the data samples in a data sequence $(x_1, x_2, \ldots, x_n)$ are processed from left to right by the arithmetic encoder, each data sample x_i is not replaced with a string of code bits as is done in conventional encoding—instead, each x_i is assigned a subinterval I_i of the unit interval [0, 1] so that

$$I_1 \supset I_2 \supset \cdots \supset I_n \quad (15)$$

and so that I_i is recursively determined from I_{i-1} and $x_i (i \ge 2)$. When the final interval I_n is determined, then the binary codeword $(b_1, b_2, \ldots, b_k)$ into which $(x_1, x_2, \ldots, x_n)$ is encoded is chosen so that the number

$$\frac{b_1}{2} + \frac{b_2}{4} + \frac{b_3}{8} + \cdots + \frac{b_k}{2^k} \quad (16)$$

is a point in I_n, where the codeword length k is approximately equal to $\log_2$ of the reciprocal of the probability assigned by the source to $(x_1, x_2, \ldots, x_n)$.

2.4.1. Precise Description.

Arithmetic codes can be constructed for any source. For simplicity, we assume a memoryless source $[n, A_n, P_n]$ in which A_n consists of all sequences of length n whose entries come from the set $\{0, 1, 2, \ldots, j-1\}$, where j is a fixed positive integer. Then, for each data sequence $(x_1, x_2, \ldots, x_n) \in A_n$, the probability assigned by the source is

$$P_n(x_1, x_2, \ldots, x_n) = \prod_{i=1}^{n} p_{x_i}$$

where $p_0, p_1, \ldots, p_{j-1}$ are given nonnegative numbers that sum to one. We specify the arithmetic code for this source by describing algorithms for encoding and decoding. Let $a_0 = 0, a_1 = 1$. Arithmetic encoding of $(x_1, x_2, \ldots, x_n) \in A_n$ takes place according to the following three-step algorithm:

Encoding Step 1. For each $i = 1, 2, \ldots, n$, a subinterval $I_i = [a_i, b_i]$ of [0, 1] is recursively determined according to the formula

$$\begin{aligned} I_i &= [a_{i-1}, a_{i-1} + (b_{i-1} - a_{i-1})p_0], x_i = 0 \\ &= [a_{i-1} + (p_0 + \cdots + p_{x_i - 1})(b_{i-1} - a_{i-1}), a_{i-1} \\ &\quad + (p_0 + \cdots + p_{x_i})(b_{i-1} - a_{i-1})], x_i > 0 \end{aligned}$$

By construction, the last interval I_n will have length equal to $P_n(x_1, x_2, \ldots, x_n)$.

Encoding Step 2. The integer

$$k = \lceil -\log_2 P_n(x_1, x_2, \ldots, x_n) \rceil + 1$$

is determined. This integer will be the length of the codeword assigned by the arithmetic encoder to $(x_1, x_2, \ldots, x_n)$.

Encoding Step 3. The midpoint M of the interval I_n is computed. The codeword $(b_1, b_2, \ldots, b_k)$ assigned to $(x_1, x_2, \ldots, x_n)$ consists of the first k digits in the binary expansion of M.

The following arithmetic decoding algorithm is applied to the codeword $(b_1, b_2, \ldots, b_k)$ in order to reclaim the data sequence $(x_1, x_2, \ldots, x_n)$ that gave rise to it:

Decoding Step 1. Compute the point $\hat{M}$ given by the expression (16). By choice of k, the point $\hat{M}$ will lie in the interval I_n.

Decoding Step 2. There are j possibilities for I_1, depending on what x_1 is. Only one of these possibilities for I_1 contains $\hat{M}$. Using this fact, the decoder is able to determine I_1. From I_1, the decoder determines x_1.

Decoding Step 3. For each $i = 2, \ldots, n$, the decoder determines from I_{i-1} (determined previously) what the j possibilities for I_i are. Since only one of these possibilities for I_i contains $\hat{M}$, the decoder is able to determine I_i. From I_i, the decoder is able to determine x_i.

Example 1. Take the source alphabet to be {0, 1}, take the probabilities defining the memoryless source to be

$p_0 = \frac{2}{5}, p_1 = \frac{3}{5}$, and take the source to be $[5, \{0, 1\}^5, P_5]$ (known to both encoder and decoder). Suppose that the data sequence to be arithmetically encoded is (1, 0, 1, 1, 0). We need to recursively determine the intervals I_1, I_2, I_3, I_4, I_5. We have

$$I_1 = \text{right } \frac{3}{5}\text{ths of } [0,1] = \left[\frac{2}{5}, 1\right]$$

$$I_2 = \text{left } \frac{2}{5}\text{ths of } I_1 = \left[\frac{2}{5}, \frac{16}{25}\right]$$

$$I_3 = \text{right } \frac{3}{5}\text{ths of } I_2 = \left[\frac{62}{125}, \frac{16}{25}\right]$$

$$I_4 = \text{right } \frac{3}{5}\text{ths of } I_3 = \left[\frac{346}{625}, \frac{16}{25}\right]$$

$$I_5 = \text{left } \frac{2}{5}\text{ths of } I_4 = \left[\frac{346}{625}, \frac{1838}{3125}\right]$$

The length of the interval I_5 is $\frac{108}{3125}$. Therefore, the length of the binary codeword must be

$$k = \left\lceil \log_2\left(\frac{3125}{108}\right)\right\rceil + 1 = 6$$

The midpoint of the interval I_5 is $M = \frac{1784}{3125}$. Expanding this number in binary, we obtain

$$\frac{1784}{3125} = .100100\ldots$$

The binary codeword is therefore $(1, 0, 0, 1, 0, 0)$. Given this codeword, how does the decoder determine the data sequence that gave rise to it? Since the decoder knows the source description, it knows that the data sequence to be found is of the form $(x_1, x_2, x_3, x_4, x_5)$, where the x_i terms are binary. To decode, the decoder first computes

$$\hat{M} = \frac{1}{2} + \frac{1}{16} = \frac{9}{16}$$

The decoder knows that

$$I_1 = \left[0, \tfrac{2}{5}\right] \quad \text{or} \quad I_1 = \left[\tfrac{2}{5}, 1\right]$$

Since $\frac{9}{16}$ is in the right interval, the decoder concludes that $I_1 = \left[\frac{2}{5}, 1\right]$ and that $x_1 = 1$. At this point, the decoder knows that

$$I_2 = \left[\frac{2}{5}, \frac{16}{25}\right] \text{ or } I_2 = \left[\frac{16}{25}, 1\right]$$

Since $\frac{9}{16}$ is in the left interval, the decoder concludes that $I_2 = \left[\frac{2}{5}, \frac{16}{25}\right]$ and that $x_2 = 0$. The decoder now knows that

$$I_3 = \left[\frac{2}{5}, \frac{62}{125}\right] \text{ or } I_3 = \left[\frac{62}{125}, \frac{16}{25}\right]$$

Since $\frac{9}{16}$ lies in the right interval, the decoder determines that $I_3 = \left[\frac{62}{125}, \frac{16}{25}\right]$, and that the third data sample is $x_3 = 1$. Similarly, the decoder can determine x_4 and x_5 by two more rounds of this procedure.

The reader sees from the preceding example that the arithmetic code as we have prescribed it requires ever greater precision as more and more data samples are processed. This creates a problem if there are a large number of data samples to be arithmetically encoded. One can give a more complicated (but less intuitive) description of the arithmetic encoder/decoder that uses only finite precision (integer arithmetic is used). A textbook [41] gives a wealth of detail on this approach.

2.4.2. Performance. In a sense that will be described here, arithmetic codes give the best possible performance, for large data length. First, we point out that for any source, an arithmetic code can be constructed. Let the source be $[n, A_n, P_n]$. For $(x_1, x_2, \ldots, x_n) \in A_n$, one can use conditional probabilities to factor the probability assigned to this sequence:

$$P_n(x_1, x_2, \ldots, x_n) = p(x_1)\prod_{i=2}^{n} p(x_i \mid x_1, x_2, \ldots, x_{i-1}).$$

The factors on the right side are explained as follows. Let $X_1, X_2, \ldots, X_n$ be the random data samples generated by the source. Then

$$p(x_1) = \Pr[X_1 = x_1]$$
$$p(x_i \mid x_1, x_2, \ldots, x_{i-1}) = \Pr[X_i = x_i \mid X_1 = x_1, X_2 = x_2, \ldots, X_{i-1} = x_{i-1}]$$

To arithmetically encode $(x_1, x_2, \ldots, x_n) \in A_n$, one constructs a decreasing sequence of intervals $I_1, I_2, \ldots, I_n$. One does this so that I_1 will have length $p(x_1)$ and for each $i = 2, \ldots, n$, the length of the interval I_i will be $p(x_i \mid x_1, x_2, \ldots, x_{i-1})$ times the length of the interval I_{i-1}. The rest of the encoding and decoding steps will be as already described for the memoryless source. Suppose a lossless data compression system S_n is built with the given source $[n, A_n, P_n]$ and the arithmetic code we have just sketched. It can be shown that the resulting compression rate satisfies

$$\frac{H_n}{n} \le R(S_n) \le \frac{H_n + 2}{n}$$

where H_n is the entropy of the source $[n, A_n, P_n]$. For large n, we therefore have $R(S_n) \approx H_n$. This is the best that one can possibly hope to do. If the source $[n, A_n, P_n]$ is memoryless or Markov, and n is large, one obtains the very good arithmetic code compression rate performance just described, but at the same time, the arithmetic code is of low complexity. As discussed in Section 2.2, for large n, the compression system S_n built using the Huffman code for the source $[n, A_n, P_n]$ will also achieve the very good compression rate performance $R(S_n) \approx H_n$, but this Huffman code will be very complex. For this reason, arithmetic codes are preferred over Huffman codes in many data compression applications. Two notable successes of arithmetic coding in practical applications are the PPM text compression algorithm [8] and IBM's Q-coder [34], used for lossless binary image compression.

2.5. Lempel–Ziv Codes

Lempel–Ziv codes are examples of *dictionary codes*. A dictionary code first partitions a data sequence into variable-length phrases (this procedure is called *parsing*). Then, each phrase in the parsing is represented by means of a pointer to that phrase in a dictionary of phrases constructed from previously processed data. The phrase dictionary changes dynamically as the data sequence is processed from left to right. A binary codeword is then assigned to the data sequence by encoding the sequence of dictionary pointers in some simple way. The most popular dictionary codes are the Lempel–Ziv codes. There are many versions of the Lempel–Ziv codes. The one we discuss here is LZ78 [55]. Another popular Lempel–Ziv code, not discussed here, is LZ77 [54]. Two widely used compression algorithms on UNIX systems are Compress and Gzip; Compress is based on LZ78 and Gzip is based on LZ77.

In the rest of this section, we discuss the parsing technique, the pointer formation technique, and the pointer encoding technique employed in Lempel–Ziv coding.

2.5.1. Lempel–Ziv Parsing.

Let $(x_1, x_2, \ldots, x_n)$ be the data sequence to be compressed. Partitioning of this sequence into variable-length blocks via *Lempel–Ziv parsing* takes place as follows. The first variable-length block arising from the Lempel–Ziv parsing of $(x_1, x_2, \ldots, x_n)$ is the single sample x_1. The second block in the parsing is the shortest prefix of $(x_2, x_3, \ldots, x_n)$ that is not equal to x_1. Suppose that this second block is $(x_2, \ldots, x_j)$. Then, the third block in Lempel–Ziv parsing will be the shortest prefix of $(x_{j+1}, x_{j+2}, \ldots, x_n)$ that is not equal to either x_1 or $(x_2, \ldots, x_j)$. In general, suppose that the Lempel–Ziv parsing procedure has produced the first k variable-length blocks $B_1, B_2, \ldots, B_k$ in the parsing, and $x^{(k)}$ is that part left of $(x_1, x_2, \ldots, x_n)$ after $B_1, B_2, \ldots, B_k$ have been removed. Then the next block B_{k+1} in the parsing is the shortest prefix of $x^{(k)}$ that is not equal to any of the preceding blocks $B_1, B_2, \ldots, B_k$. [If there is no such block, then $B_{k+1} = x^{(k)}$ and the Lempel-Ziv parsing procedure terminates.]

By construction, the sequence of variable-length blocks $B_1, B_2, \ldots, B_t$ produced by the Lempel–Ziv parsing of $(x_1, x_2, \ldots, x_n)$ are distinct, except that the last block B_t could be equal to one of the preceding ones. The following example illustrates Lempel–Ziv parsing.

Example 2. The Lempel–Ziv parsing of the data sequence

$$(1, 1, 0, 1, 1, 0, 0, 0, 1, 1, 0, 1) \tag{17}$$

is

$$B_1 = (1)$$

$$B_2 = (1, 0)$$

$$B_3 = (1, 1)$$

$$B_4 = (0)$$

$$B_5 = (0, 0)$$

$$B_6 = (1, 1, 0) \tag{18}$$

$$B_7 = (1) \tag{19}$$

2.5.2. Pointer Formation.

We suppose that the alphabet from which the data sequence $(x_1, x_2, \ldots, x_n)$ is formed is $A = \{0, 1, \ldots, k-1\}$, where k is a positive integer. After obtaining the Lempel–Ziv parsing $B_1, B_2, \ldots, B_t$ of $(x_1, x_2, \ldots, x_n)$, the next step is to represent each block in the parsing as a pair of integers. The first block in the parsing, B_1, consists of a single symbol. It is represented as the pair $(0, B_1)$. More generally, any block B_j of length one is represented as the pair $(0, B_j)$. If the block B_j is of length greater than one, then it is represented as the pair (i,s), where s is the last symbol in B_j and B_i is the unique previous block in the parsing that coincides with the block obtained by removing s from the end of B_j.

Example 3. The sequence of pairs corresponding to the parsing (19) is

$$(0, 1), (1, 0), (1, 1), (0, 0), (4, 0), (3, 0), (0, 1) \tag{20}$$

For example, $(4, 0)$ corresponds to the block $(0, 0)$ in the parsing. Since the last symbol of (0,0) is 0, the pair (4,0) ends in 0. The 4 in the first entry refers to the fact that $B_4 = (0)$ is the preceding block in the parsing, which is equal to what we get by deleting the last symbol of $(0, 0)$.

For our next step, we replace each pair (i, s) by the integer $ki + s$. Thus, the sequence of pairs (20) becomes the sequence of integers

$$I_1 = 2 * 0 + 1 = 1$$

$$I_2 = 2 * 1 + 0 = 2$$

$$I_3 = 2 * 1 + 1 = 3$$

$$I_4 = 2 * 0 + 0 = 0 \tag{21}$$

$$I_5 = 2 * 4 + 0 = 8$$

$$I_6 = 2 * 3 + 0 = 6$$

$$I_7 = 2 * 0 + 1 = 1$$

2.5.3. Encoding of Pointers.

Let $I_1, I_2, \ldots, I_t$ denote the integer pointers corresponding to the blocks $B_1, B_2, \ldots, B_t$ in the Lempel–Ziv parsing of the data sequence $(x_1, x_2, \ldots, x_n)$. To finish our description of the Lempel–Ziv encoder, we discuss how the integer pointers $I_1, I_2, \ldots, I_t$ are converted into a stream of bits. Each integer I_j is expanded to base 2, and these binary expansions are "padded" with zeros on the left so that the overall length of the string of bits assigned to I_j is $\lceil \log_2(kj) \rceil$. The reason why this many bits is necessary and sufficient is seen by examining the largest that I_j can possibly be. Let (i,s) be the pair associated with I_j. Then the largest that i can be is $j-1$ and the largest that s can be is $k-1$. Thus the largest that I_j can be is $k(j-1) + k - 1 = kj - 1$, and the number of bits in the binary expansion of $kj - 1$ is $\lceil \log_2(kj) \rceil$. Let W_j be the string of bits of length $\lceil \log_2(kj) \rceil$ assigned to I_j as described in the preceding text. Then, the Lempel–Ziv encoder output is obtained by concatenating together the strings $W_1, W_2, \ldots, W_t$.

To illustrate, suppose that the data sequence $(x_1, x_2, \ldots, x_n)$ is binary (i.e., $k = 2$), and has seven blocks $B_1, B_2, \ldots, B_7$ in its Lempel–Ziv parsing. These blocks are assigned, respectively, strings of code bits $W_1, W_2, W_3, W_4, W_5, W_6, W_7$ of lengths $\lceil \log_2(2) \rceil = 1$, $\lceil \log_2(4) \rceil = 2$, $\lceil \log_2(6) \rceil = 3$, $\lceil \log_2(8) \rceil = 3$, $\lceil \log_2(10) \rceil = 4$, $\lceil \log_2(12) \rceil = 4$, and $\lceil \log_2(14) \rceil = 4$. Therefore, any binary data sequence with seven blocks in its Lempel–Ziv parsing would result in an encoder output of length $1 + 2 + 3 + 3 + 4 + 4 + 4 = 21$ code bits. In particular, for the data sequence (2.17), the seven strings $W_1, \ldots, W_7$ are [referring to (21)]

$$W_1 = (1)$$
$$W_2 = (1, 0)$$
$$W_3 = (0, 1, 1)$$
$$W_4 = (0, 0, 0)$$
$$W_5 = (1, 0, 0, 0)$$
$$W_6 = (0, 1, 1, 0)$$
$$W_7 = (0, 0, 0, 1)$$

Concatenating, we see that the codeword assigned to data sequence (17) by the Lempel–Ziv encoder is

$$(1, 1, 0, 0, 1, 1, 0, 0, 0, 1, 0, 0, 0, 0, 1, 1, 0, 0, 0, 0, 1) \quad (22)$$

We omit a detailed description of the Lempel–Ziv decoder. However, it is easy to see what the decoder would do. For example, it would be able to break up the codeword (22) into the separate codewords for the phrases, because, from the size k of the data alphabet, it is known how many code bits are allocated to the encoding of each Lempel–Ziv phrase. From the separate codewords, the decoder recovers the integer representing each phrase; dividing each of these integers by k to obtain the quotient and remainder, the pairs representing the phrases are obtained. Finally, these pairs yield the phrases, which are concatenated together to obtain the original data sequence.

2.5.4. Performance. Let $[n, A_n, P_n]$ be any data source with alphabet of size k. Let S_n be the lossless data compression system driven by this source that employs the Lempel–Ziv code. It is known that there is a positive constant C_k (depending on k but not on n) such that

$$\frac{H_n}{n} \leq R(S_n) \leq \frac{H_n}{n} + \frac{C_k}{\log_2 n}$$

where H_n is the source entropy. Thus, the Lempel–Ziv code is not quite as good as the Huffman code or the arithmetic code, but there is an important difference. The Huffman code and arithmetic code require knowledge of the source. The preceding performance bound is valid regardless of the source. Thus, one can use the same Lempel–Ziv code for all sources — such a code is called a *universal code* [13].

In practical compression scenarios, the Lempel–Ziv code has been superseded by more efficient modern dictionary codes, such as the YK algorithm [51].

3. LOSSY COMPRESSION METHODOLOGIES

In this section, we shall be concerned with the problem of designing lossy codes. Recall from Fig. 3 that a lossy compression system consists of source, noninvertible source encoder, and source decoder. Figure 6 gives separate depictions of the source encoder and source decoder in a lossy compression system.

The same notational conventions introduced earlier are in effect here: X^n [Eq. (1)] is the random data sequence of length n generated by the source, $\hat{X}^n$ (1.3) is the reconstructed data sequence, and B^K (2) is the variable-length codeword assigned to X^n. The source decoder component of the lossy compression system in Fig. 3 is the cascade of the "quantizer" and "lossless encoder" blocks in Fig. 6a; the source decoder component in Fig. 3 is the lossless decoder of Fig. 6b. The quantizer is a many-to-one mapping that converts the data sequence X^n into its reconstruction $\hat{X}^n$. The lossless encoder is a one-to-one mapping that converts the reconstructed data sequence $\hat{X}^n$ into the binary codeword B^K from which $\hat{X}^n$ can be recovered via application of the lossless decoder to B^K.

It is the presence of the quantizer that distinguishes a lossy compression system from a lossless one. Generally speaking, the purpose of the quantizer is alphabet reduction — by dealing with a data sequence from a reduced alphabet instead of the original data sequence over the original alphabet, one can hope to perform data compression using fewer code bits. For example, the "rounding-off quantizer" is a very simple quantizer. Let the data sequence

$$X^n = (1.1, 2.6, 4.4, 2.3, 1.7) \quad (23)$$

be observed. Then, the rounding-off quantizer generates the reconstructed data sequence

$$\hat{X}^n = (1, 3, 4, 2, 2) \quad (24)$$

It takes only 10 code bits to compress (24), because its entries come from the reduced alphabet $\{1, 2, 3, 4\}$; it

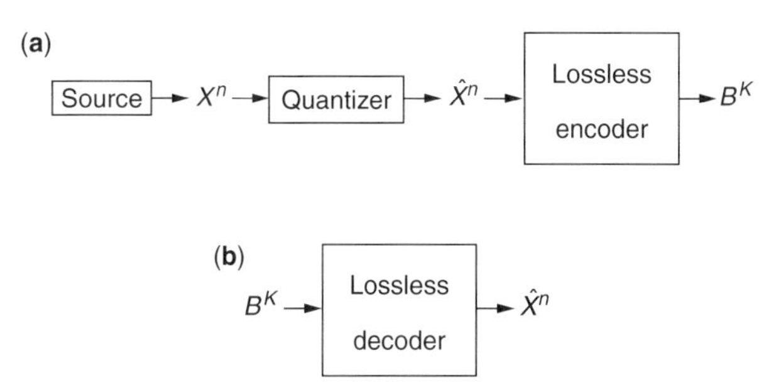

Figure 6. Source encoder (**a**) and decoder (**b**) in lossy compression system.

would take many more code bits to compress the original sequence (23).

As indicated in Fig. 6, a lossy code consists of quantizer, lossless encoder, and lossless decoder. When a lossy code is used to compress the data generated by a source $[n, A_n, P_n]$, a lossy compression system S_n results. The effectiveness of the lossy code is then evaluated by means of two figures of merit, the (compression) rate $R(S_n)$ and the distortion $D(S_n)$, defined by

$$R(S_n) \triangleq \frac{E[K]}{n}$$

$$D(S_n) \triangleq n^{-1} \sum_{i=1}^{n} E[d(X_i, \hat{X}_i)]$$

In the preceding, E denotes the expected value operator, K is the random codeword length of the codeword B^K assigned to X^n in Fig. 5a, and d is a fixed distortion function mapping pairs of source letters into nonnegative real numbers. The distortion function d is typically one of the following two types:

1. *Squared-error distortion*:

$$d(X_i, \hat{X}_i) = (X_i - \hat{X}_i)^2$$

2. *Hamming distortion*:

$$d(X_i, \hat{X}_i) = \begin{cases} 0, & X_i = \hat{X}_i \\ 1, & \text{otherwise} \end{cases}$$

Squared-error distortion is typically used for an infinite source alphabet and Hamming distortion is typically used for a finite source alphabet. When squared-error distortion is used, distortion is sometimes measured in decibels as the figure

$$[D(S_n)]_{\text{dec}} = 10 \log_{10} \left(\frac{n^{-1} \sum_{i=1}^{n} \int_{-\infty}^{\infty} x^2 f_{X_i}(x)\, dx}{D(S_n)} \right)$$

where $X_i (1 \le i \le n)$ represents the ith data sample generated by the source $[n, A_n, P_n]$ and f_{X_i} denotes the probability density function of X_i; note that small $D(S_n)$ would correspond to a large decibel measure of distortion.

Suppose that two different lossy codes have been designed to compress the random data generated by a given source $[n, A_n, P_n]$, resulting in lossy compression systems S_n^1 and S_n^2, respectively. Then, one can declare that the lossy code giving rise to system S_n^1 is better than the lossy code giving rise to system S_n^2 if $R(S_n^1) < R(S_n^2)$ and $D(S_n^1) < D(S_n^2)$. However, it may be that neither lossy code is better than the other one in this sense, since the inverse relation between rate and distortion precludes the design of a lossy code for which rate and distortion are simultaneously small. Instead, for a given source, the design goal should be to find a lossy code that yields the smallest rate for a fixed distortion, or the smallest distortion for a fixed rate. The theory detailing the rate–distortion tradeoffs that are possible in lossy code design is called *rate–distortion theory*. Section 3.1 gives an introduction to this subject.

The quantizer employed in a lossy code can be one of two types, either a scalar quantizer or vector quantizer. A *scalar quantizer* quantizes one data sample at a time, whereas for some $m > 1$, a *vector quantizer* quantizes m data samples at a time. Lossy codes that employ scalar quantizers are called "SQ-based codes" and are covered in Section 3.2; lossy codes that employ vector quantizers are called "VQ-based codes" and are covered in Section 3.3. Subsequent sections deal with two other important lossy coding techniques, trellis-based coding, and transform coding.

3.1. Distortion Bounds

In designing a lossy code for a source $[n, A_n, P_n]$ to produce a lossy compression system S_n, the usual approach is the *fixed-rate approach*, in which one attempts to find a lossy code that minimizes or approximately minimizes the distortion $D(S_n)$ among all lossy codes satisfying the rate constraint $R(S_n) \le R$, where $R > 0$ is a fixed constant. We adopt the fixed-rate approach for the rest of this article.

In this section, we will give upper and lower bounds on the distortion performance of lossy codes for a fixed source, subject to the constraint that the compression rate be no more than R code bits per data sample, on average. With these bounds, one can determine before designing a lossy code what types of performance are and are not possible for such a code to achieve. The upper and lower bounds on distortion performance that shall be developed in this section are expressible using Claude Shannon's notion of *distortion–rate function* [43], discussed next.

3.1.1. Distortion–Rate Function.

The concept of *mutual information* will be needed in order to define the distortion–rate function. Let X, Y be two random variables. Let $f(x)$ be the probability density function of X, and let $g(y \mid x)$ be the conditional probability density function of Y given $X = x$. Then, the mutual information $I(X;Y)$ of X, Y is the number

$$I(X;Y) \triangleq \int_{-\infty}^{\infty} \int_{-\infty}^{\infty} f(x) g(y \mid x) \log_2 \times \left[\frac{g(y \mid x)}{\int_{-\infty}^{\infty} f(u) g(y \mid u) du} \right] dx\, dy$$

We are now ready to define the concept of distortion–rate function. For simplicity, we restrict ourselves to the memoryless source $[n, A_n, P_n]$. We suppose that the source alphabet is a subset of the real line; let X be a random variable such that the random data samples $X_i (1 \le i \le n)$ generated according to the memoryless source $[n, A_n, P_n]$ are independent copies of X. The distortion–rate function $D(R)$ of the memoryless source $[n, A_n, P_n]$ (for a given nonnegative distortion function d such as Hamming distortion or squared-error distortion) is then defined for each $R > 0$ by

$$D(R) \triangleq \min\{E[d(X, Y)] : I(X;Y) \le R\} \tag{25}$$

where Y denotes any random variable jointly distributed with X, and we assume that the minimum in (25) exists and is finite.

Example 4. One important type of memoryless source for which the distortion–rate function has a closed-form expression is the *Gaussian memoryless source* $[n, A_n, P_n]$, which is the memoryless source whose independent random data samples are generated according to a Gaussian distribution with mean 0 and variance σ^2. For the Gaussian memoryless source with squared-error distortion

$$D(R) = \sigma^2 2^{-2R} \tag{26}$$

Sources such as this one in which $D(R)$ can be computed explicitly are rare. However, in general, one can use the Blahut algorithm [6] to approximate $D(R)$ arbitrarily closely.

3.1.2. Distortion Lower Bound.

Fix a memoryless source $[n, A_n, P_n]$ with distortion–rate function $D(R)$. Fix $R > 0$, and fix any lossy code for $[n, A_n, P_n]$ for which the resulting compression system S_n satisfies the rate constraint $R(S_n) \leq R$. Then, the following well-known [4] distortion lower bound is valid:

$$D(S_n) \geq D(R) \tag{27}$$

Example 5. Assume the nth-order Gaussian memoryless source model and squared-error distortion. Substituting the expression for $D(R)$ in (26) into (27), one concludes that any lossy code with compression rate $\leq R$ yields distortion in decibels no greater than $(20 \log_{10} 2)R \approx 6R$. Thus, a Gaussian memoryless source cannot be encoded to yield a better distortion performance than "6 decibels per bit."

3.1.3. Distortion Upper Bound.

Let f be a probability density function. For each $n = 1, 2, \ldots$, let $[n, A_n, P_n]$ be the memoryless source in which n independent random data samples are generated according to the density f. Assume finiteness of the distortion–rate function $D(R)$ for these sources [since the distortion–rate function depends only on f, all of these sources will have the same distortion-rate function $D(R)$]. Fix $R > 0$. It is well known [35, 53] that there is a positive constant C such that for every $n = 2, 3, \ldots$, there is a lossy code for $[n, A_n, P_n]$ such that the rate and distortion for the resulting compression system S_n satisfy

$$R(S_n) \leq R$$
$$D(S_n) \leq D(R) + \frac{C \log_2 n}{n} \tag{28}$$

Combining the distortion upper bound (28) with the distortion lower bound (27), if n is large, there must exist a code for an nth order memoryless source that yields rate $\leq R$ and distortion $\approx D(R)$; that is, the distortion–rate function $D(R)$ does indeed describe the distortion performance of the most efficient lossy codes. But, in general, it is not known how to find codes that are this efficient. This represents a clear difference between lossless code and lossy code design; it is known how to construct efficient lossless codes, but it is a computationally difficult problem to find efficient lossy codes [18]. For example, for large n, the nth-order Gaussian memoryless source can be encoded to yield squared-error distortion of roughly "6 decibels per bit" (the distortion–rate function performance), but only since 1990 has it been discovered how to find such codes for this simple source model [29].

3.2. SQ-Based Codes

Let R be a fixed positive integer. A 2^R*-bit scalar quantizer* for quantizing real numbers in the interval $[a, b]$ is a mapping Q from $[a, b]$ into a finite subset of $[a, b]$ of size $N = 2^R$. Let $I_1, I_2, \ldots, I_N$ be subintervals of $[a, b]$ that form a partition of $[a, b]$ (the I_j values are called the *quantization intervals* of Q). Let $L_1, L_2, \ldots, L_N$ be points in $[a, b]$ chosen so that $L_j \in I_j$ for each $j = 1, 2, \ldots, N$ (L_j is called the *quantization level* for interval I_j). The quantizer Q accepts as input any real number x in the interval $[a, b]$. The output $Q(x)$ generated by the quantizer Q in response to the input x is the quantization level L_j assigned to the subinterval I_j of $[a, b]$ containing x. In other words, the 2^R-bit quantizer Q is a nondecreasing step function taking 2^R values.

Let $[n, A_n, P_n]$ be a source (such as a memoryless source) in which each randomly generated data sample has the same probability density function f. Let the alphabet for $[n, A_n, P_n]$ be the interval of real numbers $[a, b]$. Let Q be a 2^R-bit scalar quantizer defined on $[a, b]$. We describe a lossy code for the source $[n, A_n, P_n]$ induced by Q. Referring to Fig. 6a, we must explain how the lossy code quantizes source sequences of length n and losslessly encodes the quantized sequences. The lossy code quantizes each source sequence $(x_1, x_2, \ldots, x_n) \in A^n$ into the sequence $(Q(x_1), Q(x_2), \ldots, Q(x_n))$; this makes sense because each entry of each source sequence belongs to the interval $[a, b]$ on which Q is defined. Assign each of the 2^R quantization levels of Q an R-bit binary address so that there is a one-to-one correspondence between quantization levels and their addresses; then, the lossy code losslessly encodes $(Q(x_1), Q(x_2), \ldots, Q(x_n))$ by replacing each of its entries with its R-bit address, yielding an overall binary codeword of length nR. Let S_n be the lossy compression system in Fig. 6 arising from the lossy code just described, and let d be the distortion function that is to be used. It is not hard to see that

$$R(S_n) = R$$
$$D(S_n) = \int_a^b d(x, Q(x)) f(x)\, dx$$

If in the preceding construction we let Q vary over all 2^R-bit scalar quantizers on $[a, b]$, then we obtain all possible SQ based lossy codes for the source $[n, A_n, P_n]$ with compression rate R.

Let R be a positive integer. Consider the following problem: Find the 2^R-bit scalar quantizer Q on $[a, b]$ for which

$$\int_a^b d(x, Q(x)) f(x)\, dx$$

is minimized. This quantizer Q yields the SQ based lossy code for the source $[n, A_n, P_n]$, which has compression rate R and minimal distortion. We call this scalar quantizer the *minimum distortion* 2^R-bit scalar quantizer.

3.2.1. Lloyd–Max Quantizers. We present the solution given in other papers [17,28,30] to the problem of finding the minimum distortion 2^R-bit scalar quantizer for squared-error distortion. Suppose that the probability density function f satisfies

$$\int_a^b x^2 f(x)\,dx < \infty$$

and that $-\log_2 f$ is a concave function on $[a, b]$. Let Q be a 2^R-bit scalar quantizer on $[a, b]$ with quantization levels $\{L_j\}_{j=1}^N$ and quantization intervals $\{I_j\}_{j=1}^N$, where $N = 2^R$. Let

$$y_0 < y_1 < \cdots < y_{N-1} < y_N$$

be the points such that interval I_j has left endpoint y_{j-1} and right endpoint $y_j (j = 1, \ldots, N)$. We call Q a *Lloyd–Max quantizer* if

$$y_j = (\tfrac{1}{2})[L_j + L_{j+1}],\ j = 1, \ldots, N-1 \tag{29}$$

and

$$L_j = \frac{\int_{y_{j-1}}^{y_j} xf(x)\,dx}{\int_{y_{j-1}}^{y_j} f(x)\,dx},\ j = 1, 2, \ldots, N. \tag{30}$$

There is only one 2^R-bit Lloyd–Max scalar quantizer, and it is the unique minimum distortion 2^R-bit scalar quantizer.

Example 6. One case in which it is easy to solve the Lloyd–Max equations (29) and (30) is the case in which $R = 1$ and f is an even function on the whole real line. The unique Lloyd–Max quantizer Q is then given by

$$Q(x) = \begin{cases} 2\int_0^\infty xf(x)\,dx, & x < 0 \\ -2\int_0^\infty xf(x)\,dx, & x \geq 0 \end{cases}$$

Example 7. Assume that $[a, b]$ is a finite interval. Let the source $[n, A_n, P_n]$ be memoryless with each random data sample uniformly distributed on the interval $[a, b]$. The unique 2^R-bit Lloyd–Max scalar quantizer Q for this source is obtained by partitioning $[a, b]$ into 2^R equal subintervals, and by assigning the quantization level for each interval to be the midpoint of that interval [the Lloyd–Max equations (29) and (30) are easily verified]. The SQ-based lossy code for the source $[n, A_n, P_n]$ induced by Q yields rate R and distortion in decibels equal to $(20 \log_{10} 2)R \approx 6R$. This is another "6 decibels per bit" result. Computation of the distortion–rate function for the memoryless source $[n, A_n, P_n]$ [24] reveals that for large n, $[n, A_n, P_n]$ can be encoded at rate $R = 1$ bit per sample at a distortion level of nearly 6.8 decibels. This clearly cannot be achievable using SQ-based lossy codes—more sophisticated lossy codes would be required.

In general, it may not be possible to solve the Eqs. (29) and (30) to find the Lloyd–Max quantizer explicitly. But, a numerical approximation of it can be found using the LBG algorithm covered in the next section.

3.3. VQ-Based Codes

Fix a positive integer m. Let $\mathcal{R}$ denote the set of real numbers, and let $\mathcal{R}^m$ denote m-dimensional Euclidean space, that is, the set of all sequences $(x_1, x_2, \ldots, x_m)$ in which each entry x_i belongs to $\mathcal{R}$. An *m-dimensional vector quantizer* Q is a mapping from $\mathcal{R}^m$ onto a finite subset $\mathcal{C}$ of $\mathcal{R}^m$. The set $\mathcal{C}$ is called the *codebook* of the m-dimensional vector quantizer Q; the elements of $\mathcal{C}$ are called *codevectors*, and if $x \in \mathcal{R}^m$, then $Q(x)$ is called the *codevector for* x. We define a mapping to be a *vector quantizer* if it is an m-dimensional vector quantizer for some m. The scalar quantizers can be regarded as special cases of the vector quantizers (they are the one-dimensional vector quantizers); the codebook of a scalar quantizer is just the set of its quantization levels, and a codevector for a scalar quantizer is just one of its quantization levels.

Let Q be an m-dimensional vector quantizer with codebook $\mathcal{C}$. We call Q a *nearest-neighbor quantizer* if Q quantizes each $x \in \mathcal{R}^m$ into a codevector from codebook $\mathcal{C}$ that is at least as close to x in Euclidean distance as any other codevector from $\mathcal{C}$. (We use squared-error distortion in this section; therefore, only nearest-neighbor quantizers will be of interest to us.)

Fix a memoryless source $[n, A_n, P_n]$ whose alphabet is a subset of the real line, such that n is an integer multiple of m; let f_m be the common probability density function possessed by all random vectors of m consecutive samples generated by this source. Let Q be a fixed m-dimensional nearest-neighbor vector quantizer whose codebook is of size 2^j. We describe how Q induces a lossy code for the source $[n, A_n, P_n]$ that has compression rate $R = j/m$. Referring to Fig. 6, we have to explain how the induced lossy code quantizes source sequences generated by $[n, A_n, P_n]$, and how it losslessly encodes the quantized sequences. Let $(x_1, x_2, \ldots, x_n) \in A_n$ be any source sequence. Partitioning, one obtains the following n/m blocks of length m lying in m-dimensional Euclidean space $\mathcal{R}^m$:

$$(x_1, x_2, \ldots, x_m)$$
$$(x_{m+1}, x_{m+2}, \ldots, x_{2m})$$
$$\cdots$$
$$(x_{n-m+1}, x_{n-m+2}, \ldots, x_n)$$

The induced lossy code quantizes $(x_1, x_2, \ldots, x_n)$ into $(\hat{x}_1, \hat{x}_2, \ldots, \hat{x}_n)$, where

$$(\hat{x}_1, \hat{x}_2, \ldots, \hat{x}_m) = Q(x_1, x_2, \ldots, x_m)$$
$$(\hat{x}_{m+1}, \hat{x}_{m+2}, \ldots, \hat{x}_{2m}) = Q(x_{m+1}, x_{m+2}, \ldots, x_{2m})$$
$$\cdots$$
$$(\hat{x}_{n-m+1}, \hat{x}_{n-m+2}, \ldots, \hat{x}_n) = Q(x_{n-m+1}, x_{n-m+2}, \ldots, x_n)$$

Each codevector in $\mathcal{C}$ can be uniquely represented using a j-bit binary address. The induced lossy code losslessly encodes the quantized sequence $(\hat{x}_1, \hat{x}_2, \ldots, \hat{x}_n)$ into the binary codeword obtained by concatenating together the binary addresses of the codevectors above that were used to form the quantized sequence; this binary codeword is of fixed length $(n/m)j = nR$, and so, dividing by n, the compression rate R code bits per data sample has been achieved. Let $\mathcal{S}_n$ be the lossy compression system that arises when the lossy code just described is used to compress the data sequences generated by the source $[n, A_n, P_n]$. The resulting rate and distortion performance are given by

$$R(\mathcal{S}_n) = R = \frac{j}{m}$$

$$D(\mathcal{S}_n) = m^{-1} \int_{\mathcal{R}^m} \min_{y \in \mathcal{C}} \|x - y\|^2 f_m(x)\, dx \qquad (31)$$

where $\|x - y\|^2$ denotes the square of the Euclidean distance between vectors $x = (x_i)$ and $y = (y_i)$ in $\mathcal{R}^m$:

$$\|x - y\|^2 = \sum_{i=1}^{m} (x_i - y_i)^2$$

If we let Q vary over all nearest-neightbor m-dimensional vector quantizers whose codebooks are of size 2^j, then the lossy codes for the source $[n, A_n, P_n]$ induced by these are the VQ-based codes of rate $R = j/m$. Obviously, of this large number of lossy codes, one would want to choose one of minimal distortion; however, it is typically an intractable problem to find such a minimum distortion code.

Example 8. Suppose that the source $[n, A_n, P_n]$ for which a VQ-based code is to be designed is a Gaussian memoryless source with mean 0 and variance 1. Suppose that the desired compression rate is 1.5 code bits per data sample. Since $1.5 = \frac{3}{2}$, we can use a two-dimensional vector quantizer with codebook of size $2^3 = 8$. The table below gives one possibility for such a vector quantizer. The left column gives the codevectors in the codebook; the right column gives the binary address assigned to each codevector. The codevectors in this codebook can be visualized as eight equally spaced points along the unit circle in the plane, which is of radius 1 and centered at the origin (0, 0).

Codevector	Address
$(1, 0)$	000
$(\cos(\pi/4), \sin(\pi/4))$	001
$(0, 1)$	010
$(\cos(3\pi/4), \sin(3\pi/4))$	011
$(-1, 0)$	100
$(\cos(5\pi/4), \sin(5\pi/4))$	101
$(0, -1)$	110
$(\cos(7\pi/4), \sin(7\pi/4))$	111

The lossy code induced by this 2D vector quantizer, when used to compress data generated by the source $[n, A_n, P_n]$, gives rise to a distortion figure that can be obtained using symmetry considerations. The 2D plane can be partitioned into eight congruent regions over which the integrals of the integrand in (31) are identical. One of these regions is

$$\mathcal{S} = \left\{(x_1, x_2) : x_1 \geq 0, -\tan\frac{\pi}{8}x_1 \leq x_2 \leq \tan\frac{\pi}{8}x_1\right\}$$

The region S is the set of points in the plane $\mathcal{R}^2$ that are closest in Euclidean distance to the codevector (1, 0). Therefore, the distortion is

$$4 \iint_S [(x_1 - 1)^2 + x_2^2] \left(\frac{1}{2\pi}\right) \exp\left(\frac{-(x_1^2 + x_2^2)}{2}\right) dx_1\, dx_2$$

This integral is easily evaluated via conversion to polar coordinates. Doing this, one obtains distortion of 5.55 dB. One can improve the distortion to 5.95 dB by increasing the radius of the circle around which the eight codevectors in the codebook are distributed. Examining the distortion–rate function of the Gaussian memoryless source, we see that a distortion of about 9 dB is best possible at the compression rate of 1.5 code bits per data sample. Hence, we can obtain a >3-dB improvement in distortion by designing a VQ-based code that uses a vector quantizer of dimension >2.

3.3.1. LBG Algorithm.

The LBG algorithm [27] is an iterative algorithm for vector quantizer design. It works for any dimension m, and employs a large set T of "training vectors" from $\mathcal{R}^m$. The training vectors, for example, could represent previously observed data vectors of length m. An initial codebook $\mathcal{C}_0$ contained in $\mathcal{R}^m$ of some desired size is selected (the size of the codebook is a reflection of the desired compression rate). The LBG algorithm then recursively generates new codebooks

$$\mathcal{C}_1, \mathcal{C}_2, \mathcal{C}_3, \ldots$$

as follows. Each codebook $\mathcal{C}_i (i \geq 1)$ is generated from the previous codebook $\mathcal{C}_{i-1}$ in two steps:

Step 1. For each $v \in T$, a closest vector to v in $\mathcal{C}_{i-1}$ is found (with respect to Euclidean distance), and recorded. Let x_v denote the vector that is recorded for $v \in T$.

Step 2. For each vector $y \in \{x_v : v \in T\}$ recorded in step 1, the arithmetic mean of all vectors v in T for which $x_v = y$ is computed. The set of arithmetic means forms the new codebook $\mathcal{C}_i$.

The LBG codebooks $\{\mathcal{C}_i\}$ either eventually coincide, or else eventually keep cycling periodically through a finite set of codebooks; in either case, one would stop iterations of the LBG algorithm at a final codebook $\mathcal{C}_i$ as soon as one of these two scenarios occurs. The final LBG codebook, if used to quantize the training set, will yield smaller distortion on the training set than any of the previously generated codebooks (including the initial codebook). The LBG algorithm has been used extensively since its discovery for VQ-based code design in both theoretical and practical scenarios. The one-dimensional LBG algorithm can be used to find a good approximation to the unique Lloyd–Max quantizer, if the training set is big enough. However, in higher dimensions, the LBG algorithm has some drawbacks: (1) the final LBG codebook may depend on the initial choice of codebook or (2) the final LBG codebook may not yield minimum distortion. Various stochastic relaxation and neural network–based optimization techniques have been devised to overcome these deficiencies [38, Chap. 14;52].

3.3.2. Tree-Structured Vector Quantizers. In tree-structured vector quantization [19, Chap. 12;32], the 2^j codevectors in the VQ codebook are placed as labels on the 2^j leaf vertices of a rooted binary tree of depth j. Each of the $2^j - 1$ internal vertices of the tree is also labeled with some vector. To quantize a vector x, one starts at the root of the tree and then follows a unique root-to-leaf path through the tree by seeing at each intermediate vertex of the path which of its two children has its label closer to x, whereupon the next vertex in the path is that child; x is quantized into the codevector at the terminus of this path. A tree-structured vector quantizer encodes speedily, since the time for it to quantize a vector using a codebook of size 2^j is proportional to j instead of 2^j. The minimum distortion vector quantizer for a probabilistic source or a training set is typically not implementable as a tree-structured vector quantizer, but there are some scenarios in which a tree-structured vector quantizer yields close to minimum distortion.

3.3.3. Lattice Vector Quantizers. A lattice quantizer is a vector quantizer whose codebook is formed from points in some Euclidean space lattice. Lattice quantizers are desirable because (1) there are fast algorithms for implementing them [9], and (2) for high compression rates, lattice quantizers yield nearly minimum distortion among all vector quantizers for the memoryless source in which the data samples have a uniform distribution. A monograph [10, Table 2.3] tabulates the best-known m-dimensional lattices in terms of distortion performance, for various values of m between $m = 1$ and $m = 24$. For example, for dimension $m = 2$, one should use a hexagonal lattice consisting of the centers of hexagons that tile the plane.

3.4. Trellis-Based Codes

Suppose that one has a finite-directed graph G satisfying the following properties:

- There are a fixed number of outgoing edges from each vertex of G, and this fixed number is a power of two. (Let this fixed number be 2^j.)
- Each edge of G has a label consisting of a sequence of fixed length from a given data alphabet. A. (Let this fixed length be m.)
- The graph G is connected (i.e., given any two vertices of G, there is a finite path connecting them).

Let v^* be any fixed vertex of G (since G is connected, it will not matter what v^* is). Let n be a fixed positive integer that is an integer multiple of m. Let $\mathcal{P}$ be the set of all paths in G that begin at v^* and consist of n/m edges. Then

- Each path in $\mathcal{P}$ gives rise to a sequence of length n over the alphabet A if one writes down the labels on its edges in the order in which these edges are visited.
- Let $R = j/m$. There are 2^{nR} paths in $\mathcal{P}$, and it consequently takes nR bits to uniquely identify any one of these paths.

A *trellis* is a pictorial representation of all of the paths in $\mathcal{P}$. The *trellis-based lossy code* for quantizing/encoding/decoding all the data sequences in A^n works in the following way:

Quantization Step. Given the data sequence $(x_1, x_2, \ldots, x_n)$ from A^n, the encoder finds a "minimal path" in $\mathcal{P}$, which is a path in $\mathcal{P}$ giving rise to a sequence $(y_1, y_2, \ldots, y_n)$ in A^n for which

$$\sum_{i=1}^{n} d(x_i, y_i)$$

is a minimum, where d is the distortion measure to be used. [The sequence $(y_1, y_2, \ldots, y_n)$ is the output of the quantizer in Fig. 6a.]

Encoding Step. Let $R = j/m$. Having found the minimal path in $\mathcal{P}$ for the data sequence $(x_1, x_2, \ldots, x_n)$, the encoder transmits a binary codeword of length nR in order to tell the decoder which path in $\mathcal{P}$ was the minimal path.

Decoding Step. From the received binary codeword of length nR, the decoder follows the minimal path and writes down the sequence $(y_1, y_2, \ldots, y_n)$ to which that path gives rise. That sequence is the reconstruction sequence $(\hat{x}_1, \hat{x}_2, \ldots, \hat{x}_n)$ for the data sequence $(x_1, x_2, \ldots, x_n)$.

Computation of the minimal path in the quantization step is a dynamic programming problem that can be efficiently solved using the *Viterbi algorithm*.

The trellis-based lossy code just described will work for any source $[n, A_n, P_n]$ in which $A_n \subset A^n$. The resulting compression rate is $R = j/m$; the resulting distortion depends on the source model—there is no simple formula for computing it. Trellis-based lossy coding is quite unlike SQ-based or VQ-based lossy coding because of its feedback nature. The best way to understand it is through an extended example, which follows.

Example 9. The graph G is taken as follows:

There are two vertices v_0, v_1, four edges with two outgoing edges per vertex ($j = 1$), and the edge labels are of length 2 ($m = 2$). The compression rate for the trellis-based code based on G is $R = j/m = 0.5$ code bits per data sample. Suppose that we wish to encode the data sequence (0, 1, 0, 0, −1, 0) of length $n = 6$, using squared-error distortion. Taking the vertex v_0 as our distinguished vertex v^*, the set of $2^{nR} = 8$ paths $\mathcal{P}$ in

G of length $n/m = 3$ that start at v_0 are then given the trellis representation:

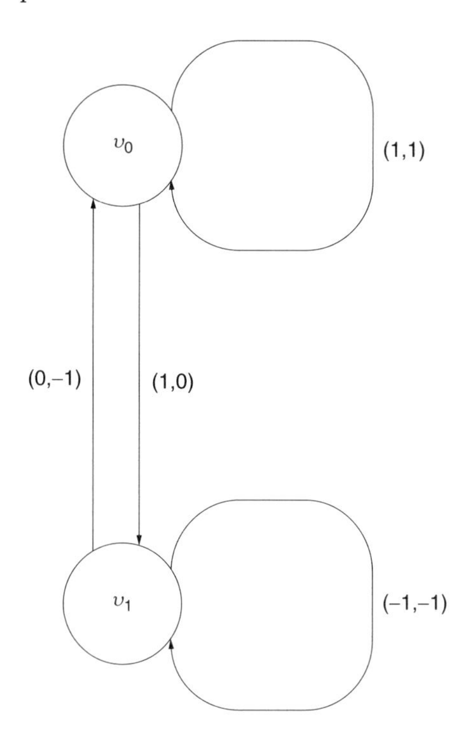

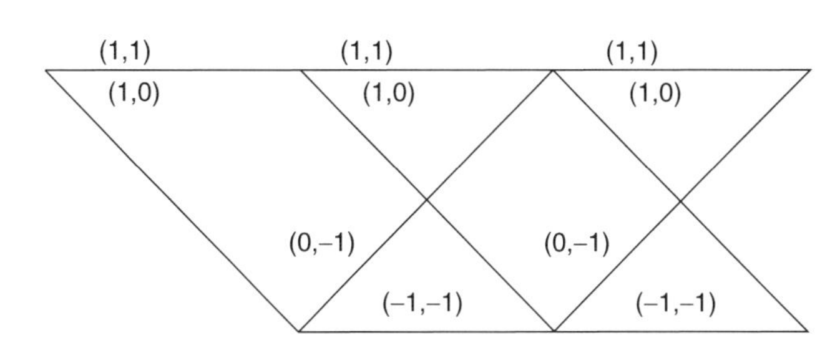

The trellis comes in three stages from left to right, which designate stage 1, stage 2, and stage 3, respectively. In stage 1, one replaces each label with the square of the Euclidean distance between that label and the pair (0,1) consisting of the first two of the six given data samples. In stage 2, one replaces each label with the square of the Euclidean distance between that label and the pair (0,0) consisting of the next two data samples. In stage 3, one replaces each label with the square of the Euclidean distance between that label and the pair $(-1, 0)$ consisting of the final two data samples. This gives us the following trellis whose edges are weighted by these squared Euclidean distances:

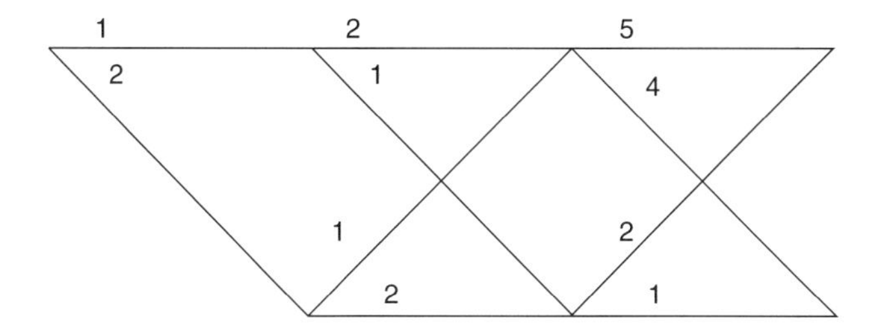

The minimal path is the path whose sum of weights is the smallest; this is the path marked in bold in the following figure:

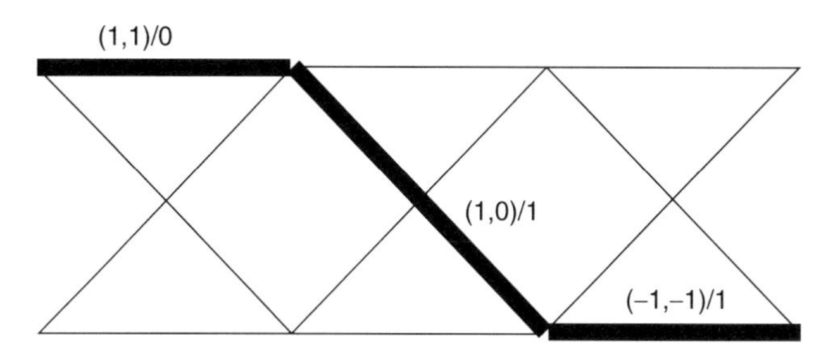

We have added an additional label to each edge of the minimal path to denote the bit sent to the decoder (bit $= 0$ means upper branch chosen; bit $= 1$ means lower branch chosen).

Example 10. Consider the memoryless source $[n, A_n, P_n]$ with alphabet $\{a, b, c, d\}$, in which each random data sample is equidistributed over this alphabet. Suppose that $[n, A_n, P_n]$ is encoded using the trellis-based code, one stage of which is the following:

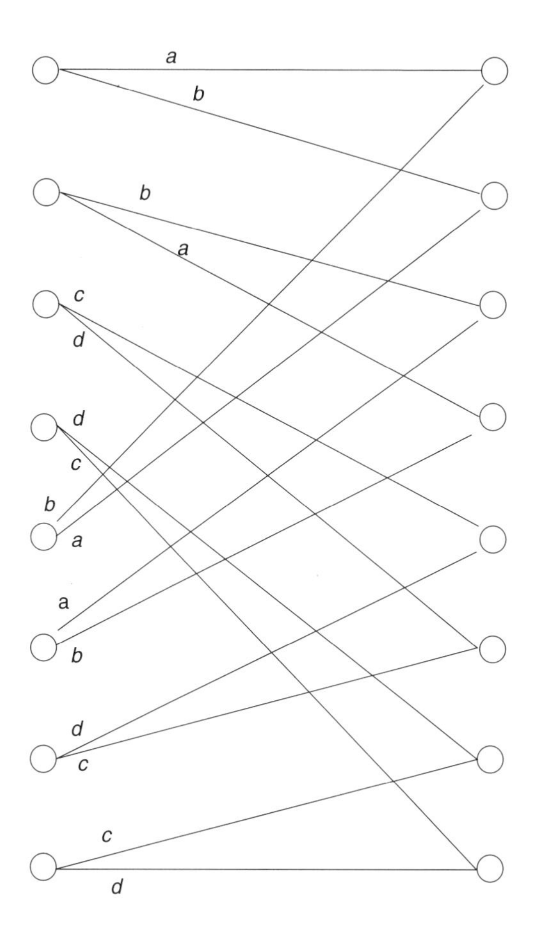

This code has a compression rate of 1 code bit per data sample. Let Hamming distortion be used; for large n, simulations show that the Hamming distortion is slightly less than 0.25. Hence, for a large number of data samples, this simple code can reconstruct over 75% of the data samples at the decoder, on average, while transmitting only half of the bits that would be necessary for perfect reconstruction (which would require a compression rate of 2 bits per sample).

3.4.1. Marcellin–Fischer Codes. Marcellin and Fischer [29] developed an important class of trellis-based codes. They use DeBrujn graphs as the basis for their codes—labeling of the edges is done using heuristic symmetry rules motivated by Ungerboeck's work on trellis-coded modulation [48]. Marcellin and Fischer report quite good distortion performance when their trellis-based codes are used to encode Gaussian and uniform memoryless sources.

3.5. Transform Codes

In certain applications, the data samples to be compressed are not one-dimensional quantities; that is, for some $m > 1$, each data sample is a point in m-dimensional Euclidean space.

Example 11. One may wish to compress a large square image by compressing the sequence of 8×8 blocks that are obtained from partitioning the image. Each data "sample" in this case could be thought of as a point in m-dimensional Euclidean space with $m = 64$. (The JPEG image compression algorithm takes this point of view.)

For simplicity, let us assume that the dimension of the data samples is $m = 2$. Let us write the data samples as random pairs

$$\begin{bmatrix} X_i^{(1)} \\ X_i^{(2)} \end{bmatrix}, \quad i = 1, 2, \ldots, n \tag{32}$$

Suppose that one is committed to lossy codes that employ only scalar quantization. The sequence

$$X_1^{(1)}, X_2^{(1)}, \ldots, X_n^{(1)}$$

might have one type of model that would dictate that some scalar quantizer Q_1 be used to quantize these samples. On the other hand, the sequence

$$X_1^{(2)}, X_2^{(2)}, \ldots, X_n^{(2)}$$

might have another type of model that would dictate the use of a different scalar quantizer Q_2. Let us assume that Q_1 is a 2^{R_1}-bit quantizer and that Q_2 is a 2^{R_2}-bit quantizer. The sequence resulting from quantization of (3.32) would then be

$$\begin{bmatrix} \hat{X}_i^{(1)} \\ \hat{X}_i^{(2)} \end{bmatrix}, \quad i = 1, 2, \ldots, n \tag{33}$$

where $\hat{X}_i^{(1)} = Q_1(X_i^{(1)})$ and $\hat{X}_i^{(2)} = Q_2(X_i^{(2)})$. The sequence (33) is transmitted by the encoder to the decoder using $nR_1 + nR_2$ bits, and is reconstructed by the decoder as the decoder's estimate of the original sequence (32). Let us use squared-error distortion. Then, the compression rate R and distortion D for this lossy code are

$$R = R_1 + R_2$$

$$D = n^{-1} \sum_{i=1}^{n} E[(X_i^{(1)} - \hat{X}_i^{(1)})^2] + n^{-1} \sum_{i=1}^{n} E[(X_i^{(2)} - \hat{X}_i^{(2)})^2]$$

Suppose that the compression rate is to be kept fixed at R. Then, to minimize D, one would optimize the lossy code just described by choosing integers R_1 and R_2 so that $R = R_1 + R_2$, and so that

$$\min_{Q_1} \sum_{i=1}^{n} E[(X_i^{(1)} - Q_1(X_i^{(1)}))^2] + \min_{Q_2} \sum_{i=2}^{n} E[(X_i^{(2)} - Q_2(X_i^{(2)}))^2]$$

is a minimum, where in the first minimization Q_1 ranges over all 2^{R_1}-bit quantizers, and in the second minimization Q_2 ranges over all 2^{R_2}-bit quantizers. Finding the best way to split up R into a sum $R = R_1 + R_2$ is called the *bit allocation problem* in lossy source coding.

We may be able to obtain a smaller distortion by *transforming* the original data pairs from $\mathcal{R}^2$ into another sequence of pairs from $\mathcal{R}^2$, and then doing the quantization and encoding of the transformed pairs in a manner similar to what we did above for the original pairs. This is the philosophy behind *transform coding*. In the following, we make this idea more precise.

Let m be a fixed positive integer. The data sequence to be compressed is

$$\mathbf{X}_1, \mathbf{X}_2, \ldots, \mathbf{X}_n \tag{34}$$

where each $\mathbf{X}_i$ is an $\mathcal{R}^m$-valued random column vector of the form

$$\mathbf{X}_i = \begin{bmatrix} X_i^{(1)} \\ X_i^{(2)} \\ \vdots \\ X_i^{(m)} \end{bmatrix}$$

We write the data sequence in more compact form as the $m \times n$ matrix $M(X)$ whose columns are the $\mathbf{X}_i$ values:

$$M(X) = [\mathbf{X}_1\ \mathbf{X}_2\ \cdots\ \mathbf{X}_n]$$

Let A be an $m \times m$ invertible matrix of real numbers. The matrix A transforms the matrix $M(X)$ into an $m \times n$ matrix $M(Y)$ as follows:

$$M(Y) = AM(X)$$

Equivalently, A can be used to transform each column of $M(X)$ as follows:

$$\mathbf{Y}_i = A\mathbf{X}_i, \quad i = 1, 2, \ldots, n$$

Then, $M(Y)$ is the matrix whose columns are the $\mathbf{Y}_i$ values:

$$M(Y) = [\mathbf{Y}_1\ \mathbf{Y}_2\ \cdots\ \mathbf{Y}_n]$$

Let $\mathbf{Y}^{(j)}$ denote the jth row of $M(Y)(j = 1, 2, \ldots, m)$. Then, the matrix $M(Y)$ can be partitioned in two different ways:

$$M(Y) = [\mathbf{Y}_1\ \mathbf{Y}_2\ \cdots\ \mathbf{Y}_n] = \begin{bmatrix} \mathbf{Y}^{(1)} \\ \mathbf{Y}^{(2)} \\ \vdots \\ \mathbf{Y}^{(m)} \end{bmatrix}$$

The row vectors $\mathbf{Y}^{(j)}(j = 1, 2, \ldots, m)$ are the *coefficient streams* generated by the transform code; these streams are separately quantized and encoded for transmission to the decoder, thereby completing the "front end" of the transform code, which is called the *analysis stage* of the transform code and is depicted in Fig. 7a.

Note in Fig. 7a that the rates at which the separate encoders operate are $R_1, R_2, \ldots, R_m$, respectively. The separate quantizers in Fig. 7a are most typically taken to be scalar quantizers (in which case the R_i terms would be integers), but vector quantizers could be used instead (in which case all R_i would be rational numbers). If we fix the overall compression rate to be the target value R, then the bit allocation must be such that $R = R_1 + R_2 + \cdots + R_m$. The m separate bit streams generated in the analysis stage are multiplexed into one rate R bit stream for transmission to the decoder, who then obtains the separate bit streams by demultiplexing—this multiplexing/demultiplexing part of the transform code, which presents no problems, has been omitted in Fig. 7a.

The decoder must decode the separate bit streams and then combine the decoded streams by means of an inverse transform to obtain the reconstructions of the original data samples; this "back end" of the system is called the *synthesis stage* of the transform code, and is depicted in Fig. 7b.

In Fig. 7b, $\hat{\mathbf{Y}}^{(j)}$ is the row vector of length n obtained by quantizing the row vector $\mathbf{Y}^{(j)}(j = 1, 2, \ldots, m)$. The matrix $M(\hat{Y})$ is the $m \times n$ matrix whose rows are the $\hat{\mathbf{Y}}^{(j)}$ terms. The matrix $M(\hat{X})$ is also $m \times n$, and is formed by

$$M(\hat{X}) = A^{-1}M(\hat{Y})$$

Let us write the columns of $M(\hat{X})$ as

$$\hat{\mathbf{X}}_1, \hat{\mathbf{X}}_2, \ldots, \hat{\mathbf{X}}_n.$$

These are the reconstructions of the original data sequence (34).

The distortion (per sample) D resulting from using the transform code to encode the source $[n, A_n, P_n]$ is

$$D = n^{-1} \sum_{i=1}^{n} E[\|\mathbf{X}_i - \hat{\mathbf{X}}_i\|^2] \tag{35}$$

With the compression rate fixed at R, to optimize the transform code, one must select $R_1, R_2, \ldots, R_m$ and quantizers at these rates so that $R = R_1 + R_2 + \cdots + R_m$

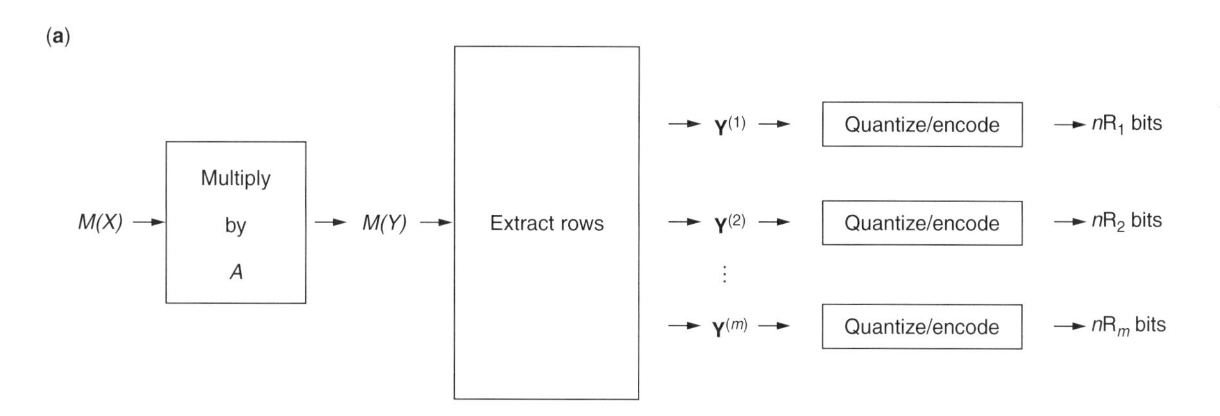

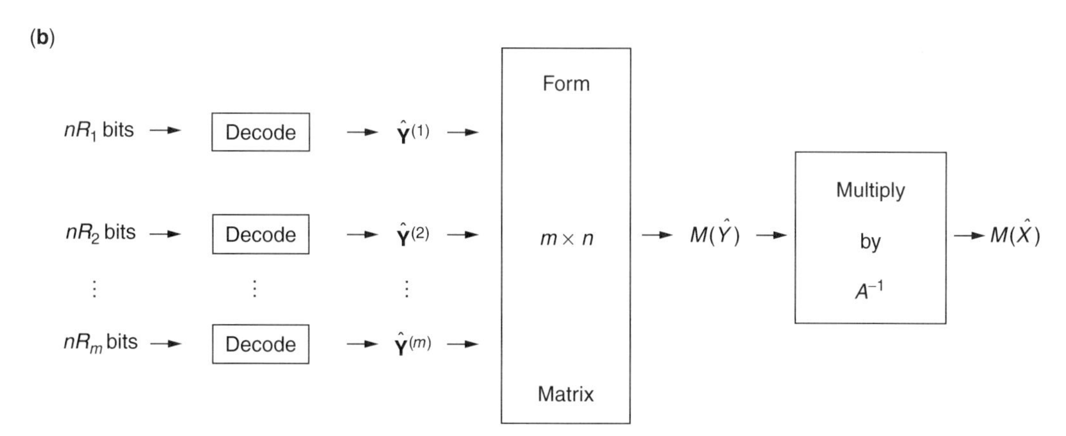

Figure 7. (**a**) Analysis stage and (**b**) synthesis stage of transform code.

and D is minimized. One can also attempt to optimize the choice of the transformation matrix A as well, although, more typically, the matrix A is fixed in advance. If A is an orthogonal matrix (i.e., the transformation is unitary), then the right side of (35) is equal to

$$n^{-1}\sum_{i=1}^{n} E[\|\mathbf{Y}_i - \hat{\mathbf{Y}}_i\|^2]$$

which makes the problem of optimally designing the transform code much easier. However, A need not be orthogonal.

Various transforms have been used in transform coding. Some commonly used transforms are the discrete cosine transform, the discrete sine transform, and the Karhunen–Loeve transform — a good account of these transforms may be found in Ref. 41, Chap. 12. Special mention should be made of the wavelet transform, which is becoming ubiquitous in state of the art compression methods. Various image compression methods employ the wavelet transform, including the EZW method [44], the SPIHT method [39], and the EBCOT method [46]; the wavelet transform is the backbone of the JPEG 2000 image compression standard [47]. The wavelet transform has also been applied to fingerprint compression, speech, audio, electrocardiogram, and video compression — a good account of these applications may be found in Ref. 45, Chap. 11.

Subband coding may be regarded as a special case of transform coding. It originally arose in the middle 1970s as an effective technique for speech compression [12]; subsequently, its applications have grown to include the compression of recorded music, image compression, and video compression. In subband coding, one filters a data sequence of length n by both a highpass filter and a lowpass filter and then throws every other filtered sample away, yielding a stream of $n/2$ highpass-filtered samples and a stream of $n/2$ lowpass-filtered samples; these two streams are separately quantized and encoded, completing the analysis stage of a subband coding system. The highpass and lowpass filters are not arbitrary; these must be complementary in the sense that if no quantization and encoding takes place, then the synthesis stage must perfectly reconstruct the original sequence of data samples. Further passes of highpass and lowpass filterings can be done after the first pass in order to yield several substreams with frequency content in different subbands; in this way, the subband coding system can become as sophisticated as one desires — one can even obtain the wavelet transform via a certain type of subband decomposition. Subband coding has received extensive coverage in several recent textbooks on multirate and multiresolution signal processing [1,49,50].

We have concentrated on lossy transform codes, but there are lossless transform codes as well. A lossless transform code is similar in concept to a lossy transform code; the main difference is that the transformed data sequence is not quantized in a lossless transform code. Good examples of lossless transform codes are (1) the transform code for text compression based on the *Burrows–Wheeler transform* [7]; and (2) grammar-based codes [25,31], which are based on *grammar transforms*.

4. NOTES ON THE LITERATURE

This article has provided an introduction to basic data compression techniques. Further material may be found in one textbook [41] that provides excellent coverage of both lossless and lossy data compression, and in another textbook [40] that provides excellent coverage of lossless data compression. There have been several excellent survey articles on data compression [5,15,20].

Various data compression standards employ the basic data compression techniques that were covered in this article. Useful material on data compression standards may be found in a book by Gibson et al. [21], which covers the JPEG and JBIG still-image compression standards, the MPEG audio and MPEG video compression standards, and multimedia conferencing standards.

4.1. Speech Compression

Speech is notoriously hard to compress — consequently, a specialized body of techniques have had to be developed for speech compression. Since these specialized techniques are not applicable to the compression of other types of data, and since this article has focused rather on general compression techniques, speech compression has been omitted; coverage of speech compression may be found in the textbook by Deller et al. [14].

4.2. Fractal Image Compression

Fractal image compression has received much attention since 1987 [2,3,16,23]. A fractal image code compresses an image by encoding parameters for finitely many contractive mappings, which, when iterated, yield an approximation of the original image. The limitations of the fractal image compression method arise from the fact that it is not clear to what extent natural images can be approximated in this way.

4.3. Differential Coding

Differential codes (such as *delta modulation* and *differential pulsecode modulation*), although not covered in this article, have received coverage elsewhere in this encyclopedia. Differential codes have become part of the standard body of material covered in a first course in communication systems [26, Chap. 6].

BIOGRAPHY

John C. Kieffer received the B.S. degree in applied mathematics in 1967 from the University of Missouri, Rolla, Missouri, and the M.S. and Ph.D. degrees in mathematics from the University of Illinois Champaign — Urbana in 1968 and 1970, respectively. He joined the University of Missouri Rolla (UMR) in 1970 as an Assistant Professor. At UMR he worked on ergodic theory and information theory. Since 1986, he has been a Professor at the University of Minneapolis Twin Cities Department of Electrical and Computer Engineering, where he has been working on data compression. Dr. Kieffer has 70 MathSciNet publications. He was named a Fellow of the IEEE in 1993. His areas of interest are grammar-based codes, trellis-coded quantization, and the interface between information theory and computer science.

BIBLIOGRAPHY

1. A. Akansu and R. Haddad, *Multiresolution Signal Decomposition: Transforms, Subbands, Wavelets*, Academic Press, San Diego, 1992.
2. M. Barnsley, *Fractals Everywhere*, Academic Press, Boston, 1988.
3. M. Barnsley and L. Hurd, *Fractal Image Compression*, Wellesley, AK Peters, Ltd., MA, 1993.
4. T. Berger, *Rate Distortion Theory: A Mathematical Basis for Data Compression*, Prentice-Hall, Englewood Cliffs, NJ, 1971.
5. T. Berger and J. Gibson, Lossy source coding, *IEEE Trans. Inform. Theory* **44**: 2693–2723 (1998).
6. R. Blahut, Computation of channel capacity and rate-distortion functions, *IEEE Trans. Inform. Theory* **18**: 460–473 (1972).
7. M. Burrows and D. Wheeler, *A block-sorting lossless data compression algorithm*, unpublished manuscript, 1994.
8. J. Cleary and I. Witten, Data compression using adaptive coding and partial string matching, *IEEE Trans. Commun.* **32**: 396–402 (1984).
9. J. Conway and N. Sloane, Fast quantizing and decoding algorithms for lattice quantizers and codes, *IEEE Trans. Inform. Theory* **28**: 227–232 (1982).
10. J. Conway and N. Sloane, *Sphere Packings, Lattices, and Groups*, 2nd ed., Springer-Verlag, New York, 1993.
11. T. Cover, Enumerative source encoding, *IEEE Trans. Inform. Theory* **19**: 73–77 (1973).
12. R. Crochiere, S. Webber and J. Flanagan, Digital coding of speech in subbands, *Bell Syst. Tech. J.* **56**: 1056–1085 (1976).
13. L. Davisson, Universal noiseless coding, *IEEE Trans. Inform. Theory* **19**: 783–795 (1973).
14. J. Deller, J. Proakis and J. Hansen, *Discrete-Time Processing of Speech Signals*, Macmillan, Englewood Cliffs, NJ, 1993.
15. D. Donoho, M. Vetterli, R. DeVore and I. Daubechies, Data compression and harmonic analysis, *IEEE Trans. Inform. Theory* **44**: 2435–2476 (1998).
16. Y. Fisher, ed., *Fractal Image Compression: Theory and Application*, Springer-Verlag, New York, 1995.
17. P. Fleischer, *Sufficient conditions for achieving minimum distortion in a quantizer*, *IEEE Int. Conv. Record*, 1964, Part I, Vol. 12, pp. 104–111.
18. M. Garey, D. Johnson and H. Witsenhausen, The complexity of the generalized Lloyd-Max problem, *IEEE Trans. Inform. Theory* **28**: 255–256 (1982).
19. A. Gersho and R. Gray, *Vector Quantization and Signal Compression*, Kluwer, Boston, 1992.
20. R. Gray and D. Neuhoff, Quantization, *IEEE Trans. Inform. Theory* **44**: 2325–2383 (1998).
21. J. Gibson et al., *Digital Compression for Multimedia: Principles and Standards*, Morgan-Kaufmann, San Francisco, 1998.
22. D. Huffman, A method for the construction of minimum redundancy codes, *Proc. IRE* **40**: 1098–1101 (1952).
23. A. Jacquin, Image coding based on a fractal theory of iterated contractive image transformations, *IEEE Trans. Image Proc.* **1**: 18–30 (1992).
24. N. Jayant and P. Noll, *Digital Coding of Waveforms*, Prentice-Hall, Englewood Cliffs, NJ, 1984.
25. J. Kieffer and E. Yang, Grammar-based codes: A new class of universal lossless source codes, *IEEE Trans. Inform. Theory* **46**: 737–754 (2000).
26. B. Lathi, *Modern Digital and Analog Communications Systems*, 3rd ed., Oxford Univ. Press, New York, 1998.
27. Y. Linde, A. Buzo and R. Gray, An algorithm for vector quantizer design, *IEEE Trans. Commun.* **28**: 84–95 (1980).
28. S. Lloyd, Least squares quantization in PCM, *IEEE Trans. Inform. Theory* **28**: 129–137 (1982).
29. M. Marcellin and T. Fischer, Trellis coded quantization of memoryless and Gauss-Markov sources, *IEEE Trans. Commun.* **38**: 82–93 (1990).
30. J. Max, Quantizing for minimum distortion, *IRE Trans. Inform. Theory* **6**: 7–12 (1960).
31. C. Nevill-Manning and I. Witten, Compression and explanation using hierarchical grammars, *Comput. J.* **40**: 103–116 (1997).
32. A. Nobel and R. Olshen, Termination and continuity of greedy growing for tree-structured vector quantizers, *IEEE Trans. Inform. Theory* **42**: 191–205 (1996).
33. R. Pasco, *Source Coding Algorithms for Fast Data Compression*, Ph.D. thesis, Stanford Univ., 1976.
34. W. Pennebaker, J. Mitchell, G. Langdon, and R. Arps, An overview of the basic principles of the Q-coder adaptive binary arithmetic coder, *IBM J. Res. Dev.* **32**: 717–726 (1988).
35. R. Pilc, The transmission distortion of a source as a function of the encoding block length, *Bell Syst. Tech. J.* **47**: 827–885 (1968).
36. J. Rissanen, Generalized Kraft inequality and arithmetic coding, *IBM J. Res. Dev.* **20**: 198–203 (1976).
37. J. Rissanen and G. Langdon, Arithmetic coding, *IBM J. Res. Dev.* **23**: 149–162 (1979).
38. H. Ritter, T. Martinez and K. Schulten, *Neural Computation and Self-Organizing Maps*, Addison-Wesley, Reading, MA, 1992.
39. A. Said and W. Pearlman, A new fast and efficient coder based on set partitioning in hierarchical trees, *IEEE Trans. Circuits Syst. Video Tech.* **6**: 243–250 (1996).
40. D. Salomon, *Data Compression: The Complete Reference*, Springer-Verlag, New York, 1998.
41. K. Sayood, *Introduction to Data Compression*, 2nd ed., Morgan Kaufmann, San Francisco, 2000.
42. C. Shannon, A mathematical theory of communication, *Bell System Tech. J.* **27**: 379–423, 623–656 (1948).
43. C. Shannon, Coding theorems for a discrete source with a fidelity criterion, *IRE Nat. Conv. Record*, 1959, Part 4, pp. 142–163.
44. J. Shapiro, Embedded image coding using zerotrees of wavelet coefficients, *IEEE Trans. Signal Process.* **41**: 3445–3462 (1993).

45. G. Strang and T. Nguyen, *Wavelets and Filter Banks*, Wellesley-Cambridge Univ. Press, Wellesley, MA, 1996.
46. D. Taubman, High performance scalable image compression with EBCOT, *IEEE Trans. Image Process.* **9**: 1158–1170 (2000).
47. D. Taubman and M. Marcellin, *JPEG 2000: Image Compression Fundamentals, Standards and Practice*, Kluwer, Hingham, MA, 2000.
48. G. Ungerboeck, Channel coding with multilevel/phase signals, *IEEE Trans. Inform. Theory* **28**: 55–67 (1982).
49. P. Vaidyanathan, *Multirate Systems and Filter Banks*, Prentice-Hall, Englewood Cliffs, NJ, 1993.
50. M. Vetterli and J. Kovačević, *Wavelets and Subband Coding*, Prentice-Hall, Englewood Cliffs, NJ, 1995.
51. E. Yang and J. Kieffer, Efficient universal lossless data compression algorithms based on a greedy sequential grammar transform. I. Without context models, *IEEE Trans. Inform. Theory* **46**: 755–777 (2000).
52. K. Zeger, J. Vaisey and A. Gersho, Globally optimal vector quantizer design by stochastic relaxation, *IEEE Trans. Signal Process.* **40**: 310–322 (1992).
53. Z. Zhang, E. Yang and V. Wei, The redundancy of source coding with a fidelity criterion. I. Known statistics, *IEEE Trans. Inform. Theory* **43**: 71–91 (1997).
54. J. Ziv and A. Lempel, A universal algorithm for data compression, *IEEE Trans. Inform. Theory* **23**: 337–343 (1977).
55. J. Ziv and A. Lempel, Compression of individual sequences via variable-rate coding, *IEEE Trans. Inform. Theory* **24**: 530–536 (1978).

DESIGN AND ANALYSIS OF A WDM CLIENT/SERVER NETWORK ARCHITECTURE

Wushao Wen
Biswanath Mukherjee
University of California, Davis
Davis, California

1. INTRODUCTION

The client/server architecture has been one of the driving forces for the development of modern data communication networks. Most of data services, such as distributed database systems, web applications, interactive multimedia services, and so on make use of client-server network architecture. In such an architecture, clients are connected to the server via data networks. The data networks can be local-area networks, such as Ethernet, or can be wide-area networks. The required bandwidth for the client–server network is increasing with the increasing penetration of interactive multimedia and World Wide Web (WWW) applications. In a traditional client–server network, the server bandwidth is limited to several hundred megabits per second (Mbps) at the maximum. Such a bandwidth is sufficient to provide conventional data service. However, with the increasing deployment of multimedia applications and services such as network games, video on demand, virtual reality applications, and similar, the system must provide more bandwidth. For example, a video-on-demand system that provides 1000 videostreams of MPEG2 quality needs about 5–6 Gbps of bandwidth at the server side. Such high-bandwidth requirement cannot be satisfied in conventional client/server systems. The most recent technological advances in optical communication, especially wavelength-division multiplexing (WDM) technology, as well as computer hardware, make the design of such a high-bandwidth server possible. However, what kind of network architecture can we use to construct such high-bandwidth network? How well is the performance of the proposed network architecture? We describe and study a client/server WDM network architecture and its performance. This architecture is suitable for multimedia services, and is unicast-, multicast-, and broadcast-capable.

The remainder of this article is organized as follows. We outline the enabling technologies in Section 2. The WDM networking technology is summarized in Section 3. In Section 4, we describe the system architecture and connection setup protocol. Section 5 describes the system performance and analysis for unicast and multicast services. We discuss the scheduling policy for user requests in Section 6. Finally, Section 7 provides concluding remarks.

2. WDM CLIENT–SERVER NETWORK-ARCHITECTURE-ENABLING TECHNOLOGIES

Optical fiber communication plays a key role to enable high-bandwidth data connection. Now, a single fiber strand has a potential bandwidth of 50 THz corresponding to the low-loss region shown in Fig. 1. To make good use of the potential bandwidth of the fiber, WDM technology is widely used. In WDM, the tremendous bandwidth of a fiber (up to 50 THz) is divided into many nonoverlapping wavelengths, called *WDM channels* [1]. Each WDM channel may operate at whatever possible speed independently, ranging from hundreds of Mbps to tens of Gbps. WDM technology is now deployed in the backbone networks [1,2]. OC-48 WDM backbone networks have already

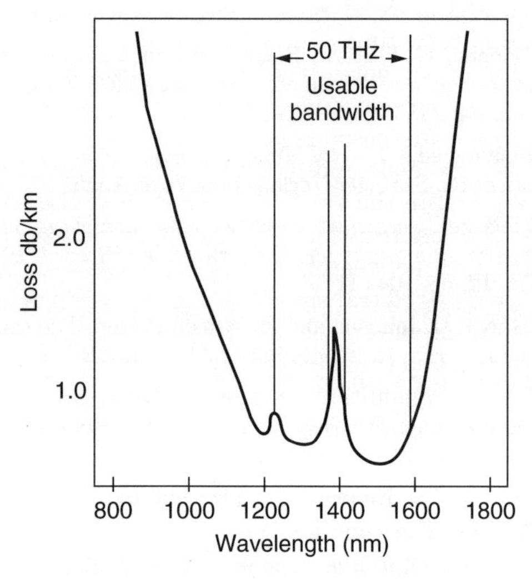

Figure 1. The low-loss region in a single-mode optical fiber.

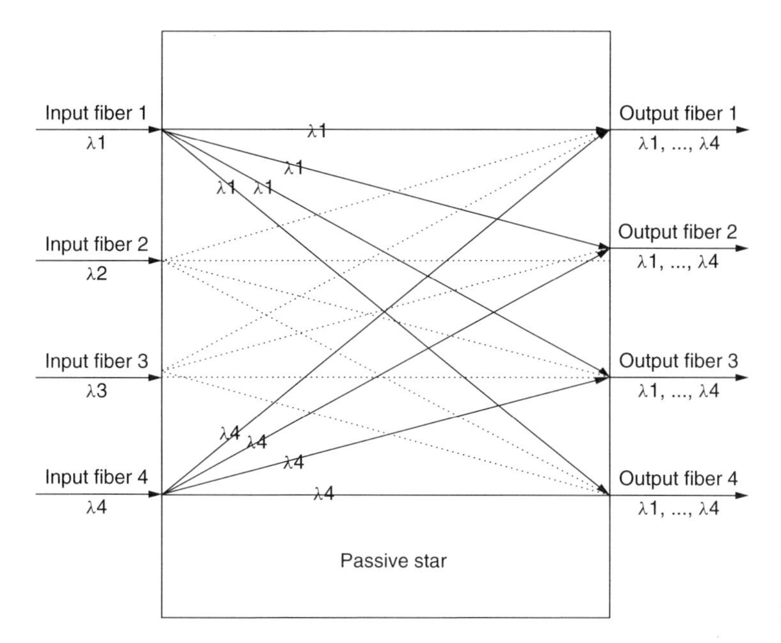

Figure 2. A 4 × 4 passive star.

been widely implemented, and the OC-192 WDM backbone network is being tested and implemented. The enabling technologies of the WDM client/server network architecture include tunable/nontunable wavelength transmitter and receiver, optical multiplexer, optical tap, and high-performance computer in additional to WDM technology. We outline the basic functions for each optical component.

- *Transmitter and Receiver*. An optical transmitter [1] is used to convert an electrical signal into an optical signal and inject it to the optical transmission media, namely, the fiber. An optical receiver actually consists of an optical filter [1] and an opticoelectrical signal transformer. The transmitter and receiver can be classified as tunable and nontunable. A nontunable transmitter/receiver transmits or receives signal only from a fixed frequency range, implying that only fixed wavelengths can be used. For a tunable transmitter/receiver, the passband can be tuned to different frequencies, which makes it possible to transmit or receive data for different wavelengths at different times. In a WDM system, the bandwidth of an optical fiber is divided into multiple channels (or wavelengths). Communication between two nodes is possible only when the transmitter of the source node and receiver of the destination node are tuned to the same channel during the period of information transfer. There are four types of configuration between two communication end nodes. They are fixed transmitter/fixed receiver, fixed transmitter/tunable receiver, tunable transmitter/fixed receiver, and tunable transmitter/tunable receiver. A tunable transmitter/receiver is more expensive than a fixed transmitter/receiver.
- *Optical Multiplexer*. An optical multiplexer combines signals from different wavelengths on its input ports (fiber) onto a common output port (fiber) so that a single outgoing fiber can carry multiple wavelengths at the same time.
- *Passive Star*. A passive star (see Fig. 2) is a "broadcast" device. A signal that is inserted on a given wavelength from an input fiber port will have its power equally divided among (and appear on the same wavelength on) all output ports. As an example, in Fig. 2, a signal on wavelength λ_1 from input fiber 1 and another on wavelength λ_4 from input fiber 4 are broadcast to all output ports. A "collision" will occur when two or more signals from the input fibers are simultaneously launched into the star on the same wavelength. Assuming as many wavelengths as there are fiber ports, an $N \times N$ passive star can route N simultaneous connections through itself.
- *Optical Tap*. An optical tap is similar to a one-input, two-output passive star. However, the output power in one port is much higher than that in the other. The output port with higher power is used to propagate a signal to the next network element. The other output port with lower power is used to connect to a local end system. The input fiber can carry more than one wavelength in an optical tap.

3. WDM NETWORKING ARCHITECTURE

3.1. Point-to-Point WDM Systems

WDM technology is being deployed by several telecommunication companies for point-to-point communications. This deployment is being driven by the increasing demands on communication bandwidth. When the demand exceeds the capacity in existing fibers, WDM is a more cost-effective alternative compared to laying more fibers.

WDM mux/demux in point-to-point links is now available in product form. Among these products, the maximum number of channels is 160 today, but this number is expected to increase.

3.2. Broadcast-and-Select (Local) Optical WDM Network

A local WDM optical network may be constructed by connecting network nodes via two-way fibers to a passive

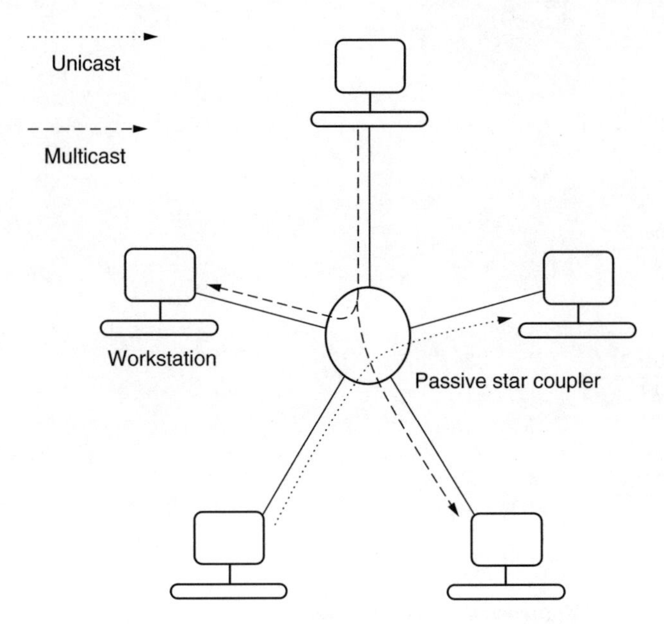

Figure 3. A passive-star-based local optical WDM network.

star, as shown in Fig. 3. A node sends its transmission to the star on one available wavelength, using a laser that produces an optical information stream. The information streams from multiple sources are optically combined by the star and the signal power of each stream is equally split and forwarded to all the nodes on their receive fibers. A node's receiver, using an optical filter, is tuned to only one of the wavelengths; hence it can receive the information stream. Communication between sources and receivers may follow one of two methods: (1) *single-hop* [3] or (2) *multihop* [4]. Also, note that, when a source transmits on a particular wavelength λ_1, more than one receiver can be tuned to wavelength λ_1, and all such receivers may pick up the information stream. Thus, the passive star can support "*multicast*" services.

Detailed, well-established discussions on these network architectures can be found elsewhere, [e.g., 1,3,4].

4. A WDM CLIENT/SERVER NETWORK ARCHITECTURE

4.1. Architecture

In our client/server architecture, the server is equipped with M fixed WDM transmitters that operate on M different wavelengths, and a conventional bidirectional Ethernet network interface that operates on a separate WDM channel is used as the control signaling channel. Every workstation (client) is equipped with a tunable WDM receiver (TR) and an Ethernet network interface. The Ethernet client and server interfaces are connected together and form the signaling channel. All control signals from the clients to the server or vice versa will be broadcast on the control channel by using the IEEE 802.3 protocol [5]. The control channel can be used as a data channel between the clients, or between a client and the server. In the normal data transmission from the server to the clients, the server uses the M data channels to provide data service to the clients. A client can tune its receiver to any of the M server channels and receive data from the server. We show in Fig. 4 a typical client/server WDM system described above. This system can be implemented using a passive star coupler, which connects the server and clients using two-way fibers [1].

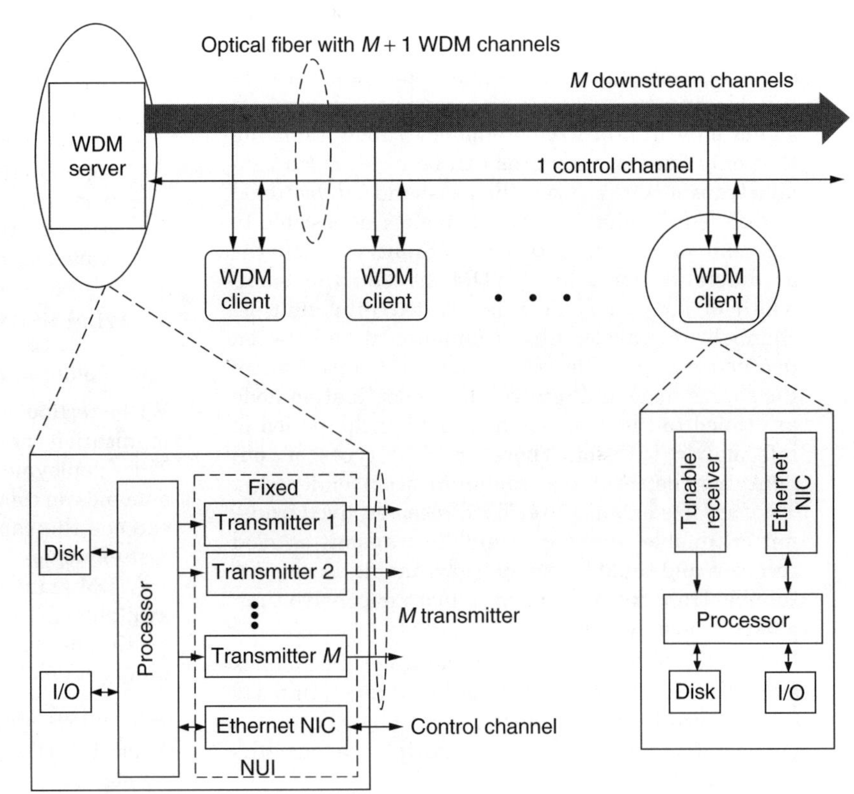

Figure 4. A typical client/server WDM network.

Since a WDM data channel can have a data rate as high as several gigabits per second, a WDM client/server network can be used for those applications that require a high volume of data transfer or those multimedia applications that require high bandwidth. However, a fiber can carry only several hundred wavelengths today, and this may not be enough if we assign a whole wavelength for a single user request. To make good use of the channel capacity, time-division multiplexing (TDM) technology is used to carve a WDM channel into multiple subchannels in time domain. The server can then assign a TDM subchannel to serve the user request. Clients must use some filtering mechanism to get the data from the subchannel. However, the client and the server must communicate with each other to negotiate on which subchannel they plan to use.

This network is also broadcast- and multicast-capable because every client can tune its receiver to any of the M data channels, and more than one client may tune their receivers to the same channel at any given time.

4.2. Operations and Signaling

In a client/server network, the server is able to provide service to clients on request. In such a network, the upstream data traffic volume (from clients to server) is generally much smaller than the downstream data traffic (from server to clients). Therefore, in a WDM client/server network, it is possible to use a single Ethernet channel to fulfill the signaling tasks between the clients and the server. All the clients' upstream traffic and the server's control traffic share the common signaling channel via statistical multiplexing. Figure 5 shows a connection setup procedure of a client's request in the WDM client/server architecture.

Figure 5 indicates that, to setup a connection for a client's request, the network must perform the following operations:

Step 1. A client sends a request message to the server via the common control channel. The message should include information such as the identification of the client, the identification of the requested-file, and the service profile. On receiving the user request, the server will respond with a request–acknowledgment message to the client. This message includes information on the request's status, the server status, and so on.

Step 2. The server processes the client's request. If no data channel or server resource is available, the request is put in the server's request queue and it waits for the server's channel scheduler (SCS) to allocate a data channel or server resource for it. If the request waits for a long time, the server should send keep-alive messages periodically to the client. As soon as the SCS knows that a channel is available, the server's scheduler will select the request (if any) with highest priority and reserve the channel for it. Then, a setup message will be sent to the client via the control channel. The setup message includes information such as the target client's identification information, request ID, the data channel that will be used, the information of the file to be sent out, and the service profile granted for the request.

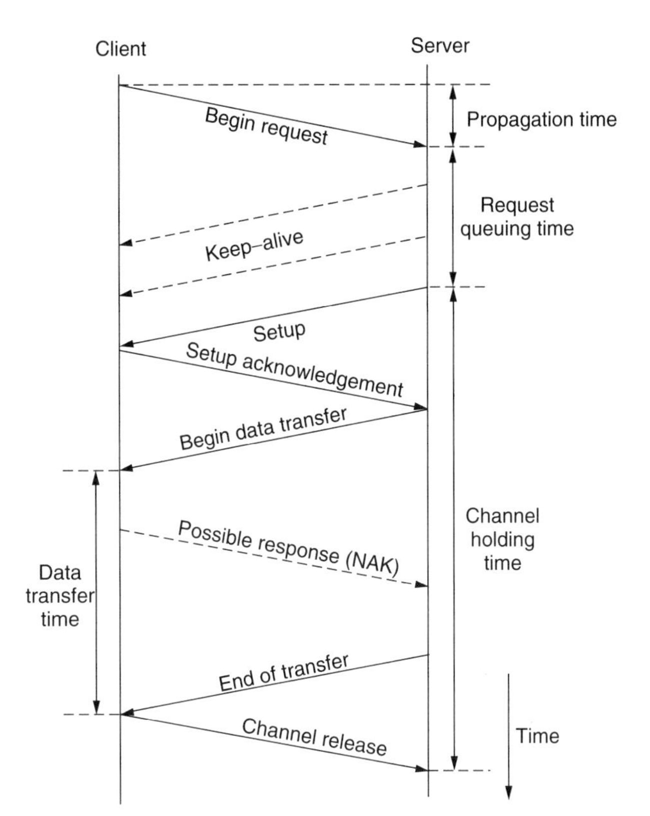

Figure 5. The request setup procedure in a WDM client/server system.

Step 3. On receiving the setup message, the target client's tunable receiver tunes to the assigned channel and sends back a setup acknowledgment message. Now, the client's receiver is waiting for the requested data.

Step 4. When the server receives the setup acknowledgment message, the connection is set up. Then, the server begins to send out the data as soon as the channel is available. If no setup acknowledgment message is received from the client within certain timeout duration, the reserved channel is released and the client's request is discarded.

Step 5. Now the connection is set up and data transfer begins. As we know, the optical channel's bit error rate (BER) is very low, on an order of $10^{-9}–10^{-12}$, so it is safe to assume that the data channels have very low packet error rates. Therefore, the system uses NAK signals to indicate any error packets. No acknowledgment signal will be sent out if a packet is correctly received. However, it is possible and helpful for a client to send out some control message to synchronize the data transmission.

Step 6. After the server finishes sending the client's requested data, it sends an end-of-connection message. The client then responds with a channel-release message to the server. Then, the server tears down the connection and frees the channel.

In this client/server WDM network, the server may receive three kinds of messages: connection request messages, acknowledgment messages, and data transmission control messages. The server should give higher priority to acknowledgment and control messages so that the system can process clients' responses promptly and minimize the channel idle-holding time.[1] It is possible that the requests from the clients come in a bursty manner. Therefore, sometimes, the server may not have enough channels to serve the requests immediately. In that case, the request is buffered. When a channel is available again, there may be more than one client waiting for a free channel. Therefore, the server should have some mechanism to allocate a free channel when it is available. The mechanism used to decide which request should be assigned with the free channel when there are multiple requests waiting for it is called the (SCS) mechanism [6].

4.3. Internetworking Technology

The WDM client/server network architecture is easy to internetwork. In this network architecture, the server has very high bandwidth; therefore, it is not cost-effective to connect the server with the outside network directly. Instead, the clients are connected to outside networks. Figure 6 shows the internetworking scheme where the clients act as proxy servers for outside networks. A client now has two major tasks: handling user requests and caching data. An end user who wants to access some data service, connects to the nearest WDM client. This can be done by a standard client/server service, which implies that no modification is necessary for the end user. On receiving a service request from the end user, the WDM client then analyzes and processes the request. Batching[2] may be used to accommodate requests for the same service from different users. If the user's requested file or program is stored at the WDM client, the request can be served directly without referring to the WDM server; otherwise, the WDM client will contact the server for the data. However, no matter whether the service is provided by the WDM client or the WDM server, it is transparent to the end users.

This internetworking architecture is suitable for multimedia service providers. A service provider can deploy the WDM clients in different cities to provide service to end users. Because the system is broadcast- and multicast-capable, it has very good scalability.

5. SYSTEM PERFORMANCE ANALYSIS

To analyze the system's performance, we consider an example system that uses fixed-length cells (k bytes/cell) to send data from the server to the clients. Every cell occupies exactly one time slot in a WDM channel. The control messages between a client and the server also use

[1] A channel's idle holding time is the time that the channel is reserved for a client but not used for data transfer.

[2] Batching makes use of multicast technology. It batches multiple requests from different users who ask for the same service and use a single stream from the server to serve the users at the same time by using multicasting [6].

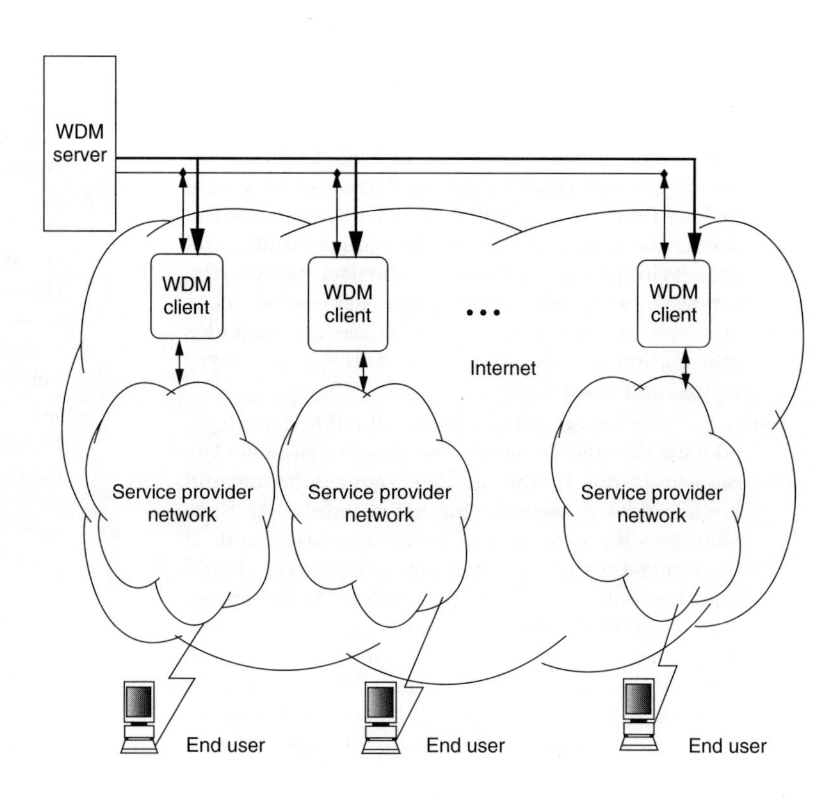

Figure 6. WDM client/server internetworking.

fixed-length cells (k_1 bytes/cell). We use E_1 to represent the bit-error rate (BER) for an optical channel.

5.1. Bottleneck Analysis

5.1.1. Unicast Service. We assume that the system uses negative acknowledgement (NAK) to ask the server to resend the packet that is received with error. It is clear that the expected number of NAK signals per second (denoted as A_1) that will be sent to the control channel is

$$A_1 = \frac{MB}{k}(1 - (1 - E_1)^k) \tag{1}$$

where M is the number of channels and B is the bandwidth for a channel. The BER for an optical channel (E_1) is very small, on the order of 10^{-11}, so Eq. (1) can be simplified as $A_1 = MBE_1$.

Now, suppose the requests from all stations form a Poisson process with average request rate λ_r requests per second. We also assume that the server responds to a user's request by sending back an acknowledgement signal after a random delay. The delay is negative exponential distributed with mean $1/\lambda_r$ minutes. As indicated by the connection setup procedure, the system exchanges five messages via the control channel for every connection in a normal situation.[3] With the additional assumption that the NAK messages are also generated as a Poisson process with average arrival rate λ_n requests/second, the number of cells sent to the Ethernet channel is a Poisson process with total arrival rate λ, ($\lambda = E_1MB + 5\lambda_r$) at the maximum.[4]

Suppose that the mean file size is $\bar{f}$. Then to serve all requests with limited delay, the system should satisfy the condition $\lambda_r * \bar{f} < MB$. So $\lambda_r < MB/\bar{f}$. Therefore, the bandwidth requirement for the control channel should satisfy

$$B_e = \lambda k_1 < \frac{K_1}{p}\left(\frac{E_1MB + 5MB}{\bar{f}}\right) = \frac{MBk_1}{p}\left(\frac{E_1 + 5}{\bar{f}}\right) \tag{2}$$

where p is the maximum offered load allowed on the control channel. In this above equation, $E_1 \ll 5/\bar{f}$ in the actual system. Figure 7 shows the plot of expected control channel bandwidth (B_e) requirement vs. the average file size ($\bar{f}$) in the server with default parameter $p = 0.5, E_1 = 10^{-8}, M = 10, B = 1$ Gbps, and $k_1 = 40$ bytes. From the plot, we notice that, when $\bar{f}$ is small, B_e must be very high. For example, when $\bar{f} = 1$ kbyte, the system requires a bandwidth of 500 Mbps for the control. However, with $f = 1$ Mbyte, the required bandwidth B_e is only about 0.5 Mbps. Clearly, with such a low bandwidth requirement, the control channel is not a bottleneck. When the average file size $\bar{f}$ is between 10 and 100 kbytes, which is a practical file size, the required bandwidth B_e is between 5 and 50 Mbps. From Fig. 7, we can conclude that the WDM client/server network architecture is useful to provide those services that need to transfer a lot of large files. If the system needs to support small files, then the control channel's bandwidth must be increased.

[3] We omit the keep-alive messages in this discussion.

[4] The aggregation of several Poisson processes is also a Poisson process. The average arrival rate of the aggregated process is the sum of the arrival rates of all the original processes.

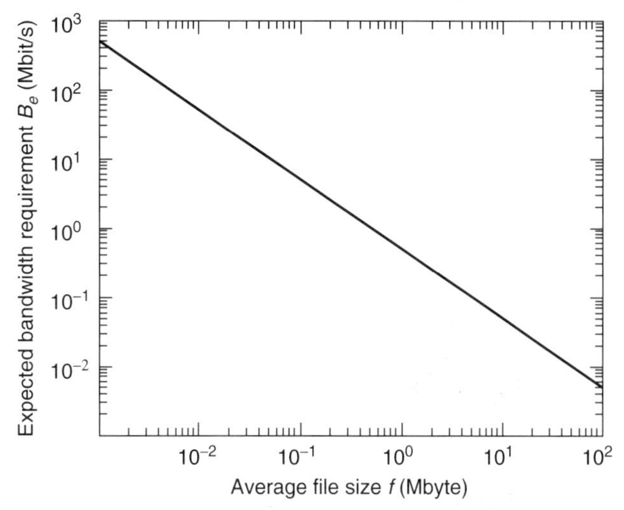

Figure 7. Expected bandwidth requirement of the control channel versus the average file size in the server.

5.1.2. Multicast Service. We have stated that the WDM client-server network is multicast- and broadcast-capable. If the system provides on-demand multicast services, then the server can group together user requests that ask for the same file coming in a certain period and use a single-server channel to multicast the requested file to all users. In an actual implementation, a WDM channel may be carved into several subchannels, and any multicast group will only use a subchannel instead of a whole WDM channel. A multicast service system scales very well even with a limited number of data channels. For example, if a server has 10-Gbps data bandwidth, it can provide about 1800 MPEG2 or 6700 MPEG1 video channels at the same time. A client/server system providing video-on-demand service, which only provides several hundred different video programs at the same time, can easily support any number of user requests by using batching. However, all the request messages and control messages are sent via the Ethernet channel, and the Ethernet channel capacity is limited. Therefore, it is possible that the Ethernet may be a bottleneck. However, in an actual system, this is unlikely to happen. Let us assume, on the average, that every request needs to exchange m 40-byte messages between a client and the server. If we assume $m = 10$ in the multicast system, a control channel with 100 Mbps can accommodate a request rate as high as $\lambda_r = 31{,}250$ requests/s. Such a high request rate is unlikely to happen in any actual system. Therefore, the control channel should not be a bottleneck.

For broadcast service, there is not a lot of interaction between clients and the server. The server notifies the clients what kind of file or program is sent in each channel. A client then chooses its interested channel to receive the data. Therefore, bottleneck analysis does not apply in this service mode.

5.2. System Performance Illustrative Study

This subsection studies the system performance in terms of delay, throughput, and average queuing length. Let us first define several keywords:

- *A Request's System Time*. A request's system time is measured from the time that a user sends out the request message to the time that the user receives all data. Let us denote the time as S_t. Then, from Fig. 5, we can express S_t as

$$S_t = 2R + t_q + t_d \tag{3}$$

  where R is the round-trip time between the client and the server; t_q is the request's queuing time, defined as the time from the instant a request reaches the server to the time the request starts to be served by the server; and t_d is the data transfer time. For every client's request, the round-trip time and the data transfer time are fixed. The only optimizable factor is the queuing time for a given system. Therefore, in order to reduce the response time, we should have a good scheduling algorithm in the server so that we can minimize the average request delay.
- *A Request's Channel Holding Time*. The channel holding time for a request, denoted as H, can be expressed by

$$H = 2R + t_d = D - t_q \tag{4}$$

  Therefore, for a given request, the channel holding time is almost nonoptimizable, because the round-trip time and data transfer time are all nonoptimizable for a given request in a specific system.
- *System Throughput*. When the system operates in the stable state, it should be able to serve all user requests. So the system throughput should equal the average request rate multiplied by the average file size. The maximum reachable throughput in the system, denoted as $\theta_{\max}$, can be express as

$$\theta_{\max} = MB\frac{\overline{f}/B}{\overline{f}/B + 2\overline{R}} \tag{5}$$

where M is total number of data channels, B is the channel capacity, $\overline{R}$ is the average round-trip time from the clients to the server, and $\overline{f}$ is the average file size. Here, $2\overline{R}$ is the average overhead conceived by the server to setup a connection.

We analyze a system's performance based on the following assumptions:

- *Clients' requests* form a Poisson process. The request rate is λ_r requests/s. So the interarrival times are negative exponentially distributed with mean $1/\lambda_r$ minutes.
- There are M data channels altogether. Each channel's capacity is B bps.
- *The file sizes* that users request from the server are exponentially distributed with average file size $\overline{f}$. We also assume the average file size to be relatively large. Therefore, the data transfer time is much larger than the round-trip time and the client's receiver tuning time, and we can omit the round-trip time and client's receiver-tuning time in our analysis.
- *The scheduler uses first-come, first-served (FCFS)* policy.

Using on the above assumptions, we analyze the system as follows. For FCFS scheduling policy, we can model the system as an $M/M/m$ queuing system. In this system, a data channel is regarded as a server in the model. The service time is equal to the channel holding time. We have assumed that the file sizes that users requested from the server are exponentially distributed with average $\overline{f}$. Therefore, the channel holding time at each channel is negative exponential distributed with mean $\mu = B/\overline{f}$. Let us define $\rho = \lambda_r/M\mu$. Clearly the system should require $0 \le \rho < 1$. If $\rho \ge 1$, then the system's service rate will be smaller than the arrival rate, which means that the waiting queue will be built up to infinity, and the system will become unstable. On the basis of these parameters and the $M/M/m$ model, we have the following results [7]:

- *The average number of busy channels* U_b in the system is given by

$$E[U_b] = M\rho = \frac{\lambda_r}{\mu} \tag{6}$$

- *The throughput* of the system is given by $\lambda_r\overline{f}$. This is because the average number of the served requests and the arrival requests should be in balance.
- *The average queue length* Q_n of the system is given by

$$Q_n = M\rho + \frac{(M\rho)^M}{M!}\frac{\rho}{(1-\rho)^2}\pi_0 \tag{7}$$

  where π_0 is the stationary zero queue-length probability and is given by

$$\pi_0 = \left[1 + \sum_{i=1}^{M-1}\frac{(M\rho)^i}{i!} + \frac{(M\rho)^M}{(M)!}\frac{1}{1-\rho}\right]^{-1} \tag{8}$$

- *The average queuing time* is given by the average system time minus the service time (we omit the round-trip time here). Therefore, we can calculate the average queuing time from the expression

$$E[t_q] = \frac{1}{\mu}\frac{(M\rho)^M}{M!}\frac{\pi_0}{M(1-\rho)^2} \qquad (9)$$

6. SCHEDULING POLICIES

In Section 4, we pointed out that the delay for a client's request can be decreased by reducing the queuing time with a proper scheduler. However, the FCFS scheduling policy does not account for a request's channel holding time in allocating an available channel. Therefore, it is possible that some long-time tasks will occupy the server channels for very long time and the short-time tasks have to wait for very long time to be served. Several scheduling policies, such as priority queues, and feedback priority queues, are available to solve this problem. We discuss these policies below.

Priority Queues (PQ) Scheduling Policy. In a priority-queue scheduling policy, the server maintains several queues with different priorities. When no channel is available, the requests for small files will be assigned a higher priority and placed in the higher-priority queue in FCFS manner, and requests for larger files will be put in lower-priority queue. When a channel is available, it is first allocated to a request in the higher-priority queue. Requests in a lower-priority queue cannot be allocated a channel unless all higher-priority queues are empty. The PQ method can greatly reduce the average queuing time. However, there is a starvation problem in this policy. When the request rate is high, requests that ask for large files may never be served because they remain in the lower-priority queue forever and the channels are always allocated to the requests in the higher-priority queue.

Priority Feedback Queues (PFQ) Scheduling Policy. To avoid the starvation problem in the PQ scheduling policy, we can use the priority feedback queues (PFQ) scheduling policy, which allows a request to move from a lower-priority queue to a higher-priority queue if it waits too long. One possible method is to use a threshold function $T_r = f(i, f, t)$. It is the function of i (the current queue it belongs to), the request's file size f, and the waiting time t. Let $T_h(i)$ be the threshold value for the current queue i. Then, we can set the scheduling policy as follows. If $T_r < T_h(i)$, then the request will remain in the same queue; otherwise, it will move up to the higher-priority queue. A scheduler will always select the request with highest priority and allocate the channel to it.

The PFQ scheduling policy solves the starvation problem, but the implementation is more complex.

7. CONCLUSION

We described a WDM client/server network architecture based on a passive-star-coupler-based broadcast-and-select network. In this architecture, all downstream data traffic uses WDM data channels and all upstream requests, upstream control messages, and downstream control messages use an Ethernet channel (called the *control channel*). This architecture is broadcast and multicast capable. We described a detailed point-to-point connection setup procedure for this architecture. An internetworking scheme was also provided. This system architecture is appropriate for an asymmetric data transmission system in which the downstream traffic volume is much higher than the upstream traffic volume. The system's performance was analyzed in terms of whether the control channel or the data channels are the bottleneck. We also analyzed the request delay, request's channel holding time, and system throughput following the model description. We concluded that the control channel is not a bottleneck in the system. When the user request rate is high, the server channel scheduler is the most important adjustable factor to reduce the user response time. An illustrative analysis of FCFS scheduling policy for the system was also provided. To explore WDM technology further, the literature please refer to [1,10,11].

BIBLIOGRAPHY

1. B. Mukherjee, *Optical Communication Networks*, McGraw-Hill, New York, 1997.
2. *http://www.ipservices.att.com/backbone/*.
3. B. Mukherjee, WDM-based local lightwave networks—Part I: Single-hop systems, *IEEE Network Mag.* **6**: 12–27 (May 1992).
4. B. Mukherjee, WDM-based local lightwave networks—Part II: Multiple-hop systems, *IEEE Network Mag.* **6**: 20–32 (July 1992).
5. Digital Equipment Corporation, INTEL corporation, and XEROX Corporation, *The Ethernet: A local Area Network Data Link Layer and Physical Layer Specification*, Sept. 1980.
6. W. Wen, G. Chan, and B. Mukherjee, Token-tray/weighted queuing-time (TT/WQT): An adaptive batching policy for near video-on-demand, *Proc. IEEE ICC 2001*, Helsinki, Finland, June 2001.
7. C. G. Cassandras, *Discrete Event Systems—Modeling and performance Analysis*, Aksen Associates, 1993.
8. P. E. Green, Optical networking update, *IEEE J. Select. Areas Commun.* **14**: 764–779 (June 1996).
9. W. J. Goralski, *SONET*, McGraw-Hill, New York, 2000.
10. R. Ramaswami and K. Sivarajan, *Optical Networks: A Practical Perspective*, Morgan-Kaufmann, San Fransciso, 1998.
11. T. E. Stern and K. Bala, *Multiplewavelength Optical Networks, a Layered Approach*, Addison-Wesley, Reading, MA, 1999.

DESIGN AND ANALYSIS OF LOW-DENSITY PARITY-CHECK CODES FOR APPLICATIONS TO PERPENDICULAR RECORDING CHANNELS

EROZAN M. KURTAS
ALEXANDER V. KUZNETSOV
Seagate Technology
Pittsburgh, Pennsylvania
BANE VASIC
University of Arizona
Tucson, Arizona

1. INTRODUCTION

Low-density parity-check (LDPC) codes, first introduced by Gallager [19], are error-correcting codes described by sparse parity-check matrices. The recent interest in LDPC codes is motivated by the impressive error performance of the Turbo decoding algorithm demonstrated by Berrou et al. [5]. Like Turbo codes, LDPC codes have been shown to achieve near-optimum performance [35] when decoded by an iterative probabilistic decoding algorithm. Hence LDPC codes fulfill the promise of Shannon's noisy channel coding theorem by performing very close to the theoretical channel capacity limit.

One of the first applications of the Gallager LDPC codes was related to attempts to prove an analog of the Shannon coding theorem [47] for memories constructed from unreliable components. An interesting scheme of memory constructed from unreliable components was proposed by M. G. Taylor, who used an iterative threshold decoder for periodic correction of errors in memory cells that degrade steadily in time [52]. The number of elements in the decoder per one memory cell was a priori limited by a capacity-like constant C as the number of memory cells approached infinity. At the time, the Gallager LDPC codes were the only class of codes with this property, and therefore were naturally used in the Taylor's memory scheme. For sufficiently small probabilities of the component faults, using Gallager LDPC codes of length N, Taylor constructed a reliable memory from unreliable components and showed that the probability of failure after T time units, $P(T,N)$, is upper-bounded by $P(T,N) < A_1TN^{-\alpha}$, where $0 < \alpha < 1$ and A is a constant. A similar bound with a better constant α was obtained [26]. In fact, Ref. 26 report even the tighter bound $P(T,N) < A_2T\exp\{-\gamma N^{\beta}\}$, with a constant $0 < \beta < 1$. Using a slightly different ensemble of LDPC and some other results [60], Kuznetsov [27] proved the existence of LDPC codes that lead to the "pure" exponential bound with a constant $\beta = 1$ and decoding complexity growing linearly with code length. A detailed analysis of the error correction capability and decoding complexity of the Gallager-type LDPC codes was done later by Pinsker and Zyablov [61].

Frey and Kschischang [18] showed that all compound codes including Turbo codes [5], classical serially concatenated codes [3,4], Gallager's low-density parity-check codes [19], and product codes [20] can be decoded by Pearl's belief propagation algorithm [41] also referred to as the *message-passing algorithm*. The iterative decoding algorithms employed in the current research literature are suboptimal, although simulations have demonstrated their performance to be near optimal (e.g., near maximum likelihood). Although suboptimal, these decoders still have very high complexity and are incapable of operating in the >1-Gbps (gigabit-per-second) regime.

The prevalent practice in designing LDPC codes is to use very large randomlike codes. Such an approach completely ignores the quadratic encoding complexity in length N of the code and the associated very large memory requirements. MacKay [35] proposed a construction of irregular LDPC codes resulting in a deficient rank parity-check matrix that enables fast encoding since the dimension of the matrix to be used in order to calculate parity bits turns out to be much smaller than the number of parity check equations. Richardson et al. [44] used the same encoding scheme in the context of their highly optimized irregular codes. Other authors have proposed equally simple codes with similar or even better performance. For example, Ping et al. [42] described concatenated tree codes, consisting of several two-state trellis codes interconnected by interleavers, that exhibit performance almost identical to turbo codes of equal block length, but with an order of magnitude less complexity.

An alternative approach to the random constructions is the algebraic construction of LDPC codes using finite geometries [24,25]. Finite geometry LDPC codes are quasicyclic, and their encoders can be implemented with linear shift registers with feedback connections based on their generator polynomials. The resulting codes have been demonstrated to have excellent performance in AWGN, although their decoder complexities are still somewhat high. Vasic [54–56] uses balanced incomplete block designs (BIBDs) [6,11] to construct regular LDPC codes. The constructions based on BIBD's are purely combinatorial and lend themselves to low complexity implementations.

There have been numerous papers on the application of turbo codes to recording channels [14–16,22] showing that high-rate Turbo codes can improve performance dramatically. More recent work shows that similar gains can be obtained by the use of random LDPC and Turbo product codes [17,28–30]. Because of their high complexity, the LDPC codes based on random constructions cannot be used for high speed applications (>1 Gbps) such as the next generation data storage channels.

In this article we describe how to construct LDPC codes based on combinatorial techniques and compare their performance with random constructions for perpendicular recording channels. In Section 2 we introduce various deterministic construction techniques for LDPC codes. In Section 3 we describe the perpendicular channel under investigation which is followed by the simulation results. Finally, we conclude in Section 4.

2. DESIGN AND ANALYSIS OF LDPC CODES

In this section we introduce several deterministic constructions of low-density parity-check codes. The constructions are based on combinatorial designs. The codes constructed in this fashion are "well-structured" (cyclic and quasicyclic) and, unlike random codes, can lend themselves to low-complexity implementations. Furthermore, the important code parameters, such as minimum distance, code rate and the graph girth, are fully determined by the underlying combinatorial object used for the construction. Several classes of codes with a wide range of code rates and minimum distances are considered First, we introduce a construction using general Steiner system (to be defined later), and then discuss projective and affine geometry codes as special cases of Steiner systems. Our main focus are regular Gallager codes [19] (LDPC codes with constant column and row weight), but as it will become clear later, the combinatorial method can be readily extended to irregular codes as well. The first construction is given by Vasic [54–56] and exploits resolvable Steiner systems to construct a class of high-rate codes. The second construction is by Kou et al. [24,25], and is based on projective and affine geometries.

The concept of combinatorial designs is well known, and their relation to coding theory is profound. The codes as well as their designs are combinatorial objects, and their relation does not come as a surprise. Many classes of codes including extended Golay codes and quadratic residue (QR) codes can be interpreted as designs (see Refs. 8 and 48 and the classical book of MacWilliams and Sloane [37]). The combinatorial designs are nice tools for code design in a similar way as bipartite graphs are helpful in visualizing the message passing decoding algorithm. Consider an (n, k) linear code C with parity-check matrix H [31]. For any vector $x = (x_v)_{1 \leq v \leq n}$ in C and any row of H

$$\sum_{v} h_{c,v} \cdot x_v = 0, 1 \leq c \leq n - k \tag{1}$$

This is called the *parity-check equation*. To visualize the decoding algorithm, the matrix of parity checks is represented as a bipartite graph with two kinds of vertices [38,58,50]. The first subset (B), consists of bits, and the second is a set of parity-check equations (V). An edge between a bit and an equation exists if the bit is involved in the check. For example, the bipartite graph corresponding to

$$H = \begin{bmatrix} 1 & 0 & 0 & 1 & 0 & 1 & 1 \\ 0 & 1 & 0 & 1 & 1 & 1 & 0 \\ 0 & 0 & 1 & 0 & 1 & 1 & 1 \end{bmatrix}$$

is shown in Fig. 1.

2.1. Balanced Incomplete Block Design (BIBD)

Now we introduce some definitions and explain the relations of 2-designs to bipartite graphs. A *balanced incomplete block design* (BIBD) is a pair (V,B), where V is a v-element set and B is a collection of $b \cdot k$-subsets of V, called blocks, such that each element of V is contained in exactly r blocks and any 2-subset of V is contained in exactly λ blocks. A design for which every block contains the same number k of points, and every point is contained in the same number r of blocks is called a *tactical configuration*. Therefore, BIBD can be considered as a special case of a tactical configuration. The notation BIBD(v, k, λ) is used for a BIBD on *v points, block size k*, and index λ. The BIBD with a block size $k = 3$ is called a *Steiner triple system*. A BIBD is called *resolvable* if there exists a nontrivial partition of its set of blocks B into *parallel classes*, each of which in turn partitions the point set V. A resolvable Steiner triple system with index $\lambda = 1$ is called a *Kirkman* system. These combinatorial objects originate from Kirkman's famous schoolgirl problem posted in 1850 [23] (see also Refs. 49 and 43). For example, collection $B = \{B_1, B_2, \ldots, B_7\}$ of blocks $B_1 = \{0, 1, 3\}$, $B_2 = \{1, 2, 4\}$, $B_3 = \{2, 3, 5\}$, $B_4 = \{3, 4, 6\}$, $B_5 = \{0, 4, 5\}$, $B_6 = \{1, 5, 6\}$ and $B_7 = \{0, 2, 6\}$ is a BIBD(7,3,1) system or a Kirkman system with $v = 7$, and $b = 7$.

We define the block-point incidence matrix of a (V, B) as a $b \times v$ matrix $A = (a_{ij})$, in which $a_{ij} = 1$ if the jth element of V occurs in the ith block of B, and $a_{ij} = 0$ otherwise. The point-block incidence matrix is A^T.

The block-point matrix for the BIBD(7,3,1) described above is

$$A = \begin{bmatrix} 1 & 1 & 0 & 1 & 0 & 0 & 0 \\ 0 & 1 & 1 & 0 & 1 & 0 & 0 \\ 0 & 0 & 1 & 1 & 0 & 1 & 0 \\ 0 & 0 & 0 & 1 & 1 & 0 & 1 \\ 1 & 0 & 0 & 0 & 1 & 1 & 0 \\ 0 & 1 & 0 & 0 & 0 & 1 & 1 \\ 1 & 0 & 1 & 0 & 0 & 0 & 1 \end{bmatrix}$$

and the corresponding bipartite graph is given in Fig. 2.

Each block is incident with the same number of points k, and every point is incident with the same number r of blocks. If $b = v$, and hence $r = k$, the BIBD is called *symmetric*. The concepts of a symmetric BIBD(v, k, λ) with $k \geq 3$ and of a finite projective plane are equivalent (see Ref. 34, Chap. 19). If one thinks of points as parity-check equations and of blocks as bits in a linear block code, then the A^T defines a matrix of parity checks H of a Gallager code [19,36]. The row weight of A is k, column weight is r, and the code rate is $R = [b - \text{rank}(H)]/b$.

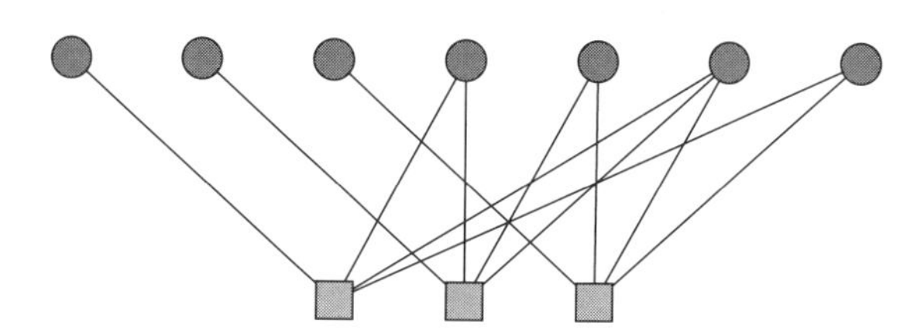

Figure 1. An example of a bipartite graph.

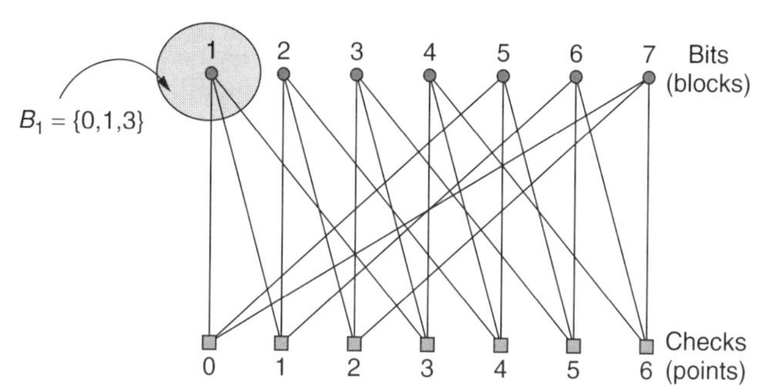

Figure 2. The bipartite graph representation of the Kirkman (7,3,1) system.

It is desirable to have each bit "checked" in as many equations as possible, but because of the iterative nature of the decoding algorithm, the bipartite graph must not contain short cycles. In other words, the graph *girth* (the length of the shortest cycle) must be large [7,51]. These two requirements are contradictory, and the tradeoff is especially difficult when we want to construct a code that is both short and has a high rate. The girth constraint is related to the constraint that every t-element subset of V is contained in as few blocks as possible. If each t-element subset is contained in exactly λ blocks, the underlying design is known as t design. An example of a 5-design is the extended ternary Golay code [8]. However, the codes based on t designs ($t > 2$) have short cycles, and therefore we will restrict ourselves to the 2-designs (i.e., BIBD) or more specifically to the designs with the index $\lambda = 1$. The $\lambda = 1$ constraint means that no more than one block contains the same pair of points or, equivalently, that there are no cycles of length four in a bipartite graph. The lower bound on a rate of a 2-(v, k, λ)-design-based code is given by

$$R \geq \frac{\lambda \dfrac{v(v-1)}{k(k-1)} - v}{\lambda \dfrac{v(v-1)}{k(k-1)}} \tag{2}$$

This bound follows from the basic properties of designs (Ref. 9, Lemma 10.2.1 and Corollary 10.2.2, pp. 344–345). By counting ones in H across the rows and across the columns, we conclude that

$$b \cdot k = v \cdot r \tag{3}$$

If we fix the point u, and find the number of pairs (u, w), $u \neq w$, we arrive to the relation

$$r \cdot (k-1) = \lambda \cdot (v-1) \tag{4}$$

Since the point-block incidence matrix (matrix of parity checks) H has v rows and b columns ($v \leq b$), the rank of A cannot be larger than v, and the code rate, which is $R = [b - \text{rank}(A^T)]/b$ cannot be smaller than $(b - v)/b$. Dividing (3) and (4) yields (2). A more precise characterization of code rate can be obtained by using the rank (and p rank) of the incidence matrix of 2-designs as explained by Hamada [21]. In the case of t-designs, $t > 1$ the number of blocks is $b = \lambda \binom{v}{t} / \binom{k}{t}$ (see Ref. 34, p. 191).

We have shown that the concept of BIBD offers a useful tool for designing codes. Let us now show a construction of Steiner triple systems using difference families of Abelian groups. Let V be an additive Abelian group of order v. Then $t \cdot k$-element subsets of V, $B_i = \{b_{i,1}, \ldots, b_{i,k}\} 1 \leq i \leq t$ form a (v, k, λ) *difference family* (DF) if every nonzero element of V can be represented exactly λ ways as a difference of two elements lying in a same member of a family, that is, it occurs λ times among the differences $b_{i,m} - b_{i,n}$, $1 \leq i \leq t, 1 \leq m, n \leq k$. The sets B_i are called *base blocks*. If V is isomorphic with Z_v, a group of integers modulo v, then a (v, k, λ) DF is called a *cyclic difference family* (CDF). For example, the block $B_1 = \{0, 1, 3\}$ is a base block of a (7,3,1) CDF. To illustrate this, we create an array $\Delta^{(1)} = (\Delta_{i,j})$, of differences $\Delta^{(1)}{}_{i,j} = b_{1,i} - b_{1j}$

$$\Delta^{(1)} = \begin{bmatrix} 0 & 6 & 4 \\ 1 & 0 & 5 \\ 3 & 2 & 0 \end{bmatrix}$$

Given base blocks B_j, $1 \leq j \leq t$, the orbits $B_j + g$ can be calculated as a set $\{b_{j,1} + g, \ldots, b_{j,k} + g\}$, where $g \in V$. A construction of a BIBD is completed by creating orbits for all base blocks. If the number of base blocks in the difference families is t, then the number of blocks in a BIBD is $b = tv$. For example, it can be easily verified (by creating the array Δ) that the blocks $B_1 = \{0, 1, 4\}$ and $B_2 = \{0, 2, 7\}$ are the base block of a (13,3,1) CDF of a group $V = Z_{13}$. The two parallel classes (or orbits) are as given below:

$$B_1 + g = \begin{Bmatrix} 0 & 1 & 2 & 3 & 4 & 5 & 6 & 7 & 8 & 9 & 10 & 11 & 12 \\ 1 & 2 & 3 & 4 & 5 & 6 & 7 & 8 & 9 & 10 & 11 & 12 & 0 \\ 4 & 5 & 6 & 7 & 8 & 9 & 10 & 11 & 12 & 0 & 1 & 2 & 3 \end{Bmatrix}$$

$$B_2 + g = \begin{Bmatrix} 0 & 1 & 2 & 3 & 4 & 5 & 6 & 7 & 8 & 9 & 10 & 11 & 12 \\ 2 & 3 & 4 & 5 & 6 & 7 & 8 & 9 & 10 & 11 & 12 & 0 & 1 \\ 7 & 8 & 9 & 10 & 11 & 12 & 0 & 1 & 2 & 3 & 4 & 5 & 6 \end{Bmatrix}$$

The corresponding matrix of parity checks is

$$H = \begin{bmatrix}
1&0&0&0&0&0&0&0&0&1&0&0&1&1&0&0&0&0&0&1&0&0&0&0&1&0 \\
1&1&0&0&0&0&0&0&0&0&1&0&0&0&1&0&0&0&0&0&1&0&0&0&0&1 \\
0&1&1&0&0&0&0&0&0&0&0&1&0&1&0&1&0&0&0&0&0&1&0&0&0&0 \\
0&0&1&1&0&0&0&0&0&0&0&0&1&0&1&0&1&0&0&0&0&0&1&0&0&0 \\
1&0&0&1&1&0&0&0&0&0&0&0&0&0&0&1&0&1&0&0&0&0&0&1&0&0 \\
0&1&0&0&1&1&0&0&0&0&0&0&0&0&0&0&1&0&1&0&0&0&0&0&1&0 \\
0&0&1&0&0&1&1&0&0&0&0&0&0&0&0&0&0&1&0&1&0&0&0&0&0&1 \\
0&0&0&1&0&0&1&1&0&0&0&0&0&1&0&0&0&0&1&0&1&0&0&0&0&0 \\
0&0&0&0&1&0&0&1&1&0&0&0&0&0&1&0&0&0&0&1&0&1&0&0&0&0 \\
0&0&0&0&0&1&0&0&1&1&0&0&0&0&0&1&0&0&0&0&1&0&1&0&0&0 \\
0&0&0&0&0&0&1&0&0&1&1&0&0&0&0&0&1&0&0&0&0&1&0&1&0&0 \\
0&0&0&0&0&0&0&1&0&0&1&1&0&0&0&0&0&1&0&0&0&0&1&0&1&0 \\
0&0&0&0&0&0&0&0&1&0&0&1&1&0&0&0&0&0&1&0&0&0&0&1&0&1
\end{bmatrix}$$

This matrix contains only columns of weight 3. Of course, since the order of the group in our example is low, the code rate is low as well ($R \geq \frac{1}{2}$). Generally, given a (v, k, λ) CDF, a $t \cdot k$-element subset of Z_v, with base blocks $B_i = \{b_{i,1}, \ldots, b_{i,k}\}$, $1 \leq i \leq t$, the matrix of parity checks can be written in the form

$$H = [H_1 \; H_2 \; \cdots \; H_t] \tag{5}$$

where each submatrix is a circulant matrix of dimension $v \times v$. Formula (5) indicates that the CDF codes have a quasicyclic structure similar to the Townsend and Weldon [53] self-orthogonal quasicyclic codes and Weldon's [57] difference set codes. Each orbit in the design corresponds to one circulant submatrix in a matrix of parity checks of quasicyclic codes.

The codes based on Z_v are particularly interesting because they are conceptually extremely simple and have a structure that can be easily implemented in hardware. Notice also that for a given constraint (v, k, λ) the CDF-based construction maximizes the code rate because for a given v the number of blocks is maximized. The code rate is independent of the underlying group. Other groups different than Z_v may lead to similar or better codes.

2.2. Constructions for Cyclic Difference Families

As we have shown, it is straightforward to construct a BIBD once the CDF is known. However, finding the CDF is a much more complicated problem and solved only for some values of v, k, and λ. Now we describe how to construct a CDF.

2.2.1. The Netto's First Construction. This scheme (see Refs. 10 and 12, p. 28) is applicable if $k = 3$, and v is a power of a prime, such that $v \equiv 1 \pmod 6$. Since v is a power of a prime, then Z_v is a Galois field [GF(v)]. Let Ψ be a multiplicative group of the field, and let ω be its generator [a primitive element in GF(v)] [39]. Write v as $v = 6t + 1, t \geq 1$, and for d, a divisor of $v - 1$, denote by Ψ^d the group of dth powers of ω in GF$(6t + 1)$, and by $\omega^i \Psi^d$ the coset of dth powers of ω^i. Then the set $\{\omega^i \Psi^{2t} \mid 1 \leq i \leq t\}$ defines a $(6t + 1, 3, 1)$ difference family [40,59]. In the literature the base blocks are typically given in the form $\{0, \omega^i(\omega^{2t} - 1), \omega^i(\omega^{4t} - 1)\}$ rather than as $\{\omega^i, \omega^{i+2t}, \omega^{i+4t}\}$.

An alternative combinatorial way of constructing a CDF is proposed by Rosa [33]. Rosa's method also generates a $(6t + 3, 3, 1)$ CDF. In Ref. 53 a list of constructions for rate-$\frac{1}{2}$ codes is given. The details are not given here, but the parameters are the same as of Netto construction. In Ref. 12 [p. 28] Colbourn noticed that this construction is perhaps wrongly associated with Netto.

2.2.2. The Netto's Second Construction. This construction [40] can be used to create a cyclic difference family when the number of points v is a power of a prime, and $v \equiv 7 \pmod{12}$. Again let ω be a generator of the multiplicative group and let $v = 6t + 1, t \geq 1$, and Ψ^d the group of dth powers in GF(v), then the set $\{\omega^{2i} \Psi^{2t} \mid 1 \leq i \leq t\}$ defines base blocks of the so called Netto triple system. The Netto systems are very interesting because of the following property. Netto triple systems on Z_v, v power of a prime and $v \equiv 19 \pmod{24}$ are Pasch-free [32,45,46] (see also Ref. 12, p. 214, Lemma 13.7), and as shown in Ref. 56 achieve the upper bound on minimum distance. For example, consider the base blocks of the Netto triple system difference family on $Z_v(\omega = 3)$ where $B_1 = \{0, 14, 2\}$, $B_2 = \{0, 40, 18\}$, $B_3 = \{0, 16, 33\}$, $B_4 = \{0, 15, 39\}$, $B_5 = \{0, 6, 7\}$, $B_6 = \{0, 11, 20\}$, and $B_7 = \{0, 13, 8\}$. The resulting code is quasi-cyclic, has $d_{\min} = 6$, length $b = 301$ and $R \geq 0.857$.

2.2.3. Burratti Construction for $k = 4$ and $k = 5$. For $k = 4$, Burratti's method gives CDFs with v points, provided that v is a prime and $v = 1 \bmod 12$. The CDF is a set $\{\omega^{6i}B: 1 \leq i \leq t\}$, where base blocks have the form $B = \{0, 1, b, b^2\}$, where ω is a primitive element in GF(v). The numbers $b \in$ GF$(12t + 1)$ for some values of v are given in Ref. 10. Similarly, for $k = 5$, the CDF is given as $\{\omega^{10i}B: 1 \leq i \leq t\}$, where $B = \{0, 1, b, b^2, b^3\}, b \in$ GF$(20t + 1)$. Some Buratti designs are given in Ref. 10.

2.2.4. Finite Euclidean and Finite Projective Geometries. The existence and construction of short designs ($b < 10^5$) is an active area of research in combinatorics. The handbook edited by Colbourn and Dinitz [12] is an excellent reference. The Abel and Greg Table 2.3 in Ref. 12 (pp. 41–42) summarizes the known results in existence of short designs. However, very often the construction of these design is somewhat heuristic or works only for a given block size. In many cases such constructions give a very small set of design with parameters of practical interests. An important subclass of BIBDs are so-called infinite families (Ref. 12, p. 67). The examples of these BIBDs include projective geometries, affine geometries, unitals, Denniston designs and some geometric equivalents of 2-designs (see Refs. 2 and 12, VI.7.12). The known infinite families of BIBD are listed in Table 1 [12,13]. Since q is a power of a prime, they can be referred as to *finite Euclidean* and *finite projective geometries*.

As we explained above, resolvable BIBDs are those BIBDs that possess parallel classes of blocks. If the blocks are viewed as lines, the analogy of BIBD and finite geometry becomes apparent. It is important to notice that not all BIBDs can be derived from finite geometries, but this discussion is beyond the scope of the discussion here. The most interesting projective and affine geometries are those constructed using Gallois fields [GF(q)]. In a m-dimensional projective geometry PG(m, q) a point $\boldsymbol{a}$ is specified by a vector $a = (a_j)_{0 \leq j \leq m}$, where $a_j \in$ GF(q). The vectors $a = (a_j)_{0 \leq j \leq m}$, and $\lambda a = (\lambda a_j)_{0 \leq j \leq m}$, [$\lambda \in$ GF(q), $\lambda \neq 0$] are considered to be representatives of the same point. There are $q^{(m+1)} - 1$ nonzero tuples, and λ can take $q - 1$ nonzero values, so that there are $(q - 1)$ tuples representing the same point. Therefore, the total number of points is $(q^{m+1} - 1)/(q - 1)$. The line through two distinct points $a = (a_j)_{0 \leq j \leq m}$ and $b = (b_j)_{0 \leq j \leq m}$ is incident with the points from the set $\{\mu a + v b\}$, where μ and v are not both zero. There are $q^2 - 1$ choices for μ and v, and each point appears $q - 1$ times in line $\{\mu a + v b\}$, so that the number of points on a line is $(q^2 - 1)/(q - 1) = q + 1$. For example, consider the nonzero elements of GF$(2^2) = \{0, 1, \omega, \omega^2\}$ that represent the points of PG(2,4). The triples $a = (1, 0, 0)$, $\omega a = (\omega, 0, 0)$, and $\omega^2 a = (\omega^2, 0, 0)$, as we ruled, all represent the same point. The line incident with a and $b = (0, 1, 0)$ is incident with all of the following distinct points: a, $b, a + b = \{1, 1, 0\}, a + \omega b = \{1, \omega, 0\}, a + \omega^2 b = (1, \omega^2, 0)$. In the expression $\{\mu a + v b\}$, there are $4^2 - 1 = 15$ combinations of μ and v that are not both zero, but each point has $4 - 1 = 3$ representations, resulting in $\frac{15}{3}$ points on each line.

A *hyperplane* is a subspace of dimension $m - 1$ in PG(m, q); that is, it consists of the points satisfying a homogenous linear equation $\sum_{0 \leq j \leq m} \lambda_j \cdot a_j = 0, \lambda_j \in$ GF(q). A

Table 1. Known Infinite Families of 2-(v, k, 1) Designs

k	v	Parameter	Name
q	q^n	$n \geq 2$, -power of a prime	Affine geometries
$q + 1$	$(q^n - 1)/(q - 1)$	$n \geq 2$, -power of a prime	Projective geometries
$q + 1$	$q^3 + 1$	q-power of a prime	Unitals
2^m	$2^m(2^s + 1) - 2^s$	$2 \leq m < s$	Denniston designs

projective plane of order q is a hyperplane of dimension $m = 2$ in PG(m, q). It is not difficult to see that the projective plane of order q is a BIBD[$(q^3 - 1)/(q - 1), q + 1, 1$]. The Euclidean (or affine) geometry EG(q, m) is obtained by deleting the points of a one hyperplane in PG(m, q). For example, if we delete a hyperplane $\lambda_0 a_0 = 0$, that is, the hyperplane with $a_0 = 0$, then the remaining q^m points can be labeled by the m-tuples $a = (a_j)_{1 \leq j \leq m}$. Euclidean or affine geometry EG(q, m) is a BIBD($q^2, q, 1$) [37]. For more details, see Kou et al. [25].

2.2.5. Lattice Construction of 2-(*v*, *k*,1) Designs. In this section we address the problem of construction of BIBDs of large block sizes. As shown, the Buratti-type CDFs and the projective geometry approach offer a quite limited set of parameters and therefore a small class of codes. In this section we give a novel construction of 2-(v, k,1) designs with arbitrary block size. The designs are lines connecting points of a rectangular integer lattice. The idea is to trade a code rate and number of blocks for the simplicity of construction and flexibility of choosing design parameters. The construction is based on integer lattices as explained below.

Consider a rectangular integer lattice $L = \{(x, y): 0 \leq x \leq k - 1, 0 \leq y \leq m - 1\}$, where m is a prime. Let $l: L \to V$ be an one-to-one mapping of the lattice L to the point set V. An example of such mapping is a simple linear mapping $l(x, y) = m \cdot x + y + 1$. The numbers $l(x, y)$ are referred to as *lattice point labels*. For example, Fig. 3 depicts the rectangular integer lattice with $m = 7$ and $k = 5$.

A *line with slope* s, $0 \leq s \leq m - 1$, *starting at* the point (x, a), contains the points $\{(x, a + sx \bmod m): 0 \leq x \leq k - 1\}$, $0 \leq a \leq m - 1$. For each slope, there are m classes of parallel lines corresponding to different values. In our example, the lines of slope 1 are the points $\{1, 9, 17, 25, 33\}$, $\{2, 10, 18, 26, 34\}$, $\{3, 11, 19, 27, 35\}$, and so on. We assume that the lattice labels are periodic in vertical (y) dimension. The slopes $2, 3, \ldots, m$ can be defined analogously.

A set $\boldsymbol{B}$ of all k-elcment sets of V obtained by taking labels of points along the lines with different slopes s, $0 \leq s \leq m - 1$ is a BIBD. Since m is a prime, for each lattice point (x, y) there is exactly one line with slope s that go through (x, y). For each pair of lattice points, there is exactly one line that passes through both points. Therefore, the set $\boldsymbol{B}$ of lines is a 2-design. The block size is k, number of blocks is $b = m^2$ and each point in the design occurs in exactly m blocks. We can also show [56] that the $(m, k = 3)$ lattice BIBD is Pasch-free. Consider a periodically extended lattice. The proof is based on the observation that it is not possible to draw the quadrilateral (no sides with infinite slope are allowed) in which each point occurs twice. Figure 4 shows one such quadrilateral. Without loss of generality, we can assume that the ting point of two lines is (0,0). The slopes of four lines in Fig. 4 are $s, p, p + a/2$, and $s - a/2$. The points (0,0), (0,a), (2,2s), and $(2, a + 2p)$ are all different, and each occupies two lines. As long as the remaining four points are concerned, they will be on two lines in one of these three cases: (1) $s = p + a/2$ and $s + a/2 = a + p$; (2) $s = s + a/2$ and $p + a/2 = p + a$; and (3) $s = p + a$ and $p + a/2 = s + a/2$ (all operations are modulo m). Case 1 implies $s - p = a/2$, which means that points (2,2s) and $(2, a + 2p)$ are identical, which is a contradiction. If both equalities in cases 2 and 3 were satisfied, then a would have to be 0, which would mean that two leftmost points are identical, again a contradiction.

By generalizing this result to $k > 3$, we can conjecture that the (m, k) lattice BIBDs contain no generalized Pasch configurations. The lattice construction can be extended to nonprime vertical dimension m, but the slopes s and m must be coprime.

Figure 5 shows the growth of the required code length with a lower bound on a minimum distance as a parameter.

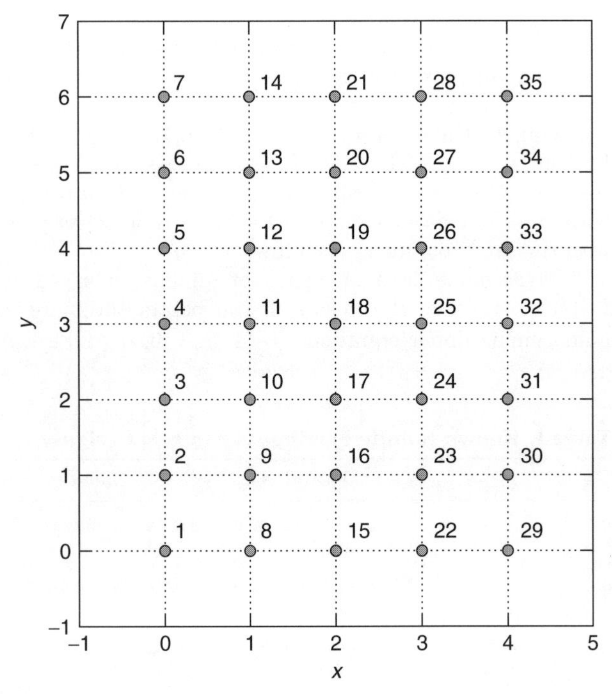

Figure 3. An example of the rectangular grid for $m = 7$ and $k = 5$.

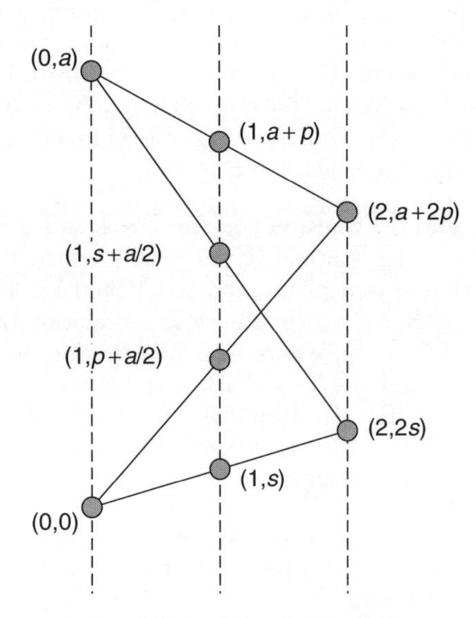

Figure 4. Quadrilateral in a lattice finite geometry.

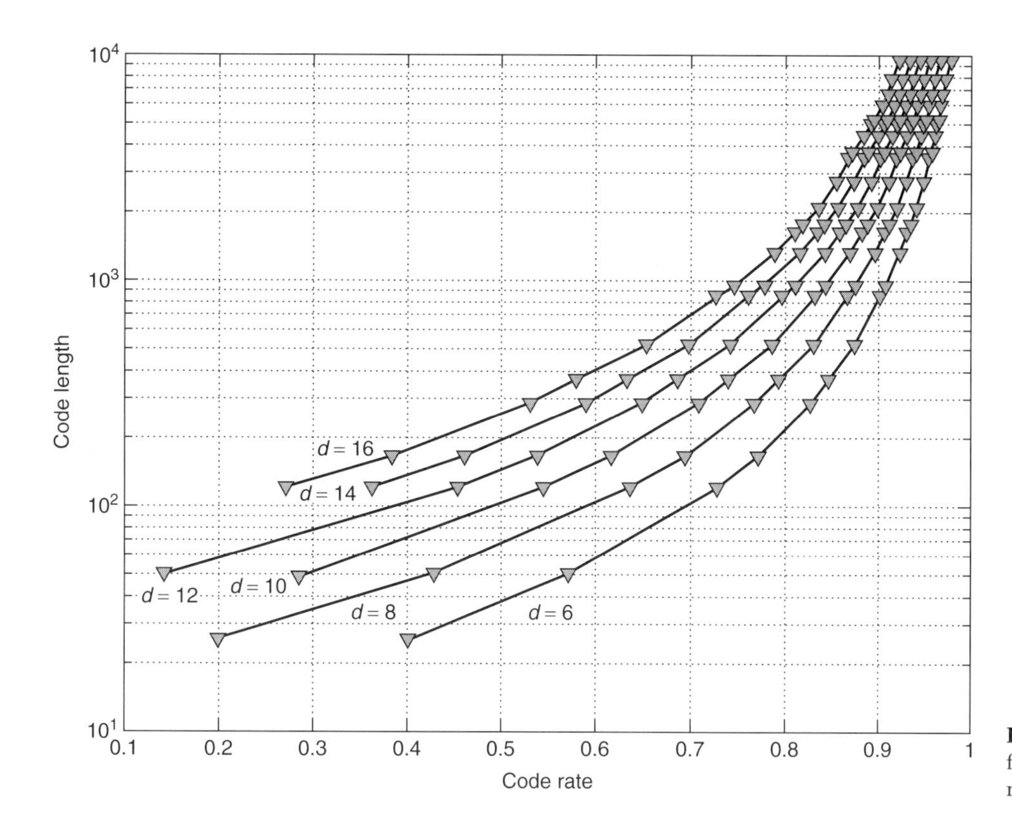

Figure 5. The rate–length curve for lattice designs with the minimum distance as a parameter.

3. APPLICATIONS OF LDPC CODES TO PERPENDICULAR RECORDING SYSTEMS

In this section we consider the performance of LDPC codes based on random and deterministic constructions in perpendicular recording systems. First we give a brief description of the perpendicular recording channel model.

3.1. Perpendicular Recording System

The write head, the magnetic medium, and the read head constitute the basic parts of a magnetic recording system. The binary data are written onto the magnetic medium by applying a positive or negative current to the inductive write head, creating a magnetic field that causes the magnetic particles on the media to align themselves in either of two directions. During playback, the alignment of the particles produces a magnetic field rising that produces a voltage in the read head when it passes over the media. This process can be approximated in the following simplified manner.

If the sequence of symbols $a_k \in C \subseteq \{\pm 1\}$ is written on the medium, then the corresponding write current can be expressed as

$$i(t) = \sum_k (a_k - a_{k-1})u(t - KT_c)$$

where $u(t)$ is a unit pulse of duration T_c seconds. Assuming the read-back process is linear, the induced read voltage, $V(t)$, can be approximated as

$$V(t) = \sum_k a_k g(t - kT_c)$$

where $g(t)$ is the read-back voltage corresponding to an isolated positive going transition of the write current (transition response). For perpendicular recording systems under consideration $g(t)$ is modeled as

$$g(t) = \frac{2}{\sqrt{\pi}} \int_0^{St} e^{-x^2}\, dx = \operatorname{erf}(St)$$

where $S = 2\sqrt{1n2}/D$ and D is the normalized density of the recording system. There are various ways one can define D. In this work we define D as $D = T_{50}/T_c$, where T_{50} represents the width of impulse response at a half of its peak value.

In Figs. 6 and 7, the transition and dibit response $[g(t + T_c/2) - g(t - T_c/2)]$ of a perpendicular recording system are presented for various normalized densities.

In our system, the received noisy read-back signal is filtered by a lowpass filter, sampled at intervals of T_c seconds, and equalized to a partial response (PR) target. Therefore, the signal at the input to the detector at the kth time instant can be expressed as

$$z_k = \sum_j a_j f_{k-j} + w_k$$

where w_k is the colored noise sequence at the output of the equalizer and $\{f_n\}$ are the coefficients of the target channel response. In this work we investigated PR targets of the form $(1 + D)^n$ with special emphasis on $n = 2$, named PR2 case. The input signal-to-noise-ratio (SNR) is defined as $\text{SNR} = 10 \times \log_{10}(V_p^2/\sigma^2)$, where V_p is the peak amplitude of the isolated transition response which is assumed to be unity.

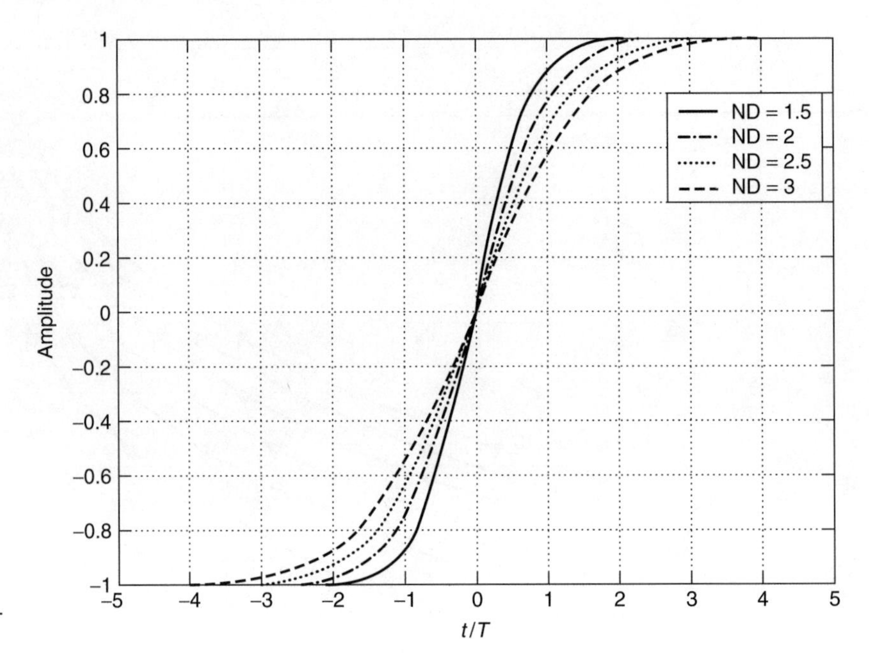

Figure 6. Transition response of a perpendicular recording channel.

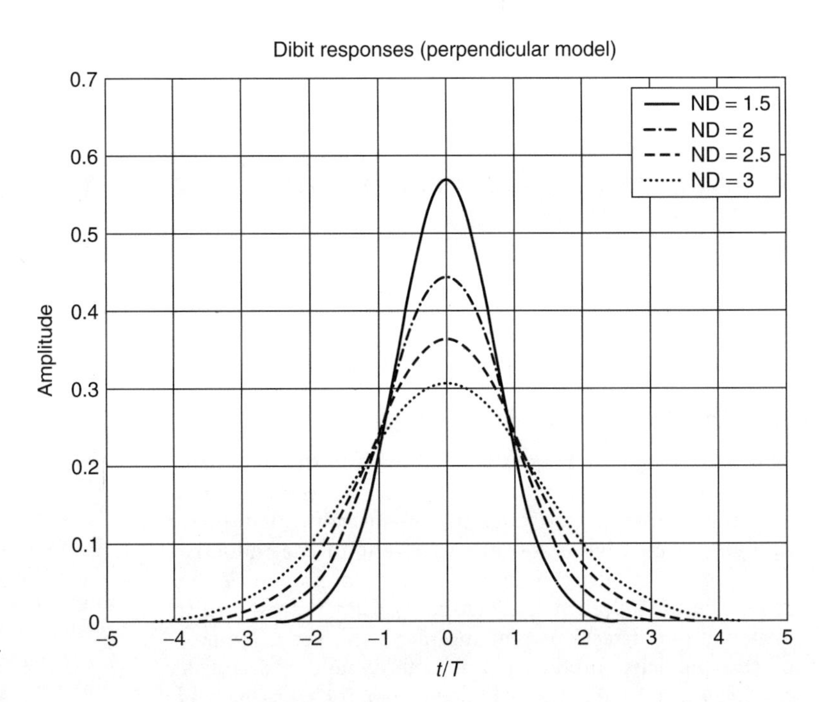

Figure 7. Dibit response of a perpendicular recording channel.

3.2. Simulation Results

We consider an iterative decoding scheme in which the BCJR algorithm [1] is used on the channel trellis, and the message-passing algorithm (MPA) is used for the decoding LDPC codes. The decoding is established by iterating between channel decoder and the outer LDPC decoder. LDPC decoder performs four inner iterations prior to supplying channel decoder with extrinsic information [17].

Throughout our simulations the normalized density for the uncoded system was 2.0 and the coded system channel density was obtained by adjusting with rate, namely, channel density $= D/R$, where R is the rate of the code.

In Fig. 8 we compare the performance of randomly constructed LDPC codes for different column weights, J, and rates. Random LDPC with $J = 3$ outperforms the other two and does not show any error flooring effect.

In Fig. 9 we compare Kirkman codes ($J = 3, v = 121, n = 2420$) and ($J = 3, v = 163, n = 4401$) and the random constructions with $J = 3$ and comparable block lengths and rates. It is clear that the LDPC codes based on Kirkman constructions show an earlier error floor than do their random counterparts. However, the lower error floor and less than 0.5 dB superior performance of random LDPC codes over Kirkman types come at a much larger implementation complexity penalty.

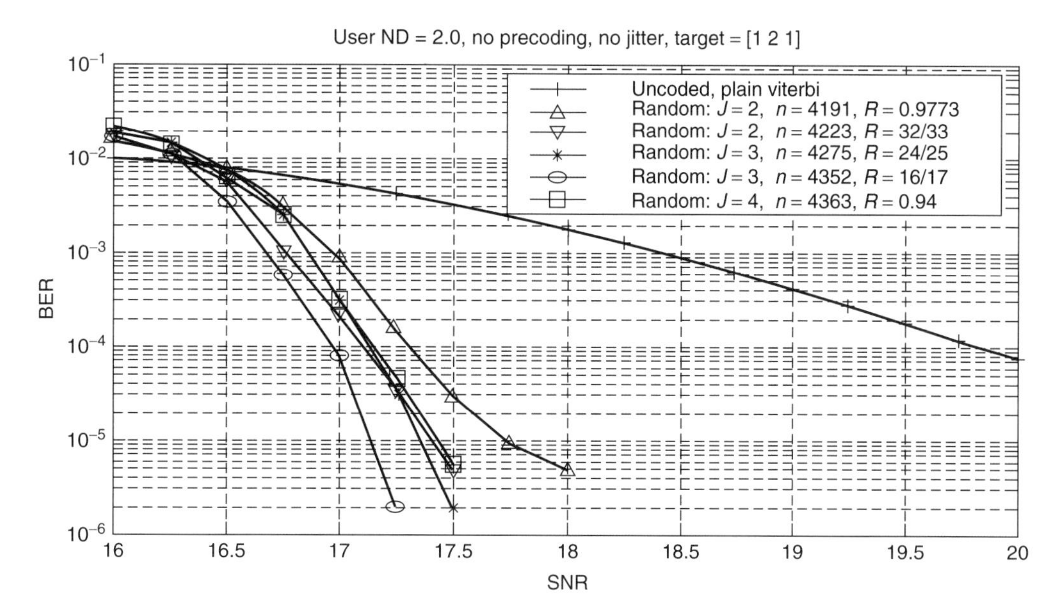

Figure 8. Random LDPC codes with different weights.

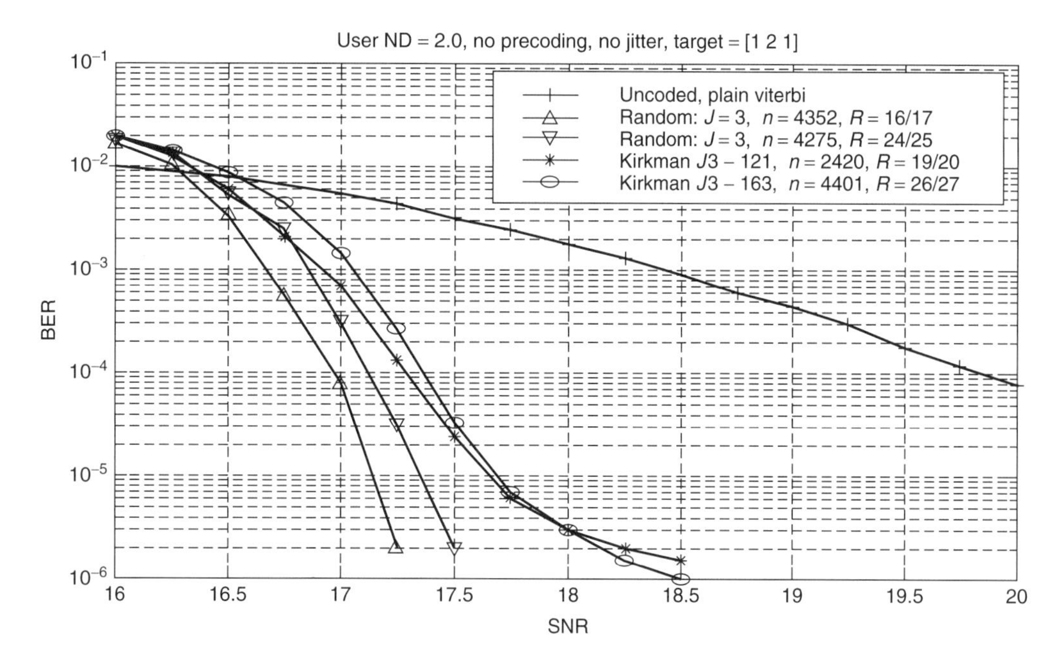

Figure 9. Random versus Kirkman LDPC codes.

Figure 10 makes a similar comparison between random LDPC codes of weights 3 and 4, and lattice designs with weights 4 and 5. As in Kirkman constructions, lattice designs have a higher error floor than do their random counterparts.

4. CONCLUSION

In this work we have presented various designs for LDPC code constructions. We have shown that LDPC codes based on BIBD designs can achieve very high rates and simple implementation compared to their random counterparts. We have analyzed the performance of these codes for a perpendicular recording channel. We have shown the tradeoff between BER performance, error floor, and complexity via extensive simulations.

Acknowledgment
The authors wish to thank our editor, John G. Proakis.

BIOGRAPHIES

Erozan M. Kurtas received his B.S. degree in electrical and electronics engineering in 1991 from Bilkent University, Ankara, Turkey, and an M.S. and Ph.D. degree in

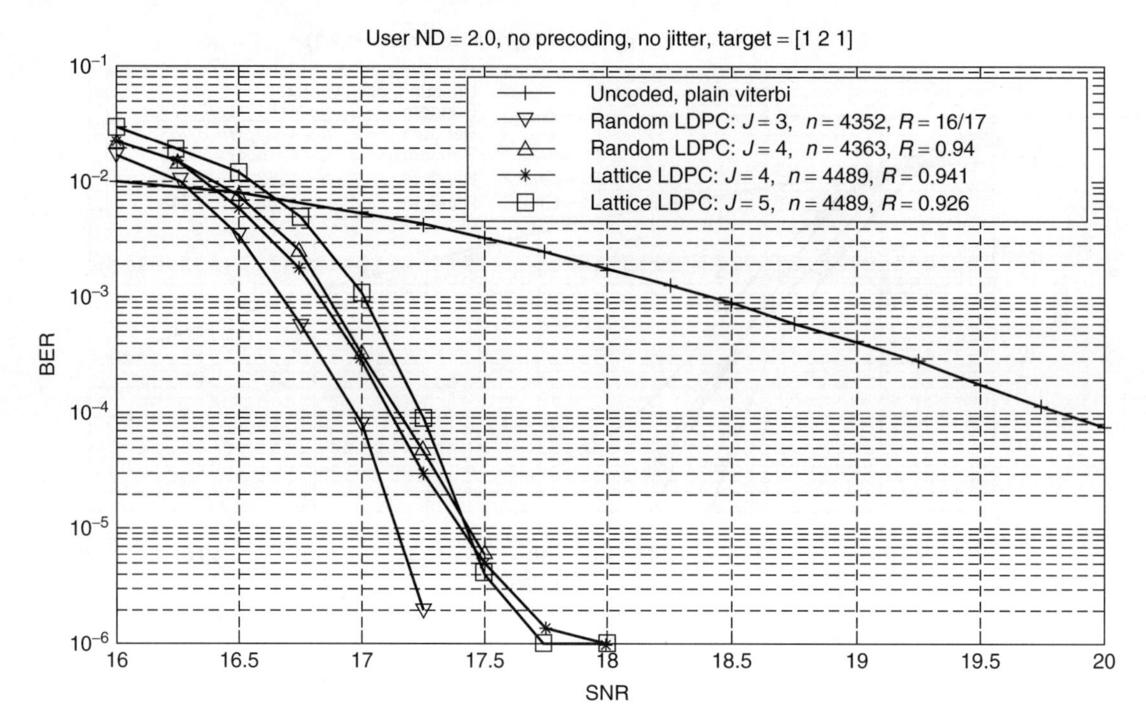

Figure 10. Random versus lattice LDPC codes.

electrical and computer engineering from the Northeastern University, Boston, Massachusetts, in 1994 and 1997, respectively. He joined Quantum Corporation (now Maxtor) in 1996 as a senior design engineer and worked on coding and signal processing for read-channel architectures employed in data storage devices. Since 1999, he has been with Seagate Technology where he is the director of channels research working on future data coding and recovery systems for storage applications. Dr. Kurtas has over 50 papers in various journals and conference proceedings in the general field of digital communications technology. His research interests span information theory, coding, detection and estimation, synchronization, and signal processing techniques. Dr. Kurtas is the recipient of the 2001 Seagate Technology Outstanding Technical Contribution and Achievement Award. Dr. Kurtas has five pending patent applications.

Alexander V. Kuznetsov received his Diploma with excellence in radio engineering and a Ph.D. degree in information theory from the Moscow Institute of Physics and Technology, Moscow, Russia, and the degree of doctor of engineering sciences from the Institute for Information Transmission Problems IPPI, Russian Academy of Sciences, in 1970, 1973, and 1988, respectively. In 1973, he joined the Russian Academy of Sciences as a scientific fellow, where for 25 years he worked on coding theory and its applications to data transmission and storage. On leave from IPPI, he held different research positions at the Royal Institute of Technology and Lund University in Sweden, Concordia University in Canada, Eindhoven Technical University in the Netherlands, Rand African University in RSA, Osaka University and Nara Advanced Institute of Sciences and Technology in Japan, Institute of Experimental Mathematics of Essen University in Germany, CRL of Hitachi, Ltd., in Tokyo, Japan. In 1998–1999, he was a member of technical staff of the Data Storage Institute (DSI) in Singapore, and worked on coding and signal processing for hard disk drives. Currently, he is a research staff member of the Seagate Technology in Pittsburgh, Philadelphia.

Dr. Vasic received his B.S., M.S., and Ph.D., all in electrical engineering from University of Nis, Serbia. From 1996 to 1997 he worked as a visiting scientist at the Rochester Institute of Technology, and Kodak Research, Rochester, New York, where he was involved in research in optical storage channels. From 1998 to 2000 he was with Lucent Technologies, Bell Laboratories. He was involved in research in read channel architectures and iterative decoding and low-density parity check codes. He was involved in development of codes and detectors for five generations of Lucent (now Agere) read channel chips. Presently, Dr. Vasic is a faculty member of the University of Arizona, Electrical and Computer Engineering Department. Dr. Vasic is an author of a more than fifteen journal articles, more than fifty conference papers, and one book chapter "Read channels for magnetic recording," in *CRC Handbook of Computer Engineering*. He is a member of the editorial board of the *IEEE Transactions on Magnetics*. He will serve as a technical program chair for IEEE Communication Theory Workshop in 2003, and as a coorganizer of the Center for Discrete Mathematics and Theoretical Computer Science Workshop on Optical/Magnetic Recording and Optical Transmission in 2003. His research interests include: coding theory, information theory, communication theory, digital communications, and recording.

BIBLIOGRAPHY

1. L. R. Bahl, J. Cocke, F. Jelinek, and J. Raviv, Optimal decoding of linear codes for minimizing symbol error rate, *IEEE Trans. Inform. Theory* **IT-20**: 284–287 (1974).
2. L. M. Batten, *Combinatorics of Finite Geometries*, London, Cambridge Univ. Press, 1997.
3. S. Benedetto, G. Montorsi, D. Divsalar, and F. Pollara, Serial concatenation of interleaved codes: Performance analysis, design, and iterative decoding, *Telecommun. Data Acquis. Progr. Rep.* **42**: 1–26 (Aug. 1996).
4. S. Benedetto and G. Montorsi, Unveiling turbo codes: Some results on parallel concatenated coding schemes, *IEEE Trans. Inform. Theory* **42**: 409–428 (March 1996).
5. G. Berrou, A. Glavieux, and P. Thitimajshima, Near Shannon limit error-correcting coding and decoding: Turbo-codes, *Proc. IEEE Int. Conf. Communications* (ICC'93), Geneva, Switzerland, May 1993, pp. 2.1064–2.1070.
6. T. Beth, D. Jungnickel, and H. Lenz, *Design Theory*, Cambridge Univ. Press, 1986.
7. R. A. Beezer, The girth of a design, *J. Combinat. Math. Combinat. Comput.* (in press) (also online, *http://buzzard.ups.edu/pubs.html)*.
8. V. K. Bhargava and J. M. Stein, (v, k, λ) configurations and self-dual codes, *Inform. Control* **28**: 352–355 (1975).
9. R. A. Brualdi, *Introductory Combinatorics*, Prentice-Hall, Upper Saddle River, NJ, 1999.
10. M. Buratti, Construction of (q, k, l) difference families with q a prime power and $k = 4, 5$, *Discrete Math.* **138**: 169–175 (1995).
11. P. J. Cameron and J. H. van Lint, *Graphs, Codes and Designs*, London Math. Soc. Lecture Note Series 23, Cambridge Univ. Press, 1980.
12. C. J. Colbourn and J. H. Dinitz, eds., *The Handbook of Combinatorial Designs*, CRC Press, Boca Raton, FL, 1996.
13. C. Colbourn and A. Rosa, *Steiner Systems*, Oxford Univ. Press (Oxford Mathematical Monographs), London, 1999.
14. T. M. Duman and E. Kurtas, Performance of Turbo codes over magnetic recording channels, *Proc. MILCOM*, Nov. 1999.
15. T. M. Duman and E. Kurtas, Comprehensive performance investigation of Turbo codes over high density magnetic recording channels, *Proc. IEEE GLOBECOM*, Dec. 1999.
16. T. M. Duman and E. Kurtas, Performance bounds for high rate linear codes over partial response channels, *Proc. IEEE Int. Symp. Information Theory* (Sorrento, Italy), IEEE, June 2000, p. 258.
17. J. Fan, A. Friedmann, E. Kurtas, and S. W. McLaughlin, Low density parity check codes for partial response channels, Allerton Conf. Communications, Control and Computing, Urbana, IL, Oct. 1999.
18. B. J. Frey, *Graphical Models for Machine Learning and Digital Communication*, MIT Press, Cambridge, MA, 1998.
19. R. G. Gallager, *Low-Density Parity-Check Codes*, MIT Press, Cambridge, MA, 1963.
20. J. Hagenauer, E. Offer, and L. Papke, Iterative decoding of binary block and convolutional codes, *IEEE Trans. Inform. Theory* **42**(2): 439–446 (March 1996).
21. N. Hamada, On the p-rank of the incidence matrix of a balanced or partially balanced incompleted block design and its applications to error correcting codes, *Hiroshima Math. J.* **3**: 153–226 (1973).
22. C. Heegard, Turbo coding for magnetic recording, *Proc. IEEE Information Theory Workshop*, (San Diego, CA), IEEE, Feb. 1998, pp. 18–19.
23. T. P. Kirkman, Note on an unanswered prize question, *Cambridge Dublin Math. J.* **5**: 255–262 (1850).
24. Y. Kou, S. Lin, and M. Fossorier, Construction of low density parity check codes: A geometric approach, *Proc. 2nd Int. Symp. Turbo Codes*, Brest, France, Sept. 4–7, 2000.
25. Y. Kou, S. Lin, and M. P. C. Fossorier, Low density parity check codes based on finite geometries: A rediscovery and new results, *IEEE Trans. Inform. Theory* **47**(7): 2711–2736 (Nov. 2001).
26. A. V. Kuznetsov and B. S. Tsybakov, On unreliable storage designed with unreliable components, *Proc. 2nd Int. Symp. Information Theory*, 1971, Tsahkadsor, Armenia (Publishing House of the Hungarian Academy of Sciences), 1973, pp. 206–217.
27. A. V. Kuznetsov, On the storage of information in memory constructed from unreliable components, *Problems Inform. Trans.* **9**(3): 100–113 (1973).
28. J. Li, E. Kurtas, K. R. Narayanan, and C. N. Georghiades, On the performance of Turbo product codes over partial response channels, *IEEE Trans. Magn.* (July 2001).
29. J. Li, E. Kurtas, K. R. Narayanan, and C. N. Georghiades, Iterative decoding for Turbo product codes over Lorentzian channels with colored noise, *Proc. GLOBECOM*, San Antonio, TX, Nov. 2001.
30. J. Li, E. Kurtas, K. R. Narayanan, and C. N. Georghiades, On the performance of Turbo product codes and LDPC codes over partial response channels, *Proc. Int. Conf. Communications*, Helsinki, Finland, June 2001.
31. S. Lin and D. J. Costello, Jr., *Error Control Coding: Fundamentals and Applications*, Englewood Cliffs, NJ, Prentice-Hall, 1983.
32. A. C. Ling, C. J. Colbourn, M. J. Grannell, and T. S. Griggs, Construction techniques for anti-pasch Steiner triple systems, *J. Lond. Math. Soc.* **61**(3): 641–657 (June 2000).
33. A. C. H. Ling and C. J. Colbourn, Rosa triple systems, in J. W. P. Hirschfeld, S. S. Magliveras, M. J. de Resmini, eds., *Geometry, Combinatorial Designs and Related Structures*, Cambridge Univ. Press, 1997, pp. 149–159.
34. J. H. van Lint and R. M. Wilson, *A Course in Combinatorics*, Cambridge Univ. Press, 1992.
35. D. J. C. MacKay, Good error-correcting codes based on very sparse matrices, *IEEE Trans. Inform. Theory* **45**: 399–431 (March 1999).
36. D. MacKay and M. Davey, Evaluation of Gallager codes for short block length and high rate applications (online), *http://www.cs.toronto.edu/~mackay/CodesRegular.html*.
37. F. J. MacWilliams and N. J. A. Sloane, *The Theory of Error-Correcting Codes*, North-Holland, Oxford, UK, 1977.
38. R. J. McEliece, D. J. C. MacKay, and J.-F. Cheng, Turbo decoding as an instance of Pearl's "Belief propagation" algorithm, *IEEE J. Select. Areas Commun.* **16**: 140–152 (Feb. 1998).
39. R. J. McEliece, *Finite Fields for Computer Scientist and Engineers*, Kluwer, Boston, 1987.

40. E. Netto, *Zur theorie der triplesysteme, Math. Ann.* **42**: 143–152 (1893).

41. J. Pearl, *Probabilistic Reasoning in Intelligent Systems: Networks of Plausible Inference*, Morgan Kaufmann, San Mateo, CA, 1988.

42. L. Ping and K. Y. Wu, Concatenated tree codes: A low-complexity, high performance approach, *IEEE Trans. Inform. Theory* (Dec. 1999).

43. D. K. Ray-Chaudhuri and R. M. Wilson, Solution of Kirkman's school-girl problem, *Proc. Symp. Pure Math.*, Amer. Mathematical Society, Providence, RI, 1971, pp. 187–203.

44. T. Richardson, A. Shokrollahi, and R. Urbanke, Design of provably good low-density parity check codes, *IEEE Trans. Inform. Theory* **47**: 619–637 (Feb. 2001).

45. R. M. Robinson, Triple systems with prescribed subsystems, *Notices Am. Math. Soc.* **18**: 637 (1971).

46. R. M. Robinson, The structure of certain triple systems, *Math. Comput.* **29**: 223–241 (1975).

47. C. E. Shannon, A mathematical theory of communication, *Bell Syst. Tech. J.* 372–423, 623–656 (1948).

48. E. Spence and V. D. Tonchev, Extremal self-dual codes from symmetric designs, *Discrete Math.* **110**: 165–268 (1992).

49. D. R. Stinson, Frames for Kirkman triple systems, *Discrete Math.* **65**: 289–300 (1988).

50. R. M. Tanner, A recursive approach to low complexity codes, *IEEE Trans. Inform. Theory* **IT-27**: 533–547 (Sept. 1981).

51. R. M. Tanner, Minimum-distance bounds by graph analysis, *IEEE Trans. Inform. Theory* **47**(2): 808–821 (Feb. 2001).

52. M. G. Taylor, Reliable information storage in memories designed from unreliable components, *Bell Syst. Tech. J.* **47**(10): 2299–2337 (1968).

53. R. Townsend and E. J. Weldon, Self-orthogonal quasi-cyclic codes, *IEEE Trans. Inform. Theory* **IT-13**(2): 183–195 (1967).

54. B. Vasic, Low density parity check codes: Theory and practice, National Storage Industry Consortium (NSIC) quarterly meeting, Monterey, CA, June 25–28, 2000.

55. B. Vasic, Structured iteratively decodable codes based on Steiner systems and their application in magnetic recording, *Proc. GLOBECOM 2001*, San Antonio, TX, Nov. 2001.

56. B. Vasic, Combinatorial constructions of structured low-density parity check codes for iterative decoding, *IEEE Trans. Inform. Theory* (in press).

57. E. J. Weldon, Jr., Difference-set cyclic codes, *Bell Syst. Tech. J.* **45**: 1045–1055 (Sept. 1966).

58. N. Wiberg, H.-A. Loeliger, and R. Kötter, Codes and iterative decoding on general graphs, *Eur. Trans. Telecommun.* **6**: 513–525 (Sept./Oct. 1995).

59. R. M. Wilson, Cyclotomy and difference families in elementary Abelian groups, *J. Number Theory* **4**: 17–47 (1972).

60. V. V. Zyablov and M. S. Pinsker, Error correction properties and decoding complexity of the low-density parity check codes, 2nd Int. Symp. Information Theory, Tsahkadsor, Armenia, 1971.

61. V. V. Zyablov and M. S. Pinsker, Estimation of the error-correction complexity for Gallager low-density codes, *Problems Inform. Trans.* **11**: 18–28 (1975) [transl. from *Problemy Peredachi Informatsii* **11**(1): 23–26].

DIFFERENTIATED SERVICES

IKJUN YEOM
Korea Advanced Institute of Science and Technology
Seoul, South Korea

A. L. NARASIMHA REDDY
Texas A&M University
College Station, Texas

1. INTRODUCTION

Providing quality of service (QoS) has been an important issue in Internet engineering, as multimedia/real-time applications requiring certain levels of QoS such as delay/jitter bounds, certain amount of bandwidth and/or loss rates are being developed. The current Internet provides only *best-effort* service, which does not provide any QoS. The Internet Engineering Task Force (IETF) has proposed several service architectures to satisfy different QoS requirements to different applications.

The Differentiated Services (Diffserv) architecture has been proposed by IETF to provide service differentiation according to customers' service profile called service-level agreement (SLA) in a scalable manner [1–4]. To provide service differentiation to different customers (or flows), the network needs to identify the flows, maintain the requested service profile and conform the incoming traffic based on the SLAs.

A simple Diffserv domain is illustrated in Fig. 1. In a Diffserv network, most of the work for service differentiation is performed at the edge of the network while minimizing the amount of work inside the network core, in order to provide a scalable solution. The routers at the edge of the network, called boundary routers or edge routers, may monitor, shape, classify, and mark *differentiated services code point* (DSCP) value assigned to specific packet treatment to packets of flows (individual or aggregated) according to the subscribed service. The routers in the network core forward packets differently to provide the subscribed service. The core routers need to provide only several forwarding schemes, called *per-hop behaviors* (PHBs) to provide service differentiation. It is expected that with appropriate network engineering, *per-domain behaviors* (PDBs), and end-to-end services can be constructed based on simple PHBs.

1.1. Differentiated Services Architecture

In the Diffserv network, we can classify routers into two types, *edge or boundary* routers and *core* routers according to their location. A boundary router may act both as an *ingress router* and as an *egress router* depending on the direction of the traffic. Traffic enters a Diffserv network at an ingress router and exits at an egress router. Each type of router performs different functions to realize service differentiation. In this section, we describe each of these routers with its functions and describe how service differentiation is provided.

1.1.1. Boundary Router. A basic function of ingress routers is to mark DSCPs to packets based on SLAs so that

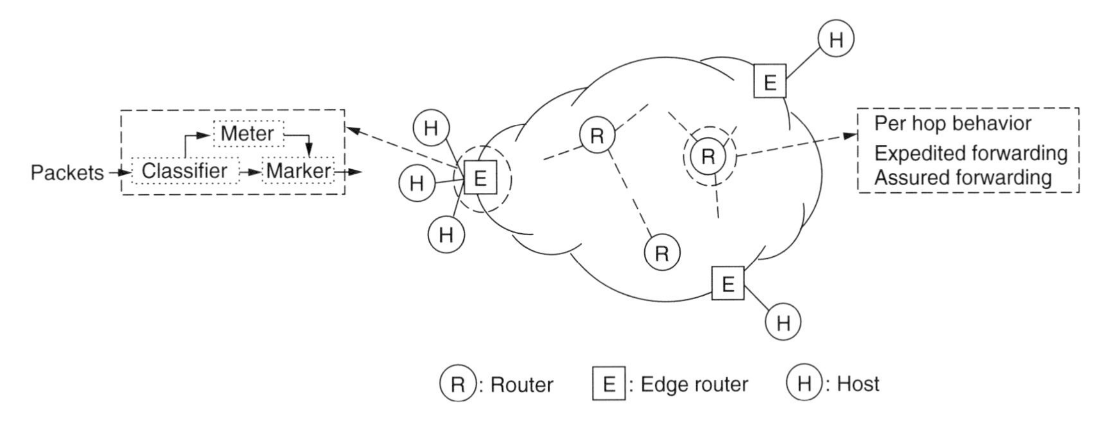

Figure 1. A Diffserv domain.

core routers can distinguish the packets and forward them differently. When a packet arrives at an ingress router, the router first identifies the flow to which the packet belongs and monitors the flow to conform to its contract. If it conforms to its contract, the packet is marked by the DSCP of the contracted service. Otherwise, the packet may be delayed, discarded, or unmarked to make injected traffic conform to its contract.

An egress router may monitor traffic forwarded to other domains and perform shaping or conditioning to follow a traffic conditioning agreement (TCA) with the other domains.

1.1.2. Core Router. In the Diffserv model, functions of core routers are minimized to achieve scalability. A core router provides requested PHB specified in the DS field of the arriving packets. Currently, *expedited forwarding* (EF) [3] and *assured forwarding* (AF) [4] PHBs have been standardized by IETF.

To provide end-to-end QoS, we may consider flows that transit over multiple network domains. While a hop-by-hop forwarding treatment in a Diffserv domain is defined by a PHB, Per-domain behavior (PDB) is used to define edge-to-edge behavior over a Diffserv domain [5]. A PDB defines metrics that will be observed by a set of packets with a particular DSCP while crossing a Diffserv domain. The set of packets subscribing to a particular PDB is classified and monitored at an ingress router. Conformant packets are marked with a DSCP for the PHB associated with the PDB. While crossing the Diffserv domain, core routers treat the packets based only on the DSCP. An egress router may measure and condition the set of packets belonging to a PDB to ensure that exiting packets follow the PDB.

1.1.3. Assured Forwarding PHB. Assured forwarding (AF) PHB has been proposed in [4,6]. In the AFPHB, the edge devices of the network monitor and mark incoming packets of either individual or aggregated flows. A packet of a flow is marked IN (in profile) if the temporal sending rate at the arrival time of the packet is within the contract profile of the flow. Otherwise, the packet is marked OUT (out of profile). Packets of a flow can be hence marked both IN and OUT. The temporal sending rate of a flow is measured using *time sliding window* (TSM) or a *token bucket* controller. IN packets are given preferential treatment at the time of congestion; thus, OUT packets are dropped first at the time of congestion.

Assured forwarding can be realized by employing RIO (RED with IN/OUT) drop policy [6] in the core routers. RIO drop policy is illustrated in Fig. 2. Each router maintains a virtual queue for IN packets and a physical queue for both IN and OUT packets. When the network is congested and the queue length exceeds minTh_OUT, the routers begin dropping OUT packets first. If the congestion persists even after dropping all incoming OUT packets and the queue length exceeds minTh_IN, IN packets are discarded. With this dropping policy, the RIO network gives preference to IN packets and provides different levels of service to users per their service contracts. Different marking policies [7–9] and correspondingly appropriate droppers have been proposed to improve the flexibility in providing service differentiation.

1.1.4. Expedited Forwarding PHB. Expedited forwarding (EF) PHB was proposed [3] as a premium service for the Diffserv network. The EFPHB can be used to guarantee low loss rate, low latency, low jitter, and assured

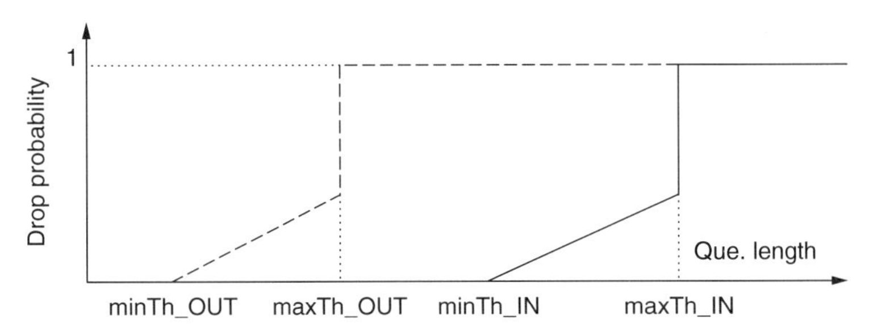

Figure 2. RED IN/OUT (RIO) drop policy.

throughput like a “virtual leased line.” According to Jacobson et al. [3], the departure rate of the EF traffic measured over any time interval equal to or longer than a packet time should equal or exceed a configurable rate independent of the intensity of any other traffic.

However, there is evidence [10] that the original EFPHB configuration in Ref. 3 is too strict and hard to implement in practice. In a network where there are many connections frequently established, closed, merged and split, it is hard to maintain each node's arrival rate to be less than its departure rate at timescales required by the EFPHB. Alternatively, it has been proposed [10–12] that the EFPHB should be redefined as “a forwarding treatment for a particular Diffserv aggregate where the node offers to the aggregate a packet scale rate guarantee R with latency E, where R is a configurable rate and E is a tolerance that depends on the particular node characteristics.” In a network providing packet scale rate guarantees, any EF packet arriving at a node at time t will leave the node no later than at time $t + Q/R + E$, where Q is the total backlogged EF traffic at time t. Here note that Q is zero in the original EF configuration since the arrival rate is less than the departure rate at any node at any given time.

Several types of queue scheduling schemes (e.g., a priority queue, a single queue with a weighted round-robin scheduler, and class-based queue [13]) may be used to implement the EFPHB.

It is important to understand the QoS delivered to individual applications by the extra support and functionality provided by a Diffserv network. In the next two sections, we present throughput analysis of AFPHB and delay analysis of EFPHBs.

2. TCP PERFORMANCE WITH AFPHB

Popular transport protocol TCP reacts to a packet loss by halving the congestion window and increases the window additively when packets are delivered successfully [14]. This AIMD congestion control policy makes the throughout of a TCP flow highly dependent on the dropping policy of the network. With AFPHB, service differentiation is realized by providing different drop precedences, and thus, TCP throughput with AFPHB is an interesting issue.

In Fig. 3, we present a simulation result using ns-2 [15] to show realized TCP throughput in a simple network with AFPHB. In this simulation, there are five TCP flows sharing one 4-Mbps bottleneck link. Each flow contracts bandwidth {0, 0.1, 0.5, 1, 2} Mbps with network provider. From the figure, it is shown that flows with higher contract rates get higher throughput than flows with lower contract rates. However, it is also shown that the flows with 2 Mbps contract rate do not reach 2 Mbps while flows with 0- and 0.1-Mbps contract rates exceed their contract rates.

Similar results have been reported in the literature [9,16,17], and it has been also shown that it is difficult to guarantee absolute bandwidth with a simple marking and dropping scheme [18]. There is a clear need to understand the performance of a TCP flow with the AFPHB.

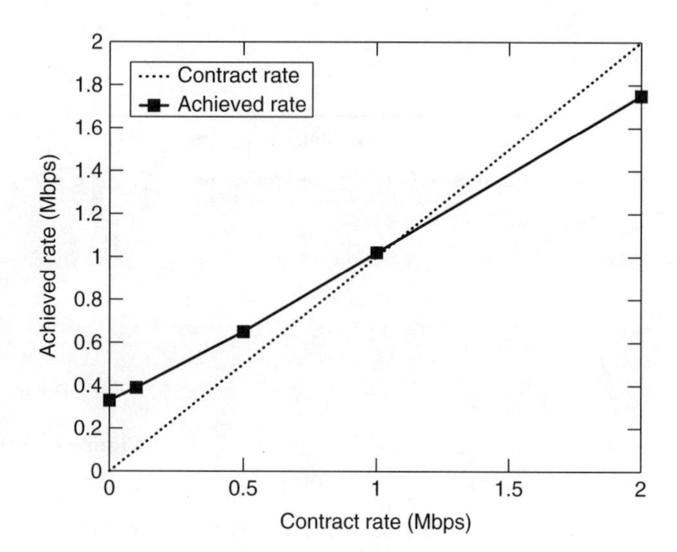

Figure 3. Realized TCP throughput with AFPHB.

In a steady state, a flow can experience different levels of congestion on the basis of its contract rate and the network dynamics. A flow that experiences no IN packet drops is said to observe an undersubscribed path. A flow that does not transmit any OUT packets either because every OUT packet is dropped or because the sending rate is less than the contract profile is said to observe an oversubscribed path. In a reasonably configured network, however, IN packets are expected to be protected, and only OUT packets may be discarded. In this section, we focus on the model for undersubscribed path.[1]

The steady-state TCP throughput, B is given by [19]

$$B = \begin{cases} \dfrac{3k}{4\text{RTT}}\left(\sqrt{\dfrac{2}{p_{\text{out}}}} + R\right) & \text{if } R \geq \dfrac{W}{2} \\ \dfrac{k}{2\text{RTT}}\left(\sqrt{R^2 + \dfrac{6}{p_{\text{out}}}} + R\right) & \text{otherwise} \end{cases} \tag{1}$$

where k is the packet size, p_{out} is the drop rate of OUT packets R is the reservation window defined as (contract rate/packet size × RTT), and W is the maximum congestion window size in steady state. From Eq. (1), B is proportional to the contract rate and inversely proportional to drop rate.

To illustrate the model described above, we present Fig. 4, where results from the model are compared with simulations. In the simulation, there are 50 TCP flows sharing a 30-Mbps link. Contract rate of each flow is randomly selected from 0 to 1 Mbps. In the figure, squares indicate simulation results, and stars indicate estimated throughput from the model. It is observed that the model can estimate TCP throughput accurately.

To discuss the interaction between contract rate and realized throughput more in depth, we define the

[1] To develop a model for a general situation where both IN and OUT packets are dropped, a model for an oversubscribed path is also required. For the general model, please refer to Ref. 19.

excess bandwidth B_e as the difference between realized throughput and its contract rate. It is given by

$$B_e = \begin{cases} \frac{k}{4\text{RTT}}\left(3\sqrt{\frac{2}{p_{\text{out}}}} - R\right) & \text{if } R \geq \frac{W}{2} \\ \frac{k}{2\text{RTT}}\left(\sqrt{R^2 + \frac{6}{p_{\text{out}}}} - R\right) & \text{otherwise} \end{cases} \quad (2)$$

If B_e of a flow is positive, this means that the flow obtains more than its contract rate. Otherwise, it does not reach its contract rate. From (2) and Fig. 4, we can observe that

- When a flow reserves relatively higher bandwidth ($R \geq \sqrt{2/p_{\text{out}}}$), B_e is decreased as the reservation rate is increased. Moreover, if R is greater than $3\sqrt{2/p_{\text{out}}}$ (see line C in Fig. 4), the flow cannot reach its reservation rate.
- When a flow reserves relatively lower bandwidth ($R < \sqrt{2/p_{\text{out}}}$; see line B in Fig. 4), it always realizes at least its reservation rate. As it reserves less bandwidth, it obtains more excess bandwidth. TCP's multiplicative decrease of sending rate after observing a packet drop results in a higher loss of bandwidth for flows with higher reservations. This explains the observed behavior.
- Equation (2) also shows that as the probability of OUT packet drop decreases, the flows with smaller reservation benefit more than do the flows with larger reservations. This points to the difficulty in providing service differentiation between flows of different reservations when there is plenty of excess bandwidth in the network.
- The realized bandwidth is observed to be inversely related to the RTT of the flow.
- For best-effort flows, $R = 0$. Hence, $B_e(= k\sqrt{6/p_{\text{out}}}/2\text{RTT}$; see line D in Fig. 4) gives the bandwidth likely to be realized by flows with no reservation.
- Comparing the above mentioned best-effort bandwidth when $R \geq \sqrt{2/p_{\text{out}}}$, we realize that the reservation rates larger than 3.5 times the best-effort bandwidth cannot be met.
- Equation (2) clearly shows that excess bandwidth cannot be equally shared by flows with different reservations without any enhancements to basic RIO scheme or to TCP's congestion avoidance mechanism.

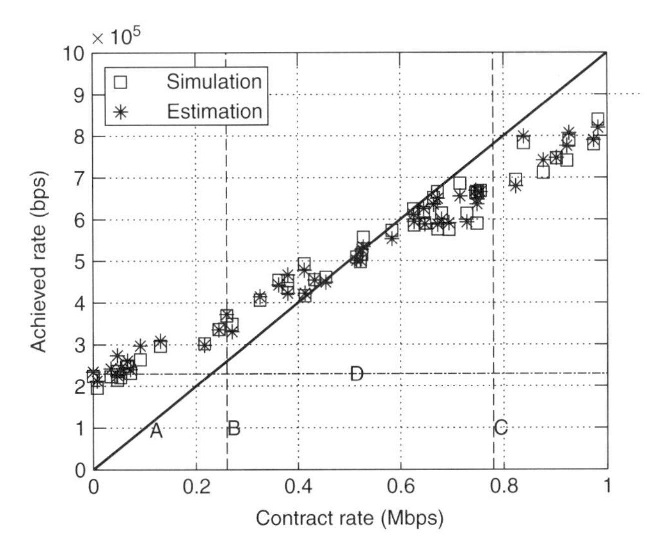

Figure 4. Observations from the model.

When several TCP flows are aggregated, the impact of an individual TCP sawtooth behavior is reduced, and the aggregated sending rate is stabilized. If the marker neither maintains per-flow state nor employs other specific methods for distinguishing individual flows, an arriving packet is marked IN with the probability, $p_m(= contract\_rate/aggregated\_sending\_rate)$. In the steady state, p_m is approximately equal for all the individual flows. A flow sending more packets then gets more IN packets, and consequently, the contract rates consumed by individual flows is roughly proportional to their sending rates. We call this marking behavior *proportional marking*.

With the assumptions that all packets of aggregated flows are of the same size, k, a receiver does not employ delayed ACK, and the network is not oversubscribed, the throughput of the ith flow and the aggregated throughput B_A of n flows are given in [19] by

$$B_i = \frac{m_i}{\sum_{j=1}^{n} m_j} \cdot \frac{3r_A}{4} + \frac{3k}{4} m_i \quad (3)$$

$$B_A = \sum_{i=1}^{n} B_i = \frac{3k}{4} \sum_{i=1}^{n} m_i \quad (4)$$

where r_A is the contract rate for the aggregated flows, and m_i is $(1/\text{RTT}_i)\sqrt{(2/p_i)}$.

Equation (3) relates the realized throughput of an individual flow to the aggregated contract rate r_A and the network conditions (RTT_i and p_i) observed by various flows within the aggregation. From (4), B_e (the excess bandwidth) of aggregated flows is calculated as follows

$$B_e = \tfrac{3}{4} r_A + B_s - r_A = B_s - \tfrac{1}{4} r_A \quad (5)$$

where $B_s = \frac{3k}{4} \sum_{i=1}^{n} m_i$, and it is approximately the throughput which the aggregated flows can achieve with zero contract rate ($r_A = 0$). According to the analysis above, the following observations can be made:

- The total throughput realized by an aggregation is impacted by the contract rate. The larger the contract rate, the smaller the excess bandwidth claimed by the aggregation.

- When the contract rate is larger than 4 times B_s, the realized throughput is smaller than the contract rate.
- The realized throughput of a flow is impacted by the other flows in the aggregation (as a result of the impact on p_m) when proportional marking is employed.
- The total realized throughput of an aggregation is impacted by the number of flows in the aggregation.

There are two possible approaches for enhancing better bandwidth differentiation with AFPHB. The first approach tries to enhance the dropping policy of the core routers [20] while the second approach tries to enhance the marking policies of the edge routers [7–9,21]. Below, we briefly outline one approach of enhancing the edge markers.

2.1. Adaptive Marking for Aggregated TCP Flows

The Diffserv architecture allows aggregated sources as well as individual sources [2]. When several individual sources are aggregated, output traffic of the aggregation is not like traffic of one big single source in the following respects: (1) each source within the aggregation responds to congestion individually, and (2) the aggregation has multiple destinations, and each source within the aggregation experiences different delays and congestion. Therefore, when we deal with aggregated sources in the Diffserv network, we need to consider not only the total throughput achieved by the aggregated sources but also the throughput achieved by individual flows.

Given individual target rates for each flow, how does one allocate a fixed aggregate contract rate among individual flows within the aggregation under dynamic network conditions? The most desirable situation is to guarantee individual target rates for all the flows. However, there exist situations in which some targets cannot be met: (1) when there is a severe congestion along the path and the current available bandwidth is less than the target and (2) when the contract rate is not enough to achieve the target. If we try to achieve the target by increasing the marking rate of an individual flow, the congestion along that flow's path may become more severe and result in wasting contracted resources of the aggregation. This is undesirable for both customers and service providers.

To solve this problem, an adaptive marking scheme has been proposed [21]. The proposed scheme achieves at least one of the following for all the flows in an aggregation:

1. Individual target rate when it is reachable.
2. Maximized throughput without IN packet loss when the current available bandwidth is less than the individual target rate.
3. Throughput achieved with a fair marking rate M/n, where M and n are the total marking rate and the number of flows within the aggregation, respectively.

These three goals correspond to (1) meeting individual flow's BW needs, (2) maximization of utility of the aggregated contract rate, and (3) fairness among the flows within the aggregation.

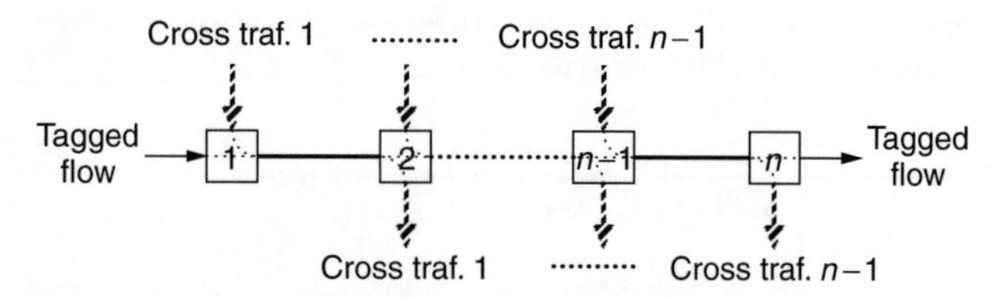

Figure 5. Multihop topology.

To observe how the marking rate is adjusted and an individual flow achieves its target rate, we conducted a set of simulations. We consider a multihop path as shown in Fig. 5. There are n routers, and cross traffic is injected to this network at the ith router and exits at the $(i+1)$th router.

In the simulation, we set the link capacity at 3 Mbps and use 10 TCP flows for cross traffic. The contract rate for each TCP flow is randomly selected from 0 to 1 Mbps, and the total contract of cross-traffic is 2.7 Mbps, so that the subscription level is 90%. The number of routers (n) is 5. For the tagged flow, we use a single TCP flow.

First, to observe path characteristics, we use a static contract rate for the tagged flow. We vary the contract rate from 0 to 0.8 Mbps. In Fig. 6, the solid line shows the achieved rate with static marking rate, and the dashed line indicates that the achieved rate is equal to 75% of the marking rate. The achieved rate increases as the marking rate increases up to 0.5 Mbps. However, the achieved rate does not increase beyond the 0.55-Mbps marking rate. In this example, the maximum achievable rate is about 0.42 Mbps.

Now the tagged flow is an individual flow within an aggregation with aggregated contract rate. The marker for the aggregation employs the adaptive marking. We vary the target rate for the tagged flow from 0.1 to 0.5 Mbps. Figure 6 shows the results. In each figure, dots indicate instantaneous marking and achieved rate, and a square shows the average. In this path, a flow gets 0.15 Mbps with zero contract rate. When the target rate is 0.1 Mbps (Fig. 6a), the marking rate stays around zero. When the target rate is achievable (<0.42 Mbps), the adaptive marking scheme finds the minimum marking rate to realize the target rate (Fig. 6a,b). In Fig. 6c, the marking rate stays below 0.55 Mbps to avoid wasting resources when the target is unachievable.

We present one other experiment to show the utility of the adaptive marker in reducing bandwidth differences due to RTTs as described by the model earlier. We use a topology in which 40 TCP flows are aggregated and compete in a 25-Mbps bottleneck link. The aggregated contract rate is 10 Mbps. The RTT (excluding queueing delay) of each flow is randomly selected from within 50–150 ms. Figure 7 shows the result. The adaptive marker effectively removes RTT bias of TCP flows and realizes QoS goals of individual flows within the aggregation.

The adaptive marker addresses the important problem of establishing a relationship between per-session

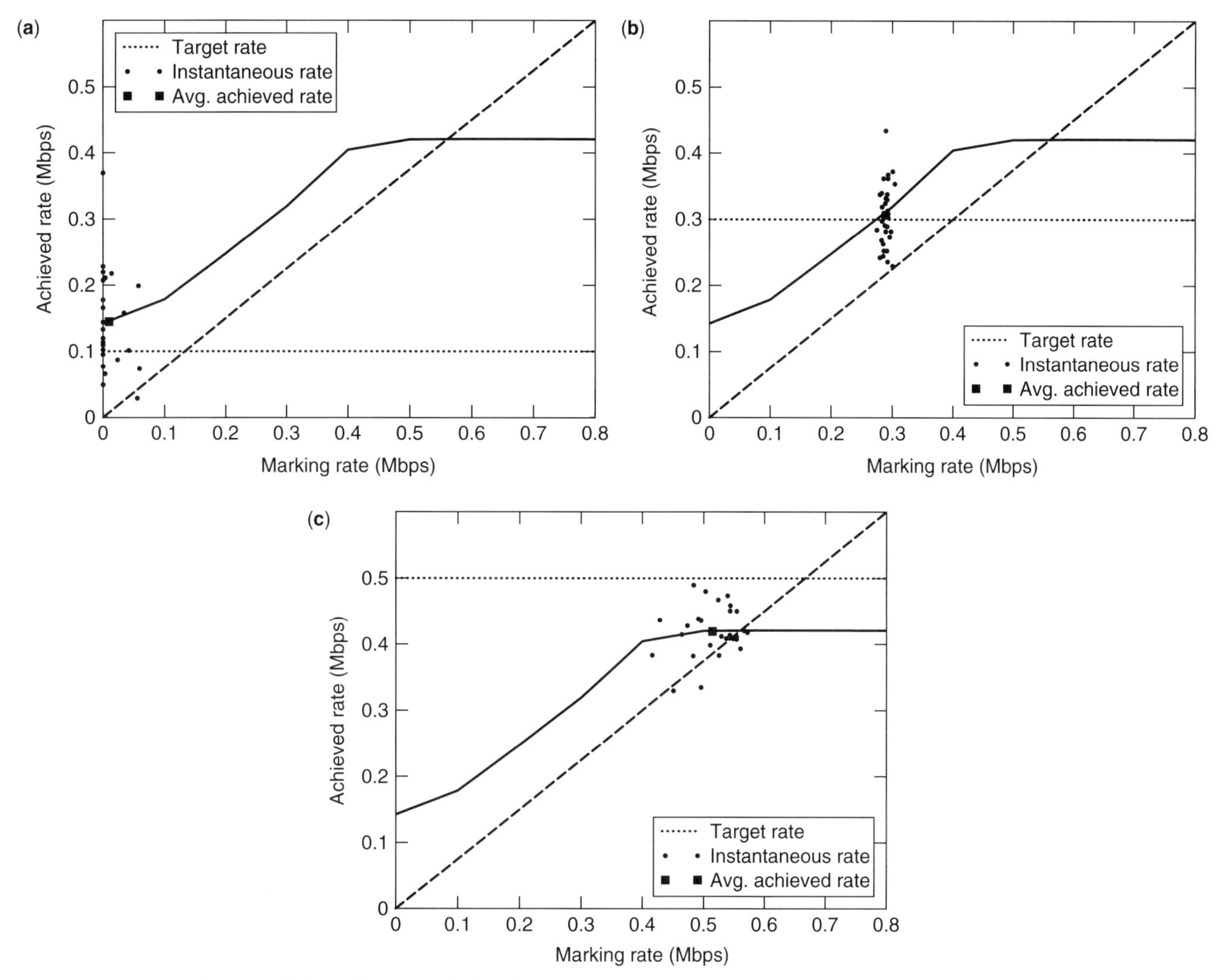

Figure 6. Achieved rates with the adaptive marking, with targets of (**a**) 0.1 Mbps, (**b**) 0.3 Mbps, and (**c**) 0.5 Mbps

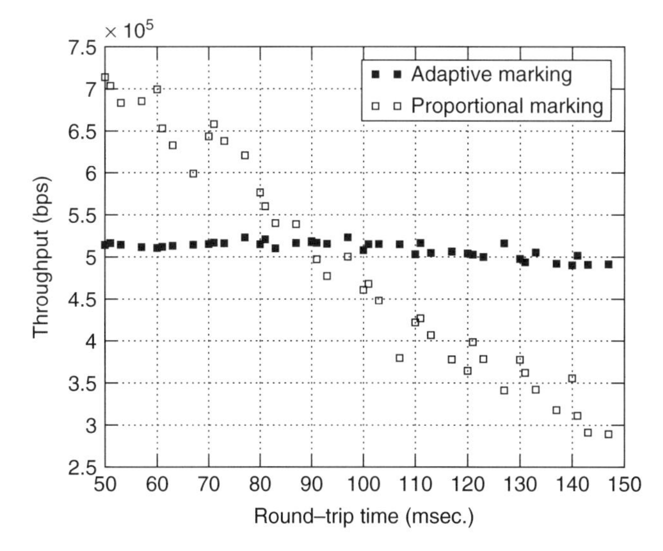

Figure 7. Throughput of flows with different round-trip times

behavior, aggregate packet marking, and packet differentiation within a differentiated services network. The adaptive marking algorithm is based on the TCP performance model within Diffserv networks. The adaptive marker enables reaching specific QoS goals of individual flows while efficiently managing the aggregate resources. Moreover, it shows that with appropriate edge devices, it is possible for applications to realize bandwidth guarantees utilizing the AFPHB.

3. DELAY ANALYSIS OF THE EFPHB

The EFPHB is proposed to provide low delay, jitter, and loss rate. Low loss rate is easily achievable with strict admission control and traffic conditioning. Regarding the EFPHB, providing low delay and jitter is an important issue and has been widely studied. In this section, we introduce studies on delay analysis of the EFPHB and present some experimental work for practical performance evaluation of the EFPHB.

3.1. Virtual Wire Service

Virtual wire (VW) service aims to provide a very low end-to-end delay jitter for flows subscribing to the EFPHB [22]. More precisely, as the term, *virtual wire* suggests, VW service is intended to "mimic, from the point of view of the originating and terminating nodes, the behavior of a hard-wired circuit of some fixed capacity [22]."

To provide VW service, the following two conditions are required: (1) each node in the domain must implement the EFPHB, and (2) conditioning the aggregate so that it's arrival rate at any node is always less than that node's departure rate. If the arrival rate is less than the virtual wire's configured rate, packets are delivered with almost no distortion in the interpacket timing. Otherwise, packets are unconditionally discarded not to disturb other traffic.

In [22], a *jitter window* is defined as the maximum time interval between two consecutive packets belonging to a flow so that the destination cannot detect delay jitter. Consider a constant-bit rate (CBR) flow with rate R. The packets of the flow arrive at an egress router through a link of bandwidth $B(=nR)$ and leave to their destination through a link of bandwidth R as shown in Fig. 8.

Let's consider two consecutive packets: P_0 and P_1. To transmit P_1 immediately after transmitting P_0 and to hide jitter, the last bit of P_1 should arrive at the node before T_1. T_1 is the time when the last bit of P_0 leaves the node and is calculated by

$$T_1 = T_0 + \frac{S}{R} \tag{6}$$

where S is the packet size. Then, the jitter window, Δ, is given by

$$\Delta = \frac{S}{R} - \frac{S}{B} = \frac{S}{R} \times \left(1 - \frac{1}{n}\right) \tag{7}$$

Generally, a packet experiences propagation, transmission, and queueing delay while traveling a network path. As long as the path is identical, propagation and transmission delay are the same. Therefore, as long as the sum of queueing delays that a packet observes in an EF domain is less than Δ, the destination does not observe jitter.

There are three possible sources of queueing delay of a VW packet: (1) non-EF packets ahead of the packet in a queue, (2) the previous VW packet of the same flow, and (3) VW packets of other flows. In a properly configured EF domain, the arrival rate of the EF traffic should be less than the departure rate at any node. Then, the queueing delay caused by the other EF traffic (cases 2 and 3) is zero.

In an EF domain, EF and non-EF packets are separately queued, and EF packets are serviced at a higher priority. Therefore, case 1 occurs only when an EF packet arrives at a node that is serving a non-EF packet. The worst-case delay occurs when the EF packet arrives immediately after the node begins to send a non-EF packet, and it is S/B, where S is the packet size. From (7), n should be at least 2 in order to satisfy the jitter window, and which means that the bandwidth assigned to EF traffic should be configured to be less than half of the link bandwidth.

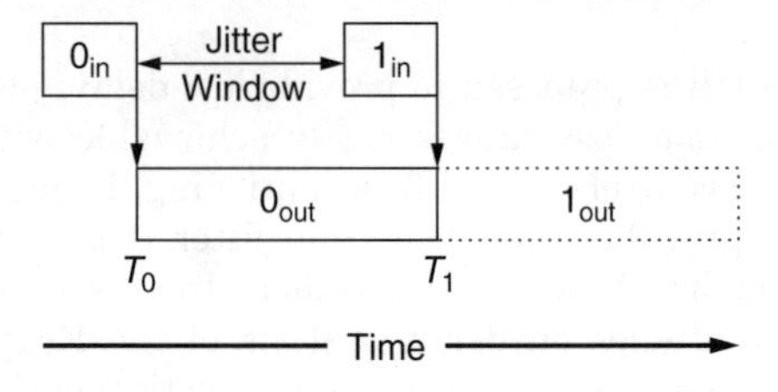

Figure 8. Jitter window.

3.2. Packet-Scale Rate Guarantee

Packet-scale rate guarantee service has been proposed to analyze and characterize the delay of the EFPHB more precisely [10–12]. A node is said to provide packet-scale rate guarantee R with latency E if, all $j \geq 0$, the jth departure time, $d(j)$, of the EF traffic is less than or equal to $F(j) + E$, where $F(j)$ is defined iteratively by

$$F(0) = 0, d(0) = 0 \tag{8a}$$

For all $j > 0$:

$$F(j) = \max[a(j), \min\{d(j-1), F(j-1)\}] + \frac{L(j)}{R} \tag{8b}$$

where $a(j)$ is the arrival time of the jth packet of length $L(j)$.

If a node provides packet-scale rate guarantee, it has been shown [10] that any EF packet arriving at a node at time t will leave that node no later than at time $t + Q/R + E$, where Q is the total backlogged EF traffic at time t. This property infers that a single-hop worst delay is bounded by $B/R + E_p$ when all input traffic is constrained by a leaky-bucket regulator with parameters (R, B), where R is the configured rate, and B is the bucket size. Note that E_p is the error term for processing individual EF packets.

If every node in a Diffserv domain regulates its EF input traffic with the maximum burst size B, then the worst case delay of each packet crossing h hops is just h times the single hop delay. However, the Diffserv architecture performs traffic conditioning only at the ingress and not in the interior. As a result, it is possible for bursty traffic larger than B to arrive at a node in the network even if the EF traffic has been regulated initially at the ingress. This problem may be solved by appropriately distributing traffic and limiting the number of hops traversed by each flow. Therefore, we need to consider topology constraints as well as ingress traffic conditioning when designing an EF service. Topology-independent utilization bounds have been obtained for providing bounded delay service that take network diameter into account [23].

3.3. QBone and Practical Performance Evaluation

QBone (Quality of Service Backbone) has been constructed by Internet2 QoS Working Group to provide an inter-domain testbed for Diffserv [24]. Currently, the EFPHB has been implemented, and the QBone Premium Service (QPS) [25] using the EFPHB is available.

To use the QPS, customers should provide their traffic information, {`source, dest, route, startTime, endTime, peakRate, MTU, jitter`}. This information is required to configure EF nodes [3]. The parameter `route` specifies a DS-domain-to-DS-domain route. The parameters `startTime` and `endTime` denote the time duration of the traffic, `peakRate` is the peak rate of the traffic, `MTU` is the maximum transmission unit, and `jitter` is the worst-case jitter required. QPS traffic is regulated at an ingress node by a token bucket profiler with token rate of `peakRate` and bucket depth of `MTU`.

The QPS provides low loss rate, low queueing delay, and low jitter as long as the traffic arriving at the ingress node conforms to its token bucket profile. It is necessary to discard the nonconformant packets or to reshape them in order to achieve the desired QoS.

QBone enables us to evaluate and analyze the Diffserv performance more practically through large scale, interdomain experiments. An interesting experiment with video streaming applications has been conducted for observing end-to-end performance of the Diffserv network [26]. Here we present a summary of the experiment and results.

3.3.1. Video Streaming Applications in Diffserv Networks. With faster computer networks and improved compression schemes, video streaming applications are becoming popular. To improve quality of video playback on a given network, the applications deploy new technologies such as FEC (forward error correction) and layered coding to adjust to network dynamics. However, there is a limit to improving the quality without any QoS support from the network. To play back video clips seamlessly, a certain amount of bandwidth, bounded jitter, and low loss rate are required. Therefore, video streaming applications can be good target applications of a Diffserv network.

Experiments with video streaming applications in a Diffserv network with EFPHB have been performed [26]. A local network with several routers configured for EFPHB connected to the QBone network was used as a testbed. *Quality index* and frame loss rate were measured with various token rates and bucket sizes at ingress router policers. *Quality index* was measured by a variant of ITS (Institute of Telecommunication Sciences) video quality measurement (VQM) tool [27]. The VQM captures features of individual and sequence of frames from both the received and original streams, and measures the difference between them. The VQM was originally developed to measure quality of television and videoconferencing systems. The variant was modified from the VQM for measuring quality of video stream transmitted over IP networks.

From the experiments and measurements, interesting relationship between video quality and token bucket parameters of EF ingress policer have been observed: (1) the quality of video stream can be controlled by the ingress policer, (2) frame loss rate itself does not reflect the quality of the video stream, (3) the token rate should be configured to be larger than the video encoding rate for achieving acceptable quality, and (4) a small increase of bucket depth may result in substantial improvement of the video quality.

The first observation confirms that the Diffserv network can provide service differentiation to real-world applications.

4. NOVEL AND OTHER QoS

The Diffserv framework is still in development, and new services are constantly being proposed. In this section, we introduce some of the proposals receiving attention.

4.1. A Bulk Handling Per-Domain Behavior

Some applications need to transfer bulk data such as movie files or backup data over the network without any timing constraints. These transfers do not need any assurance as long as they are eventually completed. A bulk-handling (BH) per-domain behavior (PDB) has been proposed [28] for supporting such traffic. It is reasonable to exclude such traffic from best-effort traffic in order to prevent competition with other valuable traffic.

BHPDB traffic has the lowest priority in a Diffserv network. A BH packet is transferred only when there is no other packet. In the presence of other traffic, it may be discarded or delayed. To implement the BHPDB, only marking for PDB is enough. Policing or conditioning is not required since there is no configured rate, reserved resources or guarantees. BHPDB traffic is different from best-effort (BE) traffic in that there are at least some resources assigned to BE traffic.

4.2. An Assured Rate Per-Domain Behavior

An assured rate (AR) PDB has been proposed [29]. The ARPDB provides assured rate for one-to-one, one-to-few, or one-to-any traffic. Here *one-to-one traffic* means traffic entering a network at one ingress router and exiting at one egress router, and *one-to-few traffic* means traffic having more than one fixed egress routers. *One-to-any traffic* refers to traffic entering at one ingress router and exiting at multiple (any) egress routers. Assured rate can be implemented using the AFPHB.

The ARPDB is suitable for traffic requiring certain amount of bandwidth but no delay bounds. A possible example service with the ARPDB is VPN (virtual private network) services with one-to-few traffic. One-to-any service of the ARPDB can be used to deliver multicast traffic with a single source, but appropriate traffic measurement should be supported since packets in a multicast are duplicated in the interior of the network and the total amount of traffic at the egress routers may be larger than the amount of traffic at the ingress router. The ARPDB stipulates how packets should be marked and treated across all the routers within a Diffserv domain.

4.3. Relative Differentiated Services

Relative differentiated service has been proposed [30] to provide service differentiation without admission control. The fundamental idea of the relative differentiated services is based on the fact that absolute service guarantees are not achievable without admission control. In such cases, only relative service differentiation can be provided since network resources are limited.

A proportional delay differentiation model, which provides delays to each class on the basis of a proportionality constraint, has been proposed [30]. *Backlog-proportional rate* schedulers and *waiting-time priority* schedulers [31] were proposed as candidate schedulers for such service. Both schedulers were shown to provide predictable and controllable delay differentiation independent of the variations of loads in each class. A proportional loss rate model has also been proposed and evaluated [32].

5. SUMMARY

We have presented the basic architecture of Diffserv networks and showed how service differentiation can be achieved in such networks. The AFPHB has been shown to enable us to realize service differentiation in TCP throughput with appropriate edge marking devices. The EFPHB has been proposed to provide low delay, jitter, and loss rate. While simple scheduling schemes such as priority queueing and class-based queueing are enough to implement the EFPHB in the network core, considerable work in traffic policing and shaping at the network edge is required. Initial experience on QBone points to the possibility that applications can realize the necessary QoS in Diffserv networks. Much research work is in progress in identifying and improving various aspects of a Diffserv network, particularly in traffic management, admission control and routing.

BIOGRAPHIES

A. L. Narasimha Reddy is currently an associate professor in the Department of Electrical Engineering at Texas A&M University, College Station. He received his Ph.D in computer Engineering from the University of Illinois at Urbana-Champaign in August 1990. He was a research staff member at IBM Almaden Research Center in San Jose from August 1990–August 1995.

Reddy's research interests are in multimedia, I/O systems, network QOS, and computer architecture. Currently, he is leading projects on building scalable multimedia storage servers and partial-state based network elements. His group is also exploring various issues related to network QOS. While at IBM, he coarchitected and designed a topology-independent routing chip operating at 100 MB/sec, designed a hierarchical storage management system, and participated in the design of video servers and disk arrays. Reddy is a member of ACM SIGARCH and is a senior member of IEEE Computer Society. He has received an NSF CAREER award in 1996. He received an Outstanding Professor Award at Texas A&M during 1997–98.

Ikjun Yeom received his B.S. degree in electronic engineering from Yonsei University, Seoul, Korea, in February 1995 and his M.S. and Ph.D. degrees in computer engineering from Texas A&M University, College Station, in August 1998 and May 2001, respectively. He worked at DACOM Company located in Seoul, Korea, between 1995 and 1996 and at Nortel Networks in 2000. After working as a research professor at Kyungpook National University in 2001, he has been an assistant professor in the Department of Computer Science at KAIST since January 2002. His research interests are in congestion control, network performance evaluation, and Internet QoS.

BIBLIOGRAPHY

1. K. Nichols, S. Blake, F. Baker, and D. Black, *Definition of the Differentiated Service Field (DS Field) in the IPv4 and IPv6 Headers*, RFC 2474, Dec. 1998.
2. S. Blake et al., *An Architecture for Differentiated Services*, RFC 2475, Dec. 1998.
3. V. Jacobson, K. Nichols, and K. Poduri, *An Expedited Forwarding PHB*, RFC 2598, June 1999.
4. J. Heinanen, F. Baker, W. Weiss, and J. Wroclawski, *Assured Forwarding PHB Group*, RFC 2597, June 1999.
5. K. Nochols and B. Carpenter, *Definition of Differentiated Services Per Domain Behaviors and Rules for Their Specification*, RFC 3086, April 2001.
6. D. Clark and W. Fang, Explicit allocation of best-effort packet delivery service, *IEEE/ACM Trans. Network.* **6**(4): 362–373 (Aug. 1998).
7. J. Heinanen, T. Finland, and R. Guerin, *A Three Color Marker*, RFC 2697, Sept. 1999.
8. J. Heinanen, T. Finland, and R. Guerin, *A Two Rate Three Color Marker*, RFC 2698, Sept. 1999.
9. F. Azeem, A. Rao, and S. Kalyanaraman, TCP-friendly traffic marker for IP differentiated services, *Proc. IWQoS'2000*, Pittsburgh, PA, June 2000, pp. 35–48.
10. J. Bennett et al., Delay jitter bounds and packet scale rate guarantee for expedited forwarding, *Proc. Infocom*, 2001.
11. B. Davie et al., *An Expedited Forwarding PHB*, Work in Progress, April 2001.
12. A. Charny et al., *Supplemental Information for the New Definition of the EFPHB*, Work in Progress, June 2001.
13. S. Floyd and V. Jacobson, Like-sharing and resource management models for packet networks, *IEEE/ACM Trans. Network.* **3**(4): 365–386 (Aug. 1995).
14. V. Jacobson and M. Karels, Congestion avoidance and control, *Proc. SIGCOMM'88*, Stanford, CA, Aug. 1998, pp. 314–329.
15. S. McCanne and S. Floyd, *ns-LBL network simulator*; see: *http://www.nrg.ee.lbl.gov/ns/*.
16. I. Yeom and A. L. N. Reddy, Realizing throughput guarantees in a differentiated services network, *Proc. ICMCS'99*, Florence, Italy, June 1999, pp. 372–376.
17. S. Sahu et al., On achievable service differentiation with token bucket marking for TCP, *Proc. SIGMETRICS'2000*, Santa Clara, CA, June 2000, pp. 23–33.
18. I. Stoica and H. Zhang, LIRA: An approach for service differentiation in the Internet, *Proc. NOSSDAV'98*, Cambridge, UK, June 1998, pp. 345–359.
19. I. Yeom and A. L. N. Reddy, Modeling TCP behavior in a differentiated services network, *IEEE/ACM Trans. Network.* **9**(1): 31–46 (Feb. 2001).
20. S. Gopalakrishnan and A. L. N. Reddy, SACRIO: An active buffer management schemes for differentiated services networks, *Proc. NOSSDAV'01*, June 2001.
21. I. Yeom and A. L. N. Reddy, Adaptive marking for aggregated flows, *Proc. Globecom*, 2001.
22. V. Jacobson, K. Nochols, and K. Poduri, *The Virtual Wire Behavior Aggregate*, March 2000, Work in Progress.
23. S. Wang, D. Xuan, R. Bettati, and W. Zhao, Providing absolute differentiated services for real-time applications in static priority scheduling networks, *Proc. IEEE Infocom*, April 2001.

24. B. Teitelbaum, *QBone Architecture (v1.0)*, Internet2 QoS Working Group Draft, Aug. 1999; see: *http://www.internet2.edu/qos/wg/qbArch/1.0/draft-i2-qbonearch-1.0.html*.

25. K. Nichols, V. Jacobson, and L. Zhang, *A Two-Bit Differentiated Services Architecture for the Internet*, Nov. 1997; see: *ftp://ftp.ee.lbl.gov/papers/dsarch.pdf*.

26. W. Ashmawi, R. Guerin, S. Wolf, and M. Pinson, On the impact of pricing and rate guarantees in Diff-Serv networks: A video streaming application perspective, *Proc. ACM Sigcomm'01*, Aug. 2001.

27. ITU-T Recommendation J. 143, *User Requirements for Objective Perceptual Video Quality Measurements in Digital Cable Television*, Recommendations of the ITU, Telecommunication Standardization Sector.

28. B. Carpenter and K. Nichols, *A Bulk Handling Per-Domain Behavior for Differentiated Services*, Internet Draft, Jan. 2001; see: *http://www.ietf.org/internet-drafts/draft-ietf-diffserv-pdb-bh-02.txt*.

29. N. Seddigh, B. Nandy, and J. Heinanen, *An Assured Rate Per-Domain Behavior for Differentiated Services*, Work in Progress, July 2001.

30. C. Dovrolis, D. Stiliadis, and P. Ramanathan, Proportional differentiated services: delay differentiation and packet scheduling, *Proc. SIGCOMM'99*, Aug. 1999, pp. 109–120.

31. L. Kleinrock, *Queueing Systems*, Vol. II, Wiley, 1976.

32. C. Dovrolis and P. Ramanathan, Proportional differentiated services, Part 2: Loss Rate Differentiation and Packet Dropping, *Proc. IWQoS'00*, Pittsburgh PA, June 2000.

DIGITAL AUDIOBROADCASTING

Rainer Bauer
Munich University of Technology (TUM)
Munich, Germany

1. INTRODUCTION

The history of broadcasting goes back to James Clerk Maxwell, who theoretically predicted electromagnetic radiation in 1861; and Heinrich Hertz, who experimentally verified their existence in 1887. The young Italian Guglielmo Marconi picked up the ideas of Hertz and began to rebuild and refine the Hertzian experiments and soon was able to carry out signaling over short distances in the garden of the family estate near Bologna. In the following years the equipment was continuously improved and transmission over long distances became possible. The first commercial application of radio transmission was point-to-point communication to ships, which was exclusively provided by Marconi's own company. Pushed by the promising possibilities of wireless communication the idea of broadcasting soon also came up. The first scheduled radio program was transmitted by a 100-W transmitter at a frequency around 900 kHz in Pittsburgh late in 1920.

The installation of more powerful transmitters and progress in receiver technology led to a rapidly increasing number of listeners. A milestone in broadcasting history was the invention of frequency modulation (FM) by Armstrong in the 1930s and the improvement in audio quality that came along with the new technology. Because FM needed higher transmission bandwidths, it was necessary to move to higher frequencies. While the first FM radio stations in the United States operated at frequencies around 40 MHz, today's FM radio spectrum is located in the internationally recognized VHF FM band from 88 to 108 MHz. Although the reach of the FM signals was below that of the former AM programs, due to the shorter wavelength of the FM band frequencies, broadcasters adopted the new technology after initial skepticism and the popularity of FM broadcasting grew rapidly after its introduction.

The next step toward increased perceived quality was the introduction of stereobroadcasting in the 1960s. Constant advances in audiobroadcasting technology resulted in today's high-quality audio reception with analog transmission technology. So, what is the reason for putting in a lot of time and effort into developing a new digital system?

In 1982 the compact disk (CD) was introduced and replaced the existing analog record technology within just a few years. The advantages of this new digital medium are constant high audio quality combined with robustness against mechanical impacts. The new technology also resulted in a new awareness of audio quality and the CD signal became a quality standard also for other audio services like broadcasting.

Using good analog FM equipment, the audio quality is fairly comparable to CD sound when the receiver is stationary. However, along with the development of broadcasting technology, the customs of radio listeners also changed. While at the beginning, almost all receivers were stationary, nowadays many people are listening to audiobroadcasting services in their cars, which creates a demand for high-quality radio reception in a mobile environment. This trend also reflects the demand for mobile communication services, which has experienced a dramatic boom.

Due to the character of the transmission channel, which leads to reflections from mountains, buildings, and cars, for example, in combination with a permanently changing environment caused by the moving receiver, a number of problems arise that cannot be handled in a satisfying way with existing analog systems. The problem here is termed multipath propagation and is illustrated in Fig. 1. The transmitted signal arrives at the receiver not only via the direct (line-of-sight) path but also as reflected and scattered components that correspond to different path delays and phase angles. This often results in severe interference and therefore signal distortion.

Another drawback is interference caused by transmission in neighboring frequency bands (adjacent-channel interference) or by transmission of different programs in the same frequency with insufficient spacial distance between the transmitters (cochannel interference).

The tremendous progresses in digital signal processing and microelectronics in the 1980s initiated a trend to supplement and even substitute existing analog systems with digital systems. Introduction of the compact disk was already mentioned above, but a change from analog

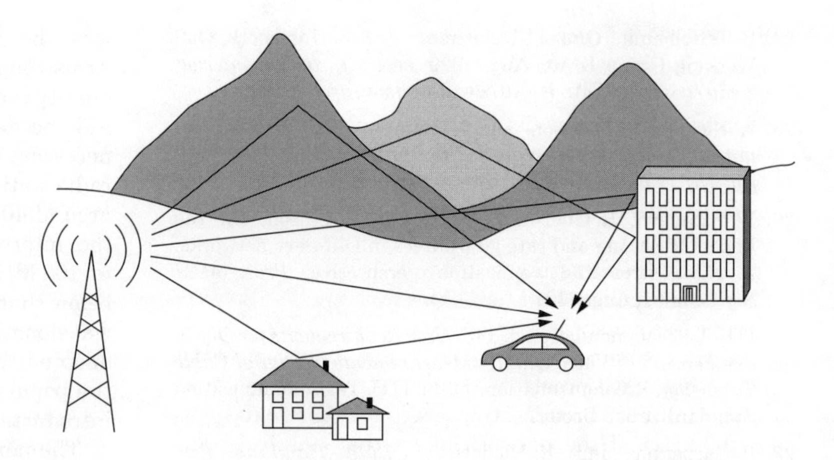

Figure 1. Scenario with multipath propagation.

to digital systems also took place in the field of personal communications. This was seen, for example, in the introduction of ISDN (Integrated Services Digital Network) in wireline communications and also in GSM (Global System for Mobile Communications) for personal mobile communications. The latter system caused a rapidly growing market for mobile communications systems and drove the development of new high-performance wireless communication techniques.

Several advantages of digital systems suggest the application of digital techniques also in broadcasting, which is still dominated by analog systems [1]. For instance, advanced digital receiver structures and transmission and detection methods are capable of eliminating the distortions caused by multipath propagation, which enables high-quality reception also in mobile environments.

To use the available frequency spectrum in an efficient way, the digitized audio signals can be compressed with powerful data reduction techniques to achieve a compression factor of up to 12 without perceptible degradation of audio quality compared to a CD signal.

Techniques such as channel coding to correct transmission errors are applicable only to digital signals. These methods allow an error free transmission even when errors occur on the transmission channel.

With advanced signal detection and channel coding techniques, a significantly lower signal to noise ratio is sufficient in producing a particular sound quality when compared to equivalent analog transmission systems. This results in lower transmission power and therefore reduced costs and also less "electromagnetic pollution," which has recently become a topic of increasing importance and relevance.

Another advantage of a digital broadcasting system is its flexibility. While conventional analog systems are restricted to audio transmission, a digital system can also provide other services besides audio transmission and is therefore open to trends and demands of the future. In combination with other digital services such as Internet or personal communications, a digital broadcasting technology further supports the convergence of information and communication techniques. Also there is a change of paradigm in the way we obtain information in everyday life. The growing importance of the Internet obliterates the boundaries between individual and mass communications. Broadcasting will have to face competition with other communication networks that have access to the consumer. To meet future demands, it is necessary to introduce digital broadcasting systems.

2. SYSTEM ASPECTS AND DIFFERENT APPROACHES

Besides technical aspects, the introduction of a digital audiobroadcasting system must also meet market issues. The transition from an analog to a digital system has to take place gradually. Users have to be convinced that digital audiobroadcasting has significant advantages compared to analog systems in order to bring themselves to buy new digital receivers instead of conventional analog ones. This can be obtained only by added value and new services. On the other hand, the situation of the broadcasters and network operators must be considered. Simultaneous transmission of digital and analog programs results in increasing costs that have to be covered. Standards have to be adopted to guarantee planning reliability to broadcasters, operators, and also manufacturers [2].

2.1. An Early Approach Starts Services Worldwide

An early approach to standardize a digital audiobroadcasting system was started in the 1980s by a European telecommunication consortium named Eureka. In 1992 the so-called Eureka-147 DAB system was recommended by the International Telecommunication Union (ITU) and became an ITU standard in 1994. The standard is aimed at terrestrial and satellite sound broadcasting to vehicular, portable and fixed receivers and is intended to replace the analog FM networks in the future. The European Telecommunications Standard Institute (ETSI) also adopted this standard in 1995 [3].

The terrestrial service uses a multicarrier modulation technique called coded orthogonal frequency-devision multiplex (COFDM), which is described later in this article along with other technical aspects of this system. With this multiplexing approach up to six high-quality audio programs (called an ensemble) are transmitted via closely spaced orthogonal carriers that allocate a total bandwidth of 1.536 MHz. The overall net data rate that can be transmitted in this ensemble is approximately 1.5 Mbps (megabits per second) [4]. Although originally a sound broadcasting system was to be developed, the multiplex signal of Eureka-147 DAB is very flexible, allowing the transmission of various data services also. Because of the spreading in frequency by OFDM combined with channel coding and time interleaving, a reliable transmission can be guaranteed with this concept. A further advantage of the Eureka-147 system is the efficient use of radio spectrum by broadcasting in a single-frequency network (SFN). This means that one program is available at the same frequency across the whole coverage area of the network. In a conventional analog audiobroadcasting environment, each program is transmitted on a separate carrier (in FM broadcasting with a bandwidth of 300–400 kHz). If the network offers the same program outside the coverage area of a particular transmitter another frequency is usually used to avoid cochannel interference. On the other hand, adjacent channels within the coverage area of one transmitter are not used in order to reduce adjacent-channel interference. This leads to a complex network and requires a sophisticated frequency management, both of which can be avoided with a single-frequency network. The basic structures of a single-frequency network and a multiple frequency network are compared in Fig. 2. A further advantage of a single-frequency network is the simple installation of additional transmitters to fill gaps in the coverage.

To operate a system like Eureka-147 DAB dedicated frequency bands have to be reserved because it cannot coexist with an analog system in the same frequency band. In 1992 the World Administrative Radio Conference (WARC) reserved almost worldwide a 40-MHz wide portion (1452–1492 MHz) of the L band for this purpose. In several countries a part of the spectrum in the VHF band III (in Germany, 223–240 MHz) is also reserved for Eureka-147 DAB.

At present, the Eureka-147 DAB system is in operation or in pilot service in many countries worldwide [5].

2.2. A System for More Flexibility

The strategy to merge several programs into one broadband multiplex signal prior to transmission has several advantages with respect to the received signal quality. However, the ensemble format also shows drawbacks that aggravate the implementation of Eureka-147 DAB in certain countries. The broadcasting market in the United States for example is highly commercial with many independent private broadcasters. In this environment it would be almost impossible to settle on a joint transition scenario from analog to digital. Also, the reserved frequency spectrum in the L band is not available in the United States [6]. Therefore, a different system was considered by the U.S. broadcasting industry. The idea was that no additional frequency spectrum (which would cause additional costs, because spectrum is not assigned but auctioned off in the United States) should be necessary, and the broadcasters should remain independent from each other and be able to decide on their own when to switch from analog to digital. A concept that fulfills all these demands is in-band/on-channel (IBOC) digital audiobroadcasting. In this approach the digital signal is transmitted in the frequency portion to the left and the right of the analog spectrum, which is left free to avoid interference from nearby signals. The basic concept is shown in Fig. 3. The spectral position and power of the digital signal is designed to meet the requirements of the spectrum mask of the analog signal. The IBOC systems will be launched in an "hybrid IBOC" mode, where the digital signal is transmitted simultaneously to the analog one. When there is sufficient market penetration of the digital services, the analog programs can be switched off and the spectrum can be filled with a digital signal to obtain an all-digital IBOC system.

After the merger of USA Digital Radio (USADR) and Lucent Digital Radio (LDR) to iBiquity Digital Corp. there is currently a single developer of IBOC technology in the United States.

2.3. Digital Sound Broadcasting in Frequency Bands Below 30 MHz

The first radio programs in the 1920s were transmitted at frequencies around 900 kHz using analog amplitude modulation (AM). Although frequency modulation (FM), which was introduced in the late 1930s and operated at higher frequencies, allows a better audio quality, there are still many radio stations, particularly in the United States, that use the AM bands for audiobroadcasting. One advantage of these bands is the large coverage area that

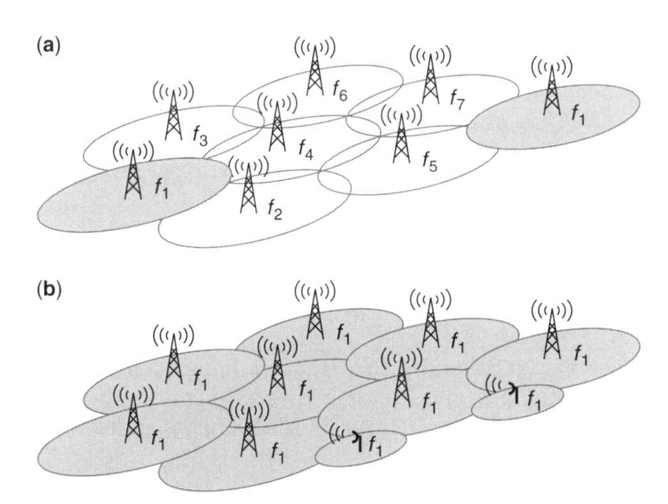

Figure 2. Conventional FM networks (**a**) and single-frequency network (**b**).

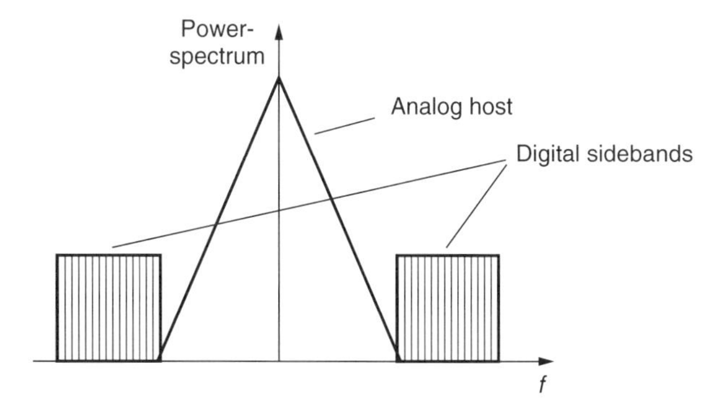

Figure 3. In-band/on-channel signaling.

can be obtained because of the propagation conditions of these frequencies. A major drawback, however, is the poor sound quality, due to the limited bandwidth available in these low-frequency bands.

On the other hand, the coverage properties still make this frequency band attractive for audiobroadcasting. With state-of-the-art audio compression technology and digital transmission methods, the sound quality can be increased to near FM quality.

In 1994 the European project (*Narrow Band Digital Broadcasting* (*NADIB*)) started to develop concepts for a digital system operating in the AM band. The international consortium Digital Radio Mondiale (DRM) [7], which was officially founded in 1998, began to develop an appropriate system. Together with a second approach proposed by iBiquity Digital Corp., the ITU finally gave a recommendation that encompasses both systems for digital sound broadcasting in the frequency bands below 30 MHz [8].

One major difference between both approaches is the strategy to deal with existing analog AM signals. While the DRM scheme is an all-digital approach that occupies all frequencies within the channels with a bandwidth of 9 or 10 kHz or multiples of these bandwidths, the iBiquity system applies the IBOC strategy described above for FM systems.

2.4. Mobile Satellite Digital Audio Radio Services (SDARS)

While the intention of terrestrial digital audio broadcasting technology is to replace existing analog radio in the AM and FM bands, new satellite platforms are emerging that are intended for mobile users. Previous satellite services such as ASTRA Digital (a proprietary standard of the satellite operator SES/ASTRA) or DVB-S, the satellite distribution channel of the European DVB (Digital Video Broadcasting) system, were designed to serve stationary users with directional satellite dishes. In 1999 the WorldSpace Corp. started its satellite service to provide digital high quality audio to emerging market regions such as Africa, Asia, and South and Central America [9]. Originally developed to serve portable receivers, the system is about to be expanded to mobile receivers as well.

Two systems that are targeting mobile users from the beginning are the two U.S. satellite systems operated by Sirius Satellite Radio Inc. and XM Satellite Radio Inc., which intended to launch their commercial service in 2001. Both systems are based on proprietary technology [10].

The three systems differ in the orbital configurations of their satellites. The complete WorldSpace network will consist of three geostationary satellites (each using three spot beams and serving one of the intended coverage areas Africa, parts of Asia, and South and Central America). The XM system uses only two geostationary satellites located to guarantee optimum coverage of the United States. A completely different approach was chosen for the Sirius system, where three satellites in a highly elliptical orbit rise and set over the coverage area every 16 h. The orbit enables the satellites to move across the coverage area at a high altitude (even higher than a geostationary orbit) and therefore also provide a high elevation angle. Two of the three satellites are always visible to provide sufficient diversity when a direct signal to one of the satellites is blocked.

To guarantee reception even in situations when the satellite signal is totally blocked, for example, in tunnels or in urban canyons, Sirius and XM use terrestrial repeaters that rebroadcast the signal.

One major advantage of a satellite system is the large coverage area. Regions with no or poor terrestrial broadcasting infrastructure can be easily supplied with high-quality audio services. On the other hand, it is difficult to provide locally oriented services with this approach.

2.5. Integrated Broadcasting Systems

Besides systems that are designed primarily to broadcast audio services, approaches emerge that provide a technical platform for general broadcasting services. One of these systems is the Japanese Integrated Services Digital Broadcasting (ISDB) approach, which covers satellite, cable, and terrestrial broadcasting as distribution channels. ISDB is intended to be a very flexible multimedia broadcasting concept that incorporates sound and television and data broadcasting in one system. The terrestrial component ISDB-T, which is based on OFDM transmission technology, will be available by 2003–2005. A second scheme that should be mentioned here is DVB-T, the terrestrial branch of the European digital video broadcasting (DVB) system, which is also capable of transmitting transparent services but is not optimized for mobile transmission.

2.6. Which Is the Best System?

To summarize the different approaches in sound broadcasting to mobile receivers, we have to consider the basic demands of the respective market (both customer and broadcaster aspects).

If the situation among the broadcasters is sufficiently homogeneous, which allows the combination of several individual programs to ensembles that are jointly transmitted in a multiplex, then the Eureka-147 system provides a framework for spectrally efficient nationwide digital audiobroadcasting. By adopting this system, a strategy for the transition from analog to digital that is supported by all parties involved must be mapped out.

A more flexible transition from analog to digital with no need for additional frequency bands is possible with the IBOC approach. Also, small local radio stations can be more easily considered using this approach.

However, if the strategy is to cover large areas with the same service, then satellite systems offer advantages over the terrestrial distribution channel.

3. THE EUREKA-147 DAB SYSTEM

Several existing and future systems have been summarized in the previous section. In this section we focus on Eureka-147 DAB as it was the first digital audiobroadcasting system in operational service.

Eureka-147 DAB is designed to be the successor of FM stereobroadcasting. It started as a proposal of a European consortium and is now in operational or pilot service in countries around the world [5]. Although several alternative approaches are under consideration, right now Eureka-147 DAB is the first all-digital sound broadcasting system that has been in operation for years. When we refer to DAB below, we mean Eureka-147 DAB. Some features of the system are

- Data compression with MPEG-Audio layer 2 (MPEG — Moving Picture Experts Group) according to the standards ISO-MPEG 11172-3 and ISO-MPEG 1318-3 (for comparison with the well-known audio compression algorithm MP3, which is MPEG-Audio layer 3).
- Unequal error protection (UEP) is provided to the compressed audio data where different modes of protection are provided to meet the requirements of different transmission channels, namely, radiofrequency transmission for a variety of scenarios or cable transmission.
- The concept of (coded orthogonal frequency-devision multiplex (COFDM)) copes very well with the problem of multipath propagation, which is one of the major problems in mobile systems. Furthermore, this transmission scheme allows the operation of a single-frequency network (SFN).
- The DAB transmission signal carries a multiplex of several sound and data services. The overall bandwidth of one ensemble is 1.536 MHz and provides a net data rate of approximately 1.5 Mbps. The services can be combined very flexibly within one ensemble. Up to six high-quality audio programs or a mixture of audio and data services can be transmitted in one ensemble.

3.1. General Concept

The basic concept of signal generation in a DAB system is shown in Fig. 4. The input to the system may either be one or several audio programs or one or several data services together with information about the multiplex structure, service information, and so on. Audio and data services form the main service channel (MSC), while service and multiplex information are combined in the fast information channel (FIC). Every input branch undergoes a specific channel coding matched to the particular protection level required by the channel. Data compression is specified only for audio signals in the DAB standard. Data services are transparent and are transmitted in either a packet mode or a stream mode.

After additional time interleaving, which is skipped in Fig. 4, audio and data services form the MSC, which consists of a sequence of so-called common interleaved frames (CIFs) assembled by the main service multiplexer. The final transmission signal is generated in the transmission frame multiplexer, which combines the MSC and the FIC. Together with the preceding synchronization information, OFDM symbols are formed and passed to the transmitter.

In the following the main components of the block diagram (flowchart) in Fig. 4 are explained in more detail.

3.2. Audio Coding

Since the bit rate of a high-quality audio signal (e.g., a CD signal with 2×706 kbps) is too high for a spectral efficient transmission, audio coding according to the MPEG Audio layer 2 standard is applied. For a sampling frequency of 48 kHz the resulting bitstream complies with the ISO/IEC 11172-3 layer 2 format, while a sampling frequency of 24 kHz corresponds to ISO/IEC 13818-3 layer 2 LSF (low sampling frequency). The main idea behind the compression algorithms is utilization of the properties of human audio perception, which are based on spectral and temporal masking phenomena. The concept is shown in Fig. 5, where the sound pressure level is plotted as a function of frequency. The solid curve indicates the threshold in quiet, which is the sound pressure level of a pure tone that is barely audible in a quiet environment. The curve shows that a tone has to show a higher level

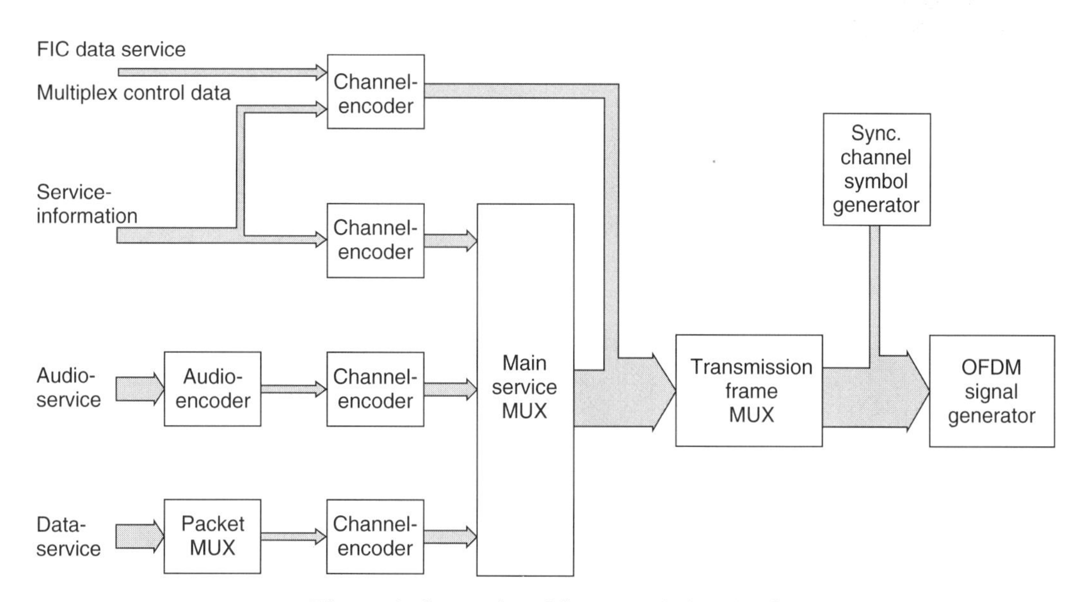

Figure 4. Generation of the transmission signal.

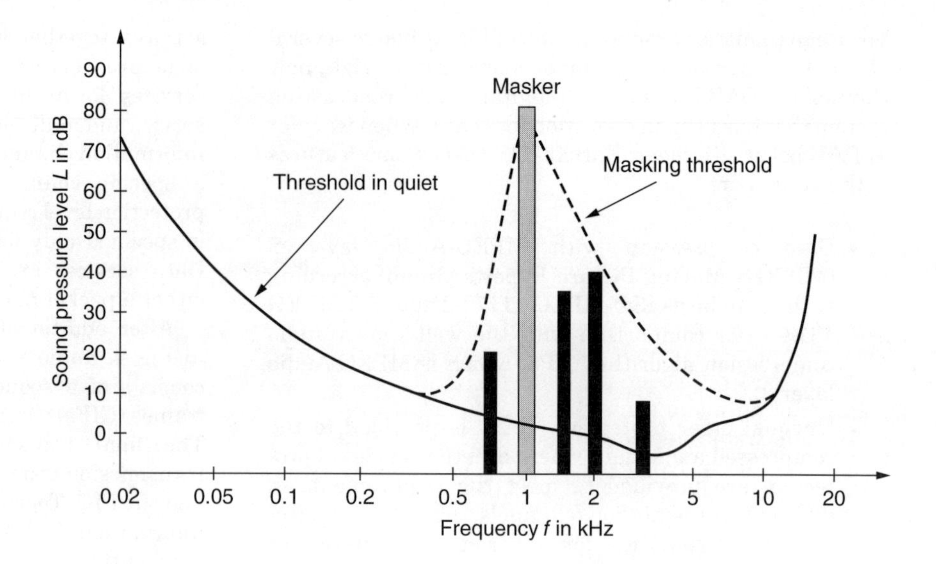

Figure 5. Example for masking.

at very low and very high frequencies to be perceivable than at medium frequencies around 3 kHz, where the human auditory system is very sensitive. Besides the threshold in quiet, each additional sound also creates a masking pattern that is also depicted in the figure. The shape of the pattern depends on the level and the frequency of the underlying sound. A general observation is that the slope of the masking pattern is steeper toward lower frequencies than in the opposite direction. All sound events that lie below this masking pattern (also indicated in Fig. 5) are not perceivable by the human ear and therefore do not have to be transmitted. Since a general audio signal consists of a more complex spectrum, the first step in audio coding is a transformation from the time domain into the frequency domain. Each frequency component creates a masking pattern by itself, and the masking pattern of the overall signal can be calculated by a superposition of the individual patterns. The signal is divided into frequency bands in the spectral domain, and each band is coded (quantized) separately in such a way that the quantization noise lies below the masking threshold in this band. By this technique the quantization noise can be shaped along the frequency axis, and the overall bit rate that is necessary to represent the signal can be reduced.

A simplified block diagram of the audio encoder is shown in Fig. 6. For processing, the sampled signal is divided into segments of length 24 or 48 ms, respectively. Each segment is then transformed from the time domain into the frequency domain by a filter bank with 32 subbands. In parallel, the masking threshold is calculated for each segment in a psychoacoustic model. The subband samples undergo a quantization process in which the number of quantization levels are controlled by the requirements given by the psychoacoustic model. Finally, the quantized samples together with the corresponding side information that is necessary to reconstruct the signal in the decoder are multiplexed into an audio frame. The frame also contains program-associated data (PAD).

The bit rates available for DAB are between 8 and 192 kbps for a monophonic channel. To achieve an audio quality that is comparable to CD quality, approximately 100 kbps are necessary per monochannel, which means a data reduction of a factor of ~7. Since the Eureka-147 DAB standard was fixed in the early 1990s, MPEG Audio layer 2 was chosen for audio coding because it combines good compression results with reasonable complexity.

With today's leading-edge audio compression algorithms such as MPEG2-AAC (advanced audiocoding) or the Lucent PAC (perceptual audiocoder) codec, compression gains of a factor of 12 can be realized without noticeable differences in a CD signal. The AAC codec, for example, is proposed for audiocoding in the Digital Radio Mondiale approach as well as in the Japanese ISDB system, while the PAC technology is applied in the Sirius and XM satellite services, for example.

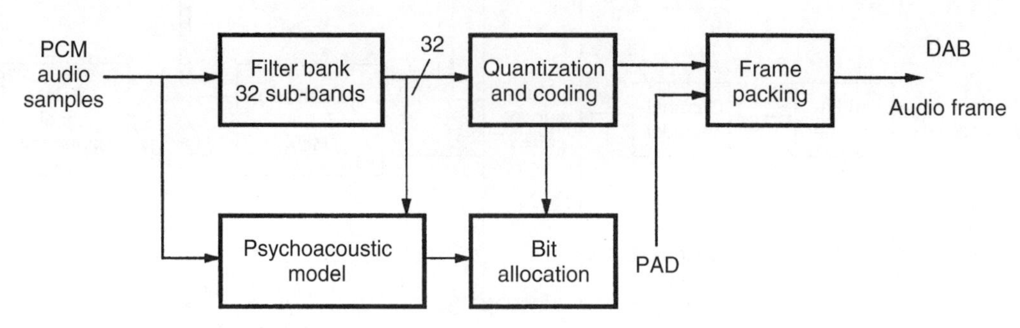

Figure 6. Simplified block diagram of an DAB audio encoder.

3.3. Channel Coding

Between the components for audioencoding and channel encoding an energy dispersal is performed by a pseudo random scrambling that reduces the probability of systematic regular bit patterns in the data stream.

The basic idea of channel coding is to add redundancy to a digital signal in such a way, that the redundancy can be exploited in the decoder to correct transmission errors. In DAB, a convolutional code with memory six and rate $\frac{1}{4}$ is applied. The encoder is depicted in Fig. 7. The code rate $\frac{1}{4}$ means that for every information bit that enters the encoder, 4 coded bits are produced. In the case of an audio signal not every part of the source coded audio frame has the same sensitivity to transmission errors. Very sensitive segments have to be protected by a strong code, while a weaker code can be applied to other parts. This concept, called unequal error protection (UEP), can easily be realized by a technique termed puncturing, which means that not every output bit of the convolutional encoder is transmitted. According to a defined rule, some of the bits are eliminated (punctured) prior to transmission. This technique is also shown in Fig. 7. A binary 1 in the puncturing pattern means that the corresponding bit at the output of the convolutional encoder is transmitted, while a binary 0 indicates that the bit is not transmitted (punctured). In the DAB specification, 24 of these puncturing patterns are defined, which allows a selection of code rates between $\frac{8}{9}$ and $\frac{8}{32}$. A code rate of $\frac{8}{9}$ means that for every 8 bits that enter the encoder, 9 bits are finally transmitted over the channel (depicted in the upper pattern in Fig. 7). On the other hand, with a code rate of $\frac{8}{32}$, all output bits are transmitted as indicated by the lower pattern in Fig. 7. To apply unequal error protection, the puncturing pattern can be changed within an audio frame. This is shown in Fig. 8. While header and side information is protected with the largest amount of redundancy, the scale factors are protected by a weaker code, and the subband samples finally have the least protection. The program-associated data (PAD) at the end of an audio frame are protected roughly by the same code rate as the scale factors. Besides the different protection classes within an audio frame, five protection levels are specified to meet the requirements of the intended transmission scenario. A cable transmission for example needs far less error protection than a very critical mobile multipath environment. Therefore, the DAB specifications contain 64 different protection profiles

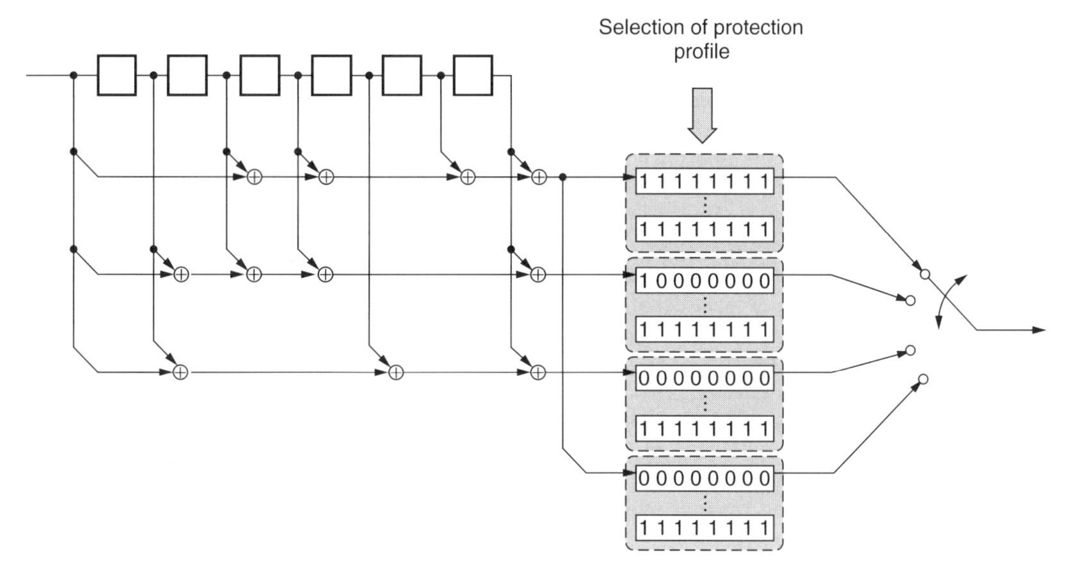

Figure 7. Channel encoder and puncturing.

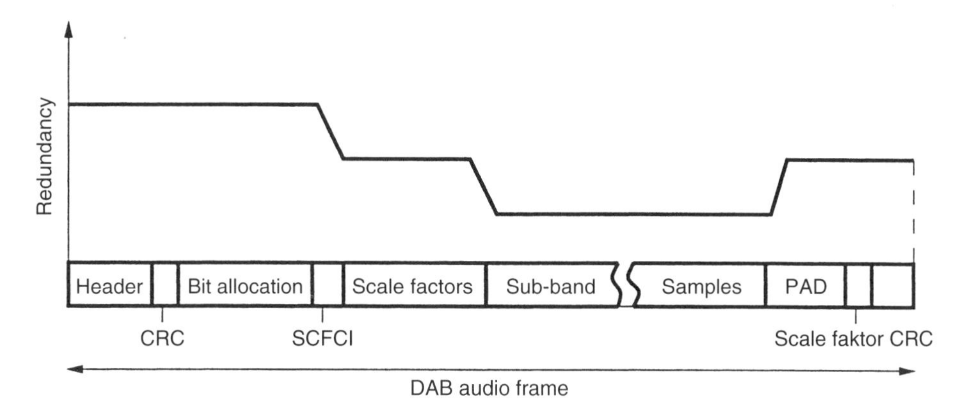

Figure 8. DAB audio frame with unequal error protection.

for different combinations of audio bit rate (32–384 kbps) and protection level.

3.4. Modulation and Transmission Format

As shown in Fig. 4, the channel coded data are multiplexed in the main service multiplexer. The structure of the frames that are finally transmitted across the air interface depends on the chosen transmission mode. DAB provides four different modes depending on the intended radiofrequency range. The duration of a transmission frame is either 24, 48, or 96 ms, and each frame consists of data from the fast information channel (FIC) and the main service channel (MSC) preceded by synchronization information.

The transmission scheme for DAB is orthogonal frequency-devision multiplex (OFDM), the motivation for which will be illustrated in a brief example.

A main problem of the mobile radio channel is multipath propagation. Let us consider a system that transmits at a symbol rate of $r = 1$ Msps (million symbols per second) (with QPSK modulation, this means a bit rate of 2 Mbps) on a single-carrier system. Hence, one symbol has a duration of $T_{sc} = 1\ \mu s$ (where subscript "sc" indicates single carrier), and the bandwidth of the system is $1/T_{sc} = 1$ MHz. We further assume a maximum channel delay τ_{max} of 80 μs. This means a difference in length between the direct path and the path with the largest delay of 24 km, which is a reasonable value for a terrestrial broadcasting channel. If we consider the relation between the symbol duration and the maximal path delay $\tau_{max}/T_{sc} = 80$, we see that this approach leads to heavy intersymbol interference since a received signal is still influenced by the 80 previously sent symbols. We can cope with this problem more easily if we do not transmit the datastream on a single carrier but multiplex the original datastream into N parallel steams (e.g., $N = 1000$) and modulate a separate carrier frequency with each individual stream. With the overall symbol rate given above, the symbol rate on each carrier is reduced to $r/N = 1000$ sps, which means a symbol duration of only $T_{mc} = 1$ ms (where subscript "mc" denotes multiple carriers). If we compare again the symbol duration with the maximal channel delay, a received symbol overlaps with only an 8% fraction of the previous symbol.

But what does this mean for the bandwidth that is necessary to transmit the large number of carriers? A very elegant way to minimize the carrier spacing without any interference between adjacent carriers is to use rectangular pulses on each subcarrier and space the resulting $\sin(x)/x$-type spectra by the inverse of the pulse (symbol) duration T_{mc}. This results in orthogonal carriers, and the overall bandwidth of the system with the parameters given above is $N \times 1/T_{mc} = 1000 \times 1\ \text{kHz} = 1$ MHz, which is the same as for the single-carrier approach. This multicarrier approach with orthogonal subcarriers, called orthogonal frequency-devision multiplex (OFDM), is used as the transmission scheme in Eureka-147 DAB. Another advantage of OFDM is the simple generation of the transmission signal by an inverse discrete Fourier transform (IDFT), which can be implemented with low complexity using the fast fourier transform (FFT). Therefore, the modulation symbols of a transmission frame are mapped to the corresponding carriers before an IFFT generates the corresponding time-domain transmission signal. The individual carriers of the DAB signal are DQPSK-modulated, where the "D" stands for differential, meaning that the information is carried in the phase difference between two successive symbols rather than in the absolute phase value. This allows an information recovery at the receiver by just comparing the phases of two successive symbols. To initialize this process, the phase reference symbol has to be evaluated, which is located at the beginning of each transmission frame.

One essential feature of OFDM we did not mention yet is the guard interval. Since each overlapping of received symbols disturbs the orthogonality of the subcarriers and leads to a rapidly decreasing system performance, this effect has to be avoided. By introduction of a guard interval at the beginning of each OFDM symbol, this interference can be avoided. This guard interval is generated by periodically repeating the tail fraction of the OFDM symbol at the beginning of the same symbol. As long as the path delay is not longer than the guard interval, only data that belong to the actual symbol fall into the receiver window. By this technique all information of delayed path components contribute constructively to the received signal. This concept is sketched in Fig. 9.

The length of the guard interval specifies the maximal allowed path delay. In order to fix the parameters of an OFDM system, a tradeoff between two elementary properties of the mobile radio channel has to be made.

On one hand, the guard interval has to be sufficiently large to avoid interference due to multipath effects. This can be obtained by a large symbol duration. Because the carrier spacing is the inverse of the symbol duration, this results in a large number of closely spaced carriers within the intended bandwidth. On the other hand, a small carrier spacing means a high sensitivity to frequency shifts caused by the Doppler effect when the receiver is moving. This, in turn, depends on the used radiofrequency and the speed of the vehicle.

In Table 1 four sets of transmission parameters are given for the different scenarios (transmission modes). Mode I is suitable for terrestrial single frequency networks in the VHF frequency range. Mode II is suitable for smaller single-frequency networks or locally oriented conventional networks because of the rather small guard interval. Mode III has an even smaller guard interval and is designed for satellite transmission on frequencies up to 3 GHz where path delay is not the dominating problem. Mode IV is designed for single-frequency networks operating at frequencies higher than those of Mode I, and the parameters are in between those of Mode I and Mode II.

3.5. DAB Network

As mentioned before, DAB allows the operation of single-frequency networks. Several transmitters synchronously broadcast the same information on the same frequency. This becomes possible because of the OFDM transmission technique. For the receiver, it makes no difference whether

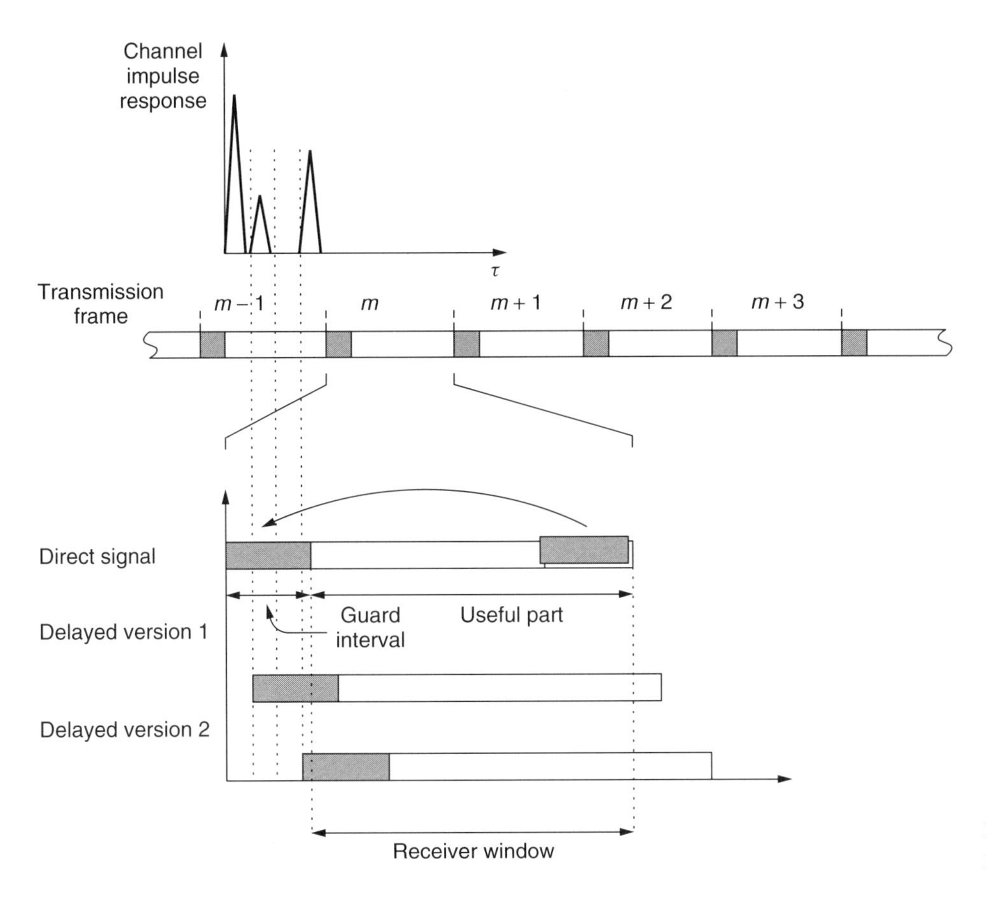

Figure 9. OFDM symbol in multipath environment.

Table 1. DAB Parameters for Transmission Modes I–IV [4]

Parameter	Mode I	Mode II	Mode III	Mode IV
Number of carriers	1536	384	192	768
Frame duration	96 ms	24 ms	24 ms	48 ms
Carrier spacing	1 kHz	4 kHz	8 kHz	2 kHz
Useful symbol duration (inverse carrier spacing)	1 ms	250 μs	125 μs	500 μs
Guard interval duration	246 μs	62 μs	31 μs	123 μs
Maximal transmitter separation	96 km	24 km	12 km	48 km
Frequency range for mobile transmission	≤375 MHz	≤1.5 GHz	≤3 GHz	≤1.5 GHz

the received signal components all come from the same transmitter or stem from different transmitters. Each component contributes constructively to the received signal as long as the path delays stay within the guard interval. Therefore, the maximal distance between the transmitters is determined by the length of the guard interval specified by the used transmission mode. To ensure synchronous transmission within the network, a time reference is necessary, which can be provided, for example, by the satellite-based Global Positioning System (GPS).

BIOGRAPHY

Rainer Bauer received his Dipl.-Ing. degree in electrical engineering and Dr.-Ing. degree from Munich University of Technology (TUM) in 1995 and 2002, respectively. Since 1995, he has been working at the Institute for Communications Engineering of TUM. His areas of interest are source-channel coding and decoding, as well as iterative decoding techniques and their application to wireless audio communications.

BIBLIOGRAPHY

1. W. Hoeg and T. Lauterbach, eds., *Digital Audio Broadcasting—Principles and Applications*, Wiley, New York, 2001.
2. WorldDAB Forum (no date), public documents (online), WorldDAB Documentation: *Thought on a transition scenario from FM to DAB*, Bayerische Medien Technik GmbH (BMT), *http://www.worlddab.org/dab/aboutdab_frame.htm* (Aug. 2001).

3. European Telecommunications Standards Institute, *Radio Broadcasting Systems; Digital Audio Broadcasting (DAB) to Mobile, Portable and Fixed Receivers*, European Standard ETSI EN 300 401 V1.3.2 (2000-09), 2000.
4. WorldDAB Forum (no date), public documents (online), *EUREKA-147—Digital Audio Broadcasting, http://www.worlddab.org/dab/aboutdab_frame.htm* (Aug. 2001).
5. WorldDAB (2001), homepage (online), *http://www.WorldDAB.org*.
6. D. Lavers, IBOC—Made in America, *Broadcast Dialogue* (Feb. 2000).
7. Digital Radio Mondiale (2001), Homepage (online), *http://www.DRM.org*.
8. International Telecommunication Union, *Systems for Digital Sound Broadcasting in the Broadcasting Bands below 30 MHz*, draft new recommendation ITU-R BS.[DOC.6/63], Oct. 2000.
9. WorldSpace (2001), homepage (online), *http://www.worldspace.com*.
10. D. H. Layer, Digital radio takes the road, *IEEE Spectrum* **38**(7): 40–46 (2001).

DIGITAL FILTERS

Hanoch Lev-Ari
Northeastern University
Boston, Massachusetts

1. INTRODUCTION

The omnipresence of noise and interference in communication systems makes it necessary to employ filtering to suppress the effects of such unwanted signal components. As the cost, size, and power consumption of digital hardware continue to drop, the superiority of digital filters over their analog counterparts in meeting the increasing demands of modern telecommunication equipment becomes evident in a wide range of applications. Moreover, the added flexibility of digital implementations makes adaptive digital filtering a preferred solution in situations where time-invariant filtering is found to be inadequate.

The design of linear time-invariant digital filters is by now a mature discipline, with methodological roots reaching back to the first half of the twentieth century. Digital filter design combines the power of modern computing with the fundamental contributions to the theory of optimized (analog) filter design, made by Chebyshev, Butterworth, Darlington, and Cauer. In contrast, the construction of adaptive digital filters is still an evolving area, although its roots can be traced back to the middle of the twentieth century, to the work of Kolmogorov, Wiener, Levinson, and Kalman on statistically optimal filtering. Adaptive filtering implementations became practical only after the early 1980s, with the introduction of dedicated digital signal processing hardware.

The practice of digital filter design and implementation relies on two fundamental factors: a well-developed mathematical theory of signals and systems and the availability of powerful digital signal processing hardware. The synergy between these two factors results in an ever-widening range of applications for digital filters, both fixed and adaptive.

1.1. Signals, Systems, and Filters

Signals are the key concept in telecommunications—they represent patterns of variation of physical quantities such as acceleration, velocity, pressure, and brightness. Communication systems transmit the information contained in a signal by converting the physical pattern of variation into its electronic analogue—hence the term "analog signal." *Systems* operate on signals to modify their shape. *Filters* are specialized systems, designed to achieve a particular type of signal shape modification. The relation between the signal that is applied to a filter and the resulting output signal is known as the *response* of the filter.

Most signals encountered in telecommunication systems are *one-dimensional* (1D), representing the variation of some physical quantity as a function of a single independent variable (usually time). Multidimensional signals are functions of several independent variables. For instance, a video signal is 3D, depending on both time and (x, y) location in the image plane. Multidimensional signals must be converted by scanning to a 1D format before transmission through a communication channel. For this reason, we shall focus here only on 1D signals that vary as a function of time.

Filters can be classified in terms of the operation they perform on the signal. Linear filters perform linear operations. Offline filters process signals after they have been acquired and stored in memory; online filters process signals instantaneously, as they evolve. The majority of filters used in telecommunication systems are online and linear. Since the input and output of an online filter are continuously varying functions of time, memory is required to store information about past values of the signal. Analog filters use capacitors and inductors as their memory elements, while digital filters rely on electronic registers.

1.2. Digital Signal Processing

Digital signal processing is concerned with the use of digital hardware to process digital representations of signals. In order to apply digital filters to real life (i.e., analog) signals, there is a need for an input interface, known as *analog-to-digital converter*, and an output interface, known as *digital-to-analog converter*. These involve *sampling* and *quantization*, both of which result in some loss of information. This loss can be reduced by increasing the speed and wordlength of a digital filter.

Digital filters have several advantages over their analog counterparts:

- *Programmability*—a time-invariant digital filter can be reprogrammed to change its configuration and response without modifying any hardware. An adaptive digital filter can be reprogrammed to change its response adaptation algorithm.
- *Flexibility*—digital filters can perform tasks that are difficult to implement with analog hardware, such as

large-scale long-term storage, linear phase response, and online response adaptation.

- *Accuracy and robustness*—the accuracy of digital filters depends mainly on the number of bits (wordlength) used, and is essentially independent of external factors such as temperature and age.
- *Reliability and security*—digital signals can be coded to overcome transmission errors, compressed to reduce their rate, and encrypted for security.
- *Cost/performance tradeoff*—the cost, size, and power consumption of digital hardware continue to drop, while its speed keeps increasing. As a result, a digital implementation often offers a better cost/performance trade-off than its analog counterpart.

Still, digital hardware is not entirely free from limitations, and the choice between digital and analog has to be made on a case-by-case basis.

1.3. Telecommunication Applications of Digital Filters

Among the many application areas of digital filters, we focus only on those applications that are directly related to communication systems. Our brief summary (Section 7) distinguishes between two main types of applications:

- *Time-invariant digital filters*—used as an alternative to analog filters in applications ranging from frequency-selective filtering and symbol detection in digital communication systems, through speech and image coding, to radar and sonar signal processing
- *Adaptive digital filters*—used in applications that require continuous adjustment of the filters response, such as channel equalization, echo cancellation in duplex communication systems, sidelobe cancellation and adaptive beamforming in antenna arrays, and linear predictive speech coding

2. FUNDAMENTALS OF DISCRETE-TIME SIGNALS AND SYSTEMS

2.1. Discrete-Time Signals

A *discrete-time signal* is a sequence of numbers (real or complex). As explained in the introduction (Section 1), such a sequence represents the variation of some physical quantity, such as voltage, pressure, brightness, and speed. Because discrete-time signals are often obtained by sampling of continuous-time signals, their elements are known as *samples*.

The samples of a discrete-time signal $x(n)$ are labeled by a discrete index n, which is an integer in the range $-\infty < n < \infty$. For instance, the signal

$$x(n) = \begin{cases} 2n+1 & 0 \le n \le 2 \\ 0 & \text{else} \end{cases}$$

consists of three nonzero samples preceded and followed by zero samples (Fig. 1).

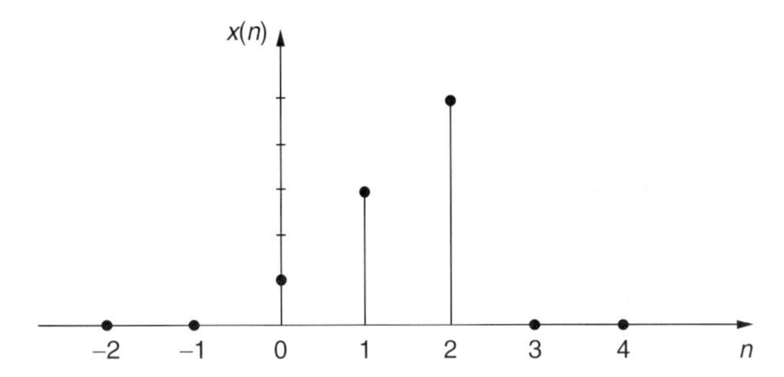

Figure 1. Graphical representation of a discrete-time signal.

Because signals represent physical quantities, they are subject to various constraints:

- A signal $x(\cdot)$ is called *bounded* if there exists a positive constant B such that $|x(n)| \le B$ for all n.
- A signal $x(\cdot)$ is said to have *finite energy* if the infinite sum $\sum_{n=-\infty}^{\infty} |x(n)|^2$ converges to a finite value.
- A signal $x(\cdot)$ is said to have *finite average power* if the limit
$$\lim_{N\to\infty} \frac{1}{2N+1} \sum_{n=-N}^{N} |x(n)|^2$$
exists and is finite.

Here, and in the sequel, we use the shorthand notation $x(\cdot)$, which is equivalent to $\{x(n); -\infty < n < \infty\}$, to refer to the entire signal sequence, rather than to a particular sample. From a mathematical standpoint, a bounded signal has finite ℓ_∞ norm, while a finite-energy signal has finite ℓ_2 norm [2].

Certain elementary signal models are essential to the characterization and analysis of discrete-time systems. These include

- The discrete-time unit impulse, also known as the Krönecker delta:[1]
$$\delta(n) = \begin{cases} 1 & n = 0 \\ 0 & n \neq 0 \end{cases}$$
- The discrete-time unit step
$$u(n) = \begin{cases} 1 & n \ge 0 \\ 0 & n < 0 \end{cases}$$
- A two-sided sinusoid, $x(\cdot) = \{\cos 2\pi f_o n; -\infty < n < \infty\}$.
- A one-sided exponential, $a^n u(n)$. A decaying exponential is obtained by choosing $|a| < 1$; in this case the parameter a is known as a "forgetting factor" [3].

We observe that the discrete-time unit impulse $\delta(n)$ satisfies all three constraints—boundedness, finite energy, and finite power—while the discrete-time step $u(n)$ and the sinusoid $\cos 2\pi f_0 n$ are both bounded and have finite power, but infinite energy. Finally, the exponential $a^n u(n)$ satisfies all three constraints when $|a| < 1$, but it violates all three when $|a| > 1$. In general, every finite-energy

[1] The Krönecker delta should not be confused with the continuous-time impulse, namely, the Dirac delta function.

signal is bounded, and every bounded signal must have finite power:

$$\text{Finite energy} \Rightarrow \text{boundedness} \Rightarrow \text{finite power}$$

Also, since finite-energy signals have the property $\lim_{n\to\pm\infty} x(n) = 0$, they represent transient phenomena. Persistent phenomena, which do not decay with time, are represented by finite-power signals.

2.2. Discrete-Time Systems

A discrete-time system is a *mapping* namely, an operation performed on a discrete-time signal $x(\cdot)$ that produces another discrete-time signal $y(\cdot)$ (Fig. 2). The signal $x(\cdot)$ is called the *input* or *excitation* of the system, while $y(\cdot)$ is called the *output* or *response*.

Most synthetic (human-made) systems are *relaxed,* in the sense that a zero input [i.e., $x(n) = 0$ for all n] produces a zero output. Nonrelaxed systems, such as a wristwatch, rely on internally stored energy to produce an output without an input. In general every system can be decomposed into a *sum* of two components: a relaxed subsystem, which responds to the system input, and an autonomous subsystem, which is responsible for the zero-input response (Fig. 3). Since every analog and every digital filter is a relaxed system, we restrict the remainder of our discussion to relaxed systems only.

Relaxed systems can be classified according to certain input–output properties:

- A relaxed system is called *memoryless* or *static* if its output $y(n)$ at any given instant n depends only on the input sample $x(n)$ at the same time instant, and does not depend on any other input sample. Such a system can be implemented without using memory; it maps every input sample, as it becomes available, into a corresponding output sample.
- A relaxed system is called *dynamic* if its output $y(n)$ at any given instant n depends on past and/or future samples of the input signal $x(\cdot)$. This means that memory is required to store each input sample as it becomes available until its processing has been completed, and it is no more needed by the system. The number of input samples that need to be kept in memory at any given instant is known as the order of the system, and it is frequently used as a measure of system complexity.
- A relaxed system is called *causal* if its output $y(n)$ at any given instant n does not depend on future samples of the input $x(\cdot)$. In particular, every memoryless system is causal. Causality is a physical constraint, applying to systems in which n indicates (discretized) physical time.
- A relaxed system is called *time-invariant* if its response to the time-shifted input $x(n - n_o)$ is a similarly shifted output $y(n - n_o)$, for every input signal $x(\cdot)$ and for every integer n_o, positive or negative. In the context of (digital) filters, time-invariance is synonymous with fixed hardware, while time variation is an essential characteristic of adaptive filters.
- A relaxed system is called *stable* if its response to any bounded input signal $x(\cdot)$ is a bounded output signal $y(\cdot)$. This is known as bounded-input/bounded-output (BIBO) stability, to distinguish it from other definitions of system stability, such as internal (or Lyapunov) stability.
- A relaxed system is called *linear* if its response to the input $x(\cdot) = a_1x_1(\cdot) + a_2x_2(\cdot)$ is $y(\cdot) = a_1y_1(\cdot) + a_2y_2(\cdot)$ for any a_1, a_2 and any $x_1(\cdot), x_2(\cdot)$. Here $y_1(\cdot)$ (resp. $y_2(\cdot)$) is the system's response to the input $x_1(\cdot)$ [resp. $x_2(\cdot)$].

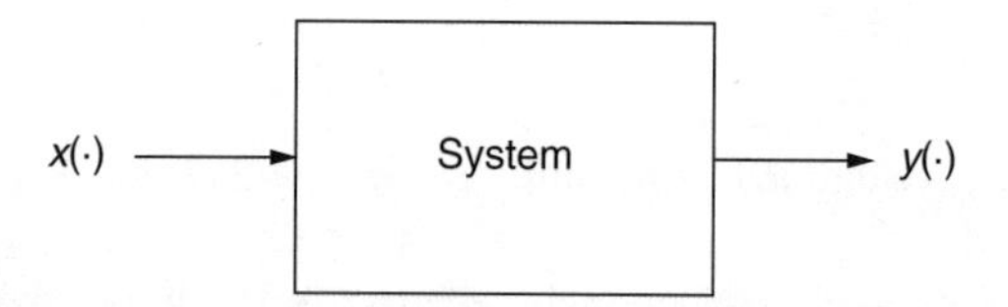

Figure 2. Block diagram representation of a discrete-time system.

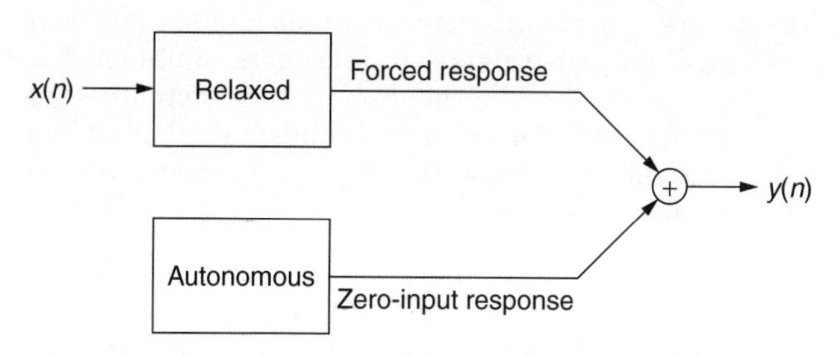

Figure 3. The decomposition of a nonrelaxed system.

As explained in the introduction, digital filters can be either fixed or adaptive. Fixed digital filters are time invariant and most often linear, while adaptive digital filters usually consist of a linear time-variant module and a nonlinear time-invariant module.

Our definitions of fundamental system properties make use of several basic building blocks of discrete-time systems in general, and digital filters in particular:

- *Unit-delay element*—delays a signal by one sample (Fig. 4).
- *Adder*—forms the sum of two signals, such that each output sample is obtained by adding the corresponding input samples (Fig. 5).
- *Signal scaler*—multiplies each sample of the input signal by a constant (Fig. 6). It is the discrete-time equivalent of an (wideband) amplifier.

These three basic building blocks—unit delay, addition, and signal scaling—are relaxed, linear, time-invariant,

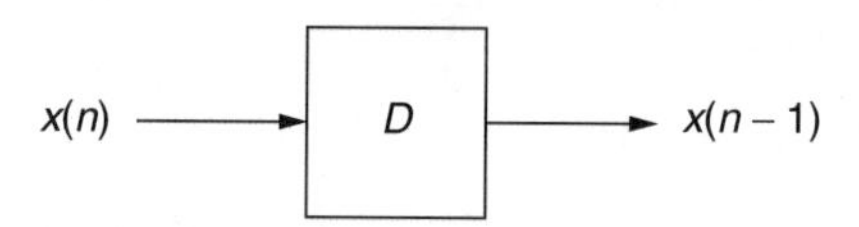

Figure 4. Block diagram representation of a unit-delay element.

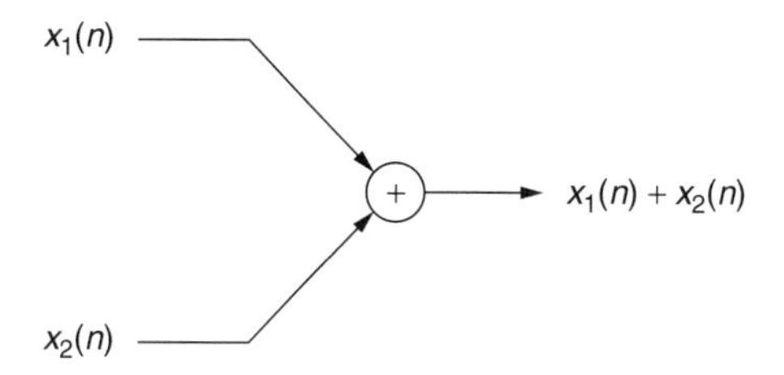

Figure 5. Block diagram representation of an adder.

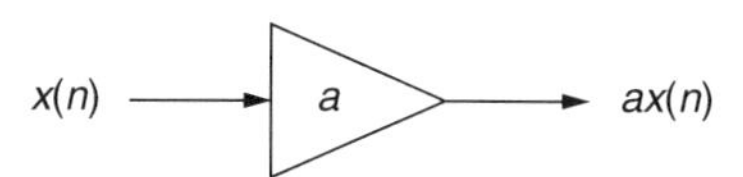

Figure 6. Block diagram representation of a signal scaler.

causal, and stable systems. In fact, every linear, time-invariant, and causal system can be implemented as a network of interconnected delays, scalers and adders (see Section 3.2). Notice that the adder and scaler are memoryless, while the unit delay is dynamic.

In reality, scalers are implemented using digital multipliers (Fig. 7). Multipliers can also be used to form the pointwise product of two signals (Fig. 8), and they serve as a fundamental building block in adaptive digital filters and other nonlinear systems.

2.3. Discrete-Time Sinusoidal Signals

A discrete-time signal of the form

$$x(n) = A\cos(2\pi f_0 n + \phi) \tag{1a}$$

is called *sinusoidal*. The *amplitude* A, *frequency* f_0, and *phase shift* ϕ are all real and, in addition

$$A > 0, \qquad 0 \le f_0 \le \tfrac{1}{2}, \qquad -\pi \le \phi \le \pi \tag{1b}$$

The *radial frequency* $\omega_0 = 2\pi f_0$ $(0 \le \omega_0 \le \pi)$ is often used as an alternative to f_0. The dimensions of f_0 are cycles per sample, and those of ω_0 are radians per sample.

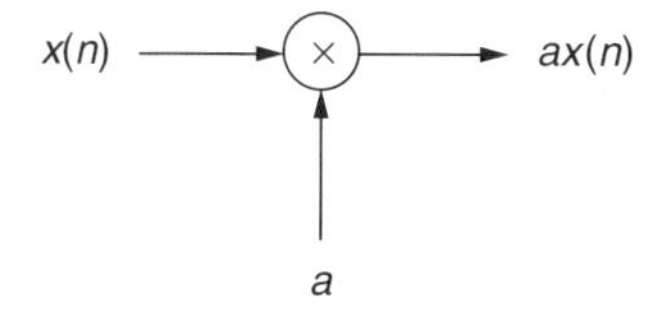

Figure 7. Implementing a signal scaler using a multiplier.

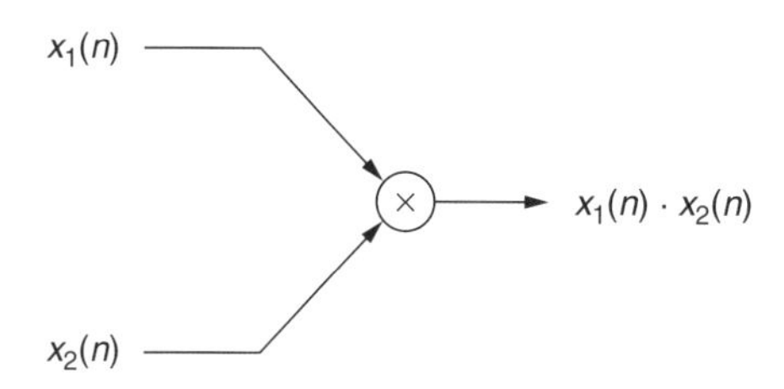

Figure 8. Block diagram representation of a signal multiplier.

The restriction imposed on the range of the frequency f_0 stands in sharp contrast to the continuous-time case, where sinusoids can have unlimited frequencies. However, in the discrete-time context this restriction is necessary in order to avoid *aliasing*.

Aliasing occurs because the two discrete-time sinusoidal signals

$$x_1(n) = \cos 2\pi f_1 n, \qquad x_2(n) = \cos 2\pi f_2 n$$

are indistinguishable when $f_1 - f_2$ is an integer. For instance, $\cos\left(\frac{3\pi}{2}n\right) = \cos\left(-\frac{\pi}{2}n\right)$, so that the frequency $f_1 = \frac{3}{4}$ cannot be distinguished from its *alias* $f_2 = -\frac{1}{4}$. From a mathematical standpoint, we observe that the sinusoidal signal of (1) is periodic in f_0 with a period of 1. In order to avoid ambiguity, we must therefore restrict f_0 to its fundamental (or principal) range $-\frac{1}{2} < f_0 \le \frac{1}{2}$. In addition, by using the symmetry property $\cos(-x) = \cos x$, we can avoid using negative frequencies altogether, which leads to the frequency range specified in Eq. (1b).

The constraint on f_0 is closely related to the well-known Nyquist condition: the range of frequencies allowed at the input of a sampler cannot exceed half the sampling rate. In particular, when the continuous-time sinusoid $\cos 2\pi f_0 t$ is sampled at a rate of F_s samples per second, the resulting sequence of samples forms a discrete-time sinusoid, $x(n) = \cos 2\pi f_0 n$, where $f_0 = f_0/F_s$. According to the Nyquist condition we must have $f_0 < F_s/2$ in order to avoid aliasing: the equivalent restriction on f_0 is $f_0 < \frac{1}{2}$. Anti-aliasing filters must be used to avoid the presence of aliased components in sampled signals.

2.4. Relaxed Linear Time-Invariant (LTI) Systems

The input–output characterization of relaxed discrete-time LTI systems relies on the notion of *impulse response*, namely the response of the system to the discrete-time unit impulse $\delta(\cdot)$ (Fig. 9).

The response of such a system to an arbitrary input $x(\cdot)$ is determined by a linear convolution between the input signal and the impulse response of the system:

$$y(n) = \sum_{k=-\infty}^{\infty} h(k)x(n-k) \tag{2}$$

This is usually expressed using the shorthand notation

$$y(\cdot) = h(\cdot) \circledast x(\cdot)$$

Thus, a relaxed LTI system is completely characterized by its impulse response. In particular, such a system is causal if, and only if

$$h(n) = 0 \quad \text{for} \quad n < 0$$

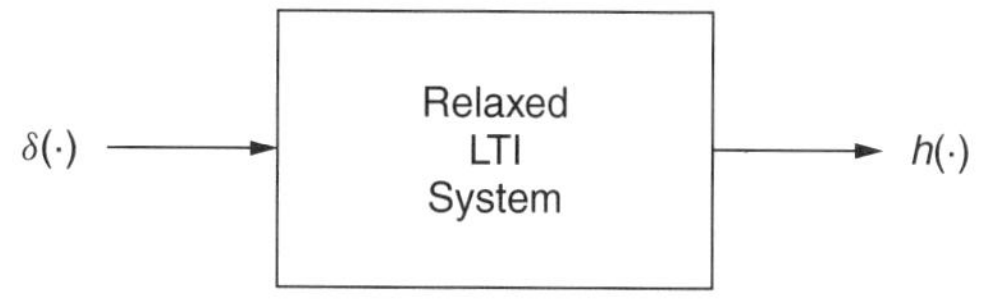

$\delta(\cdot)$ → Relaxed LTI System → $h(\cdot)$

Figure 9. The impulse response of a relaxed LTI system.

As a result the convolution sum (2) for a causal system ranges only over $0 \leq k < \infty$. Also, an LTI system is BIBO stable if, and only if

$$\sum_{k=-\infty}^{\infty} |h(k)| < \infty$$

The response of an LTI system to a sinusoidal input is of particular interest. The input signal $x(n) = \exp\{j\omega_0 n\}$ produces the output

$$y(n) = H(e^{j\omega_0})e^{j\omega_0 n} \equiv H(e^{j\omega_0})x(n) \tag{3a}$$

where

$$H(e^{j\omega}) = \sum_{k=-\infty}^{\infty} h(k)e^{-j\omega k} \tag{3b}$$

is known as the *frequency response* of the LTI system. We observe that the response to the complex sinusoid $x(n) = \exp\{j\omega_0 n\}$ is $H(e^{j\omega_0})x(n)$, namely, a scaled version of the input signal. From a mathematical standpoint, this means that complex sinusoids are eigenfunctions of LTI systems.

The infinite sum (3b) that defines the frequency response $H(e^{j\omega})$ converges for all ω if, and only if, $\sum_k |h(k)| < \infty$, which is also the necessary and sufficient conditions for BIBO stability. Thus, an unstable system does not have a frequency response—exciting it with a sinusoidal input produces an unbounded output signal. Unstable systems can be characterized by their response to exponentially decaying sinusoids, which leads us to the more general concept of a $\mathcal{Z}$ transform (see Section 3.1).

A real-life LTI system has a real-valued impulse response. The response of such a system to the real-valued sinusoid $x(n) \cos \omega_0 n$ is given by the real part of the response in (3)

$$y(n) = \text{Re}\{H(e^{j\omega_0})e^{j\omega_0 n}\} = |H(e^{j\omega_0})| \cos[\omega_0 n + \theta(\omega_0)]$$

where $\theta(\omega_0) = \arg H(e^{j\omega_0})$. The effect of an LTI system on a real sinusoid is to scale its amplitude by $|H(e^{j\omega_0})|$, and to increase its phase by $\arg H(e^{j\omega_0})$. For this reason, $|H(e^{j\omega})|$ is known as the *magnitude response* of the system, and $\arg H(e^{j\omega})$ is known as its *phase response*.

3. TRANSFORM DOMAIN ANALYSIS OF SIGNALS AND SYSTEMS

Transform-domain analysis is a powerful technique for characterization and design of (fixed) digital filters. The role of the $\mathcal{Z}$ transform in the context of discrete-time signals and systems is similar to the role of the Laplace transform in the context of continuous-time signals and systems. In fact, the two are directly related via the process of sampling and reconstruction. Recall that interpolation, specifically the process of reconstructing a continuous-time signal from its samples, produces the continuous-time signal

$$x_r(t) = \sum_{n=-\infty}^{\infty} x(n)g(t - nT)$$

where $g(\cdot)$ is the impulse response of the interpolating filter and T is the sampling interval. The Laplace transform of this signal is

$$X_r(s) = \left[\sum_{n=-\infty}^{\infty} x(n)e^{-snT}\right] G(s)$$

namely, a product of the transfer function $G(s)$ of the interpolating filter with a transform-domain characterization of the discrete-time sequence of samples $x(\cdot)$. This observation motivates the introduction of the $\mathcal{Z}$ transform

$$X(z) = \sum_{n=-\infty}^{\infty} x(n)z^{-n} \tag{4}$$

so that $X_r(s) = G(s)X(z)|_{z=e^{sT}}$.

The $\mathcal{Z}$ transform converts difference equations into algebraic equations, which makes it possible to characterize every discrete-time LTI system in terms of the poles and zeros of its *transfer function*, namely, the $\mathcal{Z}$ transform of its impulse response. This is entirely analogous to the role played by the Laplace transform in the context of continuous-time LTI systems.

3.1. The $\mathcal{Z}$ Transform and the Discrete-Time Fourier Transform

From a mathematical standpoint, the $\mathcal{Z}$ transform (4) is the sum of two power series, viz., $X(z) = \sum_{n=-\infty}^{\infty} x(n)z^{-n} = X_+(z) + X_-(z)$, where

$$X_+(z) = \sum_{n=0}^{\infty} x(n)z^{-n}, \qquad X_-(z) = \sum_{n=-1}^{-\infty} x(n)z^{-n}$$

The $\mathcal{Z}$ transform $X(z)$ is said to exist only if both $X_+(z)$ and $X_-(z)$ converge absolutely and uniformly. This implies that the *region of convergence* (RoC) of $X(z)$ is $\{z; r < |z| < R\}$, where r is the radius of convergence of $X_+(z)$ and R is the radius of convergence of $X_-(z)$. Thus a $\mathcal{Z}$ transform exists if, and only if, $r < R$. When it does exist, it is an analytic function within its RoC.

The $\mathcal{Z}$ transform converts time-domain convolutions into transform-domain products. In particular, the input–output relation of a relaxed LTI system, specifically, $y(\cdot) = h(\cdot) \circledast x(\cdot)$ transforms into $Y(z) = H(z)X(z)$, where $X(z)$, $Y(z)$, and $H(z)$ are the $\mathcal{Z}$ transforms of $x(\cdot)$, $y(\cdot)$, and $h(\cdot)$, respectively. Thus, specifying the *transfer function*

$$H(z) = \sum_{k=-\infty}^{\infty} h(k)z^{-k}$$

along with its RoC provides the same information about the input–output behavior of a relaxed LTI system as the impulse response $h(\cdot)$. In particular

- The system is causal if, and only if, the RoC of $H(z)$ is of the form $|z| > r$ for some $r \geq 0$.
- The system is (BIBO) stable if, and only if, the unit circle is within the RoC of $H(z)$.

Thus the transfer function of a stable system is always well defined on the unit circle $\mathbf{T} = \{z; |z| = 1\}$, and we recognize $H(z)|_{z=e^{j\omega}}$ as the frequency response of the system [as defined in (3)]. More generally, if $x(\cdot)$ is a discrete-time sequence, whether a signal or an impulse response, such that $\sum_{n=-\infty}^{\infty} |x(n)| < \infty$, then the unit circle $\mathbf{T}$ is included in the RoC of the $\mathcal{Z}$ transform $X(z)$, and so $X(z)|_{z=e^{j\omega}}$ is well defined. The resulting function of ω, that is

$$X(e^{j\omega}) = \sum_{n=-\infty}^{\infty} x(n)e^{-j\omega n} \tag{5a}$$

is called the *discrete-time Fourier transform* (DTFT) of $x(\cdot)$.

From a mathematical standpoint, relation (5) is a complex Fourier series representation of the periodic function $X(e^{j\omega})$, and we recognize the samples $x(n)$ as the Fourier coefficients in this representation. The standard expression for Fourier coefficients

$$x(n) = \frac{1}{2\pi}\int_{-\pi}^{\pi} X(e^{j\omega})e^{j\omega n}\,d\omega \tag{5b}$$

is known as the *inverse DTFT*. The DTFT (5) converges absolutely (and uniformly) for ℓ_1 sequences, and the resulting limit is a continuous function of ω. The DTFT can be extended to other types of sequences by relaxing the notion of convergence of the infinite sum in (5a). For instance, the DTFT of ℓ_2 sequences is defined by requiring convergence in the $\mathcal{L}^2(\mathbf{T})$ norm, and the resulting limit is a square-integrable function on the unit circle $\mathbf{T}$. A further extension to ℓ_∞ sequences (= bounded signals) results in limits that are $\mathcal{L}^1(\mathbf{T})$ functions, and thus may contain impulsive components. For instance, the DTFT of the bounded signal $x(n) = e^{j\omega_0 n}$ is $X(e^{j\omega}) = 2\pi\,\delta(\omega - \omega_0)$, where $\delta(\cdot)$ is the Dirac delta function. The convergence of (5) in this case is defined only in the distribution sense. Also, notice that this signal has no $\mathcal{Z}$ transform.

3.2. Transfer Functions of Digital Filters

A fixed digital filter implements a realizable discrete-time linear time-invariant system. Realizability restricts us to causal systems with *rational transfer functions*

$$H(z) = \frac{b(z)}{a(z)} \tag{6a}$$

where $a(z)$ and $b(z)$ are finite-order polynomials in z^{-1}:

$$a(z) = 1 + a_1 z^{-1} + a_2 z^{-2} + \cdots + a_N z^{-N} \tag{6b}$$

$$b(z) = b_o + b_1 z^{-1} + b_2 z^{-2} + \cdots + b_M z^{-M} \tag{6c}$$

The roots of the numerator polynomial $b(z)$ are known as the *zeros* of the transfer function $H(z)$, and the roots of the denominator polynomial $a(z)$ are known as the *poles* of $H(z)$. A digital filter with $N = 0$ is known as a *finite-impulse response* (FIR) filter, while one with $N > 0$ is known as an *infinite-impulse response* (IIR) filter. In view of our earlier statements about causality and stability in terms of the region of convergence, it follows that a digital filter is stable if all its poles have magnitudes strictly less than unity. Thus, FIR filters are unconditionally stable, because all of their poles are at $z = 0$.

In view of (6), the input–output relation of a digital filter can be written as $a(z)Y(z) = b(z)X(z)$, which corresponds to a *difference equation* in the time domain

$$\begin{aligned} y(n) + a_1 y(n-1) + \cdots + a_N y(n-N) \\ = b_o x(n) + b_1 x(n-1) + \cdots + b_M x(n-M) \end{aligned} \tag{7}$$

The same input–output relation can also be represented by a block diagram, using multipliers (actually signal scalers), adders, and delay elements (Fig. 10). Such a block diagram representation is called a *realization* of the transfer function $H(z) = \dfrac{b(z)}{a(z)}$. Digital filter realizations are not unique: the one shown in Fig. 10 is known as the *direct form type 2* realization. Other realizations are described in Section 5.

A hardware implementation (i.e., a digital filter) is obtained by mapping the realization into a specific platform, such as a DSP chip or an ASIC (see Section 5). A software implementation is obtained by mapping the same realization into a computer program.

3.3. The Power Spectrum of a Discrete-Time Signal

The *power spectrum* of a finite-energy signal $x(\cdot)$ is defined as the square of the magnitude of its DTFT $X(e^{j\omega})$. Alternatively, it can be obtained by applying a DTFT to the *autocorrelation sequence*

$$C_{xx}(m) = \sum_{n=-\infty}^{\infty} x(n+m)x^*(n) = x(m) \circledast x^*(-m)$$

where the asterisk ($*$) denotes complex conjugation. For finite-power signals the DTFT $X(e^{j\omega})$ may contain impulsive components, so that the square of its magnitude cannot be defined. Instead, the autocorrelation is defined in this case as

$$C_{xx}(m) = \lim_{N\to\infty} \frac{1}{2N+1}\sum_{n=-N}^{N} x(n+m)x^*(n)$$

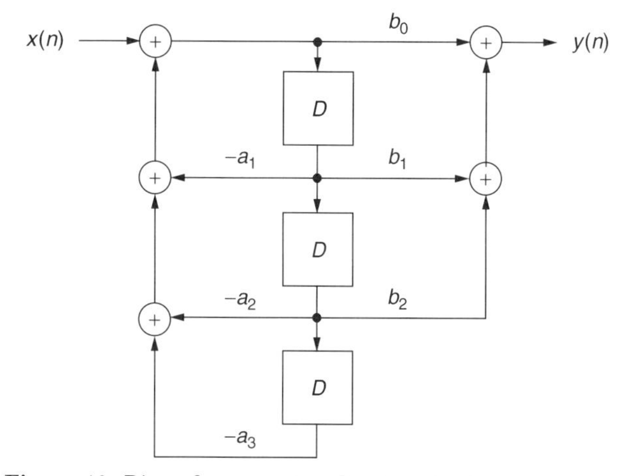

Figure 10. Direct-form type 2 realization of a digital filter with $M = 2$ and $N = 3$.

and the power spectrum is defined as the DTFT of the autocorrelation $C_{xx}(\cdot)$. Similarly, the autocorrelation of a random stationary signal is defined as

$$C_{xx}(m) = E\{x(n+m)x^*(n)\}$$

where $E\{\ \}$ denotes expectation (i.e., probabilistic mean). Thus, in all three cases the power spectrum is

$$S_{xx}(e^{j\omega}) = \sum_{m=-\infty}^{\infty} C_{xx}(m)e^{-j\omega m} \tag{8a}$$

and can be viewed as a restriction to the unit circle of the so-called *complex power spectrum*

$$S_{xx}(z) = \sum_{m=-\infty}^{\infty} C_{xx}(m)z^{-m} \tag{8b}$$

The complex power spectrum is used in the design of optimal (Wiener) filters (see Section 6.1).

Similarly, the cross-correlation of two signals is defined as

$$C_{yx}(m) = \begin{cases} \sum_{n=-\infty}^{\infty} y(n+m)x^*(n) \\ \quad \text{finite-energy signals} \\ \lim_{N\to\infty} \frac{1}{2N+1} \sum_{n=-N}^{N} y(n+m)x^*(n) \\ \quad \text{finite-power signals} \\ E\{y(n+m)x^*(n)\} \\ \quad \text{jointly stationary random signals} \end{cases} \tag{9a}$$

and the (complex) cross spectrum is the transform of the cross-correlation:

$$S_{yx}(z) = \sum_{m=-\infty}^{\infty} C_{yx}(m)z^{-m} \tag{9b}$$

When $y(\cdot)$ is the response of a stable LTI system to the input $x(\cdot)$, then

$$S_{yx}(z) = H(z)S_{xx}(z) \tag{10}$$

where $H(z)$ is the transfer function of the system, and

$$S_{yy}(z) = H(z)S_{xx}\left[H\left(\frac{1}{z^*}\right)\right]^* \tag{11}$$

In particular, by using a unit-power white noise [i.e., one with $S_{xx}(z) = 1$] input, we find that

$$S_{yy}(z) = H(z)\left[H\left(\frac{1}{z^*}\right)\right]^* \tag{12}$$

This expression is called a *spectral factorization* of the complex power spectrum $S_{yy}(z)$ and the transfer function $H(z)$ is called a *spectral factor*.

4. DESIGN OF FREQUENCY-SELECTIVE DIGITAL FILTERS

The complete process of designing a digital filter consists of seven stages:

1. *Problem analysis*—determine what the filter is supposed to accomplish.
2. *Filter specification*—select a desired (ideal) frequency response for the filter, and decide how accurately it should be approximated.
3. *Filter design*—obtain the coefficients of a realizable transfer function $H(z)$ that approximates the desired frequency response within the specified tolerance.
4. *Filter realization*—determine how to construct the filter by interconnecting basic building blocks, such as delays, adders, and signal scalers (i.e., multipliers).
5. *Implementation choices*—select the specific hardware/software platform in which the building blocks will be implemented.
6. *Performance analysis*—use the physical parameters of the selected platform (accuracy, cost, speed, etc.) to evaluate the compliance of the selected implementation with the specification of the filter.
7. *Construction*—implement the specific choices made in the design and realization stages into the selected hardware/software platform.

In the problem analysis stage the designer uses specific information about the application to determine a desired frequency response, say, $D(\omega)$, for the filter. The desired magnitude response $|D(\omega)|$ is often a classical (ideal) frequency-selective prototype—lowpass, highpass, bandpass, or bandstop—except in specialized filter designs such as Hilbert transformers, differentiators, notch filters, or allpass filters. This means that $|D(\omega)|$ is piecewise constant, so that the frequency scale decomposes into a collection of bands. The desired phase response $\arg D(\omega)$ could be linear (exactly or approximately), minimum-phase, or unconstrained.

The designer must also define the set of parameters that determine the transfer function of the filter to be designed. First a choice has to be made between FIR and IIR, considering the following facts:

- Exact linear phase is possible only with FIR filters.
- FIR filters typically require more coefficients and delay elements than do comparable IIR filters, and therefore involve higher input–output delay and higher cost.
- Currently available design procedures for FIR filters can handle arbitrary desired responses (as opposed to ideal prototypes) better than IIR design procedures.
- Finite precision effects are sometimes more pronounced in IIR filters. However, such effects can be ameliorated by choosing an appropriate realization.
- Stability of IIR filters is harder to guarantee in adaptive filtering applications (but not in fixed filtering scenarios).

Additional constraints, such as an upper bound on the overall delay (e.g., for decision feedback equalization), or a requirement for maximum flatness at particular frequencies, serve to further reduce the number of independent parameters that determine the transfer function of the filter.

On completion of the problem analysis stage, the designer can proceed to formulate a specification and determine a transfer function that meets this specification, as described in the remainder of Section 4. The remaining stages of the design process (realization, implementation, performance analysis, and construction) are discussed in Section 5.

4.1. Filter Specification

Once the desired frequency response $D(\omega)$ has been determined along with a parametric characterization of the transfer function $H(z)$ of the designed filter, the next step is to select a measure of approximation quality and a tolerance level (i.e., the highest acceptable deviation from the desired response). The most commonly used measure is the weighted Chebyshev (or $\mathcal{L}^\infty$) norm

$$\max_{\omega} W(\omega)|H(e^{j\omega}) - D(\omega)|$$

where $W(\omega)$ is a nonnegative weighting function. Alternative measures include the weighted mean-square (or $\mathcal{L}^2$) norm

$$\int_{-\pi}^{\pi} W^2(\omega)|H(e^{j\omega}) - D(\omega)|^2 d\omega$$

and the truncated time-domain mean-square norm

$$\sum_{n=0}^{L} |h(n) - d(n)|^2$$

where $h(n)$ is the impulse response of the designed filter and $d(n)$ is the inverse DTFT of the desired frequency response $D(\omega)$.

The mean-square measures are useful mainly when the desired response $D(\omega)$ is arbitrary, since in this case optimization in the Chebyshev norm can be quite demanding. However, the approximation of ideal prototypes under the Chebyshev norm produces excellent results at a reasonable computational effort, which makes it the method of choice in this case.

The approximation of ideal prototypes is usually carried out under the *modified* Chebyshev norm

$$\max_{\omega} W(\omega)||H(e^{j\omega})| - |D(\omega)||$$

The elimination of phase information from the norm expression reflects the fact that in this case phase is either completely predetermined (for linear-phase FIR filters) or completely unconstrained (for IIR filters). Furthermore, the desired magnitude response is constant in each frequency band (e.g., unity in passbands and zero in stopbands), so it makes sense to select a weighting function $W(\omega)$ that is also constant in each frequency band. As a result, constraining the Chebyshev norm to a prescribed tolerance level is equivalent to providing a separate tolerance level for each frequency band, say

$$||H(e^{j\omega})| - D_i| \leq \delta_i \quad \text{for all} \quad \omega \in B_i \tag{13}$$

where B_i is the range of frequencies for the ith band, D_i is the desired magnitude response in that band, and δ_i is the prescribed level of tolerance. For instance, the specification of a lowpass filter is

$$\begin{array}{lll} \text{Passband:} & 1 - \delta_p \leq |H(e^{j\omega})| \leq 1 + \delta_p & \text{for } |\omega| \leq \omega_p \\ \text{Stopband:} & |H(e^{j\omega})| \leq \delta_s & \text{for } \omega_p \leq |\omega| \leq \pi \end{array}$$

as illustrated by the template in Fig. 11. The magnitude response of the designed filter must fit within the unshaded area of the template. The tolerance δ_p characterizes *passband ripple*, while δ_s characterizes *stopband attenuation*.

4.2. Design of Linear-Phase FIR Filters

The transfer function of an FIR digital filter is given by (6) with $N = 0$, so that $a(z) = 1$ and $H(z) \equiv b(z) = \sum_{i=0}^{M} b_i z^{-i}$. The design problem is to determine the values of the coefficients $\{b_i\}$ so that the phase response is linear, and the magnitude response $|H(e^{j\omega})|$ approximates a specified $|D(\omega)|$. In order to ensure phase linearity, the coefficients $\{b_i\}$ must satisfy the symmetry constraint

$$b_{M-i} = sb_i \quad \text{for} \quad 0 \leq i \leq M, \quad \text{where} \quad s = \pm 1 \tag{14}$$

As a result, the frequency response satisfies the constraint $H^*(e^{j\omega}) = sH(e^{j\omega})e^{jM\omega}$, so that the phase response is indeed linear: $\arg H(e^{j\omega}) = (M\omega + s\pi)/2$. Another consequence of the symmetry constraint (14) is that the zero pattern of $H(z)$ is symmetric with respect to the unit circle: $H(z_o) = 0 \leftrightarrow H(1/z_o^*) = 0$.

The design of a linear-phase FIR filter is successfully completed when we have selected values for the filter coefficients $\{b_i\}$ such that (1) the symmetry constraint (14) is satisfied and (2) the magnitude tolerance constraints (13) are satisfied. A design is considered *optimal* when the order M of the filter $H(z)$ is as small as possible under the specified constraints. It is customary to specify the magnitude tolerances in decibels — for instance, a lowpass filter is characterized by

$$R_p = 20 \log \frac{1 + \delta_p}{1 - \delta_p}, \qquad A_s = 20 \log \frac{1}{\delta_s}$$

The most popular techniques for linear-phase FIR design are

- *Equiripple* — optimal in the weighted Chebyshev norm, uses the Remez exchange algorithm to optimize filter coefficients. The resulting magnitude response is equiripple in all frequency bands (as in Fig. 11). The weight for each band is set to be inversely proportional to the band tolerance. For instance, to design a lowpass filter with passband ripple $R_p = 1$ dB and stopband attenuation $A_s = 40$ dB, we

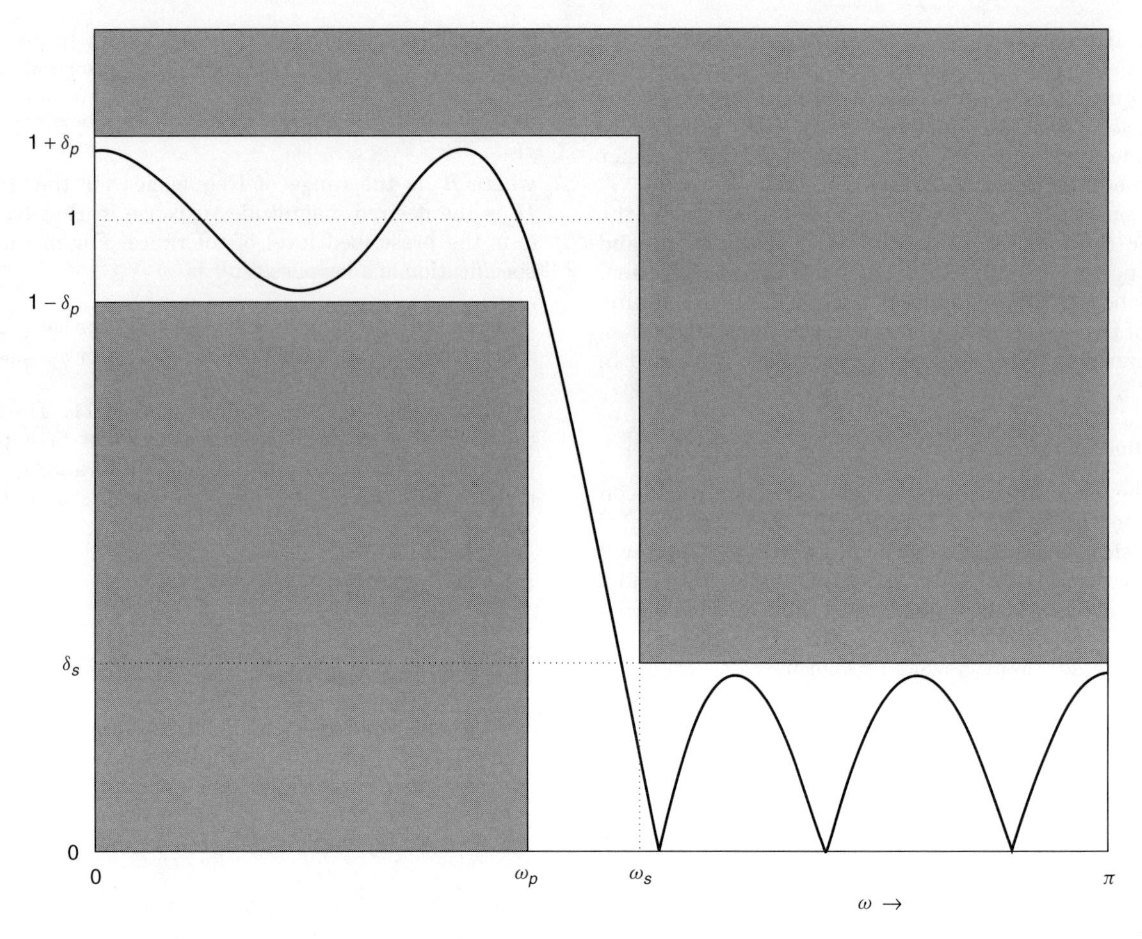

Figure 11. Specification for FIR lowpass filter design.

calculate first $\delta_p = 0.0575$ and $\delta_s = 0.01$ and then set $W_p = \delta_p^{-1} = 17.39$ and $W_s = \delta_s^{-1} = 100$. While, in principle, the Remez exchange algorithm can be applied to any desired magnitude response $|D(\omega)|$, most filter design packages (such a the *signal processing toolbox* in Matlab) accept only piecewise constant gain specifications.

- *Least squares*—optimal in the weighted mean-square norm, uses closed-form expressions for the optimal filter coefficients. As in the equiripple method, the weight for each band is set to be inversely proportional to the band tolerance. Because of the availability of a simple closed-form solution, the least-squares method is most frequently used to design filters with arbitrarily shaped magnitude responses.
- *Truncation and windowing*—optimal in the time-domain mean-square norm (when $L = M$) but not in any frequency-domain sense. The resulting filter usually has many more coefficients than the one designed by the equiripple method. The filter coefficients are obtained by (1) applying the inverse DTFT (5b) to the desired response $D(\omega)$ (which contains the appropriate linear phase term) to obtain the impulse response $d(n)$, and (2) multiplying this impulse response by a *window function* $w(n)$, which vanishes outside the range $0 \leq n \leq M$. The window function is selected to control the tradeoff between the passband attenuation of the filter, and the width of the transition band. The Kaiser window has a control parameter that allows continuous adjustment of this tradeoff. Other popular windows (e.g., Bartlett, Hamming, Von Hann, and Blackman) are not adjustable and offer a fixed tradeoff. The simplicity of the truncation and windowing method makes it particularly attractive in applications that require an arbitrarily shaped magnitude response (and possibly also an arbitrarily shaped phase response).

Specialized design methods are used to design nonstandard filters such as maximally flat or minimum-phase FIR filters, Nyquist filters, differentiators, Hilbert transformers, and notch filters [5,6].

4.3. IIR Filter Design

The transfer function of an IIR digital filter is given by (6) with $N > 0$, so that $H(z) = b(z)/a(z)$. The design problem is to determine the values of the coefficients $\{a_i\}$ and $\{b_i\}$ so that the magnitude response $|H(e^{j\omega})|$ approximates a specified $|D(\omega)|$, with no additional constraints on the phase response. The design of an IIR filter is successfully completed when we have selected values for the filter coefficients such that the magnitude tolerance constraints (13) are satisfied. A design is considered

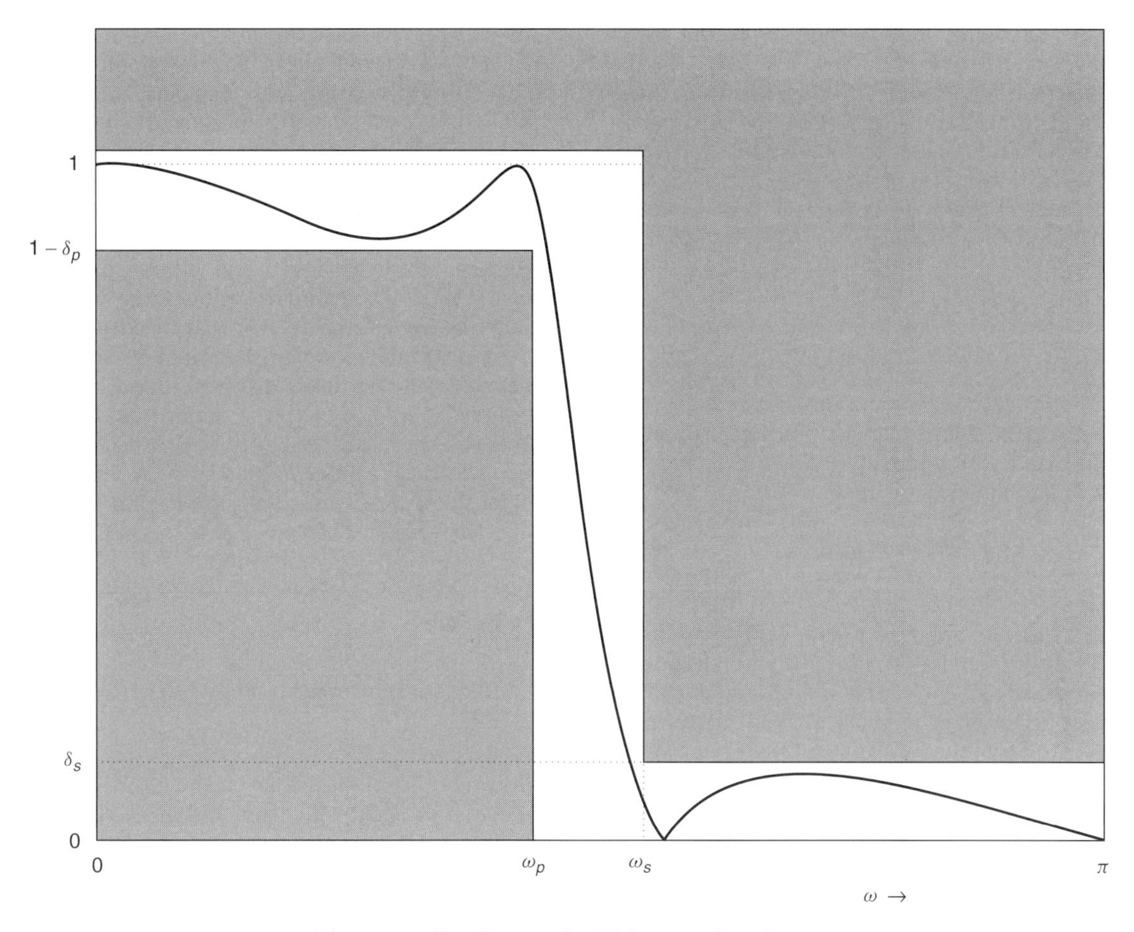

Figure 12. Specification for IIR lowpass filter design.

optimal when the filter order N is as small as possible under the specified constraints.

There exist a number of techniques for the design of IIR filters with an arbitrarily shaped $D(\omega)$, some based on the weighted (frequency-domain) $\mathcal{L}^2$ and $\mathcal{L}^\infty$ norms [5], and others on the truncated time-domain mean-square norm (with $L = M + N + 1$) [6]. Here we shall discuss only the design of classical (ideal) frequency-selective prototypes, which is the most common application for IIR digital filters. The most popular IIR design technique relies on a well-developed theory of analog filter design from closed-form (analytical) formulas. This means that the design of an IIR digital filter decomposes into three steps:

1. The specification of the desired classical frequency selective digital filter (lowpass, highpass, bandpass, or bandstop) is translated into a specification of an equivalent analog filter.
2. An analog filter $H_a(s)$ that meets the translated specification is obtained by (a) designing a lowpass analog filter $H_{\mathrm{LP}}(s)$ from closed-form formulas and (b) transforming $H_{\mathrm{LP}}(s)$ into a highpass, bandpass, or bandstop filter, as needed, by a complex variable mapping $s \to g(s)$, namely, $H_a(s) = H_{\mathrm{LP}}[g(s)]$.
3. The analog filter $H_a(s)$ is transformed into a digital filter $H(z)$ by (another) complex variable mapping.

The analog filter $H_{LP}(s)$ is one of four standard prototypes: Butterworth, Chebyshev type 1 or 2, or elliptic (also known as Chebyshev–Cauer). Since the magnitude response of such prototypes is always bounded by unity, the specification template has to be modified accordingly (see Fig. 12). As a result, the decibel scale characterization of tolerances for IIR design is also modified:

$$R_p = 20\log\frac{1}{1-\delta_p}, \qquad A_s = 20\log\frac{1}{\delta_s}$$

The variable transformation $s \to g(s)$ that converts the analog lowpass filter $H_{\mathrm{LP}}(s)$ into a highpass, bandpass, or bandstop filter $H_a(s)$ is described in Table 1. It involves $\Omega_{p,\mathrm{LP}}$, the passband edge frequency of the prototype $H_{\mathrm{LP}}(s)$,

Table 1. Frequency Transformations for Analog Filters

Type of Transformation	$g(s)$	Band-Edge Frequency of Target Filter
Lowpass to highpass	$\dfrac{\Omega_{p,\mathrm{LP}}\Omega_p}{s}$	Ω_p
Lowpass to bandpass	$\Omega_{p,\mathrm{LP}}\dfrac{s^2+\Omega_{p1}\Omega_{p2}}{s(\Omega_{p2}-\Omega_{p1})}$	Ω_{p1}, Ω_{p2}
Lowpass to bandstop	$\Omega_{p,\mathrm{LP}}\dfrac{s(\Omega_{p2}-\Omega_{p1})}{s^2+\Omega_{p1}\Omega_{p2}}$	Ω_{p1}, Ω_{p2}

as well as the passband edge frequencies of the target analog filter $H_a(s)$.

Several methods—including the impulse invariance, matched $\mathcal{Z}$, and bilinear transforms—can be used to map the analog transfer function $H_a(s)$, obtained in the second step of IIR design, into a digital transfer function $H(z)$. The most popular of these is the bilinear transform: the transfer function $H(z)$ is obtained from $H_a(s)$ by a variable substitution

$$H(z) = H_a\left(\alpha\frac{1-z^{-1}}{1+z^{-1}}\right) \tag{15a}$$

where α is an arbitrary constant. The resulting digital filter has the same degree N as the analog prototype $H_a(s)$ from which it is obtained. It also has the same frequency response, but with a warped frequency scale; as a result of the variable mapping (15a), we have

$$H(e^{j\omega}) = H_a(j\Omega)|_{\Omega=\alpha\tan(\omega/2)} \tag{15b}$$

Here we use Ω to denote the frequency scale of the analog filter $H_a(s)$ in order to distinguish it from the frequency scale ω of the digital filter $H(z)$. The frequency mapping $\Omega = \alpha\tan(\omega/2)$ governs the first step of the IIR design process—the translation of a given digital filter specification into an equivalent analog filter specification. While the tolerances R_p and A_s are left unaltered, all critical digital frequencies (such as ω_p and ω_s for a lowpass filter) are *prewarped* into the corresponding analog frequencies, using the relation

$$\Omega = \alpha\tan\frac{\omega}{2} \equiv \alpha\tan\pi f \tag{15c}$$

For instance, in order to design a lowpass digital filter with passband $|f| \le f_p = 0.2$ with ripple $R_p = 1$ dB and stopband $0.3 = f_s \le |f| \le 0.5$ with attenuation $A_s = 40$ dB, we need to design an analog filter with the same tolerances, but with passband $|\Omega| \le \Omega_p$ and stopband $|\Omega| \ge \Omega_s$, where

$$\Omega_p = \alpha\tan(0.2\pi), \qquad \Omega_s = \alpha\tan(0.3\pi)$$

Since the value of α is immaterial, the most common choice is $\alpha = 1$ (sometimes $\alpha = 2$ is used). The value of α used in the prewarping step must also be used in the last step when transforming the designed analog transfer function $H_a(s)$ into its digital equivalent $H(z)$ according to (15a).

Once the complete specification of the desired analog filter is available, we must select one of the four standard lowpass prototypes (Fig. 13):

- *Butterworth*—has a monotone decreasing passband and stopband magnitude response, which is maximally flat at $\Omega = 0$ and $\Omega = \infty$
- *Chebyshev 1*—has an equiripple passband magnitude response, and a monotone decreasing stopband response, which is maximally flat at $\Omega = \infty$
- *Chebyshev 2*—has a monotone decreasing passband magnitude response, which is maximally flat at $\Omega = 0$, and an equiripple stopband response
- *Elliptic*—has an equiripple magnitude response in both the passband and the stopband

All four prototypes have a maximum gain of unity, which is achieved either at the single frequency $\Omega = 0$ (for Butterworth and Chebyshev 2), or at several frequencies throughout the passband (for Chebyshev 1 and elliptic). When the Butterworth or Chebyshev 1 prototype is translated into its digital form, its numerator is proportional to $(1+z^{-1})^N$. Thus only $N+1$ multiplications are required to implement these two prototypes, in contrast to Chebyshev 2 and elliptic prototypes, which have nontrivial symmetric numerator polynomials, and thus require $N + \mathrm{ceil}(N+1/2)$ multiplications each.

As an example we present in Fig. 13 the magnitude response of the four distinct digital lowpass filters, designed from each of the standard analog prototypes to meet the specification $f_p = 0.23$, $f_s = 0.27$, $R_p = 1$ dB, and $A_s = 20$ dB. The order needed to meet this specification is 12 for the Butterworth, 5 for Chebyshev (both types), and 4 for the elliptic. Thus the corresponding implementation cost (=number of multipliers) is in this case 13 for Butterworth, 6 for Chebyshev 1, 11 for Chebyshev 2, and 9 for elliptic.

5. REALIZATION AND IMPLEMENTATION OF DIGITAL FILTERS

The preceding discussion of digital filters has concentrated solely on their input–output response. However, in order to build a digital filter, we must now direct our attention to the internal structure of the filter and its basic building blocks. The construction of a digital filter from a given (realizable) transfer function $H(z)$ usually consists of two stages:

- *Realization*—in this stage we construct a block diagram of the filter as a network of interconnected basic building blocks, such as unit delays, adders, and signal scalers (i.e., multipliers).
- *Implementation*–in this stage we map the filter realization into a specific hardware/software architecture, such as a general-purpose computer, a digital signal processing (DSP) chip, a field-programmable gate array (FPGA), or an application-specific integrated circuit (ASIC).

A realization provides an idealized characterization of the internal structure of a digital filter, in the sense that it ignores details such as number representation and timing. A given transfer function has infinitely many realizations, which differ in their performance, as explained in Section 5.3. The main distinguishing performance attributes are numerical accuracy and processing delay. These and other details must be addressed as part of the implementation stage, before we can actually put together a digital filter.

5.1. Realization of FIR Filters

FIR filters are always realized in the so-called transversal (also tapped-delay-line) form, which is a special case of the direct-form realization described in Section 3.2. Since

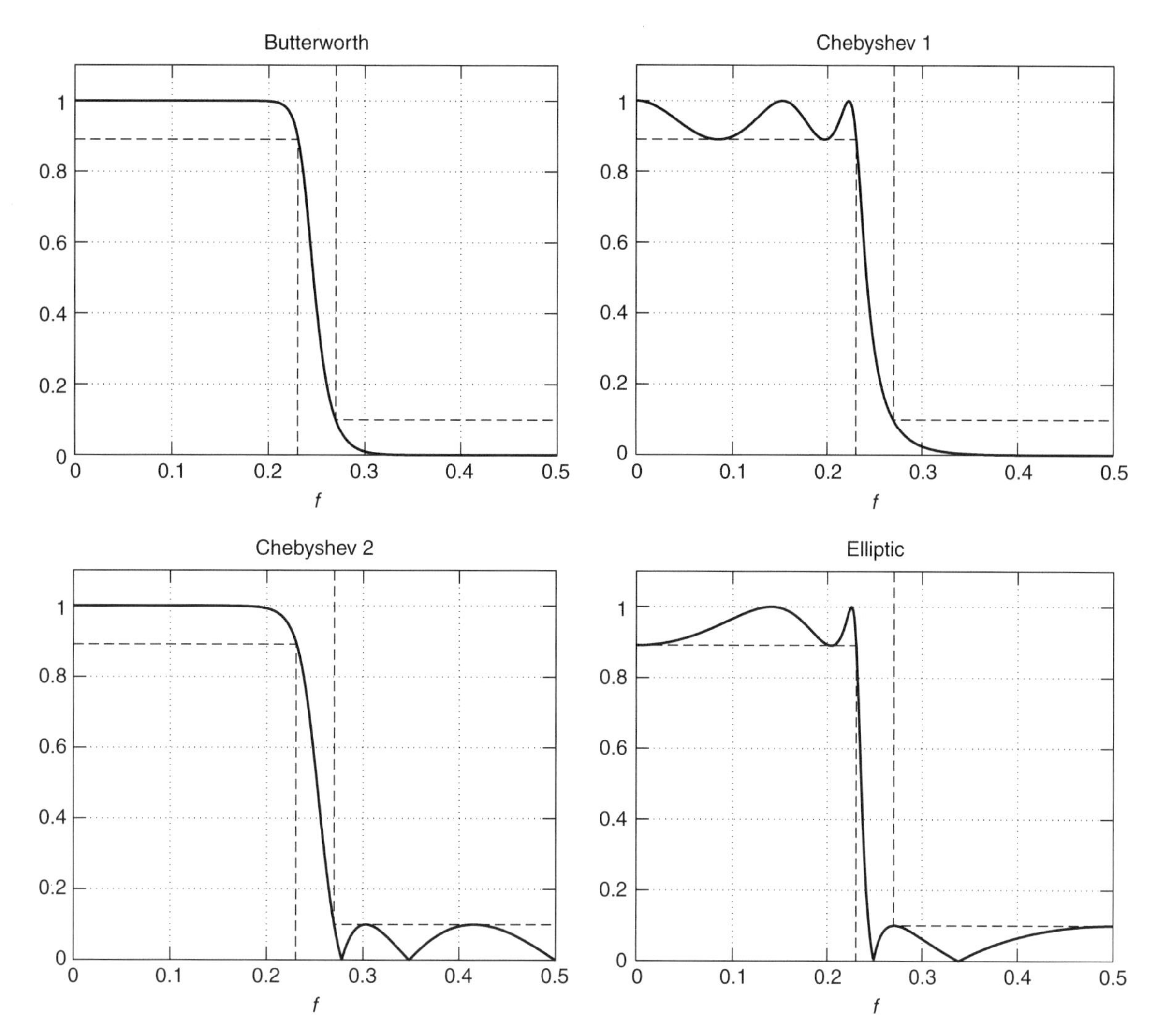

Figure 13. The four standard lowpass prototypes for IIR filter design.

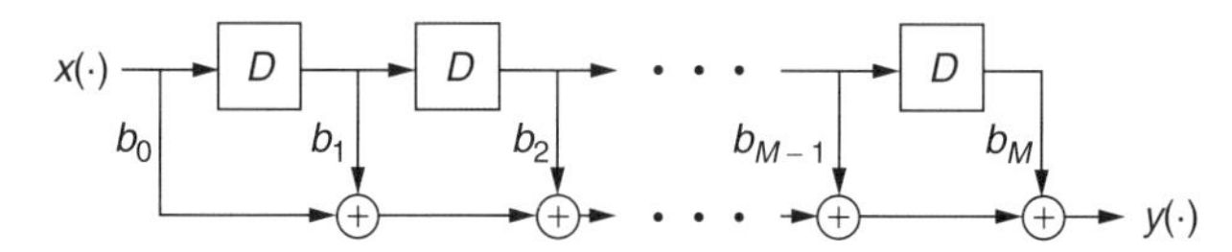

Figure 14. Transversal (direct-form) realization of an FIR digital filter.

$a(z) = 1$ for FIR filters, the direct form realization reduces to the configuration shown in Fig. 14.

The realization requires M delay elements, $M + 1$, multipliers and M adders. However, since FIR filters usually satisfy the symmetry constraint $b_i = \pm b_{M-i}$ in order to achieve linear phase, we can save about half of the number of multipliers.

5.2. Realization of IIR Filters

The transfer function of most practical IIR filters, such as those obtained from analog prototypes, have the same numerator and denominator degrees: $N \equiv \deg a(z) = \deg b(z) \equiv M$. Consequently, their direct-form realization requires N delay elements, $2N + 1$ multipliers, and N adders (see Fig. 10). We now describe two alternative realizations that require the same number of delays, multipliers, and adders as the direct-form realization.

The *parallel realization* of a rational transfer function $H(z)$ is obtained by using its partial fraction expansion

$$H(z) = A_0 + \sum_{j=1}^{N} \frac{A_j}{1 - p_j z^{-1}} \tag{16}$$

where p_j are the poles of $H(z)$ and where we assumed that (1) $M \le N$ and (2) all poles are simple. Both assumptions hold for most practical filters, and in particular for those obtained from the standard analog prototypes.

Since the denominator polynomial $a(z)$ has real coefficients, its roots (i.e., the poles p_j) are either real or conjugate pairs. In order to avoid complex filter coefficients, the two terms representing a conjugate pair of poles are combined into a single second-order term:

$$\frac{A}{1 - p_0 z^{-1}} + \frac{A^*}{1 - p_0^* z^{-1}} = 2\frac{(\mathrm{Re}A) - (\mathrm{Re}p_0)z^{-1}}{1 - (2\mathrm{Re}p_0)z^{-1} + |p_0|^2 z^{-2}}$$

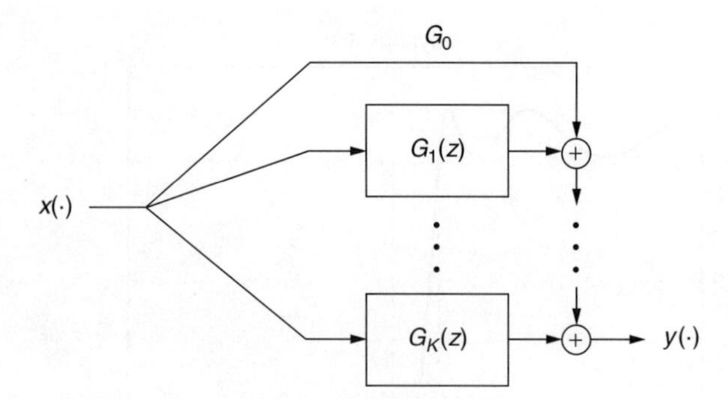

Figure 15. Parallel realization of an IIR digital filter.

Thus we obtain an additive decomposition

$$H(z) = A_0 + \sum_{k=1}^{K} G_k(z)$$

where each $G_k(z)$ is a strictly proper IIR filter or order 2 (for conjugate pairs of poles) or order 1 (for real poles). This results in a parallel connection of the subsystems $G_k(z)$ (Fig. 15). Finally, each individual $G_k(z)$ is realized in direct form.

The *cascade realization* is obtained by using the pole–zero factorization of the transfer function

$$H(z) \equiv \frac{b(z)}{a(z)} = b_0 \frac{\prod_{i=1}^{M}(1 - z_i z^{-1})}{\prod_{j=1}^{N}(1 - p_j z^{-1})} \tag{17}$$

where we recall again that the *zeros* z_i are the roots of the numerator polynomial $b(z)$ and the *poles* p_j are the roots of the denominator polynomial $a(z)$. Since these polynomials have real coefficients, their roots are either real or conjugate pairs. For each conjugate pair, the corresponding first-order factors are combined into a single second-order term:

$$(1 - z_0 z^{-1})(1 - z_0^* z^{-1}) = 1 - (2\text{Re}z_0)z^{-1} + |z_0|^2 z^{-2}$$

Thus we obtain a multiplicative factorization

$$H(z) = \prod_{k=1}^{K} H_k(z) \tag{18}$$

where each $H_k(z)$ is a proper IIR filter or order 2 (for conjugate pairs of poles) or order 1 (for real poles). This results in a cascade connection of the subsystems $H_k(z)$ (Fig. 16). Again, each individual $H_k(z)$ is realized in direct form.

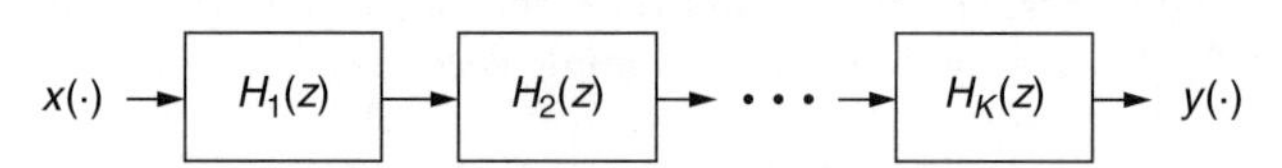

Figure 16. Cascade realization of an IIR digital filter.

There are multiple ways to form the factors $H_k(z)$ in (18), all requiring the same number of delays, multipliers, and adders as the direct-form realization. There are, however, differences in performance between these alternative factorizations, as explained in Section 5.3.

A unified algebraic framework for all possible realizations of a given finite-order transfer function $H(z)$ is provided by the theory of *factored state variable descriptions* (FSVD) [7]. The FSVD is a refinement of the well-known *state-space* description, which describes the relations between the input signal $x(\cdot)$, output signal $y(\cdot)$, and a state vector $s(n)$:

$$\begin{pmatrix} \mathbf{s}(n+1) \\ y(n) \end{pmatrix} = \begin{pmatrix} A & B \\ C & D \end{pmatrix} \begin{pmatrix} \mathbf{s}(n) \\ x(n) \end{pmatrix} \tag{19a}$$

The FSVD refines this characterization by specifying a multiplicative factorization

$$\begin{pmatrix} A & B \\ C & D \end{pmatrix} = \mathcal{F}_K \mathcal{F}_{K-1} \cdots \mathcal{F}_2 \mathcal{F}_1 \tag{19b}$$

that captures the decomposition of the realization into interconnected modules, as in the cascade or parallel realizations. Each $\mathcal{F}_k$ describes a single module, and each module is associated with a subset of the state vector $\mathbf{s}(n)$. The totality of all possible minimal realizations of a given $H(z)$ is captured by the notion of similarity transformations, combined with the added flexibility of factoring the (A, B, C, D) matrix as in (19b).

5.3. Implementation and Performance Attributes

The next stage in the design of a digital filter is *implementation*, that is, the mapping of a selected realization into a specific hardware/software architecture. Since a given transfer function has multiple implementations, which differ in their performance, the choice of a specific implementation should be based on objective performance criteria. Commonly used performance attributes for evaluating implementations include

- *Processing speed*—can be quantified in terms of two distinct attributes: (1) *throughput*, which is the number of input samples that can be processed by the digital filter in a unit of time, and (2) *input–output delay*, which is the duration between the instant a given input sample $x(n_o)$ is applied to the digital filter, and the instant the corresponding output sample $y(n_o)$ becomes available.
- *Numerical accuracy*—the level of error due to finite precision number representation.
- *Hardware cost*—the total number of each kind of basic building blocks, such as delays, multipliers, and adders.
- *Implementation effort*–the amount of work required to map a given realization into a specific hardware/software architecture. Structural attributes of the realization, such as modularity and/or regularity, contribute to the reduction of this effort.

The same attributes can be also used to evaluate the performance of realizations [6,7]. For instance, the direct realization of an IIR digital filter has much poorer numerical accuracy than does either the parallel or the cascade realizations. Similarly, the input–output delay of the parallel realization is somewhat shorter than that of the cascade or direct realizations. Such observations make it possible to optimize the selection of a realization, and to quantify the relative merits of alternative implementations.

6. OPTIMAL AND ADAPTIVE DIGITAL FILTERS

The desired response of classical frequency-selective filters is completely specified in terms of passbands, stopbands, and transition bands. In contrast, optimal filters use detailed information about the frequency content of the desired signal and the interference in achieving maximal suppression of the latter, along with minimal distortion of the former. Since the information needed to construct an optimal filter is typically not available a priori, it is usually estimated online from measurements of the available signals. When this process of frequency content estimation is carried out simultaneously with signal filtering, and the coefficients of the optimal filter are continuously updated to reflect the effect of new information, the resulting linear time-variant configuration is known as an *adaptive filter*.

Fixed frequency-selective filters are appropriate in applications where (1) the desired signal is restricted to known frequency bands (e.g., AM and FM radio, television, frequency-division multiplexing) or (2) the interfering signal is restricted to known frequency bands (e.g., stationary background in Doppler radar, fixed harmonic interference). On the other hand, in numerous applications the interference and the desired signal share the same frequency range, and thus cannot be separated by frequency selective filtering. Adaptive filters can successfully suppress interference in many such situations (see Section 7.2).

6.1. Optimal Digital Filters

Optimal filtering is traditionally formulated in a probabilistic setting, assuming that all signals of interest are *random*, and that their joint statistics are available. In particular, given the second-order moments of two discrete-time (zero-mean) random signals $x(\cdot)$ and $y(\cdot)$, the corresponding mean-square optimal filter, also known as a *Wiener filter*, is defined as the (unique) solution $h_{\text{opt}}(\cdot)$ of the quadratic minimization problem

$$\min_{h(\cdot)} E|y(n) - h(n) \circledast x(n)|^2 \tag{20}$$

where E denotes expectation. If $x(\cdot)$ and $y(\cdot)$ are jointly stationary, the Wiener filter $h_{\text{opt}}(\cdot)$ is time-invariant; otherwise it is a linear time-variant (LTV) filter. The Wiener filter can be applied in two distinct scenarios:

- *Linear mean-square estimation*—an unknown random signal $y(\cdot)$ can be estimated from observations of another random signal $x(\cdot)$. The corresponding estimate $\hat{y}(\cdot)$ is obtained by applying the observed signal $x(\cdot)$ to the input of the Wiener filter, so that $\hat{y}(\cdot) = h_{\text{opt}}(\cdot) \circledast x(\cdot)$.
- *System identification*—the impulse response of an unknown relaxed LTI system can be identified from observations of its input signal $x(\cdot)$ and output signal $y(\cdot)$. In fact, the optimal solution of the Wiener problem (20) coincides with the unknown system response, provided (1) the input signal $x(\cdot)$ is stationary, and is observed with no error and (2) the (additive) error in measuring the output signal $y(\cdot)$ is uncorrelated with $x(\cdot)$.

The unconstrained optimal solution of the Wiener problem (20) is a noncausal IIR filter, which is not realizable (see Section 3.2 for a discussion of realizability). In the case of jointly stationary $x(\cdot)$ and $y(\cdot)$ signals, realizable solutions are obtained by imposing additional constraints on the impulse response $h(\cdot)$ and/or the joint statistics of $x(\cdot)$, $y(\cdot)$:

- When both the autospectrum $S_{xx}(z)$ and the cross-spectrum $S_{yx}(z)$ are rational functions, and the Wiener filter is required to be causal, the resulting Wiener filter is realizable; specifically, it is causal and rational.
- When $x(\cdot)$ and $y(\cdot)$ are characterized jointly by a state-space model, the resulting optimal filter can be described by a state-space model of the same order. This is the celebrated *Kalman filter*.
- Realizability can be enforced regardless of structural assumptions on the joint statistics of $x(\cdot)$ and $y(\cdot)$. In particular, we may constrain the impulse response in (20) to have finite length. This approach is common in adaptive filtering.

The classical Wiener filter is time-invariant, requiring an infinitely long prior record of $x(\cdot)$ samples, and thus is optimal only in steady state. In contrast, the Kalman filter is optimal at every time instant $n \geq 0$, requiring only the finite (but growing) signal record $\{x(k); 0 \leq k \leq n\}$. Consequently, the impulse response of the Kalman filter is time-variant and its length grows with n, asymptotically converging to the classical Wiener filter response. Finally, adaptive filters continuously adjust their estimates of the joint signal statistics, so the resulting filter response remains time-variant even in steady state, randomly fluctuating around the Wiener filter response.

6.2. Adaptive Digital Filters

In practice, the statistics used to construct an optimal filter must also be estimated from observed signal samples. Thus, the construction of an optimal filter consists of two stages:

- *Training Stage*. Finite-length records of the signals $x(\cdot)$ and $y(\cdot)$ are used to estimate the joint (first- and) second-order moments of these signals.

Subsequently, the estimated moments are used to construct a realizable optimal filter.

- *Filtering Stage*. The fixed optimal filter that was designed in the training stage is now applied to new samples of $x(\cdot)$ to produce the estimate $\hat{y}(\cdot)$.

Alternatively, we may opt to continue the training stage indefinitely; as new samples of $x(\cdot)$ and $y(\cdot)$ become available, the estimated statistics and the corresponding optimal filter coefficients are continuously updated. This approach gives rise to an adaptive filter configuration — a linear time-variant digital filter whose coefficients are adjusted according to continuously updated moment estimates (Fig. 17).

Since adaptive filtering requires ongoing measurement of both $x(\cdot)$ and $y(\cdot)$, it can be applied only in applications that involve the system identification interpretation of the Wiener filter. Thus, instead of applying $h_{\text{opt}}(\cdot)$ to $x(\cdot)$ to estimate an unknown signal $y(\cdot)$, adaptive filters rely on the explicit knowledge of both $x(\cdot)$ and $y(\cdot)$ to determine $h_{\text{opt}}(\cdot)$. Once this optimal filter response is available, it is used to suppress interference and extract desired signal components from $x(\cdot), y(\cdot)$ (see Section 7.2).

The most common adaptive filters use a time-variant FIR configuration with continuously adjusting filter coefficients. Such filters take full advantage of the power of digital signal processing hardware because (1) the ongoing computation of signal statistics and optimal filter coefficients usually involves nonlinear operations that would be difficult to implement in analog hardware, and (2) modifications to the updating algorithms can be easily implemented by reprogramming. The most commonly used algorithms for updating the filter coefficients belong to the *least-mean-squares* (LMS) family and the *recursive least-squares* (RLS) family [3].

7. APPLICATIONS IN TELECOMMUNICATION SYSTEMS

Since the early 1980s, digital filtering has become a key component in a wide range in applications. We provide here a brief summary of the main telecommunication applications of digital filters, separated into two categories: (1) *time-invariant digital filters*, which are used to replace analog filters in previously known applications, and (2) *adaptive digital filters*, which enable new applications that were impossible to implement with analog hardware. A detailed discussion of numerous digital filtering applications can be found in the literature [1,4–6].

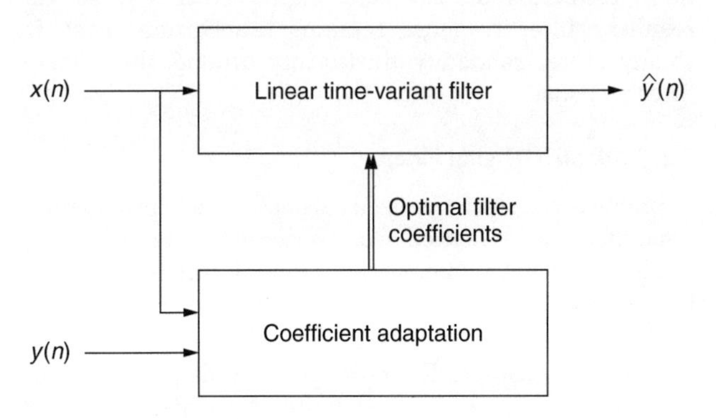

Figure 17. Adaptive filter configuration.

7.1. Time-Invariant Digital Filters

Ongoing improvements in the cost, size, speed, and power consumption of digital filters have made them an attractive alternative to analog filters in a variety of telecommunication applications that require a time-invariant filter response. The main types of such applications are:

- *Frequency-selective filters* — pass (or stop) prespecified frequency bands. Some examples include
  - RF-to-IF-to-baseband conversion in digital wireless systems
  - FDM multiplexing and demultiplexing
  - Subband decomposition for speech and image coding (e.g., MPEG)
  - Doppler radar moving-target indicator (MTI) to remove the nonmoving background
  - Digital spectrum analysis
- *Matched filters/correlators* — have an impulse response that matches a given waveform. Some examples include:
  - Matched filter for symbol detection in digital communication systems
  - Waveform-based multiplexing/demultiplexing (e.g., CDMA)
  - Front-end Doppler radar processing
- *Analog-to-digital and digital-to-analog conversion*
  - Oversampling converters (e.g., sigma–delta modulation)
  - Compact-disk recording and playing
- *Specialized filters* (such as)
  - Hilbert transformers
  - Timing recovery in digital communications (e.g., early–late gate synchronizer)

7.2. Adaptive Digital Filters

As explained in Section 6.2, adaptive filters implement the system identification scenario of the Wiener filter (Fig. 17), requiring two observed signals. Many applications use a variation of this configuration, known as the *adaptive interference canceler*, in which the received signal $y(\cdot)$ is a noisy version of an unknown desired signal $s(\cdot)$, and the so-called "reference signal" $x(\cdot)$ is a filtered version of the interference component of $y(\cdot)$ (Fig. 18).

Two assumptions must be satisfied in order for the adaptive filter to perform well as an interference canceler:

- *Lack of correlation* — the reference signal $x(\cdot)$ should be uncorrelated with the desired signal component of the received signal $y(\cdot)$.
- *FIR response* — the unknown system $H(z)$ should be FIR, and of a known order.

If these two assumptions are met, the estimated $\hat{H}(z)$ matches the true $H(z)$ perfectly, and the effect of

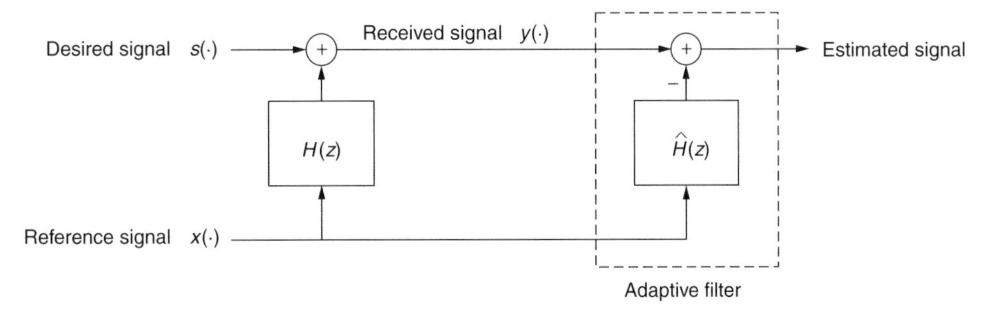

Figure 18. Adaptive interference canceller configuration.

interference is *completely cancelled*. In reality, both assumptions are met only approximately, so that only partial cancellation of the interference is achieved.

The interference cancelling configuration is useful in a variety of specific telecommunication applications, including

- *Echo cancellation*—used in duplex communication systems to suppress interference caused by signal leakage from the incoming far-end signal (the "reference signal" in Fig. 18) to the outgoing local signal (the "desired signal"). Such leakage is common in telephone lines (2W/4W converters), modems, teleconferencing systems and hands-off telephone units.
- *Decision feedback equalization*—used to reduce the effect of intersymbol interference (ISI) in a digital communication channel. Here the "reference signal" is the sequence of previously received symbols, which are assumed to be uncorrelated with the current symbol (the "desired signal"). In order for the equalizer to work correctly, the previously received symbols must be known without error. To meet this requirement, the equalizer alternates between a training phase and a tracking phase. In the training phase a prespecified "training sequence" of symbols is transmitted through the channel, and this information is used by the equalizer to determine the channel estimate $\hat{H}(z)$. In the tracking phase the equalizer uses detected previous symbols to track slow variations in the channel response: as long as the estimated response $\hat{H}(z)$ remains close to the true response $H(z)$, the cancellation of ISI is almost perfect, symbols are detected without error, and the correct "reference signal" is available to the equalizer. Since decision feedback equalization relies on previous symbols, it cannot reduce the ISI caused by the precursors of future symbols—this task is left to the feedforward equalizer.
- *Sidelobe cancellation*—used to modify the radiation pattern of a narrowbeam directional antenna (or antenna array), in order to reduce the effect of strong nearby interferers received through the sidelobes of the antenna. The "reference signal" in this case is obtained by adding an inexpensive omnidirectional antenna; if the interfering RF source is much closer than the source at which the main antenna is pointed, then the reference signal received by the omnidirectional antenna is dominated by the interference, and the lack-of-correlation assumption is (approximately) met.

The *adaptive beamformer* configuration, a variation of the sidelobe-canceling approach, is used to maintain the mainlobe of an antenna array pointed in a predetermined direction, while adaptively reducing the effect of undesired signals received through the sidelobes [3]. It can be used, for instance, to split the radiation pattern of a cellular base-station antenna into several narrow beams, each one tracking a different user.

Another common adaptive filtering configuration is the *adaptive linear predictor*, in which only one observed signal is available (Fig. 19). In this case the "reference signal" is simply a delayed version of the observed signal $y(\cdot)$, so that the Wiener problem (20) now becomes

$$\min_{h(\cdot)} E|y(n) - h(n) \circledast y(n-\Delta)|^2 \tag{21}$$

which is, a Δ-step-ahead linear prediction problem. In some applications the object of interest is the adaptively estimated linear predictor response $\hat{h}(\cdot)$, while in other applications it is the *predicted signal* $\hat{y}(n) = \hat{h}(n) \circledast y(n-\Delta)$, or the *residual signal* $y(n) - \hat{y}(n)$. Telecommunication applications of the linear predictor configuration include

- *Linear predictive coding* (LPC)—used for low-rate analog-to-digital conversion, and has become the standard speech coding method in GSM cellular systems. While the original LPC approach used only the linear predictor response $\hat{h}(\cdot)$ to represent the observed speech signal $y(\cdot)$, current LPC-based speech coders use, in addition, a compressed form of the residual signal $y(n) - \hat{y}(n)$ [6].
- *Adaptive notch filter*—used to suppress sinusoidal interference, such as narrowband jamming in spread-spectrum systems. Since spread-spectrum signals resemble broadband noise, which is almost completely unpredictable, while sinusoidal signals are

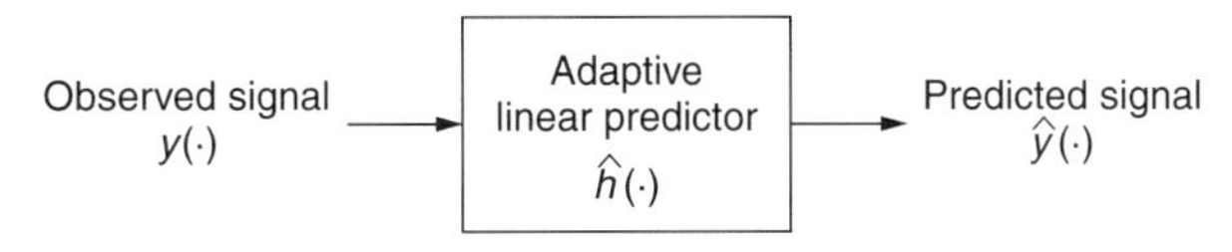

Figure 19. Adaptive linear predictor configuration.

perfectly predictable, the adaptive linear predictor response is almost entirely dominated by the sinusoidal component of the observed signal $y(\cdot)$. As a result, one can use the adaptive linear predictor configuration to suppress sinusoidal components or, with a minor modification, to enhance sinusoidal components.

- *Feedforward equalizer* — used to reduce the effect of ISI from precursors of future symbols. This is made possible by the fact that the adaptive linear predictor tends to render the equalized channel response minimum phase, thereby reducing the length of precursors. This minimum-phase property is an intrinsic characteristic of the MSE linear predictor defined by (21).

BIOGRAPHY

Hanoch Lev-Ari received the B.S. (summa cum laude) and the M.S. degrees in electrical engineering from the Technion, Israel Institute of Technology, Haifa, Israel in 1971 and 1976, respectively, and the Ph.D. degree in electrical engineering from Stanford University, Stanford, California, in 1984. During 1985 he held a joint appointment as an Adjunct Research Professor of Electrical Engineering with the Naval Postgraduate School, Monterey, California and as a Research Associate with the Information Systems Laboratory at Stanford. He stayed at Stanford as a Senior Research Associate until 1990. Since 1990 he has been an Associate Professor with the Department of Electrical and Computer Engineering at Northeastern University, Boston, Massachusetts. During 1994–1996 he was also the Director of the Communications and Digital Signal Processing (CDSP) Center at Northeastern University. Dr. Lev-Ari has over 100 journal and conference publications. He served as an Associate Editor of *Circuits, Systems and Signal Processing*, and of *Integration, the VLSI Journal*. His areas of interest include model-based spectrum analysis and estimation for nonstationary signals, scale-recursive (multirate) detection and estimation of random signals, Markov renewal models for nonstationary signals, and adaptive linear and nonlinear filtering techniques.

BIBLIOGRAPHY

1. M. E. Frerking, *Digital Signal Processing in Communication Systems*, Van Nostrand Reinhold, New York, 1994.
2. R. A. Horn and C. R. Johnson, *Matrix Analysis*, Cambridge Univ. Press, 1985.
3. S. Haykin, *Adaptive Filter Theory*, 4th ed., Prentice-Hall, Upper Saddle River, NJ, 2002.
4. V. K. Madisetti and D. B. Williams, eds., *The Digital Signal Processing Handbook*, CRC Press, 1998.
5. S. K. Mitra and J. F. Kaiser, eds., *Handbook for Digital Signal Processing*, Wiley, New York, 1993.
6. B. Porat, *A Course in Digital Signal Processing*, Wiley, New York, 1997.
7. R. Roberts and T. Mullis, *Digital Signal Processing*, Addison-Wesley, 1987.

DIGITAL OPTICAL CORRELATION FOR FIBEROPTIC COMMUNICATION SYSTEMS

John E. McGeehan
Michelle C. Hauer
Alan E. Willner
University of Southern California
Optical Communications Laboratory
Los Angeles, California

1. INTRODUCTION

As bit rates continue to rise in the optical core of telecommunication networks [toward $\geq$40 Gbps (gigabits per second)], the potential inability of electronic signal processors to handle this increase is a force driving research in optical signal processing. Correlation, or matched filtering, is an important signal processing function. The purpose of a correlator is to compare an incoming signal with one that is "stored" in the correlator. At the appropriate sample time, a maximum autocorrelation peak will be produced if the input signal is an exact match to the stored one. This function is typically used to pick a desired signal out of noise, an essential requirement for radar and wireless CDMA systems. As telecommunication systems tend toward the use of optical fiber as the preferred transmission medium, optical CDMA networks are receiving greater attention and will require the use optical correlator implementations.

In present-day fiberoptic networks, data packets are converted to electrical form at each node to process their headers and make routing decisions. As routing tables grow in size, this is becoming a predominant source of network latency. The advent of optical correlation could enable packet headers to be read at the speed of light by simply passing the packets through a bank of correlators, each configured to match a different entry in the routing table. Simple decision logic could then configure a switch to route each packet according to which correlator produced a match. Thus, with the growing demand for all-optical networking functions, there is a strong push to develop practical and inexpensive optical correlators that can identify high-speed digital bit patterns on the fly and with minimal electronic control.

2. DIGITAL OPTICAL CORRELATION

The term "correlator" is typically used to describe a hardware implementation of a matched filter. Although there exist strict definitions for matched filters versus correlators, most hardware implementations that produce the same output as a matched filter, sampled at the peak autocorrelation value, are referred to as "correlators." The detailed theory of matched filters and correlation is presented in most textbooks on communication systems [1]. The aim here is to give a brief overview of the concepts that apply to the correlation of digital binary waveforms modulated onto optical carriers. This presents a unique case in that the data bits in an optical communication system are most commonly represented by the optical power of the signal as opposed to a voltage as in electrical systems. Consequently, the digital waveforms consist of only positive values (a "unipolar" system), causing the correlation function of any two optical signals to be a completely nonnegative function. This creates some limitations for systems that transmit specially designed sets of codewords intended to have good autocorrelation properties; thus, a large autocorrelation peak when the incoming signal is synchronized with the receiver and matches the desired codeword and a low, ideally zero, level for all other codewords, for all possible shifts in time [2]. Sets of codewords with these properties are termed *orthogonal codes*. It is possible with optical phase modulation to achieve a bipolar system, but this requires a coherent receiver, which is far more complex to implement in optics than the standard direct intensity detection receivers.

Just as with electronics, prior to the development of high-speed digital signal processors, a common implementation of an optical correlator is the tapped-delay line. A basic implementation of an optical tapped-delay line is shown in Fig. 1a. The delay line is configured to match the correlation sequence 1101. Thus, the delay line requires four taps (one for each bit in the desired sequence), weighted by the factors 1, 1, 0 and 1, respectively. The weights are implemented by placing a switch in each path that is closed for weight = 1, and opened for weight = 0. The incoming optical bit stream is equally split among the four fiberoptic delay lines. Each successive delay line adds one additional bit of delay to the incoming signal before the lines are recombined, where the powers of the four signals are added together to yield the correlation output function. This function is sampled at the optimum time, T_s, and passed through a threshold detector that is set to detect a power level above 2 since the autocorrelation peak of 1101 with itself equals 3 (or more specifically, 3 times the power in each 1 bit). This threshold detection may be implemented either electronically or optically, although optical threshold detection is still a nascent research area requiring further development to become practical. Electronically, the output is detected with a photoreceiver and a simple decision circuit is used to compare the correlation output to the threshold value. The high-speed advantage of optics still prevails in this case since the correlation function is produced in the time it takes the signal to traverse the optical correlator, and the decision circuit only needs to be triggered at the sample-rate, which is often in the kilohertz–megahertz range, depending on the number of bits in each data packet or codeword. The mathematical function describing the tapped-delay-line correlator is

$$y(t) = \sum_{k=0}^{N-1} x(t - kT_{\text{bit}})h(kT_{\text{bit}}) \quad (1)$$

where N is the number of bits in the correlation sequence, T_{bit} is one bit period, $x(t - kT_{\text{bit}})$ is the input signal delayed by k bit times, and $h(kT_{\text{bit}})$ represents the k weights that multiply each of the k-bit delayed input signals. For a phase-modulated system, the same operation can be performed by replacing the switches with the appropriate optical phase shifters to match the desired codeword (e.g., $\pi\pi 0\pi$ instead of 1101). Figure 1b illustrates the delay-and-add operation of the correlator for the case when the three 4-bit words 1010, 1011, and 1101 are input to the correlator, where the second word is an exact match to the desired sequence. Since the correlation function for two 4-bit words is 7 bits long and the peak occurs during the fourth time slot, the correlation output is sampled every 4 bits and compared to a threshold as shown in Fig. 1c. As expected, the correlation output for the second word exceeds the threshold while the first and third samples produce no

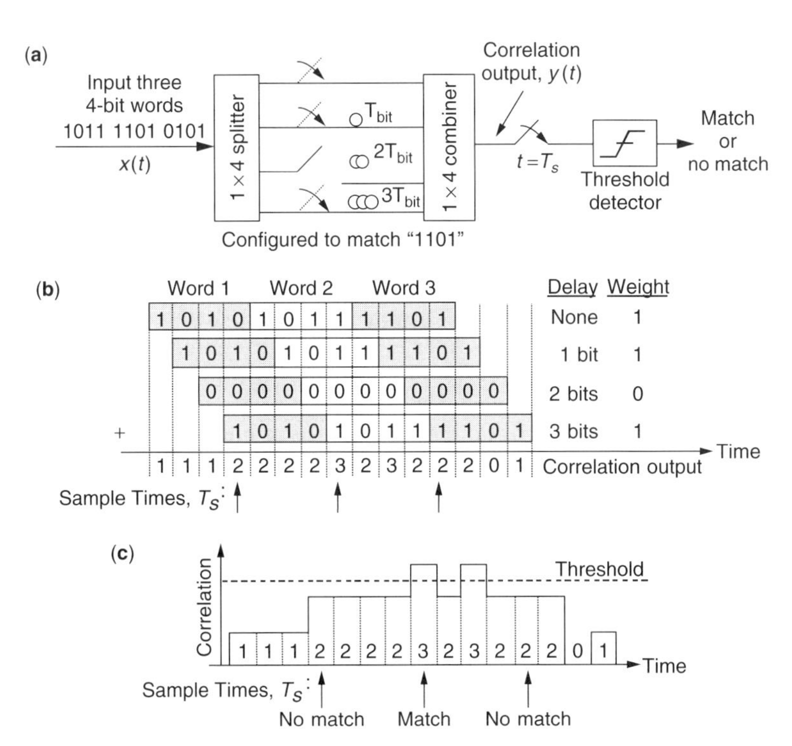

Figure 1. (**a**) A basic implementation of a fiberoptic tapped-delay-line correlator configured to produce an autocorrelation peak at the sample time, T_s, for the sequence 1101; (**b**) the weighted delay-and-add computation of the digital correlation output for three input words when correlated with the bit pattern 1101 (the three optimum sample times are labeled); (**c**) an intensity profile of the correlation output from this tapped-delay-line correlator. Only when the intensity is above the threshold at the sample time is a match signal produced.

matches. Note that for an input signal L bits long, the length of the correlation output will be $L+N-1$ bits long.

It should also be noted that the correlator as shown will also produce a level 3 peak that is above the threshold at time T_s for a 1111 input, which is not the desired codeword. This is because the open switch in the third delay line, corresponding to the third correlation bit, does not care if the third bit is a 1 or a 0 since it does not pass any light. Thus, the correlator as shown is really configured to produce a match for the sequence 11x1, where the x indicates a "don't care" bit that can either be 0 or 1. In many cases, such as in optical CDMA systems, the set of all possible codewords used in the system is specifically designed to always have a constant number of 1 bits so that this is not an issue. But, for cases in which the correlator must identify a completely unique sequence, the present correlator must be augmented with, for example, a second, parallel tapped-delay line that is configured in complement to the first one (the switches are closed for desired 0 bits and open otherwise), and produces a "match" signal when zero power is present at the sample time. Then, the incoming signal is uniquely identified only when both correlators produce a match. This configuration will be discussed later in further detail.

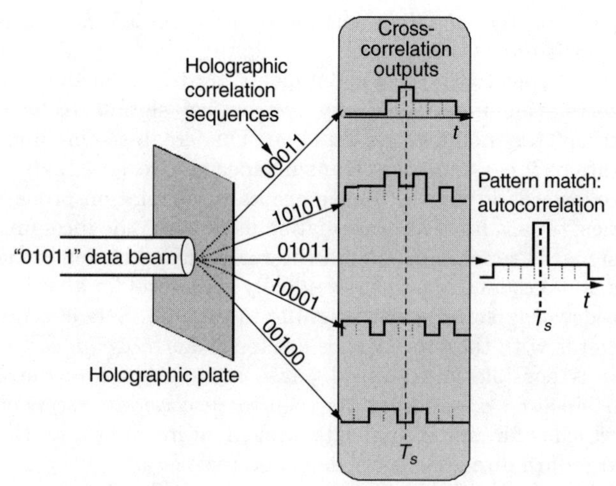

Figure 2. A collimated 01011 input beam in free space impinges on an optical angular multiplexed spectral holographic (AMSH) plate that is programmed to recognize five different correlation sequences. Each correlation sequence corresponds to a different scattering angle from the plate. Only the direction corresponding to a pattern match (01011) produces an intensity autocorrelation peak that will be above threshold at the sample time, while the rest produce cross-correlation outputs.

3. OPTICAL CORRELATOR IMPLEMENTATION

Digital optical correlators can be fabricated using free-space optics, fiber-based devices, and fiber-pigtailed crystals or semiconductors. Four varieties of optical correlators will be reviewed here, including (1) a free-space holographic correlator, (2) a fiberoptic correlator using separate fiber delay lines terminated with mirrors, (3) a single-fiber device with periodically spaced fiber Bragg gratings (FBGs) to provide time-delayed reflections, and (4) an optical phase-correlator implemented using FBGs. The first three of these correlators are explained assuming optical intensity-modulated correlation sequences. However, most can be modified to act as phase-coded correlators as well. No assumptions are made regarding any special data coding schemes.

One example of a free-space optical correlator employs spectral holography to create a correlating plate that is able to correlate an incoming bit stream with several correlation sequences simultaneously. The correlation sequences are encoded using an angular multiplexed spectral hologram (AMSH). The hologram acts upon the incoming modulated lightbeam to produce a correlation output [3]. The incoming bit pattern is spread out in the frequency domain using a grating, collimated by a lens and then sent through the AMSH plate, which is coded for a finite number of possible correlation patterns. The hologram is written using a 632.8-nm helium–neon laser and a photosensitive holographic plate and is designed to operate at signal wavelengths around 1550 nm (a standard fiberoptic communication wavelength). Each desired correlation sequence is written onto the holographic plate such that each pattern corresponds to a different diffraction angle. Figure 2 shows an input datastream of 01011 scattering off an AMSH plate. The correlation outputs corresponding to each correlation sequence are produced at different scattering angles at the output. As the number of correlation patterns increases, the difference between the output angles for each pattern decreases. At each diffraction angle, either an autocorrelation or cross-correlation output is produced, with the autocorrelation output appearing only at the diffraction angle corresponding to a matched correlation pattern, and cross-correlation outputs appearing at all other angles. A second grating can also be used to reroute any diffracted light back into an optical fiber for continued transmission. Otherwise, photodetectors may be placed at each diffraction angle to detect the correlation outputs. This free-space correlation method has a number of potential advantages over the fiber-based correlators that are detailed below, including the potential low size and cost associated with having one hologram that can correlate with many bit patterns simultaneously. However, further research is required on low-loss holographic materials at standard fiberoptic communication wavelengths, as the loss for the material used for this demonstration can exceed 80 dB due to high absorption at these frequencies.

To avoid the losses associated with exiting and reentering the fiber medium, a reconfigurable optical correlator that operates entirely within optical fibers is highly desirable for fiberoptic communication systems. In addition to the optical tapped-delay-line structure described in Fig. 1, a similar device may be constructed, except that each fiber branch is terminated with a fiber mirror (a metallic coating on the fiber end face), thereby providing a double pass through the delay lines and using the splitter also as a recombiner (see Fig. 3) [4]. The double pass requires that the relative delays be halved and an optical circulator must be added at the input to route the counterpropagating correlation output to the threshold detector. Aside from these differences, the operation of the correlator is identical to that described for Fig. 1. The

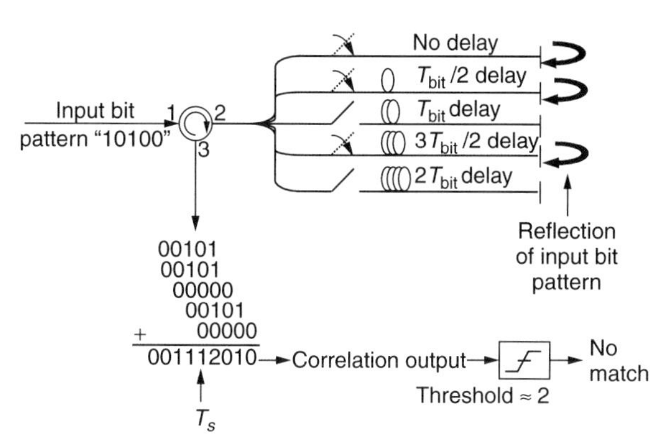

Figure 3. A fiberoptic delay line correlator in which each branch is terminated with a fiber mirror. The correlator is configured for the correlation sequence $11x1x$, where the switches are closed to represent 1 bits and opened to indicate "don't care" bits, meaning that the peak of the output correlation function at time T_s will be the same regardless of whether the "don't care" bits are ones or zeros. The correlation output for an input pattern of 10100 is 001112010. The level 1 output at the sample time does not exceed the threshold and produces no match.

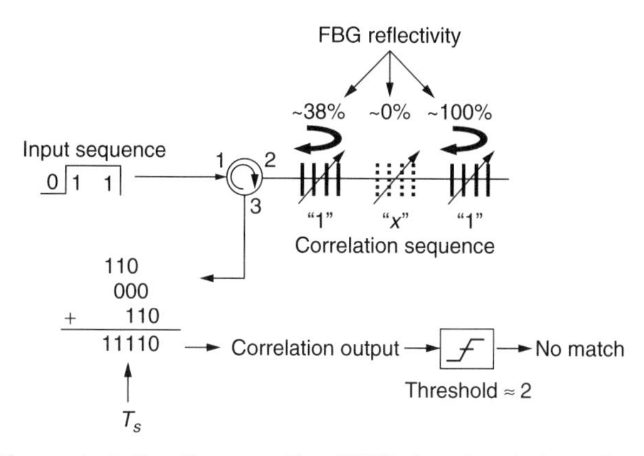

Figure 4. A fiber Bragg grating (FBG)–based optical correlator in which the correlation sequence is programmed by tuning the reflectivities of the gratings such that desired 1 bits reflect and "don't care" bits do not reflect. The cross-correlation output 11110 is the result of the correlation between the input sequence 011 and the programmed sequence $1x1$. The output is sampled at time T_s and a threshold detector produces a "no match" decision.

fiber-mirror-based correlator shown in Fig. 3 is configured to recognize the correlation sequence 11010 (or more accurately, $11x1x$, since the open switches represent "don't care" bits) by closing the first, second and fourth switches and setting the threshold just below a level 3. The figure shows the output cross-correlation function for the input sequence 10100, which will not produce a correlation peak above the threshold at the sample time.

A set of fiber Bragg gratings (FBGs) can also be used to construct an effective fiber-based optical correlator. An FBG is fabricated by creating a periodic variation in the fiber's index of refraction for a few millimeters to a centimeter of length along the fiber core [5]. Since optical fiber is photosensitive at ultraviolet frequencies, the grating can be written by illuminating the fiber from the side with the interference pattern of two ultraviolet laser beams. A conventional FBG acts as a reflective, wavelength-selective filter; that is, it reflects light at a particular wavelength that is determined by the spatial period of the index grating, and passes light at all other wavelengths. The bandwidth of the FBG filter depends largely on the magnitude of the index variation and is typically designed to be <100 GHz (0.8 nm) for communications applications. A nice feature of FBG filters is that the reflection spectrum can be adjusted by a few nanometers via heating or stretching of the grating. This allows one to alter the wavelength that the FBG reflects. The reflectivity of the grating is nearly 100% at the center of the reflected spectrum and falls off outside the grating bandwidth. By tuning the FBG so that the signal wavelength intersects with the rising or falling edge of the passband, the reflected energy at that wavelength will be reduced from 100% toward 0% as the grating's passband is tuned farther away.

Although it is possible to produce an optical correlator using FBGs to simply replace the fiber mirrors described above, the only advantage to this would be that the optical switches could be removed and the FBGs could simply be tuned *not* to reflect the incoming wavelength to represent the case of an open switch. However, the fact remains that this tapped-delay-line structure requires many separate branches of optical fiber, each precisely cut to provide delay differences as small as $\frac{1}{2}$ of a bit time, which is only 1 cm in fiber length at 10 Gbps. Using FBGs, a more compact, producible and manageable correlator can be produced [6]. The procedure for writing gratings into fibers makes it relatively simple to write several FBGs into a single fiber with precise control down to centimeter spacings (see Fig. 4). Using separate piezoelectric crystals or small heaters, each FBG can be independently stretched or heated to tune its reflection spectrum. Now the tapped-delay line may be implemented within a single piece of fiber, again with a circulator at the input to route the correlation output to the threshold detector. The FBG-based correlator shown in Fig. 4 is configured to recognize a bit pattern of 101 (or, more accurately, $1x1$). This is accomplished by writing three gratings in a row, with center-to-center spacings equal to $\frac{1}{2}$ of a bit time in fiber so that the round-trip time between gratings corresponds to a 1-bit delay. The reflectivity of the third grating is ideally 100% since it is the last "mirror" in the series. The reflection spectrum of the second grating is tuned away from the incoming signal wavelength so that it will simply transmit the light and there will be no reflection for the $2T_{\text{bit}}$ delay. This is the equivalent of the "open switch" in the previous configuration. The first grating must then be tuned to only partially reflect the incoming light so that the remaining light can pass through to reflect off the third grating. The reflectivity of the first grating must be chosen carefully to guarantee that the pulses reflecting off the first grating have power equal to that of those that reflect off the third grating. In practice this can be determined by repeatedly sending a single pulse into the FBG array and observing the powers of the two, time-delayed output pulses on an oscilloscope (the two gratings will reflect the input pulse at different times, creating two pulses at the output). The first FBG can then be tuned until the two pulses have equal power. The required reflectivities of the gratings in the array can also be calculated. Assume that there are n gratings in the array that represent 1 bits and therefore must provide some partial reflection of the incoming signal. Let the reflectivities of each grating be R_n (e.g., if $R_n = 0.4$, then the grating will reflect 40% of the incoming light and transmit the remaining 60% since $1 - R_n = 0.6$). Then the equation to determine the reflectivities needed to achieve equal output powers of all the time-delayed output pulses is

$$\frac{R_n}{(1 - R_n)^2} = R_{n+1} \quad (2)$$

Thus, the reflectivity of the last grating in the array should be set equal to 1, and then the reflectivities of all the preceding gratings can be calculated using this recursive equation. For the case shown in Fig. 4, with $R_3 = 1$, we get $R_1 = (1 - R_1)$, which results in $R_1 = 38\%$. The cross-correlation output for the case of an input sequence of 011 is depicted in Fig. 4. Limitations of this method due to the increasing attenuation of the signal as it makes a double pass through the series of gratings will be discussed in the following section.

By replacing the switches or reflectivity-tuned FBGs in the previous configurations with phase shifters or phase-encoded FBGs, the fiber-based optical intensity correlators described above can be used as phase correlators instead [7]. The holographic correlator may also be adapted to correlate with phase-encoded signals. For these cases, the correlation sequences are a series of phase shifts such as $0\pi\pi00\pi0\pi$, for an 8-bit codeword. As long as the incoming codeword contains the same phase shifts, an autocorrelation peak will result. There are several methods of generating optical codewords containing such a series of phase shifts. One all-fiber method utilizes a single FBG, 4 cm long, with seven very thin wolfram wires (diameter = 25 μm) wrapped around the grating every 5 mm [8]. The 5 mm spacing provides a 50-ps round-trip delay, corresponding to 1 bit time at 20 Gbps, and the 7 wires mean that this correlator can recognize an 8-bit phase-encoded sequence. By passing a current through one of the wires, the FBG will be heated at only that point in the grating, causing the signal being reflected by the FBG to experience a phase shift at the time delay corresponding to the location of the wire. The heat-induced phase shift occurs because the index of refraction of the glass fiber is temperature-sensitive. Note that only point heating is desired here to induce the phase shifts — it is not the aim to shift the grating's reflection spectrum, which would occur if the entire grating were heated. This device may be used to both generate phase-encoded signals at the transmitter and to correlate with them at the receiver. The most common application of optical phase correlators is to construct optical CDMA encoders and decoders. One possible advantage of phase modulation over intensity modulation is the potential for multilevel coding. While intensity modulation coding primarily uses on/off keying (OOK), where a 1 bit is represented by the presence of light and a 0 bit by the absence of it, phase coding need not utilize phase shifts of only 0 and π. For example, another article demonstrated four-level encoding, using phase shifts of 0, $\pi/2$, π, and $3\pi/2$ [9].

This is merely a sampling of the available optical correlator designs, specifically concentrating on those that are easily compatible with fiberoptic telecommunication systems and provide easy-to-understand illustrations of optical correlators. However, many other novel correlator configurations have been developed in research labs. These include correlators based on semiconductor optical amplifiers (SOA) [10], erbium-doped fibers (EDFs) [11], optical loop mirrors [12], and nonlinear optical crystals [13].

4. IMPLEMENTATION CHALLENGES

While the previous section detailed some of the more common optical correlator structures, there are a number of roadblocks facing optical correlators that keep them from seeing wide use beyond research environments. However, a number of advances not only begin tackling these roadblocks but also aim to decrease the cost and increase the commercial viability of optical correlators.

A driving motivation for research in optical signal processing is the fact that electronics may at some point present a bottleneck in telecommunication networks. Optical signals in commercial networks currently carry data at 2.5 and 10 Gbps and in research environments, at 10, 40, and even 160 Gbps. At these higher speeds, either electronic circuits will be unable to efficiently process the data traffic, or it may actually become more economical to use optical alternatives. However, optical techniques also face significant hurdles in moving to 40 Gbps and beyond. In particular, optical correlators that rely on fiberoptic delays require greater fabrication precision as the bit rate rises. At 40 Gbps, a single bit time is 25 ps, corresponding to 0.5 cm of propagation in standard optical fiber. For the fiber-mirror-based correlator, the differences in length of the fiber branches must equal a $\frac{1}{2}$ of a bit time delay, or 2.5 mm — an impractically small length. The FBG-based correlator has an additional problem in that not only must the gratings have a center-to-center spacing of 2.5 mm at 40 Gbps, but the gratings themselves are often longer than 2.5 mm. While it is possible to make FBGs that are only 100s of micrometers long, the shorter length will make it difficult, but not impossible, to achieve a high-reflectivity grating with the appropriate bandwidth to reflect the higher data-rate signal. Furthermore, there must be some method to provide precise tuning of the closely spaced individual FBGs. One report [14] demonstrated a grating-based tapped-delay line correlator created from a single FBG. A periodically spaced series of thin-film metallic heaters were deposited on the surface of the FBG to effectively create several tunable FBGs from one long one. By passing individual currents through the heaters, the reflection spectrums for the portions of the FBG beneath the heaters are tuned away. Since the heaters can be deposited with essentially lithographic precision, this resolves the issue of how to precisely fabricate and tune very closely spaced FBGs. Of course, this technique will eventually be limited by how closely spaced the heaters can be before thermal crosstalk between neighboring elements becomes a significant problem. Spectral holography is perhaps one of the more promising techniques for higher bit rates, as it is currently feasible (albeit expensive) to create holographic plates that can correlate with high-speed signals. However, the high absorption loss of current free-space holographic correlators at communication wavelengths remains a roadblock to any practical implementation.

Coherence effects can also present problems in optical correlators that utilize tapped-delay-line structures in which the light is split into several time-delayed replicas that are recombined at the correlator output. The coherence time of standard telecommunication lasers is typically tens of nanoseconds, corresponding to a coherence length in fiber of ~2 ms. When the differential time delay between two branches of the correlator is less than the coherence time of the laser (which is clearly the typical case), then the recombined signals will coherently interfere with each other, causing large power fluctuations in the correlation output function. This effect destroys the correlation output and must be mitigated or prevented in order to effectively operate the correlator. There are a number of ways that this problem can be solved. A polarization controller followed by a polarization beamsplitter (PBS) can be used at the input of each 1×2 optical splitter/combiner to ensure that the electric field polarizations between the two branches are orthogonal. This will prevent coherent interference of the recombined signals because orthogonally polarized lightbeams will not interfere with each other. Thus, for more than two branches, a tree structure of 1×2 splitters with polarization controllers and polarization beamsplitters can be used. Another, more manageable, solution is to somehow convert the coherent light into an incoherent signal before entering the correlator. One method of doing this uses cross-gain modulation in a semiconductor optical amplifier (SOA) to transfer the coherent data pattern onto incoherent light, which in this case is the amplified spontaneous emission light generated by the SOA [15].

An additional problem facing the grating-based correlator is the severe optical losses associated with multiple passes through the FBGs. This can so significantly limit the length of the correlation sequence that it can realistically correlate with before the power in the correlation output pulses are so low that they drop below the system noise floor. As explained before, the reflectivities of the gratings are determined by the need to equalize the pulses reflecting off of each grating representing a 1 bit, while recognizing that the pulses are attenuated with each pass through each grating. With four 1 bits in a correlation sequence, only ~65% of the incident light is present in the total correlation output. The rest is lost as a result of multiple reflections within the correlator, as the reflection off an internal grating can reflect again on the return trip and will no longer fall within the correlation time window. These multiple internal reflections do cause undesired replicas of the signal to add to the correlation output function, but they are so attenuated by the multiple reflections that they are not typically considered a significant impairment. As the number of 1 bits increases, the correlation sensitivity will decrease as more light is lost to these reflections. One method to mitigate this problem is to interleave the gratings between multiple fiber branches [6]. For the 4-bit sequence, two branches, each with two gratings, preceded by a splitter, can be used instead of a single row of gratings, thereby increasing the efficiency (assuming that all gratings are 1 bits) to ~75% of the incident light. This solution also alleviates the difficulty of spacing the gratings very closely together since only every other grating is on each branch, at the cost of introducing coherence effects that must be mitigated.

5. ADVANCED TOPICS

As wavelength-division-multiplexed (WDM) systems are becoming the standard in optical networks, it is becoming increasingly desirable to build modules that can act on multiple wavelength channels simultaneously. To this end, there is a growing interest in multiwavelength optical correlators. One study of a multiwavelength optical correlator uses sampled FBGs in a grating-based correlator structure. The reflection spectrum of a sampled FBG possesses multiple passbands so that it can filter several wavelength channels simultaneously [16]. When this type of FBG is stretched or heated, the entire reflection spectrum shifts, causing the reflectivity at each wavelength to experience the same variation. Thus, by replacing the standard FBGs with sampled FBGs in the correlator structure described previously, incoming signals on multiple wavelengths can simultaneously be correlated with a single correlation sequence. While it may still be necessary to demultiplex these channels prior to detection in order to check the correlation output for each channel, this technique still significantly reduces the number of components that would otherwise be required in a system that provided a separate correlator for each wavelength.

In Section 2, we mentioned that the conventional N-bit optical tapped-delay-line correlator cannot be configured to uniquely identify all 2^N possible bit patterns. This is because a 0 bit is represented by an open switch or a grating that is tuned not to reflect any light. This really means that these bit positions are considered "don't care" bits and the desired autocorrelation peak will result at the sample time whether the "don't care" bits are 1s or 0s. This is not an issue in optical CDMA systems, where the set of codewords can be specifically designed to maintain a constant number of 1 bits in each codeword. However, for applications that must be able to uniquely recognize any of the 2^N possible N-bit sequences, this situation will result in false-positive matches whenever a 1 is present where a 0 bit is desired. This is important for applications such as packet header or label recognition. A solution to this problem is to add a second correlator that is configured in complement to the first one and produces a "match" signal when *zero* power is present at the sample time. This is accomplished by placing a NOT gate at the output of the threshold detector which is set just above level zero. If the power goes above the threshold, this indicates that at least one 1 bit is present where a 0 is desired, and the NOT gate will convert the high output to a low one to indicate "no match" for this correlator. This correlator therefore correlates with the desired 0 bits in the sequence and is thus called a "zeros" correlator. Likewise, the originally described correlator is called a "ones" correlator. In the zeros correlator, the switches are closed for desired 0 bits and open otherwise (or the FBGs reflect for desired 0 bits and are tuned away otherwise). Thus, the 1 bits are "don't care" bits in a zeros correlator. By combining the outputs of the ones and zeros correlators with an AND gate, the input sequence will only produce a final "match" signal when the input pattern uniquely matches the desired correlation sequence. An illustration of how the combination of a ones and a zeros correlator can avoid false positive matches is depicted in Fig. 5. The desired correlation sequence is 1001, meaning that the ones correlator is configured to match a $1xx1$ pattern and the zeros correlator will produce a match for an $x00x$ pattern. In Fig. 5a, the incoming sequence is 1001, and so the ones and zeros correlators both produce "match" signals, resulting in a "match" signal at the output of the AND gate. In Fig. 5b, the input sequence is a 1101, causing the ones correlator to still produce a "match" signal (this would be a false-positive match if only this correlator were used). But the undesired 1 bit in the second time slot of the input sequence causes the power at the sample time in the zeros correlator to go above threshold, resulting in a "no match." The combination of the "match" and "no match" signals in the AND gate produce the correct "no match" result.

6. CONCLUSION

While optical correlators currently see limited commercial application, the frontier of all-optical networking is rapidly approaching, aided by the ever-increasing demand to transmit more bandwidth over the network core. This, in addition to the growing interest in optical CDMA networks, will push designers to develop producible optical correlators that can be dynamically adjusted to recognize very high bit-rate sequences. For applications such as header or label recognition, technologies that can implement huge arrays or banks of correlators to efficiently test the incoming signal against all possible bit sequences will be needed. Reaching these goals presents a significant engineering challenge, but research is continuing to make progress, and optical correlators, combined with the appropriate data coding techniques, offer great potential for bringing the ever-expanding field of optical signal processing to light.

(**a**) Pattern matching - "ones" and "zeros" correlators both show a match

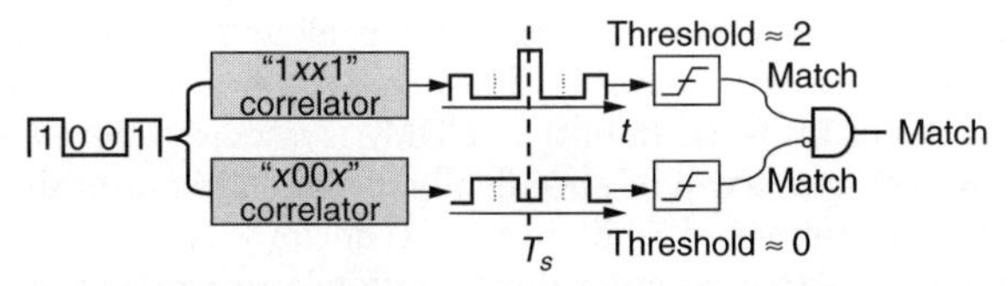

(**b**) Pattern mismatch - false match avoided by "zeros" correlator

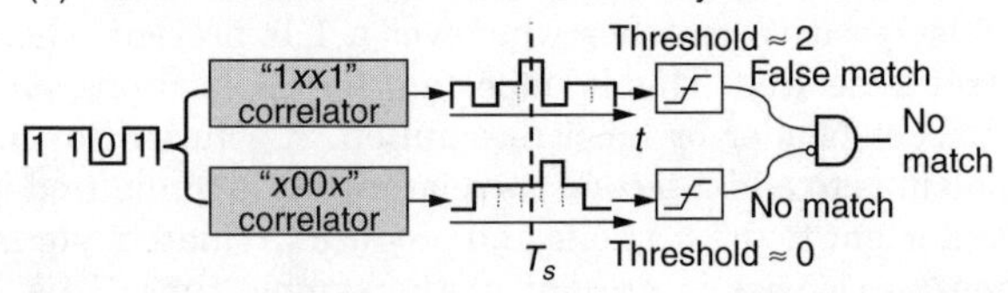

Figure 5. The concept of combining 1s and 0s correlators to uniquely recognize a bit sequence and avoid false-positive matches. The "ones" correlator tests for 1 bits in the correlation sequence and the "zeros" correlator tests for 0 bits. The correlators shown are configured to recognize a 1001 pattern when their outputs are combined in an AND gate. (**a**) The input pattern 1001 results in a match for both correlators, producing a final "match" decision at the output. (**b**) The input pattern 1101 results in a match for the "ones" correlators but a "no match" for the "zeros" correlator. The combination of these two outputs produces the correct "no match" decision at the output of the AND gate.

BIOGRAPHIES

John E. McGeehan received his B.S. and M.S. degrees in electrical engineering at the University of Southern California in Los Angeles in 1998 and 2001, respectively. He joined the Optical Communications Laboratory at the University of Southern California as a research assistant in 2001 and currently is working toward his doctorate. His research interests include the implementation of all-optical networking functions and optical signal processing as well as Raman amplification and signal monitoring. He is an author or co-author of nine technical papers.

Michelle C. Hauer received the B.S. degree in engineering physics from Loyola Marymount University, Los Angeles, California, in 1997. She currently is a research assistant in the Optical Fiber Communications Laboratory at the University of Southern California, Los Angeles, California, where she received the M.S.E.E. degree in 2000. Her doctoral research includes optical signal processing techniques for implementing all-optical networking functions. She is the author or co-author of more than 13 research papers. She is a member of the Tau Beta Pi, Eta Kappa Nu, and Sigma Pi Sigma academic honor societies in engineering, electrical engineering, and physics, respectively. She has also held a position as a systems engineer at Raytheon Company in El Segundo, California since 1997.

Alan Willner received his B.S. from Yeshiva University and his Ph.D. from Columbia University. He has worked at AT&T Bell Labs and Bellcore, and is Professor of Electrical Engineering at the University of Southern California. Professor Willner has received the following awards: the National Science Foundation (NSF) Presidential Faculty Fellows Award from the White House, the Packard Foundation Fellowship, the NSF National Young Investigator Award, the Optical Society of America (OSA) Fellow Award, the Fulbright Foundation Senior Scholars Award, the Institute of Electronic and Electrical Engineers (IEEE) Lasers and Electro-Optics Society (LEOS) Distinguished Traveling Lecturer Award, the USC/TRW Best Engineering Teacher Award, and the Armstrong Foundation Memorial Prize. His professional activities have included: editor-in-chief of the IEEE/OSA *Journal of Lightwave Technology*, editor-in-chief of the IEEE *Journal of Selected Topics in Quantum Electronics*, V.P. for Technical Affairs of the IEEE LEOS, Co-Chair of the OSA Science and Engineering Council, Elected Member of the IEEE LEOS Board of Governors, Program Co-Chair of the Conference on Lasers and Electro-Optics (CLEO), General Chair of the LEOS Annual Meeting Program, Program Co-Chair of the OSA Annual Meeting, OSA Photonics Division Chair,

General Co-Chair of the OSA Optical Amplifier Meeting, and Steering and Program Committee Member of the Conference on Optical Fiber Communications (OFC). Professor Willner has 325 publications, including one book.

BIBLIOGRAPHY

1. F. G. Stremler, *Introduction to Communications Systems*, 3rd ed., Addison-Wesley, Reading MA, 1990.
2. A. Stok and E. H. Sargent, Lighting the local area: Optical code-division multiple access and quality of service provisioning, *IEEE Network* 42–46 (Nov./Dec. 2000).
3. J. Widjaja, N. Wada, Y. Ishii, and W. Chijo, Photonic packet address processor using holographic correlator, *Electron. Lett.* **37**(11): 703–704 (2001).
4. J.-D. Shin, M.-Y. Jeon, and C.-S. Kang, Fiber-optic matched filters with metal films deposited on fiber delay-line ends for optical packet address detection, *IEEE Photon. Tech. Lett.* **8**(7): 941–943 (1996).
5. R. Kashyap, *Fiber Bragg Gratings*, Academic Press, San Diego, 1999.
6. D. B. Hunter and R. A. Minasian, Programmable high-speed optical code recognition using fibre Bragg grating arrays, *Electron. Lett.* **35**(5): 412–414 (1999).
7. A. Grunnet-Jepsen et al., Spectral phase encoding and decoding using fiber Bragg gratings, *Proc. Conf. Optical Fiber Communications (OFC) 1999*, paper PD33, 1999, pp. PD33-1–PD33-3.
8. M. R. Mokhtar, M. Ibsen, P. C. Teh, and D. J. Richardson, Simple dynamically reconfigurable OCDMA encoder/decoder based on a uniform fiber Bragg grating, *Proc. Conf. Optical Fiber Communications (OFC) 2002*, paper ThGG54, 2002, pp. 688–690.
9. P. C. Teh et al., Demonstration of a four-channel WDM/OCDMA system using 255-chip 320-Gchip/s quaternary phase coding gratings, *IEEE Photon. Tech. Lett.* **14**(2): 227–229 (2002).
10. P. Petruzzi et al., All optical pattern recognition using a segmented semiconductor optical amplifier, *Proc. Eur. Conf. Optical Communication (ECOC) 2001*, paper We.B.2.1, 2001, pp. 304–305.
11. J. S. Wey, J. Goldhar, D. L. Butler, and G. L. Burdge, Investigation of dynamic gratings in erbium-doped fiber for optical bit pattern recognition, *Proc. Conf. Lasers and Electro-optics (CLEO) 1997*, paper CThW1, 1997, pp. 443–444.
12. N. Kishi, K. Kawachi, and E. Yamashita, Auto-correlation method for weak optical short pulses using a nonlinear amplifying loop mirror, *Proc. Eur. Conf. Optical Communication (ECOC) 1997*, paper 448, 1997, pp. 215–218.
13. Z. Zheng and A. M. Weiner, Spectral phase correlation of coded femtosecond pulses by second-harmonic generation in thick nonlinear crystals, *Opt. Lett.* **25**(13): (2000).
14. M. C. Hauer et al., Dynamically reconfigurable all-optical correlators to support ultra-fast internet routing, *Proc. Conf. Optical Fiber Communications (OFC) 2002*, paper WM7, 2002, pp. 268–270.
15. P. Parolari et al., Coherent-to-incoherence light conversion for optical correlators, *J. Lightwave Technol.* **18**(9): 1284–1288 (2000).
16. J. McGeehan, M. C. Hauer, A. B. Sahin, and A. E. Willner, Reconfigurable multi-wavelength optical correlator for header-based switching and routing, *Proc. Conf. Optical Fiber Communications (OFC) 2002*, paper WM4, 2002, pp. 264–266.

DIGITAL PHASE MODULATION AND DEMODULATION

FUQIN XIONG
Cleveland State University
Cleveland, Ohio

1. INTRODUCTION

Digital signals or messages, such as binary 1 and 0, must be modulated onto a high-frequency carrier before they can be transmitted through some communication channels such as a coaxial cable or free space. The carrier is usually an electrical voltage signal such as a sine or cosine function of time t

$$s(t) = A\cos(2\pi f_c t + \theta)$$

where A is the amplitude, f_c is the carrier frequency, and θ is the initial phase. Each of them or a combination of them can be used to carry digital messages. The total phase $\Phi(t) = 2\pi f_c t + \theta$ can also be used to carry digital messages. The process that impresses a message onto a carrier by associating one or more parameters of the carrier with the message is called *modulation*; the process that extracts a message from a modulated carrier is called *demodulation*.

There are three basic digital modulation methods: amplitude shift keying (ASK), frequency shift keying (FSK), and phase shift keying (PSK). In ASK, data 1 and 0 are represented by the presence and absence of a burst of the carrier sine wave, respectively. The burst of the carrier that lasts a duration of a data period is called a *symbol*. The frequency and phase of the carrier are kept unchanged from symbol to symbol. In FSK, data 1 and 0 are represented by two different frequencies of the carrier, respectively. The amplitude and phase of the carrier are kept unchanged from symbol to symbol. In PSK, data 1 and 0 are represented by two different phases (e.g., 0 or π radians) of the carrier, respectively. The amplitude and frequency of the carrier are kept unchanged from symbol to symbol.

Modulation schemes are usually evaluated and compared using three criteria: *power efficiency, bandwidth efficiency, and system complexity*.

The bit error probability (P_b), or bit error rate (BER), as it is commonly called, of a modulation scheme is related to E_b/N_0, where E_b is the energy of the modulated carrier in a bit duration, and N_0 is the noise power spectral density. The *power efficiency* of a modulation scheme is defined as the required E_b/N_0 for a certain bit error probability (P_b), over an additive white Gaussian noise (AWGN) channel.

The *bandwidth efficiency* is defined as the number of bits per second that can be transmitted in one hertz (1 Hz) of system bandwidth. System bandwidth requirement depends on different criteria. Assuming the system uses

Nyquist (ideal rectangular) filtering at baseband, which has the minimum bandwidth required for intersymbol-interference (ISI)-free transmission of digital signals, then the bandwidth at baseband is $0.5R_s$, where R_s is the symbol rate, and the bandwidth at carrier frequency is $W = R_s$. Since $R_s = R_b/\log_2 M$, where R_b is the bit rate, for M-ary modulation, the bandwidth efficiency is

$$\eta_B = \frac{R_b}{W} = \log_2 M \quad \text{(bps/Hz) (Nyquist)}$$

This is called the Nyquist bandwidth efficiency. For modulation schemes that have power density spectral nulls such as the ones of PSK in Fig. 2, the bandwidth may be defined as the width of the main spectral lobe. Bandwidth efficiency based on this definition of bandwidth is called null-to-null bandwidth efficiency. If the spectrum of the modulated signal does not have nulls, null-to-null bandwidth no longer exists, as in the case of continuous-phase modulation (CPM). In this case, energy percentage bandwidth may be used as a definition. Usually 99% is used, even though other percentages (e.g., 90%, 95%) are also used. Bandwidth efficiency based on this definition of bandwidth is called percentage bandwidth efficiency.

System complexity refers to the circuit implementation of the modulator and demodulator. Associated with the system complexity is the cost of manufacturing, which is, of course, a major concern in choosing a modulation technique. Usually the demodulator is more complex than the modulator. A coherent demodulator is much more complex than a noncoherent demodulator since carrier recovery is required. For some demodulation methods, sophisticated algorithms, such as the Viterbi algorithm, are required. All these are basis for complexity comparison.

In comparison with ASK and FSK, PSK achieves better power efficiency and bandwidth efficiency. For example, in terms of power efficiency, binary PSK is 3 dB better than binary ASK and FSK at high signal-to-noise ratios, while BPSK is the same as BASK and better than BFSK in terms of bandwidth efficiency. Because of the advantages of PSK schemes, they are widely used in satellite communications, fixed terrestrial wireless communications, and wireless mobile communications.

A more complex, but more efficient PSK scheme is quadrature phase shift keying (QPSK), where the initial phase of the modulated carrier at the start of a symbol is any one of four evenly spaced values, say, $(0, \pi/2, \pi, 3\pi/2)$ or $(\pi/4, 3\pi/4, 5\pi/4, 7\pi/4)$. In QPSK, since there are four different phases, each symbol can represent two data bits. For example, 00, 01, 10, and 11 can be represented by $0, \pi/2, \pi$, and $3\pi/2$, respectively. In general, assuming that the PSK scheme has M initial phases, known as M-ary PSK or MPSK, the number of bits represented by a symbol is $n = \log_2 M$. Thus each 8-PSK symbol represents 3 bits, each 16-PSK symbol represents 4 bits, and so on.

As the order (M) of the PSK scheme is increased, the bandwidth efficiency is increased. In terms of null-to-null bandwidth, which is often used for PSK schemes, the efficiency is

$$\eta_B = \frac{R_b}{W} = \frac{R_b}{2R_s} = \frac{R_b}{2R_b/\log_2 M} = \frac{1}{2}\log_2 M$$

(bps/Hz) (null-to-null)

Thus the bandwidth efficiencies of BPSK, QPSK, 8-PSK, and 16-PSK are 0.5, 1, 1.5, and 2 bps/Hz.

Using the bandwidth efficiency, the bit rate that can be supported by a system bandwidth can be easily calculated. From above we have

$$R_b = \eta_B W$$

For example, for a satellite transponder that has a bandwidth of 36 MHz, assuming that Nyquist filtering is achieved in the system, the bit rate will be 36 Mbps for BPSK, 72 Mbps for QPSK, and so on. If null-to-null bandwidth is required in the system, the bit rate will be 18 Mbps for BPSK, 36 Mbps for QPSK, and so on. Practical systems may achieve bit rates somewhere between these two sets of values.

In this article, we first describe the commonly used class of PSK schemes, that is, MPSK. Then BPSK and QPSK are treated as special cases of MPSK. Next, differential PSK schemes are introduced, which are particularly useful in channels where coherent demodulation is difficult or impossible, such as fading channels. At the end of this article, advanced phase modulation schemes, such as offset QPSK (OQPSK), $\pi/4$-DQPSK, and continuous-phase modulation (CPM), including multi-h CPM, are briefly introduced. A future trend in phase modulation is in the direction of CPM and multi-h CPM.

2. PSK SCHEMES AND THEIR PERFORMANCE

2.1. Signal Waveforms

In PSK, binary data are grouped into n bits in a group, called n-tuples or symbols. There are $M = 2^n$ possible n-tuples. When each n-tuple is used to control the phase of the carrier for a period of T, *M-ary phase shift keying* (MPSK) is formed. The MPSK signal set is defined as

$$s_i(t) = A\cos(2\pi f_c t + \theta_i), \qquad 0 \le t \le T, \qquad i = 1, 2, \ldots, M \tag{1}$$

where A is the amplitude, f_c is the frequency of the carrier, and

$$\theta_i = \frac{(2i-1)\pi}{M}$$

are the (initial) phases of the signals. Note that θ_i are equally spaced. Each signal $s_i(t)$ is a burst of carrier of period T, called a *symbol waveform*. Each symbol waveform has a unique phase that corresponds to a n-tuple; or, in other words, each n-tuple is represented by a symbol waveform with a unique phase. When the PSK signal is transmitted and received, the demodulator detects the phase of each symbol. On the basis of phase information, the corresponding n-tuple is recovered. Except for the phase difference, the M symbol waveforms in MPSK have the same amplitude (A), the same frequency (f_c), and the same energy:

$$E = \int_0^T s_i^2(t)\,dt = \frac{A^2T}{2}, \qquad i = 1, 2, \ldots, M$$

The simplest PSK scheme is *binary phase shift keying* (BPSK), where binary data (1 and 0) are represented by two symbols with different phases. This is the case when $M = 2$ in MPSK. Typically these two phases are 0 and π. Then the two symbols are $\pm A\cos 2\pi f_c t$, which are said to be *antipodal*.

If binary data are grouped into 2 bits per group, called dibits, there are four possible dibits: 00, 01, 10, and 11. If each dibit is used to control the phase of the carrier, then we have *quadrature phase shift keying* (QPSK). This is the case when $M = 4$ in MPSK. Typically, the initial signal phases are $\pi/4, 3\pi/4, 5\pi/4$, and $7\pi/4$. Each phase corresponds to a dibit. QPSK is the most often used scheme since its BER performance is the same while the bandwidth efficiency is doubled in comparison with BPSK, as will be seen shortly.

2.2. Signal Constellations

It is well known that sine-wave signals can be represented by phasors. The phasor's magnitude is the amplitude of the sine-wave signal; its angle with respect to the horizontal axis is the phase of the sine-wave signal. In a similar way, signals of modulation schemes, including PSK schemes, can be represented by phasors. The graph that shows the phasors of all symbols in a modulation scheme is called a signal constellation. It is a very convenient tool for visualizing and describing all symbols and their relationship in the modulation scheme. It is also a powerful tool for analyzing the performance of the modulation scheme.

The PSK waveform can be written as

$$\begin{aligned} s_i(t) &= A\cos\theta_i \cos 2\pi f_c t - A\sin\theta_i \sin 2\pi f_c t \qquad (2) \\ &= s_{i1}\phi_1(t) + s_{i2}\phi_2(t), \qquad 0 \le t \le T, \qquad i = 1, 2, \ldots, M \end{aligned}$$

where

$$\phi_1(t) = \sqrt{\frac{2}{T}}\cos 2\pi f_c t, \qquad 0 \le t \le T \qquad (3)$$

$$\phi_2(t) = -\sqrt{\frac{2}{T}}\sin 2\pi f_c t, \qquad 0 \le t \le T \qquad (4)$$

are two orthonormal basis functions and

$$s_{i1} = \int_0^T s_i(t)\phi_1(t)\,dt = \sqrt{E}\cos\theta_i \qquad (5)$$

$$s_{i2} = \int_0^T s_i(t)\phi_2(t)\,dt = \sqrt{E}\sin\theta_i \qquad (6)$$

are the projections of the signal onto the basis functions.

The phase is related with s_{i1} and s_{i2} as

$$\theta_i = \tan^{-1}\frac{s_{i2}}{s_{i1}}$$

Thus PSK signals can be graphically represented by a signal constellation in a two-dimensional coordinate system with the two orthonormal basis functions in Eqs. (5) and (6), $\phi_1(t)$ and $\phi_2(t)$, as its horizontal and vertical axes, respectively (Fig. 1). Each signal $s_i(t)$ is represented by a point (s_{i1}, s_{i2}), or a vector from the origin to this point, in the two coordinates. The polar coordinates of the signal are $(\sqrt{E}, \theta_i)$; that is, the signal vector's magnitude is $\sqrt{E}$ and its angle with respect to the horizontal axis is θ_i. The signal points are equally spaced on a circle of radius $\sqrt{E}$ and centered at the origin. The bits-signal mapping could be arbitrary provided that the mapping is one-to-one. However, a method called Gray coding is usually used in signal assignment in MPSK. Gray coding assigns n-tuples with only one-bit difference in two adjacent signals in the constellation. When an M-ary symbol error occurs because of the presence of noise, it is most likely that the signal is detected as the adjacent signal on the constellation; thus only one of the n input bits is in error. Figure 1 shows the constellations of BPSK, QPSK, and 8-PSK, where Gray coding is used for bit assignment.

2.3. Power Spectral Density

The *power spectral density* (PSD) of a signal shows the signal power distribution as a function of frequency. It provides very important information about the relative strength of the frequency components in the signal, which, in turn, determines the bandwidth requirement of the communication system.

The power spectral density of a bandpass signal, such as a PSK signal, is centered about its carrier frequency. Moving the center frequency down to zero, we obtain the baseband signal PSD, which completely characterizes the signal [1–3]. The baseband PSD expression of MPSK can be expressed as [3]

$$\Psi_{\tilde{s}}(f) = A^2 T\left(\frac{\sin \pi f T}{\pi f T}\right)^2 = A^2 n T_b\left(\frac{\sin \pi f n T_b}{\pi f n T_b}\right)^2 \qquad (7)$$

where T_b is the bit duration, $n = \log_2 M$. Figure 2 shows the PSDs ($A = \sqrt{2}$ and $T_b = 1$ for unit bit energy: $E_b = 1$) for different values of M where the frequency axis is normalized to the bit rate (fT_b). From the figure we can see that the bandwidth decreases with M, or in other words, the bandwidth efficiency increases with M. The Nyquist bandwidth is

$$B_{\text{Nyquist}} = \frac{1}{nT_b} = \frac{R_b}{\log_2 M}$$

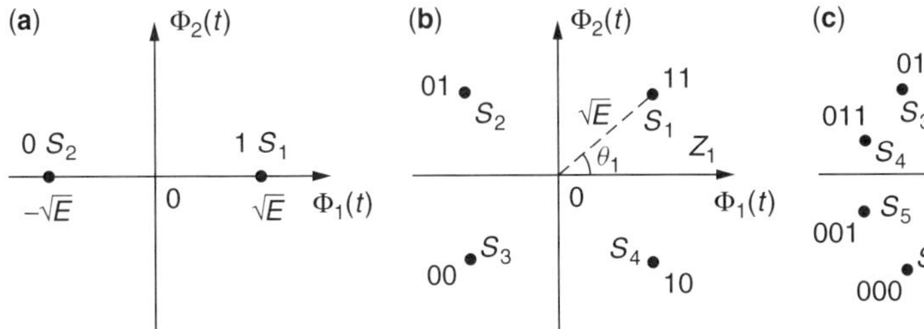

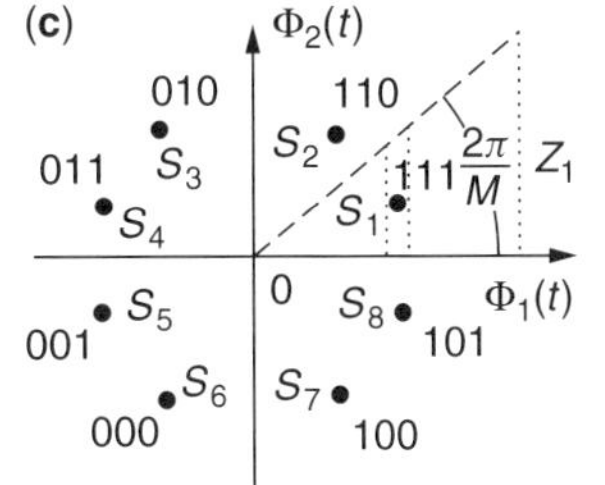

Figure 1. PSK constellations: (**a**) BPSK; (**b**) QPSK; (**c**) 8PSK.

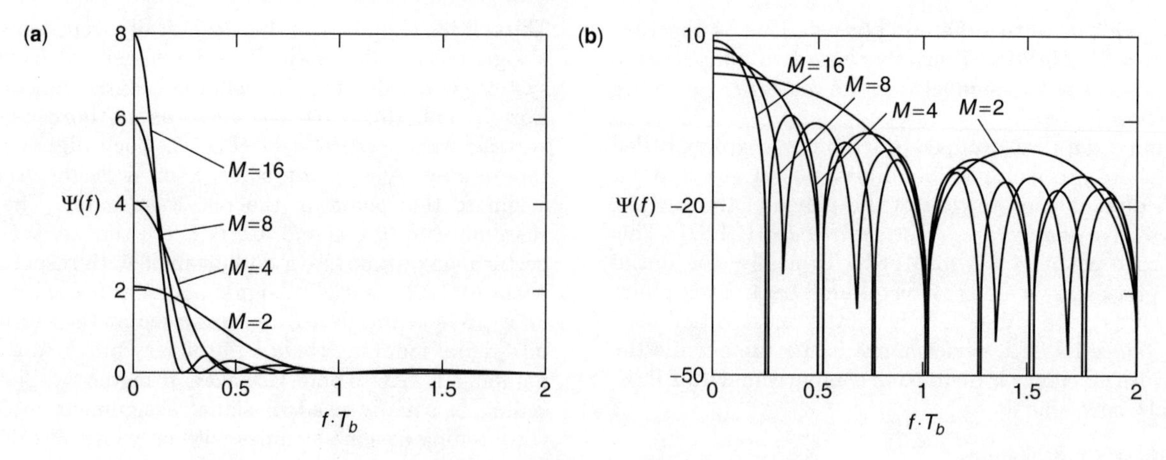

Figure 2. PSDs of MPSK: (**a**) linear; (**b**) logarithmic.

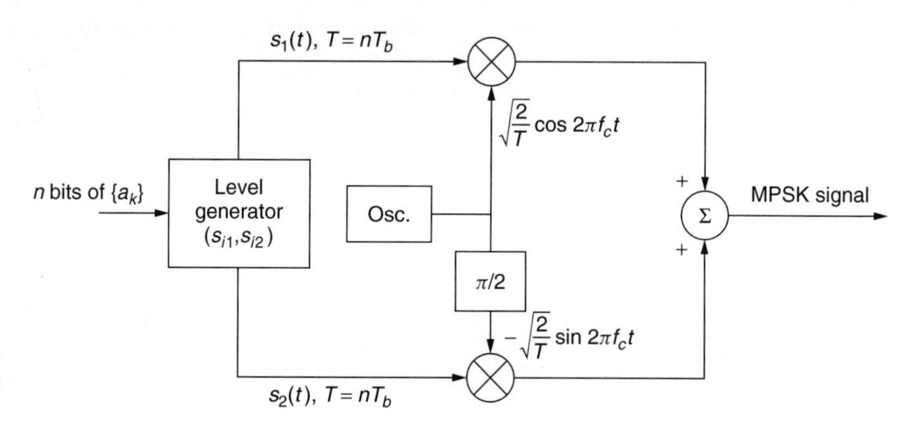

Figure 3. MPSK modulator (Osc. = oscillator). (From Ref. 3, copyright © 2000 Artech House.)

This translates to the Nyquist bandwidth efficiency of $\log_2 M$. The null-to-null bandwidth is

$$B_{null\text{-}to\text{-}null} = \frac{2}{nT_b} = \frac{2R_b}{\log_2 M}$$

This translates to a bandwidth efficiency of $0.5\log_2 M$, which is half of that of the Nyquist bandwidth efficiency. Practical systems can achieve bandwidth efficiencies between these two values.

2.4. Modulator and Demodulator

Over the entire time axis, we can write MPSK signal as

$$s(t) = s_1(t)\sqrt{\frac{\pi}{2}}\cos 2\pi f_c t - s_2(t)\sqrt{\frac{\pi}{2}}\sin 2\pi f_c t,$$
$$-\infty < t < \infty \tag{8}$$

where

$$s_1(t) = \sqrt{E}\sum_{k=-\infty}^{\infty}\cos(\theta_k)p(t-kT) \tag{9}$$

$$s_2(t) = \sqrt{E}\sum_{k=-\infty}^{\infty}\sin(\theta_k)p(t-kT) \tag{10}$$

where θ_k is one of the M phases determined by the input binary n-tuple and $p(t)$ is the rectangular pulse with unit amplitude defined on $[0, T]$. Expression (8) requires that the carrier frequency be an integer multiple of the symbol timing so that the initial phase of the signal in any symbol period is θ_k.

Since MPSK signals are two-dimensional, for $M \geq 4$, the modulator can be implemented by a quadrature *modulator* (Fig. 3), where the upper branch is called the *in-phase channel* or *I channel*, and the lower branch is called the *quadrature channel* or the *Q channel*. The only difference for different values of M is the level generator. Each n-tuple of the input bits is used to control the level generator. It provides the I and Q channels the particular sign and level for a signal's horizontal and vertical coordinates, respectively.

Modern technology intends to use completely digital devices. In such an environment, MPSK signals are digitally synthesized and fed to a D/A (digital-to-analog) converter whose output is the desired phase-modulated signal.

The coherent demodulation of MPSK is shown in Fig. 4. The input to the *demodulator* is the received signal $r(t) = s_i(t) + n(t)$, where $n(t)$ is the additive white Gaussian noise (AWGN). There are two correlators, each consisting of a multiplier and an integrator. The carrier recovery (CR) block synchronizes the reference signals for the multipliers with the transmitted carrier in frequency and phase. This makes the demodulation

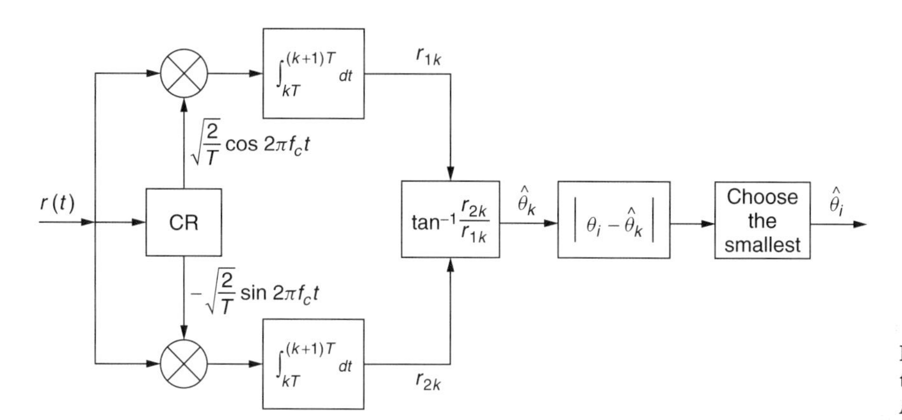

Figure 4. Coherent MPSK demodulator using two correlators. (From Ref. 3, copyright © 2000 Artech House.)

coherent and ensures the lowest bit error probability. The correlators correlate $r(t)$ with the two reference signals. Because of the orthogonality of the two components of the PSK signal, each correlator produces an output as follows:

$$r_1 = \int_0^T r(t)\phi_1(t)\,dt = \int_0^T [s(t) + n(t)]\phi_1(t)\,dt = s_{i1} + n_1$$

$$r_2 = \int_0^T r(t)\phi_2(t)\,dt = \int_0^T [s(t) + n(t)]\phi_2(t)\,dt = s_{i2} + n_2$$

where s_{i1} and s_{i2} are given as in Eqs. (5) and (6), respectively, and n_1 and n_2 are noise output. In Fig. 4 the subscript k indicates the kth symbol period.

Define

$$\widehat{\theta} \triangleq \tan^{-1}\frac{r_2}{r_1}$$

In the absence of noise, $\widehat{\theta} = \tan^{-1} r_2/r_1 = \tan^{-1} s_{i2}/s_{i1} = \theta_i$; that is, the PSK signal's phase information is completely recoverable in the absence of noise. With noise, $\widehat{\theta}$ will deviate from θ_i. To recover θ_i, the $\widehat{\theta}$ difference with all θ_i are compared and the θ_i that incurs the smallest $|\theta_i - \widehat{\theta}|$ is chosen.

The modulator and demodulator for BPSK and QPSK can be simplified from Figs. 3 and 4. For BPSK, since there is only I-channel component in the signal, the modulator is particularly simple: only the I channel is needed (Fig. 5a). The binary data (0,1) are mapped into (−1,1), which are represented by non-return-to-zero (NRZ) waveform $a(t)$, which is equivalent to Eq. (9). This NRZ waveform is multiplied with the carrier, and the result is the antipodal BPSK signal. The BPSK demodulator needs only one channel, too, as shown in Fig. 5b. In the absence of noise, the correlator output l is directly proportional to the data. Then l is compared with threshold zero, if $l \geq 0$ the data bit is 1; otherwise it is 0.

For QPSK, since it is equivalent to two parallel BPSK signals, the modulator can be simplified as shown in Fig. 6a, where $I(t)$ and $Q(t)$ are equivalent to Eqs. (9) and (10), respectively, except for a constant factor; the level generator is simply a serial-to-parallel converter. The demodulator is shown in Fig. 6b, which is simply a pair of two parallel BPSK demodulators.

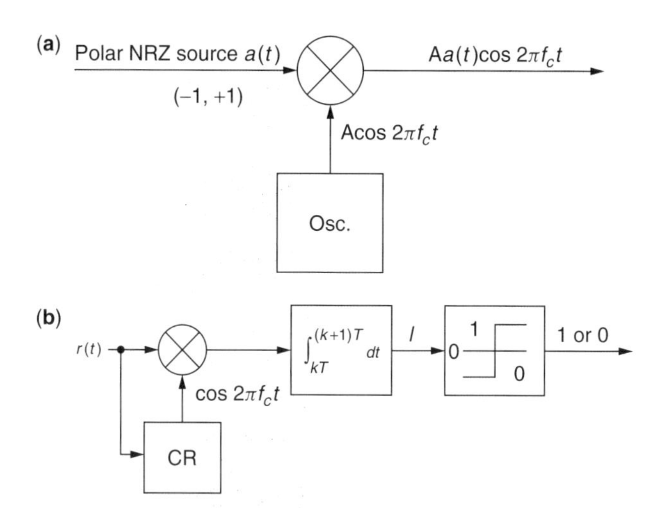

Figure 5. BPSK modulator (**a**); coherent BPSK demodulator (**b**). (From Ref. 3, copyright © 2000 Artech House.)

For $M > 4$, the general modulator and demodulator must be used.

2.5. Symbol and Bit Error Probability

In the demodulator shown in Fig. 4, a symbol error occurs when the estimated phase $\widehat{\theta}$ is such that the phase deviation $|\varphi| = |\widehat{\theta} - \theta_i| > \pi/M$ (see Fig. 1c). Thus the symbol error probability of MPSK is given by

$$P_s = 1 - \int_{-\pi/M}^{\pi/M} p(\varphi)\,d\varphi \tag{11}$$

It can be shown that $p(\varphi)$ is given by [3]

$$p(\varphi) = \frac{e^{-E/N_0}}{2\pi}\left\{1 + \sqrt{\frac{\pi E}{N_0}}(\cos\varphi)e^{(E/N_0)\cos^2\varphi} \times \left[1 + \text{erf}\left(\sqrt{\frac{E}{N_0}}\cos\varphi\right)\right]\right\} \tag{12}$$

where

$$\text{erf}\,(x) \triangleq \frac{2}{\sqrt{\pi}}\int_0^x e^{-u^2}\,du \tag{13}$$

is the error function.

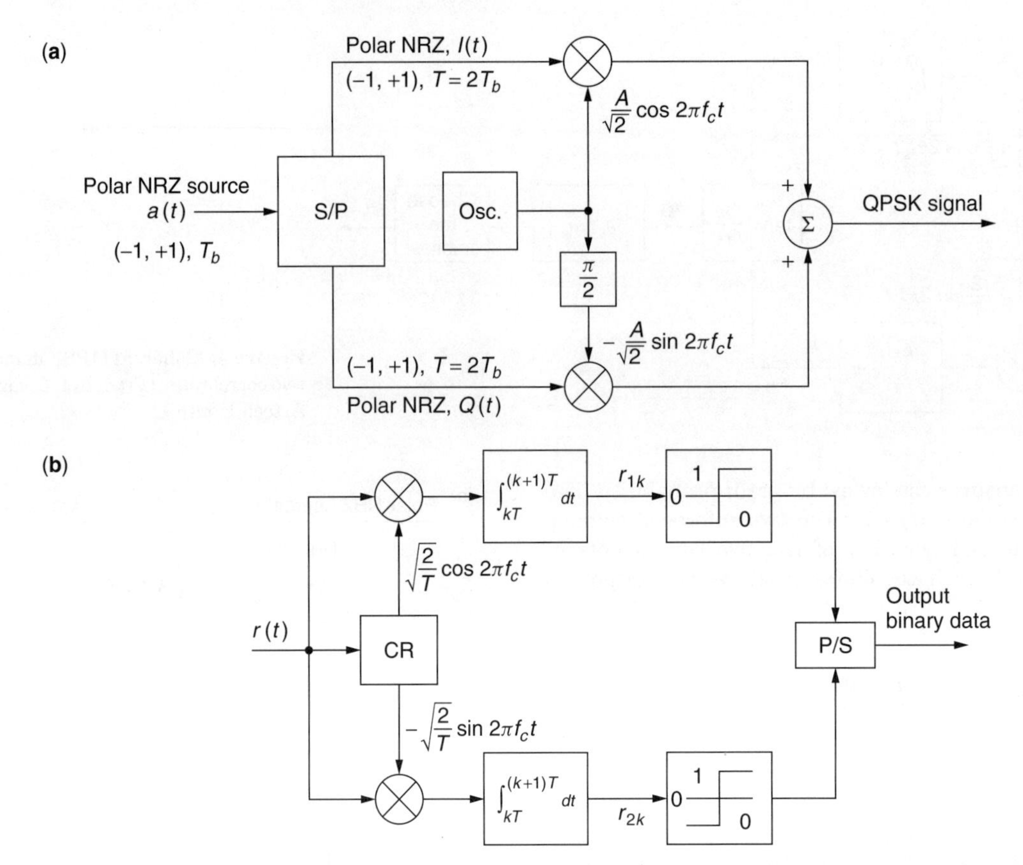

Figure 6. (**a**) QPSK modulator; (**b**) QPSK demodulator. (From Ref. 3, copyright © 2000 Artech House.)

When $M = 2$, (11) results in the formula for the symbol (and bit) error probability of BPSK:

$$P_s = P_b = Q\left(\sqrt{\frac{2E_b}{N_o}}\right) \quad \text{(BPSK)} \tag{14}$$

where the E_b is the bit energy, which is also equal to the symbol energy E since a bit is represented by a symbol in BPSK, and

$$Q(x) = \int_x^\infty \frac{1}{\sqrt{2\pi}} e^{(-u^2/2)}\, du \tag{15}$$

is called the Q function.

When $M = 4$ [Eq. (11)] results in the formula for QPSK:

$$P_s = 2Q\left(\sqrt{\frac{2E_b}{N_0}}\right) - \left[Q\left(\sqrt{\frac{2E_b}{N_0}}\right)\right]^2 \tag{16}$$

Since the demodulator of QPSK is simply a pair of two parallel BPSK demodulators, the bit error probability is the same as that of BPSK:

$$P_b = Q\left(\sqrt{\frac{2E_b}{N_0}}\right) \quad \text{(QPSK)} \tag{17}$$

Recall that QPSK's bandwidth efficiency is double that of BPSK. This makes QPSK a preferred choice over BPSK in many systems.

For $M > 4$, expression (11) cannot be evaluated in a closed form. However, the symbol error probability can be obtained by numerically integrating (11).

Figure 7 shows the graphs of P_s for $M = 2, 4, 8, 16$, and 32 given by the exact expression (11). Beyond $M = 4$, doubling the number of phases, or increasing the n-tuples represented by the phases by one bit, requires a substantial increase in E_b/N_0 [or signal-to-noise ratio (SNR)]. For example, at $P_s = 10^{-5}$, the SNR difference between $M = 4$ and $M = 8$ is approximately 4 dB, the difference between $M = 8$ and $M = 16$ is approximately 5 dB. For large values of M, doubling the number of phases requires an SNR increase of 6 dB to maintain the same performance.

For $E/N_0 \gg 1$, we can obtain an approximation of the P_s expression as

$$P_s \approx 2Q\left(\sqrt{\frac{2E}{N_0}} \sin\frac{\pi}{M}\right) \tag{18}$$

Note that only the high signal-to-noise ratio assumption is needed for the approximation. Therefore (18) is good for any values of M, even though it is not needed for $M = 2$ and 4 since precise formulas are available.

The bit error rate can be related to the symbol error rate by

$$P_b \approx \frac{P_s}{\log_2 M} \tag{19}$$

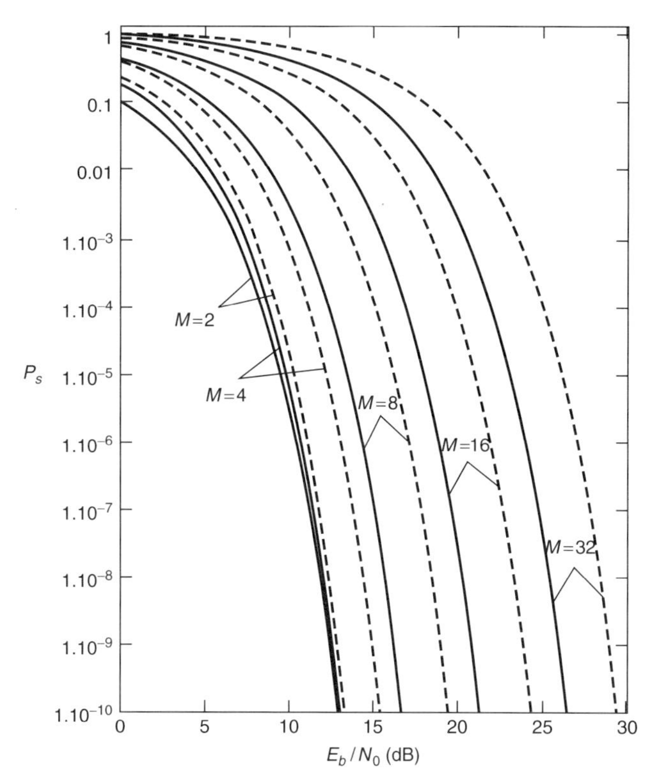

Figure 7. P_s of MPSK (solid lines) and DMPSK (dotted lines). (From Ref. 3, copyright © 2000 Artech House.)

for Gray-coded MPSK signals, since the most likely erroneous symbols are the adjacent signals that differ by only one bit.

Because of the substantial increase in SNR to maintain the same BER for larger M, higher-order PSK schemes beyond 8-PSK are not very often used. If further higher bandwidth efficiency is desired, quadrature amplitude modulation (QAM), which is a combination of phase and amplitude modulation, is a preferable choice over MPSK. For the same bandwidth efficiency, QAM delivers better power efficiency for $M > 4$ (see Section 8.7 of Ref. 3). However, QAM needs to preserve its amplitude through the transmitter stages. This can be difficult when nonlinear power amplifiers, such as traveling-wave-tube amplifiers (TWTAs) in satellite transponders, are used. QAM is widely used in telephone-line modems.

3. DIFFERENTIAL PSK SCHEMES

Differential encoding of a binary data sequence converts the original sequence into a new sequence of which each bit is determined by the difference of the current uncoded bit and the previous coded bit. Differential coding is needed in situations where coherent demodulation is difficult or phase ambiguity is a problem in carrier recovery. BPSK, QPSK, and MPSK all can be differentially coded.

3.1. Differential BPSK

We denote differentially encoded BPSK as *DEBPSK*. Figure 8a depicts the DEBPSK modulator. A DEBPSK signal can be coherently demodulated or differentially demodulated. We denote the modulation scheme that uses differential encoding and differential demodulation as *DBPSK*, which is sometimes simply called *DPSK*.

DBPSK does not require a coherent reference signal. Figure 8b shows a simple, but suboptimum, differential demodulator that uses the previous symbol as the reference for demodulating the current symbol. (This is commonly referred to as a *DPSK demodulator*. Another DPSK demodulator is the optimum differentially coherent demodulator. Differentially encoded PSK can also be coherently detected. These will be discussed shortly.) The front-end bandpass filter reduces the noise power but preserves the phase of the signal. The integrator can be replaced by a LPF (lowpass filter). The output of the integrator is

$$l = \int_{kT}^{(k+1)T} r(t)r(t-T)\,dt$$

In the absence of noise and other channel impairment

$$l = \int_{kT}^{(k+1)T} s_k(t)s_{k-1}(t)\,dt = \begin{cases} E_b & \text{if } s_k(t) = s_{k-1}(t) \\ -E_b & \text{if } s_k(t) = -s_{k-1}(t) \end{cases}$$

where $s_k(t)$ and $s_{k-1}(t)$ are the current and the previous symbols. The integrator output is positive if the current signal is the same as the previous one; otherwise the output is negative. This is to say that the demodulator makes decisions based on the difference between the two signals. Thus the information data must be encoded as the difference between adjacent signals, which is exactly what the differential encoding can accomplish. The encoding rule is

$$d_k = \overline{a_k \oplus d_{k-1}}$$

where $\oplus$ denotes modulo-2 addition. Inversely, we can recover a_k from d_k using

$$a_k = \overline{d_k \oplus d_{k-1}}$$

If d_k and d_{k-1} are the same, then they represent a 1 of a_k. If d_k and d_{k-1} are different, they represent a 0 of a_k.

The demodulator in Fig. 8 is suboptimum, since the reference signal is the previous symbol, which is noisy. The optimum noncoherent, or differentially coherent, demodulation of DEBPSK is given in Fig. 9. The derivation of this demodulator and its BER performance can be found in Ref. 3. Note that the demodulator of Fig. 9 does not require phase synchronization between the reference signals and the received signal. But it does require the reference frequency be the same as that of the received signal. Therefore the suboptimum receiver in Fig. 8b is more practical. Its error performance is slightly inferior to that of the optimum receiver.

The performance of the optimum receiver in Fig. 9 is given by

$$P_b = \frac{1}{2}e^{-E_b/N_0} \quad \text{(optimum DBPSK)} \tag{20}$$

The performance of the suboptimum receiver is given by Park [4]. It is shown that if an ideal narrowband IF

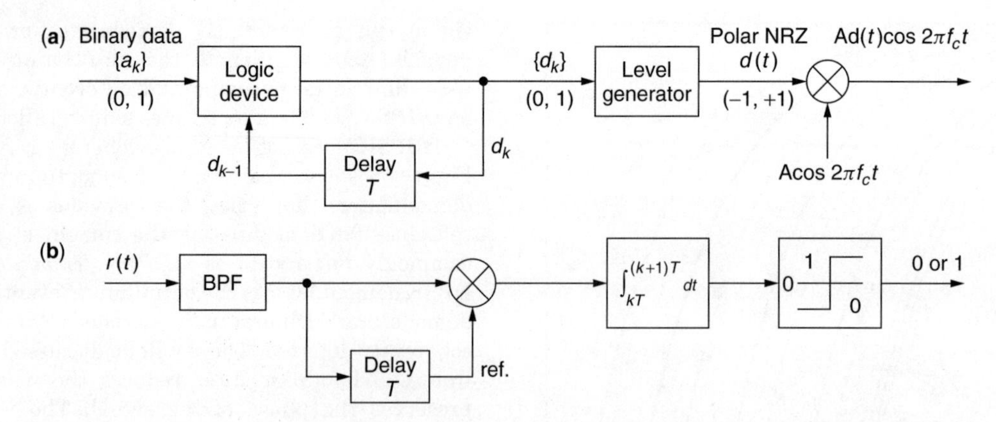

Figure 8. DBPSK modulator (**a**) and demodulator (**b**). (From Ref. 3, copyright © 2000 Artech House.)

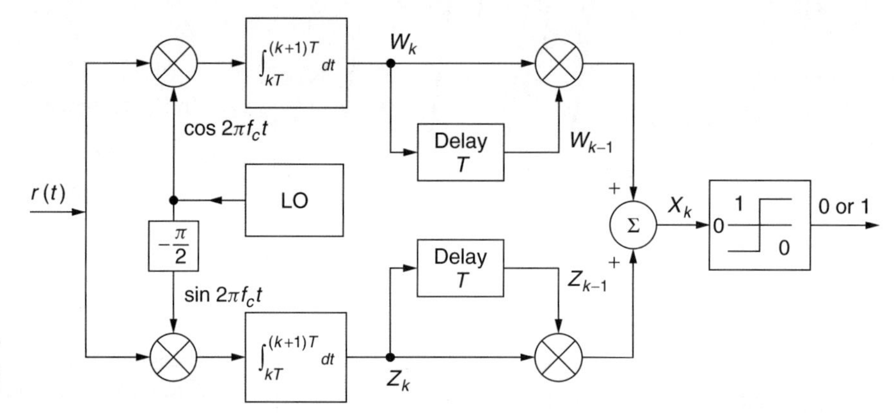

Figure 9. Optimum demodulator for DBPSK. (From Ref. 3, copyright © 2000 Artech House.)

(intermediate-frequency) filter with bandwidth W is placed before the correlator in Fig. 8b, the bit error probability is

$$P_b = \frac{1}{2}e^{-0.76E_b/N_0} \quad \text{for } W = \frac{0.5}{T} \quad \text{(suboptimum DBPSK)} \tag{21}$$

or

$$P_b = \frac{1}{2}e^{-0.8E_b/N_0} \quad \text{for } W = \frac{0.57}{T} \quad \text{(suboptimum DBPSK)} \tag{22}$$

which amounts to a loss of 1.2 and 1 dB, respectively, with respect to the optimum. If an ideal wideband IF filter is used, then

$$P_b \approx Q\left(\sqrt{\frac{E_b}{N_0}}\right) \quad \text{for } W > \frac{1}{T}$$

$$\approx \frac{1}{2\sqrt{\pi}\sqrt{E_b/2N_0}}e^{-E_b/2N_0}$$

$$\text{for } W > \frac{1}{T} \quad \text{(suboptimum DBPSK)} \tag{23}$$

The typical value of W is 1.5/T. If W is too large or too small, Eq. (23) does not hold [4]. The P_b for the wideband suboptimum receiver is about 2 dB worse than the optimum at high SNR. The bandwidth should be chosen as $0.57/T$ for the best performance. P_b curves of DBPSK are shown in Fig. 10.

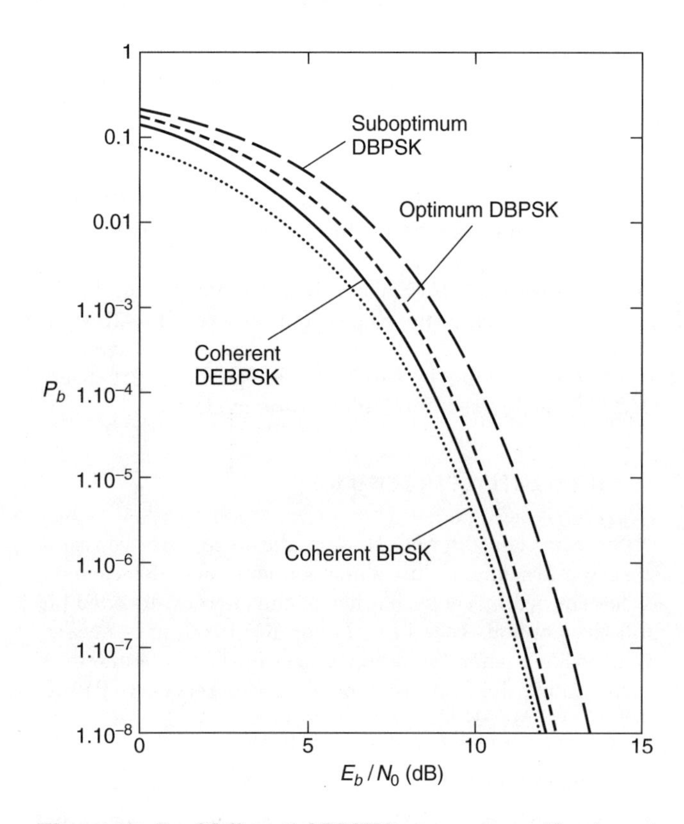

Figure 10. P_b of differential BPSK in comparison with coherent BPSK scheme. (From Ref. 3, copyright © 2000 Artech House.)

A differentially encoded BPSK signal can also be demodulated coherently (denoted as *DEBPSK*). It is used when the purpose of differential encoding is to eliminate phase ambiguity in the carrier recovery circuit for coherent PSK (see Section 4.10 of Ref. 3). This is rarely denoted by the acronym *DBPSK*. DBPSK refers to the scheme of differential encoding and differentially coherent demodulation as we have discussed above.

In the case of DEBPSK, the bit error probability (P_b) of the final decoded sequence $\{\widehat{a}_k\}$ is related to the bit error probability ($P_{b,d}$) of the demodulated encoded sequence $\{\widehat{d}_k\}$ by

$$P_b = 2P_{b,d}(1 - P_{b,d}) \tag{24}$$

(see Section 2.4.1 of Ref. 3). Substituting $P_{b,d}$ as in (14) into Eq. (24), we have

$$P_b = 2Q\left(\sqrt{\frac{2E_b}{N_0}}\right)\left[1 - Q\left(\sqrt{\frac{2E_b}{N_0}}\right)\right] \quad \text{(DEBPSK)} \tag{25}$$

for coherently detected differentially encoded PSK. For large SNR, this is just about 2 times that of coherent BPSK without differential encoding.

Finally we need to say a few words about the power spectral density of differentially encoded BPSK. Since the difference of differentially encoded BPSK from BPSK is differential encoding, which always produces an asymptotically equally likely data sequence (see Section 2.1 of Ref. 3), the PSD of the differentially encoded BPSK is the same as BPSK, in which we assume that its data sequence is equally likely. However, it is worthwhile to point out that if the data sequence is not equally likely, the PSD of the BPSK is not the one in Fig. 2, but the PSD of the differentially encoded PSK is still the one in Fig. 2.

3.2. Differential QPSK and MPSK

The principles of differential BPSK can be extended to MPSK, including QPSK. In *differential MPSK*, information bits are first differentially encoded. Then the encoded bits are used to modulate the carrier to produce the differentially encoded MPSK (DEMPSK) signal stream. In a DEMPSK signal stream, information is carried by the phase difference $\Delta\theta_i$ between two consecutive symbols. There are M different values of $\Delta\theta_i$; each represents an n-tuple ($n = \log_2 M$) of information bits.

In light of the modern digital technology, DEMPSK signals can be generated by digital frequency synthesis technique. A phase change from one symbol to the next is simply controlled by the n-tuple that is represented by the phase change. This technique is particularly suitable for large values of M.

Demodulation of DEMPSK signal is similar to that of differential BPSK [3]. The symbol error probability of the differentially coherent demodulator is approximated by

$$P_s \approx 2Q\left(\sqrt{\frac{2E}{N_0}}\sin\frac{\pi}{\sqrt{2}M}\right) \quad \text{(optimum DMPSK)} \tag{26}$$

for large SNR [6,7]. The exact curves are given as dotted lines in Fig. 7 together with those of coherent MPSK. Compared with coherent MPSK, asymptotically the DMPSK requires 3 dB more SNR to achieve the same error performance.

4. OTHER PSK SCHEMES

4.1. Offset QPSK

Offset QPSK (OQPSK) is devised to avoid the 180° phase shifts in QPSK [3,5]. OQPSK is essentially the same as QPSK except that the I- and Q-channel pulsetrains are staggered. The OQPSK signal can be written as

$$s(t) = \frac{A}{\sqrt{2}}I(t)\cos 2\pi f_c t - \frac{A}{\sqrt{2}}Q\left(t - \frac{T}{2}\right)\sin 2\pi f_c t, \quad -\infty < t < \infty \tag{27}$$

The modulator and the demodulator of OQPSK are basically identical to those of QPSK, except that an extra delay of $T/2$ seconds is inserted in the Q channel in the modulator and in the I channel in the demodulator.

Since OQPSK differs from QPSK only by a delay in the Q-channel signal, its power spectral density is the same as that of QPSK, and its error performance is also the same as that of QPSK.

In comparison to QPSK, OQPSK signals are less susceptible to spectral sidelobe restoration in satellite transmitters. In satellite transmitters, modulated signals must be band-limited by a bandpass filter in order to conform to out-of-band emission standards. The filtering degrades the constant-envelope property of QPSK, and the 180° phase shifts in QPSK cause the envelope to go to zero momentarily. When this signal is amplified by the final stage, usually a highly nonlinear power amplifier, the constant envelope will be restored. But at the same time the sidelobes will be restored. Note that arranging the bandpass filter after the power amplifier is not feasible since the bandwidth is very narrow compared with the carrier frequency. Hence the Q value of the filter must be extremely high such that it cannot be implemented by the current technology. In OQPSK, since the 180° phase shifts no longer exist, the sidelobe restoration is less severe.

4.2. π/4-DQPSK

Although OQPSK can reduce spectral restoration caused by a nonlinearity in the power amplifier, it cannot be differentially encoded and decoded. *π/4-DQPSK* is a scheme that not only has no 180° phase shifts like OQPSK but also can be differentially demodulated. These properties make it particularly suitable for mobile communications where differential demodulation can reduce the adversary effects of the fading channel. π/4-DQPSK has been adopted as the standard for the digital cellular telephone system in the United States and Japan.

π/4-DQPSK is a form of differential QPSK. The correspondence between data bits and symbol phase difference are shown as follows.

I_k Q_k	1 1	−1 1	−1 −1	1 −1
$\Delta\theta_k$	$\pi/4$	$3\pi/4$	$-3\pi/4$	$-\pi/4$

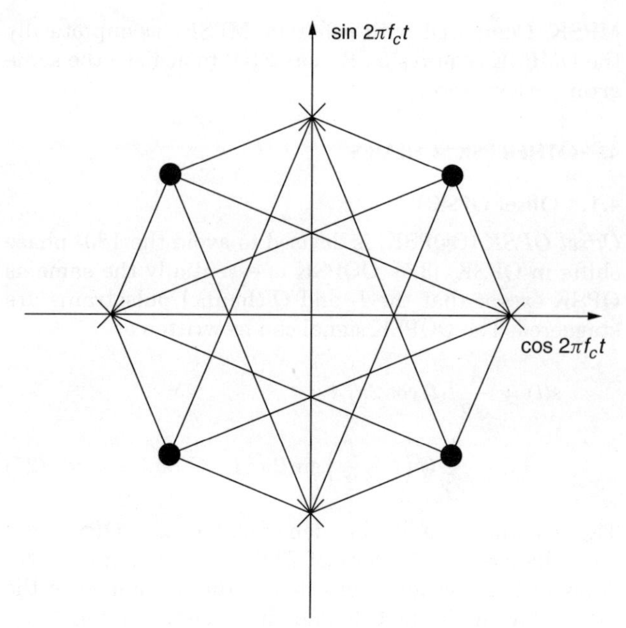

Figure 11. $\pi/4$-DQPSK signal constellation. (From Ref. 3, copyright © 2000 Artech House.)

We can see that the phase changes are confined to odd-number multiples of $\pi/4(45^\circ)$. There are no phase changes of 90° or 180°. In addition, information is carried by the phase changes $\Delta\theta_k$, not the absolute phase Φ_k. The signal constellation is shown in Fig. 11. The angle of a vector (or symbol) with respect to the positive direction of the horizontal axis is the symbol phase Φ_k. The symbols represented by • can become symbols represented only by ×, and vice versa. Transitions among themselves are not possible. The phase change from one symbol to the other is $\Delta\theta_k$.

Since information is carried by the phase changes $\Delta\theta_k$, differentially coherent demodulation can be used. However, coherent demodulation is desirable when higher power efficiency is required. There are four ways to demodulate a $\pi/4$-DQPSK signal: baseband differential detection, IF band differential detection, FM-discriminator detection, and coherent detection. The error probability of the $\pi/4$-DQPSK in the AWGN channel is about 2 dB inferior to that of coherent QPSK at high SNRs [3,8].

4.3. Continuous-Phase Modulation

The trend of phase modulation is shifting to *continuous-phase modulation* (CPM), which is a class of power- and bandwidth-efficient modulations. With proper choice of pulseshapes and other parameters, CPM schemes may achieve higher bandwidth and power efficiency than QPSK and higher-order MPSK schemes.

The CPM signal is defined by

$$s(t) = A\cos(2\pi f_c t + \Phi(t, \mathbf{a})), \quad -\infty \le t \le \infty \tag{28}$$

The signal amplitude is constant. Unlike signals of previously defined modulation schemes such as PSK, where signals are usually defined on a symbol interval, this signal is defined on the entire time axis. This is due to the continuous, time-varying phase $\Phi(t, \mathbf{a})$, which usually is influenced by more than one symbol. The transmitted M-ary symbol sequence $\mathbf{a} = \{a_k\}$ is embedded in the excess phase

$$\Phi(t, \mathbf{a}) = 2\pi h \sum_{k=-\infty}^{\infty} a_k q(t - kT) \tag{29}$$

with

$$q(t) = \int_{-\infty}^{t} g(\tau)\, d\tau \tag{30}$$

where $g(t)$ is a selected pulseshape. The M-ary data a_k may take any of the M values: $\pm 1, \pm 3, \ldots, \pm(M-1)$, where M usually is a power of 2. The phase is proportional to the parameter h which is called the modulation index. Phase function $q(t)$, together with modulation index h and input symbols a_k, determine how the phase changes with time. The derivative of $q(t)$ is function $g(t)$, which is the frequency shape pulse. The function $g(t)$ usually has a smooth pulseshape over a finite time interval $0 \le t \le LT$, and is zero outside. When $L = 1$, we have a full-response pulseshape since the entire pulse is in a symbol time T. When $L > 1$, we have a partial-response pulseshape since only part of the pulse is in a symbol time T.

The modulation index h can be any real number in principle. However, for development of practical maximum-likelihood CPM detectors, h should be chosen as a rational number.

Popular frequency shape pulses are the rectangular pulse, the raised-cosine pulse, and the Gaussian pulse. The well-known *Gaussian minimum shift keying* (GMSK) is a CPM scheme that uses the Gaussian frequency pulse:

$$g(t) = \frac{1}{2T}\left[Q\left(2\pi B_b \frac{t - \frac{T}{2}}{\sqrt{\ln 2}}\right) - Q\left(2\pi B_b \frac{t + \frac{T}{2}}{\sqrt{\ln 2}}\right)\right], \quad 0 \le B_b T \le 1 \tag{31}$$

where B_b is the 3-dB bandwidth of the premodulation Gaussian filter, which implements the Gaussian pulse effect [9]. GMSK is currently used in the U.S. cellular digital packet data system and the European GSM system.

If the modulation index h is made to change cyclically, then we obtain *multi-h CPM* (or MHPM); the phase is

$$\Phi(t, \mathbf{a}) = 2\pi \sum_{k=-\infty}^{\infty} h_k a_k q(t - kT) \tag{32}$$

where the index h_k cyclically changes from symbol to symbol with a period of K, but only one index is used during one symbol interval, that is, $h_1, h_2, \ldots, h_K, h_1, h_2, \ldots, h_K$, and so on. With proper choice of the index set and pulseshape, MHPM can be more power- and bandwidth-efficient than single-h CPM.

The major drawback of CPM and MHPM is the system complexity, especially the circuit complexity and the computational complexity of the demodulator, since optimum demodulation of CPM and MHPM requires complicated maximum-likelihood sequence estimation [3]. However, with the rapid advances in microelectronics and digital signal processing power of the electronic devices, the practical implementation and use of CPM and MHPM is quickly emerging.

Significant contributions to CPM schemes, including signal design, spectral and error performance analysis were made by Sundberg, Aulin, Svensson, and Anderson, among other authors [10–13]. Excellent treatment of CPM up to 1986 can be found in the book by Anderson, et al. [13] or the article by Sundberg [10]. A relatively concise, but up-to-date, description of CPM and MHPM can be found in Ref. 3.

BIOGRAPHY

Fuqin Xiong received his B.S. and M.S. degrees in electronics engineering from Tsinghua University, Beijing, China, and a Ph.D. degree in electrical engineering from University of Manitoba, Winnipeg, Manitoba, Canada, in 1970, 1982, and 1989, respectively. He was a faculty member at the Department of Radio and Electronics in Tsinghua University from 1970 to 1984. He joined the Department of Electrical and Computer Engineering, Cleveland State University, Ohio, as an assistant professor in 1990, where he is currently a full professor and director of the Digital Communications. Research Laboratory. He was a visiting scholar at University of Manitoba in 1984–1985, visiting fellow at City University of Hong Kong, Kowloon, in spring 1997, and visiting professor at Tsinghua University in summer 1997. His areas of research are modulation and error control coding. He is the author of the best selling book *Digital Modulation Techniques*, published by Artech House, 2000. He is an active contributor of technical papers to various IEEE and IEE technical journals and IEEE conferences. He has been a reviewer for IEEE and IEE technical journals for years. He has been the principal investigator for several NASA sponsored research projects. He is a senior member of IEEE.

BIBLIOGRAPHY

1. J. G. Proakis, *Digital Communications*, 2nd ed., McGraw-Hill, New York, 1989.
2. S. Haykin, *Digital Communications*, Wiley, New York, 1988.
3. F. Xiong, *Digital Modulation Techniques*, Artech House, Boston, 2000.
4. J. H. Park, Jr., On binary DPSK detection, *IEEE Trans. Commun.* **26**(4): 484–486 (April 1978).
5. B. Sklar, *Digital Communications, Fundamentals and Applications*, Prentice-Hall, Englewood Cliffs, NJ, 1988.
6. S. Benedetto, E. Biglieri, and V. Castellani, *Digital Transmission Theory*, Prentice-Hall, Englewood Cliffs, NJ, 1987.
7. K. M. Simon, S. M. Hinedi, and W. C. Lindsey, *Digital Communication Techniques, Signal Design and Detection*, Prentice-Hall, Englewood Cliffs, NJ, 1995.
8. C. L. Liu and K. Feher, $\pi/4$-QPSK Modems for satellite sound/data broadcast systems, *IEEE Trans. Broadcast.* (March 1991).
9. G. Stüber, *Principle of Mobile Communication*, Kluwer, Boston, 1996.
10. C-E. Sundberg, Continuous phase modulation: A class of jointly power and bandwidth efficient digital modulation schemes with constant amplitude, *IEEE Commun. Mag.* **24**(4): 25–38 (April 1986).
11. T. Aulin and C-E. Sundberg, Continuous phase modulation — part I: Full response signaling, *IEEE Trans. Commun.* **29**(3): 196–206 (March 1981).
12. T. Aulin, N. Rydbeck, and C-E. Sundberg, Continuous phase modulation — Part II: Partial response signaling, *IEEE Trans. Commun.* **29**(3): 210–225 (March 1981).
13. J. B. Anderson, T. Aulin, and C-E. Sundberg, *Digital Phase Modulation*, Plenum, New York, 1986.

DISTRIBUTED INTELLIGENT NETWORKS

Iakovos S. Venieris
Menelaos K. Perdikeas
National Technical University of Athens
Athens, Greece

1. DEFINITION AND CHARACTERISTICS OF THE DISTRIBUTED INTELLIGENT NETWORK

The distributed intelligent network represents the next stage of the evolution of the intelligent network (IN) concept. The term corresponds to no specific technology or implementation but rather is used to refer, collectively, to a family of architectural approaches for the provisioning of telecommunication services that are characterized by: use of distributed, object oriented and, potentially, mobile code technologies, and a more open platform for service provisioning.

The intelligent network [1] first emerged around 1980 when value-added services that had previously been offered only on a private branch exchange basis involving mainly corporate users, first begun to be made available on the public network as well. The problem was that the only way that new services could be introduced using the infrastructure of the time required upgrading a large number of deployed telephone switches. In the first available programmable switches, services resided in the memory space of each switch and service view was therefore local. Thus, each installed service provided a number of additional features on the switch it was installed and so could be provided only for those telephone calls that were routed through that particular switch. A certain switch could have a number of services implemented in it, while others could have different sets or none. Also, there was no guarantee that the implementation of a service would behave in the same way, uniformly across switches coming from different vendors. This was the era of the "switch-based services."

New services therefore could not be provided nationwide simultaneously and often even when all involved switches had been upgraded, different service dialects existed as a result of the heterogeneity of the equipment and the unavoidable discrepancies between the implementation of the same service in switches provided by different vendors. These problems were further aggravated by the need to guard against undesirable feature interaction problems: a process that even now cannot be automated or tackled with, in a systematic, algorithmic manner and usually involves manually checking through hundreds or even thousands of possible service feature combinations. Since this process had to also account for the different service dialects, each service that was added to the network increased the complexity in a combinatorial manner.

The result was that service introduction soon became so costly that new services could be provided only infrequently. The intelligent network idea sought to alleviate these problems. The key concept was to separate the resources (public telephony switches) from the logic that managed them. Once this was achieved, the logic could be installed in a central location called "service control point" and from there, operate on the switches under its control. Service introduction would consist simply in the installation of a new software component in the service control point and would therefore take effect immediately once this was completed and once involved switches were configured to recognize the new service prefixes. Once a switch identified an "intelligent network call," that is, a call prefixed with a number corresponding to an IN service (e.g., 0800 numbers), it would suspend call processing and request further instructions on how to proceed from the remotely located service or service logic program. No other logic therefore needed to be installed in the switches apart from a generic facility for recognizing such IN calls and participating in the interaction with the remotely located service logic programs. This interaction consisted of notifying the service logic programs of intelligent network calls or other important call events, receiving instructions from them, and putting these instructions into effect by issuing the appropriate signaling messages toward the terminals that requested the IN service or towards other resources involved in the provisioning of the service. Typical resources of the latter kind were intelligent peripherals and specialized resource functions where recorded service messages were and still are typically kept.

In order to implement a mechanism such as the one called for by the IN conception, three key artifacts are needed: (1) a finite state machine implementing an abstraction of the call resources of a public switch, (2) a remote centralized server where programs providing the algorithmic logic of a service are executed, and (3) a protocol for remote interaction between the switch and the service logic programs. Through this protocol (1) a limited aperture of visibility of the switch functionality—in terms of the said finite state machine—is presented to the service logic programs; and (2) hooks are provided allowing the latter to influence call processing in order to effect the desired behavior of any given service.

As noted above, prior to introduction of the intelligent network concept the last two of these three elements were not existent as service logic was embedded into every switch in order for a service to be provided. There was therefore no remote centralized server responsible for service execution and, hence, due to the locality of the interaction, no protocol needed to support the dialogue between the switch and the service logic program. Even the first of these three elements (the state machine abstracting call and connection processing operations in the switch) was not very formalized as switch-based services were using whatever nonstandardized programming handles a vendor's switch was exposing. This was, after all, why for a new service to be introduced, direct tampering with all affected switches was necessary, and in fact the implementation of the same service needed to be different in different switches. Also this accounted for the fact that, considering the heterogeneous nature of the switches comprising a certain operator's network, different service dialects were present, resulting in a nonuniform provision of a given service depending on the switch to which a subscriber was connected. Intelligent networks changed all that by defining a common abstraction for all switches and by centralizing service logic to a few (often one) easily administrated and managed servers. The abstraction of the switches was that of a state machine offering hooks for interest on certain call events to be registered, and supporting a protocol that allowed remote communication between the switches and the now centrally located service logic programs. It will be shown in the following paragraphs that this powerful and well-engineered abstraction is also vital in the distributed intelligent network.

2. THE NEED FOR DISTRIBUTED INTELLIGENT NETWORKS

The IN concept represents the most important evolution in telecommunications since the introduction of programmable switches that replaced the old electromechanic equipment. It allows for the introduction of new services, quickly, instantly, across large geographic areas, and at the cost of what is essentially a software development process as contrasted to the cost of directly integrating a new service in the switching matrix of each involved telephony center.

However, after 1998, a number of technical and socioeconomic developments have opened up new prospects and business opportunities and also have posed new demands, which traditional IN architectures seem able to accommodate only poorly: (1) use of mobile telephony became widespread, particularly in Europe and Japan; (2) the Internet came of age both in terms of the penetration it achieved and is projected to achieve, and also in terms of the business uses it is put to; and (3) deregulation seems to be the inevitable process globally setting network operators and carriers in fierce competition against each other. The import of the first two of these developments is that the public switched telephony network is no longer the only network used for voice communications. Cellular telephony networks and also telephony over the Internet are offering essentially the same services. Therefore, an architecture for service creation such as the IN that is entirely focused on the public telephony network falls short of providing the universal platform for service provisioning that one would ideally have wished: a single platform offering the same services over the wired, wireless and Internet components of a global integrated network for voice and data services. Deregulation, on the other hand, has two primary effects; first, as noted above, it promotes competition between operators for market share making the introduction of new and appealing services a necessity for a carrier that wishes to stay in business. Since the basic service offered by any carrier is very much the same, a large amount of differentiation can be provided in the form of value-added or intelligent services that are appealing and useful to end users. The new telecommunications landscape is no longer homogenous but instead encompasses a variety of networks with different technologies and characteristics. The prospect for innovative services that focus not only on the telephone network but also Internet and mobile networks is immense. Particularly useful would be hybrid services that span network boundaries and involve heterogeneous media and access paradigms. Typical services belonging to this genre are "click to" services whereby a user can point to a hyperlink in his or hers browsers and as a result have a phone or fax (Facsimile) call being set up in the network. Also, media conversion services whereby a user who is not able to receive a certain, urgent email can have a phone call in his mobile terminal and listen to the contents of the mail using text to speech conversion. All these services cannot be offered by relying solely on an intelligent network. Other technologies and platforms would have to be combined and this would not result in a structured approach to service creation. Essentially, traditional intelligent networks cannot fulfill the new demands on rapid service creation and deployment in a converged Internet — telecommunications environment because the whole concept of this technology had at its focus the public telephony network and revolved around its protocols, mechanisms, and business models. Internet and cellular telephony, on the other hand, have their own protocols, infrastructure, and mechanisms, and these cannot be seamlessly incorporated into an architecture designed and optimized with a different network in mind. Furthermore, the business model envisaged by the traditional intelligent network is a closed one. It has to be said that the focus of the IN standardization was not to propose and enable new business models for telephony: only to expedite the cumbersome and costly process of manual switch upgrading that made introduction of new services uneconomical to the point of impeding the further growth of the industry. Intelligent network succeeded in solving this problem by removing the service intelligence from the switches and locating that intelligence in a few centralized points inside the network. However, the same organization continued to assume the role of the network operator or carrier and that of the service provider. Deregulation and the growth of Internet that adheres to a completely different business model necessitate the separation of these two roles and so require a technological basis that would support this new model. This is the problem that the next generation of the intelligent network technologies face.

The distributed intelligent network represents the next stage of the intelligent network evolution. The three main elements of the traditional intelligent network concept survive this evolution as could be inferred by the central role they have been shown to play in delivering services to the end users in the traditional IN context. Distributed IN is characterized by the use of distributed object technologies such as the Common Object Request Broker Architecture (CORBA), Java's Remote Method Invocation (RMI) or Microsoft's Distributed Component Object Model (DCOM) to support the switch — services interaction. In a traditional IN implementation, the Intelligent Network Application Protocol (INAP) information flows are conveyed by means of static, message-based, peer-to-peer protocols executed at each functional entity. The static nature of the functional entities and of the protocols they employ means that in turn the associations between them are topologically fixed. An IN architecture as defined in Ref. 2 is inherently centralized with a small set of service control points and a larger set of service switching points engaged with it in INAP dialogs. The service control points are usually the bottleneck of the entire architecture and their processing capacity and uptime in large extent determine the number of IN calls the entire architecture can handle effectively. Distributed object technologies can help alleviate that problem by making associations between functional entities less rigid. This is a by-product of the location transparencies that use of these technologies introduces in any context. More importantly, the fact that under these technologies the physical location of an entity's communicating peer is not manifested makes service provisioning much more open. This will be explained in more detail in later paragraphs. The remainder of this encyclopedia article is structured as follows: the distributed IN's conceptual model is introduced and juxtaposed with that of the traditional IN, typical distributed IN implementation issues are presented, and finally some emerging architectures that adhere to the same distributed IN principles are identified.

3. THE DISTRIBUTED INTELLIGENT FUNCTIONAL MODEL

According to the International Telecommunication Unions (ITU) standardization of IN, an intelligent network conceptual model is defined, layered in four planes depicting different views of the intelligent network architecture from the physical plane up to the service plane. Of these planes we will use as a means of comparing the IN architecture with that of distributed IN, the distributed functional plane. The distributed functional plane identifies the main functional entities that participate in the provision of any given IN service without regard to their location or mapping to physical elements of the network (the aggregation of functional entities to physical components is reflected in the physical plane). The functional models of IN and distributed IN are quite different in certain respects. First, a number of emerging distributed IN architectures like Parlay and JAIN (discussed later on) incorporate new functional entities to account for the new types of resources that can be found in a converged Internet — Telecommunications environment and which traditional IN models could not anticipate. Apart from that, a further difference emerges in the form of the more open environment for service provisioning, the distributed or remote-object interactions that characterize distributed IN and the potential use of mobile code technologies that can change the locality of important components at runtime (and with this, the end points of control flows). Finally, the business model of distributed IN with its use of distributed object technologies explicitly enables and accommodates the separation between the roles of the network operator and the service provider.

Some of the abovementioned differences cannot be reflected in the distributed functional plane as defined in the IN's conceptual model as this is at a level of abstraction where such differences are hidden. Therefore, Fig. 1 compares the two technologies (traditional and distributed IN) by reflecting features of both their distributed and physical planes.

Figure 1a presents a limited view of the traditional IN conceptual model at the distributed and physical planes. A number of functional entities are there identified, of which the more central to this discussion will be the service creation environment function/service management function, the service control function, and the service switching function. Figure 1 also depicts a specialized resource function and an intelligent peripheral, which is were announcements are kept and digits corresponding to user's input are collected. The second part of Fig. 1 also depicts an abstract entity entitled "Internet/mobile resources," which represents functional entities present in distributed IN that correspond to resource types unique in the Internet or mobile networks components such as mail servers, media conversion servers or location servers.

Starting from the bottom up, the service switching function corresponds to the finite state machine reflecting the switch's call resources that the article drew attention to as being one of the main artifacts on which the whole IN concept is based. This functional entity is closely linked with the service control function, which is the place where service logic programs are executed. These are not depicted since they are thought of as incorporated with the latter. Finally, the service creation environment function and the service management function are the entities responsible for the creation of new service logic programs, and their subsequent injection and monitoring into the architecture and eventual withdrawal/superseding by other services or more up to date versions.

The INAP protocol is executed between the service switching and the service control functions that can be regarded as implementing a resource-listener-controller pattern. The service switching function is the resource that is monitored whereas the service control function is responsible for setting triggers corresponding to call events for which it registers an interest and also the controller that acts on these events when informed of their occurrence. A typical event for instance would be the detection of a digit pattern that should invoke an IN service (e.g., 0800). The service switching function undertakes the job of watching for these events and of suspending call processing when one is detected and delivering an event notification to the service control function for further instructions. This pattern is the same also in the distributed IN and can in fact also be identified in recent telecommunication architectures such as those articulated by the JAIN and Parlay groups.

The distributed IN conceptual model (Fig. 1b) differs by (1) explicitly enabling mobility of the service logic programs through the use of mobile code technologies and (2) replacing the message-based INAP protocol with distributed processing technologies like CORBA, RMI

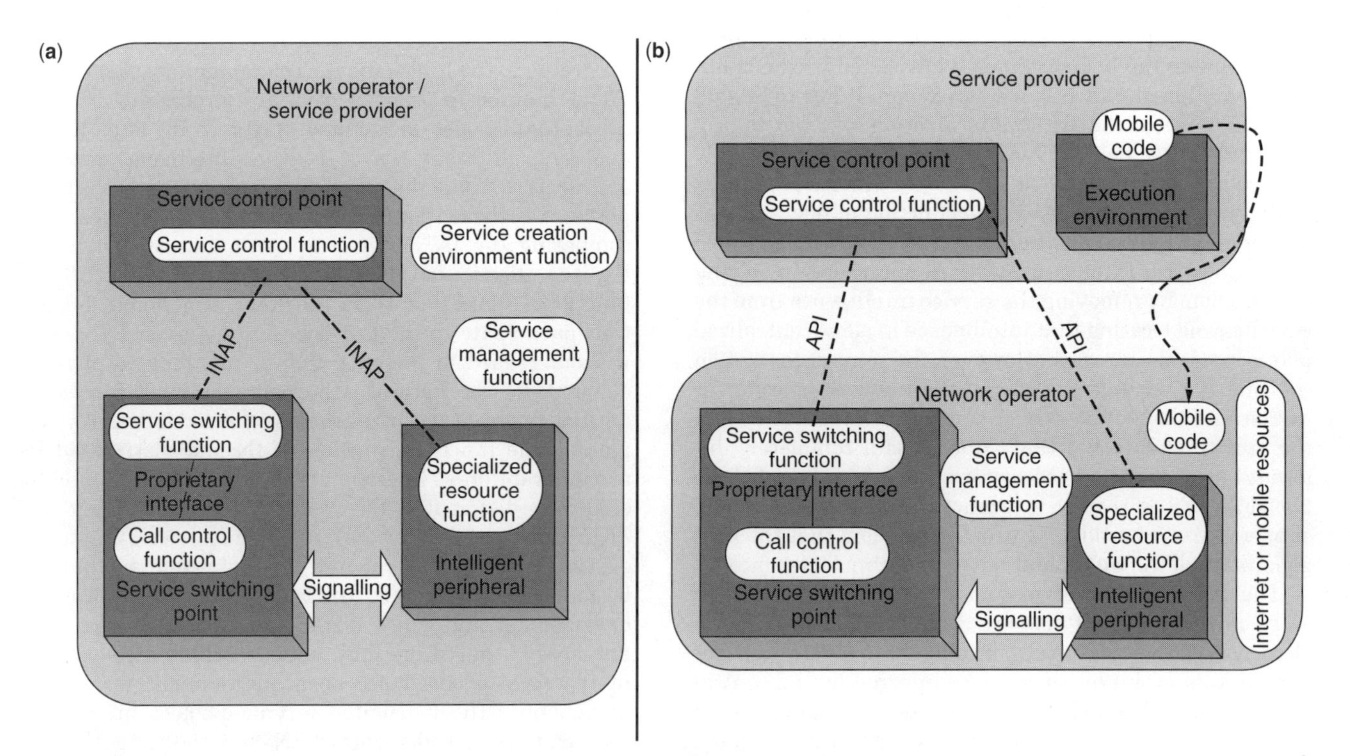

Figure 1. Traditional (**a**) and distributed (**b**) IN functional models.

or DCOM. It should be noted of course that both the traditional and the distributed IN conceptual models are populated by an additional number of functional entities which are not depicted in Fig. 1 nor enter this discussion for reasons of economy. Code mobility (often implemented using Java and/or some additional mobile code libraries) is used to allow service logic programs to reside not only in the service control function but also in the service switching function where an execution environment identical with the one that exists within the service control function can be found. Service logic programs in the form of mobile code components populate both execution environments and constitute the control part of IN service provisioning. From these execution environments, using distributed processing technologies, the switch resources, as exposed by the service switching function objects, are monitored and manipulated. This means that service logic programs in the distributed IN conceptual model are prime level entities and cannot be suitably depicted as pinned down inside the implementation of the service control function. In contrast to the functional plane of the IN conceptual model where the service control function and the service logic programs were one and the same, in the functional plane of the distributed IN conceptual model, the service control and also the service switching functions are containers (or execution environments) for service components capable of migrating to whichever environment is best suited to accommodate their execution. It is important to consider that this amount of flexibility would not be attainable were it not for the use of distributed processing technologies at the control plane. The defining characteristic of these technologies is that they abstract process, machine and network boundaries and provide to the programmer and the runtime instances of service code a view of the network as a generalized address space spanning conventional boundaries. In the same manner in which a program may hold a pointer to an object or function residing in the same local process, so it can, when distributed technologies are used, hold a pointer to an object located in a different process in the same machine, in a machine in the same local network, or in a machine located in a different network. Therefore the abstraction of a single memory space is supported and compiled code components can be mapped in an arbitrary way in processes and physical nodes without needing recompilation or rebooting of the system. This is a potent facility and it opens new capabilities which the distributed IN employs. It also has to be stated that the articulation of the traditional IN conceptual model, to an extent, anticipated a certain flexibility in the allocation of functional entities to physical ones according to the correspondences between the functional and the physical planes. This was however not materialized since the communication infrastructure used for the conveyance of the INAP information flows between the various entities of the system (and most importantly between the service switching and the service control entities) was the signaling system 7 network, which does not demonstrate the properties of a distributed processing environment. Therefore the full potential inherent in the abstract definition of the IN conceptual model was materialized only to a limited extent and mainly had to do with different configuration and aggregations of functional entities to the hardware nodes of the system. The interested reader should refer to Ref. 2 for a presentation of the various alternatives that are possible in an IN architecture. This leeway afforded to the network designer is something completely different from the ability of service logic programs to roam through the various execution environments of the system at runtime without requiring the system to suspend its operation and, under certain configuration options discussed in the following paragraph, in a manner completely automated and transparent even to the network management system.

4. ISSUES IN DISTRIBUTED IN IMPLEMENTATIONS

Given the considerations outlined above and the differences at the functional plane between traditional and distributed IN architectures, a number of approaches exist. Each of these approaches essentially answers a defining question in a different manner, and since for the most part they are orthogonal to each other, a large number of widely differing distributed IN implementations can be envisaged; all, however, share the same fundamental characteristics of increased flexibility and openness when compared with traditional IN.

The first point to examine is the set of considerations that govern the location of service logic programs inside the network. Attendant to it are the dynamics of their mobility. This is possible once employment of mobile code technologies is assumed. Service logic programs implemented as mobile components can migrate between execution environments dynamically. Considerations valid for incorporation in any given distribution algorithm are processing load, signaling load, and functionality. Given a number of execution environments present in the network's service switching and service control functions, each with its own characteristics, simple or elaborate load balancing mechanisms can be devised and implemented. These would allow service logic programs to locate themselves in such a way so that no individual execution environment's resources are strained beyond a certain point. Apparent tradeoffs exist with respect to complexity, time, and computing power spent in process or code migrations, which are in themselves costly procedures. This cost is reflected in terms of both time and processor load they temporarily create as they involve suspension or creation of new threads, allocation of objects in memory, and so on. These overheads, when also viewed in the light of the stringent performance and responsiveness demands that exist in a telecommunication system, should tilt the balance in favor of a simple and not very reactive load balancing algorithm that operates on ample time scales.

Processing load is not however the only kind of load to consider. Control load is another. Control load should not be confused with signaling load, which concerns protocol message exchanges between terminals and switches at various levels of the classic Five-layered telephony network architecture. Signaling load is for the most part transparent to the IN with the exception of those call events that invoke an IN service and for which a service switching function has been asked to suspend call processing and wait for instructions from the service control function. Control load in this discussion is about the exchange of INAP information flows using the selected distributed processing technology between the service logic programs and the switching resources of the system. The location transparency provided by distributed technologies cannot clearly be interpreted to suggest that local or remote interactions take the same time. Therefore, when service logic programs are located in an execution environment closer to the switch they control, better performance can be expected. This means that, in general, the code distribution mechanism, should locate service logic programs at the proximity of the resources with which they are engaged if not at an execution environment collocated with those resources. There is, however, a tradeoff here between processing and control load. Because of the higher level of abstraction (method calls instead of messages) offered to service logic programs when distributed processing technologies are used, it is necessary that, under the hood, the respective middleware that is used to support this abstraction enters into some heavy processing. For every method call that is issued, it is necessary that each argument's transitive closure is calculated and then the whole structure is serialized into an array of bytes for subsequent transmission in the form of packets using the more primitive mechanisms offered by the network layer. On the recipient side, the reverse procedure should take place. This set of processes is known as marshaling/demarshaling and is known to be one of the most costly operations of distributed processing. Because of the processor intensive character of marshaling/demarshaling operations, it is possible that, under certain configurations, better performance is attained when the service logic program is located to a remote execution environment than to a local, strained, one. This in spite of the fact that the time necessary for the propagation of the serialized byte arrays in the form of packets from the invoking to the invoked party will be higher in the remote case. This interrelation between processing and control load means that no clear set of rules can be used to produce an algorithm that is efficient in all cases. The complex nature of the operations that take place in the higher software layers means that this is not an optimization problem amenable to be expressed and solved analytically or even by means of simulation, and therefore an implementor should opt for simple, heuristic designs, perhaps also using feedback or historic data. Another approach would be to rely on clusters of application servers into which service logic programs can be executed. Commercially available application servers enable process migration and can implement a fair amount of load balancing operations themselves, transparently to the programmer or the runtime instances of the service components.

The second point that can lead to differentiations in architecture design in distributed intelligent network is the question of who makes the abovementioned optimization and distribution decisions. There could be a central entity that periodically polls the various execution environments receiving historic processing and signal load data, runs an optimization algorithm and instructs a number of service logic programs to change their location in the network according to the results thus produced. The merit of this approach is that the optimization is networkwide and the distribution algorithm can take into account a full snapshot of the network's condition at any given instance of time when it is invoked. The disadvantage is the single point of failure and the cost of polling for the necessary data and issuing the necessary instructions. These communications could negatively affect traffic on the control network, depending, of course, on the frequency with which they are executed. A point to consider in this respect that can lead to more efficient implementations is whether networkwide optimization is necessary or whether locally executed optimizations could serve the same purpose with similar results and while projecting a much lower burden on the communication network. Indeed, it can be shown that it is highly unlikely for the purposes of any optimization to be necessary to instruct any given service logic program to migrate to an execution environment that is very remote to the one in which it was until that time executed. To see that this is the case, one can consider that a migration operation moving a service component to a very distant location will most likely result in an unacceptably higher propagation delay that would degrade the responsiveness of the corresponding resource–service logic link. Therefore locally carried optimizations could result in comparable performance benefits at a greatly reduced communication cost when compared to a networkwide optimization. Therefore, in each subnetwork an entity (migration manager) can be responsible for performing local optimizations and instructing service components to assume different configurations accordingly. Taking this notion to its logical conclusion, a further implementation option would be to have service logic programs as autonomous entities (mobile agents) that are responsible for managing their own lifecycle and proactively migrating to where they evaluate their optimal location to be. As before, considerations for deriving such an optimal location can be signaling or control load experienced at the execution environment where they were hosted but also, more appropriately in this case, the location of mobile users. Indeed, in this last scenario one can have a large population of service code components, each instantiated to serve a specific mobile (roaming) user. See Ref. 3 for a discussion of this approach. Service logic programs could then evaluate their optimal location in the network, also taking into consideration the physical location of the user they serve. This can, for instance, lead to migration operations triggered by the roaming of a mobile user. In this manner concepts like that of the virtual home environment for both terminal and personal mobility can be readily supported enabling one user to have access to the same portfolio of intelligent services irregardless of the network into which he/she is roaming and subject only to limitations posed by the presentation capabilities of the terminal devices he/she is using. Of course, the usefulness of autonomous service logic programs is not limited to the case of roaming or personal mobility users, but this example provides an illustration of the amount of flexibility that becomes feasible when advanced software technologies such as distributed processing environments and mobile code are used in the context of an intelligent network architecture. Moreover, as more logic is implemented in the form of mobile code components, the supporting infrastructure can become much more generic, requiring fewer modifications and configuration changes and able to support different service paradigms. The generic character of the architecture means that stationary code components that require human individual or management intervention for their installation or modification are less likely to require tampering with, and that a greater part of the service logic can be deployed from a remote management station dynamically, at runtime, contributing to the robustness of the system and to an increased uptime. Naturally, the tradeoff is that making service logic programs more intelligent necessarily means increasing their size, straining both the memory resources of the execution environments into which they are executed and also requiring more time for their marshaling and demarshaling when they migrate from one functional entity to another. If their size exceeds a certain threshold, migration operations may become too costly and thus rarely triggered by the load balancing algorithm they implement, negating the advantages brought by their increased autonomy and making services more stationary and less responsive in changing network or usage conditions. Again, simpler solutions may be more appropriate and a designer should exercise caution in determining which logic will be implemented in the form of stationary code components, engraved in the infrastructure, and which will be deployed at runtime using mobile code.

Another major issue to consider in the implementation of a distributed intelligent network architecture is which distributed processing technology to use and how to support its interworking with the signaling system 7 network that interconnects the physical components hosting the functional entities in any intelligent network architecture. There are three main competing distributed technologies: CORBA, DCOM, and RMI. It is not the purpose of this article to examine or compare these three technologies, nor to identify their major strengths and weaknesses. However, for the purposes of an intelligent network implementation, CORBA is best suited because it is the only one of these three technologies to have interworking between it and the signaling system 7 network prescribed [4]. The alternative would be to bypass the native IN control network and install a private internet over which the traffic from DCOM or RMI could be conveyed. Since this is not a development that operators can be expected to implement quickly and for reasons of providing a solution that has the benefit of backward compatibility and also allows an evolutionary roadmap to be defined, CORBA should be the technology of choice in distributed IN implementations. A further reason is that CORBA, which is targeted more than the other two technologies for the telecommunications domain, has more advanced real time characteristics and is therefore better equipped to meet the demands of an operator. The interested reader is referred to Ref. 5 for an in-depth look of the implementation of a distributed IN architecture using CORBA.

5. A DISTRIBUTED IN ARCHITECTURE IN DETAIL

Figure 2 depicts a typical distributed IN implementation in more detail. The main components are (1) the network resources that provide the actual call control and media processing capabilities, (2) the remotely enabled distributed objects that wrap the functionality provided by the physical resources and expose it as a set of interfaces to the service layer, (3) the code components residing in the service layer that incorporate the business logic, and (4) the communication infrastructure that makes communication between the service logic programs and the network resources, via their wrappers, possible. Of course, in an actual architecture a lot more components would need to be identified, such as service management stations, and service creation systems. However, the purpose of the discussion presented here is to allow more insight, where appropriate, into the more technical aspects of a DIN implementation and not to provide a full and comprehensive description of a real system.

Notice, first, that terminals are not depicted in Fig. 2. This is because interaction with terminals at the signaling level is exactly the same as in the case of traditional IN. Signaling emitted by the terminals is used to trigger an IN session and signaling toward terminals or media resources of an IN system (like Intelligent peripherals or specialized resource functions) is used to carry out the execution of a service. However, service components do not directly perceive signaling as they interact with their peer entities on the network side by means of message- or method-based protocols. INAP is a typical method based protocol used at the control plane of IN and Parlay, JAIN, or a version of INAP based on remote application programming interfaces (APIs) are prime candidates for the control plane of distributed IN. In any case it is the responsibility of the network resources to examine signaling messages and initiate an IN session where appropriate, or, in the opposite direction, to receive the instructions sent to them by the service logic programs and issue the appropriate signaling messages to bring these instructions into effect. Therefore, since signaling is conceptually located 'below' the architecture presented in Fig. 2, it can safely be omitted in the discussion that follows. The reader who wishes to gain a fuller appreciation of the temporal relationships between the signaling and INAP messages that are issued in the course of the provisioning of an IN session can refer to Ref. 5.

Assuming a bottom–up approach, the first step would be to explore the programmability characteristics of the deployed equipment that forms the basis of a distributed IN architecture. We refer to deployed equipment as *distributed IN*, like IN before it, which has to be able to encompass and rely on equipment that is already deployed in the field if it is to succeed. Telecom operators have made huge investments in building their networks, and it would be uneconomical to replace all this equipment. At the very least, an evolutionary approach should

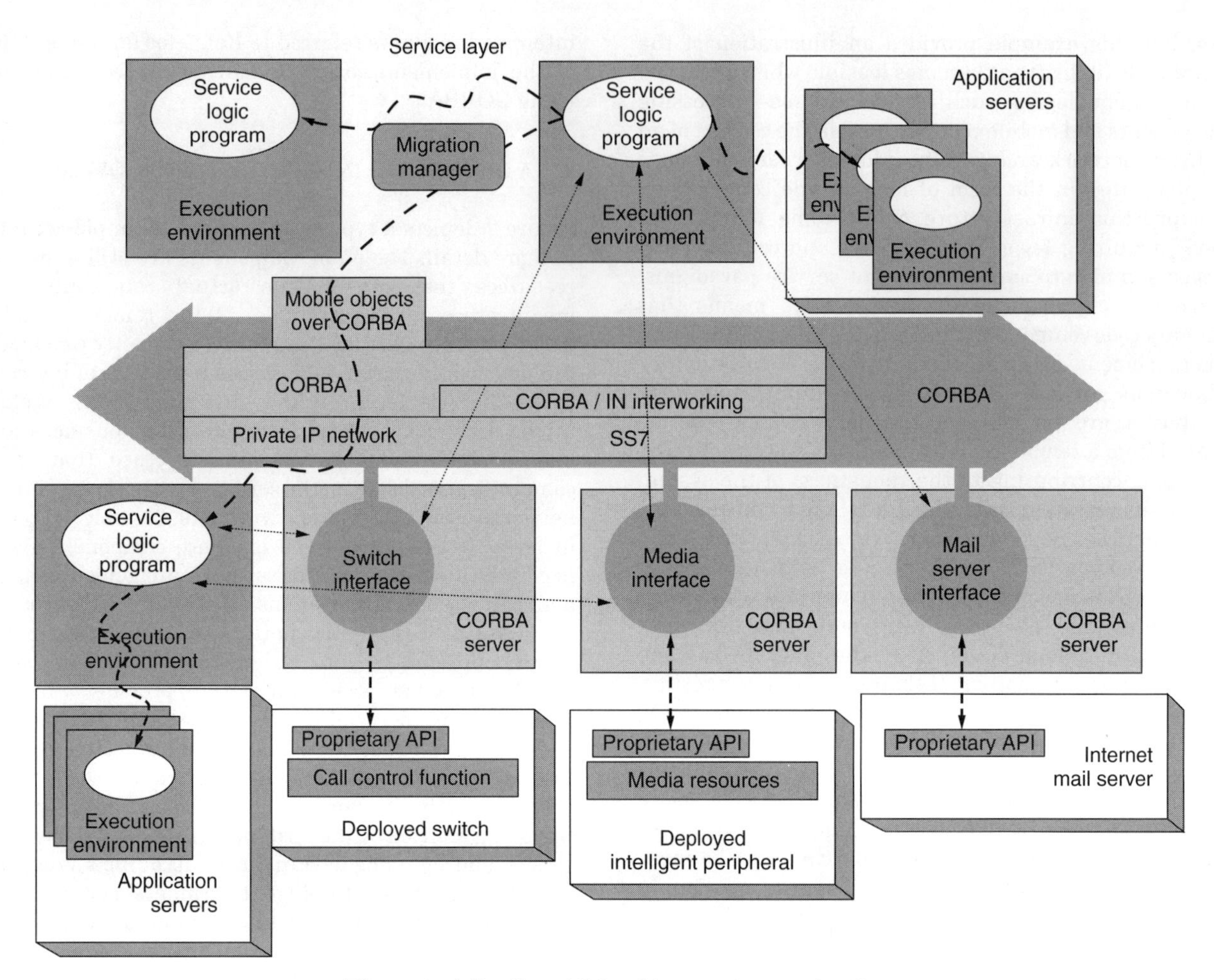

Figure 2. A distributed IN architecture in more detail.

be possible to implement where equipment would be replaced gradually over a period of years and the two technologies, IN and distributed IN, would coexist. It would be further advantageous if existing equipment could be seamlessly integrated into the new distributed intelligent network infrastructure. To address this issue we note that in general, network switches are not meant to be overly programmable. Certain proprietary APIs are always offered that provide an amount of control over how the switch responds to an incoming call, but there are few specifications defining standard interfaces that network switches should offer, and in any case there is much discrepancy between the equipment provided by different vendors. Of course, the IN itself could be regarded as a standard interface for network switches, but the purpose here is not to build the new distributed IN infrastructure over the already existing IN one. Rather, it is to build it on top of the same lower-level facilities and services that IN is using as part of its own implementation: and such lower-level facilities are accessible only to equipment vendors. An external integrator exploring the programmability features of a switch will be able to find standard interfaces only at the IN level, which, as we noted, is conceptually very elevated to be useful as the infrastructure of a distributed IN architecture (note, nevertheless, that Parlay, which is one of a group of emerging technologies that can fall under the "distributed" IN heading, also allows for an approach of using existing IN interfaces as an interim solution before it can be provided natively in switches as IN does). Below this level one, can find only proprietary APIs at different levels of abstraction.

It is therefore possible that a certain operation that a developer would wish to expose to the service components cannot be implemented because the programmability characteristics of the switch would not allow the programming of this behavior. Assuming however that the semantic gap between the operations that a network resource needs to make available for remote invocation and the programming facilities that are available for implementing these operations, can be bridged, generally accepted software engineering principles, object-oriented or otherwise could serve to provide the substrate of the distributed IN [6]. This substrate consists of set of remotely enable objects (CORBA objects "switch interface," "media interface," and "mail server interface" depicted in Fig. 2) that expose the facilities of the network resources they represent. Once this critical implementation phase has been carried out, subsequent implementation is entirely at the service layer. The "switch interface," "media interface," and "mail server interface" objects are each responsible for exposing a remotely accessible facet of the functionality of the network resource they represent. Their implementation has then to mediate the method invocations it receives remotely, to the local proprietary API that each network resource natively provides. In the opposite direction, one has events that are detected by the network resources and have, ultimately, to reach the components residing in the service layer. This is accomplished by having the CORBA objects register themselves as listeners for these events. The process of registering an external object as a listener involves identifying the events to which it is interested and providing a callback function or remote pointer that will serve to convey the notification when an event satisfying the criteria is encountered. From that point on, the implementation of the CORBA object will itself convey the notification to the higher software layers. It is interesting to note that at this second stage the same pattern of listener and controller objects is also observed, with the exception that now the listener objects are the service logic programs or, in general, the service control logic in the network and the resource is the CORBA object itself. This observation is depicted at Fig. 3.

Through this wrapping of the native control interface to remote API-based ones two things are achieved: (1) the middleware objects residing at the CORBA servers in the middle tier of Fig. 3 can be used to implement standardized interfaces. As an example, Fig. 3 indicates INAP or Parlay API. INAP is, of course, a message-based protocol, but it is relatively straightforward to derive method—based interfaces from the original protocol specification and to express the protocol semantics in terms of method calls and argument passing instead of asynchronous exchange of messages and packets with their payloads described in Abstract Syntax Notation 1. In that sense, API-based versions of INAP can be used to implement the control plane of the distributed intelligent network, and so can, for that matter, emerging technologies such as Parlay and JAIN. The next section discusses such approaches.

6. EMERGING ARCHITECTURES

A number of architectures and most notably JAIN and Parlay have emerged that, although not classified under the caption of "distributed intelligent network," have nevertheless many similarities with it. Distributed intelligent networks make telecommunications service provisioning more open by resting on distributed object technologies and utilizing software technologies such as mobile code, which make the resulting implementation more flexible and responsive to varying network conditions. Parlay and JAIN move a step further toward this direction again by leveraging on CORBA, DCOM, or RMI to implement a yet more open service model that allows the execution of the services to be undertaken by actors different than the operator's organization, in their own premises closer to the corporate data they control and manipulate [7]. In particular, Parlay and JAIN use the same approach as in Ref. 5 of defining methods corresponding more or less to actual INAP information flows and of exposing a switch's (or, for that matter, any other network resident resources)

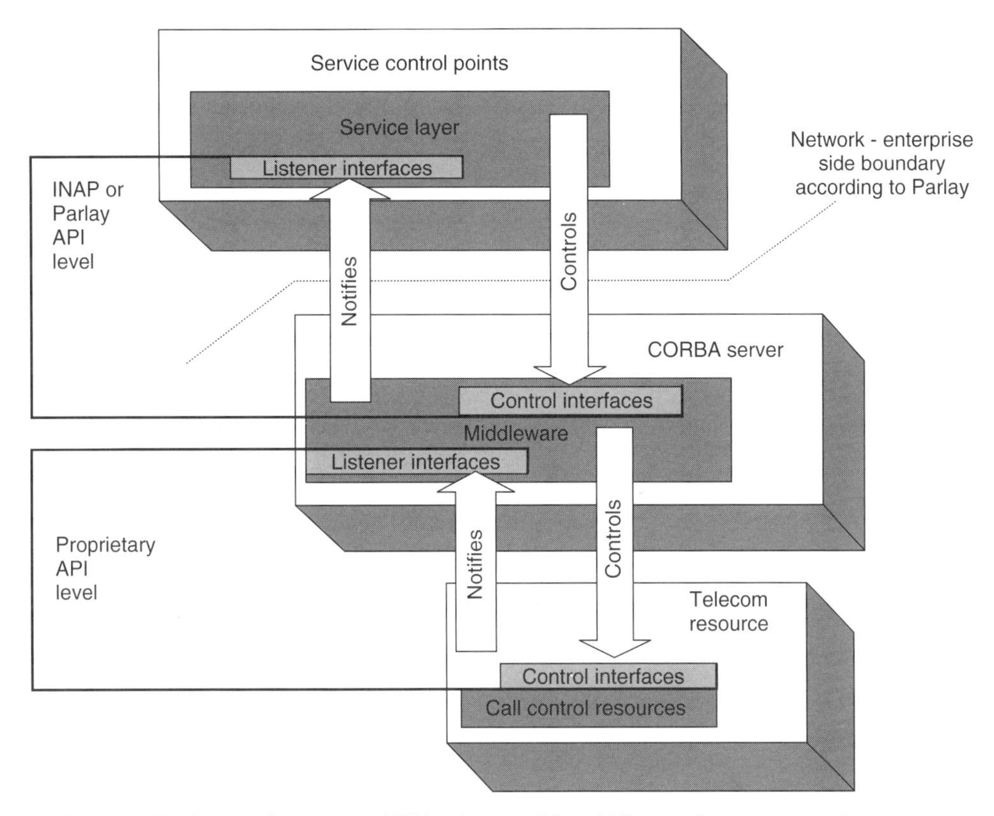

Figure 3. Parlay or other remote API implemented in middleware form over a native resource.

functionality to remotely located service code in the form of an interface of available methods. Moreover, a number of technologies such as the call processing language or telephony services expressing their interaction with the switch in the form of eXtensible Markup Language (XML) statements have appeared that can be used in the same manner as mobile code is used in a distributed IN environment. Service logic expressed in XML or Call Processing Language is inherently mobile in that it is not compiled code and can be interpreted by any appropriate execution environment or script engine. Such engines could then use INAP (message- or method-based) or Parlay and JAIN to interact with actually deployed equipment. Again, this approach enhances the open characteristics of the foreseen telecommunications environment by allowing even the service users themselves to code their services and injecting them into the network exactly in the same manner that mobile service components of the distributed IN where injected into the network. In either case, code mobility is exploited. In the call processing language or XML case, mobility accrues due to the interpreted or scripted nature of the code, in the distributed IN case, due to the semiinterpreted characteristics of the Java language and the Java virtual machine architecture. Service users could program their services themselves since the call processing language expresses telephony services logic at a very simple and abstract level relying on the actual execution engine for the translation of this logic to the appropriate set of INAP or Parlay commands. Alternatively, a graphical front end could be used where elementary building blocks are arranged in a two-dimensional canvas and then the appropriate call processing language or XML code is generated. This is not different from the service creation environments used in IN (both the traditional and the distributed ones), with the exception that such tools were very elaborate and expensive, requiring trained personnel to use them, whereas graphical front ends such as those described above can be simple and inexpensive as they need only produce a set of call processing language statements and not actual code.

7. CONCLUSIONS

The distributed intelligent network encompasses a wide range of architectures each adopting different implementation options and exhibiting different tradeoffs among the various properties that characterize such a system. The common elements in all such systems are the use of distributed processing technologies for the control plane (either INAP in the form of CORBA methods or Parlay/JAIN), mobile service code (Java components or interpreted call processing language or XML statements), and a more open service provisioning model (services executed at the premises of their users, directly managed and maintained or even coded by them).

Distributed IN thus represents the next stage in the evolution of traditional IN architectures that witnesses a new business model with the network operator concentrating on the role of the telecommunication infrastructure provider while more and more of the service provisioning role is assumed by external actors.

BIOGRAPHIES

Iakovos S. Venieris (venieris@cs.ntua.gr) was born in Naxos, Greece, in 1965. He received a Dipl.-Ing. from the University of Patras, Greece in 1988, and a Ph.D. from the National Technical University of Athens (NTUA), Greece, in 1990, all in electrical and computer engineering. In 1994 he became an assistant professor in the Electrical and Computer Engineering Department of NTUA where he is now an associate professor. His research interests are in the fields of broadband communications, Internet, mobile networks, intelligent networks, signaling, service creation and control, distributed processing, agents technology, and performance evaluation. He has over 150 publications in the above areas and has contributed to standardization bodies (ETSI, ITU-T, OMG, and IETF). He has participated in several European Union and national R&D projects. He is an associate editor of the *IEEE Communication Letters*, member of the editorial board of *Computer Communications* (Elsevier), and has been a guest editor for *IEEE Communications Magazine*. He is a reviewer for several journals and has been member of the technical program committee and session chairman of several international conferences. Dr. Venieris is the editor and coauthor of two international books on *Intelligent Broadband Networks* (Wiley 1998) and *Object oriented Software Technologies in Telecommunications* (Wiley 2000).

Dr. Ing. Menelaos Perdikeas (mperdikeas@semantix.gr) was born in Athens, Greece, in 1974 and received a Dipl.-Ing. in computer engineering and informatics (Summa Cum Laude) from the University of Patras, Greece, in 1997 and a Ph.D. in telecommunications engineering from the National Technical University of Athens in 2001. He has over 25 publications in international journals and conferences and has received a number of national and international distinctions, among which, the "D. Chorafas Award for Academic Achievement" for the year 2001. He is coauthor of the book *Object Oriented Software Technologies in Telecommunications* published by Wiley. His interests include intelligent network systems, service provisioning architectures using distributed object technologies, and the application of aspect-oriented programming methodologies and patterns in the development of telecommunications software. Dr. Perdikeas is among the founders of Semantix S. A., a Greek software company.

BIBLIOGRAPHY

1. I. S. Venieris and H. Hussman, eds., *Intelligent Broadband Networks*, Wiley, 1998.
2. ITU-T, *Distributed Functional Plane for Intelligent Network Capability Set 2*, Recommendation Q.1224.
3. M. Breugst and T. Magedanz, Impacts of mobile agent technology on mobile communications system evolution, *IEEE Pers. Commun. Mag.* **5**(4): (Aug. 1998).
4. Object Management Group Telecom Specifications, *Interworking between CORBA and Intelligent Network Systems*, *http://www.omg.org*.

5. F. Chatzipapadopoulos, M. Perdikeas, and I. Venieris, Mobile agent and CORBA technologies in the broadband intelligent network, *IEEE Commun. Mag.* **38**(6): 116–124 (June 2000).
6. I. Venieris, F. Zizza, and T. Magedanz, eds., *Object Oriented Software Technologies in Telecommunications*, Wiley, 2000.
7. M. Perdikeas and I. Venieris, Parlay-based service engineering in a converged Internet-PSTN environment, *Comput. Networks* **35**(6): 565–578 (2001).

DIVERSITY IN COMMUNICATIONS

MOHSEN KAVEHRAD
Pennsylvania State University
University Park, Pennsylvania

1. INTRODUCTION

In designing a reliable communication link, the system must be planned around the chosen transmission medium referred to as the *channel*. The disturbances of the medium must be taken into account in the process of encoding the signal at the transmitting end, and in the process of extracting the message from the received waveform at the receiving end. Once a satisfactory characterization of the anticipated channel disturbances has been made, the message encoding chosen for transmission must be designed so that the disturbances will not damage the message beyond recognition at the receiving end. With a corrupted message at hand, the receiving system must be prepared to operate continuously in the presence of the disturbances and to take maximum advantage of the basic differences between the characteristics of messages and of disturbances.

In this article, we assume that the encoded form in which the message is to be transmitted has been selected, and that the encoded form has been translated into a radiofrequency (RF) or lightwave signal by an appropriate modulation technique, such as by varying some distinguishable parameter of a sinusoidal carrier in the RF or optical frequency spectrum. Improvements in system performance can be realized only through the utilization of appropriate corrective signal processing measures. Of primary interest here will be what is widely known as *diversity techniques* [1,2] as countermeasures for combating the effects of loss of received signal energy in parts or over its entire transmission bandwidth, termed as signal fading.

In many practical situations, one seeks economical ways of either transmitting and/or receiving signals in such a way that the signal is never completely lost as a result of transmission disturbances. This has been traditionally the case, in particular, in wireless communications. Ideally, one would like to find transmission methods that are negatively correlated in the sense that the loss of signal in one channel is offset by the guaranteed presence of signal in another channel. This can occur in some diversity systems, such as those that utilize antennas at different elevations in order to minimize the received signal loss of energy. Also, in Section 4.3 of this article the same scenario (negative correlation) applies. In a way, expert investment firms, claiming to provide a diversified portfolio to investors, try to do the same. They opt for capital investment from among those economy sectors whose mutual fund return values are to some degree negatively correlated or at least fluctuate, independently from one sector to another. Consequently, over a long time period, there will be a net gain associated with the diversified portfolio.

The principles of diversity combining have been known in the radio communication field for decades; the first experiments were reported in 1927 [1]. In diversity transmission techniques, one usually settles for fluctuations of signal transmissions over each channel that are more or less uncorrelated with those in other channels and the simultaneous loss of signal will occur rarely over a number of such channels. In order to make the probability of signal loss as low as possible, an effort is made to find many channels that are either statistically independent or negatively correlated. This may be performed over the dimensions of time, frequency and space in a wireless system. For this purpose, it is occasionally possible to use two different polarizations on the receiving antennas, or receivers at several different angles of arrival for the electromagnetic wavefront, or to place antennas in several different spatial locations (spatial diversity) or to transmit the signal over several widely separated carrier frequencies or at several widely separated times (time diversity). The term "diversity improvement" or "diversity gain" is commonly employed to describe the effectiveness of various diversity configurations. There is no standard definition for the effectiveness of diversity reception techniques. One common definition is based on the significance of diversity in reducing the fraction of the time in which the signal drops below an unusable level. Thus, one may define an *outage rate* at some specified level, usually with respect to the mean output noise level of the combiner of the diversity channel outputs. In the rest of this section, the problem of correlation among such multiport channels is discussed.

Assume that a number of terminal pairs are available for different output signals $y_i(t)$ from one or more input signals $x_j(t)$. When frequency (or time) diversity is used, there is no mathematical distinction between a set of multi-terminal-pair channels, each centered on the different carriers (or different times) and a single channel whose system function encompasses all frequencies (or all times) in use. In practice, since one may use physically different receivers, the use of separate system functions to characterize the outputs from each receiver is useful. If space diversity is used, one may be concerned with the system function that depends on the spatial position as a continuous variable.

The cross-correlation function between the outputs of two diversity channels when the channels are both excited by the same signal $x_j(t)$ is fully determined by a complete knowledge of the system functions for each channel alone. In view of the random nature of the channels, the most that can be done to provide this knowledge in practice is to determine the joint statistical properties of the channels.

The signal diversity techniques mentioned above do not all lead to independent results. One must, therefore, recognize those diversity techniques that are dependent in order to avoid trying to "squeeze blood out of a bone." For example, one cannot apply both frequency *and* time diversity to the same channel in any wide sense. In fact one might argue that complete distinction between frequency and time diversity may be wholly artificial, since the signal designer usually has a given time–bandwidth product available that can be exploited in conjunction with the channel characteristics. Another pair of diversity channels that are not necessarily independent of each other are distinguished as angular diversity and space diversity channels. Consider an array of n isotropic antennas that are spaced a sufficient distance apart so that the mutual impedances between antennas can be ignored. If the transmission characteristics of the medium are measured between the transmitting-antenna terminals and the terminals of each of the n receiving antennas, then n channel system functions will result, each of which is associated with one of the spatially dispersed antennas. If the antenna outputs are added through a phase-shifting network, the resultant array will have a receiving pattern that can be adjusted by changing the phase shifting network to exhibit preferences for a variety of different angles of arrival. The problem of combining the outputs of the array through appropriate phase shifters, in order to achieve major lobes that are directed at favorable angles of arrival would be considered a problem in angular diversity, while the problem of combining the outputs of the elements in order to obtain a resultant signal whose qualities are superior to those of the individual outputs is normally considered as the problem of space diversity combining, yet both can lead to the same end result. Signal diversity techniques and methods of combining signals from a number of such channels are discussed in the next section.

2. DIVERSITY AND COMBINING TECHNIQUES

Diversity is defined here as a general technique that utilizes two or more copies of a signal with varying degrees of disturbance to achieve, by a selection or a combination scheme, a consistently higher degree of message recovery performance than is achievable from any one of the individual copies, separately. Although diversity is commonly understood to aim at improving the reliability of reception of signals that are subject to fading in the presence of random noise, the significance of the term will be extended here to cover conceptually related techniques that are intended for other channel disturbances.

The first problem in diversity is the procurement of the "diverse" copies of the disturbed signal, or, if only one copy is available, the operation on this copy to generate additional "diversified" copies. When the signal is disturbed by a combination of multiplicative and additive disturbances, as in the case of fading in the presence of additive random noise, the transmission medium can be tapped for a permanently available supply of diversity copies in any desired numbers.

Propagation media are generally time-varying in character, and this causes transmitted signals to fluctuate randomly with time. These fluctuations are usually of three types:

1. Rapid fluctuations, or fluctuations in the instantaneous signal strength, whose cause can be traced to interference among two or more slowly varying copies of the signal arriving via different paths. This may conveniently be called *multipath fading*. If the multiple paths are resolved by the receiver [2], fading is called *frequency-selective*. Otherwise, it is called *single-path* (flat) *fading*. This type of fading often leads to a complete loss of the message during time intervals that are long even when compared with the slowest components of the message. It is observed, however, that if widely spaced receiving antennas are used to pick up the same signal, then the instantaneous fluctuations in signal-to-noise ratio (SNR) at any one of the receiving sites is almost completely independent of the instantaneous fluctuations experienced at the other sites. In other words, at times when the signal at one of the locations is observed to fade to a very low level, the same signal at some other sufficiently distant site may very well be at a much higher level compared to its own ambient noise. This type of variation is also referred to as *macrodiversity*. Signals received at widely spaced time intervals or widely spaced frequencies also show almost completely independent patterns of instantaneous fading behavior. Nearly uncorrelated multipath fading has also been observed with signal waves differing only in polarization. It will be evident that by appropriate selection or combining techniques, it should be possible to obtain from such a diversity of signals a better or more reliable reception of the desired message than is possible from processing only one of the signals all the time.

2. The instantaneous fluctuations in signal strength occur about a mean value of signal amplitude that changes relatively so slowly that its values must be compared at instants separated by minutes to hours before any significant differences can be perceived. These changes in short-term (or "hourly") mean signal amplitude are usually attributable to changes in the attenuation in the medium that the signals will experience in transit between two relatively small geographic or space locations. No significant random spatial variations in the received mean signal amplitude are usually perceived in receiving localities that could be utilized for diversity protection against this attenuation fading or, as sometimes called, "fading by shadowing". However, it is possible to combat this type of fading by a feedback operation in which the receiver informs the transmitter about the level of the received mean signal amplitude, thus "instructing" it to radiate an adequate amount of power. But the usual practice is to anticipate the greatest attenuation to be expected at the design stage and counteract it by appropriate antenna design and adequate transmitter power.

3. Another type of attenuation fading is much slower than that just described. The "hourly" mean signal levels are different from day to day, just as they are from hour to hour in any single day. The mean signal level over one

day changes from day to day and from month to month. The mean signal level for a period of one month changes from month to month and from season to season, and then there are yearly variations, and so on. As in the case of the "hourly" fluctuations in paragraph 2, the long-term fluctuations are generally caused by changes in the constitution of the transmission medium, but the scale and duration of these changes for the long-term fluctuations are vastly greater than those for the "hourly" changes. Diversity techniques per se are ineffective here.

In addition to the instantaneous-signal diversity that can be achieved by seeking two or more separate channels between the transmitting and receiving antennas, certain types of useful diversity can also be achieved by appropriate design of the patterns of two or more receiving antennas placed essentially in the same location (microdiversity), or by operations in the receiver on only one of the available replicas of a disturbed signal. The usefulness of "receiver diversity" of a disturbed signal will be demonstrated in examples of the next section. The application discussed in Section 4.1 demonstrates the use of diversity (under certain circumstances) from a delayed replica of the desired signal arriving via a different path of multiple fading paths. The latter is referred to as *multipath diversity*, where the same message arrives at distinct arrival times at a receiver equipped to resolve the multipath into a number of distinct paths [3] with different path lengths. The example presented in Section 4.2 shows application of diversity for the case in which the interference from some other undesired signal source is the cause of signal distortion.

The second problem in diversity is the question of how to utilize the available disturbed copies of the signal in order to achieve the least possible loss of information in extracting the desired message. The techniques that have thus far been developed can be classified into (1) switching, (2) combining, and (3) a combination of switching and combining. These operations can be carried out either on the noisy modulated carriers (predetection) or on the noisy, extracted modulations that carry the message specifications (postdetection). In any case, if K suitable noisy waveforms described by $f_1(t), f_2(t), \ldots, f_k(t)$ are available, let the kth function be weighted by the factor a_k, and consider the sum

$$f(t) = \sum_{k=1}^{K} a_k f_k(t) \qquad (1)$$

In the switching techniques only one of the a_k values is different from zero at any given time. In one of these techniques, called *scanning diversity*, the available waveforms are tried one at a time, in a fixed sequence, until one is found whose quality exceeds a preset threshold. That one is then delivered for further processing in order to extract the desired message, until its quality falls below the preset threshold as a result of fading. It is then dropped and the next one that meets the threshold requirement in the fixed sequence is chosen. In scanning diversity, the signal chosen is often not the best one available. A technique that examines the K available signals simultaneously and selects only the best one for delivery is conceptually (although not always practically) preferable. Such a technique is referred to as *optimal selection diversity*.

In the combining techniques, all the available noisy waveforms, good and poor, are utilized simultaneously as indicated in Eq. (1); the a_k values are all nonzero all the time. Of all the possible choices of nonzero a_k values, only two are of principal interest. First, on the assumption that there is no a priori knowledge or design that suggests that some of the $f_k(t)$ values will always be poorer than the others, all the available copies are weighted equally in the summation of Eq. (1) irrespective of the fluctuations in quality that will be experienced. Thus, equal mean values of signal level and equal RMS (root-mean-square) values of noise being assumed, the choice $a_1 = a_2 = \cdots = a_k$ is made, and the technique is known as *equal-weight* or *equal-gain combining*. The second possible choice of nonzero weighting factors that is of wide interest is one in which a_k depends upon the quality of $f_k(t)$ and during any short time interval the a_k values are adjusted automatically to yield the maximum SNR for the sum $f(t)$. This is known as *maximal ratio combining*.

In the alternate switching–combining technique a number of the a_k values up to $K - 1$ can be zero during certain time intervals because some of the available signals are dropped when they become markedly noisier than the others. This approach is based on the fact that the performance of an equal-gain combiner will approximate that of the maximal ratio combiner as long as the SNRs of the various channels are nearly equal. But if any of the SNRs become significantly inferior to the others, the overall SNR can be kept closer to the maximum ratio obtainable if the inferior signals are dropped out of the sum $f(t)$.

Over a single-path fading channel, implementation of selection combining does not require any knowledge about the channel, that is, no channel state information (CSI) is necessary at the receiver, other than that needed for coherent carrier recovery, if that is employed. The receiver simply selects the diversity branch that offers the maximum SNR. For equal-gain combining some CSI estimation can be helpful in improving the combiner performance. For example, in a multipath diversity receiver, the maximum delay spread of a multipath fading channel, which is indicative of the channel memory length, can guide the integration time in equal-gain combining, such that the combiner can collect the dispersed signal energy more effectively and perhaps avoid collecting noise over low signal energy time intervals [4]. Maximal ratio combining (MRC) performs optimally, when CSI estimates on both channel phase and multiplicative path attenuation coefficient are available to the combiner. MRC, by using these estimates and proper weighting of the received signal on each branch, yields the maximum SNR ratio for the sum $f(t)$, compared to the selection and equal-gain combining. In the MRC case, diversity branches that bear a strong signal are accentuated and those that carry weak signals are suppressed such that the total sum $f(t)$ will yield the maximum SNR [2]. This is similar to the philosophy that in a free-market society, by making the rich richer, the society as a whole is better off, perhaps because of increased productivity. Needless to say, in a society as such, laws and justice must also protect the welfare of the needy in order to bring social stability.

Where there is rich scattering, for all multipath fading cases, a RAKE receiver [5] can yield MRC performance. A simplified equal-gain combiner can also be used in some cases, achieving a somewhat lesser diversity gain [4]. However, in digital transmission, if the channel maximum delay spread exceeds the duration of an information bit, intersymbol interference is introduced which needs to be dealt with.

2.1. Statistical Characterization of Fading

For analytical purposes, combined time and frequency, three-dimensional presentations of recordings of fading signals envelopes are obtained through elaborate measurements. These are usually treated as describing a random variable whose statistical properties are determined from fraction-of-time distributions and are hence intimately related to the duration of the interval of observation. The probability distribution functions of such random variables can be considered to characterize a type of stochastic process for which ergodicity theorem applies. According to this theorem, time and distribution averages of random variables described by fraction-of-time distributions are one and the same thing, and they can be used interchangeably depending on expediency.

It is important to note here that although the rate at which the envelope of a received carrier fluctuates may often appear to be high, it is usually quite slow in comparison with the slowest expected variations in the message waveform. In other words, the envelope of the carrier is usually approximately constant when observed over intervals of time that extend over the duration of the longest message element, or over a few periods of the lowest-frequency component in the message spectrum. On the timescale of the fading envelope, such time intervals are then too short for any significant changes in the envelope to occur but not so short that the details of the message waveform are perceived in averaging over the interval. The probability distribution of a fading envelope is usually determined from samples of short time duration, and the results are presented in histograms. Such histograms are invariably compared with simple mathematical curves such as the Rayleigh density and distribution functions or some other functions whose shapes resemble the appearance of the experimental presentations. The fit of the experimental distributions to the Rayleigh distribution is most often excellent for long-range SHF and UHF tropospheric transmission, quite often so for short-range UHF and for ionospheric scatter and reflection of VHF and HF. Accordingly, the Rayleigh fading model is almost always assumed in theoretical treatments, although it is well known that serious deviations from it arise in some situations. According to this model, if a sinusoid of frequency ω_c is radiated at the transmitting end, it will reach a receiver in the form:

$$R(t) = X(t)\cos[\omega_c t + \phi(t)] \tag{2}$$

where $X(t)$ is a slowly fluctuating envelope (or instantaneous amplitude) whose possible values have a probability density function (PDF)

$$p(X) = \frac{2X}{x^2}\exp\left[-\frac{X^2}{x^2}\right] \quad \text{for} \quad X \geq 0$$
$$= 0 \quad \text{otherwise} \tag{3}$$

where x^2 is the mean-square value of X during the small time interval discussed earlier.

No explicit assumptions are usually made concerning the phase $\phi(t)$ beyond the fact that its fluctuations, like those of $X(t)$, are slow compared to the slowest expected variations in the message waveform. But one possible and sometimes convenient assumption to make is that $\phi(t)$ fluctuates in a random manner and can assume all values between 0 and 2π in accordance with the probability density function:

$$p(\phi) = \frac{1}{2\pi} \quad \text{for} \quad 0 \leq \phi \leq 2\pi$$
$$= 0 \quad \text{otherwise} \tag{4}$$

The convenience that results from the assumption of a uniformly distributed phase is due to the fact that $R(t)$ of Eq. (2) can now be viewed as a sample function of a narrowband Gaussian process with zero mean and variance $\frac{x^2}{2}$. One may envision, over a multipath channel, many unresolved scattered rays combine in order to give rise to a Gaussian envelope, $R(t)$.

A Rayleigh PDF, as presented, has a single maximum that tends to occur around small values of the random variable X. Thus, in transmitting a binary modulated signal over a Rayleigh fading channel and receiving the signal in additive white Gaussian noise (AWGN), the average bit error rate (BER) for the detected bits tends to be inversely proportional to the receiver SNR, as shown in Fig. 1. This is shown for several binary modulation techniques. This is quite a slow decrease compared to the same, transmitted over an AWGN channel that only suffers with additive white Gaussian noise. For the latter, the BER drops as a function of SNR, exponentially. Naturally, the Rayleigh fading channel model demands significantly higher transmitted power for delivering the bits as reliably as over an AWGN channel. Now, consider use of diversity combining in transmitting a binary modulated signal over a Rayleigh fading channel when the signal is received in AWGN. The average BER for the detected bits now tends to decrease inversely as a function of SNR raised to the power of diversity order, L, as shown in Fig. 2. This is quite a remarkable improvement. In a way, diversity combining process modifies the Rayleigh PDF to look more like a truncated Gaussian PDF, as the order of diversity increases. Thus, loosely speaking, the performance in BER versus SNR for binary transmission over a Rayleigh fading channel starts to resemble that of transmission over an AWGN channel when diversity combining is employed.

Another case of interest in transmission over fading channels is when the received signal in AWGN has a strong deterministic (nonrandom) component. This will contribute to moving the received signal mean away from the region of small signals to a range of rather large

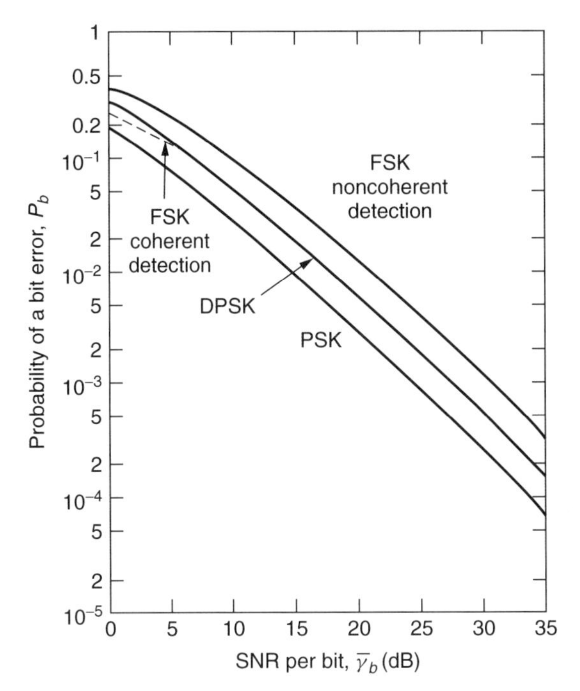

Figure 1. Performance of binary signaling on a Rayleigh fading channel (from Proakis [2]).

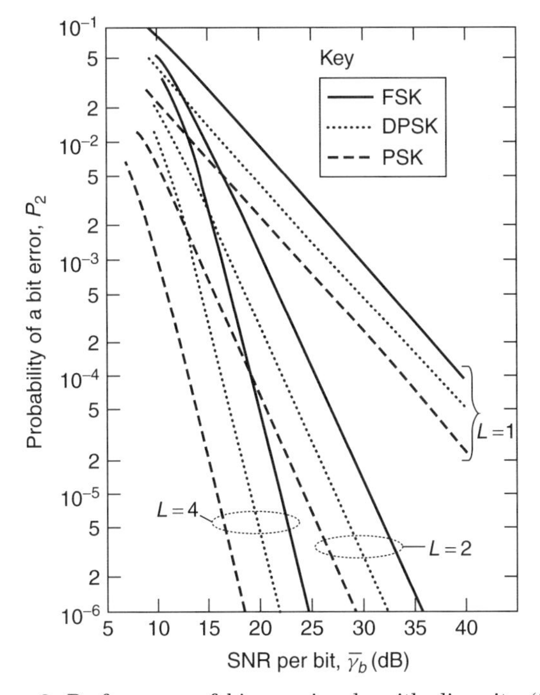

Figure 2. Performance of binary signals with diversity (from Proakis [2]).

signal values. Statistically, the Rayleigh PDF is modified to a Rice PDF [2] that is much milder in the manner by which it affects the transmitted signal. Transmission over a short (~3 km) line-of-sight microwave channel is often subject to Rice fading. Finally, another PDF that is frequently used to describe the statistical fluctuations of signals received from a multipath fading channel is the Nakagami m distribution [2]. The parameter m is defined as the ratio of moments, called the *fading figure*. By setting $m = 1$, the Nakagami PDF reduces to a Rayleigh PDF. As m increases to values above $m = 1$, transmission performance over a Nakagami channel improves. In a way, this is similar to what happens on a Rice channel. That is, receiver is provided with a stronger signal average, as m increases above the $m = 1$ value.

3. DIVERSITY THROUGH CHANNEL CODING WITH INTERLEAVING

In general, time and/or frequency diversity techniques may be viewed as a form of trivial repetition (block) coding of the information signal. The combining techniques can then be considered as soft-decision decoding of the trivial repetition codes. Intelligent (nontrivial) coding in conjunction with interleaving provides an efficient method of achieving diversity on a fading channel, as emphasized, for example, by chase [6]. The amount of diversity provided by a code is directly related to the code minimum distance, $d_{\min}$. With soft-decision decoding, the order of diversity is increased by a factor of $d_{\min}$, whereas, if hard-decision decoding is employed, the order of diversity is increased by a factor of $\frac{d_{\min}}{2}$. Note that, although coding in conjunction with interleaving can be an effective tool in achieving diversity on a fading channel, it cannot help the signal quality received through a single stationary antenna located in a deep-fade null. The interleaving will not be helpful, since in practice, the interleaving depth cannot be indefinite.

In the next section, we present some examples illustrating the benefits of diversity combining in various communications applications.

4. APPLICATION EXAMPLES

Three examples, illustrating benefits of diversity techniques in practical communications systems, are presented in this section.

4.1. Wideband Code-Division-Multiple-Access Using Direct-Sequence Spread-Spectrum (DSSS) Communications

One application [4] treats a wireless cellular scenario. This represents the up and downstream transmission mode, from user to base station (BS) and from the base to user, of a wireless local-area network (LAN) with a star architecture. Each user has a unique DSSS code and a correlator exists for each user in a channel bank structure at the base station. The output of the bank of correlators is fed to the usual BS circuitry and call setup is handled using standard BS features. Average power control is used in this system to avoid the classical near/far problem. In wireless LAN we have a severe multipath fading problem. It is really a classical Rayleigh multipath, fading channel scenario. The role of asynchronous transmission is clear — a user transmits at random. The signal arrives at the correlator bank and is detected, along with interfering signals. Because the broad bandwidth of a DSSS signal can indeed exceed the channel coherence band (this is the channel band over which all frequency components of the transmitted signal are treated in a correlated manner), there is inherent diversity in transmission that can be exploited as multipath diversity [3] by a correlation receiver. The psuedonoise (PN) sequences used as direct sequence spreading codes in this application are indeed trivial repeat codes. By exclusive-OR addition of a PN sequence to a data bit, the narrowband of data is spread out to the level of the wide bandwidth of PN sequence. A correlation receiver that knows and has available the matching PN sequence, through a correlation operation, generates correlation function peaks representing the narrowband information bits with a correlation function base, in time, twice the size of a square PN sequence pulse, called a *PN chip*. In this application, the correlator, or matched-filter output, will be a number of resolved replicas of the same transmitted information bit, displayed by several correlation peaks. The number of correlation peaks representing the same information bit corresponds to the diversity order, in this application. Many orders of diversity can be achieved this way.

The replicas obtained this way may now be presented to a combiner for diversity gain. A simple equal gain combiner has been adopted [4] that is by far simpler in implementation than a RAKE receiver [5]. The multipath diversity exploitation in conjunction with multiantenna space diversity [7] establishes a foundation for joint space and time coding.

4.2. Indoor Wireless Infrared (IR) Communications

The purpose of using infrared wireless communication systems in an indoor environment is to eliminate wiring. Utilization of IR radiation to enable wireless communication has been widely studied and remote-control units used at homes introduce the most primitive applications of this type of wireless systems. A major advantage of an IR system over an RF system is the absence of electromagnetic wave interference. Consequently, IR systems are not subject to spectral regulations as RF systems are. Infrared radiation, as a medium for short-range indoor communications, offers unique features compared to radio. Wireless infrared systems offer an inherent spatial diversity, making multipath fading much less of a problem. It is known that the dimension of the coherence area of a fully scattered light field is roughly of the order of its wavelength [8].

This is due to the fact that the receive aperture diameter of a photodiode by far exceeds the wavelength of an infrared waveform. Therefore, the random path phase of a fading channel is averaged over the photo-receiver surface. Hence, the signal strength fluctuations that are caused by phase cancellations in the RF domain are nonexistent in the optical domain. An optical receiver actually receives a large number (hundreds of thousands) of independent signal elements at different locations on the receiving aperture of the photodiode. This in fact provides spatial diversity, which is very similar to employing multiple, geographically separated antennae in an RF fading environment. In summary, because of the inherent diversity channels, the frequency-selective fading effect at the optical carrier frequency level is not a problem in an IR system.

Since infrared transmission systems use an optical medium for data transmission, they have an inherent potential for achieving a very high capacity level. However, in order to make this feasible, the communication system design has to offer solutions to the problems associated with IR propagation in a noisy environment. Various link designs may be employed in indoor wireless infrared communication systems. The classification is based on the degree of directionality and existence of a line-of-sight path between a transmitter and a receiver. In one configuration, instead of transmitting one wide beam, multi-beam transmitters are utilized [9]. These emit multiple narrow beams of equal intensity, illuminating multiple small areas (often called *diffusing spots*) on a reflecting surface. Each beam aims in a prespecified direction. Such a transmitting scheme produces multiple-line-of-sight (as seen by the receiver) diffusing spots, all of equal power, on an extended reflecting surface, such as a ceiling of a room. Each diffusing spot in this arrangement may be considered as a secondary line-of-sight light source having a Lambertian illumination pattern [10]. Compared to other configurations, this has the advantage of creating a regular grid of diffusing spots on the ceiling, thus distributing the optical signal as uniformly as possible within the communication cell as shown in Fig. 3. A direction diversity (also known as angle diversity) receiver that utilizes multiple receiving elements, with each element pointed at a different direction, is used, in order to provide a diversity scheme for optimal rejection of ambient noise power from sunlight or incandescent light, for example, and to substantially reduce the deleterious effects of time dispersion via multiple arrivals of reflecting light rays, causing intersymbol interference in digital transmission. The composite receiver consists of several narrow field-of-view (FoV) elements replacing a single element wide-FoV receiver. The receiver consists of more than one element in order to cover several diffusing spots, thus ensuring uninterrupted communication in case some of the transmitter beams are blocked. Additionally, a multiple-element receiver provides diversity, thus allowing combining of the output signals from different receiver elements, using optimum combining methods. An increased system complexity is the price that one has to pay to escape from restrictions of line-of-sight links, retaining the potential for a high capacity wireless communication system.

For indoor/outdoor wireless transmission systems, use of multiple antennas at both transmit and receive sides to achieve spatial diversity at RF has gained a significant amount of attention. This is motivated by the lack of

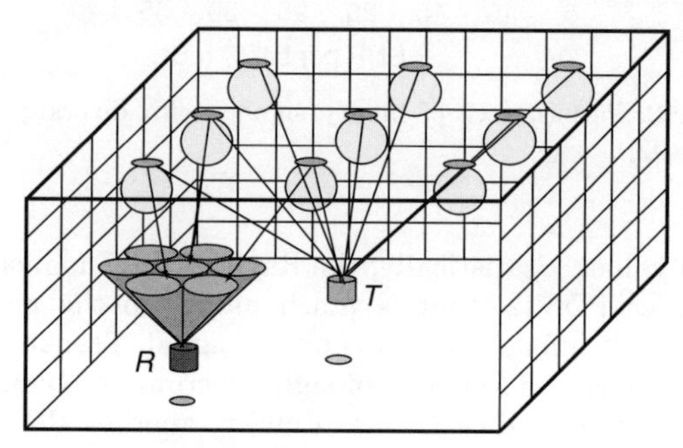

Figure 3. Multispot diffusing configuration (T—transmitter; R—receiver).

available bandwidth at the low-frequency end of the radio spectrum. The approach, in a way, is similar to the IR system described earlier. The capacity increase and spatial multiplexing for high-data-rate transmissions via multiple transmit antennas have been illustrated in Ref. 11. Earlier, transmission of orthogonally-polarized transmitted digitally modulated RF waveforms was introduced to achieve capacity increase over multipath fading channels of point-to-point high-data-rate digital radio routes [12].

4.3. Polarization-Insensitive Fiberoptic Communications

In a coherent optical receiver, receiving a modulated lightwave on a single-mode fiber, the polarization state of the received optical signal must be matched with that of the receiver local laser source signal. A mismatch reduces (perhaps drastically) the received signal strength. Unless the polarization states of the light fields are controlled carefully, receiver performance degradation is unavoidable. The problem encountered here is very similar to the flat fading on a single-path fading channel. Occasionally, the entire band of the received signal is severely attenuated. In an earlier study [13] we examined a polarization-insensitive receiver for binary frequency shift keying (FSK) transmission with discriminator demodulation. The polarization-state-insensitive discriminator receiver is shown in Fig. 4. The two branches carry horizontally and vertically polarized optical beams obtained through a polarization beamsplitter. The information-carrying optical beams are subsequently heterodyne demodulated down to FSK-modulated IF signals and are received by the discriminators for demodulation. The demodulated baseband signals are then combined; thereby a polarization-independent signal is obtained. This is yet another example of diversity combining of equal-gain type.

5. CONCLUDING REMARKS

Given proper operating condition of the equipment, the reliability of a communication system is basically determined by the properties of the signal at the receiving end. We have concentrated in this article on diversity and the effects it may have upon the signal reliability. It is established that "diversification" offers a gain in reliable signal detection. However, a wireless channel offers endless possibilities over a multiplicity of dimensions. Diversity is only one way of introducing a long-term average gain into the detection process. More recently, the availability of low-cost and powerful processors and the development of good channel estimation methods have rejuvenated an interest in adaptive transmission rate techniques with feedback. This new way of thinking is termed *opportunistic communication*, whereby dynamic rate and power allocation may be performed over the dimensions of time, frequency, and space in a wireless system. In a fading (scattering) environment, the channel can be strong sometimes, somewhere, and *opportunistic* schemes can choose to transmit in only those channel states. Obviously, some channel state information is required for an opportunistic communication approach to be successful. Otherwise, it becomes like shooting in the dark. This is in some respects similar to building a financial investment portfolio of stocks, based on some "insider's" information. Clearly, it results in more gain compared to traditional methods of building a diversified portfolio of stocks, based on long-term published trends. Similarly, one would expect a great loss, if the "insider's" information turned out to be wrong. Consequently, opportunistic communications based on wrong channel states will result in a great loss in the wireless network capacity. *Thus, in those wireless applications where reliable channel states may easily be obtained, it is possible to achieve enormous capacities, at a moderate realization complexity.*

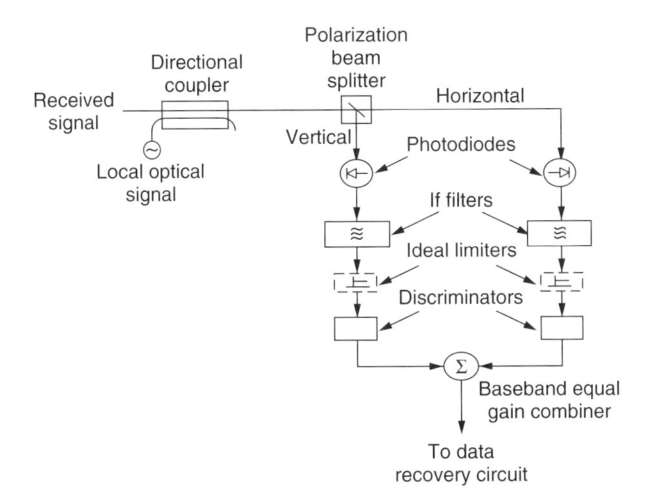

Figure 4. Basic proposed polarization-insensitive receiver.

BIOGRAPHY

Mohsen Kavehrad received his B.Sc. degree in electronics from Tehran Polytechnic, Iran, 1973, his M.Sc. degree from Worcester Polytechnic in Massachusetts, 1975, and his Ph.D. degree from Polytechnic University (Brooklyn Polytechnic), Brooklyn, New York, November 1977 in electrical engineering. Between 1978 and 1989, he worked on telecommunications problems for Fairchild Industries, GTE (Satellite and Labs.), and AT&T Bell Laboratories. In 1989, he joined the University of Ottawa, Canada, EE Department, as a full professor. Since January 1997, he has been with The Pennsylvania State University EE Department as W. L. Weiss Chair Professor and founding director of the Center for Information and Communications Technology Research. He is an IEEE fellow for his contributions to wireless communications and optical networking. He received three Bell Labs awards for his contributions to wireless communications, the 1991 TRIO feedback award for a patent on an optical interconnect, 2001 IEEE VTS Neal Shepherd Best Paper Award, 5 IEEE Lasers and Electro-Optics Society Best Paper Awards between 1991–95 and a Canada NSERC Ph.D.-thesis award in 1995, with his graduate students for contributions to wireless systems and optical networks. He has over 250 published papers, several book chapters, books, and patents in these areas. His current research interests are in wireless communications and optical networks.

BIBLIOGRAPHY

1. W. C. Jakes, Jr., *Microwave Mobile Communications*, Wiley, New York, 1974.
2. J. G. Proakis, *Digital Communications*, McGraw-Hill, New York, 1983.
3. G. L. Turin, Introduction to spread-spectrum anti-multipath techniques and their application to urban digital radio, *Proc. IEEE* **68**: 328–353 (March 1980).
4. M. Kavehrad and G. E. Bodeep, An experiment with direct-sequence spread spectrum and differential phase shift keying modulation for indoor, wireless communications, *IEEE J. Select. Areas Commun.* **SAC-5**(5): 815–823 (June 1987).
5. R. Price and P. E. Green, A communication technique for multipath channels, *Proc. IRE* **46**(3): 555–570 (1958).
6. D. Chase, Digital signal design concepts for a time-varying Ricean channel, *IEEE Trans. Commun.* **COM-24**: 164–172 (Feb. 1976).
7. M. Kavehrad and P. J. McLane, Performance of low-complexity channel-coding and diversity for spread-spectrum in indoor, wireless communication, *AT&T Tech. J.* **64**(8): 1927–1966 (Oct. 1985).
8. R. J. Collier et al., *Optical Holography*, Academic Press, New York, 1971.
9. G. Yun and M. Kavehrad, Spot diffusing and fly-eye receivers for indoor infrared wireless communications, *Proc. IEEE Int. Conf. Selected Topics in Wireless Communications*, Vancouver, Canada, 1992, pp. 262–265.
10. J. R. Barry, *Wireless Infrared Communications*, Kluwer, Boston, 1994.
11. G. J. Foschini and M. J. Gans, On limits of wireless communications in a fading environment using multiple antennas, *Wireless Pers. Commun.* 311–335 (June 1998).
12. M. Kavehrad, Baseband cross-polarization interference cancellation for *M*-quadrature amplitude modulated signals over multipath fading radio channels, *AT&T Tech. J.* **64**(8): 1913–1926 (Oct. 1985).
13. M. Kavehrad and B. Glance, Polarization-insensitive frequency shift keying optical heterodyne receiver using discriminator demodulation, *IEEE J. Lightwave Technol.* **LT-6**(9): 1388–1394 (Sept. 1988).

DMT MODULATION

ROMED SCHUR
STEPHAN PFLETSCHINGER
JOACHIM SPEIDEL
Institute of Telecommunications
University of Stuttgart
Stuttgart, Germany

1. INTRODUCTION

1.1. Outline

Discrete multitone (DMT) modulation as well as orthogonal frequency-division multiplex (OFDM) belong to the category of multicarrier schemes. The early ideas go back to the late 1960s and early 1970s [e.g., 1,2]. With the development of fast digital signal processors, the attraction of these techniques increased [e.g., 3–5] and advanced developments have been carried out [e.g., 6,7]. Meanwhile, DMT was introduced as a standard for digital communications on twisted-pair cables [digital subscriber line (DSL)] [8].

The basic idea of multicarrier modulation is to partition a high-rate datastream into a large number of low-rate data signals that are modulated onto different carrier frequencies and are transmitted simultaneously over parallel subchannels. Because of the partition of the datastream, the data rate on each subchannel is much lower than for the original signal. As low-rate signals are much less susceptible to channel impairments, the reception and reconstruction of the subchannel signals at the receiver side is simplified significantly. However, all subchannels have to be received in parallel and have to be processed simultaneously — a requirement that can be met in an economic way only with digital signal processing. Because of the large number of carriers, the subchannel signals can be well adapted to the channel characteristics. As a consequence, multicarrier schemes like DMT offer the ability to maximize the data throughput over frequency-selective channels, such as the telephone subscriber line.

As multicarrier modulation like DMT or OFDM can be interpreted as a further development of frequency-division multiplexing (FDM) [4], we begin with the explanation of the classical FDM principles.

1.2. Frequency-Division Multiplexing (FDM)

With FDM the available bandwidth of the transmission medium is separated into a number of frequency bands in order to transmit various signals simultaneously on the same medium. The principal block diagram is given in Fig. 1. Each band-limited baseband signal $x_\nu(t)$ is modulated onto a carrier $\cos(\omega_\nu t)$ with carrier frequency ω_ν, $\nu = 0, 1, \ldots, N-1$.

The received FDM signal is first bandpass (BP)-filtered with center frequency ω_μ, multiplied by $\cos(\omega_\mu t)$ and then lowpass (LP)-filtered to obtain the demodulated signal $\hat{x}_\mu(t)$.

The FDM signal at the transmitter output is

$$s(t) = \sum_{\nu=0}^{N-1} x_\nu(t)\cos(\omega_\nu t) \tag{1}$$

For the spectra of the transmitter signals, we have

$$s(t) \leftrightarrow S(\omega) \tag{2}$$

$$x_\nu(t)\cos(\omega_\nu t) \leftrightarrow \tfrac{1}{2}X_\nu(\omega-\omega_\nu) + \tfrac{1}{2}X_\nu(\omega+\omega_\nu) \tag{3}$$

$$s(t) \leftrightarrow S(\omega) = \frac{1}{2}\sum_{\nu=0}^{N-1} X_\nu(\omega-\omega_\nu) + X_\nu(\omega+\omega_\nu) \tag{4}$$

The term $S(\omega)$ is shown in Fig. 2 for the simplified case that all $X_\nu(\omega)$ are real-valued. In conventional FDM systems, the frequencies ω_ν have to be chosen in such a way that the spectra of the modulated signals do not overlap. If all baseband signals $X_\nu(\omega)$ have the same

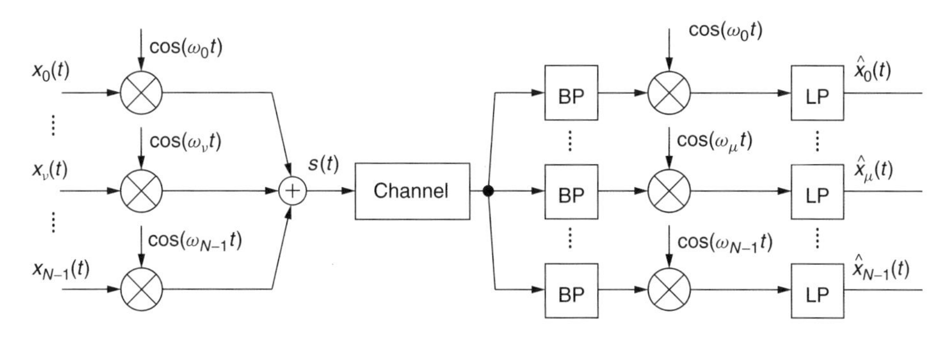

Figure 1. Analog frequency division multiplexing. The baseband signals $x_0(t), \ldots, x_{N-1}(t)$ are modulated on different carriers.

bandwidth, the carrier frequencies are normally chosen equidistantly. However, the FDM system described above wastes valuable bandwidth because (1) some space must be available for the transition between the passband and the stopband of each bandpass filter in Fig. 1; and (2), if $s(t)$ is real, $S(\omega)$ is conjugated symmetric, that is, $S(-\omega) = S^*(\omega)$, where * denotes the conjugate complex operation. As a consequence, the bandwidth of $S(\omega)$ is twice as required theoretically.

The second drawback can be solved by quadrature amplitude modulation (QAM) [9]. With QAM, two independent baseband signals are modulated onto a sine and a cosine carrier with the same frequency ω_ν which gives an unsymmetric spectrum. As a result the spectral efficiency is doubled.

The first drawback of FDM can be overcome by multicarrier modulation, such as by DMT, which allows for a certain overlap between the spectra illustrated in Fig. 2. Of course, spectral overlap in general could lead to nontolerable distortions. So the conditions for this overlap have to be carefully established in order to recover the signal perfectly at the receiver side. This will be done in the next section.

2. MULTICARRIER BASICS

2.1. Block Diagram and Elementary Impulses

Following the ideas outlined in the previous section and applying discrete-time signals $X_\nu[k]$ at the input leads us to the block diagram in Fig. 3.

The impulse modulator translates the sequence $\{\ldots, X_\nu[0], X_\nu[1], \ldots\}$ into a continuous-time function

$$x_\nu(t) = T_S \sum_{k=-\infty}^{\infty} X_\nu[k] \cdot \delta(t - kT_S), \quad \nu = 0, 1, \ldots, N-1 \tag{5}$$

where T_S is the symbol interval, $\delta(t)$ is the Dirac impulse and $\delta[k]$ is the unit impulse with

$$\delta[k] = \begin{cases} 1 & \text{for} \quad k = 0 \\ 0 & \text{for} \quad k \neq 0 \end{cases} \tag{6}$$

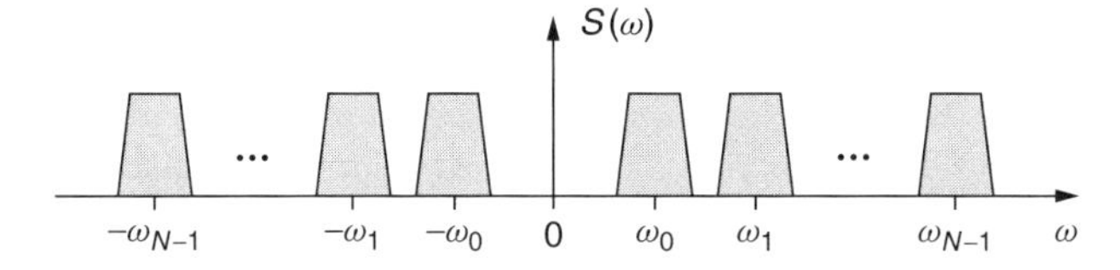

Figure 2. Spectrum $S(\omega)$ of the FDM signal $s(t)$.

The output signal $s(t)$ of the multicarrier transmitter is given by

$$s(t) = T_S \sum_{\nu=0}^{N-1} e^{j\omega_\nu t} \sum_{k=-\infty}^{\infty} X_\nu[k] g(t - kT_S) \tag{7}$$

where $g(t)$ is the impulse response of the impulse shaping filter. The carrier frequencies ω_ν are chosen as integer multiples of the carrier spacing $\Delta\omega$:

$$\omega_\nu = \nu \cdot \Delta\omega, \quad \nu = 0, 1, \ldots, N-1 \tag{8}$$

All practical multicarrier systems are realized with digital signal processing. Nevertheless, we will use the analog model of Fig. 3 in this section, because the analysis can be done more conveniently and the understanding of the principles of multicarrier modulation is easy. Section 3.1 deals with the digital implementation.

The output signal $s(t)$ can be either real or complex. If complex, $s(t)$ can be considered as the complex envelope and the channel is the equivalent baseband channel. As will be shown in Section 3.2 , DMT modulation provides a real-valued output signal $s(t)$. Nevertheless, the following derivations hold for both cases.

The receiver input signal $w(t)$ is demodulated and filtered by the receiver filters with impulse response $h(t)$, resulting in the signal $y_\mu(t)$, $\mu \in \{0, \ldots, N-1\}$. Sampling this signals at the time instants $t = kT_S$ gives the discrete-time output $Y_\mu[k]$. The signal after filtering is given by

$$\begin{aligned} y_\mu(t) &= (w(t)e^{-j\omega_\mu t}) \star h(t) \\ &= \int_{-\infty}^{\infty} w(\tau)e^{-j\omega_\mu \tau} h(t - \tau) d\tau, \quad \mu = 0, \ldots, N-1 \end{aligned} \tag{9}$$

where $\star$ denotes the convolution operation. For the moment, we assume an ideal channel. Thus $w(t) = s(t)$ and we get from (9) with (7) and (8)

$$\begin{aligned} y_\mu(t) = T_S \int_{-\infty}^{\infty} \Bigg(&\sum_{\nu=0}^{N-1} e^{j(\nu-\mu)\Delta\omega\tau} \\ &\times \sum_{k=-\infty}^{\infty} X_\nu[k] g(\tau - kT_S) h(t - \tau) \Bigg) d\tau \end{aligned} \tag{10}$$

Sampling $y_\mu(t)$ at $t = kT_S$ yields the output signal

$$\begin{aligned} Y_\mu[k] = y_\mu(kT_S) = \int_{-\infty}^{\infty} \Bigg(&\sum_{\nu=0}^{N-1} e^{j(\nu-\mu)\Delta\omega\tau} \\ &\times \sum_{l=-\infty}^{\infty} X_\nu[l] g(\tau - lT_S) h(kT_S - \tau) \Bigg) d\tau \end{aligned} \tag{11}$$

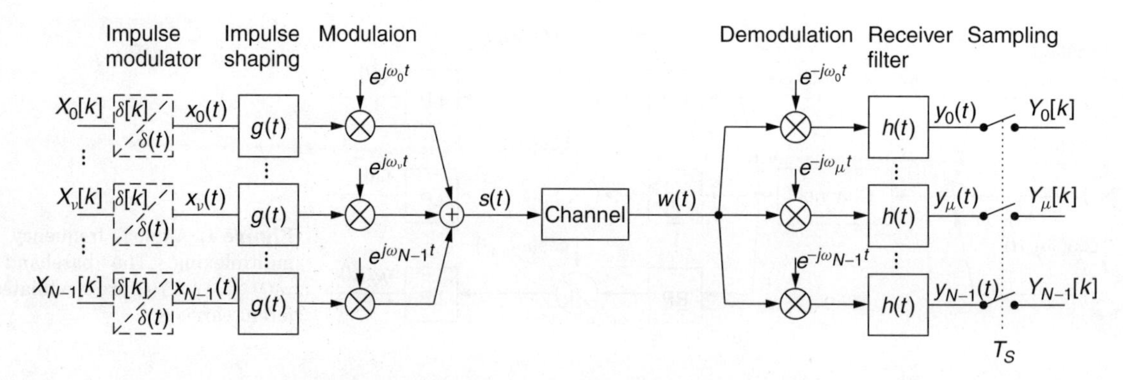

Figure 3. Block diagram of a general multicarrier system.

Now, the target is to recover the sequences $X_\nu[k]$ at the receiver side without distortion. As the given system is linear, we can restrict our analysis to the evaluation of the response to one single transmitted symbol on one subcarrier. We can then make use of the superposition theorem for the general case of arbitrary input sequences. Basically, two types of interference can be seen from Eq. (11):

1. *Intersymbol interference* (ISI), in which a symbol sent at time instant k on subcarrier μ has impact on previous or following samples $y_\mu(lT_S)$ with $l \neq k$.
2. *Intercarrier interference* (ICI), which is the result of crosstalking between different subchannels at time instants kT_S.

Without loss of generality we assume that the system model in Fig. 3 has zero delay, i.e., the filters $g(t)$, $h(t)$ are not causal. We sent one unit impulse $\delta[k]$ at the time $k = 0$ on subcarrier ν:

$$X_i[k] = \delta[\nu - i]\delta[k] \tag{12}$$

The received signal $y_\mu(t)$ is free of any interference at the sampling instants kT_S, if

$$Y_\mu[k] = \delta[\nu - \mu]\delta[k] \tag{13}$$

To gain further insight into the nature of ISI and ICI, we take a closer look at the received signals before they are being sampled. A unit impulse $X_\nu[k] = \delta[k]$ on carrier ν yields the transmitter output signal

$$s(t) = T_S\, g(t) e^{j\nu\Delta\omega t} \tag{14}$$

which is also the input signal $w(t)$ as we assume an ideal channel. We now define the elementary impulse $r_{\nu,\mu}(t)$ as the response $y_\mu(t)$ to the receiver input signal of Eq. (14).

$$\begin{aligned} r_{\nu,\mu}(t) &= T_S \left(g(t)e^{j(\nu-\mu)\Delta\omega t}\right) \star h(t) \\ &= T_S \int_{-\infty}^{\infty} g(\tau)e^{j(\nu-\mu)\Delta\omega\tau} h(t-\tau)\, d\tau \end{aligned} \tag{15}$$

Obviously, $r_{\nu,\mu}(t)$ depends only on the difference $d = \nu - \mu$. Thus we get

$$r_d(t) = T_S(g(t)e^{jd\Delta\omega t}) \star h(t) \tag{16}$$

We can now formulate the condition for zero interference as

$$r_d(kT_S) = \delta[d]\delta[k] \tag{17}$$

This can be interpreted as a more general form of the first Nyquist criterion because it forces not only zero intersymbol interference but also zero intercarrier interference [10,11]. If we set $d = 0$, Eq. (17) simplifies to the Nyquist criterion for single-carrier systems. In the context of multicarrier systems and filterbanks, (17) is also often called an *orthogonality condition* or a *criterion for perfect reconstruction*.

For DMT systems without guard interval $g(t) = h(t)$ holds and they are rectangular with duration T_S. The purpose of the guard interval will be explained in the next section. With the carrier spacing

$$\Delta\omega = \frac{2\pi}{T_S}, \tag{18}$$

we obtain from Eq. (16) the elementary impulses

$$r_0(t) = \begin{cases} 1 - \dfrac{|t|}{T_S} & |t| \leq T_S \\ 0 & \text{elsewhere} \end{cases} \tag{19}$$

$$r_d(t) = \begin{cases} \dfrac{\operatorname{sgn}(t)(-1)^d}{j2\pi d}\left(1 - e^{jd(2\pi/T_S)t}\right) & \text{for } |t| \leq T_S \quad d \neq 0 \\ 0 & \text{elsewhere} \end{cases} \tag{20}$$

These functions are shown in Fig. 4a and give us some insight into the nature of intersymbol and intercarrier interference. We clearly see that the sampling instants $t = kT_S$ do not contribute to any interference. As the elementary impulses are not always zero between the sampling instants, they cause quite some crosstalk between the subchannels. The oscillation of the crosstalk is increased with the distance d between transmitter and receiver subchannel, whereas their amplitude is decreasing proportional to $1/d$.

The impact of interference can also be seen from Fig. 4b, where the eye diagram of the real part of $y_\mu(t)$ is shown for a 16-QAM on $N = 256$ subcarriers; that is, the real and imaginary parts of the symbols $X_\nu[k]$ are taken at random out of the set $\{\pm 1, \pm 3\}$ and are modulated on all subcarriers.

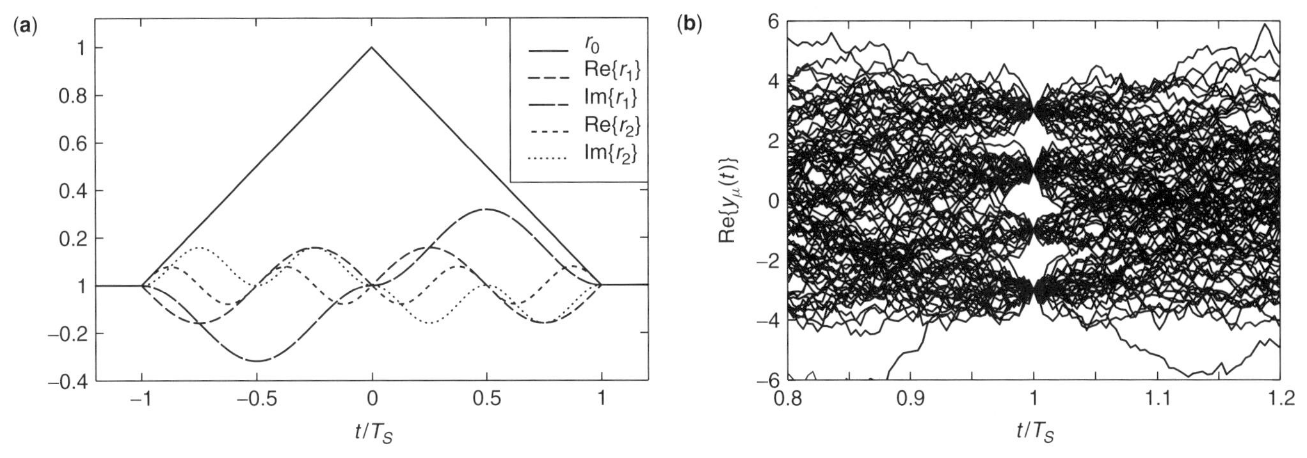

Figure 4. (**a**) Elementary impulses $r_d(t)$ of a multicarrier system with rectangular impulse shapes; (**b**) eye diagram of real part of $y_\mu(t)$ for 16-QAM and $N = 256$ carriers.

Obviously, even with an ideal channel the horizontal eye opening tends to zero, requiring very small sampling jitter and making a correct detection of the transmitted symbols extremely difficult.

2.2. Introduction of a Guard Interval

An effective solution to increase the horizontal eye opening is the introduction of a guard interval. It is introduced by choosing different impulse responses $g(t) \neq h(t)$ at transmitter and receiver. The duration of the guard interval is given as

$$T_G = T_S - T_u \geq 0 \tag{21}$$

where T_S and T_u denote the length of the impulse response of the transmitter and the receiver filter $g(t)$ and $h(t)$, respectively. Both impulse responses are rectangular and symmetric to $t = 0$.

In contrast to (18), the carrier spacing is now

$$\Delta\omega = \frac{2\pi}{T_u} \tag{22}$$

For the elementary impulses follows

$$r_0(t) = \begin{cases} \dfrac{-|t| + T_S - T_G/2}{T_S - T_G} & \text{for } \dfrac{T_G}{2} < |t| < T_S - \dfrac{T_G}{2} \\ 1 & \text{for } |t| < \dfrac{T_G}{2} \\ 0 & \text{for } |t| > T_S - \dfrac{T_G}{2} \end{cases} \tag{23}$$

$$r_d(t) = \begin{cases} \dfrac{\mathrm{sgn}(t)(-1)^d}{j2\pi d}\left(e^{j\,\mathrm{sgn}(t)\,d\pi(T_G/T_u)} - e^{jd(2\pi/T_u)t}\right) & \text{for } \dfrac{T_G}{2} < |t| < T_S - \dfrac{T_G}{2} \\ 0 & \text{elsewhere} \end{cases} \quad d \neq 0 \tag{24}$$

As can be seen from Fig. 5, there are flat regions around the sampling instants $t = kT_S$, $k = 0, \pm 1, \ldots$ which prevent any interference. In many multicarrier systems, DMT as well as OFDM, the guard interval is chosen as $T_G/T_u = \frac{1}{16}, \ldots, \frac{1}{4}$. In Fig. 5b the eye diagram for a DMT system with guard interval is shown. Obviously, the horizontal eye opening is now increased

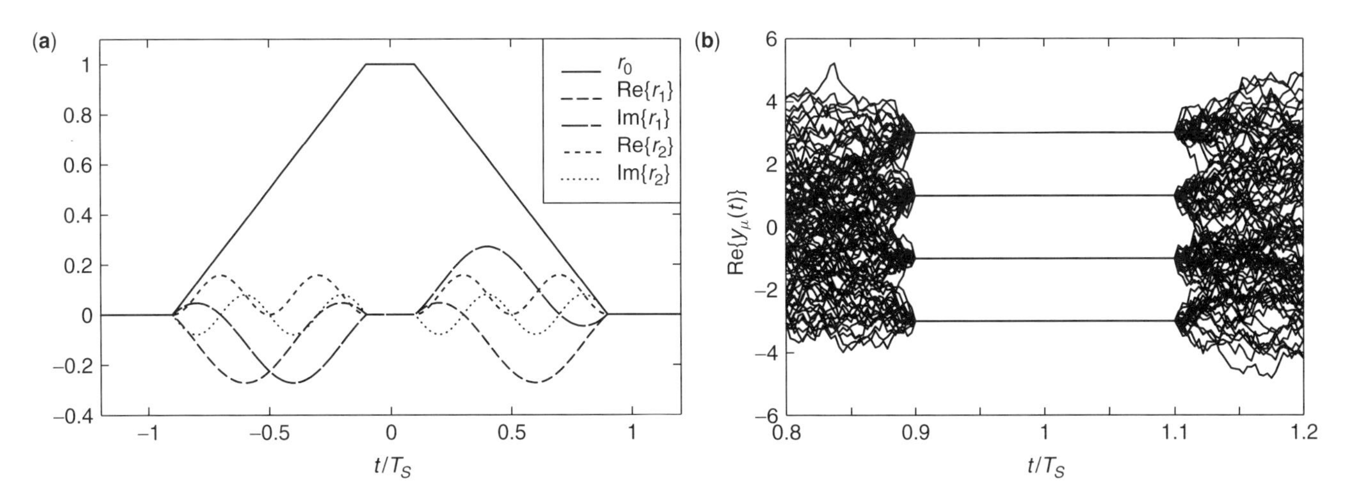

Figure 5. (**a**) Elementary impulses $r_d(t)$ with guard interval $T_G = T_u/4$; (**b**) eye diagram of real part of $y_\mu(t)$ for 16-QAM and $N = 256$ carriers and $T_G = T_u/4$.

and is as long as the guard interval. Therefore, the impact of timing jitter and interference is reduced considerably. However, DMT and OFDM systems are much more sensitive to timing jitter than are single-carrier systems [12].

During the guard interval period T_G, no information can be transmitted. Thus, the spectral efficiency is reduced by a factor T_G/T_S.

3. PRINCIPLES OF DISCRETE MULTITONE MODULATION

3.1. Implementation with Digital Signal Processing

While the continuous-time model in Fig. 3 is valuable for the analysis, the implementation of DMT systems is exclusively done using digital signal processing. Sampling the continuous-time signals in Fig. 3 with the sampling period T_A leads to the discrete-time or digital DMT system in Fig. 6.

The impulse modulators are replaced by upsamplers. They insert $N_S - 1$ zero samples between each incoming symbol $X_\nu[k]$. Thus $T_S = N_S \cdot T_A$ holds. We define

$$T_S = N_S \cdot T_A, \quad T_G = G \cdot T_A, \quad T_u = N \cdot T_A \tag{25}$$

From (21) follows

$$N_S = G + N \tag{26}$$

where G denotes the number of guard samples. For the complex carriers in Fig. 3, we obtain the following with Eqs. (8), (22), and (25) and $t = nT_A$:

$$\exp(j\omega_\nu nT_A) = \exp\left(j\frac{2\pi}{N}\nu n\right) = w^{-\nu n} \quad \text{with}$$
$$w = \exp\left(-j\frac{2\pi}{N}\right) \tag{27}$$

Of course, we have to ask whether the sampling theorem is fulfilled. As $g(t)$ and $h(t)$ are rectangular, and thus have infinite bandwidth, the sampling theorem is not met, at least not exactly. Consequently, the digital system in Fig. 6 is not an exact representation of the continuous-time system of Fig. 3. Nevertheless it is a reasonable approximation, and the concept of the elementary impulses and the generalized Nyquist criterion can be used similarly. We now adopt causal discrete-time filters:

$$g[n] = \begin{cases} \dfrac{1}{\sqrt{N}} & \text{for } n = 0, \ldots, N_S - 1 \\ 0 & \text{elsewhere} \end{cases} \tag{28}$$

The output signal of the transmitter in Fig. 6 can be written as

$$s[n] = \sum_{\nu=0}^{N-1} w^{-\nu n} \sum_{k=-\infty}^{\infty} X_\nu[k] g[n - kN_S] \tag{29}$$

We introduce the operator div for integer divisions as $n \operatorname{div} N_S = \lfloor n/N_S \rfloor$, where $\lfloor\,\rfloor$ indicates rounding toward the nearest smaller integer. With (28) we get from (29)

$$s[n] = \frac{1}{\sqrt{N}} \sum_{\nu=0}^{N-1} X_\nu[n \operatorname{div} N_S] \cdot w^{-\nu n} \tag{30}$$

Here we recognize the expression for the discrete Fourier transform (DFT; IDFT = inverse DFT) pair:

$$\text{DFT: } X_m = \frac{1}{\sqrt{N}} \sum_{n=0}^{N-1} x_n w^{nm} \quad \text{IDFT: } x_n = \frac{1}{\sqrt{N}} \sum_{m=0}^{N-1} X_m w^{-nm} \tag{31}$$

Thus, we identify the input signals $X_0[k], \ldots, X_{N-1}[k]$ as a block with index k and define the blockwise IDFT:

$$x_i[k] = \frac{1}{\sqrt{N}} \sum_{\nu=0}^{N-1} X_\nu[k] \cdot w^{-i\nu} \tag{32}$$

which allows us to express (30) as

$$s[n] = x_{n \bmod N}[n \operatorname{div} N_S] \tag{33}$$

For each block k of N input samples, $N_S = N + G$ output samples are produced. The first block $k = 0$ produces the output sequence

$$\{s[0], \ldots, s[N_S - 1]\} = \{x_0[0], \ldots, x_{N-1}[0], x_0[0], \ldots, x_{G-1}[0]\}$$

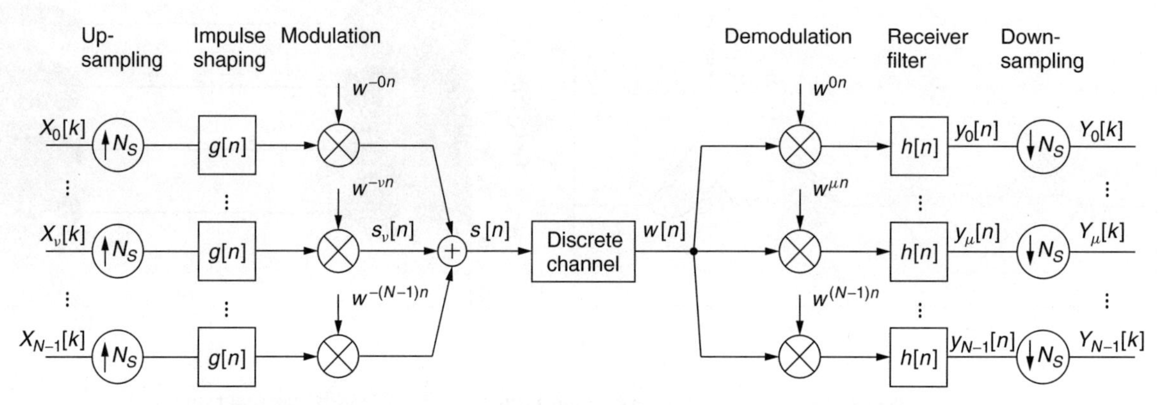

Figure 6. Transmitter and receiver for digital multicarrier transmission.

while the second block ($k = 1$) yields the output

$$\{s[N_S], \ldots, s[2N_S - 1]\}$$
$$= \{x_G[1], \ldots, x_{N-1}[1], x_0[1], \ldots, x_{2G-1}[1]\}$$

We recognize that each output block contains the symbols $x_0[k], \ldots, x_{N-1}[k]$ plus G additionally samples taken out of the same set. The calculation of $s[n]$ applying a blockwise IDFT is illustrated in Fig. 7, where a block of N input symbols is first IDFT-transformed and then parallel–serial-converted. The commutator puts out N_S symbols for each block by turning more than one revolution, resting after each block in a different position. Thus, although each block contains all transformed symbols, the ordering and the subset of doubled samples vary. To overcome this inconvenience and to facilitate the block processing at the receiver side, practically all DMT systems compute a block of N samples by inverse DFT processing and insert the additional samples at the beginning of the block. Therefore the guard interval is also called *cyclic prefix* (CP). Thus, a DMT block for index k will be ordered as follows:

$$\underbrace{x_{N-G}[k], x_{N-G+1}[k], \ldots, x_{N-1}[k]}_{G \text{ samples, cyclic prefix}}, \underbrace{x_0[k], x_1[k], \ldots, x_{N-1}[k]}_{N \text{ data samples}} \quad (34)$$

Thus, we can express the transmitter signal after insertion of the CP as

$$\tilde{x}[n] = x_{n \bmod N_S}[n \text{ div } N_S] \quad (35)$$

The discrete-time output signal with block processing is denoted as $\tilde{x}[n]$ in order to distinguish it from $s[n]$. At this point, the mathematical notation seems to be a little bit more tedious than in Eq. (30), but (32) and (35) describe the signals for independent block processing, which simplifies the operations in the transmitter and especially in the receiver considerably. Later we will derive a compact matrix notation that operates blockwise, taking advantage of the block independence.

3.2. Real-Valued Output Signal for Baseband Transmission

Because the DMT signal is transmitted over baseband channels, we must ensure that the output signal $\tilde{x}[n]$ and thus $x_i[k]$ are real-valued. Therefore we have to decompose $x_i[k]$ of (32) into real and imaginary parts

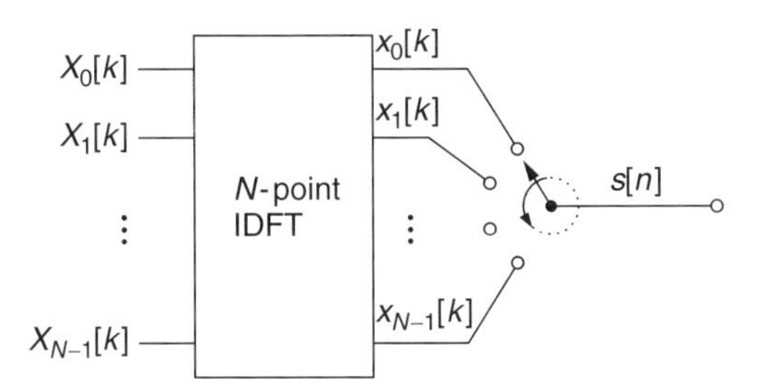

Figure 7. Inverse IDFT with parallel–serial conversion.

and set the latter to zero. After some calculations, this provides the following conditions on the input sequences $X_\nu[k]$:

$$X_0, X_{N/2} \in \mathcal{R} \quad (36)$$

$$X_\nu = X^*_{N-\nu}, \qquad \nu = 1, \ldots, \frac{N}{2} - 1 \quad (37)$$

Consequently, we get from (32) with $X_\nu[k] = X'_\nu[k] + jX''_\nu[k]$

$$x_i[k] = \frac{1}{\sqrt{N}}\left(X_0[k] + (-1)^i X_{N/2}[k] + 2 \times \sum_{\nu=1}^{\frac{N}{2}-1} X'_\nu[k]\cos\left(\frac{2\pi}{N}i\nu\right) - X''_\nu[k]\sin\left(\frac{2\pi}{N}i\nu\right)\right) \quad (38)$$

To simplify the transmitter scheme, we can choose $\tilde{X}_0 = X_0 + jX_{N/2}$ as indicated in Fig. 8. For practical implementations, this is of minor importance as usually X_0 and $X_{N/2}$ are set to zero. The reason will be discussed in Section 3.3.

It is convenient to interpret the DMT signal in (38) as a sum of $N/2$ QAM carriers where $X_\nu[k]$ modulates the νth carrier.

The detailed block diagram of a DMT transmitter as realized in practice is depicted in Fig. 8. The parallel–serial conversion and the insertion of the guard interval is symbolized by the commutator which inserts at the beginning of each block G guard samples, copied form the end of the block. A digital-to-analog converter (DAC) provides the continuous-time output $\tilde{x}(t)$. We clearly see that the guard interval adds some overhead to the signal and reduces the total data throughput by a factor $\eta = G/N_S$. Therefore, the length G of the cyclic prefix is normally chosen much smaller than the size N of the inverse DFT. We will see in Section 3.5 that the cyclic prefix allows for a rather simple equalizer.

Figure 9 depicts the corresponding DMT receiver. After analog-to-digital conversion and synchronization, the signal $\tilde{y}[n]$ is serial–parallel converted. Out of the N_S symbols of one block, the G guard samples are discarded and the remaining N symbols are fed into a DFT. Note that synchronization and timing estimation are essential receiver functions to produce reliable estimates of the transmitted data at the receiver. There have been many proposals for synchronization with DMT modulation, such as that by Pollet and Peeters [13].

Another major item to be considered with DMT modulation is the large peak-to-average power ratio (PAPR) of the output signal $\tilde{x}[n]$. Thus, a wide input range of the DAC is required. Furthermore, large PAPR can cause severe nonlinear effects in a subsequent power amplifier. The PAPR should be considered particularly for multicarrier modulation with a large numbers of subcarriers, because this ratio increases with the number of subcarriers. In order to reduce the

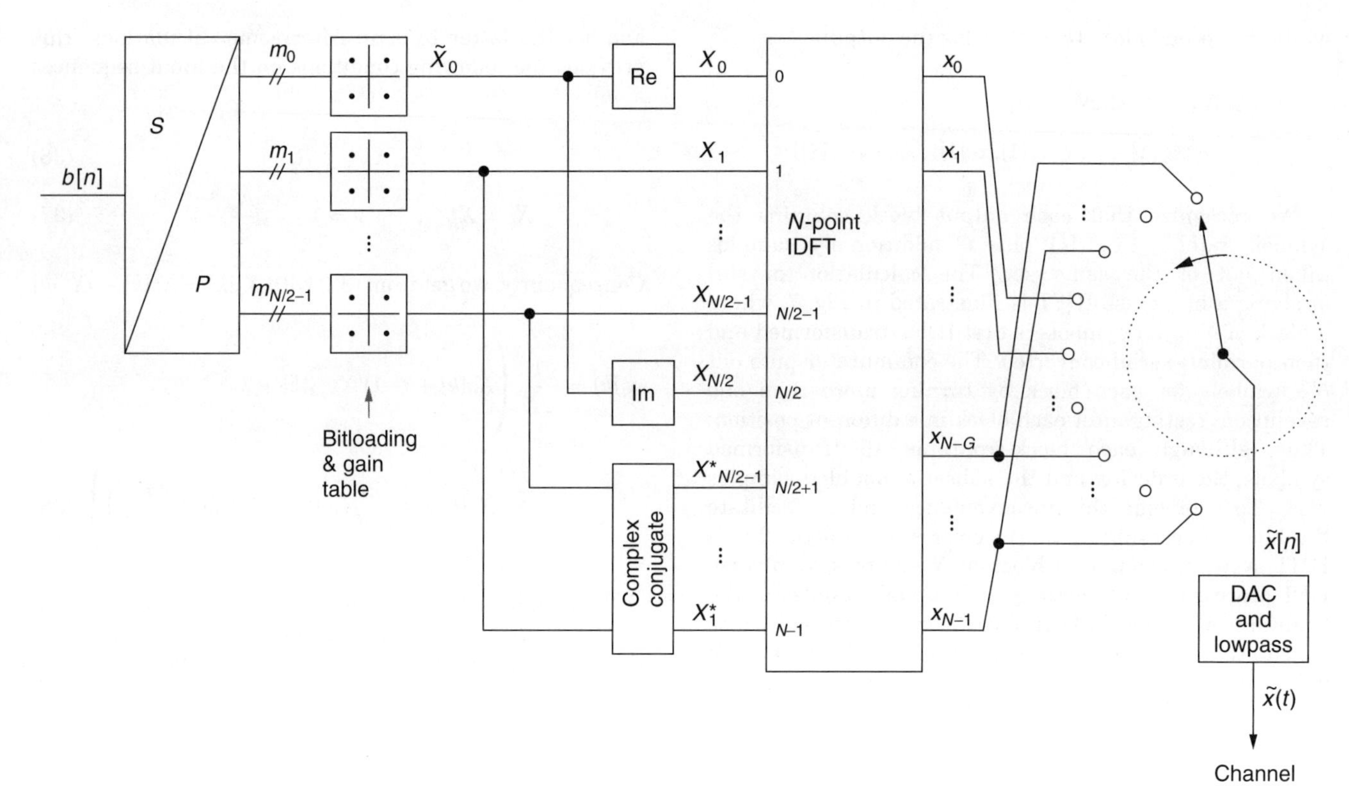

Figure 8. DMT transmitter.

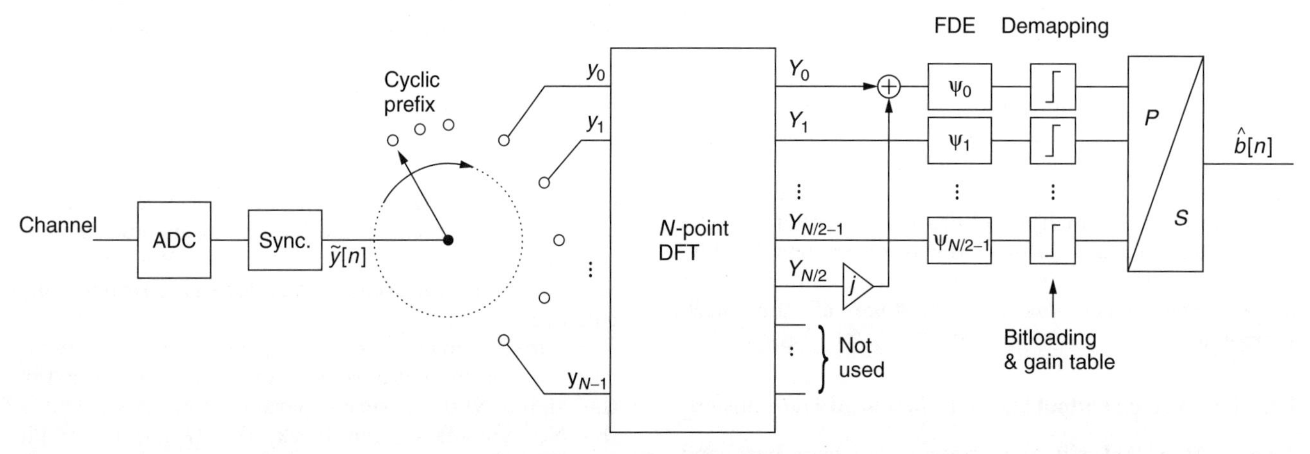

Figure 9. DMT receiver.

PAPR, several techniques have been proposed, such as "selective mapping" and "partial transmit sequence" approaches [14].

3.3. Spectral Properties

We calculate the spectrum of the output signal $s[n]$ in Fig. 6 for one modulated subchannel, namely, the input signal is given by (12). This gives

$$s[n] = g[n] \cdot w^{-\nu n} \leftrightarrow S(\omega) = \sum_{n=-\infty}^{\infty} g[n] \cdot w^{-\nu n} \cdot e^{-j\omega n T_A} \quad (39)$$

From Eqs. (22) and (25) we obtain $T_A = 2\pi/(N\Delta\omega)$. Together with (28), it follows from (39) that

$$S(\omega) = \frac{1}{\sqrt{N}} \begin{cases} \dfrac{1 - w^{(\omega/\Delta\omega - \nu)N_S}}{1 - w^{(\omega/\Delta\omega - \nu)}} & \text{for } \dfrac{\omega}{\Delta\omega} \neq \nu \\ N_S & \text{for } \dfrac{\omega}{\Delta\omega} = \nu \end{cases} \quad (40)$$

In order to obtain a real-valued output signal, according to (37), we have to sent a unit impulse on subchannel $N - \nu$, too. In Fig. 10a the magnitude of the spectrum is depicted for the case that subcarrier ν is modulated. The spectrum is quite similar to a $\sin(\omega)/\omega$-function. However,

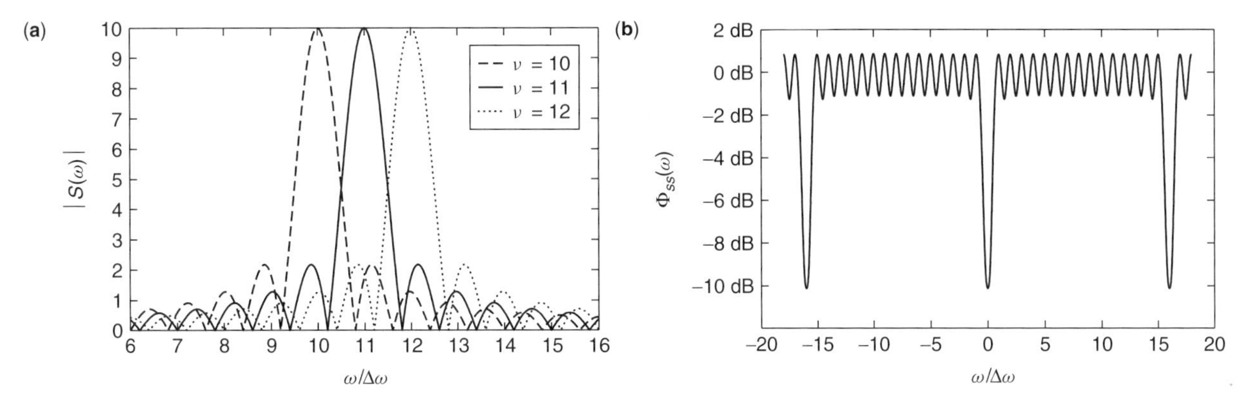

Figure 10. (**a**) Magnitude response $|S(\omega)|$ for modulated carrier ν, $G = N/8$; (**b**) power spectral density $\Phi_{ss}(\omega)$ for $N = 32$, $G = 4$. The ripple increases with the duration of the guard interval.

as Fig. 6 is a discrete-time system, $S(\omega)$ is periodic with $\omega_A = N \cdot \Delta\omega$. Note that the spectrum of a single subcarrier with carrier frequency $\nu\Delta\omega$ does *not* have zero crossings at $\omega = (\nu \pm m)\Delta\omega$, $m = 1, 2, \ldots$ if a guard interval with length $G > 0$ is used.

In order to calculate the power spectral density (PSD) of a DMT signal, we now consider stochastic input sequences $X_\nu[k]$ that are uncorrelated, have zero mean and variances σ_ν^2. Then the PSD of the output signal $s_\nu[n]$ of modulator ν in Fig. 6 becomes

$$\Phi_{s_\nu s_\nu}(\omega) = \frac{\sigma_\nu^2}{N_S} \cdot |G(\omega)|^2, \qquad \text{with}$$

$$G(\omega) = \sum_{n=-\infty}^{\infty} g[n] e^{-j\omega n T_A} = \frac{1}{\sqrt{N}} \frac{1 - w^{\frac{\omega}{\Delta\omega} N_S}}{1 - w^{\frac{\omega}{\Delta\omega}}} \tag{41}$$

From this it follows the total PSD of the transmitter

$$\Phi_{ss}(\omega) = \frac{1}{N_S} \sum_{\substack{\nu=-15 \\ \nu \neq 0}}^{15} \sigma_\nu^2 \cdot |G(\omega - \nu\Delta\omega)|^2 \tag{42}$$

which is depicted in Fig. 10b for a system with $N = 32$. All carriers are modulated with equal power $\sigma_\nu^2 = 1$, except for carriers $\nu = 0$ and $\nu = N/2$. Because of (37), the subcarriers $\nu = 17, \ldots, 31$[1] are modulated with the complex conjugate of the sequences $X_1[k], \ldots, X_{15}[k]$. Note that we can approximately add the PSD of all sequences $s_\nu[n]$ despite their pairwise correlation because their spectra overlap to only a very small extent. As the antialiasing lowpass in the DAC has a cutoff frequency of $\omega_A/2$ [15], the carriers in this frequency region cannot be modulated; therefore, at least the carrier at $\nu = N/2$ remains unused. The carrier at $\nu = 0$ is seldom used, either, because most channels do not support a DC component.

[1] Because of the periodicity of $\Phi_{ss}(\omega)$ this is equivalent to modulate the subcarriers $\nu = -15, \ldots, -1$.

3.4. System Description with Matrix Notation

We have seen that the transmitter processes the data blockwise, resulting in simple implementation without the need of saving data from previous blocks. We now consider how this block independence can be extended to the receiver when including the channel. Figure 11 shows a block diagram for the complete DMT transmission system. For the moment we focus on the signal only and consider the influence of the noise later. The receiver input signal is then given by

$$\tilde{y}[n] = \tilde{x}[n] \star c[n] = \sum_{m=-\infty}^{\infty} \tilde{x}[m]\, c[n-m] \tag{43}$$

The discrete-time channel impulse response $c[n]$ includes the influences of DAC and ADC (digital-to-analog and analog-to-digital conversion), respectively. We assume that the channel can be described by a finite-length

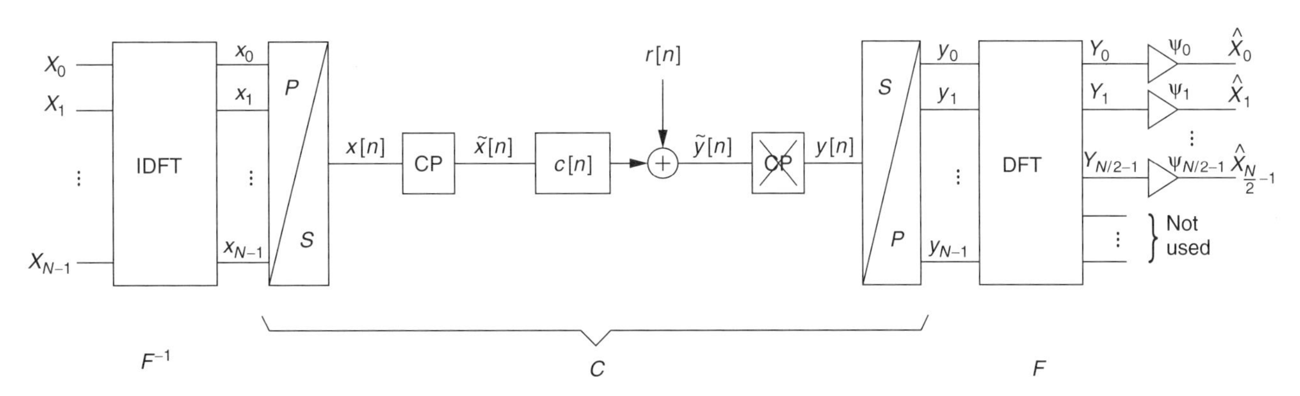

Figure 11. Discrete-time model of a DMT transmission system.

impulse response $c[n]$ with $L+1$ samples $c[0],\ldots,c[L]$. From (43) we conclude that for independence of the received blocks, it must hold that

$$L \leq G \tag{44}$$

Thus, the cyclic prefix must be at least as long as the length of the channel impulse response. If this condition is satisfied, we can adopt a vector notation for the block data:

$$\mathbf{x} = \begin{pmatrix} x_0 \\ x_1 \\ \vdots \\ x_{N-1} \end{pmatrix}, \quad \tilde{\mathbf{x}} = \begin{pmatrix} x_{N-G} \\ \vdots \\ x_{N-1} \\ x_0 \\ \vdots \\ x_{N-1} \end{pmatrix}, \quad \tilde{\mathbf{y}} = \begin{pmatrix} \tilde{y}_0 \\ \vdots \\ \tilde{y}_{G-1} \\ \tilde{y}_G \\ \vdots \\ \tilde{y}_{N+G-1} \end{pmatrix},$$

$$\mathbf{y} = \begin{pmatrix} y_0 \\ y_1 \\ \vdots \\ y_{N-1} \end{pmatrix} = \begin{pmatrix} \tilde{y}_G \\ \tilde{y}_{G+1} \\ \vdots \\ \tilde{y}_{N+G-1} \end{pmatrix} \tag{45}$$

where $\mathbf{y}$ is the input signal after removal of the cyclic prefix. It can be expressed as

$$\mathbf{y} = \begin{pmatrix} c[G] & c[G-1] & \cdots & c[0] & 0 & \cdots & 0 \\ 0 & c[G] & c[G-1] & \cdots & c[0] & \cdots & 0 \\ \cdots & & & & & \cdots & \\ 0 & 0 & \cdots & c[G] & c[G-1] & \cdots & c[0] \end{pmatrix} \cdot \tilde{\mathbf{x}} \tag{46}$$

or

$$\begin{pmatrix} y_0 \\ y_1 \\ \vdots \\ y_{G-1} \\ y_G \\ y_{G+1} \\ \vdots \\ y_{N-1} \end{pmatrix} = \underbrace{\begin{pmatrix} c[0] & 0 & \cdots & 0 & c[G] & \cdots & c[2] & c[1] \\ c[1] & c[0] & 0 & \cdots & 0 & c[G] & \cdots & c[2] \\ \cdots & \cdots & \cdots & \cdots & \cdots & \cdots & \cdots & \cdots \\ c[G-1] & \cdots & \cdots & c[0] & 0 & \cdots & 0 & c[G] \\ c[G] & c[G-1] & \cdots & \cdots & c[0] & 0 & \cdots & 0 \\ 0 & c[G] & c[G-1] & \cdots & \cdots & c[0] & 0 & \cdots \\ \cdots & \cdots & \cdots & \cdots & \cdots & \cdots & \cdots & \cdots \\ 0 & \cdots & \cdots & c[G] & \cdots & \cdots & c[1] & c[0] \end{pmatrix}}_{\mathbf{C}} \times \begin{pmatrix} x_0 \\ x_1 \\ \vdots \\ x_{G-1} \\ x_G \\ x_{G+1} \\ \vdots \\ x_{N-1} \end{pmatrix} \tag{47}$$

The latter expression is preferred as it directly relates the signals $\mathbf{x}$ and $\mathbf{y}$ and transforms the effect of the guard interval insertion and removal into the channel matrix $\mathbf{C}$. The matrix equation (47) represents the circular convolution, which is equivalent to multiplication in the frequency domain [15]:

$$y[n] = x[n] \otimes c[n] = \sum_{m=0}^{N-1} x[m] \cdot c[(n-m) \bmod N] \tag{48}$$

From this point of view, the cyclic prefix transforms the linear convolution (43) into a circular convolution. The matrix $\mathbf{C}$ is a circulant matrix whose eigenvalues λ_μ and eigenvectors $\boldsymbol{\vartheta}_\mu$ are given by

$$\lambda_\mu = \sum_{n=0}^{N-1} c[n] w^{\mu n}, \quad \boldsymbol{\vartheta}_\mu = \frac{1}{\sqrt{N}} \begin{pmatrix} w^{-0} \\ w^{-\mu} \\ w^{-2\mu} \\ \vdots \\ w^{-(N-1)\mu} \end{pmatrix}, \quad \mu = 0,\ldots,N-1 \tag{49}$$

This can be easily verified by checking the equation $\mathbf{C}\boldsymbol{\vartheta}_\mu = \lambda_\mu \boldsymbol{\vartheta}_\mu$. We identify the inverse DFT matrix as

$$\mathbf{F}^{-1} = (\boldsymbol{\vartheta}_0, \boldsymbol{\vartheta}_1, \ldots, \boldsymbol{\vartheta}_{N-1}) \tag{50}$$

This means that the eigenvectors of the channel matrix $\mathbf{C}$ are the columns of the inverse DFT matrix. As a consequence, we can diagonalize the channel matrix by multiplying with the IDFT and its inverse, the DFT matrix $\mathbf{F}$

$$\mathbf{FCF}^{-1} = \boldsymbol{\Lambda} \qquad \text{where } \boldsymbol{\Lambda} = \text{diag}(\lambda_\mu) \tag{51}$$

With $\mathbf{X} = (X_0[k],\ldots,X_{N-1}[k])^T$ and $\mathbf{Y} = (Y_0[k],\ldots,Y_{N-1}[k])^T$ we can write

$$\mathbf{x} = \mathbf{F}^{-1}\mathbf{X}, \qquad \mathbf{y} = \mathbf{Cx} = \mathbf{CF}^{-1}\mathbf{X}, \qquad \mathbf{Y} = \mathbf{Fy} = \mathbf{FCF}^{-1}\mathbf{X} \tag{52}$$

We can now describe the input-output-relation of the whole transmission system of Fig. 11 with a single diagonal matrix: $\mathbf{Y} = \boldsymbol{\Lambda}\mathbf{X}$, or $Y_\mu[k] = \lambda_\mu \cdot X_\mu[k]$. The result shows that due to the cyclic prefix the parallel subchannels are independent, and perfect reconstruction can be realized by a simple one-tap equalizer with tap weight ψ_μ per subchannel located at the output of the receiver DFT as shown in Fig. 11. As equalization is done after the DFT, this equalizer is usually referred to as a *frequency-domain equalizer* (FDE).

Following from Eq. (49), we can interpret the eigenvalues λ_μ of the channel matrix as the DFT of the channel impulse response. If we define

$$C(\omega) = \sum_{n=0}^{N-1} c[n] \cdot e^{-j\omega n T_A} \tag{53}$$

as the discrete-time Fourier transform of $c[n]$, we see that the eigenvalues are just the values of the channel transfer function at the frequencies $\omega_\mu = \mu \cdot \Delta\omega$:

$$\lambda_\mu = C(\omega_\mu) \tag{54}$$

Because DMT is a baseband transmission scheme, the channel impulse response $c[n]$ is real-valued and therefore its spectrum shows hermitian symmetry:

$$\lambda_0, \lambda_{N/2} \in \mathcal{R}; \qquad \lambda_\mu = \lambda_{N-\mu}^* \qquad \text{for } \mu = 1, \ldots, N/2 - 1 \tag{55}$$

Let us summarize the results. We have shown that the insertion of the cyclic prefix translates the linear convolution (43) into a circular convolution (48) that corresponds to a multiplication in frequency domain, as long as the impulse response of the channel is not longer than the guard interval. If the guard interval is sufficiently large, no interblock and no intercarrier interference occur and each subchannel signal is weighted only by the channel transfer function at the subchannel frequency. If the channel impulse response $c[n]$ exceeds the guard interval, the described features of DMT are not valid anymore. To overcome this problem, $c[n]$ can be compressed by a so-called time-domain equalizer (TDE) to the length of the guard interval. The TDE will be applied before the guard interval is removed at the receiver [e.g., 16–18].

3.5. Frequency-Domain Equalization (FDE)

The output signal of the DFT at the receiver in Fig. 11 is given by

$$Y_\mu[k] = \lambda_\mu \cdot X_\mu[k] + q_\mu[k] \tag{56}$$

where $q_\mu[k]$ represents the noise introduced on the subchannel after DFT processing. For many practical channels, the noise $r[n]$ on the channel will be Gaussian but not white. Therefore, the power of $q_\mu[k]$ will depend on the subchannel. If N is chosen sufficiently large, the subchannel bandwidth is very small and thus the noise in each subchannel will be approximately white with variance $\sigma_{q,\mu}^2$. Further, the noise is uncorrelated with the noise on any other subchannel [5]. The equalizer coefficients ψ_μ can be derived with or without considering the noise term in Eq. (56). If the signal-to-noise ratio (SNR) is high, we approximately neglect the noise and find the optimal FDE parameters as $\psi_\mu = 1/\lambda_\mu$. With $\hat{X}_\mu[k] = \psi_\mu Y_\mu[k]$ follows $\hat{X}_\mu[k] = X_\mu[k] + q_\mu[k]/\lambda_\mu$ for the reconstructed signal at the receiver. The SNR at the output of the μ-th subchannel is then given by

$$\text{SNR}_\mu = \frac{E\{|X_\mu|^2\}}{E\{|\hat{X}_\mu - X_\mu|^2\}} = \frac{\sigma_\mu^2 \cdot |\lambda_\mu|^2}{\sigma_{q,\mu}^2} \tag{57}$$

where $\sigma_\mu^2 = E\{|X_\mu[k]|^2\}$ is the signal power of the μth input signal at the transmitter.

However, the SNR per subchannel can be improved by considering the noise in the derivation of the equalizer coefficients ψ_μ. Considering the AWGN (additive white Gaussian noise) in the subchannels, we calculate the equalizer coefficients by minimizing the mean-square error (MSE) $E\{|\hat{X}_\mu - X_\mu|^2\}$ which can be written with $\psi_\mu = \psi_\mu' + j\psi_\mu''$ and $\lambda_\mu = \lambda_\mu' + j\lambda_\mu''$ as

$$\begin{aligned} E\{|\hat{X}_\mu - X_\mu|^2\} &= (\psi_\mu'^2 + \psi_\mu''^2)(|\lambda_\mu|^2\sigma_\mu^2 + \sigma_{q,\mu}^2) \\ &\quad - 2\sigma_\mu^2(\psi_\mu'\lambda_\mu' - \psi_\mu''\lambda_\mu'') + \sigma_\mu^2 \end{aligned} \tag{58}$$

For minimization of the MSE, we set the gradient of (58) to zero:

$$\frac{\partial}{\partial \psi_\mu'} E\{|\hat{X}_\mu - X_\mu|^2\} = 0 \quad \frac{\partial}{\partial \psi_\mu''} E\{|\hat{X}_\mu - X_\mu|^2\} = 0 \tag{59}$$

which results in

$$\begin{aligned} \psi_\mu' &= \frac{\lambda_\mu'}{|\lambda_\mu|^2 + \sigma_{q,\mu}^2/\sigma_\mu^2}, \; \psi_\mu'' = \frac{-\lambda_\mu''}{|\lambda_\mu|^2 + \sigma_{q,\mu}^2/\sigma_\mu^2} \\ \Rightarrow \psi_\mu &= \frac{\lambda_\mu^*}{|\lambda_\mu|^2 + \sigma_{q,\mu}^2/\sigma_\mu^2} \end{aligned} \tag{60}$$

The subchannel SNR can be calculated as

$$\text{SNR}_\mu = \frac{E\{|X_\mu|^2\}}{E\{|\hat{X}_\mu - X_\mu|^2\}} = \frac{\sigma_\mu^2 \cdot |\lambda_\mu|^2}{\sigma_{q,\mu}^2} + 1 \tag{61}$$

which gives a better performance than the first case, especially for low SNR.

It is this amazingly simple equalization with a one-tap equalizer per subchannel that makes DMT so popular for the transmission over frequency-selective channels like the telephone subscriber line.

4. CHANNEL CAPACITY AND BIT LOADING

The channel capacity [19], or the maximum error-free bitrate, of an ideal AWGN channel is given by

$$R_{\max} = \frac{\omega_B}{2\pi} \cdot \log_2\left(1 + \frac{\sigma_y^2}{\sigma_r^2}\right) \tag{62}$$

where $\omega_B = N \cdot \Delta\omega$ is the total channel bandwidth, σ_y^2 is the received signal power, and σ_r^2 is the noise power. The capacity of subchannel μ with very small bandwidth $\Delta\omega$ is then

$$R_\mu = \frac{\Delta\omega}{2\pi} \cdot \log_2\left(1 + \frac{\Delta\omega\Phi_{xx}(\omega_\mu)|C(\omega_\mu)|^2}{\Delta\omega\Phi_{rr}(\omega_\mu)}\right) \tag{63}$$

where $\Phi_{xx}(\omega)$ denotes the power spectral density (PSD) of the transmitter signal $\tilde{x}[n]$, $C(\omega)$ is the channel transfer function in (53) and $\Phi_{rr}(\omega)$ is the PSD of the (colored) noise $r[n]$. The total capacity of all subchannels is

$$R = \frac{\Delta\omega}{2\pi} \cdot \sum_{\mu=0}^{N-1} \log_2\left(1 + \frac{\Phi_{xx}(\omega_\mu)|C(\omega_\mu)|^2}{\Phi_{rr}(\omega_\mu)}\right) \tag{64}$$

In the limit for $\Delta\omega \to 0$ with $\Delta\omega \cdot N = \omega_B = \text{const.}$ we get

$$R = \frac{1}{2\pi} \int_0^{\omega_B} \log_2\left(1 + \frac{\Phi_{xx}(\omega)|C(\omega)|^2}{\Phi_{rr}(\omega)}\right) d\omega \tag{65}$$

We now want to maximize the channel capacity R, subject to the constraint that the transmitter power is limited:

$$\frac{1}{2\pi} \int_0^{\omega_B} \Phi_{xx}(\omega) d\omega = P_t \tag{66}$$

Thus, we search for a function $\Phi_{xx}(\omega)$ that maximizes Eq. (65) subject to the constraint (66). This can be accomplished by calculus of variations [20]. Therefore we set up the Lagrange function

$$L(\Phi_{xx}, \omega) = \log_2 \left(1 + \Phi_{xx}(\omega)\frac{|C(\omega)|^2}{\Phi_{rr}(\omega)}\right) + \lambda \cdot \Phi_{xx}(\omega) \quad (67)$$

which must fulfill the Euler–Lagrange equation

$$\frac{\partial L}{\partial \Phi_{xx}} = \frac{d}{d\omega}\frac{\partial L}{\partial \Phi'_{xx}}, \quad \text{where } \Phi'_{xx} = \frac{d\Phi_{xx}}{d\omega} \quad (68)$$

This requires that $\Phi_{xx}(\omega) + \Phi_{rr}(\omega)/|C(\omega)|^2 = \Phi_0 = \text{const}$. Thus,

$$\Phi_{xx}(\omega) = \begin{cases} \Phi_0 - \Phi_{rr}(\omega)/|C(\omega)|^2 & \text{for } |\omega| < \omega_B \\ 0 & \text{elsewhere} \end{cases} \quad (69)$$

This represents the "water-filling solution." We can interpret $\Phi_{rr}(\omega)/|C(\omega)|^2$ as the bottom of a bowl in which we fill an amount of water corresponding to P_t. The water will distribute in a way that the depth represents the wanted function $\Phi_{xx}(\omega)$, as illustrated in Fig. 12.

The optimum solution according to (69) can be achieved by appropriately adjusting the constellation sizes and the gain factors for the bit-to-symbol mapping for each carrier.

In practice, when the bit rate assignments are constrained to be integer, the "water-filling solution" has to be modified. There have been extensive studies on the allocation of power and data rate to the subchannels, known as *bitloading algorithms* [21–24]. In the following, the connection between the channel characteristics and bitloading is presented for an asymmetric digital subscriber line (ADSL) system.

5. APPLICATIONS OF DMT

The DMT modulation scheme has been selected as standard for ADSL [8], a technique for providing high-speed data services over the existing copper-based telephone network infrastructure. Typically, these phone lines consist of unshielded twisted-pair (UTP) wire, which share the same cable with multiple other pairs. Hence, the performance of ADSL transmission depends on the noise environment and cable loss [25]. When lines are bound together in a cable, they produce crosstalk from one pair to another, at levels that increase with frequency and the number of crosstalking pairs or disturbers. For this reason, crosstalk is one of the major noise sources. Additional disturbances are given by impulse noise and possibly radiofrequency interferers. Beside that, line attenuation

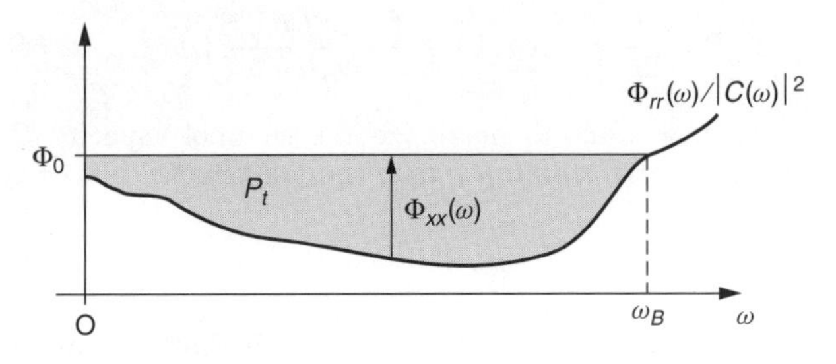

Figure 12. Optimum water-filling solution.

increases with frequency and distance. A modulation scheme that is well suited for the wide variations in the ADSL channel characteristics is provided by DMT. In particular, rate adaption is quite simple for DMT modulation to optimize the ADSL system transmission. Some aspects of an ADSL system are regarded in the remainder of this section.

The parameter of the DMT modulation scheme for downstream transmission are determined by the ADSL standard [8] as follows:

- FFT size of $N = 512$. Consequently, excluding the two carriers at $\nu = 0$ and $\nu = N/2$, 255 usable parallel subchannels result.
- Sampling rate $f_A = 1/T_A = 2.208$ MHz.
- Guard interval length $G = 32 = N/16$.
- Adaptive bitloading with a maximum number of $m_{\max} = 15$ bits per subcarrier (tone).
- A flat transmit power spectral density of about −40 dBm/Hz (dBm = decibels per milliwatt). As the spectrum ranges up to $f_A/2 = 1.104$ MHz, the total transmit power is about 20 dBm.

The achievable data throughput depends on the bit allocation that is established during initialization of the modem and is given by

$$R = \frac{1}{T_S}\sum_{\nu=1}^{N/2-1} m_\nu \quad (70)$$

where m_ν denotes the number of bits modulated on the νth subcarrier. The frequency spacing $\Delta f = \Delta\omega/2\pi$ between subcarriers is given by

$$\Delta f = \frac{f_A}{N} = 4.3125 \text{ kHz} \quad (71)$$

As ADSL services share the same line with the plain old telephone service (POTS), a set of splitter filters (POTS splitter) at the customer premises and the network node separate the different frequency bands. Usually, POTS signals occupy bandwidths of up to 4 kHz. To allow a smooth and inexpensive analog filter for signal separation, ADSL services can use a frequency range from 25 kHz up to 1.1 MHz.

Figure 13a shows the variation of the SNR in the DMT subchannels at the receiver for a subscriber line of 3 km length and 0.4 mm wire diameter. DMT modems for ADSL with rate adaption measure the SNR per subchannel during an initialization period.

Of course, the POTS splitter in an ADSL system essentially removes all signals below 25 kHz. Thus, the first few channels cannot carry any data, as can be seen in Fig. 13b, which shows the bitloading corresponding to the SNR of Fig. 13a. A minimum bitload of 2 bits per carrier is stipulated in the standard. The total bit rate is about 4 Mbps. Further, attenuation can be severe at the upper end of the ADSL frequency band. As a consequence, SNR is low and the adaptive bitloading algorithm allocates fewer bits per carrier.

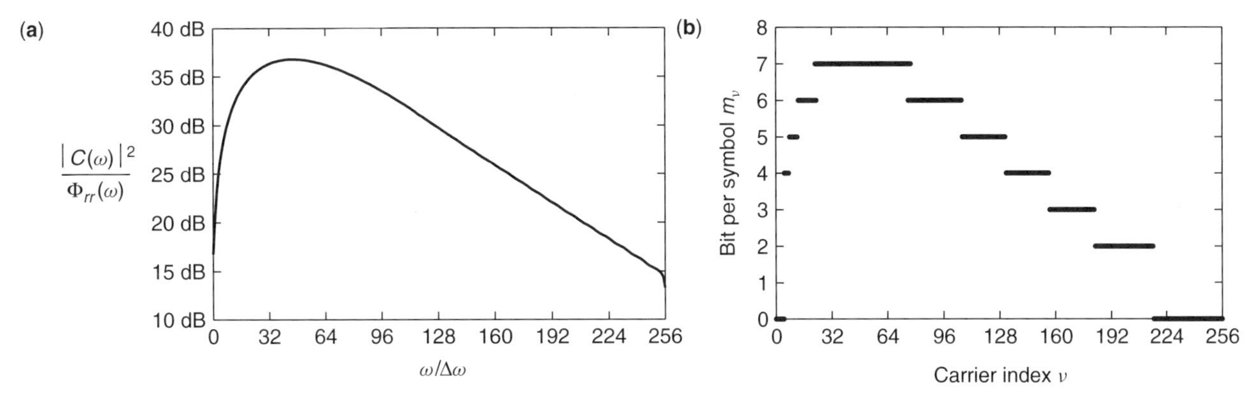

Figure 13. (**a**) SNR in the DMT subchannels at the receiver for a subscriber line of 3 km length and 0.4-mm wire diameter for an asymmetric digital subscriber line (ADSL); (**b**) adaptive bitloading for DMT modulation for transmission over a subscriber line 3 km length and 0.4 mm diameter.

BIOGRAPHIES

Romed Schur received his Dipl.-Ing. degree in electrical engineering and information technology from the University of Stuttgart, Germany, in 1997. He joined the Institute for Telecommunications at the University of Stuttgart as a research and teaching assistant in 1997 and is currently working toward his Ph.D. His interests include multicarrier modulation, xDSL transmission, as well as filterbanks and their applications to communication.

Stephan Pfletschinger studied electrical engineering and information technology at the University of Stuttgart, Germany, and received the Dipl.-Ing. degree with distinction in 1997. Since then he has been with the Institute of Telecommunications at the University of Stuttgart as a teaching and research assistant and is working toward his Ph.D. His research interests include multicarrier modulation and cable TV networks.

Joachim Speidel studied electrical engineering and information technology at the University of Stuttgart, Germany, and received his Dipl.-Ing. and Dr.-Ing. degree in 1975 and 1980, respectively, all with summa cum laude. From 1980 to 1992 he worked for Philips Communications (today Lucent Technologies Bell Labs Innovations) in the field of digital communications, ISDN, and video coding. During his industry career he has held various positions in R&D, as a member of technical staff, laboratory head, and finally as vice president. Since autumn 1992 he has been full professor at the University of Stuttgart and Director of the Institute of Telecommunications. His research areas are digital multimedia communications in mobile, optical, and electrical networks with emphasis on systems, modulation, source and channel coding.

BIBLIOGRAPHY

1. B. R. Saltzberg, Performance of an efficient parallel data transmission system, *IEEE Trans. Commun. Technol.* **COM-15**: 805–811 (Dec. 1967).
2. S. B. Weinstein and P. M. Ebert, Data transmission by frequency-division multiplexing using the discrete Fourier transform, *IEEE Trans. Commun. Technol.* **COM-19**(5): 628–634 (Oct. 1971).
3. I. Kalet, The multitone channel, *IEEE Trans. Commun.* **37**(2): 119–124 (Feb. 1989).
4. J. A. C. Bingham, Multicarrier modulation for data transmission: An idea whose time has come, *IEEE Commun. Mag.* **28**(5): 5–14 (May 1990).
5. J. S. Chow, J. C. Tu, and J. M. Cioffi, A discrete multitone transceiver system for HDSL applications, *IEEE J. Select. Areas Commun.* **9**(6): 895–908 (Aug. 1991).
6. A. N. Akansu, P. Duhamel, X. Lin, and M. de Courville, Orthogonal transmultiplexers in communications: a review, *IEEE Trans. Signal Process.* **46**(4): 979–995 (April 1998).
7. J. M. Cioffi et al., Very-high-speed digital subscriber lines, *IEEE Commun. Mag.* **37**(4): 72–79 (April 1999).
8. ITU-T Recommendation G.992.1, *Asymmetric Digital Subscriber Line (ADSL) Transceivers*, June 1999.
9. J. G. Proakis, *Digital Communications*, McGraw-Hill, New York, 2000.
10. G. Cherubini, E. Eleftheriou, S. Ölçer, and J. M. Cioffi, Filter bank modulation techniques for very high-speed digital subscriber lines, *IEEE Commun. Mag.* **38**(5): 98–104 (May 2000).
11. K. D. Kammeyer, U. Tuisel, H. Schulze, and H. Bochmann, Digital multicarrier-transmission of audio signals over mobile radio channels, *Eur. Trans. Telecommun.* **3**(3): 243–253 (May–June 1992).
12. B. R. Saltzberg, Comparison of single-carrier and multitone digital modulation for ADSL applications, *IEEE Commun. Mag.* **36**(11): 114–121 (Nov. 1998).
13. T. Pollet and M. Peeters, Synchronization with DMT modulation, *IEEE Commun. Mag.* **37**(4): 80–86 (April 1999).
14. L. J. Cimini, Jr. and N. R. Sollenberger, Peak-to-average power ratio reduction of an OFDM signal using partial transmit sequences, *Proc. IEEE ICC'99*, June 1999, Vol. 1, pp. 511–515.
15. A. V. Oppenheim and R. W. Schafer, *Discrete-Time Signal Processing*, Prentice-Hall, Englewood Cliffs, NJ, 1989.
16. P. J. W. Melsa, R. C. Younce, and C. E. Rohrs, Impulse response shortening for discrete multitone transceivers, *IEEE Trans. Commun.* **44**(12): 1662–1672 (Dec. 1996).

17. R. Schur, J. Speidel, and R. Angerbauer, Reduction of guard interval by impulse compression for DMT modulation on twisted pair cables, *Proc. IEEE Globecom'00*, Nov. 2000, Vol. 3, pp. 1632–1636.
18. W. Henkel, Maximizing the channel capacity of multicarrier transmission by suitable adaption of time-domain equalizer, *IEEE Trans. Commun.* **48**(12): 2000–2004 (Dec. 2000).
19. C. E. Shannon, A mathematical theory of communication, *Bell Syst. Tech. J.* **27**: 379–423, 623–656 (July, Oct. 1948).
20. K. F. Riley, M. P. Hobson, and S. J. Bence, *Mathematical Methods for Physics and Engineering*, Cambridge Univ. Press, Cambridge, UK, 1998.
21. U.S. Patent 4,679,227 (July, 1987), D. Hughes-Hartogs, Ensemble modem structure for imperfect transmission media.
22. P. S. Chow, J. M. Cioffi, and J. A. C. Bingham, A practical discrete multitone transceiver loading algorithm for data transmission over spectrally shaped channels, *IEEE Trans. Commun.* **43**(2–4): 773–775 (Feb.–April 1995).
23. R. F. H. Fischer and J. B. Huber, A new loading algorithm for discrete multitone transmission, *Proc. IEEE Globecom'96*, Nov. 1996, Vol. 1; pp. 724–728.
24. R. V. Sonalkar and R. R. Shively, An efficient bit-loading algorithm for DMT applications, *IEEE Commun. Lett.* **4**(3): 80–82 (March 2000).
25. S. V. Ahamed, P. L. Gruber, and J.-J. Werner, Digital subscriber line (HDSL and ADSL) capacity of the outside loop plant, *IEEE J. Select. Areas Commun.* **13**(9): 1540–1549 (Dec. 1995).

DWDM RING NETWORKS

DETLEF STOLL
JIMIN XIE
Siemens ICN, Optisphere Networks
Boca Raton, Florida

JUERGEN HEILES
Siemens Information & Communication Networks
Munich, Germany

1. INTRODUCTION

Ring networks are well known and widely used in today's Synchronous Optical Network (SONET) or Synchronous Digital Hierarchy (SDH) transport networks [1]. They are attractive because they offer reliable and cost-efficient transport. In combination with dense wavelength-division multiplexing (DWDM), ring networks provide high capacity and transparent transport for a variety of client signals. Compared to a star and bus topology, the ring topology provides two diverse routes between any two network nodes. Ring network topologies are therefore often used in combination with protection schemes for increased reliability. Additionally, network management, operation and node design are less complex in rings than in mesh topologies.

DWDM is the key technology to provide high transmission capacity by transmitting many optical channels simultaneously over the same optical fiber. Each channel uses a dedicated optical wavelength (or, equivalently, color or frequency) [2]. Multiplexing in the frequency domain has been used for many decades in the communication industry, including radio or TV broadcast, where several channels are multiplexed and transmitted over the air or cable. The information signals are modulated at the transmitter side on carrier signals of different frequencies (wavelengths), where they are then combined for transmission over the same medium. At the receiver side, the channels are selected and separated using bandpass filters. More recent advances in optical technology allow for the separation of densely spaced optical channels in the spectrum. Consequently, more than 80 optical channels can be transmitted over the same fiber, with each channel transporting payload signals of $\geq$2.5 Gbps (gigabits per second). In combination with optical amplification of multiplexed signals, DWDM technology provides cost-efficient high-capacity long-haul (several 100 km) transport systems. Because of its analog nature, DWDM transport is agnostic to the frame format and bit rate (within a certain range) of the payload signals. Therefore, it is ideal in supporting the transportation of various high–speed data and telecommunication network services, such as Ethernet, SONET, or SDH.

These advantages fit the demands of modern communication networks. Service providers are facing the merging of telecommunication and data networks and a rapid increase in traffic flow, driven mainly by Internet applications. The level of transmission quality is standardized in the network operation business, or, commonly agreed. There are few possibilities to achieve competitive advantages for network operators. But these differences, namely, better reliability of service or flexibility of service provisioning, have a major influence on the market position. This explains why the transportation of telecommunication and data traffic is a typical volume market [3, p. 288]. As a default strategy in volume markets, it is the primary goal to provide highest data throughput at the most competitive price in order to maximize market share. As a secondary goal, network operators tend to provide flexibility and reliability as competitive differentiators. DWDM ring networks fulfill these basic market demands in an almost ideal way.

In the following sections, we will approach DWDM ring networks from three different viewpoints:

- *The Network Architecture Viewpoint.* DWDM ring networks are an integral part of the worldwide communication network infrastructure. The network design and the architecture have to meet certain rules in order to provide interoperability and keep the network structured, reliable, and manageable. The DWDM ring network architecture, the network layers, and specific constrains of DWDM ring networks will be discussed in Section 2. We will also take a look at protection schemes, which are an important feature of ring networks.
- *The Network Node Viewpoint.* Optical add/drop multiplexers (OADMs) are the dominant type of network elements in DWDM ring networks. The ring features will be defined mainly by the OADM

functionality. The basic OADM concept and its functionality is discussed in Section 3.

- *The Component Viewpoint*. Different component technologies can be deployed to realize the OADM functionality. It is the goal of system suppliers, to utilize those components that minimize the cost involved in purchasing and assembling these components and maximize the functionality of the OADM and DWDM ring network. State-of-the-art technologies and new component concepts are presented in Section 4.

2. DWDM RING NETWORK ARCHITECTURE

2.1. Basic Architecture

The basic ring architecture is independent of transport technologies such as SONET, SDH, or DWDM. Two-fiber rings (Fig. 1, *left*) use a single fiber pair; one fiber for each traffic direction between any two adjacent ring nodes. Four-fiber rings (Fig. 1, *right*) use two fiber pairs between each of the nodes. The second fiber pair is dedicated to protection or low priority traffic. It allows for enhanced protection schemes as discussed in Section 2.4.

The optical channels are added and dropped by the OADM nodes. Signal processing is performed mainly in the optical domain using wavelength-division multiplexing and optical amplification techniques. All-optical processing provides transparent and cost-efficient transport of multiwavelength signals. However, it has some constraints:

- *Wavelength Blocking*. In order to establish an optical path, either the wavelength of this path has to be available in all spans between both terminating add/drop nodes or the wavelength has to be converted in intermediate nodes along the path. Otherwise, wavelength blocking occurs. It results in lower utilization of the ring resources. It can be minimized, but not avoided completely by careful and farsighted network planning and wavelength management. It has been shown that the positive effects of wavelength conversion in intermediate nodes are limited [4].
- *"Signal Quality Supervision."* As today's services are mainly digital, the bit error ratio (BER) is the preferred quality-of-service information. All-optical performance supervision does not provide a parameter that directly corresponds to the BER of the signal.
- *Optical Transparency Length*. This is the maximum distance that can be crossed by an optical channel. The transparency length is limited by optical signal impairments, such as attenuation of optical signal power (loss), dispersion, and nonlinear effects. It depends on the data rate transported by the optical channels, the ring capacity, the quality of the optical components, and the type of fiber (e.g., standard single-mode fiber, dispersion shifted fiber). For 10-Gbps signals, transparency lengths of 300 km can be achieved with reasonable effort. This makes DWDM ring networks most attractive for metropolitan-area networks (MAN; $\leq$80 km ring circumference) and regional area networks ($\leq$300 km ring circumference).

2.2. Transparent Networks and Opaque Networks

If the optical transparency length is exceeded, full regeneration of the signal, including reamplification, reshaping, and retiming (3R regeneration) is necessary. With 3R regeneration however, the transport capability is restricted to certain bit rates and signal formats. In this case, the ring is called opaque. Specific digital processing equipment must be available for each format that shall be transported. Furthermore, 3R regeneration is very costly. The typical opaque DWDM ring topology uses all-optical processing within the ring. 3R regeneration is only performed at the tributary interfaces where the signals enter and exit the ring (i.e., at the add/drop locations). This decouples the optical path design in the ring from any client signal characteristic. Furthermore, it provides digital signal quality supervision the add/drop traffic. Support of different signal formats is still possible by using the appropriate tributary interface cards without the need for reconfiguration at intermediate ring nodes.

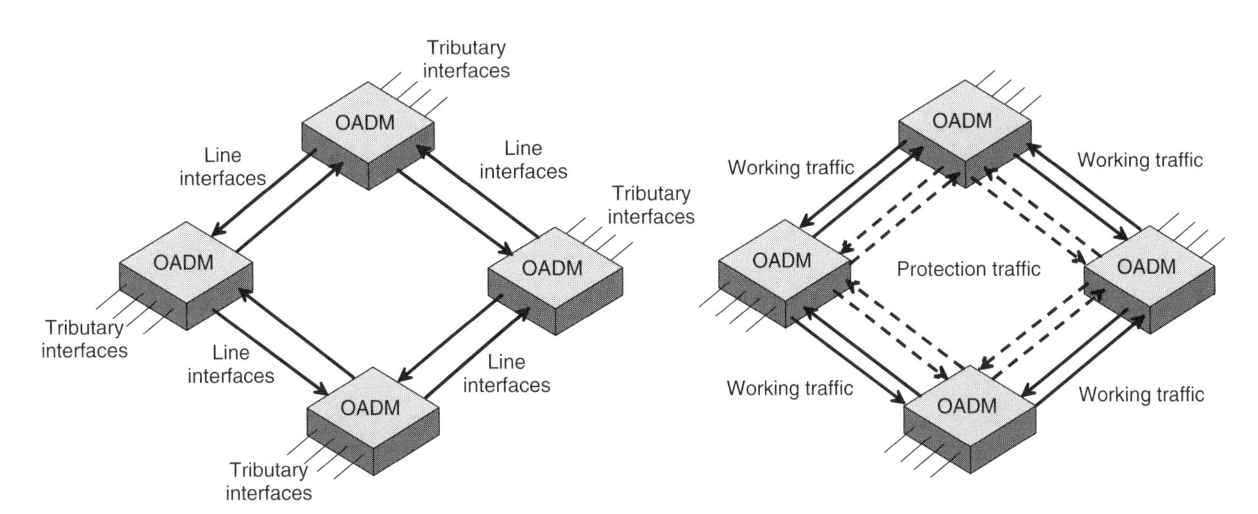

Figure 1. Basic ring architectures: *left*, two-fiber ring; *right*, four-fiber ring.

2.3. Optical-Layer Networks

The International Telecommunication Union (ITU) has standardized the architecture of the Optical Transport Network (OTN) in its recommendation ITU-T G.872 [5]. G.872 defines three optical layer networks: the optical transmission section (OTS) layer, the optical multiplex section (OMS) layer, and the optical channel (OCh) layer network. Figure 2 shows the mapping of these layer networks on a DWDM ring network.

- The OTS represents the DWDM signal transmitted between two nodes (OADM) and/or optical line amplifiers (OLA). Intermediate optical line amplifiers can be used to amplify the signal in cases of long distances between two adjacent nodes.
- The OMS represents the DWDM signal between two adjacent nodes.
- The OCh represents a single optical channel between the 3R regeneration points. For the case of the typical opaque ring architecture, the OCh is terminated at the tributary interfaces of the OADM nodes.

Supervision and management functions are defined for each network layer. Overhead that supports these functions is transported on an optical supervisory channel, which uses a dedicated wavelength for transmission between the nodes. ITU-T G.872 defines also two digital layer networks for a seamless integration of optical and digital processing. These digital layers, the optical channel transport unit (OTU) and the optical channel data unit (ODU) are further described in ITU–T recommendation G.709 [6]. They provide the following three functions: digital signal supervision, time division multiplexing for the aggregation of lower bit rate signals, and forward error correction (FEC) [7]. FEC allows for extending the optical transparency length for a given signal quality.

2.4. Protection

Ring networks support protection switching ideally, since there are always two alternate paths between any two ring nodes. Ring protection schemes are already defined for SONET and SDH [1,8]. Similar concepts can be applied to DWDM ring networks.

$1+1$ *optical channel* (*OCh*) *protection* is the most simple ring protection scheme. It belongs to the class of "dedicated protection," as it uses a dedicated protection connection for each working connection. At the add ring node, the traffic is bridged (branched) to both the working connection and the protection connection. The drop ring nodes select between the two connections based on the quality of the received signals. As a result, $1+1$ OCh protected connections always consume one wavelength per transmission direction in the entire ring, whereas unprotected connections use this wavelength on the shortest path between the terminating nodes only. The utilization of the ring capacity for working traffic can maximally reach 50%.

Optical multiplex section bidirectional self-healing ring (*OMS-BSHR*) belongs to the class of "shared protection," as it shares the protection connection between several working connections. BSHR is also known in SONET as bidirectional line switched ring (BLSR) [9]. OMS-BSHR can be used in both two- and four-fiber rings. In case of a two-fiber BSHR (Fig. 3), One-half of the wavelength channels are used for the working connections; the other half are used for protection. In order to avoid wavelength converters at the protection switches, working and protection wavelengths of opposite directions should be the same (flipped wavelength assignment). For example, in a DWDM ring of N wavelength channels, the wavelengths 1 to $N/2$ (lower band) are used for working connections whereas the wavelengths $N/2+1$ to N (upper band) are used for protection connections. This working

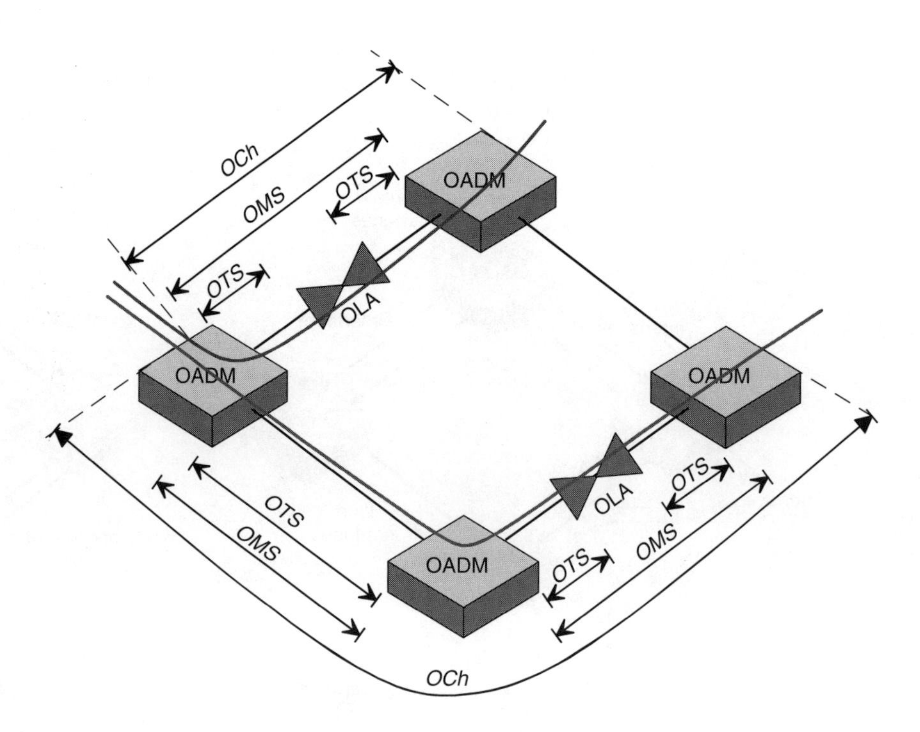

Figure 2. Optical-layer networks.

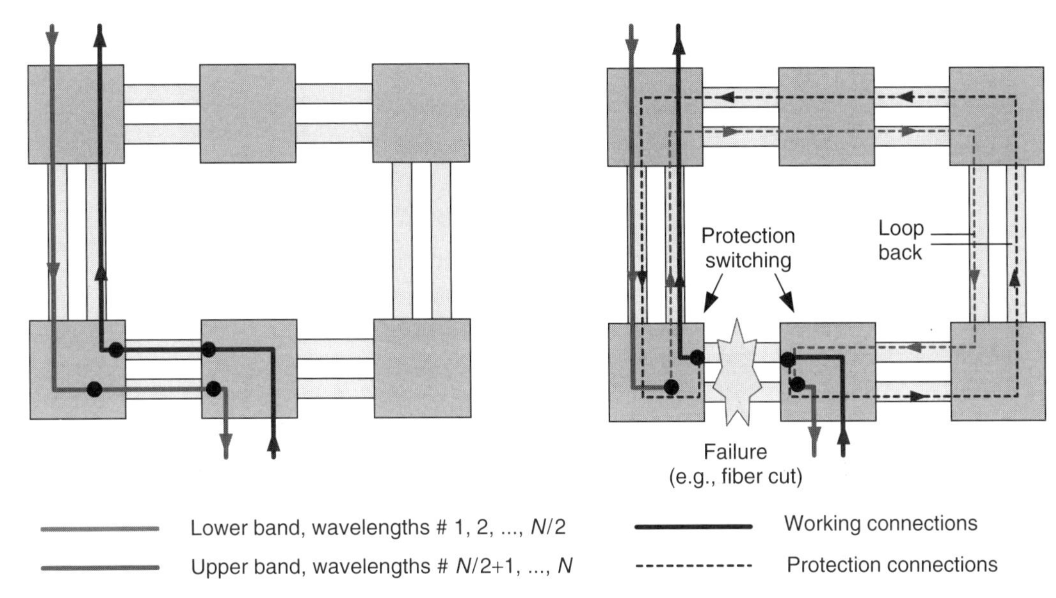

Figure 3. Two-fiber OMS-BSHR: *left*, normal operation; *right*, protection case.

and protection traffic runs in a clockwise direction. In the counter clockwise direction, the wavelengths $N/2+1$ to N are used for working connections and the wavelengths 1 to $N/2$ are used for protection. In case of a failure, the two nodes adjacent to the failure location perform ring protection switching (Fig. 3, *right*). They bridge and select the clockwise working connection to and from the counter clockwise protection connection and vise versa. The whole working traffic that passed the failure location is now looped back to the other direction of the ring.

In a four-fiber BSHR, dedicated fibers for working traffic and protection traffic are used. By this, four-fiber BSHR rings provide twice the capacity than the two-fiber BSHR rings. In addition to the ring protection of the two-fiber BSHR, the four-fiber BSHR can perform span protection. This provides protection against failures of the working fiber pair between two adjacent nodes. An "automatic protection switching" (APS) protocol is necessary for the coordination of the bridge and switch actions and for the use of the shared protection connection. Unlike in $1+1$ OCh protecting the utilization of the ring capacity for working traffic is always exactly 50%. A disadvantage of OMS-BSHR are the long pathlengths that are possible due to the loop back in the protection case. In order not to exceed the transparency length in a case of protection, the ring circumference is reduced.

The *bidirectional optical channel shared protection ring* (*OCh-SPR*) avoids the loop back problem of the OMS-BSHR. Here, ring protection is performed by the add and drop nodes of the optical channel. The add nodes switch the add channels to the working or protection connection. The drop nodes select the working or protection channels directly (Fig. 4).

OCh-SPR can be used in both the two- and four-fiber rings. An automatic protection switching protocol is required for the coordination of the bridge and switch actions; as well as, for the use of the shared protection connection. In an OCh-SPR, protection switches for each optical channel are needed, while the OMS-BSHR allows the use of a protection switch for all working/protection connections in common.

A general issue of shared protection schemes in all-optical networks is the restricted supervision capability of inactive protection connections. Failures of protection connections that are not illuminated can be detected only after activation. Usually, activation occurs due to a protection situation. In that case, the detection of failures in the protection connection comes too late. Furthermore, the dynamic activation/deactivation of protection channels on a span has an influence on the other channels of the span. It will result in bit errors for these channels if not carefully performed.

2.5. Single-Fiber Bidirectional Transmission

DWDM allows bidirectional signal transmission over a single fiber by using different wavelengths for the two signal directions. This allows for building two-fiber ring architectures by only using a single fiber between the ring nodes. It also allows for four-fiber architectures with two physical fibers. However, the two-fiber OMS-BSHR and OCh-SPR protection schemes with their sophisticated wavelength assignment are not supported.

3. OADM NODE ARCHITECTURE

The network functionalities described in the previous section are realized in the optical add/drop multiplexers. The OADM node architecture and functionality is described below.

3.1. Basic OADM Architecture

An OADM consists of three basic modules as shown in Fig. 5:

- The *DWDM line interfaces*, which physically process the ring traffic

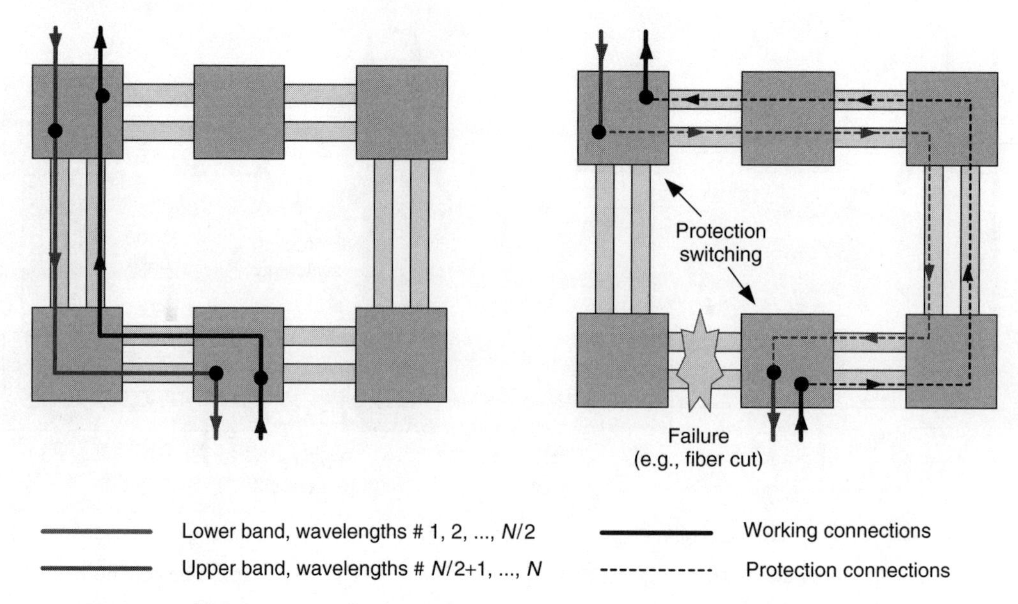

Figure 4. Two-fiber OCh shared protection: *left*, normal operation; *right*, protection case.

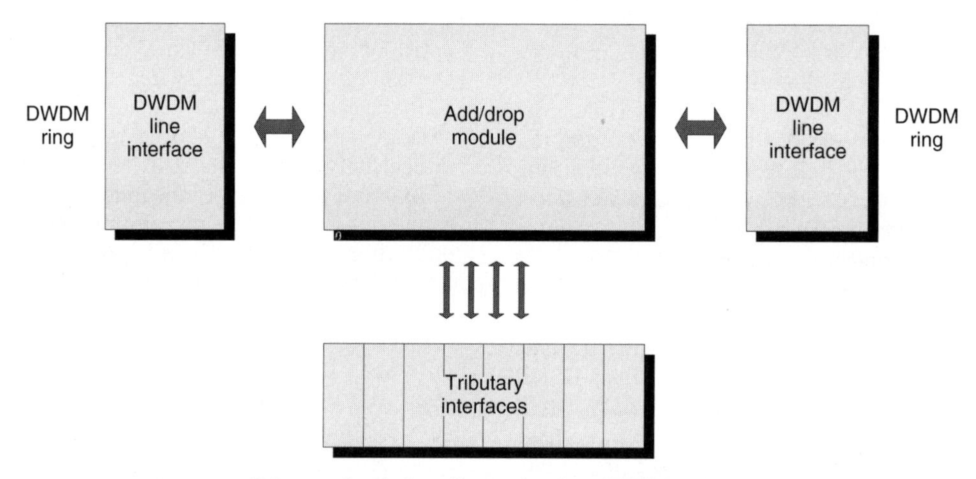

Figure 5. Basic OADM architecture.

- The *add/drop module*, which performs both through connections and the add/drop function for the optical channels
- The *tributary interfaces*, which physically process the add/drop channels

The detailed functionality of these modules depends on the OADM features.

3.2. Characteristic OADM Features

This section describes the most important features of the OADM nodes.

3.2.1. Add/Drop Capacity. The add/drop capacity is defined as the relationship between the capacity of the tributary interfaces and the capacity of both line interfaces. In an extreme case, the entire traffic entering the OADM from both sides of the ring must be dropped. Therefore, an add/drop multiplexer is considered to have 100% add/drop capacity if the capacity of the tributary interfaces equals the capacity of both line interfaces. For instance, an OADM of 100% add/drop capacity in an 80-channel DWDM ring provides 160 channels at its tributary interfaces. Designing an OADM for an appropriate add/drop capacity is an efficient approach in minimizing cost and complexity.

3.2.2. Flexibility. The selection of add/drop channels in an OADM can either be remotely configurable for channels that require frequent reconfiguration (dynamic), or it can be performed manually for "long term" connections (static). Furthermore, channels that will never be dropped can be passed through directly from one line interface to the other without add/drop capabilities (express channels). Other important OADM features are

- *Transparency* or *opacity* as discussed in Section 2.2.
- *Wavelength conversion* capability as discussed in Sections 2.

- *Protection switching* features as discussed in Section 2.4.
- *Signal supervision* capabilities can be based on optical parameters only, such as optical power and optical signal-to-noise ratio (OSNR). Additionally, in opaque rings, signal supervision can be based on bit error ratio monitoring.

3.3. The DWDM Line Interfaces

The main functionality of the DWDM line interfaces includes, primarily the compensation for physical degradation of the DWDM signal during transmission and secondarily the supervision of the DWDM signal. There are three major categories of degradation: (1) loss of optical signal power due to attenuation, (2) dispersion of the impulse shape of high-speed optical signals, and (3) impulse distortion or neighbor channel crosstalk due to nonlinear fiber-optical effects. To overcome these degradations, optical amplifiers and dispersion compensation components may be used. For metropolitan applications, moderate-gain amplifiers can be used to reduce the nonlinear fiber-optical effects, since the node distance is relatively short. Design approaches are mentioned in the references [2,10]. In the line interfaces, signal supervision is based on optical parameters (e.g., optical power, OSNR).

3.4. The Add/Drop Module

The add/drop module performs the following major functions:

- It configures the add/drop of up to 100% of the ring traffic.
- It converts the wavelengths of the add/drop and the through signals (optional).
- It protects the traffic on the OMS or OCh level (optional).

3.4.1. Optical Channel Groups. Optical channels can be classified as dynamic, static, or express traffic. In order to minimize the hardware complexity, in other words installation costs, traffic classes are assigned to optical channel groups accordingly. The optical channel groups are processed according to their specific characteristics. Figure 6 illustrates the group separation (classes) of DWDM traffic by group multiplexer and demultiplexer components. The groups are processed within the add/drop module using different techniques. Methods for defining groups and related component technologies are discussed in Section 4 based on technological boundary conditions.

For dynamic traffic, wavelength routing technologies, such as remotely controlled optical switching fabrics or tunable filters, are used. For static traffic, *manual configuration* via distribution frames (patch panels) is employed. Wavelength routing necessitates higher installation costs than manual configuration. However, it enables the reduction of operational costs due to automation. Operational costs can become prohibitive for manual configuration if frequent reconfiguration is necessary. A cost-optimal OADM processes part of the traffic via distribution frames and the other part via wavelength routing techniques. Hence, the use of costly components is focused on cases where they benefit most [11]. Designing an OADM for a specific add/drop capacity is very efficient especially in the wavelength routing part of the add/drop module. Figure 7 shows a possible architecture of a wavelength routing add/drop matrix.

The add/drop process of the dynamic traffic can be performed in two stages: the add/drop stage and the distribution stage. In order to perform wavelength conversion, wavelength converter arrays (WCA) can be used in addition. The add/drop stage is the only requirement. The distribution stage and the wavelength converters are optional.

The *add/drop-stage* filters the drop channels out of the DWDM signal and adds new channels to the ring. Single optical channels are typically processed. But it is also possible to add or to drop optical channel groups to

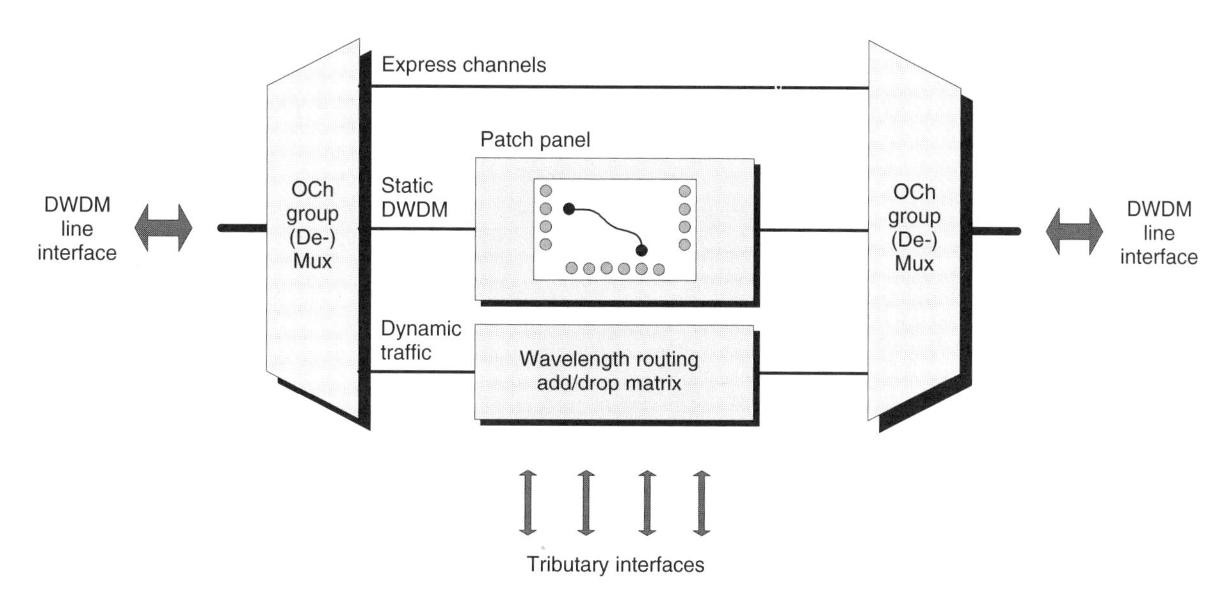

Figure 6. Add/drop module — processing according to traffic characteristics.

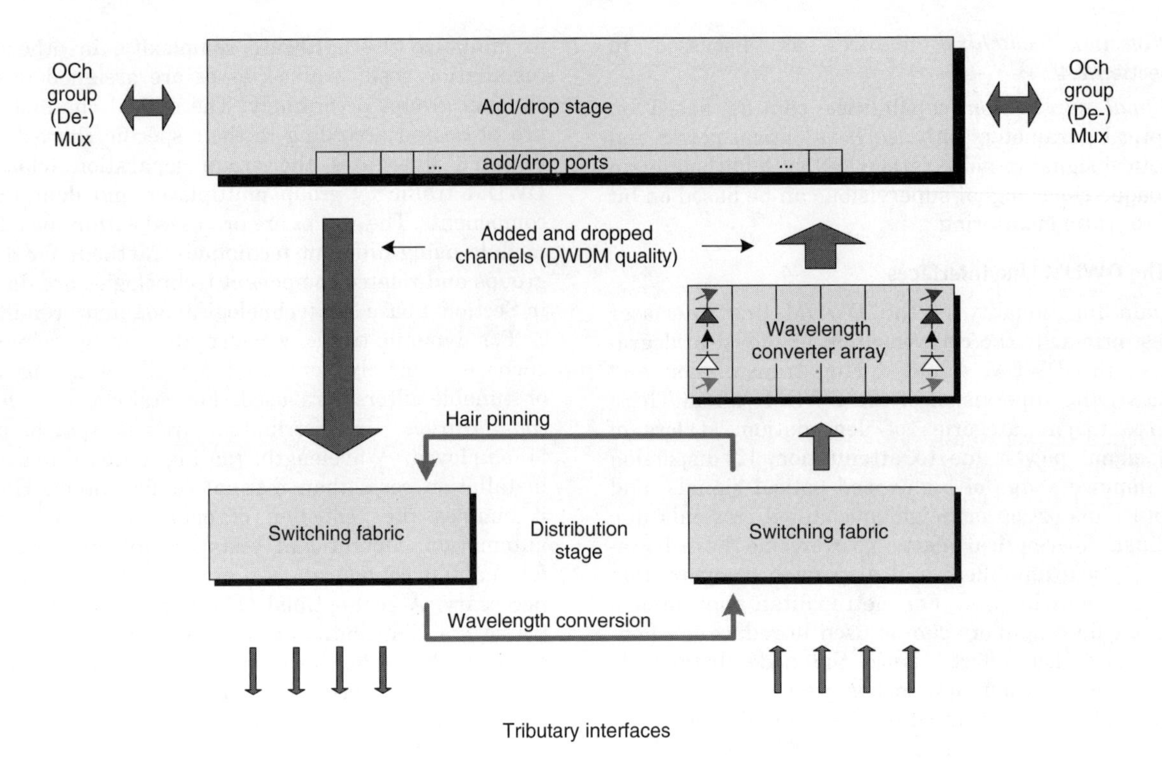

Figure 7. Wavelength routing add/drop matrix (two-stage).

a single port. Various technologies for realizing add/drop stages will be discussed in Section 4.2.

The *wavelength converter array* can be used for wavelength assignment of add channels from the tributary interfaces and for wavelength conversion of through channels. Technologies are discussed in Sections 4.3 and 4.4.

The distribution stage performs the following functions:

1. *Assignment of Channels to Specific Tributary Interfaces*. Not every tributary signal must be connected to the ring at anytime. There may be "part-time" leased lines that share the same wavelength channel such as one at business hours other one at nighttime or at weekends. For example, one customer uses the wavelength channel at the distribution stage connects tributary signals to specific add channels for insertion into the ring. For extraction, it connects them to specific drop channels.
2. *Wavelength Conversion for the Through Channels*. Dropped channels can be reinserted on another wavelength via the link "wavelength conversion" in Fig. 7. A WCA is required for this application.
3. *Hairpinning*. In some applications, add/drop multiplexers are used to route signals between tributary interfaces. This function is known as "hairpinning." It can be realized by the connection as shown in Fig. 7.
4. *Flexible Wavelength Assignment for the Add Channels*. Besides wavelength conversion, the WCA can be used to assign a specific wavelength to an incoming tributary signal. The distribution stage allows for flexible assignment of tributary signals to a WCA element of a specific wavelength. This function can be part of the tributary interfaces as well.

Techniques of switching fabrics to perform these functions are mentioned in Section 4.3.

3.5. The Tributary Interfaces

The tributary interfaces prepare the incoming signals for transmission over the DWDM ring. Incoming signals do not necessarily have DWDM quality. Furthermore, they monitor the quality of the incoming and outgoing tributary signals. It is a strong customer requirement to support a large variety of signal formats and data rates as tributary signals. In the simplest case the tributary signal is directly forwarded to the add/drop module (transparent OADM). In this case, the incoming optical signal must have the correct wavelength to fit into the DWDM line signal; otherwise, the signal has to enter the ring via wavelength converters, either as part of the tributary interface or via the WCA of the add/drop module. The simplest way of signal supervision is monitoring the optical power (loss of signal) of the incoming and outgoing tributary signals. A more advanced approach in the optical domain is the measurement of the optical signal-to-noise ratio. If digital signal quality supervision (i.e., bit error ratio measurement) is needed, 3R regeneration has to be performed. This normally requires client signal-specific processing using dedicated tributary ports for different client formats, such as SONET or Gigabit Ethernet.

4. OADM COMPONENTS

This section focuses on component technologies of the add/drop module and the tributary interfaces.

4.1. Group Filters and Multiplexer/Demultiplexer Components

A multiplexer/demultiplexer in the sense of a WDM component is a device that provides one optical fiber port for DWDM signals and multiple fiber ports for optical channels or channel groups. These multiplexers/demultiplexers are passive components that work in both directions. They can be used for either separating DWDM signals or for combining optical channels to one DWDM signal. Group filters and WDM multiplexers/demultiplexers are basically the same type of component. They just have different optical filter characteristics. A WDM multiplexer/demultiplexer combines or separates single optical channels whereas a group filter combines or separates optical channel groups. Figure 8 shows three basic realizations of wavelength multiplexer/demultiplexer filters.

A basic advantage of block channel groups as opposed to interleaved optical channel groups is the stronger suppression of neighbor group signals. Nevertheless, we will see in Section 4.2 that there are applications in which interleaved channel groups support the overall system performance better than channel blocks. Static multiplexer/demultiplexer filters are usually based on technologies such as multilayer dielectric thin-film filter (TFF), fixed fiber bragg grating (FBG), diffraction grating, arrayed waveguide grating (AWG), cascaded Mach–Zehnder interferometer, and Fabry–Perot interferometer. A detailed discussion of these technologies is given in Section 3.3 of Ref. 2. Interleaver filter technologies are described in Refs. 12 and 13.

4.2. Add/Drop Filter Technologies

Figure 9 shows two popular realization concepts of add/drop filters: (*left*) wavelength multiplexer/demultiplexer — optical switch combination and (*right*) tunable filter cascade — circulator/coupler combination.

The multiplexer/demultiplexer–optical switch solution usually demultiplexes the entire DWDM signal and provides single optical channels via switches at each add/drop port. The tunable filter cascade (Fig. 9, *right*) provides exactly the same functionality. The incoming DWDM signal is forwarded clockwise to the tunable filter cascade by a circulator. The tunable filters reflect selected channels. All other channels pass the filter cascade. The reflected channels travel back to the circulator to be forwarded to the drop port, again in a clockwise motion [14]. For adding optical channels, either a coupler

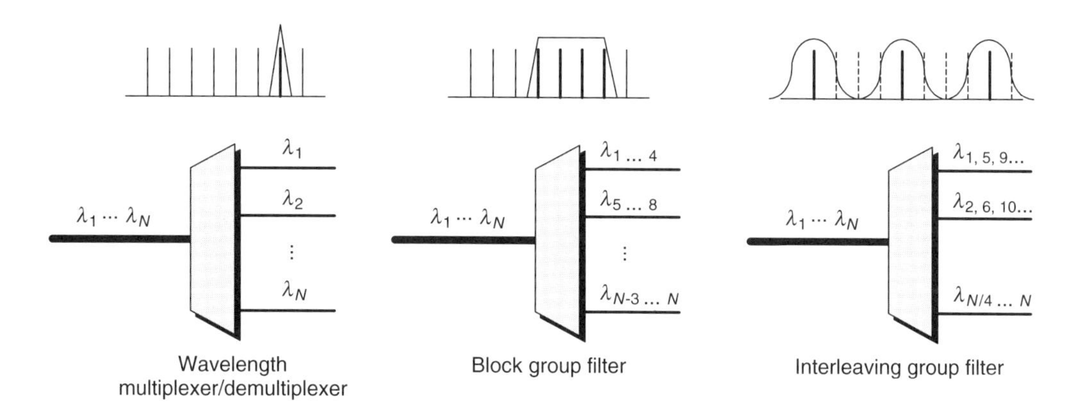

Figure 8. Static wavelength multiplexer/demultiplexer filters.

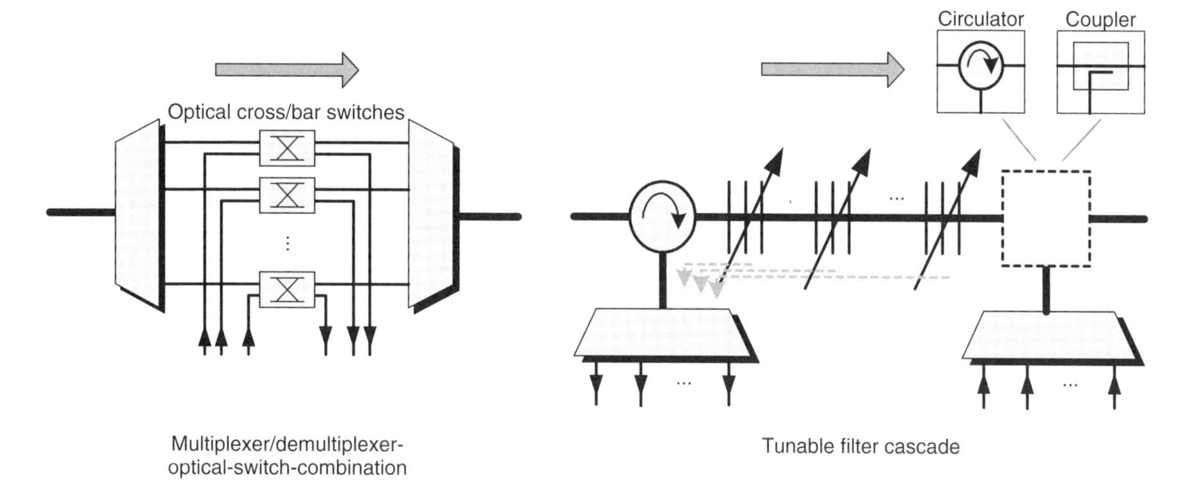

Figure 9. Add/drop filter technologies.

or a circulator can be used. A coupler provides better isolation between the add ports and the drop ports. It is also the least expensive approach. On the other hand, circulators allow for a lower transmission loss. Here, the advantage of interleaved channel groups becomes evident: the bandwidth requirements for tunable filters are lower. In tunable filter cascades, the reconfiguration process can affect uninvolved traffic. For example, if a filter is tuned from wavelength 1 to wavelength 4, it passes the wavelengths 2 and 3. Therefore, traffic running on these channels is affected by the retuning process and thus may render it unusable. The influence of the dispersion of the tunable filters is strong compared to the multiplexer/demultiplexer–optical switch solution. This is another drawback of tunable filter cascades. If tunable filters are tuned to a neutral park position between two densely spaced channels, these channels can be degraded by the dispersion of the filter. This design issue can be overcome by using interleaved channel groups. Wider spacing between the channels lowers the bandwidth requirements for the filters, and thus resulting in a lower dispersion. The basic advantage of tunable add/drop filters is the lower number of internal fiber connections and splices. However, because of the typical characteristics of fiber gratings (e.g., cladding modes), this application is limited to small and medium size channel groups. A more detailed comparison of both concepts is given in Ref. 15.

4.3. Optical Switching Fabrics

Optical switching fabrics provide cross–connection functionality. They are the technology of choice in distribution stages of optical add/drop multiplexers (see Fig. 7). From the viewpoint of OADM optical signal processing, we can distinguish between protection switching and wavelength routing applications.

Protection switching requires switching times of a few milliseconds. Switching times up to hundreds of milliseconds are sufficient for wavelength routing applications. For protection switching and wavelength routing applications, electromechanical, thermooptic, electrooptic, and acoustooptic switches can be used, see Ref. 2, Section 3.7 and Ref. 16, Chap. 10. Today, among the electromechanical switches, the MEMS technology (microelectromechanical system) is seen as the most promising technology in providing high port counts. Switches of more than 100×100 ports have been realized.

Tunable filters, as discussed in Section 4.2, are also an alternative technology to switching fabrics in order to realize cross–connection functionality. Optical circuits using tunable filters and circulators have been presented by Chen and Lee [17].

4.4. Wavelength Converter Technologies

Today, wavelength conversion is performed by O/E/O (optical/electrical/optical) conversion using photodiodes and tunable or fixed-wavelength lasers that are either modulated externally or internally (directly). Tunable lasers allow for covering a wide range of wavelengths and to minimize the number of wavelength converters that are needed. An overview about tunable laser technology is given in Ref. 18. The optical receiver and transmitter components can be connected via analog electrical amplifiers allowing for the transmission of any signal format up to a certain bit rate. If 3R regeneration is performed (digital signal processing between receiver and transmitter), a further limitation to certain signal formats may apply. In future transparent systems, the use of all-optical wavelength converters is expected. All-optical wavelength converters make use of nonlinear optical effects such as cross-gain modulation (CGM), cross phase modulation (XPM) or four-wave mixing (FWM). These effects are treated in detail in Ref. 19, Section 2.7. Semiconductor optical amplifiers (SOA) are the preferred active medium as they exhibit strong nonlinearity, wide gain bandwidth and easy integration. As a pump source, unmodulated lasers are used. Tunable pump lasers provide the capability of tunable all-optical wavelength converters.

5. SUMMARY

DWDM ring networks provide high capacity, high flexibility (multiservice integration), and high reliability (protection) at low operational costs to the operators of metropolitan and regional networks. Because of the simple ring topology, the network management is relatively less complex. The ring functions are determined mainly by the optical add/drop multiplexers. The DWDM line interface mainly determines the maximum ring circumference that can be achieved. An add/drop module, that provides manual routing capabilities, allows for low installation costs. But if frequent reconfiguration is necessary, operational costs can become prohibitive. For this type of traffic, wavelength routing capabilities that provide remotely controlled dynamic routing should be implemented. The technologies of choice for wavelength routing are integrated optical switching fabrics such as MEMS and tunable filters.

DWDM networks can either be transparent or opaque. In the transparent realization, no electronic traffic processing occurs within the ring. The transport is independent of the data format. In opaque networks, for quality-of-service supervision and management reasons, digital (electronic) processing is performed at the borders of the DWDM ring network. The transport in opaque networks may be limited to certain data formats.

The business success of network operation is driven mainly by an overall cost minimization by selling data transport in high volume. This is a general rule in markets that are characterized by low differentiation possibilities and high impact of small differences (volume markets). In the network operation business, differentiation is possible by better reliability, availability and flexibility. Therefore, DWDM ring networks are ideal in paving the way for the business success of metropolitan and regional network operators.

BIOGRAPHIES

Detlef Stoll received a Dipl.-Ing. degree (M.S.) in communication engineering in 1988 from the University of

Hannover, Germany, and a Ph.D. degree in electrical engineering in 1993 from the University of Paderborn, Germany. The subject of his Ph.D. thesis was the derivation of an integrated model for the calculation of the nonlinear transmission in single-mode optical fibers including all linear and nonlinear effects. He joined the Information & Communication Networks Division of the Siemens Corporation in 1994 as a research & development engineer. At Siemens, he worked on the advance development of broadband radio access networks and on the development of SONET/SDH systems. From 1998 to 2000 he led the advance development of a Wavelength Routing Metropolitan DWDM Ring Network for a multi-vendor field trial. Since 2000 he is managing the development of optical network solutions at Optisphere Networks, Inc. located in Boca Raton, Florida. Dr. Stoll has filed approximately 20 patents and published numerous articles in the field of optical networks and forward error correction techniques. His areas of interest are linear and nonlinear systems and wavelength routing networks.

Juergen Heiles received a Dipl.-Ing. (FH) degree in electrical engineering from the University of Rhineland-Palatinate, Koblenz, Germany, in 1986. He joined the Public Networks Division of the Siemens Corporation, Munich, Germany, in 1986 as a research & development engineer of satellite communication systems and, later on, fiber optic communication systems. Since 1998 he has been responsible for the standardization activities of the Siemens Business Unit Optical Networks. He participates in the ITU, T1, OIF and IETF on SONET/SDH, OTN, ASON, and GMPLS and is editor or coauthor of several standard documents. Juergen Heiles has over 15 years of experience in the design, development, and system engineering of optical communication systems starting from PDH systems over the first SONET/SDH systems to today's DWDM networks. He holds three patents in the areas of digital signal processing for optical communication systems. His areas of interest are network architectures and the functional modeling of optical networks.

Jimin Xie received a B.S. degree in electronics in 1986 from the Institute of Technology in Nanjing, China, and an M.S. degree in radio communications in 1988 from the Ecole Superieure d'Electricite in Paris, France, and a Ph.D. degree in opto-electronics in 1993 from the Université Paris-Sud, France. He worked on submarine soliton transmission systems for Alcatel in France from 1993 to 1995. In 1996, he worked at JDS Uniphase as a group leader in the Strategy and Research Department in Ottawa, Canada. At JDSU, he worked on the design and development of passive DWDM components, such as interleaver filters, DWDM filters, OADM modules, polarization detectors and controllers, dispersion compensators, gain flatteners, optical switches, etc. Dr. Xie has filed 12 patents in the field of optical components. Since 2001, he has been the manager of the Optical Technologies department at Optisphere Networks, Inc. located in Boca Raton, Florida. He participates in the OIF standardization activities for Siemens. His areas of interest are the new technologies of components and modules for optical networks.

BIBLIOGRAPHY

1. M. Sexton and A. Reid, *Broadband Networking: ATM, SDH, and SONET*, Artech House, Norwood, MA, 1997.
2. R. Ramaswami and K. N. Sivarajan, *Optical Networks*, Morgan Kaufmann, San Francisco, CA, 1998.
3. P. Kotler, *Marketing Management*, Prentice-Hall, Englewood Cliffs, NJ, 1999.
4. P. Arijs, M. Gryseels, and P. Demeester, Planning of WDM ring networks, *Photon. Network Commun.* **1**: 33–51 (2000).
5. ITU-T Recommendation G.872, *Architecture of the Optical Transport Networks*, International Telecommunication Union, Geneva, Switzerland, Feb. 1999.
6. ITU-T Recommendation G.709, *Interface for the Optical Transport Network (OTN)*, International Telecommunication Union, Geneva, Switzerland, Oct. 2001.
7. I. S. Reed and X. Chen, *Error-Control Coding for Data Networks*, Kluwer, Norwell, MA, 1999.
8. ITU-T Recommendation G.841, *Types and Characteristics of SDH Network Protection Architectures*, International Telecommunication Union, Geneva, Switzerland, Oct. 1998.
9. GR-1230-CORE, *SONET Bidirectional Line-Switched Ring Equipment Generic Criteria*, Telcordia, 1998.
10. P. C. Becker, N. A. Olsson, and J. R. Simpson, *Erbium-Doped Fiber Amplifiers, Fundamentals and Technology*, Academic Press, San Diego, CA, 1997.
11. D. Stoll, P. Leisching, H. Bock, and A. Richter, Best effort lambda routing by cost optimized optical add/drop multiplexers and cross-connects, *Proc. NFOEC*, Baltimore, MD, Session A-1, July 10, 2001.
12. H. van de Stadt and J. M. Muller, Multimirror Fabry-Perot interferometers, *J. Opt. Soc. Am. A* **8**: 1363–1370 (1985).
13. B. B. Dingel and M. Izutsu, Multifunction optical filter with a Michelson-Gires-Tournois interferometer for wavelength-division-multiplexed network system applications, *Opt. Lett.* **14**: 1099–1101 (1998).
14. U.S. Patent 5,748,349 (1998), V. Mizrahi, Gratings-based optical add-drop multiplexers for WDM optical communication systems.
15. D. Stoll, P. Leisching, H. Bock, and A. Richter, Metropolitan DWDM: A dynamically configurable ring for the KomNet field trial in Berlin, *IEEE Commun. Mag.* **2**: 106–113 (2001).
16. I. P. Kaminow and T. L. Koch, *Optical Fiber Telecommunications III B*, Academic Press, Boston, MA, 1997.
17. Y.-K. Chen and C.-C. Lee, Fiber Bragg grating based large nonblocking multiwavelength cross-connects, *IEEE J. Lightwave Technol.* **16**: 1746–1756 (1998).
18. E. Kapon, P. L. Kelley, I. P. Kaminow, and G. P. Agrawal, *Semiconductor Lasers II*, Academic Press, Boston, MA, 1999.
19. G. P. Agrawal, *Fiber-optic Communication Systems*, Wiley, New York, NY, 1998.

E

EXTREMELY LOW FREQUENCY (ELF) ELECTROMAGNETIC WAVE PROPAGATION

STEVEN CUMMER
Duke University
Durham, North Carolina

1. INTRODUCTION

Extremely low frequency (ELF) electromagnetic waves are currently the lowest frequency waves used routinely for wireless communication. The IEEE standard radio frequency spectrum defines the ELF band from 3 Hz to 3 kHz [1], although the acronym ELF is often used somewhat loosely with band limits near these official boundaries. Because of the low frequencies involved, the bandwidth available for ELF communication is very small, and the data rate of correspondingly low. Existing ELF communications systems with signal frequencies between 40 and 80 Hz transmit only a few bits per minute [2]. Despite this severe limitation, ELF waves have many unique and desirable properties because of their low frequencies. These properties enable communication under conditions where higher frequencies are simply not usable. Like all electromagnetic waves excited on the ground in the HF (3–30 MHz) and lower bands, ELF waves are strongly reflected by the ground and by the Earth's ionosphere, the electrically conducting portion of the atmosphere above roughly 60 km altitude [3]. The ground and ionosphere bound a spherical region commonly referred to as the *earth–ionosphere* waveguide in which ELF and VLF (very low frequency, 3–30 kHz) propagate. The combination of strongly reflecting boundaries and long ELF wavelengths (3000 km at 100 Hz) enables ELF waves to propagate extremely long distances with minimal attenuation. Measured ELF attenuation rates (defined as the signal attenuation rate in excess of the unavoidable geometric spreading of the wave energy) are typically only ~1 dB per 1000 km at 70–80 Hz [4–6]. The signal from a single ELF transmitter can, in fact, be received almost everywhere on the surface of the planet, even though only a few watts of power are radiated by existing ELF systems. Because the wavelength at ELF is so long, ELF antennas are very inefficient and it takes a very large antenna (wires tens of kilometers long) to radiate just a few watts of ELF energy. This makes ELF transmitters rather large and expensive. Only two ELF transmitters currently operate in the world, one in the United States and one in Russia. Besides their global propagation, the value of ELF waves results from their ability to penetrate effectively through electrically conducting materials, such as seawater and rock, that higher frequency waves cannot penetrate. These properties make ELF waves indispensable for applications that require long-distance propagation and penetration through conducting materials, such as communication with submarines [7]. Unlike many other radiobands, there are strong natural sources of ELF electromagnetic waves. The strongest of these in most locations on the earth is lightning, but at high latitudes natural emissions from processes in the magnetosphere (the highest portions of the earth's atmosphere) can also be strong. There are a variety of scientific and geophysical applications that rely on either natural or man-made ELF waves, including subsurface geologic exploration [8], ionospheric remote sensing [9], and lightning remote sensing [10].

2. THE NATURE OF ELF PROPAGATION

Communication system performance analysis and most geophysical applications require accurate, quantitative models of ELF propagation. Because the ELF fields are confined to a region comparable to or smaller than a wavelength, phenomena in ELF transmission and propagation are substantially different from those at higher frequencies. But before delving too deeply into the relevant mathematics, a qualitative description will give substantial insight into the physics of ELF transmission, propagation, and reception.

2.1. Transmission

A fundamental limit on essentially all antennas that radiate electromagnetic waves is that, for maximum efficiency, their physical size must be a significant fraction of the wavelength of the radiated energy [11]. At ELF, this is a major practical hurdle because of the tremendously long wavelengths (3000 km at 100 Hz) of the radiated waves. Any practical ELF transmission system will be very small relative to the radiated wavelength. Very large and very carefully located systems are required to radiate ELF energy with sufficient power for communication purposes, and even then the antenna will still be very inefficient.

Multikilometer vertical antennas are not currently practical. Thus a long horizontal wire with buried ends forms the aboveground portion of existing ELF antennas. The aboveground wire is effectively one half of a magnetic loop antenna, with the other half formed by the currents closing below ground through the earth, as shown in Fig. 1. The equivalent depth d_{eq} of the closing current with frequency ω over a homogeneous ground of conductivity σ_g is [12]

$$d_{\text{eq}} = (\omega\sigma_g\mu_0)^{-1/2} = 2^{-1/2}\delta \quad (1)$$

where δ is the skin depth [13] of the fields in the conducting ground. A grounded horizontal wire antenna of length l is thus equivalent to an electrically small magnetic loop of area $2^{-1/2}\delta l$. This effective antenna area, which is linearly related to the radiated field strength, is inversely proportional to the ground conductivity. This implies that a poorly conducting ground forces a deeper closing current and therefore increases the efficiency of the antenna. This general effect is opposite that for vertically oriented

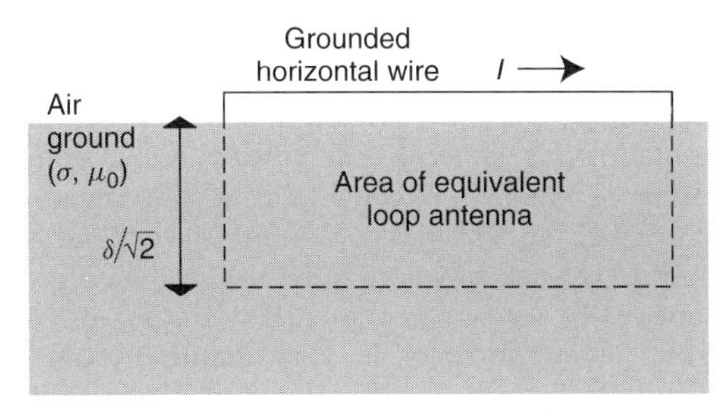

Figure 1. Typical implementation of an ELF antenna. By grounding the ends of a long, current-carrying horizontal wire, the current is forced to close through the ground at a significant depth if the ground is poorly conducting. This forms a physically large loop antenna. The U.S. Navy ELF transmitter antenna is many tens of kilometers long above ground and has an effective current closure depth of 2.6 km.

antennas, in which a good conducting ground serves to improve the radiation efficiency. Horizontal ELF antennas thus require placement over as poorly conducting ground as possible for maximum efficiency.

The U.S. Navy ELF transmitter, which transmits at frequencies between 70 and 80 Hz and is described in some detail by Friedman [2], is distributed in multiple wire antennas in Wisconsin and Michigan. This is one of only a few locations in the United States with relatively low ground conductivity; the measured conductivity is $\sim 2.4 \times 10^4$ S/m (siemens per meter) [14], giving an effective loop depth of 2.6 km. Individual horizontal wires in the antenna system range from 45 to 90 km in length. Despite their physical length, the antennas are still very short compared to the radiated wavelength, and their radiation resistance is very low. The antennas in the ELF system are driven with nearly 1 MW of total power to force 200–300 A of current through them, but the total radiated power from the two-site system is between only 2 and 8 W [15]. Such a small radiated power is still sufficient to cover the globe with a receivable ELF signal at submarine depths.

A similar 82-Hz Russian ELF transmitter became operational in the early 1990s [16]. This antenna consists of multiple 60-km wires on the Kola Peninsula in northwestern Russia. This system radiates slightly more power than does the U.S. version because of lower ground conductivity in its location.

Interestingly, lightning is a stronger radiator of ELF electromagnetic waves than are controlled artificial (human-made) sources. Cloud-to-ground lightning return strokes have typical vertical channel lengths of 5–10 km and contain current pulses that last for hundreds of microseconds with peak currents of tens of kiloamperes. An average lightning stroke radiates a peak power of 10 GW in electromagnetic waves, approximately 1% of which is spread throughout the ELF band [17]. This peak power lasts only for the duration of a lightning stroke, on the order of one millisecond. But the sum of the electromagnetic fields generated by lightning discharges over the entire globe creates a significant ELF noise background [18] that must be overcome in communications applications.

2.2. Propagation

ELF electromagnetic waves propagate along the earth's surface in a manner significantly different from waves in an unbounded medium. The main difference is that waves are bounded above and below by very efficient reflectors of electromagnetic waves. The lower boundary is either earth or water, both of which are very good electrical conductors at ELF. The atmosphere is also electrically conducting, very poorly at low altitudes but with a conductivity that increases exponentially up to ~100 km altitude, above which it continues to increase in a more complicated manner. The region above ~60 km, where the atmosphere is significantly conducting, is called the *ionosphere*. In general, the effect of electrical conductivity on electromagnetic waves is frequency-dependent; the lower the frequency, the greater the effect. The ionospheric conductivity affects ELF waves significantly above ~50–70 km, depending on the precise frequency. Higher-frequency waves can penetrate much higher, but ELF waves are strongly reflected at approximately this altitude. ELF waves are thus confined between the ground and the ionosphere, which are separated by a distance on the order of or much smaller than an ELF wavelength. This spherical shell waveguide, a section of which is shown in Fig. 2, is commonly referred to as the *earth–ionosphere waveguide*.

Because ELF electromagnetic waves are almost completely confined to this small (compared to a wavelength) region, their energy attenuates with distance from the transmitter more slowly than do higher-frequency waves. As a result of two-dimensional energy spreading, the electromagnetic fields decay with distance as $r^{-1/2}$ (producing a r^{-1} power decay), with a slight modification for long propagation distances over the spherical earth. Experimental measurements have shown that the additional attenuation with distance of ELF radiowaves due to losses or incomplete reflection in the ground or ionosphere is typically only 1 dB per 1000 km at 70–80 Hz [4–6]. In general, this attenuation through losses increases with increasing frequency. Because of their low attenuation, ELF waves can be received worldwide from a single transmitter. It is this global reach that makes ELF waves so useful in a variety of applications discussed below.

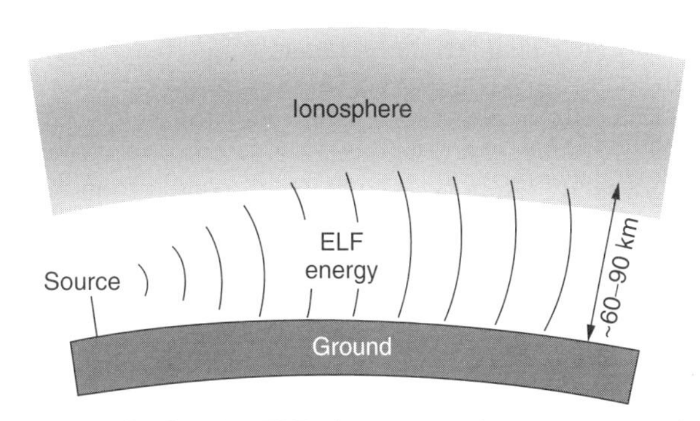

Figure 2. Qualitative ELF electromagnetic wave propagation. The ground and ionosphere (above ~60 km) are good electrical conductors that reflect ELF energy, forming the earth–ionosphere waveguide.

The low attenuation of ELF waves propagating on the surface of a sphere also produces a number of interesting propagation effects. As an ELF receiver approaches the point on the earth directly opposite the transmitter (the antipode), the signals arriving from all angles are comparable in amplitude, and therefore all contribute to the electromagnetic fields. This leads to so-called antipodal focusing, which has been theoretically investigated [12,19,20] but was experimentally verified only in 1998 [16].

At frequencies low enough that the attenuation of a fully around-the-world signal is not too severe, the multi-round-trip signals self-interfere, producing cavity resonances of the earth–ionosphere shell. These resonances are commonly called the *Schumann resonances*, after W. O. Schumann who first predicted them theoretically [21]. They can often be observed as peaks in the broadband background ELF fields in an electromagnetically quiet location and are generated primarily by steady lightning activity around the globe [22]. The frequencies of the first three resonances are approximately 8, 14, and 20 Hz [22], and in a low-noise location they can be observed up to at least 7 orders [16].

Wave propagation in the earth–ionosphere waveguide, like that in a simple parallel-plate waveguide, can be very compactly described mathematically by a sum of discrete waveguide modes that propagate independently within the boundaries. The mathematical details of the mode theory of waveguide propagation are discussed in later sections. An important consequence of this theory is that frequencies with a wavelength greater than half the waveguide height can propagate in only a single mode. For the earth–ionosphere waveguide, propagation is single mode at frequencies less than approximately 1.5–2.0 kHz, depending on the specific ionospheric conditions. This suggests a propagation-based definition of ELF, which is sometimes used in the scientific literature, as the frequencies that propagate with only a single mode in the earth–ionosphere waveguide (i.e., $f \lesssim 2$ kHz) and those for which multiple paths around the earth are not important except near the antipode (i.e., $f \gtrsim 50$ Hz, above Schumann resonance frequencies).

2.3. Reception

Receiving or detecting ELF electromagnetic waves is substantially simpler than transmitting them. Because of the long wavelength at ELF, any practical receiving antenna will be electrically small and the fields will be spatially uniform over the receiving antenna aperture. An ELF receiving antenna is thus usually made from either straight wires or loops as long or as large as possible. Both of these configurations are used as ELF receiving antennas in communication and scientific applications [12].

Again, because of the long wavelength and low transmitted field strength, the maximum practical length possible for an electric wire antenna and the maximum area and number of turns for a magnetic loop antenna are normally preferred to receive the maximum possible signal. How to deploy an antenna as big as possible is often an engineering challenge, as demonstrated by the issues involved in towing a long wire behind a moving submarine to act as an ELF antenna [23].

Designing an antenna preamplifier for an ELF system is also not trivial, especially if precise signal amplitude calibration is needed. Because the output reactance of an electrically short wire antenna is very large, the preamplifier in such a system must have a very high input impedance [12]. And even though the output impedance of a small loop antenna is very small, system noise issues usually force a design involving a stepup voltage transformer and a low-input-impedance preamplifier [24]. Another practical difficulty with ELF receivers is that in many locations, electromagnetic fields from power lines (60 Hz and harmonics in the United States, 50 Hz and harmonics in Europe and Japan) are much stronger than the desired signal. Narrowband ELF receivers must carefully filter this noise, while ELF broadband receivers, usually used for scientific purposes, need to be located far from civilization to minimize the power line noise.

The frequencies of ELF and VLF electromagnetic waves happen to overlap with the frequencies of audible acoustic waves. By simply connecting the output of an ELF sensor directly to a loudspeaker, one can "listen" to the ELF and VLF electromagnetic environment. Besides single-frequency signals from ELF transmitters and short, broadband pulses radiated by lightning, one might hear more exotic ELF emissions such as whistlers [25], which are discussed below, and chorus [26].

3. APPLICATIONS OF ELF WAVES

The unique properties of ELF electromagnetic waves propagating between the earth and the ionosphere, specifically their low attenuation with distance and their relatively deep penetration into conducting materials, make them valuable in a variety of communication and scientific applications discussed below.

3.1. Submarine Communication

The primary task of existing ELF communications systems is communication with distant submerged submarines. The ability to send information to a very distant submarine (thousands of kilometers away) without requiring it to surface (and therefore disclose its position) or even rise to a depth where its wake may be detectable on the surface is of substantial military importance. The chief difficulty in this, however, is in transmitting electromagnetic waves into highly conducting seawater. The amplitude of electromagnetic waves propagating in an electrically conducting material decays exponentially with distance. The distance over which waves attenuate by a factor of e^{-1} is commonly referred to as the "skin depth." This attenuation makes signal penetration beyond a few skin depths into any material essentially impossible. For an electromagnetic wave with angular frequency ω propagating in a material with conductivity σ and permittivity ε, if $\sigma/\omega\varepsilon \gg 1$, then the skin depth δ is very closely approximated by

$$\delta = \left(\frac{2}{\omega\sigma\mu_0}\right)^{-1/2} \qquad (2)$$

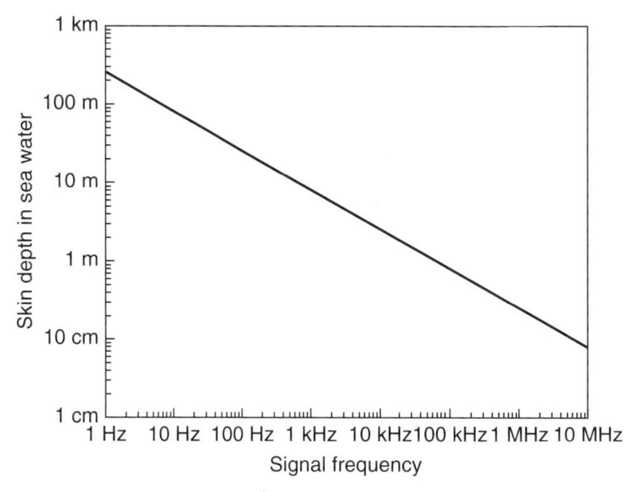

Figure 3. Skin depth (e^{-1} decay depth) of an electromagnetic wave in seawater versus frequency. Only ELF wave frequencies less than ~1 kHz can penetrate sufficiently into seawater to enable communication with submerged submarines.

This expression is usually valid in even weakly conductive materials at ELF because of the low frequency. Figure 3 shows the skin depth in seawater ($\sigma = 4$ S/m, $\varepsilon = 81\varepsilon_0$) as a function of frequency. Only ELF signals below approximately 1 kHz penetrate deeply enough to be practically useful for sending signals to submarines at a depth where they are not detectable from the surface. Because of the very low rate of data transmission on ELF (only a few bits per minute on existing systems [2]), ELF signals from transmitters are used primarily as "bell ringers" to notify the submarine to rise in order to receive more detailed information on a higher frequency. The second signal is often transmitted at ~20 kHz, a frequency low enough to/enable the submarine to remain submerged to receive the signal, using one of a number of military VLF transmitters. Submarine communication at VLF and ELF is only one-way because a transmitting system at these low frequencies is too large to install on a submarine. If higher-rate or two-way communication is needed, the submarine must extend an antenna above water to receive or transmit a signal. Because this can render the submarine detectable by radar, one-way ELF and VLF communication is often the preferred means of communication.

The same low attenuation through conducting materials that makes ELF useful for submarine communications makes it useful for communication in underground mines, where it has been used in some applications [27].

3.2. Global Communication and Navigation

Before satellites were available to relay HF and higher-frequency radio signals long distances over the horizon, global wireless communications and navigation systems depended on very long-distance coverage from a small number of transmitters. ELF and VLF signals provide a very straightforward way to do this. Besides communicating with submarines, VLF transmitters provide a one-way communication link with other military operations. These global VLF links provide a robust and jam-resistant alternative to much higher-data-rate, satellite-based communications. In the event of a global catastrophe, ordinary wireless long-distance communication channels may not be available; satellites may be disabled, and the ionosphere may be so strongly perturbed that HF signals are too strongly attenuated to be useful for long-distance communication. Even under these extreme conditions, ELF and VLF signals are still expected to propagate with low attenuation.

A VLF-based navigation system, called Omega [28], operated from 1955 until 1997. This system of eight transmitters distributed around the world and operating between 10 and 14 kHz provided global positioning accuracy to approximately 2 km. The receiver location was derived from phase differences between signals received simultaneously from multiple Omega transmitters. The satellite-based Global Positioning System (GPS) is a substantially more accurate system, leading to the decommission of the Omega system.

3.3. Geophysical Exploration

The ability of ELF electromagnetic waves to penetrate conducting materials such as rock enables their use in geophysical exploration. A standard subsurface geophysical exploration tool called *magnetotellurics* [8] relies on the fact that ELF and VLF fields on the surface of the ground are influenced by the subsurface properties because of this deep penetration. The specific source of the observed ELF and VLF fields does not matter, provided it is sufficiently far away. Both natural and artificial wave sources are commonly used. Whatever the source, the ratio of perpendicular horizontal components of the electric and magnetic field is a specific function of the subsurface electrical properties [8]. Measurements of this ratio as a function of frequency and position can be inverted into a subsurface geological map, often revealing key subsurface features such as hydrocarbon and ore-bearing deposits [29].

3.4. Scientific Applications

ELF electromagnetic waves are also used in a variety of scientific applications. Natural ELF and VLF emissions from lightning can be used very effectively for remotely sensing the ionosphere [30–32]. The lower ionosphere (~60–150 km) is one of the most inaccessible regions of the atmosphere; it is too low for satellites and too high for airplanes or balloons to probe directly, and higher-frequency incoherent scatter radar [33] used to study the higher-altitude ionosphere seldom returns a useful signal from the lower regions [34]. Because ELF and VLF waves are strongly reflected by the lower ionosphere (~60–150 km) as they propagate, they can very effectively be used to probe the lower ionosphere. And lightning is a very convenient broadband and high power source for such remote sensing.

The ELF radiation from lightning also provides a means for remotely sensing the characteristics of the source lightning discharge itself. Because this energy travels so far with minimal attenuation, a single ELF magnetic or electric field sensor can detect strong lightning discharges many thousands of kilometers away. By modeling the propagation of the ELF signal (techniques for this are described in the later sections of this article), important parameters describing the current and charge transfer in the lightning can be quantitatively measured [10]. Applications of this technique for lightning remote sensing have led to discoveries about the strength of certain lightning processes [35] and have helped us understand the kind of lightning responsible for a variety of the most recently discovered effects in the mesosphere from strong lightning [36].

A specific kind of natural ELF–VLF electromagnetic emission led to the discovery in the 1950s that near-earth space is not empty but rather filled with ionized gas, or plasma. A portion of the ELF and VLF electromagnetic wave energy launched by lightning discharges escapes the ionosphere and, under certain conditions, propagates along magnetic field lines from one hemisphere of earth to the other. These signals are called "whistlers" because when the signal is played on a loudspeaker, the sound is a frequency-descending, whistling tone that lasts around a second. Thorough reviews of the whistler phenomenon can be found in Refs. 25 and 37. L. Storey, in groundbreaking research, identified lightning as the source of whistlers and realized that the slow decrease in frequency with time in the signal could be explained if the ELF–VLF waves propagated over a long, high-altitude path through an ionized medium [38]. The presence of plasma in most of near-earth space was later confirmed with direct measurements from satellites. The study of natural ELF waves remains an area of active research, including emissions observable on the ground at high latitudes [26,39] and emissions observed directly in space on satellites [e.g., 40] and rockets [e.g., 41].

An interesting modern ELF-related research area is ionospheric heating. A high-power HF electromagnetic wave launched upward into the ionosphere can nonlinearly interact with the medium and modify the electrical characteristics of the ionosphere [42,43]. By modulating the HF wave at a frequency in the ELF band, and by doing so at high latitudes where strong ionospheric electric currents routinely flow [44], these ionospheric currents can be modulated, forming a high-altitude ELF antenna [45]. The ELF signals launched by this novel antenna could be used for any of the applications discussed above. Currently, a major research project, the High-frequency Auroral Active Research Program (HAARP), is devoted to developing such a system in Alaska [46].

4. MATHEMATICAL MODELING OF ELF PROPAGATION

Accurate mathematical modeling of ELF propagation is needed for many of the applications described in the previous sections. It is also complicated because of the influence of the inhomogeneous and anisotropic ionosphere on the propagation. Some of the best-known researchers in electromagnetic theory (most notably J. Wait and K. Budden) worked on and solved the problem of ELF and VLF propagation in the earth–ionosphere waveguide. This work is described in many scientific articles [e.g., 47–51] and technical reports [e.g., 52,53], and is also conveniently and thoroughly summarized in a few books [19,20,54]. The approaches taken by these researchers are fundamentally similar but treat the ionospheric boundary in different ways. A solution for a simplified ELF propagation problem that provides significant insight into subionospheric ELF propagation is summarized below, followed by a treatment with minimal approximations that is capable of more accurate propagation predictions.

4.1. Simplified Waveguide Boundaries

To understand the basic principles of ELF propagation, we first consider a simplified version of the problem. Consider a vertically oriented time-harmonic short electric dipole source at a height $z = z_s$ between two flat, perfectly conducting plates separated by a distance h, as shown in Fig. 4. This simple approximation gives substantial insight into the ELF propagation problem, which is harder to see in more exact treatments of the problem. This simplified problem is, in fact, a reasonably accurate representation of the true problem because at ELF the ionosphere and ground are very good electrical conductors. The main factors neglected in this treatment are the curvature of the earth and the complicated reflection from and losses generated in a realistic ionosphere.

Wait [19] solved this problem analytically and showed that the vertical electric field produced by this short electric dipole source, as a function of height z and distance r from the source, is given by

$$E_z(r,z) = \frac{\mu_0 \omega I\,dl}{4h} \sum_{n=0}^{\infty} \delta_n S_n^2 H_0^{(2)}(kS_n r) \times \cos(kC_n Z_0)\cos(kC_n z) \quad (3)$$

where $\omega = 2\pi \times$ source frequency
$k = \omega/c$
$Idl =$ source current $\times$ dipole length
$\delta_0 = \frac{1}{2}$, $\delta_n = 1$, $n \geq 1$
$C_n = n\lambda/2h =$ cosine of the eigenangle θ_n of the nth waveguide mode
$S_n = (1 - C_n^2)^{1/2} =$ sine of the eigenangle θ_n of the nth waveguide mode
$H_0^{(2)} =$ Hankel function of zero order and second kind

If $|kS_n/r| \gg 1$, the Hankel function can be replaced by its asymptotic expansion $H_0^{(2)}(kS_n r) =$

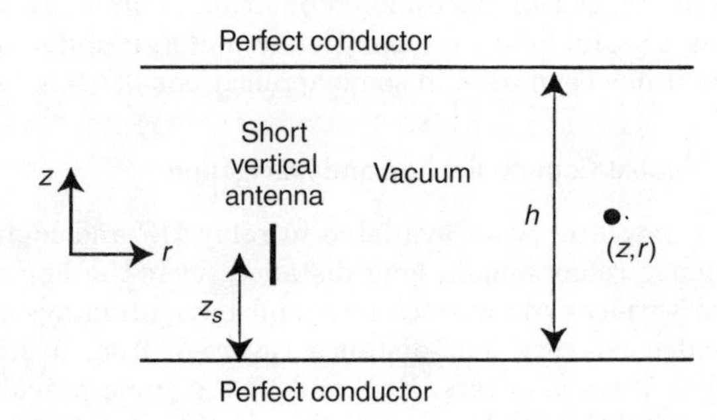

Figure 4. A simplified version of the earth–ionosphere waveguide propagation problem. A slab of free space containing a vertical antenna is bounded above and below by perfectly conducting planes.

$(2/(\pi kS_n r))^{1/2}\exp[-i(kS_n r - \pi/4)]$. This substitution, often referred to as the *far-field approximation*, is generally valid for distances more than one wavelength from the source. In this form, it is easy to see that the electric field decays with distance as $r^{-1/2}$, as it should because the propagation is confined by the waveguide to two dimensions.

This solution shows that the fields inside the perfectly conducting waveguide are the sum of the fields in an infinite number of independently propagating waveguide modes, each of which corresponds to a specific plane-wave angle θ_n of incidence on the waveguide boundaries. Each mode propagates with a different phase velocity $v_p = c/\sin(\theta_n)$, and the fields at a distance are simply the sum of the fields in each mode. This concept of waveguide modes is very general for wave propagation in bounded structures [54] and applies to light propagation in optical fibers [55], acoustic propagation in the ocean [56], and many other scenarios. A physical interpretation of the eigenangle of a waveguide mode is that the modal eigenangles are the set of incidence angles on the boundaries for which a plane wave reflected once each from the upper and lower boundaries is in phase with the incident (nonreflected) plane wave.

The summation in Eq. (3) is over an infinite number of modes. However, the equation for C_n shows that for any frequency there is a mode order $n_{\max}$ above which C_n is greater than unity. For modes of order greater than $n_{\max}$, S_n is purely imaginary, and the mode fields exponentially decay, rather than propagate, with increasing distance from the source. Such modes are called *evanescent* and, because of their exponential decay, do not contribute significantly to the total field except very close to the source. Equivalently, for a fixed mode of order n, there is a frequency below which the mode is evanescent. This frequency is called the *cutoff frequency* of the mode, which in this case is given by $f_{cn} = nc/2h$. Because at any frequency the number of propagating waves is finite, the summation can be terminated at $n_{\max}$ to a very good approximation to compute the fields beyond a significant fraction of a wavelength from the source.

The number of propagating modes $n_{\max}$ is a function of frequency. As frequency decreases, so does the number of modes required to describe the propagation of the energy. This is demonstrated in Fig. 5 with a calculation using Eq. (3). We assume a waveguide upper boundary height of 80 km, representative of the earth–ionosphere waveguide at night, and the source and receiver are on the lower boundary of the waveguide ($z = 0$). The two curves in the figure show the vertical electric field as a function of distance from a 1-A/m vertical electric dipole source at 200 Hz and 10 kHz. The difference between them is obvious and has a clear physical interpretation. At $f = 200$ Hz, only one waveguide mode propagates; all others are evanescent. This one mode is called the *transverse electromagnetic* (TEM) mode because it contains only E_z and H_ϕ fields that point transverse to the propagation direction. Thus at 200 Hz, the field decays with distance smoothly as $r^{-1/2}$ from this single mode. This represents classic, single-mode ELF propagation. But at $f = 10$ kHz, six modes propagate, each with a different phase velocity. The relative phase of these modes changes rapidly with distance, which produces a very complicated field variation with distance, as shown. The six modes interfere constructively where the amplitude is a local maximum, and interfere destructively where the amplitude is a local minimum. This complicated field pattern with distance is very typical of multimode propagation, which occurs at VLF in the earth–ionosphere waveguide. The overall amplitude difference between the two curves results from the fixed antenna length used in this calculation, which is electrically shorter (and therefore less efficient) at lower frequencies. Interesting practical issues arise in multimode propagation. The presence of deep nulls means that at certain locations and under certain waveguide conditions, the received signal level can be very low even for high transmitted power. Predicting signal coverage from military VLF transmitters and finding the locations of these nulls was one of the original motivations behind modeling the ELF–VLF propagation problem as accurately as possible.

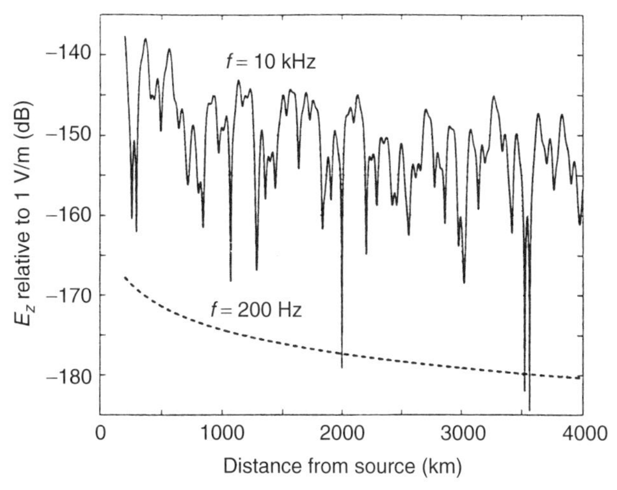

Figure 5. Vertical electric field strength versus distance from a 1-A/m electric dipole source, assuming perfectly conducting earth–ionosphere waveguide boundaries separated by 80 km, and calculated using Eq. (3). At $f = 200$ Hz, only one mode propagates, and the total field decays simply as $r^{-1/2}$. At $f = 10$ kHz, 6 modes propagate with different phase velocities. Even though the fields in each mode decay as $r^{-1/2}$, the mutual interference of all 6 modes creates the very complicated amplitude variation with distance shown here.

Equation (3) is valid for a source at any frequency. However, for frequencies high enough that the wavelength is much smaller than h, the number of propagating modes $n_{\max}$ is so great that the mode-based formulation becomes difficult to use in calculations. Ray theory [57], in which the fields at a distance are described by a sum of discrete rays that each undergo a different number of reflections from the ground and ionosphere, becomes a much more practical and compact representation of the fields at a distance. For propagation in the earth–ionosphere waveguide, this mode theory/ray theory boundary occurs around 50 kHz. At VLF and ELF frequencies, the mode theory description of the fields can very efficiently describe the propagation of the wave energy because only a few modes are propagating (i.e., are not evanescent).

There are other electromagnetic field components produced by this vertical dipole source. In the case of perfectly conducting boundaries, H_ϕ and E_r are nonzero, and expressions for these components can be found directly from the solution for E_z [19]. In general, all the field components in a single mode vary sinusoidally with altitude; thus the altitude variation of the fields is also quite complicated in the case of multimode propagation. The fields produced by a horizontal electric dipole, such as the ELF antenna described above for submarine signaling, are a similar modal series containing different field components [19].

4.2. Realistic Waveguide Boundaries

While the perfectly conducting boundary approximation demonstrates the fundamental characteristics of ELF propagation, it is rarely accurate enough for quantitatively correct simulations of earth–ionosphere waveguide propagation. The real ionosphere is not a sharp interface like that assumed above, but is a smoothly varying waveguide boundary. Nor is it a simple electrical conductor as assumed; it is a magnetized cold plasma with complicated electromagnetic properties. The key ionospheric parameters for electromagnetic wave propagation and reflection are the concentration of free electrons and ions (electron and ion density), the rate of collisions for electrons and ions with other atmospheric molecules (collision frequencies), and the strength and direction of Earth's magnetic field. Figure 6 shows representative altitude profiles of electron and ion density for midday and midnight and profiles of electron and ion collision frequencies. The index of refraction of a cold plasma like the ionosphere is a complicated function of all of these parameters [58], in which is therefore difficult to handle in analytic calculations. Also, in general, the ground is not a perfect conductor and may even be composed of layers with different electromagnetic properties, such as an ice sheet on top of ordinary ground. These realistic boundaries cannot be treated as homogeneous and sharp.

Both Budden [49] and Wait [19] derived solutions for the earth–ionosphere waveguide problem with arbitrary boundaries. Their solutions are fundamentally similar, with slight differences in the derivation and how certain complexities, such as the curved earth, are treated. We follow Wait's treatment below.

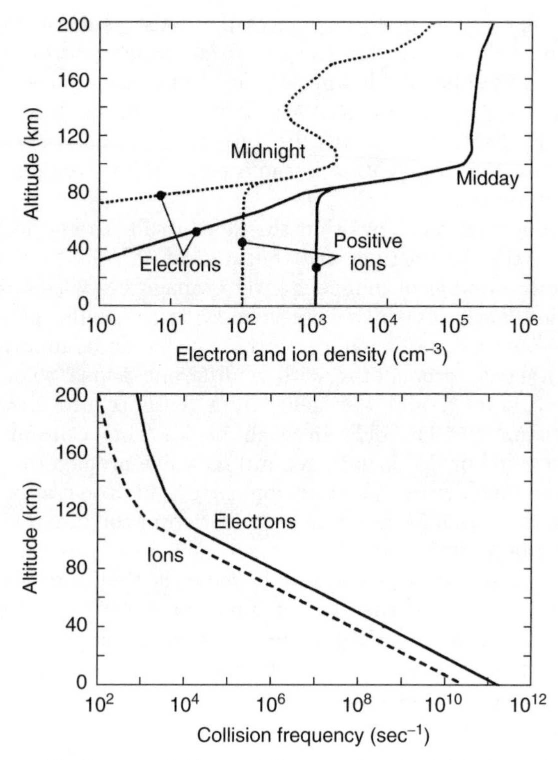

Figure 6. Representative midday and midnight ionospheric electron density, ion density, and collision frequency profiles. Negative ions are also present where needed to maintain charge neutrality.

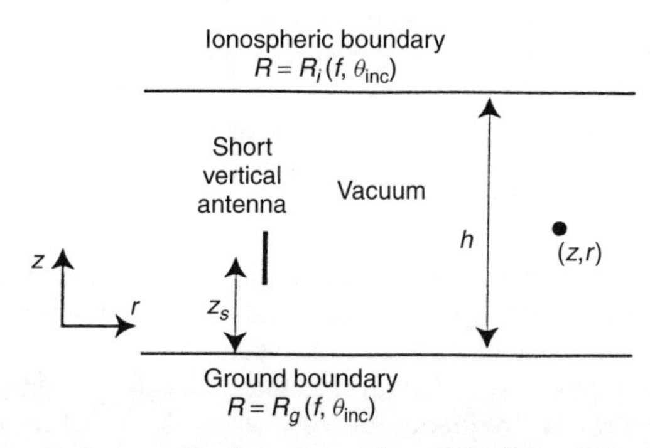

Figure 7. A generalized version of the earth–ionosphere waveguide propagation problem. A slab of free space containing a vertical antenna is bounded above and below by completely arbitrary reflecting boundaries.

4.2.1. Propagation Modeling with General Boundaries.

We again consider propagation inside a free-space region between two boundaries separated by a distance h. Now, however, the boundaries at $z = 0$ and $z = h$ are completely general and are described only by their plane-wave reflection coefficients, as shown in Fig. 7. These reflection coefficients are functions of incidence angle, incident polarization, and frequency. This important generalization makes the following formulation applicable to essentially any two-dimensional planar waveguide, regardless of size or bounding material.

Ordinarily, the electromagnetic fields can be separated into transverse electric (TE) and transverse magnetic (TM) groups that propagate independently in a two-dimensional waveguide. However, since the ionosphere is a magnetized plasma and thus is anisotropic, these field groups are coupled at the upper boundary and an incident TE or TM wave produces both TE and TM reflections. Purely TE or TM fields do not exist in the earth–ionosphere waveguide. The coupling between the fields means that the reflection coefficient from the upper boundary is not a scalar quantity but is rather a 2×2 matrix, where each matrix element is one of the four different reflection coefficients for a specific incident and reflected polarization. The lower boundary of the earth–ionosphere waveguide is the ground, which is generally isotropic so that the cross-polarized reflection coefficients in the ground reflection matrix are zero.

With this in mind, we define $\mathbf{R}_I$, the reflection matrix of the ionosphere at altitude $z = h$, and $\mathbf{R}_G$, the reflection matrix of the ground at $z = 0$, as

$$\mathbf{R}_I(\theta) = \begin{bmatrix} {}_{\parallel}R^i_{\parallel} & {}_{\parallel}R^i_{\perp} \\ {}_{\perp}R^i_{\parallel} & {}_{\perp}R^i_{\perp} \end{bmatrix} \mathbf{R}_G(\theta) = \begin{bmatrix} {}_{\parallel}R^g_{\parallel} & 0 \\ 0 & {}_{\perp}R^g_{\perp} \end{bmatrix} \tag{4}$$

These reflection coefficients are implicitly functions of the angle of incidence and the frequency. The left subscript on the matrix elements denotes the incident wave polarization (parallel or perpendicular to the plane of incidence containing the wavevector and the boundary normal), and the right subscript denotes the reflected polarization.

By using the plane-wave spectrum representation of the fields from a short vertical electric dipole, by postulating a particular solution form in the presence of the waveguide, and by enforcing continuity of the tangential field components between the free-space and boundary regions, Wait [19] shows that the fields in the waveguide, in terms of the electric and magnetic Hertz vectors $\mathbf{U}$ and $\mathbf{V}$ are given by the complex contour integral

$$\begin{bmatrix} U_z \\ V_z \end{bmatrix} = -\frac{kIdl}{8\pi\omega\varepsilon_0} \int_{\Gamma} \mathbf{F}(C) \begin{bmatrix} 1 \\ 0 \end{bmatrix} H_0^{(2)}(kSr)dC \tag{5}$$

with

$$\mathbf{F}(C) = \frac{\begin{matrix}(\exp ikCz + \mathbf{R}_G(C)\exp -ikCz) \\ (\exp ikCh + \mathbf{R}_I(C)\exp -ikCh)\end{matrix}}{\begin{matrix}\exp ikCh(1 - \mathbf{R}_G(C) \\ \mathbf{R}_I(C)\exp -2ikCh)\end{matrix}}, \tag{6}$$

where C and S are the cosine and sine of the complex angle of incidence θ of the wave on the upper and lower boundaries, respectively, as was the case for the simplified boundaries discussed above. The symbols in Eqs. (5) and (6) are consistent with those in Eq. (3). The integrand contains poles where

$$\det(1 - \mathbf{R}_G(C)\mathbf{R}_I(C)\exp -2ikCh) = 0 \tag{7}$$

and thus the integral can be evaluated as a residue series [59]. Equation (7), commonly referred to as the *mode condition*, requires that one eigenvalue of the net reflection coefficient $\mathbf{R}_G\mathbf{R}_I \exp -2ikCh$ be unity. This is equivalent to stating that the plane wave at the given incidence angle reflected once each from the upper and lower boundaries must be in phase with and equal in amplitude to the incident plane wave. In this way, the fields in the waveguide can be thought of as the sum of contributions from the angular spectrum of plane waves at angles for which propagation in the waveguide is self-reinforcing. Each angle of incidence θ_n that satisfies the mode condition is referred to as an *eigenangle* and defines a waveguide mode at the frequency ω under consideration. Budden [49] solved the same problem in a slightly different way by summing the fields produced by an infinite number of sources, each corresponding to a different multiply reflected plane wave in the waveguide. Wait's and Budden's solutions are essentially identical.

From Eqs. (5)–(7), an explicit expression for the Hertz vectors in the free-space region between the two boundaries is given by

$$\begin{bmatrix} U_z \\ V_z \end{bmatrix} = \frac{ikIdl}{4\omega\varepsilon_0} \sum_n \frac{\exp 2ikC_n h}{\partial\Delta/\partial C\Big|_{\theta=\theta_n}} \times \begin{bmatrix} (\exp 2ikC_n h - {}_{\perp}R^g_{\perp\perp}R^i_{\perp}) f_p^1(z) \\ i_{\parallel}R^g_{\parallel\parallel}R^i_{\perp} f_p^2(z) \end{bmatrix} H_0^2(kS_n r) \tag{8}$$

where $\Delta(C) = \det(\exp 2ikCh - \mathbf{R}_G\mathbf{R}_I)$. The actual electric and magnetic fields are easily derived from these Hertz vector components [19].

Each term in (8) has a physical interpretation. The leading constant is a source term that depends on the current–moment Idl of the vertical dipole source. The first term in the summation is commonly referred to as the *excitation function* for a particular mode at a given frequency, and it quantifies the efficiency with which that mode is excited by a vertical dipole on the ground. The 2×1 matrix in the summation describes the field variation with altitude, and the functions f_p^1 and f_p^2 are defined explicitly by Wait [19]. The $H_0^2(kS_n r)$ term describes the propagation of a cylindrically expanding wave, which exists because the expansion in the vertical direction is limited by the waveguide boundaries so that the mode fields spread only horizontally. For distances where $kS_n r \gtrsim 1$, we can approximate $H_0^2(kS_n r) \approx \left(\frac{2}{\pi k S_n r}\right)^{1/2} \exp[-i(kS_n r - \pi/4)]$, which more explicitly shows the $r^{-1/2}$ cylindrical spreading. In this approximation, it is also clear that each mode propagates as $\exp[-i(kS_n r)]$. The sine of the modal eigenangle thus contains all the information about the phase velocity and attenuation rate of the mode. Because the boundaries are lossy in the earth–ionosphere waveguide (due to both absorption in the boundaries and energy leakage out of the waveguide from imperfect reflection), the eigenangles and thus S_n are necessarily complex.

As was the case for the simplified, perfectly conducting waveguide, the summation in Eq. (8) is over an infinite number of modes. In practice, however, it can be limited only to the modes that contribute significantly to the fields at a distance r from the source. All modes are at least somewhat lossy with realistic waveguide boundaries, but there generally is a sharp transition between low loss and very high loss at the cutoff frequency of each mode. Often for long distances at ELF and VLF, only a few low-attenuation modes or even just one contribute significantly to the fields, leading to a very compact and efficient calculation. For very short distances, however, even highly attenuated modes can contribute to the fields and mode theory becomes less efficient and more difficult to implement.

Two factors have been neglected in this analysis of ELF propagation with realistic boundaries. One is the curvature of the earth. Treating the curvature exactly substantially increases the complexity of the modeling [19], but relatively simple corrections to Eqs. (7) and (8) can account for this curvature fairly accurately.

By introducing an artificial perturbation to the index of refraction of the free-space region between the ground and ionosphere to the flat-earth problem [19,49], the ground curvature can be handled correctly. At lower VLF ($\lesssim$5 kHz) and ELF, the flat-earth mode condition in Eq. (7) is accurate, and this correction is not needed [19]. The curved boundaries also affect the geometric decay of the fields with distance. Simply replacing the $r^{-1/2}$ term in Eq. (8) with the factor $[a\sin(r/a)]^{-1/2}$, where a is the radius of the earth, properly accounts for energy spreading over a spherical boundary [49,19].

Antipodal effects are also neglected in the analysis presented above. When the receiver approaches the antipode of the transmitter, signals that propagate around the earth in directions other than the direct great circle path are comparable in amplitude to the direct signal and thus interfere. As discussed above, this effect is most significant at ELF and has been studied theoretically [12,19,20]. Experimental results from 1998 compare favorably with predictions[16].

4.2.2. Practical Implementation. The general equations describing mode-based ELF and VLF propagation are complicated because of the complexity of the physical boundaries of the earth–ionosphere waveguide. The most difficult part of implementing calculations in Eq. (8) with realistic boundaries is solving Eq. (7) to find the modal eigenangles for a general boundary. When the reflection matrixes cannot be described analytically, such as for an arbitrary ionospheric profile, an iterative numerical procedure is needed to find the set of eigenangles that satisfy the mode condition. Such a numerical model and code was developed over a number of years by the Naval Electronics Laboratory Center (NELC), which later became the Naval Ocean Systems Center (NOSC) and is now the Space and Naval Warfare Systems Center (SSC). The details of this model are described in a series of papers [51,60] and technical reports [53,61,62]. The end result was a model called long wave propagation capability (LWPC) that calculates the eigenangles for an arbitrary ionospheric profile and arbitrary but uniform ground, and then calculates the total field at a given distance from the source. This model includes the modifications to the mode condition [Eq. (7)] and the field equation [Eq. (8)] to account for propagation over a curved earth, and it also includes the capability of accounting for sharp inhomogeneities in the ionosphere along the direction of propagation. The technique used for this calculates mode conversion coefficients at the two sides of the sharp inhomogeneity [63,64].

Using the nighttime ionospheric profiles in Fig. 6 and ground parameters representative of propagation over land ($\varepsilon_r = 15$, $\sigma = 10^{-3}$), we have used the LWPC model to calculate the vertical electric field versus distance from a 1 A m vertical electric dipole source for source frequencies of 200 Hz and 10 kHz. The results, shown in Fig. 8, are directly comparable to the simplified model results in Fig. 5. While qualitatively similar, there are significant differences. Again, the overall amplitude difference between the 200-Hz and 10-kHz signals relates to our use of a fixed antenna length in the simulation. At ELF (200 Hz), there is again only one propagating mode

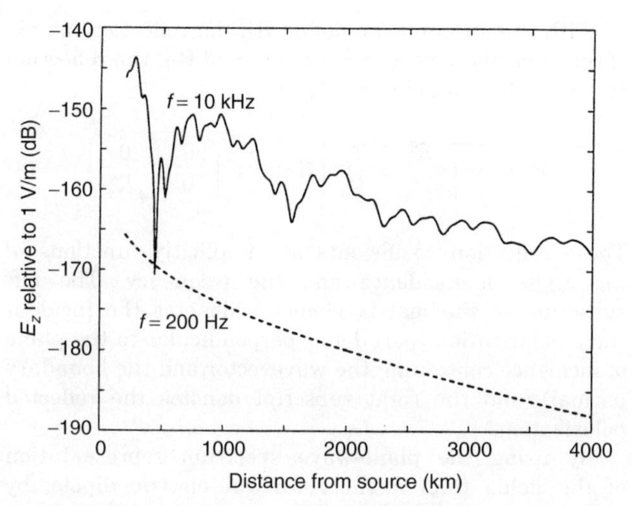

Figure 8. Model ELF and VLF calculations using LWPC with the ionospheric profiles shown in Fig. 6. At $f = 200$ Hz, only one mode propagates, and the total field decays slightly faster than $r^{-1/2}$ because of the lossy waveguide boundaries. At $f = 10$ kHz, more modes propagate and produce a more complicated amplitude variation with distance. Because the boundaries are lossy, however, the total signal is dominated by only a few modes, resulting in a simpler amplitude variation than the lossless waveguide results in Fig. 5.

that gives a smooth amplitude variation with distance. For the realistic ionosphere, however, the field attenuates faster than $r^{-1/2}$ with distance because energy is lost in the imperfectly conducting boundaries as well as through geometric spreading. At VLF (10 kHz), the amplitude varies considerably with distance, which is again a consequence of multimode propagation at the higher frequency. These variations are much less severe than in the calculation with perfectly conducting boundaries. This is also a consequence of the lossy boundaries. Certain modes are lossier than others, and these decay more rapidly with distance, so that fewer propagating modes contribute significantly to the fields at a given distance. If the calculation were extended for longer distances, there would be a point beyond which one mode dominates the fields, and the amplitude variation would become smooth. This does not happen for perfectly conducting boundaries because all modes decay with distance at the same rate. Measurements of VLF signal strength with distance made from aircraft have shown that this model can very accurately predict the variation of signal with distance, provided the correct ionosphere is used in the simulation [65].

4.3. Analytic Approximations

Analytic approximations have been developed for ELF propagation that are more realistic than the overly simplified perfectly conducting parallel-plate waveguide but that are much less complicated than the exact approach. One of the most widely used and most accurate approximations for ELF propagation was developed by Greifinger and Greifinger [66]. They recognized that because ELF wavelengths are so long, only a few key characteristics of the ionospheric altitude profile strongly influence the characteristics of ELF propagation. These approximate solutions apply only to single-mode, ELF propagation and thus to frequencies less than 1 kHz, and are most accurate at frequencies less than a few hundred hertz.

By making a few key approximations but still accounting for the anisotropy of the ionosphere, Greifinger and Greifinger [66] show that the sine of the eigenangle of the single ELF propagating mode S_0 is approximated by

$$S_0 \approx \left(\frac{h_1(h_1 + i\pi\zeta_1)}{(h_0 - \zeta_0 i\pi/2)(h_1 + \zeta_1 i\pi/2)} \right)^{1/2} \quad (9)$$

The parameter h_0 is the altitude at which the parallel ionospheric conductivity $\sigma_{\|}(h_0) = \omega\varepsilon_0$, and ζ_0 is the parallel conductivity scale height at the altitude h_0. The parameter h_1 is a higher altitude at which $4\mu_0\omega\sigma_H(h_1)\zeta_1^2 = 1$, where ζ_1 is the Hall conductivity scale height at the altitude h_1. The parallel and Hall conductivities are functions of the ionospheric electron and ion densities, collision frequencies, and magnetic field [67] and are straightforward to calculate. With this expression for S_0, the vertical electric field at ground level radiated by a vertical electric dipole is then given by [68]

$$E_z(r, 0) = \frac{\mu_0\omega Idl}{4(h_0 - i\pi\zeta_0/2)} S_0^2 H_0^{(2)}(kS_0 r) \quad (10)$$

Equations (9) and (10) give a simple, noniterative method for calculating the ELF propagation characteristics under a realistic ionosphere. This method was compared with LWPC calculations at 50 and 100 Hz and was found to be very accurate [66]. To demonstrate, we consider 200 Hz ELF propagation under the ionospheric profiles shown in Fig. 6. After calculating the associated parallel and Hall conductivity profiles, we find for this ionosphere that $h_0 = 55$ km, $\zeta_0 = 2.5$ km, $h_1 = 74$ km, and $\zeta_1 = 7.2$ km. Plugging these numbers into Eqs. (9) and (10), the vertical electric field strength versus distance from a 1-A/m source at 200 Hz is as shown in Fig. 9. The approximate method in this section is very close to the full-wave, LWPC calculation, as shown; only the modal attenuation rate is slightly underestimated. The approximate method is also much closer to the full-wave calculations than is the perfectly conducting waveguide approximation, which, because it is lossless, drastically underestimates the modal attenuation. For realistic nighttime ionospheres, there is often complicated attenuation behavior as a function of ELF frequency because of sharp ionospheric gradients between 100 and 150 km [9] that is not be captured by this approximate method. Nevertheless, this method can accurately predict ELF field strength for realistic ionospheres and is much simpler to implement than a more exact, full-wave calculation.

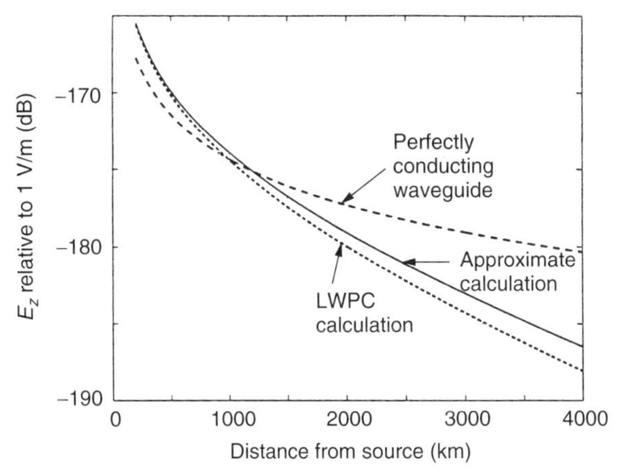

Figure 9. Comparison of 200 Hz ELF field strength with distance for a perfectly conducting waveguide, for full-wave LWPC calculations using a realistic ionosphere, and for the approximate analytical method of Greifinger and Greifinger.

4.4. Finite-Difference Methods

A completely different approach to modeling ELF propagation that has been used applies full-wave finite-difference simulations [69] to model the electromagnetic fields everywhere in the computational domain [32,70]. This brute-force but versatile approach has become usable only relatively recently because of the computing power available on modest platforms. This method is based on decomposing the volume of interest into a finite grid of discrete points at which all electromagnetic field components are computed. As a general rule, the spacing between grid points must be significantly smaller than one wavelength in order to accurately model the fields. This makes finite-difference methods well suited to low-frequency ELF and VLF propagation propagation problems in which the wavelength is quite long and the usual computational domain is a relatively small number of wavelengths in size. A major advantage of this method is that arbitrary lateral inhomogeneities, such as smooth or sharp changes in the ionospheric profile along the propagation direction, can be introduced without any increase in model complexity. Also, because the model complexity depends only on the size of the computational domain, short propagation paths are significantly easier to model than long paths. The opposite is true for mode-theory-based calculations, in which long paths are simpler because fewer modes contribute to the total fields. Mode theory and finite-difference methods are thus rather complementary. Finite-difference methods can easily handle arbitrary inhomogeneities and work especially well over short distances, while mode theory calculations can be very efficient, especially over long distances. Most importantly, the mode theory formulation of the problem also provides essential physical insight into the ELF propagation problem.

BIOGRAPHY

Steven A. Cummer is an assistant professor of the Department of Electrical and Computer Engineering at Duke University Durham, North Carolina. He received his B.S., M.S., and Ph.D. degrees in electrical engineering from Stanford University, California, in 1991, 1993, and 1997, respectively. He spent two years at NASA Goddard Space Flight Center in Greenbelt, Maryland, as an National Research Council postdoctoral research associate, and joined Duke University in 1999. He received a National Science Foundation CAREER award in 2000 and a Presidential Early Career Award for Scientists and Engineers (PECASE) in 2001. His current research is in a variety of problems in ionospheric and space physics, emphasizing electromagnetic modeling and remote sensing using ground-based and satellite instruments.

BIBLIOGRAPHY

1. J. Radatz, ed., *The IEEE Standard Dictionary of Electrical and Electronics Terms*, IEEE, New York, 1997.
2. N. Friedman, *The Naval Institute Guide to World Naval Weapons Systems*, U.S. Naval Institute, Annapolis, MD, 1997.
3. K. Davies, *Ionospheric Radio*, Peter Peregrinus, London, 1990.
4. W. L. Taylor and K. Sao, ELF attenuation rates and phase velocities observed from slow tail components of atmospherics, *Radio Sci.* **5**: 1453–1460 (1970).
5. D. P. White and D. K. Willim, Propagation measurements in the extremely low frequency (ELF) band, *IEEE Trans. Commun.* **22**(4): 457–467 (April 1974).
6. P. R. Bannister, Far-field extremely low frequency (ELF) propagation measurements, *IEEE Trans. Commun.* **22**(4): 468–473 (1974).
7. T. A. Heppenheimer, Signalling subs, *Popular Sci.* **230**(4): 44–48 (1987).
8. K. Vozoff, The magnetotelluric method, in M. Nabighian, ed., *Electromagnetic Methods in Applied Geophysics*, Vol. 2, Society of Exploration Geophysics, Tulsa, OK, 1991, pp. 641–711.
9. S. A. Cummer and U. S. Inan, Ionospheric *E* region remote sensing with ELF radio atmospherics, *Radio Sci.* **35**: 1437 (2000).
10. S. A. Cummer and U. S. Inan, Modeling ELF radio atmospheric propagation and extracting lightning currents from ELF observations, *Radio Sci.* **35**: 385–394 (2000).
11. J. D. Kraus, *Antennas*, McGraw-Hill, New York, 1988.
12. M. L. Burrows, *ELF Communications Antennas*, Peter Peregrinus, Herts, UK, 1978.
13. U. Inan and A. Inan, *Engineering Electromagnetics*, Prentice-Hall, Englewood Cliffs, NJ, 1999.
14. P. R. Bannister, Summary of the Wisconsin test facility effective earth conductivity measurements, *Radio Sci.* **11**: 405–411 (1976).
15. J. C. Kim and E. I. Muehldorf, *Naval Shipboard Communications Systems*, Prentice-Hall, Englewood Cliffs, NJ, 1995.
16. A. C. Fraser-Smith and P. R. Bannister, Reception of ELF signals at antipodal distances, *Radio Sci.* **33**: 83–88 (1998).
17. D. R. MacGorman and W. D. Rust, *The Electrical Nature of Storms*, Oxford Univ. Press, New York, 1998.
18. D. A. Chrissan and A. C. Fraser-Smith, Seasonal variations of globally measured ELF/VLF radio noise, *Radio Sci.* **31**(5): 1141–1152 (Sept. 1996).
19. J. R. Wait, *Electromagnetic Waves in Stratified Media*, Pergamon Press, Oxford, 1970.
20. J. Galejs, *Terrestrial Propagation of Long Electromagnetic Waves*, Pergamon Press, Oxford, 1972.
21. W. O. Schumann, Über die strahlungslosen eigenschingungen einer leitenden kugel, diel von einer luftschicht und einer ionosphärenhülle umgeben ist, *Z. Naturforsch.* **7a**: 149 (1952).
22. D. D. Sentman, Schumann resonances, in H. Volland, ed., *Handbook of Atmospheric Electrodynamics*, Vol. 1, CRC Press, Boca Raton, FL, 1995, pp. 267–295.
23. C. T. Fessenden and D. H. S. Cheng, Development of a trailing-wire E-field submarine antenna for extremely low frequency (ELF) reception, *IEEE Trans. Commun.* **22**: 428–437 (1974).
24. C. D. Motchenbacher and J. A. Connelly, *Low Noise Electronic System Design*, Wiley-Interscience, Englewood Cliffs NJ, 1993.
25. V. S. Sonwalkar, Whistlers, in J. G. Webster, ed., *Wiley Encyclopedia of Electrical and Electronic Engineering*, pages Wiley, New York, 1999, pp. 580–591.
26. S. S. Sazhin and M. Hayakawa, Magnetospheric chorus emissions — a review, *Planet. Space Sci.* **50**: 681–697 (May 1992).
27. J. N. Murphy and H. E. Parkinson, Underground mine communications, *Proc. IEEE* **66**: 26–50 (1978).
28. E. R. Swanson, Omega, *Proc. IEEE* **71**(10): 1140–1155 (1983).
29. G. M. Hoversten, Papua new guinea MT: Looking where seismic is blind, *Geophys. Prospect.* **44**(6): 935–961 (1996).
30. H. G. Hughes, R. J. Gallenberger, and R. A. Pappert, Evaluation of nighttime exponential ionospheric models using VLF atmospherics, *Radio Sci.* **9**: 1109 (1974).
31. S. A. Cummer, U. S. Inan, and T. F. Bell, Ionospheric *D* region remote sensing using VLF radio atmospherics, *Radio Sci.* **33**(6): 1781–1792 (Nov. 1998).
32. S. A. Cummer, Modeling electromagnetic propagation in the earth-ionosphere waveguide, *IEEE Trans. Antennas Propag.* **48**: 1420 (2000).
33. J. V. Evans, Theory and practice of ionosphere study by thomson scatter radar, *Proc. IEEE* **57**: 496 (1969).
34. J. D. Mathews, J. K. Breakall, and S. Ganguly, The measurement of diurnal variations of electron concentration in the 60–100 km, *J. Atmos. Terr. Phys.* **44**: 441 (1982).
35. S. A. Cummer and M. Füllekrug, Unusually intense continuing current in lightning causes delayed mesospheric breakdown, *Geophys. Res. Lett.* **28**: 495 (2001).
36. S. A. Cummer and M. Stanley, Submillisecond resolution lightning currents and sprite development: Observations and implications, *Geophys. Res. Lett.* **26**(20): 3205–3208 (Oct. 1999).
37. R. A. Helliwell, *Whistlers and Related Ionospheric Phenomena*, Stanford Univ. Press, Stanford, CA, 1965.
38. L. R. O. Storey, An investigation of whistling atmospherics, *Phil. Trans. Roy. Soc. London, Ser. A* **246**: 113 (1953).
39. A. J. Smith et al., Periodic and quasiperiodic ELF/VLF emissions observed by an array of antarctic stations, *J. Geophys. Res.* **103**: 23611–23622 (1998).
40. P. Song et al., Properties of ELF emissions in the dayside magnetopause, *J. Geophys. Res.* **103**: 26495–26506 (1998).
41. P. M. Kintner, J. Franz, P. Schuck, and E. Klatt, Interferometric coherency determination of wavelength or what are broadband ELF waves? *J. Geophys. Res.* **105**: 21237–21250 (Sept. 2000).

42. A. V. Gurevich, *Nonlinear Phenomena in the Ionosphere*, Springer-Verlag, Berlin, 1978.
43. K. Papadopoulos, Ionospheric modification by radio waves, in V. Stefan, ed., *Nonlinear and Relativistic Effects in Plasmas*, American Institute of Physics, New York, 1992.
44. A. D. Richmond and J. P. Thayer, Ionospheric electrodynamics: A tutorial, in S. Ohtani, ed., *Magnetospheric Current Systems*, American Geophysical Union, Washington, DC, 2000, pp. 131–146.
45. R. Barr, The generation of ELF and VLF radio waves in the ionosphere using powerful HF transmitters, *Adv. Space Res.* **21**(5): 677–687 (1998).
46. G. M. Milikh, M. J. Freeman, and L. M. Duncan, First estimates of HF-induced modifications of the D-region by the HF active auroral research program facility, *Radio Sci.* **29**(5): 1355–1362 (Sept. 1994).
47. K. G. Budden, The propagation of very-low-frequency radio waves to great distances, *Phil. Mag.* **44**: 504 (1953).
48. J. R. Wait, The mode theory of v.l.f. ionospheric propagation for finite ground conductivity, *Proc. IRE* **45**: 760 (1957).
49. K. G. Budden, The influence of the earth's magnetic field on radio propagation by wave-guide modes, *Proc. Roy. Soc. A* **265**: 538 (1962).
50. J. R. Wait, On the propagation of E.L.F. pulses in the earth-ionosphere waveguide, *Can. J. Phys.* **40**: 1360 (1962).
51. R. A. Pappert and W. F. Moler, Propagation theory and calculations at lower extremely low frequencies (ELF), *IEEE Trans. Commun.* **22**(4): 438–451 (April 1974).
52. J. R. Wait and K. P. Spies, *Characteristics of the Earth-Ionosphere Waveguide for VLF Radio Waves*, Technical Report, NBS Technical Note 300, National Bureau of Standards, 1964.
53. D. G. Morfitt and C. H. Shellman, *MODESRCH, an Improved Computer Program for Obtaining ELF/VLF/LF Mode Constants in an Earth-Ionosphere Waveguide*, Technical Report Interim Report 77T, Naval Electronic Laboratory Center, San Diego, CA, 1976.
54. K. G. Budden, *The Wave-Guide Mode Theory of Wave Propagation*, Logos Press, London, 1961.
55. G. Keiser, *Optical Fiber Communications*, McGraw-Hill, Boston, MA, 2000.
56. F. B. Jensen, W. A. Kuperman, M. B. Porter, and H. Schmidt, *Computational Ocean Acoustics*, American Institute of Physics, Woodbury, NY, 1994.
57. J. R. Wait and A. Murphy, The geometrical optics of VLF sky wave propagation, *Proc. IRE* **45**: 754 (1957).
58. K. G. Budden, *The Propagation of Radio Waves*, Cambridge Univ. Press, New York, 1985.
59. R. V. Churchill and J. W. Brown, *Complex Variables and Applications*, McGraw-Hill, New York, 1990.
60. R. A. Pappert and J. A. Ferguson, VLF/LF mode conversion model calculations for air to air transmissions in the earth-ionosphere waveguide, *Radio Sci.* **21**: 551–558 (1986).
61. J. A. Ferguson and F. P. Snyder, *Approximate VLF/LF Mode Conversion Model*, Technical Report, Technical Document 400, Naval Ocean Systems Center, San Diego, CA, 1980.
62. J. A. Ferguson, F. P. Snyder, D. G. Morfitt, and C. H. Shellman, *Long-wave Propagation Capability and Documentation*, Technical Report, Technical Document. 400, Naval Ocean Systems Center, San Diego, CA, 1989.
63. J. R. Wait, Mode conversion and refraction effects in the earth-ionosphere waveguide for VLF radio waves, *J. Geophys. Res.* **73**(11): 3537–3548 (1968).
64. R. A. Pappert and D. G. Morfitt, Theoretical and experimental sunrise mode conversion results at VLF, *Radio Sci.* **10**: 537 (1975).
65. J. E. Bickel, J. A. Ferguson, and G. V. Stanley, Experimental observation of magnetic field effects on VLF propagation at night, *Radio Sci.* **5**: 19 (1970).
66. C. Greifinger and P. Greifinger, On the ionospheric parameters which govern high-latitude ELF propagation in the earth-ionosphere waveguide, *Radio Sci.* **14**(5): 889–895 (Sept. 1979).
67. J. D. Huba and H. L. Rowland, Propagation of electromagnetic waves parallel to the magnetic field in the nightside Venus ionosphere, *J. Geophys. Res.* **98**: 5291 (1993).
68. C. Greifinger and P. Greifinger, Noniterative procedure for calculating ELF mode constants in the anisotropic Earth-ionosphere waveguide, *Radio Sci.* **21**(6): 981–990 (1986).
69. A. Taflove and S. C. Hagness, *Computational Electrodynamics: The Finite-Difference Time-Domain Method*, Artech House, Norwood, MA, 2000.
70. M. Thevenot, J. P. Berenger, T. Monediere, and F. Jecko, A FDTD scheme for the computation of VLF-LF propagation in the anisotropic earth-ionosphere waveguide, *Ann. Telecommun.* **54**: 297–310 (1999).

EM ALGORITHM IN TELECOMMUNICATIONS

Costas N. Georghiades
Texas A&M University
College Station, Texas

Predrag Spasojević
Rutgers, The State University of New Jersey
Piscataway, New Jersey

1. INTRODUCTION

Since its introduction in the late 1970s as a general iterative procedure for producing maximum-likelihood estimates in cases a direct approach is computationally or analytically intractable, the expectation-maximization (EM) algorithm has been used with increasing frequency in a wide variety of application areas. Perhaps not surprisingly, one of the areas that has seen an almost explosive use of the algorithm since the early 1990s is telecommunications. In this article we describe some of the varied uses of the EM algorithm that appeared in the literature in the area of telecommunications since its introduction and include some new results on the algorithm that have not appeared elsewhere in the literature.

The expectation-maximization (EM) algorithm was introduced in its general form by Dempster et al. in 1977 [1]. Previous to that time a number of authors had in fact proposed versions of the EM algorithm for particular applications (see, e.g., Ref. 4), but it was [1] that established the algorithm (in fact, as argued by some, a "procedure", rather than an algorithm) as a general tool for producing maximum-likelihood (ML) estimates in situations where the observed data can be viewed as "incomplete" in some sense. In addition to introducing the EM algorithm as a general tool for ML estimation, Dempster et al. [1] also dealt with the important issue of convergence, which was further studied and refined by the work of Wu [5].

Following the publication of [1], the research community experienced an almost explosive use of the EM algorithm in a wide variety of applications beyond the statistics area in which it was introduced. Early application areas for the EM algorithm outside its native area of statistics include genetics, image processing, and, in particular, positron emission tomography (PET) [6,7], (an excellent treatment of the application of the EM algorithm to the PET problem can be found in Snyder and Miller [8]). In the late 1980s there followed an increasing number of applications of the EM algorithm in the signal processing area, including in speech recognition, neural networks, and noise cancellation. Admittedly somewhat arbitrarily, we will not cite references here in the signal processing area, with the exception of one that had significant implications in the area of communications: the paper by Feder and Weistein [9]. This paper will be described in somewhat more detail later, and it is important in the area of telecommunications in that it dealt with the problem of processing a received signal that is the superposition of a number of (not individually observable) received signals. Clearly, this framework applies to a large number of telecommunications problems, including multiuser detection and detection in multipath environments. The reader can find other references (not a complete list) on the use of the EM algorithm in the signal processing (and other areas) in the tutorial paper by Moon [10]. For a general introduction to the EM algorithm and its applications in the statistics area, the reader is urged to read the original paper of Dempster et al. [1], as well as a book published in 1997 [11] on the algorithm. The appearance of a book on the EM algorithm is perhaps the best indication of its widespread use and appeal.

As mentioned above, in this manuscript we are interested in a somewhat narrow application area for the EM algorithm, that of telecommunications. Clearly, the boundaries between signal processing and telecommunications are not always well defined and in presenting the material below some seemingly arbitrary decisions were made: Only techniques that deal with some aspect of data transmission are included.

In Section 2 and for completeness we first give a brief introduction of the EM algorithm. In Section 3 we present a summary of some of the main published results on the application of the EM algorithm to telecommunications. Section 4 contains some recent results on the EM algorithm, and Section 5 concludes.

2. THE ESSENTIAL EM ALGORITHM

Let $\mathbf{b} \in \mathcal{B}$ be a set of parameters to be estimated from some observed data $\mathbf{y} \in \mathcal{Y}$. Then the ML estimate $\widehat{\mathbf{b}}$ of $\mathbf{b}$ is a solution to

$$\widehat{\mathbf{b}} = \arg\max_{\mathbf{b}\in\mathcal{B}} g(\mathbf{y} \mid \mathbf{b}) \tag{1}$$

where $g(\mathbf{y} \mid \mathbf{b})$ is the conditional density of the data given the parameter vector to be estimated. In many cases a simple explicit expression for this conditional density does not exist, or is hard to obtain. In other cases such an expression may be available, but it is one that does not lend itself to efficient maximization over the set of parameters. In such situations, the expectation–maximization algorithm may provide a solution, albeit iterative (and possibly numerical), to the ML estimation problem.

The EM-based solution proceeds as follows. Suppose that instead of the data $\mathbf{y}$ actually available one had data $\mathbf{x} \in \mathcal{X}$ from which $\mathbf{y}$ could be obtained through a many-to-one mapping $\mathbf{x} \to \mathbf{y}(\mathbf{x})$, and such that their knowledge makes the estimation problem easy (for example, the conditional density $f(\mathbf{x} \mid \mathbf{b})$ is easily obtained.) In the EM terminology, the two sets of data $\mathbf{y}$ and $\mathbf{x}$ are referred to as the incomplete and complete data, respectively.

The EM algorithm makes use of the loglikelihood function for the complete data in a two-step iterative procedure that under some conditions converges to the ML estimate given in (1) [1,5]. At each step of the EM iteration, the likelihood function can be shown to be nondecreasing [1,5]; if it is also bounded (which is mostly the case in practice), then the algorithm converges. The two-step procedure at the ith iteration is

1. *E step*: Compute $Q(\mathbf{b} \mid \mathbf{b}^i) \equiv E[\log f(\mathbf{x} \mid \mathbf{b}) \mid \mathbf{y}, \mathbf{b}^i]$.
2. *M step*: Solve $\mathbf{b}^{i+1} = \arg\max_{\mathbf{b}\in\mathcal{B}} Q(\mathbf{b} \mid \mathbf{b}^i)$.

Here $\mathbf{b}^i$ is the parameter vector estimate at the ith iteration. The two steps of the iteration are referred to as the expectation (E step) and maximization (M step) steps, respectively. Note that in the absence of the data $\mathbf{x}$, which makes $\log f(\mathbf{x} \mid \mathbf{b})$ a random variable, the algorithm maximizes its conditional expectation instead, given the incomplete data and the most recent estimate of the parameter vector to be estimated.

As mentioned earlier, the EM algorithm has been shown [1,5] to result in a monotonically nondecreasing loglikelihood. Thus, if the loglikelihood is bounded (which is the case in most practical systems), then the algorithm converges. Under some conditions, the stationary point coincides with the ML estimate.

3. AN OVERVIEW OF APPLICATIONS TO TELECOMMUNICATIONS

In this section we provide a brief overview on the use of the EM algorithm in the telecommunications area.

3.1. Parameter Estimation from Superimposed Signals

One of the earliest uses of the EM algorithm with strong direct implications in communications appeared in 1988 [9]. Feder and Weinstein [9] look at the problem of parameter estimation from a received signal, $y(t)$ (it can be a vector in general), which is a superposition of a number of signals plus Gaussian noise:

$$y(t) = \sum_{k=1}^{K} s_k(t;\theta_k) + n(t) \tag{2}$$

For simplicity in illustrating the basic idea, we will assume that $y(t)$, $n(t)$, and each $\theta_k, k = 1, 2, \ldots, K$ are scalar (Feder and Weinstein [9] handle the more general case of vectors); the objective is to estimate the parameters $\theta_k, k = 1, 2, \ldots, K$ from the data $y(t)$ observed over a T-second interval. It is assumed that the $s_k(t; \theta_k)$ are known signals given the corresponding parameters θ_k, and $n(t)$ is white Gaussian noise with $\sigma^2 = E[|n(t)|^2]$. Clearly, if instead of the superimposed data $y(t)$ one had available the individual components of $y(t)$, the problem of estimating the θ_k would become much simpler since there would be no coupling between the parameters to be estimated. Thus, the complete data $\mathbf{x}(t)$ are chosen to be a decomposition of $y(t)$: $\mathbf{x}(t) = [x_1(t), x_2(t), \ldots, x_K(t)]'$ (where the prime sign $'$ means transpose) where

$$x_k(t) = s_k(t; \theta_k) + n_k(t), \quad k = 1, 2, \ldots, K \tag{3}$$

$$\sum_{k=1}^{K} n_k(t) = n(t) \tag{4}$$

and, thus

$$y(t) = \sum_{k=1}^{K} x_k(t)$$

The $n_k(t)$ are assumed statistically independent (for analytic convenience) and have corresponding variances $\beta_k \sigma^2$, where it is assume that $\beta_k \geq 0$ and

$$\sum_{k=1}^{K} \beta_k = 1$$

Then the E step of the EM algorithm yields [9]

$$Q(\boldsymbol{\theta} \mid \hat{\boldsymbol{\theta}}) = -\sum_{k=1}^{K} \frac{1}{\beta_k} \int_T \left| \hat{x}_k(t) - s_k(t; \theta_k) \right|^2 dt \tag{5}$$

where

$$\hat{x}_k(t) = s_k(t; \hat{\theta}_k) + \beta_k \left[y(t) - \sum_{i=1}^{K} s_i(t; \hat{\theta}_i) \right] \tag{6}$$

and $\hat{\theta}_i, i = 1, 2, \ldots, K$ are the most recent estimates. At the M step, maximization of (5) with respect to the parameter vector $\boldsymbol{\theta}$ corresponds to minimizing each term in the sum in (5) with respect to the corresponding individual parameter. Thus, the desired decoupling.

Feder and Weinstein [9] went on to apply the general results presented above to the problems of estimating the multipath delays and to the problem of multiple source location. No modulation was present in the received signals, but, as mentioned above, the technique is quite general and has over the years been used in a number of applications in telecommunications. Direct applications of the results in Ref. 9 to the problem of multiuser detection can be found in Refs. 12–16.

3.2. The Multiuser Channel

In Ref. 17, also published in 1988, Poor uses the EM algorithm to estimate the amplitudes of the user signals in a multiuser environment, in the presence of modulation. In this application, the modulation symbols (binary-valued) are treated as "nuisance" parameters. The complete data are chosen as the received (incomplete) data along with the binary modulation symbols over an observation window. The E step of the EM iteration involves the computation of conditional expectations of the binary-valued symbols, which results in "soft" data estimates in the form of hyperbolic tangents. At convergence, the algorithm yields the amplitude estimates and on slicing the soft-data estimates, also estimates of the binary modulation symbols (although no optimality can be claimed). Follow-up work that relates to Ref. 17 can be found in Refs. 18–21, in which the emphasis is not on amplitude estimation (amplitudes are assumed known) but on K-user multiuser detection. In Refs. 18, 20, and 21 the complete data are taken to be the received (baud rate) matched-filter samples along with the binary modulation symbols of $(K - 1)$ users, treated as interference in detecting a particular user. In addition to the application of the standard EM algorithm, the authors in Refs. 18, 20, and 21 also study the use of the space-alternating generalized EM (SAGE) algorithm [22,23], a variant of the EM algorithm that has better convergence properties and may simplify the maximization step. Other work in the area of spread-spectrum research that uses the SAGE algorithm in jointly detecting a single user in the presence of amplitude, phase, and time-delay uncertainties has also been presented [24,25].

3.3. Channel Estimation

In Refs. 26 and 27, the authors deal with detection in an impulsive noise channel, modeled as a class A mixture. In these papers the role of the EM algorithm is not directly in detection itself, but rather in estimating the triplet of parameters of the (Gaussian) mixture model, namely, the variances of the nominal and contaminant distributions and the probability of being under one or the other distribution. The estimate of the mixture model is then used (as if it were perfect) to select the nonlinearity to be used for detection. This particular application of the EM algorithm to estimate the Gaussian mixture model is one of its original and most typical uses. Follow-up (but more in-depth) work on the use of the EM algorithm to estimating class A noise parameters can be found in Ref. 28. Further follow-up work on the problem of detection in impulsive noise can be found in Refs. 29 and 30. As in Refs. 26 and 27, the authors in Refs. 29 and 30 use the EM algorithm to estimate parameters in the impulsive noise model, which are then used for detection in a spread-spectrum, coded environment. In Ref. 31 the EM algorithm is used to estimate the noise parameters in a spatial diversity reception system when the noise is modeled as a mixture of Gaussian distributions. Also in a spatial diversity environment, Baroso et al. [32] apply the EM algorithm to estimate blindly the multiuser array channel transfer function. Data detection in the paper is considered separately.

The EM algorithm has also been used for channel estimation under discrete signaling. In a symposium paper [33] and in a follow-up journal paper [34], Vallet and Kaleh study the use of the EM algorithm for channel estimation modeled as having a finite impulse response, for both linear and nonlinear channels. In this work, the modulation symbols are considered as the "nuisance" parameters and the authors pose the problem as one of estimating the channel parameters. At convergence, symbol detection can be performed as well. Other work on the use of the EM algorithm to channel estimation/equalization in the presence of data modulation can be found in the literature [12,35–37]. Zamiri-Jafarian and Pasupathy present an algorithm for channel estimation, motivated by the EM algorithm, that is recursive in time.

3.4. The Unsynchronized Channel

In 1989 a paper dealing (for the first time) with sequence estimation using the EM algorithm appeared [38] with a follow-up journal paper appearing in 1991 [39]. In Refs. 38 and 39 the received data are "incomplete" in estimating the transmitted sequence because time synchronization is absent. The complete data in the paper were defined as the baud rate (nonsynchronized) matched-filter samples along with the correct timing phase. The E step of the algorithm was evaluated numerically, and the M step trivially corresponded to symbol-by-symbol detection in the absence of coding. The algorithm converged within 2–4 iterations, depending on the signal-to-noise ratio (SNR). The fast convergence of the algorithm can be attributed to the fact that the parameters to be estimated came from a discrete set. Follow-up work using the EM algorithm for the time unsynchronized channel appeared in 1994 and 1995 [40–42]. In one of Kaleh's papers [40], which in fact uses the Baum–Welch algorithm [4], a predecessor to EM, instead of posing the problem as sequence estimation in the presence of timing offset, the author poses the problem as one of estimating the timing offset (and additive noise variance) in the presence of random modulation. At convergence, symbol estimates can also be obtained. In Ref. 42, in addition to the absence of time synchronization, the authors assume an unknown amplitude. In the paper the authors use the SAGE algorithm [22] to perform sequence estimation.

3.5. The Random Phase Channel

Similar to Refs. 38 and 39, in a 1990 paper [43] the authors consider sequence estimation in the presence of random phase offset, for both uncoded and trellis-coded systems. We provide a brief description of the results here.

Let $\mathbf{s} = (s_1, s_2, \ldots, s_N)$ be the vector containing the complex modulation symbols, $\mathbf{D}$ be a diagonal matrix with the elements of $\mathbf{s}$ as diagonal elements, and θ be the phase offset over the observation window. Then the received (incomplete) data vector $\mathbf{r}$ can be expressed as

$$\mathbf{r} = \mathbf{D}e^{j\theta} + \mathbf{N} \tag{7}$$

where $\mathbf{N}$ is a zero mean, i.i.d., complex, Gaussian noise vector. The parameter vector to be estimated is the modulation sequence $\mathbf{s}$. Let the complete data $\mathbf{x}$ consist of the incomplete data $\mathbf{r}$ along with knowledge of the random phase vector θ, namely, $\mathbf{x} = (\mathbf{r}, \theta)$. Assuming PSK modulation (the case of QAM can also be handled easily), the E step of the algorithm at the ith iteration yields

$$Q(\mathbf{s} \mid \mathbf{s}^i) = \Re\{\mathbf{r}^\dagger \mathbf{s} \cdot (\mathbf{r}^\dagger \mathbf{s}^i)^*\} \tag{8}$$

For the maximization step of the EM algorithm, we distinguish between coded and uncoded transmission. Observe from (8) that in the absence of coding, maximizing $Q(\mathbf{s} \mid \mathbf{s}^i)$ with respect to sequences $\mathbf{s}$ is equivalent to maximizing each individual term in the sum, specifically, making symbol-by-symbol decisions. When trellis coding is used, the maximization over all trellis sequences can be done efficiently using the Viterbi algorithm. Follow-up work on sequence estimation under phase offset can be found in the literature [44–47]. The use of the EM algorithm for phase synchronization in OFDM systems has also been studied [48].

3.6. The Fading Channel

Sequence estimation for fading channels using the EM algorithm was first studied in 1993 [49], with follow-up work a bit later [45,47,50]. Here the authors look at the problem as one of sequence estimation, where the random fading are the unwanted parameters. A brief summary of the EM formulation as presented in the references cited above is given below.

Let $\mathbf{D}$, $\mathbf{s}$, and $\mathbf{N}$ be as defined above and $\mathbf{a}$ be the complex fading process modeled as a zero-mean, Gaussian vector with independent real and imaginary parts having covariance matrix $\mathbf{Q}$. Then the received data $\mathbf{r}$ are

$$\mathbf{r} = \mathbf{D}\mathbf{a} + \mathbf{N} \tag{9}$$

Let the complete data $\mathbf{x}$ be the incomplete data $\mathbf{r}$ along with the random fading vector $\mathbf{a}$: $\mathbf{x} = (\mathbf{r}, \mathbf{a})$. Then, for PSK signaling, the expectation step of the EM algorithm yields

$$Q(\mathbf{s} \mid \mathbf{s}^i) = \Re[\mathbf{r}^\dagger \mathbf{D}\hat{\mathbf{a}}_i] \tag{10}$$

where

$$\hat{\mathbf{a}}_i = E[\mathbf{a} \mid \mathbf{r}, \mathbf{s}^i] = \mathbf{Q}\left[\mathbf{Q} + \frac{\mathbf{I}}{\mathrm{SNR}}\right]^{-1} (\mathbf{D}^i)^* \mathbf{r} \tag{11}$$

When the fading is static over a number of symbols, the expectation step of the algorithm becomes $Q(\mathbf{s} \mid \mathbf{s}^i) = \Re[\mathbf{r}^\dagger \mathbf{s}\hat{a}_i]$, where $\hat{a}_i = \mathbf{r}\mathbf{s}^{i\dagger}$. Again, the maximization step can be done easily and corresponds to symbol-by-symbol detection in the uncoded case, or Viterbi decoding in the trellis-coded case.

Other work on the use of the EM algorithm for sequence estimation in multipath/fading channels can be found in the literature [51–57]. In Refs. 54 and 55 the authors introduce an approximate sequence estimation algorithm motivated by the EM algorithm.

3.7. The Interference Channel

In addition to the multiuser interference channel, a number of papers applying the EM algorithm to sequence estimation in interference and/or to interference suppression have appeared in the literature more recently. These include Refs. 58–62. An application to sequence estimation [60] is briefly presented below.

Let

$$r(t) = S(t; \mathbf{a}) + J(t) + n(t) \tag{12}$$

be the received data, where $S(t; \mathbf{a})$, $0 \leq t < T$, is the transmitted signal (assuming some arbitrary modulation) corresponding to a sequence of M information symbols $\mathbf{a}$. $J(t)$ models the interference, and $n(t)$ is zero-mean, white, Gaussian noise having spectral density $N_0/2$. For generality, we will not specify the form of the signal part, $S(t; \mathbf{a})$, but special cases of interest include direct-sequence spread-spectrum (DSSS) and multicarrier spread-spectrum (MCSS) applications.

Let the complete data be the received data $r(t)$ along with the interference $J(t)$ (which is arbitrary at this point). Application of the EM algorithm yields the following expectation step ($\mathbf{a}^k$ is the sequence estimate at the kth iteration and $\langle\cdot,\cdot\rangle$ denotes inner product):

$$Q(\mathbf{a} \mid \mathbf{a}^k) = \langle r(t) - \hat{J}^k(t), S(t; \mathbf{a})\rangle \tag{13}$$

$$\hat{J}^k(t) = E[J(t) \mid \{r(\tau) : 0 \leq \tau \leq T\}, \mathbf{a}^k] \tag{14}$$

These expressions are quite general and can be applied to arbitrary interference models, provided the conditional mean estimates (CMEs) in (14) can be computed. For a vector-space interference model

$$J(t) = \sum_{j=1}^{N} g_j \phi_j(t) \tag{15}$$

where $\{\phi_1(t), \phi_2(t), \ldots, \phi_N(t)\}$ is an orthonormal basis set and $g_j, j = 1, 2, \ldots, N$ are zero-mean, uncorrelated random variables having corresponding variances λ_j, we have

$$\hat{J}^k(t) = \sum_{i=1}^{N} \hat{g}_i^k \phi_i(t) \tag{16}$$

where

$$\hat{g}_i^k = \left(1 + \frac{N_0}{2\lambda_i}\right)^{-1} \langle r(t) - S(t; \mathbf{a}^k), \phi_i(t)\rangle$$

4. SOME RESULTS ON APPLICATION OF THE EM ALGORITHM

We present briefly next some more recent results on the use of the EM algorithm in telecommunications.

4.1. Decoding of Spacetime-Coded Systems

The results presented here are a brief summary of those reported in Ref. 63. We consider a mobile radio system where the transmitter is equipped with N transmit antennas and the mobile receiver is equipped with M receive antennas. Data blocks of length L are encoded by a spacetime encoder. After serial-to-parallel conversion, the coded symbol stream is divided into N substreams, each of which is transmitted via one transmit antenna simultaneously. The transmitted code block can be written in matrix form as

$$\mathbf{D} = \begin{pmatrix} d_1^{(1)} & d_2^{(1)} & \cdots & d_N^{(1)} \\ d_1^{(2)} & d_2^{(2)} & \cdots & d_N^{(2)} \\ \vdots & \vdots & \ddots & \vdots \\ d_1^{(L)} & d_2^{(L)} & \cdots & d_N^{(L)} \end{pmatrix} \tag{17}$$

where the superscript in $d_n^{(l)}$ represents time index and the subscript is the space index. In other words, $d_n^{(l)}$ is the complex symbol transmitted by the nth antenna during the lth symbol time. We denote the row vectors of $\mathbf{D}$ by $\mathbf{D}^{(l)}$.

Assuming quasistatic fading (the nonstatic case has been handled as well), let

$$\boldsymbol{\Gamma}_j = (\,\gamma_{1j} \quad \gamma_{2j} \quad \cdots \quad \gamma_{Nj}\,)^T, j = 1, 2, \ldots, M$$

be the fading vector whose components γ_{ij} are the fading coefficients in the channel connecting the ith transmit antenna to the jth receive antenna, for $i = 1, 2, \ldots, N$. The γ_{ij} are assumed independent. Then the received data at the jth receive antenna are

$$\mathbf{Y}_j = \mathbf{D}\boldsymbol{\Gamma}_j + \mathcal{N}_j, j = 1, 2, \ldots, M$$

Defining the complete data as $\{\mathbf{Y}_j, \boldsymbol{\Gamma}_j\}_{j=1}^M$, the E step of the EM algorithm at the kth iteration yields

$$\mathbf{Q}(\mathbf{D} \mid \mathbf{D}^k) = \sum_{l=1}^{L}\sum_{j=1}^{M}\left[\Re(\overline{y_j^{(l)}}\mathbf{D}^{(l)}\hat{\boldsymbol{\Gamma}}_j^k) - \frac{1}{2}\mathbf{D}^{(l)}(\hat{\boldsymbol{\Omega}}_j)^k(\mathbf{D}^{(l)})^*\right] \tag{18}$$

$$\hat{\boldsymbol{\Gamma}}_j^k = \left((\mathbf{D}^k)^*\mathbf{D}^k + \frac{\mathbf{I}}{\mathrm{SNR}}\right)^{-1}(\mathbf{D}^k)^*\mathbf{Y}_j \tag{19}$$

$$\hat{\boldsymbol{\Omega}}_j^k = \mathbf{I} - \left((\mathbf{D}^k)^*\mathbf{D}^k + \frac{\mathbf{I}}{\mathrm{SNR}}\right)^{-1}(\mathbf{D}^k)^*\mathbf{D}^k + \hat{\boldsymbol{\Gamma}}_j^k(\hat{\boldsymbol{\Gamma}}_j^k)^* \tag{20}$$

and the M step

$$\mathbf{D}^{k+1} = \arg\max_{\mathbf{D}} \sum_{l=1}^{L}\sum_{j=1}^{M}\left[\Re(\overline{y_j^{(l)}}\mathbf{D}^{(l)}\hat{\boldsymbol{\Gamma}}_j^k) - \frac{1}{2}\mathbf{D}^{(l)}\hat{\boldsymbol{\Omega}}_j^k(\mathbf{D}^{(l)})^*\right] \tag{21}$$

For the spacetime codes designed in Ref. 64, the Viterbi algorithm can be used to efficiently perform the required maximization. Simulation results are shown in Fig. 1 for the 8-state code over QPSK introduced in Ref. 64.

Other work using the EM algorithm in the spacetime area can be found in Ref. 65. This paper uses again Feder and Weistein's results [9] in a multitransmit antenna, multipath environment in conjunction with orthogonal frequency-division multiplexing (OFDM).

4.2. SNR Estimation

Estimation of the SNR and the received signal energy is an important function in digital communications. We present

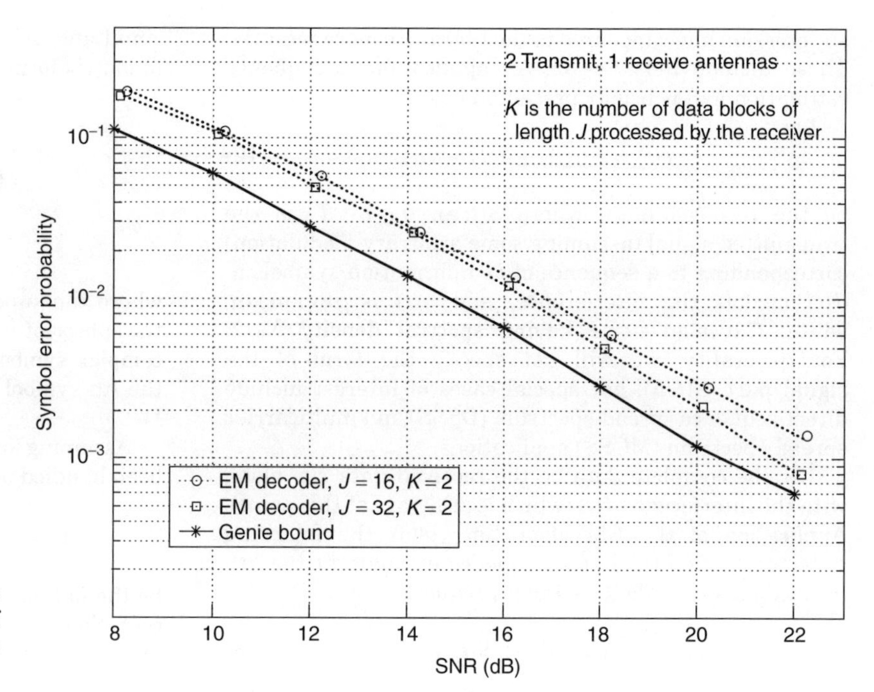

Figure 1. Symbol error probability of the "genie" and EM-based decoders as a function of block length: $N = 2$, $M = 1$, 8-state space-time code, quasistatic fading.

here an application of the EM algorithm to this problem. This is a summary of results from Ref. 66.

Let the received data be

$$r_k = \sqrt{E}d_k + \sqrt{v}n_k, k = 1, 2, \ldots, N$$

where E is the received signal energy; v is the additive noise variance; the n_k are zero-mean, unit-variance, i.i.d. Gaussian; and the d_k are binary modulation symbols (the more general case is handled as well) with $d_k \in \{1, -1\}$. We are interested in estimating the vector $\boldsymbol{\theta} = (E, v)$ from the received (incomplete) data vector $\mathbf{r}$ using the EM algorithm. Let the complete data be $\mathbf{x} = (\mathbf{r}, \mathbf{d})$. Then the E step of the EM algorithm yields

$$Q(\boldsymbol{\theta} \mid \boldsymbol{\theta}^i) = -N\ln(v) - \frac{1}{v}\sum_{k=1}^{N} r_k^2 - \frac{N \cdot E}{v} + \frac{2\sqrt{E}}{v} \cdot \sum_{k=1}^{N} r_k \hat{d}_k^i \tag{22}$$

where

$$\hat{d}_k^i = E\left[d_k \mid \mathbf{r}, \boldsymbol{\theta}^i\right] = \tanh\left(\frac{\sqrt{E^i}}{v^i} r_k\right) \tag{23}$$

Taking derivatives w.r.t. (with respect to) E and v, the M step of the EM algorithm yields the following recursions:

$$\hat{E}^{i+1} = \left(\frac{B^i}{N}\right)^2 \tag{24}$$

$$\hat{v}^{i+1} = \frac{A}{N} - \hat{E}^{i+1} \tag{25}$$

$$\widehat{\text{SNR}}^{i+1} = \frac{\hat{E}^{i+1}}{\hat{v}^{i+1}} \tag{26}$$

where

$$A = \sum_{k=1}^{N} r_k^2 \tag{27}$$

$$B^i = \sum_{k=1}^{N} r_k \tanh\left(\frac{\sqrt{E^i}}{v^i} r_k\right) \tag{28}$$

Figure 2 shows results on the bias and mean-square error in estimating the SNR and the received signal energy for the EM algorithm and a popular high-SNR approximation to the ML estimator (the true ML estimator is much too complex to implement).

4.3. On the (Non) Convergence of the EM Algorithm for Discrete Parameters

For continuous parameter estimation, under continuity and weak regularity conditions, stationary points of the EM algorithm have to be stationary points of the loglikelihood function [e.g., 11]. On the other hand, any fixed point can be a convergence point of the discrete EM algorithm even though the likelihood of such a point can be lower than the likelihood of a neighboring discrete point. In this subsection we summarize some results on the (non)convergence of the discrete EM algorithm presented in Ref. 67.

Let $\mathbf{a}$ be a M-dimensional discrete parameter with values in $\mathcal{A}$. It is assumed that, along with the discrete EM algorithm, a companion continuous EM algorithm

$$\mathbf{a}_c^{k+1} = \arg\max_{\mathbf{a}\in\mathcal{C}^{\mathbf{M}}} Q(\mathbf{a} \mid \mathbf{a}^k) \tag{29}$$

where $\mathcal{C}^M$ is the M-dimensional complex space, is well defined. The companion continuous EM mapping is the map $\mathbf{M}_c : \mathbf{a}^k \to \mathbf{a}_c^{k+1}$, for $\mathbf{a}_c^{k+1} \in \mathcal{C}^M$. The Jacobian of the continuous EM mapping $\mathbf{M}_c(\mathbf{a})$ is denoted with $\mathbf{J}_c(\mathbf{a})$, and the Hessian of its corresponding objective function $Q(\mathbf{a} \mid \mathbf{a}^k)$ is denoted with $\mathbf{H}^Q(\mathbf{a})$.

Computationally "inexpensive" maximization of the objective function $Q(\mathbf{a} \mid \mathbf{a}^k)$ over $\mathcal{A}$ can be obtained for

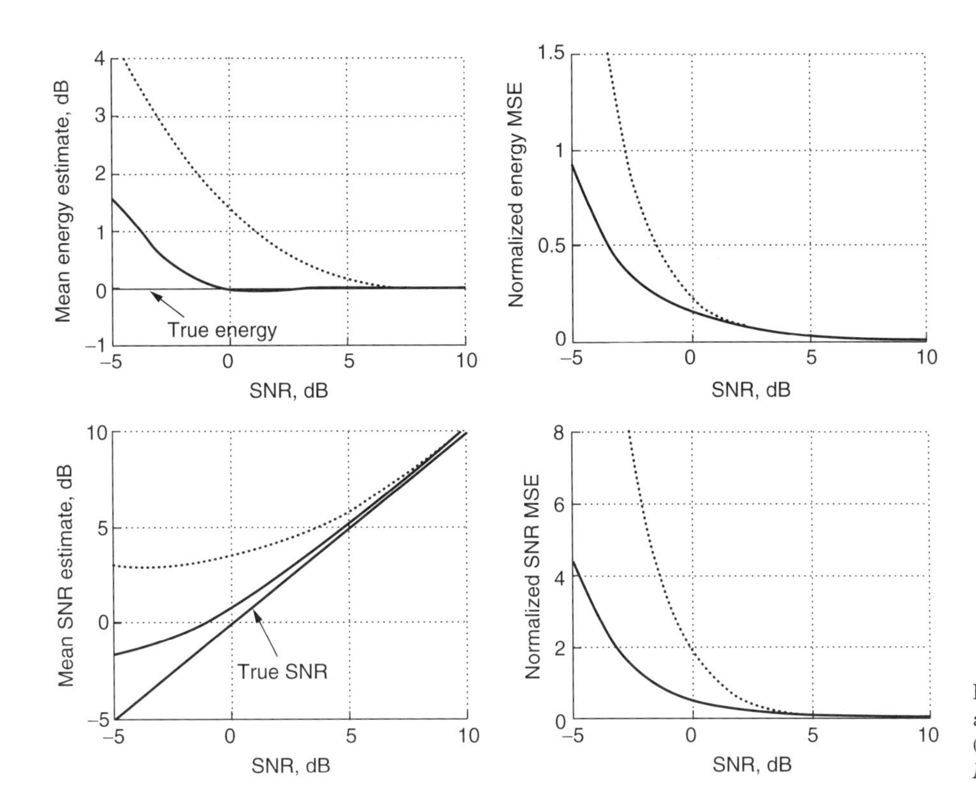

Figure 2. Comparison between the EM and high SNR ML approximation (dashed lines) for sequence length $N = 50$.

some special structures. The particular case of interest is the quadratic form

$$Q(\mathbf{a} \mid \mathbf{a}^k) = -\tfrac{1}{2}\|\mathbf{a} - \mathbf{a}_c^{k+1}\|^2 \tag{30}$$

for which the discrete maximization in the *M step* can be obtained on a parameter-by-parameter basis. Equation (30) implies that $\mathbf{a}^{k+1}$ is the discrete point closest to the continuous maximizer of $Q(\mathbf{a} \mid \mathbf{a}^k)$ over $\mathcal{C}^M$, $\mathbf{a}_c^{k+1}$. Note that $\mathbf{a}_c^{k+1}$ is (implicitly) a function of the previous iterate $\mathbf{a}^k$. As a special case of (30), $\mathbf{a}^{k+1} = \text{sign}[\mathbf{a}_c^{k+1}]$ for QPSK signaling, namely, for $a_{d,i} \triangleq [\mathbf{a}]_i \in \{\pm 1 \pm j\}$. Clearly, not only is the objective function (30) quadratic but, additionally, its Hessian is $\mathbf{H}^Q(\mathbf{a}) = -\mathbf{I}$. It is important to note that (30) holds for all the sequence estimation problems described in this overview [see Refs. 8, 10, and 13 for a linear modulation $S(t; \mathbf{a})$] except for (18), where the objective function is quadratic in the unknown symbol sequence $\{d_j^{(l)}\}$ but the Hessian is not $-\mathbf{I}$ in the general case. The arguments given below can easily be generalized for a nonidentity Hessian matrix.

Let $d_{\min}$ denote half the minimum Euclidean distance between any two discrete points from $\mathcal{A}$:

$$d_{\min} = \tfrac{1}{2} \min_{\mathbf{a}_1, \mathbf{a}_2 \in \mathcal{A}} \|\mathbf{a}_1 - \mathbf{a}_2\|. \tag{31}$$

$d_{\min} = 1$ for uncoded binary antipodal signaling. Let $\hat{\mathbf{a}}_c$ be a fixed point of the companion continuous EM algorithm (29). One of the authors [67] has been shown, based on the first two terms of the Taylor expansion of $\mathbf{M}_c(\mathbf{a})$ in a neighborhood $U(\hat{\mathbf{a}}_c)$ of $\hat{\mathbf{a}}_c$, that all $\mathbf{a} \in \mathcal{A} \cap U(\hat{\mathbf{a}}_c)$ such that

$$\|\hat{\mathbf{a}}_c - \mathbf{a}\| < \frac{d_{\min}}{1 - \hat{\lambda}_{\min}} \tag{32}$$

where $\hat{\lambda}_{\min} \triangleq \lambda_{\min}[\mathbf{J}_c(\hat{\mathbf{a}}_c)]$ is the minimum eigenvalue of the Jacobian matrix $\mathbf{J}_c(\hat{\mathbf{a}}_c)$, are stationary points of the discrete EM algorithm. Inequality (32) follows from the fact that the eigenvalues of $\mathbf{J}_c(\hat{\mathbf{a}}_c)$ are nonnegative and smaller than one (see lemma in Ref. 2). It defines a ball of radius $r^{nc} = d_{\min}/(1 - \hat{\lambda}_{\min})$ centered at a fixed point of the companion continuous EM algorithm $\hat{\mathbf{a}}_c$.

Parameter $\hat{\lambda}_{\min}$ is, in the general case, a function of the received signal. As shown by Meng and Rubin [3] (see also Ref. 11), $\mathbf{J}_c(\hat{\mathbf{a}}_c)$ and, consequently, its eigenvalues can be estimated in a reliable manner using a continuous EM algorithm. Furthermore, when the complete data of the EM algorithm are chosen in such a way that $\mathbf{H}^Q(\hat{\mathbf{a}}_c) = -\mathbf{I}$ as in (30), then matrices $\mathbf{J}_c(\hat{\mathbf{a}}_c)$ and the Hessian of the loglikelihood function, $\mathbf{H}^l(\hat{\mathbf{a}}_c)$, have the same eigenvectors. Their eigenvalues have the following relationship: $1 - \lambda_i[\mathbf{J}_c(\hat{\mathbf{a}}_c)] = \lambda_i[-\mathbf{H}^l(\hat{\mathbf{a}}_c)]$, for all i. Thus

$$r^{nc} = \frac{d_{\min}}{\lambda_{\max}[-\mathbf{H}^l(\hat{\mathbf{a}}_c)]}$$

The term $(1 - \hat{\lambda}_{\min})$ is the largest eigenvalue of the matrix $\mathbf{I} - \mathbf{J}_c(\hat{\mathbf{a}}_c)$, often referred to as the *iteration matrix* in the optimization literature [e.g., 11], since it determines the convergence iteration steps in the neighborhood $U(\hat{\mathbf{a}}_c)$. Its smallest eigenvalue is a typical measure of convergence speed of the continuous EM algorithm since it measures the convergence step along the slowest direction. Its largest eigenvalue determines the convergence step along the fastest direction. Thus, we can say that the nonconvergence radius is determined by (it is in fact inversely proportional to) the largest convergence step of the companion continuous EM algorithm.

An important implication of (32) is that $\hat{\lambda}_{\min} = 0$ is sufficient for the nonconvergence ball to hold at most one discrete stationary point. Sequence estimation, in the presence of interference when $\hat{\lambda}_{\min} = 0$ or when this identity can be forced using rank-reduction principles, has been analyzed in Ref. 67 and in part in Ref. 70, and is presented briefly next.

In the following, the received signal model given in (12) with a linear modulation

$$S(t;\mathbf{a}) = \sum_{m=0}^{M-1} a_n h(t - mT) \tag{33}$$

where pulses $\{h(t - mT), m \in [0, M-1]\}$ are orthonormal, is assumed.

For this problem, a reduced nonconvergence ball radius has been derived [67] without having to resort to the Taylor expansion of $\mathbf{M}_c(\mathbf{a})$. The reduced nonconvergence radius is a function of only $d_{\min}$ and the statistics of noise and interference. In the special case when $N = M$ and $\text{span}\{h(t - nT), n \in [0, N-1]\} = \text{span}\{\phi_n(t), n \in [0, N-1]\}$, where $\phi_n(t)$ are defined in (15), the reduced radius and r^{nc} are equal. Both are

$$r^{nc} = \left[1 + \frac{\tilde{\lambda}_{\min}}{N_0}\right] d_{\min} \tag{34}$$

where $\tilde{\lambda}_{\min} = \min_j\{\lambda_j\}$; λ_j is the variance of the coefficient g_j in (15). For example, let's assume $M = N = 2$, interference energy $J = \lambda_1 + \lambda_2 = 2\lambda_1 = 2\tilde{\lambda}_{\min}$, binary antipodal signaling with $d_{\min} = 1$, and noise and signal realizations such that $\hat{\mathbf{a}}_c \approx \mathbf{0}$ (see Ref. 67 for details). Then, if $J/N_0 > 2\sqrt{2} - 2$ all possible discrete points in $\mathcal{A}$ will be found inside the convergence ball whose radius is given by (34). Consequently, the discrete EM algorithm will generate a sequence of identical estimates $\mathbf{a}^k = \mathbf{a}^0$ for any k and any initial point $\mathbf{a}^0 \in \mathcal{A}$.

Next, we provide symbol error rate results for a binary communications example where $N > M$ and $\text{span}\{h(t - mT),\ m \in [0, M-1]\} \subset \text{span}\{\phi_n(t),\ n \in [0, N-1]\}$ holds. Figure 3 demonstrates that, in the case of an interference process whose full spectrum covers the spectrum of the useful signal, one could significantly increase the convergence probability of the EM algorithm using a rank-reduction approach. Two second-order Gaussian processes uniformly positioned in frequency within the signal spectra are used to model interference using $N > M$ eigenfunctions. The EM algorithm obtained by modeling the interference with $P < M < N$ largest eigenfunctions $\phi_n(t)$, and thus forcing $\hat{\lambda}_{\min} = 0$ in (32), is compared to the performance of the EM algorithm that uses the full-rank model. It can be observed that the reduced-rank EM algorithm has little degradation, whereas the full-rank EM algorithm does not converge and has a catastrophic performance for large SNR.

4.4. Detection in Fading Channels with Highly Correlated Multipath Components

Here we study another example where the convergence of the discrete EM algorithm can be stymied by a large nonconvergence ball.

We assume a continuous-time L-path Rayleigh fading channel model:

$$r_1(t) = \sum_{k=1}^{K}\sum_{l=1}^{L} \alpha_{l,k,1} S(t - \tau_{l,k,1}; \mathbf{a}_k) + n_1(t)$$

$$\vdots$$

$$r_{N_a}(t) = \sum_{k=1}^{K}\sum_{l=1}^{L} \alpha_{l,k,N_a} S(t - \tau_{l,k,N_a}; \mathbf{a}_k) + n_{N_a}(t) \tag{35}$$

Here N_a is the number of receive antennas and K is the number of users; $\mathbf{a}_k$ is the vector of N symbols of

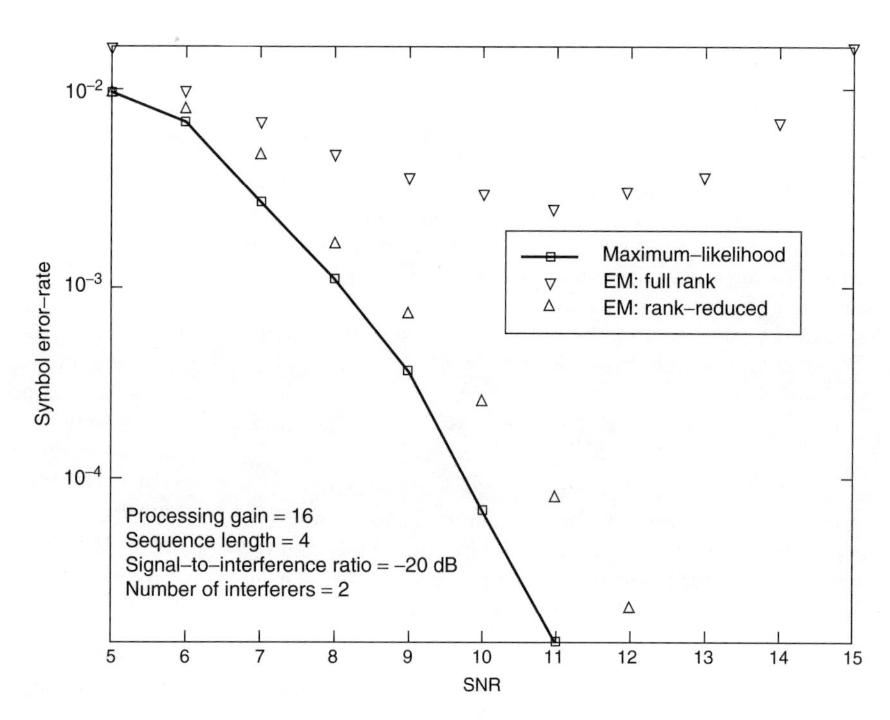

Figure 3. Comparison of ML, reduced-rank EM, and full-rank EM receivers for two second-order Gaussian interferers.

user k; $\alpha_{l,k,i}$ are zero-mean, complex circular Gaussian random variables of respective variances $\gamma_{l,k,i}$ and $\tau_{l,k,i}$ are (assumed known) path delays. We group the fading coefficients, $\alpha_{l,k,i}$, for each i into a vector $\boldsymbol{\alpha}_i$ such that $\alpha_{l,k,i} = [\boldsymbol{\alpha}_i]_{l+L\cdot(k-1)}$ is the $l + L \cdot (k-1)$th element of the vector $\boldsymbol{\alpha}_i$. The covariance matrix of $\boldsymbol{\alpha}_i$ is $\boldsymbol{\Gamma}_i$. P_s of the transmitted symbols are assumed to be known pilot symbols that allow for ambiguity resolution (67). The signal is assumed to have the following form:

$$S(t; \mathbf{a}_k) = \sum_{n=1}^{N} a_{n,k} h(t - nT)$$

where $h(t)$ is the known signaling pulse (assumed common for simplicity) modulated by the transmitted symbol sequence $\mathbf{a}$ of length N. Pulses $h(t - iT)$ are typically orthogonal to avoid ISI for frequency nonselective channels. The vector of time-delayed modulated signals is $\mathbf{s}_i(t; \mathbf{a}) = [S(t - \tau_{1,1,i}; \mathbf{a}_1), S(t - \tau_{2,1,i}; \mathbf{a}_1), \ldots, S(t - \tau_{L,K,i}; \mathbf{a}_K)]^T$, where $\mathbf{a}$ is a NK-dimensional vector of user symbols such that its subvector of length $K[\mathbf{a}]_{(k-1)\cdot N+1}^{k\cdot N}$ is the vector of N symbols of user k.

For simplicity the receiver antennas are assumed to be sufficiently separated so that any $\boldsymbol{\alpha}_i$ and $\boldsymbol{\alpha}_j$ for $i \neq j$ are independent. $\mathbf{G}_i(\mathbf{a}) = \langle \mathbf{s}_i(t; \mathbf{a}), \mathbf{s}_i^H(t; \mathbf{a})\rangle$ is the Gramian of $\mathbf{s}_i(t; \mathbf{a})$ in the L_2 space.

An EM solution is nontrivially based on the EM solution for superimposed signals introduced by Feder and Weinstein described in Section 3.1 (see, also Refs. 53 and 69). The complete data vector includes not only the path plus noise components but also the fading vectors $\boldsymbol{\alpha}_i$, as

$$(\mathbf{x}, \boldsymbol{\alpha}) \triangleq ([\mathbf{x}_1, \ldots, \mathbf{x}_{N_A}], \boldsymbol{\alpha}_1, \ldots, \boldsymbol{\alpha}_{N_a}) \quad (36)$$

$$\triangleq (\{x_{1,1,1}(t), \ldots, x_{L,K,N_a}(t), t \in [T_i, \ldots, T_f]\}, \boldsymbol{\alpha}_1, \ldots, \boldsymbol{\alpha}_{N_a}) \quad (37)$$

where $x_{l,k,i}(t) \triangleq \alpha_{l,k,i} S(t - \tau_{l,k,i}; \mathbf{a}_k) + n_{l,k,i}(t) \cdot n_{l,k,i}(t)$ is a complex zero-mean AWGN process with variance $\beta_{l,k} N_0$ such that $E\{n_{l,k,i}(t) n^*_{j,m,i}(t)\} = 0$ for all $l \neq j$ and $k \neq m$ and $n_i(t) = \sum_{l=1}^{L}\sum_{k=1}^{K} n_{l,k,i}(t)$. Clearly, $\beta_{l,k}$ has to satisfy the constraint $\sum_{l=1}^{L}\sum_{k=1}^{K} \beta_{l,k} = 1$. In the following, $\beta_{l,k} \equiv 1/(KL)$.

The terms $x_{l,k,i}(t)$ are mutually independent given the data sequence $\mathbf{a}$ and the fading vectors $\boldsymbol{\alpha}_i$, and the EM objective function, in the case of orthogonal signaling, can be represented in the following manner:

$$Q(\mathbf{a} \mid \mathbf{a}^m) = \sum_{k=1}^{K}\sum_{n=1}^{N}\left[\Re\left\{a_{n,k}\sum_{i=1}^{N_a}\sum_{l=1}^{L}\beta_{l,k}\hat{z}_{l,n,k,i}(\mathbf{a}^m)\right\} - \frac{1}{2}|a_{n,k}|^2 \sum_{i=1}^{N_a}\sum_{l=1}^{L}\beta_{l,k}\hat{\rho}_{ll,kk,i}(\mathbf{a}^m)\right] \quad (38)$$

Here $\hat{z}_{l,n,k,i}(\mathbf{a}^m) = \langle \hat{\chi}_{l,k,i}(t; \mathbf{a}^m), h(t - \tau_{l,k,i} - nT)\rangle$ is the sampled pulse-matched filter response to signal path component estimates $\hat{\chi}_{l,k,i}(t; \mathbf{a}^m)$.

The E step requires estimation of the phase and amplitude compensated signal path components

$$\hat{\chi}_{l,k,i}(t; \mathbf{a}^m) = E\{\alpha^*_{l,k,i} x_{l,k,i}(t) \mid \{r_i(t); t \in [T_i, \ldots, T_f]\}, \mathbf{a}^m\}$$

$$= \hat{\rho}_{ll,kk,i}(\mathbf{a}^m) S(t - \tau_{l,k,i}; \mathbf{a}^m) + \beta_{l,k}\left[\hat{\alpha}^*_{l,k,i}(\mathbf{a}^m) r_i(t) - \sum_{q=1}^{K}\sum_{j=1}^{L}\hat{\rho}_{lj,kq,i}(\mathbf{a}^m) S(t - \tau_{j,q,i}; \mathbf{a}^m)\right]$$

Signal path components are estimated by successive refinement. The refinement attempts to find those component estimates that explain the received signal with a smallest measurement error. $\hat{\alpha}_{l,k,i}(\mathbf{a}^m)$ and $\hat{\rho}_{lj,kq,i}(\mathbf{a}^m)$ are, respectively, the conditional mean estimates of the complex coefficients

$$\hat{\alpha}_{l,k,i}(\mathbf{a}^m) = E\{\alpha_{l,k,i} \mid \{r_i(t); t \in [T_i, \ldots, T_f]\}, \mathbf{a}^m\}$$

$$= [(N_0 \boldsymbol{\Gamma}_i^{-1} + \mathbf{G}_i(\mathbf{a}^m))^{-1}\langle \mathbf{s}_i(t; \mathbf{a}^m), r_i^H(t)\rangle]_l$$

and their cross-correlations

$$\hat{\rho}_{lj,kq,i}(\mathbf{a}^m) = E\{\alpha^*_{l,k,i}\alpha_{j,q,i} \mid \{r_i(t); t \in [T_i, \ldots, T_f]\}, \mathbf{a}^m\}$$

$$= \left[\left(\boldsymbol{\Gamma}_i^{-1} + \frac{\mathbf{G}_i(\mathbf{a}^m)}{N_0}\right)^{-1} + \hat{\boldsymbol{\alpha}}_i(\mathbf{a}^m)\hat{\boldsymbol{\alpha}}_i^H(\mathbf{a}^m)\right]_{l+(k-1)L, j+(q-1)L} \quad (39)$$

The M step of the companion continuous EM algorithm can be expressed in closed form as

$$\hat{a}^{m+1}_{cu,n,k} = \frac{\sum_{i=1}^{N_a}\sum_{l=1}^{L}\beta_{l,k}\hat{z}_{l,n,k,i}(\mathbf{a}^m)}{\sum_{i=1}^{N_a}\sum_{l=1}^{L}\beta_{l,k}\hat{\rho}_{ll,kk,i}(\mathbf{a}^m)}, \quad (n, k) \in \overline{\mathcal{J}}_{ps}$$

$$\hat{a}^{m+1}_{cu,n,k} = a_n^k, \; (n, k) \in \mathcal{J}_{ps} \quad (40)$$

where $\mathcal{J}_{ps}$ are index pairs corresponding to pilot symbols and

$$\overline{\mathcal{J}}_{ps} = \{(1, 1), \ldots, (K, N)\} \backslash \mathcal{J}_{ps}$$

It combines in an optimal manner the phase/amplitude-compensated signal path components estimated in the E step. In this way, it achieves multipath diversity combining.

The M step of the (discrete) EM algorithm requires quantization, for example, taking the sign of the components of $\mathbf{a}^{m+1}_{cu}$, in case of binary signaling.

The number of fixed points (and, consequently, convergence to the ML estimate) of the discrete EM algorithm is a function of the path Gramian matrices $\mathbf{G}_i(\mathbf{a})$ whose structure defines the size of the nonconvergence ball. For highly correlated paths the nonconvergence ball will be so large that we need to use other detection methods.

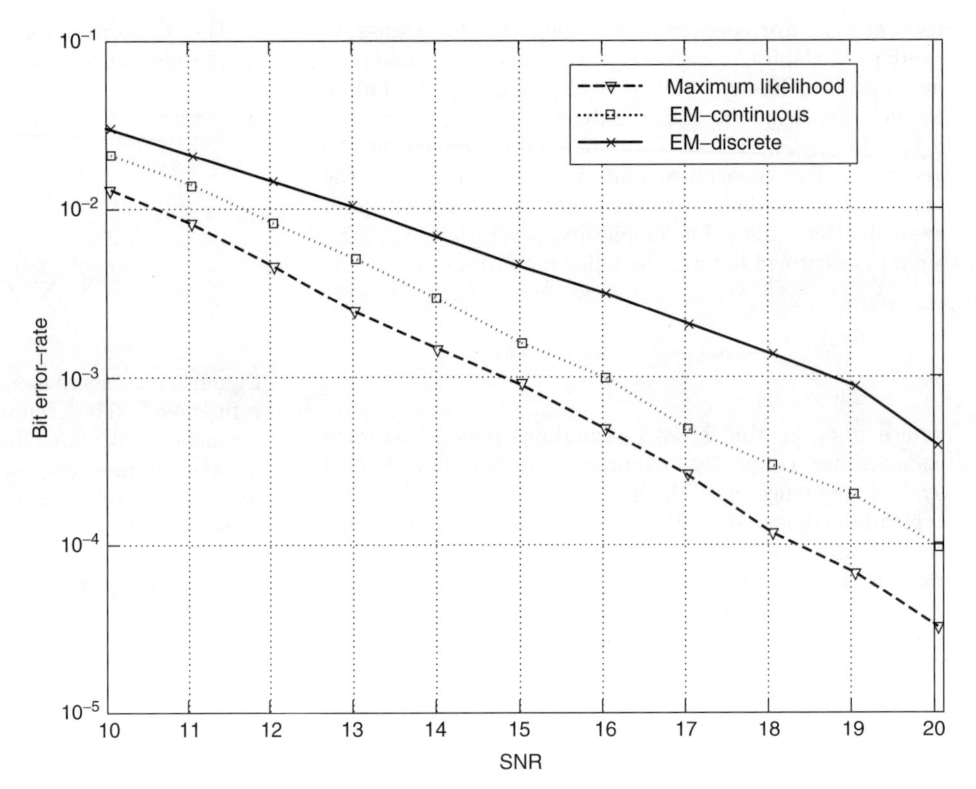

Figure 4. Performance comparison for highly correlated path components.

An alternative approach to detection termed *generalized decorrelator* (see, e.g., Refs. 18, 53, 68, and 69) allows the continuous EM algorithm to converge and quantizes the continuous maximizer. For uncoded BPSK signaling we have

$$\hat{\mathbf{a}}_c = \text{sign}[\mathbf{a}_c^{\infty}]$$

In the following example, the delays are $\tau_l/T \in \{0, \frac{1}{3}, \frac{4}{3}\}$. The number of antennas is 1, processing gain is equal to 3, the sequence length is 6 and the pilot symbol block length is 3. The paths are fading independently and their variance is $1/L$. The last two signal path components are highly correlated because of their integer symbol delay. Cross-correlation to the first component is also high as a result of the small processing gain. This particular case is detrimental to receivers that are based on the decorrelation between signal path components. The generalized decorrelator (continuous EM algorithm), as can be seen in Fig. 4, achieves a performance within 1.5–2 dB of the optimal. It manages diversity combining even for this particularly difficult situation. On the other hand, the discrete EM algorithm has a loss of 1–3 dB relative to the continuous EM detector.

5. CONCLUSION

We have provided a short overview of the use of the EM algorithm in telecommunications. Clearly, the EM algorithm has established itself as a generic tool in the design of telecommunication systems. If current work using it in various problems is any indication, the algorithm will get even more attention in the future.

BIOGRAPHIES

Costas N. Georghiades received his B.E. degree, with distinction, from the American University of Beirut in 1980, and his M.S. and D.Sc. degrees from Washington University in 1983 and 1985, respectively, all in electrical engineering. Since 1985, he has been with the Electrical Engineering Department at Texas A&M University, where he is a professor and holder of the J.W. Runyon, Jr. Endowed Professorship. He currently serves as director of the Telecommunications and Signal Processing Group in the department. Over the years he served in editorial positions with the *IEEE Transactions on Communications*, the *IEEE Transactions on Information Theory*, the *IEEE Journal on Selected Areas in Communications* and the *IEEE Communications Letters*. Dr. Georghiades is a fellow of the IEEE. His general interests are in the application of information and estimation theories to the study of communication systems, with particular interest in optimum receiver design, wireless communication, and optical systems.

Predrag Spasojevic received the Diploma of Engineering degree from the School of Electrical Engineering, University of Sarajevo, in 1990; and his M.S. and Ph.D. degrees in electrical engineering from Texas A&M University, College Station, Texas, in 1992 and 1999, respectively. From 2000 to 2001 he has been with WINLAB, Rutgers University, where he is currently an assistant professor in the Department of Electrical & Computer Engineering. His research interests are in the general areas of communication theory and signal processing.

BIBLIOGRAPHY

1. A. P. Dempster, N. M. Laird, and D. B. Rubin, Maximum-likelihood from incomplete data via the EM algorithm, *J. Roy. Stat. Soc.* **39**: 1–17 (1977).
2. X. L. Meng and D. B. Rubin, On the global and component-wise rates of convergence of the EM algorithm, *Linear Algebra Appl.* **199**: 413–425 (1994).
3. X. L. Meng and D. B. Rubin, Using EM to obtain asymptotic variance-covariance matrices: The SEM algorithm, *J. Am. Stat. Assoc.* **86**: 899–909 (1991).
4. L. E. Baum, T. Petrie, G. Soules, and N. Weiss, A maximization technique in statistical estimation for probabilistic functions of Markov chains, *Ann. Math. Stat.* **41**: 164–171 (1970).
5. C. F. Wu, On the convergence properties of the EM algorithm, *Ann. Stat.* **11**(1): 95–103 (1983).
6. L. A. Shepp and Y. Vardi, Maximum-likelihood reconstruction for emission tomography, *IEEE Trans. Med. Imag.* **1**: 113–122 (Oct. 1982).
7. D. L. Snyder and D. G. Politte, Image reconstruction from list-mode data in an emission tomography system having time-of-flight measurements, *IEEE Trans. Nucl. Sci.* **30**(3): 1843–1849 (1983).
8. D. L. Snyder and M. I. Miller, *Random Processes in Time and Space*, Springer-Verlag, New York, 1991, Chap. 3.
9. M. Feder and E. Weinstein, Parameter estimation of superimposed signals using the EM algorithm, *IEEE Trans. Acoust. Speech Signal Process.* **ASSP-36**: 477–489 (April 1988).
10. T. Moon, The expectation-maximization algorithm, *IEEE Signal Process. Mag.* (Nov. 1996).
11. G. J. McLachlan and T. Krishnan, *The EM Algorithm and Extensions*, Wiley Series in Probability and Statistics, 1997.
12. M. Feder and J. A. Catipovic, Algorithms for joint channel estimation and data recovery-application to equalization in underwater communications, *IEEE J. Ocean. Eng.* **16**(1): 42–55 (Jan. 1991).
13. J. W. Modestino, Reduced-complexity iterative maximum-likelihood sequence estimation on channels with memory, *Proc. 1993 IEEE Int. Symp. Inform. Theory* 422–422 (1993).
14. A. Radović, An iterative near-far resistant algorithm for joint parameter estimation in asynchronous CDMA systems, *5th IEEE Int. Symp. Personal, Indoor and Mobile Radio Communications*, 1994, Vol. 1, pp. 199–203.
15. M. J. Borran and M. Nasiri-Kenari, An efficient decoding technique for CDMA communication systems based on the expectation maximization algorithm, *IEEE 4th Int. Symp. Spread Spectrum Techniques and Applications Proc.*, 1996, Vol. 3, pp. 1305–1309.
16. A. Chkeif and G. K. Kaleh, Iterative multiuser detector with antenna array: An EM-based approach, *Proc. Int. Symp. Information Theory*, 1998, p. 424.
17. H. V. Poor, On parameter estimation in DS/SSMA formats, *Proc. Advances in Communications and Control Systems*, Baton Rouge, LA, Oct. 1988, pp. 59–70.
18. L. B. Nelson and H. V. Poor, Soft-decision interference cancellation for AWGN multiuser channels, *Proc. 1994 IEEE Int. Symp. Information Theory*, 1994, p. 134.
19. H. V. Poor, Adaptivity in multiple-access communications, *Proc. 34th IEEE Conf. Decision and Control*, 1995, Vol. 1, pp. 835–840.
20. L. B. Nelson and H. V. Poor, EM and SAGE algorithms for multi-user detection, *Proc. 1994 IEEE-IMS Workshop on Information Theory and Statistics*, 1994, p. 70.
21. L. B. Nelson and H. V. Poor, Iterative multiuser receivers for CDMA channels: An EM-based approach, *IEEE Trans. Commun.* **44**(12): 1700–1710 (Dec. 1996).
22. J. A. Fessler and A. O. Hero, Complete-data spaces and generalized EM algorithms, *1993 IEEE Int. Conf. Acoustics, Speech, and Signal Processing* (ICASSP-93), 1993, Vol. 4, pp. 1–4.
23. J. A. Fessler and A. O. Hero, Space-alternating generalized EM algorithm, *IEEE Trans. Signal Process.* (Oct. 1994).
24. I. Sharfer and A. O. Hero, A maximum likelihood CDMA receiver using the EM algorithm and the discrete wavelet transform, *1996 IEEE Int. Conf. Acoustics, Speech, and Signal Processing* (ICASSP-93), 1996, Vol. 5, pp. 2654–2657.
25. C. Carlemalm and A. Logothetis, Blind signal detection and parameter estimation for an asynchronous CDMA system with time-varying number of transmission paths, *Proc. 9th IEEE SP Workshop on Statistical Signal and Array Processing*, 1998, pp. 296–299.
26. D. Zeghlache and S. Soliman, Use of the EM algorithm in impulsive noise channels, *Proc. 38th IEEE Vehicular Technology Conf.*, 1988, pp. 610–615.
27. D. Zeghlache, S. Soliman, and W. R. Schucany, Adaptive detection of CPFSK signals in non-Gaussian noise, *Proc. 20th Southeastern Symp. System Theory*, 1988, pp. 114–119.
28. S. M. Zabin and H. V. Poor, Efficient estimation of class A noise parameters via the EM algorithm, *IEEE Trans. Inform. Theory* **37**(1): 60–72 (Jan. 1991).
29. A. Ansari and R. Viswanathan, Application of expectation-maximizing algorithm to the detection of direct-sequence signal in pulsed noise jamming, *IEEE Military Communications Conf., (MILCOM '92) Conference Record*, 1992, Vol. 3, pp. 811–815.
30. A. Ansari and R. Viswanathan, Application of expectation-maximization algorithm to the detection of a direct-sequence signal in pulsed noise jamming, *IEEE Trans. Commun.* **41**(8): 1151–1154 (Aug. 1993).
31. R. S. Blum, R. J. Kozick, and B. M. Sadler, An adaptive spatial diversity receiver for non-Gaussian interference and noise, *1st IEEE Signal Processing Workshop on Signal Processing Advances in Wireless Communications*, 1997, pp. 385–388.
32. V. A. N. Baroso, J. M. F. Moura, and J. Xavier, Blind array channel division multiple access (AChDMA) for mobile communications, *IEEE Trans. Signal Process.* **46**: 737–752 (March 1998).
33. R. Vallet and G. K. Kaleh, Joint channel identification and symbols detection, *Proc. Int. Symp. Information Theory*, 1991, p. 353.
34. G. K. Kaleh and R. Vallet, Joint parameter estimation and symbol detection for linear and nonlinear unknown channels, *IEEE Trans. Commun.* **42**: 2406–2413 (July 1994).
35. Min Shao and C. L. Nikias, An ML/MMSE estimation approach to blind equalization, *1994 IEEE Int. Conf. Acoustics, Speech, and Signal Processing* (ICASSP-94), 1994, Vol. 4, pp. 569–572.
36. H. Zamiri-Jafarian and S. Pasupathy, Recursive channel estimation for wireless communication via the EM algorithm,

1997 IEEE Int. Conf. Personal Wireless Communications, 1997, pp. 33–37.

37. H. Zamiri-Jafarian and S. Pasupathy, EM-Based Recursive estimation of channel parameters, *IEEE Trans. Commun.* **47**: 1297–1302 (Sept. 1999).
38. C. N. Georghiades and D. L. Snyder, An application of the expectation-maximization algorithm to sequence detection in the absence of synchronization, *Proc. Johns Hopkins Conf. Information Sciences and Systems*, Baltimore, MD, March 1989.
39. C. N. Georghiades and D. L. Snyder, The expectation maximization algorithm for symbol unsynchronized sequence detection, *IEEE Trans. Commun.* **39**(1): 54–61 (Jan. 1991).
40. G. K. Kaleh, The Baum-Welch Algorithm for detection of time-unsynchronized rectangular PAM signals, *IEEE Trans. Commun.* **42**: 260–262 (Feb.–April 1994).
41. N. Antoniadis and A. O. Hero, Time-delay estimation for filtered Poisson processes using an EM-type algorithm, *IEEE Trans. Signal Process.* **42**(8): 2112–2123 (Aug. 1994).
42. I. Sharfer and A. O. Hero, Spread spectrum sequence estimation and bit synchronization using an EM-type algorithm, *1995 Int. Conf. Acoustics, Speech, and Signal Processing* (ICASSP-95), 1995, Vol. 3, pp. 1864–1867.
43. C. N. Georghiades and J. C. Han, Optimum decoding of trellis-coded modulation in the presence of phase-errors, *Proc. 1990 Int. Symp. Its Applications* (ISITA' 90), Hawaii, Nov. 1990.
44. C. N. Georghiades, Algorithms for Joint Synchronization and Detection, in *Coded Modulation and Bandwidth Efficient Transmission*, Elsevier, Amsterdam, 1992.
45. C. N. Georghiades and J. C. Han, On the application of the EM algorithm to sequence estimation for degraded channels, *Proc. 32nd Allerton Conf. Univ. Illinois*, Sept. 1994.
46. C. R. Nassar and M. R. Soleymani, Joint sequence detection and phase estimation using the EM algorithm, *Proc. 1994 Canadian Conf. Electrical and Computer Engineering*, 1994, Vol. 1, pp. 296–299.
47. C. N. Georghiades and J. C. Han, Sequence Estimation in the presence of random parameters via the EM algorithm, *IEEE Trans. Commun.* **45**: 300–308 (March 1997).
48. E. Panayirci and C. N. Georghiades, Carrier phase synchronization of OFDM systems over frequency-selective channels via the EM algorithm, *1999 IEEE 49th Vehicular Technology Conf.*, 1999, Vol. 1, pp. 675–679.
49. J. C. Han and C. N. Georghiades, Maximum-likelihood sequence estimation for fading channels via the EM algorithm, *Proc. Communication Theory Mini Conf.*, Houston, TX, Nov. 1993.
50. J. C. Han and C. N. Georghiades, Pilot symbol initiated optimal decoder for the land mobile fading channel, *1995 IEEE Global Telecommunications Conf.* (GLOBECOM '95), 1995, pp. 42–47.
51. K. Park and J. W. Modestino, An EM-based procedure for iterative maximum-likelihood decoding and simultaneous channel state estimation on slow-fading channels, *Proc. 1994 IEEE Int. Symp. Information Theory*, 1994, p. 27.
52. L. M. Zeger and H. Kobayashi, MLSE for CPM signals in a fading multipath channel, *1999 IEEE Pacific Rim Conf. Communications, Computers and Signal Processing*, 1999, pp. 511–515.
53. P. Spasojević and C. N. Georghiades, Implicit diversity combining based on the EM algorithm, *Proc. WCNC '99*, New Orleans, LA, Sept. 1999.
54. H. Zamiri-Jafarian and S. Pasupathy, Generalized MLSDE via the EM algorithm, *1999 IEEE Int. Conf. Communication*, 1999, pp. 130–134.
55. H. Zamiri-Jafarian and S. Pasupathy, Adaptive MLSDE using the EM algorithm, *IEEE Trans. Commun.* **47**(8): 1181–1193 (Aug. 1999).
56. M. Leconte and F. Hamon, Performance of /spl pi/-constellations in Rayleigh fading with a real channel estimator, *1999 IEEE Int. Conf. Personal Wireless Communication*, 1999, pp. 183–187.
57. W. Turin, MAP decoding using the EM algorithm, *1999 IEEE 49th Vehicular Technology Conf.*, 1999, Vol. 3, pp. 1866–1870.
58. Q. Zhang and C. N. Georghiades, An application of the EM algorithm to sequence estimation in the presence of tone interference, *Proc. 5th IEEE Mediterranean Conf. Control and Systems*, Paphos, Cyprus, July 1997.
59. O. C. Park and J. F. Doherty, Generalized projection algorithm for blind interference suppression in DS/CDMA communications, *IEEE Trans. Circuits Syst. II: Analog Digital Signal Process.* **44**(6): 453–460 (June 1997).
60. C. N. Georghiades, Maximum-likelihood detection in the presence of interference, *Proc. IEEE Int. Symp. Inform. Theory* 344 (Aug. 1998).
61. C. N. Georghiades and D. Reynolds, *Interference Rejection for Spread-Spectrum Systems Using the EM Algorithm*, Springer-Verlag, London, 1998.
62. Q. Zhang and C. N. Georghiades, An interference rejection application of the EM algorithm to direct-sequence signals, *Kybernetika* (March 1999).
63. Y. Li, C. N. Georghiades, and G. Huang, Iterative maximum likelihood sequence estimation of space-time codes, *IEEE Trans. Commun.* **49**: 948–951 (June 2001).
64. V. Tarokh, N. Seshadri, and A. R. Calderbank, Space-time codes for high data rate wireless communication: Performance criterion and code construction, *IEEE Trans. Inform. Theory* **44**: 744–765 (March 1998).
65. Y. Xie and C. N. Georghiades, An EM-based channel estimation algorithm for OFDM with transmitter diversity, *Proc. Globecom 2001*, San Antonio, TX, Nov. 2001.
66. C. N. Georghiades and U. Dasgupta, On SNR and energy estimation, manuscript in preparation.
67. P. Spasojević, *Sequence and Channel Estimation for Channels with Memory*, dissertation, Texas A&M Univ., College Station, TX, Dec. 1999.
68. P. Spasojević and C. N. Georghiades, The slowest descent method and its application to sequence estimation, *IEEE Trans. Commun.* **49**: 1592–1601 (Sept. 2001).
69. P. Spasojević, Generalized decorrelators for fading channels, *Int. Conf. Information Technology: Coding and Computing*, April 2001, pp. 312–316.
70. C. N. Georghiades and P. Spasojević, Maximum-likelihood detection in the presence of interference, *IEEE Trans. Commun.* (submitted).

F

FADING CHANNELS

Alexandra Duel-Hallen
North Carolina State University
Raleigh, North Carolina

1. INTRODUCTION

Radio communication channels include shortwave ionospheric radiocommunication in the 3–30-MHz frequency band (HF), tropospheric scatter (beyond-the-horizon) radio communications in the 300–3000 MHz frequency band (UHF) and 3000–30,000-MHz frequency band (SHF), and ionospheric forward scatter in the 30–300-MHz frequency band (VHF) [1,2]. These channels usually exhibit *multipath propagation* that results from *reflection, diffraction* and *scattering* of the transmitted signal. Multipath propagation and *Doppler effects* due to the motion of the transmitter and/or receiver give rise to *multipath fading* channel characterization. The fading signal is characterized in terms of *large-scale* and *small-scale* fading.

The large-scale fading model describes the average received signal power as a function of the distance between the transmitter and the receiver. Statistical variation around this mean (on the order of 6–10 dB) resulting from *shadowing* of the signal due to large obstructions is also included in the model of large-scale fading. Large-scale fading describes the variation of the received power over large areas and is useful in estimating radio coverage of the transmitter. The large scale fading model is determined by averaging the received power over many wavelengths (e.g., 1–10 m for cellular and PCS frequencies of 1–2 GHz) [3].

The small-scale fading model describes the instantaneous variation of the received power over a few wavelengths. This variation can result in dramatic loss of power on the order of 40 dB. It is due to superposition of several reflected or scattered components coming from different directions. Figure 1 depicts small-scale fading and slower large-scale fading for a mobile radiocommunication system. The figure illustrates that small-scale variation is averaged out in the large-scale fading model.

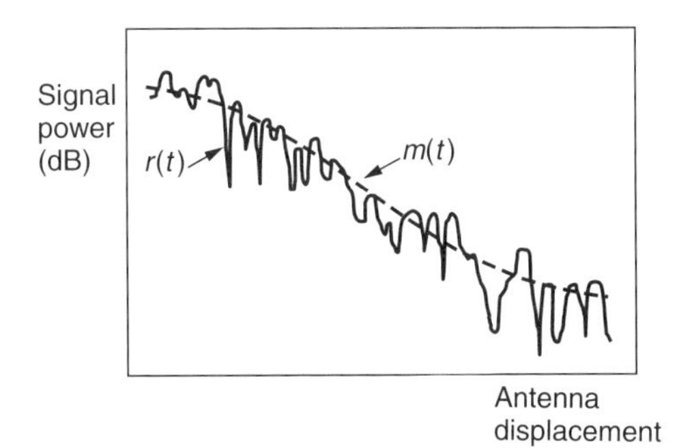

Figure 1. Large- and small-scale fading (reprinted from Ref. 2 © IEEE).

In Section 2, we review the large-scale fading models. Sections 3–5 describe small-scale fading. Multipath fading channels are characterized in Section 3. Section 4 describes useful fading models. Performance analysis for flat fading channels and diversity combining approaches are described in Section 5.

2. LARGE-SCALE FADING

Propagation path loss $L_s(d)$ is defined as the ratio of the transmitted power to the received power in a radiofrequency (RF) channel. In a free-space model, the propagation path loss is proportional to d^2, where d is the distance between the transmitter and the receiver. When the receiving antenna is isotropic, this loss is given by [2]

$$L_s(d) = \left(\frac{4\pi d}{\lambda}\right)^2 \tag{1}$$

where λ is the wavelength of the RF signal. In the mobile radio channel, the average path loss is usually more severe due to obstructions and is inversely proportional to d^n, where $2 \le n \le 6$. The average path loss is given by [2,3]

$$L_{\text{av}}(d)\ (\text{dB}) = L_s(d_0) + 10n\log_{10}\frac{d}{d_0} \tag{2}$$

where d_0 is the close-in reference distance. This distance corresponds to a point located in the far field of the antenna. For large cells, it is usually assumed to be 1 km, whereas for smaller cells and indoor environments, the values of 100 m and 1 m, respectively, are used. The value of the exponent n depends on the frequency, antenna heights, and propagation conditions. For example, in urban area cellular radio, n takes on values from 2.7 to 3.5; in building line-of-sight conditions, $1.6 \le n \le 1.8$; whereas in obstructed in building environments, $n = 4$–6 [3].

The actual path loss in a particular location can deviate significantly from its average value (2) due to *shadowing* of the signal by large obstructions. Measurements show that this variation is approximately *lognormally* distributed. Thus, the path loss is represented by the random variable

$$L_p(d)\ (\text{dB}) = L_{\text{av}}(d)\ (\text{dB}) + X_\sigma \tag{3}$$

where X_σ has Gaussian distribution (when expressed in decibels) with mean zero and standard deviation σ (also in decibels). Note that average path loss (2) corresponds to a straight line with slope $10n$ (dB) per decade when plotted on log–log scale. Thus, in practice, the values of n and σ are determined using linear regression to minimize the difference between the actual measurements (over various locations and distances between the transmitter and receiver) and the estimated average path loss in the mean-squared-error (MSE) sense. Figure 2 illustrates the actual measured data and the estimated average path loss for several transmitter-receiver separation values.

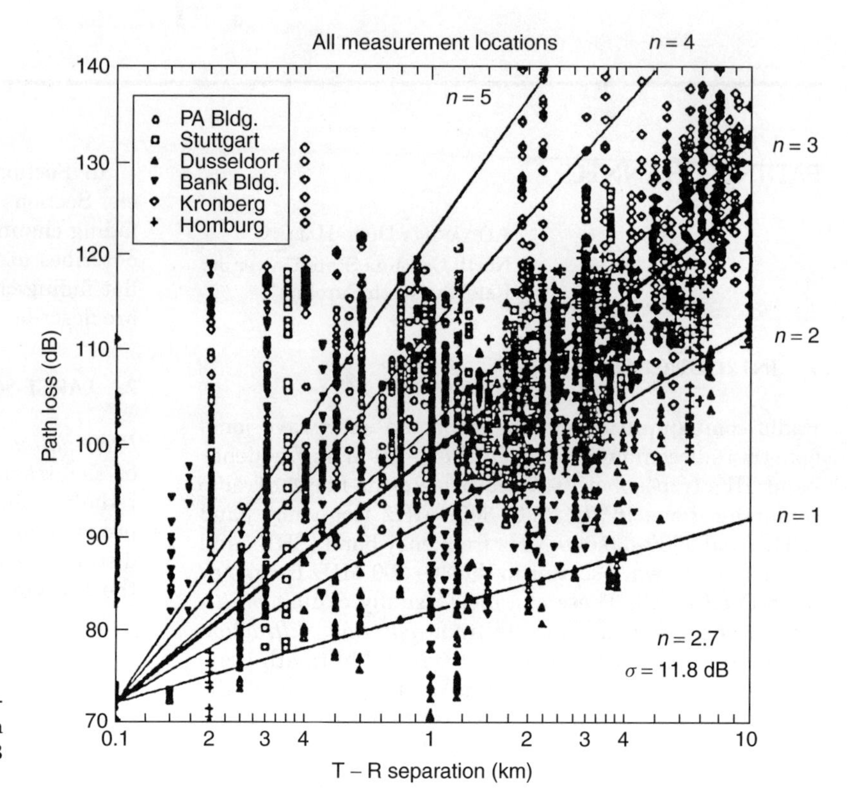

Figure 2. Scatterplot of measured data and corresponding MMSE path loss model for many cities in Germany. For these data, $n = 2.7$ and $\sigma = 11.8$ dB (reprinted from Ref. 15, © IEEE).

References 1–7 describe appropriate models and approaches to measurement and estimation of propagation path loss for various wireless channels and provide additional useful references on this subject. The rest of this article is devoted to characterization, modeling, and performance analysis of small-scale fading.

3. CHARACTERIZATION OF FADING MULTIPATH CHANNELS

A signal transmitted through a wireless channel arrives at its destination along a number of different paths (referred to as *multipath propagation*) as illustrated in Fig. 3. Multipath causes interference between *reflected or scattered* transmitter signal components. As the receiver moves through this *interference pattern*, a typical fading signal results as illustrated in Fig. 4. If an unmodulated carrier at the frequency f_c is transmitted over a fading channel, the complex envelope (the *equivalent lowpass signal*) [1] of the received fading signal is given by

$$c(t) = \sum_{n=1}^{N} A_n e^{j(2\pi f_n t + 2\pi f_c \tau_n + \phi_n)} \tag{4}$$

where N is the number of scatterers, and for the nth scatterer, A_n is the *amplitude*, f_n is the *Doppler frequency shift*, τ_n is the *excess propagation delay* (relative to the arrival of the first path), and ϕ_n is the phase. The *Doppler frequency shift* is given by [3]

$$f_n = f_c \frac{v}{c} \cos\theta = f_{dm} \cos\theta \tag{5}$$

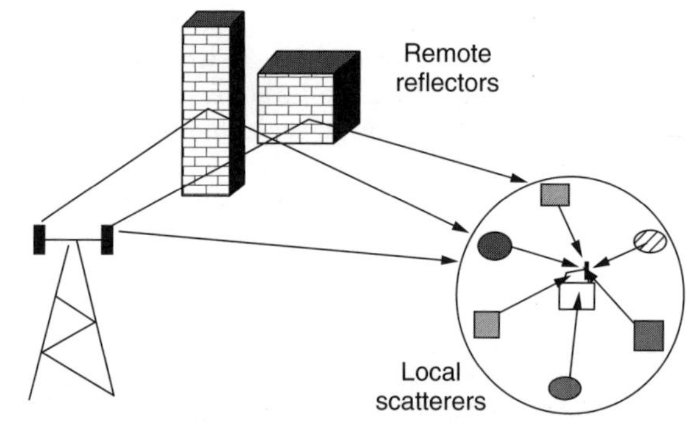

Figure 3. Typical mobile radio environment.

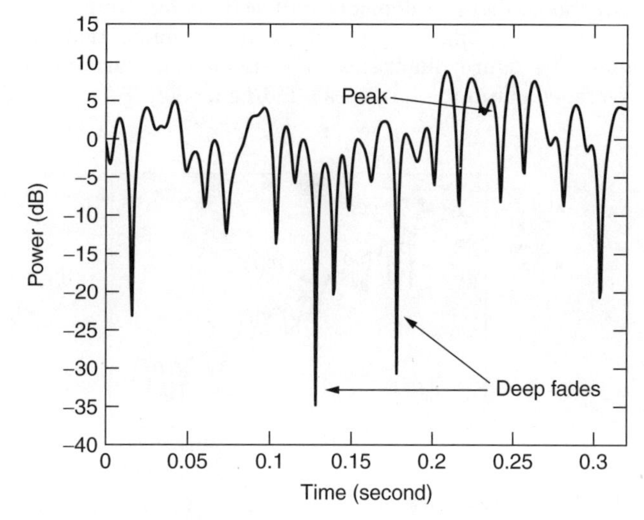

Figure 4. A typical fading signal (provided by Ericsson Inc.).

where v is the vehicle speed (assumed constant), c is the speed of light, θ is the incident radiowave angle with respect to the motion of the mobile, and f_{dm} is the *maximum Doppler frequency shift*.

The parameters A_n, f_n, τ_n, and ϕ_n are very *slowly time-variant,* and can be viewed as fixed on the timescale of a few milliseconds. Thus, the signal in (4) is a superposition of complex sinusoids with approximately constant amplitudes, frequencies, and phases, and varies in time as the mobile moves through the interference pattern. The superposition of terms in (4) can result in destructive or constructive interference, causing deep fades or peaks in the received signal, respectively, as illustrated in Fig. 4. The power of the fading signal can change dramatically, by as much as 30–40 dB. This variation can be conveniently modeled by characterizing $c(t)$ as a *stationary random process*. This statistical characterization is useful for describing time dispersion and fading rapidity of multipath fading channels [1–9].

3.1. Statistical Characterization

If we assume that the complex envelope of the transmitted signal is $s_l(t)$, the equivalent baseband signal received at the output of the fading channel is

$$r(t) = \int_{-\infty}^{\infty} c(\tau, t) s_l(t - \tau)\, d\tau \tag{6}$$

where the time-variant impulse response $c(\tau, t)$ is the response of the channel at time t to the impulse applied at time $t - \tau$ [1]. (In practice, additive Gaussian noise is also present at the output of the channel.) Expression (6) can be viewed as the superposition of delayed and attenuated copies of the transmitted signal $s_l(t)$, where the delays are given by τ_n and the corresponding complex gains have amplitudes A_n [see (4)] and time-variant phases [determined from the phase terms in (4)].

For each delay τ, the response $c(\tau, t)$ is modeled as a *wide-sense stationary* stochastic process. Typically, the random processes $c(\tau_1, t)$ and $c(\tau_2, t)$ are uncorrelated for $\tau_1 \neq \tau_2$ since different multipath components contribute to these signals (this is called *uncorrelated scattering*). For fixed delay τ, the autocorrelation function of the impulse response is defined as [1]

$$\phi_c(\tau; \Delta t) = \tfrac{1}{2} E[c^*(\tau, t) c(\tau, t + \Delta t)] \tag{7}$$

The power of $c(\tau, t)$ as a function of the delay τ is

$$\phi_c(\tau; 0) \equiv \phi_c(\tau) \tag{8}$$

This is called *multipath intensity profile* or *power delay profile* and is determined from measurements [1,3]. A "worst case" multipath intensity profile for an urban channel is shown in Fig. 5. The range of values of τ where $\phi_c(\tau)$ is nonnegligible is called the *multipath spread* of the channel and denoted as T_m. The values of multipath spread vary greatly depending on the terrain. For urban and suburban areas, typical values of multipath spread are from 1 to 10 μs, whereas in rural mountainous area, the multipath spreads are much greater with values from 10 to 30 μs [1]. The *mean excess delay* and *RMS delay spread* σ_τ are defined as the mean and the standard deviation of the excess delay, respectively, and can be determined from the multipath intensity profile [2,3]. Typical RMS delay spreads are on the order of one microsecond for urban outdoor channels, hundreds of nanoseconds for suburban channels, and tens of nanoseconds for indoor channels [3].

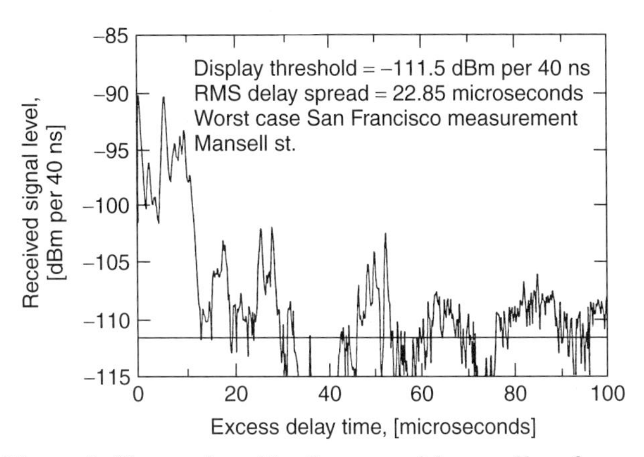

Figure 5. Measured multipath power delay profiles: from a 900-MHz cellular system in San Francisco (reprinted from Ref. 16, © IEEE).

The Fourier transform of the channel response $c(\tau, t)$ is the time-variant channel transfer function $C(f; t)$. It is also modeled as a wide-sense stationary random process. The correlation

$$\phi_c(\Delta f; \Delta t) = \tfrac{1}{2} E[C^*(f; t) C(f + \Delta f; t + \Delta t)] \tag{9}$$

is called the *spaced-frequency spaced-time autocorrelation function*. It is the Fourier transform of the autocorrelation function $\phi_c(\tau; \Delta t)$ in (7) [1]. It can be shown that this function can be factored into a product of time-domain and frequency-domain correlation functions. The latter is the *spaced-frequency correlation function* $\phi_c(\Delta f; 0) \equiv \phi_c(\Delta f)$ and is the Fourier transform of the multipath intensity profile $\phi_c(\tau)$ in (8) [1].

The complex envelope of the response $C(f; t)$ is specified by (4) with f viewed as the frequency of the unmodulated input carrier. Consider input frequency separation Δf. In the expressions for $C(f; t)$ and $C(f + \Delta f; t)$, the multipath components corresponding to the Doppler shift f_n have phase separation $\Delta\phi_n = 2\pi \Delta f \tau_n$. As Δf increases, these phase shifts result in decreased correlation between fading responses associated with two frequencies [10]. The *coherence bandwidth* $(\Delta f)_c$ provides a measure of this correlation. If the frequency separation is less than $(\Delta f)_c$, the signals $C(f; t)$ and $C(f + \Delta f; t)$ are strongly correlated, and thus fade similarly. The coherence bandwidth is inversely proportional to the multipath spread. However, the exact relationship between the coherence bandwidth and the multipath spread depends on the underlying meaning of the strong correlation of fading signals at different frequencies and varies depending on the channel model [2,3]. For example, if $|\phi_c(\Delta f)|$ is required to remain above 0.9, the corresponding coherence bandwidth is defined as $(\Delta f)_c \approx 1/(50\sigma_\tau)$, where σ_τ is the RMS delay spread. When the frequency correlation is allowed to decrease to 0.5, greater coherence bandwidth $(\Delta f)_c \approx 1/(5\sigma_\tau)$ results.

The time variation of the channel response $c(\tau, t)$ due to the Doppler shift can be statistically characterized using the *spaced-time correlation function* determined from (9) as

$$\phi_c(0; \Delta t) \equiv \phi_c(\Delta t) \tag{10}$$

or its Fourier transform $S_c(\lambda)$ [1]. The function $S_c(\lambda)$ is called the *Doppler power spectrum* of the channel. As the maximum Doppler shift increases, the channel variation becomes more rapid [see (4)], and $S_c(\lambda)$ widens, resulting in *spectral broadening* at the receiver. The shape of the autocorrelation function depends on channel characteristics. For example, the popular Rayleigh fading channel discussed in Section 4 has the autocorrelation function

$$\phi_c(\Delta t) = J_0(2\pi f_{dm} \Delta t) \tag{11}$$

where $J_0(.)$ is the zero-order Bessel function of the first kind [11]. The Doppler power spectrum for this channel is given by

$$S_c(\lambda) = \frac{1}{\pi f_{dm}} \left[1 - \left(\frac{f}{f_{dm}} \right)^2 \right]^{1/2}, |f| \leq f_{dm} \tag{12}$$

These functions are plotted in Fig. 6.

Time variation of the fading channel is characterized in terms of the *Doppler spread* and the *coherence time*. The Doppler spread B_d is defined as the range of frequencies over which the Doppler power spectrum is essentially nonzero. For example, in (12), $B_d = 2f_{dm}$. The coherence time $(\Delta t)_c$ measures the time interval over which the time variation is not significant, or the samples of the fading signal are strongly correlated when their time separation is less than the coherence time, although different interpretations are used [3]. The Doppler spread and the coherence time are inversely proportional to one another. A popular rule of thumb is $(\Delta t)_c = 0.423/f_{dm}$. This definition implies that if the time separation is greater than the coherence time, the signals will be affected differently by the channel.

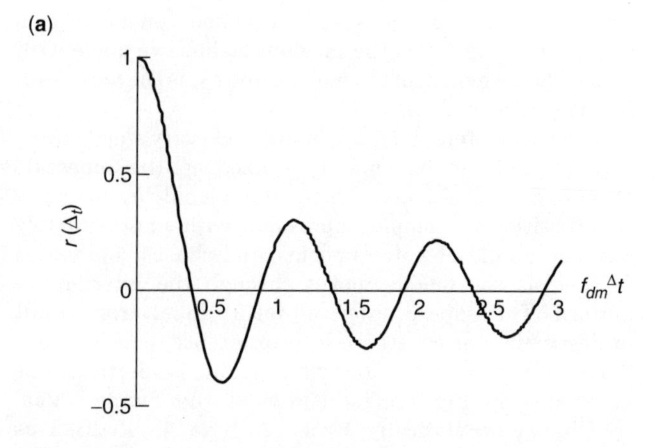

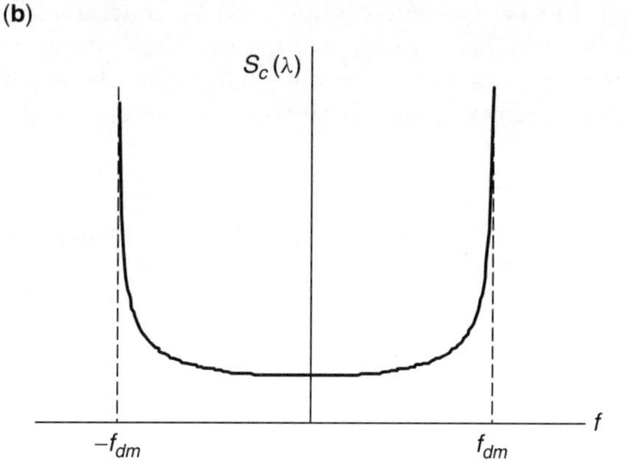

Figure 6. (**a**) The spaced-time autocorrelation function of the Rayleigh fading channel; (**b**) the Doppler power spectrum of the Rayleigh fading channel.

3.2. Relationship Between Signal Characteristics and the Fading Channel Model

Consider a linear communication system that transmits data at the symbol rate $1/T$ and employs a pulseshape with the complex envelope $s_l(t)$ and the frequency response $S_l(f)$ [1]. Assume that the bandwidth W of the transmitted signal is approximately $1/T$. The resulting output signal is characterized in terms of the response of the channel to this pulseshape as in (6), and the frequency response of the output signal is given by $C(f; t)\, S_l(f)$. The relationship between the symbol interval (or bandwidth W) and the statistical fading-channel parameters defined above dictate the choice of the underlying channel model used in analyzing performance of wireless communication systems.

First, consider the *multipath channel characterization* and *signal dispersion*. Suppose that the symbol interval is much larger than the multipath spread of the channel, $T \gg T_m$. Then all multipath components arrive at the receiver within a small fraction of the symbol interval. In this case, the coherence bandwidth significantly exceeds the bandwidth of the signal, $W \ll (\Delta f)_c$. Thus, all spectral components are affected by the channel similarly. Also, there is no *multipath-induced intersymbol interference* (ISI) in this case. This channel is modeled as complex time-varying attenuation $c(t)$ (4) [also given by $C(0; t)$], so the complex envelope of the received signal is $r(t) = c(t)\, s_l(t)$. It is called *frequency-nonselective*, or *flat fading* channel. These channels primarily occur in *narrowband* transmission systems.

If the symbol interval is smaller the multipath spread of the channel, $T < T_m$, or equivalently, the coherence bandwidth is smaller than the signal bandwidth, $W > (\Delta f)_c$, the channel becomes *frequency-selective* (it is also sometimes called the *multipath fading channel*). A common rule of thumb is that the channel is frequency-selective if $T < 10\sigma_\tau$, and flat fading if $T \geq 10\sigma_\tau$, where σ_τ is the RMS delay spread [3] With this definition of a frequency-selective channel, spectral components of the transmitted signal separated by the coherence bandwidth fade differently, resulting in frequency diversity, as discussed in Section 5. On the other hand, frequency selectivity also causes dispersion, or ISI, since delayed versions of the transmitted signal arrive at the receiver much later (relative to the symbol interval) than components associated with small delays. This channel is often modeled using several fading rays with different excess multipath delays

$$c(t) = c_1(t)\delta(t-\tau_1) + c_2(t)\delta(t-\tau_2) + \cdots + c_L(t)\delta(t-\tau_L) \tag{13}$$

where the components $c_l(t), l = 1, \ldots, L$ are uncorrelated flat fading (e.g., Rayleigh distributed) random variables. The powers associated with these rays are determined by the multipath intensity profile (8).

Now, consider *rapidity*, or *time variation* of the fading channel. The channel is considered *slowly varying* if the channel response changes much slower than the symbol rate. In this case, the symbol interval is much smaller than the coherence time, $T \ll (\Delta t)_c$, or the signal bandwidth significantly exceeds the Doppler spread, $W \gg B_d$. If the symbol interval is comparable to or greater than the coherence time, (or the coherence bandwidth is similar to or exceeds the signal bandwidth), the channel is *fast-fading*. While most mobile radio, or PCS, channels are slowly fading, as the velocity of the mobile and the carrier frequency increase, the channel becomes *rapidly time-varying* since the Doppler shift increases [see (5)]. This rapid time variation results in time selectivity (which can be exploited as time diversity), but degrades reliability of detection and channel estimation [12–14].

4. FADING-CHANNEL MODELS

The *complex Gaussian distribution* is often used to model the equivalent lowpass flat-fading channel. This model is justified since superposition of many scattered components approximates a Gaussian distribution by the central-limit theorem. Even if the number of components in (5) is modest, experiments show that the Gaussian model is often appropriate. The *Rayleigh fading* process models fading channels without strong *line of sight* (LoS). Define $c(t) = c_I(t) + jc_Q(t)$, where the in-phase (real) and quadrature (imaginary) components are independent and identically distributed (i.i.d.) zero-mean stationary real Gaussian processes with variances σ^2. The average power of this process is $\frac{1}{2}E[c^*(t)c(t)] = \sigma^2$. The amplitude of this process has a Rayleigh distribution with the probability density function (PDF):

$$p_R(r) = \frac{r}{\sigma^2}\exp\left(-\frac{r^2}{2\sigma^2}\right) \tag{14}$$

and the phase is uniformly distributed:

$$p_\theta(\theta) = \frac{1}{2\pi}, \quad |\theta| \le \pi \tag{15}$$

The Rayleigh distribution is a special case of the *Nakagami-m* distribution that provides a more flexible model of the statistics of the fading channel. The PDF of the amplitude of the Nakagami-m distribution is [1]

$$p_R(r) = \frac{2}{\Gamma(m)}\left(\frac{m}{\Omega}\right)^m r^{2m-1}\exp\left(-\frac{mr^2}{\Omega}\right) \tag{16}$$

where $\Gamma(.)$ is the Gamma function [11], $\Omega = E(R^2)$ and the parameter m is the *fading figure* given by $m = \Omega^2/E[(R^2-\Omega)^2], m \ge 1/2$. While the *Rayleigh* distribution uses a single parameter $E(R^2) = 2\sigma^2$ to match the fading statistics, the Nakagami-m distribution depends on two parameters, $E(R^2)$ and m. For $m = 1$ the density (16) reduces to Rayleigh distribution. For $\frac{1}{2} \le m < 1$, the tail of the Nakagami-m PDF decays slower than for Rayleigh fading, whereas for $m > 1$, the decay is faster. As a result, the Nakagami-m distribution can model fading conditions that are either more or less severe than Rayleigh fading. The Nakagami-m PDF for different values of m is illustrated in Fig. 7.

The Rayleigh distribution models a complex fading channel with zero mean that is appropriate for channels without the line-of-sight (LoS) propagation. When strong nonfading, or specular, components, such as LoS propagation paths, are present, a DC component needs to be added to random multipath, resulting in *Ricean* distribution. The pdf of the amplitude of Ricean fading is given by

$$p_R(r) = \frac{r}{\sigma^2}\exp-\left(\frac{r^2+s^2}{2\sigma^2}\right)I_0\left(\frac{rs}{\sigma^2}\right), \quad r \ge 0 \tag{17}$$

where s is the peak amplitude of the dominant nonfading component and $I_0(.)$ is the modified Bessel function of the first kind and zero order [11]. The *Ricean factor K*

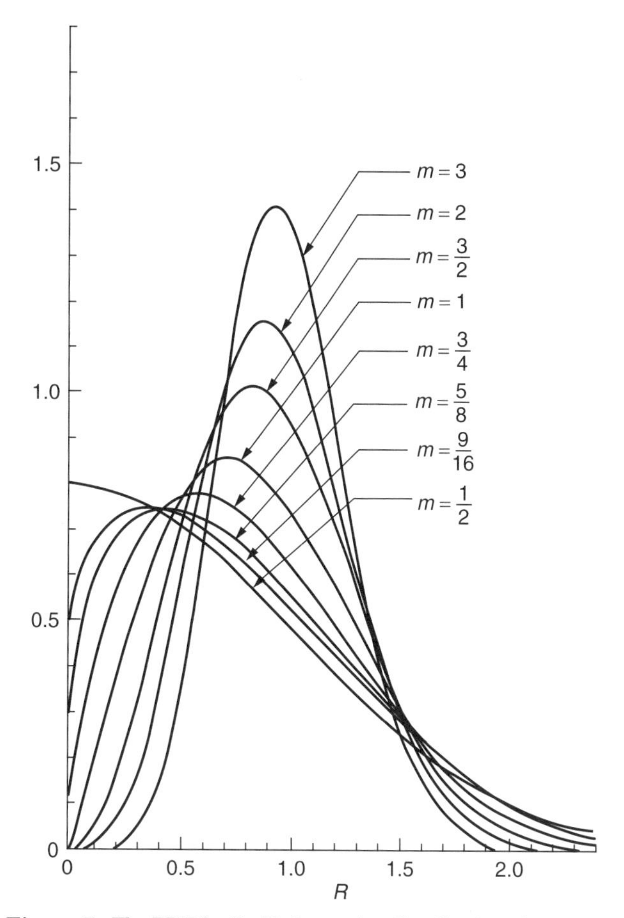

Figure 7. The PDF for the Nakagami-m distribution, shown with $\Omega = 1$, where m is the fading figure (reprinted from Ref. 17).

specifies the ratio of the deterministic signal power and the variance of the multipath:

$$K = 10\log_{10}\left(\frac{s^2}{2\sigma^2}\right) \text{ (dB)} \tag{18}$$

As $s \to 0$ ($K \to -\infty$), the power of the dominant path diminishes, and the Ricean PDF converges to the Rayleigh PDF. Examples of Ricean fading and other LoS channels include airplane to ground communication links and microwave radio channels [1].

As an alternative to modeling the Rayleigh fading as a complex Gaussian process, one can instead approximate the channel by summing a set of complex sinusoids as in (4). The number of sinusoids in the set must be large enough so that the PDF of the resulting envelope provides an accurate approximation to the Rayleigh PDF. The Jakes model is a popular simulation method based on this principle [10]. The signal generated by the model is

$$c(t) = \sqrt{\frac{2}{N}}\sum_{n=1}^{N} e^{j(\omega_d t \cos\alpha_n + \phi_n)} \tag{19}$$

where N is the total number of plane waves arriving at uniformly spaced angles α_n as shown in Fig. 8. The $c(t)$ can be further represented as

$$c(t) = \frac{E_0}{\sqrt{2N_0+1}}(c_1(t) + jc_Q(t))$$

$$c_I(t) = 2\sum_{n=1}^{N_0} \cos\phi_n \cos\omega_n t + \sqrt{2}\cos\phi_N \cos\omega_m t$$

$$c_Q(t) = 2\sum_{n=1}^{N_0} \sin\phi_n \cos\omega_n t + \sqrt{2}\sin\phi_N \cos\omega_m t$$

where $N_0 = \frac{1}{2}[(N/2) - 1)$, $\omega_m = 2\pi f_{dm}$, and $\omega_n = \omega_m \cos(2\pi n/N)$. In the Jakes model, the parameter $N_0 + 1$ is often referred as the *number of oscillators* and N is termed the *number of scatterers*. The Jakes model with as few as nine oscillators ($N_0 = 8$, and $N = 34$) closely approximates the Rayleigh fading distribution (14,15).

When a multipath fading channel with impulse response (13) needs to be modeled, the Jakes model can be extended to produce several uncorrelated fading components using the same set of oscillators [10]. The autocorrelation function and the Doppler spectrum of the signals generated by the Jakes model are characterized by Eqs. (11) and (12), respectively, and are shown in Fig. 6. This statistical characterization is appropriate for many channels where the reflectors are distributed uniformly around the mobile (*isotropic* scattering). Several other approaches to simulating Rayleigh fading channels based on *Clarke and Gans fading models* are described by Rappaport [3]. Moreover, *physical models* are useful [e.g., when the variation of amplitudes, frequencies, and phases in (4) is important to model as in *long-range fading prediction*] and *autoregressive* (*Gauss–Markov*) models are often utilized to approximate fading statistics in fading estimation algorithms since they result in rational spectral characterization [12,13].

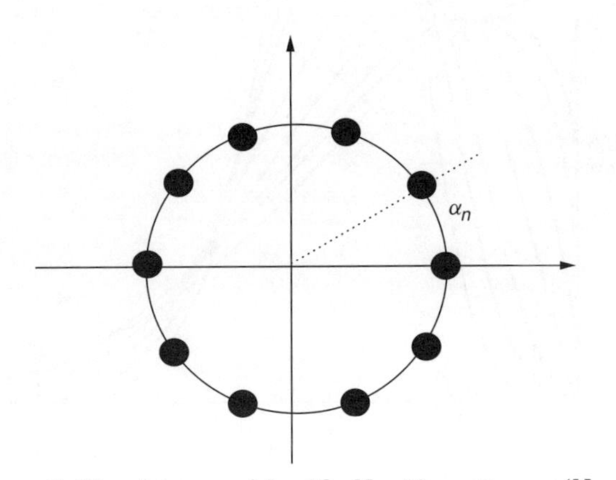

Figure 8. The Jakes model with $N = 10$ scatterers ($N_0 = 2$, 3 oscillators).

5. DIVERSITY TECHNIQUES AND PERFORMANCE ANALYSIS

5.1. Performance Analysis for Flat Fading Channels

Fading channels undergo dramatic changes in received power due to multipath and Doppler effects. When communication signals are transmitted over these channels, the bit error rate (BER) varies as a function of the signal-to-noise ratio (SNR) and is significantly degraded relative to the BER for the additive white Gaussian noise (AWGN) channel with the same average SNR. Consider the following example of transmission of binary phase-shift-keyed (BPSK) signal over the flat Rayleigh fading channel. At the output of the matched filter and sampler at the bit rate $1/T_b$ (where T_b is the bit interval), the complex envelope of the received signal is

$$r_k = c_k b_k + z_k \tag{20}$$

where c_k are the samples of the fading signal $c(t)$, b_k is the i.i.d. information sequence that takes on values $\{+1, -1\}$, and z_k is the i.i.d. complex white Gaussian noise sequence with the variance $\frac{1}{2}E[|z_k|^2] = N_0$. Since the channel is assumed to be stationary, we omit the subscript k in subsequent derivations. The equivalent passband energy is normalized as $E_b = \frac{1}{2}$, so the instantaneous SNR at the receiver is given by

$$\gamma = \frac{|c|^2}{2N_0} \tag{21}$$

Assume without loss of generality that the average power of the fading signal $E[|c_k|^2] = 1$. Then the average SNR per bit

$$\Gamma = \frac{1}{2N_0} \tag{22}$$

Suppose coherent detection of the BPSK signal is performed, and signal phase is estimated perfectly at the receiver. (This assumption is difficult to satisfy in practice for rapidly varying fading channels [12,13], so the analysis below represents a lower bound on the achievable performance.) The BER for each value of the instantaneous

SNR (21) can be computed from the BER expression for the AWGN channel [1]:

$$\text{BER}\,(\gamma) = Q[(2\gamma)^{1/2}] \tag{23}$$

where the Q function $Q(x) = 1/(2\pi)^{1/2} \int_{-\infty}^{x} \exp(-y^2/2)\,dy$. To evaluate average BER for the Rayleigh fading channel, the BER (23) has to be averaged over the distribution of the instantaneous SNR:

$$\text{BER} = \int_0^\infty Q[(2\gamma)^{1/2}]p(\gamma)\,d\gamma \tag{24}$$

Since $|c|$ is *Rayleigh* distributed, the instantaneous SNR γ has a *chi-square* distribution with the PDF $p(\gamma) = 1/\Gamma \exp(-\gamma/\Gamma)$, $\gamma \geq 0$. When this density is substituted in (24), the resulting BER for binary BPSK over flat Rayleigh fading channel is

$$\text{BER} = \frac{1}{2}\left[1 - \left(\frac{\Gamma}{1+\Gamma}\right)^{1/2}\right] \tag{25}$$

The BER for other fading distributions (e.g., Ricean or Nakagami-m) can be obtained similarly by averaging the BER for AWGN (23) over the fading statistics as in (24). Figure 9 illustrates performance of BPSK over a Nakagami-m fading channel for different values of m. Observe that as m increases, the fading becomes less severe, and the BER approaches that of the AWGN channel. Rayleigh fading ($m = 1$) results in significant SNR loss relative to the nonfading channel. In fact, for large SNR, the BER in (25) behaves asymptotically as $\frac{1}{4}\Gamma$, whereas the BER decreases exponentially with SNR for AWGN. The error rates of other modulation methods [e.g., coherent and noncoherent frequency shift keying (FSK), differential PSK (DPSK)] also *decrease only inversely with SNR*, causing very large power consumption.

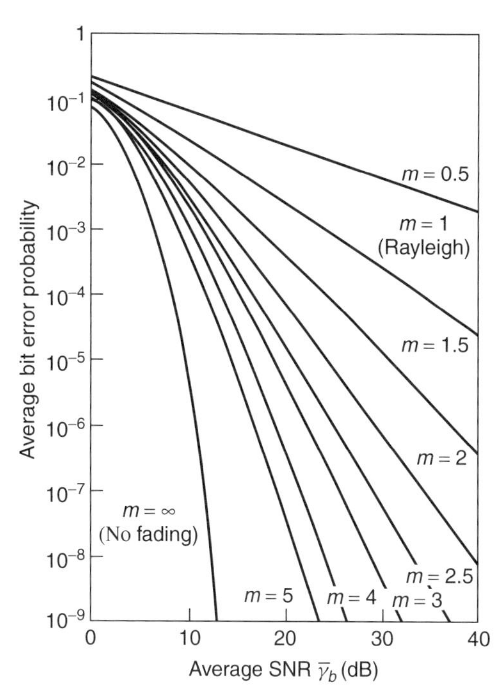

Figure 9. Average error probability for two-phase PSK with Nakagami fading (reprinted from Ref. 1).

5.2. Diversity Techniques

The poor performance of flat fading channels is due to the presence of deep fades in the received signal (Fig. 4) when the received SNR is much lower than the average SNR value. *Diversity techniques* help to combat fading by sending replicas of transmitted data over several uncorrelated (or partially correlated) fading channels. Since these channels are unlikely to go through a deep fade at the same time, higher average received SNR results when the outputs of the diversity branches are combined. Many diversity techniques are used in practice.

In *space*, or *antenna diversity* systems, several antenna elements are placed at the transmitter and/or receiver and separated sufficiently far apart to achieve desired degree of independence. (Note that the correlation function in space is analogous to that in time [10]. Typically, antenna separations of about $\lambda/2$ at the mobile station and several λ at the base station are required to achieve significant antenna decorrelation, where λ is the wavelength.)

Time diversity relies on transmitting the same information in several time slots that are separated by the coherence time $(\Delta t)_c$. Time diversity is utilized in coded systems by interleaving the outputs of the encoder.

Frequency diversity can be employed when the transmitter bandwidth is larger than the coherence bandwidth of the channel, and several fully or partially uncorrelated fading components can be resolved. The number of such uncorrelated components is determined by the ratio $W/(\Delta f)_c$ [1]. In PSK or quadrature-amplitude-modulated (QAM) systems, the transmitter bandwidth is approximately $1/T$. Because of the narrow transmission bandwidth, these channels are often frequency-nonselective. When frequency selectivity is present, it usually causes ISI since the multipath delay is significant relative to the symbol interval. Thus, *equalizers* are used to mitigate the ISI and to obtain the diversity benefit [1,12–14]. On the other hand, *direct-sequence spread-spectrum* (DSSS) systems employ waveforms with the transmission bandwidth that is much larger than the symbol rate $1/T$, and thus enjoy significant frequency diversity. DSSS signals are designed to achieve approximate orthogonality of multipath-induced components, thus eliminating the need for equalizers. Frequency diversity combining in these systems is achieved using a *RAKE correlator* [1].

In addition to the diversity techniques mentioned above, *angle of arrival* and *polarization* diversity are utilized in practice. Different diversity methods are often combined to maximize diversity gain at the receiver. To illustrate the effect of diversity on the performance of communication systems in fading channels, consider the following simple example. Suppose the transmitted BPSK symbol b_k is sent over L independent Rayleigh fading channels (we suppress the time index k below). The received equivalent lowpass samples are

$$r^i = c^i b + z^i, \qquad i = 1, \ldots, L \tag{26}$$

where c^i are i.i.d. complex Gaussian random variables with variances $E[|c^i|^2] = 1/L$ [this scaling allows us to compare performance directly with a system without diversity in (20)], b is the BPSK symbol as in Eq. (20), and z^i are i.i.d. complex Gaussian noise samples with $\frac{1}{2}E[|z_k|^2] = N_0$. Thus, the average SNR *per channel (diversity branch)* is $\Gamma_c = \Gamma/L$, where Γ is the SNR per bit.

There are several options for combining L diversity branches. For example, the branch with the highest instantaneous SNR can be chosen resulting in *selection diversity*. Alternatively, *equal gain combining* is a technique where all branches are weighted equally and cophased [3]. The maximum diversity benefit is obtained using *maximum ratio combining* (MRC) (this is also the most complex method). In MRC, the outputs r^i are weighted by the corresponding channel gains, cophased, and summed producing the decision variable (the input to the BPSK threshold detector):

$$U = (c^1) * r^1 + (c^2) * r^2 + \cdots + (c^L) * r^L$$

Note that if a signal undergoes a deep fade, it carries weaker weight than a stronger signal with higher instantaneous power. It can be shown that for large SNR, the BER for this method is approximately [1]

$$\text{BER} \approx \left(\frac{1}{(4\Gamma_c)}\right)^L \binom{2L-1}{L} \qquad (27)$$

This BER of MRC for BPSK is illustrated in Fig. 10 for different values of L (binary PSK curve). The figure also shows performance of two less complex combining methods used with orthogonal FSK (*square-law combining*) and DPSK. The latter two techniques are noncoherent (i.e., do not require amplitude or phase estimation). From Eq. (27) and Fig. 10 we observe that the *BER for all methods decreases inversely with the Lth power of the SNR*. Thus, diversity significantly reduces power consumption in fading channels.

In addition to diversity, *adaptive transmission* is an effective tool in overcoming the effects of fading. The idea is to adjust the transmitted signal (power, rate, etc.) to fading conditions to optimize average power and bandwidth requirements. Many advances in the area of communication over fading channels, and additional sources are cited in Refs. 12 and 13.

6. SUMMARY

The fading signal was characterized in terms of large-scale and small-scale fading. The large-scale fading models that describe the average and statistical variation of the received signal power were presented. The small-scale fading channel was characterized statistically in terms of its time-variant impulse response. Time and frequency-domain interpretation was provided to describe signal dispersion and fading rapidity, and it was shown how the transmitted signal affects the choice of the fading channel model. Several statistical models of fading channels were presented, and simulation techniques were discussed. Finally, performance limitations of fading channels were revealed, and fading mitigation methods using diversity techniques were reviewed.

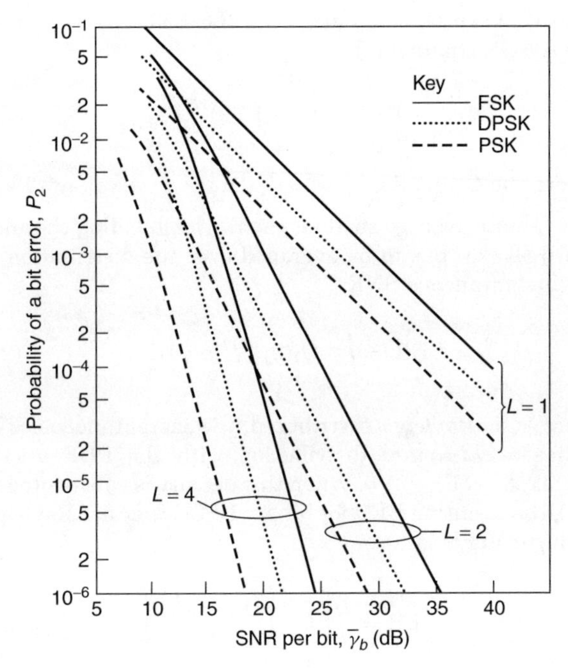

Figure 10. Performance of binary signals with diversity (reprinted from Ref. 1).

Acknowledgment

The author is grateful to Shengquan Hu, Hans Hallen, Tung-Sheng Yang, Ming Lei, Jan-Eric Berg, and Henrik Asplund for their assistance and helpful comments. This work was partially supported by NSF grant CCR-9815002 and ARO grant DAA19-01-1-0638.

BIOGRAPHY

Alexandra Duel-Hallen received the B.S. degree in Mathematics from Case Western Reserve University in 1982, the M.S. degree in Computer, Information and Control Engineering from the University of Michigan in 1983, and a Ph.D. in Electrical Engineering from Cornell University in 1987. During 1987–1990 she was a Visiting Assistant Professor at the School of Electrical Engineering, Cornell University, Ithaca, New York. In 1990–1992, she was with the Mathematical Sciences Research Center, AT&T Bell Laboratories, Murray Hill, New Jersey. She is an Associate Professor at the Department of Electrical and Computer Engineering at North Carolina State University, Raleigh, North Carolina, which she joined in January 1993. From 1990 to 1996, Dr. Duel-Hallen was Editor for Communication Theory for the *IEEE Transactions on Communications*. During 2000–2002, she has served as Guest Editor of two Special Issues on Multiuser Detection for the *IEEE Journal on Selected Areas in Communications*. Dr. Duel-Hallen's current research interests are in wireless and multiuser communications. Her 1993 paper was selected for the IEEE Communications Society 50th Anniversary Journal Collection as one of 41 key papers in physical and link layer areas, 1952–2002.

BIBLIOGRAPHY

1. J. G. Proakis, *Digital Communications*, 4th ed., McGraw-Hill, 2001.
2. B. Sklar, Rayleigh fading channels in mobile digital communication systems, Part 1: Characterization, *IEEE Commun. Mag.* **35**(7): 90–100 (July 1997).
3. T. S. Rappaport, *Wireless Communications: Principles and Practice*, 2nd ed., Prentice-Hall, 2002.
4. H. L. Bertoni, *Radio Propagation for Modern Wireless Systems*, Prentice-Hall, 2000.
5. W. C. Y. Lee, *Mobile Communications Engineering: Theory and Applications*, 2nd ed., McGraw-Hill Telecommunications, 1997.
6. R. Steele, *Mobile Radio Communications*, Pentech Press, 1992.
7. G. L. Stuber, *Principles of Mobile Communication*, Kluwer, 2001.
8. P. A. Bello, Characterization of randomly time-variant linear channels, *IEEE Trans. Commun. Syst.* **11**: 360–393 (1963).
9. S. Stein, Fading channel issues in system engineering, *IEEE J. Select. Areas Commun.* **5**(2): 68–69 (Feb. 1987).
10. W. C. Jakes, *Microwave Mobile Communications*, Wiley, New York, 1974.
11. M. Abramowitz and I. A. Stegun, *Handbook of Mathematical Functions*, National Bureau of Standards, 1981.
12. E. Biglieri, J. Proakis, and S. Shamai (Shitz), Fading channels: information-theoretic and communications aspects, *IEEE Trans. Inform. Theory* **44**(6): 2619–2692 (Oct. 1998).
13. *IEEE Signal Process. Mag.* (Special Issue on Advances in Wireless and Mobile Communications; G. B. Giannakis, Guest Editor) **17**(3): (May 2000).
14. B. Sklar, Rayleigh fading channels in mobile digital communication systems, Part 2: Mitigation, *IEEE Commun. Mag.* **35**(7): 102–109 (July 1997).
15. S. Y. Seidel et al., Path loss, scattering, and multipath delay statistics in four European cities for digital cellular and microcellular radiotelephone, *IEEE Trans. Vehic. Technol.* **40**(4): 721–730 (Nov. 1991).
16. T. S. Rappaport, S. Y. Seidel, and R. Singh, 900 MHz multipath propagation measurements for U.S. digital cellular radiotelephone, *IEEE Trans. Vehic. Technol.* 132–139 (May 1990).
17. Y. Miyagaki, N. Morinaga, and T. Namekawa, Error probability characteristics for CPFSK signal through m-distributed fading channel, *IEEE Trans. Commun.* **COM-26**: 88–100 (Jan. 1978).

FEEDBACK SHIFT REGISTER SEQUENCES

HONG-YEOP SONG
Yonsei University
Seoul, South Korea

1. INTRODUCTION

Binary random sequences are useful in many areas of engineering and science. Well-known applications are digital ranging and navigation systems because of the sharp peak in their autocorrelation functions [1], spread-spectrum modulation and synchronization using some of their correlation properties [2–6], and stream ciphers, in which the message bits are exclusive-ORed with key streams that must be as random as possible [7,8]. Several randomness tests for binary sequences have been proposed in practice [8], but no universal consensus has been made yet with regard to the true randomness of binary sequences [9].

Random binary sequences can be obtained in theory by flipping an unbiased coin successively, but this is hardly possible in most practical situations. In addition, not only must the random sequence itself be produced at some time or location but also its exact replica must also be produced at remote (in physical distance or time) locations in spread-spectrum modems. This forces us to consider the sequences that appear random but can be easily reproduced with a set of simple rules or keys. We call these *pseudorandom* or *pseudonoise* (PN) sequences. It has been known and used for many years that *feedback shift registers* (FSRs) are most useful in designing and generating such PN sequences. This is due to their simplicity of defining rules and their capability of generating sequences with much longer periods [10]. Approaches using FSR sequences solve the following two basic problems in most applications: cryptographic secrecy and ease of generating the same copy over and over.

One of the basic assumptions in conventional cryptography is that the secrecy of a system does not depend on the secrecy of how it functions but rather on the secrecy of the key, which is usually kept secret [11]. Feedback shift registers are most suitable in this situation because we do not have to keep all the terms of the sequences secret. Even though its connection is revealed and all the functionality of the system is known to the public, any unintended observer will have an hard time of locating the exact phase of the sequence in order to break the system, provided that the initial condition is kept secret. The current CDMA modem (which is used in the successful second and third generation mobile telephone systems) depends heavily on this property for its privacy [12].

One previous difficulty in employing spread-spectrum communication systems was on the effort of reproducing at the receiver the exact replica of PN sequences that were used at the transmitter [2]. Store-and-replay memory wheels to be distributed initially were once proposed, and the use of a secret and safe third channel to send the sequence to the receiver was also proposed. The theory and practice of FSR sequences have now been well developed so that by simply agreeing on the initial condition and/or connection method (which requires much smaller memory space or computing time), both ends of communicators can easily share the exact same copy.

In Section 2, the very basics of feedback shift registers (FSRs) and their operations are described, following the style of Golomb [10]. We will concentrate only on some basic terminologies, state transition diagrams, truth tables, cycle decompositions, and the like. In fact, the detailed proofs of claims and most of discussions and a lot more can be found in Golomb's study [10]. Section 3 covers mostly the *linear* FSRs. The linear FSR sequences have been studied in various mathematics literature under the term *linear recurring sequences*. Lidl and Niederreiter gave a comprehensive treatment on this subject [13]. Some other well-known textbooks on the theory of finite fields and linear recurring sequences are available [14–18]. In this article, we will discuss the condition for their output sequences to have maximum possible period. The maximum period sequences, which are known as m-sequences, are described in detail, including randomness properties. Two properties of m-sequences deserve special attention: m-sequences of period P have the two-level ideal autocorrelation, which is the *best* over all the balanced binary sequences of the same period, and they have the linear complexity of $\log_2(P)$, which is the *worst* (or the smallest) over the same set. Some related topics on these properties will also be discussed. To give some further details of the ideal two-level autocorrelation property, we describe a larger family of balanced binary sequences which come from, so called, *cyclic Hadamard difference sets*. An m-sequence can be regarded as the characteristic sequence of a cyclic Hadamard difference set of the Singer type. To give some better understanding of the linear complexity property, we describe the *Berlekamp–Massey algorithm* (BMA), which determines the shortest possible linear FSR that generates a given sequence.

In Section 4, we describe some special cases of FSRs (including nonlinear FSRs) with disjoint cycles in their state diagrams. Branchless condition and balanced logic condition will be discussed. De Bruijn sequences will briefly be described. Finally, four-stage FSRs are analyzed in detail for a complete example. Section 5 gives some concluding remarks. We will restrict our treatment to only the sequences over a binary alphabet $\{0, 1\}$ in this article.

2. FEEDBACK SHIFT REGISTER SEQUENCES, TRUTH TABLES, AND STATE DIAGRAMS

The operation of an FSR can best be described by its state transition diagram. Its output at one time instant depends only on the previous state. Figure 1 shows a generic block diagram of an FSR (linear or non-linear) with L stages. At every clock, the content of a stage is shifted to the left, and the connection logic or the boolean function f calculates a new value x_k

$$x_k = f(x_{k-L}, x_{k-L+1}, \ldots, x_{k-1}) \tag{1}$$

to be fed back to the rightmost stage. The leftmost stage gives an output sequence in which the first L terms are in fact given as an initial condition.

A *state* of this FSR at one instant k can be defined simply as the vector $(x_{k-L}, x_{k-L+1}, \ldots, x_{k-1})$, and this will be changed into $(x_{k-L+1}, x_{k-L+2}, \ldots, x_k)$ at the next instant. An FSR is called *linear* if the connection logic is a linear function on $x_{k-L}, x_{k-L+1}, \ldots, x_{k-1}$, that is, if it is of the form

$$\begin{aligned} x_k &= f(x_{k-L}, x_{k-L+1}, \ldots, x_{k-1}) \\ &= c_L x_{k-L} \oplus c_{L-1} x_{k-L+1} \oplus \cdots \oplus c_1 x_{k-1} \end{aligned} \tag{2}$$

for some fixed constants $c_1, c_2, \ldots, c_L$. Otherwise, it is called *nonlinear*. Here, the values of x_i are either 0 or 1, and hence the sequence is said to be over a *binary alphabet*, which is usually denoted as F_2, and $c_i \in F_2$ for all i. The operation $\oplus$ can easily be implemented as an *exclusive-OR* operation and $c_i x_{k-i}$ as an AND operation both using digital logic gates. In the remaining discussion, we will simply use addition and multiplication (mod 2), respectively, for these operations. Over this binary alphabet, therefore, one can add and multiply two elements, and the subtraction is the same as addition. There is only one nonzero element (which is 1), and the division by 1 is the same as the multiplication by 1.

Another method of describing an FSR is to use its truth table, in which all the 2^L states are listed on the left column and the next bits calculated from the connection logic f are listed on the right. The state change can easily be illustrated by the state diagram in which every state is a node and an arrow indicates the beginning (predecessor) and ending (successor) states. Figures 2 and 3 show examples of three-stage FSRs, including their truth tables, and state diagrams. Note that there are exactly 2^L states in total, and every state has at most two predecessors and exactly one successor. Note also that there are 2^{2^L} different L-stage FSRs, corresponding to the number of choices for the next bit column in the truth table. Finally, note that Fig. 3 has two disjoint cycles while Fig. 2 has branches and some absorbing states. Therefore, the FSR in Fig. 2 eventually will output the all-zero sequence, while that in Fig. 3 will output a sequence of period 7 unless its initial condition is 000.

In order to investigate this situation more closely, observe that any state has a unique successor, but up to two predecessors. From this, we observe that a branch occurs at a state that has two predecessors, and this

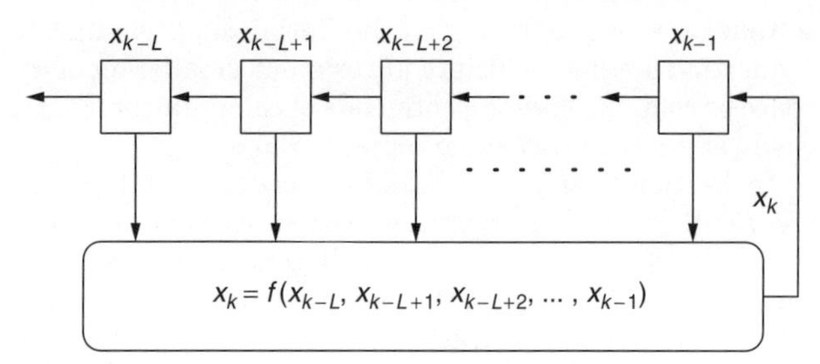

Figure 1. An L-stage FSR with a feedback (Boolean logic) function f.

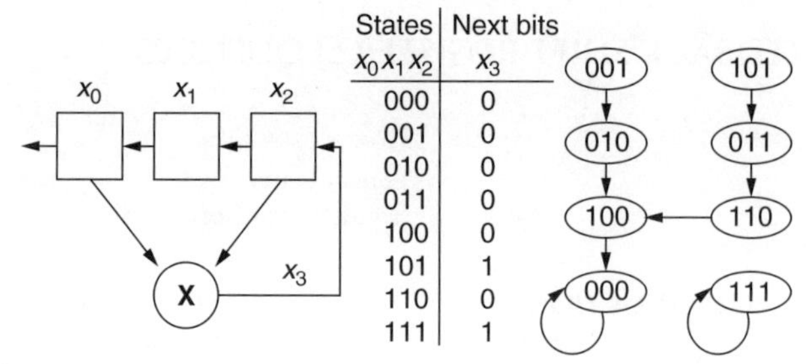

Figure 2. A three-stage nonlinear FSR with a feedback function $x_k = x_{k-1}x_{k-3}$, including its truth table and state diagram.

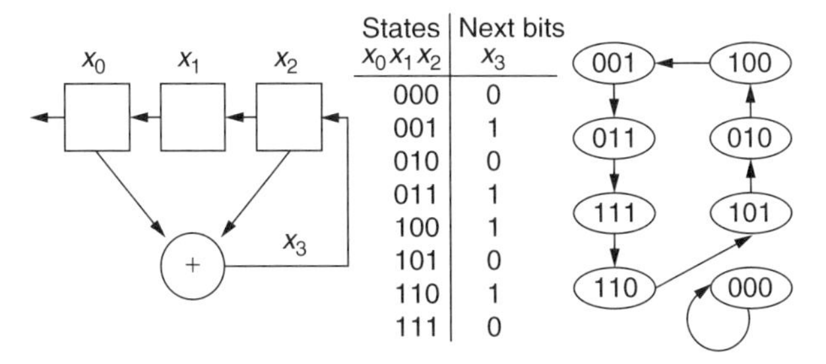

Figure 3. A three-stage linear FSR with a feedback function $x_k = x_{k-1} \oplus x_{k-3}$, including its truth table and state diagram.

happens in a state diagram if and only if there is a state that has no predecessor. A branch in a state diagram should be avoided since it will either seriously reduce the period of the output sequences or result in an ambiguous initial behavior. The necessary and sufficient condition for a branchless state diagram is, therefore, that no two states have the same successor. Consider any pair of states $(a_0, a_1, \ldots, a_{L-1})$ and $(b_0, b_1, \ldots, b_{L-1})$. If they are different in any other position except for the first, their successors $(a_1, a_2, \ldots, a_L)$ and $(b_1, b_2, \ldots, b_L)$ will still be different, because all the components except for the first will be shifted to the left and the difference remains. The remaining case is the pair of the form $(a_0, a_1, \ldots, a_{L-1})$ and $(a'_0, a_1, \ldots, a_{L-1})$, where a'_0 represents the complement of a_0. Their successors will be $(a_1, a_2, \ldots, a_{L-1}, f(a_0, a_1, \ldots, a_{L-1}))$ and $(a_1, a_2, \ldots, a_{L-1}, f(a'_0, a_1, \ldots, a_{L-1}))$. For these two states to be distinct, the rightmost component must be different:

$$f(a'_0, a_1, \ldots, a_{L-1}) = f(a_0, a_1, \ldots, a_{L-1}) \oplus 1$$
$$= f'(a_0, a_1, \ldots, a_{L-1})$$

Let $g(a_1, a_2, \ldots, a_{L-1})$ be a boolean function on $L-1$ variables such that

$$f(0, a_1, \ldots, a_{L-1}) = g(a_1, a_2, \ldots, a_{L-1})$$

Then, the relation shown above can be written as

$$f(a_0, a_1, \ldots, a_{L-1}) = a_0 \oplus g(a_1, a_2, \ldots, a_{L-1}) \quad (3)$$

This is called the *branchless condition* for an L-stage FSR. For FSRs with the branchless condition, the corresponding truth table has only 2^{L-1} independent entries, and the top half of the truth table must be the complement of the bottom half. This condition is automatically satisfied with *linear* FSRs, which are the topic of the next section.

3. LINEAR FEEDBACK SHIFT REGISTERS AND m-SEQUENCES

3.1. Basics

The output sequence $\{s(k) | k = 0, 1, 2, \ldots\}$ of a linear feedback shift register (LFSR) with L stages as shown in Fig. 4 satisfies a linear recursion of degree L. Given

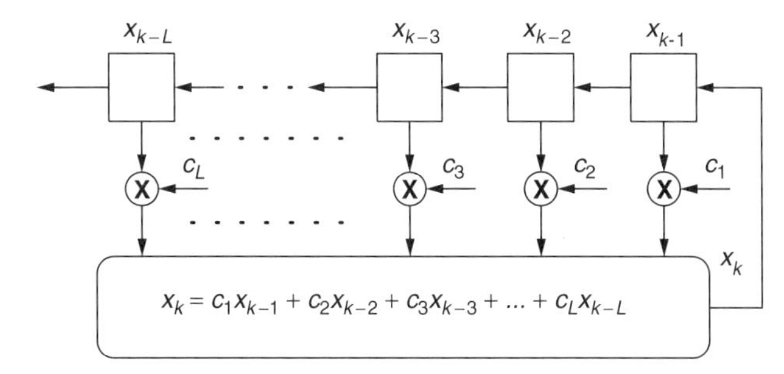

Figure 4. An L-stage LFSR with connection coefficients $c_1, c_2, \ldots, c_L$. Note that $c_L = 1$ for LFSR to have genuinely L stages.

$L + 1$ constants, $c_1, c_2, \ldots, c_L, b$, and the initial condition $s(0), s(1), \ldots, s(L-1)$, the terms $s(k)$ for $k \geq L$ satisfy

$$s(k) = c_1s(k-1) + c_2s(k-2) + \cdots + c_Ls(k-L) + b \quad (4)$$

or equivalently

$$s(k) + c_1s(k-1) + c_2s(k-2) + \cdots + c_Ls(k-L) + b = 0 \quad (5)$$

The recursion is called *homogeneous* linear if $b = 0$ and *inhomogeneous* linear if $b \neq 0$. We will assume that $b = 0$ in this section and consider mainly the homogeneous linear recursion.

The *characteristic polynomial* of the homogeneous linear recursion in Eq. (4) or (5) is defined as

$$f(x) = 1 + c_1x + c_2x^2 + \cdots + c_Lx^L \quad (6)$$

This contains all the connection coefficients, and will completely determine the operation of the LFSR provided that the initial condition is specified. Note that $c_L \neq 0$ in order for this LFSR to be genuinely with L stages. Otherwise, the recursion becomes of degree less than L, and the LFSR with less than L stages can also be used to implement the recursion.

Note that it is also the characteristic polynomial of the sequence satisfying this recursion. A given sequence may satisfy many other recursions that differ from each other. The *minimal polynomial* of a given sequence is defined as the minmum degree characteristic polynomial of the sequence. It is irreducible, it becomes unique if it is restricted to be monic, and it divides all the characteristic polynomials of the sequence. We will return to this and more later when we discuss the linear complexity of sequences.

In the state diagram of any LFSR, every state will have a unique successor and a unique predecessor, as stated at the end of the previous section. This forces the diagram to be (possibly several) disjoint cycles of states. In Fig. 3, the state diagram has two disjoint cycles, one with length 7, and the other with length 1. From this, we can easily see that the output sequence of an LFSR is *periodic*, and the period is the length of the cycle that the initial state (or the initial condition) belongs to. We can conclude, therefore, that *the output sequence of a LFSR is periodic with some period P that depends on both the initial condition and the characteristic polynomial.*

One special initial condition is the all-zero state, and this state will always form a cycle of length 1 *for any LFSR*. For any other initial state, the cycle will have a certain length ≥ 1, and this length is the period of the output sequence with the given initial condition. Certainly, the cycle with the maximum possible length must contain every not-all-zero state exactly once, and the output sequence in this case is known as the *maximal length linear feedback shift register sequence*, or the m-sequence, in short. Sometimes, PN sequences are used instead of m-sequences and vice versa, but we will make a clear distinction between these two terms. PN sequences refer to (general) pseudonoise sequences that possess some or various randomness properties, and m-sequences are a specific and very special example of PN sequences. For an L-stage LFSR, this gives the period $2^L - 1$, and Fig. 3 shows an example of an m-sequence of period 7. In fact, it shows seven different *phases* of this m-sequence depending on the seven initial conditions. Note also that the history of any stage is the same m-sequence in different phase.

The operation of an LFSR is largely determined by its characteristic polynomial. It is a polynomial of degree L over the binary alphabet F_2. How it factors over F_2 is closely related to the properties of the output sequence. In order to discuss the relation between the characteristic polynomials and the corresponding output sequences, we define some relations between sequences of the same period.

Let $\{s(k)\}$ and $\{t(k)\}$ be *arbitrary* binary sequences of period P. Then we have the following three important relations between these two sequences:

1. One is said to be a *cyclic shift* of the other if there is a constant integer τ such that $t(k - \tau) = s(k)$ for all k. Otherwise, two sequences are said to be cyclically distinct. When one is a cyclic shift of the other with $\tau \neq 0$, they are said to be in different *phases*. Therefore, there are P distinct phases of $\{s(k)\}$ that are all cyclically equivalent.
2. One is a *complement* of the other if $t(k) = s(k) + 1$ for all k.
3. Finally, one is a *decimation* (or d decimation) of the other if there are constants d and τ such that $t(k - \tau) = s(dk)$ for all k. If d is not relatively prime to the period P, then the d decimation will result in a sequence with shorter period, which is P/g, where g is the GCD of P and d.

If some combination of these three relations applies to $\{s(k)\}$ and $\{t(k)\}$, then they are called *equivalent*. Equivalent sequences share lots of common properties, and they are essentially the same sequences even if they look very different.

The necessary and sufficient condition for an L-stage LFSR to produce an m-sequence of period $2^L - 1$ is that the characteristic polynomial of degree L is *primitive* over F_2. This means simply that $f(x)$ is irreducible and $f(x)$ divides $x^{2^L-1} - 1$, but $f(x)$ does not divide $x^j - 1$ for all j from 1 to $2^L - 2$. The elementary theory of finite fields (or Galois fields) deals much more with these primitive polynomials, which we will not discuss in detail here. Instead, we refer the reader to some references for further exploration in theory [13–18]. See the article by Hansen and Mullen [19] for a list of primitive polynomials of degree up to a few hundreds, which will generally suffice for any practical application. There are $\phi(2^L - 1)/L$ primitive polynomials of degree L over F_2. Some primitive polynomials are shown in Table 1 for L up to 10. Here, $\phi(n)$ is the Euler ϕ function, and it counts the number of integers from 1 to n that are relatively prime to n.

In order to describe some properties of the output sequences of LFSRs, we include all the four-stage LFSRs: block diagrams, characteristic polynomials, truth tables, state transition diagrams, and the output sequences in Figs. 5–8. Note that there are 16 linear logics for four-stage LFSRs, and the condition $c_L = 1$ reduces it into half.

Figure 5 shows two LFSRs that generate m-sequences of period 15. The detailed properties of m-sequences will be described in Section 3.2. Here, we note that two characteristic polynomials $f_1(x) = x^4 + x^3 + 1$ and $f_2(x) = x^4 + x + 1$ are reciprocal to each other. That is, the coefficients are 11001 and 10011. This gives two m-sequences that are reciprocal to each other. In other words, one is a 14 decimation of the other. Little arrows with a dot under the output sequences indicate this fact. Note that the roots of $f_2(x)$ are the 14th power of those of $f_1(x)$. In general, if $f(x)$ and $g(x)$ are primitive of the same degree and the roots of one polynomial are dth power of the other, then the m-sequence from one polynomial is a d decimation of that from the other. The truth table shows only the top half since its bottom half is the complement of what is shown here, as a result of the branchless condition.

Figure 6 shows LFSRs with the characteristic polynomials that factor into smaller-degree polynomials. Note that $f_3(x) = x^4 + x^3 + x^2 + 1 = (x + 1)(x^3 + x + 1)$ and $f_4(x) = x^4 + x^2 + x + 1 = (x + 1)(x^3 + x^2 + 1)$. Since both of the degree 3 polynomials in their factorizations are primitive, the LFSR generates m-sequences of period 7, which could have been generated by a three-stage LFSR. Two characteristic polynomials are reciprocal, and the output

Table 1. The Number of Primitive Irreducible Polynomials of Degree L and Some Examples[a]

Degree L	$\phi(2^L - 1)/L$	Primitive Polynomial
1	1	11
2	1	111
3	2	1011
4	2	10011
5	6	100101
6	6	1000011
7	18	10000011
8	16	100011101
9	48	1000010001
10	60	10000001001

[a]The binary vector 1011 for $L = 3$ represents either $x^3 + x + 1$ or $1 + x^2 + x^3$.

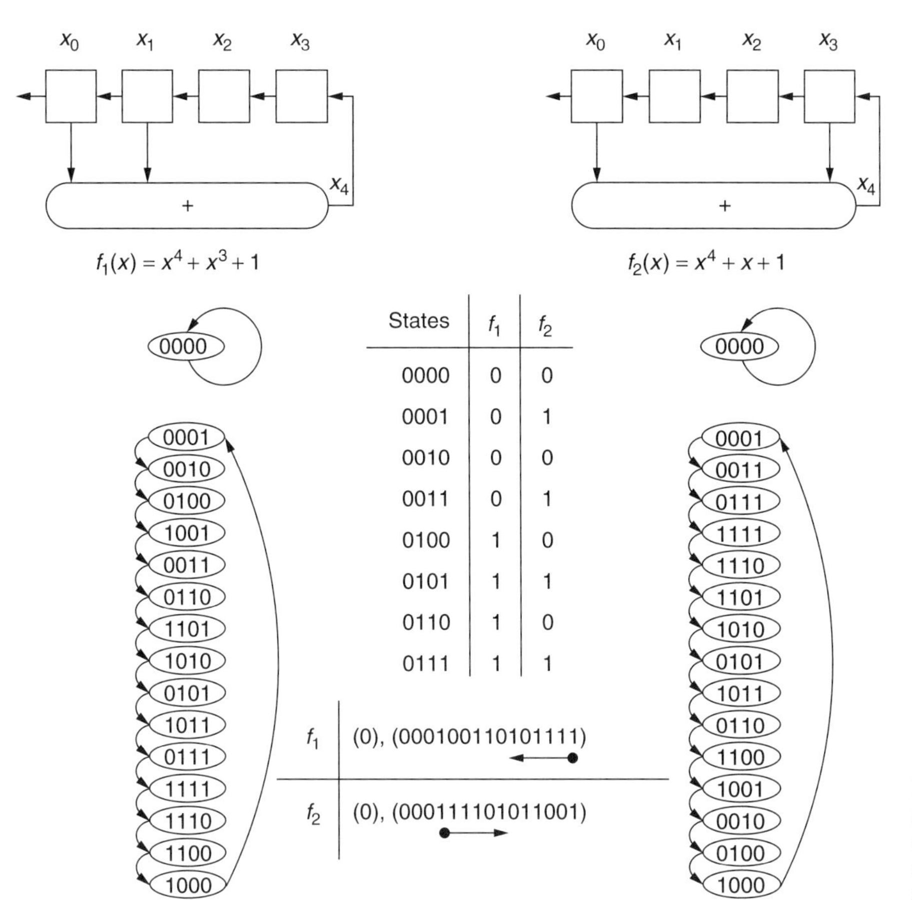

Figure 5. Block diagrams, state diagrams, truth tables, and output sequences of four-stage LFSRs that generate m-sequences: $f_1(x) = x^4 + x^3 + 1$ and $f_2(x) = x^4 + x + 1$.

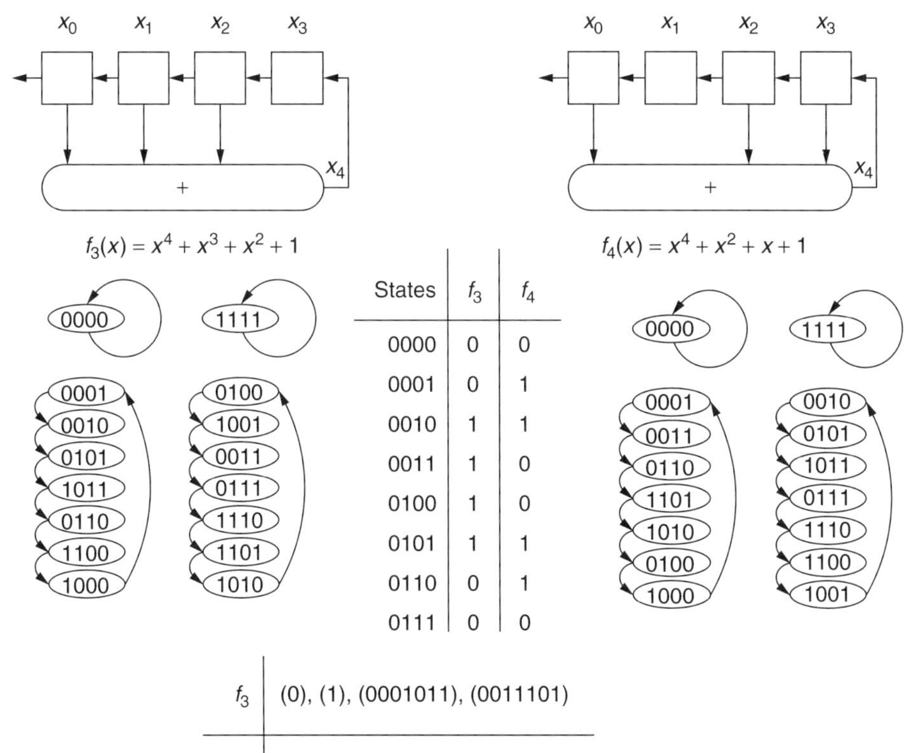

Figure 6. Block diagrams, state diagrams, truth tables, and output sequences of four-stage LFSRs with characteristic polynomials $f_3(x) = x^4 + x^3 + x^2 + 1$ and $f_4(x) = x^4 + x^2 + x + 1$.

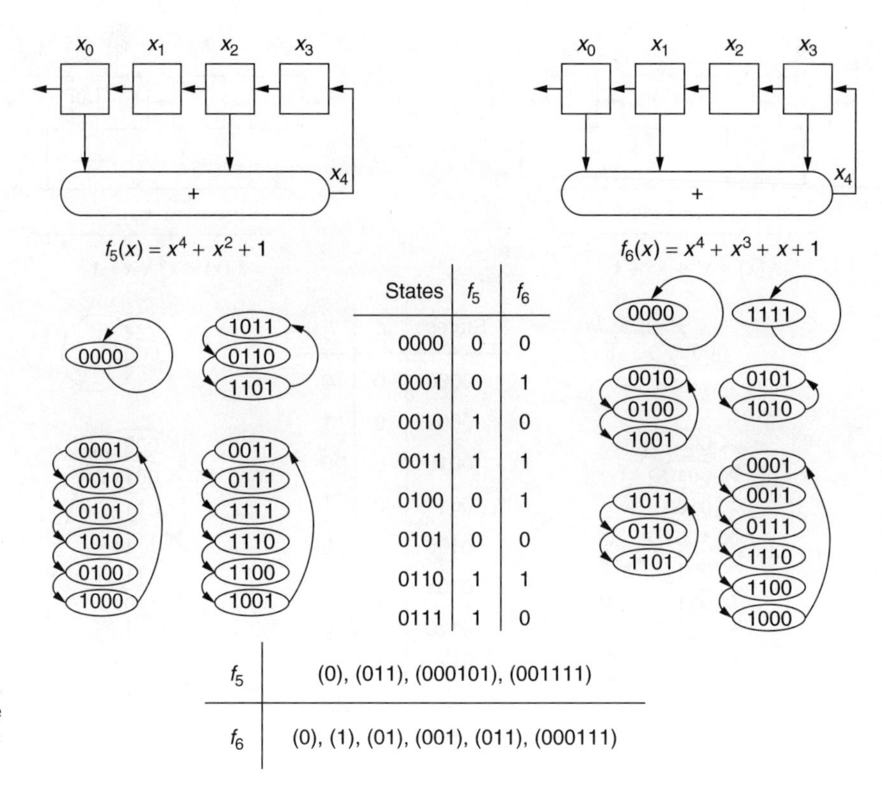

Figure 7. Block diagrams, state diagrams, truth tables, and output sequences of four-stage LFSRs with characteristic polynomials $f_5(x) = x^4 + x^2 + 1$ and $f_6(x) = x^4 + x^3 + x + 1$.

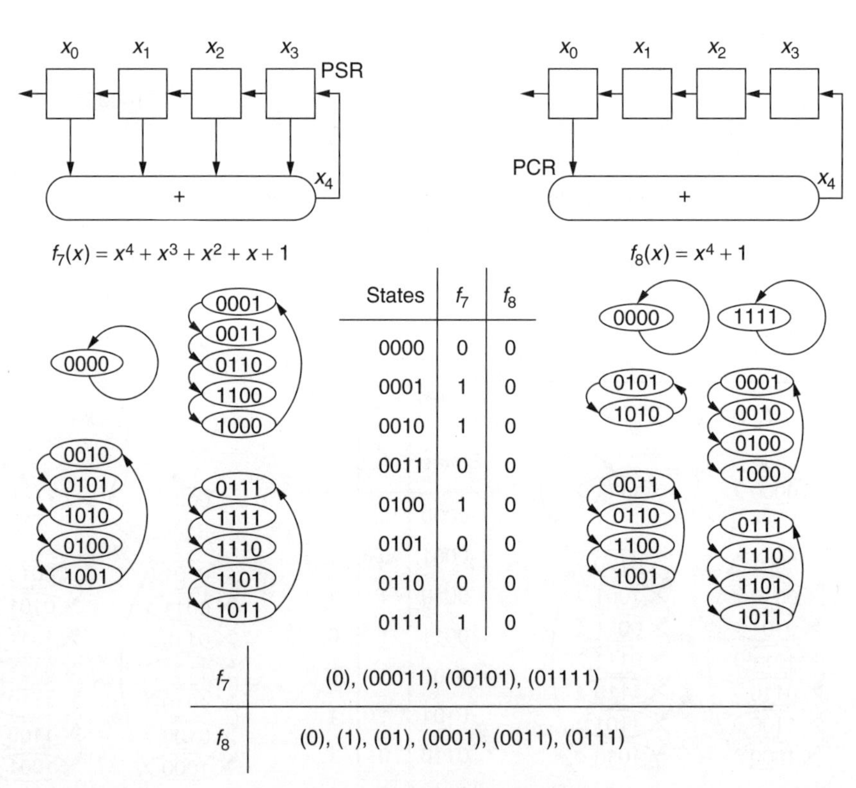

Figure 8. Block diagrams, state diagrams, truth tables, and output sequences of four-stage LFSRs with characteristic polynomials $f_7(x) = x^4 + x^3 + x^2 + x + 1$ (PSR) and $f_8(x) = x^4 + 1$ (PCR).

sequences are reciprocal. Figure 7 shows LFSRs with the self-reciprocal characteristic polynomials, and the output sequences are self-reciprocal also. This means that reading a sequence in the reverse direction gives a cyclically equivalent one to the original.

Figure 8 shows two special LFSRs: the pure summing register (PSR) and the pure cycling register (PCR). PSR has an irreducible but not primitive characteristic polynomial f_7. Observe that all the cycles except for the cycle containing (0000) have the same length, or the same period for its output sequences except for the all-zero sequence. This happens because the characteristic polynomial f_7 is irreducible. An irreducible polynomial, therefore, corresponds to a unique period, and it is called *the period of the irreducible polynomial.* A primitive polynomial of degree L is simply an irreducible polynomial with period $2^L - 1$. Possible periods of a given irreducible polynomial of degree L are the factors of the integer $2^L - 1$, which are not of the form $2^j - 1$ for $j < L$. When $2^L - 1$ is a prime, called a *Mersenne prime*, then every irreducible polynomial of degree $2^L - 1$ must be primitive.

Details on the property of PCR and some other properties of FSRs will be given at the end of Section 4.

3.2. Properties of *m*-Sequences

Now, we will describe some basic properties of m-sequences of period $2^L - 1$, mostly without proofs. The first three properties, namely, balance, run-distribution, and ideal autocorrelation are commonly known as "Golomb's postulates on random sequences" [10]. Most of the following properties can be easily checked for the examples shown in Figs. 3 and 5.

3.2.1. Balance Property. In one period of an m-sequence, the number of 1s and that of 0s are nearly the same. Since the period is an odd integer, they cannot be exactly the same, but differ by one. This is called the balance property. When a matrix of size $(2^L - 1) \times L$ is formed by listing all the states of the maximum length cycle, then the rightmost column is the m-sequence and it will contain 2^{L-1} ones and $2^{L-1} - 1$ zeros since the rows are permutations of all the vectors of length L except for the all-zero vector.

3.2.2. Run Distribution Property. A string of the same symbol of length l surrounded by different symbols at both ends is called a "run of length l." For example, a run of 1s of length 4 looks like $\ldots 0\underline{1111}0 \ldots$. The run distribution property of m-sequences refers to the fact that a shorter run appears more often than a longer run, and that the number of runs of 1s is the same as that of 0s. Specifically, it counts the number of runs of length l for $l \geq 1$ in one period as shown in Table 2. The span property of m-sequences implies this run distribution property.

3.2.3. Ideal Autocorrelation Property. A periodic unnormalized autocorrelation function $R(\tau)$ of a binary sequence $\{s(k)\}$ of period P is defined as

$$R(\tau) = \sum_{k=0}^{P-1} (-1)^{s(k)+s(k-\tau)}, \quad \tau = 0, 1, 2, \ldots,$$

Table 2. Run Distribution Property of *m*-Sequences of Period $2^L - 1$

Length	Number of Runs of 1s	Number of Runs of 0s
L	1	0
$L-1$	0	1
$L-2$	1	1
$L-3$	2	2
$L-4$	4	4
$\vdots$	$\vdots$	$\vdots$
2	2^{L-4}	2^{L-4}
1	2^{L-3}	2^{L-3}
Total	2^{L-3}	2^{L-3}

where $k - \tau$ is computed mod P. When binary phase-shift-keying is used to digitally modulate incoming bits, we are considering the incoming bits whose values are taken from the complex values $\{+1, -1\}$. The change in alphabet between $s_i \in \{0, 1\}$ and $t_i \in \{+1, -1\}$ is commonly performed by the relation $t_i = (-1)^{s_i}$. Then we have $R(\tau) = \sum_{k=0}^{P-1} t(k)t(k-\tau)$ for each τ, and this calculates *the number of agreements minus the number of disagreements* when one period of $\{s(k)\}$ is placed on top of its (cyclically) τ-shifted version. For any integer $L \geq 2$, and for any m-sequence $\{s(k)\}$ of period $P = 2^L - 1$, the ideal autocorrelation property of m-sequences refers to the following:

$$R(\tau) = \begin{cases} 2^L - 1, & \tau \equiv 0 \pmod{2^L - 1} \\ -1, & \tau \not\equiv 0 \pmod{2^L - 1} \end{cases} \tag{7}$$

The ideal two-level autocorrelation property of an m-sequence enables one to construct a Hadamard matrix of order 2^L of, so called, *cyclic* type. A Hadamard matrix of order n is an $n \times n$ matrix with entries only of ± 1 such that any two distinct rows are orthogonal to each other [20]. When the symbols of an m-sequence are mapped onto $\{\pm 1\}$ and all the cyclic shifts are arranged in a square matrix of order $(2^L - 1)$, the relation in (7) implies that the dot product of any two distinct rows is exactly -1 over the complex numbers. Therefore, adjoining a leftmost column of all $+1$s and a top row of all $+1$s will give a cyclic Hadamard matrix of order 2^L.

Cyclic-type Hadamard matrices can be constructed from a balanced binary sequence of period $P \equiv 3 \pmod 4$ that has the ideal two-level autocorrelation function. The m-sequences are one such class of sequences. Some other well-known balanced binary sequences with period $P \equiv 3 \pmod 4$ will be described later.

3.2.4. Span Property. If two vectors, $(s(i), s(i+1), \ldots, s(i+L-1))$ and $(s(j), s(j+1), \ldots, s(j+L-1))$, of length L are distinct whenever $i \neq j$, then the sequence $\{s(k)\}$ is said to have this property. The indices of terms are considered mod P. For an m-sequence of period P, in addition, all the not-all-zero vectors of length L appear exactly once on the windows of length L. This can easily be seen by observing that an m-sequence is the sequence of the rightmost bits of states in the maximum length cycle of the state diagram. Each window of length L can then easily be identified with a state in this state diagram.

If we insert an additional 0 right after the run of 0's of length $L - 1$, the sequence will have period 2^L and the span property becomes perfect in that every window of length L shows all the vectors of length L exactly once. This is an example of a *de Bruijn sequence* of order L. The above construction has been successfully adopted [12] for use in spread-spectrum modulations, for the values of $L = 14$ and $L = 41$. In general, many algorithms are currently known for de Bruijn sequences of period 2^L [21], and the modification described above can easily be implemented, as described in Section 4.

3.2.5. Constant-on-the-Coset Property. For any m-sequence of period $2^L - 1$, there are $2^L - 1$ cyclically equivalent sequences corresponding to the $2^L - 1$ starting points. The term *constant-on-the-coset property* refers to the fact that there exists exactly one among all these such that it is fixed with 2 decimation. An m-sequence in this phase is said to be in the *characteristic phase*. Therefore, for the m-sequence $\{s(k)\}$ in the characteristic phase, the following relation is satisfied:

$$s(2k) = s(k), \quad \text{for all } k \tag{8}$$

This relation deserves some special attention. It implies that every term in the $2k$th position is the same as the one in the kth position. This gives a set (or several sets) of positions in which the corresponding terms are the same. For example, $\{1, 2, 4, 8, 16, \ldots\}$ is one such set so that all the terms indexed by any number in this set are the same. Since the sequence is periodic with period $2^L - 1$, the preceding set is a finite set and called a *cyclotomic coset* mod $2^L - 1$. Starting from 3 gives another such set, $\{3, 6, 12, 24, \ldots\}$, and so on. In general, the set of integers mod $2^L - 1$ can be decomposed into some number of disjoint cyclotomic cosets, and now the constant-on-the-coset property describes itself clearly.

3.2.6. Cycle-and-Add Property. When two distinct phases of an m-sequence are added term by term, a sequence of the same period appears and it is a different phase of the same m-sequence. In other words, for any given constants $\tau_1 \not\equiv \tau_2 \pmod{2^L - 1}$, there exists yet another constant τ_3 such that

$$s(k - \tau_1) + s(k - \tau_2) = s(k - \tau_3), \quad k = 0, 1, 2, \ldots \tag{9}$$

This is the cycle-and-add property of m-sequences. On the other hand, if a balanced binary sequence of period P satisfies the cycle-and-add property, then P must be of the form $2^L - 1$ for some integer L and the sequence must be an m-sequence.

Golomb has conjectured that the span property and the ideal two-level autocorrelation property of a balanced binary sequence implies its cycle-and-add property [22]. This has been confirmed for L up to 10 by many others, but still awaits a complete solution.

3.2.7. Number of Cyclically Distinct m-Sequences of Period $2^L - 1$. For a given L, the number of cyclically distinct m-sequences of period $2^L - 1$ is equal to the number of primitive polynomials of degree L over F_2, and this is given by $\phi(2^L - 1)/L$, where $\phi(n)$ is the Euler ϕ function and counts the number of integers from 1 to n that are relatively prime to n. All these $\phi(2^L - 1)/L$ m-sequences of period $2^L - 1$ are equivalent, and they are related with some decimation of each other. Therefore, any given one m-sequence can be used to generate all the others of the same period by using some appropriate decimations.

3.2.8. Trace Function Representation of m-Sequences. Let q be a prime power, and let F_q be the finite field with q elements. Let $n = em > 1$ for some positive integers e and m. Then the trace function $\mathrm{tr}_m^n(\cdot)$ is a mapping from F_{2^n} to its subfield F_{2^m} given by

$$\mathrm{tr}_m^n(x) = \sum_{i=0}^{e-1} x^{2^{mi}}$$

It is easy to check that the trace function satisfies the following: (1) $\mathrm{tr}_m^n(ax + by) = a\,\mathrm{tr}_m^n(x) + b\,\mathrm{tr}_m^n(y)$, for all $a, b \in F_{2^m}$, $x, y \in F_{2^n}$; (2) $\mathrm{tr}_m^n(x^{2^m}) = \mathrm{tr}_m^n(x)$, for all $x \in F_{2^n}$; and (3) $\mathrm{tr}_1^n(x) = \mathrm{tr}_1^m(\mathrm{tr}_m^n(x))$, for all $x \in F_{2^n}$. See the literature [13–18] for the detailed properties of the trace function.

Let $q = 2^L$ and α be a primitive element of F_q. Then, an m-sequence $\{s(k)\}$ of period $2^L - 1$ can be represented as

$$s(k) = \mathrm{tr}_1^L(\lambda\alpha^k), \quad k = 0, 1, 2, \ldots, 2^L - 2 \tag{10}$$

where $\lambda \neq 0$ is a fixed constant in F_q. We just give a remark that λ corresponds to the initial condition and the choice of α corresponds to the choice of a primitive polynomial as a connection polynomial when this sequence is generated using an LFSR. Any such representation, on the other hand, gives an m-sequence [17]. When $\lambda = 1$, the sequence is in the characteristic phase, and the constant-on-the-coset property can easily be checked since $s(2k) = \mathrm{tr}_1^L(\alpha^{2k}) = \mathrm{tr}_1^L(\alpha^k) = s(k)$ for all k.

3.2.9. Cross-Correlation Properties of m-Sequences. No pair of m-sequences of the same period have the ideal cross-correlation. The best one can achieve is a three-level cross-correlation, and the pair of m-sequences with this property is called a *preferred pair*. Since all m-sequences of a given period are some decimations or cyclic shifts of each other, and they can all be represented as a single trace function from F_{2^L} to F_2, the cross-correlation of a pair of m-sequences can be described as

$$R_d(\tau) = \sum_{k=0}^{2^L-1} (-1)^{\mathrm{tr}_1^L(\alpha^{k+\tau}) + \mathrm{tr}_1^L(\alpha^{dk})}$$

where d indicates that the second m-sequence is a d decimation of the first, and τ represents the amount of phase offset with each other.

Many values of d have been identified that result in a preferred pair of m-sequences, but it is still unknown whether we have found them all. The most famous one that gives a Gold sequence family comes from the value $d = 2^i + 1$ or $d = 2^{2i} - 2^i + 1$ for some integer i when $L/(L, i)$ is odd. Some good references on this topic are Refs. [36–38], and also Chapter 5 of Ref. [2].

3.2.10. Linear Complexity. Given a binary periodic sequence $\{s(k)\}$ of period P, one can always construct an LFSR that outputs $\{s(k)\}$ with a suitable initial condition. One trivial solution is the *pure cycling register* as shown in Fig. 8. It has P stages, the whole period is given as its initial condition, and the characteristic polynomial (or the connection) is given by $f(x) = x^P + 1$ corresponding to $s(k) = s(k - P)$. The best one can do is to find the LFSR with the smallest number of stages, and the linear complexity of a sequence is defined as this number. Equivalently, it is the degree of the minimal polynomial of the given sequence, which is defined as the minimum degree characteristic polynomial of the sequence.

The linear complexity of a PN sequence, in general, measures cryptographically how strong it is. It is well known that the same copy (whole period) of a binary sequence can be generated whenever $2N$ consecutive terms or more are observed by a third party where N is the linear complexity of the sequence. This forces us to use those sequences with larger linear complexity in some practice. The m-sequences are the worst in this sense because an m-sequence of period $2^L - 1$ has its linear complexity L, and this number is the smallest possible over all the balanced binary sequences of period $2^L - 1$. In the following, we describe the famous Berlekamp–Massey algorithm for determining the linear complexity of a binary sequence [23].

3.3. Berlekamp–Massey Algorithm

Suppose that we are given N terms of a sequence S, which we denote as $S^N = (s(0), s(1), \ldots, s(N-1))$. It does not necessarily mean that S has period N. The goal of the Berlekamp–Massey algorithm (BMA) is to find the minimum degree recursion satisfied by S. This minimum degree $L_N(S)$ is called the *linear complexity* of S^N. This recursion can be used to form an LFSR with $L_N(S)$ stages that generates N terms of S exactly, given the initial condition of $s(0), s(1), \ldots, s(L_N(S)-1)$. We will denote this LFSR as LFSR($f^{(N)}(x), L_N(S)$), where the characteristic polynomial after the Nth iteration is given by

$$f^{(N)}(x) = 1 + c_1^{(N)}x + c_2^{(N)}x^2 + \cdots + c_{L_N(S)}^{(N)}x^{L_N(S)}$$

It is not difficult to check that (1) $L_N(S) = 0$ if and only if $s(0), s(1), \ldots, s(N-1)$ are all zeros, (2) $L_N(S) \leq N$, and (3) $L_N(S)$ must be monotonically nondecreasing with increasing N.

The BMA updates the degree $L_n(S)$ and the characteristic polynomial $f^{(n)}(x)$ for each $n = 1, 2, \ldots, N$. Assume that $f^{(1)}(x), f^{(2)}(x), \ldots, f^{(n)}(x)$ have been constructed, where the LFSR with connection $f^{(n)}(x)$ of degree $L_n(S)$ generates $s(0), s(1), \ldots, s(n-1)$. Let

$$f^{(n)}(x) = 1 + \sum_{i=1}^{L_n(S)} c_i^{(n)}x^i$$

The next discrepancy, d_n, is the difference between $s(n)$ and the $(n+1)$st bit generated by so far the minimal-length LFSR with $L_n(S)$ stages, and given as

$$d_n = s(n) + \sum_{i=1}^{L_n(S)} c_i^{(n)}s(n-i)$$

Let m be the sequence length before the last length change in the minimal-length register:

$$L_m(S) < L_n(S), \quad \text{and} \quad L_{m+1}(S) = L_n(S)$$

The LFSR with the characteristic polynomial $f^{(m)}(x)$ and length $L_m(S)$ could not have generated $s(0), s(1), \ldots, s(m-1), s(m)$. Therefore, $d_m \neq 0$.

If $d_n = 0$, then this LFSR also generates the first $n+1$ bits $s(0), s(1), \ldots, s(n)$ and therefore, $L_{n+1}(S) = L_n(S)$ and $f^{(n+1)}(x) = f^{(n)}(x)$.

If $d_n \neq 0$, a new LFSR must be found to generate the first $n+1$ bits $s(0), s(1), \ldots, s(n)$. The connection polynomial and the length of the new LFSR are updated by the following:

$$f^{(n+1)}(x) = f^{(n)}(x) - d_n d_m^{-1} x^{n-m} f^{(m)}(x)$$

$$L_{n+1}(S) = \max\,[L_n(S), n+1-L_n(S)]$$

The complete BM algorithm for implementations is as follows:

1. Initialization:

$$f(x) = 1, \quad g(x) = 1, \quad r = 1, \quad L = 0,$$
$$b = 1, \quad n = 0$$

2. If $n = N$, then stop. Otherwise compute

$$d = s(n) - \sum_{i=1}^{L} c_i s(n-i)$$

3. If $d = 0$, then $r = r + 1$, and go to step 6.
4. If $d \neq 0$ and $2L > n$, then

$$f(x) = f(x) - db^{-1}x^r g(x), \quad r = r + 1$$

   and go to step 6.
5. If $d \neq 0$ and $2L \leq n$, then

$$h(x) = f(x), \quad f(x) = f(x) - db^{-1}x^r g(x), \quad L = n + 1 - L,$$
$$g(x) = h(x), \quad b = d, \quad r = 1.$$

6. Increase n by 1 and return to step 2.

When $n = N$ and the algorithm is stopped in step (2), the quantities produced by the algorithm bear the following relations:

$$f(x) = f^{(N)}(x)$$
$$L = L_N(S)$$
$$r = N - m$$
$$d = d_{N-1}$$
$$g(x) = f^{(m)}(x)$$
$$b = d_m$$

Table 3. Example of BM Algorithm to the Sequence $(s_0, s_1, s_2, s_3, s_4, s_5, s_6) = (1, 0, 1, 0, 0, 1, 1)$ over F_2

n	L	$f(x)$	r	$g(x)$	b	s_n	d
0	0	1	1	1	1	1	1
1	1	$1+x$	1	1	1	0	1
2	1	1	2	1	1	1	1
3	2	$1+x^2$	1	1	1	0	0
4	2	$1+x^2$	2	1	1	0	1
5	3	1	1	$1+x^2$	1	1	1
6	3	$1+x+x^3$	2	$1+x^2$	1	1	0
7	3	$1+x+x^3$	3	$1+x^2$	1		

An example of BM algorithm applied to a binary sequence of length 7 is shown in Table 3.

3.4. Balanced Binary Sequences with the Ideal Two-Level Autocorrelation

In addition to m-sequences, there are some other well-known balanced binary sequences with the ideal two-level autocorrelation function. For period $P = 4n - 1$ for some positive integer n, all these are equivalent to $(4n - 1, 2n - 1, n - 1)$-cyclic difference sets [24].

In general, a (v, k, λ)-cyclic difference set (CDS) is a k-subset D of the integers mod v, Z_v, such that for each nonzero $z \in Z_v$ there are exactly λ ordered pairs (x, y), $x \in D, y \in D$ with $z = x - y \pmod{v}$ [24–28]. For example, $D = \{1, 3, 4, 5, 9\}$ is a $(11, 5, 2)$-CDS and every nonzero integer from 1 to 10 is represented by the difference $x - y \pmod{11}, x \in D, y \in D$, exactly twice.

The characteristic sequence $\{s(t)\}$ of a (v, k, λ)-cyclic difference set D is defined as $s(t) = 0$ if and only if $t \in D$.

For other values of t, the value 1 is assigned to $s(t)$. This completely characterizes binary sequences of period v with two-level autocorrelation function, and it is not difficult to check that the periodic unnormalized autocorrelation function is given as [29]

$$R(\tau) = \begin{cases} v, & \tau \equiv 0 \pmod{v} \\ v - 4(k - \lambda), & \tau \not\equiv 0 \pmod{v} \end{cases} \tag{11}$$

The out-of-phase value should be kept as low as possible in some practice, and this could happen when $v = 4(k - \lambda) - 1$, resulting in the out-of-phase value to be independent of the period v. The CDS with this parameter is called a *cyclic Hadamard difference set*, and this has been investigated by many researchers [24–28].

In the following, we will simply summarize all the known constructions for $(4n - 1, 2n - 1, n - 1)$ cyclic *Hadamard* difference sets, or equivalently, balanced binary sequences of period $v = 4n - 1$ with the two-level *ideal* autocorrelation function, which are also known as *Hadamard sequences* [30].

Three types of periods are currently known: (1) $v = 4n - 1$ is a prime, (2) $v = 4n - 1$ is a product of twin primes, and (3) $v = 4n - 1$ is one less than a power of 2. All these sequences can be represented as a sum of some decimations of an m-sequence. Song and Golomb have conjectured that a Hadamard sequence of period v exists if and only if v is one of the above three types [31], and this has been confirmed for all the values of $v = 4n - 1$ up to $v = 10000$ with 13 cases unsettled, the smallest of which is $v = 3439$ [30].

3.4.1. When $v = 4n - 1$ Is a Prime. There are two methods in this case [24,29]. The first corresponds to all such values of v, and the resulting sequences are called *Legendre sequences*. Here, D picks up integers mod v that are squares mod v. The second corresponds to some such values of v that can be represented as $4x^2 + 27$ for some integer x, and the resulting sequences are called *Hall's sextic residue sequences*. Here, D picks up integers mod v that are sixth powers mod v and some others.

3.4.2. When $v = 4n - 1$ Is a Product of Twin Primes. This is a generalization of the method for constructing Legendre sequences, and the resulting sequences are called *twin prime sequences*. Let $v = p(p + 1)$, where both p and $p + 2$ are prime. Then, D picks up the integers d that are (1) squares both mod p and mod $p + 2$, (2) nonsquares both mod p and mod $p + 2$, and (3) 0 mod $p + 2$.

3.4.3. When $v = 4n - 1 = 2^L - 1$. Currently, this is a most active area of research, and at least seven families are known. All the sequences of this case can best be described as a sum of some decimations of an m-sequence, or a sum of trace functions from F_{2^L} to F_2. The m-sequences for all the positive integers L and GMW sequences for all the composite integers L [32,33] have been known for many years. One important construction of a larger period from a given one is described by No et al. [34]. More recent discoveries were summarized by No et al. [35], most of which have now been proved by many others.

4. SOME PROPERTIES OF FSR WITH DISJOINT CYCLES

We now return to the basic block diagram of an L-stage FSR as shown in Fig. 1, and its truth table, and state transition diagram as shown in Figs. 2 and 3. Unless otherwise stated, the feedback connection logic $f(x_{k-L}, x_{k-L+1}, \ldots, x_{k-1})$ of all FSRs in this section satisfy the branchless condition given in Eq. (3).

The simplest FSR with L stages is the *pure cycling register* (PCR), as shown in Fig. 8 for $L = 4$. It is linear and has the characteristic polynomial $f(x) = x^L + 1$, or the feedback connection logic $x_k = x_{k-L}$. This obviously satisfies the branchless condition (3), and the state diagram consists only of disjoint cycles. In fact, one can prove that the number $Z(L)$ of disjoint cycles of an L-stage PCR is given as

$$Z(L) = \frac{1}{L} \sum_{d|L} \phi(d) 2^{L/d}, \tag{12}$$

where $\phi(d)$ is the Euler ϕ function and the summation is over all the divisors of L. It is not very surprising that this number is the same as the number of irreducible polynomials of degree L over the binary alphabet F_2. Golomb had conjectured that the number of disjoint cycles from an L-stage FSR with branchless condition satisfied by its connection logic is *at most* $Z(L)$ given in (12), and this was confirmed by Mykkeltveit in 1972 [39].

On the other hand, the minimum number of disjoint cycles is 1, and this corresponds to de Bruijn sequences of period 2^L. Inserting a single 0 into any m-sequence right after the run of 0s of length $L-1$ gives a de Bruijn sequence of period 2^L. This can be done by the following modification to the linear logic f_{old} that generates the m-sequence

$$f_{\text{new}} = f_{\text{old}} \oplus x'_{k-1}x'_{k-2}\cdots x'_{k-L+1}, \qquad (13)$$

where x' represents the complement of x. A de Bruijn sequence can best be described using Good's diagram. This is shown in Fig. 9 for $L=2$ and $L=3$. Note that any node in a Good diagram has two incoming edges as well as two outgoing edges. A de Bruijn sequence of period 2^L corresponds to a closed path (or a cycle) on the Good diagram of order L, which visits every node exactly once. It was shown earlier in 1946 by de Bruijn that the number of such cycles is given as $2^{2^{L-1}-L}$ [10].

The number of disjoint cycles of an L-stage FSR with (possibly) nonlinear logic connections is not completely determined. Toward this direction, we simply state a condition for the parity of this number for FSRs with branchless condition. *The number of disjoint cycles of an FSR with the branchless condition is even* (*or odd, respectively*) *if and only if the number of 1s in its truth table of* $g(x_{k-1}, x_{k-2}, \ldots, x_{k-L+1})$ *is even* (*or odd, respectively*) [10]. In other words, the parity of the top half of the truth table is the parity of the number of disjoint cycles. This implies that PCR has an even number of disjoint cycles $L > 2$.

In addition to PCR, there are three more degenerate cases: complemented cycling registers (CCRs), pure summing registers (PSRs), and complemented summing registers (CSRs). Note that PCR and PSR are homogeneous linear, but CCR and CSR are inhomogeneous linear:

$$\begin{aligned} x_k &= x_{k-L}, \quad \text{PCR} \\ &= 1 + x_{k-L}, \quad \text{CCR} \\ &= x_{k-1} + x_{k-2} + \cdots + x_{k-L} \quad \text{PSR} \\ &= 1 + x_{k-1} + x_{k-2} + \cdots + x_{k-L} \quad \text{CSR} \end{aligned}$$

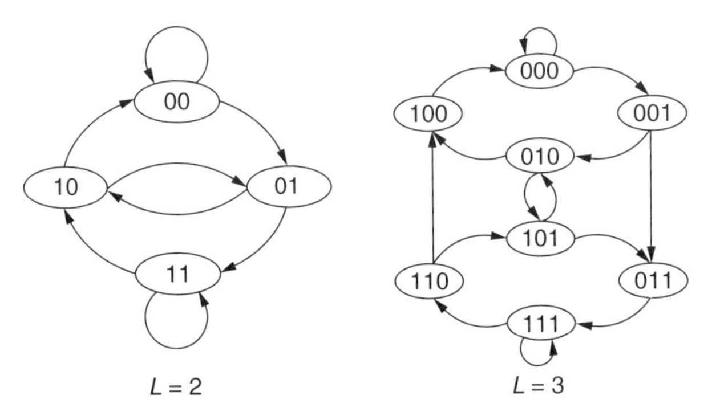

Figure 9. Good's diagrams for $L=2$ and $L=3$.

Table 4. Output Sequences from Two Degenerated FSR with $L = 4$

CCR	Period	CSR	Period
(00001111)	8	(1)	1
		(00001)	5
(01011010)	8	(00111)	5
		(01011)	5

All these satisfy the branchless condition, and the output sequences from CCR and CSR for $L=4$ are listed in Table 4.

The number of L-stage FSRs with branchless condition is given as $2^{2^{L-1}}$. Another condition for the truth table of an FSR to be practically useful is the *balanced logic* condition. A truth table of f for an L-stage FSR is said to have the balanced logic condition if f has equally many 1s and 0s in its truth table. Both the branchless condition and the balanced logic condition together imply that f has equally many 1s and 0s in its top half of the truth table. The balanced logic condition guarantees that *the autocorrelation of the output sequence of period approaching* 2^L *tends to zero for* $\tau = 1, 2, \ldots, L$.

Finally, we give a detailed analysis on the branchless four-stage FSRs with all possible connections. With the branchless condition on it, there are $2^{2^3} = 2^8 = 256$ such FSRs. Among these, 8 FSRs are linear as shown in Figs. 5–8. These are collected in Table 5. Except for PCR, all the other 7 LFSRs are balanced logic FSRs. Among the 248 nonlinear FSRs, there are $2^{2^3-4} = 16$ FSRs, which generate de Bruijn sequences of period 16, as shown in Table 6. This table also shows all the 16 output sequences in four equivalent classes, denoted as A, B, C, and D. The linear complexity of each de Bruijn sequence is also shown here. The 4th and 14th sequences are modified versions from the 6th and 8th m-sequences in Table 5, respectively. The modification is the insertion of a single 0 into the m-sequence right after the run of 0s of length 3. Note that none of the connection logic satisfies the balanced logic condition. Among the 256 FSRs with the branchless condition, there are $\binom{8}{4} = 70$ FSRs that satisfy the balanced logic condition. Table 7 shows all of them. In this table, $*$ represents that it is linear and is also shown in Table 5.

5. CONCLUDING REMARKS

There is, in fact, a large amount of literature on FSRs, on FSR sequences, and their variations, generalizations, and applications.

Analysis of LFSR and LFSR sequences can also be done using at least two other standard methods not described in this article: the generating function approach and the matrix representation approach [10,13].

There are generalizations of LFSR over a nonbinary alphabet. For this, at least two operations, addition and multiplication, must be well defined over the alphabet. The well-known example of such an alphabet is a finite field with q elements. A finite commutative ring sometimes serves as an appropriate alphabet over which an LFSR is

Table 5. Truth Tables of All Four-Stage LFSRs, Including Their Output Sequences and Characteristic Polynomials

	Truth Table	Output Sequences	Characteristic Polynomials	Figures
1	00000000 11111111	0, 1, 01, 0001, 0011, 0111	$x^4 + 1$	Fig. 8
2	00111100 11000011	0, 1, 0001011, 0011101	$x^4 + x^3 + x^2 + 1$	Fig. 6
3	01011010 10100101	0, 1, 01, 001, 011, 000111	$x^4 + x^3 + x + 1$	Fig. 7
4	01100110 10011001	0, 1, 0001101, 0010111	$x^4 + x^2 + x + 1$	Fig. 6
5	01101001 10010110	0, 00011, 00101, 01111	$x^4 + x^3 + x^2 + x + 1$	Fig. 8
6	01010101 10101010	0, 000111101011001	$x^4 + x + 1$	Fig. 5
7	00110011 11001100	0, 011, 000101, 001111	$x^4 + x^2 + 1$	Fig. 7
8	00001111 11110000	0, 000100110101111	$x^4 + x^3 + 1$	Fig. 5

Table 6. The Truth Tables of All Four-Stage FSRs That Generate de Bruijn Sequences in Four Equivalent Classes[a]

	Truth Table	Output Sequence	Equivalent Class	LC
1	11110001 00001110	0000111101100101	A	15
2	10111001 01000110	0000101001111011	C	15
3	11100101 00011010	0000110010111101	D	14
4*	11010101 00101010	0000111101011001	A	15
5	10110101 01001010	0000101100111101	D	14
6	10101101 01010010	0000101111010011	D	14
7	10011101 01100010	0000100111101011	C	15
8	11111101 00000010	0000111101001011	B	12
9	11100011 00011100	0000110111100101	C	15
10	10101011 01010100	0000101001101111	A	15
11	11000111 00111000	0000110101111001	C	15
12	10100111 01011000	0000101111001101	D	14
13	11110111 00001000	0000111100101101	B	12
14*	10001111 01110000	0000100110101111	A	15
15	11101111 00010000	0000110100101111	B	12
16	10111111 01000000	0000101101001111	B	12

[a]The linear complexity of each de Bruijn sequence is also shown. The 4th and 14th sequences are modified versions of 6th 8th m-sequences in Table 5, respectively. The asterisk denotes that the FSR is linear.

operating. Integers mod 4 is the best known in this regard, due to the application of the output sequences into QPSK modulation [13,38].

There are other directions to which FSR may be generalized. For example, one can consider LFSR with inputs. One application is to use the LFSR with input as a polynomial division circuit. These are used in the decoding/encoding of Hamming codes or other channel (block) codes. Another example is to use multiple (nonlinear) FSRs on a stack so that the stages of an FSR in one layer are used to produce inputs to upper layer FSRs on top of it. These find some important applications to generating PN sequences with larger linear complexity in streamcipher systems.

All these topics are currently under active research, and one should look at journal transactions for the most recent results and applications.

BIOGRAPHY

Hong-Yeop Song received his B.S. degree in Electronics Engineering from Yonsei University in 1984 and his M.S.E.E. and Ph.D. degrees from the University of Southern California, Los Angeles, in 1986 and 1991, respectively. After spending 2 years on the research staff in the Communication Sciences Institute at USC, he joined Qualcomm Inc., San Diego, California in 1994 as a Senior Engineer and worked in a team researching and developing North American CDMA Standards for PCS and cellular air interface. Finally, he joined the Department of Electrical and Electronics Engineering at Yonsei University, Seoul, South Korea in 1995, and is currently serving as an Associate Professor. His areas of research interest are sequence design and analysis for speed-spectrum communications and stream ciphers, theory and application of cyclic difference sets, and channel coding/decoding problems for wireless communications. He is currently visiting the University of Waterloo, Ontario, Canada, from March 2002 to Feb 2003 during his sabbatical leave.

BIBLIOGRAPHY

1. S. W. Golomb, *Digital Communications with Space Applications*, Prentice-Hall, Englewood Cliffs, NJ, 1964.

Table 7. Truth Tables and Cycle Length Distributions of All Four-Stage FSRs That Satisfy Both Branchless and Balanced Logic Conditions[a]

	Truth Table	Cycle Length Distribution		Truth Table	Cycle Length Distribution
1	1111000000001111	1:1, 15:1	36	1110000100011110	5:1, 11:1
2	1110100000010111	1:1, 4:1, 5:1, 6:1	37	1101000100101110	2:1, 14:1
3	1101100000100111	1:1, 2:1, 3:1, 10:1	38	1011000101001110	7:1, 9:1
4	1011100001000111	1:1, 15:1	39	0111000110001110	1:1, 15:1
5	0111100010000111	1:2, 5:1, 9:1	40	1100100100110110	2:1, 3:1, 5:1, 6:1
6	1110010000011011	1:1, 15:1	41	1010100101010110	5:1, 11:1
7	1101010000101011	1:1, 15:1	42*	0110100110010110	1:1, 5:3
8	1011010001001011	1:1, 15:1	43	1001100101100110	2:1, 14:1
9	0111010010001011	1:2, 6:1, 8:1	44	0101100110100110	1:1, 2:1, 3:1, 10:1
10	1100110000110011	1:1, 3:1, 6:2	45	0011100111000110	1:1, 15:1
11	1010110001010011	1:1, 15:1	46	1100010100111010	7:1, 9:1
12	0110110010010011	1:2, 5:1, 9:1	47	1010010101011010	4:1, 12:1
13	1001110001100011	1:1, 15:1	48	0110010110011010	1:1, 15:1
14	0101110010100011	1:2, 3:1, 11:1	49	1001010101101010	5:1, 11:1
15*	0011110011000011	1:2, 7:2	50*	0101010110101010	1:1, 15:1
16	1110001000011101	1:1, 15:1	51	0011010111001010	1:1, 15:1
17	1101001000101101	1:1, 2:1, 3:1, 10:1	52	1000110101110010	7:1, 9:1
18	1011001001001101	1:1, 3:1, 5:1, 7:1	53	0100110110110010	1:1, 3:1, 5:1, 7:1
19	0111001010001101	1:2, 3:1, 11:1	54	0010110111010010	1:1, 15:1
20	1100101000110101	1:1, 2:1, 3:1, 10:1	55	0001110111100010	1:1, 15:1
21	1010101001010101	1:1, 15:1	56	1100001100111100	2:1, 14:1
22	0110101010010101	1:2, 5:1, 9:1	57	1010001101011100	7:1, 9:1
23	1001101001100101	1:1, 2:1, 3:1, 10:1	58	0110001110011100	1:1, 15:1
24*	0101101010100101	1:2, 2:1, 3:2, 6:1	59	1001001101101100	2:1, 3:1, 5:1, 6:1
25	0011101011000101	1:2, 3:1, 11:1	60	0101001110101100	1:1, 2:1, 3:1, 10:1
26	1100011000111001	1:1, 15:1	61*	0011001111001100	1:1, 3:1, 6:2
27	1010011001011001	1:1, 15:1	62	1000101101110100	2:1, 14:1
28*	0110011010011001	1:2, 7:2	63	0100101110110100	1:1, 2:1, 3:1, 10:1
29	1001011001101001	1:1, 5:3	64	0010101111010100	1:1, 15:1
30	0101011010101001	1:2, 5:1, 9:1	65	0001101111100100	1:1, 2:1, 3:1, 10:1
31	0011011011001001	1:2, 5:1, 9:1	66	1000011101111000	5:1, 11:1
32	1000111001110001	1:1, 15:1	67	0100011110111000	1:1, 15:1
33	0100111010110001	1:2, 3:1, 11:1	68	0010011111011000	1:1, 15:1
34	0010111011010001	1:2, 6:1, 8:1	69	0001011111101000	1:1, 4:1, 5:1, 6:1
35	0001111011100001	1:2, 5:1, 9:1	70*	0000111111110000	1:1, 15:1

[a]Here, $a:b$ means that there are b cycles of length a, and * implies that the FSR is linear.

2. M. K. Simon, J. K. Omura, R. A. Scholtz, and B. K. Levitt, *Spread Spectrum Communications Handbook*, Computer Science Press, Rockville, MD, 1985; rev. ed., McGraw-Hill, 1994.
3. R. L. Perterson, R. E. Ziemer, and D. E. Borth, *Introduction to Spread Spectrum Communications*, Prentice-Hall, Englewood Cliffs, NJ, 1995.
4. J. G. Proakis and M Salehi, *Communication Systems Engineering*, Prentice-Hall, Englewood Cliffs, NJ, 1994.
5. A. J. Viterbi, *CDMA: Principles of Spread Spectrum Communication*, Addison-Wesley, Reading, MA, 1995.
6. J. G. Proakis, *Digital Communications*, 4th ed., McGraw-Hill, New York, 2001.
7. R. A. Rueppel, *Analysis and Design of Stream Ciphers*, Springer-Verlag, Berlin, 1986.
8. A. J. Menezes, P. C. van Oorschot, and S. A. Vanstone, *Handbook of Applied Cryptography*, CRC Press, Boca Raton, FL, 1996.
9. S. B. Volchan, What is a Random Sequence? *Am. Math. Monthly* **109**: 46–63 (2002).
10. S. W. Golomb, *Shift Register Sequences*, Holden-Day, San Francisco, CA, 1967; rev. ed., Aegean Park Press, Laguna Hills, CA, 1982.
11. D. R. Stinson, *Cryptography: Theory and Practice*, CRC Press, Boca Raton, FL, 1995.
12. TIA/EIA/IS-95, *Mobile Station–Base Station Compatibility Standard for Dual-Mode Wideband Spread Spectrum Cellular System*, published by Telecommunications Industry Association as a North American 1.5 MHz Cellular CDMA Air-Interface Standard, July 1993.
13. R. Lidl and H. Niederreiter, Finite fields, in *Encyclopedia of Mathematics and Its Applications*, Vol. 20, Addison-Wesley, Reading, MA, 1983.
14. E. R. Berlekamp, *Algebraic Coding Theory*, McGraw-Hill, New York, 1968.
15. W. W. Peterson and E. J. Weldon, *Error-Correcting Codes*, 2nd ed., MIT Press, Cambridge, MA, 1972.
16. F. MacWilliams and N. J. A. Sloane, *The Theory of Error-Correcting Codes*, North-Holland, 1977.
17. R. J. McEliece, *Finite Fields for Computer Scientists and Engineers*, Kluwer, 1987.

18. K. Ireland and M. Rosen, *A Classical Introduction to Modern Number Theory*, 2nd ed., Springer-Verlag, New York, 1991.

19. T. Hansen and G. L. Mullen, Supplement to primitive polynomials over finite fields, *Math. Comput.* **59**: S47–S50 (Oct. 1992).

20. J. H. van Lint and R. M. Wilson, *A Course in Combinatorics*, Cambridge Univ. Press, New York, 1992.

21. H. Fredricksen, A survey of full length nonlinear shift register cycle algorithms, *SIAM Rev.* **24**: 195–221 (1982).

22. S. W. Golomb, On the classification of balanced binary sequences of period $2^n - 1$, *IEEE Trans. Inform. Theory* **26**: 730–732 (1980).

23. J. L. Massey, Shift-register synthesis and BCH decoding, *IEEE Trans. Inform. Theory* **15**: 122–127 (1969).

24. L. D. Baumert, *Cylic Difference Sets, Lecture Notes in Mathematics*, Vol. 182, Springer-Verlag, New York, 1971.

25. M. Hall, Jr., A survey of difference sets, *Proc. Am. Math. Soc.* **7**: 975–986 (1956).

26. H. J. Ryser, *Combinatorial Mathematics, The Carus Mathematical Monographs* No. 14, Mathematical Association of America, 1963.

27. D. Jungnickel, Difference sets, in J. H.Dinitz and D. R. Stinson, eds., *Contemporary Design Theory*, Wiley, New York, 1992, pp. 241–324.

28. C. J. Colbourn and J. H. Dinitz, *The CRC Handbook of Combinatorial Designs*, CRC Press, New York, 1996.

29. S. W. Golomb, Construction of signals with favourable correlation properties, in A. D. Keedwell, ed., *Survey in Combinatorics, LMS Lecture Note Series*, Vol. 166, Cambridge Univ. Press, 1991, pp. 1–40.

30. J.-H. Kim, *On the Hadamard Sequences*, Ph.D. thesis, Yonsei Univ., South Korea, 2001.

31. H.-Y. Song and S. W. Golomb, On the existence of cyclic Hadamard difference sets, *IEEE Trans. Inform. Theory* **40**: 1266–1268 (1994).

32. B. Gordon, W. H. Mills, and L. R. Welch, Some new difference sets, *Can. J. Math.* **14**: 614–625 (1962).

33. R. A. Scholtz and L. R. Welch, GMW sequences, *IEEE Trans. Inform. Theory* **30**: 548–553 (1984).

34. J. S. No, H. Chung, K. Yang, and H. Y. Song, On the construction of binary sequences with ideal autocorrelation property, *Proc. IEEE Int. Symp. Information Theory and Its Application*, Victoria, BC, Canada, Sept. 1996, pp. 837–840.

35. J.-S. No et al., Binary pseudorandom sequences of period $2^n - 1$ with ideal autocorrelation, *IEEE Trans. Inform. Theory* **44**: 814–817 (1998).

36. D. V. Sarwate and M. B. Pursley, Crosscorrelation properties of pseudorandom and related sequences, *Proc. IEEE* **68**: 593–619 (1980).

37. P. Fan and M. Darnell, *Sequence Design for Communications Applications*, Research Studies Press LTD, Taunton, Somerset, England, 1995.

38. T. Helleseth and P. V. Kumar, Sequences with low correlation, in V. S. Pless and W. C. Huffman, eds., *Handbook of Coding Theory*, Elsevier Science B.V., 1998, Chap. 21.

39. J. Mykkeltveit, A proof of Golomb's conjecture on the de bruijn graph, *J. Comb. Theory Ser. B* **13**: 40–45 (1972).

FINITE-GEOMETRY CODES

M. Fossorier
University of Hawaii at Manoa
Honolulu, Hawaii

1. INTRODUCTION

Euclidean geometry (EG) and projective geometry (PG) codes belong to the class of algebraic block codes derived from finite geometries. These codes were introduced by Rudolph [1,2] and have been studied by many other researchers since [3–8].

An m-dimensional finite geometry is composed of a finite number of elements called *points*, or 1-flat. For $1 \leq \mu < m$, it is then possible to divide this set of points into subsets of identical structure defined by an equation with in general either μ, or $\mu + 1$ unknowns. Each subset is called a *μ-dimensional hyperplane* or μ-flat, and the ensemble of these μ-flats can be associated with a linear block code. In this article, we focus on codes associated with the value $\mu = 1$. These codes belong to the class of one-step majority-logic decodable codes, which have many nice structural properties that can exploited in their decoding. For example, an important but very limited class of such codes is that of difference-set cyclic (DSC) codes found independently by Weldon [9]. A DSC code has been used for error correction of a digital audio broadcasting system in Japan [10, Sect. 5-A].

The construction of EG and PG codes has long been motivated by their fast decoding with majority-logic decoding [11–13]. In addition, one-step majority logic decodable EG and PG codes can easily be decoded iteratively. Efficient iterative decoding methods have been devised for these codes, initially based on heuristic approaches [14,15], and more recently in conjunction with iterative decoding of the low-density parity-check (LDPC) codes [16,17]. On the basis of this relationship, many new classes of LDPC codes have also been constructed from finite geometries [17].

The construction concepts of EG and PG codes are summarized in Section 2 of this article. The subclasses of one-step majority logic decodable EG and PG codes are first considered, and the extension to other EG and PG codes is then discussed. The interested reader is referred to the literature [18–23] for more detailed expositions of finite geometries and their applications to error control coding. The link between EG and PG codes, and LDPC codes is finally briefly discussed. Several decoding methods are presented in Section 3.

2. CODE CONSTRUCTIONS

2.1. EG Codes

An m-dimensional Euclidean geometry over the finite Galois field GF(2^s) [denoted EG($m, 2^s$)] consists of the 2^{ms} m-tuples $\mathbf{p} = (p_0, p_1, \ldots, p_{m-1})$ (referred to as *points*), where for $0 \leq i \leq m-1$, each p_i is an element of GF(2^s). Two linearly independent points $\mathbf{p}_1$ and $\mathbf{p}_2$ [i.e., for $\alpha_1 \in$ GF(2^s) and $\alpha_2 \in$ GF(2^s), $\alpha_1\, \mathbf{p}_1 + \alpha_2\, \mathbf{p}_2 = \mathbf{0}$ if and only if

$\alpha_1 = \alpha_2 = 0]$ define a unique line $L(\mathbf{p}_1, \mathbf{p}_1 + \mathbf{p}_2)$ passing through $\mathbf{p}_1$ and $\mathbf{p}_1 + \mathbf{p}_2$. The line $L(\mathbf{p}_1, \mathbf{p}_1 + \mathbf{p}_2)$ contains a total of 2^s points $\mathbf{p}_i$ which satisfy the equation

$$\mathbf{p}_i = \mathbf{p}_1 + \alpha \mathbf{p}_2 \tag{1}$$

$\alpha \in \mathrm{GF}(2^s)$. Given a point $\mathbf{p}$, the $2^{ms} - 1$ other points in $\mathrm{EG}(m, 2^s)$ can be divided into subsets of $2^s - 1$ points, each based on Eq. (1) such that each subset corresponds to a distinct line of $\mathrm{EG}(m, 2^s)$ containing $\mathbf{p}$. As a result, there are $(2^{ms} - 1)/(2^s - 1)$ lines intersecting on each point $\mathbf{p}$ of $\mathrm{EG}(m, 2^s)$ and the total number of lines in $\mathrm{EG}(m, 2^s)$ is $2^{(m-1)s}\,(2^{ms} - 1)/(2^s - 1)$.

Let us consider the incidence matrix $H = [h_{i,j}]$ whose rows are associated with the $(2^{(m-1)s} - 1)\,(2^{ms} - 1)/(2^s - 1)$ lines not containing the all-zero point $\mathbf{0}$, and whose columns are associated with the $2^{ms} - 1$ points other than $\mathbf{0}$. Then $h_{i,j} = 1$ if the jth point of $\mathrm{EG}(m, 2^s)$ is contained in the ith line of $\mathrm{EG}(m, 2^s)$, and $h_{i,j} = 0$ otherwise. This incidence matrix H defines the dual space of a linear block EG code C denoted $C_{\mathrm{EG}}(m, s)$ of length $N = 2^{ms} - 1$. In general, the dimension K and minimum distance $d_{\min}$ of $C_{\mathrm{EG}}(m, s)$ depend on the values m and s, with $d_{\min} \geq (2^{ms} - 1)/(2^s - 1)$. From the structural properties of $\mathrm{EG}(m, 2^s)$, we conclude that (1) every row of H contains exactly 2^s "ones," (2) every column of H contains exactly $(2^{ms} - 1)/(2^s - 1) - 1$ "ones," and (3) any two rows of H have at most one "one" in common. These three properties, and especially property 3, are being used in the decoding of EG codes. Another interesting property of EG codes is that with a proper ordering of the columns of H, $C_{\mathrm{EG}}(m, s)$ is a cyclic code. This property is important for simple encoding of EG codes based on linear feedback shift registers (LFSRs).

If all 2^{ms} points of $\mathrm{EG}(m, 2^s)$ are kept to build the incidence matrix H, then we obtain an extended EG code with length $N = 2^{ms}$ and $d_{\min} \geq (2^{ms} - 1)/(2^s - 1) + 1$. In that case, properties 1 and 3 remain the same while property 2 becomes: Every column of H contains exactly $(2^{ms} - 1)/(2^s - 1)$ "ones" as each position has exactly one additional checksum associated with a line passing through the origin. This extended code is no longer cyclic, but encoding of an extended cyclic code is readily achieved from the cyclic encoder by the trivial addition of an overall parity-check bit.

As an example, let us choose $m = 2$. Then the two-dimensional EG codes $C_{\mathrm{EG}}(2, s)$ have the following parameters:

$$N = 2^{2s} - 1$$
$$K = 2^{2s} - 3^s$$
$$d_{\min} = 2^s + 1$$

For $s = 2$, $\mathrm{EG}(2, 2^2)$ has 15 lines not passing through $\mathbf{0}$, each point belonging to four different lines. Then, the incidence matrix

$$H = \begin{bmatrix}
1 & 1 & 0 & 1 & 0 & 0 & 0 & 1 & 0 & 0 & 0 & 0 & 0 & 0 & 0 \\
0 & 1 & 1 & 0 & 1 & 0 & 0 & 0 & 1 & 0 & 0 & 0 & 0 & 0 & 0 \\
0 & 0 & 1 & 1 & 0 & 1 & 0 & 0 & 0 & 1 & 0 & 0 & 0 & 0 & 0 \\
0 & 0 & 0 & 1 & 1 & 0 & 1 & 0 & 0 & 0 & 1 & 0 & 0 & 0 & 0 \\
0 & 0 & 0 & 0 & 1 & 1 & 0 & 1 & 0 & 0 & 0 & 1 & 0 & 0 & 0 \\
0 & 0 & 0 & 0 & 0 & 1 & 1 & 0 & 1 & 0 & 0 & 0 & 1 & 0 & 0 \\
0 & 0 & 0 & 0 & 0 & 0 & 1 & 1 & 0 & 1 & 0 & 0 & 0 & 1 & 0 \\
0 & 0 & 0 & 0 & 0 & 0 & 0 & 1 & 1 & 0 & 1 & 0 & 0 & 0 & 1 \\
1 & 0 & 0 & 0 & 0 & 0 & 0 & 0 & 1 & 1 & 0 & 1 & 0 & 0 & 0 \\
0 & 1 & 0 & 0 & 0 & 0 & 0 & 0 & 0 & 1 & 1 & 0 & 1 & 0 & 0 \\
0 & 0 & 1 & 0 & 0 & 0 & 0 & 0 & 0 & 0 & 1 & 1 & 0 & 1 & 0 \\
0 & 0 & 0 & 1 & 0 & 0 & 0 & 0 & 0 & 0 & 0 & 1 & 1 & 0 & 1 \\
1 & 0 & 0 & 0 & 1 & 0 & 0 & 0 & 0 & 0 & 0 & 0 & 1 & 1 & 0 \\
0 & 1 & 0 & 0 & 0 & 1 & 0 & 0 & 0 & 0 & 0 & 0 & 0 & 1 & 1 \\
1 & 0 & 1 & 0 & 0 & 0 & 1 & 0 & 0 & 0 & 0 & 0 & 0 & 0 & 1
\end{bmatrix}$$

defines the dual space of $C_{\mathrm{EG}}(2, 2)$, an $(N, K, d_{\min}) = (15, 7, 5)$ code. On the basis of H, and assuming that column i is associated with point $\mathbf{p}_i$ for $0 \leq i \leq 14$, we observe that the four lines intersecting point $\mathbf{p}_{14}$ and not passing through $\mathbf{0}$ are $\{\mathbf{p}_0, \mathbf{p}_2, \mathbf{p}_6, \mathbf{p}_{14}\}$, $\{\mathbf{p}_1, \mathbf{p}_5, \mathbf{p}_{13}, \mathbf{p}_{14}\}$, $\{\mathbf{p}_3, \mathbf{p}_{11}, \mathbf{p}_{12}, \mathbf{p}_{14}\}$, and $\{\mathbf{p}_7, \mathbf{p}_8, \mathbf{p}_{10}, \mathbf{p}_{14}\}$. Finally, the incidence matrix of the extended (16,7,6) EG code is obtained from H by adding an all-zero column in front of H and then appending to the bottom a 5×16 matrix composed of the all-one column followed by the 5×5 identity matrix repeated 3 times. The first column in the new 16×20 incidence matrix corresponds to the origin of $\mathrm{EG}(2, 2^2)$, and the five last rows correspond to the five lines of $\mathrm{EG}(2, 2^2)$ passing through $\mathbf{0}$.

The definition of a line or 1-flat given in (1) can be generalized as

$$\mathbf{p}_i = \mathbf{p}_0 + \alpha_1 \mathbf{p}_1 + \alpha_2 \mathbf{p}_2 + \cdots + \alpha_\mu \mathbf{p}_\mu \tag{2}$$

with $1 \leq \mu < m$ and for $1 \leq i \leq \mu$, $\alpha_i \in \mathrm{GF}(2^s)$. Equation (2) defines a μ-flat of $\mathrm{EG}(m, 2^s)$ that passes through the point $\mathbf{p}_0$. It is then straightforward to associate with $\mathrm{EG}(m, 2^s)$ an incidence matrix H whose rows are associated with the μ-flats of $\mathrm{EG}(m, 2^s)$ (possibly not containing $\mathbf{0}$), and whose columns are associated with the points of $\mathrm{EG}(m, 2^s)$ (possibly excluding $\mathbf{0}$). This incidence matrix H defines the dual space of a linear block EG code C. It follows that in general, an EG code is totally defined by the three parameters m, s, and μ. Further extensions of the definition of an EG code are possible by considering collections of flats. For example, twofold EG codes are obtained by considering pairs of parallel μ-flats, called "$(\mu, 2)$-frames" [6].

2.2. PG Codes

Since $\mathrm{GF}(2^{(m+1)s})$ contains $\mathrm{GF}(2^s)$ as a subfield, it is possible to divide the $2^{(m+1)s} - 1$ nonzero elements of $\mathrm{GF}(2^{(m+1)s})$ into $N(m, s) = (2^{(m+1)s} - 1)/(2^s - 1) = 2^{ms} + 2^{(m-1)s} + \cdots + 2^s + 1$ disjoint subsets of $2^s - 1$ elements each. Then each subset is regarded as a point in $\mathrm{PG}(m, 2^s)$, the m-dimensional projective geometry over the finite field $\mathrm{GF}(2^s)$.

Two distinct points $\mathbf{p}_1$ and $\mathbf{p}_2$ of PG$(m, 2^s)$ define a unique line $L(\mathbf{p}_1, \mathbf{p}_2)$ passing through $\mathbf{p}_1$ and $\mathbf{p}_2$. The line $L(\mathbf{p}_1, \mathbf{p}_2)$ contains a total of $(2^{2s} - 1)/(2^s - 1) = 2^s + 1$ points $\mathbf{p}_i$ satisfying the equation

$$\mathbf{p}_i = \alpha_1 \mathbf{p}_1 + \alpha_2 \mathbf{p}_2, \tag{3}$$

with α_1 and α_2 in GF(2^s), and not both zero. Given a point $\mathbf{p}$, the $N(m, s) - 1 = 2^s(2^{ms} - 1)/(2^s - 1)$ other points in PG$(m, 2^s)$ can be divided into subsets of 2^s points each based on Eq. (3) such that each subset corresponds to a distinct line of PG$(m, 2^s)$ containing $\mathbf{p}$. As a result, there are $(2^{ms} - 1)/(2^s - 1)$ lines intersecting on each point $\mathbf{p}$ of PG$(m, 2^s)$ and the total number of lines in PG$(m, 2^s)$ is $N(m, s)/(2^s + 1) \cdot (2^{ms} - 1)/(2^s - 1)$.

As for EG codes, we associate to PG$(m, 2^s)$ an incidence matrix H whose rows are associated with the $N(m, s)/(2^s + 1) \cdot (2^{ms} - 1)/(2^s - 1)$ lines of PG$(m, 2^s)$, and whose columns are associated with the $N(m, s)$ points of PG$(m, 2^s)$. This incidence matrix H defines the dual space of a linear block PG code C denoted $C_{PG}(m, s)$ of length $N = N(m, s)$. As for EG codes, the dimension K and minimum distance $d_{\min}$ of $C_{\text{PG}}(m, s)$ depend on the values m and s, with $d_{\min} \geq (2^{ms} - 1)/(2^s - 1) + 1$. Also, PG codes are cyclic codes and have structural properties similar to those of EG codes, namely (1) every row of H contains exactly $2^s + 1$ "ones," (2) every column of H contains exactly $(2^{ms} - 1)/(2^s - 1)$ "ones," and (3) any two rows of H have at most one "one" in common.

For $m = 2$, the 2-dimensional PG codes $C_{\text{PG}}(2, s)$ are equivalent to the DSC codes described by Weldon [9] and have the following parameters:

$$N = 2^{2s} + 2^s + 1$$
$$K = 2^{2s} + 2^s - 3^s$$
$$d_{\min} = 2^s + 2$$

For $1 \leq \mu < m$, a μ-flat of PG$(m, 2^s)$ is defined by the set of points $\mathbf{p}_i$ of the form

$$\mathbf{p}_i = \alpha_1 \mathbf{p}_1 + \alpha_2 \mathbf{p}_2 + \cdots + \alpha_{\mu+1} \mathbf{p}_{\mu+1} \tag{4}$$

with for $1 \leq i \leq \mu$, $\alpha_i \in$ GF(2^s). The dual code of a linear block PG code C is defined by the incidence matrix H whose rows and columns are associated with the μ-flats and the points of PG$(m, 2^s)$, respectively. Further extensions of this definition are also possible.

Some EG and PG codes of length $N \leq 1000$ are given in Table 1. Note that with respect to that table, for a given triplet (m, s, μ), the corresponding EG code is commonly referred to as the "$(\mu - 1, s)$th-order binary EG code," while the corresponding PG code is commonly referred to as the "(μ, s)th-order binary PG" code.

Table 1. Some EG and PG Codes of Length $N \leq 1000$

			EG Codes			PG Codes		
m	s	μ	N	K	$d_{\min}$	N	K	$d_{\min}$
2	2	1	15	7	5	21	11	6
2	3	1	63	37	9	73	45	10
3	2	1	63	13	21	85	24	22
3	2	2	63	48	5	85	68	6
2	4	1	255	175	17	273	191	18
4	2	1	255	21	85	341	45	86
4	2	2	255	127	21	341	195	22
4	2	3	255	231	5	341	315	6
3	3	1	511	139	73	585	184	74
3	3	2	511	448	9	585	520	10

2.3. EG and PG Codes Viewed as LDPC Codes

An (J, L) LDPC code is defined by as the null space of a matrix H such that (1) each column consists of J "ones," (2) each row consists of L "ones," and (3) no two rows have more than one "one" in common [24] (note that this last property is not explicitly stated in Gallager's original definition). It is readily seen that both EG and PG codes satisfy this definition. However, compared with their counterpart LDPC codes presented by Gallager [24], EG and PG codes have several fundamental differences: (1) their values J and L are in general much larger, (2) their total number of check sums defining H is larger, and (3) they usually have a cyclic structure. On the basis of these observations, new LDPC codes can be constructed from EG and PG codes by splitting the rows or columns of H or its transposed matrix [17]. If these operations are done in a systematic and uniform way, new LDPC codes are obtained. Interestingly, these codes become quasicyclic, so that fast encoding based on LFSRs is still possible.

As an example, let us consider $C_{\text{EG}}(2, 6)$, a (4095,3367) EG code with $J = L = 64$, and let us split each column of H into 16 columns of weight 4 each. Then we obtain a (65520,61425) EG extended code with $J = 4$ and $L = 64$ [17]. Note that this new code has rate $R = 0.9375 = 1 - J/L$.

3. DECODING METHODS

3.1. Decoding of EG and PG Codes with $\mu = 1$

3.1.1. One-Stage Decoding. Let us assume that an information sequence $\mathbf{u} = (u_0, u_1, \ldots, u_{K-1})$ is encoded into a codeword $\mathbf{v} = (v_0, v_1, \ldots, v_{N-1})$ using an EG, extended EG or PG (N, K) code associated with $\mu = 1$. This codeword is then transmitted over a noisy communications channel and at the receiver, either a hard-decision received vector $\mathbf{y} = (y_0, y_1, \ldots, y_{N-1})$, or a soft-decision estimate $\mathbf{r} = (r_0, r_1, \ldots, r_{N-1})$ is available.

For any codeword $\mathbf{v}$ of the code considered, each row $\mathbf{h}$ of H satisfies the checksum $\mathbf{v} \cdot \mathbf{h} = 0$. Based on properties 2 and 3 of extended EG and PG codes described in Section 2, each position i, $0 \leq i \leq N - 1$ is associated with a set $B(i)$ of $(2^{ms} - 1)/(2^s - 1)$ checksums orthogonal on that position [i.e., no other position than i appears more than once in $B(i)$]. As a result, each checksum can be used to provide uncorrelated information about bit i, and since all checksums have the same weight (see property 1), the same amount of information is provided by each of them. This suggests the following algorithm:

1. For $0 \le i \le N-1$, evaluate all checksums $\mathbf{y} \cdot \mathbf{h}$ in $B(i)$.
2. Let $B(i)^+$ and $B(i)^-$ represent the sets of satisfied and unsatisfied checksums in $B(i)$, respectively.
   If $|B(i)^+| \ge |B(i)^-|$, decode $\hat{v}_i = y_i$.
   Else, decode $\hat{v}_i = y_i \oplus 1$.

Since a majority vote is taken for each bit, this decoding method is known as "majority-logic decoding." It was first introduced by Reed [11] to decode Reed–Muller codes. Furthermore, since for the codes considered, all N bits can be decoded by this algorithm, these codes belong to the class of "one-step majority-logic decodable codes." Majority-logic decoding is a very simple algebraic decoding method that fits high-speed implementations since only binary operations are involved. It allows one to correct any error pattern of Hamming weight $(|B(i)|-1)/2 = \lfloor (2^{ms-1}-2^{s-1})/(2^s-1) \rfloor$, which corresponds to the guaranteed error-correcting capability $t = \lfloor (d_{\min}-1)/2 \rfloor$ of extended EG and PG codes. Since for EG codes, $|B(i)| = (2^{ms}-1)/(2^s-1)-1$ is even, this algorithm has to be refined in order to consider the case where $|B(i)|/2$ checksums are satisfied and $|B(i)|/2$ checksums are unsatisfied. In that case, we always decode $\hat{v}_i = y_i$, so that the guaranteed error-correcting capability $t = |B(i)|/2 = \lfloor (d_{\min}-1)/2 \rfloor$ of EG codes is also achieved.

The sets $B(i)$, $0 \le i \le N-1$ can also be used to evaluate the a posteriori probabilities q_i that the values y_i are in error, based on the a priori error probabilities p_i defined by the channel model considered. For example, $p_i = \varepsilon$ for a binary symmetric channel (BSC) with crossover probability ε and $p_i = e^{-|4r_i/N_0|}/(1+e^{-|4r_i/N_0|})$ for an additive white Gaussian noise (AWGN) channel with associated variance $N_0/2$. Since the checksums in each set $B(i)$, $0 \le i \le N-1$ are orthogonal on position i, it follows that

$$q_i = \left(1 + \left(\frac{1-p_i}{p_i}\right) \prod_{j=1}^{|B(i)|} \left(\frac{1-m_{j,i}}{m_{j,i}}\right)^{\sigma_j \oplus 1} \left(\frac{m_{j,i}}{1-m_{j,i}}\right)^{\sigma_j}\right)^{-1} \tag{5}$$

where σ_j is the result of the jth checksum in $B(i)$, and

$$m_{j,i} = \left(\frac{1}{2}\right)\left(1 - \prod_{i' \in N(j)\backslash i} (1-2p_{i'})\right)$$

$N(j)\backslash i$ representing the set of nonzero positions corresponding to checksum σ_j, but position i. Note that $m_{j,i}$ represents the probability that the sum of the bits in $N(j)\backslash i$ mismatches the transmitted bit i. The corresponding decoding algorithm is given as follows:

1. For $0 \le i \le N-1$, evaluate all checksums $\mathbf{y} \cdot \mathbf{h}$ in $B(i)$.
2. Evaluate q_i based on Eq. (5).
   If $q_i \le 0.5$, decode $\hat{v}_i = y_i$.
   Else, decode $\hat{v}_i = y_i \oplus 1$.

This decoding method, known as "a posteriori probability (APP) decoding," was introduced by Massey [25].

3.1.2. Iterative Decoding. We observe that either majority-logic decoding, or APP decoding, provides for each of the N initial inputs a new estimate of this quantity, based on exactly the same set of constraints. This is due to the fact that all EG, extended EG and PG codes presented in Section 2 are one-step majority-logic decodable codes. Consequently, this observation suggests a straightforward heuristic approach; we may use these estimates as new inputs and iterate the decoding process. If for $0 \le i \le N-1$, $x_i^{(0)}$ represents the a priori information about position i and for $l \ge 1$, and $x_i^{(l)}$ represents the a posteriori estimate of position i evaluated at iteration l, then iterative decoding is achieved on the basis of:

$$x_i^{(l)} = x_i^{(0)} + f_i(\mathbf{x}^{(l-1)}) \tag{6}$$

where $f_i()$ represents the function used to compute the a posteriori estimate of position i from the a priori inputs and $\mathbf{x}^{(l-1)}$ represents the vector of a posteriori values evaluated at iteration $(l-1)$. We notice that (6) implicitly assumes an iterative updating of the original a priori values on the basis of the latest a posteriori estimates. Iterative majority-logic decoding (also known as "iterative bit flipping" decoding) and iterative APP decoding approaches were first proposed [24] to decode LDPC codes. Refined versions of this heuristic approach for some EG, extended EG and PG codes have been proposed [14,15].

Although these iterative decoding methods allow one to achieve significant performance improvements on the first iteration, they fall short of optimum decoding. The main reason is the introduction and propagation of correlated values from the second iteration, which is readily explained as follows. Let us assume that positions i and j contribute to the same checksum. Then, according to (6), for $l \ge 2$, $x_j^{(l-1)}$ depends on $x_i^{(0)}$ and consequently, $x_i^{(0)}$ and $f_i(\mathbf{x}^{(l-1)})$ are correlated when evaluating $x_i^{(l)}$.

This problem can be overcome by computing for each position i as many a posteriori values as checksums intersecting on that position. This method, implicitly contained in the performance analysis of Ref. 24, is also known as "belief propagation" (BP) [26]. A good presentation of BP decoding can be found in Ref. 27. BP decoding of long LDPC codes has been showed to closely approach the Shannon capacity of the BSC and AWGN channel [28,29]. For $0 \le i \le N-1$, let $x_i^{(0)}$ represent the a priori information about position i and for $1 \le j \le |B(i)|$, and $l \ge 1$ let $x_{j,i}^{(l)}$ represent the a posteriori estimate of position i computed at iteration l based on the checksums other than checksum j in $B(i)$. Then at iteration l, BP decoding evaluates

$$x_{j,i}^{(l)} = x_i^{(0)} + f_{j,i}(\mathbf{x}^{(l-1)}(i)) \tag{7}$$

where $f_{j,i}()$ represents the function used to compute the a posteriori estimate of position i from the a priori inputs other than that in checksum j and $\mathbf{x}^{(l-1)}(i)$ represents the vector of a posteriori values evaluated at iteration $(l-1)$ by discarding the check sums containing position i. BP iterative decoding achieves optimum decoding if the Tanner graph representation [30] of the code considered contains no loop.

BP decoding of EG, extended EG and PG codes has been investigated [16] and its application to the LDPC codes presented in Section 2.3 was presented in a 1999 paper [17]. Despite the presence of loops in the Tanner graph of all these codes, near-optimum performance can still be achieved.

3.2. General Decoding of EG and PG Codes

One-stage decoding of EG and PG codes in general is achieved by repeating the method described in Section 3.1.1 in at most μ steps. At each step, checksums with less and less common positions are estimated until at most only one common position between any two checksums remains, as in Section 3.1.1. A good description of majority-logic decoding of EG and PG codes can be found in Chaps. 7 and 8 of Ref. 23.

On the other hand, iterative decoding of EG and PG codes in general is not as straightforwardly generalizable. In general, a multistep decoding method is also required to obtain the a posteriori error probabilities of all N bits, which inherently complicates the application of iterative decoding.

BIOGRAPHY

Marc P.C. Fossorier was born in Annemasse, France, on March 8, 1964. He received the B.E. degree from the National Institute of Applied Sciences (I.N.S.A.) Lyon, France, in 1987, and the M.S. and Ph.D. degrees from the University of Hawaii at Manoa, Honolulu, in 1991 and 1994, all in electrical engineering.

In 1996, he joined the Faculty of the University of Hawaii, Honolulu, as an Assistant Professor of Electrical Engineering. He was promoted to Associate Professor in 1999.

His research interests include decoding techniques for linear codes, communication algorithms, combining coding and equalization for ISI channels, magnetic recording and statistics. He coauthored (with S. Lin, T. Kasami, and T. Fujiwara) the book *Trellises and Trellis-Based Decoding Algorithms*, (Kluwer Academic Publishers, 1998).

Dr. Fossorier is a recipient of a 1998 NSF Career Development award. He has served as editor for the *IEEE Transactions on Communications* since 1996, as editor for the *IEEE Communications Letters* since 1999, and he currently is the treasurer of the IEEE Information Theory Society. Since 2002, he has also been a member of the Board of Governors of the IEEE Information Theory Society.

BIBLIOGRAPHY

1. L. D. Rudolph, *Geometric Configuration and Majority Logic Decodable Codes*, M.E.E. thesis, Univ. Oklahoma, Norman, 1964.
2. L. D. Rudolph, A class of majority logic decodable codes, *IEEE Trans. Inform. Theory* **13**: 305–307 (April 1967).
3. P. Delsarte, A geometric approach to a class of cyclic codes, *J. Combin. Theory* **6**: 340–358 (1969).
4. P. Delsarte, J. M. Goethals, and J. MacWilliams, On GRM and related codes, *Inform. Control* **16**: 403–442 (July 1970).
5. S. Lin, Shortened finite geometry codes, *IEEE Trans. Inform. Theory* **18**: 692–696 (July 1972).
6. S. Lin, Multifold Euclidean geometry codes, *IEEE Trans. Inform. Theory* **19**: 537–548 (July 1973).
7. C. R. P. Hartmann, J. B. Ducey, and L. D. Rudolph, On the structure of generalized finite geometry codes, *IEEE Trans. Inform. Theory* **20**: 240–252 (March 1974).
8. S. Lin and K. P. Yiu, An improvement to multifold euclidean geometry codes, *Inform. Control* **28**: (July 1975).
9. E. J. Weldon, Difference-set cyclic codes, *Bell Syst. Tech. J.* **45**: 1045–1055 (Sept. 1966).
10. D. J. Costello, Jr., J. Hagenauer, H. Imai, and S. B. Wicker, Applications of error-control coding, *Commemorative Issue, IEEE Trans. Inform. Theory* **44**: 2531–2560 (Oct. 1998).
11. I. S. Reed, A class of multiple-error-correcting codes and the decoding scheme, *IRE Trans. Inform. Theory* **4**: 38–49 (Sept. 1954).
12. C. L. Chen, On majority-logic decoding of finite geometry codes, *IEEE Trans. Inform. Theory* **17**: 332–336 (May 1971).
13. T. Kasami and S. Lin, On majority-logic decoding for duals of primitive polynomial codes, *IEEE Trans. Inform. Theory* **17**: 322–331 (May 1971).
14. K. Yamaguchi, H. Iizuka, E. Nomura, and H. Imai, Variable threshold soft decision decoding, *IEICE Trans. Elect. Commun.* **72**: 65–74 (Sept. 1989).
15. R. Lucas, M. Bossert, and M. Breitbach, On iterative soft-decision decoding of linear binary block codes and product codes, *IEEE J. Select. Areas Commun.* **16**: 276–296 (Feb. 1998).
16. R. Lucas, M. Fossorier, Y. Kou, and S. Lin, Iterative decoding of one-step majority logic decodable codes based on belief propagation, *IEEE Trans. Commun.* **48**: 931–937 (June 2000).
17. Y. Kou, S. Lin, and M. Fossorier, Low density parity check codes based on finite geometries: A rediscovery and more, *IEEE Trans. Inform. Theory* (Oct. 1999).
18. H. B. Mann, *Analysis and Design of Experiments*, Dover, New York, 1949.
19. R. D. Carmichael, *Introduction to the Theory of Groups of Finite Order*, Dover, New York, 1956.
20. W. W. Peterson and E. J. Weldon, Jr., *Error-Correcting Codes*, 2nd ed., MIT Press, Cambridge, MA, 1972.
21. I. F. Blake and R. C. Mullin, *The Mathematical Theory of Coding*, Academic Press, New York, 1975.
22. F. J. MacWilliams and N. J. A. Sloane, *The Theory of Error-Correcting Codes*, North-Holland Mathematical Library, Amsterdam, 1977.
23. S. Lin and D. J. Costello, Jr., *Error Control Coding, Fundamentals and Applications*, Prentice-Hall, Englewood Cliffs, NJ, 1983.
24. R. G. Gallager, *Low-Density Parity-Check Codes*, MIT Press, Cambridge, MA, 1963.
25. J. L. Massey, *Threshold Decoding*, MIT Press, Cambridge, MA, 1963.
26. J. Pearl, *Probabilistic Reasoning in Intelligent Systems: Networks of Plausible Inference*, Morgan Kaufmann, San Mateo, CA, 1988.
27. D. J. C. MacKay, Good error-correcting codes based on very sparse matrices, *IEEE Trans. Inform. Theory* **45**: 399–432 (March 1999).
28. T. Richardson, A. Shokrollahi, and R. Urbanke, Design of probably good low-density parity check codes, *IEEE Trans. Inform. Theory* (in press).

29. S. Y. Chung, G. D. Forney, Jr., T. Richardson, and R. Urbanke, On the design of low-density parity-check codes within 0.0051 dB of the Shannon limit, *IEEE Commun. Lett.* (in press).

30. R. M. Tanner, A recursive approach to low complexity codes, *IEEE Trans. Inform. Theory* **27**: 533–547 (Sept. 1981).

FREQUENCY AND PHASE MODULATION

MASOUD SALEHI
Northeastern University
Boston, Massachusetts

1. INTRODUCTION

Analog angle modulation methods are modulation methods in which the information is carried by the phase of a sinusoidal. Angle modulation can be carried out either by frequency modulation or by phase modulation. In frequency modulation (FM) systems, the frequency of the carrier f_c is changed by the message signal, and in phase modulation (PM) systems the phase of the carrier is changed according to the variations in the message signal. Frequency and phase modulation are obviously quite nonlinear, and very often they are jointly referred to as *angle modulation methods*. As we will see, angle modulation, due to its inherent nonlinearity, is rather difficult to analyze. In many cases only an approximate analysis can be done. Another property of angle modulation is its bandwidth expansion property. Frequency and phase modulation systems generally expand the bandwidth such that the effective bandwidth of the modulated signal is usually many times the bandwidth of the message signal.[1] With a higher implementation complexity and a higher bandwidth occupancy, one would naturally raise a question as to the usefulness of these systems. As we will show, the major benefit of these systems is their high degree of noise immunity. In fact, these systems trade off bandwidth for high noise immunity. That is the reason why FM systems are widely used in high-fidelity music broadcasting and point-to-point communication systems where the transmitter power is quite limited. Another benefit of theses systems is their constant envelop property. Unlike amplitude modulation, these systems have a constant amplitude which makes them attractive choices to use with nonlinear amplification devices (class C amplifiers or TWT devices).

FM radio broadcasting was first initiated by Edwin H. Armstrong in 1935 in New York on a frequency of 42.1 MHz. At that time when FM broadcasting started. Later with the advent of TV broadcasting the 88–108-MHz frequency band was used for FM broadcasting. Commercial stereo FM broadcasting started in Chicago in 1961.

[1] Strictly speaking, the bandwidth of the modulated signal, as will be shown later, is infinite. That is why we talk about the *effective bandwidth*.

2. REPRESENTATION OF FM AND PM SIGNALS

An angle modulated signal in general can be expressed as

$$u(t) = A_c \cos(\theta(t))$$

$\theta(t)$ is the *phase* of the signal and its *instantaneous frequency* $f_i(t)$ is given by

$$f_i(t) = \frac{1}{2\pi}\frac{d}{dt}\theta(t) \tag{1}$$

Since $u(t)$ is a bandpass signal it can be represented as

$$u(t) = A_c \cos(2\pi f_c t + \phi(t)) \tag{2}$$

and therefore

$$f_i(t) = f_c + \frac{1}{2\pi}\frac{d}{dt}\phi(t) \tag{3}$$

If $m(t)$ is the message signal, then in a PM system, we have

$$\phi(t) = k_p m(t) \tag{4}$$

and in an FM system we have

$$f_i(t) - f_c = k_f m(t) = \frac{1}{2\pi}\frac{d}{dt}\phi(t) \tag{5}$$

where k_p and k_f are phase and frequency *deviation constants*. From the above relationships we have

$$\phi(t) = \begin{cases} k_p m(t), & \text{PM} \\ 2\pi k_f \int_{-\infty}^{t} m(\tau)\,d\tau, & \text{FM} \end{cases} \tag{6}$$

This expression shows the close relation between FM and PM. This close relationship makes it possible to analyze these systems in parallel and only emphasize their main differences. The first interesting result observed from the above is that if we phase modulate the carrier with the integral of a message, it is equivalent to frequency modulation of the carrier with the original message. On the other hand, the above relation can be expressed as

$$\frac{d}{dt}\phi(t) = \begin{cases} k_p \frac{d}{dt} m(t), & \text{PM} \\ 2\pi m(t), & \text{FM} \end{cases} \tag{7}$$

which shows that if we frequency-modulate the carrier with the derivative of a message the result is equivalent to phase modulation of the carrier with the message itself. Figure 1 shows the above relation between FM and PM. Figure 2 illustrates a square-wave signal and its integral, a sawtooth signal, and their corresponding FM and PM signals.

The demodulation of an FM signal involves finding the instantaneous frequency of the modulated signal and then subtracting the carrier frequency from it. In the demodulation of PM, the demodulation process is done by finding the phase of the signal and then recovering $m(t)$. The *maximum phase deviation* in a PM system is given by

$$\Delta\phi_{\max} = k_p \max[|m(t)|] \tag{8}$$

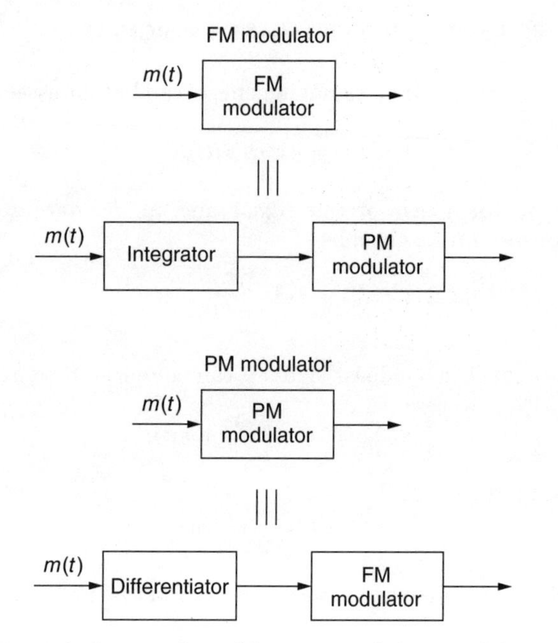

Figure 1. A comparison of frequency and phase modulators.

and the *maximum frequency deviation* in an FM system is given by

$$\Delta f_{\max} = k_f \max[|m(t)|] \tag{9}$$

We now define the *modulation index* for a general nonsinusoidal signal $m(t)$ as

$$\beta_p = k_p \max[|m(t)|] \tag{10}$$

$$\beta_f = \frac{k_f \max[|m(t)|]}{W} \tag{11}$$

where W denotes the bandwidth of the message signal $m(t)$. In terms of the maximum phase and frequency deviation $\Delta\phi_{\max}$ and $\Delta f_{\max}$, we have

$$\beta_p = \Delta\phi_{\max} \tag{12}$$

$$\beta_f = \frac{\Delta f_{\max}}{W} \tag{13}$$

2.1. Narrowband Angle Modulation

If in an angle modulation[2] system the deviation constants k_p and k_f and the message signal $m(t)$ are such that for all t we have $\phi(t) \ll 1$, then we can use a simple approximation to expand $u(t)$ as

$$\begin{aligned} u(t) &= A_c \cos 2\pi f_c t \cos\phi(t) - A_c \sin 2\pi f_c t \sin\phi(t) \\ &\approx A_c \cos 2\pi f_c t - A_c\phi(t) \sin 2\pi f_c t \end{aligned} \tag{14}$$

This last equation shows that in this case the modulated signal is very similar to a conventional AM signal of the form $A_c(1 + m(t))\cos(2\pi f_c t)$. The bandwidth of this signal is similar to the bandwidth of a conventional AM signal, which is twice the bandwidth of the message signal. Of course, this bandwidth is only an approximation to the real bandwidth of the FM signal. A phasor diagram for this signal and the comparable conventional AM signal are given in Fig. 3. Note that compared to conventional AM, the narrowband angle modulation scheme has far less amplitude variations. Of course the angle modulation system has constant amplitude and, hence, there should be no amplitude variations in the phasor diagram representation of the system. The slight variations here are due to the first-order approximation that we have used for the expansions of $\sin(\phi(t))$ and $\cos(\phi(t))$. As we will see later, the narrowband angle modulation method does not provide any better noise immunity compared to

[2] Also known as *low-index angle modulation*.

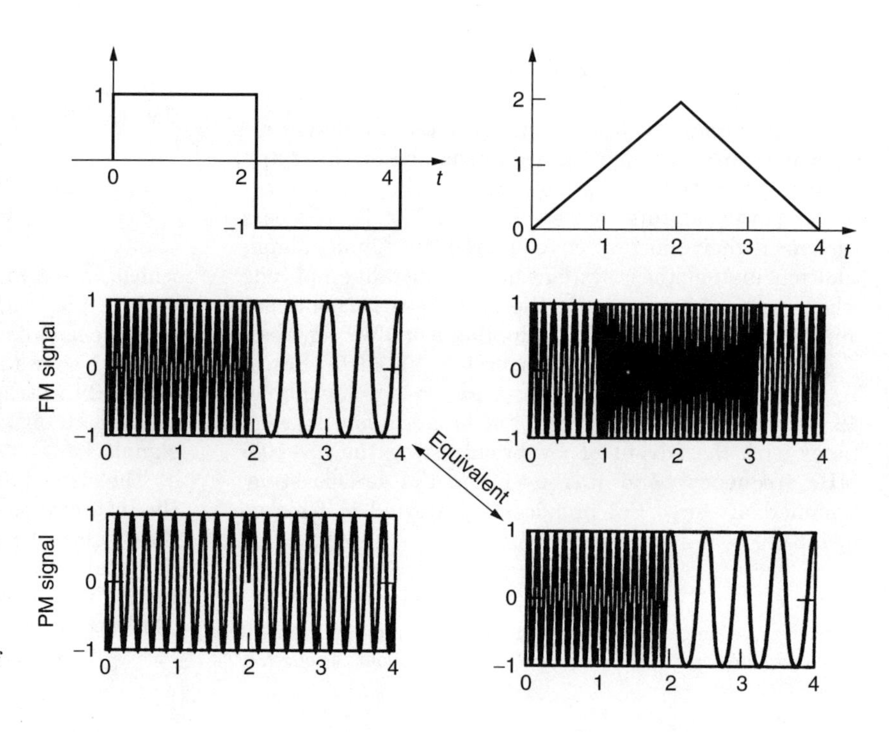

Figure 2. Frequency and phase modulation of square and sawtooth waves.

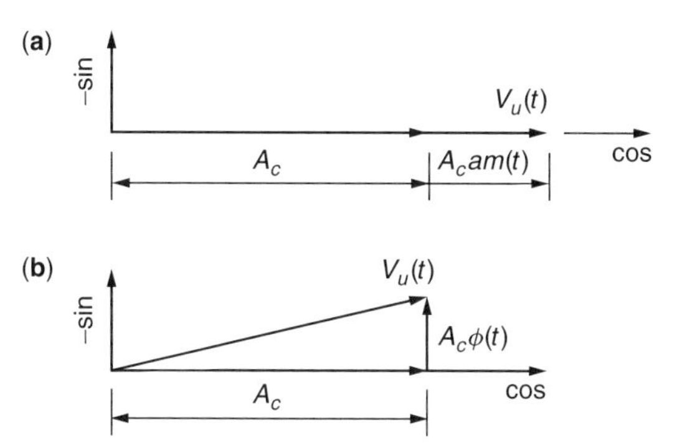

Figure 3. Phasor diagram for the conventional AM and narrowband angle modulation.

a conventional AM system, and therefore, narrowband angle modulation is not used in broadcasting. However, these systems can be used as an intermediate stage for generation of wideband angle-modulated signals, as we will discuss in Section 4.

3. SPECTRAL CHARACTERISTICS OF ANGLE-MODULATED SIGNALS

Because of the inherent nonlinearity of angle modulation systems, the precise characterization of their spectral properties, even for simple message signals, is mathematically intractable. Therefore, the derivation of the spectral characteristics of these signals usually involves the study of very simple modulating signals and certain approximations. Then, the results are generalized to the more complicated messages. We will study the spectral characteristics of an angle-modulated signal when the modulating signal is a sinusoidal signal and when the modulating signal is a general periodic signal. Then we generalize these results to an arbitrary modulating signal.

3.1. Angle Modulation by a Sinusoidal Signal

Let us begin with the case where the message signal is a sinusoidal signal $m(t) = a\cos(2\pi f_m t)$. In PM we have

$$\phi(t) = k_p m(t) = k_p a \cos(2\pi f_m t) \tag{15}$$

and in FM we have

$$\phi(t) = 2\pi k_f \int_{-\infty}^{t} m(\tau)\,d\tau = \frac{k_f a}{f_m}\sin(2\pi f_m t) \tag{16}$$

Therefore, the modulated signals will be

$$u(t) = \begin{cases} A_c\cos(2\pi f_c t + k_p a\cos(2\pi f_m t)), & \text{PM} \\ A_c\cos\left(2\pi f_c t + \dfrac{k_f a}{f_m}\sin(2\pi f_m t)\right), & \text{FM} \end{cases} \tag{17}$$

and the modulation indices are

$$\beta_p = k_p a \tag{18}$$

$$\beta_f = \frac{k_f a}{f_m} \tag{19}$$

Using β_p and β_f, we have

$$u(t) = \begin{cases} A_c\cos(2\pi f_c t + \beta_p\cos(2\pi f_m t)), & \text{PM} \\ A_c\cos(2\pi f_c t + \beta_f\sin(2\pi f_m t)), & \text{FM} \end{cases} \tag{20}$$

As shown above, the general form of an angle-modulated signal for the case of a sinusoidal message is

$$u(t) = A_c\cos(2\pi f_c t + \beta\sin 2\pi f_m t) \tag{21}$$

where β is the modulation index that can be either β_p or β_f. Therefore the modulated signal can be written as

$$u(t) = \text{Re}\,(A_c e^{j2\pi f_c t} e^{j\beta\sin 2\pi f_m t}) \tag{22}$$

Since $\sin 2\pi f_m t$ is periodic with period $T_m = \dfrac{1}{f_m}$, the same is true for the complex exponential signal

$$e^{j\beta\sin 2\pi f_m t}$$

Therefore, it can be expanded in a Fourier series representation. The Fourier series coefficients are obtained from the integral

$$\begin{aligned} c_n &= f_m \int_0^{\frac{1}{f_m}} e^{j\beta\sin 2\pi f_m t} e^{-jn2\pi f_m t}\,dt \\ &\overset{u=2\pi f_m t}{=} \frac{1}{2\pi}\int_0^{2\pi} e^{j\beta(\sin u - nu)}\,du \end{aligned} \tag{23}$$

This latter integral is a well-known integral known as the *Bessel function of the first kind of order n* and is denoted by $J_n(\beta)$. Therefore, we have the Fourier series for the complex exponential as

$$e^{j\beta\sin 2\pi f_m t} = \sum_{n=-\infty}^{\infty} J_n(\beta) e^{j2\pi n f_m t} \tag{24}$$

By substituting (24) in (22), we obtain

$$\begin{aligned} u(t) &= \text{Re}\left(A_c \sum_{n=-\infty}^{\infty} J_n(\beta) e^{j2\pi n f_m t} e^{j2\pi f_c t}\right) \\ &= \sum_{n=-\infty}^{\infty} A_c J_n(\beta)\cos(2\pi(f_c + nf_m)t) \end{aligned} \tag{25}$$

This relation shows that even in this very simple case where the modulating signal is a sinusoid of frequency f_m, the angle-modulated signal contains all frequencies of the form $f_c + nf_m$ for $n = 0, \pm 1, \pm 2, \ldots$. Therefore, the actual bandwidth of the modulated signal is infinite. However, the amplitude of the sinusoidal components of frequencies $f_c \pm nf_m$ for large n is very small and their contribution to the total power in the signal is low. Hence, we can define an finite *effective bandwidth* for the modulated signal, as the bandwidth that contains the component frequencies that contain a certain percentage (usually 98% or 99%) of the total power in the signal. A series expansion for the Bessel function is given by

$$J_n(\beta) = \sum_{k=0}^{\infty} \frac{(-1)^k \left(\frac{\beta}{2}\right)^{n+2k}}{k!(k+n)!} \tag{26}$$

The expansion here shows that for small β, we can use the approximation

$$J_n(\beta) \approx \frac{\beta^n}{2^n n!} \tag{27}$$

Thus for a small modulation index β, only the first sideband corresponding to $n = 1$ is of importance. Also, using the expansion in (26), it is easy to verify the following symmetry properties of the Bessel function.

$$J_{-n}(\beta) = \begin{cases} J_n(\beta), & n \text{ even} \\ -J_n(\beta), & n \text{ odd} \end{cases} \tag{28}$$

Plots of $J_n(\beta)$ for various values of n are given in Fig. 4, and a tabulation of the values of the Bessel function is given in Table 1. The underlined and doubly underlined entries for each β indicate the minimum value of n that guarantees the signal will contain at least 70% or 98% of the total power, respectively.

In general the effective bandwidth of an angle-modulated signal, which contains at least 98% of the signal power, is approximately given by the relation

$$B_c = 2(\beta + 1)f_m \tag{29}$$

where β is the modulation index and f_m is the frequency of the sinusoidal message signal.

It is instructive to study the effect of the amplitude and frequency of the sinusoidal message signal on the bandwidth and the number of harmonics in the modulated signal. Let the message signal be given by

$$m(t) = a\cos(2\pi f_m t) \tag{30}$$

The bandwidth[3] of the modulated signal is given by

$$B_c = 2(\beta + 1)f_m = \begin{cases} 2(k_p a + 1)f_m, & \text{PM} \\ 2\left(\dfrac{k_f a}{f_m} + 1\right)f_m, & \text{FM} \end{cases} \tag{31}$$

or

$$B_c = \begin{cases} 2(k_p a + 1)f_m, & \text{PM} \\ 2(k_f a + f_m), & \text{FM} \end{cases} \tag{32}$$

This relation shows that increasing a, the amplitude of the modulating signal, in PM and FM has almost the same effect on increasing the bandwidth B_c. On the other hand, increasing f_m, the frequency of the message signal, has a more profound effect in increasing the bandwidth of a PM signal as compared to an FM signal. In both PM and FM the bandwidth B_c increases by increasing f_m, but in PM this increase is a proportional increase and in FM this is only an additive increase which in most cases of interest, (for large β) is not substantial. Now if we look at the number of harmonics in the bandwidth (including the carrier) and denote it with M_c, we have

$$M_c = 2\lfloor\beta\rfloor + 3 = \begin{cases} 2\lfloor k_p a\rfloor + 3, & \text{PM} \\ 2\left\lfloor \dfrac{k_f a}{f_m} \right\rfloor + 3, & \text{FM} \end{cases} \tag{33}$$

Increasing the amplitude a increases the number of harmonics in the bandwidth of the modulated signal in both cases. However, increasing f_m, has no effect on the number of harmonics in the bandwidth of the PM signal and decreases the number of harmonics in the FM signal almost linearly. This explains the relative insensitivity of the bandwidth of the FM signal to the message frequency. On one hand, increasing f_m decreases the number of harmonics in the bandwidth and, at the same time, increases the spacing between the harmonics. The net effect is a slight increase in the bandwidth. In

[3] From now on, by bandwidth we mean effective bandwidth unless otherwise stated.

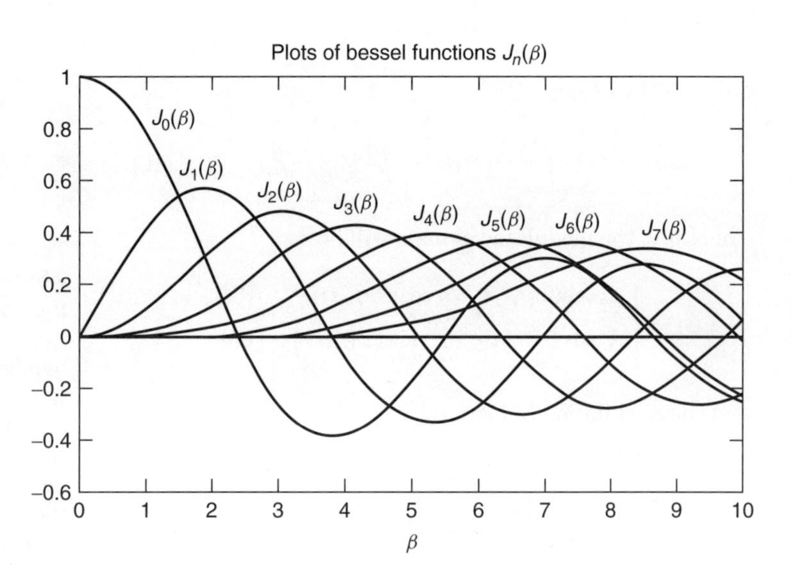

Figure 4. Bessel functions for various values of n.

Table 1. Table of Bessel Function Values

n	$\beta = 0.1$	$\beta = 0.2$	$\beta = 0.5$	$\beta = 1$	$\beta = 2$	$\beta = 5$	$\beta = 8$	$\beta = 10$
0	0.997	0.990	0.938	0.765	0.224	−0.178	0.172	−0.246
1	0.050	0.100	0.242	0.440	0.577	−0.328	0.235	0.043
2	0.001	0.005	0.031	0.115	0.353	0.047	−0.113	0.255
3	—	—	—	0.020	0.129	0.365	−0.291	0.058
4	—	—	—	0.002	0.034	0.391	−0.105	−0.220
5	—	—	—	—	0.007	0.261	0.186	−0.234
6	—	—	—	—	0.001	0.131	0.338	−0.014
7	—	—	—	—	—	0.053	0.321	0.217
8	—	—	—	—	—	0.018	0.223	0.318
9	—	—	—	—	—	0.006	0.126	0.292
10	—	—	—	—	—	0.001	0.061	0.207
11	—	—	—	—	—	—	0.026	0.123
12	—	—	—	—	—	—	0.010	0.063
13	—	—	—	—	—	—	0.003	0.029
14	—	—	—	—	—	—	0.001	0.012
15	—	—	—	—	—	—	—	0.005
16	—	—	—	—	—	—	—	0.002

Source: From Ziemer and Tranter (1990) © Houghton Mifflin, reprinted by permission.

PM, however, the number of harmonics remains constant and only the spacing between them increases. Therefore, the net effect is a linear increase in bandwidth. Figure 5 shows the effect of increasing the frequency of the message in both FM and PM.

3.2. Angle Modulation by a Periodic Message Signal

To generalize the abovementioned results, we now consider angle modulation by an arbitrary periodic message signal $m(t)$. Let us consider a PM modulated signal where

$$u(t) = A_c \cos(2\pi f_c t + \beta m(t)) \tag{34}$$

We can write this as

$$u(t) = A_c \mathrm{Re}\, [e^{j2\pi f_c t} e^{j\beta m(t)}] \tag{35}$$

We are assuming that $m(t)$ is periodic with period $T_m = \frac{1}{f_m}$. Therefore $e^{j\beta m(t)}$ will be a periodic signal with the same period and we can find its Fourier series expansion as

$$e^{j\beta m(t)} = \sum_{n=-\infty}^{\infty} c_n e^{j2\pi n f_m t} \tag{36}$$

where

$$\begin{aligned} c_n &= \frac{1}{T_m}\int_0^{T_m} e^{j\beta m(t)} e^{-j2\pi n f_m t}\, dt \\ &\overset{u=2\pi f_m t}{=} \frac{1}{2\pi}\int_0^{2\pi} e^{j\left[\beta m\left(\frac{u}{2\pi f_m}\right) - nu\right]}\, du \end{aligned} \tag{37}$$

and

$$\begin{aligned} u(t) &= A_c \mathrm{Re}\left[\sum_{n=-\infty}^{\infty} c_n e^{j2\pi f_c t} e^{j2\pi n f_m t}\right] \\ &= A_c \sum_{n=-\infty}^{\infty} |c_n| \cos(2\pi(f_c + n f_m)t + \angle c_n) \end{aligned} \tag{38}$$

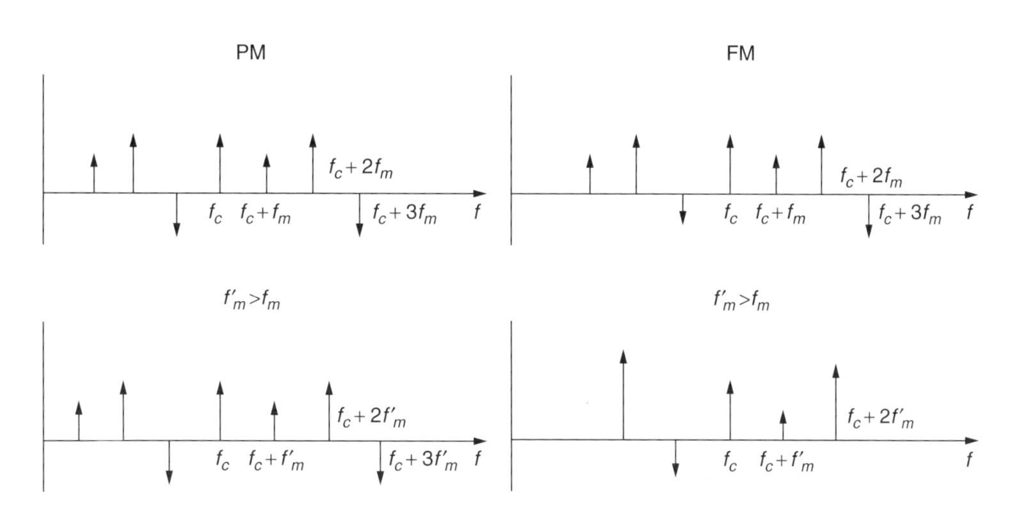

Figure 5. The effect of doubling bandwidth of the message in FM and PM.

It is seen again that the modulated signal contains all frequencies of the form $f_c + nf_m$.

The detailed treatment of the spectral characteristics of an angle-modulated signal for a general nonperiodic deterministic message signal $m(t)$ is quite involved because of the nonlinear nature of the modulation process. However, there exists an approximate relation for the effective bandwidth of the modulated signal, known as the *Carson's rule*, and given by

$$B_c = 2(\beta + 1)W \tag{39}$$

where β is the modulation index defined as

$$\beta = \begin{cases} k_p \max[|m(t)|], & \text{PM} \\ \dfrac{k_f \max[|m(t)|]}{W}, & \text{FM} \end{cases} \tag{40}$$

and W is the bandwidth of the message signal $m(t)$. Since in wideband FM the value of β is usually around ≥ 5, it is seen that the bandwidth of an angle-modulated signal is much greater than the bandwidth of various amplitude modulation schemes, which is either W (in SSB) or $2W$ (in DSB or conventional AM).

4. IMPLEMENTATION OF ANGLE MODULATORS AND DEMODULATORS

Angle modulators are in general time-varying and nonlinear systems. One method for generating an FM signal directly is to design an oscillator whose frequency changes with the input voltage. When the input voltage is zero the oscillator generates a sinusoid with frequency f_c, and when the input voltage changes, this frequency changes accordingly. There are two approaches to design such an oscillator, usually called a VCO or *voltage-controlled oscillator*. One approach is to use a *varactor diode*. A varactor diode is a capacitor whose capacitance changes with the applied voltage. Therefore, if this capacitor is used in the tuned circuit of the oscillator and the message signal is applied to it, the frequency of the tuned circuit, and the oscillator, will change in accordance with the message signal. Let us assume that the inductance of the inductor in the tuned circuit of Fig. 6 is L_0 and the capacitance of the varactor diode is given by

$$C(t) = C_0 + k_0 m(t) \tag{41}$$

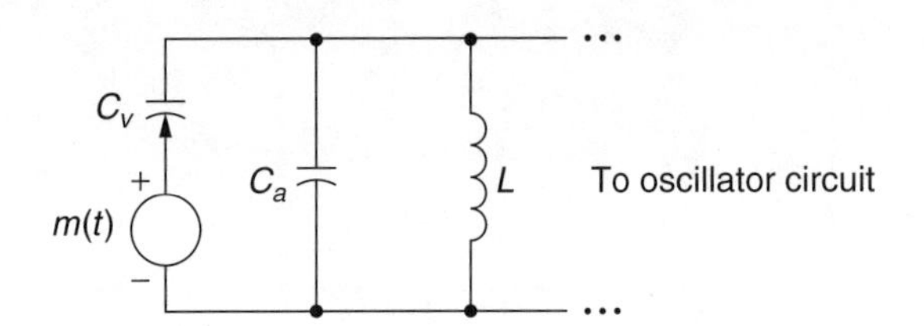

Figure 6. Varactor diode implementation of an angle modulator.

When $m(t) = 0$, the frequency of the tuned circuit is given by $f_c = \dfrac{1}{2\pi\sqrt{L_0 C_0}}$. In general, for nonzero $m(t)$, we have

$$\begin{aligned} f_i(t) &= \frac{1}{\pi\sqrt{L_0(C_0 + k_0 m(t))}} \\ &= \frac{1}{2\pi\sqrt{L_0 C_0}} \frac{1}{\sqrt{1 + \dfrac{k_0}{C_0} m(t)}} \\ &= f_c \frac{1}{\sqrt{1 + \dfrac{k_0}{C_0} m(t)}} \end{aligned} \tag{42}$$

Assuming that

$$\varepsilon = \frac{k_0}{C_0} m(t) \ll 1$$

and using the approximations

$$\sqrt{1+\varepsilon} \approx 1 + \frac{\varepsilon}{2} \tag{43}$$

$$\frac{1}{1+\varepsilon} \approx 1 - \varepsilon \tag{44}$$

we obtain

$$f_i(t) \approx f_c \left(1 - \frac{k_0}{2C_0} m(t)\right) \tag{45}$$

which is the relation for a frequency-modulated signal.

A second approach for generating an FM signal is by use of a *reactance tube*. In the reactance tube implementation an inductor whose inductance varies with the applied voltage is employed and the analysis is very similar to the analysis presented for the varactor diode. It should be noted that although we described these methods for generation of FM signals, due to the close relation between FM and PM signals, basically the same methods can be applied for generation of PM signals (see Fig. 1).

4.1. Indirect Method for Generation of Angle-Modulated Signals

Another approach for generating an angle-modulated signal is to first generate a narrowband angle-modulated signal and then change it to a wideband signal. This method is usually known as the *indirect method* for generation of FM and PM signals. Because of the similarity of conventional AM signals, generation of narrowband angle modulated signals is straightforward. In fact, any modulator for conventional AM generation can be easily modified to generate a narrowband angle modulated signal. Figure 7 is a block diagram of a narrowband angle modulator. The next step is to use the narrowband angle-modulated signal to generate a wideband angle-modulated signal. Figure 8 is a block diagram of a system that generates wideband angle-modulated signals from narrowband angle-modulated signals. The first stage of such a system is, of course, a narrowband angle modulator such as the one shown in Fig. 7. The narrowband angle-modulated signal enters a frequency multiplier that multiplies the instantaneous frequency of the input by some constant n. This is usually done by applying the

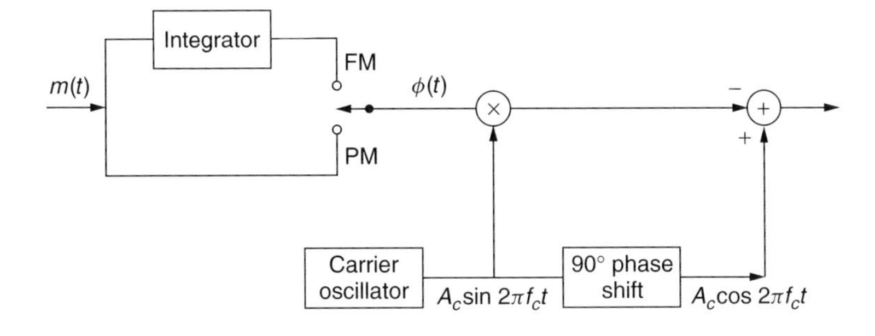

Figure 7. Generation of narrowband angle-modulated signal.

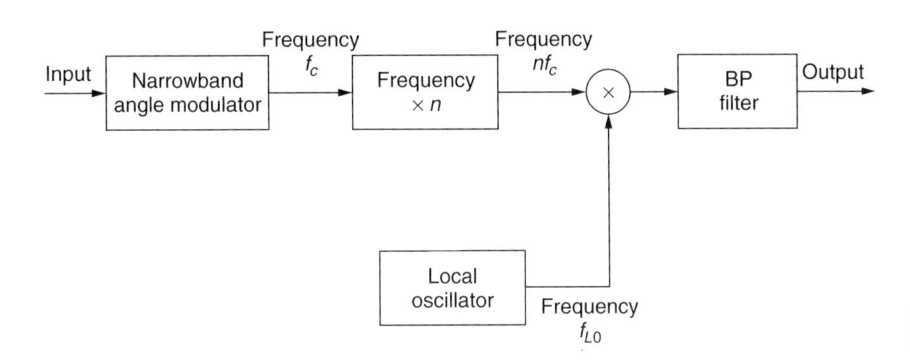

Figure 8. Indirect generation of angle-modulated signals.

input signal to a nonlinear element and then passing its output through a bandpass filter tuned to the desired central frequency. If the narrowband modulated signal is represented by

$$u_n(t) = A_c \cos(2\pi f_c t + \phi(t)) \tag{46}$$

the output of the frequency multiplier (output of the bandpass filter) is given by

$$y(t) = A_c \cos(2\pi n f_c t + n\phi(t)) \tag{47}$$

In general, this is, of course, a wideband angle-modulated signal. However, there is no guarantee that the carrier frequency of this signal, nf_c, will be the desired carrier frequency. The last stage of the modulator performs an up/down conversion to shift the modulated signal to the desired center frequency. This stage consists of a mixer and a bandpass filter. If the frequency of the local oscillator of the mixer is f_{LO} and we are using a down converter, the final wideband angle modulated signal is given by

$$u(t) = A_c \cos(2\pi(nf_c - f_{\mathrm{LO}})t + n\phi(t)) \tag{48}$$

Since we can freely choose n and f_{LO}, we can generate any modulation index at any desired carrier frequency by this method.

4.2. FM Demodulators

FM demodulators are implemented by generating an AM signal whose amplitude is proportional to the instantaneous frequency of the FM signal, and then using an AM demodulator to recover the message signal. To implement the first step, specifically, transforming the FM signal into an AM signal, it is enough to pass the FM signal through an LTI system whose frequency response is approximately a straight line in the frequency band of the FM signal. If the frequency response of such a system is given by

$$|H(f)| = V_0 + k(f - f_c) \quad \text{for } |f - f_c| < \frac{B_c}{2} \tag{49}$$

and if the input to the system is

$$u(t) = A_c \cos\left(2\pi f_c t + 2\pi k_f \int_{-\infty}^{t} m(\tau)\,d\tau\right) \tag{50}$$

then, the output will be the signal

$$v_o(t) = A_c(V_0 + kk_f m(t)) \cos\left(2\pi f_c t + 2\pi k_f \int_{-\infty}^{t} m(\tau)\,d\tau\right) \tag{51}$$

The next steps is to demodulate this signal to obtain $A_c(V_0 + kk_f m(t))$, from which the message $m(t)$ can be recovered. Figure 9 is a block diagram of these two steps.

There exist many circuits that can be used to implement the first stage of an FM demodulator, i.e., FM to AM conversion. One such candidate is a simple differentiator with

$$|H(f)| = 2\pi f \tag{52}$$

Another candidate is the rising half of the frequency characteristics of a tuned circuit as shown in Fig. 10. Such

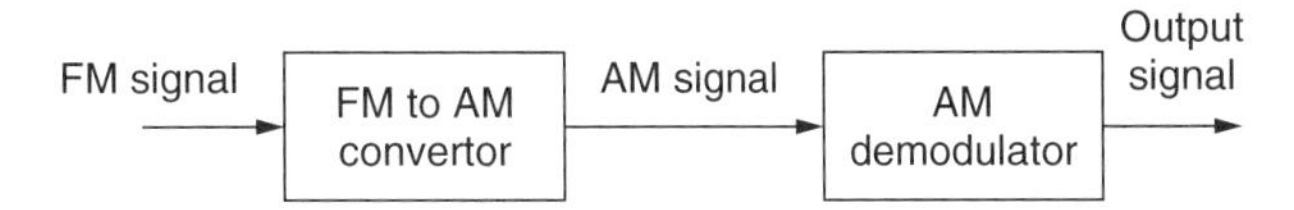

Figure 9. A general FM demodulator.

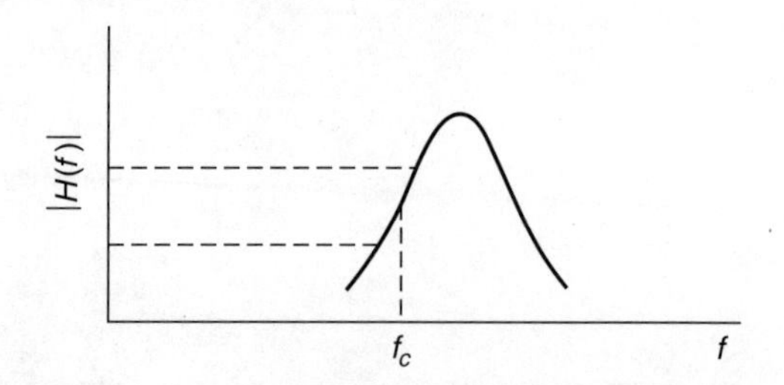

Figure 10. A tuned circuit used in an FM demodulator.

a circuit can be easily implemented but usually the linear region of the frequency characteristic may not be wide enough. To obtain a linear characteristics over a wider range of frequencies, usually two circuits tuned at two frequencies f_1 and f_2 are connected in a configuration which is known as a *balanced discriminator*. A balanced discriminator with the corresponding frequency characteristics is shown in Fig. 11.

The FM demodulation methods described above that transform the FM signal into an AM signal have a bandwidth equal to the channel bandwidth B_c occupied by the FM signal. Consequently, the noise that is passed by the demodulator is the noise contained within B_c.

4.2.1. FM Demodulation with Feedback. A totally different approach to FM signal demodulation is to use feedback in the FM demodulator to narrow the bandwidth of the FM detector and, as will be seen later, to reduce the noise power at the output of the demodulator. Figure 12 illustrates a system in which the FM discrimination is placed in the feedback branch of a feedback system that employs a voltage-controlled oscillator (VCO) path. The bandwidth of the discriminator and the subsequent lowpass filter is designed to match the bandwidth of the message signal $m(t)$. The output of the lowpass filter is the desired message signal. This type of FM demodulator is called an FM demodulator with feedback (FMFB). An alternative to FMFB demodulator is the use of a phase-locked loop (PLL), as shown in Fig. 13. The input to the PLL is the angle-modulated signal (we neglect the presence of noise in this discussion)

$$u(t) = A_c \cos[2\pi f_c t + \phi(t)] \tag{53}$$

where, for FM, we obtain

$$\phi(t) = 2\pi k_f \int_{-\infty}^{t} m(\tau)\,d\tau \tag{54}$$

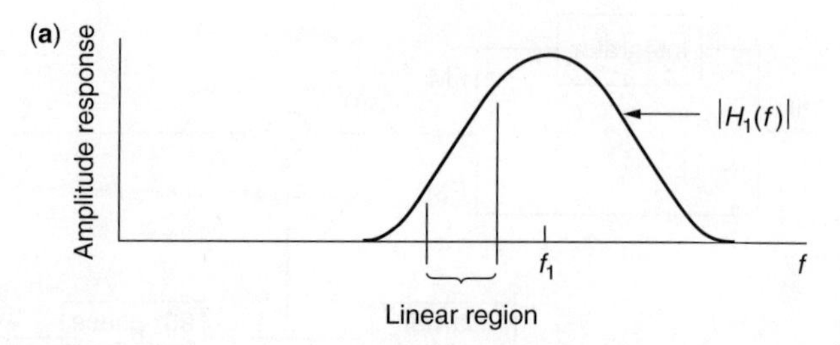

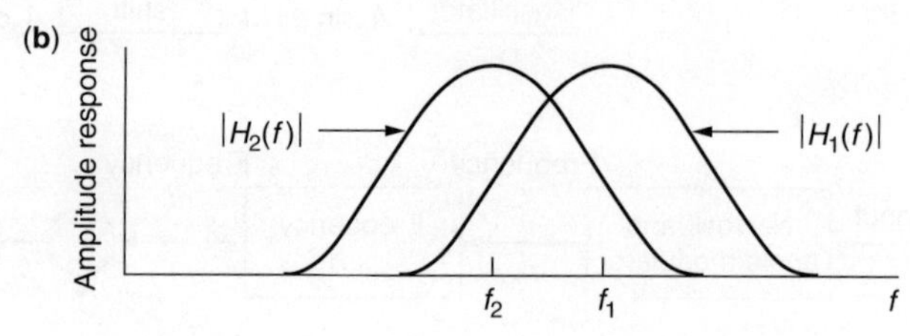

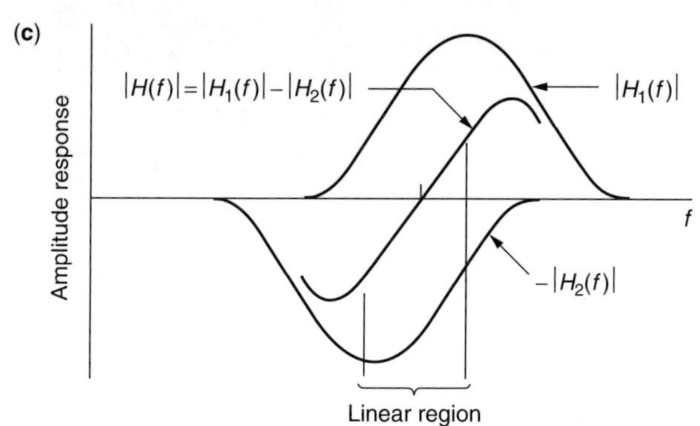

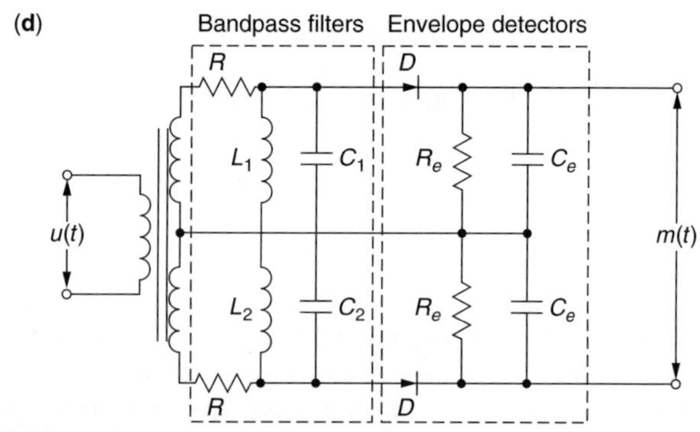

Figure 11. A ratio detector demodulator and the corresponding frequency response.

The VCO generates a sinusoid of a fixed frequency, in this case the carrier frequency f_c, in the absence of an input control voltage.

Now, suppose that the control voltage to the VCO is the output of the loop filter, denoted as $v(t)$. Then, the

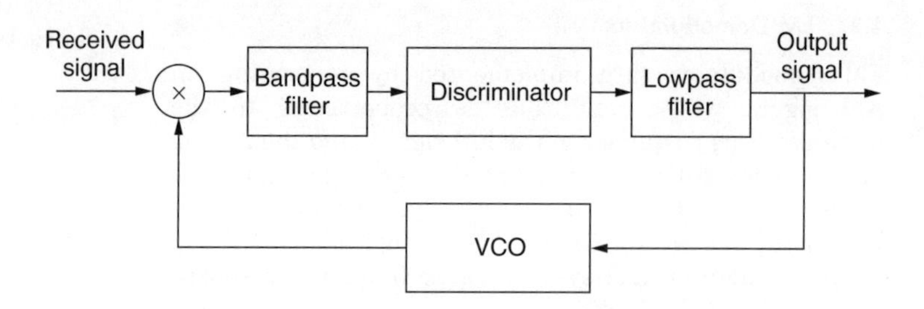

Figure 12. Block diagram of FMFB demodulator.

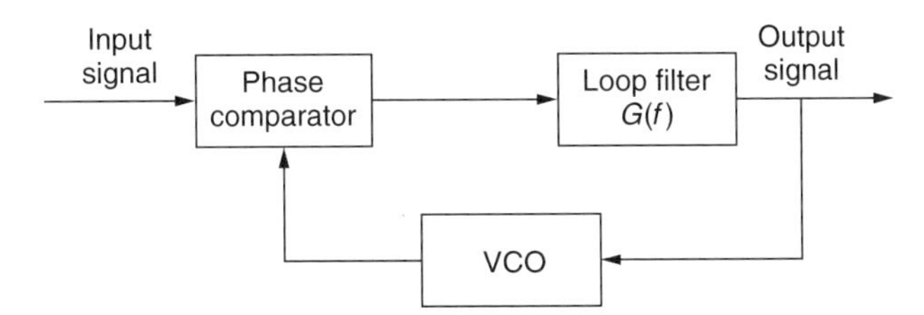

Figure 13. Block diagram of PLL FM demodulator.

instantaneous frequency of the VCO is

$$f_v(t) = f_c + k_v v(t) \tag{55}$$

where k_v is a deviation constant with units of hertz per volt (Hz/V). Consequently, the VCO output may be expressed as

$$y_v(t) = A_v \sin[2\pi f_c t + \phi_v(t)] \tag{56}$$

where

$$\phi_v(t) = 2\pi k_v \int_0^t v(\tau)\,d\tau \tag{57}$$

The phase comparator is basically a multiplier and a filter that rejects the signal component centered at $2f_c$. Hence, its output may be expressed as

$$e(t) = \tfrac{1}{2}A_v A_c \sin[\phi(t) - \phi_v(t)] \tag{58}$$

where the difference $\phi(t) - \phi_v(t) \equiv \phi_e(t)$ constitutes the phase error. The signal $e(t)$ is the input to the loop filter.

Let us assume that the PLL is in lock, so that the phase error is small. Then

$$\sin[\phi(t) - \phi_v(t)] \approx \phi(t) - \phi_v(t) = \phi_e(t) \tag{59}$$

under this condition, we may deal with the linearized model of the PLL, shown in Fig. 14. we may express the phase error as

$$\phi_e(t) = \phi(t) - 2\pi k_v \int_0^t v(\tau)\,d\tau \tag{60}$$

or, equivalently, as either

$$\frac{d}{dt}\phi_e(t) + 2\pi k_v v(t) = \frac{d}{dt}\phi(t) \tag{61}$$

or

$$\frac{d}{dt}\phi_e(t) + 2\pi k_v \int_0^\infty \phi_e(\tau) g(t-\tau)\,d\tau = \frac{d}{dt}\phi(t) \tag{62}$$

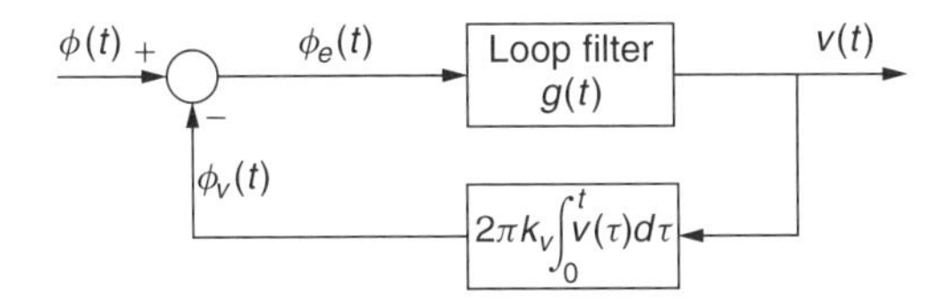

Figure 14. Linearized PLL.

The Fourier transform of the integrodifferential equation in (62) is

$$(j2\pi f)\Phi_e(f) + 2\pi k_v \Phi_e(f) G(f) = (j2\pi f)\Phi(f) \tag{63}$$

and, hence

$$\Phi_e(f) = \frac{1}{1 + \left(\dfrac{kv}{jf}\right) G(f)} \Phi(f) \tag{64}$$

The corresponding equation for the control voltage to the VCO is

$$\begin{aligned} V(f) &= \Phi_e(f) G(f) \\ &= \frac{G(f)}{1 + \left(\dfrac{k_v}{jf}\right) G(f)} \Phi(f) \end{aligned} \tag{65}$$

Now, suppose that we design $G(f)$ such that

$$\left| k_v \frac{G(f)}{jf} \right| \gg 1 \tag{66}$$

in the frequency band $|f| < W$ of the message signal. Then from (65), we have

$$V(f) = \frac{j2\pi f}{2\pi k_v} \Phi(f) \tag{67}$$

or, equivalently

$$\begin{aligned} v(t) &= \frac{1}{2\pi k_v} \frac{d}{dt} \Phi(t) \\ &= \frac{k_f}{k_v} m(t) \end{aligned} \tag{68}$$

Since the control voltage of the VCO is proportional to the message signal, $v(t)$ is the demodulated signal.

We observe that the output of the loop filter with frequency response $G(f)$ is the desired message signal. Hence, the bandwidth of $G(f)$ should be the same as the bandwidth W of the message signal. Consequently, the noise at the output of the loop filter is also limited to the bandwidth W. On the other hand, the output from the VCO is a wideband FM signal with an instantaneous frequency that follows the instantaneous frequency of the received FM signal.

The major benefit of using feedback in FM signal demodulation is to reduce the threshold effect that occurs when the input signal-to-noise-ratio to the FM demodulator drops below a critical value. We will study the threshold effect later in this article.

5. EFFECT OF NOISE ON ANGLE MODULATION

In this section we will study the performance of angle modulated signals when contaminated by additive white Gaussian noise and compare this performance with the performance of amplitude modulated signals. Recall that in amplitude modulation, the message information is contained in the amplitude of the modulated signal and since noise is additive, the noise is directly added to the signal. However, in a frequency-modulated signal, the noise is added to the amplitude and the message information is contained in the frequency of the modulated signal. Therefore the message is contaminated by the noise to the extent that the added noise changes the frequency of the modulated signal. The frequency of a signal can be described by its zero crossings. Therefore, the effect of additive noise on the demodulated FM signal can be described by the changes that it produces in the zero crossings of the modulated FM signal. Figure 15 shows the effect of additive noise on the zero crossings of two frequency-modulated signals, one with high power and the other with low power. From the above discussion and also from Fig. 15 it should be clear that the effect of noise in an FM system is less than that for an AM system. It is also observed that the effect of noise in a low-power FM system is more than in a high-power FM system. The analysis that we present in this chapter verifies our intuition based on these observations.

A block diagram of the receiver for a general angle-modulated signal is shown in Fig. 16. The angle-modulated signal is represented as[4]

$$u(t) = A_c \cos(2\pi f_c t + \phi(t)) = \begin{cases} A_c \cos(2\pi f_c t + 2\pi k_f \int_{-\infty}^{t} m(\tau)\,d\tau), & \text{FM} \\ A_c \cos(2\pi f_c t + k_p m(t)), & \text{PM} \end{cases} \tag{69}$$

The additive white Gaussian noise $n_w(t)$ is added to $u(t)$ and the result is passed through a noise limiting filter whose role is to remove the out-of-band noise. The bandwidth of this filter is equal to the bandwidth of the modulated signal and, therefore, it passes the modulated signal without distortion. However, it eliminates the out-of-band noise and, hence, the noise output of the filter is a bandpass Gaussian noise denoted by $n(t)$. The output of this filter is

$$\begin{aligned} r(t) &= u(t) + n(t) \\ &= u(t) + n_c(t)\cos 2\pi f_c t - n_s(t)\sin 2\pi f_c t \end{aligned} \tag{70}$$

where $n_c(t)$ and $n_s(t)$ denote the in-phase and the quadrature components of the bandpass noise. Because of the nonlinearity of the modulation process, an exact analysis of the performance of the system in the presence of noise is mathematically involved. Let us make the assumption that the signal power is much higher than the noise power. Then, if the bandpass noise is represented as

$$\begin{aligned} n(t) &= \sqrt{n_c^2(t) + n_s^2(t)}\cos\left(2\pi f_c t + \arctan\frac{n_s(t)}{n_c(t)}\right) \\ &= V_n(t)\cos(2\pi f_c t + \Phi_n(t)) \end{aligned} \tag{71}$$

where $V_n(t)$ and $\Phi_n(t)$ represent the envelope and the phase of the bandpass noise process, respectively, the assumption that the signal is much larger than the noise means that

$$p(V_n(t) \ll A_c) \approx 1 \tag{72}$$

Therefore, the phasor diagram of the signal and the noise are as shown in Fig. 17. From this figure it is obvious that we can write

$$\begin{aligned} r(t) &\approx (A_c + V_n(t)\cos(\Phi_n(t) - \phi(t)))\times \\ &\quad \times \cos\left(2\pi f_c t + \phi(t) + \arctan\frac{V_n(t)\sin(\Phi_n(t) - \phi(t))}{A_c + V_n(t)\cos(\Phi_n(t) - \phi(t))}\right) \\ &\approx (A_c + V_n(t)\cos(\Phi_n(t) - \phi(t))) \\ &\quad \times \cos\left(2\pi f_c t + \phi(t) + \frac{V_n(t)}{A_c}\sin(\Phi_n(t) - \phi(t))\right) \end{aligned}$$

The demodulator processes this signal and, depending whether it is a phase or a frequency demodulator, its output will be the phase or the instantaneous frequency of

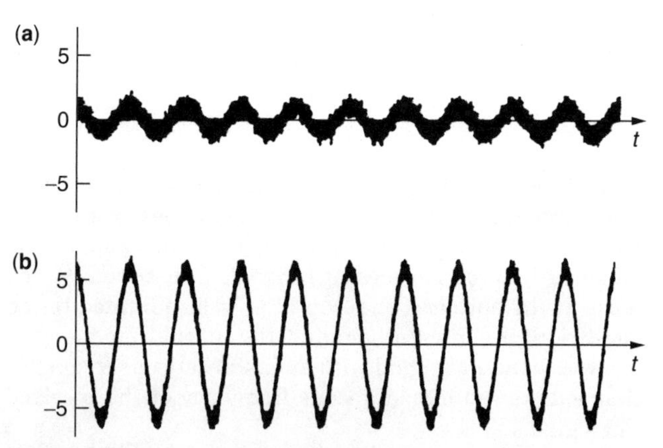

Figure 15. Effect of noise on the zero crossings of (**a**) low-power and (**b**) high-power modulated signals.

[4] When we refer to the modulated signal, we mean the signal as received by the receiver. Therefore, the signal power is the power in the received signal, not the transmitted power.

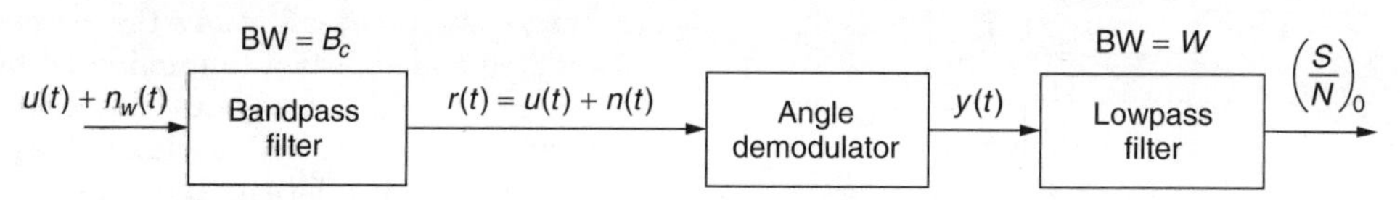

Figure 16. Block diagram of receiver for a general angle-demodulated signal.

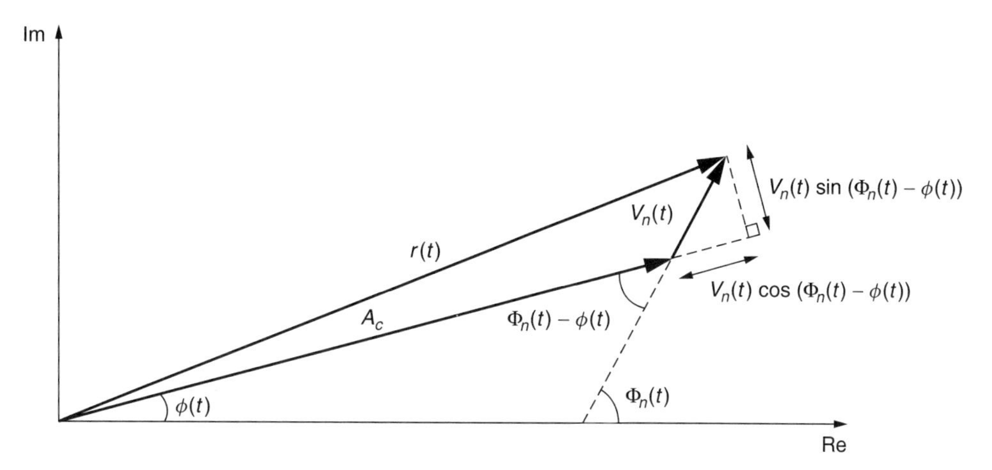

Figure 17. Phasor diagram of signal and noise in an angle-modulated system.

this signal.[5] Therefore, noting that

$$\phi(t) = \begin{cases} k_p m(t), & \text{PM} \\ 2\pi k_f \int_{-\infty}^{t} m(\tau)\, d\tau, & \text{FM} \end{cases} \tag{73}$$

the output of the demodulator is given by

$$y(t) = \begin{cases} k_p m(t) + Y_n(t), & \text{PM} \\ k_f m(t) + \dfrac{1}{2\pi}\dfrac{d}{dt}Y_n(t), & \text{FM} \end{cases}$$
$$= \begin{cases} k_p m(t) + \dfrac{V_n(t)}{A_c}\sin(\Phi_n(t) - \phi(t)), & \text{PM} \\ k_f m(t) + \dfrac{1}{2\pi}\dfrac{d}{dt}\dfrac{V_n(t)}{A_c}\sin(\Phi_n(t) - \phi(t)), & \text{FM} \end{cases} \tag{74}$$

where we have defined

$$Y_n(t) \stackrel{\text{def}}{=} \frac{V_n(t)}{A_c}\sin(\Phi_n(t) - \phi(t)). \tag{75}$$

The first term in the preceding expressions is the desired signal component, and the second term is the noise component. From this expression we observe that the noise component is inversely proportional to the signal amplitude A_c. Hence, the higher the signal level, the lower will be the noise level. This is in agreement with the intuitive reasoning presented at the beginning of this section based on Fig. 15. Note also that this is not the case with amplitude modulation. In AM systems the noise component is independent of the signal component and a scaling of the signal power does not affect the received noise power.

Let us study the properties of the noise component given by

$$Y_n(t) = \frac{V_n(t)}{A_c}\sin(\Phi_n(t) - \phi(t))$$
$$= \frac{1}{A_c}[V_n(t)\sin\Phi_n(t)\cos\phi(t) - V_n(t)\cos\Phi_n(t)\sin\phi(t)]$$
$$= \frac{1}{A_c}[n_s(t)\cos\phi(t) - n_c(t)\sin\phi(t)] \tag{76}$$

The autocorrelation function of this process is given by

$$E[Y_n(t+\tau)Y_n(t)] = \frac{1}{A_c^2}E[R_{n_s}(\tau)\cos\phi(t)\cos\phi(t+\tau) + R_{n_c}(\tau)\sin\phi(t+\tau)\sin\phi(t)]$$
$$= \frac{1}{A_c^2}R_{n_c}(\tau)E[\cos(\phi(t+\tau) - \phi(t))] \tag{77}$$

where we have used the fact that the noise process is stationary and $R_{n_c}(\tau) = R_{n_s}(\tau)$ and $R_{n_c n_s}(\tau) = 0$. Now we assume that the message $m(t)$ is a sample function of a zero mean, stationary Gaussian process $M(t)$ with the autocorrelation function $R_M(\tau)$. Then, in both PM and FM modulation, $\phi(t)$ will also be a sample function of a zero mean stationary and Gaussian process $\Phi(t)$. For PM this is obvious because

$$\Phi(t) = k_p M(t) \tag{78}$$

and in the FM case we have

$$\Phi(t) = 2\pi k_f \int_{-\infty}^{t} M(\tau)\, d\tau \tag{79}$$

Noting that $\int_{-\infty}^{t}$ represents a linear time-invariant operation, it is seen that, in this case, $\Phi(t)$ is the output of an LTI system whose input is a zero mean, stationary Gaussian process. Consequently $\Phi(t)$ will also be a zero mean, stationary Gaussian process.

At any fixed time t, the random variable $Z(t,\tau) = \Phi(t+\tau) - \Phi(t)$ is the difference between two jointly Gaussian random variables. Therefore, it is itself a Gaussian random variable with mean equal to zero and variance

$$\sigma_Z^2 = E[\Phi^2(t+\tau)] + E[\Phi^2(t)] - 2R_\Phi(\tau)$$
$$= 2[R_\Phi(0) - R_\Phi(\tau)] \tag{80}$$

[5] Of course, in the FM case the demodulator output is the instantaneous frequency deviation of $v(t)$ from the carrier frequency f_c.

Now, using this result in (5.9) we obtain

$$
\begin{aligned}
E[Y_n(t+\tau)Y_n(t)] &= \frac{1}{A_c^2} R_{n_c}(\tau) E\cos(\Phi(t+\tau)-\Phi(t)) \\
&= \frac{1}{A_c^2} R_{n_c}(\tau)\mathrm{Re}[Ee^{j(\Phi(t+\tau)-\Phi(t))}] \\
&= \frac{1}{A_c^2} R_{n_c}(\tau)\mathrm{Re}[Ee^{jZ(t,\tau)}] \\
&= \frac{1}{A_c^2} R_{n_c}(\tau)\mathrm{Re}[e^{-(1/2)\sigma_Z^2}] \\
&= \frac{1}{A_c^2} R_{n_c}(\tau)\mathrm{Re}[e^{-(R_\Phi(0)-R_\Phi(\tau))}] \\
&= \frac{1}{A_c^2} R_{n_c}(\tau) e^{-(R_\Phi(0)-R_\Phi(\tau))}
\end{aligned}
\tag{81}
$$

This result shows that under the assumption of a stationary Gaussian message, the noise process at the output of the demodulator is also a stationary process whose autocorrelation function is given above and whose power spectral density is

$$
\begin{aligned}
\mathcal{S}_Y(f) &= \mathcal{F}[R_Y(\tau)] \\
&= \mathcal{F}\left[\frac{1}{A_c^2} R_{n_c}(\tau) e^{-(R_\Phi(0)-R_\Phi(\tau))}\right] \\
&= \frac{e^{-R_\Phi(0)}}{A_c^2}\mathcal{F}[R_{n_c}(\tau)e^{R_\Phi(\tau)}] \\
&= \frac{e^{-R_\Phi(0)}}{A_c^2}\mathcal{F}[R_{n_c}(\tau)g(\tau)] \\
&= \frac{e^{-R_\Phi(0)}}{A_c^2}\mathcal{S}_{n_c}(f) * G(f)
\end{aligned}
\tag{82}
$$

where $g(\tau) = e^{R_\Phi(\tau)}$ and $G(f)$ is its Fourier transform.

It can be shown [1,2] that the bandwidth of $g(\tau)$ is $B_c/2$, that is, half of the bandwidth of the angle-modulated signal. For high-modulation indices this bandwidth is much larger than W, the message bandwidth. Since the bandwidth of the angle-modulated signal is defined as the frequencies that contain 98–99% of the signal power, $G(f)$ is very small in the neighborhood of $|f| = \frac{B_c}{2}$ and, of course,

$$
\mathcal{S}_{n_c}(f) = \begin{cases} N_0, & |f| < \dfrac{B_c}{2} \\ 0, & \text{otherwise} \end{cases}
\tag{83}
$$

A typical example of $G(f)$, $\mathcal{S}_{n_c}(f)$, and the result of their convolution is shown in Fig. 18. Because $G(f)$ is very small in the neighborhood of $|f| = \frac{B_c}{2}$, the resulting $\mathcal{S}_Y(f)$ has almost a flat spectrum for $|f| < W$, the bandwidth of the message. From Fig. 18 it is obvious that for all $|f| < W$, we have

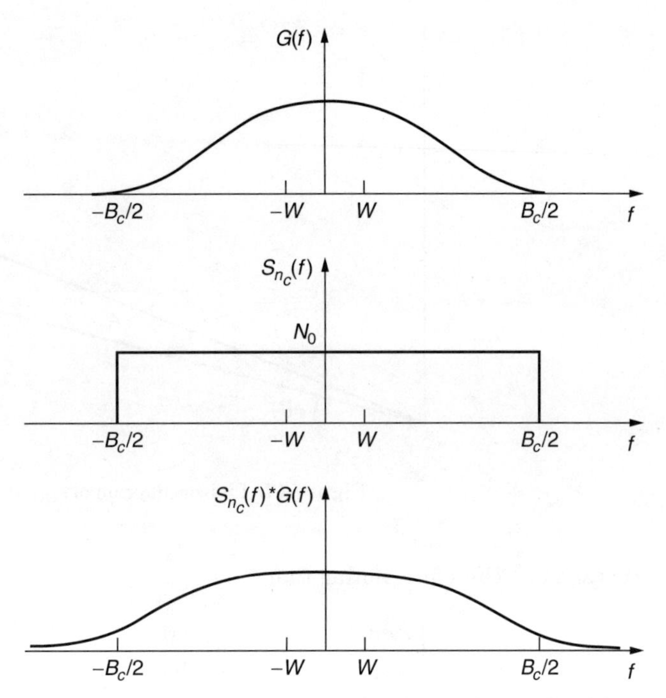

Figure 18. Typical plots of $G(f)$, $S_{n_c}(f)$ and the result of their convolution.

$$
\begin{aligned}
\mathcal{S}_Y(f) &= \frac{e^{-R_\Phi(0)}}{A_c^2}\mathcal{S}_{n_c}(f) * G(f) \\
&= \frac{e^{-R_\Phi(0)}}{A_c^2} N_0 \int_{-(B_c/2)}^{B_c/2} G(f)\,df \\
&\approx \frac{e^{-R_\Phi(0)}}{A_c^2} N_0 \int_{-\infty}^{\infty} G(f)\,df \\
&= \frac{e^{-R_\Phi(0)}}{A_c^2} N_0 g(\tau)_{|\tau=0} \\
&= \frac{e^{-R_\Phi(0)}}{A_c^2} N_0 e^{R_\Phi(0)} \\
&= \frac{N_0}{A_c^2}
\end{aligned}
\tag{84}
$$

It should be noted that this relation is a good approximation for $|f| < W$ only. This means that for $|f| < W$, the spectrum of the noise components in the PM and FM case are given by

$$
S_{n_o}(f) = \begin{cases} \dfrac{N_0}{A_c^2}, & \text{PM} \\ \dfrac{N_0}{A_c^2} f^2, & \text{FM} \end{cases}
\tag{85}
$$

where we have used the fact that in FM the noise component is given by $\frac{1}{2\pi}\frac{d}{dt}Y_n(t)$ as indicated in (74). The power spectrum of noise component at the output of the demodulator in the frequency interval $|f| < W$ for PM and FM is shown in Fig. 19. It is interesting to note that PM has a flat noise spectrum and FM has a parabolic noise spectrum. Therefore, the effect of noise in FM for higher-frequency components is much higher than the effect of noise on lower-frequency components. The noise power at the output of the lowpass filter is the noise power in the frequency range $[W, +W]$. Therefore, it is given by

$$
P_{n_o} = \int_{-W}^{+W} S_{n_o}(f)\,df
$$

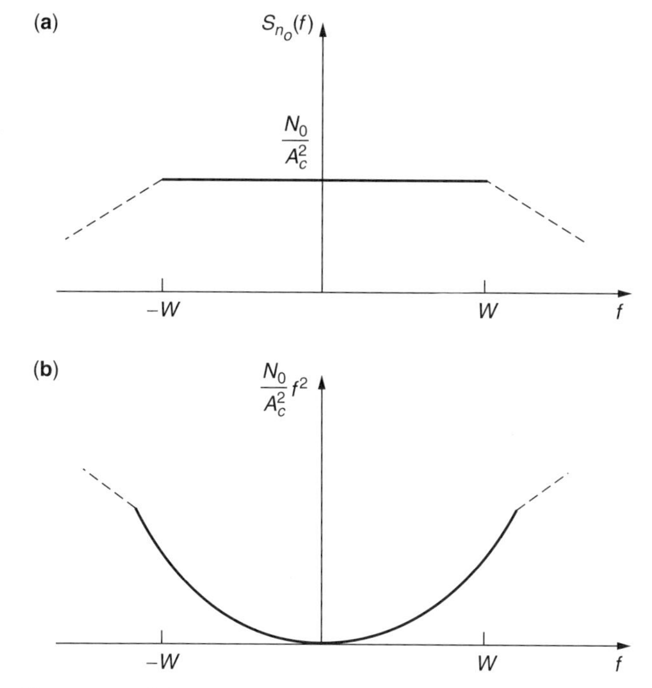

Figure 19. Noise power spectrum at demodulator output in (**a**) PM and (**b**) FM.

$$= \begin{cases} \int_{-W}^{+W} \frac{N_0}{A_c^2}\, df, & \text{PM} \\ \int_{-W}^{+W} f^2 \frac{N_0}{A_c^2}\, df, & \text{FM} \end{cases}$$

$$= \begin{cases} \frac{2WN_0}{A_c^2}, & \text{PM} \\ \frac{2N_0W^3}{3A_c^2}, & \text{FM} \end{cases} \tag{86}$$

Now we can use (74) to determine the output signal-to-noise ratio in angle modulation. First we have the output signal power

$$P_{s_o} = \begin{cases} k_p^2 P_M, & \text{PM} \\ k_f^2 P_M, & \text{FM} \end{cases} \tag{87}$$

Then, the signal-to-noise ratio, defined as

$$\left(\frac{S}{N}\right)_o = \frac{P_{s_o}}{P_{n_o}}$$

becomes

$$\left(\frac{S}{N}\right)_o = \begin{cases} \frac{k_p^2 A_c^2}{2} \frac{P_M}{N_0 W}, & \text{PM} \\ \frac{3k_f^2 A_c^2}{2W^2} \frac{P_M}{N_0 W}, & \text{FM} \end{cases} \tag{88}$$

Nothing that $\frac{A_c^2}{2}$ is the received signal power, denoted by P_R, and

$$\begin{cases} \beta_p = k_p \max|m(t)|, & \text{PM} \\ \beta_f = \frac{k_f \max|m(t)|}{W}, & \text{FM} \end{cases} \tag{89}$$

we may express the output SNR as

$$\left(\frac{S}{N}\right)_o = \begin{cases} P_R \left(\frac{\beta_p}{\max|m(t)|}\right)^2 \frac{P_M}{N_0 W}, & \text{PM} \\ 3P_R \left(\frac{\beta_f}{\max|m(t)|}\right)^2 \frac{P_M}{N_0 W}, & \text{FM} \end{cases} \tag{90}$$

If we denote $\frac{P_R}{N_0 W}$ by $\left(\frac{S}{N}\right)_b$, the signal-to-noise ratio of a baseband system with the same received power, we obtain

$$\left(\frac{S}{N}\right)_o = \begin{cases} \frac{P_M \beta_p^2}{(\max|m(t)|)^2} \left(\frac{S}{N}\right)_b, & \text{PM} \\ 3\frac{P_M \beta_f^2}{(\max|m(t|)^2} \left(\frac{S}{N}\right)_b, & \text{FM} \end{cases} \tag{91}$$

Note that in this expression $\frac{P_M}{(\max|m(t|)^2}$ is the average-to-peak-power-ratio of the message signal (or equivalently, the power content of the normalized message, P_{M_n}). Therefore

$$\left(\frac{S}{N}\right) = \begin{cases} \beta_p^2 P_{M_n} \left(\frac{S}{N}\right)_b, & \text{PM} \\ 3\beta_f^2 P_{M_n} \left(\frac{S}{N}\right)_b, & \text{FM} \end{cases} \tag{92}$$

Now using Carson's rule $B_c = 2(\beta + 1)W$, we can express the output SNR in terms of the bandwidth expansion factor, which is defined to be the ratio of the channel bandwidth to the message bandwidth and denoted by Ω:

$$\Omega = \frac{B_c}{W} = 2(\beta + 1) \tag{93}$$

From this relationship we have $\beta = \frac{\Omega}{2} - 1$. Therefore

$$\left(\frac{S}{N}\right)_o = \begin{cases} P_M \left(\frac{\frac{\Omega}{2} - 1}{\max|m(t)|}\right)^2 \left(\frac{S}{N}\right)_b, & \text{PM} \\ 3P_M \left(\frac{\frac{\Omega}{2} - 1}{\max|m(t)|}\right)^2 \left(\frac{S}{N}\right)_b, & \text{FM} \end{cases} \tag{94}$$

From (90) and (94), we observe that

1. In both PM and FM the output SNR is proportional to the square of the modulation index β. Therefore, increasing β increases the output SNR even with low received power. This is in contrast to amplitude modulation, where such an increase in the received signal-to-noise ratio is not possible.
2. The increase in the received signal-to-noise ratio is obtained by increasing the bandwidth. Therefore angle modulation provides a way to trade off bandwidth for transmitted power.
3. The relation between the output SNR and the bandwidth expansion factor, Ω, is a quadratic

relation. This is far from optimal.[6] An information-theoretic analysis shows that the optimal relation between the output SNR and the bandwidth expansion factor is an exponential relation.

4. Although we can increase the output signal-to-noise ratio by increasing β, having a large β means having a large B_c (by Carson's rule). Having a large B_c means having a large noise power at the input of the demodulator. This means that the approximation $p(V_n(t) \ll A_c) \approx 1$ will no longer apply and that the preceding analysis will not hold. In fact if we increase β such that the preceding approximation does not hold, a phenomenon known as the *threshold effect* will occur and the signal will be lost in the noise.
5. A comparison of the preceding result with the signal-to-noise ratio in amplitude modulation shows that in both cases increasing the transmitter power (or the received power), will increase the output signal-to-noise ratio, but the mechanisms are totally different. In AM, any increase in the received power directly increases the signal power at the output of the receiver. This is basically due to the fact the message is in the amplitude of the transmitted signal and an increase in the transmitted power directly affects the demodulated signal power. However, in angle modulation, the message is in the phase of the modulated signal and, consequently, increasing the transmitter power does not increase the demodulated message power. In angle modulation what increases the output signal-to-noise ratio is a *decrease in the received noise power* as seen from (86) and Fig. 15.
6. In FM the effect of noise is higher at higher frequencies. This means that signal components at higher frequencies will suffer more from noise than will the lower-frequency components. In some applications where FM is used to transmit SSB FDM signals, those channels that are modulated on higher-frequency carriers suffer from more noise. To compensate for this effect, such channels must have a higher signal level. The quadratic characteristics of the demodulated noise spectrum in FM is the basis of preemphasis and deemphasis filtering, which are discussed later in this article.

5.1. Threshold Effect in Angle Modulation

The noise analysis of angle demodulation schemes is based on the assumption that the signal-to-noise ratio at the demodulator input is high. With this crucial assumption we observed that the signal and noise components at the demodulator output are additive and we were able to carry out the analysis. This assumption of high signal-to-noise ratio is a simplifying assumption that is usually made in analysis of nonlinear modulation systems. Because of the nonlinear nature of the demodulation process, there is no reason that the additive signal and noise components at the input of the modulator result in additive signal and noise components at the output of the demodulator. In fact, this assumption is not at all correct in general, and the signal and noise processes at the output of the demodulator are completely mixed in a single process by a complicated nonlinear relation. Only under the high signal-to-noise ratio assumption is this highly nonlinear relation approximated as an additive form. Particularly at low signal-to-noise ratios, signal and noise components are so intermingled that one cannot recognize the signal from the noise and, therefore, no meaningful signal-to-noise ratio as a measure of performance can be defined. In such cases the signal is not distinguishable from the noise and a *mutilation* or *threshold effect* is present. There exists a specific signal to noise ratio at the input of the demodulator known as the *threshold SNR* beyond which signal mutilation occurs. The existence of the threshold effect places an upper limit on the tradeoff between bandwidth and power in an FM system. This limit is a practical limit in the value of the modulation index β_f.

It can be shown that at threshold the following approximate relation between $\frac{P_R}{N_0 W} = (\frac{S}{N})_b$ and β_f holds in an FM system:

$$\left(\frac{S}{N}\right)_{b,\text{th}} = 20(\beta + 1) \tag{95}$$

From this relation, given a received power P_R, we can calculate the maximum allowed β to make sure that the system works above threshold. Also, given a bandwidth allocation of B_c, we can find an appropriate β using Carson's rule $B_c = 2(\beta + 1)W$. Then, using the threshold relation given above we determine the required minimum received power to make the whole allocated bandwidth usable.

In general there are two factors that limit the value of the modulation index β. The first is the limitation on channel bandwidth that affects β through Carson's rule. The second is the limitation on the received power that limits the value of β to less than what is derived from (95). Figure 20 shows plots of the SNR in an FM system as a function of the baseband SNR. The SNR values in these curves are in decibels, and different curves correspond to different values of β as marked. The effect of threshold is apparent from the sudden drops in the output SNR. These plots are drawn for a sinusoidal message for which

$$\frac{P_M}{(\max|m(t)|)^2} = \frac{1}{2} \tag{96}$$

In such a case

$$\left(\frac{S}{N}\right)_o = \frac{3}{2}\beta^2 \left(\frac{S}{N}\right)_b \tag{97}$$

As an example, for $\beta = 5$, the preceding relation yields

$$\left(\frac{S}{N}\right)_{o|\text{dB}} = 15.7 + \left(\frac{S}{N}\right)_{b|\text{dB}} \tag{98}$$

$$\left(\frac{S}{N}\right)_{b,\text{th}} = 120 \sim 20.8 \text{ dB} \tag{99}$$

[6] By *optimal relation* we mean the maximum saving in transmitter power for a given expansion in bandwidth. An optimal system achieves the fundamental limits on communication predicted by information theory.

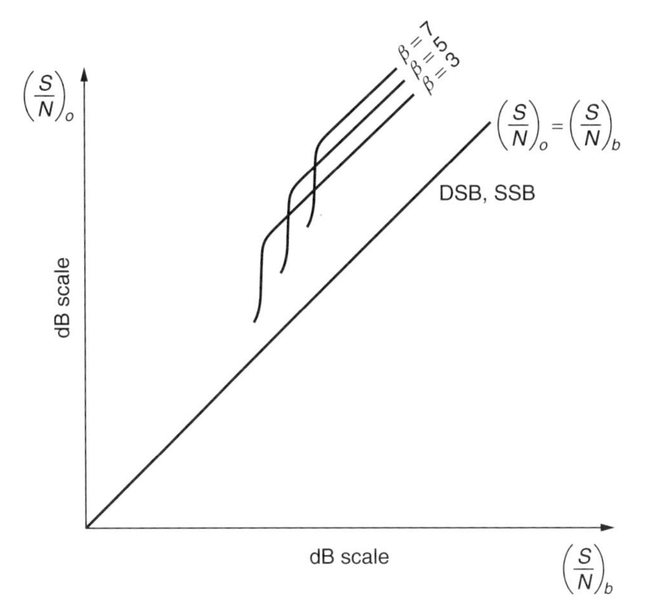

Figure 20. Output SNR versus baseband SNR in an FM system for various values of β.

On the other hand, if $\beta = 2$, we have

$$\left(\frac{S}{N}\right)_{o|_{\text{dB}}} = 7.8 + \left(\frac{S}{N}\right)_{b|_{\text{dB}}} \tag{100}$$

$$\left(\frac{S}{N}\right)_{b,\text{th}} = 60 \sim 17.8 \text{ dB} \tag{101}$$

From this discussion it is apparent that if, for example $(\frac{S}{N})_b = 20$ dB, then, regardless of the available bandwidth, we cannot use $\beta = 5$ for such a system because the demodulator will not demodulate below the threshold of 20 dB. However, $\beta = 2$ can be used, which yields an SNR equal to 27.8 dB at the output of the receiver. This is an improvement of 7.8 dB compared to a baseband system.

In general, if we want to employ the maximum available bandwidth, we must choose the largest possible β that guarantees that the system operates above threshold. This is the value of β that satisfies

$$\left(\frac{S}{N}\right)_{b,\text{th}} = 20(\beta + 1) \tag{102}$$

By substituting this value in (92), we obtain

$$\left(\frac{S}{N}\right)_o = 60\beta^2(\beta + 1)P_{M_n} \tag{103}$$

which relates a desired output SNR to the highest possible β that achieves that SNR.

5.1.1. Threshold Extension in Frequency Modulation. We have already seen that the nonlinear demodulation effect in angle modulation in general results in nonadditive signal and noise at the output of the demodulator. In high received signal-to-noise ratios, the nonlinear demodulation process can be well approximated by a linear equivalent and therefore signal and noise at the demodulator output will be additive. At high noise levels, however, this approximation is not valid anymore and the threshold effect results in signal mutilation. We have also seen that in general the modulated signal bandwidth increases with the modulation index and since the power of the noise entering the receiver is proportional to the system bandwidth, higher modulation indices cause the threshold effect to appear at higher received powers.

In order to reduce the threshold — in other words, in order to delay the threshold effect to appear at lower received signal power — it is sufficient to decrease the input noise power at the receiver. This can be done by decreasing the effective system bandwidth at the receiver.

Two approaches to FM threshold extension are to employ FMFB or PLL FM (see Figs. 12 and 13) at the receiver. We have already seen in Section 4 in the discussion following FMFB and PLL FM systems that these systems are capable of reducing the effective bandwidth of the receiver. This is exactly what is needed for extending the threshold in FM demodulation. Therefore, in applications where power is very limited and bandwidth is abundant, these systems can be employed to make it possible to use the available bandwidth more efficiently. Using FMFB the threshold can be extended approximately by 5–7 dB.

5.2. Preemphasis and Deemphasis Filtering

As observed in Fig. 19, the noise power spectral density at the output of the demodulator in PM is flat within the message bandwidth whereas for FM the noise power spectrum has a parabolic shape. This means that for low-frequency components of the message signal FM performs better and for high-frequency components, PM is a better choice. Therefore, if we can design a system that for low-frequency components of the message signal performs frequency modulation and for high-frequency components works as a phase modulator, we have a better overall performance compared to each system alone. This is the idea behind preemphasis and deemphasis filtering techniques.

The objective in preemphasis and deemphasis filtering is to design a system that behaves like an ordinary frequency modulator–demodulator pair in the low frequency band of the message signal and like a phase modulator–demodulator pair in the high-frequency band of the message signal. Since a phase modulator is nothing but the cascade connection of a differentiator and a frequency modulator, we need a filter in cascade with the modulator that at low frequencies does not affect the signal and at high frequencies acts as a differentiator. A simple highpass filter is a very good approximation to such a system. Such a filter has a constant gain for low frequencies and at higher frequencies it has a frequency characteristic approximated by $K|f|$, which is the frequency characteristic of a differentiator. At the demodulator side, for low frequencies we have a simple FM demodulator and for high-frequency components we have a phase demodulator, which is the cascade of a simple FM demodulator and an integrator. Therefore, at the demodulator we need a filter that at low frequencies has a constant gain and at high frequencies behaves as an integrator. A good approximation to such a filter is a simple lowpass filter. The modulator filter which emphasizes high frequencies is called the pre-emphasis filter and the demodulator filter which is the inverse of the modulator filter is called the *deemphasis filter*. Frequency responses of a sample preemphasis and deemphasis filter are given in Fig. 21.

Another way to look at preemphasis and deemphasis filtering is to note that due to the high level of noise in the high-frequency components of the message in FM, it is desirable to attenuate the high frequency components of the demodulated signal. This results in a reduction in the noise level but it causes the higher-frequency components of the message signal to be also attenuated. To compensate for the attenuation of the higher components of the message signal we can amplify these components at the transmitter before modulation. Therefore at the transmitter we need a highpass filter, and at the receiver we must use a lowpass filter. The net effect of these filters should be a flat frequency response. Therefore, the receiver filter should be the inverse of the transmitter filter.

The characteristics of the preemphasis and deemphasis filters depend largely on the power spectral density of the message process. In commercial FM broadcasting of music and voice, first order lowpass and highpass *RC* (resistance–capacitance) filters with a time constant of 75 microseconds (μs) are employed. In this case the frequency response of the receiver (deemphasis) filter is given by

$$H_d(f) = \frac{1}{1 + j\dfrac{f}{f_0}} \tag{104}$$

where $f_0 = \dfrac{1}{2\pi \times 75 \times 10^{-6}} \approx 2100$ Hz is the 3-dB frequency of the filter.

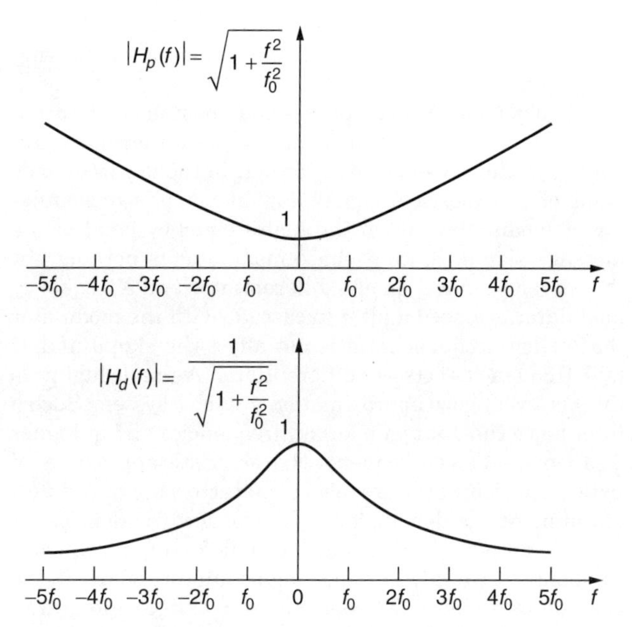

Figure 21. Preemphasis (**a**) and deemphasis (**b**) filter characteristics.

To analyze the effect of preemphasis and deemphasis filtering on the overall signal-to-noise ratio in FM broadcasting, we note that since the transmitter and the receiver filters cancel the effect of each other, the received power in the message signal remains unchanged and we only have to consider the effect of filtering on the received noise. Of course, the only filter that has an effect on the received noise is the receiver filter that shapes the power spectral density of the noise within the message bandwidth. The noise component before filtering has a parabolic power spectrum. Therefore, the noise component after the deemphasis filter has a power spectral density given by

$$\begin{aligned} S_{n_{\text{PD}}}(f) &= S_{n_o}(f)|H_d(f)|^2 \\ &= \frac{N_0}{A_c^2} f^2 \frac{1}{1 + \dfrac{f^2}{f_0^2}} \end{aligned} \tag{105}$$

where we have used (85). The noise power at the output of the demodulator now can be obtained as

$$\begin{aligned} P_{n_{\text{PD}}} &= \int_{-W}^{+W} S_{n_{\text{PD}}}(f)\, df \\ &= \frac{N_0}{A_c^2} \int_{-W}^{+W} \frac{f^2}{1 + \dfrac{f^2}{f_0^2}}\, df \\ &= \frac{2N_0 f_0^3}{A_c^2} \left[\frac{W}{f_0} - \arctan \frac{W}{f_0} \right] \end{aligned} \tag{106}$$

Because the demodulated message signal power in this case is equal to that of a simple FM system with no preemphasis and deemphasis filtering, the ratio of the output SNRs in these two cases is inversely proportional to the noise power ratios:

$$\begin{aligned} \frac{\left(\dfrac{S}{N}\right)_{o\text{PD}}}{\left(\dfrac{S}{N}\right)_o} &= \frac{P_{n_o}}{P_{n_{\text{PD}}}} \\ &= \frac{\dfrac{2N_0 W^3}{3A_c^2}}{\dfrac{2N_0 f_0^3}{A_c^2} \left[\dfrac{W}{f_0} - \arctan \dfrac{W}{f_0} \right]} \\ &= \frac{1}{3} \frac{\left(\dfrac{W}{f_0}\right)^3}{\dfrac{W}{f_0} - \arctan \dfrac{W}{f_0}} \end{aligned} \tag{107}$$

where we have used (86). Equation (107) gives the improvement obtained by employing preemphasis and deemphasis filtering.

In a broadcasting FM system with signal bandwidth $W = 15$ kHz, $f_0 = 2100$ Hz, and $\beta = 5$, using preemphasis and deemphasis filtering improves the performance of an FM system by 13.3 dB. The performance improvement of an FM system with no preemphasis and deemphasis filtering compared to a baseband system is 15–16 dB.

Thus an FM system with preemphasis and deemphasis filtering improves the performance of a baseband system by roughly 29–30 dB.

6. FM RADIO BROADCASTING

Commercial FM radio broadcasting utilizes the frequency band 88–108 MHz for transmission of voice and music signals. The carrier frequencies are separated by 200 kHz and the peak frequency deviation is fixed at 75 kHz. With a signal bandwidth of 15 kHz, this results in a modulation index of $\beta = 5$. Preemphasis filtering with $f_0 = 2100$ Hz is generally used, as described in the previous section, to improve the demodulator performance in the presence of noise in the received signal. The lower 4 MHz of the allocated bandwidth is reserved for noncommercial stations; this accounts for a total of 20 stations, and the remaining 80 stations in the 92–108-MHz bandwidth are allocated to commercial FM broadcasting.

The receiver most commonly used in FM radio broadcast is a superheterodyne type. The block diagram of such a receiver is shown in Fig. 22. As in AM radio reception, common tuning between the RF amplifier and the local oscillator allows the mixer to bring all FM radio signals to a common IF bandwidth of 200 kHz, centered at $f_{\mathrm{IF}} = 10.7$ MHz. Since the message signal $m(t)$ is embedded in the frequency of the carrier, any amplitude variations in the received signal are a result of additive noise and interference. The amplitude limiter removes any amplitude variations in the received signal at the output of the IF amplifier by band-limiting the signal. A bandpass filter centered at $f_{\mathrm{IF}} = 10.7$ MHz with a bandwidth of 200 kHz is included in the limiter to remove higher-order frequency components introduced by the nonlinearity inherent in the hard limiter.

A balanced frequency discriminator is used for frequency demodulation. The resulting message signal is then passed to the *audiofrequency amplifier*, which performs the functions of deemphasis and amplification. The output of the audio amplifier is further filtered by a lowpass filter to remove out-of-band noise and its output is used to drive a loudspeaker.

6.1. FM Stereo Broadcasting

Many FM radio stations transmit music programs in stereo by using the outputs of two microphones placed in two different parts of the stage. Figure 23 is a block diagram of an FM stereo transmitter. The signals from the left and right microphones, $m_\ell(t)$ and $m_r(t)$, are added and subtracted as shown. The sum signal $m_\ell(t) + m_r(t)$ is left as is and occupies the frequency band 0–15 kHz. The difference signal $m_\ell(t) - m_r(t)$ is used to AM modulate (DSB-SC) a 38-kHz carrier that is generated from a 19-kHz oscillator. A pilot tone at the frequency of 19 kHz is added to the signal for the purpose of demodulating the DSB SC AM signal. The reason for placing the pilot tone at 19 kHz instead of 38 kHz is that the pilot is more easily separated from the composite signal at the receiver. The combined signal is used to frequency modulate a carrier.

By configuring the baseband signal as an FDM signal, a monophonic FM receiver can recover the sum signal $m_\ell(t) + m_r(t)$ by use of a conventional FM demodulator. Hence, FM stereo broadcasting is compatible with conventional FM. The second requirement is that the resulting FM signal does not exceed the allocated 200-kHz bandwidth.

The FM demodulator for FM stereo is basically the same as a conventional FM demodulator down to the limiter/discriminator. Thus, the received signal is converted to baseband. Following the discriminator, the baseband message signal is separated into the two signals $m_\ell(t) + m_r(t)$ and $m_\ell(t) - m_r(t)$ and passed through deemphasis filters, as shown in Fig. 24. The difference signal is obtained from the DSB SC signal by means of a synchronous demodulator using the pilot tone. By taking the sum and difference of the two composite signals, we recover the two signals $m_\ell(t)$ and $m_r(t)$. These audio signals are amplified by audio band amplifiers and the two outputs drive dual loudspeakers. As indicated above, an FM receiver that is not configured to receive the

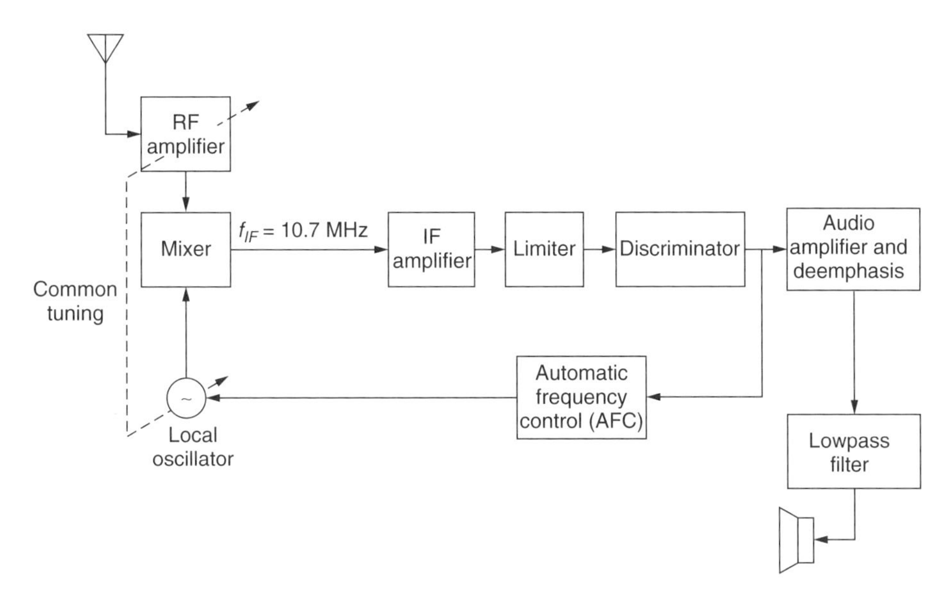

Figure 22. Block diagram of a superheterodyne FM radio receiver.

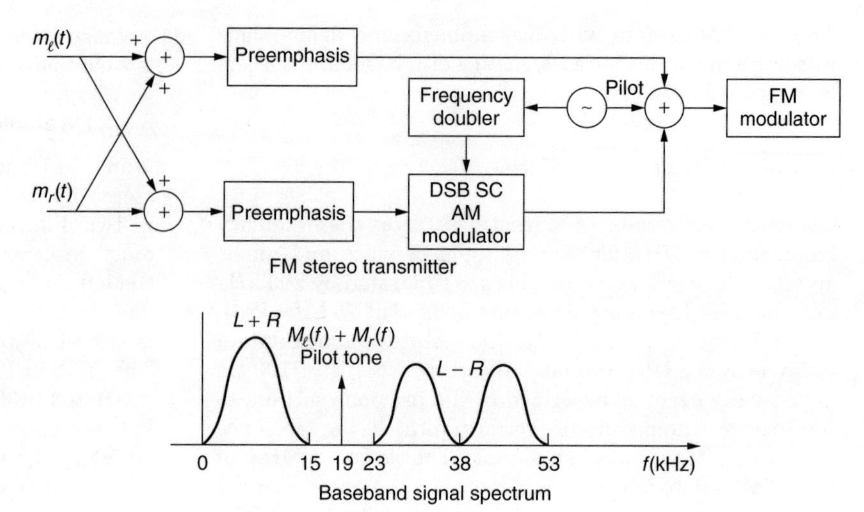

Figure 23. FM stereo transmitter and signal spacing.

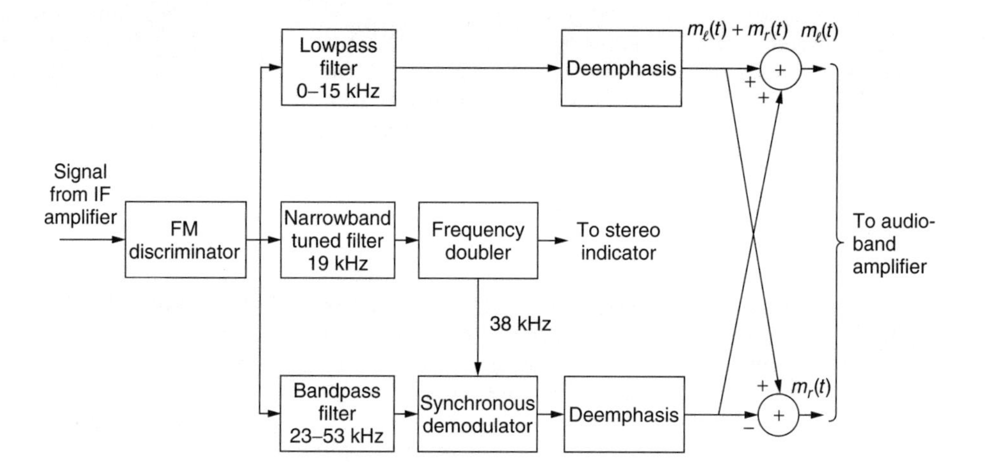

Figure 24. FM stereo receiver.

FM stereo sees only the baseband signal $m_\ell(t) + m_r(t)$ in the frequency range 0–15 kHz. Thus, it produces a monophonic output signal which consists of the sum of the signals at the two microphones.

BIOGRAPHY

Masoud Salehi received a B.S. degree from Tehran University and M.S. and Ph.D. degrees from Stanford University, all in electrical engineering. Before joining Northeastern University, he was with the Electrical and Computer Engineering Departments, Isfahan University of Technology and Tehran University.

From February 1988 until May 1989 Dr. Salehi was a visiting professor at the Information Theory Research Group, Department of Electrical Engineering, Eindhoven University of Technology, The Netherlands, where he did research in network information theory and coding for storage media.

Professor Salehi is currently with the Department of Electrical and Computer Engineering and a member of the CDSP (Communication and Digital Signal Processing) Center, Northeastern University, where he is involved in teaching and supervising graduate students in information and communication theory. His main areas of research interest are network information theory, source-channel matching problems in single and multiple user environments, data compression, channel coding, and particularly turbo codes. Professor Salehi's research has been supported by research grants from the National Science Foundation, DARPA, GTE, and Analog Devices. Professor Salehi is the coauthor of the textbooks *Communication Systems Engineering*, published by Prentice-Hall in 1994 and 2002, and *Contemporary Communication Systems Using MATLAB*, published by Brooks/Cole in 1998 and 2000.

BIBLIOGRAPHY

1. J. G. Proakis and M. Salehi, *Communication Systems Engineering*, 2nd ed., Prentice-Hall, Upper Saddle River, NJ, in press.
2. D. J. Sakrison, *Communication Theory: Transmission of Waveforms and Digital Information*, Wiley, New York, 1968.
3. A. B. Carlson, *Communication Systems*, 3rd ed., McGraw-Hill, New York, 1986.
4. L. W. Couch, II, *Digital and Analog Communication Systems*, 4th ed., Macmillan, New York, 1993.
5. J. D. Gibson, *Principles of Digital and Analog Communications*, 2nd ed., Macmillan, New York, 1993.

6. M. A. McMahon, *The Making of A Profession—Century of Electrical Engineering in America*, IEEE Press, 1984.
7. K. S. Shanmugam, *Digital and Analog Communication Systems*, Wiley, New York, 1979.
8. H. Taub and D. A. Schilling, *Principles of Communication Systems*, 2nd ed., McGraw-Hill, New York, 1986.
9. R. E. Ziemer and W. H. Tranter, *Principles of Communications*, 4th ed., Wiley, New York, 1995.
10. M. Schwartz, *Information Transmission, Modulation, and Noise*, 4th ed., McGraw-Hill, New York, 1990.
11. F. G. Stremler, *Introduction to Communication Systems*, 3rd ed., Addison-Wesley, Reading, MA, 1990.
12. S. Haykin, *Communication Systems*, 4th ed., Wiley, New York, 2001.
13. B. P. Lathi, *Modern Digital and Analog Communication Systems*, 3rd ed., Oxford Univ. Press, New York, 1998.

FREQUENCY-DIVISION MULTIPLE ACCESS (FDMA): OVERVIEW AND PERFORMANCE EVALUATION

FOTINI-NIOVI PAVLIDOU
Aristotle University of Thessaloniki
Thessaloniki, Greece

1. BASIC SYSTEM DESCRIPTION

FDMA (frequency-division multiple access) is a very basic multiple-access technique for terrestrial and satellite systems. We recall that *multiple access* is defined as the ability of a number of users to share a common transmission channel (coaxial cable, fiber, wireless transmission, etc.). Referring to the seven-layer OSI (Open System Interconnection) model, the access methods and the radio channel (frequency, time, and space) assignment are determined by the media access control (MAC) unit of the second layer.

Historically, FDMA has the highest usage and application of the various access techniques. It is one of the three major categories of fixed-assignment access methods (TDMA, FDMA, CDMA), and since it is definitely the simplest one, it has been extensively used in telephony, in commercial radio, in television broadcasting industries, in the existing cellular mobile systems, in cordless systems (CT2), and generally in many terrestrial and satellite wireless applications [1–3].

This access method is efficient if the user has a steady flow of information to send (digitized voice, video, transfer of long files) and uses the system for a long period of time, but it can be very inefficient if user data are sporadic in nature, as is the case with bursty computer data or short-message traffic. In this case it can be effectively applied only in hybrid implementations, as, for example, in FDMA/Aloha systems.

The principle of operation of FDMA is shown in Figs. 1a–d. The total common channel bandwidth is B Hz, and K users are trying to share it. In the FDMA technique each of the K users (transmitting stations) can transmit all of the time, or at least for extended periods of time but using only a portion B_i (subchannel) of the total channel bandwidth B such that $B_i = B/K$ Hz. If the users generate constantly unequal amounts of traffic, one can modify this scheme to assign bandwidth in proportion to the traffic generated by each one.

Adjacent users occupy different carriers of B_i bandwidth, with guard channels D Hz between them to avoid interference. Then the actual bandwidth available to each station for information is $B_i = \{B - (K+1)D\}/K$. The input of each source is modulated over a carrier and transmitted to the channel. So the channel transmits several carriers simultaneously at different frequencies. User separability is therefore achieved by separation in frequency. At the receiving end, the user bandpass filters select the designated channel out of the composite signal. Then a demodulator obtains the transmitted baseband signal.

Instead of transmitting one source signal on the carrier, we can feed a multiplexed signal on it [e.g., a pulse code modulation (PCM) telephone line]. Depending on the multiplexing and modulation techniques used, several transmission schemes can be considered. Although FDMA is usually considered to be built on the well-known FDM scheme, any multiplexing and modulation technique can be used for the processing of the baseband data, so several forms of FDMA are possible [4–6].

In Figs. 1b,c we give a very general implementation of the system. The first "user/station" transmits analog baseband signals, which are combined in a frequency-division multiplex (FDM) scheme. This multiplexed signal can modulate a carrier in frequency (FM) and then it is transmitted on the common channel, together with other carriers from other stations. If the technique used for all the stations is FDM/FM, the access method is called FDM/FM/FDMA (Fig. 1b).

If the stations transmit digital data, other modulation techniques can be used like PSK (phase shift keying) and TDM (time-division multiplex) can be applied. Then again this multiplexed signal can be transmitted on a frequency band on the common channel together with the carriers of other similar stations. This application is called TDM/PSK/FDMA.

The possible combinations of the form mux/mod/FDMA are numerous even if in practice the schemes shown in Fig. 1 are the most commonly used ones [3,6]. They are generally known as *multichannel-per-carrier* (MCPC) techniques.

Of course, for lower-traffic requirements the baseband signals can modulate directly a carrier in either analog or digital form (Fig. 1c). This concept, *single channel per carrier* (SCPC), has been extensively used in mobile satellites since it allows frequency reallocations according to increases of traffic and future developments in modulation schemes. The best-known application of SCPC/FDMA is the *single-channel-per-carrier pulse-code-modulation multiple-access demand assigned equipment* (SPADE) system applied in Intelsat systems.

FDMA carriers are normally assigned according to a fixed-frequency plan. However, in applications where

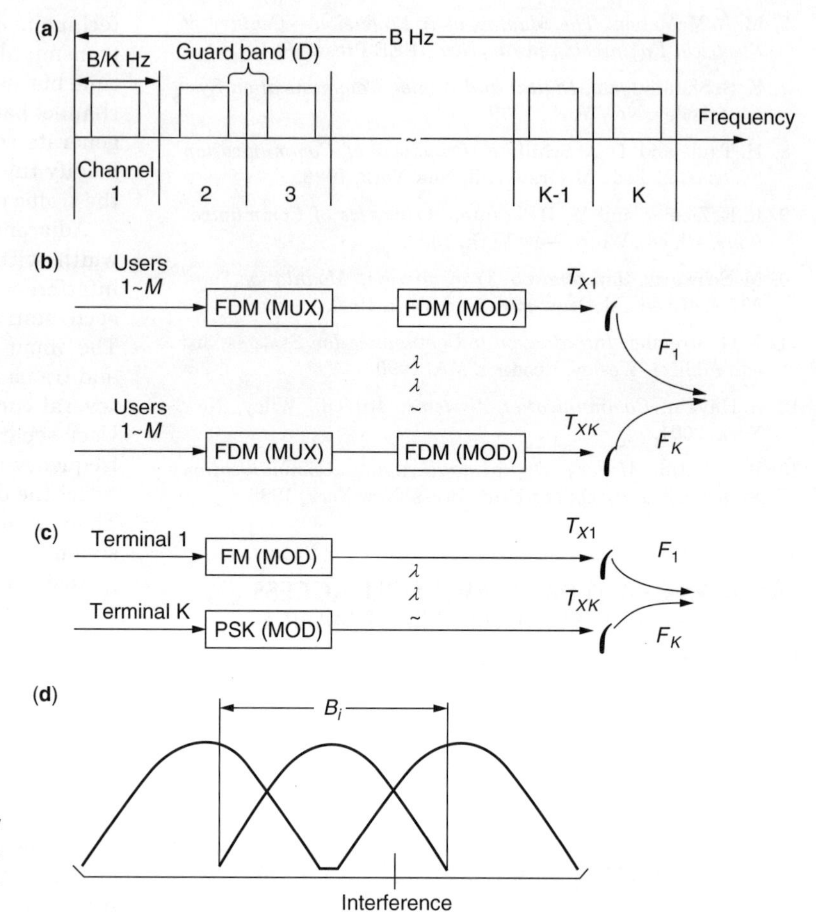

Figure 1. (**a**) FDMA frequency plan; (**b**) FDM/FM/FDMA implementation; (**c**) single-channel-per-carrier (SCPC)/FDMA implementation; (**d**) adjacent-channel interference.

limited space segment resources are to be shared by a relatively large number of users, the channel can be accessed on demand [demand assignment multiple access (DAMA)].

In wireless applications, the FDMA architecture is also known as "narrowband radio," as the bandwidth of the individual data or digitized analog signal (voice, facsimile, etc.) is relatively narrow compared with TDMA and CDMA applications.

2. SYSTEM PERFORMANCE

The major factor that determines the performance of the FDMA scheme is the received carrier-to-noise (plus interference) ratio, and for a FDM/FM/FDMA satellite system this is given by [6]

$$\frac{1}{C_T/N_T} = \frac{1}{C_u/N_u} + 1\left(\frac{C_u}{I_u}\right) + 1\left(\frac{C_d}{\text{IM}}\right) + 1\left(\frac{C_d}{N_d}\right) + \frac{1}{C_d/I_d}$$

where subscripts T, u, and d denote total, uplink, and downlink factors: N gives the white noise; I denotes the interference from other systems using the same frequency; and IM is the intermodulation distortion, explained in Sections 2.1 and 2.2.

2.1. Adjacent-Channel Interference and Intermodulation Noise

As we have noted, in FDMA applications frequency spacing is required between adjacent channels. This issue has been put forward several times by opponents of FDMA to claim that the technique is not as efficient in spectrum use as TDMA or CDMA. Indeed, excessive separation causes needless waste of the available bandwidth. Whatever filters are used to obtain a sharp frequency band for each carrier, part of the power of a carrier adjacent to the one considered will be captured by the receiver of the last one. In Fig. 1d we can see three adjacent bands of the FDMA spectrum (received at a power amplifier) composed of K carriers of equal power and identically modulated. To determine the proper spacing between FDMA carrier spectra, this adjacent channel interference (crosstalk power) must be carefully calculated. Spacings can then be selected for any acceptable crosstalk level desired. Common practice is to define the guard bands equal to around 10% of the carrier bandwidth (for carriers equal in amplitude and bandwidth). This will keep the noise level below the ITU-T (CCITT) requirements [6]. Simplified equations as well as detailed analysis for crosstalk are given in Refs. 3 and 6.

In addition, when the multiple FDMA carriers pass through nonlinear systems, like power amplifiers in satellites, two basic effects occur: (1) the nonlinear device output contains not only the original frequencies but also undesirable frequencies, that is, unwanted intermodulation (IM) products that fall into the FDMA bands as interference; and (2) the available output power decreases as a result of conversion of useful satellite power to intermodulation noise. Both of these effects depend on the type of nonlinearity and the number of simultaneous FDMA carriers present, as well as their power levels and spectral distributions. This makes it necessary to reduce the input to the amplifier from its maximum drive level in order to control intermodulation distortion. In satellite repeaters, this procedure is referred to as *input backoff* and is an important factor in maximizing the power efficiency of the repeater. So the traveling-wave tube (TWT) has to be *backed off* substantially in order to operate it as a linear amplifier. For example, Intelsat has established a series of monitoring stations to ensure that its uplink power levels are maintained. This in turn leads to inefficient usage of the available satellite power. Several nonlinear models for nonlinear power amplifiers, which have both amplitude and phase nonlinearities, have been proposed to calculate the carrier-to-IM versus input backoff.

An analysis of IM products for a satellite environment can be found elsewhere in the literature, where a detailed description and an exact calculation of intermodulation distortion are given [3,4,6–8]. Generally we have to determine the spectral distribution obtained from the mixing of the FDMA carriers. In Fig. 2a the intermodulation spectrum received by the mixing of the spectra of Fig. 1d is shown. The intermodulation power tends to be concentrated in the center of the total bandwidth B, so that the center carriers receive the most interference. Figure 2b shows the ratio of carrier power C_T to the intermodulation noise power IM, as a function of the number of carriers K and the degree of backoff.

In an ideal FDMA system the carriers have equal powers, but this is not the case in practice. When a mixture of both strong and weak carriers are present on the channel, we must ensure that the weaker carriers can maintain a communication link of acceptable quality. Some techniques have been implemented to accommodate this situation; one is known as *channelization*. In channelization carriers are grouped in *channels*, each with its own bandpass filter and power amplifier; in each channel (group) only carriers of the same power are transmitted.

(a)

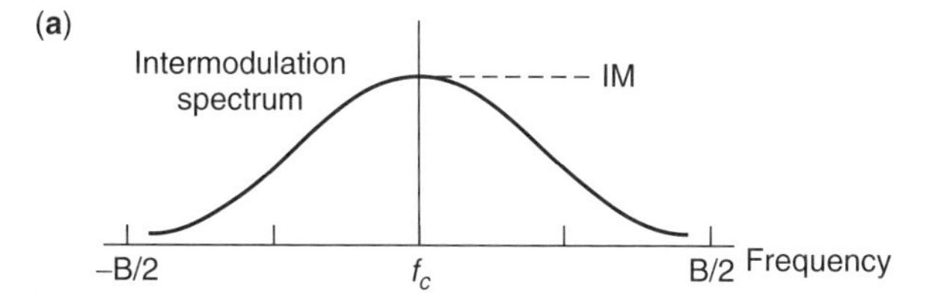

(b)

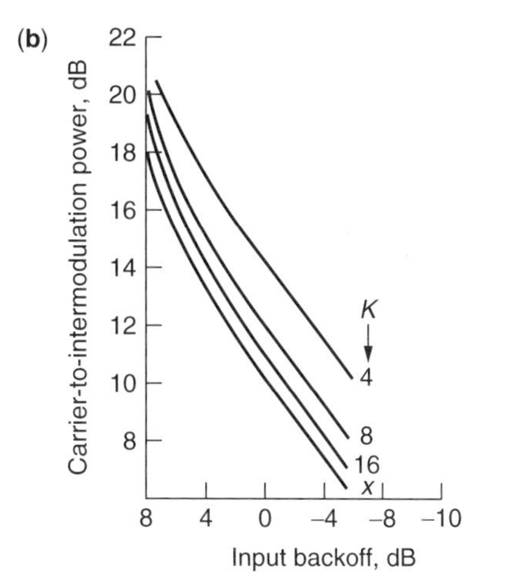

Figure 2. Intermodulation distortion.

2.2. FDMA Throughput

The throughput capability of an FDM/FM/FDMA scheme has been studied [3,6,7] as a function of the number of carriers taking into account the carrier-to-total noise (C_T/N_T) factor. The carriers are modulated by multiplexed signals of equal capacity. As the number of carriers increases, the bandwidth allocated to each carrier must decrease, and this leads to a reduction of the capacity of the modulating multiplexed signal. As the total capacity is the product of the capacity of each carrier and the total number of carriers, it could be imagined that the total capacity would remain sensibly constant. But it is not; the total capacity decreases as the number of carriers increases. This results from the fact that each carrier is subjected to a reduction in the value of C/N since the backoff is large when the number of carriers is high (extra carriers bring more IM products). Another reason is the increased need for guard bands. Figure 3 depicts the throughput of a FDMA system as a function of the number of carriers for an Intelsat transponder of 3 MHz bandwidth; it effectively shows the ratio of the total real channel capacity and the potential capacity of the channel.

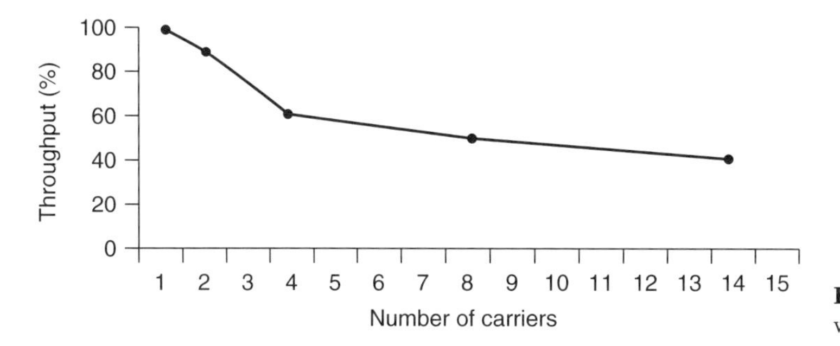

Figure 3. Throughput of FDM/FM/FDMA (channel bandwidth 36 MHz) (Dicks and Brown [4], © 1974 IEEE).

3. IMPLEMENTATION ALTERNATIVES

As we have noted, FDMA is a very basic technology that can be combined with almost all the other access techniques. Figure 4a shows FDD/FDMA for analog and digital transmission. In frequency-division duplex (FDD) there is a group of K subbands for transmission in one direction and a similar contiguous group of K subbands for transmission in the reverse direction. A band of frequencies separates the two groups. Each station is allocated a subband in both FDD bands for the duration of its call. All the first generation analog cellular systems use FDD/FDMA. FDMA can also be used with TDD (time-division duplex). Here only one band is provided for transmissions, so a timeframe structure is used allowing transmissions to be done during one-half of the frame while the other half of the frame is available to receive signals. TDD/FDMA is used in cordless communications (CT2). The TDMA/FDMA structure is used in GSM (Global System for Mobile Communication) systems (Fig. 4b), where carriers of 200 kHz bandwidth carry a frame of 8 time slots.

Further generalization of the strict fixed-assignment FDMA system is possible and has been implemented in commercial products. Frequency-hopping schemes in FDMA have been proposed for cellular systems. Hopping is based on a random sequence. In particular, slow frequency-hopped spread-spectrum (FHSS) FDMA systems, combined with power and spectrally efficient modulation techniques such as FQPSK, can have significantly increased capacity over other access methods. The GSM system supports the possibility for frequency hopping. FDMA systems in which users can randomly access each channel with an Aloha-type attempt have also been proposed.

An interesting variation of FDMA is the OFDMA (orthogonal frequency–division multiple access) scheme proposed for wideband communications. In OFDMA, multiple access is achieved by providing each user with a number of the available subcarriers, which are now orthogonal to each other. So there is no need for the relatively large guard bands necessary in conventional FDMA. An example of an OFDMA/TDMA time-frequency grid is shown in Fig. 5, where users 1, 2, and 3 each use a certain part of the available subcarriers. The part can be different for each one. Each user can have a fixed set of subcarriers, but it is relatively easy to allow hopping with different hopping patterns for each user and to result in orthogonal hopping schemes. OFDMA has been proposed for the European UMTS (Universal Mobile Telecommunications System) [9].

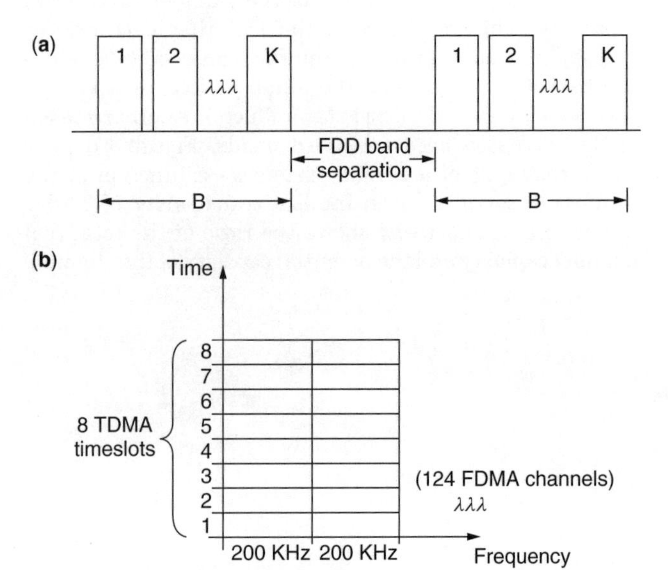

Figure 4. (**a**) FDMA/FDD arrangement; (**b**) TDMA/FDMA for GSM system.

Frequency (vertical axis) / Time (horizontal axis)

1		3		2	
	1		3		2
2		1		3	
	2		1		3
3		2		1	
	3		2		1

Figure 5. OFDMA with frequency hopping.

4. COMPARISON WITH OTHER TECHNIQUES

The comparison between the FDMA and other access methods is based on a number of performance and implementation criteria, the importance of which is varying very much depending on the type of system in which the access method is to be employed.

In digital transmission TDMA appears to be more "natural," so today most of the systems operate in this scheme or at least in combination with FDMA schemes. TDMA offers format flexibility since time-slot assignments among the multiple users are readily adjusted to provide different access rates for different users.

In some circumstances, FDMA schemes may be comparatively inefficient since they require guard bands between neighboring bands to prevent interfering phenomena, so this results in a waste of the system resources.

Another problem with FDMA systems is that they suffer from system nonlinearities; for example, they require the satellite transponder to be linear in nature, which cannot be achieved in practice. However, this is not necessarily as important in terrestrial communication systems, where the power consumption in the base station electronics is not a major design issue.

Referring to throughput performance, FDMA and TDMA should provide the same capability for carrying information over a network, and this is true with respect to bit-rate capability, if we neglect all overhead elements such as guard bands in FDMA and guard times in TDMA.

For K users generating data at a constant uniform rate and for an overall rate capability of the system equal to R bits per second, R/K bps (bits per second) is available to each user in both systems. Furthermore, both systems have the same capacity-wasting properties, because if a user has nothing to transmit, its frequency band cannot be used by another user.

However, if we examine the average delay, assuming that each user is transmitting every T seconds, the delay is found to be $D_{\mathrm{FDMA}} = T$ for the FDMA system while it is $D_{\mathrm{TDMA}} = D_{\mathrm{FDMA}} - T/2\{1 - 1/K\}$ in TDMA [1]. Therefore, for two or more users TDMA is superior to FDMA. Note that for large numbers of users the difference in packet delay is approximately $T/2$. Because of these problems, fixed assignment strategies have increasingly tended to shift from FDMA to TDMA as certain technical problems associated with the latter were overcome.

On the other hand, in transmission environments where spurious narrowband interference is a problem, the FDMA format, with a single user channel per carrier, has an advantage compared to TDMA in that a narrowband interferer can impair the performance of only one user channel.

The major advantage in the implementation of FDMA systems is that FDMA is a very mature technology and has the advantage of inexpensive terminals. Furthermore, channel assignment is simple and straightforward, and no network timing is required. This absence of synchronization problems is very attractive in fading communications channels.

Finally, in multipath fading environments, in a typical FDMA system, channel bandwidth is usually smaller than the coherence bandwidth of the transmission channel and there is no need to use an adaptive equalizer at the receiver. But at the same time, this situation removes the opportunity for the implicit frequency diversity gains that are achievable when signal bandwidth approaches the coherence bandwidth. Of course, in multipath environments CDMA is proved to be very effective. Its immunity to external interference and jamming, its low probability of intercept, and the easy integration of voice/data messages it offers establish its use in future multimedia communications systems. But a combination of FDMA/CDMA can improve the capacity of the system.

A detailed comparison of FDMA with other multiple access techniques can be found in well-known textbooks [5,6]. Schwartz et al. have provided a very fundamental comparison [10].

5. BASIC APPLICATIONS

FDMA was the first method of multiple access used for *satellite* communication systems (fixed, broadcasting, and mobile services) since around 1965 and will probably remain so for quite some time. In 1969 three satellites, Intelsat III, provided the first worldwide satellite service via analog FDM/FM/FDMA techniques for assemblies of 24, 60, or 120 telephone channels, while later, in Intelsat VI, 792 or 972 voice channels were delivered. Also, special techniques are applied for the distribution of audio and video channels on the same transponder (TV/FM/FDMA) [6]. Further, as a good example of digital operation in the FDMA mode, we can refer to Intelsat's *intermediate data rate* (IDR) carrier system described by Freeman [5].

A specific FDMA application is the single-channel-per-carrier (SCPC) system; one of the best known is the Intelsat (*SPADE*) system described above. SCPC systems are most suitable for use where each terminal is required to provide only a small number of telephone circuits, referred to as a *thin-route service*. The *satellite multiservice system* (SMS), provided by Eutelsat for business applications, has up to 1600 carriers in its 72 MHz bandwidth, which can be reallocated extremely quickly in response to changes in demand. FDMA has been used also for *VSAT* (very small-aperture terminals) satellite access, especially as DAMA method with FDMA/SCPC application. Most of today's commercial networks are based on demand assignment FDMA due to simple network control needed. Detailed VSAT implementations can be easily found in the literature.

At the present time, all *mobile satellite systems* that offer voice services have adopted a frequency–division multiple-access approach. The usual channel bandwidth is around 30 kHz, but figures of 22.5, 36, and 45 kHz are also used. The use of FDMA in conjunction with common channel signaling enables future developments in modulation and processing technologies to be easily incorporated into the system and allows a variety of communication standards to be supported. Also, as growth in the mobile terminal population occurs, incremental increases in the required system spectrum allocation can be easily assigned [6].

The modulation and multiple-access techniques used in the *IRIDIUM* system are patterned after the GSM terrestrial cellular system. A combined FDMA-TDMA access format is used along with data or vocoded voice and digital modulation techniques. Each FDMA frequency slot is 41.67 kHz wide, including guard bands, and supports 4-duplex voice channels in a TDMA arrangement [5].

All the *analog terrestrial mobile systems* were based on FDMA techniques. The band of frequencies was divided into segments, and half of the contiguous segments are assigned to outbound and the other to inbound (FDD/FDMA) cell sites with a guard band between outbound and inbound contiguous channels. A key question in these systems was the actual width of one user segment (e.g., in North American amperage, the segment width was 30 kHz).

FDMA is also applied in the contemporary cellular systems and in VHF and UHF land-mobile radio systems. The well-known *GSM* standard is based on a TDMA/FDMA combination that has been considered a very efficient scheme. The European digital cordless phone (DECT) is also based on a TDMA/TDD/FDMA format. In the United States in several FCC-authorized frequency bands, particularly those below 470 MHz, the authorized bandwidth per channel is limited to the 5–12.5-kHz range. In these narrowband mobile or cellular systems, digital FDMA could offer the most spectral- and cost-efficient solutions.

A very interesting application of the FDMA concept is found in the third-generation optical networks under the name of *wavelength-division multiple access* (WDMA) [11], meaning that the bits of the message are addressed on the basis of different wavelengths. WDMA networks have been investigated and prototyped at the laboratory level by a number of groups such as British Telecom Laboratories and AT&T Bell Laboratories.

In conclusion, we can state that FDMA as a stand-alone concept or as a basic component in hybrid implementations is being applied and will be applied in the future in most communications systems.

BIOGRAPHY

Fotini-Niovi Pavlidou received her Ph.D. degree in electrical engineering from the Aristotle University of Thessaloniki, Greece, in 1988 and the Diploma in mechanical–electrical engineering in 1979 from the same institution.

She is currently an associate professor at the Department of Electrical and Computer Engineering at the Aristotle University engaged in teaching for the under- and postgraduate program in the areas of mobile communications and telecommunications networks. Her research interests are in the field of mobile and personal communications, satellite communications, multiple access systems, routing and traffic flow in networks, and QoS studies for multimedia applications over the Internet.

She is involved with many national and international projects in these areas (Tempus, COST, Telematics, IST) and she has been chairing the European COST262 Action on "Spread Spectrum Systems and Techniques for Wired and Wireless Communications." She has served as a member of the TPC in many IEEE/IEE conferences and she has organized/chaired some conferences like the "IST Mobile Summit 2002," the 6th "International Symposium on Power Lines Communications-ISPLC2002," the "International Conference on Communications-ICT 1998," etc.

She is a permanent reviewer for many IEEE/IEE1 journals. She has published about 60 studies in refereed journals and conferences.

She is a senior member of IEEE, currently chairing the joint IEEE VT & AES Chapter in Greece.

BIBLIOGRAPHY

1. K. Pahlavan and A. Levesque, *Wireless Information Networks*, Wiley, New York, 1995.
2. V. K. Bhargava, D. Haccoun, R. Matyas, and P. P. Nuspl, *Digital Communications by Satellite-Modulation, Multiple Access and Coding*, Wiley, New York, 1981.
3. G. Maral and M. Bousquet, *Satellite Communications Systems; Systems, Techniques and Technology*, Wiley, New York, 1996.
4. R. M. Gagliardi, *Introduction to Communications Engineering*, Wiley, New York, 1988.
5. R. L. Freeman, *Telecommunications Transmission Handbook*, Wiley, New York, 1998.
6. W. L. Morgan and G. D. Gordon, *Communications Satellite Handbook*, Wiley, New York, 1989.
7. J. L. Dicks and M. P. Brown, Jr., Frequency division multiple access (FDMA) for satellite communications systems, paper presented at IEEE Electronics and Aerospace Systems Convention, Washington, DC, Oct. 7–9, 1974.
8. N. J. Muller, *Desktop Encyclopedia of Telecommunications*, McGraw-Hill, New York, 1998.
9. OFDMA evaluation report, *The Multiple Access Scheme Proposal for the UMTS Terrestrial Radio Air Interface (UTRA) System, Part 1: System Description and Performance Evaluation*, SMG2 Tdoc 362a/97, 1997.
10. J. W. Schwartz, J. M. Aein, and J. Kaiser, Modulation techniques for multiple-access to a hard limiting repeater, *Proc. IEEE* **54**: 763–777 (1966).
11. P. E. Green, *Fiber Optic Networks*, Prentice-Hall, Englewood Cliffs, NJ, 1993.

FREQUENCY SYNTHESIZERS

Ulrich L. Rohde
Synergy Microwave Corporation
Paterson, New Jersey

1. INTRODUCTION

Frequency synthesizers are found in all modern communication equipment, and signal generators, particularly wireless communication systems such as cell phones, require these building blocks [1]. On the basis of a frequency standard, the synthesizer provides a stable reference frequency for the system.

Synthesizers are used to generate frequencies with arbitrary resolution covering the frequency range from audio to millimeterwave. Today, simple frequency synthesizers consist of a variety of synthesizer chips, an external voltage-controlled oscillator (VCO), and a frequency standard. For high-volume applications, such as cell phones, cordless telephones, walkie-talkies, or systems where frequency synthesizers are required, a high degree of integration is desired. The requirements for synthesizers in cordless telephones are not as stringent as in test equipment. Synthesizers in test equipment use custom building blocks and can be modulated to be part of arbitrary waveform generators [2].

The VCO typically consists of an oscillator with a tuning diode attached. The voltage applied to the tuning diode tunes the frequency of the oscillator. Such a simple system is a phase-locked loop (PLL). The stability of the VCO is the same as the reference. There are single-loop and multiloop PLL systems. Their selection depends on the characteristic requirements. Figure 1 shows the block diagram of a single loop PLL [3].

The PLL consists of a VCO, a frequency divider, a phase detector, a frequency standard, and a loop filter [4–7].

There are limits to how high the frequency-division ratio can be. Typically, the loop where the RF frequency is divided below 1 kHz, becomes unstable. This is due to the fact that microphonic effects of the resonator will unlock the system at each occurrence of mechanical vibration. At 1 GHz, this would be a division ratio of one million. To avoid such high division ratios, either multiloop synthesizers are created, or a new breed of PLL synthesizers called *fractional-N division synthesizers* will be considered. At the same time, direct digital synthesis is being improved. Using direct digital synthesis in loops can also overcome some of the difficulties associated with high division ratios. There are also combinations of techniques, which we will refer to as *hybrid synthesizers*. They will be covered here.

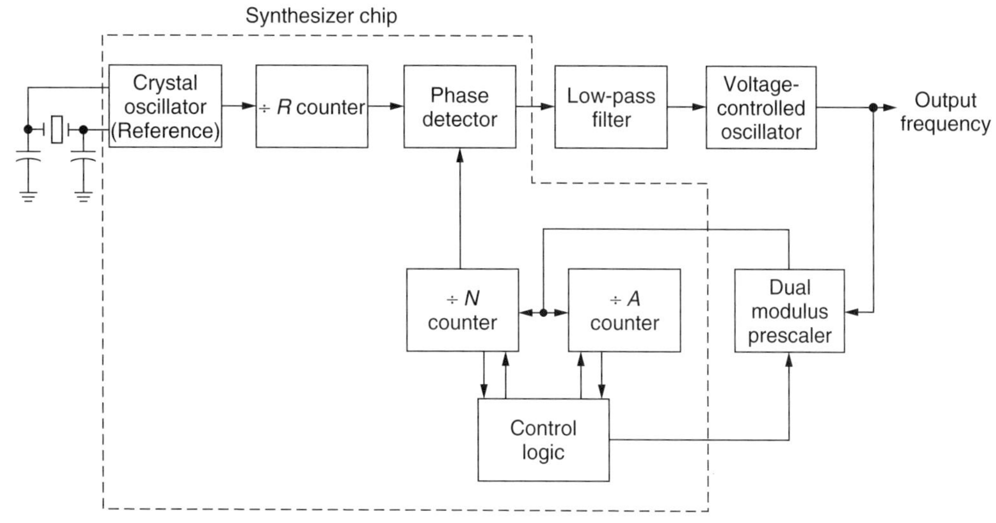

Figure 1. Block diagram of an integrated frequency synthesizer. In this case, the designer has control over the VCO and the loop filter; the reference oscillator is part of the chip. In most cases ($\leq$2.5 GHz), the dual-modulus prescaler is also inside the chip.

The quality of the system, or signal generator, is determined by the properties of the synthesizer and, of course, its building blocks. There are a variety of parameters in characterizing the synthesizer. To name a few important ones, we need to worry about the frequency stability, spurious suppression, and phase noise characteristics. The frequency resolution of the synthesizer tends to be covered by the switching speed and its spurious response. A lot of research is put into developing the ideal synthesizer, whatever ideal means. For portable applications such as cell phones, size and power consumption is a real issue, as well as the cost of the system. As mentioned, there are several competing approaches. In particular, there is a race between the fractional-N division synthesizer and the direct digital synthesis. The fractional synthesizer allows for generation of the output frequencies, which are not exact integers of the reference frequency. This results in an average frequency and problems with spurious sidebands. The direct digital synthesis uses a lookup table to construct a sine-wave, and the size of the lookup table and the sample rate determines the quality of the signal and its output frequency.

2. FREQUENCY SYNTHESIZER FUNDAMENTALS

There are several approaches to "synthesize" a frequency as we already mentioned. Probably the first and the oldest approach is called the direct frequency synthesis where a bank of crystals, as frequency standards, will be used to generate output frequencies. Such a system is called *frequency-incoherent*. There is no phase coherency between the various oscillators [8].

A simple example of direct synthesis is shown in Fig. 2. The new frequency $\frac{2}{3}f_0$ is realized from f_0 by using a divide-by-3 circuit, a mixer, and a bandpass filter. In this example $\frac{2}{3}f_0$ has been synthesized by operating directly on f_0.

Figure 3 illustrates the form of direct synthesis module most frequently used in commercial frequency

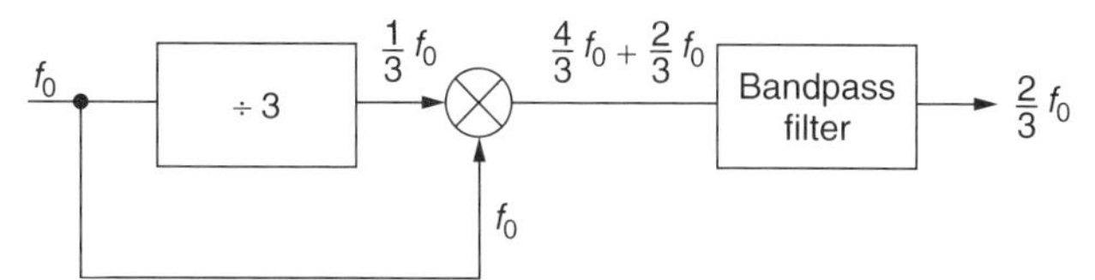

Figure 2. Direct frequency generation using the mix-and-divide principle. It requires excessive filtering.

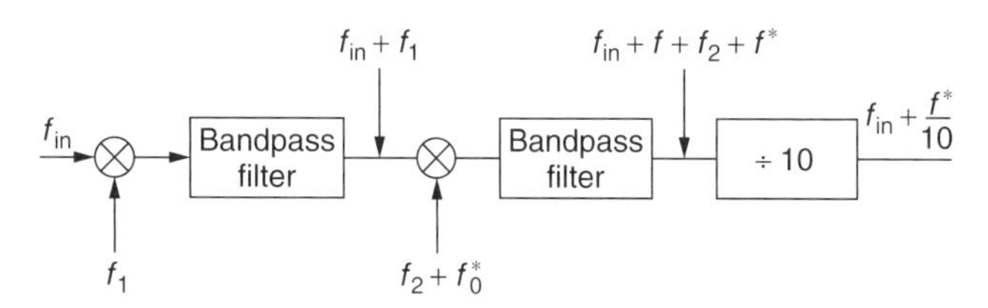

Figure 3. Direct frequency synthesizer using a mix-and-divide technique to obtain identical modules for high resolution.

synthesizers of the direct form. The method is referred to as the *double-mix-divide* approach.

An input frequency f_{in} is combined with a frequency f_1, and the upper frequency $f_1 + f_{in}$ is selected by the bandpass filter. This frequency is then mixed with a switch-selectable frequency $f_2 + f^*$ (in the following text f^* refers to any one of 10 switch-selectable frequencies). The output of the second mixer consists of the two frequencies $f_{in} + f_1 + f_2 + f^*$ and $f_{in} + f_1 - f_2 - f^*$; only the higher-frequency term appears at the output of the bandpass filter. If the frequencies f_{in}, f_1, and f_2 are selected so that

$$f_{in} + f_1 + f_2 = 10 f_{in} \tag{1}$$

then the frequency at the output of the divide by 10 will be

$$f_{out} = f_{in} + \frac{f^*}{10} \tag{2}$$

The double-mix-divide module has increased the input frequency by the switch-selectable frequency increment $f^*/10$. These double-mix-divide modules can be cascaded to form a frequency synthesizer with any degree of resolution. The double-mix-divide modular approach has the additional advantage that the frequencies f_1, f_2, and f_{in} can be the same in each module, so that all modules can contain identical components.

A direct frequency synthesizer with three digits of resolution is shown in Fig. 4. Each decade switch selects one of 10 frequencies $f_2 + f^*$. In this example the output of the third module is taken before the decade divider.

For example, it is possible to generate the frequencies between 10 and 19.99 MHz (in 10-kHz increments), using the three module synthesizer, by selecting

$$f_{in} = 1 \text{ MHz}$$

$$f_1 = 4 \text{ MHz}$$

$$f_2 = 5 \text{ MHz}$$

Since

$$f_{in} + f_1 + f_2 = 10f_{in} \tag{3}$$

the output frequency will be

$$f_0 = 10f_{in} = f_3^* + \frac{f_2^*}{10} + \frac{f_1^*}{100} \tag{4}$$

Since f^* occurs in 1-MHz increments, $f_1^*/100$ will provide the desired 10-kHz frequency increments.

Theoretically, either f_1 or f_2 could be eliminated, provided

$$f_{in} + f_1(\text{or } f_2) = 10f_{in} \tag{5}$$

but the additional frequency is used in practice to provide additional frequency separation at the mixer output. This frequency separation eases the bandpass filter requirements. For example, if f_2 is eliminated, $f_1 + f_{in}$ must equal $10f_{in}$ or 10 MHz. If an f_1^* of 1 MHz is selected, the output of the first mixer will consist of the two frequencies 9 and 11 MHz. The lower of these closely spaced frequencies must be removed by the filter. The filter required would be extremely complex. If, instead, a 5-MHz signal f_2 is also used so that $f_{in} + f_1 + f_2 - 10$ MHz, the two frequencies at the first mixer output will (for an f_1^* of 1 MHz) be 1 and 11 MHz. In this case the two frequencies will be much easier to separate with a bandpass filter. The auxiliary frequencies f_1 and f_2 can be selected in each design only after considering all possible frequency products at the mixer output.

Direct synthesis can produce fast frequency switching, almost arbitrarily fine frequency resolution, low phase noise, and the highest-frequency operation of any of the methods. Direct frequency synthesis requires considerably more hardware (oscillators, mixers, and bandpass filters) than do the two other synthesis techniques to be described. The hardware requirements result in direct synthesizers becoming larger and more expensive. Another disadvantage of the direct synthesis technique is that unwanted (spurious) frequencies can appear at the output. The wider the frequency range, the more likely that the spurious components will appear in the output. These disadvantages are offset by the versatility, speed, and flexibility of direct synthesis.

2.1. PLL Synthesizer

The most popular synthesizer is based on PLL [9]. An example of such a PLL-based synthesizer is shown in Fig. 5. In preparing ourselves for hybrid synthesizers, it needs to be noted that the frequency standard can also be replaced by a direct digital synthesizer (DDS). The number of references for synthesizers seems endless, but the most complete and relevant ones are given at the end. Its also very important to monitor the patents. If any incremental improvements are achieved there, the inventors immediately try to protect them with a patent. The following will give some insight into the major building blocks used for PLL and hybrid synthesizers. It should be noted that most designers use a combination of available integrated circuits, and most of the high-performance solutions are due to the careful design of the oscillator, the integrated/lowpass filter, and the systems architecture. All of these items will be addressed, particularly the fractional-N synthesizer, which requires a lot of handholding in the removal of spurious products.

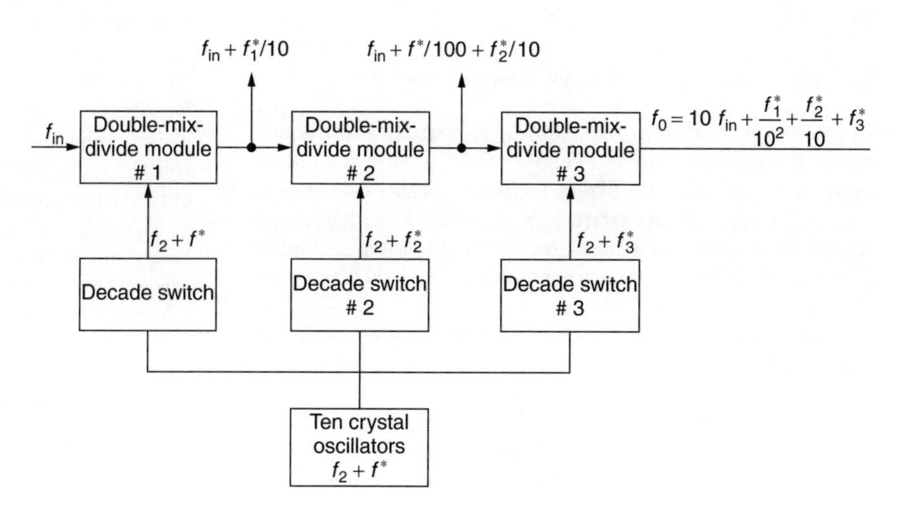

Figure 4. Phase-incoherent frequency synthesizer with three-digit resolution.

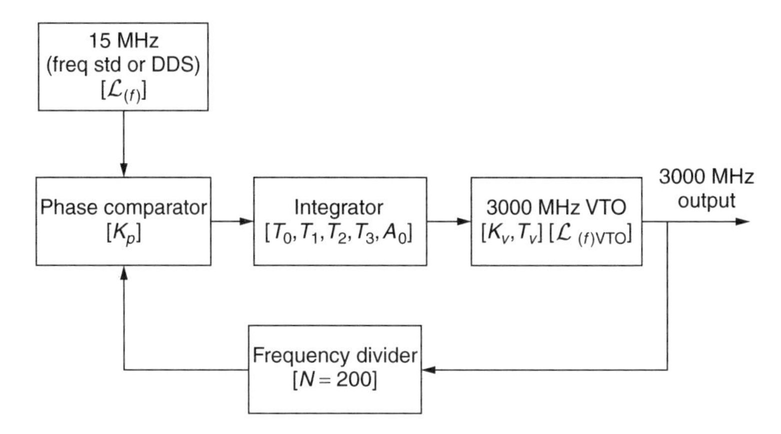

Figure 5. Block diagram of a PLL synthesizer driven by a frequency standard, DDS, or fractional-N synthesizer for high resolution at the output. The last two standards allow a relatively low division ratio and provide quasiarbitrary resolution.

How does the PLL work? According to Fig. 5, we have a free-running oscillator, which can operate anywhere from audio to millimeterwave. The output is typically sinusoidal. The VCO also is occasionally called a voltage-tuned oscillator (VTO), which drives an output stage and a pulseshaping stage to prepare the signal to drive a frequency divider chain. The frequency divider chain consists of silicon germanium or GaAs dividers to reduce the signal below 1000 MHz. At these frequencies, either silicon-based dividers or CMOS dividers can take over. The frequency divider, typically part of an integrated circuit, is a synchronous divider, which is programmable over a wide range. Division ratios as low as four and as high as one million are possible [10].

The output of the frequency divider is fed to a phase comparator, which in most cases is actually a phase-frequency detector. The phase frequency detector compares the output of the frequency divider, which typically is the same magnitude as the reference frequency, with a reference frequency, which is derived from a frequency standard. Frequency standards come as precision standards such as atomic frequency standards, followed by oven-controlled crystal oscillators, to temperature-compensated crystal oscillators. Sometimes even simple crystal oscillators will do the trick. The output from the phase comparator is a DC voltage typically between 1 and 25 V, which is applied to the tuning diode of the VTO or VCO. This tuning voltage is modulated by the differences of the two prior to lock. The frequency detector portion of the phase frequency comparator jams the voltage to one extreme charging the capacitors and integrator and acquiring frequency lock. After frequency lock is obtained, the control voltage will change the frequency at the output to have a fixed phase relationship compared to the reference frequency. The advantage of having a frequency detector in parallel to a phase detector is that the system always requires frequency lock [11,12].

2.2. Fractional-*N* Phase-Locked Loops

The principle of the fractional-N PLL synthesizer has been around for a while. In the past, implementation of this has been done in an analog system [13,14]. It would be ideal to be able to build a single-loop synthesizer with a 1.25- or 50-MHz reference and yet obtain the desired step-size resolution, such as 25 kHz. This would lead to the much smaller division ratio and much better phase noise performance. Figure 6 shows the block of an analog fractional-N synthesizer.

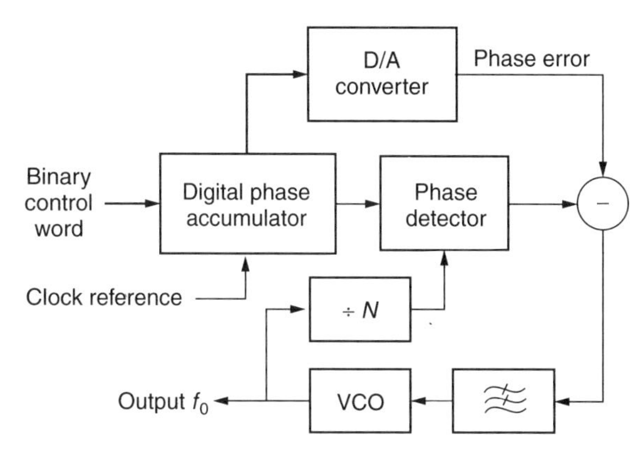

Figure 6. Simplified block diagram of an analog fractional-N synthesizer.

An alternative would be for N to take on fractional values. The output frequency could then be changed in fractional increments of the reference frequency. Although a digital divider cannot provide a fractional division ratio, ways can be found to accomplish the same task effectively.

The most frequently used method is to divide the output frequency by $N + 1$ every M cycles and to divide by N the rest of the time. The effective division ratio is then $N + 1/M$, and the average output frequency is given by

$$f_0 = \left(N + \frac{1}{M}\right) f_r \tag{6}$$

This expression shows that f_0 can be varied in fractional increments of the reference frequency by varying M. The technique is equivalent to constructing a fractional divider, but the fractional part of the division is actually implemented using a phase accumulator. The phase accumulator approach is illustrated in Section 6. This method can be expanded to frequencies much higher than 6 GHz using the appropriate synchronous dividers. For more details, see Section 6 [15–53].

2.3. Digital Direct Frequency Synthesizer

The digital direct frequency uses sampled data methods to produce waveforms [54–69]. The digital hardware block provides a datastream of k bits per clock cycle for digital-to-analog conversion (DAC). Ideally, the DAC is a linear device with glitch-free performance. The practical limits of the DAC will be discussed later in this article. The DAC output is the desired signal plus replications of it around the clock frequency and all of the clock's harmonics. Also present in the DAC output signal is a small amount of quantization noise from the effects of finite math in the hardware block. Figure 7 shows the frequency spectrum of an ideal DAC output with a digitally sampled sine-wave datastream at its input. Note that the desired signal, f_0 (a single line in the frequency domain), is replicated around all clock terms. Figure 8 shows the same signal in the time domain.

The DAC performs a sample-and-hold operation as well as converting digital values to analog voltages. The sample occurs on each rising edge of the clock; the hold occurs during the clock period. The transfer function of a sample-and-hold operator is a $(\sin x)/x$ envelope response with linear phase. In this case, $x = (\pi F/F_{\text{clock}})$. It should be noted that the sinc function rolloff affects the passband flatness. A 2.4 dB roll-off should be expected at 40% of F_{clock}.

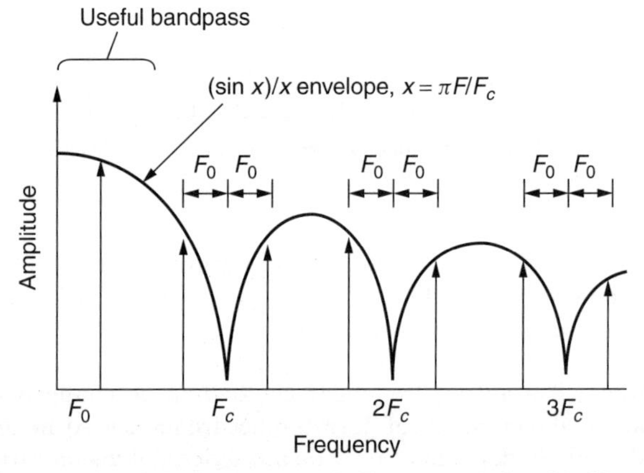

Figure 7. Ideal DAC output with F_0, a sampled-and-held sine wave, at its output. Notice the $(\sin x)/x$ envelope rolloff. As F_0 moves up in frequency, an aliased component $F_C - F_0$ moves down into the passband.

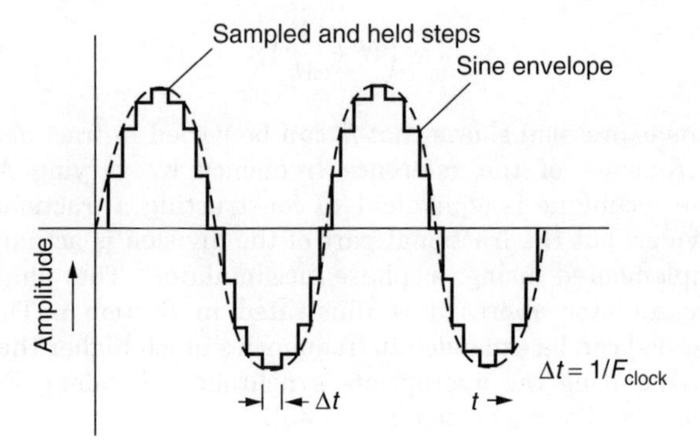

Figure 8. Samples per cycle sine wave. This is typical of a single-tone DAC output. $F_0 = F_{\text{clock}}/16$ after low pass filtering; only the sine envelope is present. The lowpass filter removes the sampling energy. Each amplitude step is held for a clock period.

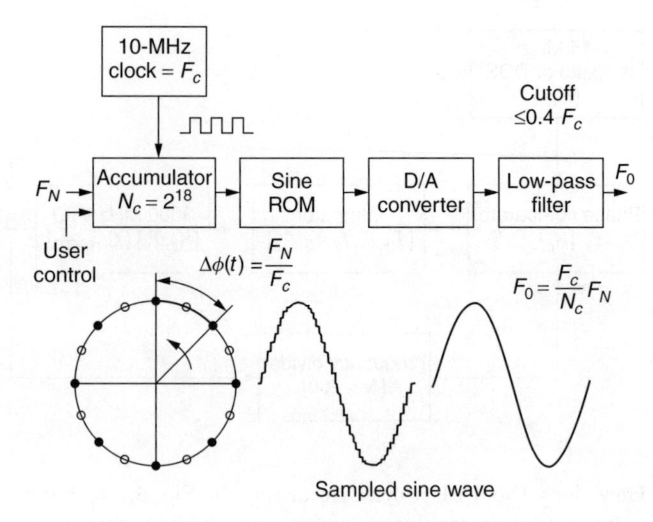

Figure 9. Direct digital frequency synthesizer.

Referring to Fig. 9, the output of the DAC is passed through a lowpass filter (LPF). With proper attention to design, an LPF may be realized that has linear phase in a flat passband with a width of $0.4F_{\text{clock}}$. With this design, the maximum available bandwidth is achieved. For example, with $F_{\text{clock}} = 125$ MHz, the useful synthesized bandwidth of about 50 MHz is attained. The LPF output is the desired signal without any sampling artifacts. Viewing the LPF strictly as a device to remove sampling energy, it is obvious why the output contains only the desired signal. It is also instructive to view the LPF from the time domain. From this point, the LPF may be seen as the perfect interpolator. It fills the space between time samples with a smooth curve to reconstruct perfectly the desired signal.

In the design of a DDS, the following guidelines apply:

- The desired frequency resolution determines the lowest output frequency f_L.
- The number of D/A conversions used to generate f_L is $N = 4k = 4f_U/f_L$, provided four conversions are used to generate $f_U (P = 4)$.
- The maximum output frequency f_U is limited by the maximum sampling rate of the DDS, $f_U \leq 1/4T$. Conversely, $T \leq 1/4f_U$.

To generate nf_L, the integer n addresses the register, and each clock cycle kn is added to the content of the accumulator so that the content of the memory address register is increased by kn. Each knth point of the memory is addressed, and the content of this memory location is transferred to the D/A converter to produce the output sampled waveform.

To complete the DDS, the memory size and length (number of bits) of the memory word must be determined. The word length is determined by system noise requirements. The amplitude of the D/A output is that of an exact sinusoid corrupted with the deterministic noise due to truncation caused by the finite length of the digital words (quantization noise). If an $(n + 1)$-bit word length (including one sign bit) is used and the output of the A/D converter varies between ±1, the mean noise from the quantization will be

$$\rho^2 = \frac{1}{12}\left(\frac{1}{2}\right)^{2n} = \frac{1}{3}\left(\frac{1}{2}\right)^{2(n+1)} \quad (7)$$

The mean noise is averaged over all possible waveforms. For a worst-case waveform, the noise is a square wave with amplitude $\frac{1}{2}(\frac{1}{2})^n$ and $\rho^2 = \frac{1}{4}(\frac{1}{2})^{2n}$ for each bit added to the word length, the spectral purity improves by 6 dB.

The main drawback of a low-power DDS is that it is limited to relatively low frequencies. The upper frequency is directly related to the maximum usable clock frequency; today, the limit is about 1 GHz. DDS tends to be noisier than other methods, but adequate spectral purity can be obtained if sufficient lowpass filtering is used at the output. DDS systems are easily constructed using readily available microprocessors. The combination of DDS for fine frequency resolution plus other synthesis techniques to obtain higher-frequency output can provide high resolution with very rapid setting time after a frequency change. This is especially valuable for frequency-hopping spread-spectrum systems.

In analyzing both the resolution and signal-to-noise ratio (or rather signal-to-spurious performance) of the DDS, one has to know the resolution and input frequencies. As an example, if the input frequency is approximately 35 MHz and the implementation is for a 32-bit device, the frequency resolution compared to the input frequency is $35 \times 10^6 \div 2^{32} = 35 \times 10^6 \div 4.294967296 \times 10^9$ or 0.00815 Hz ≈0.01 Hz. Given the fact that modern shortwave radios with a first IF of about 75 MHz will have an oscillator between 75 and 105 MHz, the resolution at the output range is more than adequate. In practice, one would use the microprocessor to round it to the next increment of 1 Hz relative to the output frequency.

As to the spurious response, the worst-case spurious response is approximately $20 \log 2^R$, where R is the resolution of the digital/analog converter. For an 8-bit A/D converter, this would mean approximately 48 dB down (worst case), as the output loop would have an analog filter to suppress close-in spurious noise. Modern devices have a 14-bit resolution. Fourteen bits of resolution can translate into $20 \log 2^{14}$ or 80 dB, worse case, of suppression. The actual spurious response would be much better. The current production designs for communication applications, such as shortwave transceivers, despite the fact that they are resorting to a combination of PLLs and DDSs, still end up somewhat complicated. By using 10 MHz from the DDS and using a single-loop PLL system, one can easily extend the operation to above 1 GHz but with higher complexity and power consumption. This was shown in Fig 5. Figure 10 shows a multiple-loop synthesizer using a DDS for fine resolution.

3. IMPORTANT CHARACTERISTICS OF SYNTHESIZERS

The following is a list of parameters that are used to describe the performance of the synthesizer. These are referred to as figures of merit.

3.1. Frequency Range

The output frequency of a synthesizer can vary over a wide range. A synthesizer signal generator typically offers output frequencies from as low as 100 kHz to as high as several gigahertz. The frequency range is determined by the architecture of the signal generator as the system frequently uses complex schemes of combining frequencies

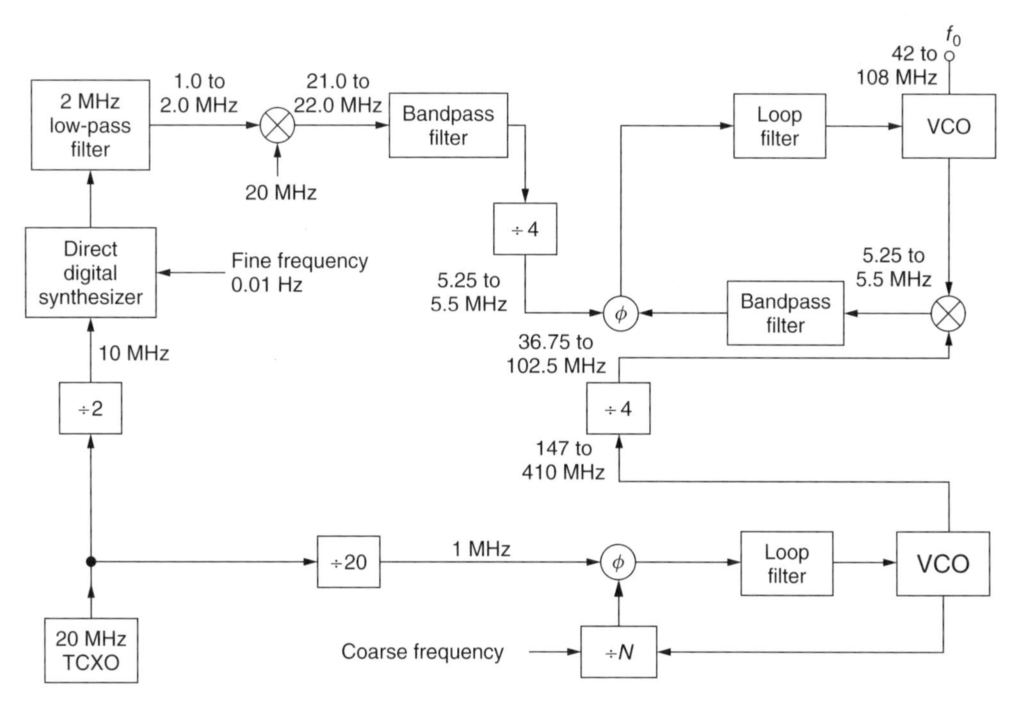

Figure 10. A multiple-loop synthesizer using a DDS for fine resolution.

in various loops. A standard loop-based synthesizer has a frequency range typically less than 1-2, as an example, 925–1650 MHz.

3.2. Phase Noise

Oscillators unfortunately are not clean, but the various noise sources in and outside of the transistor modulate the VCO, resulting in energy or spectral distribution on both sides of the carrier. This occurs via modulation and conversion. The noise, or better, FM noise is expressed as the ratio of output power divided by the noise power relative to 1 Hz bandwidth measured at an offset of the carrier. Figure 11 shows a typical phase noise plot of a synthesizer. Inside the loop bandwidth, the carrier signal is cleaned up, and outside the loop bandwidth, the measurement shows the performance of the VCO itself.

3.3. Output Power

The output power is measured at the designated output port of the frequency synthesizer. Practical designs require an isolation stage. Typical designs require one or more isolation stages between the oscillator and the output. The output power needs to be flat. While the synthesized generator typically is flat with only 0.1 dB +/− deviation, the VCO itself can vary as much as +/− 2 dB over the frequency range.

3.4. Harmonic Suppression

The VCO inside a synthesizer has a typical harmonic suppression of better than 15 dB. For high-performance applications, a set of lowpass filters at the output will reduce the harmonic contents to a desired level. Figure 12 shows a typical output power plot of a VCO.

3.5. Output Power as a Function of Temperature

All active circuits vary in performance as a function of temperature. The output power of an oscillator over a temperature range should vary less than a specified value, such as 1 dB.

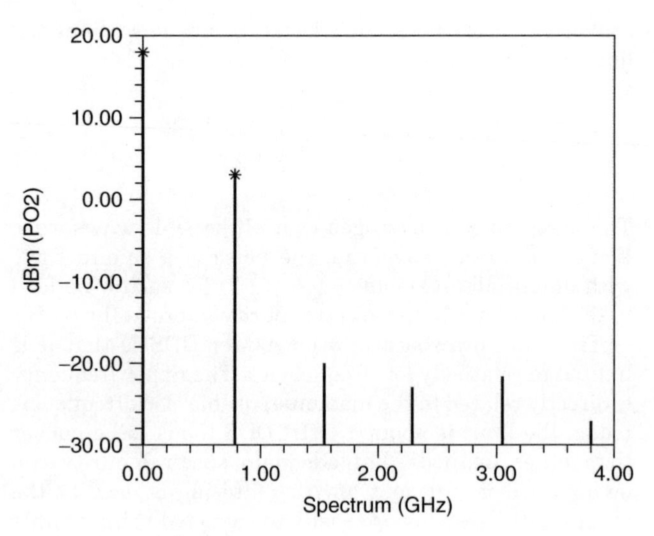

Figure 12. Predicted harmonics at the output of a VCO.

3.6. Spurious Response

Spurious outputs are signals found around the carrier of a synthesizer that are not harmonically related. Good, clean synthesizers need to have a spurious-free range of 90 dB, but these requirements make them expensive. While oscillators typically have no spurious frequencies besides possibly 60- and 120-Hz pickup, the digital electronics in a synthesizer generates a lot of signals, and when modulated on the VCO, are responsible for these unwanted output products. (See also Fig. 11.)

3.7. Step Size

The resolution, or step size, is determined by the architecture.

3.8. Frequency Pushing

Frequency pushing characterizes the degree to which an oscillator's frequency is affected by its supply voltage. For

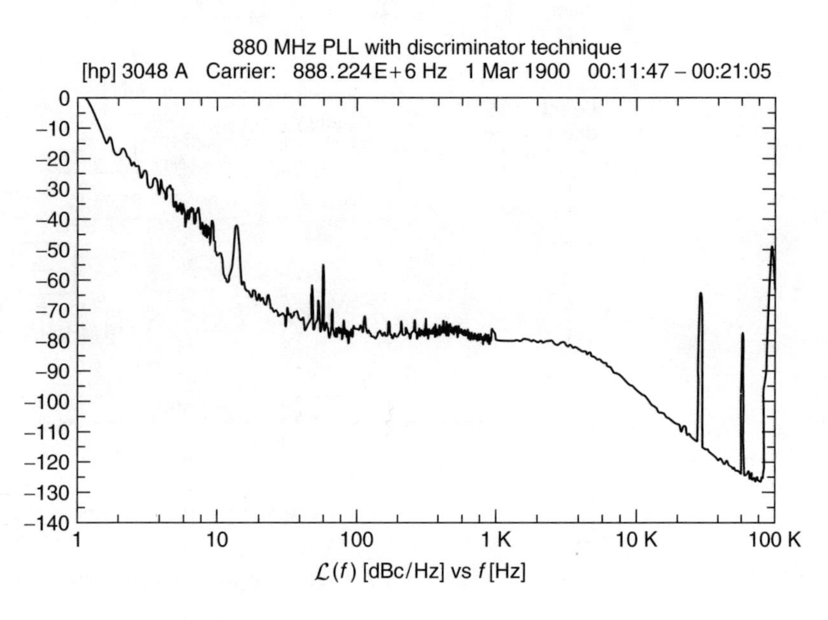

Figure 11. Measured phase noise of a 880-MHz synthesizer using a conventional synthesizer chip.

example, a sudden current surge caused by activating a transceiver's RF power amplifier may produce a spike on the VCO's DC power supply and a consequent frequency jump. Frequency pushing is specified in frequency/voltage form and is tested by varying the VCO's DC supply voltage (typically ± 1 V) with its tuning voltage held constant.

3.9. Sensitivity to Load Changes

To keep manufacturing costs down, many wireless applications use a VCO alone, without the buffering action of a high reverse-isolation amplifier stage. In such applications, frequency pulling, the change of frequency resulting from partially reactive loads, is an important oscillator characteristic. Pulling is commonly specified in terms of the frequency shift that occurs when the oscillator is connected to a load that exhibits a nonunity VSWR (such as 1.75, usually referenced to 50 Ω), compared to the frequency that results with unity-VSWR load (usually 50 Ω). Frequency pulling must be minimized, especially in cases where power stages are close to the VCO unit and short pulses may affect the output frequency. Such poor isolation can make phase locking impossible.

3.10. Tuning Sensitivity

This is a VCO parameter also expressed in frequency/voltage and is not part of a synthesizer specification.

3.11. Posttuning Drift

After a voltage step is applied to the tuning diode input, the oscillator frequency may continue to change until it settles to a final value. The posttuning drift is one of the parameters that limits the bandwidth of the VCO input.

3.12. Tuning Characteristic

This specification shows the relationship, depicted as a graph, between the VCO operating frequency and the tuning voltage applied. Ideally, the correspondence between operating frequency and tuning voltage is linear.

3.13. Tuning Linearity

For stable synthesizers, a constant deviation of frequency versus tuning voltage is desirable. It is also important to make sure that there are no breaks in tuning range, for example, that the oscillator does not stop operating with a tuning voltage of 0 V.

3.14. Tuning Sensitivity and Tuning Performance

This datum, typically expressed in megahertz per volt (MHz/V), characterizes how much the frequency of a VCO changes per unit of tuning voltage change.

3.15. Tuning Speed

This characteristic is defined as the time necessary for the VCO to reach 90% of its final frequency upon the application of a tuning voltage step. Tuning speed depends on the internal components between the input pin and the tuning diode, including the capacitance present at the input port. The input port's parasitic elements determine the VCO's maximum possible modulation bandwidth.

3.16. Power Consumption

This characteristic conveys the DC power, usually specified in milliwatts and sometimes qualified by operating voltage, required by the oscillator to function properly [70].

4. BUILDING BLOCKS OF SYNTHESIZERS

4.1. Oscillator

An oscillator is essentially an amplifier with sufficient feedback so the amplifier becomes unstable and begins oscillation. The oscillator can be divided into an amplifier, a resonator, and a feedback system [71]. One of the most simple equations describes this. It describes the input admittance of an amplifier with a tuned circuit at the output described by the term Y_L.

$$Y_{11}^{*} = Y_{11} - \frac{Y_{12} \times Y_{21}}{Y_{22} + Y_L} \tag{8}$$

The feedback is determined by the term Y_{12}, or in practical terms, by the feedback capacitor. The transistor oscillator (Fig. 13a) shows a grounded base circuit. For this type of oscillator, using microwave transistors, the emitter and collector currents are in phase. That means that the input circuit of the transistor needs to adjust the phase of the oscillation. The transistor stage forms the amplifier, the tuned resonator at the output determines the frequency of oscillation, and the feedback capacitor provides enough energy back into the transistor so that oscillation can occur. To start oscillation, the condition, relative to the output, can be derived in a similar fashion from the input:

$$Y_{22}^{*} = (Y_{22} + Y_L) - \frac{Y_{12} \times Y_{21}}{Y_{11} + Y_T} \tag{9}$$

If the second term of the equation on the right side is larger than the first term on the right of the = sign, then Real (Y_{22}^{*}) is negative and oscillation at the resonant frequency occurs.

The oscillator circuit shown in Fig 13c is good for low-frequency applications, but the Colpitts oscillator is preferred for higher frequencies. The Colpitts oscillator works by rotating this circuit and grounds the collector instead of the base. The advantage of the Colpitts oscillator is the fact that it is an emitter–follower amplifier with feedback and shows a more uniform gain of a wider frequency range (see Fig. 14). Oscillators with a grounded emitter or source have more stability difficulties over wider frequency ranges.

Depending on the frequency range, there are several resonators available. For low frequencies up to about 500 MHz, lumped elements such as inductors are useful. Above these frequencies, transmission line–based resonators are microwave resonators, which are better. Very

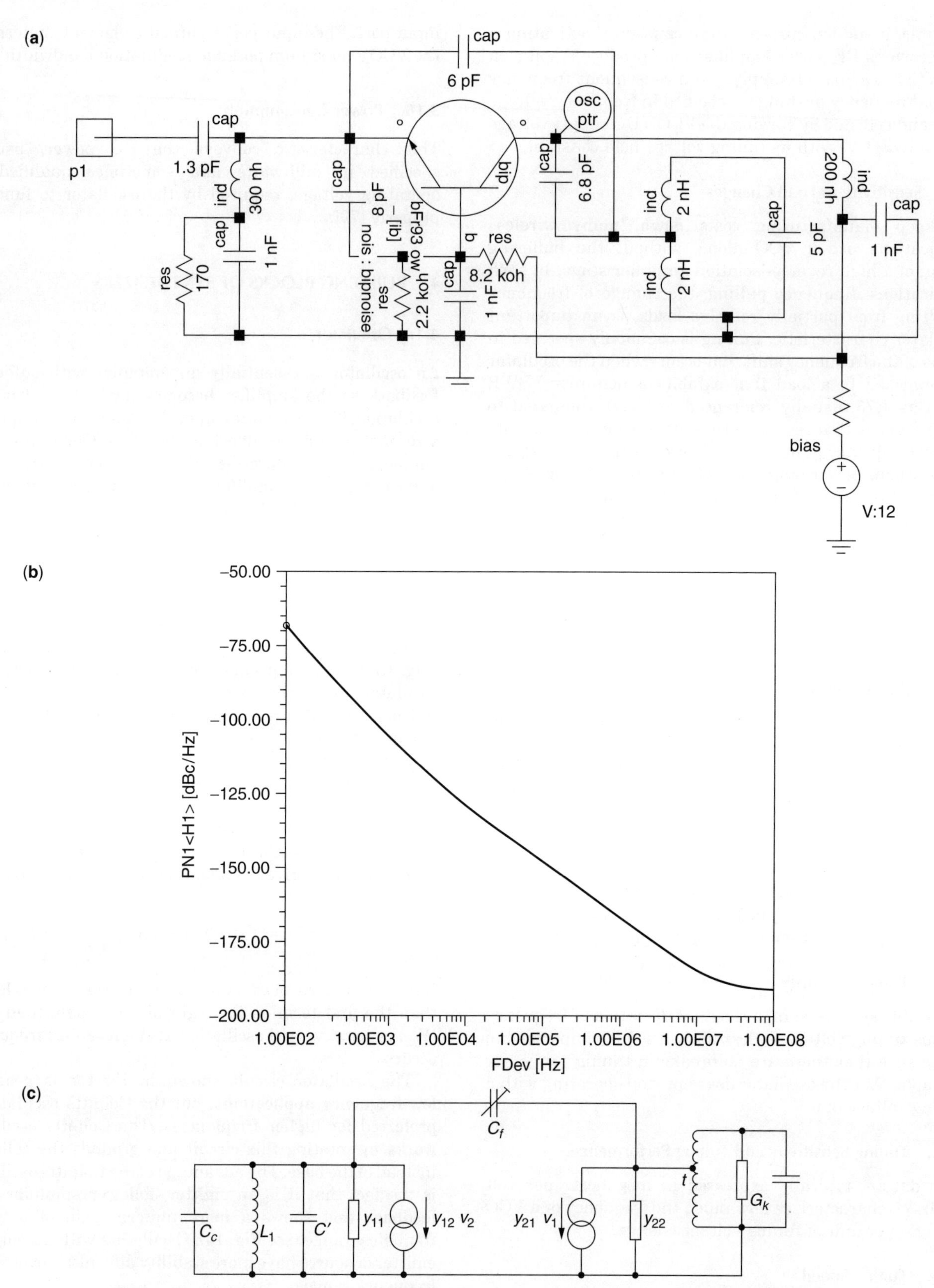

Figure 13. (**a**) A grounded base VHF/UHF oscillator with a tuned resonator tapped in the middle to improve operational Q; (**b**) the predicted phase noise of the oscillator in (**a**); **c** the equivalent circuit of the oscillator configuration of (**a**). For the grounded base configuration, emitter and collector currents have to be in phase. The phase shift introduced by C_f is compensated by the input tuned circuit.

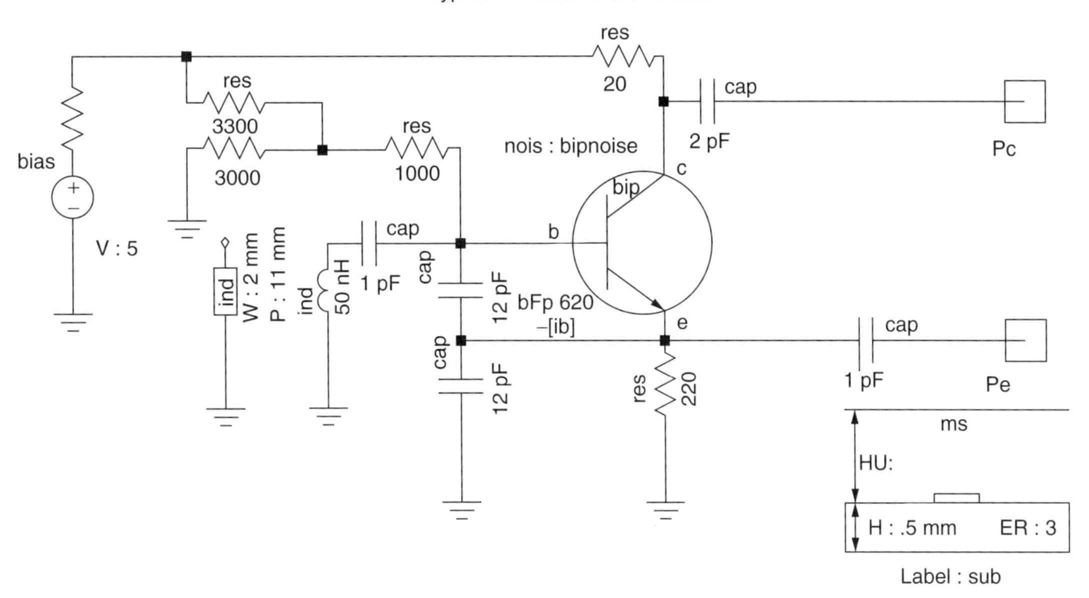

Figure 14. Colpitts oscillator using a coil inductor or a transmission line as a resonator. In this case all the physical parameters of the transmission line have to be provided.

small oscillators use printed transmission lines. High-performance oscillators use ceramic resonators (CROs), dielectric resonators (DROs), and surface acoustic wave resonators (SAWs), to name only a few. The frequency standard, if not derived from an atomic frequency standard is typically a crystal oscillator. Even the atomic frequency standards synchronize a crystal against an atomic resonance to compensate for the aging of the crystal. Figure 15a shows a circuit of a crystal oscillator with typical component values that are based on a 10-MHz third overtone AT cut crystal. Q is the ratio of stored energy divided by dissipated energy. The Q of the crystal can be as high as one million. The figure of merit Q is defined by $\omega L/R$. In our case $\omega L/R$ is

$$\frac{2\pi \times 1Hy \times 10 \times 10^6}{50} = 1.25 \text{ million}$$

Therefore, the resulting Q is 1.25 million. Typical resonator Q values for LC oscillators are 200; for structure-based resonators such as ceramic resonators or dielectric resonators, Q values of 400 are not uncommon.

An oscillator operating at 700 MHz is shown in Fig. 15a, including its schematic. CAD simulation was used to determine the resonance frequency phase noise and the harmonic contents, shown in Fig. 15b. The output power can be taken off either the emitter, at which provides better harmonic filtering, or from the collector, which provides smaller interaction between the oscillator frequency and the load. This effect is defined as *frequency pulling*. The tuned capacitor can now be replaced by a voltage-dependent capacitor. A tuning diode is a diode operated in reverse. Its PN junction capacitance changes as a function of applied voltage.

A two-port oscillator analysis will now be presented. It is based on the fact that an ideal tuned circuit (infinite Q), once excited, will oscillate infinitely because there is no resistance element present to dissipate the energy. In the actual case where the inductor Q is finite, the oscillations die out because energy is dissipated in the resistance. It is the function of the amplifier to maintain oscillations by supplying an amount of energy equal to that dissipated. This source of energy can be interpreted as a negative resistor in series with the tuned circuit. If the total resistance is positive, the oscillations will die out, while the oscillation amplitude will increase if the total resistance is negative. To maintain oscillations, the two resistors must be of equal magnitude. To see how a negative resistance is realized, the input impedance of the circuit in Fig. 16 will be derived.

If Y_{22} is sufficiently small ($Y_{22} \ll 1/R_L$), the equivalent circuit is as shown in Fig. 16. The steady-state loop equations are

$$V_{\text{in}} = I_{\text{in}}(X_{C_1} + X_{C_2}) - I_b(X_{C_1} - \beta X_{C_2}) \tag{10}$$

$$0 = -I_{\text{in}}(X_{C_1}) + I_b(X_{C_{1+h_{ie}}}) \tag{11}$$

After I_b is eliminated from these two equations, Z_{in} is obtained as

$$Z_{\text{in}} = \frac{V_{\text{in}}}{I_{\text{in}}} = \frac{(1+\beta)X_{C_1}X_{C_2} + h_{ie}(X_{C_1} + X_{C_2})}{X_{C_1} + h_{ie}} \tag{12}$$

If $X_{C_1} \ll h_{ie}$, the input impedance is approximately equal to

$$Z_{\text{in}} \approx \frac{1+\beta}{h_{ie}} X_{C_1}X_{C_2} + (X_{C_1} + X_{C_2}) \tag{13}$$

$$Z_{\text{in}} \approx \frac{-g_m}{\omega^2 C_1 C_2} + \frac{1}{j\omega[C_1C_2/(C_1+C_2)]} \tag{14}$$

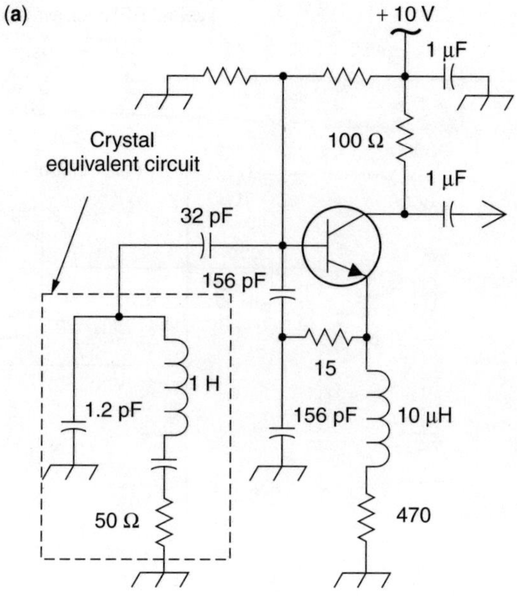

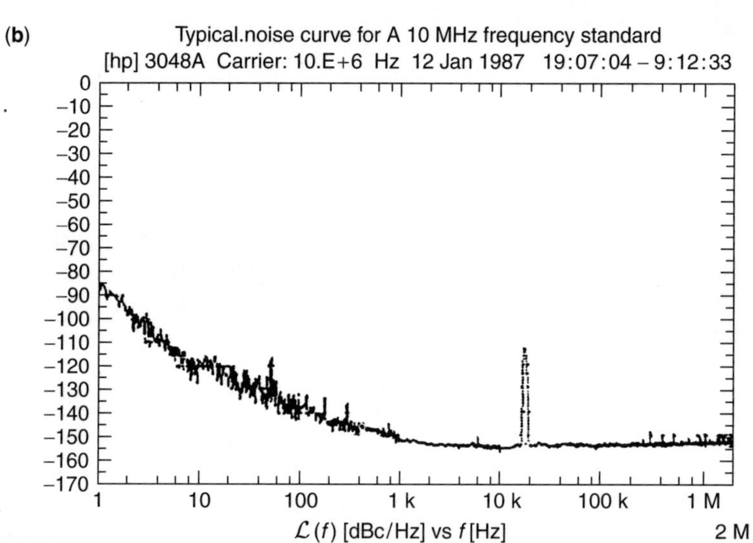

Figure 15. (**a**) Abbreviated circuit of a 10-MHz crystal oscillator; (**b**) measured phase noise for this frequency standard by HP of (**a**).

Derivation:

I_{in} C_1 R_L V_{in} Z_{in} C_2

Figure 16. Calculation of input impedance of the negative-resistance oscillator.

That is, the input impedance of the circuit shown in Fig. 17 is a negative resistor

$$R = \frac{-g_m}{\omega^2 C_1 C_2} \tag{15}$$

in series with a capacitor

$$C_{\text{in}} = \frac{C_1 C_2}{C_1 + C_2} \tag{16}$$

which is the series combination of the two capacitors.

With an inductor L (with the series resistance R_S) connected across the input, it is clear that the condition for sustained oscillation is

$$R_S = \frac{g_m}{\omega^2 C_1 C_2} \tag{17}$$

and the frequency of oscillation

$$f_o = \frac{1}{2\pi\sqrt{L[C_1 C_2/(C_1 + C_2)]}} \tag{18}$$

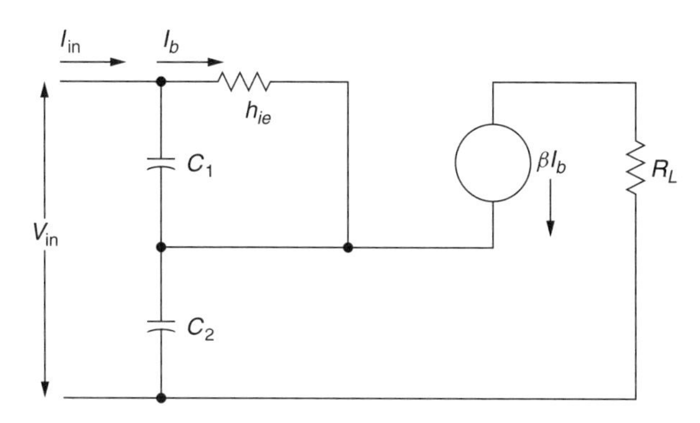

Figure 17. Equivalent small-signal circuit of Fig. 16.

This interpretation of the oscillator readily provides several guidelines that can be used in the design. First, C_1 should be as large as possible so that

$$X_{C_1} \ll h_{ie} \tag{19}$$

and C_2 is to be large so that

$$X_{C_2} \ll \frac{1}{Y_{22}} \tag{20}$$

When these two capacitors are large, the transistor base-to-emitter and collector-to-emitter capacitances will have a negligible effect on the circuit's performance. However, there is a maximum value of the capacitances since

$$r \leq \frac{g_m}{\omega^2 C_1 C_2} \leq \frac{G}{\omega^2 C_1 C_2} \tag{21}$$

where G is the maximum value of g_m. For a given product of C_1 and C_2, the series capacitance is at maximum when $C_1 = C_2 = C_m$. Thus

$$\frac{1}{\omega C_m} > \sqrt{\frac{r}{G}} \tag{22}$$

The design rule is

$$C_2 = C_1 \times \left|\frac{Y_{21}}{Y_{11}}\right| \tag{23}$$

This equation is important in that it shows that for oscillations to be maintained, the minimum permissible reactance $1/\omega C_m$ is a function of the resistance of the inductor and the transistor's mutual conductance, g_m. Figure 18 shows the resonant and oscillation condition for optimum performance. The negative real value should occur at $X = 0$!

An oscillator circuit known as the *Clapp circuit* or *Clapp–Gouriet circuit* is shown in Fig. 19. This oscillator is equivalent to the one just discussed, but it has the practical advantage of being able to provide another degree of design freedom by making C_0 much smaller than C_1 and C_2.

It is possible to use C_1 and C_2 to satisfy the condition of Eq. (20) and then adjust C_o for the desired frequency of

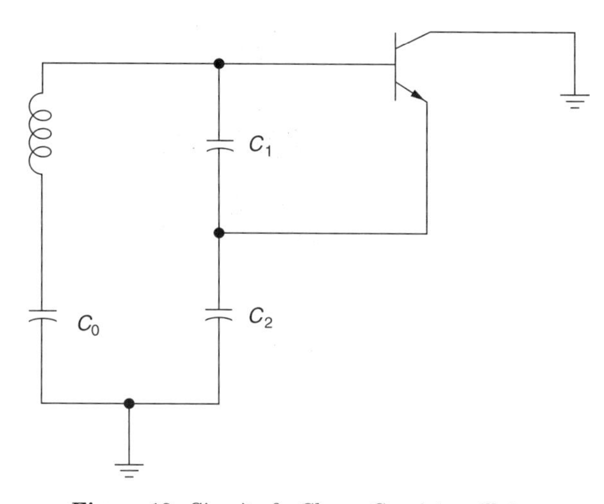

Figure 19. Circuit of a Clapp–Gouriet oscillator.

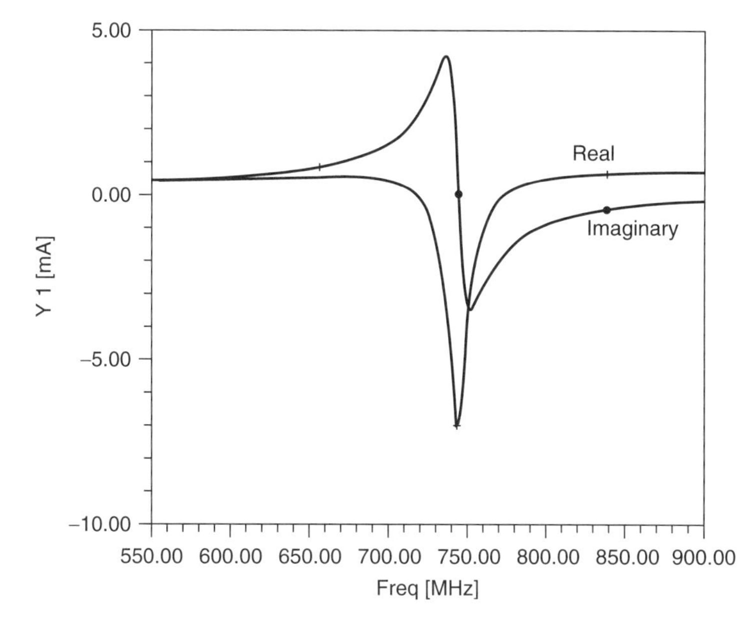

Figure 18. The CAD-based linearized currents. Conditions for oscillation are $X = 0$ and $R < 0$.

oscillation ω_o, which is determined from

$$\omega_o L - \frac{1}{\omega_o C_o} - \frac{1}{\omega_o C_1} - \frac{1}{\omega_o C_2} = 0 \tag{24}$$

Figure 20 shows the Clapp–Gouriet oscillator. Like the Colpitts, the Clapp–Gouriet obtains its feedback via a capacitive voltage divider; unlike the Colpitts, an additional capacitor series-tunes the resonator. The Pierce oscillator, a configuration used only with crystals, is a rotation of the Clapp–Gouriet oscillator in which the emitter is at RF ground [72–75].

4.1.1. Phase Noise. An estimate of the noise performance of an oscillator is as follows:

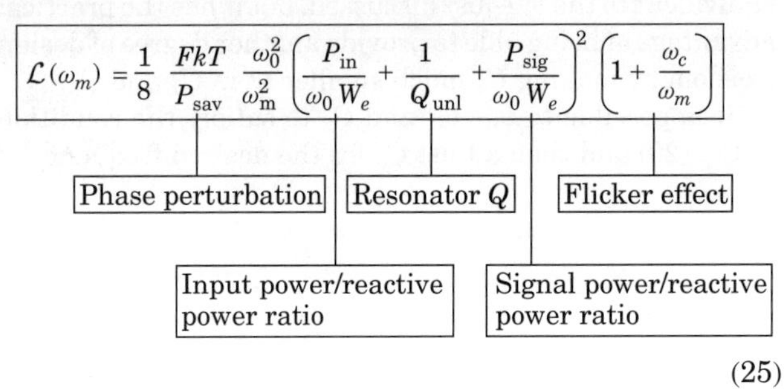

$$\mathcal{L}(\omega_m) = \frac{1}{8}\frac{FkT}{P_{sav}}\frac{\omega_0^2}{\omega_m^2}\left(\frac{P_{in}}{\omega_0 W_e} + \frac{1}{Q_{unl}} + \frac{P_{sig}}{\omega_0 W_e}\right)^2\left(1 + \frac{\omega_c}{\omega_m}\right) \tag{25}$$

Equation (25) is based on work done by Dieter Scherer of Hewlett-Packard about 1978. He was the first to introduce the flicker effect to the Leeson equation by adding the AM-to-PM conversion effect, which is caused by the nonlinear capacitance of the active devices [76]. Figure 21 shows details of the noise contribution. This equation must be further expanded:

$$£(f_m) = 10 \log \left\{\left[1 + \frac{f_0^2}{(2f_m Q_{load})^2}\right]\left(1 + \frac{f_c}{f_m}\right) \times \frac{FkT}{2P_{sav}} + \frac{2kTRK_0^2}{f_m^2}\right\} \tag{26}$$

where $£(f_m)$ = ratio of sideband power in 1-Hz bandwidth at f_m to total power in dB
f_m = frequency offset
f_0 = center frequency
f_c = flicker frequency
Q_{load} = loaded Q of the tuned circuit
F = noise factor
$kT = 4.1 \times 10^{-21}$ at 300 K_0 (room temperature)
P_{sav} = average power at oscillator output
R = equivalent noise resistance of tuning diode (typically 200 Ω–10 kΩ)
K = oscillator voltage gain

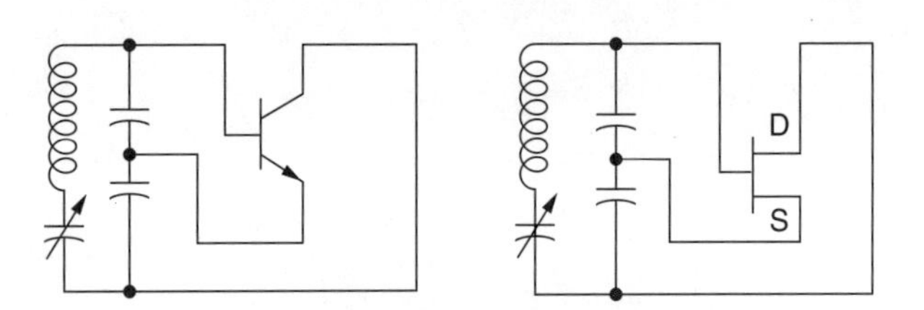

Figure 20. Clapp–Gouriet oscillator.

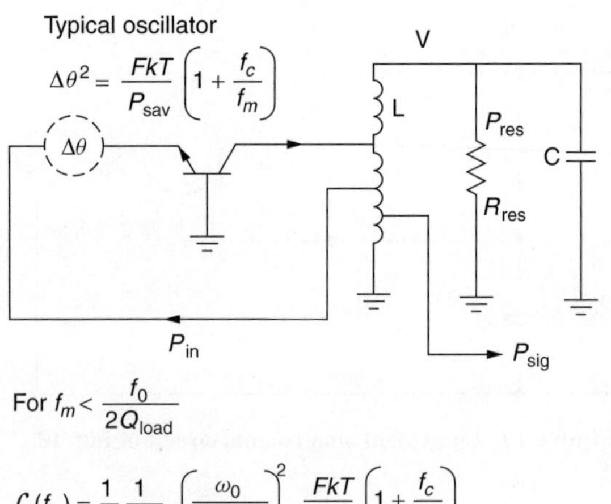

$$\mathcal{L}(f_m) = \frac{1}{2}\frac{1}{\omega_m^2}\left(\frac{\omega_0}{2Q_{load}}\right)^2 \frac{FkT}{P_{sav}}\left(1 + \frac{f_c}{f_m}\right)$$

$$Q_{load} = \frac{\omega_0 W_e}{P_{diss.\,total}} = \frac{\omega_0 W_e}{P_{in} + P_{res} + P_{sig}} = \frac{\text{Reactive power}}{\text{Total dissipated power}}$$

Maximum energy in C or L: $W_e = \frac{1}{2}CV^2$

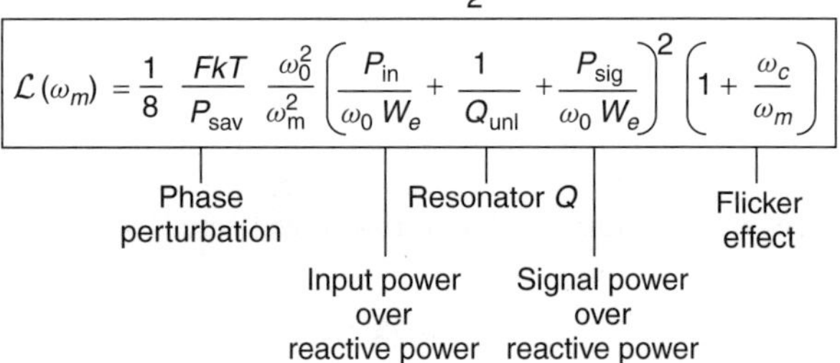

Figure 21. Diagram for a feedback oscillator showing the key components considered in the phase noise calculation and its contribution.

Table 1. Flicker Corner Frequency F_C as a Function of I_C

I_C (mA)	I_C (mA)
0.25	1
0.5	2.74
1	4.3
2	6.27
5	9.3

Source: Motorola

Table 1 shows the flicker corner frequency f_C as a function of I_C for a typical small-signal microwave BJT. $I_{C(max)}$ of this transistor is about 10 mA.

Note that f_C, which is defined by AF and KF in the SPICE model, increases with I_C. This gives us a clue about how f_C changes when a transistor oscillates. As a result of the bias-point shift that occurs during oscillation, an oscillating BJT's average I_C is higher than its small-signal

I_C. KF is therefore higher for a given BJT operating as an oscillator than for the same transistor operating as a small-signal amplifier. This must be kept in mind when considering published f_C data, which are usually determined under small-signal conditions without being qualified as such. Determining a transistor's oscillating f_C is best done through measurement; operate the device as a high-Q UHF oscillator (we suggest using a ceramic-resonator-based tank in the vicinity of 1 GHz), and measure its close-in (10 Hz–10 kHz) phase noise–offset from the carrier. f_C will correspond to a slight decrease in the slope of the phase noise–offset curve. Generally, f_C varies with device type as follows: silicon JFETs, 50 Hz and higher; microwave RF BJTs, 1–10 kHz (as above); MOSFETs, 10–100 kHz; GaAs FETs, 10–100 MHz. Figure 22 shows the phase noise of oscillators using different semiconductors and resonators.

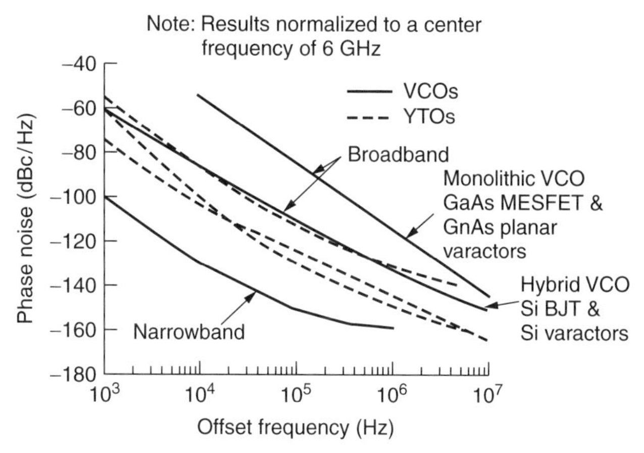

Figure 22. Phase noise of oscillators using different semiconductors and resonators.

The additional term introduces a distinction between a conventional oscillator and a VCO. Whether the voltage- or current-dependent capacitance is internal or external makes no difference; it simply affects the frequency.

For a more complete expression for a resonator oscillator's phase noise spectrum, we can write

$$s_\phi(f_m) = \frac{\alpha_R F_0^4 + \alpha_E \left(\dfrac{F_0}{2Q_L}\right)^2}{f_m^3} + \frac{\left(\dfrac{2GFkT}{P_0}\right)\left(\dfrac{F_0}{2Q_L}\right)^2}{f_m^2} + \frac{2\alpha_R Q_L F_0^3}{f_m^2} + \frac{\alpha_E}{f_m} + \frac{2GFkT}{P_0} \tag{27}$$

where
G = compressed power gain of the loop amplifier
F = noise factor of the loop amplifier
k = Boltzmann's constant
T = temperature in kelvins
P_0 = carrier power level (in watts) at the output of the loop amplifier
F_0 = carrier frequency in hertz
f_m = carrier offset frequency in hertz
$Q_L(=\pi F_0 \tau_g)$ = loaded Q of the resonator in the feedback loop
α_R, α_E = flicker-noise constants for the resonator and loop amplifier, respectively

[77–86].

4.2. Frequency Divider

The output from the VCO has to be divided down to the reference frequency. The reference frequency can vary from a few kilohertz to more than 100 MHz [87–89]. A smaller division ratio provides better phase noise. Most of the frequency dividers are either off-the-shelf devices or custom devices. A typical frequency divider consists of a CMOS synchronous divider that can handle division ratios as low as 5 and as high as 1 million. The division ratio is determined by the number of dividers. Typical CMOS dividers end at 250 MHz. To extend the frequency range by using an asynchronous divider means extending the frequency range up to several gigahertz, but then the frequency resolution is compromised. This prescaler has to be a synchronized counter that has to be clocked by the main divider, but because of propagation delays, this can become difficult to achieve and can introduce phase jitter. A way around this is to use a dual-modulus prescaler, which toggles between two stages, dividing by N and dividing by $N+1$. Dual-modulus counters are available in numbers such as $\frac{5}{6}$, $\frac{10}{11}$, and $\frac{20}{21}$.

Consider the system shown in Fig. 23. If the $P/(P+1)$ is a $\frac{10}{11}$ divider, the A counter counts the units and the M counter counts the tens. The mode of operation depends on the type of programmable counter used, but the system might operate as follows. If the number loaded into A is greater than zero, then the $P/(P+1)$ divider is set to divide by $P+1$ at the start of the cycle. The output from the $P/(P+1)$ divider clocks both A and M. When A is full, it ceases count and sets the $P/(P+1)$ divider into the P mode. Only M is then clocked, and when it is full, it resets both A and M, and the cycle repeats:

$$(M-A)P + A(P+1) = MP + A \tag{28}$$

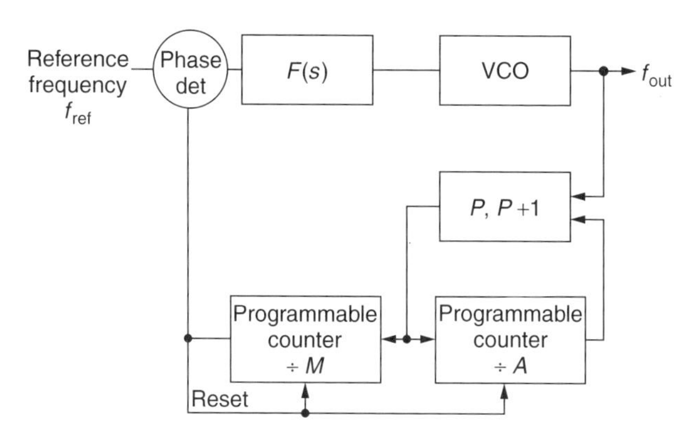

Figure 23. System using dual-modulus counter arrangement.

Therefore

$$f_{\text{out}} = (MP + A)f_{\text{ref}} \tag{29}$$

If A is incremented by one, the output frequency changes by f_{ref}. In other words, the channel spacing is equal to f_{ref}. This is the channel spacing that would be obtained with a fully programmable divider operating at the same frequency as the $P/(P+1)$ divider. For this system to work, the A counter must underflow before the M counter does; otherwise, $P/(P+1)$ will remain permanently in the $P+1$ mode. Thus, there is a minimum system division ratio, M_{min}, below which the $P/(P+1)$ system will not function. To find that minimum ratio, consider the following. The A counter must be capable of counting all numbers up to and including $P-1$ if every division ratio is to be possible, or

$$A_{\text{max}} = P - 1 \tag{30}$$

$$M_{\text{min}} = P \text{ since } M > A \tag{31}$$

The divider chain divides by $MP + A$; therefore, the minimum systems division ratio is

$$M_{\text{min}} = M_{\text{min}}(P + A_{\text{min}}) \tag{32}$$

$$= P(P + 0) = p^2$$

Using a $\frac{10}{11}$ ratio, the minimum practical division ratio of the system is 100.

In the system shown in Fig. 23, the fully programmable counter, A, must be quite fast. With a 350-MHz clock to the $\frac{10}{11}$ divider, only about 23 ns is available for counter A to control the $\frac{10}{11}$ divider. For cost reasons it would be desirable to use a TTL fully programmable counter, but when the delays through the ECL-to-TTL translators have been taken into account, very little time remains for the fully programmable counter. The $\frac{10}{11}$ function can be extended easily, however, to give a $+N(N+1)$ counter with a longer control time for a given input frequency, as shown in Figs. 24 and 25. Using the $\frac{20}{21}$ system shown in Fig. 24, the time available to control $\frac{20}{21}$ is typically 87 ns at 200 MHz and 44 ns at 350 MHz. The time available to control the $\frac{40}{41}$ (Fig. 25) is approximately 180 ns at 200 MHz and 95 ns at 350 MHz.

Figure 26 is a block diagram of an advanced digital synthesizer block produced by analog devices. There are numerous manufacturers of such chips on the market. Figure 25 gives some insight into the frequency divider system. The top accepts input from a frequency standard, also referred to as a *reference signal*, which is reduced to a number between 5 kHz and 20 MHz. The use of a high-frequency reference requires a higher division ratio, but typically these reference frequencies are also used for some other mixing processes. The 24-bit input register controls both the reference divider and the frequency divider. The frequency divider uses a prescaler like $\frac{5}{6}$ or $\frac{10}{11}$, and its output is applied to the phase frequency detector. The multiplex unit on the right is doing all the housekeeping and providing information such as block and, detect. The divider typically has very little control over this portion of the synthesizer, and it's a constant battle to find better parts. Very few high-end synthesizers use custom ICs. To

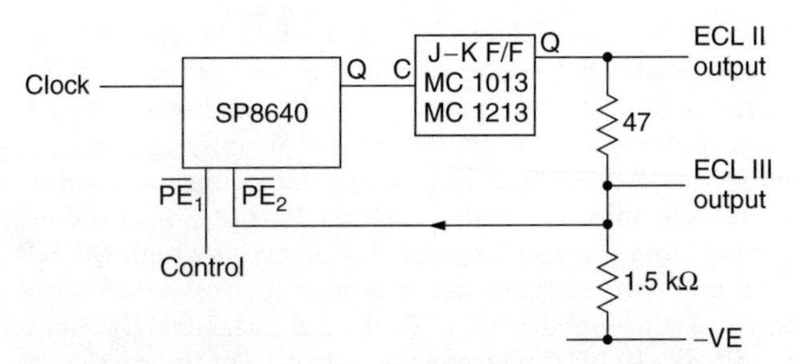

Figure 24. Level shifting information for connecting the various ECL2 and ECL3 stages.

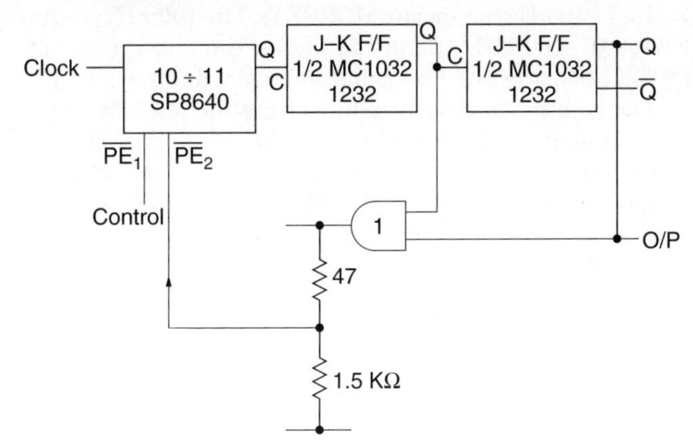

Figure 25. Level shifter diagram to drive from ECL2 and ECL3 levels.

have these built is typically very costly and is cost-effective only if the frequency resolution required is not provided by any other chip on the market. Most of the high-end fractional-N synthesizer chips fall in this category.

4.3. Phase Detector

The phase detector at minimum consists of a phase-sensitive circuit such as a double balanced mixer [90–93]. Such a simple circuit has two disadvantages: (1) it's not sensitive to frequency changes, and (2) the DC output level is only 0.7 V per diode in the ring; therefore, a poor signal-to-noise ratio can be expected. Today, modern circuits use a phase discriminator with a charge pump output. The phase frequency is edge-triggered and sensitive to both phase and frequency changes. Figure 27 shows a digital phase frequency discriminator with a programmable delay. Under locked condition, the charge pump does not supply current. Under unlocked condition, the current at the point CP charges or discharges a capacitor, which is part of the integrated system and smooths the output voltage to become ripple-free. The output from the reference divider is a pulsetrain with a small duty cycle, and the input from the frequency divider(/N) is a pulsetrain with a very small duty cycle. The duty cycle typically is as short as a division ratio is high. So, for a division ratio of 1000, the duty cycle is 0.1%, but the repetition frequency is equal to the output. The charge output, therefore, is also a very narrow train of pulses that are fed to the loop filter. The phase detector has to deal with complicated issues such as zero crossings causing instability. The best phase frequency detectors are

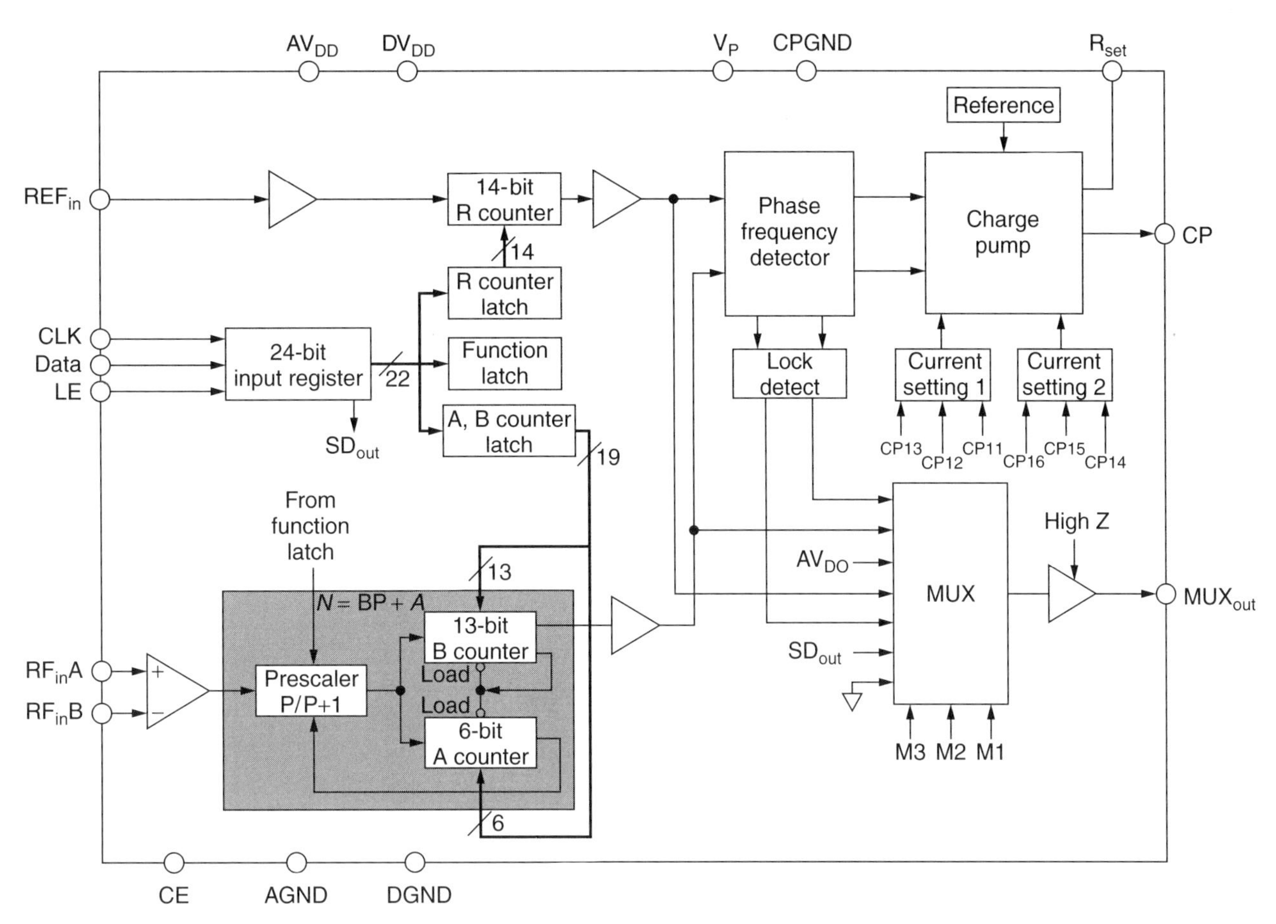

Figure 26. Block diagram of an advanced digital fractional-N synthesizer.

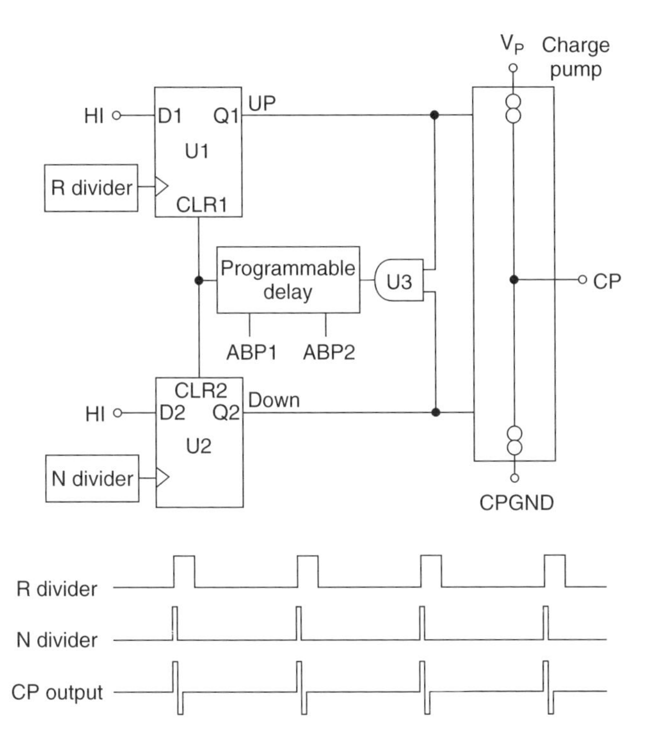

Figure 27. A digital phase frequency discriminator with a programmable delay.

either TTL or CMOS because they have a larger voltage output swing. In some cases, the charge pump is built from discrete components if a very high performance is required.

4.4. Loop Filter

Loop filters range from a simple lowpass filter to a complex arrangement of active filters. Figure 28 shows the configuration and frequency response of passive and active loop filters. Figure 29 shows an arrangement of more complex filters, including their calculations. The charge pumps can frequently lift with a purely passive filter, as seen in Fig. 30; however, the DC gain of the active filters provides better close-in phase noise and tracking. There may be a penalty if the active circuit is noise; however, the latest available operation amplifiers have sufficient performance [94].

5. PHASE-LOCKED LOOP DESIGNS

5.1. The Type 2, Second-Order Loop

The following is a derivation of the properties of the type 2, second-order loop. This means that the loop has two integrators, one being the diode and the other the operational amplifier, and is built with the order

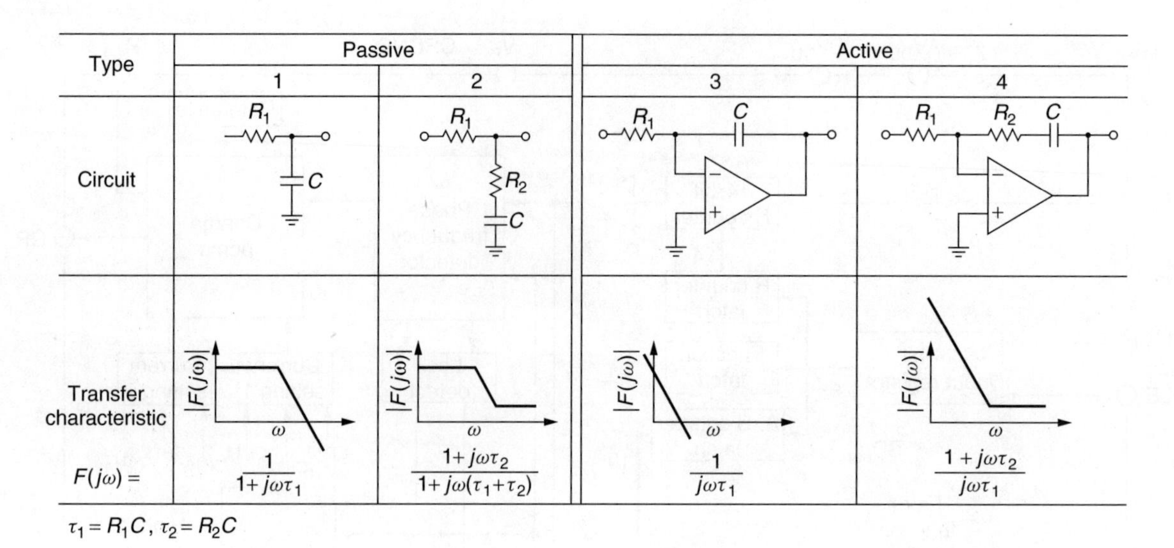

Figure 28. Circuit and transfer characteristics of several PLL filters.

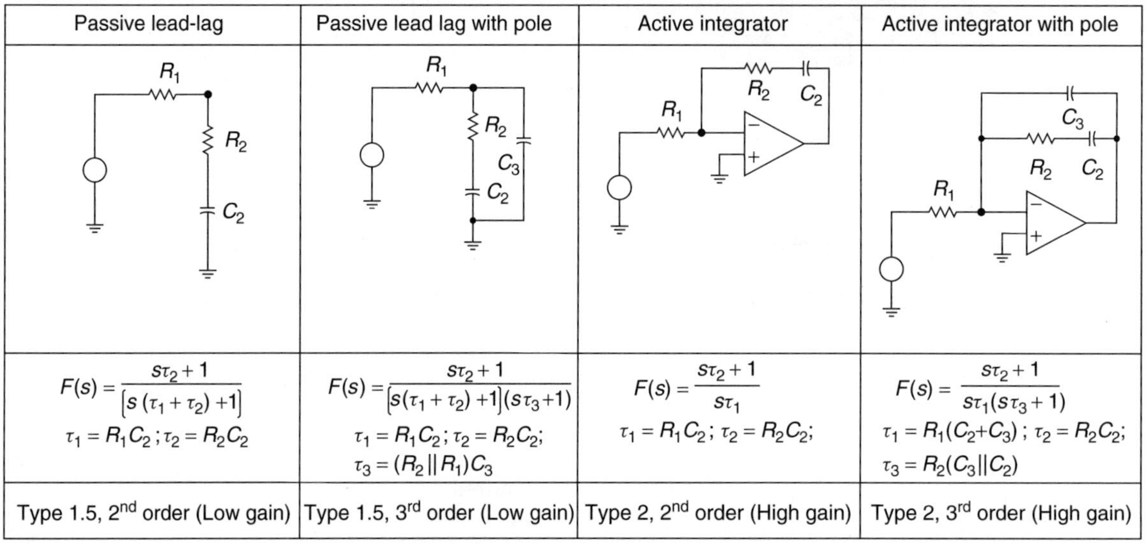

Passive lead-lag	Passive lead lag with pole	Active integrator	Active integrator with pole
$F(s) = \frac{s\tau_2 + 1}{[s(\tau_1 + \tau_2) + 1]}$	$F(s) = \frac{s\tau_2 + 1}{[s(\tau_1 + \tau_2) + 1](s\tau_3 + 1)}$	$F(s) = \frac{s\tau_2 + 1}{s\tau_1}$	$F(s) = \frac{s\tau_2 + 1}{s\tau_1(s\tau_3 + 1)}$
$\tau_1 = R_1C_2; \tau_2 = R_2C_2$	$\tau_1 = R_1C_2; \tau_2 = R_2C_2;$ $\tau_3 = (R_2 \| R_1)C_3$	$\tau_1 = R_1C_2; \tau_2 = R_2C_2;$	$\tau_1 = R_1(C_2 + C_3); \tau_2 = R_2C_2;$ $\tau_3 = R_2(C_3 \| C_2)$
Type 1.5, 2nd order (Low gain)	Type 1.5, 3rd order (Low gain)	Type 2, 2nd order (High gain)	Type 2, 3rd order (High gain)

Figure 29. Implementation of different loop filters.

of 2 as can be seen from the pictures above. The basic principle to derive the performance for higher-order loops follows the same principle, although the derivation is more complicated. Following the math section, we will show some typical responses [95].

The type 2, second-order loop uses a loop filter in the form

$$F(s) = \frac{1}{s}\frac{\tau_2 s + 1}{\tau_1} \tag{33}$$

The multiplier $1/s$ indicates a second integrator, which is generated by the active amplifier. In Table 1, this is the type 3 filter. The type 4 filter is mentioned there as a possible configuration but is not recommended because, as stated previously, the addition of the pole of the origin creates difficulties with loop stability and, in most cases, requires a change from the type 4 to the type 3 filter. One can consider the type 4 filter as a special case of the type 3 filter, and therefore it does not have to be treated separately. Another possible transfer function is

$$F(s) = \frac{1}{R_1 C}\frac{1 + \tau_2 s}{s} \tag{34}$$

with

$$\tau_2 = R_2 C \tag{35}$$

Under these conditions, the magnitude of the transfer function is

$$|F(j\omega)| = \frac{1}{R_1 C\omega}\sqrt{1 + (\omega R_2 C)^2} \tag{36}$$

and the phase is

$$\theta = \arctan(\omega\tau_2) - 90^\circ \tag{37}$$

Integrator	Integrator with poles	Integrator with 2 poles
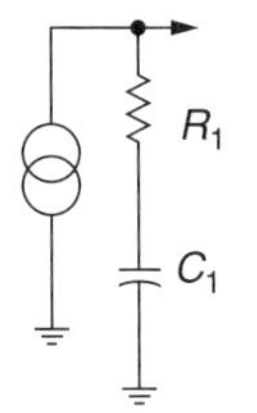	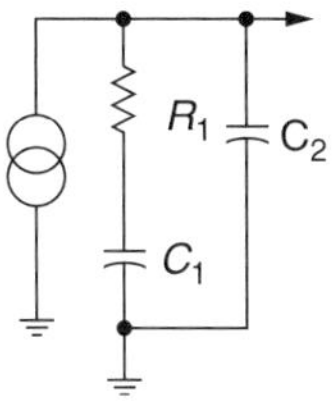	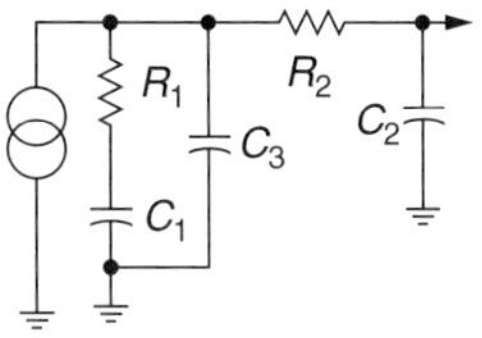
$F(s) = R_1 \frac{s\tau_1 + 1}{s\tau_1}$ $\tau_1 = R_1 C_1$	$F(s) = R_1 \frac{s\tau_1+1}{s\tau_1(s\tau_2+1)}$ $\tau_1 = R_1 C_1 ;\ \tau_2 = R_2 \left(\frac{C_1 C_2}{C_1 + C_2}\right)$	$F(s) = R_1 \frac{s\tau_1 + 1}{s\tau_1(s\tau_2 + 1)(s\tau_3 + 1)}$ $\tau_1 = R_1 C_1 ;\ \tau_2 = R_1 \frac{C_1 C_3}{C_1 + C_3};$ $\tau_3 = R_2 C_2$
Type 2, 2nd order	Type 2, 3rd order	Type 2, 4th order

Figure 30. Recommended passive filters for charge pumps.

Again, as if for a practical case, we start off with the design values ω_n and ξ, and we have to determine τ_1 and τ_2. Taking an approach similar to that for the type 1, second-order loop, the results are

$$\tau_1 = \frac{K}{\omega_n} \tag{38}$$

and

$$\tau_2 = \frac{2\zeta}{\omega_n} \tag{39}$$

and

$$R_1 = \frac{\tau_1}{C} \tag{40}$$

and

$$R_2 = \frac{\tau_2}{C} \tag{41}$$

The closed-loop transfer function of a type 2, second-order PLL with a perfect integrator is

$$B(s) = \frac{K(R_2/R_1)[s + (1/\tau_2)]}{s^2 + K(R_2/R_1)s + (K/\tau_2)(R_2/R_1)} \tag{42}$$

By introducing the terms ξ and ω_n, the transfer function now becomes

$$B(s) = \frac{2\zeta\omega_n s + \omega_n^2}{s^2 + 2\zeta\omega_n s + \omega_n^2} \tag{43}$$

with the abbreviations

$$\omega_n = \left(\frac{K}{\tau_2}\frac{R_2}{R_1}\right)^{1/2} \text{ rad/s} \tag{44}$$

and

$$\zeta = \frac{1}{2}\left(K\tau_2 \frac{R_2}{R_1}\right)^{1/2} \tag{45}$$

and $K = K_\theta K_o/N$.

The 3-dB bandwidth of the type 2, second-order loop is

$$B_{3\text{ dB}} = \frac{\omega_n}{2\pi}\left[2\zeta^2 + 1 + \sqrt{(2\zeta^2+1)^2 + 1}\right]^{1/2} \text{ Hz} \tag{46}$$

and the noise bandwidth is

$$B_n = \frac{K(R_2/R_1) + 1/\tau_2}{4} \text{ Hz} \tag{47}$$

Again, we ask the question of the final error and use the previous error function

$$E(s) = \frac{s\theta(s)}{s + K(R_2/R_1)\{[s + (1/\tau_2)]/s\}} \tag{48}$$

or

$$E(s) = \frac{s^2\theta(s)}{s^2 + K(R_2/R_1)s + (K/\tau_2)(R_2/R_1)} \tag{49}$$

As a result of the perfect integrator, the steady-state error resulting from a step change in input phase or change of magnitude of frequency is zero.

If the input frequency is swept with a constant range change of input frequency ($\Delta\omega/dt$), for $\theta(s) = (2\Delta\omega/dt)/s^3$, the steady-state phase error is

$$E(s) = \frac{R_1}{R_2}\frac{\tau_2(2\Delta\omega/dt)}{K} \text{ rad} \tag{50}$$

The maximum rate at which the VCO frequency can be swept for maintaining lock is

$$\frac{2\Delta\omega}{dt} = \frac{N}{2\tau_2}\left(4B_n - \frac{1}{\tau_2}\right) \text{ rad/s} \tag{51}$$

The introduction of N indicates that this is referred to the VCO rather than to the phase/frequency comparator. In the previous example of the type 1, first-order loop, we referred it only to the phase/frequency comparator rather than the VCO.

Figure 31 shows the closed-loop response of a type 2, third-order loop having a phase margin of 10° and with the optimal 45°.

A phase margin of 10° results in overshoot, which in the frequency domain would be seen as peaks in the oscillator noise–sideband spectrum. Needless to say, this is a totally undesirable effect, and since the operational amplifiers and other active and passive elements add to

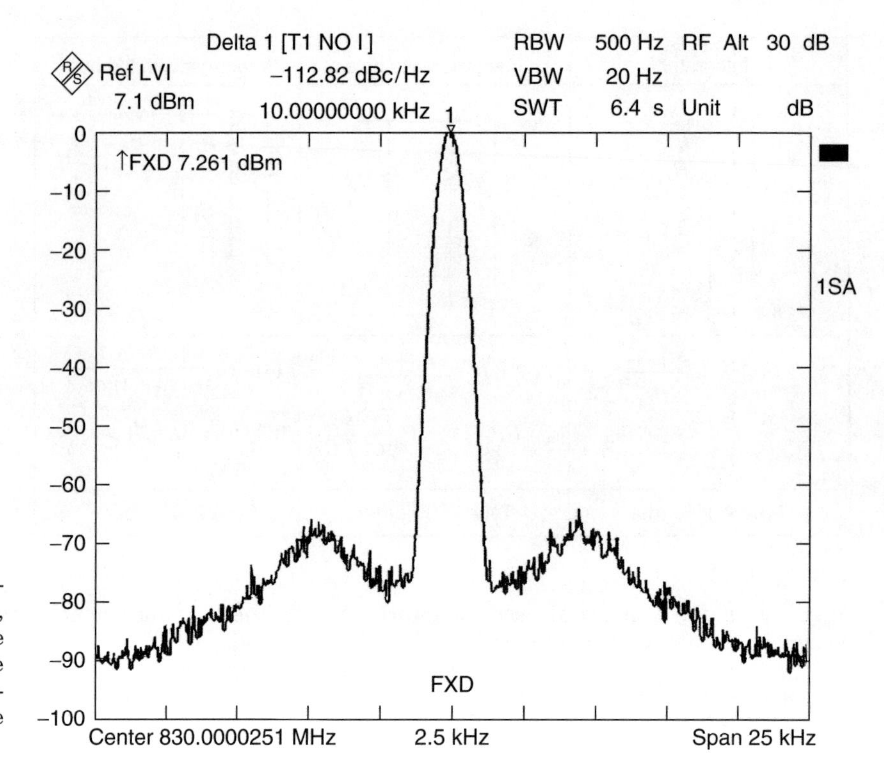

Figure 31. Measured spectrum of a synthesizer where the loop filter is underdamped, resulting in ≈10-dB increase of the phase noise at the loop filter bandwidth. In this case, we either don't meet the 45° phase margin criterion, or the filter is too wide, so it shows the effect of the up-converted reference frequency.

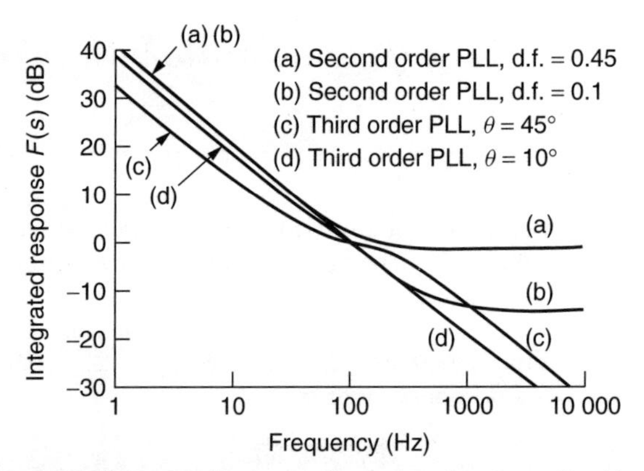

Figure 32. Integrated response for various loops as a function of the phase margin.

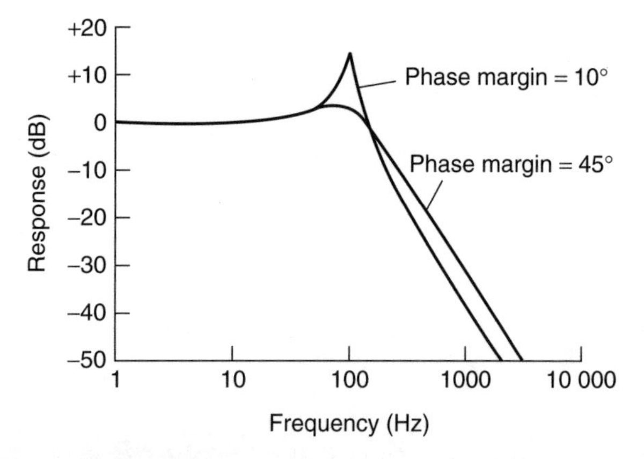

Figure 33. Closed-loop response of a type 2, third-order PLL having a phase margin of 10°.

this, the loop filter has to be adjusted after the design is finalized to accommodate the proper resulting phase margin (35°–45°). The open-loop gain for different loops can be seen in Figs. 32 and 33.

5.2. Transient Behavior of Digital Loops Using Tristate Phase Detectors

5.2.1. Pullin Characteristic. The type 2, second-order loop is used with either a sample/hold comparator or a tristate phase/frequency comparator.

We will now determine the transient behavior of this loop. Figure 34 shows the block diagram.

Very rarely in the literature is a clear distinction between pullin and lockin characteristics or frequency and phase acquisition made as a function of the digital phase/frequency detector. Somehow, all the

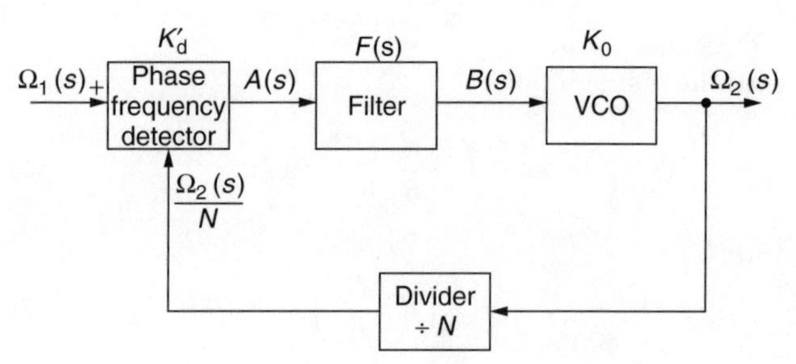

Figure 34. Block diagram of a digital PLL before lock is acquired.

approximations or linearizations refer to a sinusoidal phase/frequency comparator or its digital equivalent, the exclusive-OR gate.

The tristate phase/frequency comparator follows slightly different mathematical principles. The phase detector gain is

$$K_d' = \frac{V_d}{\omega_0} = \frac{\text{phase detector supply voltage}}{\text{loop idling frequency}}$$

and is valid only in the out-of-lock state and is a somewhat coarse approximation to the real gain which, due to nonlinear differential equations, is very difficult to calculate. However, practical tests show that this approximation is still fairly accurate.

Definitions are

$$\Omega_1(s) = \mathcal{L}[\Delta\omega_1(t)] \quad \text{(reference input to } \delta/\omega \text{ detector)}$$

$$\Omega_2(s) = \mathcal{L}[\Delta\omega_2(t)] \quad \text{(signal VCO output frequency)}$$

$$\Omega_e(s) = \mathcal{L}[\omega_e(t)] \quad \text{(error frequency at } \delta/\omega \text{ detector)}$$

$$\Omega_e(s) = \Omega_1(s) - \frac{\Omega_2(s)}{N}$$

$$\Omega_2(s) = [\Omega_1(s) - \Omega_e(s)]N$$

From the circuit described above

$$A(s) = \Omega_e(s)K_d'$$

$$B(s) = A(s)F(s)$$

$$\Omega_2(s) = B(s)K_o$$

The error frequency at the detector is

$$\Omega_e(s) = \Omega_1(s)N\frac{1}{N + K_oK_d'F(s)} \tag{52}$$

The signal is stepped in frequency:

$$\Omega_1(s) = \frac{\Delta\omega_1}{s} \quad (\Delta\omega_1 = \text{magnitude of frequency step}) \tag{53}$$

5.2.1.1. *Active Filter of First Order.* If we use an active filter

$$F(s) = \frac{1 + s\tau_2}{s\tau_1} \tag{54}$$

and insert this in (51), the error frequency is

$$\Omega_e(s) = \Delta\omega_1 N\frac{1}{s\left(N + K_oK_d'\dfrac{\tau_2}{\tau_1}\right) + \dfrac{K_oK_d'}{\tau_1}} \tag{55}$$

Utilizing the Laplace transformation, we obtain

$$\omega_e(t) = \Delta\omega_1\frac{1}{1 + K_oK_d'(\tau_2/\tau_1)(1/N)}\exp\left[-\frac{t}{(\tau_1N/K_oK_d') + \tau_2}\right] \tag{56}$$

and

$$\lim_{t\to 0}\omega_e(t) = \frac{\Delta\omega_1 N}{N + K_oK_d'(\tau_2/\tau_1)} \tag{57}$$

$$\lim_{t\to\infty}\omega_e(t) = 0 \tag{58}$$

5.2.1.2. *Passive Filter of First Order.* If we use a passive filter

$$\lim_{t\to\infty}\omega_e(t) = 0 \tag{59}$$

for the frequency step

$$\Omega_1(s) = \frac{\Delta\omega_1}{s} \tag{60}$$

the error frequency at the input becomes

$$\Omega_e(s) = \Delta\omega_1 N\left\{\frac{1}{s}\frac{1}{s[N(\tau_1+\tau_2) + K_oK_d'\tau_2] + (N + K_oK_d')} + \frac{\tau_1+\tau_2}{s[N(\tau_1+\tau_2) + K_oK_d'\tau_2] + (N + K_oK_d')}\right\} \tag{61}$$

For the first term we will use the abbreviation A, and for the second term we will use the abbreviation B:

$$A = \frac{1/[N(\tau_1+\tau_2) + K_oK_d'\tau_2]}{s\left[s + \dfrac{N + K_oK_d'}{N(\tau_1+\tau_2) + K_oK_d'\tau_2}\right]} \tag{62}$$

$$B = \frac{\dfrac{\tau_1+\tau_2}{N(\tau_1+\tau_2) + K_oK_d'\tau_2}}{s + \dfrac{N + K_oK_d'}{N(\tau_1+\tau_2) + K_oK_d'\tau_2}} \tag{63}$$

After the inverse Laplace transformation, our final result becomes

$$\mathcal{L}^{-1}(A) = \frac{1}{N + K_oK_d'} \times \left\{1 - \exp\left[-t\frac{N + K_oK_d'}{N(\tau_1+\tau_2) + K_oK_d'\tau_2}\right]\right\} \tag{64}$$

$$\mathcal{L}^{-1}(B) = \frac{\tau_1+\tau_2}{N(\tau_1+\tau_2) + K_oK_d'\tau_2} \times \exp\left(-t\frac{N + K_oK_d'}{N(\tau_1+\tau_2) + K_oK_d'\tau_2}\right) \tag{65}$$

and finally

$$\omega_e(t) = \Delta\omega_1 N[\mathcal{L}^{-1}(A) + (\tau_1+\tau_2)\mathcal{L}^{-1}(B)] \tag{66}$$

What does the equation mean? We really want to know how long it takes to pull the VCO frequency to the reference. Therefore, we want to know the value of t, the time it takes to be within 2π or less of lockin range.

The PLL can, at the beginning, have a phase error from -2π to $+2\pi$, and the loop, by accomplishing lock, then takes care of this phase error. We can make the reverse assumption for a moment and ask ourselves, as we have done earlier, how long the loop stays in phase lock. This is called the *pullout range*. Again, we apply signals to the input of the PLL as long as the loop can follow and the phase error does not become larger than 2π. Once the error is larger than 2π, the loop jumps out of lock. When the loop is out of lock, a beat note occurs at the output of the loop filter following the phase/frequency detector.

The tristate state phase/frequency comparator, however, works on a different principle, and the pulses generated and supplied to the charge pump do not allow the generation of an AC voltage. The output of such a phase/frequency detector is always unipolar, but relative to the value of $V_{\text{batt}}/2$, the integrator voltage can be either positive or negative. If we assume for a moment that this voltage should be the final voltage under a locked condition, we will observe that the resulting DC voltage is either more negative or more positive relative to this value, and because of this, the VCO will be "pulled in" to this final frequency rather than swept in. The swept-in technique applies only in cases of phase/frequency comparators, where this beat note is being generated. A typical case would be the exclusive-OR gate or even a sample/hold comparator. This phenomenon is rarely covered in the literature and is probably discussed in detail for the first time in the book by Roland Best [9].

Let us assume now that the VCO has been pulled in to final frequency to be within 2π of the final frequency, and the time t is known. The next step is to determine the lockin characteristic.

5.2.2. Lockin Characteristic. We will now determine the lockin characteristic, and this requires the use of a different block diagram. Figure 5 shows the familiar block diagram of the PLL, and we will use the following definitions:

$$\theta_1(s) = \mathcal{L}[\Delta\delta_1(t)] \quad \text{(reference input to } \delta/\omega \text{ detector)}$$

$$\theta_2(s) = \mathcal{L}[\Delta\delta_2(t)] \quad \text{(signal VCO output phase)}$$

$$\theta_e(s) = \mathcal{L}[\delta_e(t)] \quad \text{(phase error at } \delta/\omega \text{ detector)}$$

$$\theta_e(s) = \theta_1(s) - \frac{\theta_2(s)}{N}$$

From the block diagram, the following is apparent:

$$A(s) = \theta_e(s)K_d$$

$$B(s) = A(s)F(s)$$

$$\theta_2(s) = B(s)\frac{K_o}{s}$$

The phase error at the detector is

$$\theta_e(s) = \theta_1(s)\frac{sN}{K_oK_dF(s) + sN} \tag{67}$$

A step in phase at the input, where the worst-case error is 2π, results in

$$\theta_1(s) = 2\pi\frac{1}{s} \tag{68}$$

We will now treat the two cases using an active or passive filter.

5.2.2.1. ***Active Filter.*** The transfer characteristic of the active filter is

$$F(s) = \frac{1 + s\tau_2}{s\tau_1} \tag{69}$$

This results in the formula for the phase error at the detector:

$$\theta_e(s) = 2\pi\frac{s}{s^2 + (sK_oK_d\tau_2/\tau_1)/N + (K_oK_d/\tau_1)/N} \tag{70}$$

The polynomial coefficients for the denominator are

$$a_2 = 1$$

$$a_1 = \frac{K_oK_d\tau_2/\tau_1}{N}$$

$$a_0 = \frac{K_oK_d/\tau_1}{N}$$

and we have to find the roots W_1 and W_2. Expressed in the form of a polynomial coefficient, the phase error is

$$\theta_e(s) = 2\pi\frac{s}{(s + W_1)(s + W_2)} \tag{71}$$

After the inverse Laplace transformation has been performed, the result can be written in the form

$$\delta_e(t) = 2\pi\frac{W_1e^{-W_1t} - W_2e^{-W_2t}}{W_1 - W_2} \tag{72}$$

with

$$\lim_{t\to 0}\delta_e(t) = 2\pi$$

and

$$\lim_{t\to\infty}\delta_e(t) = 0$$

The same can be done using a passive filter.

5.2.2.2. ***Passive Filter.*** The transfer function of the passive filter is

$$F(s) = \frac{1 + s\tau_2}{1 + s(\tau_1 + \tau_2)} \tag{73}$$

If we apply the same phase step of 2π as before, the resulting phase error is

$$\theta_e(s) = 2\pi\frac{[1/(\tau_1 + \tau_2)] + s}{s^2 + s\dfrac{N + K_oK_d\tau_2}{N(\tau_1 + \tau_2)} + \dfrac{K_oK_d}{N(\tau_1 + \tau_2)}} \tag{74}$$

Again, we have to find the polynomial coefficients, which are

$$a_2 = 1$$

$$a_1 = \frac{N + K_oK_d\tau_2}{N(\tau_1 + \tau_2)}$$

$$a_0 = \frac{K_oK_d}{N(\tau_1 + \tau_2)}$$

and finally find the roots for W_1 and W_2. This can be written in the form

$$\theta_e(s) = 2\pi\left[\frac{1}{\tau_1+\tau_2}\frac{1}{(s+W_1)(s+W_2)} + \frac{s}{(s+W_1)(s+W_2)}\right] \quad (75)$$

Now we perform the inverse Laplace transformation and obtain our result:

$$\delta_e(t) = 2\pi\left(\frac{1}{\tau_1+\tau_2}\frac{e^{-W_1t}-e^{-W_2t}}{W_2-W_1} + \frac{W_1e^{-W_1t}-W_2e^{-W_2t}}{W_1-W_2}\right) \quad (76)$$

with

$$\lim_{t\to 0}\delta_e(t) = 2\pi$$

with

$$\lim_{t\to\infty}\delta_e(t) = 0$$

When analyzing the frequency response for the various types and orders of PLLs, the phase margin played an important role. For the transient time, the type 2, second-order loop can be represented with a damping factor or, for higher orders, with the phase margin. Figure 35 shows the normalized output response for a damping factor of 0.1 and 0.47. The ideal Butterworth response would be a damping factor of 0.7, which correlates with a phase margin of 45°.

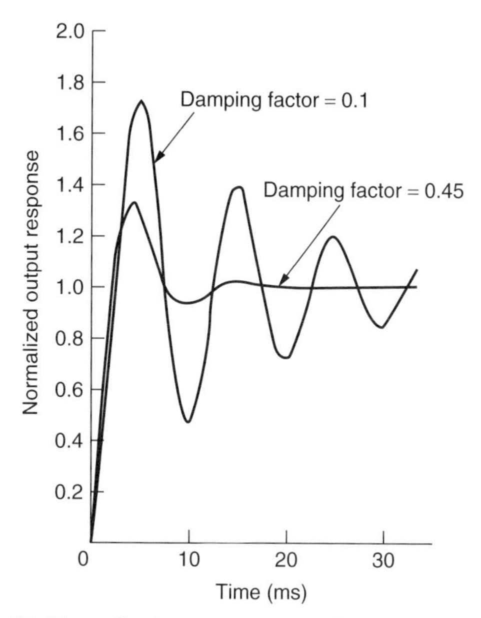

Figure 35. Normalized output response of a type 2, second-order loop with a damping factor of 0.1 and 0.05 for $\Omega_n = 0.631$.

5.3. Loop Gain/Transient Response Examples

Given the simple filter shown in Fig. 36 and the parameters as listed, the Bode plot is shown in Fig. 37. This approach can also be translated from a type 1 into a type 2 filter as shown in Fig. 38 and its frequency response is shown in Fig. 39. The lockin function for this type 2, second-order loop with an ideal damping factor of 0.707 (Butterworth response) is shown in Fig. 40. Figure 41 shows an actual settling time measurement. Any deviation from ideal damping, as we'll soon see, results in ringing (in an underdamped system) or, in an overdamped system, the voltage will crawl to its final value. This system can be increased in its order by selecting a type 2, third-order loop using the filter shown in Fig. 42. For an ideal synthesis of the values, the Bode diagram looks as shown in Fig. 43, and its resulting response is given in Fig. 44.

The order can be increased by adding an additional lowpass filter after the standard loop filter. The resulting system is called a *type 2, fifth-order loop*. Figure 45 shows the Bode diagram or open-loop diagram, and Fig. 46 shows the locking function. By using a very

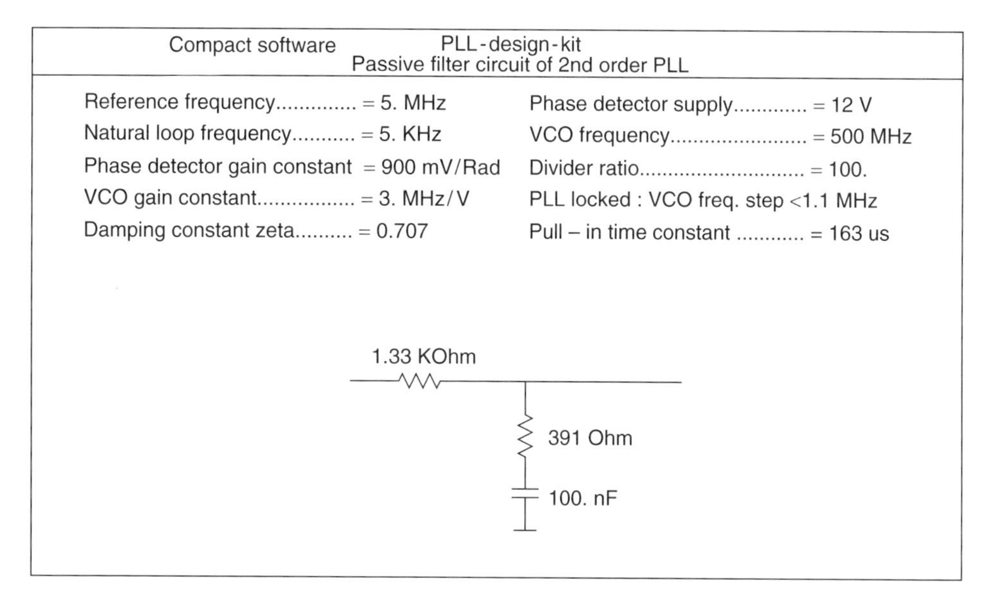

Figure 36. Loop filter for a type 1, second-order synthesizer.

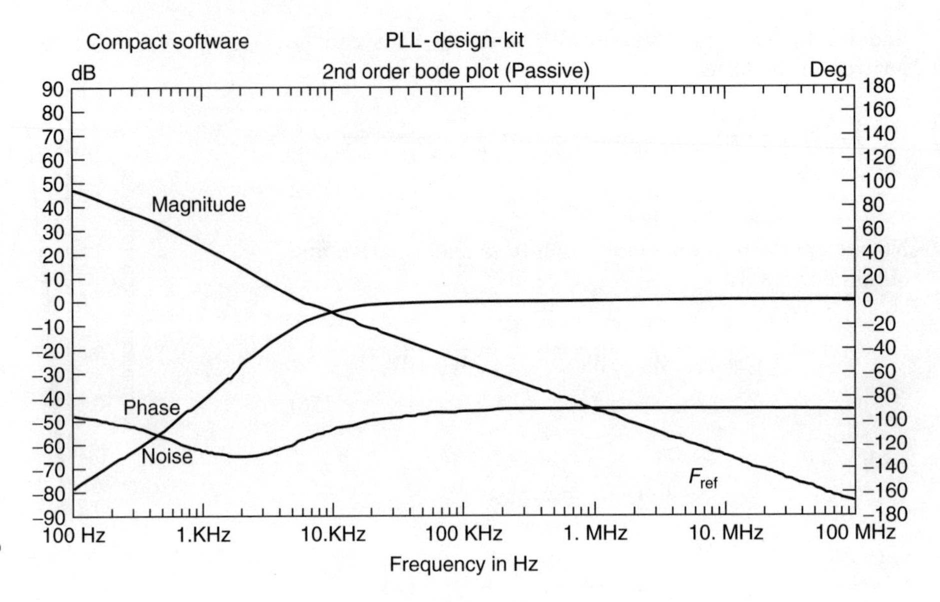

Figure 37. Type 1, second-order loop response.

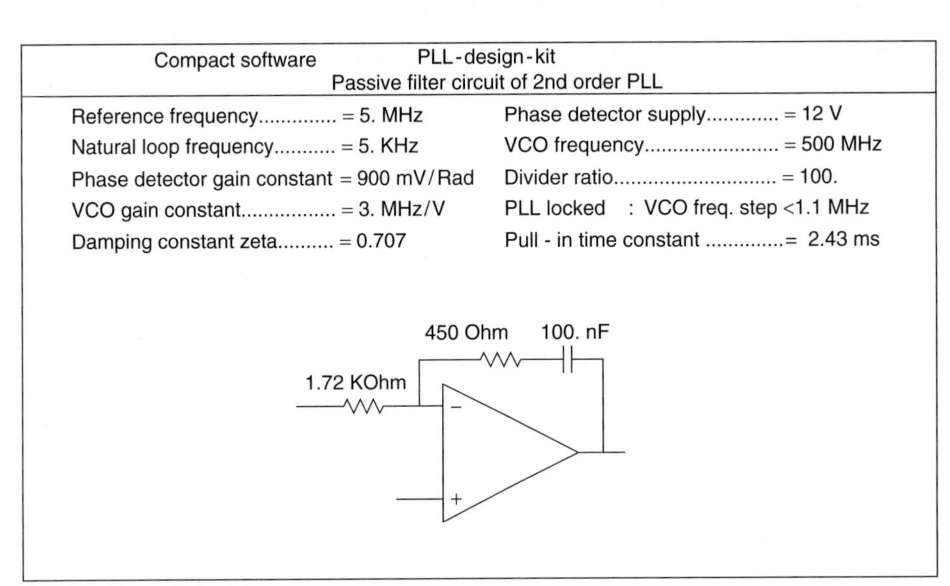

Figure 38. Loop filter for a type 2, second-order synthesizer.

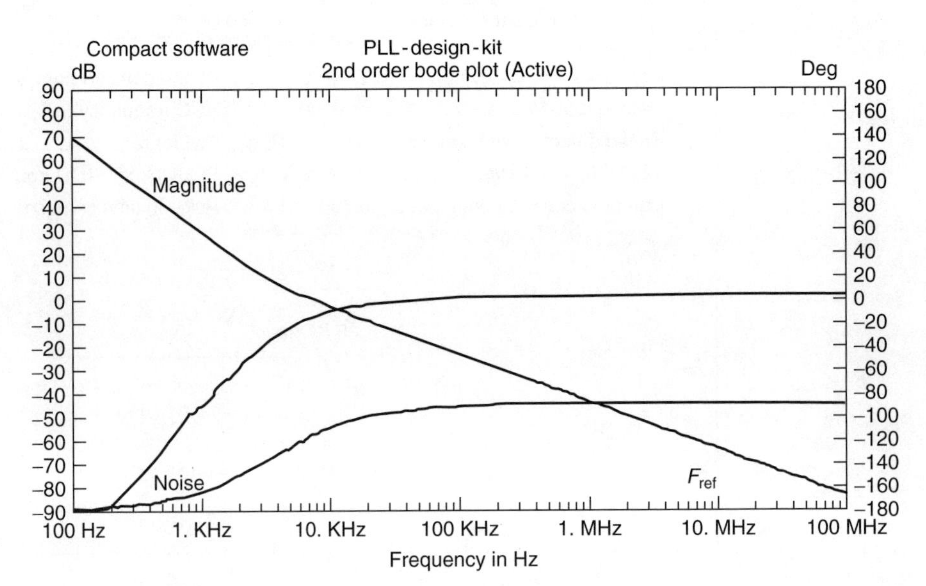

Figure 39. Response of the type 2, second-order loop.

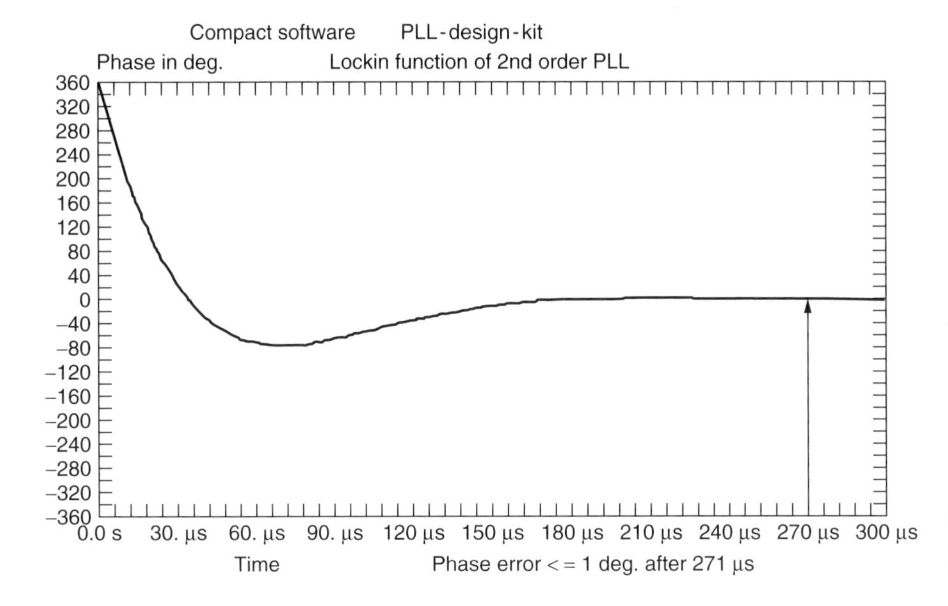

Figure 40. Lock-in function of the type 2, second-order PLL, indicating a lock time of 271 μs and an ideal response.

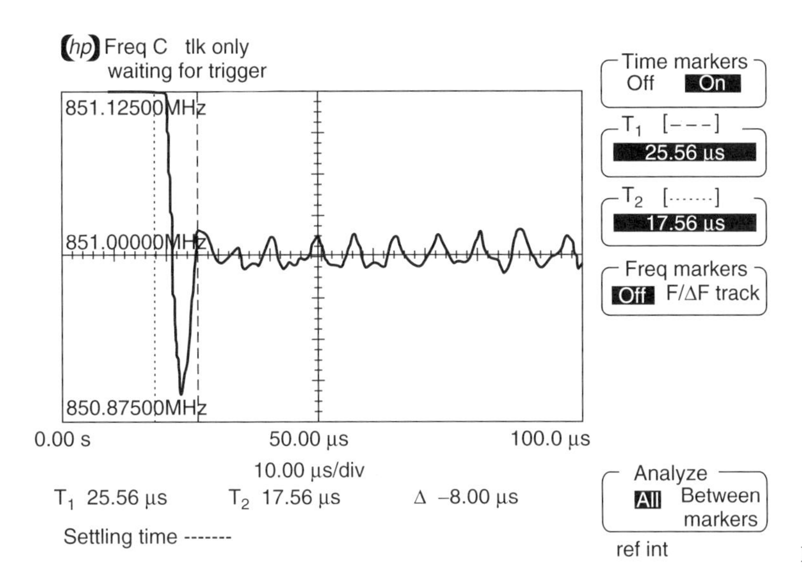

Figure 41. Example of settling time measurement.

wide loop bandwidth, this can be used to clean up microwave oscillators with inherent comparatively poor phase noise. This cleanup, has a dramatic influence on the performance.

By deviating from the ideal 45° to a phase margin of 33°, one obtains the above mentioned ringing, as is evident from Fig. 47. The time to settle has increased from 13.3 to 62 μs.

To more fully illustrate the effects of nonideal phase margin, Figs. 48, 49, 50, and 51 show the lockin function of a different type 2, fifth-order loop configured for phase margins of 25°, 35°, 45°, and 55°, respectively.

I have already mentioned that the loop should avoid "ears" (Fig. 31) with poorly designed loop filters. Another interesting phenomenon is the tradeoff between loop bandwidth and phase noise. In Fig. 52 the loop bandwidth has been made too wide, resulting in a degradation of the phase noise but provides faster settling time. By reducing the loop bandwidth from about 1 kHz to 300 Hz, only a very slight overshoot remains, improving the phase noise significantly. This is shown in Fig. 53.

5.4. Practical Circuits

Figure 54 shows a passive filter used for a synthesizer chip with constant current output. This chip has a charge pump output, which explains the need for the first capacitor.

Figure 55 shows an active integrator operating at a reference frequency of several megahertz. The notch filter at the output reduces the reference frequency considerably. The notch is about 4.5 MHz.

Figure 56 shows the combination of a phase/frequency discriminator and a higher-order loop filter as used in more complicated systems, such as fractional-division synthesizers.

Figure 57 shows a custom-built phase detector with a noise floor of better than — 168 dBc/Hz.

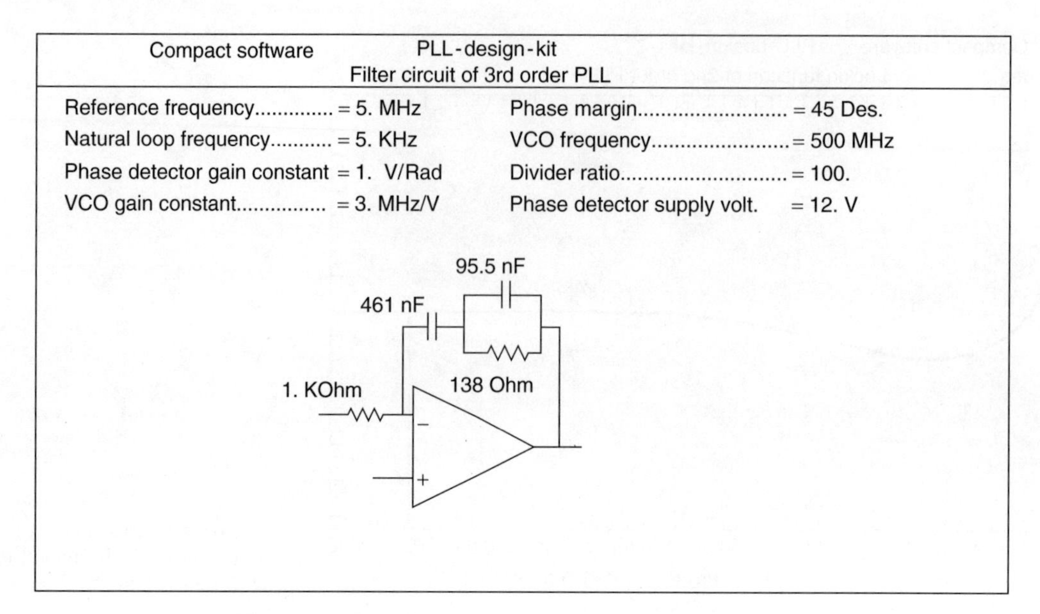

Figure 42. Loop filter for a type 2, third-order synthesizer.

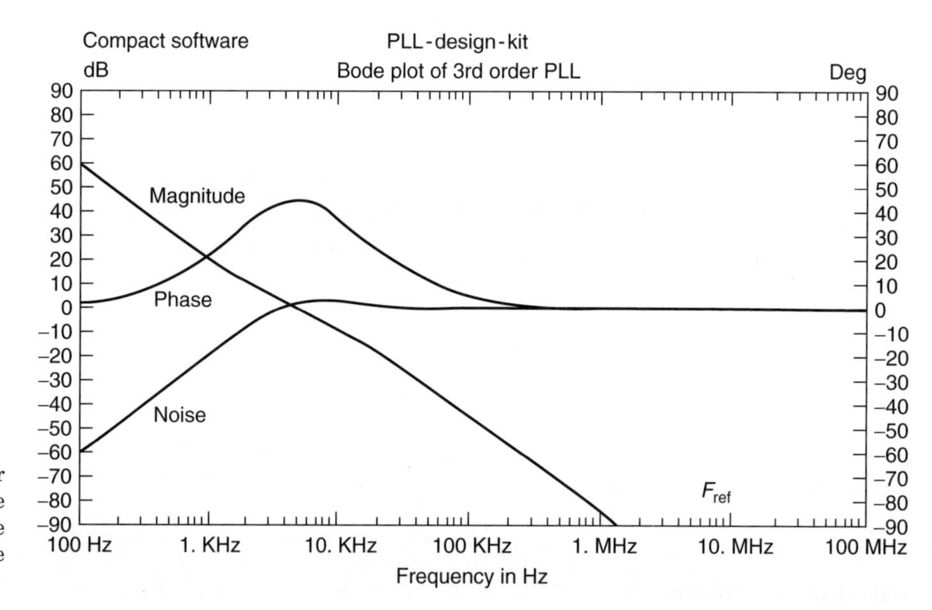

Figure 43. Open-loop Bode diagram for the type 2, third-order loop. It fulfills the requirement of 45° phase margin at the 0-dB crossover point, and corrects the slope down to −10 dB gain.

6. THE FRACTIONAL-*N* PRINCIPLE

The principle of the fractional-N PLL synthesizer was briefly mentioned in Section 2. The following is a numerical example for better understanding.

Example 1. Considering the problem of generating 899.8 MHz using a fractional-N loop with a 50-MHz reference frequency, we obtain

$$899.8\ \text{MHz} = 50\ \text{MHz}\left(N + \frac{K}{F}\right)$$

The integral part of the division N has to be set to 17 and the fractional part K/F needs to be $\frac{996}{1000}$; (the fractional part $\frac{K}{F}$ is not a integer) and the VCO output has to be divided by 996× every 1000 cycles. This can easily be implemented by adding the number 0.996 to the contents of an accumulator every cycle. Every time the accumulator overflows, the divider divides by 18 rather than by 17. Only the fractional value of the addition is retained in the phase accumulator. If we move to the lower band or try to generate 850.2 MHz, N remains 17, and K/F becomes $\frac{4}{1000}$. This method of using fractional division was first introduced by using analog implementation and noise cancellation, but today it is implemented totally as a digital approach. The necessary resolution is obtained from the dual-modulus prescaling, which allows for a well-established method for achieving a high-performance frequency synthesizer operating at UHF and higher frequencies. Dual-modulus prescaling avoids the loss of resolution in a system compared to a simple prescaler; it allows a VCO step

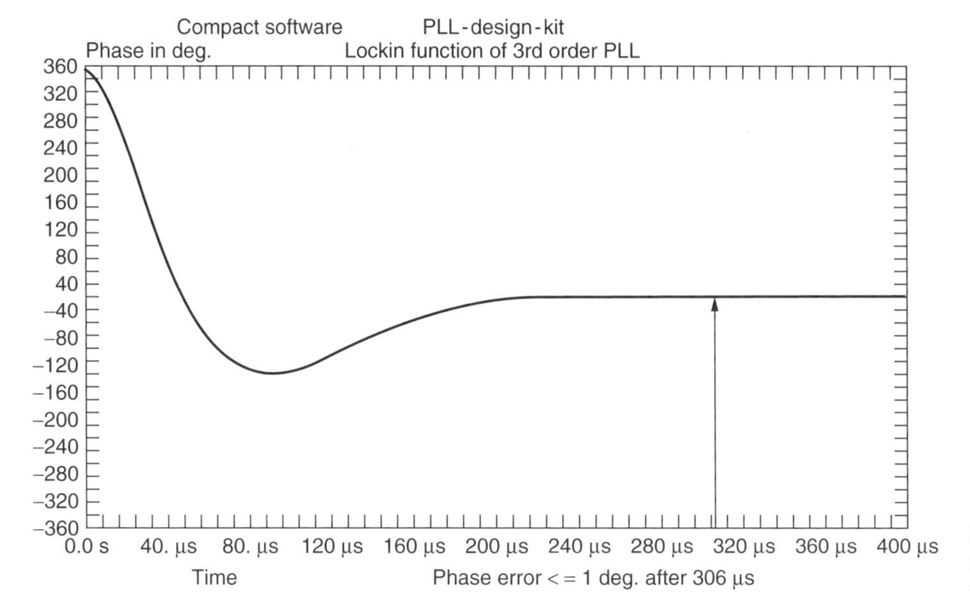

Figure 44. Lockin function of the type 2, third-order loop for an ideal 45° phase margin.

equal to the value of the reference frequency to be obtained. This method needs an additional counter and the dual modulus prescaler then divides one or two values depending on the state of its control. The only drawback of prescalers is the minimum division ratio of the prescaler for approximately N^2. The dual-modulus divider is the key to implementing the fractional-N synthesizer principle. Although the fractional-N technique appears to have a good potential of solving the resolution limitation, it is not free of having its own complications. Typically, an overflow from the phase accumulator, which is the adder with the output feedback to the input after being latched, is used to change the instantaneous division ratio. Each overflow produces a jitter at the output frequency, caused by the fractional division, and is limited to the fractional portion of the desired division ratio.

In our case, we had chosen a step size of 200 kHz, and yet the discrete sidebands vary from 200 kHz for $K/F = \frac{4}{1000}$ to 49.8 MHz for $K/F = \frac{996}{1000}$. It will become the task of the loop filter to remove those discrete spurious components. While in the past the removal of the discrete spurs has been accomplished by using analog techniques, various digital methods are now available. The microprocessor has to solve the following equation:

$$N* = \left(N + \frac{K}{F}\right) = [N(F - K) + (N + 1)K]$$

Example 2. For $F_O = 850.2$ MHz, we obtain

$$N* = \frac{850.2\text{ MHz}}{50\text{ MHz}} = 17.004$$

Following the formula above we have

$$N* = \left(N + \frac{K}{F}\right) = \frac{[17(1000 - 4) + (17 + 1) \times 4]}{1000}$$
$$= \frac{[16932 + 72]}{1000} = 17.004$$

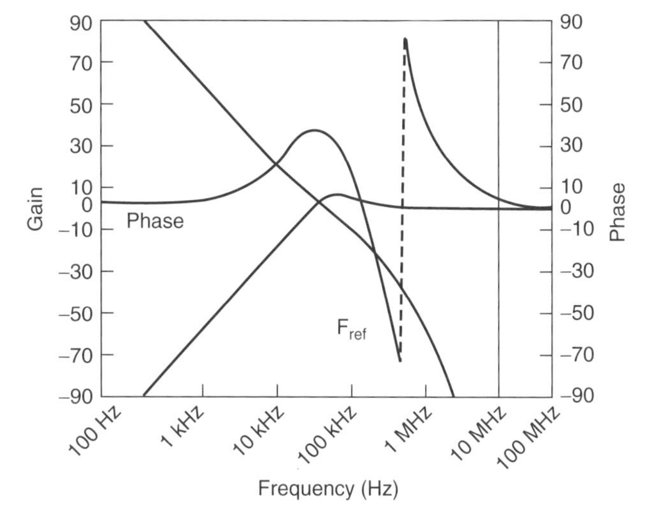

Figure 45. Bode plot of the fifth-order PLL system for a microwave synthesizer. The theoretical reference suppression is better than 90 dB.

$$F_{\text{out}} = 50\text{ MHz} \times \frac{[16932 + 72]}{1000}$$
$$= 846.6\text{ MHz} + 3.6\text{ MHz}$$
$$= 850.2\text{ MHz}$$

By increasing the number of accumulators, frequency resolution much below 1-Hz step size is possible with the same switching speed.

There is an interesting, generic problem associated with *all* fractional-N synthesizers. Assume for a moment that we use our 50-MHz reference and generate a 550-MHz output frequency. This means that our division factor is 11. Aside from reference frequency sidebands (±50 MHz) and harmonics, there will be no unwanted spurious frequencies. Of course, the reference sidebands

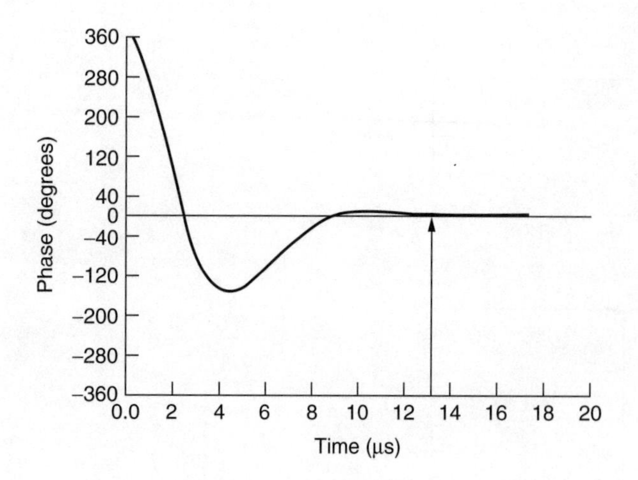

Figure 46. Lockin function of the fifth-order PLL. Note that the phase lock time is approximately 13.3 μs.

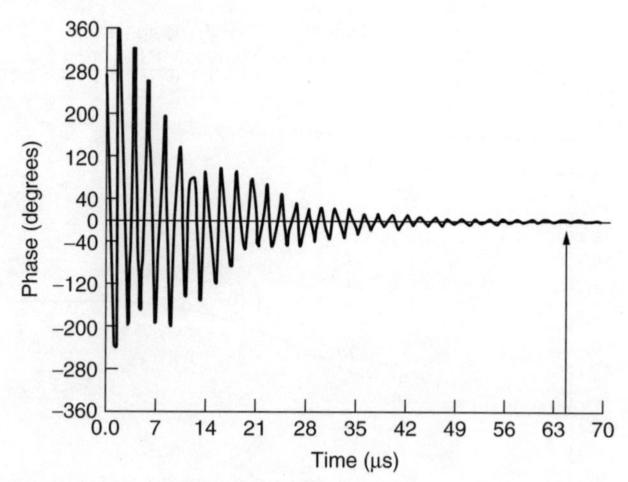

Figure 47. Lockin function of the fifth-order PLL. Note that the phase margin has been reduced from the ideal 45°. This results in a much longer settling time of 62 μs.

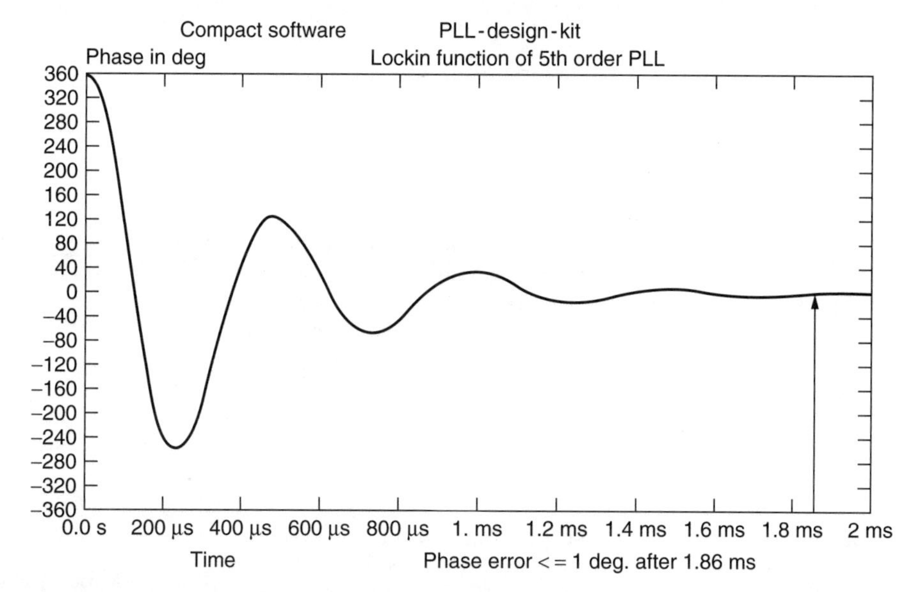

Figure 48. Lockin function of another type 2, fifth-order loop with a 25° phase margin. Noticeable ringing occurs, lengthening the lockin time to 1.86 ms.

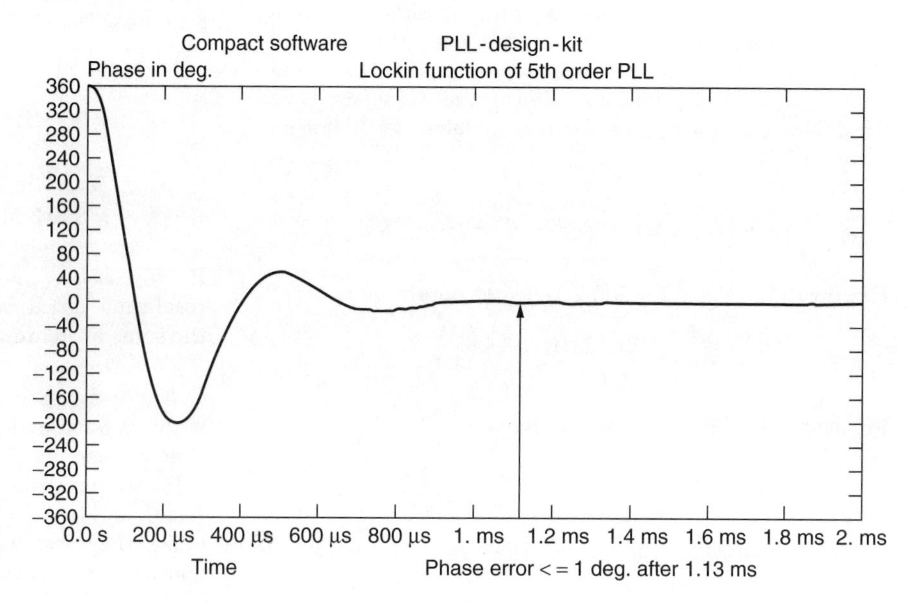

Figure 49. Lockin function of the type 2, fifth-order loop with a 35° phase margin. Ringing still occurs, but the lockin time has decreased to 1.13 ms.

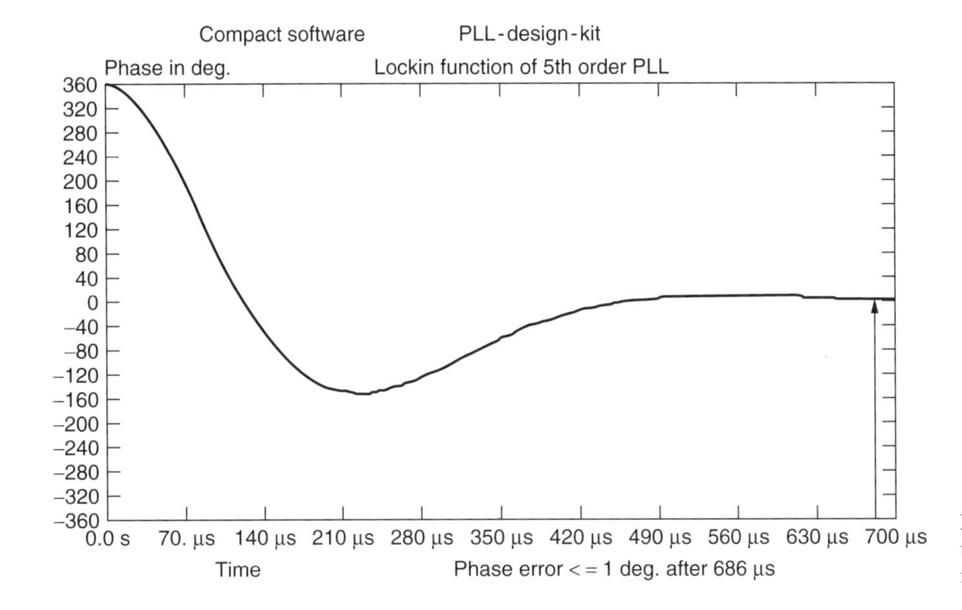

Figure 50. Lockin function of the type 2, third-order loop with an ideal 45° phase margin. The lockin time is 686 µs.

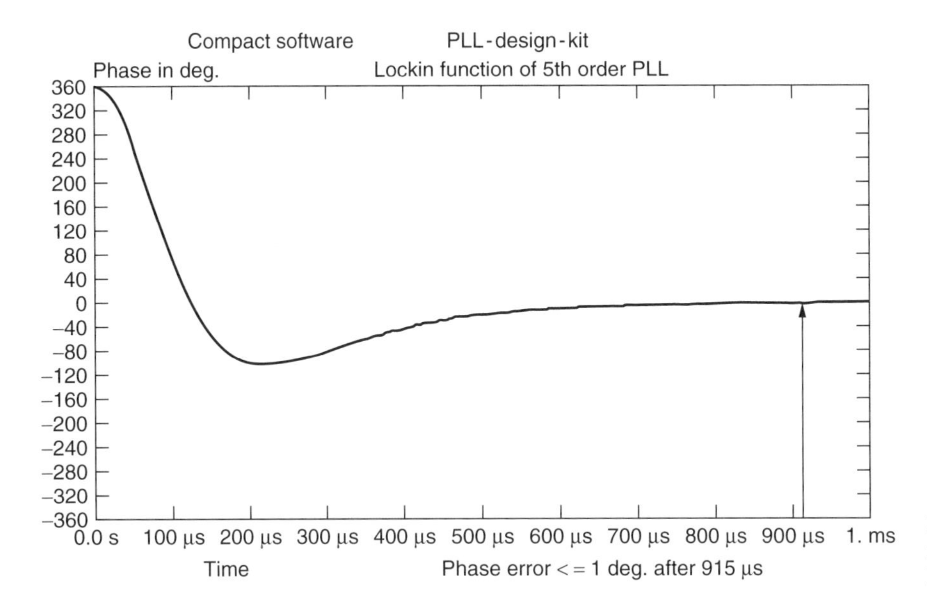

Figure 51. Lockin function of the type 2, fifth-order loop, for a 55° phase margin. The lockin time has increased to 915 µs.

will be suppressed by the loop filter by more than 90 dB. For reasons of phase noise and switching speed, a loop bandwidth of 100 kHz has been considered. Now, taking advantage of the fractional-N principle, say that we want to operate at an offset of 30 kHz (550.03 MHz). With this new output frequency, the inherent spurious signal reduction mechanism in the fractional-N chip limits the reduction to about 55 dB. Part of the reason why the spurious signal suppression is less in this case is that the phase frequency detector acts as a mixer, collecting both the 50-MHz reference (and its harmonics) and 550.03 MHz. Mixing the 11th reference harmonic (550 MHz) and the output frequency (550.03 MHz) results in output at 30 kHz; since the loop bandwidth is 100 kHz, it adds nothing to the suppression of this signal. To solve this, we could consider narrowing the loop bandwidth to 10% of the offset. A 30-kHz offset would equate to a loop bandwidth of 3 kHz, at which the loop speed might still be acceptable, but for a 1-kHz offset, the necessary loop bandwidth of 100 Hz would make the loop too slow. A better way is to use a different reference frequency—one that would place the resulting spurious product considerably outside the 100-kHz loop filter window. If, for instance, we used a 49-MHz reference, multiplication by 11 would result in 539 MHz. Mixing this with 550.03 MHz would result in spurious signals at ±11.03 MHz, a frequency so far outside the loop bandwidth that it would essentially disappear. Starting with a VHF, low-phase-noise crystal oscillator, such as 130 MHz, one can implement an intelligent reference frequency selection to avoid these discrete spurious signals. An additional method of reducing the spurious contents is maintaining a division ratio greater than 12 in all cases. Actual tests have shown that these reference-based spurious frequencies can be repeatedly suppressed by 80 to 90 dB.

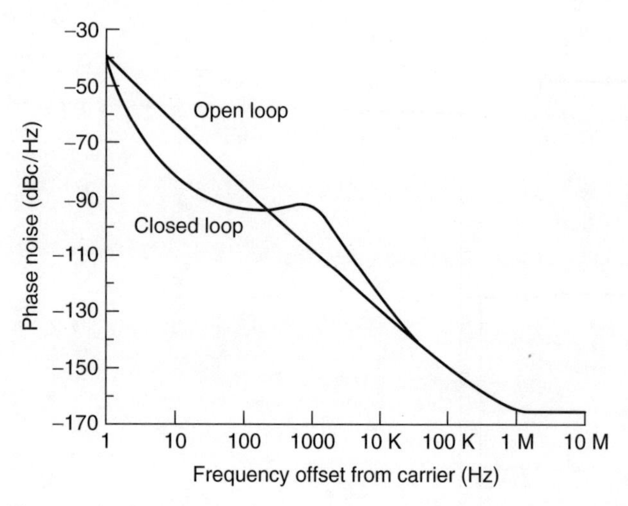

Figure 52. Comparison between open- and closed-loop noise prediction. Note the overshoot of around 1 kHz off the carrier.

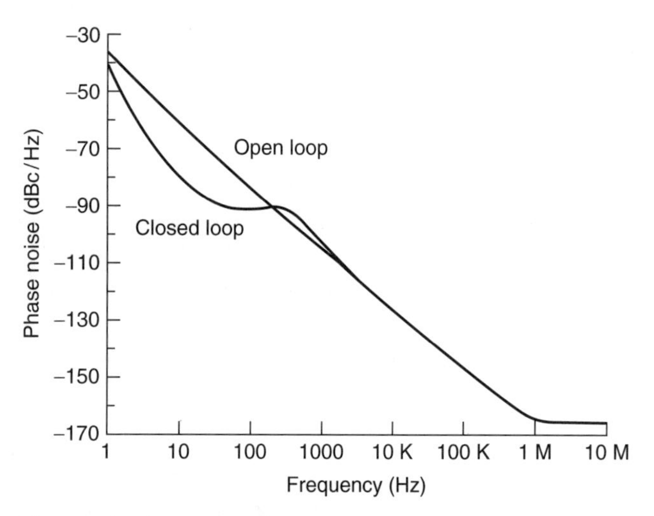

Figure 53. Comparison between open- and closed-loop noise prediction. Note the overshoot at around 300 Hz off the carrier.

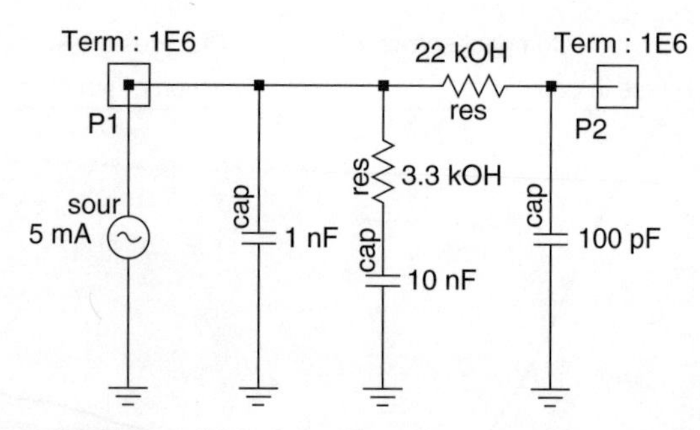

Figure 54. Type 1 high-order loop filter used for passive filter evaluation. The 1-nF capacitor is used for spike suppression as explained in the text. The filter consists of a lag portion and an additional lowpass section.

6.1. Spur Suppression Techniques

While several methods have been proposed in the literature, the method of reducing the noise by using a sigma–delta modulator has shown to be most promising. The concept is to get rid of the low-frequency phase error by rapidly switching the division ratio to eliminate the gradual phase error at the discriminator input. By changing the division ratio rapidly between different values, the phase errors occur in both polarities, positive as well as negative, and at an accelerated rate that explains the phenomenon of high-frequency noise pushup. This noise, which is converted to a voltage by the phase/frequency discriminator and loop filter, is filtered out by the lowpass filter. The main problem associated with this noise shaping technique is that the noise power rises rapidly with frequency. Figure 58 shows noise contributions with such a sigma–delta modulator in place.

On the other hand, we can now, for the first time, build a single-loop synthesizer with switching times as fast as 6 μs and very little phase noise deterioration inside the loop bandwidth, as seen in Fig. 58. Since this system maintains the good phase noise of the ceramic-resonator-based oscillator, the resulting performance is significantly better than the phase noise expected from high-end signal

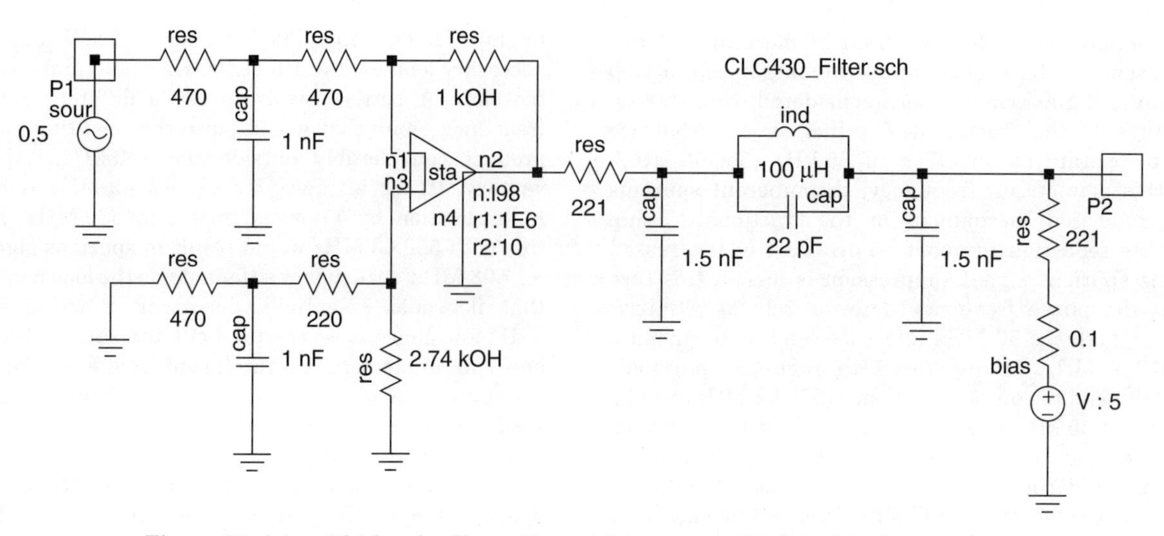

Figure 55. A type 2 high-order filter with a notch to suppress the discrete reference spurs.

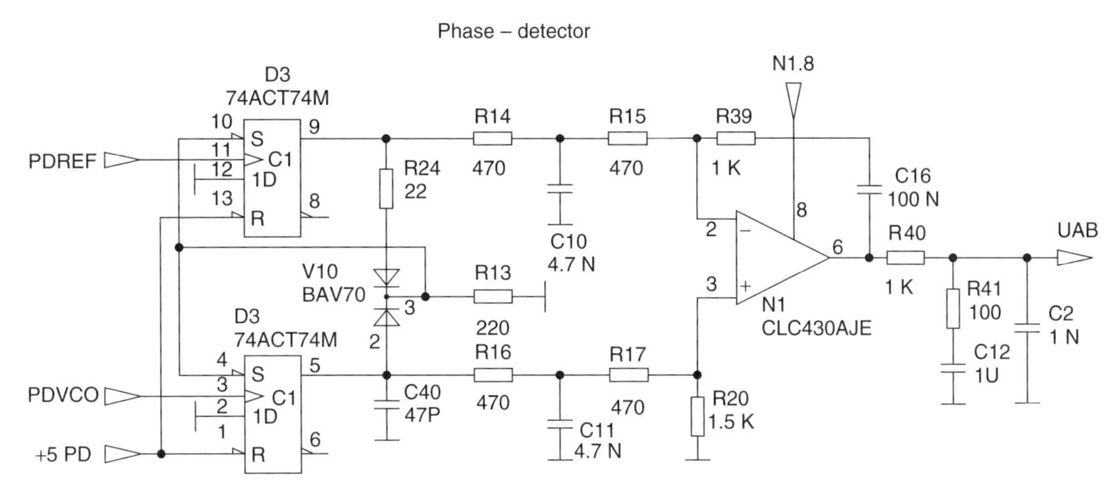

Figure 56. Phase/frequency discriminator including an active loop filter capable of operating up to 100 MHz.

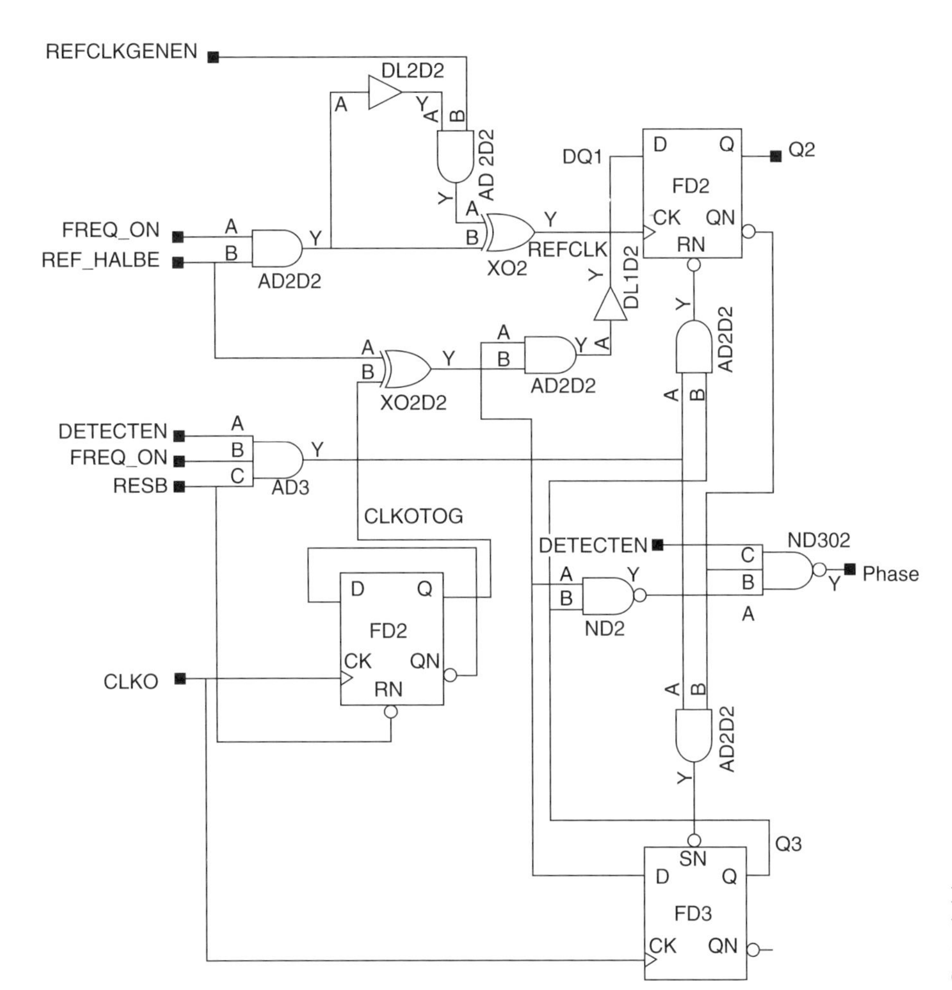

Figure 57. Custom-built phase detector with a noise floor of better than −168 dBc/Hz. This phase detector shows extremely low phase jitter.

generators. However, this method does not allow us to increase the loop bandwidth beyond the 100-kHz limit, where the noise contribution of the sigma–delta modulator takes over.

Table 2 shows some of the modern spur suppression methods. These three-stage sigma–delta methods with larger accumulators have the most potential.

The power spectral response of the phase noise for the three-stage sigma–delta modulator is calculated from

$$\mathcal{L}(f) = \frac{(2\pi)^2}{12 \times f_{\mathrm{ref}}} \times \left[2\sin\left(\frac{\pi f}{f_{\mathrm{ref}}}\right)\right]^{2(n-1)} \ \mathrm{rad}^2/\mathrm{Hz} \qquad (78)$$

where n is the number of the stage of the cascaded sigma–delta modulator. Equation (78) shows that the

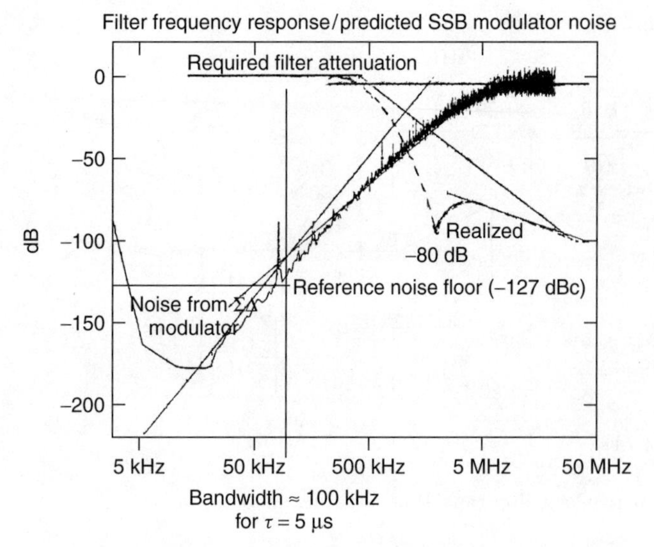

Figure 58. This filter frequency response/phase noise analysis graph shows the required attenuation for the reference frequency of 50 MHz and the noise generated by the sigma–delta converter (three steps) as a function of the offset frequency. It becomes apparent that the sigma–delta converter noise dominates above 80 kHz unless attenuated.

phase noise resulting from the fractional controller is attenuated to negligible levels close to the center frequency, and further from the center frequency, the phase noise is increased rapidly and must be filtered out prior to the tuning input of the VCO to prevent unacceptable degradation of spectral purity. A loop filter

Table 2. Spur Suppression Methods

Technique	Feature	Problem
DAC phase estimation	Cancel spur by DAC	Analog mismatch
Pulse generation	Insert pulses	Interpolation jitter
Phase interpolation	Inherent fractional divider	Interpolation jitter
Random jittering	Randomize divider	Frequency jitter
Sigma–delta modulation	Modulate division ratio	Quantization noise

must be used to filter the noise in the PLL loop. Figure 58 shows the plot of the phase noise versus the offset frequency from the center frequency. A fractional-N synthesizer with a three-stage sigma–delta modulator as shown in Fig. 59 has been built. The synthesizer consists of a phase/frequency detector, an active lowpass filter (LPF), a voltage-controlled oscillator (VCO), a dual-modulus prescaler, a three-stage sigma–delta modulator, and a buffer. Figure 60 shows the inner workings of the chip in greater detail.

After designing, building, and predicting the phase noise performance of this synthesizer, it becomes clear that measuring the phase noise of such a system becomes tricky. Standard measurement techniques that use a reference synthesizer will not provide enough resolution because there are no synthesized signal generators on the market sufficiently good to measure such low values of

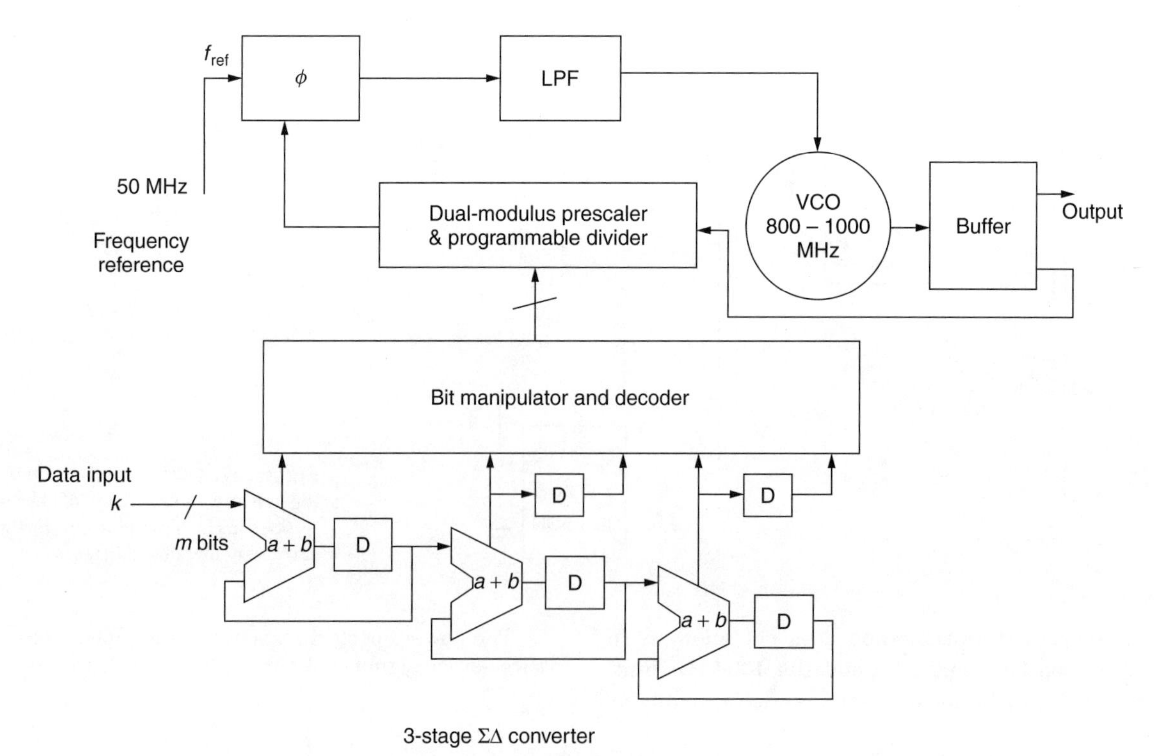

Figure 59. Block diagram of the fractional-N synthesizer built using a custom IC capable of operation at reference frequencies up to 150 MHz. The frequency is extensible up to 3 GHz using binary (÷2, ÷4, ÷8, etc.) and fixed-division counters.

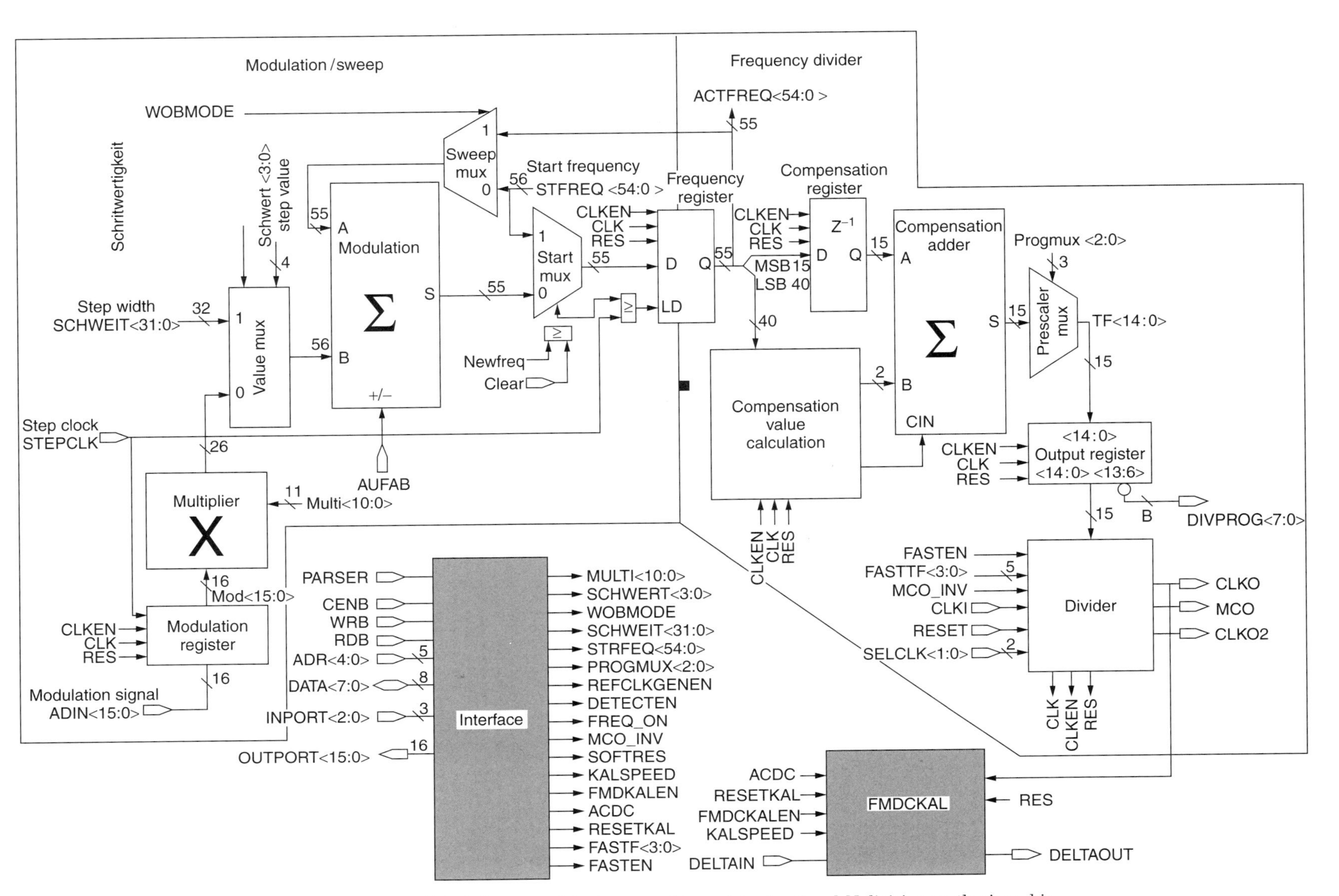

Figure 60. Detailed block diagram of the inner workings of the fractional-N-division synthesizer chip.

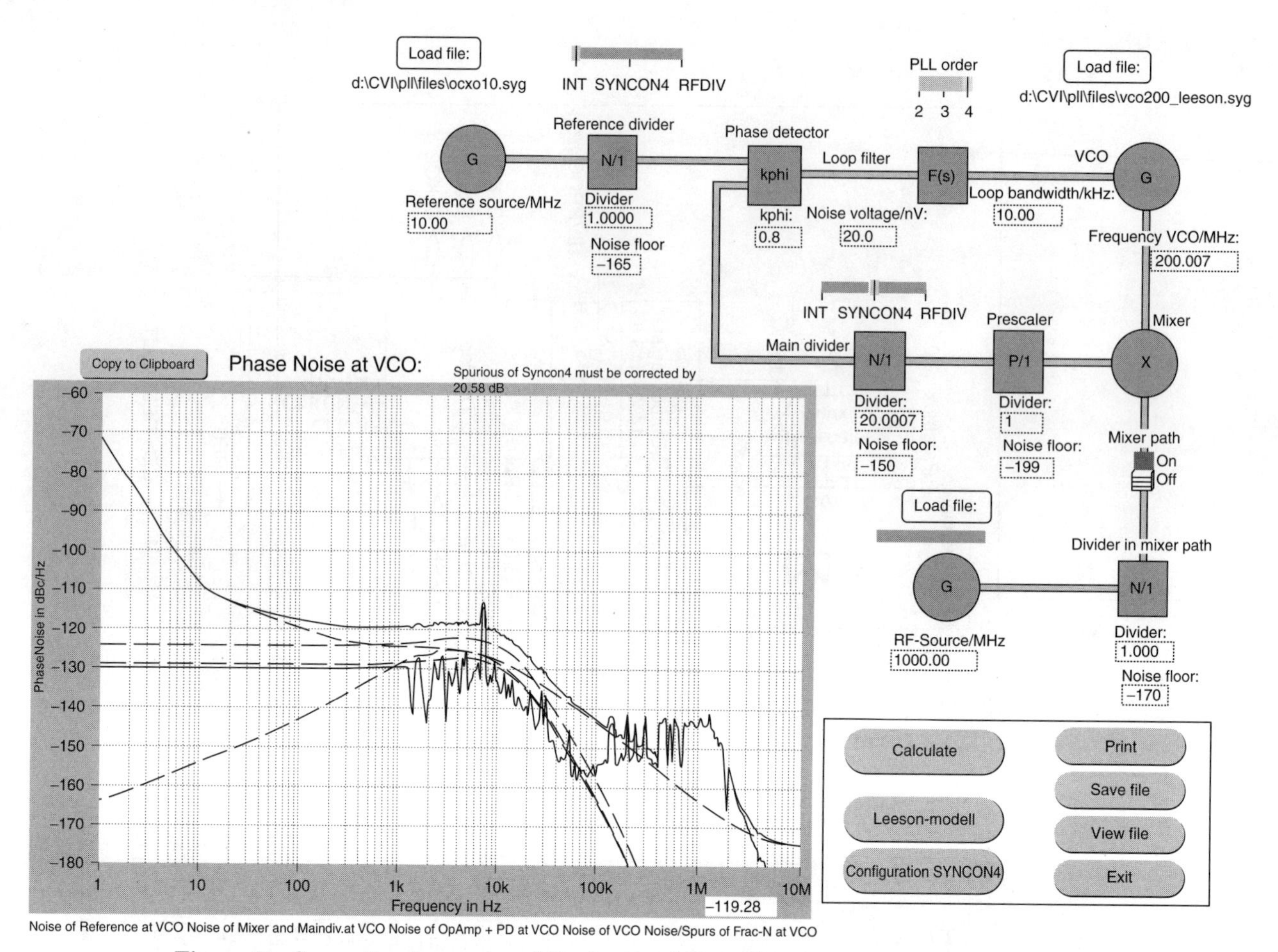

Figure 61. Composite phase noise of the fractional-N synthesizer system, including all noise and spurious signals generated within the system. The discrete spurious of 7 kHz is due to the nonlinearity of the phase detector. Its value needs to be corrected by 20.58 dB to a lesser value because of the bandwidth of the FFT analyzer.

phase noise. Therefore, a comb generator is needed that would take the output of the oscillator and multiply this up 10–20 times.

Figure 61 shows a simulated phase noise and termination of spurious outputs for the fraction-N-division synthesizer with $\Sigma\Delta$ converter. At the moment, it is unclear if the PLL with DDS or the fractional-N-division synthesizer principle with $\Sigma\Delta$ converter is the winning approach. The DDS typically requires two or three loops and is much more expensive, while the fractional-N approach requires only one loop and is a very intelligent spurious removal circuit, and high-end solutions are typically custom-built [96–111].

BIBLIOGRAPHY

1. U. L. Rohde and J. Whitaker, *Communications Receivers*, 3rd ed., McGraw-Hill, New York, Dec. 2000.
2. L. E. Larson, *RF and Microwave Circuit Design for Wireless Communications*, Artech House, Norwood, MA, 1996.
3. U. L. Rohde, *Microwave and Wireless Synthesizers: Theory and Design*, Wiley, New York, Aug. 1997.
4. W. F. Egan, *Frequency Synthesis by Phase Lock*, Wiley-Interscience, New York, 1981.
5. F. Gardner, *Phaselock Techniques*, 2nd ed., Wiley-Interscience, New York, 1979.
6. J. Gorski-Popiel, *Frequency Synthesis: Techniques and Applications*, IEEE, New York, 1975.
7. V. F. Kroupa, *Frequency Synthesis: Theory, Design, and Applications*, Wiley, New York, 1973.
8. V. Manassewitsch, *Frequency Synthesizers: Theory and Design*, Wiley, New York, 1976.
9. R. E. Best, *Phase-Locked Loops: Theory, Design, and Applications*, McGraw-Hill, New York, 1989.
10. P. Danzer, ed., *The ARRL Handbook for Radio Amateurs*, 75th ed., ARRL, Newington, 1997, Chap. 14, AC/RF sources (Oscillators and Synthesizers).
11. C. R. Chang and U. L. Rohde, The accurate simulation of oscillator and PLL phase noise in RF sources, *Wireless '94 Symp.*, Santa Clara, CA Feb. 15–18, 1994.
12. M. M. Driscoll, Low noise oscillator design using acoustic and other high Q resonators, *44th Annual Symp. Frequency Control*, Baltimore, MD, May 1990.
13. U. L. Rohde, Low-noise frequency synthesizers fractional N phase-locked loops, *Proc. SOUTHCON/81*, Jan. 1981.
14. W. C. Lindsey and C. M. Chie, eds., *Phase-Locked Loops*, IEEE Press, New York, 1986.

15. U.S. Patent 4,492,936 (Jan. 8, 1985), A. Albarello, A. Roullet, and A. Pimentel, Fractional-division frequency synthesizer for digital angle-modulation (Thomson-CSF).

16. U.S. Patent 4,686,488 (Aug. 11, 1987), C. Attenborough, Fractional-*N* frequency synthesizer with modulation compensation, (Plessey Overseas Ltd., Ilford, UK).

17. U.S. Patent 3,913,928 (Oct., 1975), R. J. Bosselaers, PLL including an arithmetic unit.

18. U.S. Patent 2,976,945, R. G. Cox, Frequency synthesizer (Hewlett-Packard).

19. U.S. Patent 4,586,005 (April 29, 1986), J. A. Crawford, Enhanced analog phase interpolation for fractional-*N* frequency synthesis (Hughes Aircraft Co., Los Angeles).

20. U.S. Patent 4,468,632 (Aug. 28, 1984), A. T. Crowley, PLL frequency synthesizer including fractional digital frequency divider (RCA Corp., New York).

21. U.S. Patent 4,763,083 (Aug. 9, 1988), A. P. Edwards, Low phase noise RF synthesizer (Hewlett-Packard), Palo Alto, CA.

22. U.S. Patent 4,752,902 (June 21, 1988), B. G. Goldberg, Digital frequency synthesizer (Sciteq Electronics, Inc., San Diego, CA).

23. U.S. Patent 4,958,310 (Sept. 18, 1990), B. G. Goldberg, Digital frequency synthesizer having multiple processing paths.

24. U.S. Patent 5,224,132 (June 29, 1993), B. G. Goldberg, Programmable fractional-*N* frequency synthesizer (Sciteq Electronics, Inc., San Diego, CA).

25. B. G. Goldberg, Analog and digital fractional-*n* PLL frequency synthesis: A survey and update, *Appl. Microwave Wireless* 32–42 (June 1999).

26. U.S. Patent 3,882,403, W. G. Greken, Digital frequency synthesizer (General Dynamics).

27. U.S. Patent 5,093,632 (March 3, 1992), A. W. Hietala and D. C. Rabe, Latched accumulator fractional *N* synthesis with residual error reduction (Motorola, Inc., schaumburg, IL).

28. U.S. Patent 3,734,269 (May, 1973), L. Jackson, Digital frequency synthesizer.

29. U.S. Patent 4,758,802 (July 19, 1988), T. Jackson, Fractional-*N* synthesizer (Plessey Overseas Ltd., Ilford, UK).

30. U.S. Patent 4,800,342 (Jan. 24, 1989), T. Jackson, Frequency synthesizer of the fractional type (Plessey Overseas Ltd., Ilford, UK).

31. Eur. Patent 0214217B1 (June 6, 1996), T. Jackson, Improvement in or relating to synthesizers (Plessey Overseas Ltd., Ilford, Essex, UK).

32. Eur. Patent WO86/05046 (Aug. 28, 1996), T. Jackson, Improvement in or relating to synthesizers (Plessey Overseas Ltd., Ilford, Essex, UK).

33. U.S. Patent 4,204,174 (May 20, 1980), N. J. R. King, Phase locked loop variable frequency generator (Racal Communications Equipment Ltd., England).

34. U.S. Patent 4,179,670 (Dec. 18, 1979), N. G. Kingsbury, Frequency synthesizer with fractional division ratio and jitter compensation (Marconi Co. Ltd., Chelmsford, UK).

35. U.S. Patent 3,928,813 (Dec. 23, 1975), C. A. Kingsford-Smith, Device for synthesizing frequencies which are rational multiples of a fundamental frequency (Hewlett-Packard, Palo Alto, CA).

36. U.S. Patent 4,918,403 (April 17, 1990), F. L. Martin, Frequency synthesizer with spur compensation (Motorola, Inc., Schaumburg, IL).

37. U.S. Patent 4,816,774 (March 28, 1989), F. L. Martin, Frequency synthesizer with spur compensation (Motorola).

38. U.S. Patent 4,599,579 (July 8, 1986), K. D. McCann, Frequency synthesizer having jitter compensation (U.S. Philips Corp., New York).

39. U.S. Patent 5,038,117 (Aug. 6, 1991), B. M. Miller, Multiple-modulator fractional-*N* divider (Hewlett-Packard).

40. U.S. Patent 4,206,425 (June 3, 1980), E. J. Nossen, Digitized frequency synthesizer (RCA Corp., New York).

41. O. Peña, SPICE tools provide behavioral modeling of PLLs, *Microwaves RF Mag.* 71–80 (Nov. 1997).

42. U.S. Patent 4,815,018 (March 21, 1989), V. S. Reinhardt and I. Shahriary, Spurless fractional divider direct digital frequency synthesizer and method (Hughes Aircraft Co., Los Angeles).

43. U.S. Patent 4,458,329 (July 3, 1984), J. Remy, Frequency synthesizer including a fractional multiplier, (Adret Electronique, Paris).

44. U.S. Patent 4,965,531 (Oct. 23, 1990), T. A. D. Riley, Frequency synthesizers having dividing ratio controlled sigma-delta modulator (Carleton Univ., Ottawa, Canada).

45. U.S. Patent 5,021,754 (June 4, 1991), W. P. Shepherd, D. E. Davis, and W. F. Tay, Fractional-*N* synthesizer having modulation spur compensation (Motorola, Inc., Schaumburg, IL).

46. U.S. Patent 3,959,737 (May 25, 1976), W. J. Tanis, Frequency synthesizer having fractional-*N* frequency divider in PLL (Engelman Microwave).

47. Eur. Patent 0125790B2 (July 5, 1995), J. N. Wells, Frequency synthesizers (Marconi Instruments, St. Albans, Hertfordshire, UK).

48. U.S. Patent 4,609,881 (Sept. 2, 1986), J. N. Wells, Frequency synthesizers (Marconi Instruments Ltd., St. Albans, UK).

49. U.S. Patent 4,410,954 (Oct. 18, 1983), C. E. Wheatley, III, Digital frequency synthesizer with random jittering for reducing discrete spectral spurs (Rockwell International Corp., El Segundo, CA).

50. U.S. Patent 5,038,120 (Aug. 6, 1991), M. A. Wheatley, L. A. Lepper, and N. K. Webb, Frequency modulated phase locked loop with fractional divider and jitter compensation (Racal-Dana Instruments Ltd., Berkshire, UK).

51. U.S. Patent 4,573,176 (Feb. 25, 1986), R. O. Yaeger, Fractional frequency divider, (RCA Corp., Princeton, NJ).

52. U. L. Rohde, A high performance synthesizer for base stations based on the fractional-*N* synthesis principle, *Microwaves RF Mag.* (April 1998).

53. U. L. Rohde and G. Klage, Analyze VCOs and fractional-*N* synthesizers, *Microwaves RF Mag.* (Aug. 2000).

54. R. Hassun and A. Kovalic, An arbitrary waveform synthesizer for DC to 50 MHz, *Hewlett-Packard J.* (Palo Alto, CA) (April 1988).

55. L. R. Rabiner and B. Gold, *Theory and Application of Digital Signal Processing*, Prentice-Hall, Englewood Cliffs, NJ, 1975, Chap. 2.

56. H. T. Nicholas and H. Samueli, An analysis of the output spectrum of direct digital frequency synthesizers in the

presence of phase-accumulator truncation, *41st Annual Frequency Control Symp.*, IEEE Press, New York, 1987.

57. L. B. Jackson, Roundoff noise for fixed point digital filters realized in cascade or parallel form, *IEEE Trans. Audio Electroacous.* **AU-18**: 107–122 (June 1970).
58. Technical Staff of Bell Laboratories, *Transmission Systems for Communication*, Bell Labs, Inc., 1970, Chap. 25.
59. U.S. Patent 4,482,974, A. Kovalick, Apparatus and method of phase to amplitude conversion in a SIN function generator.
60. C. J. Paull and W. A. Evans, Waveform shaping techniques for the design of signal sources, *Radio Electron. Eng.* **44**(10): (Oct. 1974).
61. L. Barnes, Linear-segment approximations to a sinewave, *Electron. Eng.* **40**: (Sept. 1968).
62. U.S. Patent 4,454,486, R. Hassun and A. Kovalic, Waveform synthesis using multiplexed parallel synthesizers.
63. D. K. Kikuchi, R. F. Miranda, and P. A. Thysel, A waveform generation language for arbitrary waveform synthesis, *Hewlett-Packard J.* (Palo Alto, CA) (April 1988).
64. H. M. Stark, *An Introduction to Number Theory*, MIT Press, Cambridge, MA, 1978, Chap. 7.
65. W. Sagun, Generate complex waveforms at very high frequencies, *Electron. Design* (Jan. 26 1989).
66. G. Lowitz and R. Armitano, Predistortion improves digital synthesizer accuracy, *Electron. Design* (March 31 1988).
67. A. Kovalic, Digital synthesizer aids in testing of complex waveforms, *EDN Mag.* (Sept. 1 1988).
68. G. Lowitz and C. Pederson, RF testing with complex waveforms, *RF Design* (Nov. 1988).
69. C. M. Merigold, in Kamilo Fehrer, ed., *Telecommunications Measurement Analysis and Instrumentation*, Prentice-Hall, Englewood Cliffs, NJ, 1987.
70. Synergy Microwave Corporation Designer Handbook, *Specifications of Synthesizers*, 2001.
71. G. Vendelin, A. M. Pavio, and U. L. Rohde, *Microwave Circuit Design Using Linear and Nonlinear Techniques*, Wiley, New York, 1990.
72. J. A. Crawford, *Frequency Synthesizer Design Handbook*, Artech House, Norwood, MA, 1994.
73. D. B. Leeson, Short-term stable microwave sources, *Microwave J.* 59–69 (June 1970).
74. J. Smith, *Modern Communication Circuits*, McGraw-Hill, New York, 1986, pp. 252–261.
75. J. S. Yuan, Modeling the bipolar oscillator phase noise, *Solid State Electron.* **37**(10): 1765–1767 (Oct. 1994).
76. D. Scherer, Design principles and test methods for low phase noise RF and microwave sources, *RF & Microwave Measurement Symp. Exhibition*, Hewlett-Packard.
77. S. Alechno, Analysis method characterizes microwave oscillators, *Microwaves RF Mag.* 82–86 (Nov. 1997).
78. W. Anzill, F. X. Kärtner, and P. Russer, Simulation of the single-sideband phase noise of oscillators, *2nd Int. Workshop of Integrated Nonlinear Microwave and Millimeterwave Circuits*, 1992.
79. N. Boutin, RF oscillator analysis and design by the loop gain method, *Appl. Microwave Wireless* 32–48 (Aug. 1999).
80. P. Braun, B. Roth, and A. Beyer, A measurement setup for the analysis of integrated microwave oscillators, *2nd Int. Workshop of Integrated Nonlinear Microwave and Millimeterwave Circuits*, 1992.
81. C. R. Chang et al., Computer-aided analysis of free-running microwave oscillators, *IEEE Trans. Microwave Theory Tech.* **39**: 1735–1745 (Oct. 1991).
82. P. Davis et al., Silicon-on-silicon integration of a GSM transceiver with VCO resonator, *1998 IEEE Int. Solid-State Circuits Conf. Digest of Technical Papers*, pp. 248–249.
83. P. J. Garner, M. H. Howes, and C. M. Snowden, *Ka*-band and MMIC pHEMT-basd VCO's with low phase-noise properties, *IEEE Trans. Microwave Theory Tech.* **46**: 1531–1536 (Oct. 1998).
84. A. V. Grebennikov and V. V. Nikiforov, An analytic method of microwave transistor oscillator design, *Int. J. Electron.* **83**: 849–858 (Dec. 1997).
85. A. Hajimiri and T. H. Lee, A general theory of phase noise in electrical oscillators, *IEEE J. Solid-State Circuits* **33**: 179–194 (Feb. 1998).
86. Q. Huang, On the exact design of RF oscillators, *Proc. IEEE 1998 Custom Integrated Circuits Conf.*, pp. 41–44.
87. Fairchild Data Sheet, *Phase/Frequency Detector, 11C44*, Fairchild Semiconductor, Mountain View, CA.
88. Fairchild Preliminary Data Sheet, *SH8096 Programmable Divider-Fairchild Integrated Microsystems*, April 1970.
89. U. L. Rohde, *Digital PLL Frequency Synthesizers — Theory and Design*, Prentice-Hall, Englewood Cliffs, NJ, Jan. 1983.
90. W. Egan and E. Clark, Test your charge-pump phase detectors, *Electron. Design* **26**(12): 134–137 (June 7 1978).
91. S. Krishnan, Diode phase detectors, *Electron. Radio Eng.* 45–50 (Feb. 1959).
92. Motorola Data Sheet, *MC12012*, Motorola Semiconductor Products, Inc. Phoenix, AZ, 1973.
93. Motorola Data Sheet, *Phase-Frequency Detector, MC4344, MC4044*.
94. U. L. Rohde and D. P. Newkirk, *RF/Microwave Circuit Design for Wireless Applications*, Wiley, 2000.
95. W. C. Lindsey and M. K. Simon, eds., *Phase-Locked Loops & Their Application*, IEEE Press, New York, 1978.
96. W. Z. Chen and J. T. Wu, A 2 V 1.6 GHz BJT phase-locked loop, *Proc. IEEE 1998 Custom Integrated Circuits Conf.*, pp. 563–566.
97. J. Craninckx and M. Steyaert, A fully integrated CMOS DCS-1800 frequency synthesizer, 1998 *IEEE Int. Solid-State Circuits Conf. Digest of Technical Papers*, pp. 372–373.
98. B. De Smedt and G. Gielen, Nonlinear behavioral modeling and phase noise evaluation in phase locked loops, *Proc. IEEE 1998 Custom Integrated Circuits Conf.*, pp. 53–56.
99. N. M. Filiol et al., An agile ISM band frequency synthesizer with built-in GMSK data modulation, *IEEE J. Solid-State Circuits* **33**(7): 998–1007 (July 1998).
100. Fujitsu Microelectronics, Inc., *Super PLL Application Guide*, 1998.
101. Hewlett-Packard Application Note 164-3, *New Technique for Analyzing Phase-Locked Loops*, June 1975.
102. V. F. Kroupa, ed., *Direct Digital Frequency Synthesizers*, IEEE Press, New York, 1999.

103. S. Lo, C. Olgaard, and D. Rose, A 1.8V/3.5 mA 1.1 GHz/300 MHz CMOS dual PLL frequency synthesizer IC for RF communications, *Proc. IEEE 1998 Custom Integrated Circuits Conf.*, pp. 571–574.

104. G. Palmisano et al., Noise in fully-integrated PLLs, *Proc. 6th AACD 97*, Como, Italy, 1997.

105. B. H. Park and P. E. Allen, A 1 GHz, low-phase-noise CMOS frequency synthesizer with integrated LC VCO for wireless communications, *Proc. IEEE 1998 Custom Integrated Circuits Conf.*, pp. 567–570.

106. B. Sam, Hybrid frequency synthesizer combines octave tuning range and millihertz steps, *Appl. Microwave Wireless* 76–84 (May 1999).

107. M. Smith, *Phase Noise Measurement Using the Phase Lock Technique*, Wireless Subscriber Systems Group (WSSG) AN1639, Motorola.

108. V. von Kaenel et al., A 600 MHz CMOS PLL microprocessor clock generator with a 1.2 GHz VCO, *1998 IEEE Int. Solid-State Circuits Conf. Digest of Technical Papers*, pp. 396–397.

109. Motorola MECL Data Book, Chapter 6, *Phase-Locked Loops*, 1993.

110. U. L. Rohde, Low noise microwave synthesizers, WFFDS: Advances in microwave and millimeter-wave synthesizer technology, *IEEE-MTT-Symp.*, Orlando, FL, May 19, 1995.

111. U. L. Rohde, Oscillator design for lowest phase noise, *Microwave Eng. Eur.* 35–40 (May 1994).

G

GENERAL PACKET RADIO SERVICE (GPRS)

Christian Bettstetter
Christian Hartmann
Technische Universität München
Institute of Communication Networks
Munich, Germany

1. INTRODUCTION

The *General Packet Radio Service* (GPRS) is a data bearer service for GSM and IS136 cellular networks. Its packet-oriented radio transmission technology enables efficient and simplified wireless access to Internet protocol-based networks and services. With GPRS-enabled mobile devices, users benefit from higher data rates [up to ~50 kbps (kilobits per second)], shorter access times, an "always on" wireless connectivity, and volume-based billing.

In conventional GSM networks without GPRS, access to external data networks from GSM mobile devices has already been standardized in GSM phase 2; however, on the air interface, such access occupied a complete circuit-switched traffic channel for the entire call period. In case of bursty traffic (e.g., Internet traffic), it is obvious that packet-switched bearer services result in a much better utilization of the traffic channels. A packet channel will be allocated only when needed and will be released after the transmission of the packets. With this principle, multiple users can share one physical channel (statistical multiplexing). Moreover, in conventional GSM exactly one channel is assigned for both uplink and downlink. This entails two disadvantages: (1) only symmetric connections are supported and (2) the maximum data rate is restricted since the use of multiple channels in parallel is not possible. In order to address the inefficiencies of GSM phase 2 data services, the General Packet Radio Service has been developed in GSM phase 2+ by the *European Telecommunications Standards Institute* (ETSI). It offers a genuine packet-switched transmission technology also over the air interface, and provides access to networks based on the *Internet Protocol* (IP) (e.g., the global Internet or private/corporate intranets) and X.25. GPRS enables asymmetric data rates as well as simultaneous transmission on several channels (multislot operation). Initial work on the GPRS standardization began in 1994, and the main set of specifications was approved in 1997 and completed in 1999. Market introduction took place in the year 2000. GPRS was also standardized for IS136 [1]; however, in this description we focus on GPRS for GSM.

Data transmission in conventional circuit switched GSM is restricted to 14.4 kbps, and the connection setup takes several seconds. GPRS offers almost ISDN-like data rates up to ~50 kbps and session establishment times below one second. Furthermore, GPRS supports a more user-friendly billing than that offered by circuit-switched data services. In circuit-switched services, billing is based on the duration of the connection. This is unsuitable for applications with bursty traffic, since the user must pay for the entire airtime even for idle periods when no packets are sent (e.g., when the user reads a Webpage). In contrast to this, packet-switched services allow charging based on the amount of transmitted data (e.g., in kilobytes) and the *quality of service* (QoS) rather than connection time. The advantage for the user is that he/she can be online over a long period of time ("always online") but will be billed only when data are actually transmitted. The network operators can utilize their radio resources in a more efficient way and simplify the access to external data networks.

Typical scenarios for GPRS are the wireless access to the *World Wide Web* (WWW) and corporate local-area networks. Here, GPRS provides an end-to-end IP connectivity; that is, users can access the Internet without first requiring to dial in to an Internet service provider. Examples for applications that can be offered over GPRS are mobile e-commerce services, location-based tourist guides, and applications in the telemetry field. GPRS can also be used as a bearer for the *Wireless Application Protocol* (WAP) and the *Short Message Service* (SMS).

Considering the network evolution of second generation cellular networks to third generation, GPRS enables a smooth transition path from GSM/TDMA networks toward the *Universal Mobile Telecommunication System* (UMTS). Especially, the IP backbone network of GPRS forms the basis for the UMTS core network. With the introduction of *Enhanced Data Rates for GSM Evolution* (EDGE), which will use 8-PSK modulation, GPRS will offer approximately a 3 times higher data rate and a higher spectral efficiency. This mode will be called *Enhanced GPRS* (EGPRS).

2. SYSTEM ARCHITECTURE

To incorporate GPRS into existing GSM networks, several modifications and enhancements have been made in the GSM network infrastructure as well as in the mobile stations. On the network side, a new class of nodes has been introduced, namely, the *GPRS support nodes* (GSNs). They are responsible for routing and delivery of data packets between the mobile stations and external packet data networks. Figure 1 shows the resulting GSM/GPRS system architecture [2,3]. A mobile user carries a *mobile station* (MS), which can communicate over a wireless link with a *base transceiver station* (BTS). Several BTSs are controlled by a *base station controller* (BSC). The BTSs and BSC together form a *base station subsystem* (BSS).

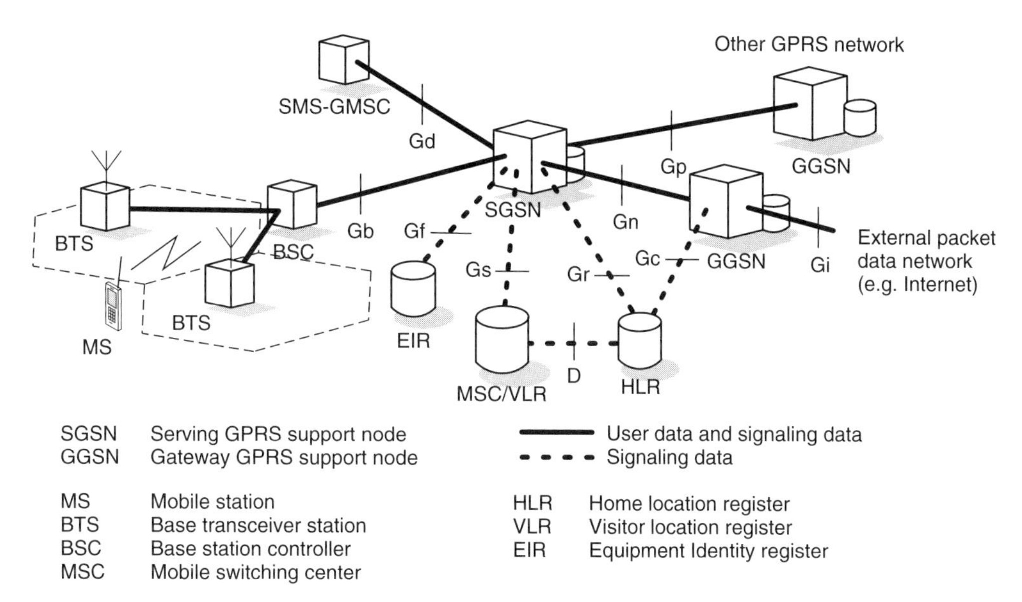

Figure 1. GPRS system architecture and interfaces.

For an detailed explanation of the GSM elements and the basic addresses, see the → GSM entry.

A *serving GPRS support node* (SGSN) delivers packets from and to the MSs within its service area. Its tasks include packet routing and transfer, functions for attach/detach of MSs and their authentication, mobility management, radio resource management, and logical link management. The location register of the SGSN stores location information of all GPRS users registered with this SGSN [e.g., current cell, current *visitor location register* (VLR)] and their user profiles [e.g., *international mobile subscriber identity* (IMSI), address used in the packet data network].

A *gateway GPRS support node* (GGSN) acts as an interface to external packet data networks (e.g., to the Internet). It converts GPRS packets coming from the SGSN into the appropriate *packet data protocol* (PDP) format (i.e., IP) and sends them out on the corresponding external network. In the other direction, the PDP address of incoming data packets (e.g., the IP destination address) is mapped to the GSM address of the destination user. The readdressed packets are sent to the responsible SGSN. For this purpose, the GGSN stores the current SGSN addresses and profiles of registered users in its location register.

The functionality of the SGSN and GGSN can be implemented in a single physical unit or in separate units. All GSNs are connected via an IP-based backbone network. Within this GPRS backbone, the GSNs encapsulate the external packets and transmit (tunnel) them using the so-called *GPRS tunneling protocol* (GTP). If there is a roaming agreement between two operators, they may install an inter-operator backbone between their networks (see Fig. 2). The *border gateways* (BGs) perform security functions in order to protect the private GPRS networks against attacks and unauthorized users.

Via the Gn and the Gp interfaces (see Figs. 1 and 2), user payload and signaling data are transmitted between the GSNs. The Gn interface will be used, if SGSN and GGSN are located in the same network, whereas the Gp interface will be used, if they are in different networks. These interfaces are also defined between two SGSNs. This allows the SGSNs to exchange user profiles when a mobile station moves from one SGSN area to another. The Gi interface connects the GGSN with external networks. From an external network's point of view, the GGSN looks like a usual IP router, and the GPRS network looks like any other IP subnetwork.

Figure 1 also shows the signaling interfaces between the GSNs and the conventional GSM network entities [2]. Across the Gf interface, the SGSN may query and verify the *international mobile equipment identity* (IMEI) of a mobile station trying to attach to the network. GPRS also adds some entries to the GSM registers. For mobility management, a user's entry in the *home location register* (HLR) is extended with a link to its current SGSN. Moreover, his/her GPRS-specific profile and current PDP address(es) are stored. The Gr interface is used to exchange this information between HLR and SGSN. For example, the SGSN informs the HLR about the current location of the mobile station, and when a mobile station registers with a new SGSN, the HLR will send the user profile to the new SGSN. In a similar manner, the signaling interface between GGSN and HLR (Gc interface) may be used by the GGSN to query the location and profile of a user who is unknown to the GGSN. In addition, the MSC/VLR may be extended with functions and register entries that allow efficient coordination between packet-switched and conventional circuit-switched GSM services. Examples for this optional feature are combined GPRS and GSM location updates and combined attachment procedures. Moreover, paging requests of circuit-switched GSM calls can be performed via the SGSN. For this purpose, the Gs interface connects the registers of SGSN and MSC/VLR.

In order to exchange messages of the Short Message Service via GPRS, the Gd interface interconnects the *SMS gateway MSC* (SMS-GMSC) with the SGSN.

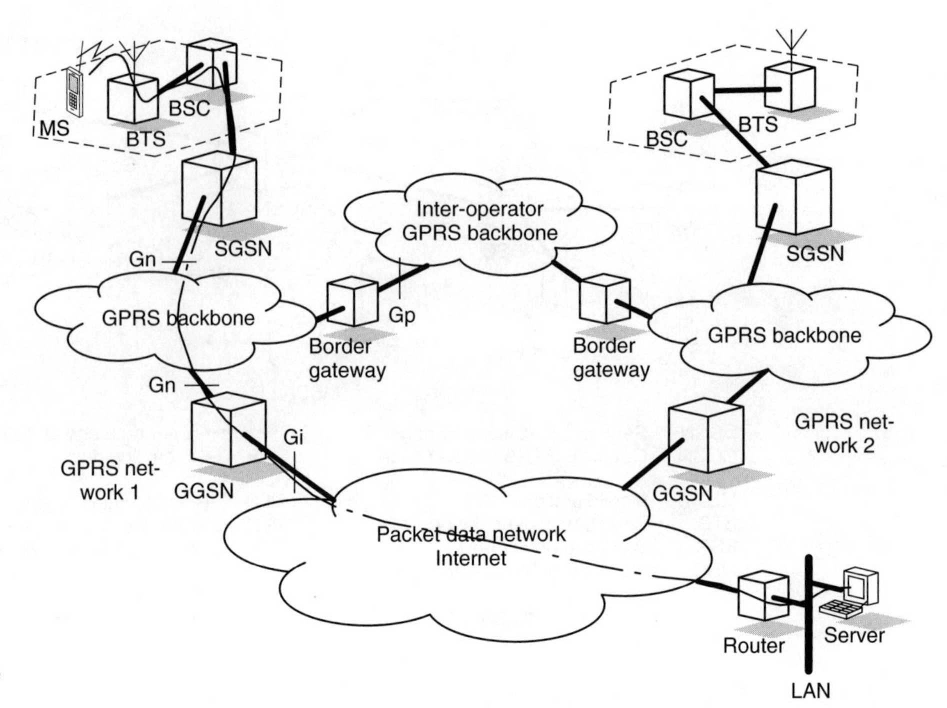

Figure 2. GPRS system architecture and routing example.

3. SERVICES

The bearer services of GPRS offer end-to-end packet-switched data transfer to mobile subscribers. A *point-to-point* (PTP) service is specified, which comes in two variants [4]: a connectionless mode for IP and a connection-oriented mode for X.25. Furthermore, SMS messages can be sent and received over GPRS.

It is planned to implement a *point-to-multipoint* (PTM) service, which offers transfer of data packets to several mobile stations. For example, IP multicast routing protocols can be employed over GPRS for this purpose [5]. Packets addressed to an IP multicast group will then be routed to all group members. Also supplementary services can be implemented, such as reverse charging and barring.

Based on these standardized bearer and supplementary services, a huge variety of nonstandardized services can be offered over GPRS. The most important application scenario is the wireless access to the Internet and to corporate intranets (for e-mail communication, database access, Web browsing). Also WAP services can be accessed in a more efficient manner than with circuit-switched GSM. Especially mobility-related services that require an "always on" connectivity but only infrequent data transmissions (e.g., interactive city guides) benefit from the packet-oriented transmission and billing method of GPRS.

3.1. Quality of Service (QoS)

Support of different QoS classes is an important feature in order to support a broad variety of applications but still preserve radio and network resources in an efficient way. Moreover, QoS classes enable providers to offer different billing options. The billing can be based on the amount of transmitted data, the service type itself, and the QoS profile. The GPRS standard [4] defines four QoS parameters: service precedence, reliability, delay, and throughput. Using these parameters, QoS profiles can be negotiated between the mobile user and the network for each session, depending on the QoS demand and the currently available resources.

The service precedence is the priority of a service (in relation to other services). There are three levels of priority defined in GPRS. In case of heavy traffic load, for example, packets of low priority will be discarded first.

The reliability indicates the transmission characteristics required by an application. Three reliability classes are defined, which guarantee certain maximum values for the probability of packet loss, packet duplication, missequencing, and packet corruption (i.e., undetected errors in a packet); see Table 1.

As shown in Table 2, the delay parameters define maximum values for the mean delay and the 95-percentile delay. The latter is the maximum delay guaranteed in 95% of all transfers. Here, delay is defined as the end-to-end transfer time between two communicating mobile stations or between a mobile station and the Gi interface to an external network, respectively. This includes all delays within the GPRS network, such as the delay for request and assignment of radio resources, transmission over the air interface, and the transit delay in the GPRS backbone network. Delays outside the GPRS network, for example in external transit networks, are not taken into account.

Table 1. QoS Reliability Classes

Class	Loss	Duplication	Missequencing	Corruption
1	10^{-9}	10^{-9}	10^{-9}	10^{-9}
2	10^{-4}	10^{-5}	10^{-5}	10^{-6}
3	10^{-2}	10^{-5}	10^{-5}	10^{-2}

Table 2. QoS Delay Classes (in seconds)

	128 Byte		1024 Byte	
Class	Mean	95%	Mean	95%
1	<0.5	<1.5	<2	<7
2	<5	<25	<15	<75
3	<50	<250	<75	<375
4	Best effort			

Finally, the throughput parameter specifies the maximum and mean bit rate.

3.2. Simultaneous Usage of Packet-Switched and Circuit-Switched Services

GPRS services can be used in parallel to circuit-switched GSM services. The GPRS standard defines three classes of mobile stations [4]. Mobile stations of class A fully support simultaneous operation of GPRS and conventional GSM services. Class B mobile stations are able to register with the network for both GPRS and conventional GSM services simultaneously and listen to both types of signaling messages, but they can use only one of the service types at a given time. Finally, class C mobile stations can attach for either GPRS or conventional GSM services at a given time. Simultaneous registration (and usage) is not possible, except for SMS messages, which can be received and sent at any time.

4. SESSION MANAGEMENT, MOBILITY MANAGEMENT, AND ROUTING

In this section we describe how a mobile station registers with the GPRS network and becomes known to an external packet data network. We show how packets are routed to or from mobile stations, and how the network keeps track of the user's current location [2].

4.1. Attachment and Detachment Procedure

Before a mobile station can use GPRS services, it must attach to the network (similar to the IMSI attach used for circuit-switched GSM services). The mobile station's ATTACH REQUEST message is sent to the SGSN. The network then checks if the user is authorized, copies the user profile from the HLR to the SGSN, and assigns a *packet temporary mobile subscriber identity* (P-TMSI) to the user. This procedure is called *GPRS attach*. It establishes a logical link between the mobile station and the SGSN, such that the SGSN can perform paging of the mobile station and deliver SMS messages. For mobile stations using circuit- and packet-switched services, it is possible to implement combined GPRS/IMSI attach procedures. The disconnection from the GPRS network is called *GPRS detach*. It can be initiated by the mobile station or by the network.

4.2. Session Management and PDP Context

To exchange data packets with external packet data networks after a successful GPRS attach, a mobile station must apply for an address to be used in the external network. In general, this address is called *PDP address* (packet data protocol address). In case the external network is an IP network, the PDP address is an IP address.

For each session, a so-called PDP context is created [2], which describes the characteristics of the session. It includes the PDP type (e.g., IPv6), the PDP address assigned to the mobile station (e.g., an IP address), the requested QoS class, and the address of a GGSN that serves as the access point to the external network. This context is stored in the MS, the SGSN, and the GGSN. Once a mobile station has an active PDP context, it is visible for the external network and can send and receive data packets. The mapping between the two addresses (PDP ↔ GSM address) makes the transfer of packets between MS and GGSN possible. In the following we assume that access to an IP-based network is intended.

The allocation of an IP address can be static or dynamic. In the first case, the mobile station permanently owns an IP address. In the second case, using a dynamic addressing concept, an IP address is assigned on activation of a PDP context. In other words, the network provider has reserved a certain number of IP addresses, and each time a mobile station attaches to the GPRS network, it will obtain an IP address. After its GPRS detach, this IP address will be available to other users again. The IP address can be assigned either by the user's home network operator or by the operator of the visited network. The GGSN is responsible for the allocation and deactivation of addresses. Thus, the GGSN should also include DHCP (*Dynamic Host Configuration Protocol* [6]) functionality, which automatically manages the available IP address space.

A basic PDP context activation procedure initialized by an MS is as follows [2]: Using the message ACTIVATE PDP CONTEXT REQUEST, the MS informs the SGSN about the requested PDP context. Afterward, the usual GSM security functions are performed. If access is granted, the SGSN will send a CREATE PDP CONTEXT REQUEST to the affected GGSN. The GGSN creates a new entry in its PDP context table, which enables the GGSN to route data packets between the SGSN and the external network. It confirms this to the SGSN and transmits the dynamic PDP address (if needed). Finally, the SGSN updates its PDP context table and confirms the activation of the new PDP context to the MS.

In case the GGSN receives packets from the external network that are addressed to a known static PDP address of an MS, it can perform a network-initiated PDP context activation procedure.

4.3. Routing

In Fig. 2 we give an example of how IP packets are routed to an external IP-based data network. A GPRS mobile station located in the GPRS network 1 addresses IP packets to a Web server connected to the Internet. The SGSN to which the mobile station is attached encapsulates the IP packets coming from the mobile station, examines the PDP context, and routes them through the GPRS backbone to the appropriate GGSN.

The GGSN decapsulates the IP packets and sends them out on the IP network, where IP routing mechanisms transfer the packets to the access router of the destination network. The latter delivers the IP packets to the Web server.

In the other direction, the Web server addresses its IP packets to the mobile station. They are routed to the GGSN from which the mobile station has its IP address (e.g., its home GGSN). The GGSN queries the HLR and obtains information about the current location of the user. In the following, it encapsulates the incoming IP packets and tunnels them to the appropriate SGSN in the current network of the user. The SGSN decapsulates the packets and delivers them to the mobile station.

4.4. Location Management

As in circuit-switched GSM, the main task of location management is to keep track of the user's current location, so that incoming packets can be routed to his/her MS. For this purpose, the MS frequently sends location update messages to its SGSN.

In order to use the radio resources occupied for mobility-related signaling traffic in an efficient way, a state model for GPRS mobile stations with three states has been defined [2]. In IDLE state the MS is not reachable. Performing a GPRS attach, it turns into READY state. With a GPRS detach it may deregister from the network and fall back to IDLE state, and all PDP contexts will be deleted. The STANDBY state will be reached when an MS in READY state does not send any packets for a long period of time, and therefore the READY timer (which was started at GPRS attach and is reset for each incoming and outgoing transmission) expires.

The location update frequency depends on the state in which the MS currently is. In IDLE state, no location updating is performed; that is, the current location of the MS is unknown. If an MS is in READY state, it will inform its SGSN of every movement to a new cell. For the location management of an MS in STANDBY state, a GSM location area (→ see GSM entry) is divided into so-called *routing areas*. In general, a routing area consists of several cells. The SGSN will be informed when an MS moves to a new routing area; cell changes will not be indicated. In addition to these event-triggered routing area updates, periodic routing area updating is also standardized. To determine the current cell of an MS that is in STANDBY state, paging of the MS within a certain routing area must be performed. For MSs in READY state, no paging is necessary.

Whenever a mobile user moves to a new routing area, it sends a ROUTING AREA UPDATE REQUEST to its assigned SGSN [2]. The message contains the *routing area identity* (RAI) of its old routing area. The BSS adds the *cell identifier* (CI) of the new cell to the request, from which the SGSN can derive the new RAI. Two different scenarios are possible: intra-SGSN routing area updates and inter-SGSN routing area updates.

In the first case, the mobile user has moved to a routing area that is assigned to the same SGSN as the old routing area. The SGSN has already stored the necessary user profile and can immediately assign a new P-TMSI. Since the routing context does not change, there is no need to inform other network elements, such as GGSN or HLR.

In the second case, the new routing area is administered by a SGSN different from the old routing area. The new SGSN realizes that the MS has entered its area. It requests the PDP context(s) of the user from the old SGSN and informs the involved GGSNs about the user's new routing context. In addition, the HLR and (if needed) the MSC/VLR are informed about the user's new SGSN number.

Besides pure routing updates, there also exist combined routing/location area updates. They are performed whenever an MS using GPRS as well as conventional GSM services moves to a new location area.

To sum up, we can say that GPRS mobility management consists of two levels: (1) micromobility management tracks the current routing area or cell of the user and (2) macromobility management keeps track of the user's current SGSN and stores it in the HLR, VLR, and GGSN.

5. PROTOCOL ARCHITECTURE

The protocol architecture of GPRS comprises transmission and signaling protocols. This includes standard GSM protocols (with slight modifications), standard protocols of the Internet Protocol suite, and protocols that have specifically been developed for GPRS. Figure 3 illustrates the protocol architecture of the transmission plane [2]. The architecture of the signaling plane includes functions for the execution of GPRS attach and detach, mobility management, PDP context activation, and the allocation of network resources.

5.1. GPRS Backbone: SGSN-GGSN

As mentioned earlier, the GPRS tunneling protocol (GTP) carries the user's IP or X.25 packets in an encapsulated manner within the GPRS backbone (see Fig. 3). GTP is defined both between GSNs within the same network (Gn interface) and between GSNs of different networks (Gp interface).

The signaling part of GTP specifies a tunneling control and management protocol. The signaling is used to create, modify, and delete tunnels. A *tunnel identifier* (TID), which is composed of the IMSI of the user and a *network-layer service access point identifier* (NSAPI), uniquely indicates a PDP context. Below GTP, one of the standard Internet protocols of the transport layer, *Transmission Control Protocol* (TCP) or *User Datagram Protocol* (UDP), are employed to transport the GTP packets within the backbone network. TCP is used for X.25 (since X.25 expects a reliable end-to-end connection), and UDP is used for access to IP-based networks (which do not expect reliability in the network layer or below). In the network layer, IP is employed to route the packets through the backbone. Ethernet, ISDN, or ATM-based protocols may be used below IP. To summarize, in the GPRS backbone we have an IP/X.25-over-GTP-over-UDP/TCP-over-IP protocol architecture.

For signaling between SGSN and the registers HLR, VLR, and EIR, protocols known from conventional GSM are employed, which have been partly extended with

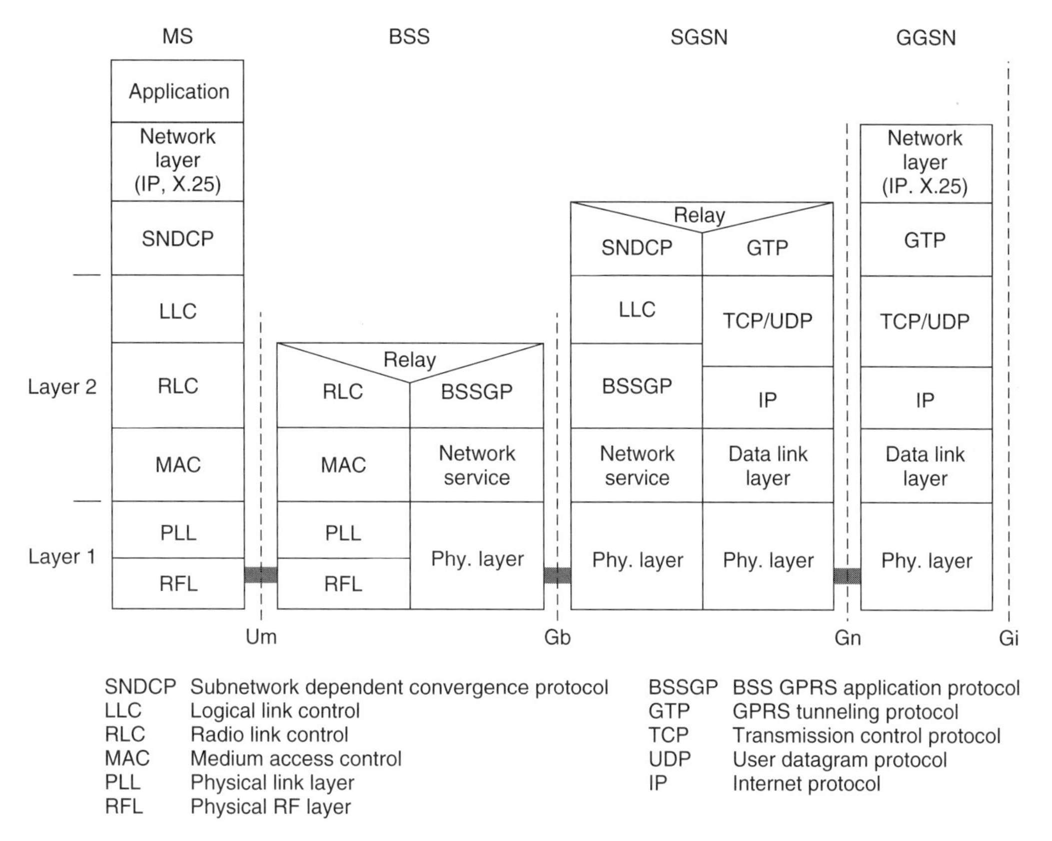

Figure 3. Protocol architecture: transmission plane.

GPRS-specific functionality. Between SGSN and HLR as well as between SGSN and EIR, an enhanced *mobile application part* (MAP) is used. The exchange of MAP messages is accomplished over the *transaction capabilities application part* (TCAP), the *signaling connection control part* (SCCP), and the *message transfer part* (MTP). The *BSS application part* (BSSAP+), which is based on GSM's BSSAP, is applied to transfer signaling information between SGSN and VLR (Gs interface). This includes, in particular, signaling of the mobility management when coordination of GPRS and conventional GSM functions is necessary (e.g., for combined GPRS and non-GPRS location updates, combined GPRS/IMSI attach, or paging of a user via GPRS for an incoming GSM call).

5.2. Air Interface

In the following we consider the transport and signaling protocols at the air interface (Um) between the mobile station and the BSS or SGSN, respectively.

5.2.1. Subnetwork-Dependent Convergence Protocol. An application running in the GPRS mobile station (e.g., a Web browser) uses IP or X.25, respectively, in the network layer. The *subnetwork-dependent convergence protocol* (SNDCP) is used to transfer these packets between the MSs and their SGSN. Its functionality includes multiplexing of several PDP contexts of the network layer onto one virtual logical connection of the underlying *logical link control* (LLC) layer and segmentation of network-layer packets onto LLC frames and reassembly on the receiver side. Moreover, SNDCP offers compression and decompression of user data and redundant header information (e.g., TCP/IP header compression).

5.2.2. GPRS Mobility Management and Session Management. For signaling between an MS and the SGSN, the *GPRS mobility management and session management* (GMM/SM) protocol is employed above the LLC layer. It includes functions for GPRS attach/detach, PDP context activation, routing area updates, and security procedures.

5.2.3. Data-Link Layer. The data-link layer is divided into two sublayers [7]:

- Logical link control (LLC) layer (between MS and SGSN)
- *Radio-link control/medium access control* (RLC/MAC) layer (between MS and BSS)

The LLC layer provides a reliable logical link between an MS and its assigned SGSN. Its functionality is based on the *link access procedure D mobile* (LAPDm) protocol, which is a protocol similar to *high-level data-link control* (HDLC). LLC includes in-order delivery, flow control, error detection, and retransmission of packets [*automatic repeat request* (ARQ)], and ciphering functions. It supports variable frame lengths and different QoS classes, and besides point-to-point, point-to-multipoint transfer is also possible. A logical link is uniquely addressed with a *temporary logical link identifier* (TLLI). The mapping between TLLI and IMSI is unique within a routing area. However, the user's identity remains confidential, since the TLLI is derived from the P-TMSI of the user.

The RLC/MAC layer has two functions. The purpose of the radio-link control (RLC) layer is to establish a reliable link between the MS and the BSS. This includes the segmentation and reassembly of LLC frames into RLC data blocks and ARQ of uncorrectable blocks. The MAC layer employs algorithms for contention resolution of random access attempts of mobile stations on the radio channel (slotted ALOHA), statistical multiplexing of channels, and a scheduling and prioritizing scheme, which takes into account the negotiated QoS. On one hand, the MAC protocol allows for a single MS to simultaneously use several physical channels (several time slots of the same TDMA frame). On the other hand, it also controls the statistical multiplexing; that is, it controls how several MSs can access the same physical channel (the same time slot of successive TDMA frames). This will be explained in more detail in Section 6.

5.2.4. Physical Layer. The physical layer between MS and BSS can be divided into the two sublayers: *physical link layer* (PLL) and *physical radiofrequency layer* (RFL). The PLL provides a physical channel between the MS and the BSS. Its tasks include channel coding (i.e., detection of transmission errors, forward error correction, and indication of uncorrectable codewords), interleaving, and detection of physical link congestion. The RFL operates below the PLL and includes modulation and demodulation.

5.2.5. Data Flow. To conclude this section, Fig. 4 illustrates the data flow between the protocol layers in the mobile station. Packets of the network layer (e.g., IP packets) are passed down to the SNDCP layer, where they are segmented to LLC frames. After adding header information and a *framecheck sequence* (FCS) for error protection, these frames are segmented into one or several RLC data blocks. Those are then passed down to the MAC layer. One RLC/MAC block contains a MAC and RLC header, the RLC payload (information bits), and a *block-check sequence* (BCS) at the end. The channel coding of RLC/MAC blocks and the mapping to a burst in the physical layer will be explained in Section 6.3.

5.3. BSS-SGSN Interface

At the Gb interface, the *BSS GPRS application protocol* (BSSGP) is defined on layer 3. It delivers routing and QoS-related information between BSS and SGSN. The underlying *network service* (NS) protocol is based on the Frame Relay protocol.

5.4. Routing and Conversion of Addresses

We now explain the routing of incoming IP packets in more detail. Figure 5 illustrates how a packet arrives at the GGSN and is then routed through the GPRS backbone to the responsible SGSN and finally to the MS. Using the PDP context, the GGSN determines from the IP destination address a TID and the IP address of the relevant SGSN. Between GGSN and the SGSN, the GPRS tunneling protocol is employed. The SGSN derives the TLLI from the TID and finally transfers the IP packet to the MS. The NSAPI, which is part of the TID, maps a given IP address to the corresponding PDP context. An NSAPI/TLLI pair is unique within one routing area.

6. AIR INTERFACE

The packet-oriented air interface [7] is one of the key aspects of GPRS. Mobile stations with multislot capability can transmit on several time slots of a TDMA frame, uplink and downlink are allocated separately, and physical channels are assigned only for the duration of the transmission, which leads to a statistical multiplexing gain. This flexibility in the channel allocation results in a more efficient utilization of the radio resources. On top of the physical channels, a number of logical packet channels have been standardized. A special packet traffic channel is used for payload transmission. The GPRS signaling channels are used, such as for broadcast of system information, multiple access control, and paging. GPRS channel coding defines four different coding schemes, which allow one to adjust the tradeoff between the level of error protection and data rate.

6.1. Logical Channels

Several GPRS logical channels are defined in addition to the logical channels of GSM. As with logical channels in conventional GSM, they can be grouped into two categories: traffic channels and signaling (control) channels. The signaling channels can be further divided into packet broadcast control, packet common control, and packet dedicated control channels.

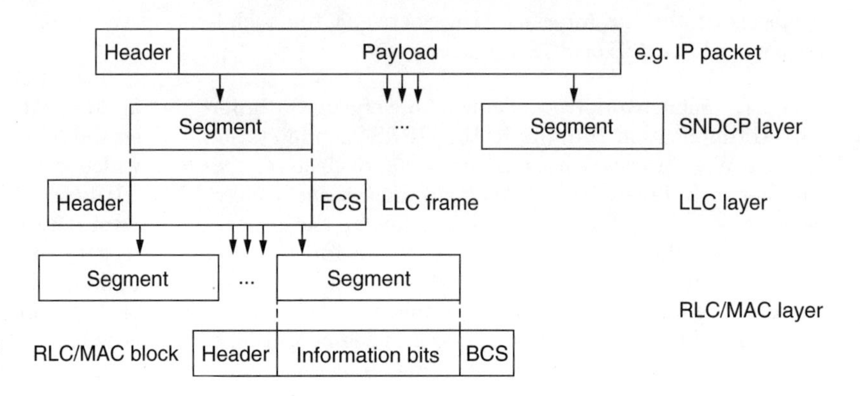

Figure 4. Data flow and segmentation between the protocol layers in the MS.

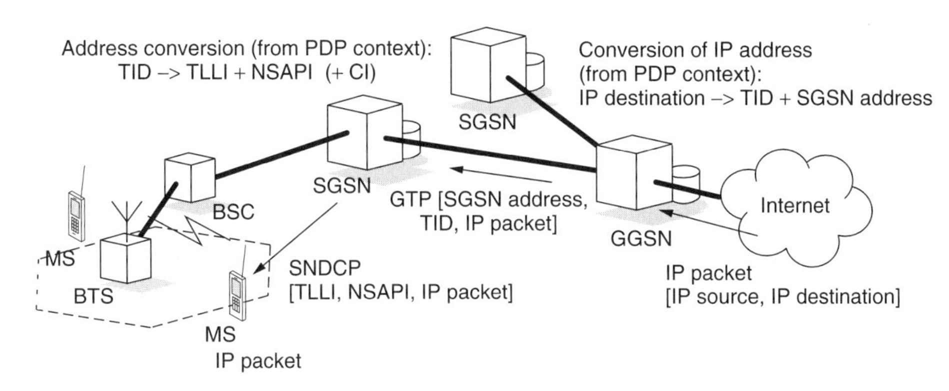

Figure 5. Routing and address conversion: incoming IP packet.

The *packet data traffic channel* (PDTCH) is employed for the transfer of user data. It is assigned to one mobile station (or in case of PTM to multiple mobile stations). One mobile station can use several PDTCHs simultaneously.

The *packet broadcast control channel* (PBCCH) is a unidirectional point-to-multipoint signaling channel from the BSS to the mobile stations. It is used to broadcast information about the organization of the GPRS radio network to all GPRS mobile stations of a cell. Besides system information about GPRS, the PBCCH should also broadcast important system information about circuit-switched services, so that a GSM/GPRS mobile station does not need to listen to the GSM *broadcast control channel* (BCCH).

The *packet common control channels* (PCCCHs) transport signaling information for functions of the network access management, specifically, for allocation of radio channels, medium access control, and paging. The group of PCCCHs comprises the following channels:

- The *packet random access channel* (PRACH) is used by the mobile stations to request one or more PDTCH.
- The *packet access grant channel* (PAGCH) is used to allocate one or more PDTCH to a mobile station.
- The *packet paging channel* (PPCH) is used by the BSS to determine the location of a mobile station (paging) prior to downlink packet transmission.
- The *packet notification channel* (PNCH) is used to inform mobile stations of incoming PTM messages.

The packet dedicated control channels are bidirectional point-to-point signaling channels. This group consists of the following two channels:

- The *packet-associated control channel* (PACCH) is always allocated in combination with one or more PDTCH. It transports signaling information related to one specific mobile station (e.g., power control information).
- The *packet timing-advance control channel* (PTCCH) is used for adaptive frame synchronization. The MS sends over the uplink part of the PTCCH, the PTCCH/U, access bursts to the BTS. From the delay of these bursts, the correct value for the timing advance can be derived. This value is then transmitted in the downlink part, the PTCCH/D, to inform the MS.

Coordination between GPRS and GSM logical channels is also possible here to save radio resources. If the PCCCH is not available in a cell, a GPRS mobile station can use the *common control channel* (CCCH) of circuit-switched GSM to initiate the packet transfer. Moreover, if the PBCCH is not available, it can obtain the necessary system information via the BCCH.

Four different coding schemes (CS1–CS4) are defined for data transmission on the PDTCH. Depending on the used coding scheme, the net data throughput on the PDTCH can be 9.05, 13.4, 15.6, or 21.4 kbps. The respective coding schemes are described in Section 6.3. Theoretically, a mobile station can be assigned up to eight PDTCHs, each of which can be either unidirectional or bidirectional.

6.2. Multiple Access and Radio Resource Management

On the physical layer, GPRS uses the GSM combination of FDMA and TDMA with eight time slots per TDMA frame. However, several new methods are used for channel allocation and multiple access [7]. They have significant impact on the performance of GPRS. In circuit-switched GSM, a physical channel (i.e., one time slot of successive TDMA frames) is permanently allocated for a particular MS during the entire call period (regardless of whether data are transmitted). Moreover, a GSM connection is always symmetric; that is, exactly one time slot is assigned to uplink and downlink.

GPRS enables a far more flexible resource allocation scheme for packet transmission. A GPRS mobile station can transmit on several of the eight time slots within the same TDMA frame (multislot operation). The number of time slots that an MS is able to use is called *multislot class*. In addition, uplink and downlink are allocated separately, which saves radio resources for asymmetric traffic (e.g., Web browsing).

The radio resources of a cell are shared by all GSM and GPRS users located in this cell. A cell supporting GPRS must allocate physical channels for GPRS traffic. A physical channel that has been allocated for GPRS transmission is denoted as *packet data channel* (PDCH). The number of PDCHs can be adjusted according to the current traffic demand ("capacity on demand" principle). For example, physical channels not currently in use for GSM calls can be allocated as PDCHs for GPRS to increase the quality of service for GPRS. When there is a resource demand for GSM calls, PDCHs may be de-allocated.

As already mentioned, physical channels for packet-switched transmission (PDCHs) are allocated only for a particular MS when this MS sends or receives data packets, and they are released after the transmission. With this dynamic channel allocation principle, multiple MSs can share one physical channel. For bursty traffic this results in a much more efficient use of the radio resources.

The channel allocation is controlled by the BSC. To prevent collisions, the network indicates in the downlink which channels are currently available. An *uplink state flag* (USF) in the header of downlink packets shows which MS is allowed to use this channel in the uplink. The allocation of PDCHs to an MS also depends on its multislot class and the QoS of its current session.

In the following we describe the procedure of uplink channel allocation (mobile originated packet transfer). A mobile station requests a channel by sending a PACKET CHANNEL REQUEST message on the PRACH or RACH [the GSM equivalent of the PRACH (→ GSM entry)]. The BSS answers on the PAGCH or AGCH [the GSM equivalent of the PAGCH (→ GSM entry)], respectively. Once the PACKET CHANNEL REQUEST is successful, a *temporary block flow* (TBF) is established. With that, resources (e.g., PDTCH and buffers) are allocated for the mobile station, and data transmission can start. During transfer, the USF in the header of downlink blocks indicates to other MSs that this uplink PDTCH is already in use. On the receiver side, a *temporary flow identifier* (TFI) helps to reassemble the packets. Once all data have been transmitted, the TBF and the resources are released again.

The downlink channel allocation (mobile terminated packet transfer) is performed in a similar fashion. Here, the BSS sends a PACKET PAGING REQUEST message on the PPCH or PCH [the GSM equivalent of the PPCH (→ GSM entry)] to the mobile station. The mobile station replies with a PACKET CHANNEL REQUEST message on the PRACH or RACH, and the further channel allocation procedure is similar to the uplink case.

6.3. Channel Coding

Channel coding is used to protect the transmitted data packets against errors and perform forward error correction. The channel coding technique in GPRS is quite similar to the one employed in conventional GSM (→ see GSM entry). An outer block coding, an inner convolutional coding, and an interleaving scheme is used. Figure 6 shows how a block of the RLC/MAC layer is encoded and mapped onto four bursts.

As shown in Table 3, four coding schemes (CS1, CS2, CS3, and CS4) with different code rates are defined [8]. For each scheme, a block of 456 bits results after encoding; however, different data rates are obtained depending on the used coding scheme, due to the different code rates of the convolutional encoder and to different numbers of parity bits. Figure 6 illustrates the encoding process, which will be briefly explained in the following.

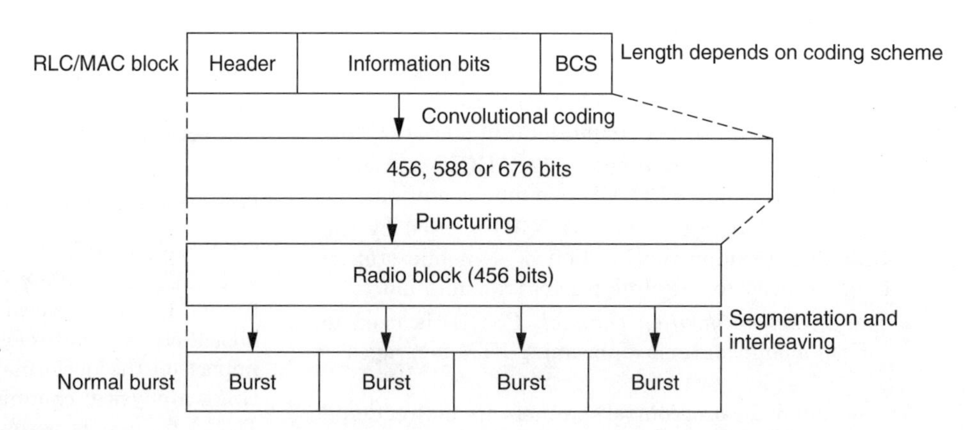

Figure 6. Physical layer at the air interface: channel coding, interleaving, and formation of bursts.

Table 3. Channel Coding Schemes for the Traffic Channels

CS	Preencoded USF	Info Bits without USF	Parity BCS	Tail Bits	Output Convolutional Encoder	Punctured Bits	Code Rate	Data Rate (kbps)
1	3	181	40	4	456	0	$\frac{1}{2}$	9.05
2	6	268	16	4	588	132	$\approx\frac{2}{3}$	13.4
3	6	312	16	4	676	220	$\approx\frac{3}{4}$	15.6
4	12	428	16	—	456	—	1	21.4

As an example we choose coding scheme CS2. First of all, the 271 information bits of an RLC/MAC block (268 bits plus 3 bits USF) are mapped to 287 bits using a systematic block encoder; thus 16 parity bits are added. These parity bits are denoted as *block-check sequence* (BCS). The USF pre-encoding maps the first 3 bits of the block (i.e., the USF) to 6 bits in a systematic way. Afterward, 4 zero bits (tail bits) are added at the end of the entire block. The tail bits are needed for termination of the subsequent convolutional coding. For the convolutional coding, a nonsystematic rate-$\frac{1}{2}$ encoder with memory 4 is used. This is the same encoder as used in conventional GSM for full-rate speech coding (→ see GSM entry). At the output of the convolutional encoder a codeword of length 588 bits results. Afterward, 132 bits are punctured (deleted), resulting in a radio block of length 456 bits. Thus, we obtain a code rate of the convolutional encoder (taking puncturing into account) of $\frac{2}{3}$. After encoding, the codewords are finally fed into a block interleaver of depth 4.

For the encoding of the traffic channels (PDTCH), one of the four coding schemes is chosen, depending on the quality of the signal. The two stealing flags in a normal burst (→ GSM entry) are used to indicate which coding scheme is applied. The signaling channels are encoded using CS1 (an exception is the PRACH).

For CS1 a systematic "fire" code is used for block coding and there is no precoding of the USF bits. The convolutional coding is done with the known rate-$\frac{1}{2}$ encoder, however, this time the output sequence is not punctured. Using CS4, the 3 USF bits are mapped to 12 bits, and no convolutional coding is applied. We achieve a data rate of 21.4 kbps per time slot, and thus obtain a theoretical maximum data rate of 171.2 kbps per TDMA frame. In practice, multiple users share the time slots, and, thus, a much lower bit rate is available to the individual user. Moreover, the quality of the radio channel will not always allow the use of CS4. The data rate available to the user depends (among other things) on the current total traffic load in the cell (i.e., the number of users and their traffic characteristics), the coding scheme used, and the multislot class of the MS.

7. SECURITY ASPECTS

The security principles of GSM (→ GSM entry) have been extended for GPRS [9]. As in GSM, they protect against unauthorized use of services (by authentication and service request validation), provide data confidentiality (using ciphering), and keep the subscriber identity confidential. The standard GSM algorithms are employed to generate security data. Moreover, the two keys known from GSM, the *subscriber authentication key* (Ki) and the *cipher key* (Kc), are used. The main difference is that not the MSC but the SGSN handles authentication. In addition, a special GPRS ciphering algorithm has been defined, which is optimized for encryption of packet data.

7.1. User Authentication

In order to authenticate a user, the SGSN offers a random number to the MS. Using the key Ki, the MS calculates a *signature response* (SRES) and transmits it back to the SGSN. If the mobile station's SRES is equal to the SRES calculated (or maintained) by the SGSN, the user is authenticated and is allowed to use GPRS services. If the SGSN does not have authentication sets for a user (i.e., Kc, random number, SRES), it requests them from the HLR.

7.2. Ciphering

The ciphering functionality is performed in the LLC layer between MS and SGSN (see Fig. 3). Thus, the ciphering scope reaches from the MS all the way to the SGSN (and vice versa), whereas in conventional GSM the scope is only between MS and BTS/BSC. The standard GSM algorithm generates the key Kc from the key Ki and a random number. Kc is then used by the GPRS encryption algorithm for encryption of user data and signaling. The key Kc that is handled by the SGSN is independent of the key Kc handled by the MSC for conventional GSM services. An MS may thus have more than one Kc key.

7.3. Confidentiality of User Identity

As in GSM, the identity of the subscriber (i.e., his/her IMSI) is held confidential. This is done by using temporary identities on the radio channel. In particular, a packet temporary mobile subscriber identity (P-TMSI) is assigned to each user by the SGSN. This address is valid and unique only in the service area of this SGSN. From the P-TMSI, a temporary logical link identity (TLLI) can be derived. The mapping between these temporary identities and the IMSI is stored only in the MS and in the SGSN.

BIOGRAPHIES

Christian Bettstetter is a research and teaching staff member at the Institute of Communication Networks at Technische Universität München TUM, Germany. He graduated from TUM in electrical engineering and information technology Dipl.-Ing. in 1998 and then joined the Institute of Communication Networks, where is he working toward his Ph.D. degree. Christian's interests are in the area of mobile communication networks, where his current main research area is wireless ad-hoc networking. His interests also include 2G and 3G cellular systems and protocols for a mobile Internet. He is coauthor of the book *GSM–Switching, Services and Protocols* (Wiley/Teubner) and a number of articles in journals, books, and conferences.

Christian Hartmann studied electrical engineering at the University of Karlsruhe (TH), Germany, where he received the Dipl.-Ing. degree in 1996. Since 1997, he has been with the Institute of Communication Networks at the Technische Universität München, Germany, as a member of the research and teaching staff, pursuing a doctoral degree. His main research interests are in the area of mobile and wireless networks including capacity and performance evaluation, radio resource management, modeling, and simulation. Christian Hartmann is a student member of the IEEE.

BIBLIOGRAPHY

1. S. Faccin et al., GPRS and IS-136 integration for flexible network and services evolution, *IEEE Pers. Commun.* **6**: (June 1999).
2. ETSI/3GPP, *GSM 03.60: GPRS Service Description; Stage 2*, technical specification, Mar. 2001.
3. C. Bettstetter, H.-J. Vögel, and J. Eberspächer, GSM phase 2+ General Packet Radio Service GPRS: Architecture, protocols, and air interface, *IEEE Commun. Surv.* **2**(3): (1999).
4. ETSI/3GPP, *GSM 02.60: GPRS Service Description; Stage 1*, technical specification, July 2000.
5. 3GPP, *3G 22.060: GPRS: Service Description; Stage 1*, technical specification, Oct. 2000.
6. R. Droms, Automated configuration of TCP/IP with DHCP, *IEEE Internet Comput.* **3**: (July 1999).
7. ETSI/3GPP, *GSM 03.64: GPRS Overall Description of the Air Interface; Stage 2*, technical specification, Feb. 2001.
8. ETSI, *GSM 05.03: GSM Phase 2+: Channel Coding*, technical specification, April 1999.
9. ETSI, *GSM 03.10: GSM phase 2+: Security Related Network Functions*, technical specification, July 2001.
10. J. Eberspächer, H.-J. Vögel, and C. Bettstetter, *GSM—Switching, Services, and Protocols*, 2nd ed., Wiley, 2001.
11. Y.-B. Lin, H. C.-H. Rao, and I. Chlamtac, General packet radio service (GPRS): Architecture, interfaces, and deployment, *Wiley Wireless Commun. Mobile Comput.* **1**: (Jan. 2001).
12. B. Walke, *Mobile Radio Networks*, 2nd ed., Wiley, 2002.
13. G. Brasche and B. Walke, Concepts, services, and protocols of the new GSM phase 2+ General Packet Radio Service, *IEEE Commun.* (Aug. 1997).
14. H. Granbohm and J. Wiklund, GPRS: General Packet Radio Service, *Ericsson Rev.* (2): (1999).
15. R. Kalden, I. Meirick, and M. Meyer, Wireless Internet access based on GPRS, *IEEE Pers. Commun.* (April 2000).
16. D. Staehle, K. Leibnitz, and K. Tsipotis, QoS of Internet access with GPRS, *Proc. 4th ACM Int. Workshop on Modeling, Analysis, and Simulation of Wireless and Mobile Systems (MSWiM'01)*, Rome, Italy, July 2001.
17. J. Korhonen, O. Aalto, A. Gurtov, and H. Laamanen, Measured performance of GSM HSCSD and GPRS, *Proc. IEEE Int. Conf. Commun. (ICC)*, Helsinki, Finland, June 2001.

FURTHER READING

This article is based on the author's survey paper [3] (© 1999 IEEE) and the GPRS chapter of the Wiley book [10]. Both contain more detailed descriptions of the GPRS architecture, protocols, and air interface. An extensive description of the signaling protocols can be found in Ref. 11. The book [12] also contains a GPRS chapter. The paper [13] gives an overview of GPRS as of 1997 and proposes and analyzes a MAC protocol. The authors of Ref. 14 give an equipment manufacturer's view on the GPRS architecture. A simulative study on GPRS performance was done, for example, in Refs. 15 and 16. Measured performance analyzes of GPRS can be found in Ref. 17.

GEOSYNCHRONOUS SATELLITE COMMUNICATIONS

Lin-Nan Lee
Khalid Karimullah
Hughes Network Systems
Germantown, Maryland

1. THE DEVELOPMENT OF GEOSYNCHRONOUS SATELLITE COMMUNICATIONS

The origin of the geosynchronous satellite concept can be traced back to Arthur C. Clarke. In his 1948 article, Clarke demonstrated the feasibility of providing worldwide communication by placing three radio repeaters as artificial satellites in space, each orbiting the earth with a period of 24 h. The geosynchronous orbit is about 22,300 m above the equator. Satellites on this orbit rotate around the earth at the same rate as the earth spins. They therefore appear as a fixed point in the sky to a fixed point on earth. Clarke's concept is visionary in its use of artificial satellites to bounce the radiowave for worldwide communication. It is also powerful, because a directional antenna at practically any fixed point on earth can be assured to maintain communications with at least one of these three satellites once accurate pointing is accomplished the first time.

In 1962, the United States created the Communications Satellite Corporation (COMSAT) to realize Clarke's vision under a Congressional Act. The first commercial application of geosynchronous satellites was transoceanic communications. In 1964, the International Satellite consortium, or INTELSAT, was formed by COMSAT and the Post Telephone and Telegraph entities (PTTs) around the world to operate a fleet of geosynchronous satellites and provide service to themselves as the carriers' carrier. In 1965, the *Early Bird*, also known as *INTELSAT I*, was successfully launched into orbit to demonstrate the feasibility with limited operational capability. Generations of INTELSAT satellites that followed have changed international telecommunications ever since.

In 1976, the launch of MARISAT satellites by COMSAT marked the beginning of a new era of maritime, mobile communications for ships at sea. The International Maritime Satellite Communications consortium, or INMARSAT, was formed by the PTTs in the following year to provide international maritime communications service. Generations of INMARSAT satellites have since played a very important role in maritime and aeronautical communications worldwide.

Following the success of the INTELSAT system, governments or private organizations in many countries since the 1970s have also launched many national and regional satellite systems. As the purpose of these national and regional systems is to facilitate radio communications within a country or region, a single satellite is typically capable of covering the entire service area, even though multiple satellites may be employed to provide additional traffic-carrying capability and redundancy. The birth of these national and regional systems has resulted in many forms of new applications, including network television distribution, private data networking, and rural telephony services.

In the late 1980s and early 1990s, satellite direct broadcast of television to homes emerged as cost for receiving equipment came down. These direct broadcast satellites provide tens of millions of rural as well as urban and suburban consumers with access to a large number of channels of television programming using small, inexpensive satellite receivers within their coverage areas. New geosynchronous satellite systems capable of public switched telephone network (PSTN) access from consumer handheld mobile terminals have also been developed. At this writing, two audio broadcast satellite systems are in the process of launching their services in the United States to provide digital compact-disk (CD) quality audio programming to automobiles. A new generation of onboard processing satellites is being developed for broadband Internet access. It is clear that new geosynchronous satellite systems and applications continue to evolve well into the twenty-first century as the required technologies continue to become more available.

2. FREQUENCY BANDS AND ORBITAL SLOTS

Like all radio systems, bandwidth is a resource being shared by all communications. Coordination is required to prevent systems from interfering one another. Frequency bands are allocated by the International Telecommunications Union (ITU), based on the intended use. Satellite bands are typically designated as Fixed Satellite Services (FSS), Mobile Satellite Services (MSS), and Broadcast Satellite Services (BSS). Actual assignment of frequency bands may vary, however, from country to country because of nationalistic considerations as well as requirements to protect legacy systems. As device technology for the lower frequency bands are more available than the higher frequencies, lower-frequency bands are often more desirable and being used up first. Equipment for higher-frequency bands, however, can be constructed with much smaller physical dimension, due to their shorter wavelength. Higher-frequency bands are also less congested. The common frequency bands for geosynchronous satellite communications are C band and Ku band for FSS, L band for MSS, Ku band for BSS. After decades of intensive research and development, Ka-band technology is just becoming commercially viable at the turn of this century, and it has become the new frontier for intensive activities.

For FSS and BSS, highly directional antennas are used to transmit and receive signals at the ground terminal. It is therefore possible to reuse the same frequency band by other satellites a few degrees away. Orbital slots are allocated by the ITU along the geosynchronous ring around the world. The spacing between satellites is 2° for FSS. It is extended to 9° for BSS to allow the use of smaller receive antennas. To accommodate the use of low-gain handheld antennas, however, the MSS satellites need much wider orbit spacing.

To maximize reliability, early geosynchronous satellites were designed as simple repeaters. The signals are transmitted from the ground and received by the satellite in one frequency band. They are frequency translated to another frequency band, then amplified and transmitted back to the ground terminals. This approach generally fits well with the peer-to-peer network architecture common for the FSS. The communications link that ground terminals transmit and the satellite receives is referred to as *uplink*, and that satellite transmits and ground terminals receive is referred to as *downlink*.

For MSS, however, the mobile terminals generally need to communicate with a PSTN through a gateway station, which is fixed and much more capable. Also, the MSS frequency bands are generally quite limited. Therefore, signal from the mobile terminals in the MSS band is translated to a feeder link frequency, then sent down to the gateway stations. Similarly, signals from the gateway station are transmitted on a feeder link frequency to the satellite, and then amplified and send to the MSS downlink. In this manner, the more valuable MSS spectrum is more efficiently used.

3. SATELLITE ANTENNA BEAM PATTERNS

As the spectrum and orbital slots are fixed resources that cannot expand with traffic increase, frequency reuse on the same satellite is essential to increase the capacity of each satellite. One way to accomplish it is to use spatial separation between multiple antenna beams.

As a primary application for the INTELSAT series of satellites is transoceanic trunking, it becomes obvious that the same frequency band can be reused at both continents across the ocean with two separate, focused beams. The uplink from the one beam is connected to the downlink to the other beam and vice versa. This leads to the use of so called "hemibeam" and "spot beams". Most of the modern INTELSAT satellites include a mix of "global beam," hemibeams, and spot beams. Hemibeams and spot beams often use different polarization from the global beam, allowing more frequency reuse.

A narrower, more focused beam also provides higher gain, which helps reduce the size of the ground terminals. Therefore, most national and regional satellite systems employ shaped antenna beams matching their intended service area. For example, the a CONUS beam covers the lower 48 of the continental United States. Spot beams are often added to provide coverage of Hawaii and parts of Alaska in a number of domestic satellite systems.

As more spot beams are placed on a satellite, interconnection between these antenna beams becomes a significant problem. Beam interconnection evolves from a fixed cross-connect to an onboard RF switch, and eventually to an onboard baseband packet switch. (See Sections 10 and 11.)

Multibeam satellite antennas are typically based either on offset multiple feeds or phased-array technology. Higher antenna gain, however, implies larger physical dimensions. Given the limited launching vehicle alternatives, they are relatively easier to implement for higher-frequency bands. At lower frequencies, designs based on foldable reflectors that are deployed after launch are often used to get around the problem.

4. THE SATELLITE TRANSPONDER

Most of the geosynchronous satellite systems assume a simple, wideband repeater architecture. The frequency band is partitioned into sections of a reasonable bandwidth, typically in the tens of megahertz (MHz). At the receive antenna output, the individual sections are filtered, frequency-converted and then amplified. The amplified signal is then filtered, multiplexed, and transmitted to the ground via the transmit antenna. The repeater chain of such a section is often called a *transponder*. A typical satellite may carry tens of transponders. The amplitude-to-amplitude (AM/AM) and amplitude-to-phase (AM/PM) response of its power amplifier, and the overall amplitude and group delay responses of the filter and multiplexer chain usually characterize the behavior of a satellite transponder. Solid-state power amplifiers are gradually replacing traveling-wave tube (TWT) amplifiers for low and medium power satellites. For further discussion, see Section 11.

Maximum power is delivered by a satellite transponder when its power amplifier is operated at saturation. The amplitude and phase response of the transponder is highly nonlinear when the power amplifier is saturated. Carriers of different frequencies may generate significant intermodulation as a result. Small signals may be suppressed by higher power signals when the transponder is operated in the nonlinear region. However, when a satellite transponder is used to transmit a single wideband carrier, such as a full-transponder analog television (TV) or a high-speed digital bitstream, operating the transponder at or close to saturation minimizes the size of receiving antennas on the ground. When the transponder is used to send many signals with different center frequencies, however, the common practice is to operate the transponder with sufficient input backoff so that the amplifier is operating at its linear region to minimize the impairments due to amplifier nonlinearity.

5. TRANSMISSION TECHNIQUES FOR GEOSYNCHRONOUS SATELLITES

As communications technology has evolved from applications to applications since the mid-1960s, the transmission techniques for geosynchronous satellites also changed. The most significant changes are certainly caused by the conversion from analog to digital in all forms of communications. For example, early transoceanic links carried groups of frequency-division multiplexed (FDM) analog telephony. The power amplifiers at the transmitting earth stations as well as the satellite transponder were required to operate with substantial backoff to ensure their linearity. This technique has very much been replaced by digital trunking, for which a single high-speed digital carrier containing hundreds of telephone circuits may be transmitted by the earth terminal. In this way the earth station power amplifier can be operated much more efficiently.

Because of the relatively higher bandwidth of television signals, most of the analog television signals have been transmitted with "full-transponder TV." The baseband video signal is frequency modulated (FM) to occupy a significant portion of the satellite transponder bandwidth, typically in the range of 24–30 MHz. The satellite transponder is often operated at saturation, because the FM/TV is the only signal in the transponder. In the INTELSAT system, a "half-transponder TV" scheme have also been used. In such case, there are two video signals, each FM modulated (with overdeviation) to ~17.5–20 MHz, depending on the width of the transponder. The satellite transponder is operated with moderate output backoff, typically 2 dB. The half-transponder TV is a tradeoff between transponder bandwidth efficiency and signal quality. Since network television distribution has been a large application segment for national and regional satellite systems, full-transponder FM/TV are still widely used throughout the world.

Advances in solid-state power amplifier technology and use of digital transmission techniques have brought dramatic cost reduction to satellite communications in terms of both space and ground segments. The single-channel per carrier (SCPC) mode of transmission coupled with power efficient forward error correction (FEC) coding allows small ground terminals to transmit a single digital circuit at speeds from a few kilo bits per second (kbps) up to hundreds of kbps with very small antenna and power amplifier. Digital transmission with power efficient FEC coding also allows direct broadcasting satellites to transmit tens of megabits per second of digitally encoded and multiplexed television signals to very small receive antennas with the satellite transponder operating at very close to saturation. Technology advancements in digital modulation, FEC coding, satellite power amplifiers, and antennas have enabled millions of consumers to watch direct satellite broadcasts of hundreds of television programming and to Web-surf via the geosynchronous satellites at speeds comparable to or higher than cable modems or digital subscriber lines (DSLs).

6. MULTIPLE-ACCESS TECHNIQUES FOR GEOSYNCHRONOUS SATELLITES

With the exception of television broadcast and distribution, most of the user applications do not require a single ground terminal to use the entire capacity of a satellite transponder. To use the transponder resource effectively, individual ground terminals may be assigned a different section of the frequency band within the transponder so that they can share the satellite transponder without interfering with one another. Similarly, terminals may use the same frequency, but each is assigned a different time slot periodically, so that only one terminal uses this frequency at any given time instant. Sharing of the same satellite transponder by frequency separation is known as *frequency-division multiple access* (FDMA), sharing by time separation is known as *time-division multiple access* (TDMA). Generally, FDMA requires less transmit power for each ground terminal, but the satellite transponder must be operated at the linear region, resulting in less efficiency in the downlink. For TDMA, each ground station must send information in a small time slot at higher speed and not transmitting during the rest of the time. Therefore, higher power is needed at the ground transmitter, but the satellite transponder can be operated at saturation. When sending packetized data, TDMA also has the advantage of inherently high burst rate. Many practical systems use a combination of FDMA and TDMA to obtain the best engineering tradeoff.

The ground stations can also modulate their digital transmission using different code sequences with very low, or no cross-correlation, and transmit the resulting signal at the same frequency. The signals can then be separated at the receiver by correlating the received signal with the same sequence. This technique is known as *code-division multiple access* (CDMA). It is possible to combine CDMA with TDMA, FDMA, or TDMA and FDMA.

To achieve higher efficiency, a ground terminal is assigned a transmission resource, such as a frequency, a time slot, or a code sequence, only when they need it. This is accomplished by setting aside a separate communication channel for the ground terminals to request the resources. The transmission resource is then allocated based on the requests received from all the terminals. This general approach is known as *demand assignment multiple access* (DAMA). Most DAMA systems use centralized control, and the resource assignments are sent to the terminals via another separate channel. In some case, a distributed control approach has also been implemented in the INTELSAT system, since all national gateway stations in the INTELSAT network are considered to be equals. For distributed control, requests are received and used by all ground terminals. They then execute the same demand assignment algorithm to reach identical assignments. No separate assignment messages need to be sent. DAMA has been adopted by terrestrial cellular radio later, and is now commonly known as part of the media access control (MAC) protocol.

7. ROUND-TRIP PROPAGATION DELAY AND SERVICE QUALITY

Depending on the location of the ground terminals with respect to the geosynchronous satellites, it takes about 240–270 ms for the transmitted signal to propagate to a geosynchronous satellite and back down to the receive ground terminal. This round-trip propagation delay presents service quality issues for two-way communications. PSTN handsets are connected to the subscriber lines via 4-wire–2-wire 2/4-wire hybrid. As a result of limited isolation at the receiving end hybrid, an attenuated echo of one's own speech can be heard two round trips later. Because of the long round-trip delay of the geosynchronous satellites, the quality of conversational speech can be degraded significantly by the echo. This problem is overcome by the introduction of echo cancelers. An echo canceler stores a digital copy of the speech at the transmit end. It estimates the amount of the round-trip delay and the strength of the echo at the receive end. The properly delayed and attenuated replica of the stored signal is reproduced and subtracted from the received signal, thus canceling the echo. Although long propagation delay also introduces response delay in conversational speech, subjective tests have demonstrated that $\leq$400-ms round-trip delay is tolerable when echo cancelers are employed. The quality of conversational speech degrades considerably even with echo cancellation, however, if two hops of geosynchronous satellite links are in between.

The round-trip propagation delay must also be considered in designing data links using automatic repeat request (ARQ) for error recovery. For high-speed links, a very sizable amount of data can be transmitted during the two round-trip times it takes for an acknowledgment (ack) for the initial data packet to arrive at the transmit terminal. Well-known protocols such as Go-back-N can perform poorly in this situation. The round-trip delay also adversely affects the Internet end-to-end Transmit Control Protocol (TCP). The TCP transmission window is opened gradually on successful transmission of successive packets. The window is cut back quickly when a packet needs to be retransmitted. The longer propagation delay can cause the TCP to operate at a very small transmission window most of the time. Since the end user devices generally are unaware of the presence of a satellite link in between, such issues are best resolved by *performance enhancement proxies* (PEPs). PEP is a software proxy installed at the ground terminal. The proxy at the transmitting terminal emulates the destination device for the transmitting end user, whereas the proxy at the receiving terminal emulates the source device for the receiving end user. The proxy also implements a protocol optimized for the satellite link between the transmit and receive ground terminals. By breaking the communications into three segments, PEP is able to maximize the performance for the satellite link while maintaining the compatibility to the standard data communications protocols with the end user devices at both ends.

8. VERY SMALL APERTURE TERMINALS (VSAT) FOR GEOSYNCHRONOUS SATELLITES

Geosynchronous satellites were initially conceived as "repeaters in the sky." Every ground terminal is capable of communicating to any other ground terminals within a satellite's coverage area on a peer-to-peer basis. This fully connected mesh-shaped network is ideal for transoceanic trunking in the INTELSAT system. When maritime communications via satellite emerged, it became clear that most of the communications are between individual ship-terminals and their home country through their "home" coastal earth station. The INMARSAT system is essentially made up of many star-shaped networks with coastal earth stations as their hubs. Ship-to-ship communications must be relayed through the hub. By trading off the direct peer-to-peer communication capability, total ground segment cost is drastically reduced. Since there are far more remote terminals than hub terminals, minimizing the remote terminal cost tends to minimize the overall system cost. Very small aperture terminals (VSATs) exploit this property to create inexpensive private networks with geosynchronous FSS satellites.

In private networks, the remote VSAT terminals typically need to send data at speeds up to low hundreds of kbps (kilobits per second) to the hub. The remote-to-hub communication is often referred to as "inroute". At these rates, reliable data communications can typically be established via Ku-band FSS satellites using low-cost antennas with diameter less than a meter. Such antennas have reasonable beamwidth, requiring no constant tracking of the satellite position. Typically, VSATs use near-constant envelope digital modulation along with powerful FEC codes, and a small power amplifier operated at near saturation. With SCPC transmission, a large number of remote terminals share a part or a whole satellite transponder on a demand assigned TDMA/FDMA basis. Demand-assigned TDMA/FDMA takes advantage of the low-duty factor, bursty nature of the data traffic, allowing a number of terminals to share a single frequency. To minimize the size of the remote terminal antenna and power amplifier, the hub terminal typically uses a much larger antenna so that the downlink contributes very little to the overall link noise. Signal from each individual remote terminal is demodulated and decoded separately, and then routed to the appropriate terrestrial network interfaces.

For the "outroute," the hub station typically time multiplexes all the traffic toward the remote terminals into a single high-speed time-division multiplex (TDM) carrier, and transmits it to the entire network of terminals via either a part of the transponder at a different frequency, or through a separate transponder. Similarly, because of the much larger transmit antenna at the hub, the overall outroute link noise is dominated by the downlink to the remote terminal. The speed of the TDM data carrier is typically in the range from submegabits per second to tens of megabits to second, scaled according to network size.

In the rare cases for which data must be exchanged between two remote terminals, they are routed through the hub. The extra delay does not cause problem for most private network applications. But double hopping through the synchronous satellite creates too much delay for conversational telephony. Alternative full-mesh VSATs have also been developed for voice applications. In such private networks, a hub is responsible for interface to the PSTN and demand assign control. The remote terminals typically utilize low-bit-rate voice coders such as the ITU G.729 8-kbps voice coder. When a remote-to-remote call is initiated, the hub assigns a pair of frequencies for the two remote terminals to transmit on. They in turn tune their transmitters and receivers to the respective frequencies accordingly, and the circuit is established. The remote-to-remote communications is possible because the SCPC signals transmitted by these remotes are at lower bit rate than normally used for data communications.

9. DIGITAL VOICE AND TELEVISION FOR GEOSYNCHRONOUS SATELLITES

As just demonstrated by the example of one-hop voice communications between VSATs, low-bit-rate voice coders are instrumental for voice communications via geosynchronous satellites using small terminals. At the beginning of the digital conversion, 64-kbps pulse-coded modulation (PCM) and 56-kbps PCM were used by the INTELSAT system. In the late 1980s, 32-kbps adaptive differential PCM (ADPCM) combined with digital speech interpolation (DSI) was used to increase INTELSAT trunking capacity by a factor of four. DSI detects the silence periods of conversational speech and transmits only the active periods of the speech. By statistically aggregating a large number of circuits, DSI provides twofold capacity after taking into account the control overhead required.

Most of the early national and regional FSS networks use 64- or 56-kbps PCM. Newer networks have adopted the ITU G-728 16-kbps voice coder and the G.729 8-kbps voice coder. The lower bit rates not only increase the number of circuits carried by a satellite transponder, but also minimize the size of the remote terminal by reducing the transmit power needed to support a fixed number of voice circuits.

INMARSAT used low-bit-rate voice coders such as 16 and 9.6 kbps in their early digital terminals. It selected a 4-kbps voice coder for their smallest terminals in the early 1990s. A similar 4-kbps voice coder is also used by a geosynchronous mobile satellite system that supports handheld cell-phone-like terminals. In fact, with its low-gain antennas, handheld, high-quality, voice communications via satellite can only be made possible with the success of these very low bit rate voice coders.

Digital television signals based on ADPCM were demonstrated via satellites at 45 Mbps in the late 1970s. 1.5-Mbps and 384-kbps coders have been used for teleconference applications in the late 1980s. Not until digital direct satellite broadcast services were launched in the early 1990s, has digital television coding been widely used. Based on enhanced Motion Pictures Experts Group (MPEG) or MPEG-2 standards, these video coders are capable of compressing broadcast quality television signals to about 3–5 Mbps, depending on motion contents. The direct broadcast satellites typically time-multiplex up to 10 such signals into a single satellite transponder. Digital video compression is truly the underlining technology that makes satellite direct broadcast a commercial success.

10. ONBOARD PROCESSING FOR GEOSYNCHRONOUS SATELLITES

Advanced geosynchronous satellites use multibeam antennas to divide their coverage areas into cells like a cellular network to increase their traffic carrying capacity. Similar to cellular networks, the frequency band is reused by cell clusters. The adjacent cells within a cluster use different frequency bands or different polarization to provide the needed isolation in addition to the spatial discrimination provided by the antenna pattern. As the number of antenna beams increases, cross-connection between uplink beams and downlink beams becomes increasingly complicated. When the number of beams are more than a few tens on both the uplink and downlink, the task can no longer be handled with simple RF switches. Telephone switch technology has evolved since the 1970s from crossbar to digital, and again from circuit-based "hard" switch to packet-based "soft" switch. With each step of the evolution, the capacity of the switch has increased and the cost reduced by orders of magnitude. Beam cross-connection on the latest high-capacity geosynchronous satellites requires a digital switch solution.

Onboard processing satellites typically demodulate the uplink in each antenna beam into digital signals. Data packets are then FEC decoded, and put into an input buffer. Based on the header information in each packet, an onboard switch routes the packets to the output queue of each individual downlink beam. Header information may also provide traffic classification and quality of service (QoS) requirements. The onboard packet switch can also assign priority and exercise flow control in case of traffic congestion. For each downlink beam, data in the output queue are FEC-encoded, modulated, and upconverted to the RF frequency. They are then amplified and transmitted.

Onboard processing, in addition, offers the opportunity to optimize the uplink and downlink independently. Satellite transponders are most efficiently used when operated in TDM or TDMA mode at saturation. The size of ground terminals is minimized, however, if the satellite transponder is accessed via FDMA or CDMA. With onboard processing, both links can be operated with their most efficient access approach. When a large number of downlink beams are needed to cover an area with uneven traffic load, a common practice is to switch a much smaller number of active downlink transmitters to all the beams by phase-array technology. The dwell time on each individual beam can be directly proportional to the downlink traffic to its corresponding cell, thus optimizing the utilization of the satellite resources.

Also, the bit error rate (BER) or the packet error rate (PER) of the overall link equals to the sum of those incurred in the two links in a processing satellite, whereas the thermal noise and interference for the overall link is the sum of those incurred in the two links with a classic "bent pipe" satellite. Since the slope of BER or PER versus noise and interference is very steep when the link is protected by the FEC, the overall link can be much better optimized for an onboard processing satellite.

As a satellite system generally needs to accommodate different sizes of terminals, one of the challenges for cost-effective onboard processing is to implement the capability of dynamically assigning demodulator and FEC decoder resources for different-sized uplinks. This is accomplished with digital signal processing techniques that are scalable within limits.

11. SATELLITE COMMUNICATIONS SYSTEM EQUIPMENT

11.1. Gateway Architecture

There are several configurations of the satellite communications payload. It would be difficult to develop a generic architecture for the system since satellite payloads and Gateways are designed based on the traffic requirements it is meant to serve. In Fig. 1, we illustrate a simplified gateway architecture that supports digitized voice and data for TDMA application.

The gateway will transmit, via a satellite transponder, to multiple subscriber terminals (STs) on a single wideband TDM signal occupying one transponder bandwidth (e.g., 30 MHz.). This outbound signal will be in the 6- or 14-GHz bands for C-band or K band systems, respectively. Similarly, the gateway will receive FDM/TDMA bursts from multiple STs via another satellite transponder. As an example, the composite 30-MHz inbound signal could be 100 FDM channels, each 300 kHz wide, carrying 200 ksymbol/sec QPSK modulated carrier time-shared by multiple users in a TDMA protocol.

The gateway transmit traffic is converted to data blocks (slots), CRC and FEC encoded, and multiplexed into a TDM frame structure. The wideband TDM baseband signal is applied to a QPSK modulator, which converts it to a fixed IF. The IF is up converted (U/C) to C or K band and transmitted to the satellite as the outbound signal, via the RF Subsystem, which includes the power amplifier (HPA) and the antenna.

The gateway inbound FDM/TDMA signal at the C or K band, which is composed of transmission from multiple STs, is received by the same antenna, amplified by the LNA and downconverted (D/C) to IF. The IF is demodulated by the TDMA burst demodulators, each operating on a separate 300-kHz channel in this example. Each demodulator decodes its traffic slot by slot which is subsequently routed to appropriate destination by the TDMA Rx controller.

The gateway transmit power requirements are high since it has to transmit a wideband TDM signal. The high effective isotropic radiated power (EIRP) requirement is adjusted by suitable choice of HPA power rating and the antenna gain. Typically, gateway antenna beamwidths are narrow because of the HPA output power limitations, and thus a tracking antenna is required at most Gateways stations to track the satellite motion.

11.2. Satellite Communications Payload

Satellite communications payload is also very specific to applications. Conventional designs evolved from a simple bent pipe, global beam approach that acted as a simple repeater in the sky, to spot beam, regenerative repeaters with traffic switched between spots at baseband. There are far more advanced designs expected to operate in the near future. The fundamental objective is to support as much simultaneously active traffic as possible within the allocated bandwidth and the satellite power limitation.

In general, noninterfering spot beams and polarization are used to achieve frequency reuse. Spot beams with high gain, also provide higher EIRP that may illuminate areas with dense traffic, thus providing bandwidth reuse and power efficiency. Further, hopping spot beams, at the expense of complex control, provide further improvement in power efficiency. In this approach many spots are arranged over the coverage area, but the available transmit power is dynamically assigned only to a small sub-set of spots, at any given time. A mix of configurations is also possible.

In Fig. 2 we show a simple fixed spot beam payload system with two spot beams where traffic can be routed between the East and West spots by a RF switch matrix.

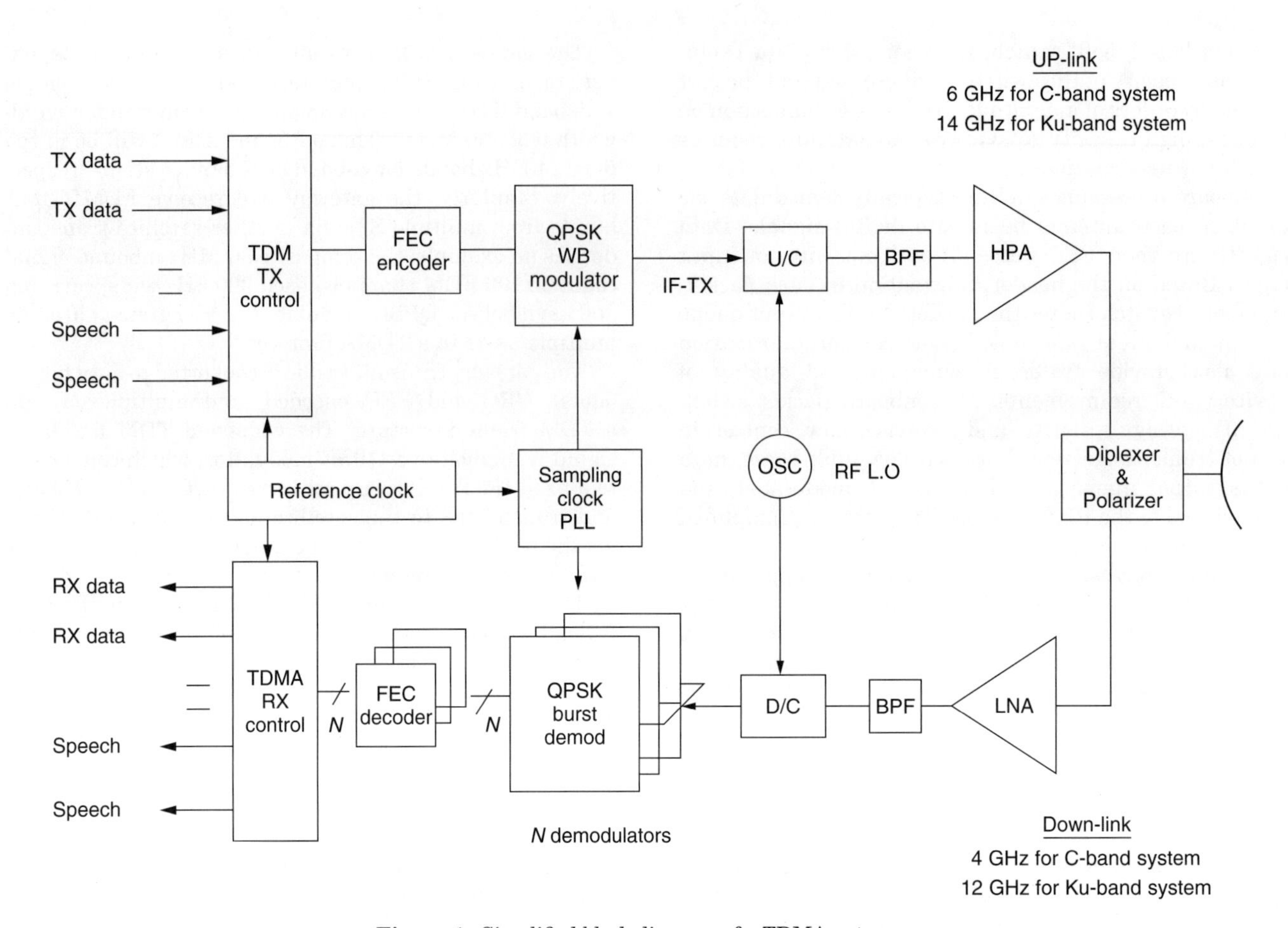

Figure 1. Simplified block diagram of a TDMA gateway.

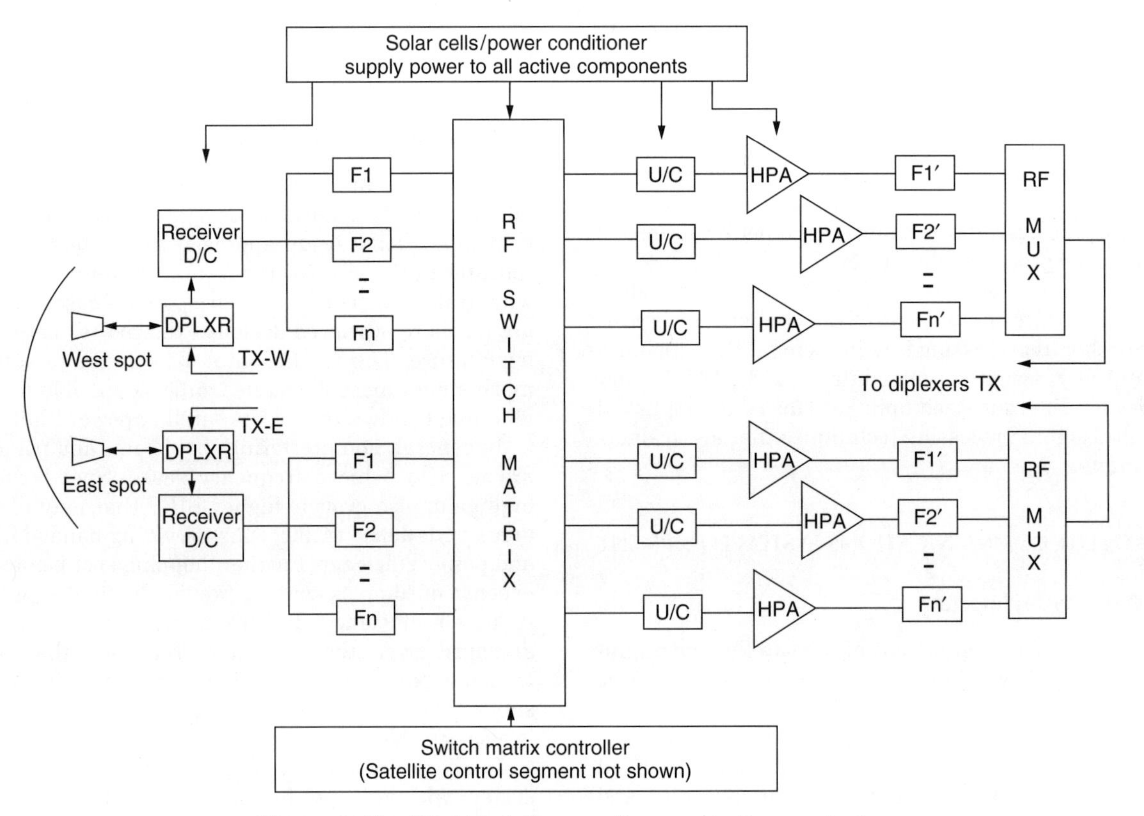

Figure 2. Simplified block diagram of communications payload.

The switch configuration is programmable by the control equipment. The satellite payload is controlled from ground by a dedicated command and control link with a control station.

Signals received from the ground are transmitted from either gateways or STs. The received signals are amplified by the LNA and demultiplexed by bandpass filters F. The bandwidth of these filters sets the transponder bandwidth. In our example F1 could be assigned to outbound transmission from a gateway, while F2 could be assigned to Inbound transmissions from a group of STs. The HPA for outbound may operate near saturation since there is only one (wideband TDMA) signal present. The HPA for Inbound traffic has to be backed off (typically output backoff of 4 dB) to support FDM/ TDMA carriers. The RF switch provides a simple return path to the same region or to a different region, based on the switch configuration. Typically this configuration would support traffic between hotspots on two ends of the coverage area for example New York and Los Angeles. Actual payload designs are more complicated than the example here. There could be a mix of CONUS beams and many spot beams, and a mix of C- and K-band transponders all interconnected via the RF switch matrix or matrices. It all depends on the system requirements and payload cost.

12. GEO SATELLITE SYSTEM LINK BUDGETS

As in any radio communications system design, link budgets are needed to size the system components to achieve performance objectives. For geosynchronous satellites, the uplink and downlink microwave path losses should be overcome by the antenna gain and transmit power to overcome the inherent noise and interference in the receivers. This analysis provides the basis of designing the hardware for the gateway and the ST, constrained by the cost and data rate capability tradeoffs. The following terminology is used for the various parameters of a link budget calculation.

Transmit Power. Actual power delivered to the antenna subsystem. If there are no losses between the HPA and the antenna, this will be the power output of the HPA.

Antenna Gain. The actual gain of the Tx (transmitting) antenna, i.e., directivity minus losses from illumination efficiency and other reflector and/or radome losses. The gain of a parabolic reflector fed by a feedhorn is given by $G = 4\pi A\eta/\lambda^2$, where A is the planar area of the aperture, λ is the wavelength, and η is the aperture efficiency of the antenna (typically 65%).

EIRP. Effective isotropic radiated power is the product of the antenna gain and transmit power (or the sum if these are in decibels).

Edge-of-Beam Loss. The satellite antenna boresight points to one location on earth. The coverage area is defined by constant power contours decreasing in power as one moves away from the boresight. If the edge of a satellite antenna beam is defined by −3 dB contour, then this loss will be 3 dB.

Antenna Pointing Loss. The ground equipment antenna boresight may not point exactly at the satellite. To account for this pointing error, typically 0.5 dB pointing loss is assumed.

Free-Space Path Loss. The square-law propagation loss factor normalized with the wavelength. This definition simplifies calculation of received carrier power based on transmit EIRP propagation loss and receive antenna gain. Loss $= (4\pi d/\lambda)^2$, where d is the distance to the satellite, a function of elevation angle. A value of 40,000 km is typically used.

Atmospheric Loss. The atmospheric loss is due to the gaseous content of the atmosphere and higher layers, encountered during wave propagation even in clear-sky conditions. Typical values range from 0.5 (C band) to 0.8 dB (Ka band).

Receive Flux Density. The receive flux density at the satellite is a measure of received power per unit area assuming isotropic radiation from the transmit antenna. It can be obtained from the EIRP and the distance d (after accounting for other fixed losses such as pointing error, edge-of-beam loss, and atmospheric loss but excluding free-space path loss).

$$\text{Rx flux density}\left[\frac{(\text{dBW})}{\text{m}^2}\right] = 10\log\frac{\text{EIRP}}{4\pi d^2} - \text{losses (dB)}$$

Typically the satellite repeater function is specified in terms of the Rx flux density and transmit EIRP at saturation. Actual EIRP is computed by first computing the incident flux density, hence, the backoff from the saturation point, then finding the output backoff from saturation EIRP, based on the power amplifier nonlinear gain characteristics.

Receiver Noise Figure. The receiver noise figure (NF) is a measure of the thermal noise created in the receive chain referenced to the antenna port (LNA input). In satellite applications this can be assumed to come entirely from the LNA. Gateway LNA noise figures are very low (<1 dB), while satellite LNA NF could be >3 dB.

Equivalent Noise Temperature. The equivalent noise temperature (T_e), a more practical parameter, is directly related to the NF by the relationship $T_e = (\text{NF} - 1)T_0$, where T_0 is ambient temperature (290 K is assumed for room temp).

Antenna Noise Temperature. The antenna noise temperature (T_a) is the noise picked up by the antenna environment, which includes galactic, atmospheric and spillover noise. For a gateway looking at the satellite, this parameter is only 40–60 K. For the satellite looking at the earth, this noise temperature could be >500 K.

System Noise Temperature (T_s). This is the total noise temperature at antenna port: $T_s = T_e + T_a$.

Typically, system noise temperature of satellites is around 1000 K.

Receive G/T. This ratio establishes the figure of merit for the receiver subsystem. It is the ratio of Rx antenna gain to the system noise temperature.

Received C/N_0. The carrier-to-noise density ratio is computed directly from the G/T and EIRP after accounting for all the losses. It is independent of the bit rate. (In fact, it gives an easy way to determine what bit rates can be supported by the system given the threshold E_b/N_0).

$$C/N_0(\text{dB}-\text{Hz}) = \text{EIRP (dBW)} + G/T\ (\text{dB/K}) - \text{losses (dB)} + 228.6\ \text{dB/K/Hz}\ \text{(Boltzmann)}$$

In a bent-pipe repeater model the uplink and downlink system noise temperatures should be accounted for using the relationship

$$\left(\frac{C}{N_0}\right)^{-1} = \left(\frac{C}{N_{\text{ou}}}\right)^{-1} + \left(\frac{C}{N_{\text{od}}}\right)^{-1}$$

Received C/I_0. The carrier-to-interference density is computed by finding the ratio of the received carrier power to the interference density. The interference density is obtained by the interference power in the channel divided by the channel bandwidth. In a bent-pipe repeater model the uplink and downlink interference should be accounted for by a similar relationship as given above:

$$\left(\frac{C}{I_0}\right)^{-1} = \left(\frac{C}{I_{\text{ou}}}\right)^{-1} + \left(\frac{C}{I_{\text{od}}}\right)^{-1}$$

Received $E_{\text{bi}}/(N_0+I_0)$. The $E_{\text{bi}}/(N_0+I_0)$ is the net information bit energy-to-noise plus interference density received by the demodulator. This includes the effects of thermal noise and interference, and for bent-pipe satellites, includes effects of the uplink and the downlink. This quantity should be compared with the threshold E_{bi}/N_0 to determine the link margin:

$$\frac{E_{\text{bi}}}{N_0+I_0}\ \text{dB} = \frac{C}{N_0+I_0}\ \text{dB} - 10\log \times \text{(information bit rate)}$$

where the quantity $(C/(N_0+I_0))^{-1} = (C/N_0)^{-1} + (C/I_0)^{-1}$, to account for system noise temperature and interference.

E_b/N_0 Threshold. The modem is guaranteed to deliver a target information bit error rate at this threshold. Usually this threshold is derived by system simulation E_b/N_0 and ideal modem assumption, based on a chosen FEC approach and then a modem implementation margin is added.

Available Fade Margin. Typically the link budget is set up for clear-sky conditions. The difference between the received $E_{\text{bi}}/(N_0+I_0)$ and the threshold E_{bi}/N_0 gives the fade margin. The fade margin reflects the availability of the system during rain fades. System availability as a function of rain statistics is beyond the scope of this paper. For the C band, a 1.5 dB margin is adequate to achieve reasonable availability for typical locations.

12.1. Example of a Link Budget

A hypothetical system is chosen as an example to be used with an INTELSAT standard B, C-band (6/4-GHz) gateway (GW) earth station. The outbound is a 20-Mbps TDM, transmitted with antenna gain of 58 dBi and HPA power 200 W. The G/T of the GW is 33.5 dB/K.

The satellite is a bent-pipe model with saturation flux density of -83 dBW/m^2 corresponding to an EIRP of 37 dBW, giving a gain of 120 dB/m^2 at saturation or 125 dB/m^2 in linear region, based on a standard TWTA nonlinear transfer characteristics. The system noise temperature is 1000 K. The G/T is -6.0 dB, based on a gain of 24 dBi, which gives an 8° half-power CONUS beam. The inbound traffic is FDM/TDMA with 250 kbps information per channel, 120 channels spaced 300 khz apart using one transponder. The subscriber terminal (ST) contains a 2-W power amplifier and 1.8 m dish antenna. Its receiver noise figure is 1.5 dB.

Threshold E_b/N_0 of 3.0 dB (including 1 dB modem implementation loss) and fade margin of 1.5 dB are assumed in both directions. Other parameters are in the link budgets shown in Tables 1 and 2 for outbound and inbound time-division multiplexing, respectively.

BIOGRAPHY

Lin-Nan Lee is vice president of engineering at Hughes Network Systems (HNS), Germantown, Maryland, responsible for advanced technology development. Dr. Lee received his B.S. degree from National Taiwan University and his M.S. and Ph.D. from University of Notre Dame, Indiana, all in electrical engineering. He started his career at Linkabit Corporation, where he was a senior scientist working on packet communications over satellites at the dawn of the Internet age. He then worked at Communication Satellite Corporation (COMSAT) in various research and development capacities with emphasis on source and channel coding technology development for satellite transmission and eventually assumed the position of chief scientist, COMSAT systems division. After joining HNS in late 1992, he has contributed to HNS' effort in wireless and satellite communications areas. Dr. Lee is a fellow of IEEE. He was the corecipient of the COMSAT Exceptional Invention Award, and the 1985 and 1988 COMSAT Research Award. He has authored or coauthored more than 20 U.S. patents.

Khalid Karimullah received his Ph.D in electrical engineering from Michigan State University, East Lansing, Michigan in 1980. He started his professional career at

Table 1. Outbound Link Budget for 20-Mbps TDM

	Units	Clear Sky
6-GHz Uplink		
Outbound bit rate	Mbps	20.0
Gateway antenna gain	dB	58.0
HPA Tx power	dBW	23.0
Transmit EIRP	dBW	81.0
Uplink frequency	GHz	6.0
Wavelength	m	0.05
Distance to satellite (max)	km	40,000
Free-space path loss at 6 GHz	dB	200
Atmospheric loss	dB	0.5
Edge-of-beam loss	dB	3.0
Gateway antenna pointing loss	dB	0.5
Satellite G/T ($G = 24$ dB; $T_s = 1000$ K)	dB/K	−6.0
Carrier noise density $(C/N_0)_u$	dB−Hz	99.6
Rx flux density (saturation = −83)	(dBW)/m^2	−86
4-GHz Downlink		
Satellite EIRP (0.5 dB backoff)	dBW	36.5
Downlink Frequency	GHz	4.0
Wavelength	m	0.075
Free-space path loss at 4 GHz	dB	196.5
Atmospheric loss	dB	0.5
Edge-of-beam loss	dB	3.0
ST antenna diameter	m	1.8
ST antenna Rx gain	dB	36.0
ST antenna pointing loss	dB	0.5
ST noise figure	dB	1.5
Equivalent noise temp (NF = 1.5 dB)	K	119.6
Antenna noise temperature	K	52.0
System noise temperature	dBK	22.3
ST G/T	dB/K	13.7
Carrier noise density $(C/N_0)_d$	dB−Hz	78.2
Outbound Overall		
(C/N_0) received	dB−Hz	78.2
(C/I_0) assumed	dB−Hz	86.0
$C/(N_0 + I_0)$	dB−Hz	77.5
$E_b/(N_0 + I_0)$	dB	4.5
Threshold E_b/N_0	dB	3.0
Fade margin	dB	1.5

COMSAT Laboratories, Maryland, where he worked on regenerative satellite transponder modem designs. He later joined MA-Com Linkabit, San Diego, California, in April 1987, where he developed his expertise in the areas of communications and signal processing. He joined Hughes Network Systems, Germantown, Maryland, in 1989 and has since worked on CDMA technology. Currently, he works at HNS as a senior Director, engineering, involved in R.F. and CDMA technology related activities. He has been active in the TIA/EIA TR45.5 cdma2000 standards development, participating in the physical layer and enhanced access procedures development. He chaired the TR45.5 cdma2000 enhanced access ad-hoc group. He has authored/ coauthored several patents in the CDMA physical layer and enhanced access procedures. In 1998, he received Hughes Electronics Patent Award and was the corecipient of the 1998 CDMA Technical Achievement Award.

Table 2. Inbound Link Budget for 250-kbps/s FDMA/TDMA (120 Channels)

	Units	Clear Sky
6 GHz Uplink		
Inbound bit rate	Mbps	0.25
ST Tx antenna gain	dB	39.5
ST SSPA Tx power	dBW	3.0
Transmit EIRP	dBW	42.5
Uplink frequency	GHz	6.0
Wavelength	m	0.05
Distance to satellite	km	40,000
Free-space path loss at 6 GHz	dB	200
Atmospheric loss	dB	0.5
Edge-of-beam loss	dB	3.0
ST antenna pointing loss	dB	0.5
Satellite G/T ($G = 24$ dB; $T_s = 1000$ K)	dB/K	−6.0
Carrier noise density $(C/N_0)_u$	dB−Hz	61.1
Rx flux density (saturation = −83)	(dBW)/m^2	−124.5
4 GHz Downlink		
Downlink EIRP/ ST (linear)	dBW	0.5
Downlink frequency	GHz	4.0
Wavelength	m	0.075
Free-space path loss at 4 GHz	dB	196.5
Atmospheric loss	dB	0.5
Edge-of-beam loss	dB	3.0
Gateway antenna diameter	m	15.2
Gateway antenna Rx gain	dB	54.5
GW antenna pointing loss	dB	0.5
GW noise figure	dB	1.0
Equivalent noise temperature (NF = 1.0 dB)	K	75.1
Antenna noise temperature	K	52.0
System noise temperature	dBK	21.0
GW G/T	dB/K	33.5
Carrier noise Density $(C/N_0)_d$	dB−Hz	62.0
Inbound Overall		
(C/N_0) Received	dB−Hz	58.5
(C/I_0) Assumed	dB−Hz	86.0
$C/(N_0 + I_0)$	dB−Hz	58.5
$E_b/(N_0 + I_0)$	dB	4.5
Threshold E_b/N_0	dB	3.0
Fade margin	dB	1.5

GOLAY CODES

Marcus Greferath
San Diego State University
San Diego, California

1. INTRODUCTION

Among the various codes and code families that have been enjoying the attention of coding theorists, two sporadic examples of (extended) cyclic codes have continuously attracted the interest of many scholars. These two codes, named after their discoverer M. Golay, have been known since the very first days of algebraic coding theory in the late 1940s. Their enduring role in contemporary coding theory results from their simplicity, structural beauty, depth of mathematical background, and connection to other fields of discrete mathematics, such as finite geometry and the theory of lattices.

This presentation is devoted to a description of the basics about these codes. The reader should be aware that the issues that we have chosen for a more detailed discussion form only a narrow selection out of the material that is available about these codes. So the article at hand does not claim in any way to be a comprehensive or complete treatment of the issue. For the many aspects that we are not discussing here, the reader is referred to the Bibliography, and in particular to the treatise by Conway and Sloane [6]. This article is organized as follows. In Section 1 we will introduce the Golay codes as (extended) quadratic residue codes. We will discuss their parameters and immediate properties. In Section 2 we will address alternative descriptions of the Golay codes. Besides a collection of nice generator matrices, we will briefly mention the miracle octad generator (MOG) construction and also Pasquier's description of the binary Golay code. Section 3 is devoted to the decoding of these codes. Among the various algorithms from the literature we have picked a particular one due to M. Elia, which is an algebraic decoder.

We will then briefly come to the relationship of these codes to finite geometry (design theory) and the theory of lattices. In particular, we will mention an important lattice called the *Leech lattice*, which closely connected to the binary Golay code, more precisely, with a $\mathbb{Z}_4$-linear version of this code. This gives us a natural bridge to the final discussion of the article, the role of the Golay codes in the investigation of ring-linear codes.

1.1. Basic Notions

We define the Hamming distance d_H on a finite set $\mathbb{F}$ (usually a field or a ring) defined by

$$d_H\colon \mathbb{F}\times\mathbb{F}\to\mathbb{N},\quad (x,y)\mapsto\begin{cases}0, & x=y\\ 1, & \text{else}\end{cases}$$

This function is usually extended additively to the space $\mathbb{F}^n$ for every $n\in\mathbb{N}$. A block code is a subset C of $\mathbb{F}^n$, and its Hamming minimum distance is

$$d_{\min}(C) := \{d_H(x,y)\mid x,y\in C, x\neq y\}$$

A block code of length n and minimum distance d is referred to as (n,M,d) code, where M is the number of its words.

If $\mathbb{F}$ is a (finite) field, then we call a code C-*linear* if it is a subspace of F^n. A linear code of dimension k is usually denoted as an $[n,k]$ code. If it has minimum distance d, we also speak of an $[n,k,d]$ code. Note that for a linear code C, the minimum Hamming weight, namely, $\min\{d_H(c,0)\mid c\in C-\{0\}\}$, coincides with its minimum distance.

A linear code $C\leq\mathbb{F}^n$ is called *cyclic* if is invariant under a cyclic shift of its coordinates. Algebraically, cyclic linear codes are exactly the ideals of the (residual) polynomial ring $\mathbb{F}[x]/(x^n-1)$, and for a cyclic code C there exists a unique monic divisor g of x^n-1 in $\mathbb{F}[x]$ such that $C=\mathbb{F}[x]g/(x^n-1)$.

Given a linear code C, we define the dual code

$$C^{\perp} := \{x\in\mathbb{F}^n\mid c\cdot x=0 \text{ for all } c\in C\}$$

and we call C *self-dual*, if $C=C^{\perp}$.

2. THE GOLAY CODES AS (EXTENDED) QUADRATIC RESIDUE CODES

M. J. E. Golay (1902–1989) was a Swiss physicist known for his work in infrared spectroscopy, among other things. He was one of the founding fathers of coding theory, discovering the two binary Golay codes in 1949 [11] and the ternary Golay codes in 1954. These codes, which have been called *Golay codes* since then, have been important not only because of theoretical but also practical reasons—the extended binary Golay code, for example, has frequently been applied in the U.S. space program, most notably with the *Voyager I and II* spacecraft that transmitted clear color pictures of Jupiter and Saturn.

Let p and q be prime numbers such that p is odd and q is a quadratic residue modulo p, and let ω be a primitive pth root in a suitable extension field of $\mathbb{F}_q$. Let $\mathrm{QR} := \{i^2\mid i\in\mathbb{F}_p^{\times}\}$ denote the set of quadratic residues modulo p and set $\mathrm{NQR} := \mathbb{F}_p^{\times}\setminus\mathrm{QR}$. Both of these sets have size $(p-1)/2$, the polynomials

$$f_{\mathrm{QR}} = \prod_{i\in\mathrm{QR}}(x-\omega^i) \text{ and } f_{\mathrm{NQR}} = \prod_{i\in\mathrm{NQR}}(x-\omega^i)$$

have coefficients in $\mathbb{F}_q$, and $x^p-1=f_{\mathrm{QR}}\cdot f_{\mathrm{NQR}}\cdot(x-1)$.

Example 1

(a) For $p=23$ and $q=2$ and choosing a suitable primitive 23rd root in $\mathbb{F}_{2^{11}}$, we obtain

$$f_{\mathrm{QR}} = x^{11}+x^9+x^7+x^6+x^5+x+1 \text{ and}$$
$$f_{\mathrm{NQR}} = x^{11}+x^{10}+x^6+x^5+x^4+x^2+1.$$

(b) For $p=11$ and $q=3$ and a suitable primitive 11th root in $\mathbb{F}_{3^5}$, we obtain

$$f_{\mathrm{QR}} = x^5+x^4-x^3+x^2-1 \text{ and}$$
$$f_{\mathrm{NQR}} = x^5-x^3+x^2-x-1$$

Definition 1

(a) The cyclic binary code of length 23 generated by the polynomial $x^{11}+x^9+x^7+x^6+x^5+x+1\in\mathbb{F}_2[x]$ is called the *binary Golay code*. This code is a [23,12,7] code; its extension by a parity check will be denoted by $\overline{G}_2$ and is a [24,12,8] code.

(b) The cyclic ternary code G_3 of length 11 generated by the polynomial $x^5+x^4-x^3+x^2-1\in\mathbb{F}_3[x]$ is called the ternary Golay code. This code is a [11,6,5] code;

its extension by a parity check will be denoted by $\overline{G}_3$ and is a [12,6,6] code.

The sphere packing bound says that if $|F| = q$ and $d_{\min}(C) = 2t + 1$, then

$$|C| \sum_{i=0}^{t} \binom{n}{i} (q-1)^i \leq q^n$$

Codes meeting this bound with equality are called *perfect codes*. Given their minimum distance, it is indeed easily verified that G_2 as well as G_3 are perfect codes, since

$$2^{12}\left(\binom{23}{0} + \binom{23}{1} + \binom{23}{2} + \binom{23}{3}\right) = 2^{23} \text{ and}$$

$$3^6\left(\binom{11}{0} + \binom{11}{1}2 + \binom{11}{2}4\right) = 3^{11}$$

The codes $\overline{G}_2$ and $\overline{G}_3$ are, of course, not perfect, but they are still what is called *quasiperfect*: the spheres of radius $t = 3$ (or $t = 2$, respectively) are still disjoint, whereas the spheres of radius $t + 1$ already cover the ambient space.

2.1. Weight Enumerators

Let C be a linear $[n, k, d]$ code. Defining the Hamming weight enumerator of C as the integer polynomial

$$W_C(t) := \sum_{c \in C} t^{w_H(c)} = \sum_{i=0}^{n} A_i t^i$$

where $A_i := |\{c \in C \mid w_H(c) = i\}|$, we obtain the Hamming weight enumerators for the four Golay codes as listed in Table 1.

Call a linear code C *self-dual*, if it coincides with its dual code

$$C^{\perp} := \{x \in \mathbb{F}^n \mid c \cdot x = 0 \text{ for all } c \in C\}$$

It can be shown that the codes $\overline{G}_2$ and $\overline{G}_3$ are self-dual codes. Moreover $\overline{G}_2$ is called a doubly even code because all of its weights are divisible by 4. Note that all weights of $\overline{G}_3$ are divisible by 3.

2.2. Uniqueness of the Golay codes

Two block codes C and D of length n over the alphabet $\mathbb{F}$ are called *equivalent* if there is a coordinate permutation π and a set $\sigma_1, \ldots, \sigma_n$ of permutation on $\mathbb{F}$ such that C is mapped to D under the bijection

$$\mathbb{F}^n \to \mathbb{F}^n, (c_1, \ldots, c_n) \mapsto (\sigma_1(c_\pi(1)), \ldots, \sigma_n(c_\pi(n)))$$

It has been proved by Pless, Goethals, and Delsarte that the Golay codes are uniquely determined up to equivalence by their parameters. This is the content of the following theorem. Proofs can be found in the literature [8,21].

Theorem 1

(a) Every binary $(23, 2^{12}, 7)$ code is equivalent to the Golay code G_2, and every binary $(24, 2^{12}, 8)$ code is equivalent to $\overline{G}_2$.

(b) Every ternary $(11, 3^6, 5)$ code is equivalent to the Golay code G_3, and every ternary $(12, 3^6, 6)$ code is equivalent to $\overline{G}_3$.

Table 1. Weight Enumerators for the Golay Codes

Code	Weight Enumerator
G_2	$t^{23} + 253\,t^{16} + 506\,t^{15} + 1288\,t^{12} + 1288\,t^{11} + 506\,t^8 + 253\,t^7 + 1$
$\overline{G}_2$	$t^{24} + 759\,t^{16} + 2576\,t^{12} + 759\,t^8 + 1$
G_3	$t^{11} + 132\,t^6 + 132\,t^5 + 330\,t^3 + 110\,t^2 + 24$
$\overline{G}_3$	$t^{12} + 264\,t^6 + 440\,t^3 + 24$

3. ALTERNATIVE CONSTRUCTIONS

In this section we will discuss alternative constructions of the Golay codes. First, we will find generator matrices of rather beautiful form for equivalent versions of these.

In the binary case consider the matrix

$$[I \mid A] := \left[\begin{array}{cccccccccccc|cccccccccccc}
1&0&0&0&0&0&0&0&0&0&0&0&0&1&1&1&1&1&1&1&1&1&1&1\\
0&1&0&0&0&0&0&0&0&0&0&0&1&1&1&0&1&1&1&0&0&0&1&0\\
0&0&1&0&0&0&0&0&0&0&0&0&1&1&0&1&1&1&0&0&0&1&0&1\\
0&0&0&1&0&0&0&0&0&0&0&0&1&0&1&1&1&0&0&0&1&0&1&1\\
0&0&0&0&1&0&0&0&0&0&0&0&1&1&1&1&0&0&0&1&0&1&1&0\\
0&0&0&0&0&1&0&0&0&0&0&0&1&1&1&0&0&0&1&0&1&1&0&1\\
0&0&0&0&0&0&1&0&0&0&0&0&1&1&0&0&0&1&0&1&1&0&1&1\\
0&0&0&0&0&0&0&1&0&0&0&0&1&0&0&0&1&0&1&1&0&1&1&1\\
0&0&0&0&0&0&0&0&1&0&0&0&1&0&0&1&0&1&1&0&1&1&1&0\\
0&0&0&0&0&0&0&0&0&1&0&0&1&0&1&0&1&1&0&1&1&1&0&0\\
0&0&0&0&0&0&0&0&0&0&1&0&1&1&0&1&1&0&1&1&1&0&0&0\\
0&0&0&0&0&0&0&0&0&0&0&1&1&0&1&1&0&1&1&1&0&0&0&1
\end{array}\right]$$

Then consider the ternary matrix

$$[I \mid B] = \left[\begin{array}{cccccc|cccccc}
1&0&0&0&0&0&0&1&1&1&1&1\\
0&1&0&0&0&0&1&0&1&2&2&1\\
0&0&1&0&0&0&1&1&0&1&2&2\\
0&0&0&1&0&0&1&2&1&0&1&2\\
0&0&0&0&1&0&1&2&2&1&0&1\\
0&0&0&0&0&1&1&1&2&2&1&0
\end{array}\right]$$

These matrices have the following properties.

1. Both A and B are symmetric and satisfy $A^2 = I$ and $B^2 = -I$.
2. Each row of A has exactly 7 or exactly 11 ones. Each row of B has exactly 5 nonzero entries.
3. Each pair of rows of A differs in exactly 6 places. For B the sum and the difference of each pair of rows have at least 4 nonzero entries.

It can be seen that the code generated by $[I \mid A]$ has the parameters of the extended binary Golay code, and that the code generated by $[I \mid B]$ has those of the ternary Golay code. By the uniqueness theorem (Theorem 1, above) we therefore see that they are (up to equivalence) the binary and ternary Golay codes.

3.1. The MOG Construction

We now discuss a construction of the binary Golay code using the $\mathbb{F}_4$-linear hexacode [6, Chap. 11].

The *hexacode* is the [6,3,4] code H generated by the matrix

$$\begin{bmatrix} 1 & 0 & 0 & \omega^2 & 1 & \omega \\ 0 & 1 & 0 & 1 & 1 & 1 \\ 0 & 0 & 1 & \omega & 1 & \omega^2 \end{bmatrix}$$

Its Hamming weight enumerator is given by $1 + 45t^4 + 18t^6$, and it is clear that any word of the hexacode has 0, 2, or 6 zeros.

Let $\mathbb{F}_2^{4\times 6}$ denote the space of all 4×6 matrices with binary entries. As an $\mathbb{F}_2$-vector space it can clearly be identified with $\mathbb{F}_2^{24}$. We are going to isolate a subset of $\mathbb{F}_2^{4\times 6}$ that will in a natural way turn out to be equivalent to the Golay code.

For this we first define a mapping $\varphi\colon \mathbb{F}_2^{4\times 6} \to \mathbb{F}_4^6, G \mapsto (0, 1, \omega, \omega^2)G$. Now define C to be the subset of all matrices $G \in \mathbb{F}_2^{4\times 6}$ having the following two properties:

1. For every column $j \in \{1, \ldots, 6\}$ there holds $\sum_{i=1}^{4} g_{ij} = \sum_{j=1}^{6} g_{1j}$; thus, the parity of each column is that of the first row of G.
2. $\varphi(G) \in H$.

As both of these conditions are preserved under addition of matrices that satisfy these conditions, we have C to be a subspace of $\mathbb{F}_2^{4\times 6}$. The first condition imposes 6 restrictions on the matrix, and the second (at most) another 6. For this reason $\dim(C) \geq 24 - (6 + 6) = 12$. To find that this code is equivalent to the Golay code, we only have to check if its minimum weight is given by at least 8, because this forces its dimension down to 12 and hence makes it a [24,12,8] code, and we are finished by Theorem 1.

Let G be a nonzero element of C. First, assume that $\sum_{j=1}^{6} g_{1j} = 0$, which by the preceding description means that $\sum_{i=1}^{4} g_{ij} = 0$ for all $j \in \{1, \ldots, 6\}$. If $\varphi(G) \neq 0$, then $\varphi(G)$ has at least 4 nonzero entries, and hence G must have at least 4 nonzero columns. As each of these columns has an even number of ones, the weight of G is at least 8. If $\varphi(G) = 0$ then each column of G must be the all-zero or the all-one vector because $0, 1, \omega$ and ω^2 can be combined to zero under even parity only in these two ways. Hence, again the weight of G is at least 8 because we have an even number of nonzero columns.

Now assume that the parity of the first row of G is 1. Then each column of G has either 1 or 3 nonzero entries. Its weight will therefore be at least 8 unless each column has exactly 1 nonzero entry. Exactly the zero-entries of $\varphi(G)$ correspond to those columns of G having their nonzero entry in the first row. As $\varphi(G) \in H$ we conclude that there is an even number of zero entries in $\varphi(G)$ contradicting the fact that the parity of the first row of G is 1. Hence, this case cannot happen, and we know that the weight of G must be at least 8.

3.2. Pasquier's Construction

Pasquier observed [20] that the extended binary Golay code can be obtained from a Reed-Solomon code over $\mathbb{F}_8$. In order to see this let

$$\text{tr}\colon \mathbb{F}_8 \to \mathbb{F}_2, x \mapsto x + x^2 + x^4$$

be the trace map and let $\alpha \in \mathbb{F}_8$ be an element satisfying the equation $\alpha^3 + \alpha^2 + 1 = 0$. It is easy to see that $\text{tr}(\alpha) = 1$, and that $B := \{\alpha, \alpha^2, \alpha^4\}$ forms a trace-orthogonal basis of $\mathbb{F}_8$ over $\mathbb{F}_2$. This means that

$$\text{tr}(xy) = \begin{cases} 1, & x = y \\ 0, & x \neq y \end{cases}$$

for all $x, y \in B$.

Now consider the [7,4,4] Reed–Solomon code generated by the polynomial $\Pi_{i=1}^{3}(x + \alpha^i)$. Its extension by a parity check yields the self-dual [8,4,5] extended Reed–Solomon code generated by the matrix

$$R := \begin{bmatrix} 1 & \alpha^5 & 1 & \alpha^6 & 0 & 0 & 0 & \alpha^3 \\ 0 & 1 & \alpha^5 & 1 & \alpha^6 & 0 & 0 & \alpha^3 \\ 0 & 0 & 1 & \alpha^5 & 1 & \alpha^6 & 0 & \alpha^3 \\ 0 & 0 & 0 & 1 & \alpha^5 & 1 & \alpha^6 & \alpha^3 \end{bmatrix}$$

We now consider the $\mathbb{F}_2$-linear mapping

$$\mathbb{F}_8 \to \mathbb{F}_2^3, \quad a_0\alpha + a_2\alpha^2 + a_1\alpha^4 \mapsto (a_0, a_1, a_2)$$

and extend it componentwise to an $\mathbb{F}_2$-linear mapping $\varphi\colon \mathbb{F}_8^8 \to \mathbb{F}_2^{24}$. The image of the extended Reed–Solomon code described above is clearly a binary [24,12] code, because the vectors $\{\alpha v, \alpha^2 v, \alpha^4 v\}$ are independent over $\mathbb{F}_2$ for each of the rows v of the preceding matrix R. Hence this binary image is a [24,12] code.

Observing $\text{tr}(xy) = \varphi(x)\varphi(x)$ for all vectors $x, y \in \mathbb{F}_8^8$, we see that our [24,12] code is self-dual because the Reed–Solomon code that we started with is so. Even more is true: this code is "doubly even," which means that the Hamming weight of its vectors is always a multiple of 4. We can easily check this by finding a basis consisting of doubly even words and keeping in mind that under self-duality this property inherits to linear combinations of vectors that have this property. Hence the minimum weight of the above code is a multiple of 4. However the underlying Reed–Solomon code has already minimum weight 5, and so our code must have minimum weight ≥ 8. Finally we can apply Theorem 1 and find that it is the binary Golay code.

Remark 1. It is worth noting that Goldberg [12] has constructed the ternary Golay code as an image of an $\mathbb{F}_9$-linear [6,3,4] code in a similar manner. Again an obvious mapping between $\mathbb{F}_9^6$ and $\mathbb{F}_3^{12}$ is used to map the [6,3,4] code into a [12,6] code, which finally turns out to be equivalent to the ternary Golay code.

4. DECODING THE GOLAY CODES

There are various decoders for the (cyclic) binary and ternary Golay codes; the most common are probably that by Kasami and the systematic search decoder (both have been nicely described in Ref. 16). Focusing on the binary Golay code G_2, we prefer to discuss an algebraic decoder that has been developed by M. Elia [9]. Even though this decoder is not the fastest known, it is of interest because it makes use of the algebraic structure of the code in question. Furthermore in a modified form it has been used in order to decode ring-linear versions of the Golay code [14].

4.1. Decoding the Binary [23,12,7] Code

Let $\alpha \in \mathbb{F}_{2^{11}}$ be a root of the generator polynomial $g = x^{11} + x^9 + x^7 + x^6 + x^5 + x + 1$ of G_2. It clearly satisfies $\alpha^{23} = 1$, and its associated cyclotomic coset is given by $B = \{1, 2, 4, 8, 16, 9, 18, 13, 3, 6, 12\}$. This shows that g has roots α, α^3 and α^9.

As we are dealing with a cyclic code, we will represent its words by polynomials in the sequel. Assume the word $r = fg + e$ has been received, where the Hamming weight of e is at most 3. We compute the syndromes

$$s_1 := r(\alpha) = e(\alpha), \quad s_3 := r(\alpha^3) = e(\alpha^3), \text{ and}$$
$$s_9 := r(\alpha^9) = e(\alpha^9)$$

and our plan is to recover e and hence fg and f from these. This can be done in a quite simple way. As we are in a binary situation, we are interested only in what is called the *error locator polynomial*

$$L(z) := \prod\{(z + \alpha^i) \mid i \in \{0, \ldots, 22\}, e_i \neq 0\}$$

Once we can express this in terms of the syndromes described above, we only have to find its roots, and we know the error locations. Let us distinguish the following four cases:

1. There is no error, which means that $e = 0$. Then $L(z) = 1$.
2. There is one error in the position i, which means that $L(z) := z + \sigma_1$ where we have set $\sigma_1 = \alpha^i$.
3. There are two errors in position i and j. Then $L(z) = z^2 + \sigma_1 z + \sigma_2$ where

$$\sigma_1 = \alpha^i + \alpha^j \text{ and } \sigma_2 = \alpha^i \alpha^j$$

4. There are three errors in position i,j and k. Then $L(z) = z^3 + \sigma_1 z^2 + \sigma_2 z + \sigma_3$, where

$$\sigma_1 = \alpha^i + \alpha^j + \alpha^k$$
$$\sigma_2 = \alpha^i \alpha^j + \alpha_j \alpha^k + \alpha^i \alpha^k$$
$$\sigma_3 = \alpha^i \alpha^j \alpha^k$$

Case 1 is easily recognized by the fact that here $s_1 = s_3 = s_9 = 0$. Case 2 occurs exactly if $s_1^3 = s_3$ and $s_3^3 = s_9$, which is easily verified considering the above mentioned definitions. For the remaining cases we observe first that $s_1^3 \neq s_3$. Furthermore we still have $\sigma_1 = s_1$ by definition. Setting

$$D = (s_1^3 + s_3)^2 + \frac{s_1^9 + s_9}{s_1^3 + s_3}$$

and verifying that $D = (\sigma_2 + s_1^2)^3$, we easily get

$$\sigma_2 = s_1^2 + \sqrt[3]{D} \text{ and } \sigma_3 = s_3 + s_1 \sqrt[3]{D}$$

where in the finite field $\mathbb{F}_{2^{11}}$ the (unique) third root of D can is also computed as D^{1365}.

All in all, the knowledge of s_1, s_3, and s_9 can be used to compute the error locator polynomial $L(z)$ and by finding its roots, we are able to solve for e and f.

Remark 1. It is worth noting that Elia and Viterbo have developed an algebraic decoder also for G_3 [10]. This decoder works in a similar fashion and makes use of two syndromes. Even though it has to consider error values in addition to error locations, it treats the entire problem just by determining one polynomial.

5. GOLAY CODES AND FINITE GEOMETRY—AUTOMORPHISM GROUPS

Given a finite set S of v elements, we recall that a subset $B \subseteq \binom{S}{k}$ is called a $t-(v, k, \lambda)$ *block design*, provided that every t-element subset of S is contained in exactly λ elements of B. The elements of B are called *blocks*, and B is often referred to simply as a t design.

Now let C be a binary code. For a word $c \in C$, we call the set $\{i \mid c_i \neq 0\}$ the *support* of c. Furthermore, we say that the word C covers the word c' if $\text{Supp}(C) \supseteq \text{Supp}(c')$. Let C_d be the set of codewords that have weight d. We say that C_d holds a $t-(n, d, \lambda)$ design, if the supports of the words in C_d form the blocks of such a design.

Theorem 2. If C is a perfect binary (n, M, d) code (containing the all-zero word), then the set C_d of all codewords of minimum weight d hold a $t-(n, d, 1)$ design, where $t = (d+1)/2$.

Proof The spheres of radius t are disjoint and cover $\mathbb{F}_2^n$. Hence for every binary word x of weight t there exists exactly one codeword $c \in C$ such that x is contained in the sphere of radius $t-1$ centered in c. By $d_H(c, x) \leq t-1$ we immediately get

$$w_H(c) \leq d_H(c, x) + w_H(x) \leq 2t - 1 = d$$

and so $d_H(c, x) = t - 1$ and $c \in C_d$. All in all, we now have

$$2|\text{Supp}(x) \cap \text{Supp}(c)| = w_H(x) + w_H(c)$$
$$- d_H(x, c) \geq t + 2t - 1 - t = 2t$$

and therefore C covers x, which finishes our proof.

Corollary 1. The words of weight 7 of the binary Golay code hold a $4-(23,7,1)$ design.

For the extended binary Golay code, there is another interesting result.

Theorem 3. The codewords of weight 8 in the extended binary Golay code hold a $5-(24,8,1)$ design.

Proof If a word of weight 5 in $\mathbb{F}_2^{24}$ were covered by two codewords of minimum weight, then the distance between these words would be at most 6, which contradicts the minimum distance of binary Golay code. Hence every word of weight 5 is covered by at most one minimum weight codeword of G_2. We now count the number of words of weight 5 in $\mathbb{F}_2^{24}$. On one hand, this is clearly given by $\binom{24}{5}$; on the other hand, it is by the foregoing arguments clear that it is at least $|C_8|\binom{8}{5}$. We have seen however that $|C_8| = 759$, and hence by $759\binom{8}{5} = \binom{24}{5}$ every word of weight 5 is covered by at least one word in C_8. This completes the proof.

Remark 2. Using a modified technique of proof it can be shown that the supports of minimum weight words of G_3 hold a $4-(11,5,1)$ design, and that the supports of words of weight 6 of the extended ternary Golay code $\overline{G}_3$ form the blocks of a $5-(12,6,1)$ design.

The *automorphism group* of a design is the permutation group acting on its points that maps the set of blocks into itself. The automorphism group of a binary linear code is the set of all coordinate permutations that map the code into itself. In the ternary case we need to consider coordinate permutations and sign flips (monomial transformations) that map the code in question into itself. Without proof, we state the following basic facts.

Theorem 4

(a) The automorphism group of the $5-(24,8,1)$ design held by the words of weight 8 in the extended binary Golay code is given by the (simple) Mathieu group M_{24} of order $24\cdot 23\cdot 22\cdot 21\cdot 20\cdot 48$. This group is also the automorphism group of the extended binary Golay code $\overline{G}_2$.

(b) The automorphism group of the $5-(12,6,1)$ design held by the words of weight 6 in the extended ternary Golay code is given by the (simple) Mathieu group M_{12} of order $12\cdot 11\cdot 10\cdot 9\cdot 8$. There is a normal subgroup N of the automorphism group $\mathrm{Aut}(\overline{G}_3)$. This group has order 2, and $\mathrm{Aut}(\overline{G}_3)/N$ is isomorphic to M_{12}.

6. GOLAY CODES AND RING-LINEAR CODES

One very important observation in algebraic coding theory in the early 1990s was the discovery of the $\mathbb{Z}_4$-linearity of the Preparata codes, the Kerdock codes, and related families [15]. These had previously been known as notoriously nonlinear binary codes that had more codewords than any known linear code of the same length and minimum distance.

Defining the *Lee weight* on $\mathbb{Z}_4$ as $w_{\mathrm{Lee}}\colon \mathbb{Z}_4 \to \mathbb{N}, r \mapsto \min\{|r|, |4-r|\}$, we obtain what is called the *Gray isometry* of $\mathbb{Z}_4$ into $\mathbb{F}_2^2$ as

$$\gamma\colon (\mathbb{Z}_4, w_{\mathrm{Lee}}) \to (\mathbb{Z}_2^2, w_H)$$
$$r \mapsto r_0(0,1) + r_1(1,1)$$

where $r_i \in \mathbb{Z}_2$ are the coefficients of the binary representation of $r \in \mathbb{Z}_4$, i.e., $r = r_0 + 2r_1$. The image of γ is the full space $\mathbb{F}_2^2$, and by abuse of notation we denote by γ also its componentwise extension to $\mathbb{Z}_4^n \to \mathbb{Z}_2^{2n}$. A (linear or nonlinear) binary code C of length $2n$ is said to have a $\mathbb{Z}_4$-linear representation if there is a $\mathbb{Z}_4$-linear code D of length n such that C is equivalent to $\gamma(D)$.

The results in Ref. 15 show that the Nordstrom–Robinson code has a $\mathbb{Z}_4$-linear representation by the [8,4] octacode, which is a lift of the Reed–Muller code RM(2,3). Long before there had been known another interesting way of constructing the Nordstrom–Robinson code. We present this code in the following paragraphs.

6.1. The Extended Binary Golay Code and the Nordstrom–Robinson Code

As the extended binary Golay code has minimum weight 8, we can assume (after column permutations) that it contains the word $(1^8 0^{16})$, where x^n is an abbreviation for n (consecutive) occurrences of the element x. If A is a generator matrix for this version of the Golay code, then it is clear that the first 7 columns of A must be linearly independent since A also serves as a check matrix for the code because of self-duality. On the other hand, the first 8 columns of A are linearly dependent, and hence we have the 8th column as the sum of the foregoing 7 columns. By elementary row operations we can finally achieve that the last 5 rows of this matrix have zeros in their first 8 positions. Hence, we end up with a generator matrix of the form

$$\left[\begin{array}{ccccccc|c|ccc}
1 & 0 & 0 & 0 & 0 & 0 & 0 & 1 & * & \cdots & * \\
0 & 1 & 0 & 0 & 0 & 0 & 0 & 1 & * & \cdots & * \\
0 & 0 & 1 & 0 & 0 & 0 & 0 & 1 & * & \cdots & * \\
0 & 0 & 0 & 1 & 0 & 0 & 0 & 1 & * & \cdots & * \\
0 & 0 & 0 & 0 & 1 & 0 & 0 & 1 & * & \cdots & * \\
0 & 0 & 0 & 0 & 0 & 1 & 0 & 1 & * & \cdots & * \\
0 & 0 & 0 & 0 & 0 & 0 & 1 & 1 & * & \cdots & * \\
\hline
0 & 0 & 0 & 0 & 0 & 0 & 0 & 0 & * & \cdots & * \\
\hline
0 & 0 & 0 & 0 & 0 & 0 & 0 & 0 & * & \cdots & * \\
0 & 0 & 0 & 0 & 0 & 0 & 0 & 0 & * & \cdots & * \\
0 & 0 & 0 & 0 & 0 & 0 & 0 & 0 & * & \cdots & * \\
0 & 0 & 0 & 0 & 0 & 0 & 0 & 0 & * & \cdots & *
\end{array}\right]$$

where the asterisks represent some binary entries.

From this matrix it can be seen that there are $32 = 2^{12-7}$ codewords that have zeros in their first 8 positions. Furthermore for every $i \in \{1,\ldots,7\}$ there are 32 codewords that have a 1 in the ith and 8th positions. We define N to be the (nonlinear) binary code to consist of the union of all these $8\cdot 32$ words, where we have cut off the first 8 entries. This is certainly a code of length 16. To determine its minimum distance we observe that any word in this code results from truncation of a word of the Golay code starting with either $(0, \ldots, 1 \ldots, 0, 1)$ or $(0, \ldots, 0)$. Any two words of these words therefore differ in at most two entries in the first 8 positions, and since the minimum distance of the Golay code is 8, the initially given words must differ in at least 6 positions. Applying the sphere packing bound we see that the minimum distance is at most 6, and overall this proves it to be given by 6.

Remark 3. In light of the statements at the beginning of this section we have questioned whether the binary Golay code itself might enjoy a $\mathbb{Z}_4$-linear representation. This has been answered to the negative in Ref. 15.

6.2. The $\mathbb{Z}_4$-Linear Golay Code and the Leech Lattice

Bonnecaze et al. investigated $\mathbb{Z}_4$-linear lifts of binary codes in connection with lattices. To explain how this works for the binary Golay code, we first Hensel-lift its generator polynomial $f = x^{11} + x^9 + x^7 + x^6 + x^5 + x + 1 \in \mathbb{F}_2[x]$ to the divisor

$$F = x^{11} + 2x^{10} - x^9 - x^7 - x^6 - x^5 + 2x^4 + x - 1$$

of $x^{23} - 1$ in $\mathbb{Z}_4[x]$. (Recall that according to Hensel's lemma [18, Sect. XIII.4] the polynomial F is the unique polynomial that reduces to f modulo 2 and divides $x^{23} - 1$ in $\mathbb{Z}_4[x]$.)

Extending the cyclic [23, 12] code G_4 that is generated by F, we obtain a [24,12] code $\overline{G}_4$ of minimal *Lee weight* 12. Defining the *Euclidean weight* on $\mathbb{Z}_4$ as the squared Lee weight and extending it additively, we obtain the minimum Euclidean weight as 16.

The so-called construction A provides a means to construct a lattice Λ from G_4, by setting

$$\Lambda := \nu^{-1}(\overline{G}_4)$$

where ν denotes the natural map $\mathbb{Z}^{24} \to \mathbb{Z}_4^{24}$. Surprisingly, this lattice is one of the best studied lattices so far. It has maximal density that is achievable in 24 dimensions. Referring to the proof in Ref. 2 we state the following theorem.

Theorem 5. Construction A of the quaternary Golay code $\overline{G}_4$ yields up to lattice equivalence the Leech lattice.

We should mention that there are several different constructions for the Leech lattice involving the binary Golay code, but not purely via construction A. This construction, however, is one of the most natural ways to construct a lattice from a code, and hence inspired us to mention it here.

6.3. Higher Lifts of Golay Codes

It is possible to define a weight on rings (e.g., $\mathbb{Z}_8$ or $\mathbb{Z}_9$) for which a generalized version of the above Gray map exists. The only difference to keep in mind is that these maps are not necessarily subjective anymore.

Specifically, the normalized homogeneous weight (defined in [4]) on $\mathbb{Z}_8$ as

$$w_{\text{hom}}: \mathbb{Z}_8 \to \mathbb{N},\ r \mapsto \begin{cases} 0, & \text{r} = 0 \\ 2, & \text{r} = 4 \\ 1, & \text{else} \end{cases}$$

is such a weight. The according Gray map of this ring into $\mathbb{Z}_2^4$ is given by

$$\gamma: (\mathbb{Z}_8, 2w_{\text{hom}}) \to (\mathbb{Z}_2^4, w_H)$$
$$r \mapsto r_0(0,0,1,1) + r_1(0,1,0,1) + r_2(1,1,1,1)$$

where $r_i \in \mathbb{Z}_2$ are the coefficients of the binary representation of $r \in \mathbb{Z}_8$, namely, $r = r_0 + 2r_1 + 4r_2$. Again we also denote by γ its componentwise extension to $\mathbb{Z}_8^n \to \mathbb{Z}_2^{4n}$.

Following the presentation in Ref. 7 Hensel lifting the generator polynomial of G_2 to $\mathbb{Z}_8[x]$ results in the polynomial $x^{11} + 2x^{10} - x^9 + 4x^8 + 3x^7 + 3x^6 - x^5 + 2x^4 + 4x^3 + 4x^2 + x - 1$, which generates a free [23,12] code G_8 over $\mathbb{Z}_8$. Extending the latter code by a parity check, we obtain a free $\mathbb{Z}_8$-linear self-dual [24,12] code $\overline{G}_8$.

Looking into Brouwer's and Litsyn's tables [3,17], it is remarkable that this code has more codewords than does any presently known binary code of length 96 and minimum distance 24. Hence we have an outperforming example of a nonlinear code that is constructed using the binary Golay code.

Remark 4. Using a similar technique it has been shown that a non-linear ternary $(36, 3^{12}, 15)$ code can be constructed as the image of a $\mathbb{Z}_9$-linear lift of the ternary Golay code. This code is not outperforming, but so far no ternary $(36, 3^{12})$ codes with a better minimum distance is known. The details can be found in Ref. 13.

BIOGRAPHY

Marcus Greferath received his Diploma and Ph.D. degrees in mathematics in 1990 and 1993, respectively, from Mainz University (Germany). He joined the Department of Mathematics of Duisburg University 1992 as Research Assistant. In Duisburg he obtained the position of an Assistant Professor in 1994 and finished his Habilitation in 2000. Dr. Greferath has published more than 20 papers in the areas of ring geometry, coding theory, and cryptography. In 1997 he started a 2-year research stay at AT&T Shannon Laboratory in New Jersey. In 1999 he held a one-year visiting professorship at Ohio University in Athens (Ohio) and was appointed as Assistant Professor at the Department of Mathematics of San Diego State University in 2001. His areas of interest are finite geometry, coding theory, and cryptography with a focus on the role of rings and modules in these disciplines.

BIBLIOGRAPHY

1. A. Barg, At the dawn of the theory of codes, *Math. Intelligencer* **15**(1): 20–26 (1993).

2. A. Bonnecaze, P. Sole and A. R. Calderbank, Quaternary quadratic residue codes and unimodular lattices, *IEEE Trans. Inform. Theory* **41**(2): 366–377 (1995).
3. A. E. Brouwer, Bounds on the minimum distance of linear codes, *http://www.win.tue.nl/math/dw/voorlincod.html*, 2002.
4. I. Constantinescu and W. Heise, A metric for codes over residue class rings of integers, *Problemy Peredachi Informatsii* **33**(3): 22–28 (1997).
5. J. H. Conway et al., M12, *Atlas of Finite Groups*, Clarendon Press, Oxford, UK, 1985.
6. J. H. Conway and N. J. A. Sloane, *Sphere Packings, Lattices and Groups*, 3rd ed., Grundlehren der Mathematischen Wissenschaften (Fundamental Principles of Mathematical Sciences), Springer-Verlag, New York, 1999.
7. I. M. Duursma, M. Greferath, S. N. Litsyn, and S. E. Schmidt, A $\mathbb{Z}_8$-linear lift of the binary Golay code and a nonlinear binary (96, 2^{37}, 24)-code, *IEEE Trans. Inform. Theory* **47**(4): 1596–1598 (2001).
8. P. Delsarte and J.-M. Goethals, Unrestricted codes with the Golay parameters are unique, *Discrete Math.* **12**: 211–224 (1975)
9. M. Elia, Algebraic decoding of the (23, 12, 7) Golay code, *IEEE Trans. Inform. Theory* **33**(1): 150–151 (1987).
10. M. Elia and E. Viterbo, Algebraic decoding of the ternary (11, 6, 5) Golay code, *Electron. Lett.* **28**(21): 2021–2022 (1992).
11. M. J. E. Golay, Notes on digital coding, *Proc. IRE* **37**: 657 (1949).
12. D. Y. Goldberg, Reconstructing the ternary Golay code, *J. Combin. Theory S A* **42**(2): 296–299 (1986).
13. M. Greferath and S. E. Schmidt, Gray isometries for finite chain rings and a nonlinear ternary (36, 3^{12}, 15) code, *IEEE Trans. Inform. Theory* **45**(7): 2522–2524 (1999).
14. M. Greferath and E. Viterbo, On $\mathbb{Z}_4$- and $\mathbb{Z}_9$-linear lifts of the Golay codes, *IEEE Trans. Inform. Theory* **45**(7): 2524–2527 (1999).
15. A. R. Hammons et al., The $\mathbb{Z}_4$-linearity of Kerdock, Preparata, Goethals, and related codes, *IEEE Trans. Inform. Theory* **40**(2): 301–319 (1994).
16. S. Lin and D. J. Costello, *Error Control Coding: Fundamentals and Applications*, Prentice-Hall, Englewood Cliffs, NJ, Inc. 1983.
17. S. Litsyn, An updated table of the best binary codes known, in *Handbook of Coding Theory*, North-Holland, Amsterdam, 1998, Vols. I and II, pp. 463–498.
18. B. R. McDonald, *Finite Rings with Identity*, Marcel Dekker, New York, 1974.
19. F. J. MacWilliams and N. J. A. Sloane, *The Theory of Error-Correcting Codes*, North-Holland, Amsterdam, N, 1977.
20. G. Pasquier, The binary Golay code obtained from an extended cyclic code over $\mathbb{F}_8$, *Eur. J. Combin. Theory* **1**(4): 369–370 (1980).
21. V. Pless, On the uniqueness of the Golay codes, *J. Combin. Theory* **5**: 215–228 (1968).
22. V. S. Pless, W. C. Huffman, and R. A. Brualdi, *Handbook of Coding Theory*, North-Holland, Amsterdam, 1998, Vols. I and II.
23. S. Roman, *Coding and Information Theory*, Springer, New York, 1992.

GOLAY COMPLEMENTARY SEQUENCES

MATTHEW G. PARKER
University of Bergen
Bergen, Norway

KENNETH G. PATERSON
University of London
Egham, Surrey, United Kingdom

CHINTHA TELLAMBURA
University of Alberta
Edmonton, Alberta, Canada

1. INTRODUCTION

Complementary sequences were introduced by Marcel Golay [1] in the context of infrared spectrometry. A *complementary pair* of sequences (CS pair) satisfies the useful property that their out-of-phase *aperiodic autocorrelation* coefficients sum to zero [1,2]. Let $\mathbf{a} = (a_0, a_1, \ldots, a_{N-1})$ be a sequence of length N such that $a_i \in \{+1, -1\}$ (we say that $\mathbf{a}$ is bipolar). Define the aperiodic autocorrelation function (AACF) of $\mathbf{a}$ by

$$\rho_{\mathbf{a}}(k) = \sum_{i=0}^{N-k-1} a_i a_{i+k}, \quad 0 \le k \le N-1 \tag{1}$$

Let $\mathbf{b}$ be defined similarly to $\mathbf{a}$. The pair $(\mathbf{a}, \mathbf{b})$ is called a Golay complementary pair (GCP) if

$$\rho_{\mathbf{a}}(k) + \rho_{\mathbf{b}}(k) = 0, \quad k \neq 0 \tag{2}$$

Each member of a GCP is called a Golay complementary sequence (GCS, or simply Golay sequence). Note that this definition (2) can be generalized to nonbinary sequences. For example, a_i and b_i can be selected from the set $\{\zeta^0, \zeta^1, \ldots, \zeta^{2^h-1}\}$ where ζ is a primitive q-th root of unity, which yields so-called polyphase Golay sequences. In this survey, however, we emphasize binary GCPs.

It is helpful to view (2) in polynomial form. A sequence $\mathbf{a}$ can be associated with the polynomial $a(z) = a_{N-1}z^{N-1} + a_{N-2}z^{N-2} + \cdots + a_1 z + a_0$ in indeterminate z with coefficients ± 1. The pair $(\mathbf{a}, \mathbf{b})$ is then a GCP if the associated polynomials $(a(z), b(z))$ satisfy

$$a(z)a(z^{-1}) + b(z)b(z^{-1}) = 2N \tag{3}$$

Equations (2) and (3) are equivalent expressions because $a(z)a(z^{-1}) = \rho_{\mathbf{a}}(0) + \sum_{1}^{N-1} \rho_{\mathbf{a}}(k)(z^k + z^{-k})$. A further condition can be obtained by restricting z to lie on the unit circle in the complex plane, i.e., $z \in \{e^{2\pi jt} \mid j^2 = -1, 0 \le t < 1\}$. Then $|a(z)|^2 = a(z)a(z^{-1})$ and we have

$$|a(z)|^2 + |b(z)|^2 = 2N, \quad |z| = 1 \tag{4}$$

This means that the absolute value of each polynomial on the unit circle is bounded by $\sqrt{2N}$.

Golay complementary pairs and sequences have found application in physics (Ising spin systems), combinatorics (*orthogonal designs* and *Hadamard matrices*) and telecommunications (e.g., to surface-acoustic wave design, the Loran C precision navigation system, channel-measurement, optical time-domain reflectometry [3], synchronization, spread-spectrum communications, and, recently, *orthogonal frequency division multiplexing* (OFDM) systems [4–7]. Initially, the properties of the *pair* were primarily exploited [1] in a two-channel setting, and periodic GCPs have lately been proposed for two-sided *channel-estimation*, where the two sequences in the pair form a preamble and postamble training sequence, respectively [8]. In recent years, the spectral spread properties of each individual sequence in the pair have also been used. As an example of this, we briefly describe the application of Golay sequences in OFDM. Here, given a data sequence, $\mathbf{a} = (a_0, a_1, \ldots, a_{N-1})$, the transmitted signal $s_{\mathbf{a}}(t)$ as a function of time t is essentially the real part of a discrete Fourier transform (DFT) of $\mathbf{a}$:

$$s_{\mathbf{a}}(t) = \sum_{i=0}^{N-1} a_i e^{2\pi j(i\Delta f + f_0)t} \tag{5}$$

where Δf the frequency separation between adjacent subcarrier pairs and f_0 is the base frequency. Notice that $|s_{\mathbf{a}}(t)| = |a(e^{2\pi j i \Delta f t})|$ where $a(z)$ is the polynomial corresponding to $\mathbf{a}$. Thus, the power characteristics of the OFDM signal can be studied by examining the behavior of an associated polynomial on $|z| = 1$. In particular, if $\mathbf{a}$ is a GCS, then we have that $|s_{\mathbf{a}}(t)|^2 \leq 2N$ so that the *peak-to-mean envelope power ratio* (PMEPR) of the signal is at most 2.0. Having such tightly bounded OFDM signals eases amplifier specification at the OFDM transmitter.

Let $\mathbf{A} = (A_0, A_1, \ldots, A_{N'-1})$ be the N'-point oversampled DFT of $\mathbf{a}$, where $N' \geq N$, i.e.,

$$A_k = \sum_{i=0}^{N-1} a_i \omega^{ik} = a(\omega^k) \quad 0 \leq k < N',$$

where $\omega = e^{2\pi j/N'}$ is a complex N'-th root of unity. For N' large, the values of the N'-point oversampled DFT of $\mathbf{a}$ can be used to approximate the values $a(z)$, $|z| = 1$, and thus the complex OFDM signal in (5).

Example 1. Let $\mathbf{a} = -++-+-+++-$, $\mathbf{b} = -++++++--+$, where '+' and '−' mean 1 and −1, respectively. The AACFs of $\mathbf{a}$ and $\mathbf{b}$ are:

$$\rho_{\mathbf{a}}(k) = (10, -3, 0, -1, 0, 1, 2, -1, -2, 1),$$

$$\rho_{\mathbf{b}}(k) = (10, 3, 0, 1, 0, -1, -2, 1, 2, -1).$$

It is evident that the AACFs of $\mathbf{a}$ and $\mathbf{b}$ sum to a δ-function, as required by (2) and (3), so $(\mathbf{a}, \mathbf{b})$ is a GCP. The absolute squared values of the 20-point oversampled DFT of $\mathbf{a}$ and $\mathbf{b}$ are:

$$\mathbf{A} = 10 \cdot (0.40, 0.44, 0.15, 0.73, 1.85, 0.20, 1.05, 1.67, 0.95, 1.96, 1.60, 1.96, 0.95, 1.67, 1.05, 0.20, 1.85, 0.73, 0.15, 0.44)$$

$$\mathbf{B} = 10 \cdot (1.60, 1.56, 1.85, 1.27, 0.15, 1.80, 0.95, 0.33, 1.05, 0.04, 0.40, 0.04, 1.05, 0.33, 0.95, 1.80, 0.15, 1.27, 1.85, 1.56)$$

At every point these two power spectra add to 20, as required by (4).

It should be stressed that the bound of $\sqrt{2N}$ on the amplitude of $a(z)$ on $|z| = 1$ is extremely low for any bipolar sequence of length greater or equal to about 16. One would not find such sequences by chance, and the complementary sequence/aperiodic correlation approach is currently the only construction method known that tightly upper bounds these values for bipolar sequences. There is also the *Rudin-Shapiro* (RuS) construction [9], which appeared soon after Golay's initial work, but the RuS construction can be viewed as a basic recursive Golay construction technique. This construction is described in Section 2.2. Indeed, research on the uniformity of polynomials on the unit circle has continued largely independently in the mathematical community for many years [10–13], and this work indicates that sequences with good AACFs and flat DFT spectra, or equivalently, polynomials that are approximately uniform on $|z| = 1$, are rather difficult to construct. For example, the celebrated conjecture of Littlewood on *flat polynomials* on the unit circle is still open:

Conjecture 1 [12]. There exist a pair of constants C_0, C_1 and a series of degree $N-1$ polynomials $a(z)$ with ± 1 coefficients such that, as $N \to \infty$,

$$C_0\sqrt{N} \leq |a(z)| \leq C_1\sqrt{N}, \quad |z| = 1 \tag{6}$$

There is no known construction that produces polynomials satisfying the lower bound of Conjecture 1, and the complementary sequence approach is the only one known that gives polynomials satisfying the upper bound.

2. EXISTENCE AND CONSTRUCTION

2.1. Necessary Conditions

As we shall see in the next section, GCPs are known to exist for all lengths $N = 2^{\alpha}10^{\beta}26^{\gamma}$, $\alpha, \beta, \gamma \geq 0$, [14]. GCPs are not known for any other lengths. Golay showed that the length N of a Golay sequence must be the sum of two squares (where one square may be 0) [2]. More recently, it has been shown that GCPs of length N do not exist if there is a prime p with $p = 3 (\mathrm{mod}\, 4)$ such that $p \mid N$, [16]. This generalized earlier, weaker nonexistence results. Therefore, the admissible lengths <100 are

$$1, 2, 4, 8, 10, 16, 20, 26, 32, 34^*, 40, 50^*, 52, 58^*, 64, 68^*, 74^*, 80, 82^*$$

Moreover, various computer searches have eliminated lengths marked by "$*$" in the preceding list. Therefore, the lengths, $N < 100$, for which GCPs exist are:

$$1, 2, 4, 8, 10, 16, 20, 26, 32, 40, 52, 64, 80 \tag{7}$$

In the next sections, we provide constructions covering all these lengths.

2.2. Recursive Constructions

Using Eq. (3), many recursive constructions for GCPs can be obtained via simple algebraic manipulation. For example, if $a(z)$ and $b(z)$ are a Golay pair of length N, then simple algebraic manipulation shows that $a(z) + z^N b(z)$ and $a(z) - z^N b(z)$ also satisfy Eq. (3) with $2N$ being replaced by $4N$. This is in fact the well-known Golay–Rudin–Shapiro recursion, generating a length $2N$ GCP from a length N GCP [13]. We may write this more simply in terms of sequences as

$$(\mathbf{a}, \mathbf{b}) \to (\mathbf{a} \mid \mathbf{b}, \mathbf{a} \mid \overline{\mathbf{b}}) \tag{8}$$

where '|' means concatenation.

The following are a few other recursive constructions:

- The construction of Turyn [14] can be stated as follows. Let $(\mathbf{a}, \mathbf{b})$ and $(\mathbf{c}, \mathbf{d})$ be GCPs of length M and N, respectively. Then

$$\begin{aligned} &a(z^N)(c(z) + d(z))/2 + z^{N(M-1)} b(z^{-N})(c(z) - d(z))/2, \\ &b(z^N)(c(z) + d(z))/2 - z^{N(M-1)} a(z^{-N})(c(z) - d(z))/2 \end{aligned} \tag{9}$$

  is a GCP of length MN.
- The constructions of Golay in [2] are obtained as follows. Let $(\mathbf{a}, \tilde{\mathbf{b}})$ and $(\mathbf{c}, \mathbf{d})$ be GCPs of lengths M and N, respectively, where $\tilde{\mathbf{b}}$ means reversal of $\mathbf{b}$.

  Golay's *concatenation* construction can be stated as

$$a(z^N)c(z) + b(z^N)d(z)z^{MN}, \quad \tilde{b}(z^N)c(z) - \tilde{a}(z^N)d(z)z^{MN} \tag{10}$$

  is a GCP of length $2MN$.

  Golay's *interleaving* construction can be stated as

$$\begin{aligned} &a(z^{2N})c(z^2) + b(z^{2N})d(z^2)z, \\ &\tilde{b}(z^{2N})c(z^2) - \tilde{a}(z^{2N})d(z^2)z \end{aligned} \tag{11}$$

  is a GCP of length $2MN$.

Repeated application of Turyn's construction, beginning with pairs of lengths 2, 10, and 26 given in Section 2.4, can be used to construct GCPs for all lengths $N = 2^{\alpha}10^{\beta}26^{\gamma}$, $\alpha, \beta, \gamma \geq 0$.

2.3. Direct Constructions

In Ref. 2, Golay gave a direct construction for GCPs of length $N = 2^m$. Reference [4] gave a particularly compact description of this construction by using algebraic normal forms (ANFs). With a Boolean function $a(\mathbf{x}) = \mathbf{a}(\mathbf{x_0}, \mathbf{x_1}, \ldots, \mathbf{x_{m-1}})$ in m variables, we associate a length 2^m sequence $\mathbf{a} = (a_0, a_1, \ldots, a_{2^m-1})$, where

$$a_i = (-1)^{a(i_0, i_1, \ldots, i_{m-1})}, \quad i = \sum_{k=0}^{m-1} i_k 2^k$$

Thus the i-th term of the sequence $\mathbf{a}$ is obtained by evaluating the function a at the 2-adic decomposition of i. Then Ref. 4 showed that, for any permutation π of $\{0, 1, \ldots, m-1\}$, and any choice of constants $c_j, c, c' \in \mathbb{Z}_2$, the pair of functions

$$\begin{aligned} a(\mathbf{x}) &= \sum_{i=0}^{m-2} x_{\pi(i)} x_{\pi(i+1)} + \left(\sum_{j=0}^{m-1} c_j x_j \right) + c \\ b(\mathbf{x}) &= a(\mathbf{x}) + x_{\pi(0)} + c' \end{aligned} \tag{12}$$

yields a length 2^m GCP $(\mathbf{a}, \mathbf{b})$.

It is simple, given this representation, to show that this construction gives a set of $m!2^m$ distinct Golay sequences of length 2^m, each of which occurs in at least 4 GCPs. Perhaps more important, by expressing this set in the form Eq. (12), Ref. 4 identified a large set of GCS occurring as a subset of the binary Reed–Muller code RM$(2, m)$. Consequently, each sequence in the set has PMEPR at most 2, and the Hamming distance between any two sequences in the set is at least 2^{m-2}. This set therefore has a very attractive combination of PMEPR and error-correcting properties making it applicable in OFDM applications. For further details, see Ref. 4.

It was shown in Ref. 7 that the direct construction of Golay described above and Golay's recursive constructions Ref. 2 described in Section 2.2 in fact result in the same set of Golay sequences.

2.4. Symmetry and Primitivity

We consider simple symmetry operations that leave complementary properties and sequence length of a GCP invariant.

A length N GCP $(\mathbf{a}, \mathbf{b})$ remains a GCP under the operations [2]:

- Swapping the sequences in the pair.
- Negation (element multiplication by -1) of either or both sequences.
- Reversal of either or both sequences.
- The "linear offset" transformation $a_i \to (-1)^i a_i$, $b_i \to (-1)^i b_i$.

The action of these symmetries on a GCP generates a set of GCPs of size at most 64, which we call the *conjugates* of the original GCP. These symmetries can aid in computer searches for new GCPs.

We define a *primitive GCP* to be one that cannot be constructed from any shorter GCP by means of the recursive constructions described in Section 2.2. Primitive GCPs are only known to exist for lengths 2, 10, and 26. There is one primitive pair for lengths 2 and 26, and two primitive pairs for length 10, up to equivalence via the above symmetry operations. These primitive pairs are as follows:

(++, +−);

(−++−+−+++−, −++++++−−+);

(+−+−++++−−, ++++−++−−+);

(+−++−−+−−−−+−+−−−−++−−−+−+,

−+−−++−++++−−−−−−−++−−−+−+).

Golay points out that the two GCPs of length 10 are equivalent under decimation [2]. Specifically, the second pair of length 10 above is obtained from the first pair by taking successive 3rd sequence elements, cyclically. There is no proof that more primitive pairs cannot exist for $N > 100$, but none have been discovered for 40 years and it is conjectured that all GCPs arise from the four primitive pairs of lengths 2, 10, and 26, as given above.[1]

2.5. Enumeration

Two main types of enumeration are possible. One can enumerate the number of GCPs of a given length. Second, one can enumerate the number of Golay *sequences* of a given length. Since a Golay sequence is present in more than one GCP, the number of the former is greater than the number of the latter. As we have already seen in our brief discussion of OFDM, the enumeration of Golay sequences is of some practical importance. Table 1 provides a complete enumeration of GCPs for all possible lengths up to 100.

From Table 1 it is evident that the largest sets occur for lengths that have a large power of 2 as a factor. Here is a useful enumeration theorem:

Theorem 1 [2]. There are exactly $2^{m+2}m!$ GCPs of length 2^m that can be derived from the primitive pair $\{++, +-\}$ by repeated application of the symmetry operations and Golay's recursive constructions.

Next we consider the enumeration of Golay sequences. We have:

Theorem 2 [4,18]. Golay's direct construction produces exactly $m!2^m$ Golay sequences of length 2^m.

It was shown in Ref. 7 that the set of sequences in this theorem is identical to that which can be obtained from the primitive pair $(++, +-)$ by repeated application of Golay's recursive constructions. Theorem 2 accounts for all Golay sequences of lengths 2^m when $1 \le m \le 6$. It is not known if every Golay sequence of length 2^m must arise from Golay's direct construction when $m \ge 7$.

3. THE MERIT FACTOR OF COMPLEMENTARY SEQUENCES

The *merit factor* is a useful measure of the quality of sequences in certain applications where aperiodic correlations are important. It was introduced by Golay in Ref. 18. Let $\mathbf{a}$ be any length N sequence. Then the merit factor of $\mathbf{a}$ is defined to be

$$F(\mathbf{a}) = \frac{N^2}{2\sum_{k=1}^{N-1} |\rho_{\mathbf{a}}(k)|^2} \tag{13}$$

where $\rho_{\mathbf{a}}(k)$ is the AACF of $\mathbf{a}$.

The merit factor is, in fact, a spectral measure; it measures the mean-square deviation from the flat Fourier spectrum. Specifically,

$$1/F(\mathbf{a}) = \frac{1}{N^2}\int_0^1 (|a(e^{j2\pi t})|^2 - N)^2\, dt \tag{14}$$

where $a(z)$ is a polynomial whose values on $|z| = 1$ gives the Fourier transform of $\mathbf{a}$.

It is desirable to find sequences with high merit factor. A random sequence has merit factor around 1. It has been established that the asymptotic merit factor of a length 2^m Golay–Rudin–Shapiro (RuS) sequence is 3.0, which is high [19]. This is not the best possible; for example, shifted-Legendre sequences attain an asymptotic merit factor of 6.0 [20], and computer searches up to length 200 have revealed ± 1-sequences of merit factor around 8.5. There is also the celebrated Barker sequence of length 13, which has merit factor 14.08. However the length $N = 2^m$ RuS sequences $\mathbf{a}_m$ are notable as the quantities $\sigma_m = \sum_{k=0}^{2^m-1} |\rho_{\mathbf{a}_m}(k)|^2$ obey a simple *generalized Fibonacci recursion*, namely,

$$\sigma_m = 2\sigma_{m-1} + 8\sigma_{m-2} \tag{15}$$

with initial conditions $\sigma_1 = 1$, $\sigma_2 = 2$ [19].

This recursion immediately gives an asymptotic merit factor for the RuS sequences of 3.0 and is significant because it demonstrates the existence of a sequence family with large merit factor for which the merit factors do not need to be computed explicitly. This asymptotic value of 3.0 also holds for any Golay sequence obtained by applying symmetry operations to the RuS sequences [19]. Taking any other non-Golay pair as a starting seed always gives an asymptotic merit factor of K, for some constant, $K < 3.0$ [21].

4. LOW-COMPLEXITY CORRELATION

The pairwise property of GCPs has been exploited for channel measurement [1], but until 1990 or thereabouts the properties of individual sequences of a GCP were not wholly exploited [22], although Shapiro had stated Eq. (4) in his master's thesis of 1951 [9]. Budisin argues that Golay sequences are as good as, if not better than,

[1] However, a very recent paper by Borwein and Ferguson [17] also regards a length 20 pair as primitive, specifically: $\{+ + + + - + - - - + + - - + + - + - - +,\ + + + + - + + + + + - - - + - + - + + -\}$.

Table 1. The Number of Golay Complementary Pairs for All Lengths, $N < 100$

N	1	2	4	8	10	16	20	26	32	40	52	64	80
#GCPs [17]	4	8	32	192	128	1536	1088	64	15360	9728	512	184320	102912

m-sequences for application as pseudo-noise sequences due to their superior aperiodic spectral properties [22]. Also, there are more Golay sequences than m-sequences. Figure 1 lends some support for this view, where the Fourier spectra (from left to right) of a length 127 m-sequence and a length 127 shifted-Legendre sequence are compared with that of a length 128 RuS sequence.

Budisin [22] proposed a highly efficient method to perform correlation of an incoming data stream with a Golay sequence of length N, which achieves a complexity of $2\log_2(N)$ operations per sample, as opposed to N operations per sample for direct correlation. To do this, he interpreted the Golay construction for length $N = 2^m$ sequences using delay to implement concatenation, i.e., $\mathbf{a} \mid \mathbf{b}$ can be implemented as $\mathbf{a}[k] + \mathbf{b}[k + D]$, where $[k]$ indicates the starting time of $\mathbf{a}$ and D is the length (duration) of $\mathbf{a}$. Implementing the recursion of (8) is then achieved by serially combining delay elements, D_i, of duration 2^i, as shown in Fig. 2. Now we commence our Rudin–Shapiro recursion with $\mathbf{a} = \mathbf{b} = 1$. In other words, we input the δ function to the left-hand side of Fig. 2, and output our GCP, $(\mathbf{a}', \mathbf{b}')$, on the right. So the pair of Golay sequences realized by Fig. 2 are two impulse responses. We can therefore reinterpret Fig. 2 as a filter that correlates a received sequence, input from the left, with the reversals of $\mathbf{a}$ and $\mathbf{b}$. By choosing the ω_i in Fig. 2 from $\{1, -1\}$, we can choose to correlate with different length 2^m GCPs.

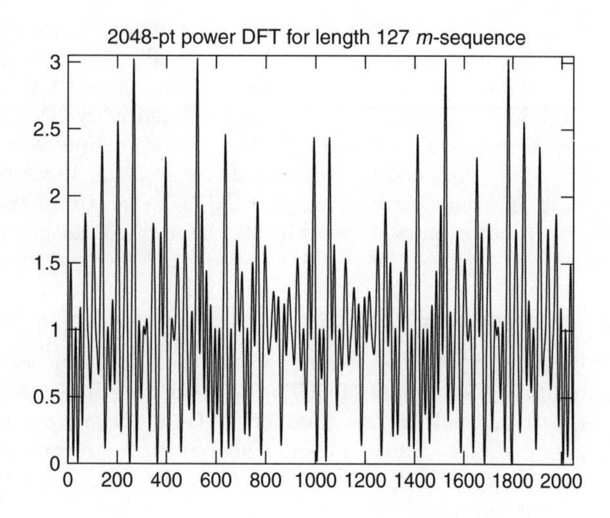

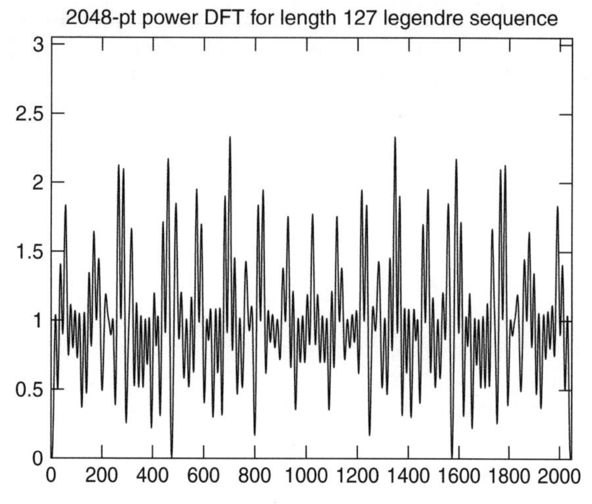

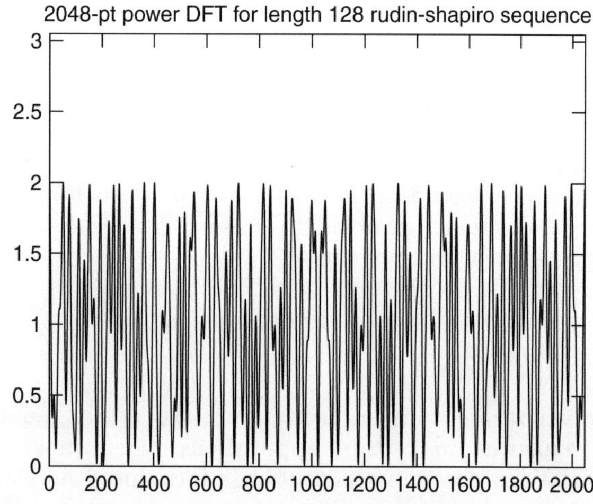

Figure 1. Power spectra for length 127 m-sequence, length 127 shifted-Legendre, and length 128 Rudin-Shapiro sequences, (power on y-axis, spectral index on x-axis).

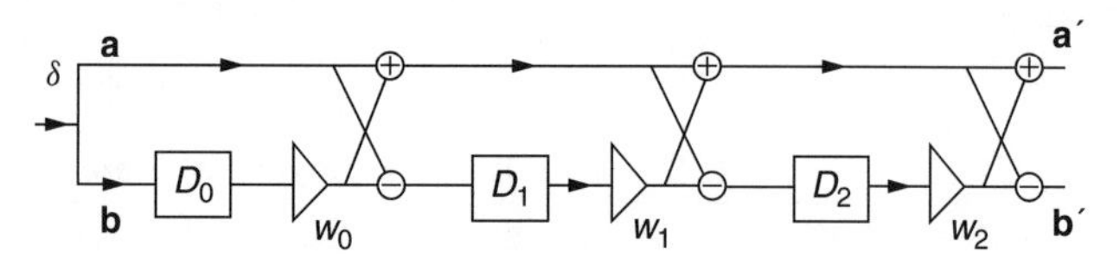

Figure 2. Fast Golay correlator.

5. COMPLEMENTARY SETS AND ORTHOGONAL MATRICES

5.1. Complementarity with Respect to a Set of Sequences

Golay complementary pairs can be generalized to sets containing two or more sequences [23]. Analogously to Eq. (2), we say that a set of T bipolar sequences of length N $(\mathbf{a}_1, \mathbf{a}_2, \ldots, \mathbf{a}_T)$ form a complementary sequence set of size T (a T-CSS) if

$$\sum_{i=1}^{T} \rho_{\mathbf{a}_i}(k) = 0 \quad k \neq 0 \tag{16}$$

In terms of polynomials, this is equivalent to

$$\sum_{i=1}^{T} |a_i(z)|^2 = TN, \quad |z| = 1 \tag{17}$$

Thus, all the DFT components of a sequence $\mathbf{a}$ that lies in a T-CSS are of size at most $\sqrt{TN}$. Of course, a 2-CSS is just a Golay complementary pair.

To date, little work has been done to formally establish primitivity conditions for T-CSS, $T > 2$, although Golay already found 4-CSS in Ref. 1. It can be shown that CSS only occur for T even, and Turyn showed that 4-CSS are only admissible at lengths N if N is a sum of at most *three* squares [14]. Dokovic later showed that 4-CSS exist for all even $N < 66$ [24]. Tseng and Liu [23] showed that for a

Table 2. Number of Possibly Primitive Quadriphase Golay Pairs

Length	2	3	4	5	6	7	8	9	10	11	12	13
#Inequivalent Pairs [29]	—	1	—	1	2	—	4	—	14	1	32	1

CSS of odd length N, T must be a multiple of 4. By way of example, here are 4-CSS of lengths 3, 5, 7:

$\{+++,\ -++,\ +-+,\ ++-\}$

$\{+----,\ -++-+,\ +---+,\ ---+-\}$

$\{+++-+++,\ +-+++--,\ +--+-++,\ ++-+---\}$

Thus, 4-CSS can exist at lengths N where 2-CSS cannot. Turyn presented constructions for 4-CSS for all odd lengths $N \leq 33$, and $N = 59$ [14].

An orthogonal matrix is defined as a matrix whose columns are pairwise orthogonal. The following theorem is straightforward.

Theorem 3 [23]. Let $\mathbf{P}$ be a $T \times N$ orthogonal matrix, $N \leq T$. Then the rows of P form a T-CSS of length N.

The primitive GCP $(++, +-)$, is an example of Theorem 3. When the elements of $\mathbf{P}$ are ± 1 and $N = T$, then $\mathbf{P}$ is a Hadamard matrix, so a subset of T-CSS is provided by the set of Hadamard matrices.

5.2. Symmetries and Constructions

The symmetries and constructions for GCPs given in Sections 2.2 and 2.3 generalize to give constructions for T-CSS. One can also construct T'-CSS by combining T-CSS, $T' > T$. As one example, we have:

- Let $(\mathbf{u}, \mathbf{v})$ and $(\mathbf{x}, \mathbf{y})$ be GCPs of length N_0 and N_1, respectively. Then, $(\mathbf{a}, \mathbf{b}, \mathbf{c}, \mathbf{d})$ is a 4-CSS of length $N_0 + N_1$, where

$$\mathbf{a} = (\mathbf{u} \mid \mathbf{x}),\ \mathbf{b} = (\mathbf{u} \mid -\mathbf{x}),\ \mathbf{c} = (\mathbf{v} \mid \mathbf{y}),\ \mathbf{d} = (\mathbf{v} \mid -\mathbf{y})$$

A fundamental recursive construction generalizing Golay's recursive constructions and relating CSS to orthogonal matrices is given in Ref. 23.

Theorem 4 [23]. Let $(\mathbf{a_0}, \mathbf{a_1}, \ldots, \mathbf{a_{T-1}})$ be a T-CSS of length N, represented by a $T \times N$ matrix, $\{\mathbf{F}\}$, with rows $\mathbf{a_j}$. Let $\mathbf{O} = (o_{ik})$ be an $S \times T$ orthogonal matrix (so $S \geq T$). Define

$$\mathbf{F}' = \mathbf{F} \odot \mathbf{O} = \begin{pmatrix} o_{00}\mathbf{a_0} & o_{01}\mathbf{a_1} & \cdots & o_{0(T-1)}\mathbf{a_{T-1}} \\ o_{10}\mathbf{a_0} & o_{11}\mathbf{a_1} & \cdots & o_{1(T-1)}\mathbf{a}_{T-1} \\ \cdots & \cdots & \cdots & \cdots \\ o_{(S-1)0}\mathbf{a_0} & o_{(S-1)1}\mathbf{a_1} & \cdots & o_{(S-1)(T-1)}\mathbf{a_{T-1}} \end{pmatrix} \tag{18}$$

Then $\mathbf{F}'$ is an $S \times TN$ matrix whose rows form an S-CCS of length TN.

Taking $(\mathbf{a}_0, \mathbf{a}_1)$ to be a GCP and $O = \begin{pmatrix} 1 & 1 \\ 1 & -1 \end{pmatrix}$, we recover Golay's concatenation construction. The basic symmetry operations can be interpreted as row/column permutations of $\mathbf{O}$ and as point-multiplication of rows of $\mathbf{O}$ by a constant vector (these operations maintain the orthogonality of $\mathbf{O}$).

The following theorem combines T-CSS of different lengths to build T'-CSS, where $T' > T$.

Theorem 5 [25]. Suppose there exist T_0-CSS, T_1-CSS, $\ldots$, T_{t-1}-CSS, of lengths $N_0, N_1, \ldots, N_{t-1}$, respectively. Let $T = \text{lcm}(T_0, T_1, \ldots, T_{t-1})$. Suppose there also exists an $S \times T$ orthogonal matrix with ± 1 entries. Then there exists an ST-CSS of length $N' = N_0 + N_1 + \ldots + N_{t-1}$.

As with GCPs, CSS also have applications to OFDM: a primary drawback with the proposal to use the set of length 2^m GCPs as a codeset for OFDM is that the code rate of the set rapidly decreases as m increases. To obtain a larger codeset, one can consider sequences that lie in T-CSS for some $T > 2$. The resulting codeset will have PMEPR at most T.

5.3. Complementarity with Respect to a Larger Set of Transforms

Although CS pairs and, more generally, CSS, are usually defined to be complementary with respect to their AACFs (with a corresponding property on power spectra under the one-dimensional DFT), one can more generally define and discover sets that are complementary with respect to *any* specified transform. It can be shown, Ref. 26, that Golay CSS of length 2^m, as constructed using Theorem 5, have a very strong property:

Theorem 6. Let $\mathbf{U}$ be a 2×2 complex-valued matrix such that $\mathbf{U}\mathbf{U}^{\dagger} = 2\mathbf{I}$, where $\mathbf{I}$ is the 2×2 identity matrix, '$\dagger$' means transpose-conjugate, and the elements of $\mathbf{U}$, u_{ij}, satisfy $|u_{00}| = |u_{01}| = |u_{10}| = |u_{11}|$. Let $\{\mathbf{U}_k, 0 \leq k < m\}$ be a set of any m of these matrices. Define $\mathbf{M} = \mathbf{U}_0 \otimes \mathbf{U}_1 \otimes \cdots \otimes \mathbf{U}_{m-1}$. Let $N = 2^m$ and let $(\mathbf{a}_0, \mathbf{a}_1, \ldots, \mathbf{a}_{T-1})$ be any T-CSS of length N constructed by Theorem 5. Finally, let $\mathbf{A}_i^M = \mathbf{M}\mathbf{a_i}$ be the N-point spectrum of $\mathbf{a}$ with respect to $\mathbf{Ma}$ with elements, $A_{k,i}^M$, $0 \leq k < N$. Then:

$$\sum_{i=0}^{T-1} |A_{k,i}^M|^2 = TN \tag{19}$$

Theorem 6 implies that

$$|A_{k,i}^M|^2 \leq TN \quad \forall k, i, T, \mathbf{M} \tag{20}$$

The combined set of all rows of all possible transform matrices, $\mathbf{M}$, includes the one-dimensional DFT and the *Walsh–Hadamard Transform* (WHT), along with infinitely many other transforms.

As an example of the application of this result, recall that cryptographers typically perform linear cryptanalysis of cipher components by looking for peaks in the WHT spectrum, (see, e.g., Ref. 27). It is known that for m even, length 2^m GCPs are bent, that is, have a completely flat WHT spectrum [28]. Reference [26] shows that Golay constructions generate a large set of GCPs and CSS that also have a relatively flat spectrum with respect to the WHT, among other transforms. So Golay CSS may have applications to cryptography.

5.4. CSS Mates

Let $\mathbf{A} = (\mathbf{a}_0, \mathbf{a}_1, \ldots, \mathbf{a}_{T-1})$ and $\mathbf{B} = (\mathbf{b}_0, \mathbf{b}_1, \ldots, \mathbf{b}_{T-1})$ be two T-CSS. Then $\mathbf{A}$ and $\mathbf{B}$ are called "mates" if $\mathbf{a}_i$ and $\mathbf{b}_i$ are orthogonal as vectors for each i. We say that $\mathbf{A}$ and $\mathbf{B}$ are "mutually orthogonal CSS" (although, in general, $\mathbf{a}_i$ is not orthogonal to $\mathbf{b}_j$, $i \neq j$). Sets $(\mathbf{A}_0, \mathbf{A}_1, \ldots, \mathbf{A}_{U-1})$ of pairwise mutually orthogonal CSS can be recursively constructed in a similar way to CSS [23].

6. COMPLEMENTARY SEQUENCES OVER LARGER ALPHABETS

Virtually all the symmetries and constructions mentioned so far for bipolar sequences can be generalized to sequences over other alphabets, but note that autocorrelation is now modified to include conjugacy, i.e., Eq. (1) is modified to,

$$\rho_{\mathbf{a}}(k) = \sum_{i=0}^{N-k-1} a_i a_{i+k}^*, \quad 0 \le k \le N-1 \tag{21}$$

where * means "complex conjugate."

For quadriphase CSS, the symmetry operations generate an equivalence class of up to 1,024 sequences [29]. For polyphase pairs, unlike the GCP case, there is no restriction that the length must be the sum of two squares. Sivaswamy and Frank investigated and discovered many polyphase CSS, including those of odd length [30,31]. The simplest polyphase T-CSS of length T is formed from the rows of the T-point DFT matrix. In fact, from Theorem 3, the rows of any $T \times T$ orthogonal polyphase matrix form a T-CSS of length T. Sivaswamy [30] identified a length 3 quadriphase primitive pair, (002, 010) (where $0, 1, 2, 3$ mean i^0, i^1, i^2, i^3, respectively), derived quadriphase versions of GCPs and synthesized sequence pairs of lengths 3.2^k. Frank [31] further presented the following primitive quadriphase Golay pairs, of lengths 5 and 13: (01321, 00013), and (0001200302031, 0122212003203). Note that the lengths here, 5 and 13, are half the length of the lengths 10 and 26 primitive bipolar GCPs, but no transform is known between the sets.

Davis and Jedwab [4] and Paterson [7] constructed many CSS with phase alphabet 2^h and any even phase alphabet, respectively. To do so, they worked with nonbinary generalizations of the Reed–Muller codes. The resulting sequences have application to OFDM with nonbinary modulation.

Many polyphase 3-CSS exist. For example, Frank [31] presented the triphase 3-CSS (01110,11210,00201) and provided a (possibly nonexhaustive) list of lengths, N, for which a CSS exists, N up to 100: Polyphase 3-CSS: 1–22,24–27,30,32,33,36,37,39–42,45,48,49,51–54,57,58, 60,61,63–66,72,73,75,78,80,81,90,96,97,100 Polyphase 4-CSS: All lengths except 71,89 $T > 4$: All lengths.

A recent exhaustive search [29] found *all* quadriphase Golay pairs up to length $N = 13$. These are summarized below where only those for which no construction is known have been counted. These are the *possible* primitive pairs. The figures also omit the GCPs that are a subset of quadriphase pairs:

Golay pairs over the alphabet $\{0, 1, -1\}$ have been found for all lengths, N. For such a set, the weight of the set becomes an important extra parameter. The weight W is the sum of the in-phase AACF coefficients, i.e., for a set $(\mathbf{a}_j)$, we have $W = \sum_j \rho_0(\mathbf{a}_j)$. For example, here is a 4-CSS of weight 7 and length 7 over the alphabet $\{0, 1, -1\}$:

$$\{+000+00,\ 0+0+0-0,\ 00+0000,\ 000000+\}$$

The larger W, the closer is the CSS to one over a bipolar alphabet.

Yet more CSS can be found by considering multilevel and QAM alphabets.

7. HADAMARD MATRICES FROM COMPLEMENTARY SEQUENCES

In Section 5, we showed that Hadamard matrices can be used to construct CSS. The converse is also true: CSS can be used to construct Hadamard matrices. Here, we present the two best-known constructions, where Theorems 7 and 8 use the *periodic* complementary property of a complementary set (see Section 8).

Theorem 7 [32]. Let $(\mathbf{a}, \mathbf{b})$ be a GCP of length N. Let $\mathbf{A}$ and $\mathbf{B}$ be $N \times N$ circulant matrices with first rows $\mathbf{a}$ and $\mathbf{b}$, respectively. Then,

$$\begin{pmatrix} \mathbf{A} & -\mathbf{B} \\ \mathbf{B}^{\mathrm{T}} & \mathbf{A}^{\mathrm{T}} \end{pmatrix} \text{ is a } 2N \times 2N \text{ Hadamard matrix}$$

Theorem 7 can be generalized to quadriphase Hadamard matrices by making $(\mathbf{a}, \mathbf{b})$ a quadriphase Golay pair and by substituting conjugation for transpose.

Theorem 8 [14,15]. Let $(\mathbf{u}, \mathbf{v})$ and $(\mathbf{x}, \mathbf{y})$ be GCPs of lengths N_0 and N_1, respectively. Then, $\mathbf{a} = \mathbf{u} \mid \mathbf{x}$, $\mathbf{b} = \mathbf{u} \mid -\mathbf{x}$, $\mathbf{c} = \mathbf{v} \mid \mathbf{y}$, and $\mathbf{d} = \mathbf{v} \mid -\mathbf{y}$ form a length $N = N_0 + N_1$4-CSS. Let $\mathbf{A}, \mathbf{B}, \mathbf{C}, \mathbf{D}$ be $N \times N$ circulant matrices with first rows $\mathbf{a}, \mathbf{b}, \mathbf{c}, \mathbf{d}$, respectively. Let $\mathbf{R}$ be a back-circulant $N \times N$ permutation matrix. Then,

$$\begin{pmatrix} \mathbf{A} & -\mathbf{BR} & -\mathbf{CR} & -\mathbf{DR} \\ \mathbf{BR} & \mathbf{A} & -\mathbf{D}^{\mathrm{T}}\mathbf{R} & \mathbf{C}^{\mathrm{T}}\mathbf{R} \\ \mathbf{CR} & \mathbf{D}^{\mathrm{T}}\mathbf{R} & \mathbf{A} & -\mathbf{B}^{\mathrm{T}}\mathbf{R} \\ \mathbf{DR} & -\mathbf{C}^{\mathrm{T}}\mathbf{R} & \mathbf{B}^{\mathrm{T}}\mathbf{R} & \mathbf{A} \end{pmatrix} \text{ is a } 4N \times 4N \text{ Goethals–Seidel (Hadamard) matrix.}$$

8. PERIODIC AND ODD-PERIODIC (NEGAPERIODIC) COMPLEMENTARY SEQUENCES

Researchers have recently become interested in constructing *periodic* and/or *odd-periodic* CSS. Periodic CSS were considered in Ref. 33.

Using polynomial form, periodic autocorrelation (PACF) of the length N sequence **a** can be expressed as

$$\text{PACF}(a(x)) = \langle a(x)a(x^{-1})\rangle_{x^{N}-1}$$

where '$\langle * \rangle_M$' reduces $*$ mod M.

Similarly, negaperiodic (odd-periodic) autocorrelation (NACF) can be expressed as

$$\text{NACF}(a(x)) = \langle a(x)a(x^{-1})\rangle_{x^{N}+1}$$

There is a simple relationship relating aperiodic, periodic, and odd-periodic AACF, as follows:

Let $a(x)$ represent a length N sequence. Then, the AACF of **a** can be computed, via the Chinese remainder theorem, as

$$a(x)a(x^{-1}) = \frac{x^N+1}{2}\langle a(x)a(x^{-1})\rangle_{x^{N}-1} - \frac{x^N-1}{2}\langle a(x)a(x^{-1})\rangle_{x^{N}+1} \quad (22)$$

Equation (22) expresses AACF in terms of PACF and NACF. It follows that

- A T-CSS is also a periodic and a negaperiodic T-CSS.
- A set of length T sequences is only a T-CSS if it is both a periodic and negaperiodic T-CSS.

Periodic and negaperiodic CSS can be used instead of, say, m-sequences, for their desirable correlation and spectral properties. They are much easier to find than aperiodic CSS due to their algebraic structure via embedding in a finite polynomial ring. Moreover, in a search for aperiodic CSS, an initial sieve can be undertaken by first identifying periodic and negaperiodic CSS and then looking for the intersection of the two sets.

It is known that periodic GCP do not exist for lengths 36 and 18, respectively.

We have the following theorem.

Theorem 9 [34]. If a length N periodic GCP exists, such that $N = p^{2l}u, p \neq u, p$ prime, $p = 3(\text{mod}\,4)$, then $u \geq 2p^l$.

Dokovic [24] discovered a periodic GCP of length 34. This is significant because no aperiodic GCP exists at that length. The next unresolved case for periodic GCPs is at length 50. References 33 and 24 also present many T-CSS for $T > 2$.

Lüke [35] has found many odd-periodic GCPs for even lengths N where an aperiodic GCP cannot exist, e.g., $N = \frac{p^{u+1}}{2}$, p an odd prime, and also for all even $N, N < 50$. He did not find any odd-length pairs.

BIOGRAPHIES

Matthew G. Parker received a B.Sc. in electrical and electronic engineering in 1982 from University of Manchester Institute of Science and Technology, U.K. and, in 1995, a Ph.D. in residue and polynomial residue number systems from University of Huddersfield, U.K. From 1995 to 1998 he was a postdoctoral researcher in the Telecommunications Research Group at the University of Bradford, U.K., researching into coding for peak factor reduction in OFDM systems. Since 1998 he has been working as a postdoctoral researcher with the Coding and Cryptography Group at the University of Bergen, Norway. He has published on residue number systems, number-theoretic transforms, complementary sequences, sequence design, quantum entanglement, coding for peak power reduction, factor graphs, linear cryptanalysis, and VLSI implementation of Modular arithmetic.

Kenneth G. Paterson received a B.Sc. (Hons) degree in mathematics in 1990 from the University of Glasgow and a Ph.D. degree, also in mathematics, from the University of London in 1993. He was a Royal Society Fellow at Institute for Signal and Information Processing at the Swiss Federal Institute of Technology, Zurich, from 1993 to 1994, investigating algebraic properties of block chipers. He was then a Lloyd's of London Tercentenary Foundation Fellow at Royal Holloway, University of London from 1994 to 1996, working on digital signatures. He joined the mathematics group at Hewlett-Packard Laboratories Bristol in 1996, becoming project manager of the Mathematics, Cryptography and Security Group in 1999. While at Hewlett-Packard, he worked on a wide variety of pure and applied problems in cryptography and communications theory. In 2001, he joined the Information Security Group at Royal Holloway, University of London, becoming a Reader in 2002. His areas of research are sequences, the mathematics of communications, and cryptography and cryptographic protocols.

C. Tellambura received his B.Sc. degree with honors from the University of Moratuwa, Sri Lanka, in 1986; his M.Sc. in electronics from the King's College, U.K., in 1988; and his Ph.D. in electrical engineering from the University of Victoria, Canada, in 1993. He was a postdoctoral research fellow with the University of Victoria and the University of Bradford. Currently, he is a senior lecturer at Monash University, Australia. He is an editor for the *IEEE Transactions on Communications* and the *IEEE Journal on Selected Areas in Communications* (Wireless Communications Series). His research interests include coding, communications theory, modulation, equalization, and wireless communications.

BIBLIOGRAPHY

1. M. J. E. Golay, Multislit spectroscopy, *J. Opt. Soc. Amer.* **39**: 437–444 (1949).
2. M. J. E. Golay, Complementary series, *IRE Trans. Inform. Theory* **IT-7**: 82–87 (1961).
3. M. Nazarathy et al., Jr., Real-time long range complementary correlation optical time domain reflectometer, *IEEE J. Lightwave Technol.* **7**: 24–38 (1989).
4. J. A. Davis and J. Jedwab, Peak-to-mean power control in OFDM, Golay complementary sequences and Reed-Muller codes, *IEEE Trans. Inform. Theory* **IT-45**: 2397–2417 (1999).

5. R. D. J. van Nee, OFDM codes for peak-to-average power reduction and error correction, in *IEEE Globecomm 1996* (London, U.K., Nov. 1996), pp. 740–744.
6. H. Ochiai and H. Imai, Block coding scheme based on complementary sequences for multicarrier signals, *IEICE Trans. Fundamentals* **E80-A**: 2136–2143, (1997).
7. K. G. Paterson, Generalized Reed-Muller codes and power control in OFDM modulation, *IEEE Trans. Inform. Theory* **IT-46**: 104–120 (2000).
8. P. Spasojevic and C. N. Georghiades, Complementary sequences for ISI channel estimation, *IEEE Trans. Inform. Theory* **IT-47**: 1145–1152 (2001).
9. H. S. Shapiro, Extremal Problems for Polynomials, M.S. Thesis, M.I.T., 1951.
10. J. Beck, Flat polynomials on the unit circle — note on a problem of Littlewood, *Bull. London Math. Soc.* **23**: 269–277 (1991).
11. J. Kahane, Sur les polynomes à coefficients unimodulaires, *Bull. London Math. Soc.* **12**: 321–342 (1980).
12. J. E. Littlewood, On polynomials $\sum \pm z^m$, $\sum \exp(\alpha_m) z^m$, $z = e^{i\theta}$, *J. London Math. Soc.* **41**: 367–376 (1966).
13. W. Rudin, Some theorems on Fourier coefficients, *Proc. Amer. Math. Soc.* **10**: 855–859 (1959).
14. R. Turyn, Hadamard matrices, Baumert-Hall units, four-symbol sequences, pulse compression, and surface wave encodings, *J. Comb. Theory Ser. A* **16**: 313–333 (1974).
15. C. H. Yang, Hadamard matrices, finite sequences, and polynomials defined on the unit circle, *Math. Comp.* **33**: 688–693 (1979).
16. S. Eliahou, M. Kervaire, and B. Saffari, A new restriction on the lengths of Golay complementary sequences, *J. Comb. Theory Ser. A* **55**: 49–59 (1990).
17. P. Borwein and R. A. Ferguson, A complete description of Golay pairs for lengths up to 100, *in preparation*, preprint available [Online]. Simon Fraser University, *http://www.cecm.sfu.ca/~pborwein/* [May, 2002].
18. M. J. E. Golay, Sieves for low autocorrelation binary sequences, *IEEE Trans. Inform. Theory* **IT-23**: 43–51 (1977).
19. T. Høholdt, H. E. Jensen, and J. Justesen, Aperiodic correlations and merit factor of a class of binary sequences, *IEEE Trans. Inform. Theory* **IT-31**: 549–552 (1985).
20. T. Høholdt and H. E. Jensen, Determination of the merit factor of Legendre sequences, *IEEE Trans. Inform. Theory* **IT-34**: 161–164 (1988).
21. P. Borwein and M. Mossinghoff, Rudin-Shapiro like polynomials in L_4, *Math. Comp.* **69**: 1157–1166 (2000).
22. S. Z. Budisin, Efficient pulse compressor for Golay complementary sequences, *Elec. Lett.* **27**: 219–220 (1991).
23. C.-C. Tseng and C. L. Liu, Complementary sets of sequences, *IEEE Trans. Inform. Theory* **IT-18**: 644–651 (1972).
24. D. Z. Dokovic, Note on periodic complementary sets of binary sequences, *Designs, Codes and Cryptography* **13**: 251–256 (1998).
25. K. Feng, P. J.-S. Shiue, and Q. Xiang, On aperiodic and periodic complementary binary sequences, *IEEE Trans. Inf. Theory* **45**: 296–303 (1999).
26. M. G. Parker and C. Tellambura, A construction for binary sequence sets with low peak-to-average power ratio, Int. Symp. Inform. Theory, Lausanne, Switzerland, June 30–July 5, 2002.
27. M. Matsui, Linear cryptanalysis method for DES, in *Advances in Cryptology — Eurocrypt93*, Lecture Notes in Computer Science, Vol. 765, pp. 386–397, Springer, 1993.
28. F. J. MacWilliams and N. J. A. Sloane, *The Theory of Error-Correcting Codes*, Amsterdam, North-Holland, 1977.
29. W. H. Holzmann and H. Kharaghani, A computer search for complex Golay sequences, *Aust. J. Comb.* **10**: 251–258 (1994).
30. R. Sivaswamy, Multiphase complementary codes, *IEEE Trans. Inform. Theory* **IT-24**: 546–552 (1978).
31. R. L. Frank, Polyphase complementary codes, *IEEE Trans. Inform. Theory* **IT-26**: 641–647 (1980).
32. C. H. Yang, On Hadamard matrices constructible by circulant submatrices, *Math. Comp.* **25**: 181–186 (1971).
33. L. Bomer and M. Antweiler, Periodic complementary binary sequences, *IEEE Trans. Inform. Theory* **IT-36**: 1487–1494 (1990).
34. K. T. Arasu and Q. Xiang, On the existence of periodic complementary binary sequences, *Designs, Codes and Cryptography* **2**: 257–262 (1992).
35. H. D. Lüke, Binary odd-periodic complementary sequences, *IEEE Trans. Inform. Theory* **IT-43**: 365–367 (1997).

GOLD SEQUENCES

HABONG CHUNG
Hongik University
Seoul, Korea

1. INTRODUCTION

In such applications as CDMA communications, ranging, and synchronization, sequences having good correlation properties have played an important role in signal designs. When a single sequence is used for the purpose of synchronization, it must possess a good autocorrelation property so that it can be easily distinguished from its time-delayed versions. Similarly, when a set of sequences are used for CDMA communications systems or multitarget ranging systems, the sequences should exhibit good cross-correlation properties so that each sequence is easy to distinguish from every other sequence in the set. Many individual sequences as well as families of sequences with desirable correlation properties have been found and reported. The Gold sequence family [1] is one of the oldest and best-known families of binary sequences with optimal correlation properties. This article focuses on the binary Gold sequences and their relatives. After briefly reviewing basic definitions, some of the well-known bounds on the correlation magnitude, and cross-correlation properties of m sequences in Sections 2 and 3, we discuss the Gold sequences and Gold-like sequences in Section 4.

A longer and more detailed overview on the subject of the well-correlated sequences in general can be found in Chapter 21 of the book by Pless and Huffman [11]. The reader can also find the article by Sarwate and Pursley [3] to be an excellent survey.

2. PRELIMINARIES

Given two complex-valued sequences $a(t)$ and $b(t)$, $t = 0, 1, \ldots, N-1$, of length N, the (periodic) cross-correlation function $R_{a,b}(\tau)$ of the sequences $a(t)$ and $b(t)$ is defined as follows

$$R_{a,b}(\tau) = \sum_{t=0}^{N-1} a(t+\tau)b^*(t) \tag{1}$$

where the asterisk $(*)$ denotes complex conjugation and the sum $t+\tau$ is computed modulo n. When $a(t) = b(t)$, $R_{a,a}(\tau)$ is called the *autocorrelation function* of the sequence $a(t)$, and will be denoted by $R_a(\tau)$.

Here, for the sake of simplicity and for practical reasons, we will restrict our discussion mainly on sequences whose symbols are qth roots of unity for some integer q. In this situation, it is manifest that $R_a(0) = N$ for a sequence $a(t)$ of length N. In applications such as ranging systems, radar systems, and CDMA systems, one may need a set of sequences such that the magnitude of the cross-correlation function between any two sequences in the set as well as that of the out-of-phase autocorrelation function of each sequence in the set must be small compared to N. More precisely, let $\mathcal{S}$ be a set of M cyclically distinct sequences of length N whose symbols are qth roots of unity given by

$$\mathcal{S} = \{s_0(t), s_1(t), \ldots, s_{M-1}(t)\}$$

Then for the given set $\mathcal{S}$, we can define the peak out-of-phase autocorrelation magnitude θ_a and the peak cross-correlation magnitude θ_c as follows

$$\theta_a = \max_i \max_{1\le l\le N-1} |R_{s_i}(l)|$$

and

$$\theta_c = \max_{i\ne j} \max_{0\le l\le N-1} |R_{s_i,s_j}(l)|$$

The maximum of θ_a and θ_c is called the *maximum correlation parameter* $\theta_{\max}$ of the set $\mathcal{S}$:

$$\theta_{\max} = \max\{\theta_a, \theta_c\}$$

Conventionally, by the term "set of sequences with optimal correlation property," we imply the set $\mathcal{S}$, $\theta_{\max}$ of which is the smallest possible for a given length N and the set size M. Certainly, $\theta_{\max}$ of a given set must be the function of sequence length N and the set size M. For example, Sarwate [2] shows that

$$\left(\frac{\theta_c^2}{N}\right) + \frac{N-1}{N(M-1)}\left(\frac{\theta_a^2}{N}\right) \ge 1 \tag{2}$$

From Eq. (2), one can obtain a simple lower bound on $\theta_{\max}$ as

$$\theta_{\max} \ge N\sqrt{\frac{M-1}{NM-1}} \tag{3}$$

Other than the bound in (3), various lower bounds on $\theta_{\max}$ in terms of N and M are known. The following bound is due to Welch [8].

Theorem 1: Welch Bound. Given a set of M complex-valued sequences of length N whose in-phase autocorrelation magnitude is N, and for an integer $k(\ge 1)$, we obtain

$$(\theta_{\max})^{2k} \ge \frac{1}{(MN-1)}\left\{\frac{MN^{2k+1}}{\binom{k+N-1}{N-1}} - N^{2k}\right\} \tag{4}$$

Especially when the sequence symbols are complex qth roots of unity, the Sidelnikov [9] bound given as the following theorem is known.

Theorem 2: Sidelnikov Bound. In the case $q = 2$, then

$$\theta_{\max}^2 > (2k+1)(N-k) + \frac{k(k+1)}{2} - \frac{2^k N^{2k+1}}{M(2k)!\binom{N}{k}},$$
$$0 \le k < \frac{2}{5}N \tag{5}$$

In the case $q > 2$, then

$$\theta_{\max}^2 > \left(\frac{k+1}{2}\right)(2N-k) - \frac{2^k N^{2k+1}}{M(k!)^2\binom{2N}{k}}, \quad k \ge 0 \tag{6}$$

3. CROSS-CORRELATION OF *m* SEQUENCES

A maximal-length linear feedback shift register sequence (m sequence) may be the best-known sequence with an ideal autocorrelation property. Like other finite field sequences, m sequences are best described in terms of the trace function over a finite field.

Let F_{q^n} be the finite field with q^n elements. Then the trace function from F_{q^n} to F_{q^m} is defined as

$$\mathrm{Tr}_m^n(x) = \sum_{i=0}^{n/m-1} x^{q^{m\cdot i}}$$

where $x \in F_{q^n}$ and $m \mid n$. The trace function satisfies the following:

1. $\mathrm{Tr}_m^n(ax+by) = a\mathrm{Tr}_m^n(x) + b\mathrm{Tr}_m^n(y)$, for all $a, b \in F_{q^m}$, $x, y \in F_{q^n}$.
2. $\mathrm{Tr}_m^n(x^{q^m}) = \mathrm{Tr}_m^n(x)$, for all $x \in F_{q^n}$.
3. Let k be an integer such that $m|k|n$. Then

$$\mathrm{Tr}_m^n(x) = \mathrm{Tr}_m^k(\mathrm{Tr}_k^n(x)) \qquad \text{forall} \quad x \in F_{q^n}$$

A q-ary m sequence $s(t)$ of length $q^n - 1$ can be expressed as

$$s(t) = \mathrm{Tr}_1^n(a\alpha^t) \tag{7}$$

where a is some nonzero element in F_{q^n} and α is a primitive element of F_{q^n}. Note that the m sequence in (7) is not complex-valued. Its symbols are the elements of the finite field F_q. When q is a prime, the natural way of converting this finite-field sequence $s(t)$ into a complex-valued sequence is taking $\omega^{s(t)}$, where ω is the primitive

qth root of unity, $e^{j2\pi/q}$. For example, an m sequence $s(t)$ of length 7 is given by

$$s(t) = \mathrm{Tr}_1^3(\alpha^t) = 1001011$$

when α is the primitive element of F_8 having minimal polynomial $x^3 + x + 1$. This m sequence $s(t)$ is easily converted to its complex-valued counterpart:

$$(-1)^{s(t)} = - + + - + - -$$

An m sequence possesses many desirable properties such as balance property, run property, shift-and-add property, and ideal autocorrelation property [15]. Given a finite field F_{q^n}, there are $\phi(q^n - 1)/n$ cyclically distinct m sequences whose symbols are drawn from F_q. Each of them corresponds to different primitive element α values with different minimal polynomials. Thus, in other words, each of the cyclically distinct m sequences of given length can be viewed as the decimation $s(dt)$ of a given m sequence $s(t)$ by some d relatively prime to $q^n - 1$. The cross-correlation properties between an m sequence and its decimation are very important, since many sequence families, including the Gold sequence family, having optimal cross-correlation properties, are constructed from pairs of m sequences.

Now, consider the case of $q = 2$, that is, of binary m sequences. Without loss of generality, we can assume that an m sequence $s(t)$ is given by

$$s(t) = \mathrm{Tr}_1^n(\alpha^t)$$

Let $\theta_{1,d}(\tau)$ be the cross-correlation function between $s(t)$ and its d-decimation $s(dt)$, where d is some integer relatively prime to $2^n - 1$:

$$\theta_{1,d}(\tau) = \sum_{t=0}^{N-1} (-1)^{s(t+\tau)+s(dt)} \tag{8}$$

From previous research, the values $\theta_{1,d}(\tau)$ have been known for various d [11], although the complete evaluation of $\theta_{1,d}(\tau)$ for each possible d is still ongoing. One well-known result on $\theta_{1,d}(\tau)$ is that $\theta_{1,d}(\tau)$ takes on at least three distinct values as τ varies from 0 to $2^n - 2$, as long as the decimation $s(dt)$ is cyclically distinct to $s(t)$. One can obtain two examples of such decimation d from the following theorem which is in part due to Gold [1], Kasami [4], and Welch [10].

Theorem 3. Let $e = \gcd(n, k)$ and $\frac{n}{e}$ be odd. Let $d = 2^k + 1$ or $d = 2^{2k} - 2^k + 1$. Then the cross-correlation $\theta_{1,d}(\tau)$ of m sequence $\mathrm{Tr}_1^n(\alpha^t)$ and its decimated sequence $\mathrm{Tr}_1^n(\alpha^{dt})$ by d takes on the following three values:

$$\begin{cases} -1 + 2^{(n+e)/2}, & 2^{n-e-1} + 2^{(n-e-2)/2} \text{ times} \\ -1, & 2^n - 2^{n-e} - 1 \text{ times} \\ -1 - 2^{(n+e)/2}, & 2^{n-e-1} - 2^{(n-e-2)/2} \text{ times} \end{cases} \tag{9}$$

When $\theta_{1,d}(\tau)$ takes on the following three values

$$-1, -1 + 2^{\lfloor (n+2)/2 \rfloor}, -1 - 2^{\lfloor (n+2)/2 \rfloor}$$

the pair of m sequences $s(t)$ and $s(dt)$ is called a preferred pair. Note that either when $n = 2m$ or $n = 2m + 1$, the above three values become -1, $-1 + 2^{m+1}$, and $-1 - 2^{m+1}$. Theorem 3 can be applied to obtain a preferred pair as long as $n \not\equiv 0 \pmod 4$. In the case when n is odd, selecting k relatively prime to n yields a preferred pair, and in the case when $n \equiv 2 \pmod 4$, making $e = 2$ also yields a preferred pair. When $n \equiv 0 \pmod 4$, Calderbank and McGuire [12] proved the nonexistence of a preferred pair.

4. GOLD SEQUENCES AND GOLD-LIKE SEQUENCES

Consider the set $\mathcal{G}$ of $(N + 2)$ sequences constructed from two binary sequences $u(t)$ and $v(t)$ of length N given as follows:

$$\mathcal{G} = \{u(t), v(t), u(t) + v(t + i) \mid i = 0, 1, \ldots, N - 1\} \tag{10}$$

Especially when both $u(t)$ and $v(t)$ are m sequences, the cross-correlation function between any two members in the set becomes either the cross-correlation function between $u(t)$ and $v(t)$, or simply the autocorrelation function of m sequence $u(t)$ or $v(t)$, due to the shift-and-add property of m sequences. In the late 1960s, Gold used this method to construct the set called *Gold sequences family*. *Gold sequences family* is defined as the set $\mathcal{G}$ when $u(t)$ and $v(t)$ are preferred pair of m sequences of length $2^n - 1$. The cross-correlation values of the Gold sequence family can be directly computed from Theorem 3. Applying the set construction method above with $u(t) = \mathrm{Tr}_1^n(\alpha^{dt})$ and $v(t) = \mathrm{Tr}_1^n(\alpha^t)$, the pair of m sequences in Theorem 3, one can easily construct the set $\mathcal{W}$ (referred to here as the *sequences family*) of size $2^n + 1$ as follows

$$\mathcal{W} = \{w_i(t) \mid 0 \le i \le 2^n\}$$

where

$$w_i(t) = \begin{cases} \mathrm{Tr}_1^n(\alpha^{(t+i)}) + \mathrm{Tr}_1^n(\alpha^{dt}), & \text{for } 0 \le i \le 2^n - 2 \\ \mathrm{Tr}_1^n(\alpha^{dt}), & \text{for } i = 2^n - 1 \\ \mathrm{Tr}_1^n(\alpha^t), & \text{for } i = 2^n \end{cases}$$

As mentioned in the previous section, when n is odd and $e = 1$, the two m sequences $\mathrm{Tr}_1^n(\alpha^t)$ and $\mathrm{Tr}_1^n(\alpha^{dt})$ in $\mathcal{W}$ are preferred pair, and in this case, the family $\mathcal{W}$ becomes the Gold sequence family. Then $\theta_{\max}$ is given by

$$\theta_{\max} = 2^{(n+1)/2} + 1$$

When we apply the Sidelnikov bound in Eq. (5) with $k = 1$, $N = 2^n - 1$, and $M = 2^n + 1$, we have $\theta_{\max}^2 > 2^{n+1} - 2$, specifically, $\theta_{\max} \ge 2^{(n+1)/2}$. But, since N is odd, $\theta_{\max}$ must be odd. Therefore, the Sidelnikov bound tells us that

$$\theta_{\max} \ge 2^{(n+1)/2} + 1 \tag{11}$$

which, in turn, implies that the Gold sequence family is optimal with respect to the Sidelnikov bound when n is odd. On the other hand, the Gold sequence family in the case when $n \equiv 2 \pmod 4$ is not optimal, since the actual $\theta_{\max}$ in this case is $2^{(n/2)+1}$, which is roughly $\sqrt{2}$ times the bound in (11). Finally, when $n \equiv 0 \pmod 4$, no Gold sequence family exists, since no preferred pair exists.

The following example shows the Gold sequence family of length 31.

Example 1: Gold Sequence of Length 31. There are 32 sequences of length 31 in the set. By taking α as the primitive element in F_{2^5} having minimal polynomial $x^5 + x^2 + 1$, we have

$$w_{32}(t) = \mathrm{Tr}_1^5(\alpha^t) = 1001011001111100011011101010000$$

Setting $k = 1$ and $d = 2^k + 1 = 3$, we have

$$w_{31}(t) = \mathrm{Tr}_1^5(\alpha^{3t}) = 1111101111000101011010000110 0100$$

and $w_i(t) = w_{32}(t + i) + w_{31}(t), 0 \le i \le 30$.

In the literature, the term *Gold-like* has been used in at least two different contexts. Sarwate and Pursley [3] used this term to introduce the set $\mathcal{H}$ in the following example.

Example 2: Gold-Like Sequences in Ref. 3. Let $n = 2m \equiv 0 \pmod 4$, $N = 2^n - 1$, α be a primitive element of F_{2^n}, and $d = 1 + 2^{m+1}$. Let $s(t) = \mathrm{Tr}_1^n(\alpha^t)$ and $s_j(t) = s(dt + j)$ for $j = 0, 1, 2$. Then, the set $\mathcal{H}$ is given as follows:

$$\mathcal{H} = \left\{ s(t), s(t) + s_j(t+i) \mid j = 0, 1, 2, i = 0, 1, \ldots, \frac{N}{3} - 1 \right\}$$

Certainly, there are 2^n sequences in the set $\mathcal{H}$, and all except one are the sums of the shifted m sequence and the decimated sequence just like the members in $\mathcal{W}$. The major distinction of this set $\mathcal{H}$ with the Gold family $\mathcal{W}$ is that the decimation of the m sequence $s(t)$ by d results in three distinct subsequences $s_j(t), j = 0, 1, 2$ of period $N/3$ according to the initial decimation position j, since $\gcd(N, d) = \gcd(2^{2m} - 1, 2^{m+1} + 1) = 3$ (where gcd = greatest common divisor). Kasami [4] showed that the cross-correlation function between any two sequences in $\mathcal{H}$ takes on values in the set

$$\{-1, -1 - 2^m, -1 + 2^m, -1 - 2^{m+1}, -1 + 2^{m+1}\}$$

The set $\mathcal{H}$ in the preceding example has parameters very similar to those of the Gold sequence family for the case when $n \equiv 2 \pmod 4$. The size of the set $\mathcal{H}$ is $N + 1$ while that of Gold sequence family is $N + 2$, and $\theta_{\max}$ for both family is the same in terms of N. If the term *Gold-like* was used in this context, as it seems, then there are at least two better candidates that are entitled by this term when $n \equiv 0 \pmod 4$. Niho [13] found the following family of binary sequences.

Theorem 4. Let $n \equiv 0 \pmod 4$ and $n = 2m$. Let $d = 2^{m+1} - 1$ and $s(t) = \mathrm{Tr}_1^n(\alpha^t)$. Let the set

$$\mathcal{N} = \{s(t), s(dt), s(t+i) + s(dt) \mid i = 0, 1, \ldots, 2^n - 2\}$$

be a family of $2^n + 1$ binary sequences of length $N = 2^n - 1$. Then the cross-correlation function of the sequences in $\mathcal{N}$ takes on the following four values:

$$\{-1 + 2^{m+1}, -1 + 2^m, -1, -1 - 2^m\}$$

Note that compared to the Gold-like sequences in Example 2, the Niho family has one more sequence in the set and slightly smaller $\theta_{\max}$.

Udaya [14] introduced the family of binary sequences for even n with five-valued cross-correlation property as in the following definition and theorem.

Definition 1. For even $n = 2m$, a family $\mathcal{G}_e$ of $2^n + 1$ sequences is defined as

$$\mathcal{G}_e = \{g_e(t) \mid 0 \le i \le 2^n, 0 \le t \le 2^n - 2\}$$

where

$$g_e(t) = \begin{cases} \mathrm{Tr}_1^n(\alpha^{(t+i)}) + \sum_{k=1}^{m-1} \mathrm{Tr}_1^n(\alpha^{(2^k+1)t}) + \mathrm{Tr}_1^m(\alpha^{(2^m+1)t}), & \text{for } 0 \le i \le 2^n - 2 \\ \sum_{k=1}^{m-1} \mathrm{Tr}_1^n(\alpha^{(2^k+1)t}) + \mathrm{Tr}_1^m(\alpha^{(2^m+1)t}), & \text{for } i = 2^n - 1 \\ \mathrm{Tr}_1^n(\alpha^t), \text{ for } i = 2^n \end{cases}$$

Theorem 5. This theorem was proposed by Udaya [14]. For the family of sequences in (12), the cross-correlation function takes on the following values:

$$\{-1, -1 + 2^m, -1 - 2^m, -1 + 2^{m+1}, -1 - 2^{m+1}\}$$

The term *Gold-like* appeared much later, in 1994, when Boztas and Kumar [7] introduced the family of binary sequences with the three-valued cross-correlation property. They called this family *Gold-like sequences* since they are identical to Gold sequences in terms of family size, correlation parameter $\theta_{\max}$, and even the range of symbol imbalance. The following give their definition and cross-correlation values.

Definition 2. For odd $n = 2m + 1$, a family $\mathcal{G}_o$ of Gold-like sequences of period $2^n - 1$ is defined as

$$\mathcal{G}_o = \{g_i(t) \mid 0 \le i \le 2^n, 0 \le t \le 2^n - 2\}$$

where

$$g_i(t) = \begin{cases} \mathrm{Tr}_1^n(\alpha^{(t+i)}) + \sum_{k=1}^{m} \mathrm{Tr}_1^n(\alpha^{(2^k+1)t}), & \text{for } 0 \le i \le 2^n - 2 \\ \sum_{k=1}^{m} \mathrm{Tr}_1^n(\alpha^{(2^k+1)t}), & \text{for } i = 2^n - 1 \\ \mathrm{Tr}_1^n(\alpha^t), & \text{for } i = 2^n \end{cases}$$

Theorem 6. This theorem was proposed by Boztas and Kumar [7]. The cross-correlation function of the sequences in the family $\mathcal{G}_o$ of Gold-like sequences defined in (13) takes on the following three values:

$$-1, -1 + 2^{m+1}, -1 - 2^{m+1}$$

The set construction method for the families $\mathcal{G}_e$ and $\mathcal{G}_o$ is identical to that of the Gold sequence family. In other words, the sets $\mathcal{G}_e$ and $\mathcal{G}_o$ are of the same type as $\mathcal{G}$ in Eq. (10). The difference is that in $\mathcal{G}_e$ and $\mathcal{G}_o$, the sequence $u(t)$ is the sum of many m sequences, whereas it is a single m sequence in the Gold sequence family. For this reason, the linear span of the sequences in the families $\mathcal{G}_e$ and $\mathcal{G}_o$ is much larger than that of the Gold sequences.

The Gold sequences and Gold-like sequences we have reviewed are not optimal when n is even. Here, we end this article by introducing two examples of optimal families in the case when n is even.

Example 3: Small Set of Kasami Sequences. Let $n = 2m, m \geq 2, \alpha$ be a primitive element of F_{2^n}, $d = 2^m + 1$ and the set

$$\mathcal{K} = \{s_b(t) = \mathrm{Tr}_1^n(\alpha^t) + \mathrm{Tr}_1^m(b\alpha^{dt}) \mid b \in F_{2^m}\}$$

be a family of 2^m binary sequences of length $N = 2^n - 1$. The cross-correlation function of the sequences in the family $\mathcal{K}$ takes on the following three values:

$$-1, -1 + 2^m, -1 - 2^m$$

The set $\mathcal{K}$ is called the *small set of Kasami sequences*. The Welch bound in (4) with $k = 1$ gives us

$$\theta_{\max} > 2^m - 1$$

when $N = 2^{2m} - 1$ and $M = 2^m$. But, since $\theta_{\max}$ in this case must be an odd integer, we have

$$\theta_{\max} \geq 2^m + 1$$

which implies that the small set of Kasami sequences is optimal with respect to the Welch bound.

"No sequence family" [6] is another example of an optimal family when n is even. It has the same size and correlation distribution as the small set of Kasami sequences.

Example 4: No Sequences. Let $n = 2m$, $m \geq 2$, α be a primitive element of F_{2^n} and $d = 2^m + 1$. Let the integer $r, 1 \leq r \leq 2^m - 1, r \neq 2^i, 1 \leq i \leq m$, satisfy $\gcd(r, 2^m - 1) = 1$. Then the set

$$\mathcal{F} = \{\mathrm{Tr}_1^m[\mathrm{Tr}_m^n(\alpha^t + b\alpha^{dt})]^r \mid b \in F_{2^m}\}$$

is called a "no sequence family." The cross-correlation function of the sequences in the family $\mathcal{F}$ takes on the following three values:

$$-1, -1 + 2^m, -1 - 2^m$$

BIOGRAPHY

Habong Chung was born in Seoul, Korea. He received the B.S. degree in 1981 in electronics from Seoul National University, Seoul, Korea, and the M.S. and Ph.D. degrees in electrical engineering from the University of Southern California in 1985 and 1988, respectively. From 1988 to 1991, he was an Assistant Professor in the Department of Electrical and Computer Engineering, the State University of New York at Buffalo. Since 1991, he has been with the School of Electronic and Electrical Engineering, Hongik University, Seoul, Korea, where he is a professor. His research interests include coding theory, combinatorics, and sequence design.

BIBLIOGRAPHY

1. R. Gold, Maximal recursive sequences with 3-valued recursive cross-correlation functions, *IEEE Trans. Inform. Theory* **14**: 154–156 (Jan. 1968).
2. D. V. Sarwate, Bounds on cross-correlation and autocorrelation of sequences, *IEEE Trans. Inform. Theory* **IT-25**: 720–724 (1979).
3. D. V. Sarwate and M. B. Pursley, Crosscorrelation properties of pseudorandom and related sequences, *Proc. IEEE* **68**: 593–619 (1980).
4. T. Kasami, *Weight Distribution Formula for Some Class of Cyclic Codes*, Technical Report R-285 (AD 632574), Coordinated Science Laboratory, Univ. Illinois, Urbana, April 1966.
5. T. Kasami, Weight distribution of Bose-Chaudhuri-Hocquenghem codes, in *Combinatorial Mathematics and Its Applications*, Univ. North Carolina Press, Chapel Hill, NC, 1969.
6. J. S. No and P. V. Kumar, A new family of binary pseudorandom sequences having optimal correlation properties and large linear span, *IEEE Trans. Inform. Theory* **35**: 371–379 (March 1989).
7. S. Boztas and P. V. Kumar, Binary sequences with Gold-like correlation but larger linear span, *IEEE Trans. Inform. Theory* **40**: 532–537 (March 1994).
8. L. R. Welch, Lower bounds on the maximum cross correlation of signals, *IEEE Trans. Inform. Theory* **IT-20**: 397–399 (1974).
9. V. M. Sidelnikov, On mutual correlation of sequences, *Soviet Math. Dokl.* **12**: 480–483 (1979).
10. H. M. Trachtenberg, *On the Crosscorrelation Functions of Maximal Linear Recurring Sequences*, Ph.D. thesis, Univ. Southern California, 1970.
11. V. S. Pless and W. C. Huffman, *Handbook of Coding Theory*, North-Holland, 1998.
12. G. McGuire and A. R. Calderbank, Proof of a conjecture of Sarwate and Pursley regarding pairs of binary m-sequences, *IEEE Trans. Inform. Theory* **IT-41**: 1153–1155 (1995).
13. Y. Niho, *Multi-valued Cross-correlation Functions between Two Maximal Linear Recursive Sequences*, Ph.D. dissertation, Univ. of Southern California, 1972.
14. P. Udaya, *Polyphase and Frequency Hopping Sequences Obtained from Finite Rings*, Ph.D. dissertation, Dept. Electrical Engineering, Indian Institute of Technology (IIT), Kanpur, India, 1992.
15. M. K. Simon, J. K. Omura, R. A. Scholtz, and B. K. Levitt, *Spread-Spectrum Communications Handbook*, McGraw-Hill, New York, 1994.
16. F. J. Macwilliams and N. J. A. Sloane, *The Theory of Error-Correcting Codes*, North-Holland, New York, 1977.

17. J. S. No, K. Yang, H. Chung, and H. Y. Song, New construction for families of binary sequences with optimal correlation properties, *IEEE Trans. Inform. Theory* **43**: 1596–1602 (Sept. 1997).

GSM DIGITAL CELLULAR COMMUNICATION SYSTEM

Christian Bettstetter
Christian Hartmann
Technische Universität München
Institute of Communication Networks
Munich, Germany

1. INTRODUCTION AND OVERVIEW

The *Global System for Mobile Communication* (GSM) is an international standard for wireless, cellular, digital telecommunication networks. GSM subscribers can use their mobile phones almost worldwide for high-quality voice telephony and low-rate data applications. International roaming and automatic handover functions make GSM a system that supports seamless connectivity and mobility.

Work on GSM was started by the Groupe Spécial Mobile of the European CEPT (Conférence Européenne des Administrations des Postes et des Télécommunications) in 1982. The aim of this working group was to develop and standardize a new pan-European mobile digital communication system to replace the multitude of incompatible analog cellular systems existing at that time. The acronym GSM was derived from the name of this group; later it was changed to Global System for Mobile Communication.

In 1987 the prospective network operators and the national administrations signed a common memorandum of understanding, which confirmed their commitment to introducing the new system based on a comprehensive set of GSM guidelines. This was an important step for international operation of the new system. In 1989 the GSM group became a technical committee of the newly founded European Telecommunications Standards Institute (ETSI). The first set of GSM technical specifications was published in 1990, and in 1991 the first GSM networks started operation. After 2 years, more than one million users made phone calls in GSM networks. The GSM standard soon received recognition also outside Europe: At the end of 1993, networks were installed for example in Australia, Hong Kong, and New Zealand. In the following years the number of subscribers increased rapidly, and GSM was deployed in many countries on all continents. Figure 1 shows the development of the GSM subscribers worldwide and the number of networks and countries on the air.

The aim of this article is to give an overview of the technical aspects of a GSM network. We first explain the functionality of the GSM components and their interworking. Next, in Section 3, we describe the services that GSM offers to its subscribers. Section 4 explains how data is transmitted over the radio interface (frequencies, modulation, channels, multiple access, coding). Section 5 covers networking-related topics, such as mobility management and handover. Section 6 discusses security-related aspects.

2. SYSTEM ARCHITECTURE

A GSM network consists of several components, whose tasks, functions, and interfaces are defined in the standard [1]. Figure 2 shows the fundamental components of a typical GSM network. A mobile user carries a *mobile station* (MS) that can communicate over the radio interface with a *base transceiver station* (BTS). The BTS contains transmitter and receiver equipment as well as a few components for signal and protocol processing. The radio range of a BTS in Fig. 2 forms one cell. In practice, a BTS with sectorized antennas can supply several cells (typically three). Also, transcoding and rate adaption of speech, error protection coding, and link control are performed in the BTS. The essential control and protocol functions reside in the *base station controller* (BSC). It handles the allocation of radio channels, channel setup, frequency hopping, and management of handovers. Typically, one BSC controls several BTSs. The BTSs and BSC together form a *base station subsystem* (BSS).

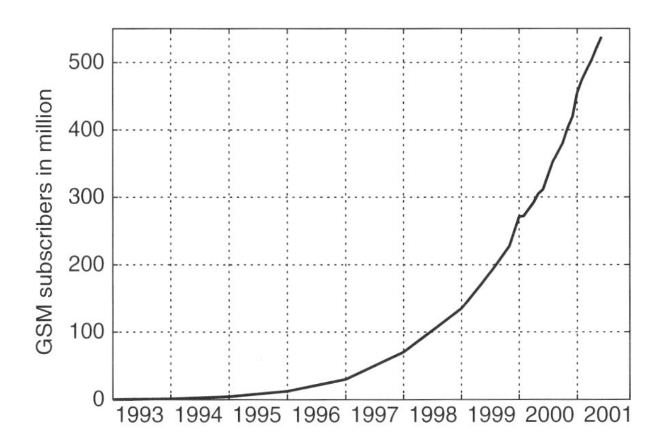

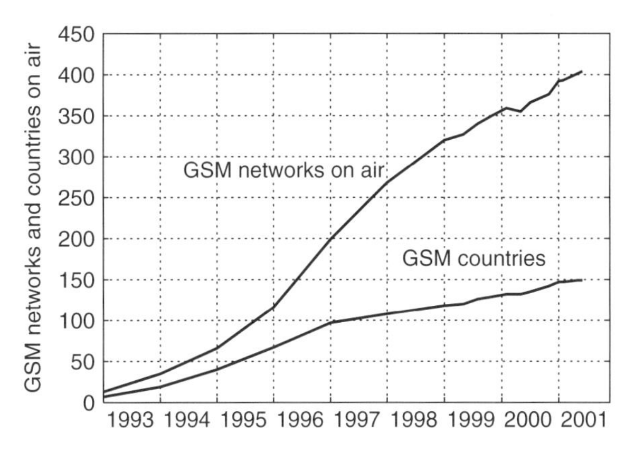

Figure 1. GSM subscriber and network statistics. (*Source*: GSM Association.)

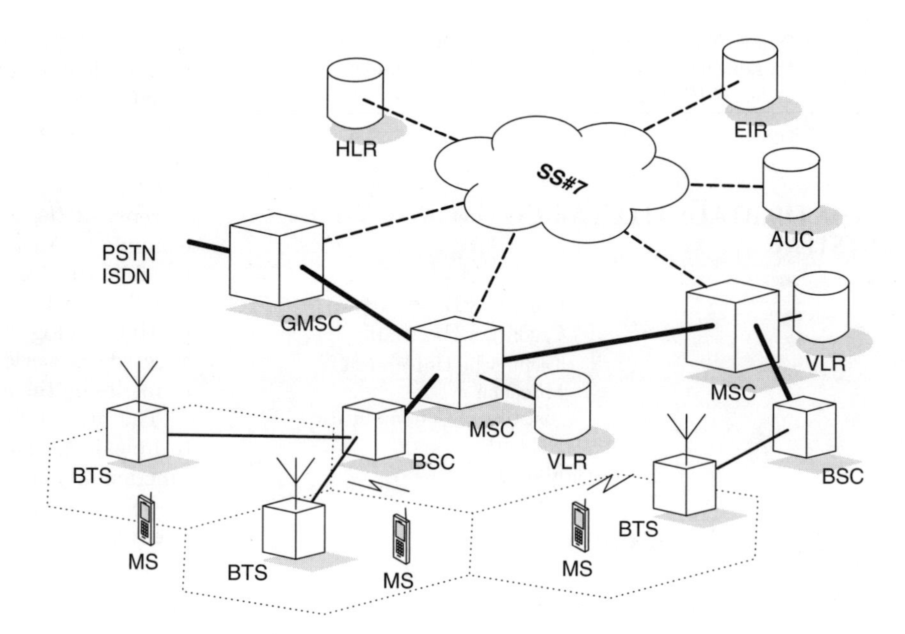

Figure 2. GSM system architecture.

The *mobile switching center* (MSC) performs the switching functions needed to route a phone call toward its destination user. Usually one MSC is allocated to several BSCs. In addition to the functionality known from usual switches in fixed ISDN networks, it must also handle the mobility of users. Such functions include the authentication of users, location updating, handover, and call routing to roaming users. Traffic between the GSM network and the fixed network [e.g., PSTN (public switched telephone network) and ISDN] is handled by a dedicated *gateway MSC* (GMSC). GSM was designed to be compatible with ISDN systems using standardized interfaces.

Two types of databases, namely, the *home location register* (HLR) and the *visitor location registers* (VLRs), are responsible for storage of profiles and location information of mobile users. There is one central HLR per network operator and typically one VLR for each MSC. The specific configuration is left to the network operator.

The HLR has a record for all subscribers registered with a network operator. It stores, for example, each user's telephone number, subscription profile, and authentication data. Besides this permanent administrative data, it also contains temporary data, such as the current location of a user. In case of incoming traffic to a mobile user, the HLR is queried in order to determine the user's current location. This allows for routing the traffic to the appropriate MSC. The mobile station must periodically inform the network about its current location. To assist this process, several cells are combined to a so-called *location area*. Whenever a mobile station changes its location area, it sends a location update to the network, indicating its current location.

A VLR is responsible for a group of location areas and stores data of all users that are currently located in this area. The data include part of the permanent subscriber data, which have been copied from the HLR to the VLR for faster access. In addition to this, the VLR may also assign and store local data, such as temporary identifiers. A user may be registered either with a VLR of his/her home network or with a VLR of a "foreign" network. On a location update, the MSC forwards the identity of the user and his/her current location area to the VLR, which subsequently updates its database. If the user has not been registered with this VLR before, the HLR is informed about the current VLR of the user.

Each user is identified by a so-called *international mobile subscriber identity* (IMSI). Together with all other personal information about the subscriber, the IMSI is stored on a chip card. This card is denoted as the *subscriber identity module* (SIM) in GSM and must be inserted into the mobile terminal in order to access the network and use the services. The IMSI is also stored in the HLR. In addition to this worldwide unique address, a mobile user receives a temporary identifier, denoted as the *temporary mobile subscriber identity* (TMSI). It is assigned by the VLR currently responsible for the user and has only local validity. The TMSI is used instead of the IMSI for transmissions over the air interface. This way, nobody can determine the identity of the subscriber.

The actual "telephone number" of a user is denoted as the *mobile subscriber ISDN number* (MSISDN). It is stored in the SIM card and in the HLR. In general, one user can have several MSISDNs.

Two further databases are responsible for various aspects of security (verification of equipment and subscriber identities, ciphering). The *authentication center* (AUC) generates and stores keys employed for user authentication and encryption over the radio channel. The *equipment identity register* (EIR) contains a list of all serial numbers of the mobile terminals, denoted as *international mobile equipment identities* (IMEI). This register allows the network to identify stolen or faulty terminals and deny network access.

As shown in Fig. 2, signaling between the GSM components in the mobile switching network is based on the Signaling System Number 7 (SS#7). For mobility-specific signaling, the MSC, HLR, and VLR hold extensions of SS#7, the so-called *mobile application part* (MAP).

Operation and maintenance of a GSM network are organized from a central *operation and maintenance center* (OMC), which is not shown in Fig. 2. Its functions include network configuration, operation, and performance management, as well as administration of subscribers and terminals.

To summarize, the entire GSM network can be divided into three major subsystems: the radio network (BSS), the switching network (including MSCs, databases, and wired core network), and the *operation and maintenance subsystem* (OSS).

3. SERVICES AND EVOLUTION

The first GSM networks mainly offered basic telecommunication services — in the first place, mobile voice telephony — and a few supplementary services. This step in the GSM evolution is called phase 1. The supplementary services of phase 1 include features for call forwarding (e.g., call forwarding on mobile subscriber busy, call forwarding on mobile subscriber not reachable) and call restriction (e.g., barring of all outgoing/incoming calls, barring of all outgoing international calls, and barring of incoming international calls when roaming outside the home network). All these services had to be implemented as mandatory features by all network operators.

Besides mobile voice telephony, GSM also offers services for data transmission, such as fax and circuit-switched access to data networks (e.g., X.25 networks) with data rates up to 9.6 kbps (Kilobits per second).

Of particular importance is the *short message service* (SMS). It allows users to exchange short text messages in a store-and-forward fashion. The network operator establishes a service center that accepts and stores text messages. Later, some value-added services, such as SMS cell broadcast and conversion of SMS messages from/to email and to speech, have been implemented.

The standardization of phase 2 basically added further supplementary services, such as call waiting, call holding, conference calling, call transfer, and calling line identification. Many parts of the GSM technical specifications had to be reworked, but all networks and terminals retained compatibility to phase 1. Phase 2 was completed in 1995, and market introduction followed in 1996.

In the following years, a broad number of additional services and improvements have been developed in independent standardization units (GSM phase 2+) [2]. These topics affect almost all aspects of GSM, and enable a smooth transition from GSM to the *Universal Mobile Telecommunication System* (UMTS); see Fig. 3. For example, the GSM voice codecs have been improved to achieve a much better speech quality (see Section 4.2). Furthermore, a set of group call and push-to-talk speech services with fast connection setup has been

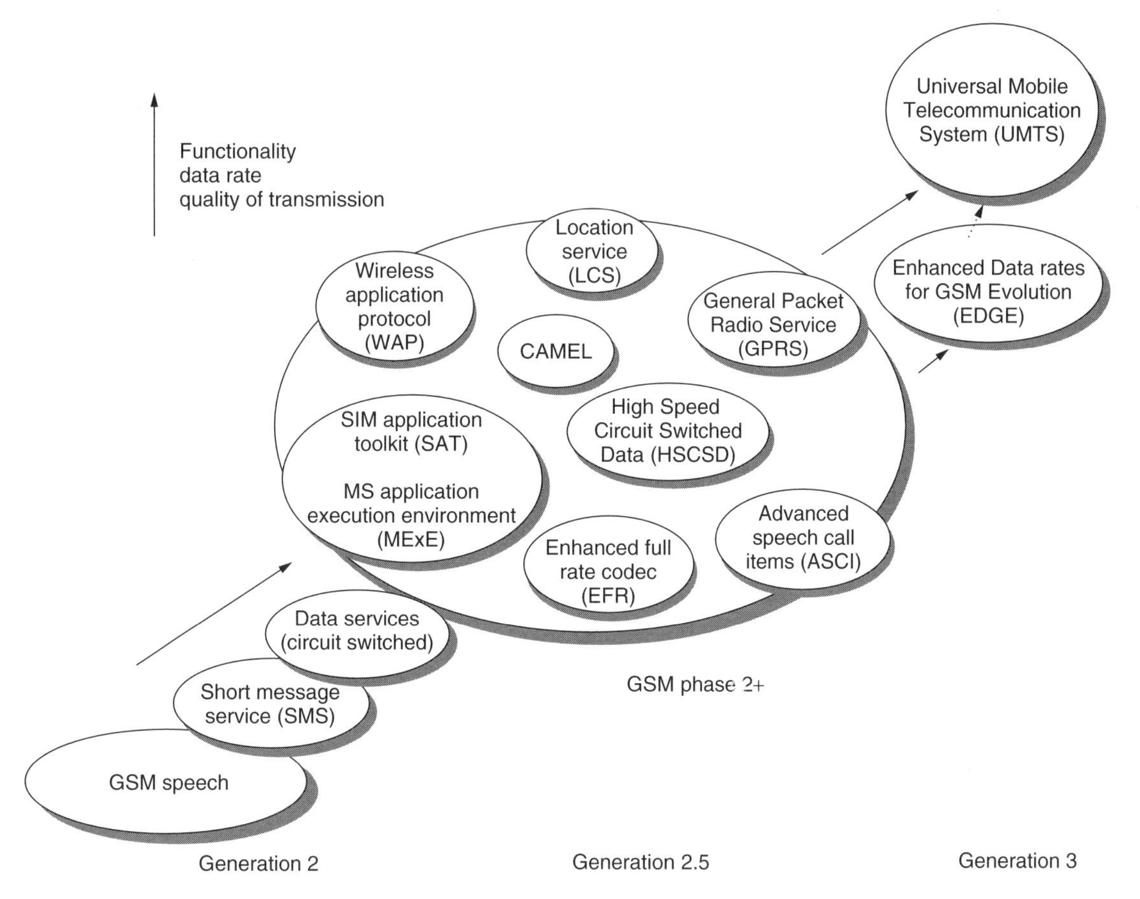

Figure 3. GSM evolution.

standardized under the name *advanced speech call items* (ASCIs). These services are especially important for closed user groups, such as police and railroad operators.

Another important aspect is the definition of service platforms. Instead of standardizing services, only features to create services and mechanisms that enable their introduction are specified. This allows the network providers to introduce new operator-specific services in a faster way. *Customized application for mobile network enhanced logic* (CAMEL) represents the integration of intelligent network (IN) principles into GSM. It enables use of operator-specific services also in foreign networks. For example, a subscriber roaming in a different country can easily check his/her voicebox with the usual short number. Other features include roaming services for prepaid subscriptions and speed dial for closed user groups. Also on the mobile station side, service platforms have been developed: the SIM application toolkit and the *MS application execution environment* (MExE). The SIM application toolkit allows the operator to run specific applications on the SIM card. With the toolkit, the SIM card is able to display new operator-specific items and logos and to play sounds. For example, users can download new ringing tones to their SIM card. The new applications can be transmitted, for example, via SMS to the mobile station. The most important components of MExE are a virtual machine for execution of Java code and the *Wireless Application Protocol* (WAP). With a virtual machine running on the mobile station, applications can be downloaded and executed. The WAP defines a system architecture, a protocol family, and an application environment for transmission and display of Web-like pages for mobile devices. WAP has been developed by the WAP forum; services and terminals have been available since 1999. Using a WAP-enabled mobile GSM phone, subscribers can download information pages, such as news, weather forecasts, stock reports, and local city information. Furthermore, mobile e-commerce services (e.g., ticket reservation, mobile banking) are offered.

If information about the current physical location of a user is provided by the GSM network or by GPS (Global Positioning System), location-aware applications are possible. A typical example is a location-aware city guide that informs mobile users about nearby sightseeing attractions, restaurants, hotels, and public transportation.

Development also continued with improved bearer services for data transmission. The *High-Speed Circuit-Switched Data* (HSCSD) service achieves higher data rates by transmitting in parallel on several traffic channels (multislot operation). The *General Packet Radio Service* (GPRS) offers a packet-switched transmission at the air interface. It improves and simplifies wireless access to the Internet. Users of GPRS benefit from shorter access times, higher data rates, volume-based billing, and an "always on" wireless connectivity. (For further details, see the GPRS entry of this encyclopedia.) The *Enhanced Data Rates for GSM Evolution* (EDGE) service achieves even higher data rates and a better spectral efficiency. It replaces the original GSM modulation by an 8-PSK (8-phase shift keying) modulation scheme.

4. AIR INTERFACE: PHYSICAL LAYER

Figure 4 gives a schematic overview of the basic elements of the GSM transmission chain. The stream of sampled speech is fed into a source encoder, which compresses the data. The resulting bit sequence is passed to the channel encoder. Its purpose is to add, in a controlled manner, some redundancy to the bit sequence. This redundancy serves to protect the data against the negative effects of noise and interference encountered in the transmission over the radio channel. On the receiver side, the introduced redundancy allows the channel decoder to detect and correct transmission errors. Without channel coding, the achieved bit error rate would be insufficient, not only for speech but also for data communication. Reasonable bit error rates are on the order of 10^{-5} to 10^{-6}. To achieve these rates, GSM uses a combination of block and convolutional coding. Moreover, an interleaving scheme is used to deal with burst errors that occur over multipath and fading channels. After coding and interleaving, the data are encrypted to guarantee secure and confident data transmission. The encryption technique as well as the methods for subscriber authentication and secrecy of the subscriber identity are explained in Section 6. The encrypted data are subsequently mapped to bursts, which are then multiplexed. Finally, the stream of bits is differentially coded, modulated, and transmitted on the respective carrier frequency over the mobile radio channel.

After transmission, the demodulator processes the signal, which was corrupted by the noisy channel. It

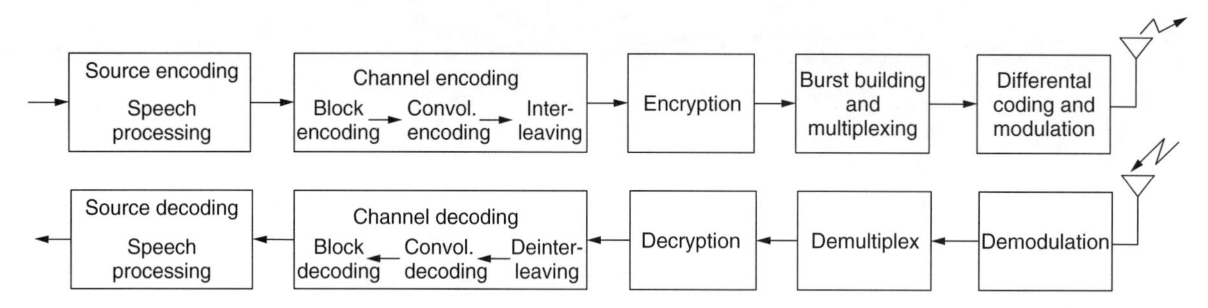

Figure 4. Basic function of the GSM transmission chain on the physical layer at the air interface.

attempts to recover the actual signal from the received signal. The next steps are demultiplexing and decryption. The channel decoder attempts to reconstruct the original bit sequence, and, as a final step, the source decoder tries to rebuild the original source signal.

4.1. Logical Channels

GSM defines a set of logical channels [3], which are divided into two categories: traffic channels and signaling channels. Some logical channels are assigned in a dedicated manner to a specific user; others are common channels that are of interest for all users in a cell.

The traffic channels (TCHs) are used for the transmission of user data (e.g., speech, fax). A TCH may either be fully used (full-rate TCH; TCH/F) or split into two half-rate channels that can be allocated to different subscribers (half-rate TCH; TCH/H). A TCH/F carries either 13 kbps of coded speech or datastreams at 14.5, 12, 6, or 3.6 kbps. A TCH/H transmits 5.6 kbps of half-rate coded speech or datastreams at 6 or 3.6 kbps.

In addition to the traffic channels, GSM specifies various signaling channels. They are grouped into broadcast channels (BCHs), common control channels (CCCHs), and dedicated control channels (DCCHs).

The BCHs are unidirectional and used to continually distribute information to all MSs in a cell. The following three broadcast channels are defined:

- The broadcast control channel (BCCH) broadcasts configuration information, including the network and BTS identity, the frequency allocations, and availability of optional features such as voice activity detection (Section 4.2) and frequency hopping (Section 4.5).
- The frequency correction channel (FCCH) is used to distribute information about correction of the transmission frequency.
- The synchronization channel (SCH) broadcasts data for frame synchronization of an MS.

The group of CCCHs consists of four unidirectional channels that are used for radio access control:

- The random-access channel (RACH) is an uplink channel that is used by the MSs in a slotted ALOHA fashion for the purpose of requesting network access.
- The access grant channel (AGCH) is a downlink channel used to assign a TCH or a DCCH to an MS.
- The paging channel (PCH), which is also a downlink channel, is employed to locate an MS in order to inform it about an incoming call.
- The notification channel (NCH) serves to inform MSs about incoming group and broadcast calls.

Finally, GSM uses three different DCCHs:

- The stand-alone dedicated control channel (SDCCH), which is applied for signaling between BSS and MS when there is no active TCH connection. This is necessary to update location information, for instance.
- The slow associated control channel (SACCH), which carries information for synchronization, power control, and channel measurements. It is always assigned in conjunction with a TCH or an SDCCH.
- The fast associated control channel (FACCH), which is used for short time signaling. It can be made available by stealing bursts from a TCH.

GSM also defines a set of logical channels for the General Packet Radio Service (GPRS). They are treated in the GPRS entry of this encyclopedia.

4.2. Speech Processing Functions and Codecs

Transmission of voice is one of the most important services in GSM. The user's analog speech signal is sampled at the transmitter at a rate of 8000 samples/s, and these samples are quantized with a resolution of 13 bits. At the input of the speech encoder, a speech frame containing 160 samples, each 13 bits long, arrives every 20 ms. This corresponds to a bit rate of 104 kbps for the speech signal. The compression of this speech signal is performed in the speech encoder. The functions of encoder and decoder are typically combined in a single building block, called a *codec*.

An optional speech processing function is *discontinuous transmission* (DTX). It allows the radio transmitter to be switched off during speech pauses. This saves battery power of the MS and reduces the overall interference level at the air interface. Voice pauses are recognized by the *voice activity detection* (VAD). During pauses, the missing speech frames are replaced by a synthetic background noise generated by the comfort noise synthesizer.

Another function on the receiver side is the replacement of bad frames. If a transmitted speech frame cannot be corrected by the channel coding mechanism, it is discarded and replaced by a frame that is predictively calculated from the preceding frame.

In the following paragraphs, the speech codecs used in GSM are briefly characterized. The channel coding is described in Section 4.3.

4.2.1. Full-Rate Speech Codec. The first set of GSM standards defined a full-rate speech codec for transmission via the TCH/F. It is an RPE-LTP (regular pulse excitation–long-term prediction) codec [4], which is based on linear predictive coding (LPC). The RPE-LTP codec has a compression rate of 1/8 and thus produces a data rate of 13 kbps at its output.

With the further development of GSM the speech codecs have also been improved. Two competing objectives have been considered: (1) the improvement of speech quality toward the quality offered by fixed ISDN networks and (2) better utilization of the frequency bands assigned to GSM, in order to increase the network capacity.

4.2.2. Half-Rate Speech Codec. The half-rate speech codec has been developed to improve bandwidth utilization. It produces a bit stream of 5.6 kbps and is used for speech transmission over the TCH/H. Instead of using the RPE-LTP coding scheme, the algorithm is based on code-excited linear prediction (CELP), in which the excitation signal is an entry in a very large stochastically populated codebook. The codec has a higher complexity and higher latency. Under normal channel conditions, it achieves — in spite of half the bit rate — almost the same speech quality as the full-rate codec. However, quality loss occurs for mobile-to-mobile communication, since in this case (due to the ISDN architecture) one has to go twice through the GSM speech coding/decoding process. A method to avoid multiple transcoding has been passed under the name tandem free operation in GSM Release 98.

4.2.3. EFR Speech Codec. The *enhanced full-rate* (EFR) speech codec [5] was standardized by ETSI in 1996 and has been implemented in GSM since 1998. It improves the speech quality compared to the full- and half-rate speech codecs without using more system capacity than the full-rate codec. The EFR codec produces a bitstream of 12.2 kbps (244 code bits for each 20-ms frame) and is based on the algebraic code excitation linear prediction (ACELP) technique [6].

A detailed explanation of the full-rate, half-rate, and EFR codecs can be found in Ref. 4.

4.2.4. AMR Codec. The speech codecs mentioned before all use a fixed source bit rate, which has been optimized for typical radio channel conditions. The problem with this approach is its inflexibility; whenever the channel conditions are much worse than usual, very poor speech quality will result, since the channel capacity assigned to the mobile station is too small for error-free transmission. On the other hand, radio resources will be wasted for unneeded error protection if the radio conditions are better than usual. To overcome these problems, a much more flexible codec has been developed and standardized: the *adaptive multirate* (AMR) codec. It can improve speech quality by adaptively switching between different speech coding schemes (with different levels of error protection) according to the current channel quality.

To be more precise, AMR has two principles of adaptability [7]: channel mode adaptation and codec mode adaptation. Channel mode adaptation dynamically selects the type of traffic channel that a connection should be assigned to: either a full-rate (TCH/F) or a half-rate traffic channel (TCH/H). The basic idea here is to adapt a user's gross bit rate in order to optimize the usage of radio resources. The task of codec mode adaptation is to adapt the coding rate (i.e., the tradeoff between the level of error protection versus the source bit rate) according to the current channel conditions. When the radio channel is bad, the encoder operates at low source bit rates at its input and uses more bits for forward error protection. When the quality of the channel is good, less error protection is employed.

The AMR codec consists of eight different modes with source bit rates ranging from 12.2 to 4.75 kbps. All modes are scaled versions of a common ACELP basis codec; the 12.2-kbps mode is equivalent to the EFR codec.

4.3. Channel Coding

GSM uses a combination of block coding (as external error protection) and convolutional coding (as internal error protection). Additionally, interleaving of the encoded bits is performed in order to spread the channel symbol errors. The channel encoding and decoding chain is depicted in Fig. 4. The bits coming from the source encoder are first block-encoded; that is, parity and tail bits are appended to the blocks of bits. The resulting stream of coded bits is then fed into the convolutional encoder, where further redundancy is added for error correction. Finally, the blocks are interleaved. This is done because the mobile radio channel frequently causes burst errors (a sequence of erroneous bits), due to long and deep fading events. Spreading the channel bit errors by means of interleaving diminishes this statistical dependence and transforms the mobile radio channel into a memoryless binary channel. On the receiver side, the sequence of received channel symbols — after demodulation, demultiplexing, and deciphering — is deinterleaved before it is fed into the convolutional decoder for error correction. Finally, a parity check is performed based on the respective block encoding.

It depends on the channel type which specific channel coding scheme is employed. Different codes are used, for example, for speech traffic channels, data traffic channels, and signaling channels. The block coding used for external error protection is either a *cyclic redundancy check* (CRC) code or a fire code. In some cases, just some tail bits are added. For the convolutional coding, the memory of the used codes is either 4 or 6. The basic code rates are $r = \frac{1}{2}$ and $r = \frac{1}{3}$; however, other code rates can be obtained by means of puncturing (deleting some bits after encoding). In the following, the basic coding process is explained for some channels. The detailed channel coding and interleaving procedures of all channels are described in Ref. 8.

4.3.1. Full-Rate Speech Codec. Let us explain how the speech of the full-rate codec is protected. One speech block coming from the source encoder consists of 260 bits. These bits are graded into different classes, which have different impact on speech quality. The 182 bits of class 1 have more impact on speech quality and thus must be better protected. The block coding stage calculates three parity bits for the most important 50 bits of class 1 (known as class 1a bits). Next, four tail bits are added, and the resulting 189 bits are fed into a rate-$\frac{1}{2}$ convolutional encoder of memory 4 (see Fig. 5). The 78 bits of class 2 are not encoded. Finally, the resulting 456 bits are interleaved.

A frame in which the bits of class 1 have been recognized as erroneous in the receiver is reported to the speech codec using the *bad-frame indication* (BFI). In order to maintain a good speech quality, these frames are discarded, and the last frame received correctly is repeated instead, or an extrapolation of received speech data is performed.

4.3.2. EFR Speech Codec. Using the EFR codec, a special preliminary channel coding is employed for the most significant bits; 8 parity bits (generated by a CRC

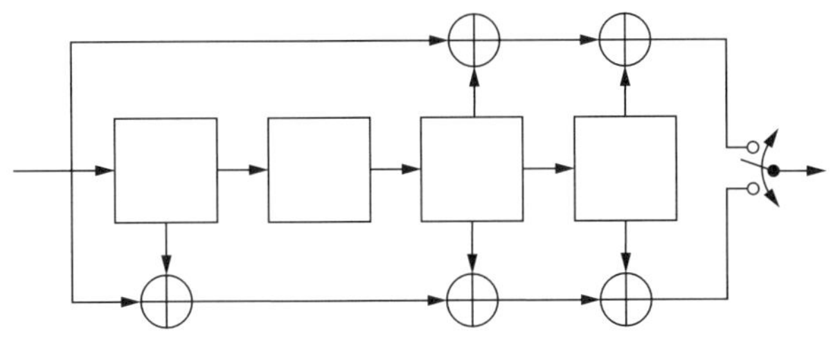

Figure 5. Convolutional encoder (used for full-rate speech, EFR speech, some data channels, and signaling channels).

encoder) and 8 repetition bits are added to the 244 bits at the output of the source encoder for additional error detection. The coding process for the resulting 260 bits is the same as for the full-rate codec.

4.3.3. AMR Speech Codec. The error protection using the AMR codec is more sophisticated. From the results of link quality measures, an adaptation unit selects the most appropriate codec mode. Channel coding is performed using a punctured recursive systematic convolutional code. Since not all bits of the voice data are equally important for audibility, AMR also employs an unequal error protection structure. The most important bits are additionally protected by a CRC code. Also, the degree of puncturing depends on the importance of the bits.

4.3.4. Data Traffic. For data traffic channels, no actual block coding is performed. Instead, a tail of all-zero bits is appended to the data blocks in order to obtain block sizes that are suitable for the convolutional encoder. In the convolutional coding stage, the channels TCH/F14.4, TCH/F9.6, TCH/H4.8 use a punctured version of the rate-$\frac{1}{2}$ encoder depicted in Fig. 5. For the TCH/F9.6 and TCH/H4.8, the 244 bit blocks at the input of the encoder are mapped to 488 bits. These blocks are reduced to 456 bits by puncturing 32 bits. Using TCH/F14.4, the 294 bits are mapped to 588 bits, followed by a puncturing of 132 bits. The channels TCH/F4.8, TCH/F2.4, and TCH/H2.4 use a rate-$\frac{1}{3}$ channel encoder.

4.3.5. Signaling Traffic. The majority of the signaling channels (SACCH, FACCH, SDCCH, BCCH, PCH, AGCH) use a fire code for error detection. It is a powerful cyclic block code that appends 40 redundancy bits to a 184-bit data block. A different approach is taken for error detection on the RACH. The very short random access burst of the RACH allows a data block length of only 8 bits, which is supplemented by six redundancy bits using a cyclic code. The SCH, as an important synchronization channel, uses a somewhat more elaborate error protection than the RACH channel. The SCH data blocks have a length of 25 bits and receive another 10 bits of redundancy for error detection through a cyclic code. The convolutional coding is performed using the encoder in Fig. 5 for all signaling channels.

4.4. Multiple Access

As illustrated in Fig. 6, GSM uses the frequency bands 890–915 MHz for transmission from the MS to the BTS (uplink) and 935–960 MHz for transmission from the BTS to the MS (downlink). Hence, the duplexing method used in GSM is *frequency-division duplex* (FDD). In addition, GSM systems developed later have been assigned frequency ranges around 1800 and 1900 MHz.

The multiple-access method uses a combination of *frequency-division multiple access* (FDMA) and *time-division multiple access* (TDMA). The entire 25-MHz frequency range is divided into 124 carrier frequencies of 200 kHz bandwidth. Each of the resulting 200 kHz radiofrequency channels is further divided into eight time slots, namely, eight TDMA channels (see Fig. 6). One timeslot carries one data burst and lasts 576.9 μs (156.25 bits with a data rate of 270.833 kbps). It can be

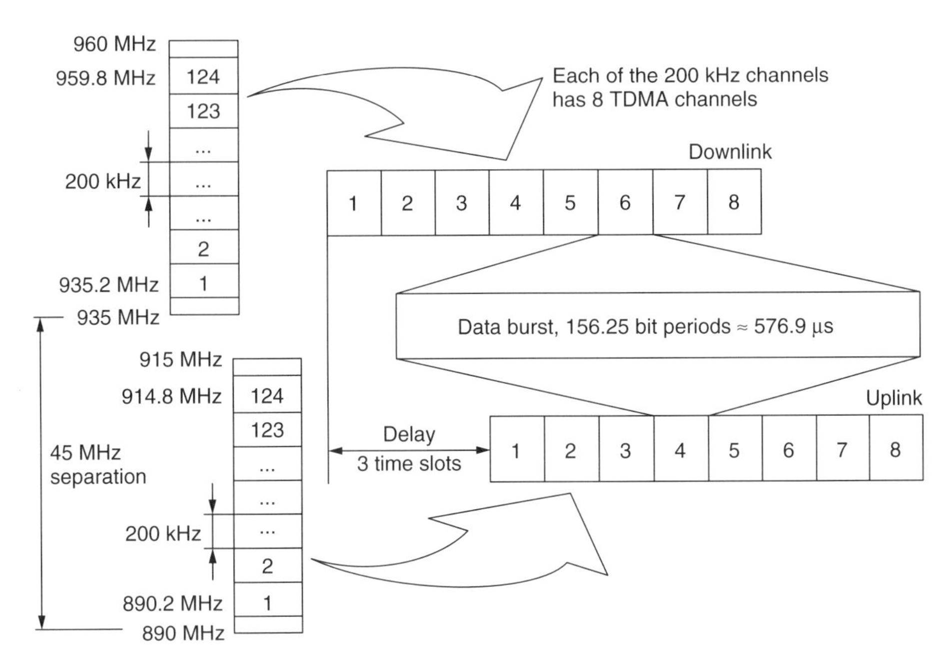

Figure 6. Carrier frequencies, duplexing, and TDMA frames.

considered the basic unit of a TDMA frame. A TDMA frame contains eight time slots and thus lasts 4.615 ms. Consequently, each TDMA channel is able to transmit at a data rate of 33.9 kbps.

Five different data bursts are distinguished in GSM [3]. A normal burst consists of two sequences of 57 data bits that are separated by a 26-bit midamble containing a training sequence. The training sequence is used for channel estimation and equalization. Furthermore, a normal burst contains two signaling bits, called "stealing flags," which indicate whether the burst contains traffic data or signaling data. Three tail bits, which are always set to logical 0, mark the beginning as well as the end of each burst. A guard period lasting the equivalent of 8.25 bits separates consecutive bursts. The other burst types are the frequency correction burst (used for frequency synchronization of an MS), the synchronization burst (used for time synchronization of the MS with the BTS), the dummy burst (used to avoid empty bursts on the BCCH in order to ensure continuous transmission on the BCCH carrier), and the access burst (used by MSs for random access on the RACH).

In order to efficiently use the bandwidth assigned to a GSM network, frequency channels have to be reused in a certain spatial distance (cellular principle). Determining a good allocation of frequencies is a complicated optimization problem, called *frequency planning*. The goal is to provide each cell with as many channels as possible, while securing that frequencies are only reused in cells that are sufficiently far apart to avoid severe cochannel interference. Assuming idealistic regular hexagonal cell patterns, frequency planning leads to dividing the system area into clusters that contain a certain number of cells (see Fig. 7). Each cluster can use all available channels of the network. Each channel can be allocated to exactly one cell of the cluster (i.e., channel reuse within a cluster is not possible). The set of frequency channels assigned to a cell is denoted as cell allocation. If each cluster of the network has the same assignment pattern, that is, if cells that have the same relative position within their respective clusters receive the same cell allocation throughout the network, a minimal cochannel reuse distance is secured in the whole network if the cluster size is reasonably chosen. Figure 7 gives an example in which 14 frequencies ($f_1 \cdots f_{14}$) have been allocated to clusters with seven cells each.

In real systems, the cluster size should be chosen high enough to avoid severe cochannel interference but not too high to obtain as many frequency channels per cell as possible. Given realistic propagation conditions and BTS positions, a typical cluster size in real GSM networks is on the order of 12. The actual frequency planning process becomes even more complicated since traffic conditions are seldom uniform and some cells will need a higher number of channels than others.

One channel of the cell allocation is used for broadcasting the configuration data on the BCCH as well as synchronization data (FCCH and SCH). This channel is called the BCCH carrier.

4.5. Frequency Hopping

As a result of multipath propagation, mobile radio channels experience frequency selective fading; that is, the

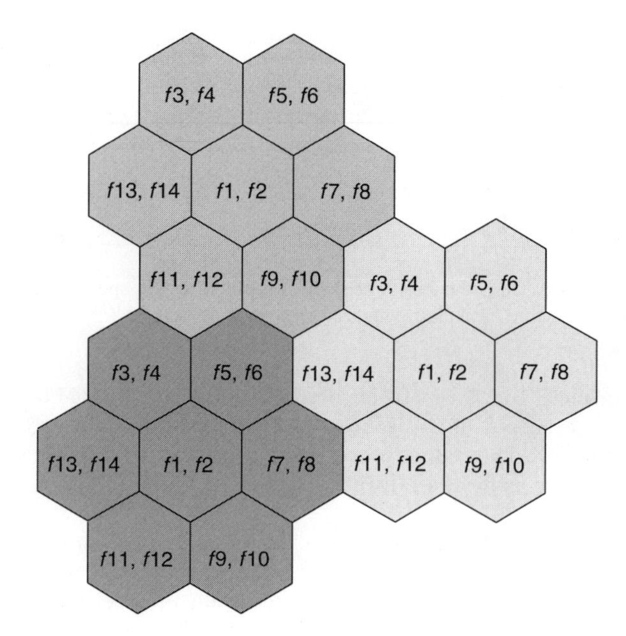

Figure 7. Cellular principle (clusters with seven cells).

instantaneous fading values that an MS experiences at a certain position are frequency-dependent. Therefore, GSM provides an optional frequency hopping procedure [9] that changes the transmission frequency periodically in order to average the interference over the frequencies in one cell. The frequency is changed with each burst, which is considered slow frequency hopping with a resulting hopping rate of about 217 changes per second. The use of frequency hopping is an option left to the network operator, which can be decided on an individual cell basis.

4.6. Synchronization

For successful operation of a mobile radio system, synchronization between MSs and BTSs is necessary [10]. Two kinds of synchronization are distinguished: frequency synchronization and time synchronization.

Frequency synchronization is necessary so that the transmitter and receiver operate on the same frequency (in both up- and downlink). By periodically monitoring the FCCH, the MS can synchronize its oscillators with the BTS.

Time synchronization is important to adjust propagation time differences of signals from different MSs, in order to achieve synchronous reception of time slots at the BTS. This way, adjacent time slots do not overlap and interfere with each other. Furthermore, synchrony is needed for the frame structure, since there is a higher-level frame structure superimposed on the TDMA frames for multiplexing logical signaling channels onto one physical channel. To keep track of the frame structure, the MSs monitor the synchronization bursts on the SCH, which contain information on the frame number. To synchronize the BTS's reception of data bursts from all MSs within the cell, a parameter called *timing advance* is used. The mobile station receives the timing advance value that it must use from the BTS on the SACCH downlink, and it reports the currently used value on the SACCH uplink.

There are 64 steps for the timing advance, where one step corresponds to 1-bit duration. The value tells the MS when it must transmit relative to the downlink. Thus, the required adjustment always corresponds to the round-trip delay between the respective mobile station and the base station. Therefore, the maximum timing advance value defined in GSM also determines the maximum cell size (radius 35 km).

4.7. Modulation

The modulation technique used in GSM is *Gaussian minimum shift keying* (GMSK). GMSK belongs to the family of continuous-phase modulation, which have the advantages of a narrow transmitter power spectrum with low adjacent channel interference and a constant amplitude envelope. This allows for use of simple and inexpensive amplifiers in the transmitters without stringent linearity requirements. In order to facilitate demodulation, each burst is encoded differentially before modulation is performed.

4.8. Power Control

The main purpose of power control in a mobile communication system is to minimize the overall transmitted power in order to keep interference levels low, while providing sufficient signal quality for all ongoing communications.

In GSM, the transmit power of the BTS and each MS can be controlled adaptively. As part of the radio subsystem link control [11], the transmit power is controlled in steps of 2 dBm. For the uplink power control, 16 control steps are defined from step 0 (43 dBm = 20 W) to step 15 (13 dBm) with a gap of 2 dBm between neighboring values. Similarly, the downlink can be controlled in steps of 2 dBm. However, the number of downlink power control steps depends on the power class of the BTS, which defines the maximum transmission power of a BTS (up to 320 W). It should be noted that downlink power control is not applied to the BCCH carrier, which must maintain constant power to allow comparative measurements of neighboring BCCH carriers by the mobile stations.

GSM uses two parameters to describe the quality of a connection: the *received signal level* (RXLEV, measured in dBm) and the *received signal quality* (RXQUAL, measured as bit error ratio before error correction). Power control as well as handover decisions are based on these parameters. The received signal power is measured continuously by mobile and base stations in each received burst within a range of −110 to −48 dBm. The respective RXLEV values are obtained by averaging.

For power control, upper and lower threshold values for uplink and downlink are defined for the parameters RXLEV and RXQUAL. If a certain number of consecutive values of RXLEV or RXQUAL are above or below the respective threshold value, the BSS can adjust the transmitter power. For example, if the upper threshold value for RXLEV on the uplink is exceeded, the transmission power of the MS will be reduced. In the other case, if RXLEV on the uplink falls below the respective lower threshold, the MS will be ordered to increase its transmission power. If the criteria for the downlink are exceeded, the transmitter power of the BTS will be adjusted. Equivalent procedures will be performed if the values for RXQUAL violate the respective range.

5. NETWORKING ASPECTS AND PROTOCOLS

The GSM standard defines a complex protocol architecture that includes protocols for transport of user data as well as for signaling. The fact that users can roam within a network from cell to cell and also internationally to other networks, requires the GSM system to handle signaling functions for registration, user authentication, location updating, routing, handover, allocation of channels, and so on. Some signaling protocols for these tasks are explained in the following; a more comprehensive explanation can be found in Ref. 2.

Figure 8 illustrates the signaling protocol architecture between MS and MSC. Layer 1 at the air interface

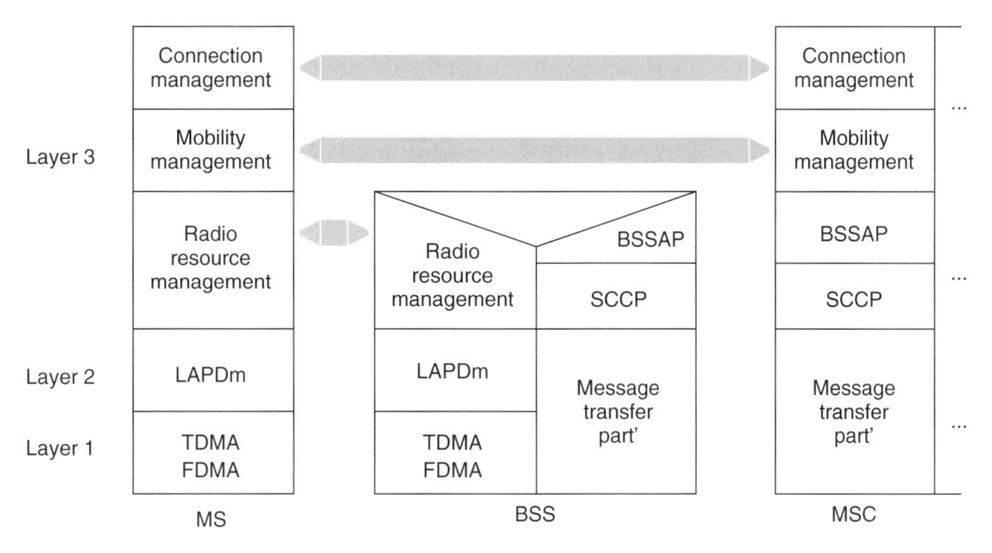

Figure 8. Signaling protocols.

(between MS and BTS) is the physical layer as described in Section 4. It implements the logical signaling channels. The data-link layer (layer 2) at the air interface employs a modified version of the LAPD (*link access procedure D*) protocol used in ISDN. It is denoted as LAPDm (LAPD mobile). The signaling protocols in Layer 3 between mobile stations and the network are divided into three sublayers: (1) radio resource management, (2) mobility management, and (3) connection management [12]. They are explained in the following subsections. In the fixed network part, a slightly modified version of the *message transfer part* (MTP) of SS#7 is used to transport signaling messages between BSC and MSC. Above this protocol, the *signaling connection control part* (SCCP) and the *BSS application part* (BSSAP) have been defined. The latter consists of the *direct transfer application part* (DTAP) and the *BSS mobile application part* (BSSMAP).

5.1. Radio Resource Management

The objective of radio resource management is to set up, maintain, and release traffic and signaling channels between a mobile station and the network. This also includes cell selection and handover procedures.

A mobile station is continuously monitoring the BCCH and CCCH on the downlink. It measures the signaling strength of the BCCHs broadcasted by the nearby BTSs in order to select an appropriate cell. At the same time, the MS periodically monitors the PCH for incoming calls.

A radio resource management connection establishment can be initiated either by the network or by the MS. In either case, the MS sends a channel request on the RACH in order to get a channel assigned on the AGCH. In case of a network-initiated connection, this procedure is preceded by a paging call to be answered by the MS.

Once a radio resource management connection has been set up, the MS has either an SDCCH or a TCH with associated SACCH/FACCH available for exclusive bidirectional use. On the SACCH the MS continuously sends channel measurements if no other messages need to be sent. These measurements include the values RXLEV and RXQUAL (see Section 4.8) of the serving cell and RXLEV of up to six neighboring cells. The system information sent by the BSS on the SACCH downlink contains information about the current and neighboring cells and their BCCH carriers.

In order to change the configuration of the physical channel in use, a channel change within the cell can be performed. The channel change can be requested by the radio resource management sublayer or by higher protocol layers. However, it is always initiated by the network and reported to the MS by means of an assignment command. A second signaling procedure to change the physical channel configuration of an established radio resource management connection is a handover, which is described in Section 5.4.

The release of radio resource management connections is always initiated by the network. Reasons for the channel release could be the end of the signaling transaction, insufficient signal quality, removal of the channel in favor of a higher priority call (e.g., emergency call), or the end of a call. After receiving the channel release command, the MS changes back to idle state.

Another important procedure of GSM radio resource management is the activation of ciphering, which is initiated by the BSS by means of the cipher mode command.

5.2. Mobility Management and Call Routing

Mobility management includes tasks that are related to the mobility of a user. It keeps track of a user's current location (location management) and performs attach and detach procedures, including user authentication.

Before a subscriber can make a phone call or use other GSM services, his/her mobile station must attach to the network and register its current location. Usually, the subscriber will attach to its home network, that is, the network with which he/she has a contract. Attachment to other networks is possible if there is a roaming agreement between the providers. To perform an attach, the MS sends the IMSI of the user to the current network. The MSC/VLR queries the HLR to check whether the user is allowed to access the network. If authentication is successful, the subscriber will be assigned a TMSI and an MSRN (*mobile station roaming number*) for further use. The TMSI is only valid within a location area. The MSRN contains routing information, so that incoming traffic can be routed to the appropriate MSC of the user.

After a successful attach, GSM's location management functions are responsible for keeping track of a user's current location, so that incoming calls can be routed to the user. Whenever a powered-on MS crosses the boundary of a location area, it sends a location update request message to its MSC. The VLR issues a new MSRN and informs the HLR about the new location. As opposed to the location registration procedure, the MS has already got a valid TMSI in this case. In addition to this event-triggered location update procedure, GSM also supports periodic location updating.

The telephone number dialed to reach a user (MSISDN) gives no information about the current location of this user. Clearly, we always dial the same number no matter whether our communication partner is attached to its home network or he/she is currently located in another country. To establish a connection, the GSM network must determine the cell in which the user resides.

Figure 9 gives an example for an incoming call from the fixed ISDN network to a mobile user. An ISDN switch realizes that the dialed number corresponds to a subscriber in a mobile GSM network. From the country and network code of the MSISDN it can determine the home network of the called user and forward the call to the appropriate GMSC. On arrival of the call, the GMSC queries the HLR in order to obtain the current MSRN of the user. With this number, the call can be forwarded to the responsible MSC. Subsequently, the MSC obtains the user's TMSI from its VLR and sends out a paging message for the user in the cells of the relevant location area. The MS responds, and the call can be switched through.

It is important to note that a subscriber can attach to a GSM network irrespective of a specific mobile terminal. By inserting the SIM card into another terminal, he/she

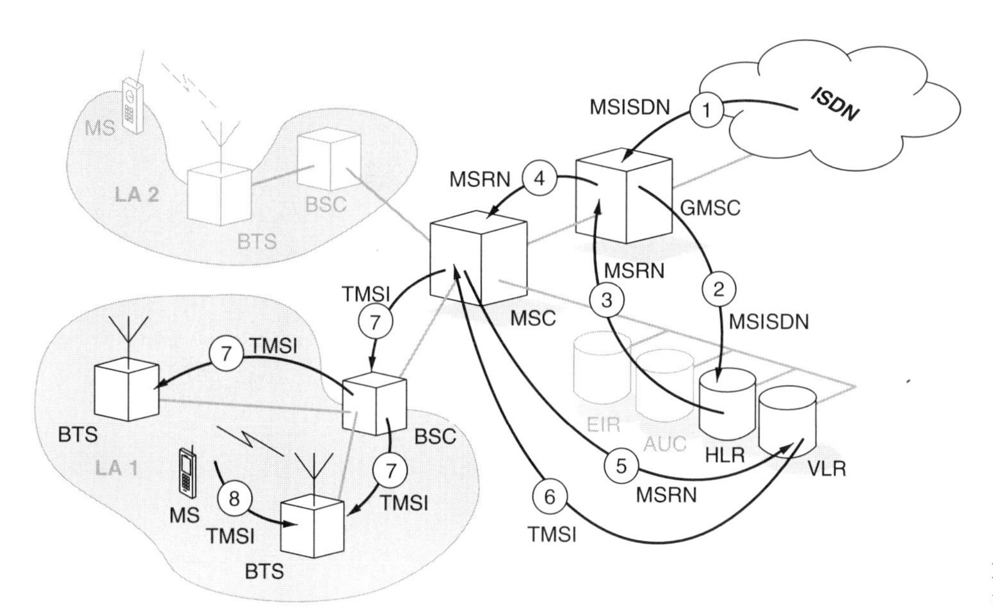

Figure 9. Routing calls to mobile users.

can make or receive calls at that terminal and use other subscribed services. This is possible because the IMEI (identifying the terminal) and the IMSI (identifying the user, stored in the SIM) are independent of each other. Consequently, GSM offers personal mobility in addition to terminal mobility.

5.3. Connection Management

The GSM connection management consists of three entities: call control, supplementary services, and short message service (SMS). The call control handles the tasks related to setting up, maintaining, and taking down calls. The services of call control include the establishment of normal calls (MS-originating and -terminating) and emergency calls (MS-originating only), the termination of calls, dual-tone multifrequency (DTMF) signaling, and incall modifications. The latter allows for changing the bearer service during a connection, for example, from speech to data transmission. Essentially, call control signaling in GSM corresponds to the call setup according to Q.931 in ISDN. Some additional features are incorporated to account for the limited resources in a wireless system. For example, it is possible to queue call requests if no traffic channel is immediately available. Furthermore, the network operator has the option to choose between early and late assignment procedures. With early assignment, the TCH is assigned immediately after acknowledging the call request. In case of late assignment, the call is first processed completely, and the TCH assignment occurs only after the destination subscriber has been called. This variant avoids unnecessary allocation of radio resources if the called subscriber is not available. Thus, the call blocking probability can be reduced.

5.4. Roaming and Handover

GSM supports roaming of mobile users within a network as well as between different GSM networks (as long as a roaming agreement exists between the respective network providers). While roaming within a single network supports the continuation of ongoing connections, roaming from and to other networks terminates the connection and requires a new attachment.

When a mobile user moves within a network and crosses cell borders, the ongoing connection is switched to a different channel through the neighboring BTS (see Fig. 10). This procedure is called *intercell handover* [13]. Handovers may even take place within the same cell, when the signal quality on the current channel becomes insufficient and an alternative channel of the same BTS can offer improved reception. This event is called *intracell handover*.

In case of an intercell handover, GSM distinguishes between intra-MSC handover (if both old and new BTS are connected to the same MSC) and inter-MSC handover (in case the new BTS is connected to a different MSC than the old BTS). In the latter case, the connection is rerouted from the old MSC (MSC-A) and through the new MSC (MSC-B). If during the same connection another inter-MSC handover takes place, either back to MSC-A or to a third one (MSC-C), the connection between MSC-A and MSC-B is taken down and the connection is newly routed from MSC-A to MSC-C (unless MSC-C equals MSC-A). This repeated inter-MSC handover event is called a *subsequent handover* (see Fig. 10).

Since handovers not only induce additional signaling traffic load but also temporarily reduce the speech quality, the importance of a well-dimensioned handover decision algorithm, which should even be locally optimized, is obvious. This is one reason why GSM does not have a standardized uniform algorithm for the determination of the moment of a handover. Network operators can develop and deploy their own algorithms that are optimally tuned for their networks. This is made possible through standardizing only the signaling interface that defines the processing of the handover and through transferring the handover decision to the BSS. The GSM handover is thus a network-originated handover as opposed to a mobile-originated handover, where the handover decision is made by the mobile station. An advantage of the GSM handover

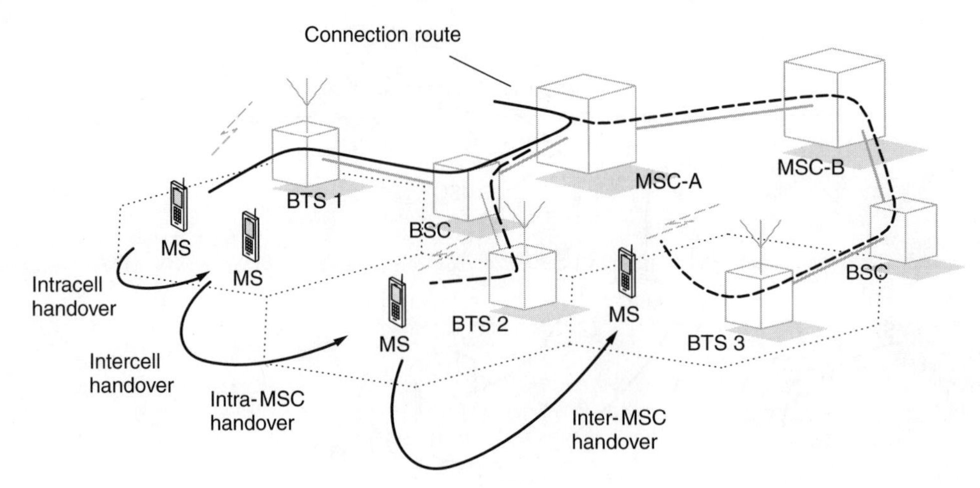

Figure 10. Handover types.

approach is that the software of the MS does not need to be changed when the handover strategy or the handover decision algorithm is changed in the network. Even though the GSM standard does not prescribe a mandatory decision algorithm, a simple algorithm is proposed.

The GSM handover decision is based on measurement data transmitted on the SACCH, most importantly the values RXLEV and RXQUAL. At least 32 values of RXLEV and RXQUAL must be averaged and the resulting mean values are continuously compared with threshold values. Additional handover decision criteria are the distance to the current BTS, an indication of unsuccessful power control, and the pathloss values to neighboring BTSs. The latter are obtained by monitoring the BCCH carriers of nearby cells. Thus, either of the following handover causes can be distinguished:

- Received signal level too low (uplink or downlink)
- Bit error ratio too high (uplink or downlink)
- Power control range exceeded
- Distance between MS and BTS too high
- Pathloss to neighboring BTS is lower than pathloss to the current BTS

A special handover situation exists if the bit error ratio (uplink or downlink) is too high but at the same time the uplink and downlink received signal levels are sufficiently high. This strongly hints at severe cochannel interference. This problem can be solved with an intracell handover, which the BSS can perform on its own without support from the MSC.

Once the decision on an intercell handover is made, a target cell must be chosen. Therefore it is determined which neighboring cell's BCCH is received with sufficient signal level. All potential target channels with lower path loss than the current channel are then reported to the MSC with the message HANDOVER REQUIRED. Once the target cell has been determined, the BSS sends the signaling message HANDOVER COMMAND to the MS on the FACCH. The HANDOVER COMMAND contains the new channel configuration and also information about the new cell (e.g., BTS identity and BCCH frequency) and a handover reference number. On reception of the HANDOVER COMMAND, the mobile station interrupts the current connection, deactivates the old physical channel, and switches over to the channel newly assigned in the HANDOVER COMMAND. On the FACCH, the mobile sends the message HANDOVER ACCESS in an access burst to the new base station. In case both BTSs have synchronized their TDMA transmission, the access burst is sent in exactly four subsequent time slots of the FACCH. Thereafter, the mobile station activates the new physical channel in both directions, activates encryption, and sends the message HANDOVER COMPLETE to the BSS. In case the BTSs are not synchronized, the mobile station repeats the access burst until either a timer expires (handover failure) or the BTS answers with a message PHYSICAL INFORMATION that contains the currently needed timing advance to enable the mobile station to activate the new physical channel.

To avoid a series of repeated handover events at the cell boundary where varying channel conditions can lead to frequent changes of the pathloss to both base stations, a hysteresis is used in GSM. This is done by defining a handover margin for each cell. A handover will be performed only if the path-loss difference of both base stations is higher than the handover margin of the potential new cell.

6. SECURITY ASPECTS

The wireless transmission over the air interface leads to the danger that unauthorized people may eavesdrop on the communication of a subscriber or that they use radio resources at the cost of the network provider. The security functions of GSM protect against unauthorized use of services (by authentication), provide data confidentiality (by encryption), and ensure the confidentiality of the subscriber identity [14].

Authentication is needed to verify that users are who they claim to be. Each subscriber has a secret key (*subscriber authentication key* Ki), which is stored in the SIM as well as in the AUC. The SIM is protected by a *personal identity number* (PIN) that is known only by the subscriber. To authenticate a user, the network sends a random number to the mobile station. Using the key Ki and a specified algorithm, the mobile station calculates a *signature response* (SRES) from the random number and sends it back to the network. If the SRES value received from the MS matches the value calculated at the network side, the user is authenticated. In addition to user authentication, the mobile terminal must also be authenticated. On the basis of the IMEI, the network can restrict access for terminals that are listed in the EIR as stolen or corrupted.

The encryption of speech and data is a feature of GSM that is not supported in analog cellular and fixed ISDN networks. It is used to protect the transmission over the air interface against eavesdropping. As illustrated in the transmission chain of Fig. 4, data coming from the channel encoder is encrypted before it is passed to the multiplexer. The initial random number and the key Ki are used on both sides, the network and the MS, to calculate a *ciphering key* Kc. This key is then employed by the encryption algorithm for the symmetric ciphering of user data and signaling information. On the receiver side, the key Kc can be used to decipher the data.

Another security feature in GSM is that a user's identity remains undisclosed to listeners on the radio channel. The IMSI uniquely identifies a subscriber worldwide and should thus not be transmitted over the air interface. Thus, in order to protect his/her identity on network attach, the subscriber obtains a temporary identifier, the TMSI, from the network. The TMSI is used instead of the IMSI for transmission over the air interface. The association between the two values is stored in the VLR. The TMSI changes frequently and has only local validity within a location area. Thus, an attacker listening on the radio channel cannot deduce the subscriber's identity from the TMSI.

7. BIBLIOGRAPHIC NOTES

Article 15 is an early survey article mainly describing the system architecture and signaling protocols. The GSM book [2] covers both network aspects (system architecture, protocols, services, roaming) and transmission aspects. It also includes a detailed description of the GSM phase 2+ services. The book [16] also contains a detailed GSM part. The GSM chapter of Ref. 4 covers the physical layer at the air interface in detail.

BIOGRAPHIES

Christian Bettstetter is a research and teaching staff member at the Institute of Communication Networks at Technische Universität München TUM, Germany. He graduated from TUM in electrical engineering and information technology Dipl.-Ing. in 1998 and then joined the Institute of Communication Networks, where is he working toward his Ph.D. degree. Christian's interests are in the area of mobile communication networks, where his current main research area is wireless ad-hoc networking. His interests also include 2G and 3G cellular systems and protocols for a mobile Internet. He is coauthor of the book *GSM–Switching, Services and Protocols* (Wiley/Teubner) and a number of articles in journals, books, and conferences.

Christian Hartmann studied electrical engineering at the University of Karlsruhe (TH), Germany, where he received the Dipl.-Ing. degree in 1996. Since 1997, he has been with the Institute of Communication Networks at the Technische Universität München, Germany, as a member of the research and teaching staff, pursuing a doctoral degree. His main research interests are in the area of mobile and wireless networks including capacity and performance evaluation, radio resource management, modeling, and simulation. Christian Hartmann is a student member of the IEEE.

BIBLIOGRAPHY

1. ETSI, *GSM 03.02: Network Architecture*, Technical Specification, 2000.
2. J. Eberspächer, H.-J. Vögel, and C. Bettstetter, *GSM — Switching, Services, and Protocols*, 2nd ed., Wiley, Chichester, March 2001.
3. ETSI, *GSM 05.02: Multiplexing and Multiple Access on the Radio Path*, Technical Specification, 2000.
4. R. Steele and L. Hanzo, eds., *Mobile Radio Communications*, 2nd ed., Wiley, 1999.
5. R. Salami et al., Description of the GSM enhanced full rate speech codec, in *Proc. IEEE Int. Conf. Communications (ICC'97)*, Montreal, Canada, June 1997, 725–729.
6. J. Adoul, P. Mabilleau, M. Delprat, and S. Morisette, Fast CELP coding based on algebraic codes, *Proc. ICASSP*, April 1987, pp. 1957–1960.
7. D. Bruhn, E. Ekudden, and K. Hellwig, Adaptive multi-rate: a new speech service for GSM and beyond, *Proc. 3rd ITG Conf. Source and Channel Coding*, Munich, Germany, Jan. 2000.
8. ETSI, *GSM 05.03: Channel Coding*, Technical Specification, 1999.
9. ETSI, *GSM 05.01: Physical Layer on the Radio Path; General Description*, Technical Specification, 2000.
10. ETSI, *GSM 05.10: Radio Subsystem Synchronization*, Technical Specification, 2001.
11. ETSI, *GSM 05.08: Radio Subsystem Link Control*, Technical Specification, 2000.
12. ETSI, *GSM 04.08: Mobile Radio Interface Layer 3 Specification*, Technical Specification, 2000.
13. ETSI, *GSM 03.09: Handover Procedures*, Technical Specification, 1999.
14. ETSI, *GSM 03.20: Security Related Network Functions*, Technical Specification, 2001.
15. M. Rahnema, Overview of the GSM system and protocol architecture, *IEEE Commun. Mag.* 92–100 (April 1993).
16. B. Walke, *Mobile Radio Networks*, 2nd ed., Wiley, Chichester, New York, 2002.

H.324: VIDEOTELEPHONY AND MULTIMEDIA FOR CIRCUIT-SWITCHED AND WIRELESS NETWORKS*

DAVE LINDBERGH
Polycom, Inc.
Andover, Massachusetts

BERNHARD WIMMER
Siemens AG
Munich, Germany

1. INTRODUCTION

ITU-T Recommendation H.324 [1] is the international standard for videotelephony and real-time multimedia communication systems on low-bit-rate circuit-switched networks. The basic H.324 protocol can be used over almost any circuit-switched network, including modems on the PSTN (public switched telephone network, often known as POTS — "plain old telephone service"), on ISDN networks, and on wireless digital cellular networks.

H.324 is most commonly used for dialup videotelephony service over modems, and as the basis for videotelephony service in the Universal Mobile Telecommunications System (UMTS) of the Third Generation Partnership Project (3GPP). The standard enables interoperability among a diverse variety of terminal devices, including PC-based multimedia videoconferencing systems, inexpensive voice/data modems, encrypted telephones, and remote security cameras, as well as standalone videophones.

H.324 is a "toolkit" standard that gives implementers flexibility to decide which media types (audio, data, video, etc.) and features are needed in a given product, but ensures interoperability by specifying a common baseline mode of operation for all systems that support a given feature. In addition to the baseline modes, H.324 allows other optional modes, standard or nonstandard, which may be better in various ways, to be used automatically if both ends have the capability to do so.

Above all, H.324 is designed to provide the best performance possible (video and audio quality, delay, etc.) on low-bit-rate networks. This is achieved primarily by reducing protocol overhead to the minimum extent possible, and by designing the multiplexer and channel aggregation protocols to allow different media channels to interleave frequently and flexibly, minimizing latency. These protocol optimizations mean that H.324 is not suitable for use on packet-switched networks, including IP networks, because H.324 depends on reasonably constant end-to-end latency in the connection, which can't be guaranteed in packet-routed networks.

The design of the H.324 standard benefits from industry's earlier experience with ITU-T H.320, the widespread international standard for ISDN videoconferencing, approved in 1990. H.324 shares H.320's basic architecture, consisting of a multiplexer that mixes the various media types into a single bitstream (H.223), audio and video compression algorithms (G.723.1 and H.263), and a control protocol that performs automatic capability negotiation and logical channel control (H.245). Other parts of the H.324 set, such as H.261 video compression, H.233/234 encryption, and H.224/281 far-end camera control, come directly from H.320. One of the considerations in the development of H.324 was practical interworking, through a gateway, with the installed base of H.320 ISDN systems, as well as with the ITU-T standards for multimedia on LANs and ATM networks, H.323 and H.310 respectively, and the ITU-T T.120 series of data conferencing standards. Table 1 compares H.324 with H.320 and the other major ITU-T multimedia conference standards.

As a second-generation standard, the design of H.324 was able to avoid the limitations of H.320. As a result, H.324 has features missing from (or retrofitted to) H.320 such as receiver-controlled mode preferences, the ability to support multiple channels of each media type, and dynamic assignment of bandwidth to different channels.

1.1. Variations

The original version of the H.324 standard was approved in 1995, exclusively for use on PSTN connections over

* This article is based on the article entitled The H.324 Multimedia Communication Standard, by Dave Lindbergh, which appeared in *IEEE Communications Magazine*, Dec. 1996, Vol. 24, No. 12, pp. 46–51. © 1996 IEEE.

Table 1. Major ITU-T Multimedia Conferencing Standards (Basic Modes)

Standard	Initial Approval	Networks Supported	Baseline Video	Baseline Audio	Multiplex	Control
H.324	1995	PSTN, ISDN, wireless	H.263	G.723.1	H.223	H.245
H.320	1990	ISDN	H.261	G.711	H.221	H.242
H.323	1996	Internet, LANs, Intranets	H.261	G.711	H.225.0	H.245
H.310	1996	ATM/B-ISDN	H.262	MPEG-1	H.222	H.245

dialup V.34 modems, at rates of up to 28,800 bps. Since then, modems have increased in speed and H.324 has been extended to support not only PSTN connections but also ISDN connections and wireless mobile radio links.

The original PSTN version of H.324 is called "H.324." The ISDN version is called "H.324/I," and the mobile version "H.324/M." The specific variant of H.324/M used in UMTS is called "3G-324" (see Table 2).

The requirements, capabilities, and protocols during a call are essentially the same for all variants of H.324. The main differences are the network interfaces and call setup procedures, and in the case of H.324/M, the use of a variant of the H.223 multiplex that is more resilient to the very high bit error rates encountered on wireless networks.

We begin by describing the base PSTN version of H.324, and describe the differences in the other variations afterwards.

2. SYSTEM OVERVIEW

Figure 1 illustrates the major elements of the H.324 system. The mandatory components are the V.34 modem (for PSTN use), H.223 multiplexer, and H.245 control protocol. The data, video, and audio streams are optional, and several of each kind may be present, set up dynamically during a call. H.324 terminals negotiate a common set of capabilities with the far end, independently for each direction of transmission (terminals with asymmetric capabilities are allowed). Multipoint operation in H.324, in which three or more sites can join in the same call, is possible through the use of a multipoint control unit (MCU, also known as a "bridge").

Table 2. Variants of H.324

Name	Description	Main Differences
H.324	For PSTN over modems	—
H.324/I	For ISDN	Network interface, call setup
H.324/M	For mobile wireless networks	Network interface, error-robust multiplex
3G-324M	Variant of H.324/M for 3GPP UMTS wireless video telephony	Network interface, error-robust multiplex, AMR audio codec instead of G.723.1

3. MODEM

When operating on PSTN circuits, H.324 requires support for the full-duplex V.34 modem [2], which runs at rates of up to 33,600 bps. The modem can operate at lower rates, in steps of 2400 bps, as line conditions require. H.324 is intended to work at rates down to 9600 bps. The mandatory V.8 or optional V.8bis protocol is used at call startup to identify the modem type and operation mode. In most implementations, the preferred V.8bis protocol allows a normal voice telephone call to switch into a multimedia call at any time. ITU-T V.25ter (the "AT" command set) is used to control local modem functions such as dialing. The V.34 modulation/demodulation is quite complex, and typically adds 30–40 ms of end-to-end delay.

An important V.34 characteristic for H.324 system design is its error burst behavior. Because of the V.34 trellis coder and other factors, bit errors rarely occur singly, but instead in bursts. The rate of these error bursts is also highly dependent on the bit rate of the modem. A single "step" of 2400 bps changes the error rate by more than an order of magnitude. This influences the proper "tuning" of the modem for use with H.324: a modem used for simple V.42 data transfer is often more efficient at the "next higher" rate, since retransmissions will consume less bandwidth than will the loss from stepping down. But a H.324 system should be run at the "next lower" rate,

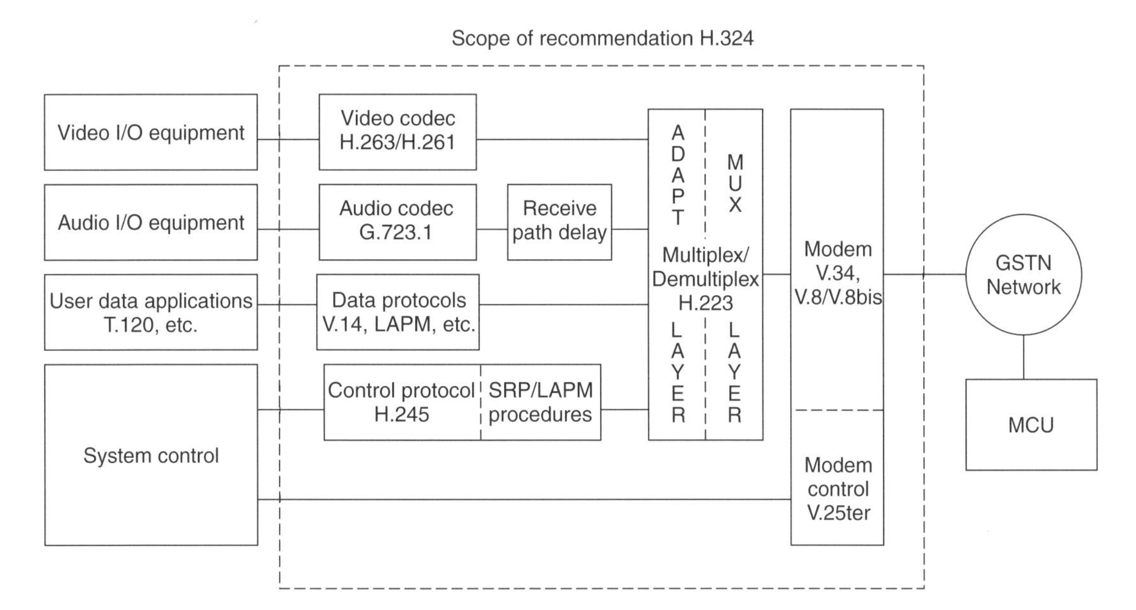

Figure 1. H.324 block diagram.

since the real-time requirements make retransmission inappropriate in many cases, requiring a lower error rate.

H.324 uses the V.34 modem directly as a synchronous data pump, to send and receive the bitstream generated by the H.223 multiplexer. Data compression such as V.42bis, and retransmission protocols such as V.42 link-access procedure for modems (LAPM) or the Microcom network protocol (MNP) are not used at the modem level, although these same protocols can be used at a higher layer for some types of data streams.

Like the V.42 error correction protocol supported in most modems, H.324 requires a synchronous interface to the V.34 data pump. Since most PCs have only asynchronous RS-232 modem interfaces, PC-based H.324 terminals require either synchronous interface hardware or some other method to pass the synchronous bitstream to the modem over the PC's asynchronous interface. Many modems support ITU-T V.80, which provides a standard sync-over-async "tunneling" protocol between the PC and the modem for this purpose.

Faster modems such as V.90 can also be used, but in most cases these modems offer asymmetric bit rates, where the modem is able to receive data faster than it can transmit. When two such modems interconnect, both sides transmit at the lower rate, yielding little or no improvement over V.34 speeds.

When operating on circuit-switched networks other than the PSTN, the modem is replaced by an equivalent interface that allows the output from the H.223 multiplex to be carried over the network in use. H.324 Annex C specifies operation of H.324 on wireless (digital cellular and satellite) networks, and H.324 Annex D specifies H.324 operation on ISDN.

4. CALL SETUP AND TEARDOWN

The setup and teardown of a H.324 call proceeds in seven phases, phase A–phase G.

In *phase A*, an ordinary telephone call is setup using the normal procedures for dialing, ringing, and answering. This can be completely manual, with the user dialing and answering by hand; or automatic, with the V.25ter protocol (the "AT" command set) used to control the modem's dialing and answering functions.

Once the call is connected, the H.324 terminal can immediately start the multimedia call, or, if both terminals support the optional V.8bis protocol, the two users can choose to have an ordinary voice conversation first. This is called *phase B*, which can continue indefinitely, until one of the users decides to switch the call into H.324 multimedia mode. V.8bis allows this "late start" mode, in which an ordinary phone call can be switched into multimedia mode at any time. On the ISDN network, the V.140 protocol substitutes for V.8bis.

In *phase C*, digital communication between the terminals is set up. For ISDN and digital wireless networks, this is inherent in the network connection. For PSTN calls, either terminal starts the modem negotiation by sending V.8 or V.8bis messages, using V.23 300 bps FSK modulation (which doesn't require any "training" time). The modems exchange capabilities and select V.34 modulation and H.324 protocol. A PC or other external DTE can control this V.8 or V.8bis negotiation using the procedures of V.25ter Annex A. If V.8bis is used, the terminal's gross ability to support H.324 video, audio, data, or encryption is also communicated, to quickly determine if the desired mode is available. The V.34 startup procedure then takes about 10 seconds, after which time end-to-end digital communication is established at the full V.34 data rate allowed by the quality of the telephone connection, up to 33,600 bps.

Phase D then starts, in which the H.324 terminals first communicate with each other on the H.245 control channel. The terminals are initialized, and detailed terminal capabilities are exchanged using H.245. This happens very quickly (less than 1 second), as the full connection bit rate is available for transfer of this control information. Logical channels can then be opened for the various media types, and table entries can be downloaded to support those channels in the H.223 multiplexer.

At this point the terminal enters *phase E*, normal multimedia communication mode. The number and type of logical channels can be changed during the call, and each terminal can change its capabilities dynamically.

Phase F is entered when either user wants to end the call. All logical channels are closed, and an H.245 message tells the far-end terminal to terminate the H.324 portion of the call, and specifies the new mode (disconnect, back to voice mode, or other digital mode) to use.

Finally, in *phase G*, the terminals disconnect, return to voice telephone mode, or go into whatever new mode (fax, V.34 data, etc.) was specified.

5. MULTIPLEX

H.324 uses a new multiplexer standard, H.223, to mix the various streams of video, audio, data, and the control channel, together into a single bitstream for transmission. The H.324 application required the development of a new multiplexing method, as the goal was to combine low multiplexer delay with high efficiency and the ability to handle bursty data traffic from a variable number of logical channels.

Time-division multiplexers (TDMs), such as H.221, were considered unsuitable because they can't easily adapt to dynamically changing modem and payload data rates, and are difficult to implement in software because of complex frame synchronization and bit-oriented channel allocation.

Packet multiplexers, such as V.42 (LAPM) and Q.922 (LAPF), avoid these problems but suffer from "blocking delay," where transmission of urgent data, such as audio, must wait for the completion of a large packet already started. This problem occurs when the underlying channel bit rate is low enough to make the transmission time of a single packet significant. In a packet multiplexer, this delay can be reduced only by limiting the maximum packet size or aborting large packets, both of which reduce efficiency.

The H.223 multiplexer combines the best features of TDM and packet multiplexers, along with some new ideas. It incurs less delay than TDM and packet multiplexers, has low overhead, and is extensible to multiple channels of each data type. H.223 is byte-oriented for ease of implementation, able to byte-fill with flags to match differing data rates, and uses a unique synchronization character that cannot occur in valid data. In H.223, each HDLC (*h*igh-level *d*ata *l*ink *c*ontrol) framed [3] multiplex–protocol data unit (MUX-PDU) can carry a mix of different data streams in different proportions, allowing fully dynamic allocation of bandwidth to the different channels.

H.223 consists of a lower multiplex layer, which actually mixes the different media streams, and a set of adaptation layers that perform logical framing, sequence numbering, error detection, and error correction by retransmission, as appropriate to each media type.

5.1. Multiplex Layer

The multiplex layer uses normal HDLC zero insertion, which inserts a 0 bit after every sequence of five 1 bits, making the HDLC flag (01111110) unique. The multiplex consists of one or more flags followed by a one-byte header, followed by a variable number of bytes in the information field. This sequence repeats. Each sequence of header byte and information field is called a MUX-PDU, as shown in Fig. 2.

The header byte includes a multiplex code that specifies, by reference to a multiplex table, the mapping of the information field bytes to various logical channels. Each MUX-PDU may contain a different multiplex code, and therefore a different mix of logical channels.

The selected multiplex table entry specifies a pattern of bytes and corresponding logical channels contained in the information field, which repeats until the closing flag. Each MUX-PDU may contain bytes from several different logical channels, and may select a different multiplex table entry, so bandwidth allocation may change with each MUX-PDU. This allows many logical channels, low-overhead switching of bandwidth allocation, and many different types of channel interleave and priority. All of this is under control of the transmitter, which may choose any appropriate multiplex for the application, and change multiplex table entries as needed. Many syntactically compliant multiplexing algorithms, optimized for different applications, are possible [4].

Figure 3 illustrates only four logical channels (audio, video, data, and control), but there may be any number of channels, as specified by the multiplex table. This allows multiple audio, video, and data channels for different data protocols, multiple languages, continuous presence, or other uses.

A slightly different, but equally valid, way of thinking about H.223 is as a continuous stream of bit-stuffed bytes carrying information in a given pattern of logical channels. This pattern (again refer to Fig. 3) is occasionally interrupted by an "escape sequence" consisting of an HDLC flag followed by a single "header" byte. The header byte contains a multiplex code that indicates a change to a new pattern of logical channel bytes.

Generally, an HDLC flag may be inserted on any byte boundary, terminating the previous MUX-PDU and beginning a new one. Unlike traditional packet multiplexers, H.223 can interrupt a adaptation-layer variable-length packet at any byte boundary to reallocate bandwidth, such as when urgent real-time data must be sent. The reallocation can occur within 16–24 bit times (16 bits for the flag and header, plus 0 to 8 bits for the byte already being transmitted), less than 1 ms at 28,800 bps. This compares very favorably with both TDM muxes and packet multiplexes which suffer from "blocking delay."

H.223 maintains a 16-entry multiplex table at all times, selected by 4 bits of the MUX-PDU header byte. Entries in the multiplex table may be changed during the call, to meet requirements as various logical channels are opened or closed.

Of the remaining 4 bits in the header, 3 are used as a CRC (cyclic redundancy code) on the multiplex code. The final bit is the packet marker (PM) bit, which is used for marking the end of some types of higher-level packets.

5.2. Adaptation Layer

Different media types (audio, video, data) require different levels of error protection. For example, T.120 application data are relatively delay-insensitive, but require full error correction. Real-time audio is extremely delay sensitive, but may be able to accept occasional errors with only minor degradation of performance. Video falls between these two extremes. Ideally, each media type should use an error-handling scheme appropriate to the requirements of the application.

The H.223 adaptation layers provide this function, working above the multiplex layer on the unmultiplexed sequence of logical channel bytes. H.223 defines three adaptation layers, AL1, AL2, and AL3. AL1 is intended primarily for variable-rate framed information, such as HDLC protocols and the H.245 control channel. AL2 is intended primarily for digital audio such as G.723.1, and includes an 8-bit CRC and optional sequence numbers. AL3 is intended primarily for digital video such as H.261 and H.263, and includes provision for retransmission using sequence numbers and a 16-bit CRC.

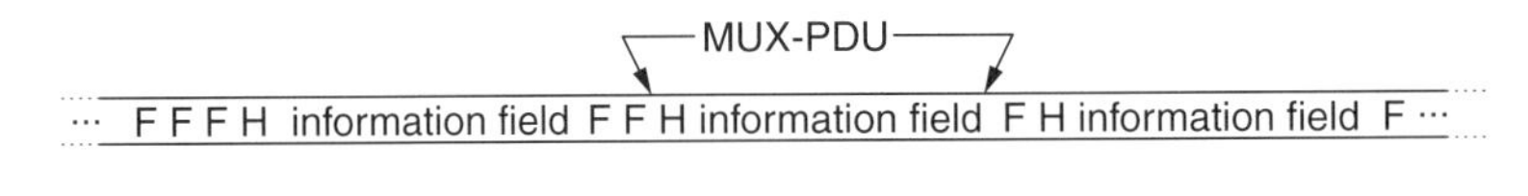

Figure 2. MUX-PDU in an H.223 multiplex stream (F = HDLC flag, H = header byte).

...VVVCVVVCV **FH** AAAAVVDCAAAAVVDCAAAAVV **FH** VDVDVDVDVDVDV **FH** CCCCCC **FH** AAAAVVDCAAA...

(Each letter represents one byte, F = Flag, H = Header, A = Audio, V = Video, D = Data, C = Control)

Figure 3. Example of an H.223 multiplex bitstream.

5.3. Encryption

Encryption is an option in H.324, and makes use of ITU-T H.233 and H.234, both of which were originally developed for use with H.320. H.233 covers encryption procedures and algorithm selection, in which FEAL, B-CRYPT, and DES may be used as well as other standardized and nonstandardized algorithms. H.234 covers key exchange and authentication using the ISO 8732, Diffie-Hellman, or RSA algorithms.

In H.324, encryption is applied at the H.223 multiplexer, where the encryptor produces a pseudorandom bitstream (cipher stream), which, prior to flag insertion and HDLC zero insertion, is exclusive-ORed with the output of the H.223 multiplexer. The exclusive-OR procedure is not applied to the H.223 MUX-PDU header and the H.245 control channel. The receiver reverses this process.

6. CONTROL

H.324 uses the H.245 multimedia system control protocol, which is also used by H.310 (for ATM networks) and H.323 (for packet-switched LANs and corporate intranets). Many of the extensions added to H.245 to support H.323 on IP networks are not used by H.324.

The control model of H.245 is based on "logical channels," independent unidirectional bitstreams with defined content, identified by unique numbers arbitrarily chosen by the transmitter. There may be up to 65,535 logical channels.

The H.245 control channel carries end-to-end control messages governing operation of the H.324 system, including capabilities exchange, opening and closing of logical channels, mode preference requests, multiplex table entry transmission, flow control messages, and general commands and indications. The H.245 structure allows future expansion to additional capabilities, as well as manufacturer-defined nonstandard extensions to support additional features.

H.245 messages are defined using ASN.1 syntax, coded according to the packed encoding rules (PERs) [5] of ITU-T X.691, which provide both clarity of definition and flexibility of specification. In H.324, the H.245 control channel runs over logical channel (LC) 0, a separate channel out of band from the various media streams. LC 0 is considered to be already open when the H.324 call starts up.

Within LC 0, H.245 is carried on a numbered simplified retransmission protocol (NSRP) based on V.42 exchange identification (XID) procedures. Optionally, the full V.42 protocol may be used instead. H.245 itself assumes that this link layer guarantees correct, in-order delivery of messages.

6.1. Capabilities Exchange

The large set of optional features in H.324 necessitates a method for the exchange of capabilities, so that terminals can become aware of the common subset of capabilities supported by both ends.

H.245 capabilities exchange provides for separate receive and transmit capabilities, as well as a system by which the terminal may describe its ability (or inability) to operate in various combinations of modes simultaneously, as some implementations are limited in processing cycles or memory availability.

Receive capabilities describe the terminal's ability to receive and process incoming information streams. Transmitters are required to limit the content of their transmitted information to that which the receiver has indicated it is capable of receiving. The absence of a receive capability indicates that the terminal cannot receive (is a transmitter only).

Transmit capabilities describe the terminal's ability to transmit information streams. They serve to offer receivers a choice of possible modes of operation, so the receiver can request the mode it prefers to receive using the H.245 RequestMode message. This is an important feature, as local terminals directly control only what they transmit, but users care about controlling what they receive. The absence of a transmit capability indicates that the terminal is not offering a choice of preferred modes to the receiver (but it may still transmit anything within the capability of the receiver).

Terminals may dynamically add or remove capabilities during a call. Since many H.324 implementations are on general-purpose PCs, other application activity on the machine can result in changing resource levels available to H.324. H.245 is flexible enough to handle such a scenario.

Nonstandard capabilities and control messages may be issued using the NonStandardParameter structure defined in H.245. This allows nonstandardized features to be supported automatically, in the same way as standardized features.

6.2. Logical Channel Signaling

H.324 terminals transmit media information from transmitter to receiver over unidirectional streams called *logical channels* (LCs). Each LC carries exactly one channel of one media type, and is identified by a logical channel number (LCN) arbitrarily chosen by the transmitter. Since transmitters completely control allocation of LCNs, there is no need for end-to-end negotiation of LCNs.

When a transmitter opens a logical channel, the H.245 OpenLogicalChannel message fully describes the content of the logical channel, including media type, codec in use, H.223 adaptation layer and any options, and all other information needed for the receiver to interpret the content of the logical channel. Logical channels may be closed when no longer needed. Open logical channels may be inactive, if the information source has nothing to send.

Logical channels in H.324 are unidirectional, so asymmetrical operation, in which the number and type of information streams is different in each direction of transmission, is allowed. However, if a terminal is capable only of certain symmetrical modes of operation, it may send a capability set that reflects its limitations.

6.3. Bidirectional Logical Channels

Certain media types, including data protocols such as T.120 and LAPM, and video carried over AL3, inherently require a bidirectional channel for their operation. In such cases a pair of unidirectional logical channels, one in each direction, may be opened and associated together to form a bidirectional channel. Such pairs of associated channels need not share the same logical channel number, since logical channel numbers are independent in each direction of transmission. To avoid race conditions that could cause duplicate sets of bidirectional channels to be opened, a symmetry-breaking procedure is used, in which master and slave terminals are chosen on the basis of random numbers. There is no advantage to being master or slave.

7. AUDIO CHANNELS

The baseline audio mode for H.324 is the G.723.1 speech coder/decoder (codec), which runs at 5.3 or 6.4 kbps. The 3G-324M variant of H.324 for 3G wireless networks uses the adaptive multirate (AMR) codec as the baseline.

G.723.1 provides near-toll-quality speech, using a 30 ms frame size and 7.5 ms lookahead. A G.723.1 implementation is estimated to require 18–20 fixed-point (MIPS) (million instructions per second) in a general-purpose DSP (digital signal processor). Transmitters may use either of the two rates, and can change rates for each transmitted frame, as the coder rate is sent as part of the syntax of each frame. The average audio bit rate can be lowered further by using silence suppression, in which silence frames are not transmitted, or are replaced with smaller frames carrying background noise information. In typical conversations both ends rarely speak at the same time, so this can save significant bandwidth for use by video or data channels.

Receivers can use a H.245 message to signal their preference for low- or high-rate audio. The audio channel uses H.223 adaptation layer AL2, which includes an 8-bit cyclic redundancy code (CRC) on each audio frame or group of frames.

The G.723.1 codec imposes about 97.5 ms of end-to-end audio delay, which, together with modem, jitter buffer, transmission time, multiplexer, and other system delays, results in about 150 ms of total end-to-end audio delay (exclusive of propagation delay) [6]. On ISDN connections, modem latency is eliminated, leading to about 115 ms end-to-end delay. These audio delays are generally less than that of the video codec, so additional delay needs to be added in the audio path if lip synchronization is desired. This is achieved by adding audio delay in the receiver only, as shown in Fig. 1. H.245 is used to send a message containing the time skew between the transmitted video and audio signals. Since the receiver knows its decoding delay for each stream, the time skew message allows the receiver to insert the correct audio delay, or alternatively, to bypass lip synchronization, and present the audio with minimal delay. While multiple channels of audio can be transmitted, H.324 does not currently provide for the exact sample-level channel synchronization needed for stereo audio.

Many H.324 applications do not require lip synchronization, or do not require video at all. For these applications, optional H.324 audio codecs such as G.729, an 8-kbps speech codec which can reduce the total end-to-end audio delay to about 85 ms for PSTN use, or 50 ms on ISDN, can be important.

Other optional audio modes can also be used, such as the wideband ITU-T G.722.1 codec, which offers 7-kHz audio bandwidth (approximately double that of conventional telephone lines) at rates of 24 or 32 kbps. Nonstandard audio modes can be also used in the same manner as standardized codecs.

8. VIDEO CHANNELS

H.324 can send color motion video over any desired fraction of the available modem bandwidth. H.324 supports both H.263 and H.261 for video coding. H.263 is the preferred method, with H.261 available to allow interworking with older ISDN H.320 videoconferencing systems without the need to convert video formats, which would add an unacceptable delay.

H.263 is based on the same video compression techniques as H.261, but includes many enhancements, including much improved motion compensation, which result in H.324 video quality estimated equivalent to H.261 at a 50–100% higher bitrate. This dramatic improvement is most apparent at the low bit rates used by H.324; the difference is less when H.263 is used at higher bitrates. H.263 also includes a broader range of picture formats, as shown in Table 3.

H.324 video can range from 5 to 30 frames/second, depending on bitrate and picture format, H.263 options in use, and the amount of complexity and movement in the scene. A H.245 control message, the videoTemporalSpatialTradeOff, feature allows the receiver to specify a preference for the tradeoff between frame rate and picture resolution.

Video channels use H.223 adaptation layer AL3, which includes a 16-bit CRC and sequence numbering, and provision for retransmission of errored video data, at the option of the receiver.

Since H.324 systems can support multiple channels of video (although bandwidth constraints may make this impractical for many applications), continuous-presence multipoint operation, in which separate images of each transmitting site are presented at the receiver "Hollywood Squares" style, can be easily implemented via multiple logical channels of video. It is up to the receiver to locally arrange an appropriate set of the channels for display.

9. DATA CHANNELS

The H.324 multimedia system can carry data channels as well as video and audio channels. These can be used for any

Table 3. Video Picture Formats

Picture Format	Luminance Pixels	H.324 Video Decoder Requirements	
—	—	H.261	H.263
SQCIF	128 × 96 for H.263[a]	Optional[a]	Required
QCIF	176 × 144	Required	Required
CIF	352 × 288	Optional	Optional
4CIF	704 × 576	Not defined	Optional
16CIF	1408 × 1152	Not defined	Optional
Custom formats	≤2048 × 1152	Not defined	Optional

[a]H.261 SQCIF is any active size less than QCIF, filled out by a black border, coded in QCIF format.

data protocol or application in the same way as an ordinary modem. Standardized protocols include T.120 data [7] for conferencing applications such as electronic whiteboards and computer application sharing, user data via V.14 or V.42 (with retransmission), T.84 still image transfer, T.434 binary file transfers, H.224/H.281 for remote control of far-end cameras with pan, tilt, and zoom functions, and ISO/IEC TR9577 network-layer protocols [8] such as IP (Internet protocol) and IETF PPP (point-to-point protocol) [9], which can be used to provide Internet access over H.324. As with other media types, H.324 provides for extension to non-standard protocols, which can be negotiated automatically and used just like standardized protocols.

The same capability exchange and logical channel signaling procedures defined for video and audio channels are also used for data channels, so automatic negotiation of data capabilities can be performed. This represents a major step forward from current practice on data-only PSTN modems, where data protocols and applications to be used must generally be arranged manually by users before a call.

As with all other media types, data channels are carried as distinct logical channels over the H.223 multiplexer, which can accommodate bursty data traffic by dynamically altering the allocation of bandwidth among the different channels in use. For instance, a video channel can be reduced in rate (or stopped altogether) when a data channel requires additional bandwidth, such as during a file or still image transfer.

10. MULTIPOINT OPERATION AND H.320 ISDN INTEROPERABILITY

H.324 terminals can directly participate in multipoint calls, in which three or more participants join the same call, through a central bridge called a multipoint control unit (MCU). Since the connections on each link in a multipoint call may be operating at different rates, MCUs can send H.245 FlowControlCommand messages to limit the overall bit rate of one or more logical channels, or the entire multiplex, to a supportable common mode.

A similar situation arises when a H.324 terminal interoperates with an H.320 terminal on the ISDN. For interworking with H.320 terminals, an H.324/H.320 gateway can transcode the H.223 and H.221 multiplexes, and the content of control, audio, and data logical channels between the H.324 and H.320 protocols. If the H.320 terminal doesn't support H.263, then H.261 (QCIF) Quarter CIF video can be used to avoid the delay of video transcoding. In this case the gateway, like the MCU in the multipoint case, can send the H.245 FlowControlCommand to force the transmitted H.324 video bit rate to match the H.320 video bit rate in use by the H.221 multiplexer.

One way for a dual-mode (H.320 ISDN and H.324 PSTN) terminal on the ISDN to work directly with H.324 terminals on the PSTN is by using a "virtual modem" on the ISDN, which generates a G.711 audio bitstream representing the V.34 analog modem signal.

11. CHANNEL AGGREGATION FOR MULTILINK OPERATION

H.324 Annex F specifies the use of the H.226 channel aggregation protocol, which allows H.324 PSTN and ISDN calls to aggregate the capacity of multiple separate connections into one faster channel. This can provide improved video quality, compared to what can be delivered by a single V.34 or V.90 connection (at no more than 56 kbps) or by a single ISDN B-channel (at 64 kbps).

Although H.226 is a general-purpose channel aggregation protocol, it was designed specifically to meet the requirements of H.324. Modems like V.34 vary their bit rate as telephone-line noise conditions change. When multiple modems are used on separate lines, each modem changes its bit rate independently, so the bitrate available on each channel varies over time without channel-to-channel coordination.

This characteristic rules out the use of conventional synchronous channel aggregation protocols like H.221 and ISO/IEC CD 13871 "BONDING" [10]. These synchronous protocols work by distributing units of data to each channel in a fixed pattern, "spreading out" a higher-rate data stream over a number of lower-rate channels. The fixed pattern avoids the need to send overhead data to tell the receiver to how to reconstruct the original data. With no overhead, the units of distributed data can be arbitrarily small, resulting in very little latency added by the channel aggregation protocol, as well as high efficiency. However, because the distribution pattern is fixed, these synchronous protocols cannot make use of channels with arbitrary or varying bitrates, such as those provided by V.34 modems.

Packet-oriented channel aggregation protocols avoid this problem, but cannot simultaneously provide both very low latency and very low overhead. In a packet-oriented channel aggregation protocol, such as the "PPP Multilink Protocol" of RFC 1990 [11], data are repeatedly divided into packets, and each packet is transmitted on the channel whose transmit queue is least full. This proportionally distributes packets on all channels regardless of their rate. Because the receiver doesn't know which packets were sent on which channels, the transmitter must include a header to delineate packet boundaries and identify the original order of the packets. The overhead of such a protocol is inversely proportional the size of the packets — on small packets, the header overhead forms a larger proportion of the total rate. The latency of transmission, however, is proportional to the size of the packets — larger packets mean greater buffering, and thus more latency. This results in an unavoidable tradeoff between latency and overhead in a packet-based channel aggregation protocol. H.226 provides the ability to operate on channels with arbitrary or varying bitrates, while still achieving low latency and efficiency close to that of synchronous channel aggregation protocols.

11.1. H.226 Operation

The H.226 channel aggregation protocol operates between the H.223 multiplexer and multiple physical network connections, as shown in Fig. 4.

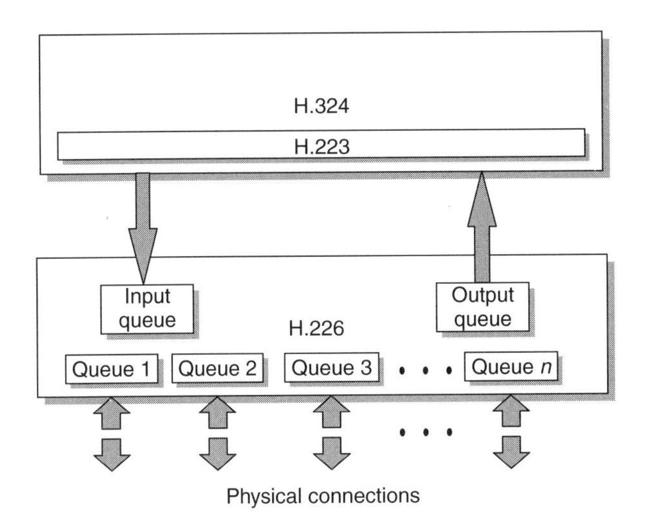

Figure 4. H.226/H.324 protocol stack.

H.226 operates by maintaining a separate transmit (and receive) queue for each individual transmission channel. The full-rate bitstream is divided into 8-bit samples, and each sample is placed on a transmit queue corresponding to one of the channels. The channel used to transmit each sample is determined not by actual queue fullness, but by the output of a finite-state machine called the *channel reference model* (CRM). The CRM simulates queue fullness for each channel using a simple algorithm that is also used identically and in synchrony at the receiver, so that the transmitter and receiver choose the same channel to transmit and receive each sample.

The CRM state machine in H.226 distributes the samples among the different channels in proportion to the bitrate of each channel, so that the full capacity of all the channels is used. For example, if there are two channels, and the first operates at double the bitrate of the other, the first channel will send two samples for each one that the other channel sends.

For each available transmission channel, the CRM maintains a state variable which represents the modeled (not actual) "queue emptiness" of that channel. It also maintains a set of "channel proportion" values, which represent the ratio of the bitrate of that channel to the total combined bitrate of all available channels. As each sample is processed, the state variable of each channel is incremented by the amount of the "channel proportion" value, so that the state variables reflect each channel's capacity to transmit. The CRM selects the channel having the largest state variable to transmit the sample. The state variable for that channel is then decreased by the sum of the "channel proportion" values for all channels, which compensates for the transmission of the sample.

Periodically, and when the transmitter detects an excessive divergence between the actual transmit queues and their expected fullness (possibly as a result of changing channel bitrates), the transmitter sends an updated set of "channel proportion" values to the receiver, and the CRM on each side is reset. This allows H.226 to track changing channel rates and correct for any other divergences between the Channel Reference Model and the actual transmit queues.

11.2. Call Setup for Multilink

Use of H.226 is negotiated during *phase C* of the H.324 call setup, using the procedures of V.8bis (for PSTN) or V.140 (for ISDN). If both terminals indicate in phase C that they wish to use the multilink mode, H.226 is used as a layer between the H.223 multiplex and the physical network connection on the channel that started the call.

The setup of additional channels is coordinated by H.245 message exchange as specified in H.324 Annex F. To avoid the need for the end user to dial additional telephone numbers for each extra channel, a special procedure has the answering terminal automatically provide for the answering terminal to automatically provide telephone numbers for additional connections. This information is sent in a H.245 DialingInformation message that contains only the rightmost digits of the telephone number that differ from the original number dialed by the end user.

For example, if the initial connection was established by dialing "0019786234349," and the DialingInformation message contains "51," the number to be dialed for the additional connection is "0019786234351." This differential digit method is used instead of the full E.164 [12] international number because the first few digits of the number to be dialed can vary according to the geographic location of the two terminals, such as whether they are located in the same building, city, or country.

12. H.324/I FOR ISDN

Operation of H.324 on ISDN is covered by Annex D of H.324, and is essentially identical to the PSTN use of H.324, except for call setup. For ISDN, the V.34 modem, which is used on PSTN lines to provide a digital bitstream, is replaced by an I.400 series ISDN user–network interface [13], which provides a direct digital connection to the ISDN network.

In *phase A* of the call setup, the end-to-end connection is made using normal ISDN procedures that make use of D-channel signaling. When this is complete, the V.140 ISDN call setup protocol is started.

12.1. V.140 ISDN Call Setup

V.140 is used to characterize the end-to-end digital connection before use, and has an automatic mode selection system that allows the terminals to automatically determine if a H.324/I, H.320, or ordinary G.711 voice telephone call is desired when the call first connects. V.140 proceeds in three phases: *phases 1–3*.

In *phase 1*, a V.140 "signature" pattern is sent in the low-order bits of each byte. The signature indicates that the terminal supports V.140, and is capable of proceeding to *phases 2* and *3*. The signature is designed to avoid conflict with a similar pattern called the *frame alignment sequence* (FAS) used to start H.320. By transmitting in the low-order bits, these patterns create minimal disruption to G.711 coded audio signals, creating only low-level background noise. Therefore G.711 audio, which can contain speech or tones from PSTN modems, the H.320 FAS signal, and the V.140 signature, can all be sent simultaneously. This allows the terminal to negotiate and interoperate correctly if the far-end terminal supports any of these protocols, even if it doesn't support H.324/I or V.140. Once both terminals detect the V.140 signature, *phase 2* is entered.

Phase 2 of V.140 probes the network connection to determine which of a variety of national ISDN and ISDN-like digital network types is use. Most ISDN networks supply an 8.0-kHz clock that times each 8-bit byte transmitted over the network. The clock indicates the boundaries between bytes. Some networks transmit only 7 bits out of each byte end-to-end (providing 56 kbps), while others carry all 8 bits (for 64 kbps). Other network types don't provide any byte timing at all. Even between two terminals with full 64-kbps connections with byte timing, sometimes there can be an intervening network that drops 1 out of each 8 bits. Phase 2 of V.140 sorts this out by exchanging a series of test patterns across the connections. Each terminal can then determine which bits are usable for transmission, and in what order. This process can make H.324/I calls possible on connections that would be mismatched otherwise.

Finally, in *phase 3*, the two terminals exchange a list of modes that they are capable of using, and choose one to be used for the call. This capability exchange process can also be used as a substitute for the V.8bis "late start" procedure in *phase B* of the PSTN H.324 call startup.

After V.140 is complete, call setup continues from phase D. Operation of H.324/I after that is identical to H.324 on the PSTN.

13. H.324/M FOR MOBILE (WIRELESS) NETWORKS

The success of digital cellular telephone systems led to a strong industry desire for a real-time videophone service for $2\frac{1}{2}$ and 3G cellular systems offering higher data rates, such as GPRS (general packet radio service), EDGE (enhanced data rates for GSM evolution), cmda2000, and UMTS.

The H.324 variant called H.324/M supports this application, and is described in H.324 Annex C. The main difference from H.324 is the addition of robust error-detection and error-correction features in the H.223 multiplexer, and the use of G.723.1 Annex C scalable error-protection coding for audio. These allow H.324/M to cope with the high bit error rates (as poor as 5×10^{-2}) of these wireless networks.

The 3rd Generation Partnership Project (3GPP) videophone standard for UMTS, 3G-324M [14], uses the structure of H.324/M with a slightly different set of codecs and requirements.

13.1. Level Structure of H.324/M

H.324/M supports five different error-detection/correction schemes, called "levels," organized in a hierarchical structure. Level 0 is the same as the PSTN and ISDN variants of H.324. Levels 1 and 2 (in H.223 Annexes A and B) add robustness to the H.223 multiplex layer, while levels 3 and 3a add more error-robust adaptation layers in H.223 Annexes C and D. Each higher-numbered level is more error-robust than the previous level, and also requires more overhead and processing cycles. Each level supports the features of all lower levels. Table 4 summarizes these levels.

13.1.1. Level 0. Level 0 is identical to the base H.324 and H.223.

13.1.2. Level 1. Defined in H.223 Annex A, Level 1 replaces the 8-bit HDLC synchronization flag "01111110" with a 16-bit pseudo-noise (PN) flag "1011001010000111."

Longer flags allow more robust detection of MUX-PDU boundaries in the case of high bit error rates, because implementations can use correlation thresholds for flag detection. For example, if 15 out of 16 bits are correct, a flag may be detected. In addition, implementations should also consider the correctness of the header byte to verify the correct detection of the flag. In an optional extension, two consecutive PN flags may be used (double flag mode), permitting even more robust flag detection.

At level 1, there is no equivalent to the HDLC zero-insertion process, so emulation of the PN flag in the data stream is not prevented. If possible, transmitters should prevent the appearance of a PN flag pattern in the information field by terminating the MUX-PDU and starting a new one when a PN flag emulation occurs.

Avoiding HDLC zero-insertion ensures that bit errors cannot unsynchronize the zero-removal process, which could lead to error propagation through the entire MUX-PDU [15].

13.1.3. Level 2. Level 2, defined in H.223 Annex B, uses the 16-bit PN flag from level 1 and additionally

Table 4. Level Structure of H.324/M

Level	Multiplex Summary	Ref.
3a	16-bit PN flag	H.324, Annex C
	24-bit multiplex header error-robust AL1M′, AL2M′, AL3M′	H.223, Annexes A, B, C, D
3	16-bit PN flag	H.324, Annex C
	24-bit multiplex header error-robust AL1M, AL2M, AL3M	H.223, Annexes A, B, C
2	16-bit PN flag	H.324, Annex C
	24-bit multiplex header AL1, AL2, AL3	H.223, Annexes A, B
1	16-bit PN flag	H.324, Annex C
	8-bit multiplex header AL1, AL2, AL3	H.223, Annex A
0	8-bit HDLC flag	H.324
	8-bit multiplex header AL1, AL2, AL3 (same as base PSTN H.324)	H.223

replaces the 8-bit H.223 header with an extended 24-bit level 2 header.

The level 2 header includes a 4-bit multiplex code (the same as in base H.324), an 8-bit multiplex payload length (MPL), and a 12-bit extended Golay code.

The MPL field indicates the count of bytes in the information field. The combination of PN flag synchronization and the MPL makes it possible to overcome critical bit error or packet-loss situations that are likely in the mobile environment.

The extended Golay code is used for error detection. It can also be used for a combination of error detection and error correction, where error-detection capability decreases as error-correction increases.

The packet marker (PM) bit is not included in the MUX-PDU header, as in levels 0 and 1. Instead, the PM is signaled by inverting the following PN flag.

An optional extension provides even more redundancy by re-sending the previous MUX-PDU header, in the 8-bit level 0 header format, immediately after each level 2 header.

When the transmitter has no information to send, the "stuffing sequence," sent repeatedly to fill the channel, is the 2-byte PN flag followed by the 3-byte level 2 header, containing a multiplex code of "0000."

13.1.4. Level 3.

H.223 Annex C defines level 3, which, in addition to the level 2 multiplex enhancements, replaces H.223 adaptation layers AL1, AL2, and AL3 with more error-robust "mobile" versions: AL1M, AL2M, and AL3M.

The level 3 adaptation layers can protect data with forward error correction (FEC) using a rate-compatible punctured convolutional code (RCPC) [16], and with automatic repeat request (ARQ) schemes, to allow incorrectly received data to be automatically retransmitted. A single higher-level packet can also be split into several smaller parts, to reduce the amount of data that can be damaged by uncorrected bit errors.

An optional interleaving scheme, which scrambles the order of the payload data, can be used to make the FEC more effective by spreading out the effect of a short burst of errors over a longer period. The headers of the higher-level packets are also protected with FEC, using either a systematic extended BCH code, or an extended Golay code.

The level 3 stuffing sequence is the 2-byte PN flag followed by the 3-byte level 2 header, containing a multiplex code of "1111."

13.1.5. Level 3a.

Level 3a, defined in H.323 Annex D, is identical to level 3, except that the RCPC FEC codes are replaced by shortened Reed–Solomon (RS) codes. Depending on the error characteristics of the transmission channel, either RS or RCPC codes can produce a lower residual error rate.

The choice between levels 3 and 3a is selected by the H.245 protocol.

13.2. Call Setup

The H.324/M call setup is essentially the same as for the PSTN and ISDN variants of H.324. Similar to H.324/I, the V.34 modem is replaced by a suitable wireless interface, and a digital end-to-end connection is established.

The main difference is the negotiation used to select which of the different H.223 levels, just described, will be used. The "stuffing sequence," sent when the transmitter is has no information to send, is unique for each level and is therefore used to signal the sending terminal's highest supported level. Table 5 shows the stuffing sequences sent by each level.

Each terminal starts by sending consecutive stuffing sequences for the highest level it supports. Each receiver tries to detect stuffing sequences that correspond to one of the levels supported by the receiver. If either terminal receives stuffing corresponding to a level lower than the one it is sending, it changes its transmitted level to match the far-end terminal. Once both terminals detect the same stuffing sequence, the terminals operate at the same level and proceed with phase D of the H.324 call setup. (The choice between levels 3 and 3a is signaled using H.245.) This procedure guarantees that the terminals initially start at the highest commonly supported level.

H.324/M also defines an optional procedure to change levels dynamically during the call (in *phase E*). This procedure is signaled via H.245 and applies separately to each direction of transmission. For example, it is possible to operate at level 1 in one direction and at level 2 in the opposite direction. This procedure is useful if the transmission characteristics (primarily bit error rate) change, and if this doesn't happen too frequently.

13.3. Control Channel Segmentation and Reassembly Layer (CCSRL)

At levels 1 and higher a *control channel segmentation and reassembly layer* (CCSRL) is added between the NSRP and H.245 layers for logical channel 0. H.324/M has to guarantee error-free delivery for H.245 messages. In general, longer NSRP frames are more likely to be affected by errors than are shorter ones. Therefore the CCSRL layer segments long H.245 messages into smaller NSRP frames. This reduces the amount of NSRP data that needs to be retransmitted due to errors.

13.4. Videotelephony in UMTS: 3G-324M

3G-324M [14] is the videotelephony standard for the 3GPP's UMTS. H.324/M was taken as the baseline, but with a few changes, including

- Support for H.324/M level 2 is mandatory.
- The Adaptive Multirate Codec (AMR) [17], the mandatory speech codec in UMTS, is required; support for G.723.1 is also encouraged.

Table 5. Stuffing Sequences for Each Level

Level	Stuffing Sequence (Repeated)
0	HDLC flag
1	PN flag
2	PN flag + level 2 header, multiplex code 0000
3 and 3a	PN flag + level 2 header, multiplex code 1111

Figure 5. Prototype H.324/M mobile videotelephone.

- The H.263 video codec is mandatory, but support for the ISO/IEC 14496-1 (MPEG-4) [18] codec is strongly encouraged; it is expected that this will be widely supported. H.324 Annex G defines the use of MPEG-4 in H.324.

The North American and Asian-based 3GPP2 project also plans to use a variant of H.324/M for video telephony in cdma2000 [19].

14. CONCLUSION

The H.324 standard makes possible multimedia terminal devices capable of handling a variety of media types, including audio, video, and data, over switched circuit networks of all types. (Useful reference material for implementers of H.324 is available at *http://www.packetizer.com.*)

BIOGRAPHIES

Dave Lindbergh is an engineer with Polycom Inc. Since 1993, he has been active in U.S. and international standards organizations, including Committee T1, TIA, IETF, and ITU, where he was a principal contributor to ITU-T Recommendations, H.223, H.224, H.281, and V.140; served as editor for Recs. H.226 and H.324; and was chairman of the ITU-T H.324 Systems Experts Group. He currently is chairman of the IMTC Requirements Work Group and cochairman of the IMTC Media Processing Activity Group. In 2002 he received an IMTC leadership award for his leadership role in the standards community. Dave was a coauthor of *Digital Compression for Multimedia: Principles and Standards* (Morgan Kaufmann, 1998) and a contributor to *Multimedia Communications* (Academic Press, 2001). In 1990 Dave cofounded CD Atlas Company, a multimedia mapping start-up, and as a consulting engineer, he designed modem protocols and software and developed the APT (Asynchronous Performance Tester) data communications measurement tool used in the modem industry. In 1981 he founded Lindbergh Systems, maker of OMBITERM data communications software. He is credited with two U.S. patents in the field of data compression and data communications.

Bernhard G. Wimmer received the diploma degree in electrical engineering from Technical University Munich, Germany, in 1995. His diploma thesis was honored by the Siemens Information Technology User Group and by IEEE (Student Paper Award) in 1996. He joined the Corporate Research and Development of Siemens AG in 1996 as research engineer for multimedia technologies. In 1999 he changed to the division of Siemens Mobile Phones. Since 2002 he has headed the Multimedia Technology Group for 3G terminals. Since 1996 he has been active in standardization, including ETSI, IETF, ITU-T, and 3GPP, where he contributed to H.223, H.324M, H.263, 3G-324M, 3GPP PSS, and TS26.140 (MMS codecs). From 1997to 1999 he served as chairman of the ITU-T H.324 Systems Experts Group and currently as coeditor for codecs for MMS in 3GPP SA4 group. He presently is chairman of the IMTC H.324 Activity Group. His areas of interest are video/image coding algorithms, multimedia protocols, and applications for mobile devices. Wimmer is holding several patents in the area of multimedia technologies and applications.

BIBLIOGRAPHY

1. ITU-T Study Group 16, Recommendations of the H Series: H.324, H.223, H.245, G.723.1, H.263, H.226, ITU, Geneva, 1995–2001. *http://www.itu.int.*
2. ITU-T Study Group 16, Recommendations of the V Series: V.8, V.8bis, V.14, V.23, V.25, V.25ter, V.34, V.42, V.42bis, V.80, V.90, V.140, ITU, Geneva, 1990–2001, *http://www.itu.int.*
3. ISO/IEC 3309, *Information Technology — Telecommunications and Information Exchange between Systems — High-Level Data Link Control (HDLC) Procedures — Frame Structure*, 1991, *http://www.iso.ch.*
4. D. Lindbergh and H. Malvar, Multimedia teleconferencing with H.324, in K. R. Rao, ed., *Standards and Common Interfaces for Video Information Systems*, SPIE, Philadelphia, Oct. 1995, pp. 206–232.
5. ITU-T Recommendation X.691, *Information Technology; ASN.1 Encoding Rules — Specification of Packed Encoding Rules (PER)*, ITU, Geneva, 1995, *http://www.itu.int.*
6. J. Gibson et al., *Digital Compression for Multimedia*, Morgan Kaufmann, San Francisco, 1998, pp. 356–361.
7. ITU-T Study Group 8, Recommendations of the T Series: T.84, T.434, T.120, T.122–T.128, ITU, Geneva, 1992–1998. *http://www.itu.int.*
8. ISO/IEC TR9577, *Information Technology — Telecommunications Information Exchange between Systems — Protocol Identification in the Network Layer*, ISO, Geneva, 1990 *http://www.iso.ch.*
9. W. Simpson, ed., *The Point to Point Protocol*, IETF RFC 1661, July 1994, *http://www.ietf.org/rfc.*
10. ISO/IEC 13871: 1995, *Information Technology — Telecommunications and Information Exchange between*

Systems — Private Telecommunications Networks — Digital Channel Aggregation, ISO, Geneva, 1995, *http://www.iso.ch*.

11. K. Sklower et al., *The PPP Multilink Protocol (MP)*, IETF RFC 1990, Aug. 1996. *http://www.ietf.org/rfc*.
12. ITU-T Study Group 2, *Recommendation E.164*, ITU, Geneva, 1997, *http://www.itu.int*.
13. ITU-T Study Group 15, *Recommendations of the I Series*, ITU, Geneva, 1988–2000, *http://www.itu.int*.
14. 3rd Generation Partnership Project (3GPP), TSG-SA Codec Working Group, 3G TS 26.110, *Codec(s) for Circuit Switched Multimedia Telephony Service: General Description*, V.4.0.0, *http://www.3gpp.org*.
15. J. Hagenauer, E. Hundt, T. Stockhammer, and B. Wimmer, Error robust multiplexing for multimedia applications, *Signal Process. Image Commun.* **14**: 585–597 (1999).
16. J. Hagenauer, Rate-compatible punctured convolutional codes (RCPC codes) and their applications, *IEEE Trans. Commun.* **36**(4): 389–400 (April 1988).
17. 3GPP TS 26.071: *Mandatory Speech Codec; General Description*, *http://www.3gpp.org*.
18. ISO/IEC 14496-1: 1999, *Information Technology — Coding of Audio-visual Objects, Part 1: Systems*, ISO, Geneva, 1999, *http://www.iso.ch*.
19. 3GPP2, *Video Conferencing Service in cdma2000*, S.R0022, V.1.0, July 2000, *http://www.3gpp2.org*.

HADAMARD MATRICES AND CODES

Martin Bossert
University of Ulm
Ulm, Germany

1. INTRODUCTION

Hadamard matrices were born as a theoretical topic in combinatorics and have found various applications in different fields. They consist of n rows and n columns, where n is called the *order* and the matrix elements are either $+1$ or -1. The rows are pairwise orthogonal.

Their main property, the orthogonality, is among others exploited in communication systems for the separation of users. If different users simultaneously send orthogonal signals, each user can be extracted from the signal sum at the receiver. Examples therefore are the European standard UMTS (Universal Mobile Telecommunications System), the U.S. American standards IS95 (Interim Standard), GPS (Global Positioning System), and many more. Furthermore, the rows of specific Hadamard matrices possess a good periodic autocorrelation property that can be used for the resolution of multipath propagation in mobile communication.

Hadamard codes is a general name for binary codes that can be constructed on the basis of Hadamard matrices. Historically, most such codes were constructed independently and, only later, relationships to Hadamard matrices were discovered. Famous representatives of such code constructions are the class of simplex codes and the class of first- order Reed–Muller codes. The unique Golay code can also be derived based on Hadamard matrices.

In signal processing the concept of orthogonal transforms have a wide range of applications. Hadamard matrices can be used as an orthogonal transform and, as in the case of the Fourier transform, a fast Hadamard transform can be derived.

First we will give the definition and two possible constructions of Hadamard matrices, namely, the Sylvester and the Payley constructions. For $n \geq 4$ only orders n divisible by 4 may exist. By these two constructions, all but six possible Hadamard matrices of order up to 256 will be constructed. Namely, the orders 156, 172, 184, 188, 232, and 236 cannot be constructed; however, there exist constructions for these orders, for example, in Refs. 5 and 3. Of course, orders larger than 256 may be obtained by the two constructions presented.

Afterward we will describe the connections to coding theory and then derive the fast Hadamard transform for signal processing. Finally, we comment on several applications.

2. DEFINITIONS AND CONSTRUCTIONS OF HADAMARD MATRICES

Definition 1. *A Hadamard matrix, $\boldsymbol{H} = \boldsymbol{H}(n)$, of order n, is an $n \times n$ matrix with entries $+1$ and -1, such that $\mathbf{H}\mathbf{H}^T = n\boldsymbol{I}$, where $\boldsymbol{I}$ is the $n \times n$ identity matrix and $\boldsymbol{H}^T$ is the transposed matrix of $\boldsymbol{H}$.*

Example 1. As examples we give possible Hadamard matrices of orders 1, 2, and 4:

$$\boldsymbol{H}(1) = 1; \quad \boldsymbol{H}(2) = \begin{pmatrix} +1 & +1 \\ +1 & -1 \end{pmatrix};$$

$$\boldsymbol{H}(4) = \begin{pmatrix} +1 & -1 & +1 & +1 \\ +1 & -1 & -1 & -1 \\ -1 & -1 & +1 & -1 \\ +1 & +1 & +1 & -1 \end{pmatrix}.$$

Properties. The rows of $\boldsymbol{H}(n)$ are pairwise orthogonal; that is, the scalar product of any two distinct rows is zero.

Permutations of rows or columns or inversion (multiplication by -1) of some rows and columns do not change this property. A matrix that can be derived by row or column operations from another is called *equivalent*. For any given Hadamard matrix there exists an equivalent one for which the first row and the first column consist entirely of $+1$s. Such a Hadamard matrix is called *normalized*.

Except for the cases $n = 1$ and $n = 2$, a Hadamard matrix may exist if $n = 4s$ for some integer $s > 0$. It is conjectured that a Hadamard matrix exists for all such integers n. At present, the smallest n for which no Hadamard matrix is known is 428.

2.1. The Sylvester Construction

The *Kronecker product* of matrices $\mathbf{A} = (a_{ij}), i, j \in \{1, 2, \ldots, n\}$ and $\mathbf{B}$ is defined as a blockwise matrix

$$\mathbf{A} \otimes \mathbf{B} = (a_{ij}\mathbf{B})$$

Let $\boldsymbol{H}(n)$ and $\boldsymbol{H}(m)$ be two Hadamard matrices. Then $\boldsymbol{H}(nm) := \boldsymbol{H}(n) \otimes \mathbf{H}(m)$ is a Hadamard matrix of order nm.

This construction was given by Sylvester [1]. Note that often only constructions with $n = 2$ are called Sylvester type, however his original definition was more general.

Example 2. Let $\boldsymbol{H}(2) = \begin{pmatrix} +1 & +1 \\ +1 & -1 \end{pmatrix}$ and $\boldsymbol{H}(n)$ be Hadamard matrices. Then the Hadamard matrix of order $2n$, using Sylvester's construction, is as follows:

$$\boldsymbol{H}(2n) = \begin{pmatrix} \boldsymbol{H}(n) & \boldsymbol{H}(n) \\ \boldsymbol{H}(n) & -\boldsymbol{H}(n) \end{pmatrix}$$

Walsh–Hadamard. In particular, the mth Kronecker powers of the normalized matrix $\boldsymbol{S}_1 := \boldsymbol{H}(2)$ gives a sequence of Hadamard matrices of the Sylvester type

$$\boldsymbol{S}_m := \boldsymbol{H}(2^m) = \boldsymbol{S}_1 \otimes \boldsymbol{S}_{m-1}, m = 2, 3, \ldots$$

which are of special interest in communications. The rows are often called Walsh or Walsh–Hadamard sequences.

2.2. The Paley Construction

A *conference matrix* $\boldsymbol{C}(n)$ of order n is an $n \times n$ matrix with diagonal entries 0 and other entries $+1$ or -1, which satisfies $\mathbf{C}\mathbf{C}^T = (n-1)\boldsymbol{I}$.

Example 3. As an example, we give a conference matrix of order 4:

$$\boldsymbol{C}(4) = \begin{pmatrix} 0 & -1 & -1 & -1 \\ 1 & 0 & 1 & -1 \\ 1 & -1 & 0 & 1 \\ 1 & 1 & -1 & 0 \end{pmatrix}.$$

The property $\mathbf{C}\mathbf{C}^T = (n-1)\boldsymbol{I}$ remains unchanged for permutations of rows and columns that do not change the diagonal elements or negations of rows and columns. Again, matrices constructed by such operations are called *equivalent*.

Conference matrices $\boldsymbol{C}(n)$ exist only for even n. Two special cases are distinguished: (1) if $\boldsymbol{C}(n) = -\boldsymbol{C}^T(n)$ this matrix is called a *skew-symmetric conference matrix*, which is possible only for $n \equiv 0 \pmod 4$; and (2) if $\boldsymbol{C}(n) = \boldsymbol{C}^T(n)$, this is called a *symmetric conference matrix* and $n \equiv 2 \pmod 4$.

We denote by $\boldsymbol{I}$ an identity matrix of order n. With both, symmetric and skew-symmetric conference matrices, Hadamard matrices can be constructed in the following way.

Payley Construction P1 (Skew-Symmetric Conference Matrices). If $n \equiv 0 \pmod 4$ and there exists a skew-symmetric conference matrix $\boldsymbol{C}(n)$, then

$$\boldsymbol{H}(n) = \boldsymbol{I} + \boldsymbol{C}(n) \tag{1}$$

is a Hadamard matrix.

Payley Construction P2 (Symmetric Conference Matrices). If $n \equiv 2 \pmod 4$ and there exists a symmetric conference matrix $\boldsymbol{C}(n)$, then

$$\boldsymbol{H}(2n) = \begin{pmatrix} \boldsymbol{I} + \boldsymbol{C}(n) & -\boldsymbol{I} + \boldsymbol{C}(n) \\ -\boldsymbol{I} + \boldsymbol{C}(n) & -\boldsymbol{I} - \boldsymbol{C}(n) \end{pmatrix} \tag{2}$$

is a Hadamard matrix.

Construction of Conference Matrices. Now we need to construct conference matrices $\boldsymbol{C}(n)$, and for this we will describe Paley's constructions based on quadratic residues in finite fields [2]. Note that in any field, addition and multiplication are defined and, in the following, we assume that the operations with field elements are done accordingly. Let $\mathbb{F}_q$ be a finite field, q odd. A non-zero-element $x \in \mathbb{F}_q$ is called a *quadratic residue* if an element $y \in \mathbb{F}_q$ exists such that $x = y^2$. Among the $q-1$ nonzero elements of $\mathbb{F}_q$ there exist exactly $(q-1)/2$ quadratic residues and $(q-1)/2$ quadratic nonresidues. The *Legendre* function $\chi\colon \mathbb{F}_q \Rightarrow \{0, \pm 1\}$ is defined by.

$$\chi(x) = \begin{cases} 0 & \text{if } x = 0 \\ +1 & \text{if } x \text{ is a quadratic residue} \\ -1 & \text{if } x \text{ is a quadratic nonresidue} \end{cases}$$

Clearly the Legendre symbol is an indicator for quadratic residues and nonresidues. A $q \times q$ matrix $\boldsymbol{Q} = (Q_{x,y})$ can be defined where the indices x,y of the entries $Q_{x,y}$ are the elements of the field, $x, y \in \mathbb{F}_q$ and their value is $Q_{x,y} := \chi(x-y)$. We denote by $\mathbf{1}$ the all one vector.

Skew-Symmetric Payley Conference Matrix. If $q \equiv 3 \pmod 4$, then

$$\boldsymbol{C}(q+1) = \begin{pmatrix} 0 & \mathbf{1} \\ -\mathbf{1}^T & \boldsymbol{Q} \end{pmatrix} \tag{3}$$

is a skew-symmetric Paley conference matrix of order $q+1$.

Symmetric Payley Conference Matrix. If $q \equiv 1 \pmod 4$, then

$$\boldsymbol{C}(q+1) = \begin{pmatrix} 0 & \mathbf{1} \\ \mathbf{1}^T & \boldsymbol{Q} \end{pmatrix} \tag{4}$$

is a symmetric Paley conference matrix of order $q+1$.

Clearly Eq. (1) gives Paley construction P1 for a Hadamard matrix of order $q+1$ if $q \equiv 3 \pmod 4$, and Eq. (2) gives Paley construction P2 for a Hadamard matrix of order $2(q+1)$ if $q \equiv 1 \pmod 4$. For both cases we will give an example.

Example 4. Let $q = 7$, $\mathbb{F}_q = \mathbb{F}_7 = \{0, 1, 2, 3, 4, 5, 6\}$. The elements $\{1, 2, 4\}$ are quadratic residues while $\{3, 5, 6\}$ are quadratic nonresidues. We have $q \equiv 3 \bmod 4$, and thus we will get a Hadamard matrix of order $q+1 = 8$ using construction P1 in Eq. (1). First we construct the skew-symmetric conference matrix with the help of matrix $\boldsymbol{Q}$ defined by the Legendre symbols. The rows and columns of $\boldsymbol{Q}$ are labeled by the elements of the field $\mathbb{F}_7 = \{0, 1, 2, 3, 4, 5, 6\}$, respectively.

$$\boldsymbol{Q} = \begin{pmatrix} 0 & 1 & 1 & -1 & 1 & -1 & -1 \\ -1 & 0 & 1 & 1 & -1 & 1 & -1 \\ -1 & -1 & 0 & 1 & 1 & -1 & 1 \\ 1 & -1 & -1 & 0 & 1 & 1 & -1 \\ -1 & 1 & -1 & -1 & 0 & 1 & 1 \\ 1 & -1 & 1 & -1 & -1 & 0 & 1 \\ 1 & 1 & -1 & 1 & -1 & -1 & 0 \end{pmatrix},$$

$$\boldsymbol{C}(8) = \begin{pmatrix} 0 & 1 & 1 & 1 & 1 & 1 & 1 & 1 \\ -1 & 0 & 1 & 1 & -1 & 1 & -1 & -1 \\ -1 & -1 & 0 & 1 & 1 & -1 & 1 & -1 \\ -1 & -1 & -1 & 0 & 1 & 1 & -1 & 1 \\ -1 & 1 & -1 & -1 & 0 & 1 & 1 & -1 \\ -1 & -1 & 1 & -1 & -1 & 0 & 1 & 1 \\ -1 & 1 & -1 & 1 & -1 & -1 & 0 & 1 \\ -1 & 1 & 1 & -1 & 1 & -1 & -1 & 0 \end{pmatrix}$$

With $\boldsymbol{C}(8)$ we get the Hadamard matrix as

$$\boldsymbol{H}(8) = \boldsymbol{C}(8) + \boldsymbol{I}$$

$$= \begin{pmatrix} 1 & 1 & 1 & 1 & 1 & 1 & 1 & 1 \\ -1 & 1 & 1 & 1 & -1 & 1 & -1 & -1 \\ -1 & -1 & 1 & 1 & 1 & -1 & 1 & -1 \\ -1 & -1 & -1 & 1 & 1 & 1 & -1 & 1 \\ -1 & 1 & -1 & -1 & 1 & 1 & 1 & -1 \\ -1 & -1 & 1 & -1 & -1 & 1 & 1 & 1 \\ -1 & 1 & -1 & 1 & -1 & -1 & 1 & 1 \\ -1 & 1 & 1 & -1 & 1 & -1 & -1 & 1 \end{pmatrix}$$

Example 5. Let $q = 3^2 = 9$, $\mathbb{F}_q = \mathbb{F}_9 = \{0, 1, \alpha, \alpha^2, \alpha^3, \alpha^4, \alpha^5, \alpha^6, \alpha^7\}$, where α is a root of the polynomial $x^2 - x - 1$. Elements $\{1, \alpha^2, \alpha^4, \alpha^6\}$ are quadratic residues, while $\{\alpha, \alpha^3, \alpha^5, \alpha^7\}$ are quadratic nonresidues. With this, we again can construct the matrix $\boldsymbol{Q}$ as

$$\boldsymbol{Q} = \begin{pmatrix} 0 & 1 & -1 & 1 & -1 & 1 & -1 & 1 & -1 \\ 1 & 0 & -1 & -1 & -1 & 1 & 1 & -1 & 1 \\ -1 & -1 & 0 & 1 & 1 & 1 & -1 & -1 & 1 \\ 1 & -1 & 1 & 0 & -1 & -1 & -1 & 1 & 1 \\ -1 & -1 & 1 & -1 & 0 & 1 & 1 & 1 & -1 \\ 1 & 1 & 1 & -1 & 1 & 0 & -1 & -1 & -1 \\ -1 & 1 & -1 & -1 & 1 & -1 & 0 & 1 & 1 \\ 1 & -1 & -1 & 1 & 1 & -1 & 1 & 0 & -1 \\ -1 & 1 & 1 & 1 & -1 & -1 & 1 & -1 & 0 \end{pmatrix}$$

The symmetric conference matrix is

$$\boldsymbol{C}(10)$$

$$= \begin{pmatrix} 0 & 1 & 1 & 1 & 1 & 1 & 1 & 1 & 1 & 1 \\ 1 & 0 & 1 & -1 & 1 & -1 & 1 & -1 & 1 & -1 \\ 1 & 1 & 0 & -1 & -1 & -1 & 1 & 1 & -1 & 1 \\ 1 & -1 & -1 & 0 & 1 & 1 & 1 & -1 & -1 & 1 \\ 1 & 1 & -1 & 1 & 0 & -1 & -1 & -1 & 1 & 1 \\ 1 & -1 & -1 & 1 & -1 & 0 & 1 & 1 & 1 & -1 \\ 1 & 1 & 1 & 1 & -1 & 1 & 0 & -1 & -1 & -1 \\ 1 & -1 & 1 & -1 & -1 & 1 & -1 & 0 & 1 & 1 \\ 1 & 1 & -1 & -1 & 1 & 1 & -1 & 1 & 0 & -1 \\ 1 & -1 & 1 & 1 & 1 & -1 & -1 & 1 & -1 & 0 \end{pmatrix}$$

and the Hadamard matrix of order 20 is given by:

$$\boldsymbol{H}(20) = \begin{pmatrix} \mathbf{I} + \mathbf{C}(10) & -\mathbf{I} + \mathbf{C}(10) \\ -\mathbf{I} + \mathbf{C}(10) & -\mathbf{I} - \mathbf{C}(10) \end{pmatrix}$$

There exist many other constructions of Hadamard matrices (see, e.g., constructions and tables in Refs. 3–5). However, the two constructions we have presented give almost all Hadamard matrices of order ≤ 256 as shown in Table 1. The matrices are constructed by combining the Sylvester and Paley constructions. $\boldsymbol{S}_m$ denotes the Hadamard matrix $\boldsymbol{H}(2^m)$ of order 2^m constructed by recursion with the Hadamard matrix $\boldsymbol{H}(2)$. P1(q) and P2(q) denote Hadamard matrices of order $q + 1$ and $2(q + 1)$, respectively, given by the Paley constructions P1 and the P2.

3. HADAMARD CODES

A binary code C of length n, of cardinality (number of code words) M, and of minimal Hamming distance d is called an $[n, M, d]$ code.

3.1. Binary Codes from a Hadamard Matrix

By replacing +1s by 0s and −1s by 1s in a normalized Hadamard matrix $\mathbf{H}(n)$ of order n, a *base* binary code

Table 1. List of Hadamard Matrices

n	Construction	n	Construction
4	$\mathbf{S}_2$	132	P1(131)
8	$\mathbf{S}_3$	136	$\boldsymbol{S}_1 \otimes \boldsymbol{H}(68)$
12	P1(11)	140	P1(139)
16	$\mathbf{S}_4$, **P2**(7)	144	$\boldsymbol{S}_2 \otimes \boldsymbol{H}(36)$
20	P1(19), P2(3^2)	148	P2(73)
24	P1(23), $\boldsymbol{S}_1 \otimes \boldsymbol{H}(12)$	152	$\boldsymbol{S}_1 \otimes \boldsymbol{H}(76)$
28	P2(13)	156	—
32	$\mathbf{S}_5$	160	$\boldsymbol{S}_2 \otimes \boldsymbol{H}(40)$
36	P2(17)	164	P1(163), P2(3^4)
40	$\boldsymbol{S}_1 \otimes \boldsymbol{H}(20)$	168	P1(167)
44	P1(23)	172	—
48	P1(23), $\boldsymbol{S}_2 \otimes \boldsymbol{H}(12)$	176	$\boldsymbol{S}_2 \otimes \boldsymbol{H}(44)$
52	P2(5^2)	180	P1(179), P2(89)
56	$\boldsymbol{S}_1 \otimes \boldsymbol{H}(28)$	184	—
60	P1(59)	188	—
64	$\boldsymbol{S}_6$	192	$\boldsymbol{S}_4 \otimes \boldsymbol{H}(12)$
68	P1(67)	196	P2(97)
72	$\boldsymbol{S}_1 \otimes \boldsymbol{H}(36)$	200	P1(199)
76	P2(37)	204	P2(101)
80	*P*1(67), $\boldsymbol{S}_2 \otimes \boldsymbol{H}(20)$	208	$\boldsymbol{S}_2 \otimes \boldsymbol{H}(52)$
84	P1(83)	212	P1(211)
88	P1(87), $\boldsymbol{S}_1 \otimes \boldsymbol{H}(44)$	216	—
92	—	220	P2(109)
96	$\boldsymbol{S}_1 \otimes \boldsymbol{H}(48)$, $\boldsymbol{S}_2 \otimes \boldsymbol{H}(24)$	224	$\boldsymbol{S}_3 \otimes \boldsymbol{H}(28)$
100	P2(7^2)	228	P2(113)
104	P1(103)	232	—
108	P2(53)	236	—
112	$\boldsymbol{S}_2 \otimes \boldsymbol{H}(28)$	240	$\boldsymbol{S}_2 \otimes \boldsymbol{H}(60)$
116	—	244	P1(3^5)
120	$\boldsymbol{S}_1 \otimes \boldsymbol{H}(60)$	248	$\boldsymbol{S}_1 \otimes \boldsymbol{H}(124)$
124	P2(61)	252	P1(251), P2(5^3)
128	$\boldsymbol{S}_7$	256	$\mathbf{S}_8$

$C(=[n,n,d])$ of length n and cardinality $M=n$ is obtained. Since the rows of $\mathbf{H}(n)$ are orthogonal, any two codewords of C differ exactly in $n/2$ coordinates, hence the minimal (Hamming) distance of C is $d=n/2$. This code is called the *Walsh–Hadamard* code. With small manipulations we can get three types of code from the base code C.

1. $C1[n-1, M=n, d=n/2]$. Note that the first column of $\mathbf{H}(n)$ consists of $+1$s, since we assumed a normalized Hadamard matrix. Correspondingly, the first coordinate of any codeword of C is equal to 0. So it can be deleted without changing the cardinality and minimal distance. Thus one gets an $[n-1, M=n, d=n/2]$ code $C1$.
2. $C2[n, M=2n, d=n/2]$. Let $C2$ be a code consisting of all codewords of the base code C and all their complements. Then $C2$ is an $[n, M=2n, d=n/2]$ code.
3. $C3[n-1, M=2n, d=n/2-1]$. Let $C3$ be a code obtained from the code $C2$ by deleting first coordinates. Then $C3$ is an $[n-1, M=2n, d=n/2-1]$ code.

3.2. The Plotkin Bound and Hadamard Matrices

The Plotkin bound states that for an $[n, M, d]$ binary code

$$\begin{array}{ll} M \le 2\left[\dfrac{d}{[2d-n]}\right] & \text{if } d \text{ is even, } 2d > n \\ M \le 4d & \text{if } d \text{ is even, } n = 2d \\ M \le 2\left[\dfrac{d+1}{[2d+1-n]}\right] & \text{if } d \text{ is odd, } 2d+1 > n \\ M \le 4d+4 & \text{if } d \text{ is odd, } n = 2d+1. \end{array} \tag{5}$$

There exists a conjecture that for any integer $s > 0$ there exists a Hadamard matrix $\boldsymbol{H}(4s)$. Levenstein proved that if this conjecture is true then codes exist meeting the Plotkin bound (5) (for details, see Ref. 6).

3.3. Simplex Codes

The class of codes $C1$ is known as simplex codes, because the Hamming distance of every pair of codewords is the same. Codewords can be geometrically interpreted as vertices of a unit cube in $n-1$ dimensions and form a regular simplex. If $n=2^m$ and $C1$ is obtained from the Hadamard matrix $\boldsymbol{S}_m = \boldsymbol{S}_1 \otimes \boldsymbol{S}_{m-1}$, then $C1$ is a *linear* $(2^m-1, m, 2^{m-1})$ code[1] known also as a *maximal-length feedback shift register code*. Any nonzero codeword is known also as an *m-sequence*, or as a *pseudonoise* (PN) *sequence*. If $\mathbf{c}=(c_0, c_1, \ldots, c_{2^m-2})$ is a nonzero codeword, then

$$c_s = c_{s-1}h_{m-1} + c_{s-2}h_{m-2} + \cdots + c_{s-m+1}h_1 + c_{s-m},$$
$$s = m, m+1, \ldots,$$

[1] In case of linear codes one can give the dimension k instead of the number of codewords M. It is $M=2^k$ for binary linear codes.

where subscripts are calculated (mod 2^m-1), and

$$1 + h_1x + \cdots + h_{m-1}x^{m-1} + x^m$$

is a primitive irreducible polynomial.

Simplex codes have many applications, such as range-finding, synchronizing, modulation, scrambling, or pseudonoise sequences. The dual code of a simplex code is the Hamming code, which is a perfect code with minimal distance 3. In the application section below, we will describe the use of m sequences. For further information, see also Ref. 7.

3.4. Reed–Muller Codes of First Order

Codes $C2$ obtained from the Hadamard matrix $\boldsymbol{S}_m = \boldsymbol{S}_1 \otimes \boldsymbol{S}_{m-1}$ are *linear* $(2^m, m+1, 2^{m-1})$ codes. They are known also as first-order Reed–Muller codes and are denoted by $\mathcal{R}(1,m)$. The all-zero vector $\mathbf{0}$ and the all-one vector $\mathbf{1}$ are valid codewords. All the other codewords have Hamming weight 2^{m-1}.

The encoding is the mapping between an information vector into a codeword. This mapping can be a multiplication of the information vector by a so-called generator matrix. One possible encoding of Reed–Muller codes is based on a special representation of the generator matrix $\mathbf{G}_m$ consisting of $m+1$ rows and 2^m columns. The first row $\mathbf{v}_0$ of $\mathbf{G}$ is the all-one vector $\mathbf{1}$. Consider this vector as a run of 1s of length 2^m. The second vector $\mathbf{v}_1$ is constructed by replacing this run by two runs of identical length 2^{m-1} where the first run consists of 0s and the second of 1s. In the third stage ($\mathbf{v}_2$), each run of length 2^{m-1} is replaced by two runs of 0s and 1s with lengths 2^{m-2}. The procedure is continued in a similar manner until one gets the last row $\mathbf{v}_m$ of the form $0101\cdots0101$.

Example 6. *Let* $m=4, n=2^m=16$. *Then the generator matrix of* $\mathcal{R}(1,m)$ *is as follows*:

$$\mathbf{G}_m = \begin{pmatrix} \mathbf{v}_0 \\ \mathbf{v}_1 \\ \mathbf{v}_2 \\ \mathbf{v}_3 \\ \mathbf{v}_4 \end{pmatrix}$$
$$= \begin{pmatrix} 1&1&1&1&1&1&1&1&1&1&1&1&1&1&1&1 \\ 0&0&0&0&0&0&0&0&1&1&1&1&1&1&1&1 \\ 0&0&0&0&1&1&1&1&0&0&0&0&1&1&1&1 \\ 0&0&1&1&0&0&1&1&0&0&1&1&0&0&1&1 \\ 0&1&0&1&0&1&0&1&0&1&0&1&0&1&0&1 \end{pmatrix}$$

If the information bits are $u_0, u_1, \ldots, u_m$, then the corresponding codeword is

$$\mathbf{c} = u_0\mathbf{v}_0 + u_1\mathbf{v}_1 + \cdots + u_m\mathbf{v}_m$$

For the code $\mathcal{R}(1,m)$, a systematic encoding is also possible as an extended cyclic code [7].

The decoding can be performed by the fast Hadamard transform described below. However, there exist several decoding methods; so-called soft decoding, which uses reliability information on the code symbols (if available) is also possible [7]. In the following we describe the multistep majority-logic decoding algorithm (the Reed algorithm), using Example 6. Let u_0, u_1, u_2, u_3, u_4 be information bits and $c_0, c_1, \ldots, c_{15}$ be the corresponding code symbols. Note that

$$u_1 = \begin{cases} c_0 + c_8 \text{ or} \\ c_1 + c_9 \text{ or} \\ c_2 + c_{10} \text{ or} \\ c_3 + c_{11} \text{ or} \\ c_4 + c_{12} \text{ or} \\ c_5 + c_{13} \text{ or} \\ c_6 + c_{14} \text{ or} \\ c_7 + c_{15} \end{cases}, \quad u_2 = \begin{cases} c_0 + c_4 \text{ or} \\ c_1 + c_5 \text{ or} \\ c_2 + c_6 \text{ or} \\ c_3 + c_7 \text{ or} \\ c_8 + c_{12} \text{ or} \\ c_9 + c_{13} \text{ or} \\ c_{10} + c_{14} \text{ or} \\ c_{11} + c_{15} \end{cases},$$

$$u_3 = \begin{cases} c_0 + c_2 \text{ or} \\ c_1 + c_3 \text{ or} \\ c_4 + c_6 \text{ or} \\ c_5 + c_7 \text{ or} \\ c_8 + c_{10} \text{ or} \\ c_9 + c_{11} \text{ or} \\ c_{12} + c_{14} \text{ or} \\ c_{13} + c_{15} \end{cases}, \quad u_4 = \begin{cases} c_0 + c_1 \text{ or} \\ c_2 + c_3 \text{ or} \\ c_4 + c_5 \text{ or} \\ c_6 + c_7 \text{ or} \\ c_8 + c_9 \text{ or} \\ c_{10} + c_{11} \text{ or} \\ c_{12} + c_{13} \text{ or} \\ c_{14} + c_{15} \end{cases}.$$

These equations give 8 votes for each information bit u_1, u_2, u_3, u_4 and the majority of the votes determines the value of each information bit. If 3 or less errors occurred during transmission then all those bits will be decoded correctly. It remains to decode u_0. Since $\mathbf{c} - (u_1\mathbf{v}_1 + u_2\mathbf{v}_2 + u_3\mathbf{v}_3 + u_4\mathbf{v}) = u_0\mathbf{v}_0$, we have an equation to determine u_0 by the majority rule.

One of the first applications of coding was the first order Reed–Muller code of length 32, which was used in the beginning of the 1970s for data transmission in the *Mariner* 9 mission.

3.5. Quadratic Residue Codes

Binary codes obtained from Hadamard matrices of the Paley type are nonlinear for $n > 8$. The linear span of these codes results in *linear quadratic residue codes*. Most known codes are linear spans of codes $C1$ and $C3$ obtained from Paley constructions $P1$ for prime q. Such linear codes have length q, dimension $k = \frac{1}{2}(q \pm 1)$, and minimum distance satisfying the inequality $d^2 - d + 1 \geq q$. The linear quadratic residue code obtained from $\boldsymbol{H}(24)$ is the famous perfect Golay (23,12,7) code.

4. THE HADAMARD TRANSFORM

Let $\mathbf{x} = (x_1, x_2, \ldots, x_n)$ be a real-valued vector. If a Hadamard matrix $\boldsymbol{H}(n)$ exists, then the vector $\mathbf{y} = \mathbf{x}\boldsymbol{H}(n)$ is said to be its *Hadamard* transform (or *Walsh–Hadamard* transform). The Hadamard transform is used in communications and physics, mostly for $n = 2^m$ and $\boldsymbol{H}(n) = \boldsymbol{S}_m$. In general, the Hadamard transform with $\boldsymbol{S}_m$ requires $2^m \times 2^m$ additions and subtractions. A computationally more efficient implementation is the *fast Hadamard transform* based on the representation of $\boldsymbol{S}_m$ as a product of sparse matrices. Specifically, define for $1 \leq i \leq m$ matrices by

$$\mathbf{M}_n^{(i)} := \boldsymbol{I}_{2^{m-i}} \otimes \boldsymbol{S}_1 \otimes \boldsymbol{I}_{2^{i-1}}$$

where $\boldsymbol{I}_s$ denotes an identity matrix of order s. Such a matrix has exactly two nonzero entries ($+1$ or -1) in each row and in each column.

Example 7. Let $m = 3, n = 8$. Then

$$\mathbf{M}_8^{(1)} = \begin{pmatrix} +1 & +1 & & & & & & \\ +1 & -1 & & & & & & \\ & & +1 & +1 & & & & \\ & & +1 & -1 & & & & \\ & & & & +1 & +1 & & \\ & & & & +1 & -1 & & \\ & & & & & & +1 & +1 \\ & & & & & & +1 & -1 \end{pmatrix},$$

$$\mathbf{M}_8^{(2)} = \begin{pmatrix} +1 & & +1 & & & & & \\ & +1 & & +1 & & & & \\ +1 & & -1 & & & & & \\ & +1 & & -1 & & & & \\ & & & & +1 & & +1 & \\ & & & & & +1 & & +1 \\ & & & & +1 & & -1 & \\ & & & & & +1 & & -1 \end{pmatrix},$$

$$\mathbf{M}_8^{(3)} = \begin{pmatrix} +1 & & & & +1 & & & \\ & +1 & & & & +1 & & \\ & & +1 & & & & +1 & \\ & & & +1 & & & & +1 \\ +1 & & & & -1 & & & \\ & +1 & & & & -1 & & \\ & & +1 & & & & -1 & \\ & & & +1 & & & & -1 \end{pmatrix}.$$

The matrix $\boldsymbol{S}_m$ can be written as

$$\boldsymbol{S}_m = \mathbf{M}_n^{(1)}\mathbf{M}_n^{(2)} \cdots \mathbf{M}_n^{(m)}$$

Thus, to evaluate the Hadamard transform $\mathbf{y} = \mathbf{x}\mathbf{S_m} = \mathbf{x}\mathbf{M_n^{(1)}}\mathbf{M_n^{(2)}} \cdots \mathbf{M_n^{(m)}}$, one calculates in the first stage $\mathbf{y}_1 = \mathbf{x}\mathbf{M_n^{(1)}}$. This requires 2^m additions and subtractions. Next, $\mathbf{y}_2 = \mathbf{y}_1\mathbf{M_n^{(2)}}$ is obtained with the same number of calculations and so on. In the mth stage, one obtains $\mathbf{y}$ with a total number of $m2^m$ calculations.

5. COMMUNICATION APPLICATIONS

In this section we describe two main principles as examples of the various applications of Hadamard matrices and related codes and sequences. Example 8 is for user separation in communication systems; Example 9 is for correlation in distance measurement or to measure the channel impulse response.

5.1. Code-Division Multiple Access (CDMA)

In CDMA communication systems, such as the standards IS95 and UMTS, the user separation is done with Walsh–Hadamard sequences. Hereby, the link from the mobiles to the base station uses other means of user separation but the so-called downlink (base station to mobiles) uses Walsh–Hadamard sequences. First we consider a small example.

Example 8. Each user is assigned one row of the Hadamard matrix. In the case of $\boldsymbol{H}(4)$ we can have four users:

$$Alice\text{: } +1, +1, +1, +1 \quad Bob\text{: } +1, -1, +1, -1$$

$$Charlie\text{: } +1, +1, -1, -1 \quad Dan\text{: } +1, -1, -1, +1$$

Now for each user, information bits can be sent by using her/his sequence or the inverted (negative) sequence, for example, in the case of Charlie $+1, +1, -1, -1$ for bit zero and $-1, -1, +1, +1$ for bit one. We assume that there is a synchronous transmission of information for all users. For Alice, Bob, and Dan, bit 0 is send and for Charlie a one. Each of the four user's receiver gets the sum of all signals

$$(+2, -2, +2, +2) = (+1, +1, +1, +1) + (+1, -1, +1, -1) + (-1, -1, +1, +1) + (+1, -1, -1, +1)$$

Now the receiver can detect which bit was send by so called correlation of the user's sequence with the received sequence as follows. For example, Bob computes

$$(+1) \cdot (+2) + (-1) \cdot (-2) + (+1) \cdot (+2) + (-1) \cdot (+2) = +4$$

From the result $+4$, Bob concludes that a zero (the sequence itself) was transmitted.

For Charlie we get

$$(+1) \cdot (+2) + (+1) \cdot (-2) + (-1) \cdot (+2) + (-1) \cdot (+2) = -4$$

From the result -4, Charlie concludes that the transmission was a one (the inverted sequence).

For the description of the CDMA principle, we restrict ourselves to real-valued sequences. Then the crosscorrelation $\Phi_{\mathbf{x,y}}$ of two sequences $\boldsymbol{x}$ and $\boldsymbol{y}$ of length n is defined by

$$\Phi_{x,y} = \frac{1}{n} \sum_{i=1}^{n} x_i y_i$$

Clearly, with this definition, the crosscorrelation $\Phi_{\boldsymbol{x,y}} = 0$, if $\boldsymbol{x}$ and $\boldsymbol{y}$ are orthogonal and $\Phi_{\boldsymbol{x,y}} = 1$ if $\boldsymbol{x} = \boldsymbol{y}$. Let $\mathbf{r} = u_1\boldsymbol{x}_1 + u_2\boldsymbol{x}_2 + \cdots + u_n\boldsymbol{x}_n$ be a received signal, where $u_1, u_2, \ldots, u_n$ are the information bits $\{+1, -1\}$ and the $\boldsymbol{x}_i$ are the rows of a Hadamard matrix. Then the CDMA principle gives

$$\Phi_{\mathbf{r},\boldsymbol{x}_i} = u_i, \quad i = 1, 2, \ldots, n$$

because of the orthogonality of the sequences.

This shows that the users do not mutually interfere. This is a direct consequence of the use of orthogonal sequences. However, in practice there is noise and multipath propagation, which introduces difficulties within this concept. The IS95 standard in the downlink uses Walsh–Hadamard sequences of order 64, while in the uplink the same set of sequences are used as for a Reed–Muller code. The UMTS standard uses Walsh–Hadamard sequences with variable orders, namely, between 4 and 256 in order to adapt to variable data rates and transmission conditions.

5.2. Measurement of the Channel Impulse Response and Scrambling

For user separation, the receiver evaluates the crosscorrelation function between the received sequence and each user sequence. Measurement of distances or the channel impulse response exploits the autocorrelation function. Again we restrict ourself to real-valued sequences. Let $\boldsymbol{x}^{(k)}$ be the cyclic shift of $\boldsymbol{x}$ by k positions. Then, the autocorrelation is defined as

$$\Phi_{\boldsymbol{x}}(k) = \frac{1}{n} \sum_{i=1}^{n} x_i x_i^{(k)}$$

The autocorrelation can take values between -1 and $+1$. In the case of PN sequences the autocorrelation has either the value 1 for $k = 0$ and $k = n$ or

$$\Phi_{\boldsymbol{x}}(k) = \begin{cases} -\dfrac{1}{n} & 0 < k < n, n \text{ odd}, \\ -\dfrac{1}{n-1} & 0 < k < n, n \text{ even}, \end{cases}$$

The mobile communication channel suffers from multipath propagation when the receiver gets the sum of delayed and attenuated versions of the transmitted signal. Usually it is assumed that the channel is time-invariant for a small period and can be described by the impulse response of a linear time-invariant system. We will explain the measurement of the channel impulse response using an example.

Example 9. We assume that the sender transmits periodically the PN sequence $\boldsymbol{pn} = +1, +1, -1, -1, -1, +1, -1$ of length 7. Therefore the transmitted signal is (";" is used to mark the period):

$$\boldsymbol{x} = \ldots; +1, +1, -1, -1, -1, +1, -1; +1, +1, -1, -1, -1, +1, -1; \ldots$$

Suppose the receiver gets

$$y_i = x_i + \tfrac{1}{2}x_{i-2} = \ldots; \tfrac{3}{2}, \tfrac{1}{2}, -\tfrac{1}{2}, -\tfrac{1}{2}, -\tfrac{3}{2}, \tfrac{1}{2}, -\tfrac{3}{2}; \tfrac{3}{2}, \tfrac{1}{2}, -\tfrac{1}{2}, -\tfrac{1}{2}, \ldots$$

If we correlate the received sequence with the PN sequence, we get the following results:

$(\ldots; \tfrac{3}{2}, \tfrac{1}{2}, -\tfrac{1}{2}, -\tfrac{1}{2}, -\tfrac{3}{2}, \tfrac{1}{2}, -\tfrac{3}{2}; \ldots)$ *correlated with* $\boldsymbol{pn}$ *is* $\tfrac{13}{14}$

$(\ldots, \tfrac{1}{2}, -\tfrac{1}{2}, -\tfrac{1}{2}, -\tfrac{3}{2}, \tfrac{1}{2}, -\tfrac{3}{2}; \tfrac{3}{2}, \ldots)$ *correlated with* $\boldsymbol{pn}$ *is* $-\tfrac{3}{14}$

$(\ldots, -\tfrac{1}{2}, -\tfrac{1}{2}, -\tfrac{3}{2}, \tfrac{1}{2}, -\tfrac{3}{2}; \tfrac{3}{2}, \tfrac{1}{2}, \ldots)$ *correlated with* $\boldsymbol{pn}$ *is* $\tfrac{5}{14}$

$(\ldots, -\tfrac{1}{2}, -\tfrac{3}{2}, \tfrac{1}{2}, -\tfrac{3}{2}; \tfrac{3}{2}, \tfrac{1}{2}, -\tfrac{1}{2}, \ldots)$ *correlated with* $\boldsymbol{pn}$ *is* $-\tfrac{3}{14}$

The other three values of the period will also be $-\tfrac{3}{14}$. Thus the influence of the channel can be measured.

In many CDMA systems, base stations transmit pilot tones consisting of PN sequences. Using the measurement method presented above, the mobile terminals can determine the channel impulse response and use it to adjust a so-called RAKE receiver. Because of the good autocorrelation and good cross-correlation properties of PN sequences, they are also used to scramble the transmitted signals. Clearly the autocorrelation property reduces the influence of the paths on each other and the crosscorrelation property guarantees the user separation. Thus, if we know the channel paths at the receiver, we can use a correlator for each path and add their results. This is the concept of a RAKE receiver. In IS95, a PN sequence of length $2^{42} - 1$ is used for this purpose, and in UMTS, scrambling codes are used on the basis of the PN sequences of the polynomials $x^{41} + x^3 + 1$ and $x^{41} + x^{20} + 1$.

The GPS system exploits both the good autocorrelation and crosscorrelation properties of PN sequences. Several satellites synchronously send their PN sequences. A GPS receiver exploits the crosscorrelation to separate the signals from different satellites and the autocorrelation to determine the timing offsets between different satellites caused by signal propagation. Combining these timing offsets with information on the actual position of the satellites allows the precise calculations of the receiver's position. The public PN sequences of length 1023 used in GPS are generated by the polynomials $x^{10} + x^3 + 1$ or $x^{10} + x^9 + x^8 + x^6 = x^3 + x^2 + 1$.

Acknowledgment
The author would like to acknowledge the valuable hints and help from Professor Ernst Gabidulin from the Institute of Physics and Technology, Moscow, Russia.

BIOGRAPHY

Martin Bossert received the Dipl.-Ing. degree in electrical engineering from the Technical University of Karlsruhe, Germany, in 1981, and a Ph.D. degree form the Technical University of Darmstadt, Germany, in 1987. After a DFG scholarship at Lingkoeping University, Sweden, he joined AEG Mobile Communication, Ulm, Germany, where he was, among others, involved in the specification and developement of the GSM System. Since 1993, he has been a professor with the University of Ulm, Germany, where he is currently head of the Departement of Telecommunications and Applied Information Theory. His main research areas concern secure and reliable data transmission, especially generalized concatenation/coded modulation, code constructions for block and convolutional codes, and soft-decision decoding.

BIBLIOGRAPHY

1. J. J. Sylvester, Thoughts on orthogonal matrices, simultaneous sign-successions, and tessellated pavements in two or more colours, with applications to Newton's rule, ornamental tile-work, and the theory of numbers, *Phil. Mag.* **34**: 461–475 (1867).
2. R. E. A. C. Paley, On orthogonal matrices, *J. Math. Phys.* **12**: 311–320 (1933).
3. W. D. Wallis, A. P. Street, and J. Seberry Wallis, Combinatorics: Room squares, sum-free sets, Hadamard matrices, in *Lecture Notes in Mathematics*, Springer, Berlin, 1972, p. 292.
4. J. Seberry and M. Yamada, Hadamard matrices, sequences, and block designs, in J. H. Dinitz and D. R. Stinson, eds., *Contemporary Design Theory: A Collection of Surveys*, Wiley, New York, 1992, pp. 431–560.
5. R. Craigen, Hadamard matrices and designs, in C. J. Colbourn and J. H. Dinitz, eds., *The CRC Handbook of Combinatorial Designs*, CRC Press, New York, 1996, pp. 370–377.
6. F. J. MacWilliams and N. J. A. Sloan, *The Theory of Error-Correcting Codes*, (3rd printing) North-Holland, 1993.
7. M. Bossert, *Channel Coding for Telecommunications*, Wiley, New York, 1999.

HELICAL AND SPIRAL ANTENNAS

HISAMATSU NAKANO
Hosei University
Koganei, Tokyo, Japan

1. INTRODUCTION

This article discusses the radiation characteristics of various helical and spiral antennas. For this, numerical techniques applied to these antennas are summarized in Section 2, where fundamental formulas to evaluate the radiation characteristics are presented. Section 3 on helical antennas, is composed of three subsections, which present the radiation characteristics of normal-mode, axial-mode, and conical-mode helical antennas, respectively. Section 4, on spiral antennas, is composed of five subsections; Sections 4.2 and 4.3 *qualitatively* describe the radiation mechanism of spiral antennas, and Sections 4.4 and 4.5 *quantitatively* refer to the radiation characteristics of the spirals. Finally, Section 5 presents additional information on helical and spiral antennas: a backfire-mode helix and techniques for changing the beam direction of a spiral antenna.

2. NUMERICAL ANALYSIS TECHNIQUES

This section summarizes numerical analysis techniques for helical and spiral antennas. The analysis is based on an electric field integral equation [1,2]. Using the current distribution obtained by solving the integral equation, the radiation characteristics, including the radiation field, axial ratio, input impedance, and gain, are formulated.

2.1. Current on a Wire

Figure 1 shows an arbitrarily shaped wire with a length of L_{arm} (from S_0 to S_E) in free space. It is assumed that the wire is thin relative to an operating wavelength and the current flows only in the wire axis direction. It is also assumed that the wire is perfectly conducting and hence the tangential component of the electric field on the wire

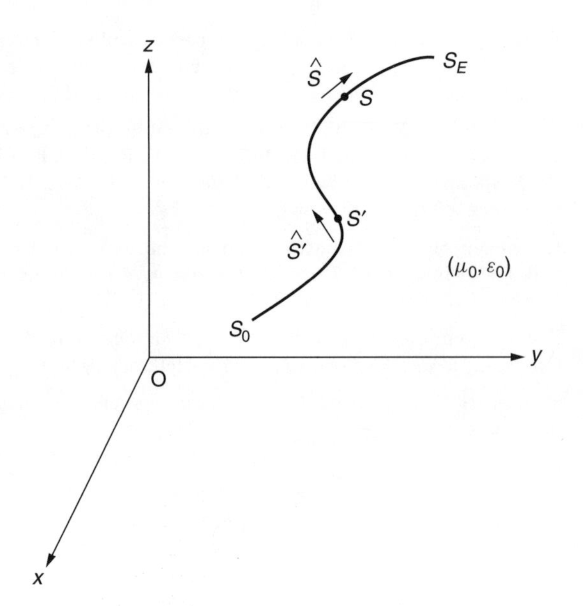

Figure 1. Arbitrarily shaped wire.

surface is zero. This boundary condition of the electric field leads to an *integral equation*.

$$\frac{1}{j\omega\varepsilon_0}\int_{S_0}^{S_E} I(s')\left[-\frac{\partial^2 G(s,s')}{\partial s\,\partial s'}+\beta^2 G(s,s')\hat{s}\cdot\hat{s}'\right]ds' = -E_s^i(s) \tag{1}$$

where $j^2 = -1$; ω is the angular frequency ($= 2\pi f$, $f =$ frequency); ε_0 is the permittivity of free space; s and s' are the distances measured along the wire from the origin (that can arbitrarily be chosen on the wire) to an observation point and a source point, respectively; $I(s')$ is the current at the source point; β is the phase constant ($= 2\pi/\lambda$, $\lambda =$ wavelength); $\hat{s}$ and $\hat{s}'$ are unit vectors, parallel to the wire axis, at the observation and source points, respectively; $E_s^i(s)$ is the tangential component of an incident electric field on the wire; and $G(s, s')$ is Green's function, which is defined as

$$G(s,s') = \frac{1}{4\pi}\frac{e^{-j\beta r_{o,s}(s,s')}}{r_{o,s}(s,s')} \tag{2}$$

where $r_{o,s}(s, s')$ is the distance between the observation and source points.

The *method of moments* (MoM) [3] is adopted to obtain the current $I(s')$ in Eq. (1). For this, the current is expanded as

$$I(s') = \sum_{n=1}^{N} I_n J_n(s') \tag{3}$$

where $J_n(s')$ and I_n $(n = 1, 2, \ldots, N)$ are called the "expansion functions" and "unknown coefficients of the expansion functions," respectively. Note that one can arbitrarily choose $J_n(s')$. Therefore, $J_n(s')$ are known functions in Eq. (3).

Substituting Eq. (3) into Eq. (1), one obtains

$$\sum_{n=1}^{N} I_n e_n(s) = -E_s^i(s) \tag{4}$$

where

$$e_n(s) = \frac{1}{j\omega\varepsilon_0}\int_{S_0}^{S_E} J_n(s')\left[-\frac{\partial^2 G(s,s')}{\partial s\,\partial s'}+\beta^2 G(s,s')\hat{s}\cdot\hat{s}'\right]ds' \tag{5}$$

Multiplying both sides of Eq. (1) by functions $W_m(s)(m = 1, 2, \ldots, N)$ and integrating the multiplied results over the wire length from S_0 to S_E, one obtains

$$\sum_{n=1}^{N} I_n Z_{mn} = V_m, \qquad m = 1, 2, \ldots, N \tag{6}$$

where

$$Z_{mn} = \int_{S_0}^{S_E} e_n(s) W_m(s) ds \tag{7}$$

$$V_m = -\int_{S_0}^{S_E} E_s^i(s) W_m(s) ds \tag{8}$$

Note that one can arbitrarily choose $W_m(s)$, which are called "weighting functions." When the $W_m(s)$ have the same form as the expansion functions $J_m(s')$, the MoM is called the "Galerkin method."

Equation (6) is written in matrix form:

$$[Z_{mn}][I_n] = [V_m] \tag{9}$$

where $[Z_{mn}]$, $[I_n]$, and $[V_m]$ are called the "impedance, current, and voltage matrices," respectively. The unknown coefficients are obtained as

$$[I_n] = [Z_{mn}]^{-1}[V_m] \tag{10}$$

Substituting the obtained $I_n(n = 1, 2, \ldots, N)$ into Eq. (3), one can determine the current distributed along the wire.

2.2. Radiation Field, Axial Ratio, Input Impedance, and Gain

The electric field $\mathbf{E}$ at a far-field point, radiated from the current $I(s')$ and called the "radiation field," is calculated to be

$$\mathbf{E}(r,\theta,\phi) = E_\theta(r,\theta,\phi)\hat{\theta} + E_\phi(r,\theta,\phi)\hat{\phi} \tag{11}$$

where

$$E_\theta(r,\theta,\phi) = -\frac{j\omega\mu_0}{4\pi r}e^{-j\beta r}\hat{\theta}\cdot\int_{S_0}^{S_E}\hat{s}' I(s')e^{j\beta\hat{\mathbf{r}}\cdot\mathbf{r}'}ds' \tag{12}$$

$$E_\phi(r,\theta,\phi) = -\frac{j\omega\mu_0}{4\pi r}e^{-j\beta r}\hat{\phi}\cdot\int_{S_0}^{S_E}\hat{s}' I(s')e^{j\beta\hat{\mathbf{r}}\cdot\mathbf{r}'}ds' \tag{13}$$

in which (r, θ, ϕ) and $(\hat{r}, \hat{\theta}, \hat{\phi})$ are the spherical coordinates and their unit vectors, respectively; μ_0 is the permeability of free space; and the vector $\mathbf{r}'$ is the position vector where current $I(s')$ exists. Other notations are defined in Section 2.1.

The radiation field of Eq. (11) is decomposed into two circularly polarized (CP) wave components:

$$\mathbf{E}(r,\theta,\phi) = E_R(r,\theta,\phi)(\hat{\theta} - j\hat{\phi}) + E_L(r,\theta,\phi)(\hat{\theta} + j\hat{\phi}) \tag{14}$$

where the first term represents *a right-hand CP wave component* and the second represents *a left-hand CP wave component*. Using these two components, the axial ratio (AR) is defined as $\text{AR} = \{|E_R| + |E_L|\}/\{||E_R| - |E_L||\}$. The AR is an indicator of the uniformity of a CP wave. Note that $\text{AR} = 1$ (0 dB) when the radiation is *perfectly* circularly polarized.

The input impedance Z_{in} is defined as $Z_{\text{in}} = R_{\text{in}} + jX_{\text{in}} = V_{\text{in}}/I_{\text{in}}$, where V_{in} and I_{in} are the voltage and current at antenna feed terminals, respectively. The power input to the antenna is given as $P_{\text{in}} = R_{\text{in}}|I_{\text{in}}/\sqrt{2}|^2$. The gain G is defined as

$$G(\theta, \phi) = \frac{|\mathbf{E}(r, \theta, \phi)/\sqrt{2}|^2/Z_0}{P_{\text{in}}/4\pi r^2}$$

$$G(\theta, \phi) = \frac{|\mathbf{D}(\theta, \phi)|^2}{60P_{\text{in}}} \tag{15}$$

where Z_0 is the intrinsic impedance of free space ($Z_0 = 120\ \Omega$) and $\mathbf{D}(\theta, \phi)$ is defined as

$$\mathbf{D}(\theta, \phi) = \left(\frac{r}{e^{-j\beta r}}\right)\mathbf{E}(r, \theta, \phi) \tag{16}$$

$\mathbf{D}(\theta, \phi)$ is decomposed into two components as $\mathbf{E}(r, \theta, \phi)$ in Eq. (14):

$$\mathbf{D}(\theta, \phi) = D_R(\theta, \phi)(\hat{\theta} - j\hat{\phi}) + D_L(\theta, \phi)(\hat{\theta} + j\hat{\phi}) \tag{17}$$

Therefore, the gains for right- and left-hand CP waves are calculated by $G_R(\theta, \phi) = |D_R(\theta, \phi)|^2/30P_{\text{in}}$ and $G_L(\theta, \phi) = |D_L(\theta, \phi)|^2/30P_{\text{in}}$, respectively.

3. HELICAL ANTENNAS

Figure 2 shows a helical arm, which is specified by the pitch angle α, the number of helical turns n, and the circumference of the helix C. Helical antennas are classified into three groups in terms of the circumference C relative to a given wavelength λ [4]: a normal-mode helical antenna ($C \ll \lambda$), an axial-mode helical antenna ($C \approx \lambda$), and a conical-mode helical antenna ($C \approx 2\lambda$). The normal-mode helical antenna radiates a linearly polarized wave. The axial-mode and conical-mode helical antennas radiate CP waves. The beam direction for each mode is illustrated using arrows in Fig. 3.

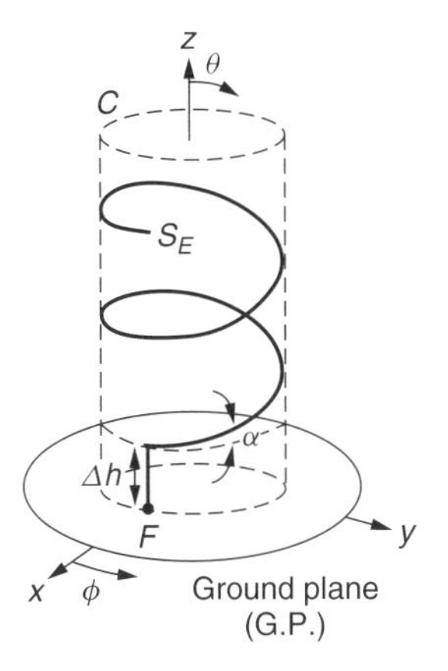

Figure 2. Helical arm.

The ground plane in Fig. 3, which backs each helical arm, is very large and assumed to be of infinite extent in the following analysis. This assumption allows the use of image theory, where the ground plane is removed. The removal of the ground plane enables one to use the techniques described in Section 2.

3.1. Normal-Mode Helical Antenna

The radiation when the circumference C is very small relative to a given wavelength is investigated in this subsection. For this, a circumference of $C = 0.0422\lambda_{0.5}$ (25.3 mm) is chosen for the antenna structure shown in Fig. 3a, together with a pitch angle of $\alpha = 40^\circ$ and number of helical turns $n = 4.5$, where $\lambda_{0.5}$ is the wavelength at a test frequency of 0.5 GHz = 500 MHz. The total arm length L_{arm}, including an initial wire length of $\Delta h = 1.5$ mm, is $\frac{1}{4}\lambda_{0.5}$. Therefore, the antenna characteristics are expected to be similar to those of a monopole antenna of one-quarter wavelength.

Figure 4 shows the current $I(= I_r + jI_i)$ distributed along the helical arm at 0.5 GHz. The helical arm is chosen to be thin: wire radius $\rho = 0.001\lambda_{0.5}$. It is found that the current distribution is a standing wave, as seen from the phase distribution. Note that the phase is calculated from $\tan^{-1}(I_i/I_r)$.

The radiation field from the current distribution at 0.5 GHz is shown in Fig. 5, where parts (a) and (b) are the radiation patterns in the x–z and y–z planes, respectively, and part (c) is the azimuth radiation pattern in the horizontal plane ($\theta = 90^\circ$, $0^\circ \le \phi \le 360^\circ$). It is clearly seen that the helical antenna radiates a linearly polarized wave: $E_\theta \ne 0$ and $E_\phi = 0$. The maximum radiation is in the horizontal direction ($\theta = 90^\circ$), where the polarization is in the antenna axis (z-axis) direction. The radiation field component E_θ in the horizontal plane is omnidirectional. The gain in the x direction is calculated to be approximately 4.9 dB.

Additionally, Fig. 6 shows the radiation patterns at 0.5 GHz as a function of the number of helical turns n, where the pitch angle and circumference remain unchanged: $\alpha = 40^\circ$ and $C = 0.0422\lambda_{0.5}$. It is found that, as n increases, the radiation beam becomes sharper.

3.2. Axial-Mode Helical Antenna

When the frequency is chosen to be 11.85 GHz, the physical circumference of the helix, $C = 25.3$ mm used in Section 3.1, corresponds to a length of one wavelength (1λ). An antenna having this circumference is analyzed in this subsection, using helical configuration parameters of $\alpha = 12.5^\circ$ and $n = 15$. The total arm length, including an initial wire length of $\Delta h = 1$ mm, is $L_{\text{arm}} = 15.4\lambda_{11.85}$, where $\lambda_{11.85}$ is the wavelength at a test frequency of 11.85 GHz.

Figure 7 shows the current $I(= I_r + jI_i)$ at 11.85 GHz, where the helical arm is chosen to be thin: wire radius $\rho = 0.001\lambda_{11.85}$. It is found that the current distribution has three distinct regions: a region from the feed point F to point P_1, a region from point P_1 to point P_2, and a region from point P_2 to the arm end S_E. The amplitude of the current in each region shows a different form.

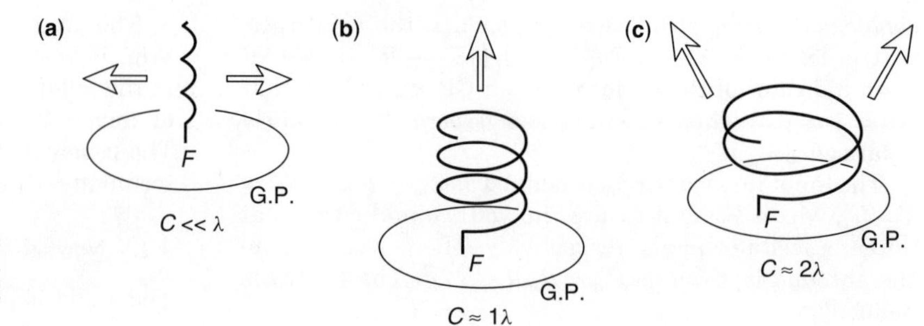

Figure 3. Helical antennas: (**a**) a normal-mode helical antenna ($C \ll \lambda$); (**b**) an axial-mode helical antenna ($C \approx \lambda$); and (**c**) a conical-mode helical antenna ($C \approx 2\lambda$).

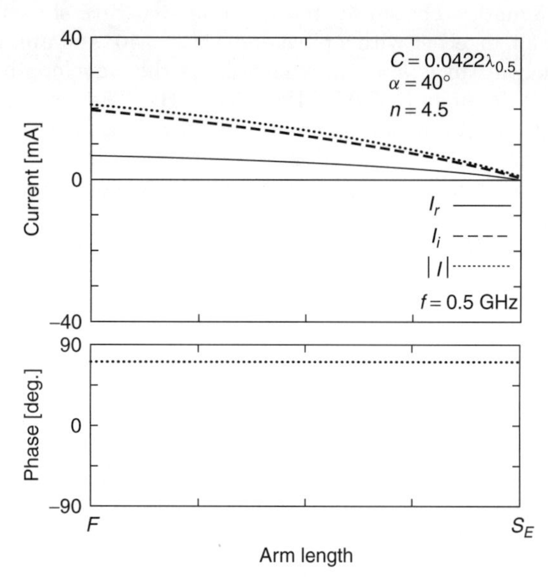

Figure 4. Current distribution of a normal-mode helical antenna.

The amplitude $|I|$ in the first region F–P_1 shows a rapid decay. Point P_1 is the position where the current first becomes minimal. The current in this region F–P_1 is a traveling wave, as seen from the phase progression. It is noted that the first region acts as an *exciter* for the remaining helical turns. This is proved by the fact that the current distribution remains almost the same even when the helical wire is cut at point P_1 [5].

The second region P_1–P_2 is the region called the "director," to which part of the power in the first region F–P_1 is guided. The amplitude of the current, $|I|$, in the second region is relatively constant. This is obviously different from the amplitude of the current in the first region. Detailed calculations reveal that the phase velocity of the current in the second region is such that the field in the z direction from each turn adds nearly in phase over a wide frequency bandwidth [6].

The outgoing current flowing along the director (the forward current) reaches the arm end S_E and is reflected. As a result, the current forms a standing wave near the arm end. The third region P_2–S_E reflects this fact. The reflected current from the arm end S_E is not desirable for a CP antenna, because it radiates a CP wave whose rotational sense is opposite that of the forward current, resulting in degradation of the axial ratio.

Figure 8 shows radiation patterns at 11.85 GHz, where the electric field at a far-field point is decomposed into right- and left-hand CP wave components. The forward current traveling toward the arm end generates the copolarization component. The copolarization component for this helical antenna is a right-hand CP wave component E_R. The component E_L for this helical antenna is called the "cross-polarization component." The cross-polarization component is generated by the undesired reflected current. The axial ratio and gain in the z direction are calculated to be approximately 0.8 and 11.5 dB, respectively.

Additionally, Fig. 9 shows the gains for a right-hand CP wave at 11.85 GHz as a function of the number of helical turns, n, for various pitch angles α, where the

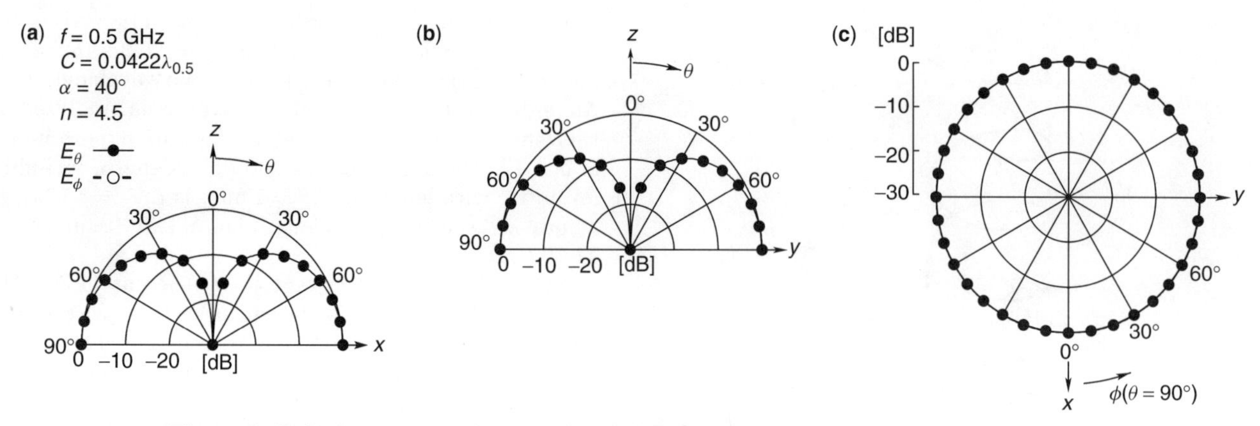

Figure 5. Radiation patterns of a normal-mode helical antenna: (**a**) in the x–z plane; (**b**) in the y–z plane; (**c**) in the horizontal plane ($\theta = 90°$, $0° \leq \phi \leq 360°$).

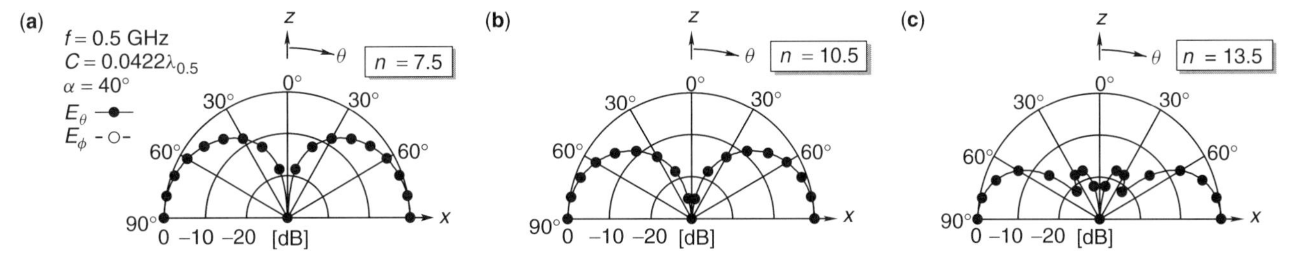

Figure 6. Radiation pattern of a normal-mode helical antenna as a function of number of helical turns: (**a**) $n = 7.5$; (**b**) $n = 10.5$; (**c**) $n = 13.5$.

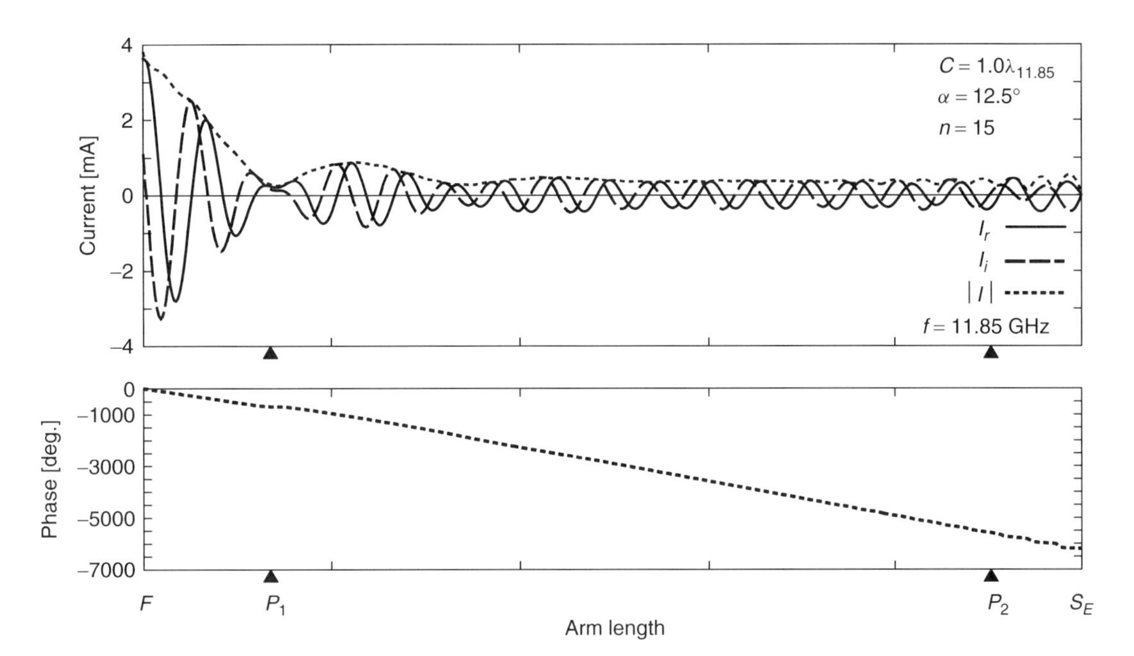

Figure 7. Current distribution of an axial-mode helical antenna.

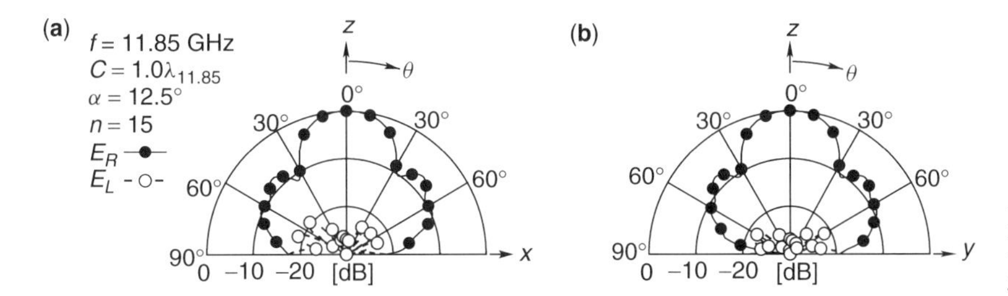

Figure 8. Radiation patterns of an axial-mode helical antenna: (**a**) in the x–z plane; (**b**) in the y–z plane.

circumference is kept constant: $C = 1.0\lambda_{11.85}$. Note that each gain increases as n increases. However, the gain increase is bounded; that is, there is a maximum gain value for a given pitch angle α.

3.3. Conical-Mode Helical Antenna

When the helical arm has a circumference of approximately two wavelengths (2λ), the radiation from the helix forms a conical beam [7,8]. To reflect this fact, a frequency of 23.7 GHz is used for an antenna with a circumference of $C = 25.3$ mm $= 2.0\lambda_{23.7}$, where $\lambda_{23.7}$ is the wavelength at 23.7 GHz. Other configuration parameters are arbitrarily chosen as follows: pitch angle $\alpha = 4^\circ$ and number of helical turns $n = 2$. The total arm length, including the initial wire length $\Delta h = 1$ mm, is $L_{\text{arm}} = 4.09\lambda_{23.7}$.

Figure 10 shows the current distribution along the helical arm, whose wire radius is $\rho = 0.001\lambda_{23.7}$. As observed in the first region F–P_1 of the axial-mode helical antenna (see Fig. 7), the current is a decaying traveling wave, which radiates a CP wave.

It is obvious that local radiation from the helical arm forms the total radiation. If each turn of the helix is a local radiation element and approximated by a loop whose circumference is two wavelengths, the currents along n local loops produce a zero field in the z direction and a maximum radiation off the z axis; that is, the total

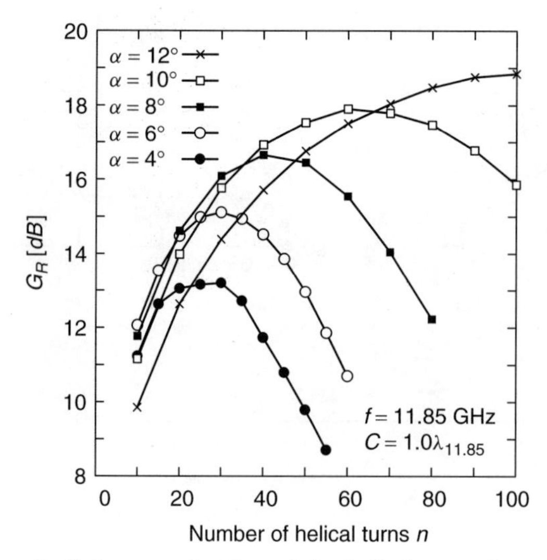

Figure 9. Gains as a function of the helical turns for various pitch angles.

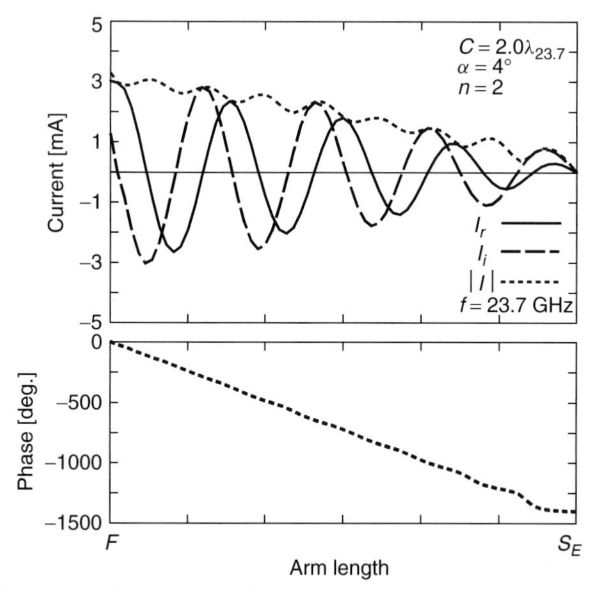

Figure 10. Current distribution of a conical-mode helical antenna.

radiation forms a conical beam. Figure 11 reflects this fact, where the radiation in the beam direction ($\theta = 34°$) is circularly polarized. Note that the azimuth pattern of Fig. 11c is calculated in a plane of ($\theta = 34°, 0° \leq \phi \leq 360°$), where the copolarization component E_R shows an omnidirectional pattern.

4. SPIRAL ANTENNA

A spiral antenna is an element that radiates a CP wave over a wide frequency bandwidth [9–12]. The mechanism of the CP wave radiation is investigated for two cases of excitation: (1) antiphase and (2) in-phase. First, the CP radiation is qualitatively explained in terms of *current band theory* [9], which is based on transmission line theory; second, the numerical results (quantitative results) obtained by the MoM are presented and discussed.

4.1. Configuration

Figure 12 shows the configuration of a two-arm spiral antenna. The two arms, A and B, are symmetrically wound with respect to the centerpoint o. The radial distance from the centerpoint o to a point on arm A is defined as $r_A = a_{\rm sp}\phi_{\rm wn}$, where $a_{\rm sp}$ and $\phi_{\rm wn}$ are the spiral constant and winding angle, respectively. The winding angle starts at $\phi_{\rm st}$ and ends at $\phi_{\rm end}$. Similarly, the radial distance from the centerpoint o to a point on arm B is defined as $r_B = a_{\rm sp}(\phi_{\rm wn} - \pi)$, with a starting angle of $\phi_{\rm st} + \pi$ and an ending angle of $\phi_{\rm end} + \pi$. It is noted that the spiral specified by r_A and r_B is an *Archimedean spiral antenna*.

The spiral antenna is fed from terminals T_A and T_B in the center region. The mechanism of CP wave radiation is qualitatively explained by current band theory [9], in which it is assumed that the two arms A and B are tightly wound. It is also assumed that the currents along the two arms gradually decrease, radiating electromagnetic power into free space.

To apply current band theory, four points P_A, P'_A, P_B, and P'_B are defined as follows. P_A and P'_A are points on arm A, and P_B and P'_B are points on arm B. Points P_A and P_B are symmetric with respect to the centerpoint o. P'_A is a point on arm A and a neighboring point of P_B. P'_B is a point on arm B and a neighboring point of P_A.

4.2. Antiphase Excitation

Discussion is devoted to the radiation from the spiral when terminals T_A and T_B are fed with the same amplitude and a phase difference of 180°. The excitation is called "antiphase excitation," which is realized by inserting a voltage source between terminals T_A and T_B.

4.2.1. First-Mode Radiation. The current along arm A travels through point P_A and reaches arm end E_A. Similarly, the current along arm B travels through point P_B and reaches arm end E_B. The phase of the current at point P_B always differs from that at point P_A by 180°, because of the antiphase excitation at terminals T_A and T_B. This is illustrated using arrows at P_A and P_B in Fig. 13. Note that the direction of the arrow at P_B is opposite that of the arm growth, and the two arrows at P_A and P_B are in the same direction.

An interesting phenomenon is found when points P_A and P_B are located on a ring region, with a center circumference of one wavelength (λ), in the spiral plane. Points P_A and P_B in this case are separated by approximately one-half wavelength (0.5λ) along the circumference. The current at P_A and the current at its neighboring point P'_B on arm B are approximately in phase, because the current traveling from point P_B to point P'_B along arm B (traveling by approximately one-half wavelength, since the spiral arms are tightly wounded) experiences a phase change of approximately 180°. Similarly, the current at P_B and the current at its neighboring point P'_A on arm A are approximately in phase. Figure 13 illustrates these four currents at points P_A, P'_B,

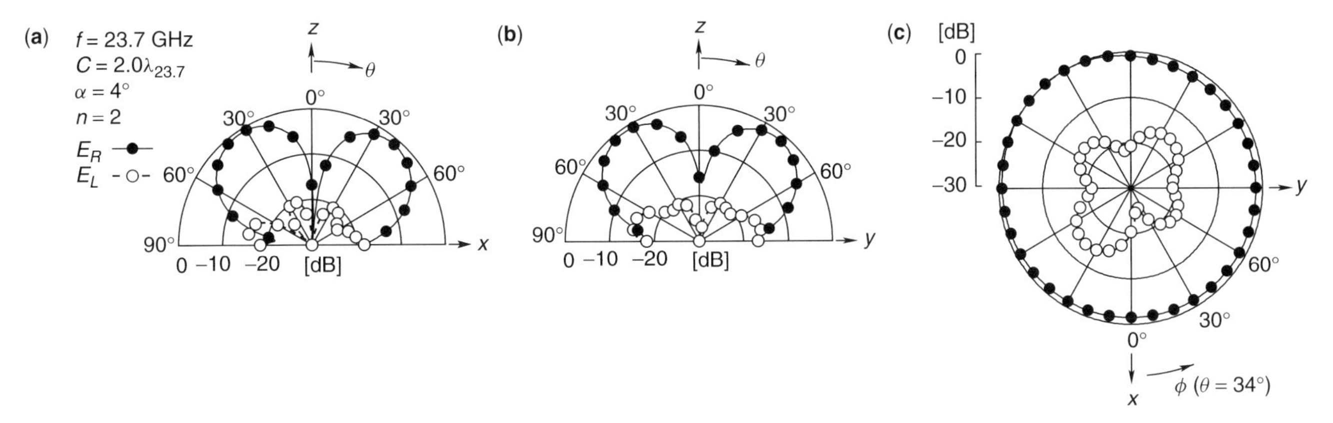

Figure 11. Radiation patterns of a conical-mode helical antenna: (**a**) in the x–z plane; (**b**) in the y–z plane; (**c**) in a plane of ($\theta = 34^\circ, 0^\circ \leq \phi \leq 360^\circ$).

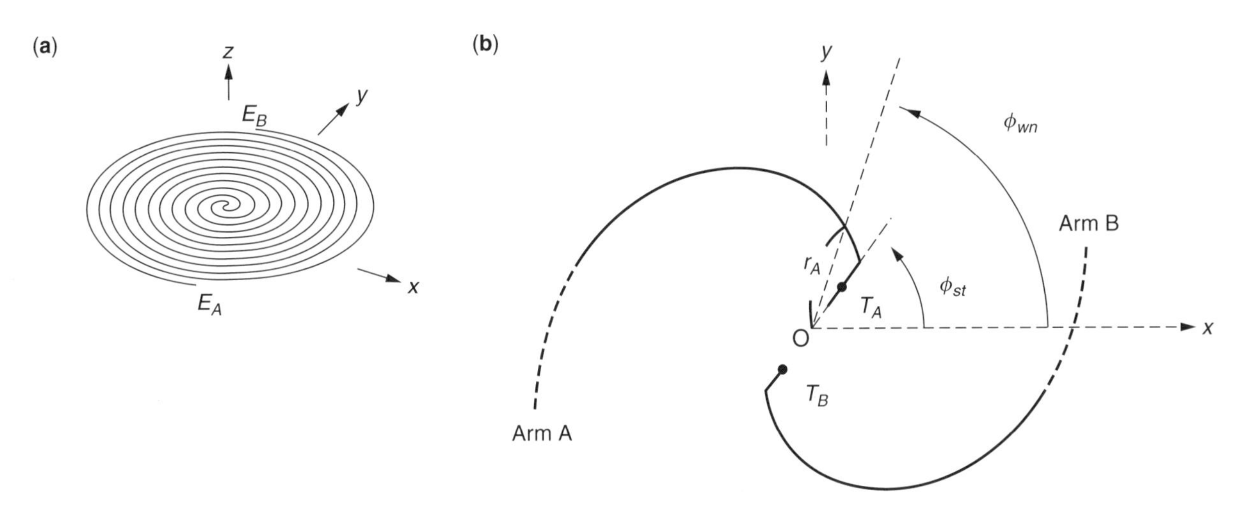

Figure 12. A two-wire spiral antenna: (**a**) perspective view; (**b**) top view.

P_B, and P'_A, where each pair of currents forms a band of current.

The two current bands in Fig. 13 rotate around the centerpoint o with time. This means that the electric field radiated from each current band also rotates. In other words, the radiation field is circularly polarized. The two circularly polarized waves radiated from the two current bands are in phase on the z axis, resulting in maximum radiation on the z axis. This radiation is called "first-mode radiation."

4.2.2. Third-Mode Radiation. When points P_A and P_B are located on a ring region of three-wavelength (3λ) circumference in the spiral plane (see Fig. 14), points P_A and P_B are separated by 1.5 wavelengths along this 3λ circumference. Therefore, the current along arm B experiences a phase change of approximately $360^\circ + 180^\circ$ from point P_B to point P'_B. As a result, the currents at points P_A and P'_B are approximately in phase; that is, a current band is formed. Similarly, the currents at points P_B and P'_A are approximately in phase, and a current band is formed.

The currents on arms A and B become in-phase with a period of one-half wavelength along the 3λ circumference.

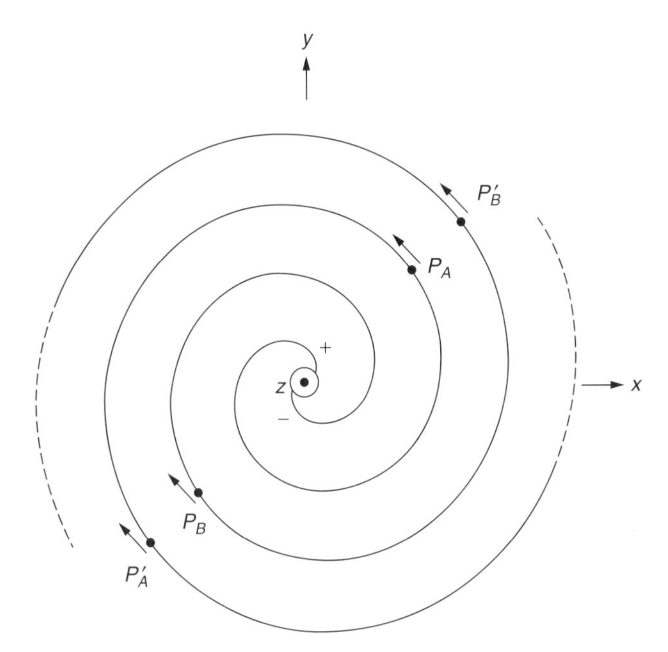

Figure 13. Current bands for first-mode radiation.

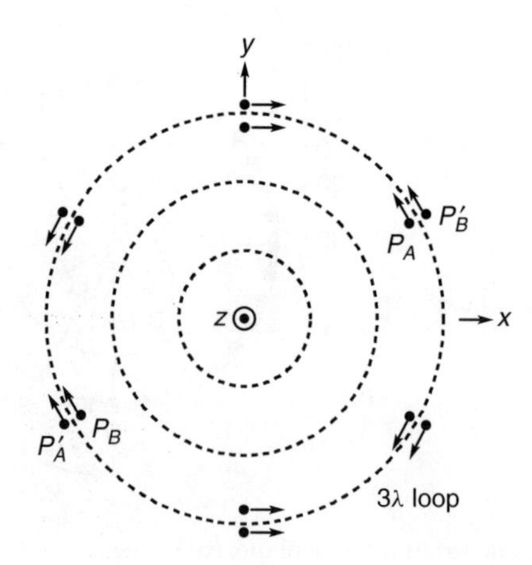

Figure 14. Current bands for third-mode radiation.

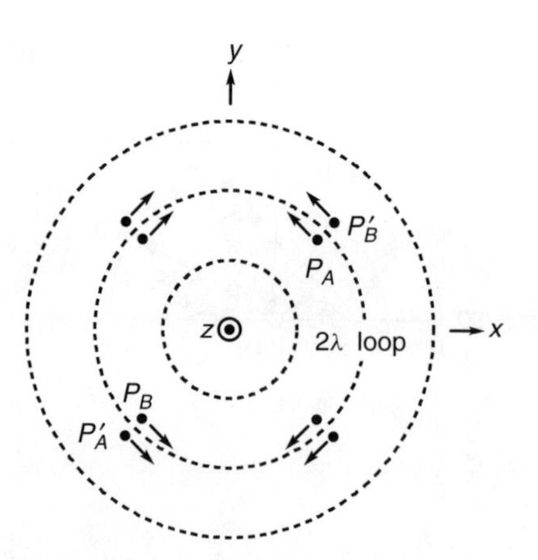

Figure 15. Current bands for second-mode radiation.

As a result, four more current bands are formed between points P_A and P_B, as shown in Fig. 14. The directions of all current bands along the 3λ circumference cause the electric fields, rotating with time, to cancel on the z axis and add off the z axis. This radiation is called "third-mode radiation."

4.2.3. Odd-Mode Radiation. So far, first-mode radiation and third-mode radiation have been discussed under the condition that the excitation at terminals T_A and T_B is antiphase. The first- and third-mode radiation components result from the current bands formed over two regions of 1λ and 3λ circumferences in the spiral plane, respectively. The mechanism described in the previous section leads to the formation of higher odd mth-mode radiation ($m = 5, 7, \ldots$) as long as the currents exist over regions of circumference of $m\lambda$ in the spiral plane. Note that each higher odd mth-mode radiation component becomes maximal off the z axis.

As the currents leave the regions of circumference of $m\lambda (m = 1, 3, 5, \ldots)$ in the spiral plane, the in-phase condition of the neighboring currents on arms A and B becomes destructive. The destructive currents contribute little to the radiation. The radiation, therefore, is characterized by a sum of odd-mode radiation components that the spiral supports.

4.3. In-Phase Excitation

When terminals T_A and T_B are excited in antiphase, the spiral has odd-mode radiation components, as discussed in the previous subsection. Now the radiation when terminals T_A and T_B are excited in phase (terminals T_A and T_B are excited with the same amplitude and the same phase) is considered. Realization of in-phase excitation is found in Section 4.5.

4.3.1. Second-Mode Radiation. A situation where points P_A and P_B are located on a ring region of 2λ circumference in the spiral plane is investigated. The phases of the currents at points P_A and P_B are the same due to two facts; terminals T_A and T_B are excited in phase, and the distance from terminal T_A to point P_A along arm A is equal to that from terminal T_B to point P_B along arm B. Arrows P_A and P_B in Fig. 15 illustrate the phases of the currents at points P_A and P_B. The directions of the arrows at points P_A and P_B are in the arm-growth direction and opposite each other.

The current traveling from points P_B to P'_B experiences a phase change of approximately 360°, because the distance from points P_B to P'_B along arm B is approximately one wavelength. It follows that the direction of the arrow for the current at point P'_B is in the arm-growth direction, as shown in Fig. 15. The arrows at points P_A and P'_B have the same direction, and indicate the formation of a current band.

Similarly, the current traveling from points P_A to P'_A experiences a phase change of approximately 360°, due to an approximately one-wavelength difference between these points. An arrow at point P'_A in Fig. 15 shows this fact. The arrows at points P'_A and P_B have the same direction, and indicate the formation of a current band.

The two middle points between points P_A and P_B along the 2λ circumference are separated from P_A (and P_B) by one-half wavelength. A current band is formed at each middle point. It follows that four current bands are formed over a ring region of 2λ circumference in the spiral plane. As seen from the direction of the arrows in Fig. 15, the radiation from these four current bands is zero on the z axis and maximal off the z axis. This radiation is called "second-mode radiation."

4.3.2. Even-Mode Radiation. One can conclude from the previous observation that the spiral with in-phase excitation does not have current bands in the ring region whose circumference is 3λ in the spiral plane. The phase relationship of the currents over the 3λ ring region is destructive (out of phase), thereby not forming current bands. However, as the currents on arms A and B further travel toward their arm ends, the phase relationship of the currents gradually becomes constructive. When the currents reach a ring region whose circumference is four wavelengths in the spiral plane, the phases of the neighboring currents become in-phase. Again current bands are formed. This radiation is called "fourth-mode radiation." Similarly, higher even-mode radiation occurs until the currents die out. It is noted that even mth-mode radiation ($m = 2, 4, \ldots$) have zero intensity on the z axis and maximal off the z axis.

4.4. Numerical Results of a Spiral Antenna with Antiphase Excitation

The radiation mechanisms of a spiral antenna have been *qualitatively* discussed in Sections 4.2 and 4.3. In this subsection, the radiation characteristics of a spiral antenna with antiphase excitation are *quantitatively* obtained on the basis of the numerical techniques presented in Section 2.

The configuration parameters of the spiral are chosen as follows: spiral constant $a_{sp} = 0.0764$ cm/rad and winding angle ϕ_{wn} ranging from $\phi_{st} = 2.60$ rad to $\phi_{end} = 36.46$ rad. These configuration parameters lead to an antenna diameter of $2\pi a_{sp}\phi_{end} = 17.5$ cm $= 3.5\lambda_6$, where λ_6 is the wavelength at a frequency of 6 GHz. In other words, the spiral at a frequency of 6 GHz includes a ring region of three-wavelength circumference in the spiral plane. Note that the wire radius of the spiral arm is $\rho = 0.012\lambda_6$.

Figure 16 shows the current $I(= I_r + jI_i)$ distributed along arm A at a frequency of 6 GHz. Since the currents on arms A and B are symmetric with respect to the centerpoint o (note that the distance between terminals T_A and T_B is assumed to be infinitesimal), this figure shows only the current along arm A. It is found that the current decreases, traveling toward the arm end. The phase progression of the current close to that in free space ($= -2\pi s/\lambda_6$, where s is the distance measured along the spiral arm from the centerpoint o to an observation point). This means that the wavelength of the current along the arm, called the "guided wavelength λ_g," is close to the wavelength propagating in free space, λ_6.

Figure 17 shows the radiation patterns in the x–z plane and y–z plane. The spiral equally radiates waves in the $\pm z$ hemispheres. The radiation has a maximum value in the $\pm z$ directions and is circularly polarized. The axial ratio in the $\pm z$ directions is approximately 0.1 dB, and the gain is approximately 5.5 dB.

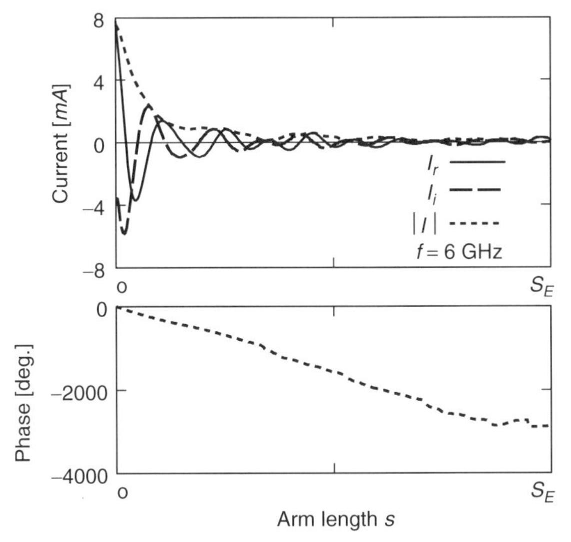

Figure 16. Current distribution of a spiral antenna with antiphase excitation.

So far, the antenna characteristics at a frequency of 6 GHz have been discussed. Now, the frequency responses of the radiation characteristics are investigated. Figure 18 shows the axial ratio (AR) in the positive z direction as a function of frequency, together with the gain for a right-hand CP wave, G_R, in the positive z direction. It is observed that, as the frequency decreases, the axial ratio deteriorates. This is due to the fact that, as the frequency decreases, the ring region of one-wavelength circumference in the spiral plane moves toward the periphery of the spiral and finally disappears. With the movement of the one-wavelength circumference ring, the polarization of the radiation becomes elliptical, that is, the axial ratio increases.

Figure 19 shows the input impedance $Z_{in}(= R_{in} + jX_{in})$ as a function of frequency. The input impedance is relatively constant over a wide frequency bandwidth. Note that the input impedance is always $Z_{in} = 60\pi\,\Omega$ when the following two conditions are satisfied: (1) arms A and B, each made of a *strip* conductor, are infinitely wound; and (2) the spacing between the two arms equals the width of the strip conductor. The antenna satisfying these conditions is called the "self-complementary antenna" [13].

4.5. Numerical Results of a Spiral Antenna with In-Phase Excitation

Figure 20 shows a spiral antenna with in-phase excitation, where a round conducting disk, approximated by wires for analysis (see Fig. 20c), is used for exciting the spiral. A voltage source is inserted between the spiral and the conducting disk. The spiral is a backed by a conducting plane of infinite extent.

The configuration parameters are as follows: spiral constant $a_{sp} = 0.04817$ cm/rad, winding angle ϕ_{wn} ranging from $\phi_{st} = 8\pi$ rad to $\phi_{end} = 37.5$ rad, wire radius $\rho = 0.00314\lambda_6$, disk diameter $D_{disk} = 0.49\lambda_6$, spacing between spiral and disk $H = 0.046\lambda_6$, and spacing between spiral and conducting plane $H_r = \frac{1}{4}\lambda_6$. The spiral at 6 GHz includes a ring region of two-wavelength circumference in the spiral plane. Note that a ring region of one-wavelength circumference does not contribute to the radiation, and hence the arms inside the one–wavelength circumference are deleted, as shown in Fig. 20a.

Figure 21 shows the radiation at a frequency of 6 GHz. The radiation occurs in the positive z hemisphere because the conducting plane is of infinite extent. As expected from current band theory, the maximum value of the radiation is off the z axis, as shown in Fig. 21a. Figure 21b shows the azimuth radiation pattern at $\theta = 40^\circ$ (beam direction angle from the z axis). The variation in the azimuth radiation component E_R is very small. The axial ratio in a plane of ($\theta = 40^\circ$, $0^\circ \leq \phi \leq 360^\circ$) is presented in Fig. 21c. It is concluded that the spiral forms a circularly polarized conical beam.

5. ADDITIONAL INFORMATION

The axial-mode helix in Section 3.2 has been analyzed under the condition that its ground plane is of infinite extent. An interesting phenomenon is observed when

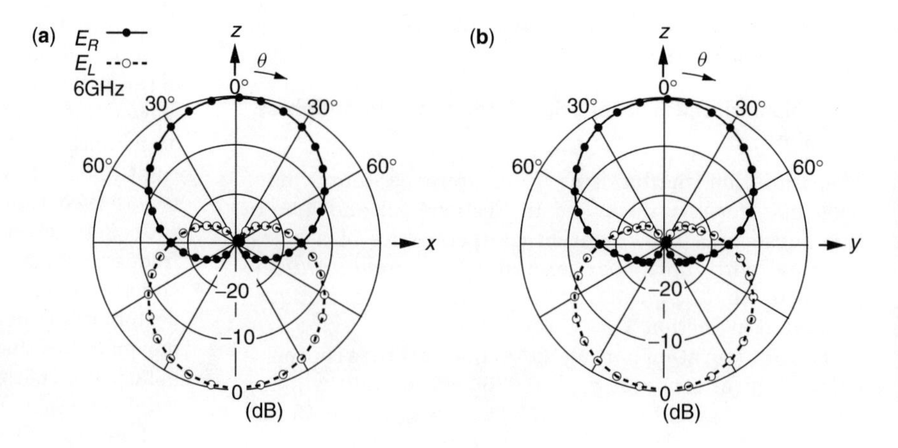

Figure 17. Radiation patterns of a spiral antenna with antiphase excitation: (**a**) in the x–z plane; (**b**) in the y–z plane.

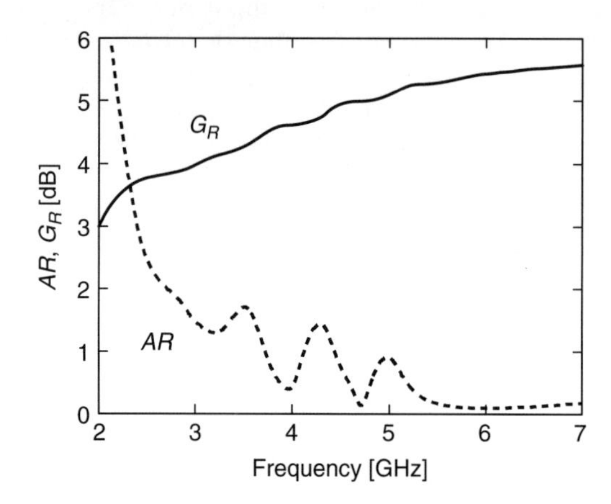

Figure 18. Axial ratio and gain as a function of frequency.

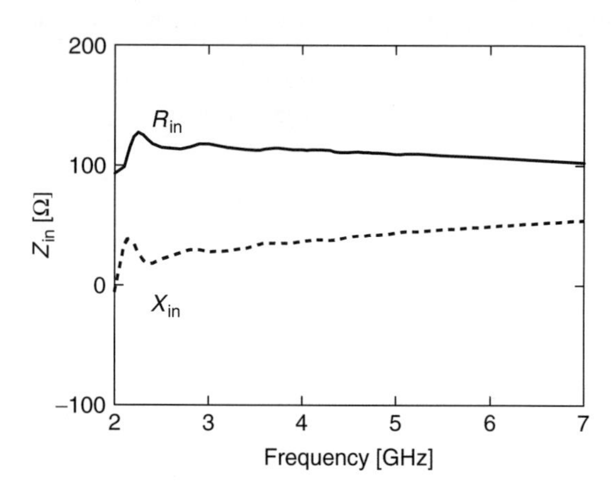

Figure 19. Input impedance as a function of frequency.

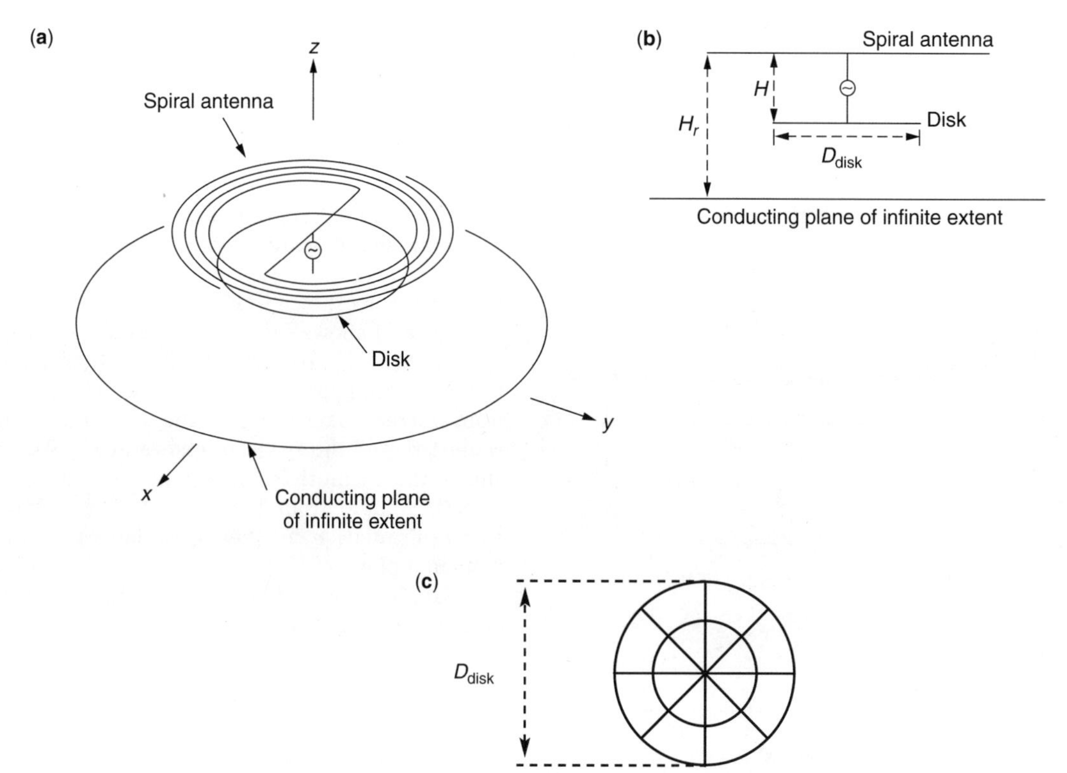

Figure 20. A spiral antenna with in-phase excitation: (**a**) perspective view; (**b**) side view; (**c**) disk approximated by wires.

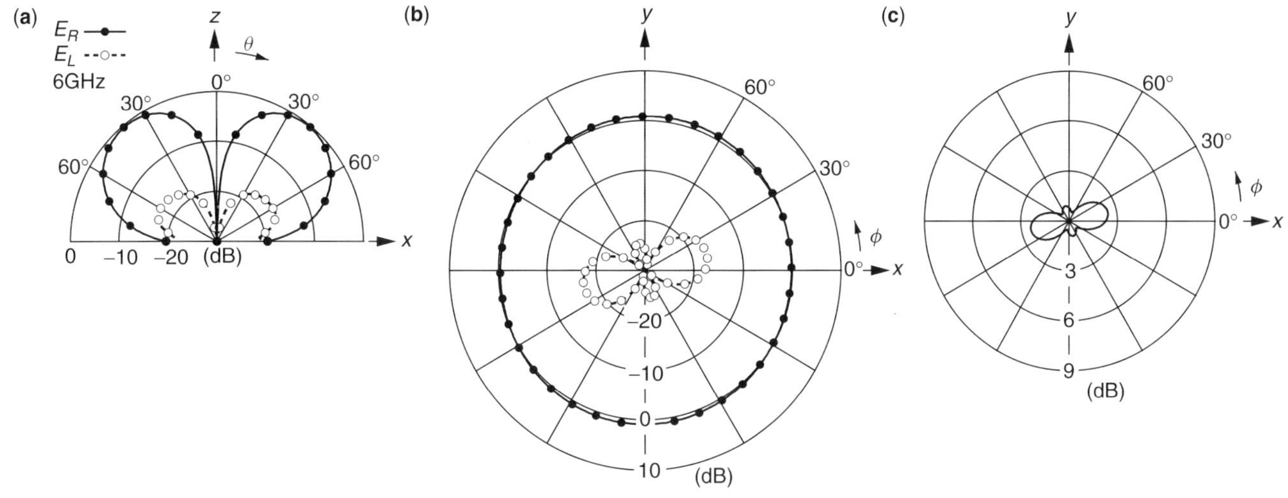

Figure 21. Radiation from a spiral antenna with in-phase excitation: (**a**) radiation pattern in the x–z plane; (**b**) radiation pattern in a plane of ($\theta = 40°$, $0° \leq \phi \leq 360°$); and (**c**) axial ratio pattern in a plane of ($\theta = 40°$, $0 \leq \phi \leq 360°$).

the size of the ground plane is made comparable to the circumference of the helix — the helix radiates a CP wave in the *backward* direction (in the negative z direction). This is called the "backfiremode." An application of the backfiremode is found in Ref. 14, where the helix is used as a CP primary feed for a parabolic reflector. This reflector antenna has a high gain of more than 30 dB with an aperture efficiency of more than 70%. It is widely used for *direct broadcasting satellite* (DBS) signal reception. For more recent research on helical antennas, including the analysis of a helix wound on a dielectric rod and four helices with a cavity, readers are directed to Refs. 15 and 16.

The spiral antenna discussed in Section 4.4 has a, "bidirectional beam"; that is, the radiation occurs in the $\pm z$-hemispheres, as shown in Fig. 17. It is possible to change the bi-directional beam to a "unidirectional beam" (radiation in the $+z$ hemisphere) using two techniques. One is to use a conducting plane, as shown in Fig. 22a, and the other is to use a cavity, as shown in Fig. 22b.

The conducting plane in Fig. 22a is put behind the spiral usually with a spacing of one-quarter wavelength (0.25λ) at the operating center frequency. The gain is increased because the rear radiation is reflected by the conducting plane and added to the front radiation. However, the conductor plane affects the wideband characteristics of the spiral. To keep the antenna characteristics over a wide frequency bandwidth, it is recommended that the spacing between the spiral and the conducting plane be small and absorbing material be inserted between the outermost arms and the conducting plane [12]. A theoretical analysis for this case using a finite-difference time-domain method [17] is found in Ref. 18.

The inside of the cavity in Fig. 22b is filled with absorbing material. The rear radiation is absorbed and does not contribute to the front radiation. Only one-half of the power input to the spiral is used for the radiation. Therefore, the gain does not increase, unlike the gain for the spiral with a conducting plane. However, the antenna characteristics are stable over a wide frequency band, for example, ranging from 1 to 10 GHz.

BIOGRAPHY

Hisamatsu Nakano received his B.E., M.E., and Dr. E. degrees in electrical engineering from Hosei University, Tokyo, Japan, in 1968, 1970, and 1974, respectively. Since 1973 he has been a member of the faculty of Hosei University, where he is now a professor of electronic informatics. His research topics include numerical methods for antennas, electromagnetic wave scattering problems, and light wave problems. He has published more than 170 refereed journal papers and 140 international symposium papers on antenna and relevant problems. He is the author of *Helical and Spiral Antennas* (Research Studies Press, England, Wiley, 1987). He published the chapter "Antenna analysis using integral equations," in *Analysis Methods of Electromagnetic Wave Problems*, vol. 2 (Norwood, MA: Artech House, 1996).

He was a visiting associate professor at Syracuse University, New York, during March–September 1981, a visiting professor at University of Manitoba, Canada, during March–September 1986, and a visiting professor at the University of California, Los Angeles, during September 1986–March 1987.

Dr. Nakano received the Best Paper Award from the IEEE 5th International Conference on antennas and propagation in 1987. In 1994, he received the IEEE AP-S Best Application Paper Award (H. A. Wheeler Award).

BIBLIOGRAPHY

1. K. K. Mei, On the integral equations of thin wire antennas, *IEEE Trans. Antennas Propag.* **13**(3): 374–378 (May 1965).
2. E. Yamashita, ed., *Analysis Methods for Electromagnetic Wave Problems*, Artech House, Boston, 1996, Chap. 3.
3. R. F. Harrington, *Field Computation by Moment Methods*, Macmillan, New York, 1968.

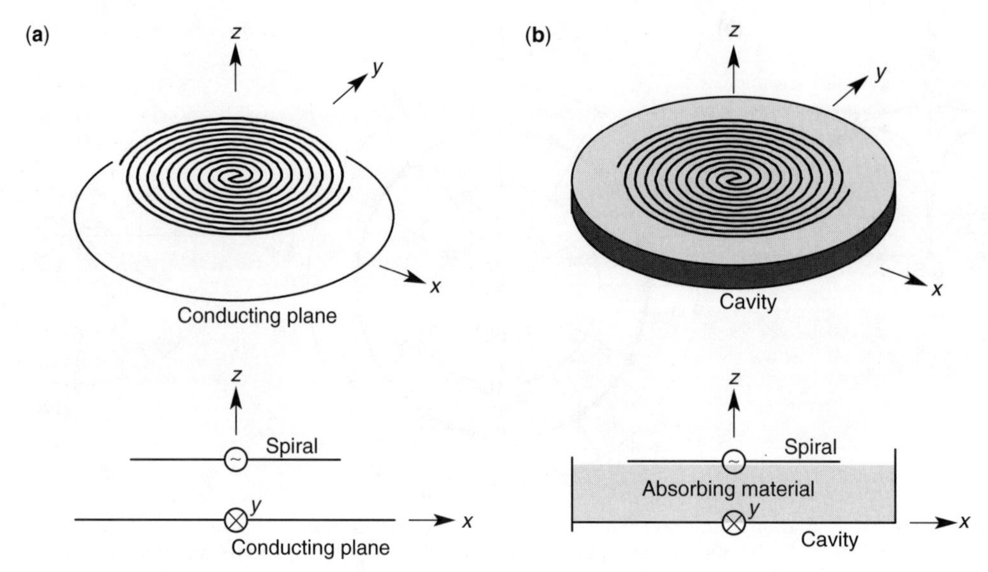

Figure 22. Techniques for unidirectional beam: (**a**) a conducting plane; (**b**) a cavity filled with absorbing material.

4. J. D. Kraus, *Antennas*, 2nd ed., McGraw-Hill, New York, 1988, Chap. 7.

5. H. Nakano and J. Yamauchi, Radiation characteristics of helix antenna with parasitic elements, *Electron. Lett.* **16**(18): 687–688 (Aug. 1980).

6. H. Nakano and J. Yamauchi, The balanced helices radiating in the axial mode, *IEEE AP-S Int. Symp. Digest*, Seattle, WA, 1979, pp. 404–407.

7. R. C. Johnson, *Antenna Engineering Handbook*, 3rd ed., McGraw-Hill, New York, 1993, Chap. 13.

8. H. Mimaki and H. Nakano, A small pitch helical antenna radiating a circularly polarized conical beam, *Proc. 2000 Communications Society Conf. IECE*, Nagoya, Japan, Sept. 2000, p. B-1-82.

9. J. A. Kaiser, The Archimedean two-wire spiral antenna, *IRE Trans. Antennas Propag.* **AP-8**(3): 312–323 (May 1960).

10. H. Nakano and J. Yamauchi, Characteristics of modified spiral and helical antennas, *IEE Proc.* **129**(5)(Pt. H): 232–237 (Oct. 1982).

11. H. Nakano et al., A spiral antenna backed by a conducting plane reflector, *IEEE Trans. Antennas Propag.* **AP-34**(6): 791–796 (June 1986).

12. J. J. H. Wang and V. K. Tripp, Design of multioctave spiral-mode microstrip antennas, *IEEE Trans. Antennas Propag.* **39**(3): 332–335 (March 1991).

13. Y. Mushiake, Self-complementary antennas, *IEEE Antennas Propag. Mag.* **34**(6): 23–29 (Dec. 1992).

14. H. Nakano, J. Yamauchi, and H. Mimaki, Backfire radiation from a monofilar helix with a small ground plane, *IEEE Trans. Antennas Propag.* **36**(10): 1359–1364 (Oct. 1988).

15. H. Nakano, M. Sakai, and J. Yamauchi, A quadrifilar helical antenna printed on a dielectric prism, Proc. *2000 Asia-Pacific Microwave Conf.*, Sydney, Australia, (Dec. 2000), Vol. 1, pp. 1432–1435.

16. M. Ikeda, J. Yamauchi, and H. Nakano, A quadrifilar helical antenna with a conducting wall, *Proc. 2001 Communications Society Conf.* IEICE, Choufu, Japan, (Sept. 2001), p. B-1-70.

17. A. Taflove, *Computational Electrodynamics: The Finite-Difference Time Domain Method*, Artech House, Norwood, MA, 1995.

18. Y. Okabe, J. Yamauchi, and H. Nakano, A strip spiral antenna with absorbing layers inside a cavity, *Proc. 2001 Communications Society Conf. IEICE*, Choufu, Japan, Sept. 2001, p. B-1-107.

HF COMMUNICATIONS

Julian J. Bussgang
Steen A. Parl
Signatron Technology Corporation
Concord, Massachusetts

1. INTRODUCTION TO HF COMMUNICATION

High-frequency (HF) *communications* is defined by the International Telecommunication Union (ITU) as radio transmission in the frequency range 3–30 MHz. However, it is common to consider the lower end of the HF band as extending to 2 MHz.

HF radio communications originally became popular because of its long-range capabilities and low cost. For high-rate digital communications, the HF channel is, however, a rather difficult communications medium. It subjects the transmitted signal to large variations in signal level and multiple propagation paths with delay-time differences large enough to cause signal overlap, and hence self-interference. A relatively high rate of fading can result and hinder the ability of receivers to adapt to the channel. High-data-rate communications over HF have also been hampered by having to contend with narrowband channel allocations, a large number of legacy users, and restrictive regulatory requirements.

Communications in the HF band rely on either ground-wave propagation or sky-wave propagation, which

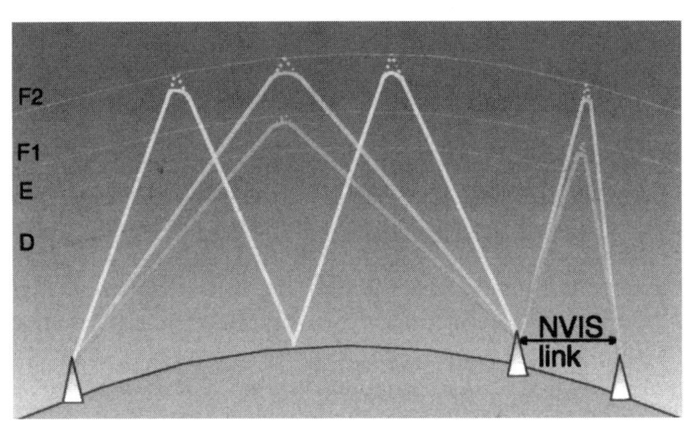

Figure 1. HF sky-wave propagation examples, showing multipath from different layer reflections and multihop propagation.

involves reflections from the ionosphere (Fig. 1). Ground-wave propagation is composed of a direct ray, a ground reflected ray, and a surface wave due to ground currents. The lower the frequency is, the farther ground-wave signals propagate. At the upper end of HF frequencies, the ground-wave attenuates rather rapidly with distance and plays less of a role for long-distance communications. The other mode of HF propagation, *sky-wave*, relies on the ionosphere, which is a series of ionized layers above the earth. Ionization is induced by solar radiation on the day side of the earth and by cosmic rays and meteors on the night side. This causes rays to be reflected, or rather *refracted*, from the ionosphere.

Refraction in the ionosphere is stronger at lower frequencies. As a result, frequencies below HF are refracted from lower layers. At frequencies above HF rays do not refract enough, even from the higher layers. Thus there is a highest frequency supported by HF sky-wave propagation. D-layer absorption determines the lowest frequency in the HF band supporting a sky-wave mode.

Well below HF, in the very low frequency (VLF) band ground-wave and sky-wave propagation combine to form the earth–ionosphere waveguide, which can become nearly lossless, allowing worldwide communications. HF is normally the highest frequency band that can propagate over long distances within this waveguide.

The importance of HF derives from the fact that HF radiowaves are capable of long-distance transmission with relatively little power, which was crucial before satellites and long-range cables became available. The ability to communicate using radio transmission at long distances was discovered by Guglielmo Marconi when he demonstrated the first transatlantic wireless telegraphy in December 1901. He used radio transmission to send Morse code signals from England to Newfoundland. The existence of the ionosphere was not known at that time. A year later Oliver Heaviside and independently Arthur Edwin Kennelly suggested the existence of an ionized layer to explain long-range propagation. It was not until 1924 that Edward V. Appleton confirmed this.

More recently the role of HF has to some extent been taken over by communications satellites, which hover over the earth in geostationary positions and permit over-the-horizon communication independent of day or night. However, satellite communications is more costly than HF, is less convenient for stations on a moving platform, and is not available without prior arrangements and usage fees. Thus, military services, commercial airplanes and ships, and emergency services continue to use HF radio. The lower cost of HF communications and advances in technology are the major driving forces toward continued use and further enhancements.

Today HF is increasingly used for data transmission. With advances in electronic signal processing, the interest in reviving HF is growing. The popularity of the Internet has created a demand for higher data rates. However, significant moves in that direction are stifled by the fact that the existing HF frequency allocations are very narrowband.

2. HF RADIO APPLICATIONS

HF radio was initially used for Morse code and voice transmission. Accordingly, under international radio conventions, much of the spectrum was organized into a series of narrowband channels, with voice channel allotted a nominal 3-kHz band for single sideband (SSB) transmission. Either the upper (USB) or the lower (LSB) sideband could be selected. For voice communications, HF radio has sometimes been combined with vocoders (speech bandwidth compression devices), particularly when double-sideband modulation of speech had to be accommodated within the narrow bandwidth of a typical HF channel.

Nowadays even voice is transmitted digitally, so that the principal interest has shifted to various digital modulation schemes and modems tailored to be effective over HF channels.

As digital transmission was introduced, frequency shift keying (FSK) and phase shift keying (PSK) became increasingly applied as modulation techniques. In addition, frequency hopping has been used for military digital transmission, when it is desirable to avoid detection of the transmissions so as to prevent a jammer from locating and concentrating on a particular transmission frequency.

A typical application of HF radio has been marine radio from coastal stations to ships and back, and between naval vessels. The frequency is selected according to the time of day, season of the year, and distance to be covered. Some of the marine communications have switched to satellite links, but HF is still commonly used because of its lower cost. HF voice links continue to be very much in use for emergency or distress calls.

A new digital service has been introduced by ARINC, augmenting existing HF voice radios to provide data communications between commercial aircraft and ground. This service is intended primarily for commercial aircraft flying over the oceans, out of reach of other radio communications. The service, HF Data Link (HFDL) [1], and 11 HF shore stations around the world under control of two long-distance operational control (LDOC) centers, one in New York and the other in San Francisco. HF links operated by ARINC are used by the Federal Aviation Agency (FAA) for air traffic control (ATC) communications. Above 80° north, HFDL is the only communications medium available to commercial aviation. Elsewhere, the HFDL service competes with satellite services primarily on cost. HF is still used in Australia for the Royal Flying Doctor Service (RFDS), but its role has shifted to function more as an emergency backup to the telephone network and to satellite systems.

As HF data links become more sophisticated and better able to cope with link variability, they find more applications in digital networking, data transmission, and email. Indeed, email is a growing marine application. The shipboard HF radio stays in contact with one of several shore-based HF stations, which act as mail servers. Several commercial HF email networks have been established as an outgrowth of marine amateur radio operations [2]. The Red Cross has used the CLOVER protocol. Another network protocol is called PACTOR, which is an FSK based scheme developed in Germany in the late 1980s by a group of ham operators. A newer proprietary protocol is PACTOR-II, which uses a two-tone differential phase shift keying (DPSK) [3]. Both use a parallel-tone modulation scheme. The data rates for PACTOR-II range from 100 to 800 baud depending on conditions. The throughput is up to 140 characters per second.

HF radios serve as important links for data communications for all military services. The U.S. Air Force operates Scope Command, a network of HF stations. The U.S. Army uses HF for long-range handheld radio communications. The U.S. Navy uses HF for ship-to-ship and shore-to-ship communications, mostly at short range using ground waves. The U.S. Army and the U.S. Marine Corps often use HF in near-vertical incidence sky-wave (NVIS) mode to communicate short range over mountains and other obstacles (Fig. 1). The National Communications System (NCS) links a large network of HF stations (SHARES) and a large number of allocated frequencies for national emergency communications. The NATO Standardization Agreement (STANAG) 5066 data-link protocol is used in HF email gateways by the NCS and by U.S. and European military services, and has been adopted by NATO for the Broadcast and Ship-Shore (BRASS) system.

Other important HF radio applications include encrypted communications with diplomatic posts in various countries and private networks linking remote outposts around the world, such as International Red Cross field sites and oil exploration stations.

The allocation of HF frequencies is strictly regulated by licensing in individual countries, and is coordinated worldwide by the ITU and in the United States by the Federal Communications Commission (FCC). Some of the frequencies are allocated to citizens band (CB), some to amateur fixed public use, and some to marine and aeronautical communications. Government groups, including the military, have their own allocations. The full FCC table of HF allocations is available on the Internet [4].

Wideband HF has been an active topic of research, but it is still considered impractical by many because of the legacy of narrowband frequency allocations.

3. HF CHANNEL PROPAGATION

HF frequencies propagate by ground-wave signals along the conducting earth or by skywave signals refracted from the layers of the ionosphere, as summarized in Table 1. Many additional details about HF propagation and communication methods may be found in the literature [5–7].

Ground-wave propagation is quite predictable and steady. Signal strength drops off inversely with distance, but depends strongly on the polarization, the ground conductivity, the dielectric constant, and the transmission frequency. Saltwater provides excellent conductivity, and therefore transmission over seawater is subject to the least attenuation. At frequencies below HF, ground-wave communications reach several thousand kilometers, whereas at HF communications can range from tens of kilometers over dry ground to hundreds of kilometers over seawater.

Skywave signals depend primarily on the ionospheric layer from which they reflect. The layers can often support several rays, or paths, between two terminals. The transmissions at lower frequencies tend to reflect from lower layers, while the transmissions at higher frequency penetrate deeper and reflect from higher layers. Transmissions at frequencies above the HF band tend to

Table 1. Earth–Ionosphere Propagation at HF and Lower Frequencies

	Propagation Medium			
Feature	Ground Wave	D Layer	E Layer	F Layer
Height above earth	0	50–90 km	90–140 km	140–250 km (day) 250–400 km (night)
Variability	Varies with surface characteristics and frequency	Daytime only	Reduced or disappeared at night	Split into F1 and F2 during day Highly dependent on solar activity
Approximate communications range	Depends on frequency, ground conductivity, and noise	2100 km (single-hop)	2800 km (single-hop)	3500 km (day) 4500 km (night)

go right through the ionosphere, as they are not reflected back to earth.

We use the term *reflection* to indicate that a ray returns to earth. Actually as a radiowave hits the ionosphere, it is not reflected as if from a smooth surface boundary; rather, it is *refracted*, or somewhat bent, so that it appears as if it had been reflected from a mirrorlike surface at a greater height. This apparent height of reflection is called the *virtual height*.

The lowest ionospheric layer is the D layer, which thins out at night. The E layer above it also thins out at night. During the day it can refract frequencies in the lower HF band. The highest layer, F, is split during the day into the F1 layer and the higher F2 layer. By refracting from a higher layer, the radiowaves can reach out further. The F2 layer is the most important layer for daytime propagation of long-range HF rays; it permits communications in the higher end of the HF band.

The ionization in the F2 layer, so important for long-range HF, is highly variable because of its dependence on the sun. These variations have cycles of 1 day (due to the rotation of the earth), about 27 days (due to the rotation of the sun), seasonal (due to the movement of the earth around the sun), and about 11 years (due to the observed period of sunspot activity).

The reflecting property of the ionosphere is often characterized by the *critical frequency*, which is the highest frequency that can be reflected from the ionosphere at vertical incidence. The critical frequency f_c is determined by the electron density N (the number of electrons per cm^3) and it varies with time and location, but has a typical value around 12 MHz. The critical frequency can be determined by vertical sounding or predicted from

$$f_c = 9 \cdot 10^{-3}\sqrt{N}$$

The *maximum usable frequency* (MUF) is the highest frequency that connects transmitter and receiver on a given link:

$$\text{MUF} = \frac{kf_c}{\sin\varphi} = \frac{kf_c}{\cos(\theta/2)}$$

where φ is the angle of incidence, θ is the angle by which the ray is refracted, and $k \cong 1$ is a correction factor accounting for ray bending in lower layers. This equation can also be used to determine the *critical angle*, the maximum angle of incidence at which reflection can occur at a particular frequency.

When the critical frequency is high enough, it is possible to establish HF sky-wave communications at short ranges by tilting the HF antennas so that enough energy is radiated straight up. This unique mode of HF propagation, called near-vertical incidence sky-wave (NVIS) communication, is used for short-range valley-to-valley-type communication links. The NVIS transmission mode is possible only at the lower HF frequencies.

At the low end of the HF band D-layer absorption increases, making daytime communications difficult. *The lowest usable frequency* (LUF) is the frequency below which the signal becomes too weak. Whereas the MUF is defined entirely by propagation effects, the LUF depends on system parameters (transmitter power, antenna, modulation, and noise) as well.

The E layer sometimes contains patches of denser ionization that generate an additional reflection called "sporadic E." Sporadic E is strong and usually nonfading. It can degrade long-range HF communications by preventing signals from reaching the F2 layer. On the other hand, it can make possible medium-range communications at frequencies well above the HF band.

A ray reflected from the ionosphere can exhibit fading and time-delay dispersion due to absorption and small variations of the propagation medium along the path. The fading of a single ray is generally uniform across a wide bandwidth and is called *flat fading*.

Multiple rays can be created by reflections from different layers (see Fig. 1). The different rays arrive at the receiver with different delays and combine to cause a very-frequency-dependent fading called *frequency-selective fading*.

A sky-wave signal can be rereflected by the ground (especially seawater) to create a ray with multiple ionospheric reflections, usually referred to as *hops*, as also illustrated in Fig. 1. When this happens, delay spreads can be especially large.

The decomposition of a transmitted ray into multiple rays can also be caused by the influence of the earth's magnetic field. This can be explained by noting that a transmitted vertically polarized electromagnetic wave can be considered as a superposition of two waves of opposite circular polarizations. Because of the interaction of the magnetic field with electrons in the ionosphere, each of these component waves is subjected to different refraction levels and therefore follows a slightly different ray path between transmitter and receiver. These two rays, which are termed the *ordinary ray* (O-mode) and the *extraordinary ray* (X-mode), can arrive with different delays, also generating multipath.

Sky-wave signals exhibit a considerable amount of variability. The ionosphere is not static and has variations in the electron density distribution. Solar flares and sunspots cause ionospheric turbulence and atmospheric phenomena such as sudden ionospheric disturbances (SIDs), and polar cap absorption (PCA) can disrupt and disturb transmission.

The initial frequency can be selected from ionospheric predictions based on historical data, vertical sounding, or oblique ionospheric sounding probes. It can be updated by monitoring the transmitted signal directly and switching frequencies as necessary.

External noise is often a limiting factor on HF links. Its three components, *atmospheric noise*, *galactic noise*, and *man-made (human-generated) noise*, are important sources of noise in the HF band. We note that below HF only atmospheric noise and synthetic noise are significant. Atmospheric noise is due mainly to lightning and is therefore highly variable. Atmospheric noise levels tend to be more severe in equatorial regions and are 30–40 dB weaker near the poles of the earth. Atmospheric noise can be dominant at nighttime at frequencies below 16 MHz. Like the other external noise sources, it decreases with frequency. Galactic noise is fairly predictable and can be dominant in radio-quiet areas at frequencies above 4 MHz. Synthetic noise is highly variable and depends on transmission activity in adjoining frequency bands.

In summary, the properties of HF sky-wave transmission depend on the height and density of the ionospheric layers and on ambient noise sources. Because the ionosphere has several layers and sublayers, several refracted ray paths are possible. Thus multipath and fading effects are common. Typical time spread within each path is 20–40 μs, with individual paths separated by 2–8 ms. Some of the fades last just a fraction of a second and some a few minutes. Fading and the limited bandwidth impose critical constraints on the achievable data rates, which can be quantified using Shannon's channel capacity [8].

4. KEY ELEMENTS OF THE HF DIGITAL TERMINAL

A typical HF digital terminal consists of an antenna, a radio (transmitter, receiver), control equipment (frequency scanning, frequency selection, link establishment, synchronization, selective calling, and link quality control), a modem, and computer or network interfaces as illustrated in Fig. 2. These functions are elements of the ISO layered protocol model illustrated in Fig. 3.[1] The terminals can range in complexity from fairly simple "ham" (amateur) radios to sophisticated automated radio network terminals.

HF voice terminals traditionally operate in a one-way (simplex) mode using manual or automated "push to talk" (PTT). Modern data modems operate in a full-duplex mode, requiring separate frequency allocations for each direction of communications. Typically each terminal is allocated several frequencies on which it can receive, so as to increase the likelihood that one of them will permit a good connection.

More recent advances in electronic hardware, radiofrequency circuits, programmable chips, and software technology have permitted the building of HF radio terminals that are miniaturized and relatively inexpensive. These

[1] The International Standards Organization (ISO) is a worldwide federation of national standards bodies from some 140 countries that publishes international standards.

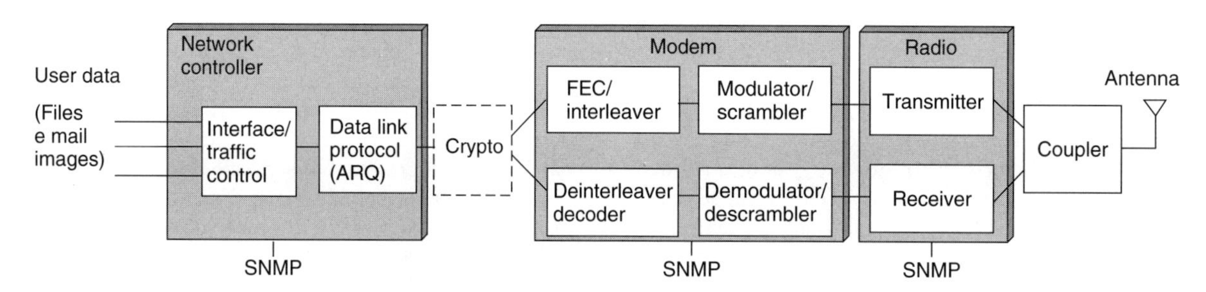

Figure 2. Block diagram of digital HF terminal.

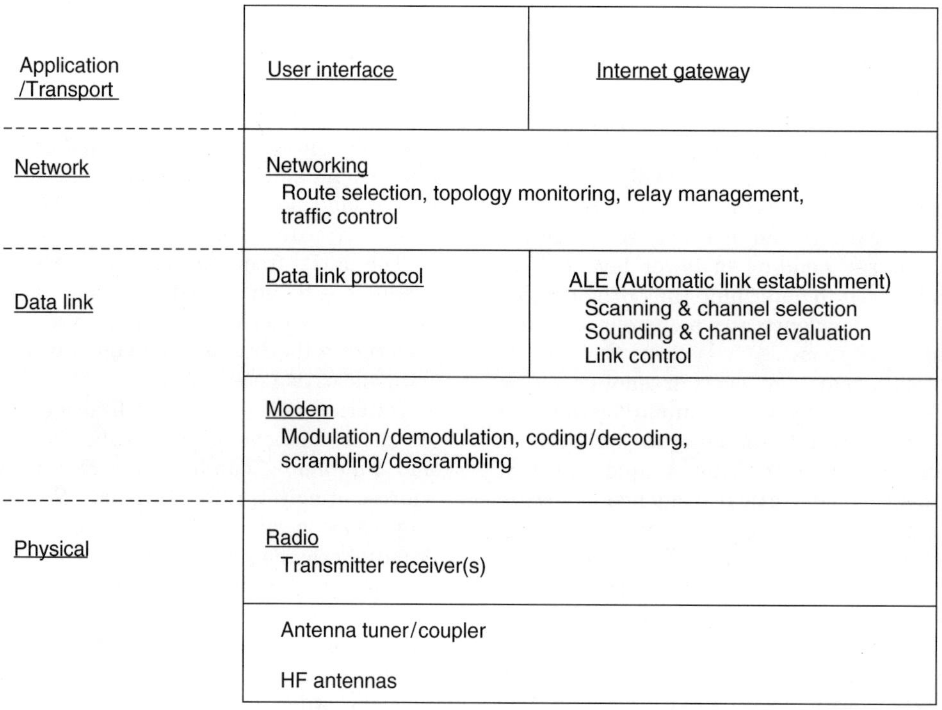

Layer	Functions	
Application /Transport	User interface	Internet gateway
Network	Networking Route selection, topology monitoring, relay management, traffic control	
Data link	Data link protocol	ALE (Automatic link establishment) Scanning & channel selection Sounding & channel evaluation Link control
	Modem Modulation/demodulation, coding/decoding, scrambling/descrambling	
Physical	Radio Transmitter receiver(s)	
	Antenna tuner/coupler HF antennas	

Figure 3. HF terminal functions.

radios can be versatile. They operate in different modes, use a range of data rates, and interface with different networks.

With this background, we describe the components shown in Fig. 2, starting with antennas (the bottom layer in Fig. 3).

4.1. Antennas and Couplers

Antennas are characterized by directivity, gain, bandwidth, radiation efficiency, physical size, and polarization. HF antennas [5] on the ground often require a ground screen to improve ground conductivity. Directivity is measured by assessing the proportion of the total radiated power propagating in a given direction. Gain expresses the combined directivity and radiation efficiency and thus measures the portion of the total transmitted power radiated in a given direction. For ground-wave propagation, the transmitter and receiver must have the same polarization. For sky-wave propagation, the received polarization is generally elliptic and can vary greatly from the transmitted polarization due to different propagation of the ordinary and extraordinary waves. Therefore, some receiver terminals use diversity, whereby the received signals from two orthogonally polarized antennas are combined to get the best signal.

Vertically polarized antennas, such as the vertical whip and high-power towers, tend to be narrowband. These antennas are best suited for ground-wave propagation or for long-range sky-wave propagation. Their antenna patterns are omnidirectional in azimuth, making such antennas ideal for point-to-multipoint communications. A vertical whip can be tilted or bent for intermediate- or short-range NVIS skywave. Horizontal polarization requires that the antenna be raised above the ground. Another common HF antenna type is the half-wave horizontal dipole on a mast, which can be used at low frequencies and medium range. The horizontal Yagi antenna is also used at HF, but becomes large at low frequencies (wavelength at 3 MHz is 50 m or 150 ft) and is very narrowband. An omnidirectional pattern can be achieved with a crossed-dipole antenna. Rhombic antennas can achieve significantly better antenna gain, but when designed for the low end of the HF band (2–10 MHz) can be very large, requiring several acres.

Narrowband antennas require an antenna *tuner*, which is a circuit of capacitors and inductors that allows adjustment of the antenna impedance to maximize radiation efficiency at the desired frequency. An antenna tuner automated under microprocessor control is called a *coupler*.

Wideband antennas that do not require a tuner or coupler are usually a variation of the logperiodic antenna, which can be either vertically or horizontally polarized. The logperiodic antenna is a combination of several narrowband elements, each tuned to a different frequency. Properly designed, they can cover the entire 2–30 MHz HF band.

Small wideband antennas are used primarily for receive-only functions, as they have low radiation efficiency due to the difficulty of matching the impedance over a wide bandwidth. Low-efficiency antennas do not degrade the signal-to-noise ratio when external noise dominates, as is the case in the HF band.

4.2. HF Radio

The radio transmitter generates the radiated power. The required amount of power (hence, the size of the power supply) depends, of course, on the intended transmission distance. Power can be emitted in bursts or continuously, according to the selected type of modulation. Transmitted power can range from tens of watts for handheld units to several kilowatts for point-to-multipoint (broadcast) ground stations. Automatic level control (ALC) or transmit level control (TLC) is used to moderate the increase in transmitted RF (radiofrequency) power when audio input increases.

A single HF radio channel accommodates transmission of voice in the range of 300–3000 Hz. HF radios generally use single-sideband (SSB) modulation, transmitting only one of the two sidebands generated by modulating the voiceband signal onto a carrier frequency. An SSB HF radio is typically designed for voice and supports only low-data-rate transmission.

To introduce higher data throughput, some modern radios operate over two channels (6-kHz band) or four channels (12-kHz band) at once. Two adjoining channels use both sidebands in an independent sideband (ISB) modulation. In a four-channel mode two sidebands on either side of the carrier are used [9].

At the receiving end, it is possible to improve performance by using polarization and space diversity. This works best if the antennas can be spaced several wavelengths apart, which may not always be practical at HF frequencies. However, some diversity improvement can still be achieved with smaller antenna spacing [10,11].

4.3. HF Modem

While the earliest radios used analog amplitude modulation (AM) or frequency modulation (FM), modern digital HF systems achieve significant information throughput by using a modem (modulator/demodulator) at each end of the link. The modem modulates the data onto a transmitted carrier, and inversely demodulates the received carrier waveform to get the transmitted data.

The modem must be designed to effectively cope with the severe multipath and fading found on the HF channel. A well-designed modem can take advantage of the redundancies generated by encoding and by propagation through the different paths spaced in time or space. It can reassemble the signal that may have been spread out, taking advantage of all propagation paths.

The modem is the heart of the digital HF terminal and is discussed in more detail in a later section.

4.4. Link Establishment and Maintenance

The station desiring to transmit has to select from a number of preassigned frequencies. While in the past link establishment was a difficult manual task for the operator, more and more modern HF radios use automatic link establishment (ALE) [12]. At the receiving station the ALE receiver scans the allocated frequencies to detect the transmission and to make the best frequency selection for establishing the link. The selection may be based on what is best for one particular link or for connecting a transmitting station to a number of users at the same time.

An ALE-equipped radio can passively or actively evaluate the channel for the transmission frequency best suited for the intended receiver. Active evaluation of what is the proper frequency might be the result of a broadband sounder identifying the ionospheric layers. It can involve feedback from an ALE receiver to report when a carrier frequency change is needed. Passive evaluation, such as link prediction based on stored or broadcast ionospheric data, can be reasonably effective at estimating gross ionospheric effects.

The ALE module coordinates communications for automatic HF link establishment (handshake) and status information between the two nodes, controlling both half- and full-duplex modes. Another function of ALE is *selective calling*. This feature allows a radio to mute the received signal until the transmission intended for that particular radio is received. Selective calling requires that the receiving radio have a unique address and that this address be included as part of the transmission.

ALE functions international rely on standards to make radios from different manufacturers as compatible as possible, such as FED-STD-1045 [13] and MIL-STD-188-141B [9]. Combined with availability of adaptive equalization in the modem, ALE enhances the likelihood of link establishment and good communications.

Reliable HF communication also requires monitoring of the connectivity between the transmitter and the receiver to enable automatic switching to another frequency when the current frequency has faded out or has become less effective for the desired range of transmission.

4.5. Network Interface and Traffic Management

A network controller couples the modem to the network. It controls network traffic using the link and manages changing conditions over the HF link. Interfaces may be able to adapt the modem to more than one local-area network (LAN) standard (e.g., ATM or Ethernet). The interface performs appropriate handshake operations with the network equipment and may also implement a traffic management function, such as congestion control.

As an example, consider a marine email application. The modem at the user's end of the link is connected through a network interface to the computer running the local email software, while the modem at the shore station is connected to the Internet.

4.6. Data-Link Protocol

Digital transmission over the HF link must be controlled by a well-defined protocol. For instance, military radios use the data-link protocol in MIL-STD-188-110B, Appendix E [14]. The link data frame usually includes cyclic redundancy code (CRC) check bits, which allow detection of errors in a frame. When errors are detected, the link protocol may use a feedback channel to initiate retransmissions. Since such a feature is usually automatic, it is known as automatic repeat request (ARQ).

ARQ can be considered an alternative, or supplement, to interleaving and coding, which are discussed below. Essentially error-free communications can be achieved this way at the cost of significant delay variations. Combining interleaving and coding with an ARQ method can get the benefits of both approaches. ARQ is commonly used by higher layer protocols, such as TCP/IP, to provide end-to-end reliability. It can also be very useful at the link level to minimize end-to-end retransmissions.

The most common ARQ scheme is "go-Back-N" [15], where retransmissions start over with the frame that contains errors. The most advanced ARQ method is selective ARQ, which allows the terminals to select individual data frames for retransmission. One advantage of selective ARQ over coding and interleaving is that the delay is determined by the actual channel characteristics, while an interleaver generally has to be designed for the worst-case scenario (long fades).

5. HF MODEMS

HF modems perform several functions other than modulation and demodulation. These include interleaving and deinterleaving, coding and decoding, and security. Modems incorporate circuits for initial acquisition, adaptive equalizers for automatic compensation of multipath dispersion, and circuits for performance monitoring and for controlling the switching of carrier frequencies. This section describes some applicable standards and key functions of the modem.

5.1. Standards

The proliferation of different modulations, link protocols, and data rates has led to the establishment of a series of standards promoting interoperability.

The ITU, the Department of Defense (DoD), NATO, Telecommunications Industries Association (TIA), and others publish carefully crafted standards. The DoD, in particular, has issued a comprehensive series of mandatory and recommended standards for military communications, many of which have been adopted for commercial use. Military HF modems follow the MIL-STD-188-110B [14], which specifies a number of modulation and coding waveforms for various data rates and for both single- and multichannel modems. The HF terminals follow the MIL-STD-188-141B [9], which covers transmitters, receivers, link establishment procedures, data-link protocols, security, antijam technology, and network maintenance (using SNMP). This standard is coordinated with FED-STD-1045 [13]. Military standards are available on the Internet [16]. International standards, and many others, are available from the American National Standards Institute (ANSI) [17].

NATO has established its own Standardization Agreements (STANAGs). The U.S., NATO, and ITU standards are highly coordinated. For instance, MIL-188-STD-110B specifies STANAG 5066 as an optional data-link protocol.

Table 2 lists representative standardized HF modulations, several of which are discussed further below.

Table 2. Characteristics of Selected HF Modem Standards

	16-Tone DPSK Modem, MIL-STD-188-110B, Appendix A	39-Tone DPSK Modem, MIL-STD-188-110B, Appendix B	ALE Single Modem, MIL-STD-188-141B, Appendix A	PSK Single-Tone Modem, MIL-STD-188-110B, Fixed Frequency	PSK Single-Tone Modem, MIL-STD-188-110B, Appendix C
Data rates (bps)	75–2400	75–2400	61.2	75–4800	3200–12,800
Symbol rate/period	75 baud/13.3 ms	44.44 baud/22.5 ms	125 baud/8 ms	2400 baud/416.6 μs	2400 baud/416.6 μs
Known (training) transmissions	Tone for Doppler correction	Tone for Doppler correction; 1 framing bit per 31.5 data bits	1 stuff bit per 49 bits transmitted	2400 or 4800 bps: 16 training symbols following 32 data symbols 150–1200 bps: 20 training symbols following 20 data symbols 75 bps no training symbols	31 training symbols every 256 data symbols 72 preamble symbols reinserted after every 72nd training sequence
Modulation	DQPSK (1200, 2400 bps) DBPSK (75–600 bps) Parallel tones, spaced 110 Hz	DQPSK Parallel tones, spaced 56.25 Hz	8-FSK	8-PSK (2400, 4800 bps) QPSK (1200 bps) BPSK (150–600 bps) QPSK spread by 32 (75 bps)	QPSK (3200 bps) 8-PSK (4800 bps) 16-QAM (6400 bps) 32-QAM (8000 bps) 64-QAM (9600, 12,800 bps)
Coding and interleaving	Bits repeated at rates below 2400 bps	Shortened Reed-Solomon (15,11) : (14,10) at 2400 bps, (7,3) otherwise; repeat coded bits at rates below 2400	Rate-$\frac{1}{2}$ Golay code, triple redundancy	4800 bps: no coding 75, 600–2400 bps: rate $\frac{1}{2}$ 150–300 bps: rate $\frac{1}{2}$ plus repetitions	Rate $\frac{3}{4}$ convolutional code (with tail-biting to match interleaver length)
Interleaving	None	Selectable	48-bit interleaving	0, 0.6, or 4.8 s	0.12-8.64 s (6 steps)
Preamble	2 tones for 66.6 ms followed by all tones for 13.3 ms	4 tones over 14 or 27 periods followed by 3 tones over 8 or 27 periods followed by all tones over 1 or 12 periods (second number for optional extended preamble)	No fixed preamble; transmission is an asynchronous sequence of 24-bit words; each word has three 7-bit characters and a 3-bit preamble identifying word type	0.6 or 4.8 s of a fixed BPSK sequence	0-7 blocks of 184 8PSK symbols followed by 184 sync symbols followed by 103 symbols with information on data rate and interleaver settings

5.2. Signal Acquisition

HF modems must be able to achieve signal acquisition in the presence of several channel disturbances: severe fading, multipath, non-Gaussian noise, and interference from other transmitters. Transmissions normally include a known preamble designed to permit the modem to adjust as necessary to achieve signal acquisition.

After the initial acquisition, transmissions may incorporate a known signal to permit continuous monitoring of channel conditions. Such a signal may be in the form of a pilot tone in a parallel-tone system or a periodically inserted training sequence in a single-tone system. Equalization without a training sequence is known as "blind" equalization [18] and has been studied for HF [19].

5.3. Types of Modulation

The easiest way to communicate over a channel with multipath is to modulate tones with symbols of duration much longer than the multipath spread. By ignoring the first part of each received tone, intersymbol interference can be avoided. This principle is used in older frequency shift keying (FSK) modems, which transmit one or more of several possible tones at a time, and in the parallel-tone phase shift keying (PSK) modems commonly employed by military and commercial users. Table 2 shows an example of two parallel-tone PSK modems (16-tone modem in column 1 and 39-tone modem in column 2) and a robust FSK modem used for ALE (column 3). Parallel tone modems are also used with marine single-sideband (SSB) radios for commercial HF email and HF Web applications [3].

Modems with a narrow HF bandwidth allocation can benefit from using multiphase PSK (e.g., 4, 8, or 16 phases), to achieve higher data rates. Differential PSK (DPSK), which compares the phase received in each time slot to the phase received in the preceding slot, offers a simple implementation that does not require accurate phase tracking. However, since the phase in the previous slot is also noisy, DPSK incurs a performance loss relative to coherent PSK.

Parallel-tone modems with narrow frequency bands for each tone and low keying rates have several advantages: (1) they are easy to implement, (2) they can be quite bandwidth-efficient, and (3) the multipath spread is usually a fraction of the tone duration. The main disadvantage is the high peak-to-average power ratio resulting from the superposition of several parallel tones. This leads to the need for a linear power amplifier with a dynamic range much larger than the average transmitted power. Another disadvantage is the fact that Doppler spread can cause interference between the parallel tones. Coding is used to provide an in-band diversity gain, compensating for the selective fading caused by multipath.

In serial transmission, with single-tone modems, the modulation can have virtually constant amplitude, so the peak-to-average power ratio is close to unity. A higher average power is then achieved with a given peak power. This means that a more efficient class C amplifier can be used, and the transmitter can be made more compact. A single-tone modem requires faster keying to match the data rate. This means that intersymbol interference cannot be ignored, except perhaps at the lowest data rates. Single-tone modems use adaptive equalization to undo the intersymbol interference. By using the entire received signal, a serial modem can get better performance than a modem that ignores the part of the signal containing intersymbol interference.

Thanks to more recent advances in signal processing and linear amplification technologies, single-tone modems can use amplitude modulation to achieve even higher data rates. For example, the MIL-STD-188-110B includes 16-, 32-, and 64-quadrature amplitude modulation (QAM) modems (see the last column in Table 2). These new modulations can have rates up to 12,800 bps within a 3-kHz band. Although QAM spaces signal points as far apart as possible, it requires significantly higher signal-to-noise ratio (SNR), due the smaller spacing between signal constellation points. In general, one result of this "compressed" signal constellation is the *constellation loss*, meaning that the required power grows faster than the increase in data rate. Higher-data-rate modes can be used in short-range ground-wave applications where the signal is sufficiently strong. With sufficient transmitter power, they may also be appropriate for short- or medium-range sky-wave communications.

5.4. Coding

Coding can be used to correct errors by introducing redundancy in the transmitted data [20]. It can be used with both parallel- and serial-tone modulations. The selection of a proper code is one of the key design issues. Forward error correction (FEC) block codes are simple to use because they are inherently compatible with the data frame structure. However, short-constraint-length convolutional codes are often applied and can offer performance advantages, particularly as they can easily be decoded using the actual received samples (soft decision), as opposed to only using demodulated symbols (hard decision). Table 2 lists several types of codes used in standard HF systems. All these coding techniques add redundancy, thus increasing the required channel data rate. With conventional modulations it is then necessary to expand the bandwidth.

Trellis modulation coding [21] is a way of combining coding and modulation that can offer coding gain without expanding bandwidth. It has seen few applications in HF in spite of the band-limited nature of HF channels. A newer form of coding, Turbo coding [22], promises performance even closer to Shannon's capacity limit by interleaving the coded data over relatively long time periods.

5.5. Interleaving

Fades cause errors to appear in clusters. The objective of interleaving is to randomize error occurrences by forming the encoded block, not out of consecutive bits, but out of bits that are selected in a pattern spread out over several blocks. When the inverse process is implemented at the receiver, the received clusters of error are converted into randomly distributed errors. Coding then permits the random errors to be corrected at the receiver.

To be effective, the interleaving period needs to be longer than the duration of a fade, which can be several seconds on the HF channel. The resulting delay can be undesirable, and many modems using interleaving include means for selecting the interleaving period.

5.6. Link Protection

A number of methods may be used to protect the data transmitted over the link. These include data scrambling and encryption. Similarly, ALE transmissions may be protected, for instance, as described in Appendix B of the military standard for automatic link establishment [9].

5.7. Adaptive Equalization

Adaptive equalization is a modem feature that reassembles a signal that has been dispersed in time by the channel propagation effects. It was first introduced in telephony [23] and later applied to fading channels [24]. Adaptive equalization is an effective technique for HF links to overcome multipath.

When the delay spread is small compared to the symbol duration, a simpler technique called adaptive matched filtering, or a RAKE receiver, may be used [20]. A RAKE receiver estimates individual channel gains and uses these estimates to adjust the gains of a tapped delay line to best match the channel. This approach is most often used with spread-spectrum systems, where bandwidth is large enough to resolve the individual paths, and symbols are long enough to permit neglecting intersymbol interference.

When intersymbol interference cannot be neglected, adaptive equalization is needed. Many forms of adaptive equalization may be used to undo the channel multipath [20]. The main concept is to combine different delayed versions of the received signal in order to maximize the signal-to-interference ratio in each symbol being demodulated. The interference meant here includes that from adjoining symbols. The equalizer applies proper amplitude and phase to each delayed tap and combines the resulting signals in such a way as to best reconstruct the transmitted symbol.

Two common adaptive equalizer techniques, the decision-feedback equalizer (DFE) and the maximum-likelihood sequence estimation (MLSE) equalizer, have not been widely used with HF transmission. The MLSE equalizer is an optimal receiver when the channel is known, but it has rarely been used because of its complexity and sensitivity to channel perturbations. It has been used mostly at HF in experimental modems but could become more practical as the processing power of new digital signal processing (DSP) chips increases. Because of its simplicity, the DFE is popular for modems with data rates that are high relative to the fade rate, but it has not been used much at HF because of the low data rates involved. As the trend to increase HF modem data rates continues, however, the DFE may become more common.

Since the channel is continuously changing as a result of fading, a training sequence is periodically inserted to effectively measure the individual path gains and phase shifts. The MIL-STD-188-110B serial-mode waveform uses 33–50% of the transmissions for training sequences (see Table 2, column 4). The extra reference permits using other equalization methods offering performance slightly better than a DFE, such as the data decision estimation method [25].

6. LINK PREDICTION AND SIMULATION

Performance prediction, based on HF channel models, is used to plan and operate HF communications links. Channel simulators are used to test modem performance using synthesized model channels.

Multipath characterization of the HF communications channel is traditionally based on Watterson's narrowband HF channel model [26]. The Watterson model represents the channel as an ideal tapped-delay line with tap spacings selected to match the relative propagation delays of the resolvable multipath components. Each tap has a complex gain multiplier, and summation elements combine the delayed and weighted signal components (see Fig. 4). The transmitted signal feeds the tapped-delay-line channel model. The tap gains of each path are modeled as mutually independent complex-valued Gaussian random processes, producing Rayleigh fading. Each tap gain fades in accordance with a Gaussian spectrum.

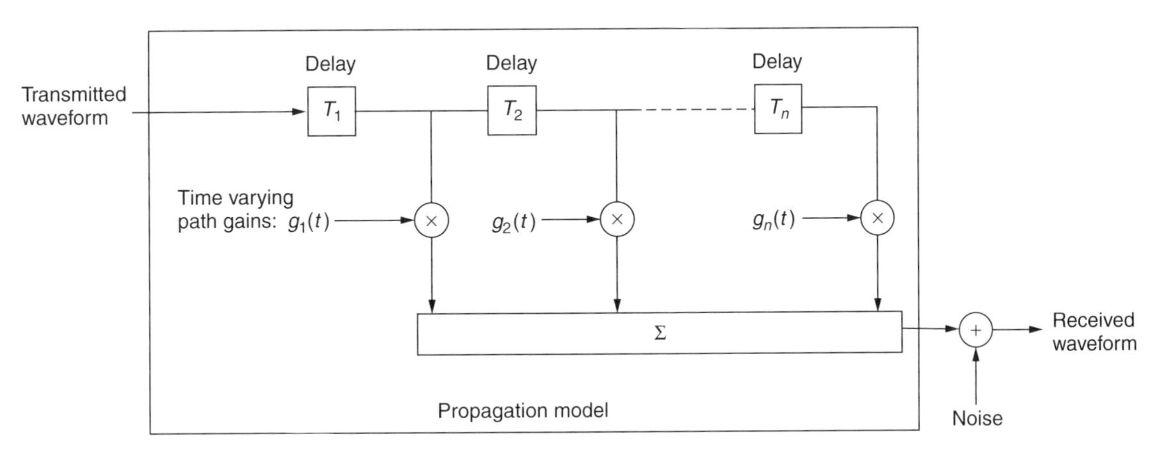

Figure 4. HF channel modeled as a topped delay line.

The Watterson model emulates the propagation channel with the addition of random noise. Atmospheric noise, galactic noise, and synthetic noise are important sources of noise in the HF band, and a proper noise model should combine all three.

Galactic noise is generally Gaussian, whereas atmospheric noise is due to lightning and therefore impulsive in nature. Atmospheric noise values have been measured on a worldwide basis, and typical ranges can be predicted based on geographic location, season, and time of day using charts calculated by the International Telecommunications Union (ITU) [27].

A number of models of synthetic noise have been developed; however, because they are often based on data that are decades old, they should be used with caution. The ITU man-made noise model recommends a distinction between several classes of areas [27]. The most important of these are in business, residential, and rural areas, where the median noise falls off by 27.7 dB per decade in frequency. Synthetic noise can dominate for transmissions at frequencies as low as the low-frequency (LF) band and as high as 4 MHz in the HF band.

The high variability of HF propagation means that the best frequency to transmit depends strongly on ionospheric parameters. Path loss prediction is an important component of link planning. Aided by tabulated data or ionosonde data, it can help select the best frequency. Several propagation models are available [28], most of them derived from the Ionospheric Communications Analysis and Prediction (IONCAP) model developed by the National Telecommunications and Information Administration — Institute for Telecommunications Sciences (NTIA/ITS). One such model is the ICEPAC model by the NTIA/ITS, which adds a newer ionospheric electron density model that provides better worldwide coverage. Another model is the Voice of America Communications Analysis and Prediction (VOACAP) program [28], which extends the IONCAP program with several features and adds a graphical user interface. The ITU skywave model [29] is available as both a table-based model and a computer program, as is the ITU ground-wave model [30]. NTIA/ITS offers a computer program (HFWIN32) combining all three models [31].

For digital communications it is important to model the delay spreads and Doppler spreads, especially when different modems are to be tested and compared. One such prediction model [32] generates the parameters of the Watterson model directly for a hardware simulator.

Hardware channel simulators are essential laboratory instruments for testing the performance of actual modems and network equipment under controlled repeatable HF channel conditions [33]. Such simulators are most valuable in that they model not just the channel but also additive Gaussian noise, impulse noise, and typical types of interference. Hardware channel simulators are available with interfaces at baseband, intermediate frequency (IF), a fixed radiofrequency (RF), and at a frequency-hopping RF. Comparisons of link performance and of simulator test data have validated that simulators are particularly useful for quantitative and comparative performance evaluation.

7. FUTURE TRENDS

Future HF radios and modems are likely to develop in several directions: higher data rates, different modulations, software-controlled multifunction radios, more intelligent signal processing, and improved networking and traffic management.

Apart from the propagation effects on the HF channel, the greatest impediment to achieving higher data rates is the current narrowband channel allocations, which allow only about 3 kHz of spectrum per channel. The simplest design to increase the data rate is to modulate separately each of several parallel channels. Military standards have been developed for combining two or four channels (see [14], Appendix F). However, if the channels could be lumped together and guard bands eliminated, then data rates could be increased and a wider choice of modulations would be available.

Radio users are assigned a group of frequencies to select from, depending on channel conditions. Combining the current 3-kHz channels into wider bands would mean a paradigm shift in HF frequency allocations and radio design. A wideband HF radio would have to use fewer frequencies with wider allocations, but would achieve higher data rates without necessarily increasing the total allocated bandwidth. Experimental HF modems have been studied by DARPA, aiming at data rates as high as 64 or 96 kbps.

Another approach to achieving higher data rates is to combine noncontiguous narrowband frequency channels, so as to be compatible with the current frequency allocations. However, in that case the use of the spectrum would be less efficient, because of the multiple guard bands required on the edges of each transmission band. Amplifier linearity would be a critical problem that might have to be overcome by introducing multiple amplifiers to service each individual frequency band. Another problem would be antennas, which would need to be sufficiently wideband. Having separate antennas for individual bands is probably impractical.

Wideband HF has also been proposed for secure military applications [34]. In that situation the idea is not necessarily to increase the data rate, but to use code-division multiple access (CDMA) with a spread-spectrum technique to reduce power density across the band. The signal is spread so much that interference to narrowband users in the same band is negligible and the transmissions are not readily detected.

Bandwidth-on-demand protocols are being used increasingly in wireless and satellite communications systems. Demand assignment multiple access (DAMA) is standardized for satellite systems [35]. It is possible that the principles of DAMA will be applied to improve the bandwidth utilization of HF in the future.

The so-called third-generation (3G) radios will exhibit enhanced automation and capability to respond to circumstances. Modems of the future will be increasingly software-controlled, and consequently may be reprogrammable depending on the link situation and the network into which they are being interfaced.

BIOGRAPHIES

Steen A. Parl received the degree of Civ. Ing. from the Technical University of Denmark in 1970 and the doctor of philosophy degree from the Massachusetts Institute of Technology (MIT) in 1974. He joined Signatron, Inc., Concord, Massachusetts, in 1972, conducting research in detection and estimation, microwave propagation prediction, underwater acoustic communications, adaptive modems, wideband interference cancellation, adaptive arrays for radio communications, coding, and statistical methods of channel simulation. Since 1993, he has been president and chief scientist of Signatron Technology Corporation, where he has worked on Over-the-Horizon (OTH) systems for applications such as providing Internet access for schools in rural areas, and on nonGPS position-location systems designed to find or track tagged objects. Dr. Parl holds seven patents in the areas of radio communications, interference suppression, channel simulation, and position location. He is a fellow of the IEEE. His current technical interests include wireless telecommunications, efficient data link protocols, digital HF and troposcatter communication systems, antenna nulling techniques, and radiolocation techniques.

Julian J. Bussgang received his B.Sc. (Engineering) degree in 1949 from University of London, England, M.S.E.E. degree from M.I.T. in 1951, and Ph.D. degree in applied physics from Harvard University, Cambridge, Massachusetts, in 1955. From 1951 to 1955 he was a staff member and then a consultant at the M.I.T. Lincoln Laboratory. He joined RCA in 1955 and became manager in Radar Development, and later manager in Applied Research at the Aerospace Division in Burlington, Massachusetts. In 1962 he founded Signatron, Inc. and served as President till 1987. He now consults to Signatron Inc., Concord, Massachusetts.

He is a patentee in the field and has published and presented over forty technical papers, some reprinted in the collections of significant publications in the field. He was visiting lecturer at Harvard University in the Division of Engineering and Applied Science one year and taught a graduate communications course at Northeastern University, Boston, Massachusetts.

Dr. Bussgang is fellow of the IEEE, served on the board of the Information Theory Group, and as chair of the Boston section of the IEEE.

His areas of interest have been correlation functions, sequential detection, radar system design, radio communications over multipath channels, and coding/decoding algorithms.

BIBLIOGRAPHY

1. *HF Data Link Protocols*, ARINC Specification 635, 2000.
2. Technical descriptions CLOVER, CLOVER-2000, G-TOR, PACTOR, PACTOR II, & PSK 31, American Radio Relay League (ARRL), #6982, 2000.
3. J. Corenman, PACTOR primer (Jan. 16, 1998) (online): *http://www.airmail2000.com/pprimer.htm* (Oct. 2001).
4. FCC Office of Engineering and Technology (Sep. 28, 2001), FCC Radio Spectrum Homepage (online): *http://www.fcc.gov/oet/spectrum* (Oct. 2001).
5. J. M. Goodman, *HF Communications*, Van Nostrand-Reinhold, New York, 1992.
6. K. Davies, *Ionospheric Radio*, Peregrinus, London (IEE), 1990.
7. L. Wiesner, *Telegraph and Data Transmission over Shortwave Radio Links: Fundamental Principles and Networks*, Wiley, New York, 1984.
8. E. Biglieri, J. Proakis, and S. Shamai, Fading channels: Information-theoretic and communications aspects, *IEEE Trans. Inform. Theory* **IT-44**(6): 2619–2692 (1998).
9. *Interoperability and Performance Standards for Medium and High Frequency Radio Systems*, U.S. Department of Defense MIL-STD-188-141B, March 1, 1999.
10. W. C.-Y. Lee and Y. S. Yeh, Polarization diversity system for mobile radio, *IEEE Trans. Commun.* **COM-20**: 912–913 (Oct. 1972).
11. W. C.-Y. Lee, *Mobile Communication, Design Fundamentals*, Wiley, New York, 1993.
12. E. E. Johnson et al., *Advanced High-Frequency Radio Communications*, Artech House, Boston, 1997.
13. *High Frequency (HF) Radio Automatic Link Establishment*, National Communications System, Office of Technology and Standards, FED-STD-1045A, 1993.
14. *Interoperability and Performance Standards for Data Modems*, U.S. Dept. Defense MIL-STD-188-110B, April 27, 2000.
15. W. Stallings, *Data and Computer Communications*, 6th ed., Prentice-Hall, Englewood Cliffs, NJ, 1999.
16. Defense Technical Information Center (Oct. 9, 2001), *Scientific and Technical Information Network* (online): *http://stinet.dtic.mil/* (Oct. 2001).
17. American National Standards Institute (no date), *Electronic Standards Store* (online): *http://webstore.ansi.org* (Oct. 2001).
18. S. Haykin, *Adaptive Filter Theory*, Prentice-Hall, Englewood Cliffs, NJ, 1996.
19. J. Q. Bao and L. Tong, Protocol-aided channel equalization for HF ATM networks, *IEEE J. Select. Areas Commun.* **18**(3): 418–435 (2000).
20. J. G. Proakis, *Digital Communications*, 4th ed., McGraw-Hill Higher Education, New York, 2000.
21. C. Schlegel and L. Perez, *Trellis Coding*, IEEE Press, Piscataway, NJ, 1997.
22. B. Vucetic and J. Yuan, *Turbo Codes: Principles and Applications*, Kluwer, Boston, 2000.
23. R. W. Lucky, Automatic equalization for digital communications, *Bell Syst. Tech. J.* (April 1965).
24. U.S. Patent 3,879,664 (April, 1975), P. Monsen (Signatron, Inc.), High speed digital communications receiver.
25. F. M. Hsu, Data directed estimation techniques for single tone HF modems, *Proc. IEEE MILCOM85*, 1985, pp. 12.4.1–12.4.10.

26. C. C. Watterson, J. R. Juroshek, and W. D. Bensema, Experimental confirmation of an HF channel model, *IEEE Trans. Commun. Technol.* **COM-18**(6): 792–803 (1970).
27. *Radio Noise*, International Telecommunications Union Recommendation ITU-R P.372-7, 2001.
28. J. Coleman (no date), *Propagation Theory and Software, Part II* (online): *http://www.n2hos.com/digital/prop2.html* (Oct. 2001).
29. *HF Propagation Prediction Method*, International Telecommunications Union Recommendation ITU-R, 2001, pp. 533–537.
30. *Ground-Wave Propagation Curves for Frequencies between 10 kHz and 30 MHz*, International Telecommunications Union Recommendation ITU-R, 2000, pp. 368–377.
31. G. R. Hand (no date), *NTIA/ITS High Frequency Propagation Models* (online): *http://elbert.its.bldrdoc.gov/hf.html* (Oct. 2001).
32. A. Malaga, A characterization and prediction of wideband HF skywave propagation, *Proc IEEE MILCOM85*: 1985, pp. 281–288.
33. L. Ehrman, L. Bates, J. Eschle, and J. Kates, Simulation of the HF channel, *IEEE Trans. Commun.* **COM-30**(8): 1809–1817 (1982).
34. B. D. Perry, A new wideband HF technique for MHz-bandwidth spread spectrum radio communications, *IEEE Commun. Mag.* **21**(6): 28–36 (1983).
35. *DAMA Demand Assignment Multiple Access*, National Communications System, Office of Technology and Standards, FED-STD-1037C, 2000.

HIDDEN MARKOV MODELS

BIING-HWANG JUANG
Bell Laboratories
Lucent Technologies
Holmdel, New Jersey

1. INTRODUCTION

Many real-world processes or systems change their characteristics over time. For example, the traffic condition at the Lincoln Tunnel connecting New York City and New Jersey displays drastically different volume and congestion situations several times a day—morning rush hours, midday flow, evening rush hours, night shifts, and perhaps occasional congestions due to construction. Telephony traffic bears a similar resemblance. As another example, the speech sound carries the so-called linguistic codes for a language in an acoustic wave with varying characteristics in terms of its energy distribution across time and frequency. In order to properly characterize these processes or systems, one has to employ models of measurement beyond the simple long-term average. The average number of daily phone calls going in and out of a city does not allow efficient, dynamic resource management to meet the telephony–traffic needs during peak hours. And the average power spectrum of a spoken utterance does not convey the linguistic content of the speech. The need to have a model that permits characterization of the average behavior as well as the nature of behavioral changes of the system gives rise to the mathematical formalism called the *hidden Markov model* (HMM).

A hidden Markov model is a doubly stochastic process with an underlying stochastic process that is not readily observable but can only be observed through another set of stochastic processes that produce the sequence of observations [1–3]. By combining two levels of stochastic processes in a hierarchy, one is able to address the short(er) term or instantaneous randomness in the observed event via one set of probabilistic measures while coping with the long(er) term, characteristic variation with another stochastic process, namely, the Markov chain. This formalism underwent rigorous theoretical developments in the 1960s and 1970s and reached some important milestones in the 1980s [1–7]. Today, it has found widespread applications in stock market prediction [6], ecological modeling [7], cryptoanalysis [8], computer vision [9], and most notably, automatic speech recognition [10]. Most, if not all, of the modern speech recognition systems are based on the HMM methodology.

2. DEFINITION OF HIDDEN MARKOV MODEL

Let $\boldsymbol{X}$ be a sequence of observations, $\boldsymbol{X} = (\boldsymbol{x}_1, \boldsymbol{x}_2, \ldots, \boldsymbol{x}_T)$, where $\mathbf{x}_t$ denotes an observation or measurement, possibly vector-valued. Further consider a first-order N-state Markov chain governed by a *state transition probability* matrix $A = [a_{ij}]$, where a_{ij} is the probability of the Markov system making a transition from state i to state j; that is

$$a_{ij} = P(q_t = j | q_{t-1} = i) \qquad (1 \le i, j \le N) \tag{1}$$

where q_t denotes the system state at time t. Note that

$$a_{ij} \ge 0 \qquad \forall \; i \text{ and } j \tag{2a}$$

$$\sum_{j=1}^{N} a_{ij} = 1 \qquad \forall \; i \tag{2b}$$

Assume that at $t = 0$ the state of the system q_0 is specified by the *initial-state probability* vector $\pi = [\pi_i]_{i=1}^{N}$, where

$$\pi_i = P(q_0 = i)$$
$$\sum_{i=1}^{N} \pi_i = 1 \tag{3}$$

Then, for any state sequence $\boldsymbol{q} = (q_0, q_1, \ldots, q_T)$, the probability of $\boldsymbol{q}$ being generated by the Markov chain is

$$P(\boldsymbol{q}|A, \pi) = \pi_{q_0} a_{q_0 q_1} a_{q_1 q_2} \cdots a_{q_{T-1} q_T} \tag{4}$$

Suppose that the system, when in state q_t, puts out an observation $\boldsymbol{x}_t$ according to a probability density function $b_{q_t}(\boldsymbol{x}_t) = P(\boldsymbol{x}_t|q_t)$, $q_t \in \{1, 2, \ldots, N\}$. The hidden Markov model thus defines a density function for the observation sequence $\boldsymbol{X}$ as follows:

$$
\begin{aligned}
P(\boldsymbol{X}|\pi, A, \{b_j\}_{j=1}^N) &= P(\boldsymbol{X}|\Lambda) \\
&= \sum_{\boldsymbol{q}} P(\boldsymbol{X}, \boldsymbol{q}|\Lambda) \\
&= \sum_{\boldsymbol{q}} P(\boldsymbol{X}|\boldsymbol{q}, \Lambda)P(\boldsymbol{q}|\Lambda) \qquad (5) \\
&= \sum_{\boldsymbol{q}} \pi_{q_0} \prod_{t=1}^{T} a_{q_{t-1}q_t} b_{q_t}(\boldsymbol{x}_t)
\end{aligned}
$$

where $\Lambda = (\pi, A, \{b_j\}_{j=1}^N)$ is the parameter set for the model.

In this model, $\{b_{q_t}\}$ defines the distribution for short-time observations $\boldsymbol{x}_t$ and A characterizes the behavior and interrelationship between different states of the system. In other words, the structure of a hidden Markov model provides a reasonable means for defining the distribution of a signal in which characteristic changes take place from one state to another in a stochastic manner. Normally N, the total number of states in the system, is much smaller than T, the time duration of the observation sequence. The state sequence $\boldsymbol{q}$ displays a certain degree of stability among adjacent q_ts if the rate of change of state is slow compared to the rate of change in observations. The use of HMM has been shown to be practically effective for many real-world processes such as a speech signal.

2.1. An Example of HMM: Drawing Colored Balls in Urns

To fix the idea of an HMM model, let us try to analyze an observation sequence consisting of a series of colors, say, {G(reen), G, B(lue), R(ed), R, G, Y(ellow), B, ...} (see [3]). The scenario may be as depicted in Fig. 1, in which N urns each with a large quantity of colored balls are shown. We assume there are M distinct colors of the balls. We choose an urn, according to some random procedure, and then pick out a ball from the chosen urn again randomly. The color of the ball is recorded as the observation. The ball is replaced back in the urn from which it was selected and the procedure repeats itself—a new urn is chosen followed by a random selection of a colored ball. The entire process generates a sequence of colors (i.e., observations) that can be a candidate for hidden Markov modeling.

It should be obvious that the number of states, N, defines the "resolution" or "complexity" of the model, which is intended to explain the generation of the color sequence as accurately as possible. One could attempt to solve the modeling problem using a single-state machine, namely, $N = 1$, with no possibility of addressing the potential distinction among various urns in their composition of colored balls. Alternatively, one can construct a model with a large N, such that detailed distinctions in the collection of colored balls among urns can be analyzed and the sequence of urns that led to the color observations can be hypothesized. This is the essence of the hidden Markov model.

2.2. Elements of HMM

A hidden Markov model is parametrically defined by the triplet $\{\pi, A, B = \{b_j\}_{j=1}^N\}$. The significance of each of these elements is as follows:

1. N, the number of states. The number of states defines the resolution and complexity of the model. Although the states are hidden, for many applications there is often some physical significance attached to the states or to sets of states of the model. In the urn-and-ball model, the states correspond to the urns. If in the application it is important to recover at any time the state the system is in, some prior knowledge on the meaning or the utility of the states has to be assumed. In other applications, this prior assumption may not be necessary or desirable. In any event, the choice of N implies our assumed knowledge of the source in terms of the number of distinct states in the observation.
2. $A = [a_{ij}]$, the state-transition probability matrix. The states of the Markov model are generally interconnected according to a certain *topology*. An ergodic model [3] is one in which each state can be reached from any other state (in a nonperiodic fashion). If the process is progressive in nature, say, from state $1, 2, \ldots$ to state N as time goes on and observations are made, then a so-called left-to-right topology may be appropriate. Figure 2 shows examples of an ergodic model and two left-to-right models. The topological interconnections of the

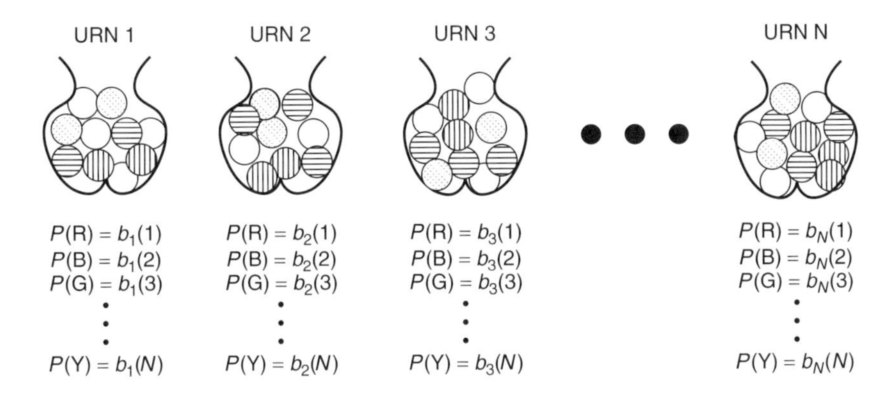

Figure 1. An N-state urn-and-ball model illustrating a case of discrete HMM.

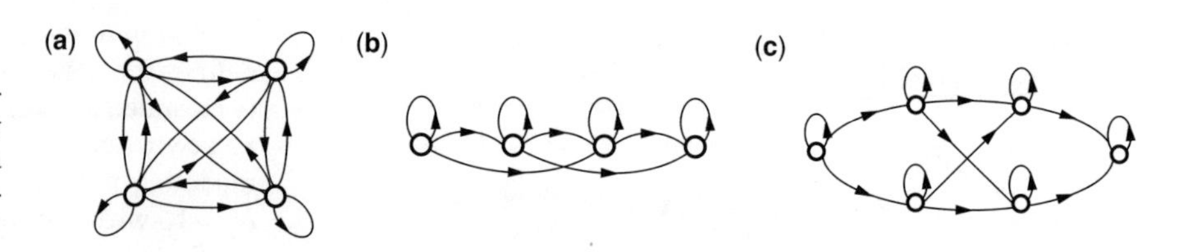

Figure 2. Three types of HMM differentiated according to the topology of the Markov chain: (**a**) a 4-state ergodic model; (**b**) a 4-state left–right model; (**c**) a 6-state parallel left–right model.

Markov chain can be expressed in a state-transition probability matrix. The state-transition probability is defined by Eq. (1) and satisfies the probability constraints of Eq. (2) For an ergodic model, one can calculate the "stationary" or "equilibrium" state probability, the probability that the system is in a particular state at an arbitrary time, by finding the eigenvalues of the state-transition probability matrix [11].

3. $B = \{b_j\}_{j=1}^{N}$, the set of observation probability distributions. Each function in the set defines the distribution of the observation in the corresponding state. These functions can take the form of a discrete density when the observation assumes one of a finite set of values, or a probability density function for observations that are real-valued over a continuous range. The former is often referred to as a *discrete* HMM; the latter, a *continuous-density* HMM.
4. $\pi = \{\pi_i\}_{i=1}^{N}$, the initial-state distribution. The matrix is defined according to Eq. (3).

3. THREE BASIC PROBLEMS OF HMM

Three basic problems associated with the HMM must be solved for the model to be useful in real-world applications. The first, the evaluation problem, relates to the computation of the probability of an observed sequence evaluated on a given model. This is important because a given model represents our knowledge of an information source, and the evaluated probability indicates how likely it is that the observed sequence came from the information source. The second, the decoding problem, is to uncover the sequence of states that is most likely to have led to the generation of the observed sequence in some optimal sense. We have mentioned that in many real-world problems a state often carries a certain physical meaning, such as realization of a phoneme in a speech utterance or a letter in handwritten script. Decoding aims at making the associated meaning explicit. The third, the estimation problem, is about obtaining the values of the model parameters, given an observation sequence or a set of observation sequences. A model can be viewed as a characterization of the regularity in the "random" event, which forms the basis of our knowledge of the source. The regularity is encapsulated by the model parameters, which have to be estimated from a set of given observation sequences, known to have come from the source.

3.1. The Evaluation Problem

We wish to calculate the probability of the observation sequence $\boldsymbol{X} = (\boldsymbol{x}_1, \boldsymbol{x}_2, \ldots, \boldsymbol{x}_T)$, given the model Λ, specifically, to compute $P(\boldsymbol{X}|\Lambda)$. Obviously, this can be accomplished by enumerating every possible state sequence of length T together with the observation sequence as defined in Eqs. (4) and (5):

$$P(\boldsymbol{X}|\Lambda) = \sum_{\boldsymbol{q}} P(\boldsymbol{X}, \boldsymbol{q}|\Lambda)$$
$$= \sum_{q_0, q_1, \ldots, q_T} \pi_{q_0} a_{q_0 q_1} b_{q_1}(\boldsymbol{x}_1) a_{q_1 q_2} b_{q_2}(\boldsymbol{x}_2) \times \cdots a_{q_{T-1} q_T} b_{q_T}(\boldsymbol{x}_T) \quad (6)$$

This direct calculation, however, is not particularly efficient as its computational complexity is exponential in time. The total number of state sequences is N^T and approximately $2T$ essential calculations are required for each state sequence. It thus involves on the order of $2TN^T$ calculations. This is computationally infeasible even for small values of N and T. For example, for $N = 5$ and $T = 100$, the total number of calculations would be on the order of 10^{72}!

An efficient procedure, called the *forward algorithm*, exists that evaluates the HMM probability in linear time. Define the forward probability $\alpha_t(i)$ as

$$\alpha_t(i) = P(\boldsymbol{x}_1, \boldsymbol{x}_2, \ldots, \boldsymbol{x}_t, q_t = i|\Lambda) \quad (7)$$

that is, the probability of the partial observation sequence $(\boldsymbol{x}_1, \boldsymbol{x}_2, \ldots, \boldsymbol{x}_t)$ and state i at time t. Calculation of the forward probabilities can be realized inductively as follows:

1. *Initialization:*
$$\alpha_0(i) = \pi_i \quad (8)$$
2. *Induction:*
$$\alpha_{t+1}(j) = \left[\sum_{i=1}^{N} \alpha_t(i) a_{ij}\right] b_j(\boldsymbol{x}_{t+1}) \quad (0 \le t \le T-1, \quad 1 \le j \le N) \quad (9)$$
3. *Termination:*
$$P(\boldsymbol{X}|\Lambda) = \sum_{i=1}^{N} \alpha_T(i). \quad (10)$$

This procedure requires on the order of N^2T calculations. Using the same example of $N = 5$ and $T = 100$, we see that this procedure entails about 3000 calculations instead of 10^{72} as required in the direct procedure.

In a similar manner, one can define the backward probability $\beta_t(i)$ as the probability of the partial observation sequence from $t+1$ to the end, given the system state i at time t:

$$\beta_t(i) = P(\boldsymbol{x}_{t+1}, \boldsymbol{x}_{t+2}, \ldots, \boldsymbol{x}_T | q_t = i, \Lambda) \qquad (11)$$

$$\beta_T(i) = 1 \quad (1 \leq i \leq N) \qquad (12)$$

An induction procedure similar to Eq. (9) can be employed to compute the backward probability for all i and t. The forward and the backward probabilities are very useful in solving other fundamental problems in hidden Markov modeling.

3.2. The Decoding Problem

When the state of the system carries information of interest, it may be necessary to "uncover" the state sequence that led to the observation event $\boldsymbol{X}$. There are several ways of solving this problem, depending on the definition of the "optimal" state sequence and the associated criterion that one may choose to optimize. For example, one possible optimality criterion is to choose the states q_t^* that are individually most likely at each time t:

$$q_t^* = \arg \max_{1 \leq i \leq N} P(q_t = i | \boldsymbol{X}, \Lambda) \qquad (1 \leq t \leq T) \qquad (13)$$

This optimality criterion maximizes the expected number of correct individual states. Other criteria are obviously possible, such as one that solves for the state sequence that maximizes the expected number of correct pairs of states (q_t, q_{t+1}) or triples of states (q_t, q_{t+1}, q_{t+2}).

The most widely used criterion is to find the *single* best state sequence to maximize $P(\boldsymbol{q}|\boldsymbol{X}, \Lambda)$, which is equivalent to maximizing $P(\boldsymbol{q}, \boldsymbol{X}|\Lambda)$, the joint state-observation probability. A dynamic programming based algorithm, called the Viterbi algorithm [12], can be used to efficiently find the best single-state sequence.

3.2.1. The Viterbi Algorithm.

The Viterbi algorithm can be used to find the single best state sequence $\boldsymbol{q}^*$ defined as

$$\boldsymbol{q}^* = \arg \max_{\boldsymbol{q}} P(\boldsymbol{q}, \boldsymbol{X} | \Lambda) \qquad (14)$$

for a given observation sequence $\boldsymbol{X}$. To accomplish the goal of maximization, define the following best partial score

$$\delta_t(i) = \max_{q_1, q_2, \ldots, q_{t-1}} P(q_1, q_2, \ldots, q_{t-1}, q_t = i, \boldsymbol{x}_1, \boldsymbol{x}_2, \ldots, \boldsymbol{x}_t | \Lambda) \qquad (15)$$

that is, the maximum probability at time t along a single path that accounts for the first t observations and ends in state i. The best partial score can be computed by induction as follows:

1. *Initialization:*

$$\delta_0(i) = \pi_i \qquad (16)$$

$$\psi_0(i) = 0 \qquad (17)$$

2. *Recursion:*

$$\delta_t(j) = \max_{1 \leq i \leq N} [\delta_{t-1}(i) a_{ij}] b_j(\boldsymbol{x}_t) \qquad (1 \leq t \leq T, \quad 1 \leq j \leq N) \qquad (18)$$

$$\psi_t(j) = \arg \max_{1 \leq i \leq N} [\delta_{t-1}(i) a_{ij}] \qquad (1 \leq t \leq T, \quad 1 \leq j \leq N) \qquad (19)$$

3. *Termination:*

$$P^* = \max_{1 \leq i \leq N} [\delta_T(i)] \qquad (20)$$

$$q_T^* = \arg \max_{1 \leq i \leq N} [\delta_T(i)] \qquad (21)$$

4. *Backtracking:*

$$q_t^* = \psi_{t+1}(q_{t+1}^*) \qquad (t = T-1, \quad T-2, \ldots, 0) \qquad (22)$$

The array $\psi_t(j), t = T, \; T-1, \ldots, 1$ and $j = 1, 2, \ldots, N$ records the partial optimal state sequence, and is necessary in producing the single best state sequence $\boldsymbol{q}^*$. The Viterbi algorithm has been widely used in many applications such as speech recognition and data communication.

3.3. The Estimation Problem

A signal-modeling task involves estimating the parameter values from a given set of observations, say, $\Omega_X = \{\boldsymbol{X}_1, \boldsymbol{X}_2, \ldots, \boldsymbol{X}_L\}$. The HMM parameters to be estimated is the set $\Lambda = (\pi, A, B = \{b_j\}_{j=1}^N)$. (In the following, we shall speak of the model and the model parameter set interchangeably without ambiguity.) Parameter estimation is normally carried out according to some well-known optimization criterion; "maximum likelihood" (ML) is one of the most prevalent.

3.3.1. Maximum Likelihood.

The ML estimate of the model is obtained as

$$\Lambda_{\mathrm{ML}} = \arg \max_{\Lambda} P(\Omega_X | \Lambda) \qquad (23)$$

For HMM, unfortunately, there is no known way to analytically solve for the model parameter optimization problem in closed form. Instead, a general hill-climbing algorithm is used to iteratively improve the model parameter set until the procedure reaches a fixed-point solution, which is at least locally optimal. The algorithm is called the *Baum–Welch algorithm* [13] [or the expectation–maximization (EM) algorithm [14] in other statistical contexts].

The Baum–Welch algorithm accomplishes likelihood maximization in a two-step procedure, known as "reestimation." On the basis of an existing model parameter set Λ, the first step of the algorithm is to transform the objective function $P(\Omega_X|\Lambda)$ into an auxiliary function $Q(\Lambda, \Lambda')$, which measures a divergence between the model Λ and another model Λ', a variable to be optimized. The auxiliary function is defined, for the simplest case with a single observation sequence $\boldsymbol{X}$, as

$$Q(\Lambda, \Lambda') = \sum_{\boldsymbol{q}} P(\boldsymbol{X}, \boldsymbol{q}|\Lambda) \log P(\boldsymbol{X}, \boldsymbol{q}|\Lambda') \tag{24}$$

where $P(\boldsymbol{X}, \boldsymbol{q}|\Lambda)$ can be found in Eq. (6). It can be shown that $Q(\Lambda, \Lambda') \geq Q(\Lambda, \Lambda)$ for a certain Λ' implies $P(\boldsymbol{X}, \boldsymbol{q}|\Lambda') \geq P(\boldsymbol{X}, \boldsymbol{q}|\Lambda)$. Therefore, the second step of the algorithm involves maximizing $Q(\Lambda, \Lambda')$ as a function of Λ' to obtain a higher, improved likelihood. These two steps iterate interleavingly until the likelihood reaches a fixed point. Detailed derivation of the reestimation formulas can be found in three papers [1–3].

3.3.2. Other Optimization Criteria. The need to consider other optimization criteria comes primarily from the potential inconsistency between the form of the chosen model (i.e., an HMM) and that of the true distribution of the data sequence. If inconsistency or model mismatch exists, the optimality achieved in ML (maximum likelihood) estimation may not represent its real significance, either in the sense of data fitting or in indirect applications such as statistical pattern recognition. In statistical pattern recognition, one needs to evaluate and compare the a posteriori probabilities of all the classes, on receipt of an unknown observation, in order to achieve the theoretically optimal performance of the so-called Bayes risk [15]. A model mismatch prevents one from achieving the goal of Bayes risk. In these situations, application of optimization criteria other than maximum likelihood may be advisable.

One proposal to deal with the potential problem of a model mismatch is the method of minimum discrimination information (MDI) [16]. Let the observation sequence $\boldsymbol{X} = (\boldsymbol{x}_1, \boldsymbol{x}_2, \ldots, \boldsymbol{x}_T)$ be associated with a constraint $\boldsymbol{R}$, for example, $\boldsymbol{R} = (\boldsymbol{r}_1, \boldsymbol{r}_2, \ldots, \boldsymbol{r}_T)$, in which each $\boldsymbol{r}_t$ is an autocorrelation vector corresponding to a short-time observation $\boldsymbol{x}_t$. Note that $\boldsymbol{X}$ is just a realization of a random event, which might be governed by a set of possibly uncountably many distributions that satisfy the constraint $\boldsymbol{R}$. Let's denote this set of distributions by $\Theta(\boldsymbol{R})$. The minimum discrimination information is a measure of closeness between two probability measures under the given constraint $\boldsymbol{R}$ and is defined by

$$v(\boldsymbol{R}, P(\boldsymbol{X}|\Lambda)) \equiv \inf_{G \in \Theta(R)} I(G : P(\boldsymbol{X}|\Lambda)) \tag{25}$$

where

$$I(G : P(\boldsymbol{X}|\Lambda)) \equiv \int g(\boldsymbol{X}) \log \frac{g(\boldsymbol{X})}{p(\boldsymbol{X}|\Lambda)} d\boldsymbol{X} \tag{26}$$

where $g(\cdot)$ and $p(\cdot|\Lambda)$ denote the probability density functions corresponding to G and $P(\boldsymbol{X}|\Lambda)$, respectively. The MDI criterion tries to choose a model parameter set such that $v(R, P(\boldsymbol{X}|\Lambda))$ is minimized. An interpretation of MDI is that it attempts to find not just the HMM parameter set to fit the data $\boldsymbol{X}$ but also an HMM that is as close as it can be to a member of the distribution set $\Theta(\boldsymbol{R})$, of which $\boldsymbol{X}$ could have been a true realization. If $\boldsymbol{X}$ is indeed from a hidden Markov source (of the right order), the attainable minimum discrimination information is zero. While the MDI criterion does not fundamentally change the problem in model selection, it provides a way to deemphasize the potentially acute mismatch between the data and the chosen model.

Another concern about the choice of the HMM optimization criterion arises in pattern (e.g., speech) recognition problems in which one needs to estimate a number of distributions, each corresponding to a class of events to be recognized. Let V be the number of classes of events, each of which is characterized by an HMM Λ_v, $1 \leq v \leq V$. Also let $P(v)$ be the prior distribution of the classes of events. The set of HMMs and the prior distribution thus define a probability measure for an arbitrary observation sequence $\boldsymbol{X}$:

$$P(\boldsymbol{X}) = \sum_{v=1}^{V} P(\boldsymbol{X}|\Lambda_v)P(v) \tag{27}$$

A measure of mutual information between class v and a realized class v observation $\boldsymbol{X}^v$ can be defined, in the context of the composite probability distribution $\{P(\boldsymbol{X}^v|\Lambda_v)\}_{v=1}^{V}$, as

$$\begin{aligned} I(\boldsymbol{X}^v, v; \{\Lambda_v\}_{v=1}^{V}) &= \log \frac{P(\boldsymbol{X}^v, v|\{\Lambda_v\}_{v=1}^{V})}{P(\boldsymbol{X})P(v)} \\ &= \log P(\boldsymbol{X}^v|\Lambda_v) - \log \sum_{w=1}^{V} P(\boldsymbol{X}^v|\Lambda_w)P(w) \end{aligned} \tag{28}$$

The maximum mutual information (MMI) criterion [17] is to find the entire model parameter set such that the mutual information is maximized:

$$(\{\Lambda_v\}_{v=1}^{V})_{\text{MMI}} = \arg \max_{\{\Lambda_v\}_{v=1}^{V}} \left\{ \sum_{v=1}^{V} I(\boldsymbol{X}^v, v; \{\Lambda_v\}_{v=1}^{V}) \right\} \tag{29}$$

The key difference between an MMI and an ML estimate is that optimization of model parameters is carried out on all the models in the set for any given observation sequence. Equation (28) also indicates that the mutual information is a measure of the class likelihood evaluated in contrast to an overall "probability background" (the second term in the equation). Since all the distributions are involved in the optimization process for the purpose of comparison rather than individual class data fitting, it in some way mitigates the model mismatch problem. For direct minimization of recognition errors in pattern recognition problems, the minimum classification error (MCE) criterion provides another alternative [18].

Optimization of these alternative criteria is usually more difficult than the ML hill-climbing method and requires general optimization procedures such as the descent algorithm to obtain the parameter values.

4. TYPES OF HIDDEN MARKOV MODELS

Hidden Markov models can be classified according to the Markov chain topology and the form of the in-state observation distribution functions. Examples of the

Markov chain topology have been shown in Fig. 2. The in-state observation distribution warrants further discussion as it significantly affects the re-estimation procedure.

In many applications, the observation is discrete, assuming a value from a finite set or alphabet, with cardinality, say, M, and thus warrants the use of a discrete HMM. Associated with a discrete HMM is the set of observation symbols $\boldsymbol{V} = \{\boldsymbol{v}_1, \boldsymbol{v}_2, \ldots, \boldsymbol{v}_M\}$, where M is the number of distinct observation symbols in each state. The observation probability distribution is defined, for state j, as

$$b_j(\boldsymbol{x_t} = \boldsymbol{v}_k) = b_j(k) = P(\boldsymbol{x_t} = \boldsymbol{v}_k | q_t = j) = b_{jk} \tag{30}$$

forming a matrix $B = [b_{jk}], 1 \leq j \leq N, 1 \leq k \leq M$. To obtain an ML estimate, the Baum–Welch algorithm involves maximization of the auxiliary function $Q(\Lambda, \Lambda')$ defined in Eq. (24). Note that

$$\log P(\boldsymbol{X}, \boldsymbol{q} | \Lambda') = \log \pi'_{q_0} + \sum_{t=1}^{T} \log a'_{q_{t-1}q_t} + \sum_{t=1}^{T} \log b'_{q_t}(\boldsymbol{x}_t) \tag{31}$$

which allows us to write the auxiliary function in the following form

$$Q(\Lambda, \Lambda') = Q_\pi(\Lambda, \pi') + \sum_{i=1}^{N} Q_a(\Lambda, \boldsymbol{a}'_i) + \sum_{i=1}^{N} Q_b(\Lambda, \boldsymbol{b}'_i) \tag{32}$$

where $\pi' = [\pi'_1, \pi'_2, \ldots, \pi'_N]$, $\boldsymbol{a}'_i = [a'_{i1}, a'_{i2}, \ldots, a'_{iN}]$, $\boldsymbol{b}'_i = [b'_{i1}, b'_{i2}, \ldots, b'_{iM}]$, and

$$Q_\pi(\Lambda, \pi') = \sum_{i=1}^{N} P(\boldsymbol{X}, q_0 = i | \Lambda) \log \pi'_i \tag{33}$$

$$Q_a(\Lambda, \boldsymbol{a}'_i) = \sum_{j=1}^{N} \sum_{t=1}^{T} P(\boldsymbol{X}, q_{t-1} = i, q_t = j | \Lambda) \log a'_{ij} \tag{34}$$

$$Q_b(\Lambda, \boldsymbol{b}'_i) = \sum_{t=1}^{T} P(\boldsymbol{X}, q_t = i | \Lambda) \log b'_i(\boldsymbol{x}_t) \tag{35a}$$

$$= \sum_{m=1}^{M} \sum_{t=1}^{T} P(\boldsymbol{X}, q_t = i | \Lambda) \delta(\boldsymbol{x}_t, \boldsymbol{v_m}) \log b'_{im} \tag{35b}$$

Note that in Eq. 35b, which is for a discrete HMM specifically, we define

$$\delta(\boldsymbol{x}_t, \boldsymbol{v_m}) = 1 \quad \text{if} \quad \boldsymbol{x}_t = \boldsymbol{v_m}; \quad = 0, \qquad \text{otherwise.}$$

These individual terms can be maximized separately over π', $\{\boldsymbol{a}'_i\}_{i=1}^N$, $\{\boldsymbol{b}'_i\}_{i=1}^N$. They share an identical form of an optimization problem. Find y_i, $i = 1, 2, \ldots, N$, (or M in 35b) to maximize $\sum_{i=1}^{N} w_i \log y_i$ subject to $\sum_{i=1}^{N} y_i = 1$ and $y_i \geq 0$. This problem attains a global maximum at the single point

$$y_i = \frac{w_i}{\sum_{k=1}^{N} w_k}, i = 1, 2, \ldots, N \tag{36}$$

This leads to the following reestimation transformation for the HMM parameters

$$\overline{\pi}_i = \frac{P(\boldsymbol{X}, q_0 = i | \Lambda)}{\sum_{j=1}^{N} P(\boldsymbol{X}, q_0 = j | \Lambda)} = \frac{P(\boldsymbol{X}, q_0 = i | \Lambda)}{P(\boldsymbol{X} | \Lambda)} \tag{37}$$

$$\overline{a}_{ij} = \frac{\sum_{t=1}^{T} P(\boldsymbol{X}, q_{t-1} = i, q_t = j | \Lambda)}{\sum_{k=1}^{N} \sum_{t=1}^{T} P(\boldsymbol{X}, q_{t-1} = i, q_t = k | \Lambda)} = \frac{\sum_{t=1}^{T} P(\boldsymbol{X}, q_{t-1} = i, q_t = j | \Lambda)}{\sum_{t=1}^{T} P(\boldsymbol{X}, q_{t-1} = i | \Lambda)} \tag{38}$$

$$\overline{b}_{im} = \frac{\sum_{t=1}^{T} P(\boldsymbol{X}, q_t = i | \Lambda) \delta(\boldsymbol{x}_t, \boldsymbol{v_m})}{\sum_{k=1}^{M} \sum_{t=1}^{T} P(\boldsymbol{X}, q_t = i | \Lambda) \delta(\boldsymbol{x}_t, \boldsymbol{v}_k)} = \frac{\sum_{t=1}^{T} P(\boldsymbol{X}, q_t = i | \Lambda) \delta(\boldsymbol{x}_t, \boldsymbol{v_m})}{\sum_{t=1}^{T} P(\boldsymbol{X}, q_t = i | \Lambda)} \tag{39}$$

where $\{\overline{\pi}\}$, $\{\overline{a}\}$, and $\{\overline{b}\}$ achieve $\max Q(\Lambda, \Lambda')$ as a function of Λ'.

For continuous-density HMMs, the same iterative procedure applies, although one needs to construct the right form of the auxiliary function in order to be able to carry out the maximization step. During the course of development of the HMM theory, Baum et al. [6] first showed an algorithm to accomplish ML estimation for an HMM with continuous in-state observation densities that are log-concave. The Cauchy distribution, which is not log-concave, was cited as one that the algorithm would have difficulty with. Later, Liporace [4], by invoking a representation theorem, relaxed this restriction and extended the algorithm to enable it to cope with the family of elliptically symmetric density functions. The algorithm was further enhanced at Bell Laboratories in the early 1980s [5] and extended to the case of general mixture densities taking the form, for state i

$$b_i(\boldsymbol{x}) = \sum_{k=1}^{M} c_{ik} f(\boldsymbol{x}; \boldsymbol{b}_{ik}) \tag{40}$$

where M is the number of mixture components, c_{ik} is the weight for mixture component k, and $f(\boldsymbol{x}; \boldsymbol{b}_{ik})$ is a continuous probability density function, log-concave or elliptically symmetric, parameterized by $\boldsymbol{b}_{ik}$. The significance of this development is that the modeling technique now has the capacity to deal with virtually any distribution function since the mixture density can be

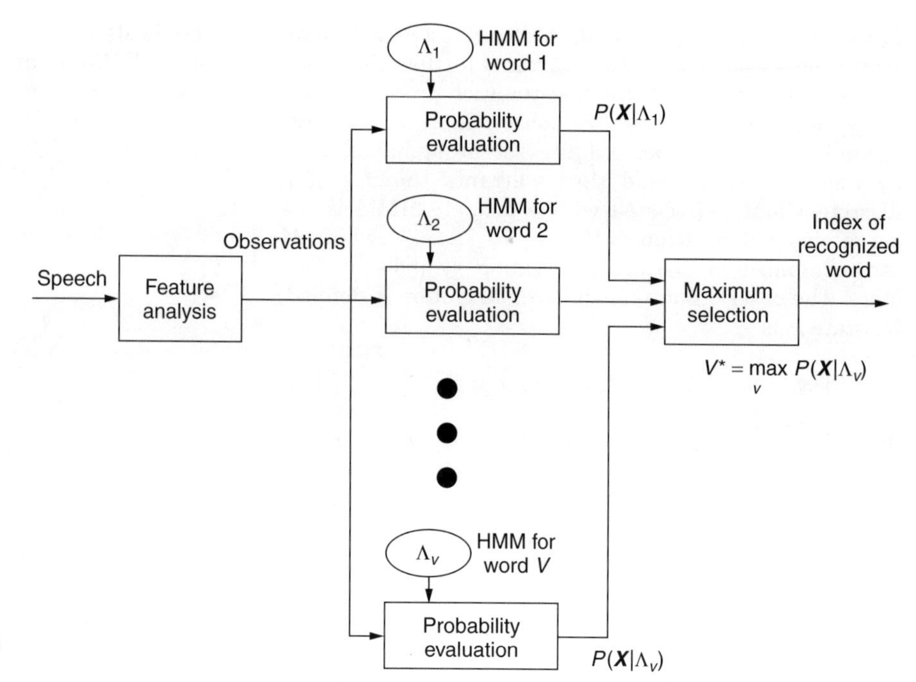

Figure 3. Block diagram of an isolated word HMM recognizer (after Rabiner [3]).

used to approximate a distribution with arbitrary accuracy (by increasing M). This is important because many real-world processes such as speech parameters do not follow the usual symmetric or single modal distribution form. For detailed derivations of the reestimation formulas for various kinds of density functions embedded in a hidden Markov model, consult Rabiner [3] and Rabiner and Juang [10].

5. APPLICATIONS OF HMM

Since its inception in the late 1960s and through its development in the 1970s and 1980s, the HMM has found widespread use in many problem areas, such as stockmarket analysis, cryptoanalysis, telecommunication modeling and management, and most notably automatic speech recognition, which we highlight here.

Automatic speech recognition (ASR) was traditionally treated as a speech analysis problem because of the complex nature of the signal. The main impact of the HMM methodology was to formalize the statistical approach as a feasible paradigm in dealing with the vast variability in the speech signal. Major sources of variability in a speech signal include the inherent uncertainty in speaking rate variation, physiological variation such as speaker-specific articulatory characteristics, changes in speaking conditions (e.g., the Lombard effect [19]), uncertainty and coarticulation in speech production (e.g., underarticulated phonemes), regional accents, and use of language. The HMM has an intrinsic ability to cope with the speaking rate variation due to the Markov chain, but the broad nature of variability calls for the use of a mixture-density HMM, particularly in the design of a system that is supposed to perform equally well for a large population of users (i.e., a speaker-independent speech recognition system [10]). Today, most high-performance large-vocabulary ASR systems are based on mixture-density HMMs.

Figure 3 illustrates the use of the HMM in an isolated word recognition task. A recognizer requires the knowledge of the a posteriori probabilities $P(v|\boldsymbol{X})$ for all the words/classes, v, $1 \leq v \leq V$, evaluated on an observed pattern or sequence, $\boldsymbol{X}$. For simplicity, we assume that the prior probabilities for all the classes are equal and the recognition decision is thus based on the likelihood functions $P(\boldsymbol{X}|\Lambda_v)$, $1 \leq v \leq V$, with each model Λ_v corresponding to a class v. The models $P(\boldsymbol{X}|\Lambda_v)$, $1 \leq v \leq V$, need to be "trained" on the basis of a set of known samples during the design phase. "Training" is the task of model parameter estimation using one or more observation sequences or patterns. Section 3.3 outlines the procedure for parameter estimation. Once these models are trained, they are stored in the system for likelihood evaluation on receipt of an unknown observation sequence. The solution based on the forward probability calculation as discussed in Section 3.1 is an efficient way to obtain the likelihood for each word in the vocabulary. The word model that achieves the highest likelihood determines the recognized word. This simple design principle has become the foundation of many ASR systems.

For more sophisticated speech recognition tasks, such as large-vocabulary, continuous speech recognition, the design principle described above has to be modified to cope with the increased complexity. In continuous speech recognition, event classes (whether they are based on "lexical words" or "phonemes") are observed consecutively without clear demarcation (or pause), making it necessary to consider composite HMMs, constructed from elementary HMMs. For example, an elementary HMM may be chosen to represent a phoneme, and a sequence of words would thus be modeled by a concatenation of the corresponding phoneme models. The concatenated model is then used

Table 1. Performance Benchmark of HMM-Based Automatic Speech Recognition Systems for Various Tasks or Applications

Task/Application	Vocabulary Size	Perplexity[a]	Word Accuracy (%)
Isolated word/digits	10	N/A	~100
Connected digit strings	10	10	~99
Isolated spell letters and digits	37	N/A	95
Navy resource management	991	<60	97
Air travel information system	1,800	<25	97
Wall St. Journal transcription	64,000	<140	94
Broadcast news transcription	64,000	<140	86

[a]Perplexity in this context is defined as the average number of words that can follow another word and is a rough measure of the difficulty or complexity of the task.

in the likelihood calculation for making the recognition decision. The procedure is rather complex, and dynamic programming techniques together with heuristics and combinatorics (to form search algorithms) are usually employed to obtain the result. Also, because of the much-increased variability in the realization of a continuous speech signal, an elementary model is often qualified by its context (e.g., an "e-**l**-i" model for the phoneme /l/ to be used in the word "element") in order to achieve the needed accuracy in modeling. These context-dependent models greatly increase the complexity of the system design.

Today, HMM-based ASR systems are deployed in telecommunication networks for automatic call routing and information services, and in personal computers for dictation and word-processing. Table 1 provides a gist of the performance, in terms of word accuracy of various systems and tasks. Voice-enabled portals that bring information services to the user over the Internet are also emerging and are expected to gain popular acceptance in the near future. The HMM is the underpinning technology of these modern-day communication services.

BIOGRAPHY

Dr. Biing-Hwang Juang received his Ph.D. in electrical and computer engineering from the University of California, Santa Barbara in 1981 and is currently Director of Multimedia Technologies Research at AVAYA Labs Research, a spin-off of Bell Laboratories. Before joining AVAYA Labs Research, he was Director of Acoustics and Speech Research at Bell Laboratories. He is engaged in and responsible for a wide range of research activities, from speech coding, speech recognition, and intelligent systems to multimedia and broadband communications. He has published extensively and holds a number of patents in the area of pattern recognition, speech communication and telecommunication services. He is co-authr of the book *Fundamentals of Speech Recognition* published by Prentice-Hall. He received the Technical Achievement Award from the IEEE Signal Processing Society in 1998 and the IEEE Third Millennium Medal in 2000. He is an IEEE Fellow and a Bell Labs Fellow.

BIBLIOGRAPHY

1. L. E. Baum, An inequality and associated maximization techniques in statistical estimation for probabilistic functions of Markov processes, *Inequalities* **3**: 1–8 (1972).
2. L. R. Rabiner and B. H. Juang, An introduction to hidden Markov models, *IEEE ASSP Mag.* **3**(1): 4–16 (Jan. 1986).
3. L. R. Rabiner, A tutorial on hidden Markov models and selected applications in speech recognition, *Proc. IEEE* **77**(2): 257–286 (Feb. 1989).
4. L. R. Liporace, Maximum likelihood estimation for multivariate observations of Markov sources, *IEEE Trans. Inform. Theory* **IT-28**: 729–734 (Sept. 1982).
5. B. H. Juang, Maximum likelihood estimation for mixture multivariate stochastic observations of Markov chains, *AT&T Tech. J.* **64**: 1235–1249 (1985).
6. L. E. Baum, T. Petri, G. Soules, and N. Weiss, A maximization technique occurring in the statistical analysis of probabilistic functions of Markov chains, *Ann. Math. Stat.* **41**: 164–171 (1970).
7. L. E. Baum and J. A. Eagon, An inequality with applications to statistical estimation for probabilistic functions of Markov processes and to a model for ecology, *Bull. Am. Math. Soc.* **73**: 360–363 (1967).
8. D. Andelman and J. Reeds, On the cryptoanalysis of rotor machines and substitution-permutation networks, *IEEE Trans. Inform. Theory* **IT-28**(4): 578–584 (July 1982).
9. H. Bunke and T. Caelli, *Hidden Markov Model in Vision,* a special issue of the *International Journal of Pattern Recognition and Artificial Intelligence*, World Scientific, Singapore, Vol. 45, June 2001.
10. L. R. Rabiner and B. H. Juang, *Fundamentals of Speech Recognition*, Prentice-Hall, Englewood Cliffs, NJ, 1993.
11. A. T. Bharucha-Reid, *Elements of the Theory of Markov Processes and Their Applications*, McGraw-Hill, New York, 1960.
12. G. D. Forney, The Viterbi algorithm, *Proc. IEEE* **61**: 268–278 (March 1973).
13. S. E. Levinson, L. R. Rabiner, and M. M. Sondhi, An introduction to the application of the theory of probabilistic functions of a Markov process to automatic speech recognition, *Bell Syst. Tech. J.* **62**(4): 1035–1074 (April 1983).
14. A. P. Dempster, N. M. Laird, and D. B. Rubin, Maximum likelihood from incomplete data via the EM algorithm, *J. Roy. Stat. Soc.* **39**(1): 1–38 (1977).
15. R. Duda and P. Hart, *Pattern Classification and Scene Analysis*, Wiley, New York, 1973.
16. Y. Ephraim, A. Dembo, and L. R. Rabiner, A minimum discrimination information approach for hidden Markov modeling, *IEEE. Trans. Inform. Theory* **IT-35**(5): 1001–1003 (Sept. 1989).

17. A. Nadas, D. Nahamoo, and M. A. Picheny, On a model-robust training method for speech recognition, *IEEE Trans. Acoust. Speech Signal Process.* **ASSP-36**(9): 1432–1436 (Sept. 1988).

18. B. H. Juang, Wu Chou, and C. H. Lee, Minimum classification error rate methods for speech recognition, *IEEE Trans. Speech Audio Process.* **SA-5**(3): 257–265 (May 1997).

19. R. P. Lippmann, E. A. Martin, and D. B. Paul, Multi-style training for robust isolated-word speech recognition, *IEEE ICASSP-87 Proc.*, April 1987, pp. 705–708.

HIGH-DEFINITION TELEVISION*

Jae S. Lim
Massachusetts Institute of Technology
Cambridge, Massachusetts

A high-definition television (HDTV) system is a television system whose performance is significantly better than a conventional television system. An HDTV system delivers spectacular video and multichannel CD (compact disk) quality sound. The system also has many features that are absent in conventional systems, such as auxiliary data channels and easy interoperability with computers and telecommunications networks.

Conventional television systems were developed during the 1940s and 1950s. Examples of conventional systems are NTSC (National Television Systems Committee), SECAM (Sequential Couleur a Memoire), and PAL (phase-alternating line). These systems have comparable performance in video quality, audio quality, and transmission robustness. The NTSC system, used in North America, will be used as a reference for conventional television systems when discussing HDTV in this article.

For many decades, conventional television systems have been quite successful. However, they were developed on the basis of the technology that was available during the 1940s and 1950s. Advances in technologies such as communications, signal processing, and very-large-scale integration (VLSI) has enabled a major redesign of a television system with substantial improvements over conventional television systems. An HDTV system is one result of this technological revolution.

1. CHARACTERISTICS OF HDTV

Many characteristics of an HDTV system markedly differ from a conventional television system. These characteristics are described in this section.

1.1. High Resolution

An HDTV system can deliver video with spatial resolution much higher than a conventional television system. Typically, video with a spatial resolution of at least four times that of a conventional television system is called *high-resolution video*. Resolution represents the amount of detail contained within the video, which can also be called "definition." This is the basis for high-definition television. An NTSC system delivers video at a resolution of approximately 480 lines in an interlaced format at an approximate rate of 60 fields/s (it is actually 59.94 Hz, but we will not make a distinction between 59.94 and 60). Each line of resolution contains approximately 420 pixels or picture elements. The number of lines represents the vertical spatial resolution in the picture, and the number of pixels per line represents the horizontal spatial resolution. *Interlaced scanning* refers to the scanning format. All conventional television systems use this format. Television systems deliver pictures that are snapshots of a scene recorded at a certain number of times per second. In interlaced scanning, a single snapshot consists of only odd lines, the next snapshot consists of only even lines, and this sequence repeats. A snapshot in interlaced scanning is called a *field*. In the NTSC system, 60 fields are used per second. Although only snapshots of a scene are shown, the human visual system perceives it as continuous motion, as long as the snapshots are shown at a sufficiently high rate. In this way, the video provides accurate motion rendition.

More lines and more pixels per line in a field provide more spatial details than the field can retain. An HDTV system may have 1080 lines and 1920-pixel/line resolution in an interlaced format of 60 fields/sec. In this case, the spatial resolution of an HDTV system would be almost 10 times that of an NTSC system. This high spatial resolution is capable of showing details in the picture much more clearly, and the resultant video appears much sharper. It is particularly useful for sports events, graphic material, written letters, and movies.

The high spatial resolution in an HDTV system enables a large-screen display and increased realism. For an NTSC system, the spatial resolution is not high. To avoid the visibility of a line structure in an NTSC system, the recommended viewing distance is approximately 7 times the picture height. For a 2-ft-high display screen the recommended viewing distance from the screen is 14 ft, 7 times the picture height. This makes it difficult to have a large-screen television receiver in many homes. Because of the long viewing distance, the viewing angle is approximately 10°, which limits realism. For an HDTV system with more than twice the number of lines, the recommended viewing distance is typically 3 times the picture height. For a 2-ft-high display, the recommended viewing distance would be 6 ft. This can accommodate a large-screen display in many environments. Because of a short viewing distance and wider aspect (width-to-height) ratio, the viewing angle for an HDTV system is approximately 30°, which significantly increases realism.

An HDTV system can also deliver higher temporal resolution by using progressive scanning. Unlike interlaced scanning, where a snapshot (field) consists of only even lines or only odd lines, all the lines in progressive scanning are scanned for each snapshot. The snapshot in progressive scanning is called a *frame*. Both progressive scanning and interlaced scanning have their own merits, and the choice between the two generated much discussion during the digital television standardization process in the

*This article is a modified version of High Definition Television, published in the *Wiley Encyclopedia of Electrical and Electronics Engineering*; 1999, Vol. 8, pp. 725–739.

United States. An HDTV system can have only interlaced scanning, or only progressive scanning, or a combination of the two.

An HDTV system delivers video with substantially higher spatial and temporal resolution than does a conventional television system. In addition to its superior resolution, an HDTV system typically has other important features, discussed below.

1.2. Wide Aspect Ratio

An NTSC television receiver has a display area with an aspect ratio of 4 : 3. The aspect ratio is a width-to-height ratio. The 4 : 3 aspect ratio was chosen because movies were made with a 4 : 3 aspect ratio when the NTSC system was first developed. Since then, movies have been made with a wider aspect ratio. To reflect this change, an HDTV system typically has a wider aspect ratio, such as 16 : 9. The difference in spatial resolution and aspect ratio between an NTSC system and an HDTV system is illustrated in Fig. 1. Figure 1a is a frame with an aspect ratio of 4 : 3, and Fig. 1b is a frame with an aspect ratio of 16 : 9. The difference in spatial detail between the two pictures is approximately the difference in spatial resolution between a conventional television and HDTV.

(a)

(b)

Figure 1. Resolution and aspect ratio of a conventional and a high-definition television system: (**a**) a segment of a conventional television video frame — 4 : 3 aspect ratio; (**b**) the corresponding segment of a high-definition television video frame — 16 : 9 aspect ratio with higher spatial resolution than in part (a).

1.3. Digital Representation and Transmission

In a conventional television system, the video signal is represented in an analog format, and the analog representation is transmitted. However, the analog representation is highly susceptible to channel transmission degradations such as multipath effects or random noise. In a conventional television system, video received through the air (terrestrial broadcasting) often has visible degradations such as ghosts and snowlike noise.

In an HDTV system, the video signal is represented digitally and transmitted. An HDTV system can be developed using an analog transmission system. However, transmission of the digital representation of the video signal is much more efficient in the bandwidth utilization. Digital transmission can utilize such modern technologies as digital video compression and digital communications. The effects of channel degradation manifest themselves differently in a digital transmission system. In an HDTV system that is broadcast digitally over the air, the video received is essentially perfect within a certain coverage area (within a certain level of channel degradation). Outside that area, the video is not viewable. Unlike an analog NTSC system, where the video degrades more as the channel degradation increases, a digital HDTV system will deliver either an essentially perfect picture or no picture at all. This is referred to as the "cliff effect" or "digitally clean" video.

1.4. Multichannel Digital Audio

An HDTV system has the capability to deliver multichannel sound. The number of audio channels that can accompany a video program may be as many as one desires. Multiple audio channels can be used to produce the effect of surround sound, which is often used in movie theaters. They can also be used for transmitting different languages in the same video program. In addition to multichannel sound, the reproduced sound achieves the quality of an audio CD (compact disk).

A television system is often considered as primarily a video service. However, the audio service is particularly important for HDTV applications. Generally, people will not watch video with poor-quality audio, even when the video quality may be comparable to HDTV. In addition, high-quality audio enhances our visual experience. The same video, when accompanied by higher-quality audio, gives the impression of higher-quality video than when it is accompanied by low-quality audio. An HDTV system delivers multichannel audio with CD-quality sound. In addition to a superb listening experience, it enhances our visual experience beyond what is possible with high-resolution video alone.

1.5. Data Channel

A conventional television system is a standalone system whose primary objective is entertainment. A digital HDTV system utilizes a data transmission channel. Its data can represent not only high-resolution video and audio but also any digital data like computer data, newspapers, telephone books, and stockmarket quotes. The digital HDTV system can be integrated easily to operate with computers and telecommunication networks.

2. HISTORY OF HDTV IN THE UNITED STATES

The NTSC system was developed for the terrestrial transmission of television signals. Since the NTSC system requires 6 MHz of bandwidth, the available VHF (very-high-frequency) and UHF (ultra-high-frequency) bands, which are suitable for the terrestrial broadcasting of television signals, were divided into 6-MHz channels. Initially, there was plenty of spectrum. The NTSC system, however, utilizes its given 6 MHz of spectrum quite inefficiently. This inefficiency generates interference among the different NTSC signals. As the number of NTSC signals that were broadcast terrestrially increased, the interference problem became serious. The solution was to avoid using certain channels. These unused channels are known as "taboo channels." In a typical highly populated geographic location in the United States, only one of two VHF channels is used and only one of six UHF channels is used. In addition, in the 1980s, other services such as mobile radio requested the use of the UHF band spectrum. As a result, an HDTV system that requires a large amount of bandwidth, such as Japan's MUSE system, was not an acceptable solution for terrestrial broadcasting in the United States.

At the request of the broadcast organizations, the United States Federal Communications Commission (FCC) created the Advisory Committee on Advanced Television Service (ACATS) in September 1987. ACATS was chartered to advise the FCC on matters related to the standardization of the advanced television service in the United States, including establishment of a technical standard. At the request of ACATS, industries, universities, and research laboratories submitted proposals for the ATV (advanced television) technical standard in 1988.

While the ACATS screened the proposals and prepared testing laboratories for their formal technical evaluation, the FCC made a key decision. In March 1990, the FCC selected the simulcast approach for advanced television service rather than the receiver-compatible approach, in which existing NTSC television receivers can receive an HDTV signal and generate a viewable picture. This was the approach taken when the NTSC introduced color. A black-and-white television receiver can receive a color television signal and display it as a viewable black-and-white picture. In this way, the then-existing black-and-white television receivers would not become obsolete. It was possible to use the receiver-compatible approach in the case of color introduction. This is because color information did not require a large amount of bandwidth and a small portion of the 6-MHz channel used for a black-and-white picture could be used to insert the color information without seriously affecting the black-and-white picture.

In the case of HDTV, the additional information needed was much more than the original NTSC signal and the receiver-compatibility requirement would require additional spectrum to carry the HDTV signal. Among the proposals received, the receiver-compatible approaches typically required a 6-MHz augmentation channel that carried the enhancement information, which was the difference between the HDTV signal and the NTSC signal. Even though the augmentation approach solves the receiver-compatibility problem, it has several major problems. The approach requires an NTSC channel as a basis to transmit an HDTV signal. This means that the highly spectrum-inefficient NTSC system cannot be converted to a more efficient technical system. In addition, the introduction of HDTV would permanently require a new channel for each existing NTSC channel. The FCC rejected this spectrum-inefficient augmentation channel approach.

Although the FCC's decision did not require receiver compatibility, it did require transmission of an entire HDTV signal within a single 6-MHz channel. In the simulcast approach adopted by the FCC, an HDTV signal that can be transmitted in a single 6-MHz channel can be designed independently of the NTSC signal. An NTSC television receiver cannot receive an HDTV signal. In order to receive an HDTV signal, a new television receiver would be needed. To ensure that existing television receivers do not become obsolete when HDTV service is introduced, the FCC would give one new channel for HDTV service to each NTSC station that requested it. During the transition period, both NTSC and HDTV services coexist. After sufficient penetration of HDTV service, NTSC service will be discontinued. The spectrum previously occupied by NTSC services will be used for additional HDTV channels or for other services. Initially, the FCC envisioned that the new HDTV channel and the existing NTSC channel would carry the same programs, so as not to disadvantage NTSC receivers during the transition period. This is the basis for the term "simulcasting." Later, this requirement was removed. The simulcast approach is illustrated in Fig. 2.

The simulcast approach provides several major advantages. It presents the possibility of designing a new spectrum-efficient HDTV signal that requires significantly less power and that does not interfere with other signals, including the NTSC signal. This allows the use of the taboo channels, which could not be used for additional NTSC service because of the strong interference characteristics of the NTSC signals. Without the taboo channels, it would not have been possible to give an additional channel to each existing NTSC broadcaster for HDTV service. In addition, it eliminates the spectrum-inefficient NTSC channels following the transition period. The elimination

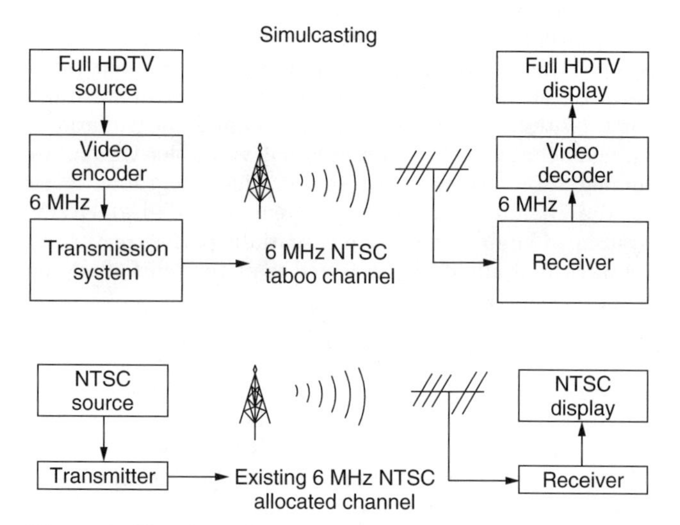

Figure 2. Flowchart of simulcasting approach for transition from an NTSC system to a digital high-definition television system.

of NTSC broadcasting vacates the spectrum that it occupied. Furthermore, by removing the NTSC signals that have strong interference characteristics, other channels could be used more efficiently. The 1990 FCC ruling was a key decision in the process to standardize the HDTV system in the United States.

The 1990 decision also created several technical challenges. The HDTV signal had to be transmitted in a single 6-MHz channel. In addition, the signal was required to produce minimal interference with NTSC signals and other HDTV signals. At the time of the FCC's decision in 1990, it was not clear whether such a system could be developed within a reasonable period of time. Later events proved that developing such a system at a reasonable cost to broadcasters and consumers was possible using modern communications, signal processing, and VLSI technologies.

Before the formal technical evaluation of the initial HDTV proposals began, some were eliminated, others were substantially modified, and still others combined their efforts. Five HDTV system proposals were ultimately approved for formal evaluation. One proposed an analog system while four others proposed all-digital systems. The five systems were evaluated in laboratory tests at the Advanced Television Testing Center (ATTC) in Alexandria, Virginia. Subjective evaluation of picture quality was performed at the Advanced Television Evaluation Laboratory (ATEL) in Ottawa, Canada. In February 1993, a special panel of experts reviewed the test results of the five HDTV system proposals and made a recommendation to the ACATS.

The panel concluded that the four digital systems performed substantially better than the analog system. The panel also concluded that each of the four digital systems excelled in different aspects. Therefore, the panel could not recommend one particular system. The panel recommended that each digital system be retested after improvements were made by the proponents. The four digital proponents had stated earlier that substantial improvements could be made to their respective system. The ACATS accepted the panel's recommendation and decided to retest the four systems after improvements were made. As an alternative to the retest, the ACATS encouraged the four proponents to combine the best elements of the different systems and submit one single system for evaluation.

The four digital system proponents evaluated their options and decided to submit a single system. In May 1993, they formed a consortium called the Grand Alliance to design and construct an HDTV prototype system. The Grand Alliance consisted of seven organizations who were members at the inception of the Grand Alliance. Later, some member organizations changed their names: (1) *General Instrument* first proposed digital transmission of an HDTV signal and submitted one of the four initial systems; *Massachusetts Institute of Technology* submitted a system together with General Instrument; *AT&T and Zenith* submitted one system together; and *Philips*, the *David Sarnoff Research Center*, and *Thomson Consumer Electronics* submitted one system together. Between the years 1993 and 1995, the Grand Alliance chose the best technical elements from the four systems and made further improvements on them. The Grand Alliance HDTV system was submitted to the ATTC and ATEL for performance verification. Test results verified that the Grand Alliance system performed better than the previous four digital systems. A technical standard based on the Grand Alliance HDTV prototype system was documented by the Advanced Television System Committee (ATSC), an industry consortium.

The HDTV prototype proposed by the Grand Alliance was a flexible system that carried approximately 20 million bits per second (20 Mbps). Even though it used the available bit capacity to transmit one HDTV program, the bit capacity could also be used to transmit several programs of standard definition television (SDTV) or other digital data such as stock quotes. SDTV resolution is comparable to that of the NTSC, but it is substantially less than the HDTV. The documented technical standard (known as the ATSC standard) allowed the transmission of SDTV programs as well as HDTV programs.

In November 1995, the ACATS recommended the ATSC standard as the United States advanced television standard to the FCC. The ATSC standard had allowed a set of only 18 video resolution formats for HDTV and SDTV programs. The FCC eased this restriction in December 1996, and decided that the ATSC standard with a relaxation of the requirements for video resolution format would be the United States digital television standard. In early 1997, the FCC made additional rulings to support the new technical standard, such as channel allocation for digital television service.

3. AN HDTV SYSTEM

A block diagram of a typical HDTV system is shown in Fig. 3. The information transmitted includes video, audio, and other auxiliary data such as stock quotes. The input video source may have a format (spatial resolution, temporal resolution, or scanning format) different from the formats used or preferred by the video encoder. In this case, the input video format is converted to a format used or preferred by the video encoder. The video is then compressed by a video encoder. Compression is needed because the bit rate supported by the modulation system is typically much less than the bit rate needed for digital video without compression. The audio, which may be multichannel for one video program, is also compressed. Since the bit rate required for audio is much less than that for video, the need to compress the audio is not as crucial. Any bit-rate savings, however, can be used for additional bits for video or other auxiliary data. The data may represent any digital data, including additional information for video and audio. The compressed video data, compressed audio data, and any other data are multiplexed by a transport system. The resulting bitstream is modulated. The modulated signal is then transmitted over a communication channel.

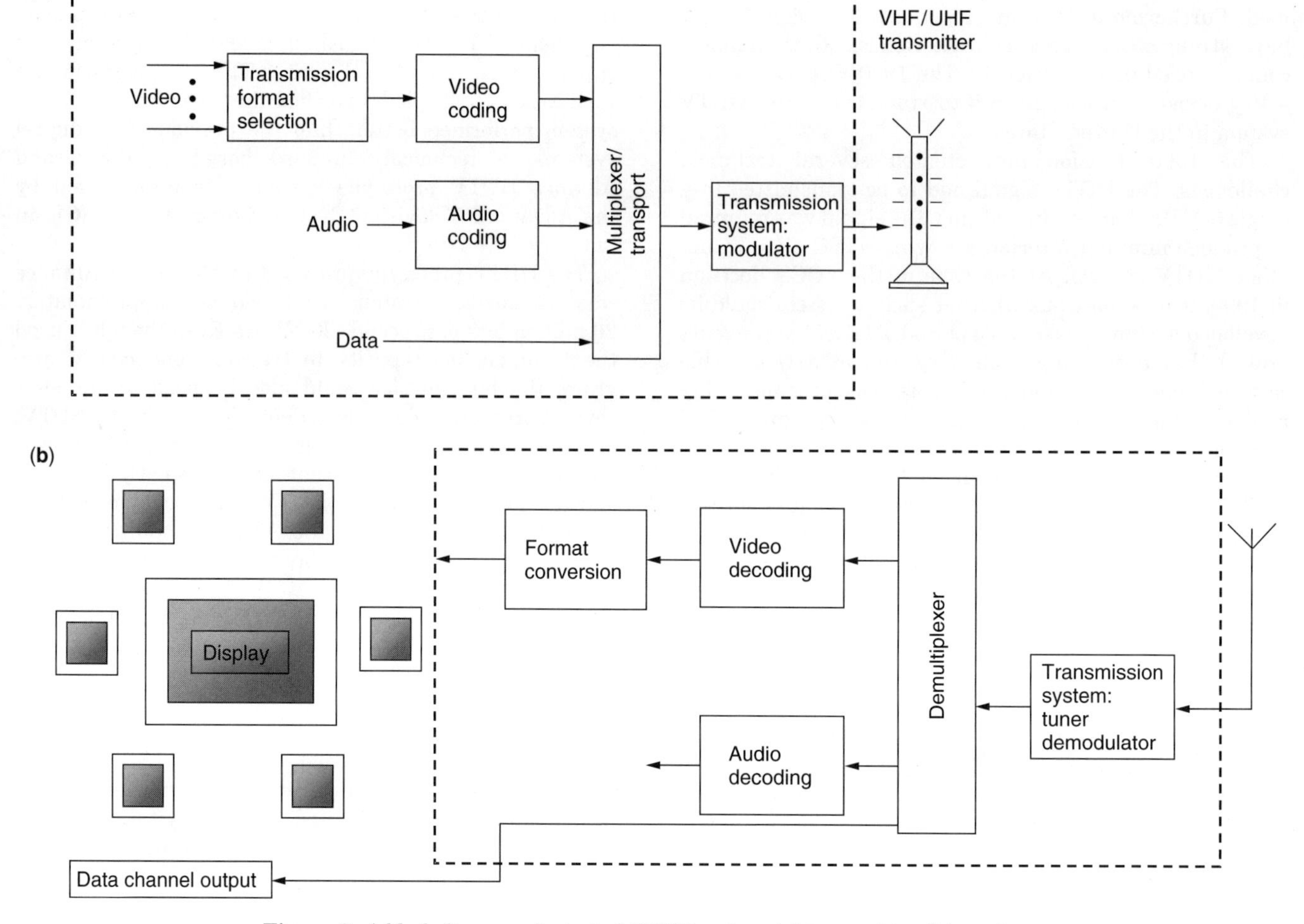

Figure 3. A block diagram of a typical HDTV system: (**a**) transmitter; (**b**) receiver.

At the receiver, the received signal is demodulated to generate a bitstream, which is demultiplexed to produce compressed video, compressed audio, and other data. The compressed video is then decompressed. The video format received may not be the same as the format used in the display. In this case, the received video format is converted to the proper display format. The compressed audio, which may be multichannel, is decompressed and distributed to different speakers. The use of the data received depends on the type of information the data contains. The communication channel in Fig. 3 may represent a storage device such as digital video disk. If the available bit rate can support more than one video program, multiple video programs can be transmitted.

There are many different possibilities for the design of an HDTV system. For example, various methods can be used for video compression, audio compression, and modulation. Some modulation methods may be more suitable for terrestrial transmission, while others may be more suitable for satellite transmission. Among the many possibilities, this article will focus on the Grand Alliance HDTV system. This system was designed over many years of industry competition and cooperation. The system's performance was carefully evaluated by laboratory and field tests, and was judged to be acceptable for its intended application. The system was the basis for the United States digital television standard. Even though this article focuses on one system, many issues and design considerations encountered in the Grand Alliance HDTV system could be applied to any HDTV system.

The overall Grand Alliance HDTV system consists of five elements: transmission format selection, video coding, audio coding, multiplexing, and modulation. These are described in the following sections.

3.1. Transmission Format Selection

A television system accommodates many video input sources such as videocameras, film, magnetic and optical media, and synthetic imagery. Even though these different input sources have different video formats, a conventional television system such as the NTSC uses only one single transmission format. This means the various input sources are converted to one format and then transmitted. Using one format simplifies the receiver design because a receiver can eliminate format conversion by designating the display format to be the same as the transmission format. This is shown in Fig. 4. When the NTSC system was standardized in the 1940s and 1950s, format conversion would have been costly.

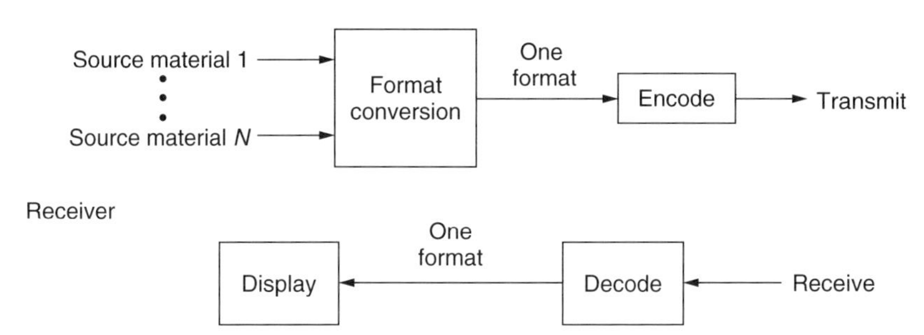

Figure 4. A television system with one single video transmission format.

The disadvantage of using one transmission format is the inefficient use of the available spectrum, since all the video input sources must be converted to one format and then transmitted in that format. For example, in the NTSC system, film (whose native format is 24 frames/s with progressive scanning) is converted to 60 fields/sec with interlaced scanning. It is then transmitted in the NTSC format. Transmission of video in a format other than its native format is an inefficient use of the spectrum.

The Grand Alliance HDTV system utilizes multiple transmission formats. This allows the use of a video transmission format that is identical to or approximates the native video source format. In addition, the system allows the use of different formats for various applications. From the viewpoint of spectrum efficiency, allowing all possible video formats would be ideal. Since a display (such as a CRT) has typically one display format, the different formats received must be converted to one display format, as shown in Fig. 5. Allowing for too many formats can complicate the format conversion operation. In addition, most of the benefits derived from multiple formats can be obtained by carefully selecting a small set of formats. For HDTV applications, the Grand Alliance system utilizes six transmission formats as shown in Table 1. In the table, the spatial resolution of $C \times D$ means C lines of vertical resolution with D pixels of horizontal resolution. The scanning format is either a progressive scan or an interlaced scan format. The "frame/field rate" refers to the number of frames/s for progressive scan and the number of fields/s for interlaced scan.

Table 1. HDTV Transmission Formats Used in the Grand Alliance System

Spatial Resolution	Scanning Format	Frame/Field Rate (Frames/s)
720 × 1280	Progressive scanning	60
720 × 1280	Progressive scanning	30
720 × 1280	Progressive scanning	24
1080 × 1920	Progressive scanning	30
1080 × 1920	Progressive scanning	24
1080 × 1920	Interlaced scanning	60

The Grand Alliance system utilizes both 720 lines and 1080 lines. The number of pixels per line was chosen so that the aspect ratio (width-to-height ratio) is 16×9 with square pixels. When the spatial vertical dimension that corresponds to one line equals the spatial horizontal dimension that corresponds to one pixel, it is called "square pixel." For 720 lines, the scanning format is progressive. The highest frame rate is 60 frames/s. The pixel rate is approximately 55 Mpixels/s (million pixels per second). For the video compression and modulation technologies used, a substantial increase in the pixel rate above 60–70 Mpixels/s may result in a noticeable degradation in video quality. At 60 frames/s with progressive scanning, the temporal resolution is very high, and smooth motion rendition results. This format is useful for sports events and commercials. The 720-line format also allows the temporal resolution of 30 frames/s and 24 frames/s. These frame rates were

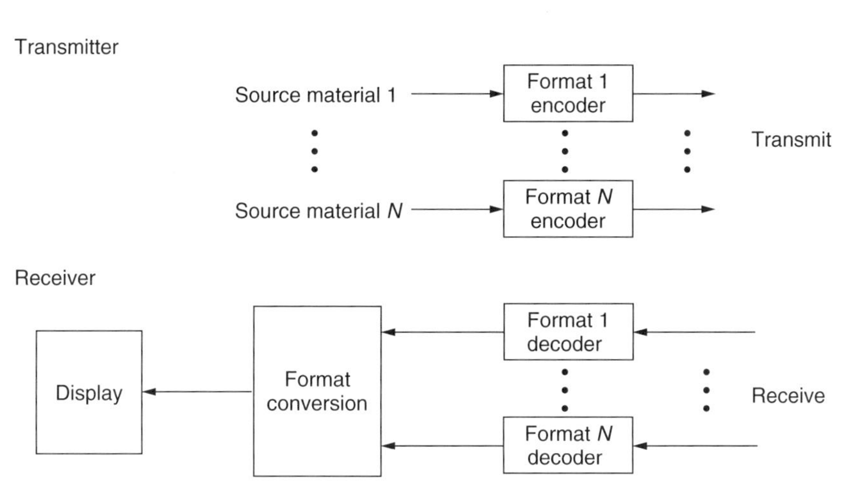

Figure 5. A television system with multiple video transmission formats.

chosen to accommodate film and graphics. For film, whose native format is 24 frames/s, conversion to 60 frames/s and then compressing it will result in a substantial inefficiency in spectrum utilization. For 720 lines at 24 frames/s, it is possible to simultaneously transmit two high-resolution video programs within a 6-MHz channel because of the lower pixel rate (approximately 22 Mpixels/s each).

In the 1080-line format, two temporal rates in progressive scan are 30 and 24 frames/s. These temporal rates were chosen for film and graphics with the highest spatial resolution. Another temporal rate used is the 1080-line interlaced scan at 60 fields/s. This is the only interlaced scan HDTV format used in the Grand Alliance system. It is useful for scenes obtained with a 1080-line interlaced-scan camera. The pixel rate for 1080-line progressive scan at 60 frames/s would be more than 120 Mpixels/s. The video encoded for such a high pixel rate can result in a substantial degradation in video quality for the compression and modulation technologies used in the Grand Alliance system. Therefore, there is no 1080-line, 60-frame/s progressive scan format in the system.

All conventional television systems, such as the NTSC, utilize only interlaced scanning. In such systems, a display format is matched to the single-transmission format. The display requires at least a 50–60 Hz rate with a reasonable amount of spatial resolution (approximately 480 active lines for the NTSC system). An alternative strategy in the NTSC system would be to preserve 480 lines with progressive scan, but at 30 frames/s. To avoid display flicker, each frame can be repeated twice at the display, making the display rate 60 Hz. Repetition of a frame at the receiver would require frame memory, which was not possible with the technologies available when the NTSC system was standardized. Because of the exclusive use of interlaced scanning in conventional television systems, early HDTV video equipment, such as videocameras, was developed for interlaced scanning.

An interlaced display has video artifacts such as interline flicker. Consider a sharp horizontal line that is in the odd field, but not in the even field. Even though the overall large area flicker rate is 60 Hz, the flicker rate for the sharp horizontal line is only 30 Hz. As a result, the line flickers in a phenomenon called *interline flicker*. Interline flickers are particularly troublesome for computer graphics or written material that contains many sharp lines. Partly for this reason, almost all computer monitors use progressive scanning.

When a television system is used as a standalone entertainment device, its interoperability with computers is not a serious issue. For a digital HDTV system, however, it is no longer a standalone entertainment device, and its interoperability with computers and telecommunications networks is useful. When a display device uses progressive scan and an interlaced transmission format is used, a conversion process called *deinterlacing* must convert the interlaced scan format to a progressive scan format before it is displayed. A high-performance deinterlacer requires complex signal processing. Even when a high-performance deinterlacer is used, a progressive transmission format yields better performance than an interlaced transmission format for graphics, animation, and written material. For this and other reasons like simple processing, the computer industry preferred only the progressive transmission format for television. Other industries, such as television manufacturers and broadcasters, preferred interlaced scanning. This is because they were accustomed to interlaced scanning. Interlaced scanning worked well for entertainment material. Early video equipment, such as videocameras, was developed only for interlaced scanning. The Grand Alliance HDTV system used five progressive scan formats and one interlaced scan format. The FCC decision in December 1996 removed most of the restrictions on transmission formats and allowed both progressive and interlaced formats. This decision left the choice of transmission format to free market forces.

A multiple transmission format system utilizes the available spectrum more efficiently than does a single transmission format system by better accommodating video source materials with different native formats. In addition, multiple transmission formats can be used for various applications. Table 2 shows possible applications for the six Grand Alliance HDTV formats. For a multiple transmission format system, one of the allowed transmission formats is chosen for a given video program prior to video encoding. The specific choice depends on the native format of the input video material and its intended application. The same video program within a given format may be assigned to a different transmission format, depending on the time of broadcast. If the transmission format chosen is different from the native format of the video material, format conversion occurs.

3.2. Video Coding

The Grand Alliance HDTV system transmits at least one HDTV program in a single 6-MHz channel. For the modulation technology used in the Grand Alliance system the maximum bit rate available for video is approximately 19 Mbps. For a typical HDTV video input, the incompressed bit rate is on the order of 1 Gbps. This means the input video must be compressed by a factor of >50.

For example, consider an HDTV video input of 720×1280 with progressive scan at 60 frames/s. The pixel rate is 55.296 Mpixels/s. A color picture consists of three monochrome images: red, green, and blue. The red, green, and blue colors are three primary colors of an

Table 2. Applications of HDTV Transmission Formats Used in the Grand Alliance System

Format	Applications
720×1280, PS, 60 frames/s	Sports, concerts, animation, graphics, upconverted NTSC, commercials
720×1280, PS, 24 frames/s or 30 frames/s	Complex film scenes, graphics, animation
1080×1920, PS, 24 frames/s or 30 frames/s	Films with highest spatial resolution
1080×1920, IS, 60 fields/s	Scenes shot with an interlaced scan camera

additive color system. By mixing the appropriate amounts of red, green, and blue lights, many different color lights can be generated. By mixing a red light and a green light, for example, a yellow light can be generated. A pixel of a color picture consists of the red, green, and blue components. Each component is typically represented by 8 bits (256 levels) of quantization. For many video applications, such as television, 8 bits of quantization are considered sufficient to avoid video quality degradation by quantization. Each pixel is then represented by 24 bits. The bit rate for the video input of 720×1280 with progressive scan at 60 frames/s is approximately 1.3 Gbps. In this example, reducing the data rate to 19 Mbps requires video compression by a factor of 70.

Video compression is achieved by exploiting the redundancy in the video data and the limitations of the human visual system. For typical video, there is a considerable amount of redundancy. For example, much of the change between two consecutive frames is due to the motion of an object or the camera. Therefore, a considerable amount of similarity exists between two consecutive frames. Even within the same frame, the pixels in a neighborhood region typically do not vary randomly. By removing the redundancy, the same (redundant) information is not transmitted.

For television applications, the video is displayed for human viewers. Even though the human visual system has enormous capabilities, it has many limitations. For example, the human visual system does not perceive well the spatial details of fast-changing regions. The high spatial resolution in such cases does not need to be preserved. By removing the redundancy in the data and exploiting the limitations of the human visual system, many methods of digital video compression were developed. A digital video encoder usually consists of the three basic elements shown in Fig. 6. The first element is representation. This element maps the input video to a domain more suitable for subsequent quantization and codeword assignment. The quantization element assigns reconstruction (quantization) levels to the output of the representation element. The codeword assignment selects specific codewords (a string of zeros and ones) to the reconstruction levels. The three elements work together to reduce the required bit rate by removing the redundancy in the data and exploiting the limitations of the human visual system.

Many different methods exist for each of the three elements in the image coder. The Grand Alliance system utilizes a combination of video compression techniques that conform to the specifications of the MPEG-2 (Moving Pictures Expert Group) video compression standard. This is one of many possible approaches to video compression.

MPEG-2 Standard. The International Standard Organization (ISO) established the Moving Pictures Expert Group (MPEG) in 1988. Its mission was to develop video coding standards for moving pictures and associated audio. In 1991, the group developed the ISO standard 11172, called *coding of moving pictures and associated audio*. This standard, known as MPEG-1, is used for digital storage media at up to ~1.5 Mbps.

In 1996, the group developed the ISO standard 13818 called *Generic Coding of Moving Pictures and Associated Audio*. This standard, known as MPEG-2, is an extension of MPEG-1 that allows flexibility in the input format and bit rates. The MPEG-2 standard specifies only the syntax of the coded bitstream and the decoding process. This means that there is some flexibility in the encoder. As long as the encoder generates a bitstream that is consistent with the MPEG-2 bitstream syntax and the MPEG-2 decoding process, it is considered a "valid" encoder. Since many methods of generating the coded bitstream are consistent with the syntax and the decoding process, some optimizations and improvements can be made without changing the standard. The Grand Alliance HDTV system uses some compression methods included in MPEG-2 to generate a bitstream that conforms to the MPEG-2 syntax. An MPEG-2 decoder can decode a video bitstream generated by the Grand Alliance video coder.

3.3. Audio Processing. The Grand Alliance HDTV system compresses the audio signal to efficiently use the available spectrum. To reconstruct CD-quality audio after compression, the compression factor for audio is substantially less than that of the video. High-quality audio is important for HDTV viewing. The data rate for audio is inherently much lower than that for video, and the additional bit-rate efficiency that can be achieved at the expense of audio quality is not worthwhile for HDTV applications.

Consider one channel of audio. The human auditory system is not sensitive to frequencies above 20 kHz. The audio signal sampled at a 48-kHz rate is sufficient to ensure that the audio information up to 20 kHz is preserved. Each audio sample is typically quantized at 16 bits/sample. The total bit rate for one channel of audio input is 0.768 Mbps. Exploiting the limitations of the human auditory system, the bit-rate requirement is reduced to 0.128 Mbps, with the reproduced audio quality almost indistinguishable from that of the input audio. The compression factor achieved is 6, which is substantially less than the video's compression factor of >50. In the case of video, it is necessary to obtain a very high compression factor, even at the expense of some noticeable quality degradation for difficult scenes because of the very high-input video bit rate (>1 Gbps). In the case of audio, additional bit-rate savings from 0.128 Mbps at the expense of possible audio quality degradation is not considered worthwhile for HDTV applications. Similar to video compression, audio compression utilizes reduction of redundancy in the audio data and the limitations of the human auditory system.

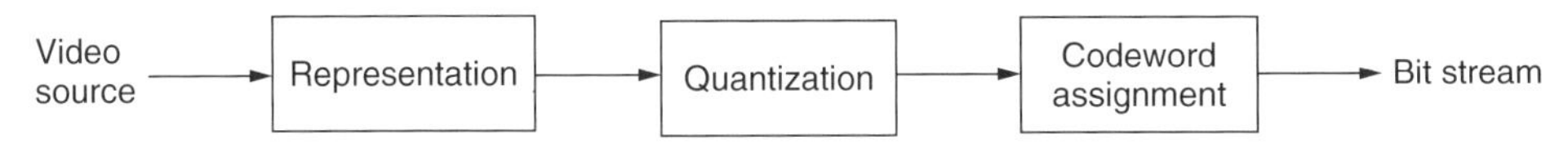

Figure 6. Three basic elements of a video encoder.

The Grand Alliance HDTV system uses a modular approach to the overall system design. Various technologies needed for a complete HDTV system can be chosen independently from each other. The audio compression method, for example, can be chosen independent from the video compression method. Even though the MPEG-2 standard used for video compression in the Grand Alliance system includes an audio compression method, the Grand Alliance selected the Audio Coder 3 (AC-3) standard on the basis of several factors including performance, bit-rate requirement, and cost.

The Grand Alliance system can encode a maximum of six audio channels per audio program. The channelization follows the ITU-R recommendation BS-775: *Multichannel Stereophonic Sound System with and without Accompanying Picture*. The six audio channels are left, center, right, left surround, right surround, and low-frequency enhancement. The bandwidth of the low-frequency enhancement channel extends to 120 Hz, while the other five channels extend to 20 kHz. The six audio channels are also referred to as "5.1 channels." Since the six channels are not completely independent for a given audio program, this dependence can be exploited to reduce the bit rate. The Grand Alliance system encodes 5.1 channel audio at a bit rate of 384 kbps with audio quality essentially the same as that of the original.

3.4. Transport System

The bitstreams generated by the video and audio encoders and the data channel must be multiplexed in an organized manner so that the receiver can demultiplex them efficiently. This is the main function of the transport system. The Grand Alliance system uses a transport format that conforms to the MPEG-2 system standard, but it does not utilize all of its capabilities. This means that the Grand Alliance decoder cannot decode an arbitrary MPEG-2 systems bitstream, but all MPEG-2 decoders can decode the Grand Alliance bitstream.

The bitstream that results from a particular application such as video, audio, or data is called an *elementary bitstream*. The elementary bitstreams transmitted in a 6-MHz channel are multiplexed to form the program transport bitstream. Each elementary bitstream has a unique program identification (PID) number, and all the elementary bitstreams within a program transport bitstream have a common timebase.

An example of the multiplex function used to form a program transport stream is shown in Fig. 7. The first two elementary streams are from one television program. The next two elementary streams are from another television program. As long as the available bit rate for the channel can accommodate more than one television program, the transport system will accommodate them. The fifth elementary stream is only an audio program without the corresponding video. The next two elementary streams are from two different datastreams. The last elementary stream contains the control information, which includes a program table that lists all the elementary bit streams, their PIDs, and the applications such as video, audio, or data. All eight elementary streams that form the program transport bitstream have the same timebase.

The Grand Alliance transport system uses the fixed-length packet structure shown in Fig. 8. Each packet consists of 188 bytes, which is divided into a header field and payload. The header field contains overhead information, and the payload contains the actual data that must be transmitted. The size of the packet is chosen to ensure that the actual payload-to-overhead ratio is sufficiently high and that a packet lost during

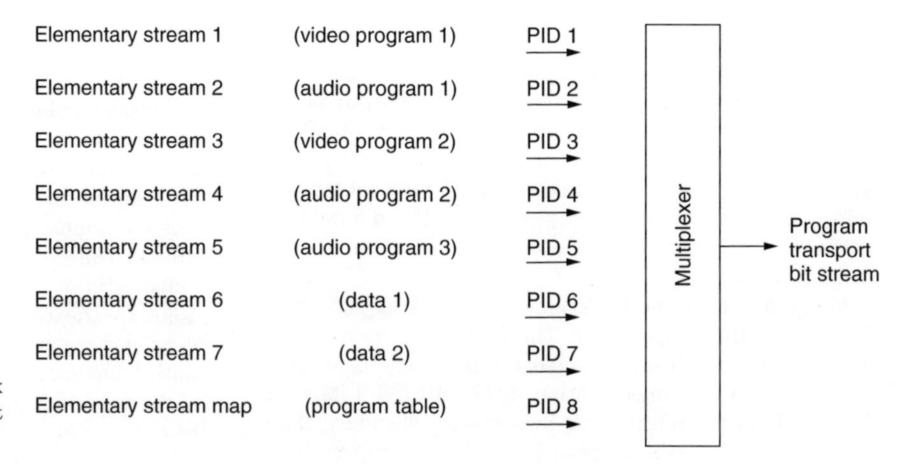

Figure 7. An example of the multiplex function used to form a program transport bitstream.

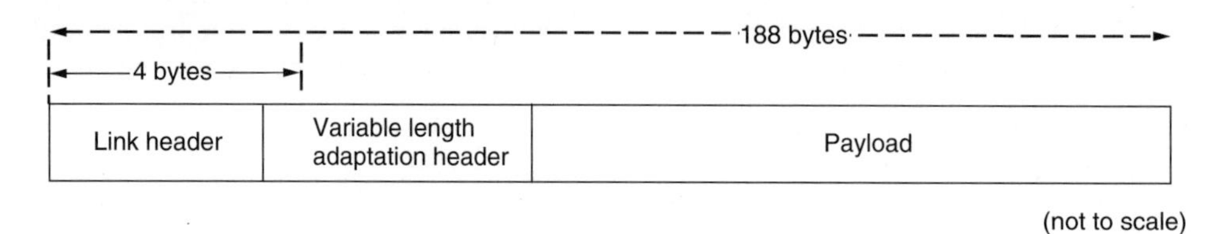

Figure 8. Fixed-length packet structure used in the Grand Alliance transport system.

transmission will not have serious consequences on the received video, audio, and data. The payload of each packet contains bits from only one particular elementary bitstream. For example, a packet cannot have bits from both a video elementary bitstream and an audio elementary bitstream.

Information that is not actual data but is important or useful for decoding the received bitstream is contained in the header. The header of each packet includes a 4-byte link header and may also include a variable-length adaptation header when needed. The link header includes 1-byte synchronization to indicate the beginning of each packet, a 13-bit PID to identify which elementary stream is contained in the packet, and information as to whether the adaptation header is included in the packet. The adaptation header contains timing information to synchronize decoding and a presentation of applications such as video and audio. This information can be inserted into a selected set of packets. The adaptation header also contains information that facilitates random entry into application bitstreams in order to support functions such as program acquisition and change.

On the transmitter side, the bits from each elementary stream are divided into packets that contain the same PID. The packets are multiplexed to form the program transport bitstream. An example is shown in Fig. 9. At the receiver, the synchronization byte, which is the first byte in each packet, is used to identify the beginning of each packet. From the program table contained in the control packet, information can be obtained on which elementary streams are in the received program transport bitstream. This information, together with the PID in each packet, is used to separate the packets into different elementary bitstreams. The information contained in the adaptation header in a selected set of packets is used for the timing and synchronization of the decoding and for the presentation of different applications (video, audio, data, etc.).

The transport system used in the Grand Alliance system has many advantages. The system is very flexible in dynamically allocating the available channel capacity to video, audio, and data. The system can devote all available bits to video, audio, or data, or any combination thereof. The system also can allocate available bits to more than one television program. If video resolution is not high, several standard definition television programs (comparable to the NTSC resolution) can be transmitted. This is in sharp contrast with the NTSC system, where a fixed bandwidth is allocated to one video program and a fixed bandwidth is allocated to audio. The capability to dynamically allocate bits as the need arises is a major feature of the transport system.

The transport system is also scalable. If a higher-bit-rate channel is available, the same transport system can be used by simply adding elementary bitstreams. The system is also extensible. If future services become available, such as 3D television, they can be added as an elementary stream with a new PID. Existing receivers that do not recognize the new PID will ignore the new elementary stream. New receivers will recognize the new PID.

The transport system is also robust in terms of transmission errors, and is amenable to cost-effective implementation. The detection and correction of transmission errors can be synchronized easily because of the fixed-length packet structure; this structure also facilitates simple demultiplex designs for low-cost, high-speed implementation.

3.5. Transmission System

The bitstream generated by the transport system must be processed in preparation for modulation, and then modulated for transmission. Choosing from the many modulation methods depends on several factors, including the transmission medium and the specific application. The best method for terrestrial broadcasting may not be the best for satellite broadcasting. Even for terrestrial broadcasting, the use of taboo channels means that interference with existing NTSC channels must be considered to determine the specific modulation technology. Considering several factors, such as coverage area, available bit rate, and complexity, the Grand Alliance system uses an 8-VSB system for terrestrial broadcasting. A block diagram of the 8-VSB system is shown in Fig. 10.

3.5.1. Data Processing. The data processing part of the 8-VSB system consists of a data randomizer, a Reed–Solomon encoder, a data interleaver, and trellis encoder. Data packets of 188 bytes per packet are received from the transport system and randomized. Portions of the bitstream from the transport system

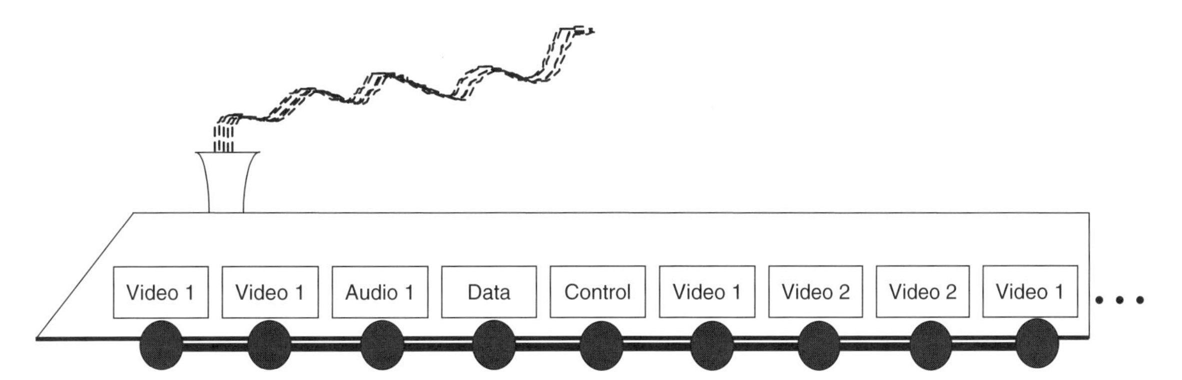

Figure 9. Example of packet multiplexing to form the program transport bitstream.

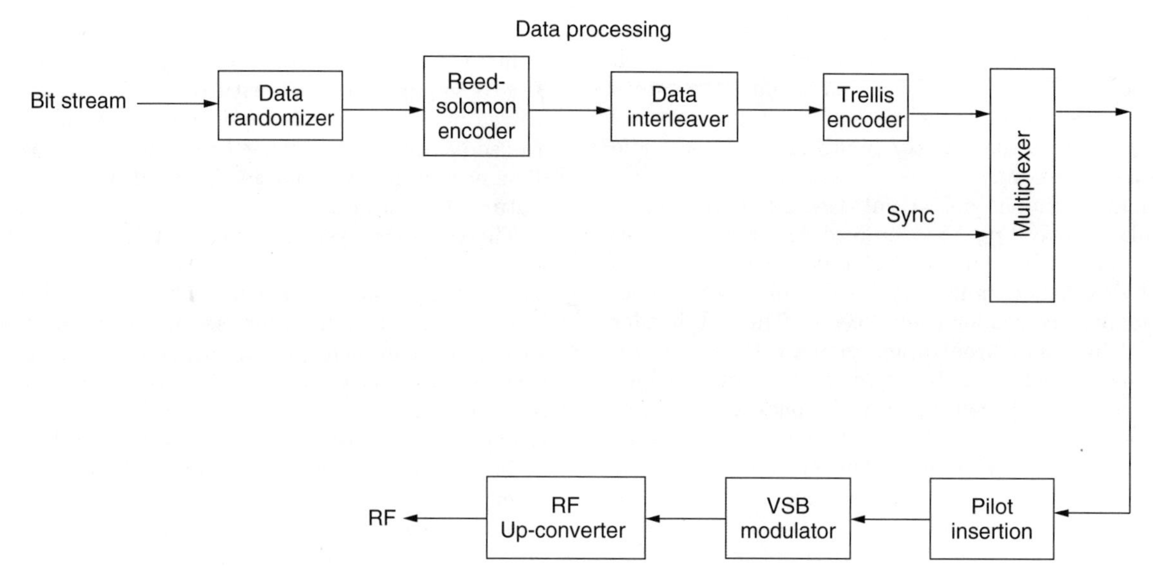

Figure 10. A block diagram of an 8-VSB system used in the Grand Alliance HDTV system.

may have some pattern and may not be completely random. Randomization ensures that the spectrum of the transmitted signal is flat and is used efficiently. In addition, the receiver exhibits optimal performance when the spectrum is flat.

The transmission channel introduces various types of noise such as random noise and multipath. They manifest themselves as random and bursty bit errors. To handle these bit errors, two forward error correction (FEC) schemes are used. These FEC schemes add redundancy to the bitstream. Errors are then detected and corrected by exploiting this added redundancy. Even though error correction bits use some of the available bit rate, they increase overall performance of the system by detecting and correcting errors.

The first error correction method is the Reed–Solomon code, known for its burst noise correction capability and efficiency in overhead bits. The burst noise correction capability is particularly useful when the Reed–Solomon code is used with the trellis code. The trellis code is effective in combating random and short impulsive noise, but it tends to generate burst errors in the presence of strong interference and burst noise. The Reed–Solomon code used in the 8-VSB system adds approximately 10% of overhead to each packet of data. The result is a data segment that consists of a packet from the transport system and the Reed–Solomon code bits. The resulting bytes are convolutionally interleaved over many data segments. The convolutional interleaving that is useful in combating the effects of burst noise and interference is part of the trellis encoding. The trellis encoder, which is a powerful technique for correcting random and short-burst bit errors, adds additional redundancy. In the 8-VSB system, the trellis coder creates a 3-bit symbol (eight levels) from a 2-bit data symbol.

At the transmitter, the Reed–Solomon encoder precedes the trellis encoder. At the receiver, the trellis decoder precedes the Reed–Solomon decoder. The trellis decoder is effective in combating the random and short-burst bit errors, but it can create long-burst bit errors in the presence of strong interference and bursts. The Reed–Solomon code is effective in combating long-burst bit errors.

The 8-VSB system transmits approximately 10.76 million symbols per second (SPS), with each symbol representing 3 bits (eight levels). When accounting for the overhead associated with the trellis coder, the Reed–Solomon coder, and additional synchronization bytes, the bit rate available to the transport system's decoder is approximately 19.4 Mbps. The bit rate is not only for applications such as video, audio, and data, but also for overhead information (link header, etc.) in the 188-byte packets. The actual bit rate available for the applications is less than 19.4 Mbps. The VSB system can deliver a higher bit rate, for example, by reducing the redundancy in the error correction codes and by increasing the number of levels per symbol. However, the result is loss of performance in other aspects such as the coverage area. The 8-VSB system delivers the 19.4-Mbps bit rate to ensure that an HDTV program can be delivered within a 6-MHz channel with a coverage area at least comparable to an NTSC system, and with an average power level below the NTSC power level in order to reduce interference with the NTSC signals.

The trellis encoding results in data segments. At the beginning of each set of 312 data segments, a data segment is inserted that contains the synchronization information for the set of 312 data segments. This data segment also contains a training sequence that can be used for channel equalization at the receiver. Linear distortion in the channel can be accounted for by an equalizer at the receiver. At the beginning of each data segment, a 4-symbol synchronization signal is inserted. The data segment sync and the 312-segment set sync are not affected by the trellis encoder and can provide synchronization independent of the data.

3.5.2. **Pilot Insertion.** Prior to modulation, a small pilot is inserted in the lower band within the 6-MHz band. The location of the pilot is on the Nyquist slope of NTSC receivers. This ensures that the pilot does not seriously impair existing NTSC service. The channel assigned for the HDTV service may be a taboo channel that is currently unused because of cochannel interference with an existing NTSC service located some distance away. The HDTV signal must be designed to ensure that its effect on existing service is minimal.

The NTSC system is rugged and reliable. The main reason for this is the use of additional signals for synchronization that do not depend on the video signals. The NTSC receiver reliably synchronizes at noise levels well below the loss of pictures. In the 8-VSB system, a similar approach is taken. The pilot signal that does not depend on the data is used for carrier acquisition. In addition, the data segment sync is used to synchronize the data clock for both frequency and phase. The 312-segment-set sync is used to synchronize the 312 segment-set and equalizer training. Reliance on additional signals for carrier acquisition and clock recovery is very useful. Even when occasional noise in the field causes a temporary loss of data, a quick recovery is possible as long as the carrier acquisition and clock recovery remain locked during the data loss. The 8-VSB system ensures that carrier acquisition and clock recovery remain intact well below the threshold level of data loss by relying on additional signals for such functions.

3.5.3. **Modulation.** For transmission of the prepared bitstream (message) over the air, the bitstream must be mapped to a bandpass signal that occupies a 6-MHz channel allocated to a station's HDTV service. A modulator modulates a carrier wave according to the prepared bitstream. The result is a bandpass signal that occupies a given 6-MHz channel. The 8-VSB system modulates the signal onto an IF (intermediate-frequency) carrier, which is the same frequency for all channels. It is followed by an upconversion to the desired HDTV channel.

3.5.4. **Receiver.** At the receiver, the signal is processed in order to obtain the data segments. One feature of the receiver is the NTSC rejection filter, which is useful because of the way HDTV service is being introduced in the United States. In some locations, an HDTV channel occupies the same frequency band as an existing NTSC channel located some distance away. The interference of the HDTV signal with the NTSC channel is minimized by a very-low-power level of the HDTV signal. The low-power level was made possible for HDTV service because of its efficient use of the spectrum. To reduce interference between the NTSC signal and the HDTV channel, the 8-VSB receiver contains an NTSC rejection filter. This is a simple comb filter whose rejection null frequencies are close to the video carrier, chroma carrier, and audio carrier frequencies of the NTSC signal. The comb filter, which reduces the overall performance of the system, can be activated only when a strong cochannel NTSC signal interferes with the HDTV channel.

3.5.5. **Cliff Effect.** Although additive white Gaussian noise does not represent typical channel noise, it is often used to characterize the robustness of a digital communication system. The segment error probability for the 8-VSB system in the presence of additive white Gaussian noise is shown in Fig. 11. At the signal-to-noise ratio (S/N) of 14.9 dB, the segment error probability is 1.93×10^{-4} or 2.5 segment errors/sec. At this threshold the segment errors become visible. Thus, up to an S/N of 14.9 dB, the system is perfect. At an S/N of 14 dB, which is just 0.9 dB less than the threshold of visibility (ToV), practically all segments are in error. This means a system that operates perfectly above the threshold of visibility becomes unusable when the signal level decreases by 1 dB or when the noise level increases by 1 dB. This is known as the "cliff effect" in a digital communication system. In the case of the NTSC system, the picture quality decreases gradually as the S/N decreases. To avoid operating near the cliff region, a digital system is designed to operate well above the threshold region within the intended coverage area. Both laboratory and field tests have indicated that the coverage area of the HDTV channel is comparable to or greater than the NTSC

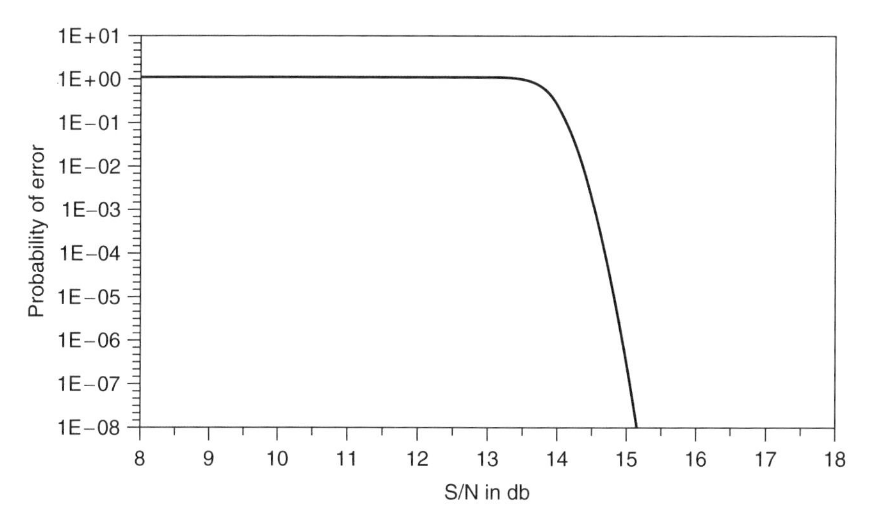

Figure 11. Segment error probability for the 8-VSB system in the presence of additive white Gaussian noise.

channel, despite the substantially lower power level used for the HDTV signal.

3.5.6. Cable Mode. A cable environment introduces substantially less channel impairment than does terrestrial broadcasting for a variety of reasons. In the case of cable, for example, cochannel interference from an existing NTSC station is not an issue. This can be exploited to deliver a higher bit rate for applications in a 6-MHz cable channel. Although the Grand Alliance HDTV system was designed for terrestrial broadcasting, the system includes a cable mode that doubles the available bit rate using a small modification. Doubling the bit rate means the ability to transmit two HDTV programs within a single 6-MHz cable channel.

In order to double the bit rate for cable mode, the Grand Alliance System uses 16-VSB rather than 8-VSB. All other aspects of the system such as video compression, audio compression, and transport remain the same. In the 8-VSB system, a symbol is represented by eight levels or 3 bits. One of the 3 bits is due to the redundancy created by trellis encoding. For a cable environment, error correction from the powerful trellis coding is no longer needed. In addition, the higher available S/N for cable means that a symbol can be represented by 16 levels or 4 bits. Since the symbol rate remains the same between the 8-VSB and 16-VSB systems, the available bit rate for the 16-VSB system doubles in comparison with the 8-VSB system. For the 16-VSB system without trellis coding, the S/N ratio for the segment error rate that corresponds to the threshold of visibility in the environment of additive white Gaussian noise is approximately 28 dB. This is 13 dB higher than the 8-VSB system with the trellis coding. This increase in the S/N is acceptable in a cable environment that has substantially less channel impairments than a typical terrestrial environment.

4. HDTV AND INTEROPERABILITY

The Grand Alliance HDTV system has served as the basis for the digital television standard in the United States. The standard itself defines a significant number of technical elements. The technologies involved, however, will continue to develop without the need to modify the standard. For example, the video compression system that was adopted defines syntax for only the decoder. There is much room for improvement and advances in the encoder. Technical elements can also be added in a backward-compatible manner. The transmission of a very high-definition (VHD) television format, which was not provided in the initial standard, can be accomplished in a backward-compatible manner. This can be achieved by standardizing a method to transmit enhancement data. This, in turn can be combined with an allowed video transmission format to deliver the VHD format.

The Grand Alliance system was designed for HDTV delivery in terrestrial environments in the United States. For other delivery environments such as satellite, cable, and environments in other parts of the world, other standards will emerge. Depending on the degree of common elements, interoperability among different standards will become an important issue. In terms of program exchange for different delivery media and throughout the world, technologies that convert one standard to another will continue to be developed. Efforts to facilitate the conversion process by adopting common elements among the different standards also will continue.

Interoperability will be an issue not only among the different HDTV standards, but among telecommunication services and computers as well. A traditional television system has been a standalone device whose primary purpose was entertainment. Although an HDTV system is used for entertainment, it can be an integral part of a home center for entertainment, telecommunications, and information. The HDTV display can be used as a videophone, a newspaper service, or a computer display. Interoperability between an HDTV system and other services that will be integrated in the future is an important consideration.

BIBLIOGRAPHY

1. ATSC, *Digital Audio Compression (AC-3)*, Dec. 20, 1995.
2. ATSC, *Digital Television Standard*, Sept. 16, 1995.
3. ATSC, *Guide to the Use of the ATSC Digital Television Standard*, Oct. 4, 1995.
4. V. Bhaskaran and K. Konstaninides, *Image and Video Compression Standards and Architectures*, Kluwer Academic Publishers, 1995.
5. W. Bretl, G. Sgrignoli, and P. Snopko, *VSB Modem Subsystem Design for Grand Alliance Digital Television Receivers*, ICCE, 1995.
6. A. B. Carlson, *Communication Systems*, 3rd ed., McGraw-Hill, 1986.
7. The Grand Alliance Members, The U.S. HDTV standard. The Grand Alliance, *IEEE Spectrum* 36–45 (April 1995).
8. ISO/IEC JTC1 CD 11172, *Coding of Moving Pictures and Associated Audio for Digital Storage Media up to 1.5 Mbits/s*, International Organization for Standardization (ISO), 1992.
9. ISO/IEC JTC1 CD 13818, *Generic Coding of Moving Pictures and Associated Audio*, International Organization for Standardization (ISO), 1994.
10. N. S. Jayant and P. Noll, *Digital Coding of Waveforms*, Prentice-Hall, Englewood Cliffs, NJ, 1984.
11. J. S. Lim, *Two-Dimensional Signal and Image Processing*, Prentice-Hall, Englewood Cliffs, NJ, 1990.
12. A. N. Netravali and B. G. Haskell, *Digital Pictures: Representation and Compression*, Plenum Press, New York, 1988.
13. C. C. Todd et al., *AC-3: Flexible Perceptual Coding for Audio Transmission and Storage*, Audio Engineering Society Convention, Amsterdam, Feb. 28, 1994.
14. J. S. Lim, Digital television: Here at last, *Sci. Am.* 78–83 (May 1998).
15. B. G. Haskell, A. Puri, and A. N. Netravali, *Digital Video: An Introduction to MPEG-2*, Digital Multimedia Standard Series, Chapman and Hill, New York, 1997.

16. J. L. Mitchell, W. B. Pennebaker, C. E. Fogg, and D. J. LeGall, *MPEG Video Compression Standard*, Digital Multimedia Standard Series, Chapman and Hill, New York, 1997.

17. T. Sikora, MPEG digital video-coding standards, *IEEE Signal Process. Mag.* **14**(5): 82–100 (Sept. 1997).

HIGH-RATE PUNCTURED CONVOLUTIONAL CODES

David Haccoun
Ecole Polytechnique de Montréal
Montréal, Quebec, Canada

1. INTRODUCTION

For discrete memoryless channels where the noise is essentially white (such as the space and satellite channels), error control coding systems using convolutional encoding and probabilistic decoding are among the most attractive means of approaching the reliability of communication predicted by the Shannon theory; these systems provide substantial coding gains while being readily implementable [1–3].

By far, error control techniques using convolutional codes have been dominated by low-rate $R = 1/v$ codes [1–3,11,16,19]. Optimal low-rate codes providing large coding gains are available in the literature [1,3,11,16,19], and practical implementations of powerful decoders using Viterbi, sequential, iterative, or Turbo decoding schemes exist for high data rates in tens of Mbits/s. However, as the trend for ever-increasing data transmission rates and high error performance continues while conserving bandwidth, the needs arise for good high-rate $R = b/v$ convolutional codes as well as practical encoding and decoding techniques for these codes. Unfortunately, a straightforward application of the usual decoding techniques for rates $R = 1/v$ codes to high-rate $R = b/v$ codes becomes very rapidly impractical as the coding rate increases. Furthermore, a conspicuous lack of good nonsystematic long-memory ($M > 9$) convolutional codes with rates R larger than 2/3 prevails in the literature.

With the advent of high-rate punctured convolutional codes [6], the inherent difficulties of coding and decoding of high-rate codes can be almost entirely circumvented. Decoding of rate $R = b/v$ punctured convolutional codes is hardly more complex than for rate $R = 1/v$ codes; furthermore, puncturing facilitates the implementation of rate-adaptive, variable-rate, and rate-compatible coding-decoding [3,6–9].

In this article, we present high-rate punctured convolutional codes especially suitable for Viterbi and sequential decoding. We provide the weight spectra and upper bounds on the error probability of the best-known punctured codes having memory $2 \leq M \leq 8$ and coding rates $2/3 \leq R \leq 7/8$ together with rate-2/3 and 3/4 long-memory punctured convolutional codes having $9 \leq M \leq 23$. All these codes are derived from the best-known rate-1/2 codes of the same memory lengths. The short memory codes are useful for Viterbi decoding, whereas the long memory codes are provided for archival purposes in addition to being suitable for sequential decoding.

We assume that the reader is familiar with the basic notions of convolutional encoding and the tree, trellis, and state diagram representations of convolutional codes. Without loss of generality, we consider binary convolutional codes of coding rate $R = b/v$, b and v integers with $1 \leq b \leq v$. A rate $R = b/v$ convolutional code produced by a b-input/ v-output encoder may also be denoted as a (v, b) convolutional code. The encoder is specified by a generating matrix $G(D)$ of dimension $b.v$ whose elements are the generator polynomials

$$g_{ij}(D) = \sum_{k=0}^{m_i} g_{ij}^k D^k = g_{ij}^0 + g_{ij}^1 D + \ldots + g_{ij}^{m_i} D^{m_i} \quad (1)$$

where $i = 1, \ldots, b; j = 1, \ldots, b$.

The total memory of the encoder is $M = \sum_i m_i$. Hence, for low rate $R = 1/v$ codes, there are two branches emerging from each node in all the representations of the code, with two branches or paths remerging at each node or state of the encoder. For usual high rate $R = b/v$ codes, 2^b branches enter and leave each encoder state. As a consequence, compared to $R = 1/v$ codes, for $R = b/v$ codes the encoding complexity is multiplied by b, whereas the Viterbi decoding complexity is multiplied by 2^b. By using the notion of puncturing, these difficulties can be entirely circumvented, since regardless of the coding rate $R = (v - 1)/v$, the encoding or decoding procedure is hardly more complex than for the rate $R = 1/v$ codes.

The article is structured as follows. Section 2 introduces the basic concepts of encoding for punctured codes and their perforation patterns. Section 3 presents Viterbi decoding for punctured codes and their error performances. The search for good punctured codes is the objective of Section 4. This section presents extensive lists of the best-known punctured codes of coding rates varying from $R = 2/3$ to $R = 7/8$ together with their weight spectra and bit error probability bounds. Finally, the problem of generating punctured codes equivalent to the best-known usual nonpunctured high-rate codes is presented in Section 5. Again, extensive lists of long memory punctured equivalent to the best usual codes of the same rate are provided.

2. BASIC CONCEPTS OF PUNCTURED CONVOLUTIONAL CODES

2.1. Encoding of Punctured Codes

A punctured convolutional code is a high-rate code obtained by the periodic elimination (i.e., puncturing) of specific code symbols from the output of a low-rate encoder. The resulting high-rate code depends on both the low-rate code, called the *original code* or *mother code*, and the number and specific positions of the punctured symbols. The pattern of punctured symbols is called the perforation pattern of the punctured code, and it is conveniently described in a matrix form called the *perforation matrix*.

Consider constructing a high-rate $R = b/v$ punctured convolutional code from a given original code of any low-rate $R = 1/v_o$. From every $v_o b$ code symbols corresponding to the encoding of b information bits by the original low-rate encoder, a number $S = (v_o b - v)$ symbols are deleted according to some specific perforation pattern. The resulting rate is then $R = b/(v_o b - S)$, which is equal to the desired target rate $R = b/v$. By a judicious choice of the original low-rate code and perforation pattern, any rate code may thus be obtained [6–9,21–24].

For example, Fig. 1 shows the trellis diagram of a rate-1/2, memory $M = 2$ code where every fourth symbol is punctured (indicated by X on every second branch on the diagram). Reading this trellis two branches (i.e., two information bits) at a time and redrawing it as in Fig. 2, we see that it corresponds to a rate-2/3, memory $M = 2$ code. A punctured rate-2/3 code has therefore been obtained from an original rate-1/2 encoder. The procedure can be generalized as described below.

2.2. Perforation Patterns

As shown in Fig. 3, an encoder for rate $R = b/v$ punctured codes may be visualized as consisting of an original low-rate $R = 1/v_o$ convolutional encoder followed by a symbol selector or sampler that deletes specific code symbols according to a given perforation pattern. The perforation pattern may be expressed as a perforation matrix $\mathbf{P}$ having v_o rows and b columns, with only binary elements 0s and 1s, corresponding to the deleting or keeping of the corresponding code symbol delivered by the original encoder, that is, for $i \in \{1, \ldots, v_o\}$, $j \in \{1, 2, \ldots, b\}$ the elements of $\mathbf{P}$ are

$$p_{ij} = \begin{cases} 0 & \text{if symbol } i \text{ of every } j\text{th branch is punctured} \\ 1 & \text{if symbol } i \text{ of every } j\text{th branch is not punctured} \end{cases} \quad (2)$$

Clearly, both the punctured code and its rate can be varied by suitably modifying the elements of the perforation matrix. For example, starting from an original rate-1/2 code, the perforation matrix of the rate-2/3 punctured code of Fig. 1 is given by:

$$\mathbf{P}_1 = \begin{pmatrix} 1 & 1 \\ 1 & 0 \end{pmatrix}$$

whereas a rate-4/5 code could be obtained using the perforation matrix

$$\mathbf{P}_2 = \begin{pmatrix} 1 & 1 & 0 & 0 \\ 1 & 0 & 1 & 1 \end{pmatrix}$$

Likewise, using an original $R = 1/3$ code, perforation matrix $\mathbf{P}_3$ also yields a punctured $R = 2/3$ code

$$\mathbf{P}_3 = \begin{pmatrix} 1 & 0 \\ 1 & 0 \\ 0 & 1 \end{pmatrix}$$

Variable-rate coding may be readily obtained if all punctured rates of interest are obtained from the same low-rate encoder where only the perforation matrices are modified accordingly, as illustrated by $\mathbf{P}_1$ and $\mathbf{P}_2$.

Variable-rate coding may be further specialized by adding the restriction that all the code symbols of the higher rate punctured codes are required by the lower rate punctured codes. This restriction implies minimal modifications of the perforation matrix as the coding rates vary. Punctured codes satisfying this restriction are said to be *rate-compatible*.

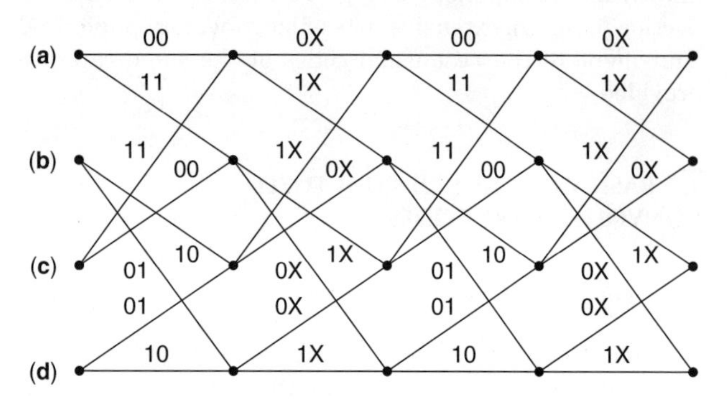

Figure 1. Trellis for $M = 2$, $R = 1/2$ convolutional code.

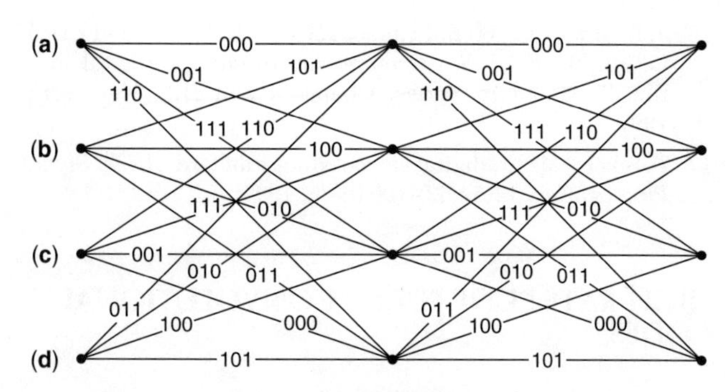

Figure 2. Trellis for punctured $M = 2$, $R = 2/3$ code.

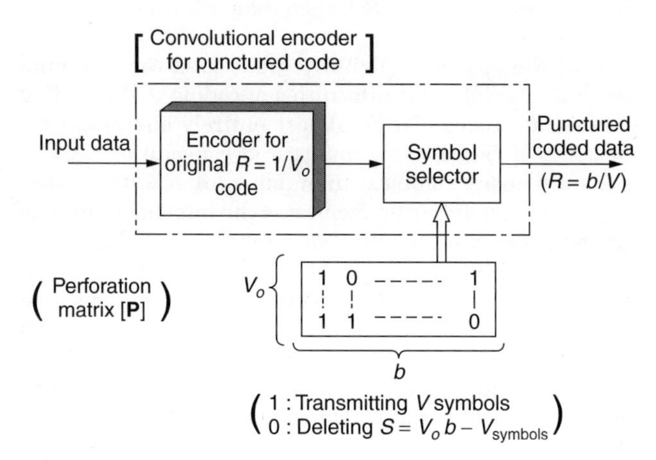

Figure 3. Schematic of an encoder for high rate punctured codes.

For example, using a mother code of rate $R = 1/2$, the sequence of perforation matrices $\mathbf{P}_4$, $\mathbf{P}_5$, and $\mathbf{P}_6$, yields the rate-compatible punctured codes with coding rates $R = 4/5$, 4/6, and 4/7, respectively,

$$\mathbf{P}_4 = \begin{pmatrix} 1 & 1 & 1 & 0 \\ 1 & 0 & 0 & 1 \end{pmatrix}$$

$$\mathbf{P}_5 = \begin{pmatrix} 1 & 1 & 1 & 0 \\ 1 & 1 & 0 & 1 \end{pmatrix}$$

$$\mathbf{P}_6 = \begin{pmatrix} 1 & 1 & 1 & 1 \\ 1 & 1 & 0 & 1 \end{pmatrix}$$

Rate-compatible punctured codes are especially useful in some rate-adaptive ARQ/FEC applications since only the incremental redundancy must be transmitted as the coding rate is decreased. Families of good noncatastrophic short-memory rate-compatible punctured codes with rates varying from 8/9 to 1/4 have been found by Hagenauer [9].

Finally, another class of perforation patterns called *orthogonal perforation patterns* plays an important part in the generation of specific punctured codes [21,22]. An orthogonal perforation pattern is a pattern in which any code symbol that is not punctured on one of the b branches is punctured on all the other $(b - 1)$ branches of the resulting rate-b/v punctured code. With an orthogonal perforation pattern, the perforation matrix has $v_o = v$ rows and b columns, with each row containing only one element 1.

For example the perforation patterns $\mathbf{P}_7$ is orthogonal

$$\mathbf{P}_7 = \begin{pmatrix} 1 & 0 & 0 \\ 1 & 0 & 0 \\ 0 & 1 & 0 \\ 0 & 0 & 1 \end{pmatrix}$$

and generates a $R = 3/4$ punctured code from a $R = 1/4$ original code.

Orthogonal perforation patterns ensure that all the different generators of the original low-rate $1/v_o$ code are used in deriving a desired punctured rate-b/v code. In particular, it can be shown that any punctured code can be obtained by means of an orthogonal perforation pattern. Using this concept, punctured codes strictly identical to the best-known usual rate-2/3 and 3/4 codes have been obtained [22]. These notions of puncturing may be generalized for obtaining any coding rate $R = b/v$ or $R = 1/v$, $v < v_o$.

For example, starting from a low rate code, $R = 1/v_o$, one could obtain a series of low-rate codes $R = 1/v$, $v < v_o$ using degenerate perforation matrices (or perforation vectors) having 1 column and v_o rows. Clearly, then, for a rate $R = 1/v$ code, the perforation vector will have v 1s and $(v_o - v)$ 0s. For example, starting from a mother code of rate $R = 1/6$, the perforation vectors $\mathbf{P}_8$ to $\mathbf{P}_{11}$ yield punctured codes of rates $R = 1/5$, 1/4, 1/3, and 1/2, respectively,

$$\mathbf{P}_8 = \begin{pmatrix} 1 \\ 0 \\ 1 \\ 1 \\ 1 \\ 1 \end{pmatrix} \mathbf{P}_9 = \begin{pmatrix} 1 \\ 0 \\ 0 \\ 1 \\ 1 \\ 1 \end{pmatrix} \mathbf{P}_{10} = \begin{pmatrix} 1 \\ 0 \\ 0 \\ 1 \\ 0 \\ 1 \end{pmatrix} \mathbf{P}_{11} = \begin{pmatrix} 1 \\ 0 \\ 0 \\ 1 \\ 0 \\ 0 \end{pmatrix} \quad (3)$$

Clearly, all the above codes are rate-compatible, and in principle one could further extend the procedure to the usual perforation patterns and obtain punctured high-rate codes $R = b/v$, which would all be issued from the same very low-rate original $R = 1/v_o$ code.

Finally, instead of puncturing some of the code symbols at the output of the original encoder, using a *repetition matrix* one could perform instead a repetition of some code symbols leading to a decrease of the coding rate. Starting with a coding rate $R = \dfrac{b}{bv_o}$, a repetition matrix will have v_o rows and b columns, and repetition of $(n - bv_o)$ code symbols will yield a coding rate $R = b/n$, $n > bv_o$, thus allowing one to obtain practically any coding rate. In a repetition matrix, the code symbol being repeated is identified by an integer equal to the number of repetition. For example, starting from a coding rate $R = 5/10$, the repetition matrices $\mathbf{Q}_1$, $\mathbf{Q}_2$, and $\mathbf{Q}_3$ yield the coding rates $R = 5/11$, 5/13, and 5/16, respectively,

$$\mathbf{Q}_1 = \begin{pmatrix} 1 & 1 & 1 & 1 & 1 \\ 2 & 1 & 1 & 1 & 1 \end{pmatrix}$$

$$\mathbf{Q}_2 = \begin{pmatrix} 1 & 1 & 1 & 1 & 1 \\ 2 & 1 & 2 & 1 & 2 \end{pmatrix}$$

$$\mathbf{Q}_3 = \begin{pmatrix} 3 & 1 & 2 & 1 & 2 \\ 1 & 2 & 1 & 2 & 1 \end{pmatrix}$$

The basic notions of encoding punctured codes having been established, the problem of their decoding by either the Viterbi algorithm or sequential decoding is examined in Section 3.

3. DECODING OF PUNCTURES CODES

3.1. Viterbi Decoding

Given the received sequence from the channel, Viterbi decoding consists essentially of computing for every distinct encoder state the likelihood (also called the metric) that a particular sequence has been transmitted. For rate $R = b/v$ codes, there are 2^b paths merging at every trellis state, and only the path with the largest metric is selected at each state. The process is repeated at each of the 2^M encoder states and for each trellis depth so that, clearly, for a given M as b increases, the complexity of decoding also increases very rapidly.

For punctured high-rate b/v codes, Viterbi decoding is hardly more complex than for the original low-rate $1/v_o$ code from which the punctured codes are derived. The decoding is performed on the trellis of the original low-rate code, where the only modification consists of discarding the metric increments corresponding to the punctured code symbols. Given the perforation pattern of the punctured code, this can be readily performed by inserting dummy data into the positions corresponding to the deleted code symbols. In the decoding process, the dummy data are discarded by assigning them the same metric value (usually zero) regardless of the code symbol, 0 or 1. For either hard- or soft-quantized channels, this procedure in effect inhibits the conventional metric calculation for the punctured symbols. In addition to that metric inhibition, the only coding rate-dependent modification in a variable-rate codec is the truncation path length, which must be increased with the coding rate. All other operations of the decoder remain essentially unchanged [6–9,21].

Therefore, Viterbi codecs for high-rate punctured codes involve none of the complexity required for the straightforward decoding of high-rate b/v codes. They can be implemented by adding relatively simple circuitry to the codecs of the original low-rate $1/v_o$ code. Furthermore, since a given low-rate $1/v_o$ code can give rise to a large number of high-rate punctured codes, the punctured approach leads to very attractive realizations of variable-rate Viterbi decoding. In fact, virtually all hardware implementations of convolutional codecs for high-rate codes use the punctured approach.

The punctured approach to high-rate codes can just as easily be applied to sequential decoding, where the decoding of the received message is performed one tree-branch at a time, without searching the entire tree. Here again, decoding is performed on the tree of the original low-rate code rather than on the tree on the high-rate code where, like for Viterbi decoding, the metric of the punctured symbols is inhibited. Therefore, either the Fano [10] or the Ziganzirov–Jelinek stack decoding algorithm could be used for the decoding of high-rate punctured codes with minimal modifications [13–15,19].

Finally, the same approach as described above can also be applied for iterative Turbo decoding using the MAP algorithm or its variants [4,5].

3.2. Error Performance

For discrete memoryless channels, upper bounds on both the sequence and bit error probabilities of a convolutional code can be obtained. The derivation of the bounds is based on a union bound argument on the transfer function $T(D, B)$ of the code that describes the weight distribution, or weight spectrum, of the incorrect codewords and the number of bit errors on these codewords or paths [1,16]. Except for short memory codes, the entire transfer function of the code is rarely known in closed form, but upper bounds on the error performances can still be calculated using only the first few terms of two series expansions related to the transfer function $T(D, B)$, that is

$$T(D,B)\mid_{B=1} = \sum_{j=d_{\text{free}}}^{\infty} a_j D^j \tag{4}$$

and

$$\frac{dT(D,B)}{dB}\mid_{B=1} = \sum_{j=d_{\text{free}}}^{\infty} c_j D^j \tag{5}$$

In these expressions, d_{free} is the free distance of the code and a_j is the number of incorrect paths or adversaries of Hamming weight $j, j \geq d_{\text{free}}$, that diverge from the correct path and remerge with it sometime later. As for c_j, it is simply the total number of bit errors in all the adversaries having a given Hamming weight j.

Using the weight spectrum, upper bounds on the sequence error probability P_E of a code of rate $R = b/v$ is given by

$$P_E \leq \sum_{j=d_{\text{free}}}^{\infty} a_j P_j \tag{6}$$

where P_j is the pair-wise error probability between two codewords having Hamming distance j. As for the upper bound on the bit error probability P_B, it is easily obtained by weighing each erroneous path by the number of information bits 1 contained in that path.

Since the coding rate is $R = b/v$, P_B is bounded by

$$P_B \leq \frac{1}{b} \sum_{j=d_{\text{free}}}^{\infty} c_j P_j \tag{7}$$

Evaluation of this bound depends on the actual expression of the pair-wise error probability P_j, which in turn depends on the type of modulation and channel parameters used [1,16]. For coherent PSK modulation and unquantized additive white Gaussian noise channels, the error probability between two codewords that differ over j symbols is given by

$$P_j = Q(\sqrt{2jRE_b/No}) \tag{8}$$

where

$$Q(x) = \int_x^{\infty} \frac{1}{\sqrt{2\pi}} \exp\left(-\frac{1}{2}z^2\right) dz \tag{9}$$

and where E_b/No is the energy per bit-to-noise density ratio.

For binary symmetric memoryless channels having transition probability p, $p < 0.5$, the pair-wise error probabilities P_j may be bounded by

$$P_j \leq 2^j \, (p(1-p))^{j/2} \tag{10}$$

However, using the exact expressions for P_j yields tighter bounds for P_B and P_E that closely match performance simulation results, that is

$$P_j = \begin{cases} \displaystyle\sum_{i=(j+1)/(2)}^{j} C_i^j p^i (1-p)^{j-i}, & j \text{ odd} \\ \displaystyle\sum_{i=(j)/(2)+1}^{j} C_i^j p^i (1-p)^{j-i} + \frac{1}{2} C_{j/2}^j \, (p(1-p))^{j/2}, & j \text{ even} \end{cases}$$

where $C_i^j = \dfrac{j!}{i!(j-i)!}, j > i$.

A good evaluation of the bounds on either P_E or P_B requires knowledge of the transfer functions (4) or (5). However, for the vast majority of codes, only the first few terms of either functions are known, and sometimes only the leading coefficients $a_{d_{\text{free}}}$ and $c_{d_{\text{free}}}$ are available. However, for channels with large E_b/No values such as those usually used with high-rate codes, the bounds on P_E and P_B are dominated by the very first terms $a_{d_{\text{free}}}$ and $c_{d_{\text{free}}}$.

Naturally, bounds (6) and (7) are also applicable for punctured codes. Therefore, in deriving the error performances of punctured codes, the free distances and at least the first few terms of the weight spectra of these codes must be known. Partial weight spectra are provided here for both the best known short-memory and for long-memory codes.

4. SEARCH FOR GOOD PUNCTURED CODES

Since punctured coding was originally devised for Viterbi decoding, the criterion of goodness for these codes was the free distance, and hence the best free distance punctured codes that first appeared in the literature were all short-memory codes [6–9]. For sequential decoding, good long-memory punctured codes should have, in addition to a large free distance, a good distance profile.

In searching for good punctured codes of rate b/v and memory M, one is confronted with the problem of determining both an original code of memory M and rate $R = 1/v_o$, and its accompanying perforation pattern. Not unlike the search for usual convolutional codes, the search for punctured codes is often based on intuition and trial and error rather than on a strict mathematical construction [21–23]. An approach that yielded good results is based on the notion that "good codes generate good codes." Consequently, one could choose as the mother code a known good (even the best) code of memory M and rate $R = 1/v_o$, (e.g., R = 1/2, 1/3, 1/4, ...) and exhaustively try out all possible perforation patterns to generate the best possible punctured codes of rates $R = b/v$ and same memory M.

For each perforation pattern, the error probability must be evaluated, which in turn implies determining the corresponding weight spectrum. For a punctured code of rate $R = b/v$ obtained from a mother code of rate $R = 1/v_o$, the number of possible perforation matrices to consider is equal to $C_v^{bv_o}$. For example, using a mother code of rate $R = 1/2$, there are $C_8^{14} = 3003$ different perforation matrices to consider for determining the best punctured codes of rate $R = 7/8$. This number may be substantially reduced if the code must satisfy some specific properties. For example, if the code is systematic, that is the information sequence appears unaltered in the output code sequence, then clearly the first row of the perforation matrix is all composed of 1s, and the puncturing may take place only in the other rows. For example, for $R = 7/8$ systematic punctured codes, the puncturing matrices are of the following form

$$\mathbf{P} = \begin{pmatrix} 1 & 1 & 1 & 1 & 1 & 1 & 1 \\ 1 & 0 & 0 & 0 & 0 & 0 & 0 \end{pmatrix}$$

There are only seven possible alternatives, and in general the mother codes for $R = b/v$ systematic punctured codes must also be systematic.

Naturally, if families of variable-rate punctured codes are desired, then all the required perforation patterns must be applied to the same low-rate original code. Furthermore, if the codes are to be rate-compatible, then the perforation patterns must be selected accordingly [9].

Obviously, puncturing a code reduces its free distance, and hence a punctured code cannot achieve the free distance of its original code. Although the free distance of a code increases as its rate decreases, using original codes with rates $1/v_o$ lower than 1/2 does not always warrant punctured codes with larger free distances since, for a given b and v, the proportion of deleted symbols, $S/v_ob = 1 - (v/v_ob)$, also increases with v_o. Consequently, good results and ease of implementation tend to favor the use of rate-1/2 original codes for generating good punctured codes with coding rates of the form $R = (v - 1)/v$. Results on both short-($M \leq 8$) and long-memory punctured codes ($M \geq 9$) that are derived from the best-known $R = 1/2$ codes are provided in this article [21–23].

Although one could select the punctured code on the basis of its free distance only, a finer method consists of determining the weight spectrum of the punctured code according to Eqs. (4) and (5) and then calculating the error probability bounds Eqs. (6) and (7). The code yielding the best error performance may thus be selected as the best punctured code, provided that it is not catastrophic.

Clearly, then, starting from a known optimal low-rate code, a successful search for good punctured codes hinges on the ability for determining the weight spectrum corresponding to each possible perforation pattern. Although seemingly simple, finding the weight spectrum of a punctured code is a difficult task. This is due to the fact that even if the spectrum of the low-rate original code were available, the spectrum of the punctured code cannot easily be derived from it. One has to go back exploring the tree or trellis of the low-rate original code and apply the perforation pattern to each path of interest. For the well-known short-memory codes, the procedure is at best a rediscovery of their weight spectra, whereas for long-memory codes where often only the free distance is known, it is a novel determination of their spectra. The problem is further compounded by the fact that since puncturing a path reduces its Hamming weight, in obtaining spectral terms up to some target Hamming distance, a larger number of paths must be explored over much longer lengths for a punctured than for a usual non-punctured low-rate code.

For each perforation pattern, the corresponding weight spectrum consists essentially in the triplet $\{D^l, a_l, c_l\}$ from which the bounds on the error probabilities P_E and P_B can be calculated. Of course, the best punctured code will correspond to the perforation pattern yielding the best error probability bounds. In the search for the best punctured codes, the difficulty is compounded by the fact that in determining the weight spectrum of a code of a memory length M up to some Hamming weight value L, the computational effort is exponential in both M and L.

The best punctured codes presented here have been obtained using very efficient trellis-search algorithms for determining the weight spectra of convolutional codes [17,18]. Each algorithm has been designed to be especially efficient within a given range of memory lengths, making it possible to obtain substantial extensions of the known spectra of the best convolutional codes. The spectral terms obtained by these exploration methods have been shown to match exactly the corresponding terms derived from the transfer function of the code [21], thus validating the search procedure.

4.1. Short-Memory Punctured Codes

A number of short-memory lengths ($M \leq 8$) punctured codes of rates $R = (v-1)/v$ varying from 2/3 to 16/17 have been determined first by Cain et al. [6], Yasuda et al. [7] and [8], and then by Hagenauer [9] for rate-compatible codes. In particular, in Ref. [7] all the memory-6 punctured codes of rates varying from 2/3 to 16/17 have been derived from the same original memory-6, rate-1/2, optimal convolutional code due to Odenwalder [16]. A more complete list of rate 2/3 to 13/14 punctured codes has been derived from the optimal rate-1/2 codes with memory M varying from 2 to 8 and compiled by Yasuda et al. [8]. In this list, the perforation matrix for each code is provided, but the weight spectrum is limited to the first term only, that is, the term $c_{d_{\text{free}}}$ corresponding to d_{free}.

Using the given perforation patterns, Haccoun et al. [21–23] have extended Yasuda's results by determining the first 10 spectral terms a_n and c_n for all the codes having memory lengths $2 \leq M \leq 8$ and coding rates 2/3, 3/4, 4/5, 5/6, 6/7, and 7/8. These results are given in Tables 1 to 6, respectively. Each table lists the generators of the original low-rate code, the perforation matrix, the free distance of the resulting punctured code, and the coefficients a_n and c_n of the series expansions of the corresponding weight spectra up to the 10 spectral terms. Beyond 10 spectral coefficients the required computer time becomes rather prohibitive, and hardly any further precision of the bounds is gained by considering more than 10 spectral terms.

To verify the accuracy and validity of these results, the entire transfer functions $T(D, B)$ and $\frac{D(T,B)}{dB}$ have been analytically derived for both rate-2/3 and rate-4/5 memory-2 punctured codes. The transfer functions of the codes have been determined by solving the linear equations describing the transitions between the different states of the encoder. As expected, all the spectral terms obtained by the search algorithms have been shown to match exactly the corresponding terms of the transfer functions [22].

The bit error probability upper bound P_B over both unquantized and hard quantized additive white Gaussian noise channels has been evaluated for all the codes listed in Tables 1 to 6 [21–23], using all the listed weight spectra terms. For all these codes, the error performance improves as the coding rate decreases, indicating well-chosen perforation patterns. A notable exception concerns the memory $M = 3$ code, where the bit error performance turned out to be slightly better at rate 4/5 than at rate 3/4. This anomaly may be explained by an examination of the spectra of these two codes. As shown in Tables 2 and 3, although the free distances of the rates 3/4 and 4/5 codes are $d_f = 4$ and $d_f = 3$, respectively, the number of bit errors c_n on the various spectral terms are far larger for the rate 3/4 code than for the rate 4/5 code. Clearly, it is the coefficients c_n that adversely affect the error performance of the rate 3/4 code. This anomaly illustrates the fact that selecting a code according to only the free distance is good in general but may sometimes be insufficient. Knowledge of further terms of the spectrum will always provide more insight on the code performance.

The theoretical bound on P_B for the original $M = 6$, $R = 1/2$ optimal-free-distance code is plotted in Fig. 4, together with those of all the best punctured codes with rates $R = 2/3$, 3/4, 4/5, 5/6, 6/7, and 7/8 derived from it. For comparison purposes, Fig. 4 also includes the performance curve for the uncoded coherent PSK modulation. It shows that the performance degradation from the original rate-1/2 code is rather gentle as the coding rate increases from 1/2 to 7/8. For example, at $P_B = 10^{-5}$, the coding gains for the punctured rate-2/3 and 3/4 codes are 5.1 dB and 4.6 dB, respectively. These results indicate that these codes are

Table 1. Weight Spectra of Yasuda et al. Punctured Codes with $R = 2/3$, $2 \leq M \leq 8$, [21]

Original Code				Punctured Code		
M	G_1	G_2	d_f	[P]	d_f	$(a_n, n = d_f, d_{f+1}, d_{f+2}, \ldots)$ $[c_n, n = d_f, d_{f+1}, d_{f+2}, \ldots]$
2	5	7	5	1 0 1 1	3	(1, 4, 14, 40, 116, 339, 991, 2897, 8468, 24752) [1, 10, 54, 226, 856, 3072, 10647, 35998, 119478, 390904]
3	15	17	6	1 1 1 0	4	(3, 11, 35, 114, 381, 1276, 4257, 14208, 47413, 158245) [10, 43, 200, 826, 3336, 13032, 49836, 187480, 696290, 2559521]
4	23	35	7	1 1 1 0	4	(1, 0, 27, 0, 345, 0, 4528, 0, 59435, 0) [1, 0, 124, 0, 2721, 0, 50738, 0, 862127, 0]
5	53	75	8	1 0 1 1	6	(19, 0, 220, 0, 3089, 0, 42790, 0, 588022, 0) [96, 0, 1904, 0, 35936, 0, 638393, 0, 10657411]
6	133	171	10	1 1 1 0	6	(1, 16, 48, 158, 642, 2435, 9174, 34705, 131585, 499608) [3, 70, 285, 1276, 6160, 27128, 117019, 498860, 2103891, 8784123]
7	247	371	10	1 0 1 1	7	(9, 35, 104, 372, 1552, 5905, 22148, 85189, 323823, 1232139) [47, 237, 835, 3637, 17770, 76162, 322120, 1374323, 5730015, 23763275]
8	561	753	12	1 1 1 0	7	(3, 9, 50, 190, 641, 2507, 9745, 37121, 142226, 545002) [11, 46, 324, 1594, 6425, 29069, 127923, 544616, 2313272, 9721227]

Table 2. Weight Spectra of Yasuda et al. Punctured Codes with $R = 3/4$, $2 \leq M \leq 8$, [21]

Original Code				Punctured Code		
M	G_1	G_2	d_f	[P]	d_f	$(a_n, n = d_f, d_{f+1}, d_{f+2}, \ldots)$ $[c_n, n = d_f, d_{f+1}, d_{f+2}, \ldots]$
2	5	7	5	1 0 1 1 1 0	3	(6, 23, 80, 290, 1050, 3804, 13782, 49930, 180890, 655342) [15, 104, 540, 2557, 11441, 49340, 207335, 854699, 3471621, 13936381]
3	15	17	6	1 1 0 1 0 1	4	(29, 0, 532, 0, 10059, 0, 190112, 0, 3593147, 0) [124, 0, 4504, 0, 126049, 0, 3156062, 0, 74273624, 0]
4	23	35	7	1 0 1 1 1 0	3	(1, 2, 23, 124, 576, 2852, 14192, 70301, 348427, 1726620) [1, 7, 125, 936, 5915, 36608, 216972, 1250139, 7064198, 39308779]
5	53	75	8	1 0 0 1 1 1	4	(1, 15, 65, 321, 1661, 8396, 42626, 216131, 1095495, 5557252) [3, 85, 490, 3198, 20557, 123384, 725389, 4184444, 23776067, 133597207]
6	133	171	10	1 1 0 1 0 1	5	(8, 31, 160, 892, 4512, 23307, 121077, 625059, 3234886, 16753077) [42, 201, 1492, 10469, 62935, 379644, 2253373, 13073811, 75152755, 428005675]
7	247	371	10	1 1 0 1 0 1	6	(36, 0, 990, 0, 26668, 0, 726863, 0, 19778653, 0) [239, 0, 11165, 0, 422030, 0, 14812557, 0, 493081189, 0]
8	561	753	12	1 1 1 1 0 1	6	(10, 77, 303, 1599, 8565, 44820, 236294, 1236990, 6488527, 34056195) [52, 659, 3265, 21442, 133697, 805582, 4812492, 28107867, 162840763, 935232173]

Table 3. Weight Spectra of Yasuda et al. Punctured Codes with $R = 4/5$, $2 \leq M \leq 8$, [21]

Original Code				Punctured Code		
M	G_1	G_2	d_f	[P]	d_f	$(a_n, n = d_f, d_{f+1}, d_{f+2}, \ldots)$ $[c_n, n = d_f, d_{f+1}, d_{f+2}, \ldots]$
2	5	7	5	1 0 1 1 1 1 0 0	2	(1, 12, 53, 238, 1091, 4947, 22459, 102030, 463451) [1, 36, 309, 2060, 12320, 69343, 375784, 1983150, 10262827]
3	15	17	6	1 0 1 1 1 1 0 0	3	(5, 36, 200, 1070, 5919, 32721, 180476, 995885, 5495386, 30323667) [14, 194, 1579, 11313, 77947, 514705, 3305113, 20808587, 129003699, 790098445]
4	23	35	7	1 0 1 0 1 1 0 1	3	(3, 16, 103, 675, 3969, 24328, 147313, 897523, 5447618, 33133398) [11, 78, 753, 6901, 51737, 386465, 2746036, 19259760, 132078031, 896198879]
5	53	75	8	1 0 0 0 1 1 1 1	4	(7, 54, 307, 2005, 12970, 83276, 534556, 3431703, 22040110) [40, 381, 3251, 27123, 213451, 1621873, 12011339, 87380826, 627189942]
6	133	171	10	1 1 1 1 1 0 0 0	4	(3, 24, 172, 1158, 7409, 48729, 319861, 2097971, 13765538, 90315667) [12, 188, 1732, 15256, 121372, 945645, 7171532, 53399130, 392137968, 2846810288]
7	247	371	10	1 0 1 0 1 1 0 1	5	(20, 115, 694, 4816, 32027, 210909, 1392866, 9223171, 61013236) [168, 1232, 9120, 78715, 626483, 4758850, 35623239, 263865149, 1930228800]
8	561	753	12	1 1 0 1 1 0 1 0	5	(7, 49, 351, 2259, 14749, 99602, 663936, 4431049, 29536078, 197041141) [31, 469, 4205, 34011, 268650, 2113955, 16118309, 121208809, 898282660, 2301585211]

very good indeed, even though their free distances, which are equal to 6 and 5, respectively, are slightly smaller than the free distances of the best-known usual rate-2/3 and 3/4 codes, which are equal to 7 and 6, respectively [11].

The error performance of these codes has been verified using an actual punctured Viterbi codec [7,8], and independently, using computer simulation [21–23]. Both evaluations have been performed using 8-level soft decision Viterbi decoding with truncation path lengths equal to 50, 56, 96, and 240 bits for the coding rates 2/3, 3/4, 7/8, and 15/16, respectively. Both hardware and software evaluations have yielded identical error performances that closely match the theoretical upper bounds.

Even for the rate 15/16, the coding gain of the $M = 6$ code has been shown to reach a substantial 3 dB at $P_B = 10^{-6}$ [3]. The fact that such a coding gain can be achieved with only a 7% redundancy and a Viterbi decoder that is hardly more complex than for a rate-1/2 code makes the punctured coding technique very attractive indeed for short-memory codes. For longer codes and larger coding gains, Viterbi decoding becomes impractical and other decoding techniques such as sequential decoding should be considered instead. Long-memory length punctured codes and their error performance are presented next.

4.2. Long-Memory Punctured Codes

Following essentially the same approach as for the short-memory codes, one could choose a known optimal long-memory code of rate 1/2 and exhaustively try out all possible perforation patterns to generate all punctured codes of rate $R = b/v$. The selection of the best punctured code is again based on its bit error performance, which is calculated from the series expansion of its transfer function. Here, one of the difficulties is that for the original

Table 4. Weight Spectra of Yasuda et al. Punctured Codes with $R = 5/6$, $2 \le M \le 8$, [21]

Original Code				Punctured Code		
M	G_1	G_2	d_f	[P]	d_f	$(a_n, n = d_f, d_{f+1}, d_{f+2}, \ldots)$ $[c_n, n = d_f, d_{f+1}, d_{f+2}, \ldots]$
2	5	7	5	1 0 1 1 1 1 1 0 0 0	2	(2, 26, 129, 633, 3316, 17194, 88800, 459295, 2375897, 12288610) [2, 111, 974, 6857, 45555, 288020, 1758617, 10487425, 61445892, 355061333]
3	15	17	6	1 0 1 0 0 1 1 0 1 1	3	(15, 96, 601, 3918, 25391, 164481, 1065835, 6906182, 44749517) [63, 697, 6367, 53574, 426471, 3277878, 24573195, 180823448, 1311630186]
4	23	35	7	1 0 1 1 1 1 1 0 0 0	3	(5, 37, 309, 2282, 16614, 122308, 900991, 6634698, 48853474) [20, 265, 3248, 32328, 297825, 2638257, 22710170, 191432589, 1587788458]
5	53	75	8	1 0 0 0 0 1 1 1 1 1	4	(19, 171, 1251, 9573, 75167, 585675, 4558463, 35513472) [100, 1592, 17441, 166331, 1591841, 14627480, 131090525, 1155743839]
6	133	171	10	1 1 0 1 0 1 0 1 0 1	4	(14, 69, 654, 4996, 39699, 315371, 2507890, 19921920, 158275483) [92, 528, 8694, 79453, 792114, 7375573, 67884974, 610875423, 1132308080]
7	247	371	10	1 1 1 0 0 1 0 0 1 1	4	(2, 51, 415, 3044, 25530, 200878, 1628427, 12995292, 104837990) [7, 426, 5244, 49920, 514857, 4779338, 44929071, 406470311, 3672580016]
8	561	753	12	1 0 1 1 0 1 1 0 0 1	5	(19, 187, 1499, 11809, 95407, 775775, 6281882, 50851245) [168, 2469, 25174, 242850, 2320429, 21768364, 199755735, 1807353406]

Table 5. Weight Spectra of Yasuda et al. Punctured Codes with $R = 6/7$, $2 \le M \le 8$, [21]

Original Code				Punctured Code		
M	G_1	G_2	d_f	[P]	d_f	$(a_n, n = d_f, d_{f+1}, d_{f+2}, \ldots)$ $[c_n, n = d_f, d_{f+1}, d_{f+2}, \ldots]$
2	5	7	5	1 0 1 1 1 1 1 1 0 0 0 0	2	(4, 39, 221, 1330, 8190, 49754, 302405, 1840129, 11194714, 68101647) [5, 186, 1942, 16642, 131415, 981578, 7076932, 49784878, 343825123, 2340813323]
3	15	17	6	1 0 0 0 1 1 1 1 1 1 0 0	2	(1, 25, 188, 1416, 10757, 81593, 619023, 4697330, 35643844) [2, 134, 1696, 18284, 179989, 1676667, 15082912, 132368246, 1140378555]
4	23	35	7	1 0 1 0 1 0 1 1 0 1 0 1	3	(14, 100, 828, 7198, 60847, 513573, 4344769, 36751720) [69, 779, 9770, 113537, 1203746, 12217198, 120704682, 1167799637]
5	53	75	8	1 1 0 1 1 0 1 0 1 0 0 1	3	(5, 55, 517, 4523, 40476, 362074, 3232848, 28872572) [25, 475, 6302, 73704, 823440, 8816634, 91722717, 935227325]
6	133	171	10	1 1 1 0 1 0 1 0 0 1 0 1	3	(1, 20, 223, 1961, 18093, 169175, 1576108, 14656816, 136394365) [5, 169, 2725, 32233, 370861, 4169788, 45417406, 483171499, 768072194]
7	247	371	10	1 0 1 0 0 1 1 1 0 1 1 0	4	(11, 155, 1399, 13018, 122560, 1154067, 10875198, 102494819, 965649475) [85, 1979, 24038, 282998, 3224456, 35514447, 383469825, 4075982541, 4092715598]
8	561	753	12	1 1 0 1 1 0 1 0 1 0 0 1	4	(2, 48, 427, 4153, 39645, 377500, 3600650, 34334182) [9, 447, 5954, 76660, 912140, 10399543, 115459173, 1256388707]

low-rate and long-memory codes of interest, only very partial knowledge of their weight spectra is available. In fact, beyond memory length $M = 15$, very often only the free distances of these codes are available in the literature [12].

Results of computer search for the best rate-2/3 and 3/4 punctured codes with memory lengths ranging from 9 to 23 that are derived from the best-known rate-1/2 are provided in Refs. 21–23, where for each code the first few terms of the weight spectrum have been obtained for each possible distinct perforation pattern. The final selection of the best punctured codes was based on the evaluation of the upper bound on the bit error probability. Naturally, the codes obtained with this approach are suitable for variable-rate codecs using an appropriate decoding technique such as sequential decoding. Only these two rates have been considered because a definite comparison of the resulting punctured codes with the best-known nonsystematic high-rate codes is limited to the rate-2/3 and 3/4 codes, since with very few exceptions, optimal long memory codes suitable for sequential decoding are known only for rates 2/3 and 3/4.

Tables 7 and 8 list the characteristics of the best punctured codes of rate 2/3 and 3/4, respectively, with memory lengths M varying from 9 to 23, that are derived from the best nonsystematic rate-1/2 codes [21]. From $M = 9$ to $M = 13$ the original codes are the maximal free distance codes discovered by Larsen [19], whereas for $14 \le M \le 23$ the original codes are those of Johannesson and Paaske [12]. In both of these tables, for each memory length the generators of the original code and its perforation matrix are given, together with the free distances of both the original and resulting punctured code. Just as with short-memory codes, the first few terms

Table 6. Weight Spectra of Yasuda et al. Punctured Codes with $R = 7/8$, $2 \le M \le 8$, [21]

Original Code				Punctured Code		
M	G_1	G_2	d_f	[P]	d_f	$(a_n, n = d_f, d_{f+1}, d_{f+2}, \ldots)$ $[c_n, n = d_f, d_{f+1}, d_{f+2}, \ldots]$
2	5	7	5	1 0 1 1 1 1 1 1 1 0 0 0 0 0	2	(6, 66, 408, 2636, 17844, 119144, 793483, 5293846, 35318216) [8, 393, 4248, 38142, 325739, 2647528, 20794494, 159653495, 1204812440]
3	15	17	6	1 0 0 0 0 1 0 1 1 1 1 1 0 1	2	(2, 38, 346, 2772, 23958, 201842, 1717289, 14547758, 123478503) [4, 219, 3456, 38973, 437072, 4492304, 45303102, 442940668, 4265246076]
4	23	35	7	1 0 1 0 0 1 1 1 1 0 1 1 0 0	3	(13, 145, 1471, 14473, 143110, 1416407, 14019214, 138760394) [49, 1414, 21358, 284324, 3544716, 42278392, 489726840, 1257797047]
5	53	75	8	1 0 1 1 1 0 1 1 1 0 0 0 1 0	3	(9, 122, 1195, 12139, 123889, 1259682, 12834712, 130730637, 1331513258) [60, 1360, 18971, 252751, 3165885, 38226720, 450898174, 923001734, 3683554219]
6	133	171	10	1 1 1 1 0 1 0 1 0 0 0 1 0 1	3	(2, 46, 499, 5291, 56179, 599557, 6387194, 68117821) [9, 500, 7437, 105707, 1402743, 17909268, 222292299, 2706822556]
7	247	371	10	1 0 1 0 1 0 0 1 1 0 1 0 1 1	4	(26, 264, 2732, 30389, 328927, 3571607, 38799203) [258, 3652, 52824, 746564, 9825110, 125472545, 1567656165]
8	561	753	12	1 1 0 1 0 1 1 1 0 1 0 1 0 0	4	(6, 132, 1289, 13986, 154839, 1694634, 18532566) [70, 1842, 24096, 337514, 4548454, 58634237, 738611595]

a_n and c_n, $n = d_{\text{free}}, d_{\text{free}} + 1, d_{\text{free}} + 2, \ldots$ of the weight spectra are also given for each punctured codes. In deriving these spectral coefficients, the tree exploration of the original code had to be performed over a considerable length [21].

In the search for the best punctured codes, the perforation patterns were chosen as to yield both a maximal free distance and a good distance profile. Although all perforation patterns were exhaustively examined, the search was somewhat reduced by exploiting

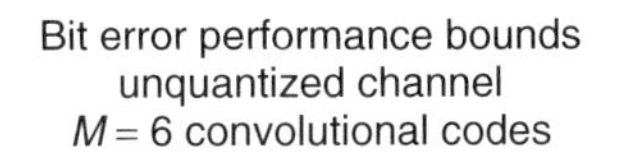

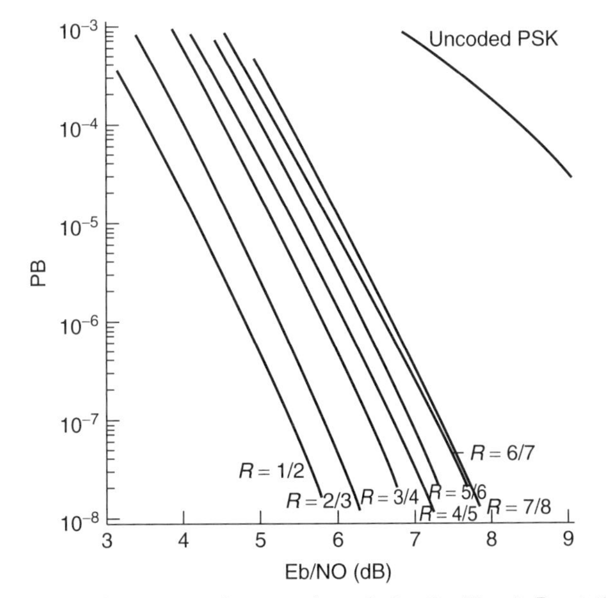

Figure 4. Bit error performance bounds for the $M = 6$, $R = 1/2$ original and punctured rates 2/3, 3/4, 4/5, 5/6, 6/7, and 7/8 codes, [21].

equivalences of the perforation patterns under cyclical shifts of their rows [21–23]. Among all the codes that were found, Tables 7 and 8 list only those having the smallest number of bit errors $c_{d_{\text{free}}}$ at the free distance d_{free}, and obviously all the codes listed are noncatastrophic.

Table 7 lists two $M = 19$, $R = 2/3$ punctured codes derived from two distinct good $M = 19$, $R = 1/2$ original codes. The free distances of these two punctured codes are equal to 12 and 13, respectively, but the coefficients c_n are larger for the $d_{\text{free}} = 13$ code than they are for the $d_{\text{free}} = 12$ code. Consequently, as confirmed by the calculation of the bit error bound P_B, the code with $d_{\text{free}} = 13$ turned out to be slightly worse by approximately 0.35 dB than the code with $d_{\text{free}} = 12$. This anomaly again confirms the need to determine at least the first few terms of the weight spectra when searching for the best punctured codes.

Figure 5 plots the free distances of the original rate 1/2 codes and the punctured rate-2/3 and 3/4 codes as a function of the memory length M, $2 \le M \le 22$. Except for the two anomalies with the $M = 19$ mentioned above, the behavior of the free distances is as expected: the free distance of the punctured codes of a given rate is generally nondecreasing with the memory length, and for a given memory length the free distance decreases with increasing coding rates.

When the punctured codes of rate b/v are determined from the best original low-rate $1/v$ code, an upper bound on the free distance of the punctured code can be derived [21,22]. This derivation, which is based on an analysis of the effect of the different perforation patterns on the spectrum of the original code, yields the bound

$$d_{\text{free}(p)} \le (1/b)\, d_{\text{free}(0)} \tag{12}$$

where $d_{\text{free}(p)}$ and $d_{\text{free}(o)}$ are the free distances of the punctured and best codes of rate $1/v$, respectively. This bound is also plotted on Fig. 5 for the rate-2/3 and 3/4 codes with memory $M \le 13$.

Table 7. Best Rate-2/3, $9 \le M \le 23$ Punctured Codes with Their Weight Spectra, Perforation Matrix, and Original $R = 1/2$ Codes, [21]

Original Code				Punctured Code		
M	G_1	G_2	d_f	$[P]$	d_f	$(a_n, n = d_f, d_{f+1}, d_{f+2}, \ldots)$ $[c_n, n = d_f, d_{f+1}, d_{f+2}, \ldots]$
9	1167	1545	12	1 1 1 0	7	(1, 10, 29, 94, 415, 1589, 5956) [3, 70, 207, 836, 4411, 19580, 82154]
10	2335	3661	14	1 0 1 1	8	(1, 21, 65, 226, 907, 3397, 13223) [8, 165, 560, 2321, 10932, 46921, 204372]
11	4335	5723	15	1 1 1 0	9	(10, 38, 137, 518, 1990, 7495, 28907) [86, 326, 1379, 6350, 27194, 114590, 492275]
12	10533	17661	16	1 1 1 0	9	(4, 8, 45, 193, 604, 2383, 9412) [25, 65, 413, 1991, 6925, 31304, 139555]
13	21675	27123	16	1 1 1 0	10	(5, 30, 104, 380, 1486) [46, 268, 1066, 4344, 19992]
14	55367	63121	18	1 1 1 0	10	(2, 6, 37, 153, 582) [13, 62, 334, 1606, 7321]
15	111653	145665	19	1 1 1 0	10	(3, 0, 46, 0, 683, 0) [28, 0, 397, 0, 7735, 0]
16	347241	246277	20	1 1 1 0	12	(8, 45, 145, 567, 2182) [68, 495, 1569, 7112, 31556]
17	506477	673711	20	1 0 1 1	12	(2, 24, 79, 320, 1251) [11, 253, 889, 3978, 18056]
18	1352755	1771563	21	1 1 1 0	12	(2, 11, 27, 137) [18, 105, 276, 1679]
19	2451321	3546713	22	1 1 1 0	12	(1, 3, 14, 71) [9, 21, 139, 715]
19	2142513	3276177	22	1 1 1 0	13	(10, 34, 101, 417, 1539) [99, 425, 1425, 6158, 25037]
20	6567413	5322305	22	1 1 1 0	12	(1, 0, 18, 0, 333, 0) [8, 0, 210, 0, 4290, 0]
21	15724153	12076311	24	1 1 1 0	13	(1, 1, 20, 62) [11, 5, 231, 736]
22	33455341	24247063	24	1 0 1 1	14	(1, 12, 67) [17, 163, 927]
23	55076157	75501351	25	1 0 1 1	15	(2, 14, 71) [39, 170, 852]

The upper bounds on the bit error probability over unquantized white Gaussian channels have been evaluated for all the punctured codes of rates $R = 2/3$ and 3/4 and are shown for even values of memory lengths in Figs. 6 and 7. These bounds indicate a normal behavior for all the punctured codes listed in Tables 7 and 8. All the bit error performances improve as the coding rate decreases and/or as the memory increases, with approximately 0.4 dB improvement for each unit increase of the memory length. At $P_B = 10^{-5}$, the $M = 22$, $R = 2/3$, and $R = 3/4$ punctured codes can yield substantial coding gains of 8.3 dB and 7.7 dB, respectively.

The selection of the best punctured codes listed in Tables 7 and 8 has been based on both the maximal free distance and the calculated bit error probability bound. However, the choice of the best punctured code is not always clear-cut as different puncturing patterns may yield only marginally different error performances. In some cases, the performance curves may even be undistinguishable, leading to several "best" punctured codes having the same coding rate and memory length. However, since these long codes are typically for sequential decoding applications, the final selection of the code should also be based on the distance profile and computational performance. Short of analyzing the computational behavior, when in doubt, the codes finally selected and listed in the tables had the fastest-growing column distance function.

In the search for punctured codes, the above approach will produce good but not necessarily optimal codes since the original low-rate code is imposed at the outset. A measure of the discrepancy between optimal and punctured codes of the same rate and memory length is provided in Fig. 8, which shows the bit error performance bound for both the best punctured and maximal free distance codes of memory length 9 and rates 2/3 and 3/4. These bounds have been computed using only the term at d_{free} for the maximum free distance (MFD) codes and using both the terms at d_{free} and $d_{\text{free}} + 1$ for the punctured codes. Based on these terms only, the two MFD codes appear to be only slightly better than the punctured codes. Therefore, it may be concluded that, although not optimal, the error performances of the rate-2/3 and 3/4 punctured codes of memory 9 closely match those of the MFD codes of the same rates and memory lengths. The same general

Table 8. Best Rate-3/4, $9 \leq M \leq 23$ Punctured Codes with Their Weight Spectra, Perforation Matrix, and Original $R = 1/2$ Codes, [21]

Original Code				Punctured Code		
M	G_1	G_2	d_f	$[P]$	d_f	$(a_n, n = d_f, d_{f+1}, d_{f+2}, \ldots)$ $[c_n, n = d_f, d_{f+1}, d_{f+2}, \ldots]$
9	1167	1545	12	1 0 0 1 1 1	6	(4, 31, 151, 774, 3967, 21140) [38, 270, 1640, 10554, 63601, 387227]
10	2335	3661	14	1 0 1 1 1 0	6	(2, 7, 59, 338, 1646) [9, 40, 517, 3731, 22869]
11	4335	5723	15	1 0 0 1 1 1	7	(12, 55, 236, 1271, 6853) [107, 628, 3365, 20655, 126960]
12	10533	17661	16	1 1 0 1 0 1	7	(4, 18, 90, 476, 2466) [34, 182, 965, 6294, 38461]
13	21675	27123	16	1 1 0 1 0 1	7	(1, 11, 41, 202, 1334) [12, 109, 387, 2711, 20403]
14	55367	63121	18	1 0 1 1 1 0	8	(3, 19, 95, 529) [28, 159, 1186, 7461]
15	111653	145665	19	1 0 0 1 1 1	8	(1, 14, 47, 259) [9, 143, 512, 3571]
16	347241	246277	20	1 1 0 1 0 1	8	(1, 5, 28, 167) [5, 49, 311, 2266]
17	506477	673711	20	1 0 0 1 1 1	9	(1, 13, 101, 427) [5, 142, 1375, 6842]
18	1352755	1771563	21	1 1 1 1 0 0	10	(6, 51, 217, 1014) [104, 735, 3368, 18736]
19	2451321	3546713	22	1 1 0 1 0 1	10	(4, 18, 81, 429) [40, 240, 1219, 6934]
19	2142513	3276177	22	1 1 0 1 0 1	10	(4, 18, 89, 461, 2529) [48, 202, 1248, 7445, 46981]
20	6567413	5322305	22	1 1 1 1 0 0	10	(4, 19, 82, 436, 2443) [40, 249, 1510, 8120, 53164]
21	15724153	12076311	24	1 1 1 1 0 0	11	(8, 19, 120) [143, 266, 2038]
22	33455341	24247063	24	1 0 0 1 1 1	12	(7, 68, 298) [79, 1275, 5279]
23	55076157	75501351	25	1 0 0 1 1 1	13	(21, 141, 707) [292, 2340, 13196]

conclusions may be made about the slight suboptimality of the other punctured codes with different memory lengths [21–23]. However, as mentioned earlier, the small error performance degradation of the punctured codes is compensated for by a far simpler practical implementation.

Given an optimal usual high-rate code of rate $R = b/v$ and memory length M, one could attempt to determine the low-rate $1/v$ code that, after puncturing, will yield a punctured code that is equivalent to that optimal code. This approach, which is the converse of the usual code searching method, can be used to find the punctured code equivalent to any known usual high-rate code. Based on this approach and using the notion of orthogonal perforation patterns, a systematic construction technique has been developed by Begin and Haccoun [22]. Using this technique, punctured codes strictly equivalent to the best-known nonsystematic rate-2/3 codes with memory lengths up to $M = 24$ have been found. Likewise, punctured codes equivalent to the best-known rate-3/4 codes with memory lengths up to $M = 9$ have also been determined [22]. Therefore, optimal high-rate codes may be obtained as punctured codes, but it should

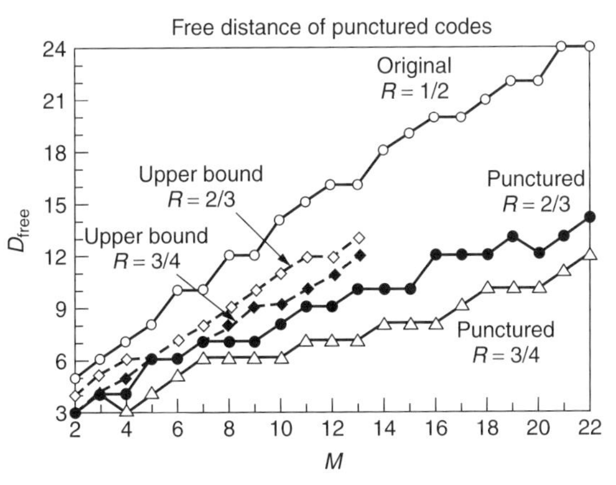

Figure 5. Free distances of original rate-1/2 codes and punctured $R = 2/3$ and 3/4 codes derived from them as a function of M, $2 \leq M \leq 22$, and upper bounds, [21].

be pointed out that the punctured codes generated by this latter approach are not suitable for variable-rate applications since each punctured code has its own distinct low-rate original code and orthogonal perforation pattern.

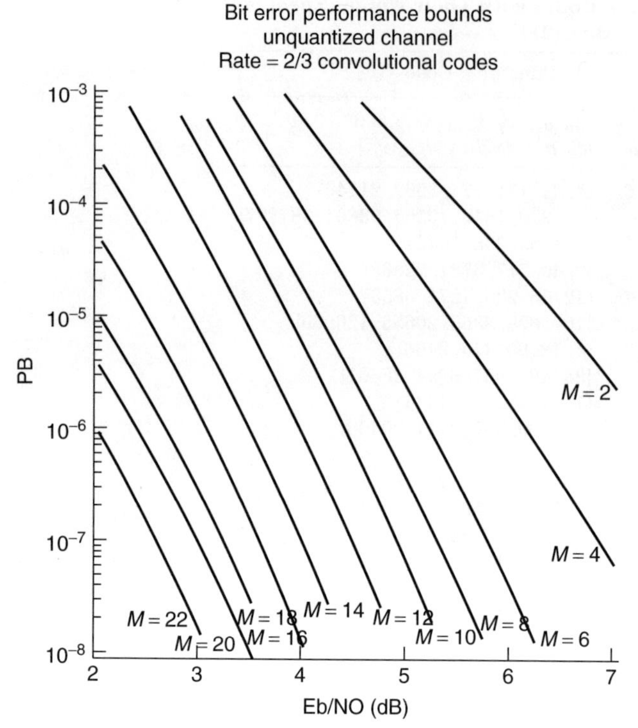

Figure 6. Upper bounds on the bit error probability over unquantized white Gaussian noise channels for $R = 2/3$ punctured codes with $2 \leq M \leq 22$, M even, [21].

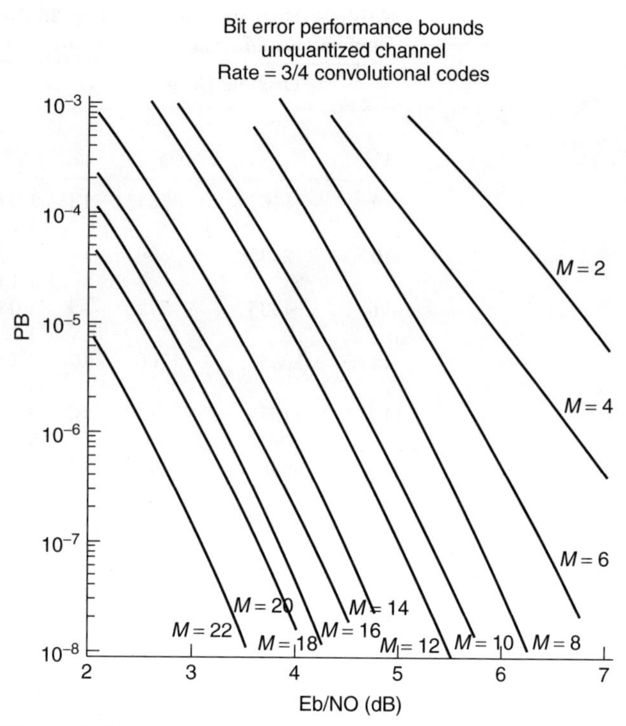

Figure 7. Upper bounds on the bit error probability over unquantized white Gaussian noise channels for $R = 3/4$ punctured codes with $2 \leq M \leq 22$, M even, [21].

5. PUNCTURED CODES EQUIVALENT TO THE BEST USUAL HIGH-RATE CODES

5.1. Nonsystematic Codes

Given a best-known high-rate usual nonsystematic code, Haccoun and Begin have devised a construction method for generating a low-rate original code that, when punctured by an orthogonal perforation pattern, yields a punctured code having both the same rate and same weight spectrum as the best-known usual code [22]. These codes are listed in Tables 9 to 12. Tables 9 to 11 use the same orthogonal perforation matrix $\mathbf{P}_{01}$ whereas Table 12 uses the orthogonal perforation matrix $\mathbf{P}_{02}$:

$$\mathbf{P}_{01} = \begin{pmatrix} 1 & 0 \\ 0 & 1 \\ 0 & 1 \end{pmatrix} \quad \mathbf{P}_{02} = \begin{pmatrix} 1 & 0 & 0 \\ 0 & 1 & 0 \\ 0 & 0 & 1 \\ 0 & 0 & 1 \end{pmatrix} \tag{13}$$

The $R = 2/3$ codes are obtained by puncturing $R = 1/3$ original codes, and the $R = 3/4$ punctured codes are obtained from $R = 1/4$ original codes. Tables 9 to 12 provide the memory lengths and generators of the original low-rate codes and the resulting punctured codes equivalent to their respective best usual codes.

We can observe from the tables that, for most of the cases, the memory of the required original code is larger than that of the resultant equivalent punctured code. The difference in memory length is quite small, usually one or two units, and this difference appears to be independent of the actual memory length of the code. Therefore, its relative importance decreases as the memory length of the code increases. For large memory codes, this memory

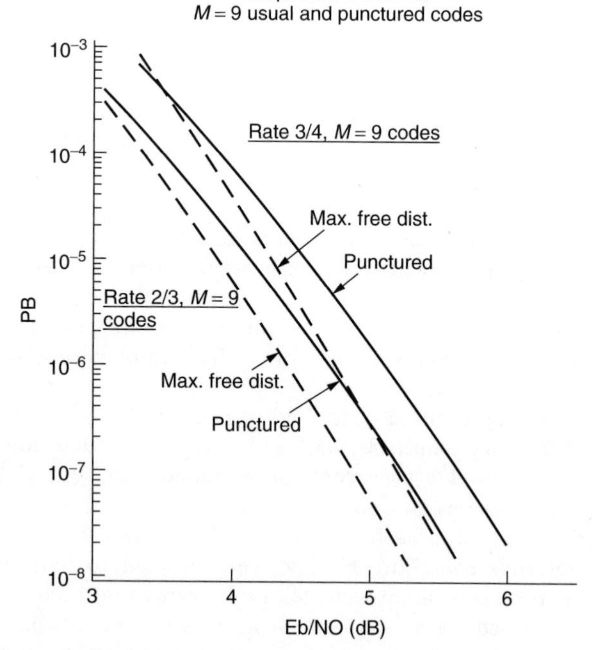

Figure 8. Bit error performance bounds for maximal free distance (MFD) codes and punctured codes of rates 2/3 and 3/4 with memory $M = 9$, [21].

increment is of no consequence whatsoever since the codes certainly would be decoded by sequential decoding methods.

As mentioned earlier, the punctured codes listed in Tables 9 to 12 are not suitable for variable-rate or rate-compatible decoding applications. However, they do provide a practical method for coding and decoding the best-known long memory rate 2/3 and 3/4 codes, especially using sequential decoding. Decoding these codes by the normal (nonpunctured) approach is cumbersome because of the large number (2^b) of nodes stemming from each node in the tree of the high-rate $R = b/v$ code. With the punctured approach, the decoding proceeds on the low-rate tree structure, so that the number of nodes stemming from a single node is always 2, regardless of the actual coding rate.

5.2. Systematic Codes

Using a similar procedure, all high-rate systematic codes, whether known or yet to be discovered, may be obtained by puncturing some low-rate original (systematic) code. Furthermore, since for any $R = b/(b+1)$ target code the branches of the high-rate code consist of b information digits and a single parity symbol, the original code need only be a $R = 1/2$ systematic code: the b information digits are directly issued from the information digits of the

Table 9. Original Codes that Yield the $R = 2/3$ Codes of Johannesson and Paaske, Perforation Pattern P_{01}, [22]

Original $R = 1/3$				Punctured $R = 2/3$			
M_0	G_1	G_2	G_3	M	G_{11} G_{12}	G_{21} G_{22}	G_{31} G_{32}
4	26	22	35	3	6 1	2 4	4 7
5	54	47	67	4	6 1	3 5	7 5
6	172	137	152	5	14 07	06 17	16 10
7	314	271	317	6	12 05	05 16	13 13
8	424	455	747	7	26 00	14 23	32 33
9	1634	1233	1431	8	32 13	05 33	25 22
10	3162	2553	3612	9	54 25	16 71	66 60
11	6732	4617	7153	10	53 36	23 53	51 67
12	17444	11051	17457	11	162 064	054 101	156 163

Table 10. Original Codes that Yield the $R = 2/3$ Codes of Johannesson and Paaske, Perforation Pattern P_{01}, [22]

Original $R = 1/3$				Punctured $R = 2/3$			
M_0	G_1	G_2	G_3	M	G_{11} G_{12}	G_{21} G_{22}	G_{31} G_{32}
16	377052	221320	314321	12	740 367	260 414	520 515
16	274100	233221	331745	13	710 140	260 545	670 533
16	163376	101657	153006	14	337 127	023 237	342 221
16	370414	203175	321523	15	722 302	054 457	642 435
18	1277142	1144571	1526370	16	1750 0165	0514 1235	1734 1054
18	1066424	1373146	1471265	17	1266 0140	0652 1752	1270 1307
19	2667576	2153625	3502436	18	1567 0337	0367 1230	1066 1603
20	4600614	4773271	6275153	19	2422 0412	1674 2745	2356 2711
20	12400344	13365473	15646505	20	3414 0005	1625 3367	3673 2440
22	24613606	22226172	35045621	21	6562 0431	2316 4454	4160 7225
24	117356622	126100341	151373474	22	13764 03251	02430 16011	14654 11766
24	106172264	130463065	141102467	23	12346 01314	05250 14247	10412 11067

Table 11. Original Codes that Yield the $R = 2/3$ Codes of Johannesson and Paaske, Perforation Pattern P_{01}, [22]

Original $R = 1/3$				Punctured $R = 2/3$			
M_0	G_1	G_2	G_3	M	G_{11} G_{12}	G_{21} G_{22}	G_{31} G_{32}
3	16	11	15	2	3 1	1 2	3 2
4	22	22	37	3	4 1	2 4	6 7
6	72	43	72	4	7 2	1 5	4 7
7	132	112	177	5	14 03	06 10	16 17
8	362	266	373	6	15 06	06 15	15 17
9	552	457	736	7	30 07	16 23	26 36
10	2146	2512	3355	9	52 05	06 70	74 53
11	7432	5163	7026	10	63 32	15 65	46 61

Table 12. Original Codes that Yield the $R = 3/4$ Codes of Johannesson and Paaske, Perforation Pattern P_{02}, [22]

Original $R = 1/4$					Punctured $R = 3/4$				
M_0	G_1	G_2	G_3	G_4	M	G_{11} G_{12} G_{13}	G_{21} G_{22} G_{23}	G_{31} G_{32} G_{33}	G_{41} G_{42} G_{43}
6	100	170	125	161	3	4 0 0	4 6 2	4 2 5	4 4 5
7	224	270	206	357	5	6 1 0	2 6 2	2 0 5	6 7 5
8	750	512	446	731	6	6 3 2	1 4 3	0 1 7	7 6 4
10	2274	2170	3262	3411	8	16 03 01	06 12 02	04 00 17	10 13 10
11	6230	4426	4711	7724	9	10 01 07	03 15 00	07 04 14	14 16 15

original code and the single parity symbol is obtained by puncturing all but one of the b remaining code symbols at the output of the original encoder. Equivalently, the perforation matrix has two rows, whereby the first one is filled with 1s and the second one is filled with 0s except at one position.

Using this procedure, Haccoun and Begin [22] have obtained the original codes and associated perforation matrices allowing to duplicate all the systematic codes of Hagenauer [24]. These codes are listed in Table 13. The possibility of generating these codes by perforation allows once again their easy and practical decoding by any sequential decoding algorithm. However, just like the nonsystematic codes, the systematic punctured codes found here are optimal but do not readily lend themselves to variable-rate or rate-compatible decoding.

Table 13. Systematic Punctured Codes that Duplicate the Codes of Hagenauer. G2 is Given in Octal [22]

R	P	G_2
2/3	1 1 0 1	33275606556377737
3/4	1 1 1 0 0 1	756730246717030774725
4/5	1 1 1 1 0 0 0 1	7475464466521133456725475223
5/6	1 1 1 1 1 0 0 0 0 1	17175113117122772233670106777
7/8	1 1 1 1 1 1 1 0 0 0 0 0 0 1	1773634453774014541375437553121

6. CONCLUSION

In this article, we have presented the basic notions, properties and error performances of high-rate punctured convolutional codes. These high-rate codes, which are derived from well-known optimal low-rate codes and a judicious choice of the perforation patterns, are no more complex to encode and decode than low-rate codes, yielding easy implementations of high coding rate codecs as well as variable-rate and rate-compatible coding and decoding. Extensive lists of both short- and long-memory length punctured codes have been provided together with up to the first 10 terms of their weight spectra and their bit error probability bounds. The substantial advantages of using high-rate punctured codes over the usual high-rate codes open the way for powerful, versatile, and yet practical implementations of variable-rate codecs extending from very low to very high coding rates.

BIOGRAPHY

David Haccoun received the Engineer and B.Sc.Ap. degrees (Magna Cum Laude) in engineering physics from École Polytechnique de Montréal, Canada, in 1965; the S.M. degree in electrical engineering from the Massachusetts Institute of Technology, Cambridge, Massachusetts, in 1966; and the Ph.D. degree in electrical engineering from McGill University, Montréal, Canada, in 1974.

Since 1966 he has been with the Department of Electrical Engineering, École Polytechnique de Montréal, where he has been Professor of Electrical Engineering since 1980 and was the (founding) Head of the Communication and Computer Section from 1980 to 1996. He was a Research Visiting Professor at several universities in Canada and in France. Dr. Haccoun is involved in teaching at both undergraduate and graduate levels, conducting research, and performing consulting work for both government agencies and industries.

His current research interests include the theory and applications of error-control coding, mobile and personal communications, and digital communications systems by satellite. He is the author or coauthor of a large number of journal and conference papers in these areas. He holds a patent on an error control technique and is a coauthor of the books *The communications Handbook* (CRC press and IEEE Press, 1997 and 2001), and *Digital Communications by Satellite: Modulation, Multiple-Access and Coding* (New York : Wiley, 1981). A Japanese translation of that book was published in 1984.

BIBLIOGRAPHY

1. A. J. Viterbi, Convolutional codes and their performance in communications systems, *IEEE Trans. Commun. Technol.* **COM-19** (1971).
2. I. M. Jacobs, Practical applications of coding, *IEEE Trans. Inform. Theory* **IT-20**: 305–310 (1974).
3. W. W. Wu, D. Haccoun, R. Peile, and Y. Hirata, Coding for satellite communication, *IEEE J. Select. Areas. Commun.* **SAC-5**: 724–748 (1987).
4. C. Berrou, A. Glavieux, and P. Thitimasjhima, Near Shannon Limit Error Correcting Coding and Decoding: Turbo Codes, Proceedings of ICC'93, pp. 1064–1070 (1993).
5. D. Divsalar and F. Pollara, Turbo codes for deep-space communications, *TDA Progress Report* **42–120**: 29–39 (1995).
6. J. B. Cain, G. C. Clark, and J. Geist, Punctured convolutional codes of rate $(n-1)/n$ and simplified maximum likelihood decoding, *IEEE Trans. Inform. Theory* **IT-25**: 97–100 (1979).
7. Y. Yasuda, Y. Hirata, K. Nakamura, and S. Otani, Development of a variable-rate Viterbi decoder and its performance characteristics, 6th Int. Conf. Digital Satellite Commun., Phoenix, Arizona, Sept. 1983.
8. Y. Yasuda, K. Kashiki, and Y. Hirata, High-rate punctured convolutional codes for soft decision Viterbi decoding, *IEEE Trans. Commun.* **COM-32**: 315–319 (1984).
9. J. Hagenauer, Rate compatible punctured convolutional codes and their applications, *IEEE Trans. Commun.* **36**: 389–400 (1988).
10. R. M. Fano, A heuristic discussion of probabilistic decoding, *IEEE Trans. Inform. Theory* **IT-9** (1962).
11. E. Paaske, Short binary convolutional codes with maximal free distance for rates 2/3 and 3/4, *IEEE Trans. Inform. Theory* **IT-20**: 683–686 (1974).
12. R. Johannesson and E. Paaske, Further results on binary convolutional codes with an optimum distance profile, *IEEE Trans. Inform. Theory* **IT-24**: 264–268 (1978).
13. F. Jelinek, A fast sequential decoding algorithm using a stack, *IBM J. Res. Develop.* **13**: 675–685 (1969).
14. K. Zigangirov, Some sequential decoding procedures, *Problemii Peradachi Informatsii* **2**: 13–15 (1966).
15. D. Haccoun and M. J. Ferguson, Generalized stack algorithms for decoding convolutional codes, *IEEE Trans. Inform. Theory* **IT-21**: 638–651 (1975).
16. J. P. Odenwalder, Optimal decoding of convolutional codes, Ph.D. dissertation, Dept. Elect. Eng., U.C.L.A., Los Angeles, 1970.
17. P. Montreuil, Algorithmes de determination de spectres des codes convolutionnels, M.Sc.A. thesis, Dep. Elect. Eng., Ecole Polytechnique de Montreal, 1987.
18. D. Haccoun and P. Montreuil, Weight spectrum determination of convolutional codes, to be submitted to *IEEE Trans. Commun.*, Book of Abstracts, 1988 Int. Symp. Inform. Theory, Kobe, Japan, June 1988, 49–50.
19. K. Larsen, Short convolutional codes with maximal free distance for rates 1/2, 1/3, and 1/4, *IEEE Trans. Inform. Theory* **IT-19**: 371–372 (1973).
20. G. Begin and D. Haccoun, Performance of sequential decoding of high rate punctured convolutional codes, *IEEE Trans. Commun.* **42**: 996–978 (1994).
21. D. Haccoun and G. Begin, High-rate punctured convolutional codes for Viterbi and sequential decoding, *IEEE Trans. Commun.* **37**: 1113–1125 (1989).
22. G. Begin and D. Haccoun, High-rate punctured convolutional codes: structure properties and construction technique, *IEEE Trans. Commun.* **37**: 1381–1385 (1989).
23. G. Begin, D. Haccoun, and C. Paquin, Further results on high-rate punctures convolutional codes for Viterbi and sequential decoding, *IEEE Trans. Commun.* **38**: 1922–1928 (1990).
24. J. Hagenauer, High rate convolutional codes with good profiles, *IEEE Trans. Inform. Theory* **IT-23**: 615–618 (1977).

HIGH-SPEED PHOTODETECTORS FOR OPTICAL COMMUNICATIONS

M. Selim Ünlü
Olufemi Dosunmu
Matthew Emsley
Boston University
Boston, Massachusetts

1. INTRODUCTION

The capability to detect, quantify, and analyze an optical signal is the first requirement of any optical system. In optical communication systems, the detector is a crucial element whose function is to convert the optical signal at the receiver end into an electrical signal, which is then amplified and processed to extract the information content. The performance characteristics of the photodetector, therefore, determines the requirements for the received optical power and dictates the overall system performance along with other system parameters such as the allowed optical attenuation, and thus the length of the transmission channel.

Progress in optical communications and processing requires simultaneous development in light sources, interconnects, and photodetectors. While optical fibers have been developed for nearly ideal links for optical signal transmission, semiconductors have become the material of choice for optical sources and detectors, primarily because of their well-established technology, fast electrical response, and optical generation and absorption properties. Many years of research on semiconductor devices have led to the development of high-performance photodetectors in all of the relevant wavelengths throughout the visible and near-infrared spectra. The choice of wavelengths in optical communications is driven by limitations relating to the availability of suitable transmission media and, to a lesser extent, light-emitting devices (LEDs). In general, photodetectors are not the limiting factors. As a result, discussion of photodetectors is usually limited to one or two chapters in books on optoelectronic devices [1–4] or optical communications [5,6] with only a few dedicated books on photodetectors [7–9].

1.1. Optical Communication Wavelengths

The early development work on optical fiber waveguides focused on the 0.8–0.9-μm wavelength range because the first semiconductor optical sources, based in GaAs/AlAs alloys, operated in this region. As silica fibers were further refined, however, it was discovered that transmission at longer wavelengths (1.3–1.6 μm) would result in lower losses [6]. The rapid development of long wavelength fibers has led to attenuation values as low as 0.2 dB/km, which is very close to the theoretical limit for silicate glass fiber [10]. In addition to the intrinsic absorption due to electronic transitions at short wavelengths (high photon energies) and interaction of photons with molecular vibrations within the glass at long (infrared) wavelengths, absorption due to impurities in the silica host, most notably water incorporated into the glass as hydroxyl or OH ions, limits the transmission properties of the optical fiber. Furthermore, Rayleigh scattering resulting from the unavoidable inhomogeneities in glass density, manifesting themselves as subwavelength refractive-index fluctuations, represents the dominant loss mechanism at short wavelengths. The overall typical silica fiber loss has the well-known form given in Fig. 1. Two of the important communication wavelength regions are clearly identifiable as 1.3 and 1.55 μm as a direct result of the fiber attenuation characteristics. The third wavelength region we will consider is at 0.85 μm, due to the advent of GaAs-based light emitters and detectors.

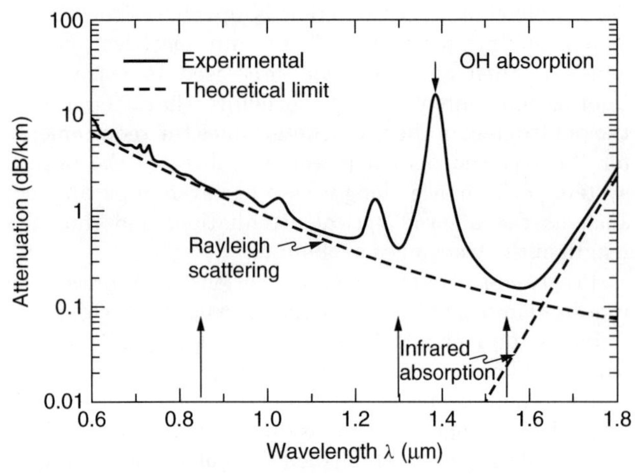

Figure 1. Measured attenuation and theoretical limits in silica fibers (see Ref. 1, for example). The three communication wavelengths we consider are indicated by arrows.

1.2. Basic Performance Requirements

Before we continue with specific photodetector structures, we must first discuss some of the performance requirements. Since the relevant wavelengths for optical communications are in the near-infrared region of the spectrum, both external and internal photoemission of electrons can be utilized to convert the optical signal to electrical signal. External photoemission devices such as vacuum tubes not only are very bulky but also require high voltages, and therefore cannot be used for optical communication applications. Internal photoemission devices, especially semiconductor photodiodes, provide high performance in compact and relatively inexpensive structures. Semiconductor photodetectors are made in a variety of materials such as silicon, germanium and alloys of III–V compounds, and satisfy many of the important performance and compatibility requirements:

1. Compact size for efficient coupling with fibers and easy packaging
2. High responsivity (efficiency) to produce a maximum electrical signal for given optical power
3. Wide spectral coverage through the use of different materials to allow for photodetectors in all communication wavelengths
4. Short response time to operate at high bit rates (wide bandwidth)
5. Low operating (bias) voltage to be compatible with electronic circuits
6. Low noise operation to minimize the received power requirements
7. Low cost to reduce the overall cost of the communication link
8. Reliability (long mean time to failure) to prevent failure of the system
9. Stability of performance characteristics working within a variety of ambient conditions, especially over a wide range of temperatures
10. Good uniformity of performance parameters to allow for batch production

2. FUNDAMENTAL PROPERTIES AND DEFINITIONS

In this section, we describe various definitions relating to the performance of photodetectors, including detection efficiency and noise, and identify suitable materials for a given operation wavelength. We will consider the high-speed properties in the following section. A typical photodetection process in semiconductors can be summarized in three steps:

1. Photons are absorbed in the material, resulting in the generation of mobile charge carriers (electron–hole pairs).
2. The charge carriers drift under an internal electric field.
3. The carriers are collected at the contacts and detection is completed with an electrical response in the external circuit.

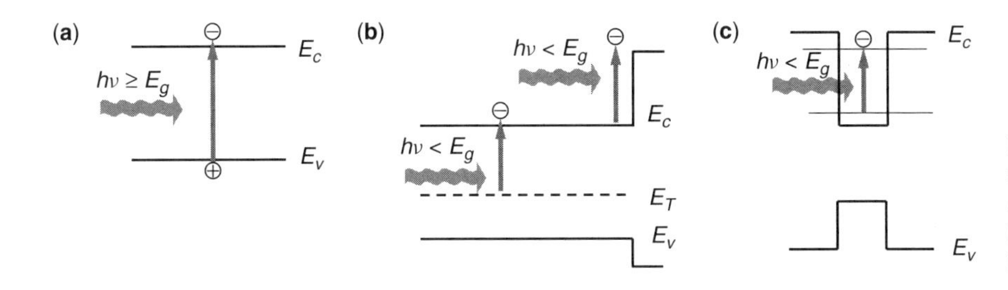

Figure 2. Various absorption mechanisms for photodetection: (**a**) intrinsic (band-to -band); (**b**) extrinsic utilizing an impurity level or free-carrier absorption; (**c**) intersub-band transition in a quantum well.

2.1. Optical Absorption and Quantum Efficiency

The first requirement of photodetection — absorption of light — implies that photons have sufficient energy to excite a charge carrier inside the semiconductor. The most common absorption event results in the generation of an electron–hole pair in typical intrinsic photodetectors. In contrast, an *extrinsic* photodetector responds to photons of energy less than the bandgap energy. In these photodetectors, the absorption of photons result in transitions to or from impurity levels, or between the subband energies in quantum wells as is depicted in Fig. 2.

As will be shown below, thin semiconductor devices for high-speed operation will be required and, therefore, it is crucial to have materials that absorb the incident light very quickly. The absorption of photons at a particular wavelength is dependent on the absorption coefficient α, which is the measure of how fast the photon flux Φ decays in the material:

$$\frac{d\Phi}{dx} = -\alpha\Phi \Rightarrow \Phi(x) = \Phi(0)\exp(-\alpha x) \tag{1}$$

Therefore, the amount of photon flux that is absorbed in a material of thickness (or length) L and absorption coefficient α, can be expressed as [11]

$$\Delta\Phi = \Phi(0)\cdot(1-R)\cdot[1-\exp(-\alpha L)] \tag{2}$$

where R is the power reflectivity at the incidence surface.

The *quantum efficiency* ($0 \leq \eta \leq 1$) of a photodetector is the probability that an incident photon will create or excite a charge carrier that contributes to the detected photocurrent. When many photons are present, we consider the ratio of the detected flux to the incident flux of photons. Assuming that all photoexcited carriers contribute to the photocurrent, we obtain

$$\eta = \frac{\Delta\Phi}{\Phi(0)} \tag{3}$$

$$\eta = (1-R)\cdot[1-\exp(-\alpha L)] \tag{4}$$

2.2. Material Selection

As can be deduced from the equations above, it is necessary to have a large α to realize high-efficiency photodetectors. To capture most of the photons for single-pass absorption, a semiconductor with a thickness of several absorption depth lengths ($L_{abs} = 1/\alpha$) is needed. Figure 3 shows the wavelength dependence of

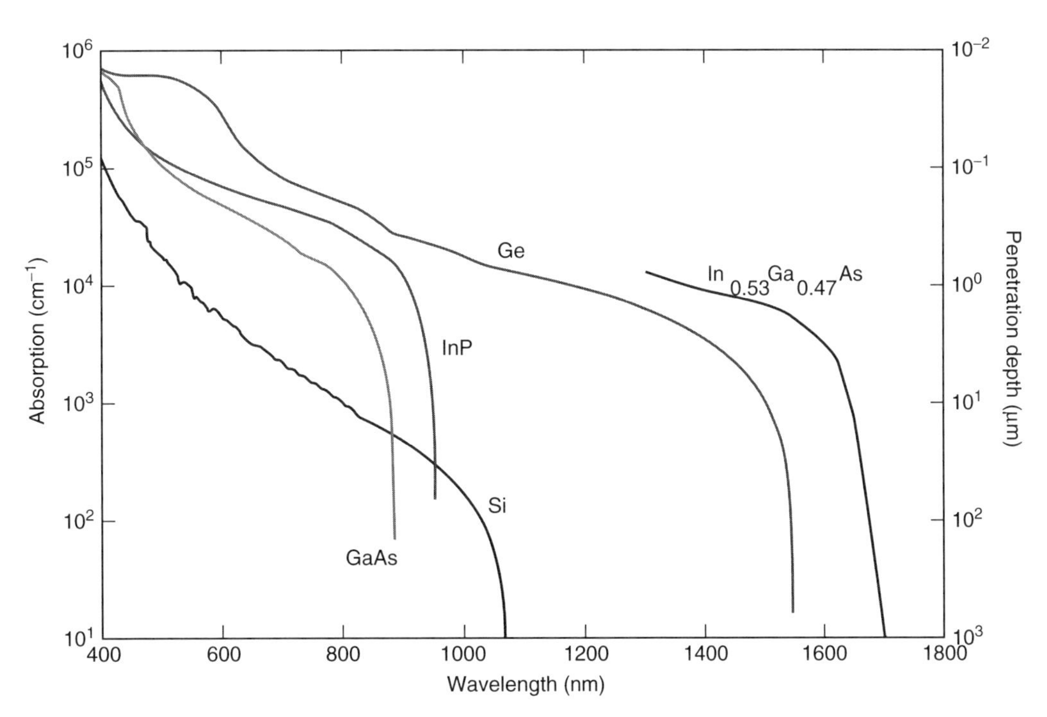

Figure 3. Absorption coefficient for various semiconductors.

α and absorption (or penetration) depth for a variety of semiconductor materials. For very small α, L_{abs} is very large; this results not only in impractical detector structures but also in detectors with a necessarily slow response. At the other extreme, L_{abs} is very small; that is, all the radiation is absorbed in the immediate vicinity of the surface. For typical high-speed semiconductor photodetectors, it is desirable to have an absorption region between a fraction of a micron and a few micrometers in length; specifically, α should be about 10^4 cm^{-1}. Therefore, the high-speed application of the most commonly used detector material, silicon, is limited to the visible wavelengths (0.4 μm $< \lambda <$ 0.7 μm). While the absorption spectrum of an indirect bandgap semiconductor like germanium covers all the optical communication wavelengths, direct bandgap semiconductors such as III–V compounds are more commonly used for high-speed applications. For example, GaAs has a cutoff wavelength of $\lambda_c = 0.87$ μm and is ideal for optical communications at 0.85 μm. At longer wavelengths, ternary compounds such as InGa(Al)As and quaternary materials such as InGa(Al)AsP and a variety of their combinations are used. An important consideration when various semiconductor materials are used together to form heterostructures is the lattice matching of the different constituents and the availability of a suitable substrate. Figure 4 shows the lattice constants and bandgap energies of various compound semiconductors. For most of the photodetector structures designed to operate at 1.3 and 1.55 μm wavelengths, compound semiconductors lattice-matched to InP are used.

2.3. Responsivity

In an analog photodetection system η is recast into a new variable, namely, *responsivity*, or the ratio of the photocurrent (in amperes) to the incident light (in watts):

$$\Re = \frac{\text{photocurrent}}{\text{incident optical power}} = \frac{I_p}{P_0} \quad \text{(A/W)} \tag{5}$$

From here, a simple relationship between quantum efficiency and responsivity can be developed:

$$\{P_0 = \Phi(0)h\nu \text{ and } I_p = q \cdot \Delta\Phi\} \Rightarrow \Re = \frac{q\eta}{h\nu} \tag{6}$$

Responsivity ($\Re$) can also be expressed in terms of wavelength (λ):

$$\nu = \frac{c}{\lambda} \Rightarrow \Re = \frac{q\lambda\eta}{hc} \simeq \frac{\eta \cdot \lambda(\mu\text{ m})}{1.24} \tag{7}$$

It should be noted that the responsivity is directly proportional to η at a given wavelength. The ideal responsivity versus λ is illustrated in Fig. 5 together with the response of a generic InGaAs photodiode showing long- and short-wavelength cutoff and deviation from the ideal behavior.

2.4. Noise Performance and Detectivity

The responsivity equation $I_p = \Re P_0$ suggests that the transformation from optical power to photocurrent is deterministic or noise-free. In reality, even in an ideal photoreceiver, two fundamental noise mechanisms [12]—*shot noise* [13] and *thermal noise* [14,15]—lead to fluctuations in the current even with constant optical input. The overall sensitivity of the photodetector is determined by these random fluctuations of current that occur in the presence and absence of an incident optical signal. Various figures of merit have been developed to assess the noise performance for photodetectors. Although these are not always

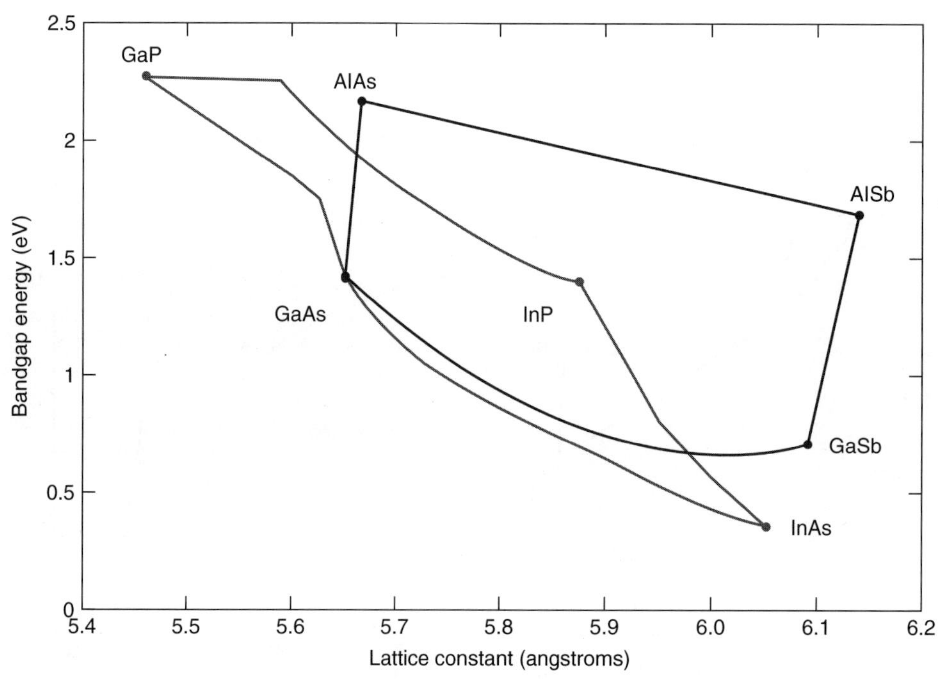

Figure 4. Correlation between lattice constant and bandgap energy for various semiconductor materials.

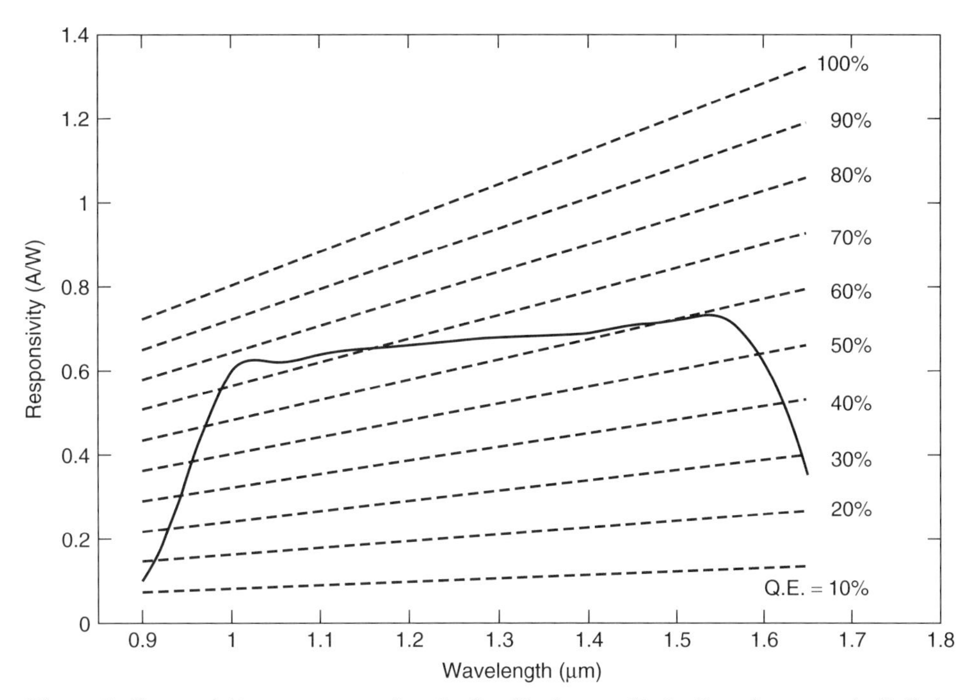

Figure 5. Responsivity versus wavelength for ideal case (dashed) and a generic InGaAs photodiode (solid).

appropriate for high-speed photodetectors for optical communications, it is instructive to discuss the most commonly used definitions.

The *noise equivalent power* (NEP) is the input optical power that results in unity signal-to-noise ratio (SNR = 1). The *detectivity D* is defined as the inverse of NEP ($D = 1/\mathrm{NEP}$), while *specific detectivity* D^* (*D*-star) incorporates the area of the photodetector A and the bandwidth Δf of the signal:

$$D^* = \frac{(A \times \Delta f)^{1/2}}{\mathrm{NEP}} \tag{8}$$

For thermal-noise-limited photodetectors (most semiconductor detectors without internal gain), the noise power (rms) is given by

$$\sigma_T^2 = \langle i_T^2(t) \rangle = \left(\frac{4k_B T}{R_L} \right) \Delta f \tag{9}$$

where R_L is the load resistance, T is temperature in degrees Kelvin and k_B is the Boltzmann constant. Therefore, the NEP has the units of W/Hz$^{1/2}$ as shown below:

$$\mathrm{SNR} = \frac{I_p^2}{\sigma^2} = 1 = \frac{R_L \Re^2}{4k_B T F_n \Delta f} P_{\mathrm{in}}^2 \tag{10}$$

$$\mathrm{NEP} = \frac{P_{\mathrm{in}}}{\sqrt{\Delta f}} = \frac{1}{\Re} \sqrt{\frac{4k_B T F_n}{R_L}} = \frac{h\nu}{\eta q} \sqrt{\frac{4k_B T F_n}{R_L}} \tag{11}$$

where P_{in} is the input optical power and F_n represents the noise figure. The NEP or detectivity can be used to estimate the optical power needed to obtain a specific value of SNR if the bandwidth Δf is known. In the case of shot noise limit, the noise power scales with the total current ($I_p + I_d$, photocurrent + dark current) and bandwidth Δf:

$$\sigma_s^2 = \langle i_s^2(t) \rangle = 2q(I_p + I_d)\Delta f \tag{12}$$

where q is the electronic charge. For high-gain detectors such as avalanche photodetectors, or at high incident power limit, shot noise is much greater than thermal noise:

$$\mathrm{SNR} = \frac{\Re P_{\mathrm{in}}}{2q\Delta f} = \frac{\eta P_{\mathrm{in}}}{2h\nu \Delta f} \tag{13}$$

This further neglects the dark current:

$$\mathrm{SNR} = \frac{I_p^2}{\sigma^2} = \frac{(\Re P_{\mathrm{in}})^2}{2q(\Re P_{\mathrm{in}} + I_d)\Delta f} \tag{14}$$

In this case, the SNR scales linearly with the input optical power and the NEP would not have the traditional units of W/Hz$^{1/2}$ and is not traditionally used. Instead, one can calculate and refer to the power (or number of photons) in a "one" bit at given bit rate and SNR values.

3. HIGH-SPEED PERFORMANCE AND LIMITATIONS

The most common high-speed photodetector response limitations include drift, diffusion, capacitance, and charge-trapping limitation. The type of detector used, or even changes in the detector's geometry, can result in one or more of these limitations, severely degrading the overall bandwidth of the photodetector. In addition, as will be discussed later in this section, the direct relationship between the detector responsivity and bandwidth often serves as an obstacle in designing detectors that are both fast and efficient.

3.1. Transit-Time Limitation

Drift, or transit-time, limitation is directly related to the time needed for photogenerated carriers to traverse the depletion region. To fully understand the mechanism behind this limitation, one must first examine the transient response of photogenerated carriers within a typical PN junction photodiode. A cross-sectional view of a basic PN junction photodiode under reverse bias is illustrated in Fig. 6. In this basic design, photogeneration occurs throughout the photodiode structure. In particular, those carriers photogenerated within the depletion region are swept across at or close to their saturation velocities due to the high electric field within the depletion region.

Assuming that no absorption takes place within the neutral P and N regions of the photodiode, an incident optical pulse generates carriers only within the depletion region. The length of the depletion region, L, can be determined from the following expression:

$$L = \sqrt{\frac{2\varepsilon_s(V_0 - V_a)}{q}\left(\frac{1}{N_a} + \frac{1}{N_d}\right)} \tag{15}$$

Here, ε_s and V_0 represent the semiconductor permittivity and contact potential at the junction, while N_a and N_d represent the doping concentration in the P and N regions, respectively. The external bias, V_a, is negative when reverse-biased, and positive when forward-biased. As is illustrated in Fig. 6, these carriers are then swept to the opposite ends of the depletion region both by the built-in potential induced by the ionized dopants as well as any externally applied potential. At the same time, current at the photodiode contacts is "induced" by the movement of charge within the depletion region. The induced current can be seen as displacement current, where any delay between the movement of charge through the depletion region and the induced current at the contacts is set by the electromagnetic propagation time through the device, which will always be less than 10 fs [16].

Figure 7 shows that as an incremental sheet of photogenerated carriers moves from one particular point within the depletion region to its edge, the ideal current waveform would be a step function whose magnitude is determined by the electron and hole velocities, as well as the width of the depletion region:

$$I_{\text{tot}} = I_e(t) + I_h(t) \tag{16}$$

$$I_e(t) = \frac{q}{L}v_e, \text{ for } 0 < t < t_e \tag{17}$$

$$I_h(t) = \frac{q}{L}v_h, \text{ for } 0 < t < t_h \tag{18}$$

where q represents the electron charge and $v_e(v_h)$ and $t_e(t_h)$ represent the electron (hole) terminal velocities and transport times within the depletion region, respectively. In this figure, it is assumed that both carriers immediately reach their terminal velocities, and the electron velocity is higher than that of the holes. Now, taking into account carriers generated throughout the depletion region, the resulting current waveforms induced by all the photogenerated electrons and holes would be as shown in Fig. 8. Here, $t_e(t_h)$ represents the time required for the electron (hole) generated furthest from its respective depletion edge to reach that edge.

The response speed of any photodiode is limited, in part, by the speed of the slower carrier; in this case, holes. The temporal response of a typical high-speed photodiode, illustrated in Fig. 9, represents the combination of responses by both electrons and holes. Because current at the detector contacts is induced from the time the carriers are photogenerated to the time they reach the depletion region boundaries, the response current does not end until both carriers reach their respective depletion edges. The 3-dB bandwidth of a transit-time-limited photodetector can be expressed as [17]

$$f_{tr} = 0.4\frac{v}{L} \tag{19}$$

where v represents the speed of the slower carrier.

One way to reduce the effects of this transit-time limitation is to make only part of the depletion region

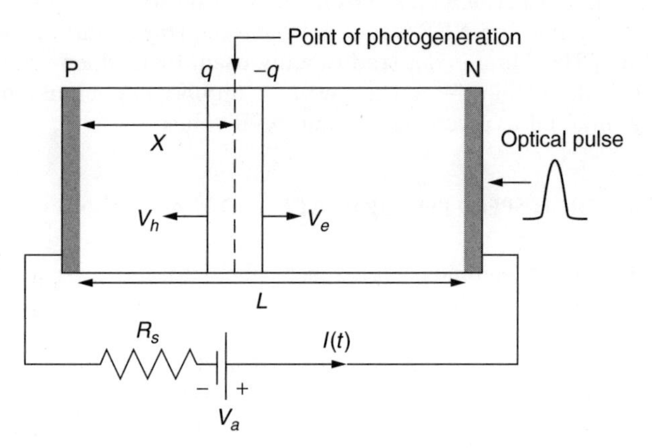

Figure 6. Movement of an incremental sheet of photogenerated carriers within a depletion region.

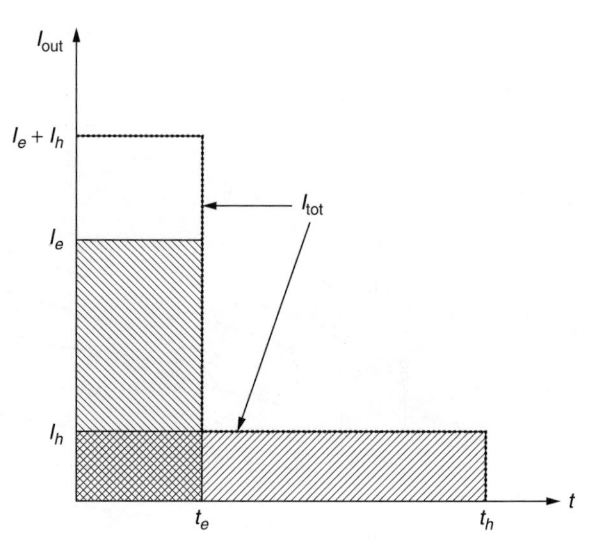

Figure 7. Current induced by incremental sheet of electrons and holes.

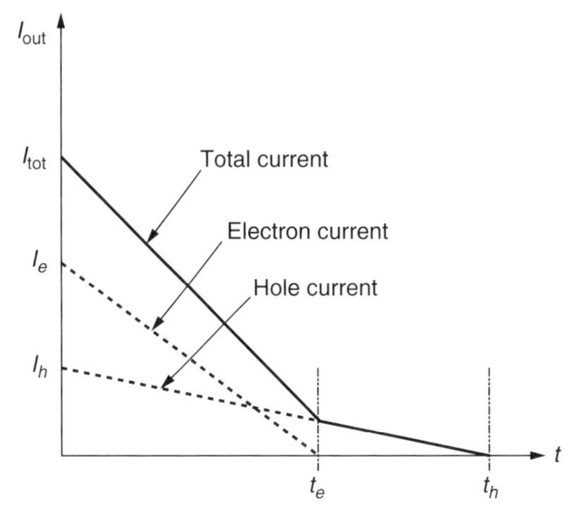

Figure 8. Current induced by all photogenerated electrons and holes.

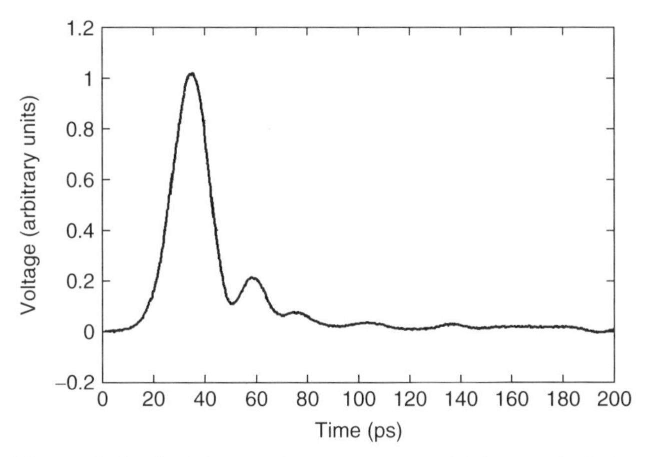

Figure 9. Typical temporal response of a high speed photodetector.

absorptive over the wavelength range of interest. As a result, photogeneration does not occur at the extremes of the depletion region, thus reducing the duration of induced current at the contacts. In addition, the absorption layer should be optimally positioned such that the average electron and hole transit times are equal.

3.2. Capacitance Limitation

Another possible way to reduce the effects of transit-time limitation would be to reduce the length of the depletion region itself. However, a reduction of the depletion length will also result in an increase of the junction capacitance. The bandwidth of a junction-capacitance-limited photodetector can be formulated by the following expression:

$$f_{RC} = \frac{L}{2\pi R_L \varepsilon A} \qquad (20)$$

where R_L is the load resistance, ε is the semiconductor permittivity, and A represents the detector's cross-sectional area. To minimize this limitation, the photodetector should be designed such that the depletion length is not so small that capacitance limitations are dominant. At the same time, however, the length should not be increased to the point where transit-time limitations become an overwhelming limiting factor. Basically, there exists an optimum point where the combined effect of the transit-time and capacitance limitations is at a minimum. In addition to junction capacitance, other capacitances relating to the detector contact pad and other external elements can also serve to degrade the overall detector bandwidth. Called *parasitic capacitance*, they can be minimized by using an airbridge directly from the detector to the contact pad [18]. A top-view image of such a photodetector can be seen in Fig. 10.

3.3. Diffusion Limitation

If a photodetector is designed such that all the semiconductor regions, including the nondepleted or neutral regions, are absorptive over the wavelength range of interest, then limitations due to carrier diffusion will dominate any bandwidth limitations imposed on the detector. Unlike the depletion region, there is no potential drop across the neutral regions to drive the photogenerated minority carriers in any particular direction. Instead, the movement of carriers within these regions is dictated by the diffusion process, where a net current induced by the photogenerated minority carriers in the neutral regions does not occur until some of these carriers reach the depletion edge and are swept across to the other end by the high electric field. Here, the average amount of time needed for the photogenerated carriers within the nondepleted regions to reach the depletion edge is dictated in part by the thickness of the region, which can lead to a time period on the order of 10 ns or more [5]. As a result, the temporal response of a diffusion-limited photodetector will exhibit a "tail" at the end of the pulse as long as the diffusion time period. Limitations caused by carrier diffusion can be minimized, even totally eliminated, simply by making the neutral regions nonabsorptive.

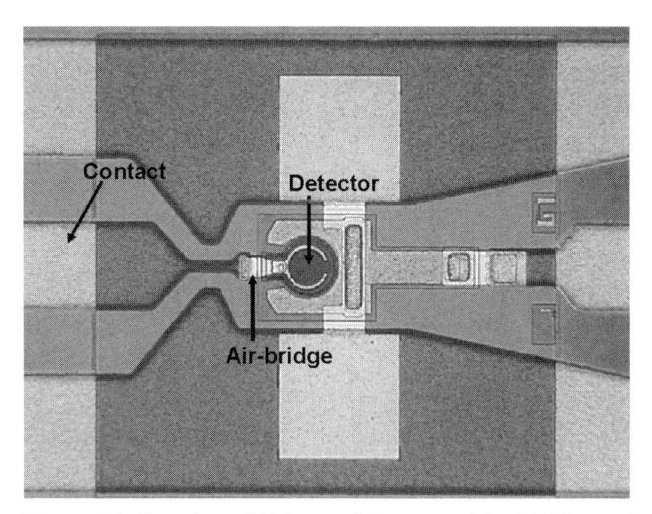

Figure 10. Top view of high-speed detector with airbridge and contacts for high-speed probe.

3.4. Charge-Trapping Limitation

Making only certain regions within a photodetector absorptive over a given wavelength range involves forming heterojunctions, or junctions between two different semiconductor materials, where one possesses a much higher absorption coefficient than the other over the wavelength range of interest. When two different semiconductors are brought abruptly into contact, however, the differences in their respective valence and conduction bands will result in the temporary storage of charge at the interface. This effect, referred to as *charge trapping*, ultimately impedes the transport of charge through the depletion region, reducing the device bandwidth as a result. One effective way to reduce the amount of charge trapping in heterojunction-based devices is to compositionally grade the region around the heterojunction itself, thereby smoothing out any abruptness in the energy bands between the two semiconductor materials.

3.5. Bandwidth–Efficiency Product

Designing photodetectors for high-speed optical communications requires the optimization of both the bandwidth, or speed, as well as the quantum efficiency of the detection system. Unfortunately, there is often a trade-off between bandwidth and efficiency, especially when carries are collected along the same direction as the photons being absorbed. The bandwidth–efficiency (BWE) product serves as an important figure of merit for high-speed detectors.

Assuming that all other speed limitations are minimized, the BWE product for a transit-time-limited photodetector can be expressed as

$$\mathrm{BWE} = f_{tr} * \eta = 0.4\frac{v}{L}(1-R)(1-e^{-\alpha L}) \tag{21}$$

For a thin photodetector, the transit-time-limited BWE product can be approximated as

$$\mathrm{BWE} = f_{tr} * \eta = 0.4v\alpha(1-R) \tag{22}$$

which is independent of the depletion region length and depends only on material properties, presenting a fundamental limitation.

4. SEMICONDUCTOR PHOTODETECTORS

4.1. Classification and Structures

Semiconductor photodetectors can be classified into two major groups:

1. *Photovoltaics*. These detectors produce a voltage drop across its terminals when illuminated by an external optical source. A subcategory of photovoltaics is the photodiode, the most commonly used detector in high-speed optical communications. Photodiodes detect light through the photogeneration of carriers within the depletion region of a diode structure, typically under reverse bias. Below, we will discuss the most common types of photodiodes, including PIN, Schottky, and avalanche photodetectors (APDs).
2. *Photoconductors*. These detectors respond to an external optical stimulus through a change in their conductivity. The optical signal is detected by applying an external bias voltage and observing the change in current due to a change in resistivity. Unlike other types of photodetectors, photoconductors can exhibit gain, since the photoexcited carriers may contribute to the external current multiple times when the recombination lifetimes of the carriers are greater than their transit times [19].

It is also possible to categorize semiconductor photodetectors according to a variety of other properties. Below we emphasize different photodetector structures based on illumination geometry and collection of photoexcited carriers. It is important to distinguish between structures where photon absorption and carrier collection occur along the same direction, and those that do not. In the first case, there is a direct trade-off between the efficiency and speed of the photodetector, as discussed in the previous section.

Most common photodetectors are vertically illuminated and carriers are collected along the same direction as schematically shown in Fig. 11a. Alternatively, carrier collection can be in the lateral direction (Fig. 11b). In a conventional (one-pass) structure, the efficiency is limited by the thickness and absorption coefficient of the detector material. To increase the efficiency without requiring a thick absorption region in vertically illuminated photodetectors, a resonant cavity enhanced (RCE) structure can be utilized, effectively resulting in multiple passes through the detector (Fig. 11c). Photodetectors can be formed as optical waveguides in an edge-illuminated configuration as shown in Fig. 11d. In waveguide photodetectors (WGPD), light is absorbed along the waveguide and the carriers are collected in the transverse direction, permitting nearly independent

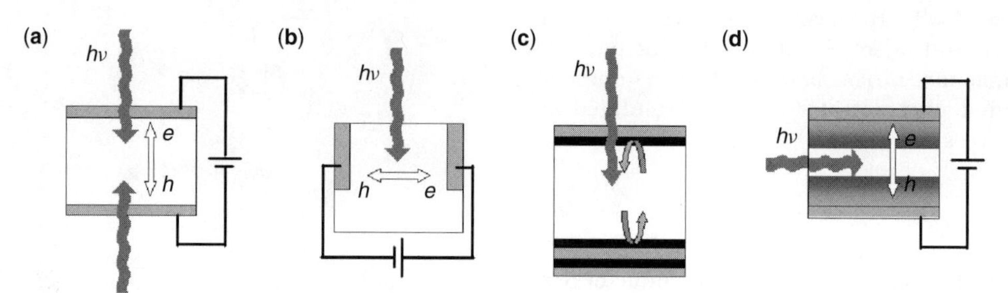

Figure 11. Photodetector structures: (**a**) vertically illuminated (top or bottom); (**b**) vertically illuminated with lateral carrier collections; (**c**) resonant cavity enhanced (RCE); (**d**) waveguide photodetector (edge-illuminated).

optimization of both the bandwidth and efficiency. A further refinement of the WGPD is the traveling-wave photodetector (TWPD), in which the optical signal and resulting electrical signal travel at the same speed, overcoming the capacitance limitations [20].

4.2. PIN Detectors

PN junctions are formed simply by bringing into contact n- (donors) and p-doped (acceptors) semiconductor regions [21]. When a junction between these two regions is formed, diffusion of the electrons from the n region and holes from the p region will cause a charge imbalance to form as a result of the uncompensated ions that are left behind. This charge imbalance will create an opposing electric field that will prevent further diffusion of electrons and holes, resulting in a region depleted of carriers at the junction. The length of the depletion was stated earlier in Eq. (15).

As was described earlier, photons absorbed in the depletion region will create electron–hole pairs that are separated by the built-in or externally applied field and drift across the depletion region, inducing current at the contacts until reaching the depletion edge. Conversely, photons that are absorbed in the neutral regions can contribute to the photocurrent if the minority carrier is sufficiently close to the depletion region, or if the diffusion length is long enough that recombination does not occur before it reaches the depletion region, where it will be swept across while inducing current at the contacts. The minority carrier moves by diffusion through the neutral region, which is an inherently slow process. Therefore, great care must be taken to inhibit this process by making the highly doped P and N regions extremely thin.

In general, the depletion region is designed to be as large as possible to increase the absorption of photons. For PN junctions, applying a reverse bias increases the depletion region. An alternative approach is a one-sided abrupt junction, formed between a highly doped $p(n)$ region and lightly doped $n(p)$ region. The depletion length is given by [22]

$$L = \sqrt{\frac{2\varepsilon_s(V_{bi} - V)}{qN_B}} \tag{23}$$

where N_B is the doping concentration on the lightly doped side. However, one-sided PN junctions suffer from high series resistance, which is detrimental to high-speed photodiodes. Increasing the background doping, N_B, in a PN junction decreases the transit time τ_d for the carriers by increasing the maximum field strength in the depletion region. Since the field is triangular, however, the mean electric field will be half the maximum E field:

$$\tau_d \approx \frac{L^2}{2V\mu} \tag{24}$$

where μ represents the carrier mobility.

PIN photodiodes, where "I" denotes intrinsic, improve on the conventional PN photodiode by having a large intrinsic region sandwiched between two heavily doped regions. Since the intrinsic region is depleted and is much larger than the depletion length in the highly doped p and n regions, it is sufficient to say that the field is at a maximum across the entire depletion length. Therefore, the mean electric field is twice that of a conventional PN junction and, correspondingly, the transit time is cut in half [7], assuming the carriers have not reached their saturation velocities:

$$\tau_d \approx \frac{L^2}{4V\mu} \tag{25}$$

An added benefit of the PIN photodiode is that one does not need to apply an external bias to deplete the intrinsic region, as it is usually fully depleted by the built-in potential. Also, having L much greater in length than the neutral regions results in a greatly reduced diffusion contribution to the photocurrent. PIN photodiodes can also have very highly doped p and n regions, thereby reducing access resistance as compared to the one-sided PN junction photodiode.

An alternative approach to making the neutral regions thin is the use of heterojunctions, or junctions between two different semiconductors. As was discussed earlier, the neutral regions can be made of materials that do not absorb at the operating wavelength. Therefore, carriers will be photogenerated only within the depletion region, thereby eliminating any diffusion current. For example, InGaAs-InP heterojunction PIN photodiodes operating at speeds in excess of 100 GHz have been reported [23,24].

4.3. MSM Photodetectors

Unlike most photodetectors used in high-speed optical communications, metal–semiconductor–metal (MSM) photodetectors are photoconductors. Photoconductors operate by illuminating a biased semiconductor material layer, which results in the creation of electron–hole pairs that raise the carrier concentration and, in turn, increase the conductivity as given by

$$\sigma = q(\mu_n n + \mu_p p) \tag{26}$$

where μ_n and μ_p are the electron and hole mobility, and n and p represent the electron and hole carrier concentrations, respectively. This increase in carrier concentration and subsequent increase in conductivity results in an increase of the photocurrent given by [4]

$$I_{ph} = \frac{q\eta GP}{h\nu} \tag{27}$$

where q is the electron charge, G is the photoconductor gain, and P represents the optical input power. The photoconductor gain G, which results from the difference between the transit time for the majority carrier and the recombination lifetime of the minority carrier, is given by

$$G = \frac{\tau}{\tau_{tr}} \tag{28}$$

where τ is the minority carrier lifetime and τ_{tr} is the majority carrier transit time. It is usually advantageous to have a high defect density in the photoconductor absorption area so that the carrier lifetime is reduced, resulting in a reduction of the impulse response [9].

MSMs are important devices because they can be easily monolithically integrated into FET-based circuit technologies. MSMs, for example, have been demonstrated on InGaAs with bandwidths in excess of 40 GHz operating at 1.3 μm [25], and it has been shown that monolithic integration with InP-based devices is possible [26].

4.4. Schottky Photodetectors

A Schottky detector consists of a conductor in contact with a semiconductor, which results in a rectifying contact. As a result, a depletion region forms at the conductor/semiconductor junction, entirely on the semiconductor side. The conductor can be metal or silicide, while the semiconductor region is usually moderately doped ($\sim 10^{17}$ cm^{-3}) to prevent tunneling through a thin barrier. A generic energy band diagram of a Schottky junction is illustrated in Fig. 12.

In a basic Schottky detector, the depletion length W and semiconductor work function Φ_s are determined by the semiconductor doping, while the metal-to-semiconductor barrier height Φ_b is set by the high density of surface states formed at the metal/semiconductor interface, where the Fermi level is "pinned" at a certain position, regardless of doping. The semiconductor-to-metal barrier height Φ_v, on the other hand, can be controlled through doping, as is described through the following equation [21]:

$$\Phi_v = \Phi_m - \Phi_s \tag{29}$$

where Φ_m represents the metal work function.

In the case where the photon energy is greater than the semiconductor bandgap, the Schottky barrier behaves much like a one-sided PN junction; for photons incident on the metal side of the Schottky barrier, however, the metal must be thin such that it is effectively transparent. For energies less than the semiconductor bandgap, but greater than the barrier formed at the metal–semiconductor junction ($\Phi_v < h\nu < E_g$), photons incident on the semiconductor side will cross the region

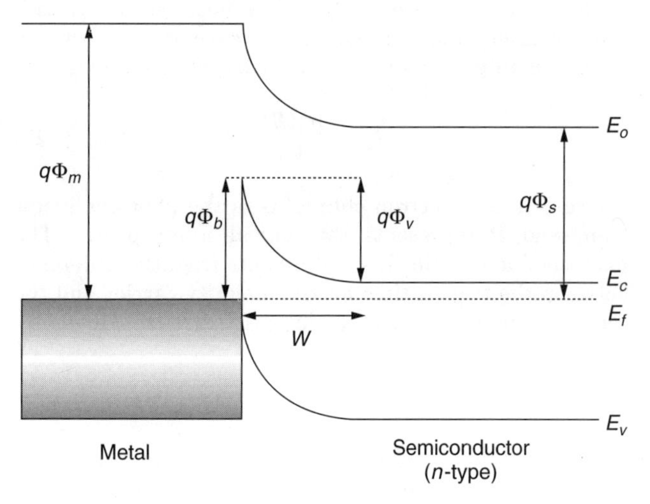

Figure 12. Energy band diagram of Schottky junction.

without any absorption and will excite carriers within the metal over the metal–semiconductor barrier.

One advantage that Schottky detectors have over their PIN counterparts is their reduced contact resistance, resulting in a faster response. For example, Schottky detectors with measured 3-dB bandwidths well above 100 GHz have been reported [9,27]. An additional advantage is their simple material structure and relative ease in fabrication [18]; however, detector illumination is a common design issue. Because metals and silicides are very absorptive because of the effects of free-carrier absorption, one cannot design a traditional Schottky detector to be illuminated from the metal side, unless the metal layer is very thin such that reflection is minimized. Ignoring this issue would result in a photodiode with poor quantum efficiency.

4.5. Avalanche Photodetectors

Avalanche photodetectors are PN junctions with a large field applied across the depletion region so that electron–hole pairs created in this region will have enough energy to cause impact ionization, and in turn avalanche multiplication, while they are swept across the depletion length. Under this condition, a carrier that is generated by photon absorption is accelerated across the depletion region, lifting it to a kinetic energy state that is large enough to ionize the neighboring crystal lattice, resulting in new electron–hole pair generation. Likewise, these new pairs are accelerated and multiply with the production of more electron–hole pairs, until finally these carriers reach the contacts. This avalanche effect results in a multiplication of the photogenerated carriers and, in turn, an increase in the collected photocurrent. The multiplication factor M is used to refer to the total number of pairs produced by the initial photoelectron. This factor represents the internal gain of the photodiode; as a result, APDs can have quantum efficiencies greater than unity

$$M = \frac{1}{1 - (V/V_{\mathrm{BR}})^r} \tag{30}$$

where V_{BR} represents the breakdown voltage and r is a material dependent coefficient. Since APDs exhibit gain, they are highly sensitive to incident light and therefore are limited by shot noise, as a single incident photon has the potential to create many carrier pairs. Also, there is a finite time associated with the avalanche process, resulting in a slower impulse response. Regardless, APDs for optical communications have been demonstrated exhibiting internal gains in excess of 200 at 9 V reverse bias [28], as well as devices with BWE product of 90 GHz [29].

4.6. RCE Photodetectors

As discussed earlier, the important figure of merit for high speed photodetectors is the BWE product. That is, the transit time of the photogenerated carriers must be kept to a minimum while the absorption length must be sufficiently long so that a reasonable number of photons are absorbed and, in turn, carriers generated. Because these two quantities are inversely related, an increase in efficiency will result in the reduction of

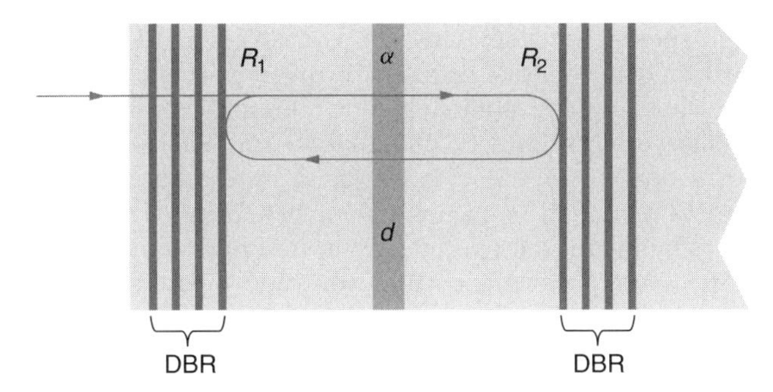

Figure 13. Simplified cross-sectional view of a typical RCE detector.

bandwidth. One way to circumvent this problem involves fabricating the photodetector within a Fabry–Perot cavity. Fabry–Perot cavities are formed by sandwiching a region between two reflecting surfaces, where light incident perpendicular to the cavity results in spectrally dependent constructive and destructive interference of the electric field inside the cavity. A cross-sectional illustration of a typical RCE photodetector can be seen in Fig. 13. The constructive interference results in an increase of the electric field amplitude at specific points within the cavity over a limited spectral range. Since high-speed optical communications typically rely on narrow-linewidth laser sources, fabricating photodetectors within a Fabry–Perot cavity is an ideal solution. For an RCE photodetector, a conventional PIN diode is placed within a Fabry–Perot cavity so that the electric field maximum occurs over the absorption region, thereby increasing the number of absorbed photons and, in turn, increasing the quantum efficiency. For a given absorption length, the quantum efficiency of the detector over a specific wavelength range is increased, while the transit time and bandwidth remain constant as compared to a conventional photodiode. This results in an overall increase in the bandwidth–efficiency product for the RCE photodiode. The peak quantum efficiency for an RCE photodetector is given by

$$\eta_{\max} = \left\{ \frac{(1 + R_2 e^{-\alpha d})}{(1 - \sqrt{R_1 R_2} e^{-\alpha d})^2} \right\} \times (1 - R_1) \cdot (1 - e^{-\alpha d}) \quad (31)$$

RCE structures can be fabricated in semiconductors using distributed Bragg reflectors (DBRs), which are alternating layers of different refractive index materials. DBRs can yield reflectivities in excess of 90% when many periods are used. Typical resonant cavity photodetectors are fabricated from GaAs-based materials by molecular beam epitaxy (MBE). AlAs/GaAs distributed Bragg reflectors are lattice matched to GaAs and achieve high reflectivity using greater than 10 periods. Figure 14 illustrates the cross section of a GaAs-based RCE photodiode optimized for 900 nm.

Figure 15 shows the quantum efficiency of a GaAs PIN photodetector [30]. The RCE photodiode exhibits greatly improved efficiency at 850 nm over a conventional single-pass photodiode. Additionally, RCE Schottky GaAs-based photodetectors have been demonstrated with bandwidths in excess of 50 GHz and peak quantum

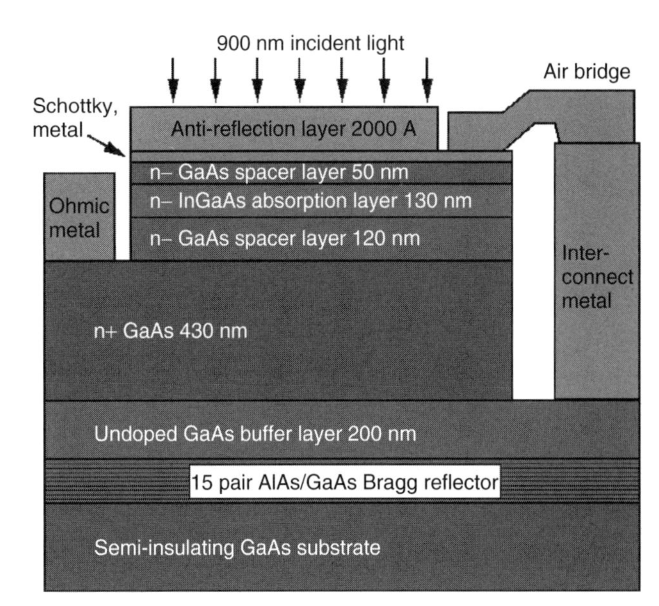

Figure 14. Cross-sectional view of a GaAs-based RCE photodiode optimized for 900 nm.

efficiency of 75% [31,32]. Making use of silicon on insulator technology (SOI), silicon-based RCE photodiodes have also been developed [33]. Conventional silicon photodiodes provide cost-efficient alternatives to GaAs- or InP-based semiconductors, due mainly to the ubiquitous silicon processing industry, but typically suffer from poor response at the communication wavelengths compared to its more expensive counterparts. The use of an RCE structure greatly improves the efficiency of silicon photodiodes, making them perform on par with GaAs- and InP-based photodiodes operating at 850 nm. Figure 16 shows the cross section of a silicon RCE photodiode.

4.7. Waveguide Photodetectors

One characteristic common to all the photodetectors described so far is the fact that the propagation direction

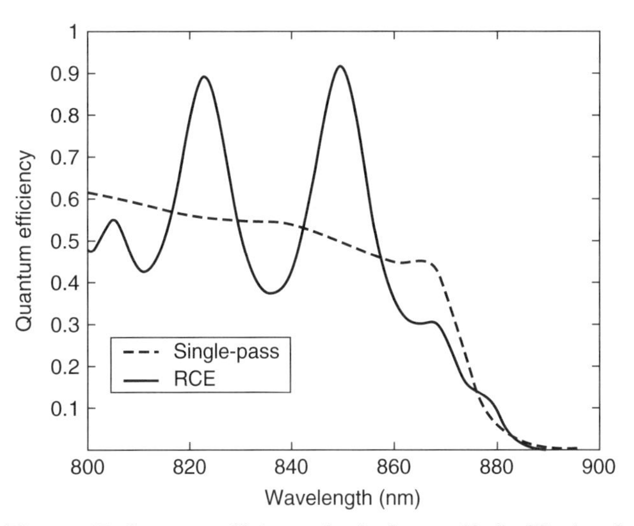

Figure 15. Quantum efficiency of a single-pass (dashed line) and a RCE (solid line) PIN photodetector.

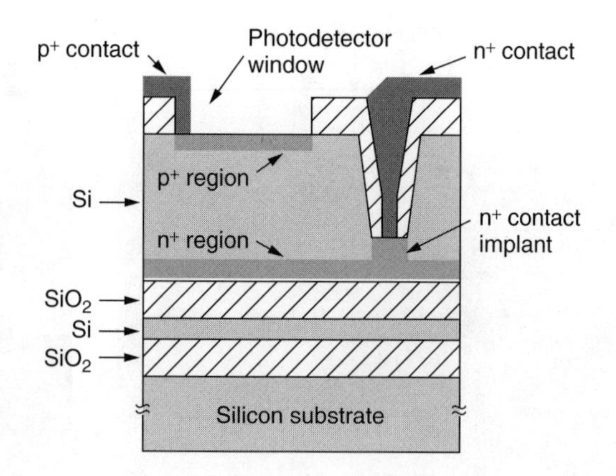

Figure 16. Cross-section of a silicon RCE PIN photodetector.

of both the incident light and the photogenerated carriers are parallel to one another. This characteristic is at the very heart of the bandwidth/efficiency "tug of war." Another interesting solution to this detector design issue involves making the propagation of the incident light and photogenerated carriers perpendicular to one another. This particular characteristic can be seen in waveguide detectors where, instead of top or bottom illumination, the photodetector is *side*-illuminated through a waveguide-like structure [34]. The guiding region doubles as the absorbing region and, under biasing, the photogenerated carriers within the guiding region are swept across the depletion region in a direction perpendicular to the propagation of the guided light. As a result, the photodetector quantum efficiency and bandwidth are not coupled to one another; because the quantum efficiency here depends on the waveguide length, absorption coefficient and confinement factor, while the bandwidth is related to the waveguide thickness, one can design a waveguide photodetector that has both high quantum efficiency and wide bandwidth. For example, a multimode waveguide PIN photodetector with a 3-dB bandwidth of 110 GHz and 50% quantum efficiency was demonstrated at 1.55 μm [35,36].

One design issue encountered with waveguide photodetectors involves saturation caused by photocarrier screening. This particular effect occurs when the absorption coefficient of the guiding region is high enough that a high carrier density is photogenerated near the entrance of the waveguide. As a result, the density of empty energy states for other photogenerated carriers are dramatically reduced, thereby preventing or *screening* further carrier photogeneration. This problem is often solved by reducing the absorption coefficient, in effect diluting and distributing the photogeneration of carriers along the guiding layer. However, an even more severe limitation of the waveguide photodetector involves its low coupling efficiency with respect to the incident light source.

5. CONCLUSIONS

In this section, we discussed the basic operation principles and properties of photodetectors, focusing on requirements relevant to high-speed optical communications. High-speed photodetectors are available for all communication wavelengths, utilizing a variety of semiconductor materials. Silicon photodetectors dominate applications for short wavelengths (visible to 0.85 μm) where low-cost, high-volume production is the most important consideration. Compound semiconductors, benefiting from their direct bandgap and availability of heterostructures, provide higher performance and coverage over the longer communication wavelengths (1.3–1.6 μm). The performance of conventional photodetectors has been refined to the point of fundamental material limitations. To meet the increasing demand for higher bit rates, more recent photodetector development has focused on innovative designs such as avalanche photodiodes (APDs), resonant cavity enhanced (RCE), and waveguide photodetectors, to name only a few.

Optical communications offering higher bandwidths have replaced their electrical counterparts, starting with long-distance applications. In these applications, the cost of a photodetector is not significant in the context of the overall system, and thus the choice for a receiver is performance-driven. Current trends indicate that optical communications will dominate medium- to short-distance applications in the near future. While the research and development efforts for faster and more efficient photodetectors will continue, the development of photodetectors with low-cost manufacturing potential will be increasingly important. We expect that there will be a greater effort behind the integration of high-speed photodetectors with electronic circuits and photonic components, leading not only to improved performance and functionality but also ultimately to the reduction of cost.

BIOGRAPHIES

Professor M. Selim Ünlü was born in Sinop, Turkey, in 1964. He received the B.S. degree in electrical engineering from Middle East Technical University, Ankara, Turkey, in 1986, and the M.S.E.E. and Ph.D. in electrical engineering from the University of Illinois, Urbana-Champaign, in 1988 and 1992, respectively. His dissertation topic dealt with resonant cavity enhanced (RCE) photodetectors and optoelectronic switches. In 1992, he joined the Department of Electrical and Computer Engineering, Boston University, as an Assistant Professor, and he has been an Associate Professor since 1998. From January to July 2000, he worked as a visiting professor at University of Ulm, Germany.

Dr. Ünlü's career interest is in *research and development of photonic materials, devices and systems* focusing on the design, processing, characterization, and modeling of semiconductor optoelectronic devices, especially photodetectors.

During 1994/95, Dr. Ünlü served as the Chair of IEEE Laser and Electro-Optics Society, Boston Chapter, winning the LEOS Chapter-of-the-Year Award. He served as the vice president of SPIE New England Chapter in 1998/99. He was awarded National Science Foundation Research Initiation Award in 1993, United Nations TOKTEN award in 1995 and 1996, and both the National Science Foundation CAREER and Office of Naval Research

Young Investigator Awards in 1996. Dr. Ünlü has authored and co-authored more than 150 technical articles and several book chapters and magazine articles; edited one book; holds one U.S. patent; and has several patents pending. During 1999–2001, he served as the chair of the IEEE/LEOS technical subcommittee on photodetectors and imaging, and he is currently an associate editor for *IEEE Journal of Quantum Electronics*.

Olufemi Dosunmu was born in Bronx, New York, in 1977. He graduated as Salutatorian from Lakewood High School in Lakewood, New Jersey in 1995, and received both his B.S. and M.S. degrees in Electrical Engineering with his thesis entitled *Modeling and Simulation of Intrinsic and Measured Response of High Speed Photodiodes*, along with a minor in Physics from Boston University in May 1999. While at Boston University, he was awarded the 4-year Trustee Scholarship in 1995, the Golden Key National Honor Society award in 1998, as well as the National Defense Science & Engineering Graduate (NDSEG) Fellowship in 2001. In 1997, he was a summer intern at AT&T Labs in Red Bank, New Jersey and, in 1998, at Princeton Plasma Physics Laboratories in Princeton, New Jersey. Between 1999 and 2000, he was employed at Lucent Technologies in the area of ASIC design for high-speed telecommunication systems. Mr. Dosunmu is expected to complete his PhD in January 2004.

Matthew K. Emsley was born in 1975, in the northern suburbs of Wilmington, Delaware. He graduated from Brandywine High School in 1993 and received his B.S. degree in Electrical Engineering from The Pennsylvania State University in December 1996, and a M.S. degree in Electrical Engineering from Boston University in May 2000 for his thesis entitled *Reflecting Silicon-on-Insulator (SOI) Substrates for Optoelectronic Applications*. While at Boston University Mr. Emsley was awarded the Electrical and Computer Engineering Chair Fellowship in 1997 and the Outstanding Graduate Teaching Fellow award for 1997/98. In 2001 Matthew was awarded a LEOS Travel Award as well as the H.J. Berman "Future of Light" Prize in Photonics for his poster entitled *Silicon Resonant-Cavity-Enhanced Photodetectors Using Reflecting Silicon on Insulator Substrates* at the annual Boston University Science Day. Mr. Emsley is expected to complete his Ph.D. in May 2003.

BIBLIOGRAPHY

1. D. Wood, *Optoelectronic Semiconductor Devices*, Prentice-Hall, New York, 1994.
2. J. Singh, *Semiconductor Optoelectronics*, McGraw Hill, New York, 1995.
3. K. J. Ebeling, *Integrated Opto-electronics*, Springer-Verlag, New York, 1992.
4. P. Bhattacharya, *Semiconductor Optoelectronic Devices*, 2nd ed., Prentice-Hall, Englewood Cliffs, NJ, 1997.
5. G. P. Agrawal, *Fiber-Optic Communication Systems*, Wiley, New York, 1992.
6. J. M. Senior, *Optical Fiber Communications*, Prentice-Hall, New York, 1992.
7. S. Donati, *Photodetectors: Devices, Circuits, and Applications*, Prentice-Hall, Englewood Cliffs, NJ, 2000.
8. E. L. Dereniak and G. D. Boreman, *Infrared Detectors and Systems*, Wiley, New York, 1996.
9. H. S. Nalwa, ed., *Photodetectors and Fiber-Optics*, Academic Press, New York, 2001.
10. T. Miya, Y. Terunuma, T. Hosaka, and T. Miyashita, Ultimate low-loss single-mode fibre at 1.55 μm, *Electron. Lett.* **15**(4): 106–108 (1979).
11. T. P. Lee and T. Li, *Photodetectors*, in S. E. Miller and A. G. Chynoweth, eds., *Optical Fiber Telecommunications*, Academic Press, New York, 1979, pp. 593–626.
12. D. K. C. MacDonald, *Noise and Fluctuations in Electronic Devices and Circuits*, Oxford Univ. Press, Oxford, 1962.
13. W. Schottky, *Ann. Phys.* **57**: 541 (1918).
14. J. B. Johnson, *Phys. Rev.* **32**: 97 (1928).
15. H. Nyquist, *Phys. Rev.* **32**: 110 (1928).
16. D. G. Parker, The theory, fabrication and assessment of ultra high speed photodiodes, *Gec. J. Res.* **6**(2): 106–117 (1988).
17. S. M. Sze, *Semiconductor Device Physics and Technology*, Wiley, New York, 1985.
18. M. Gökkavas et al., Design and optimization of high-speed resonant cavity enhanced schottky photodiodes, *IEEE J. Quant. Electron.* **35**(2): 208–215 (1999).
19. B. E. A. Saleh and M. C. Teich, *Fundamentals of Photonics*, Wiley, New York, 1991.
20. K. S. Giboney, M. J. W. Rodwell, and J. E. Bowers, Traveling-wave photodetectors, *IEEE Photon. Technol. Lett.* **4**(12): 1363–1365 (1992).
21. B. G. Streetman, *Solid State Electronic Devices*, 4th ed., Prentice-Hall, Englewood Cliffs, NJ, 1995.
22. S. M. Sze, *Physics of Semiconductor Devices*, 2nd ed., Wiley, New York, 1981.
23. D. L. Crawford et al., High speed InGaAs-InP p-i-n photodiodes fabricated on a semi-insulating substrate, *IEEE Photon. Technol. Lett.* **2**(9): 647–649 (1990).
24. Y. Wey et al., 108-GHz GaInAs/InP p-i-n photodiodes with integrated bias tees and matched resistors, *IEEE Photon. Tech. Lett.* **5**(11): 1310–1312 (1993).
25. E. H. Böttcher et al., Ultra-wide-band (>40 GHz) submicron InGaAs metal-semiconductor-metal photodetectors, *IEEE Photon. Technol. Lett.* **8**(9): 1226–1228 (1996).
26. J. B. D. Soole and H. Schumacher, InGaAs metal-semiconductor-metal photodetectors for long wavelength optical communications, *IEEE J. Quantum Electron.* **27**(3): (1991).
27. E. Özbay, K. D. Li, and D. M. Bloom, 2.0 ps, 150 GHz GaAs monolithic photodiode and all-electronic sampler, *IEEE Photon. Technol. Lett.* **3**(6): 570–572 (1991).
28. R. Kuchibhotla et al., Low-voltage high-gain resonant-cavity avalanche photodiode, *IEEE Photon. Technol. Lett.* **3**(4): 354–356 (1991).
29. H. Kuwatsuka et al., An $Al_xGa_{1-x}Sb$ avalanche photodiode with a gain bandwidth product of 90 GHz, *IEEE Photon. Technol. Lett.* **2**(1): 54–55 (1990).
30. M. Gökkavas et al., High-speed high-efficiency large-area resonant cavity enhanced p-i-n photodiodes for multimode

fiber communications, *IEEE Photon. Technol. Lett.* **13**(12): 1349–1351 (2001).

31. M. S. Ünlü et al., High bandwidth-efficiency resonant cavity enhanced schottky photodiodes for 800–850 nm wavelength operation, *Appl. Phys. Lett.* **72**(21): 2727–2729 (1998).
32. N. Biyikli et al., 45 GHz bandwidth-efficiency resonant cavity enhanced ITO-schottky photodiodes, *IEEE Photon. Technol. Lett.* **13**(7): 705–707 (2001).
33. M. K. Emsley, O. I. Dosunmu, and M. S. Ünlü, High-speed resonant-cavity-enhanced silicon photodetectors on reflecting silicon-on-insulator substrates, *IEEE Photon. Technol. Lett.* **14**(4): 519–521 (2002).
34. D. Dragoman and M. Dragoman, *Advanced Optoelectronic Devices*, Springer, 1999.
35. K. Kato et al., 110 GHz, 50% efficiency mushroom-mesa waveguide p-i-n photodiode for a 1.55 μm wavelength, *IEEE Photon. Technol. Lett.* **6**(6): 719–721 (1994).
36. K. Kato, Ultrawide-band/high-frequency photodetectors, *IEEE Trans. Microwave Theory Tech.* **47**(7): 1265–1281 (1999).

HORN ANTENNAS

Edward V. Jull
University of British Columbia
Vancouver, British Columbia,
Canada

A "horn" antenna is a length of conducting tube, flared at one end, and used for the transmission and reception of electromagnetic waves. For an efficient transition between guided and radiated waves, the horn dimensions must be comparable to the wavelength. Consequently horns are used mostly at centimeter and millimeter wavelengths. At lower or higher frequencies they are inconveniently large or small, respectively. They are most popular at microwave frequencies (3–30 GHz), as antennas of moderate directivity or as feeds for reflectors or elements of arrays.

Since acoustic horns have been in use since prehistoric times, the design of horns as musical instruments was a highly developed art well before the appearance of the first electromagnetic horns. This occurred shortly after Hertz in 1888 first demonstrated the existence of electromagnetic waves. Experimenters placed their spark-gap sources in hollow copper tubes (Figs. 1a,5a). These tubes acted as highpass filters for microwave and millimeter wave radiation from the open end. In London in 1897 Chunder Bose used rectangular conducting tubes with "collecting funnels," or pyramidal horns (Fig. 1d) in his demonstrations at 5 and 25 mm wavelengths [1]. Thus the electromagnetic horn antenna was introduced but this early beginning of microwave invention closed with Marconi's demonstration that longer wavelengths could be received at greater distances. Horns were too large to be practical at those wavelengths, and it was almost 40 years before microwave horns reappeared with the need for directive antennas for communications and radar. Horns alone were often not sufficiently directive but combined in an array or with a lens (Fig. 4a), or more often a parabolic reflector (Fig. 4b,c) highly directive antenna beams are obtained.

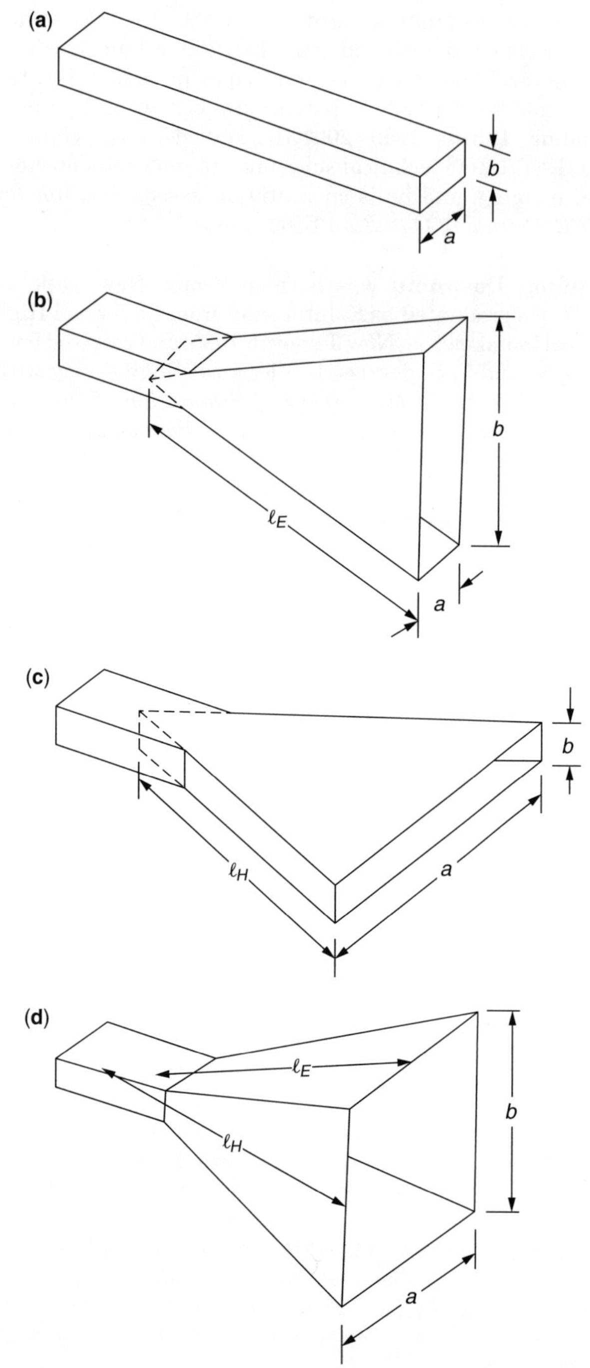

Figure 1. (**a**) Open-ended rectangular waveguide; (**b**) E-plane sectoral horn; (**c**) H-plane sectoral horn; (**d**) Pyramidal horn.

1. RADIATING WAVEGUIDES AND HORNS

Horns are normally fed by waveguides supporting only the dominant waveguide mode. For a rectangular waveguide (Fig. 1a) with TE_{01} mode propagation only, these dimensions in wavelengths λ are $\lambda/2 < a < \lambda$ and $b \approx a/2$. Open-ended waveguides have broad radiation

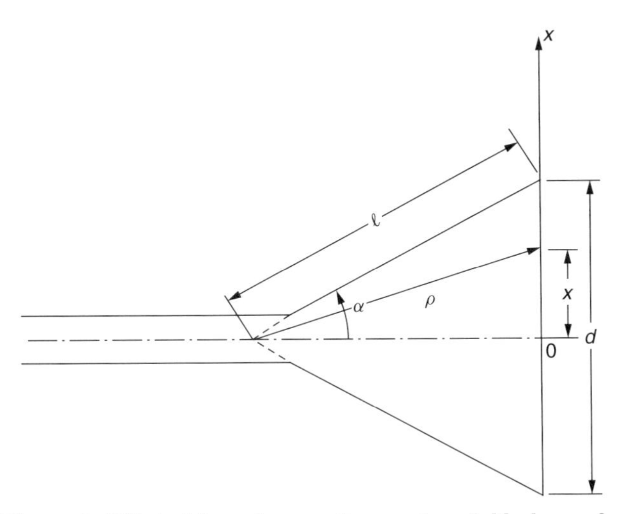

Figure 2. Effect of horn flare on the aperture field phase of a horn.

patterns, so when used as a feed for a reflector, there is substantial spillover, or radiation missing the reflector and radiation directly backward from the feed. To increase the directivity of a radiating waveguide and its efficiency as a reflector feed, for example, its aperture dimensions must be enlarged, for the beamwidth of an aperture of width $a \gg \lambda$ is proportional to λ/a radians.

This waveguide enlargement by a flare characterizes horns. The aperture fields of a horn are spherical waves originating at the horn apex (Fig. 2). The path from the horn apex to the aperture plane at a distance x from the aperture center of a horn of slant length ℓ is

$$\rho = ((\ell\cos\alpha)^2 + x^2)^{1/2} \approx \ell\cos\alpha + \frac{x^2}{2\ell\cos\alpha} \tag{1}$$

when $x \ll \ell\cos\alpha$. Thus the phase variation in radians across the aperture for small flare angles α is approximately $kx^2/(2\ell)$, where $k = 2\pi/\lambda$ is the propagation constant. This quadratic phase variation increases with increasing flare angle, thus reducing directivity increase due to the enlarged aperture dimension. It is convenient to quantify aperture phase variation by the parameter

$$s = \frac{\ell(1-\cos\alpha)}{\lambda} \approx \frac{d^2}{8\lambda\ell},\ d \ll \ell \tag{2}$$

which is the approximate difference in wavelengths between the distance from the apex to the edge ($x = d/2$) and the center ($x = 0$) of the aperture. The radiation patterns of Fig. 3a,b [2] show the effect of increasing s on the E- and H-plane radiation patterns of sectoral and pyramidal horns. The main beam is broadened, the pattern nulls are filled, and the sidelobe levels are raised over those for an in-phase aperture field ($s = 0$). With large flare angles radiation from the extremities of the aperture can be so out of phase with that from the center that the horn directivity decreases with increasing aperture width.

The adverse effects of the flare can be compensated by a lens in the aperture (Fig. 4a), but because that adds to the weight and cost and because bandwidth limitations

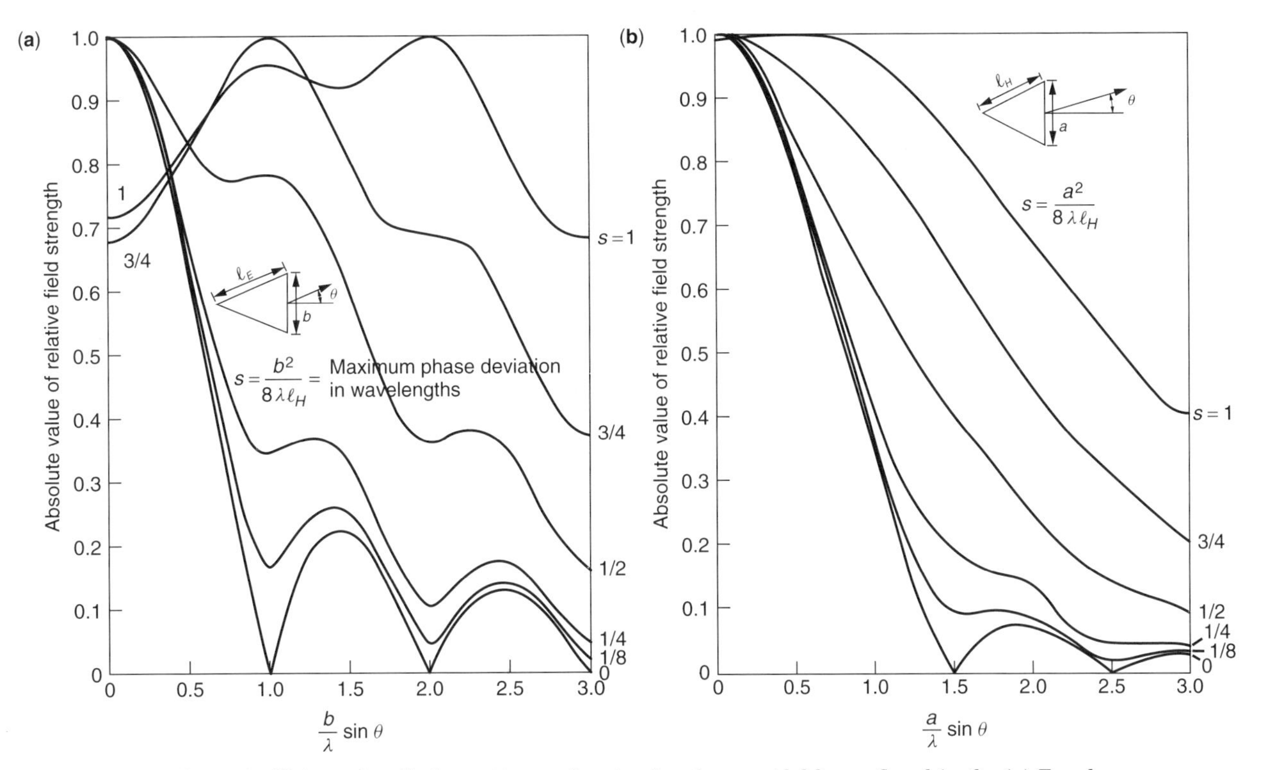

Figure 3. Universal radiation patterns of sectoral and pyramidal horns flared in the (**a**) E and (**b**) H-planes. The parameter $s = b^2/8\lambda\ell_E$ in (a) and $a^2/8\lambda\ell_H$ in (b); $2\pi s/\lambda$ is the maximum phase difference between the fields at the center and the edge of the aperture. (Copyright 1984, McGraw-Hill, Inc. from Love [2].)

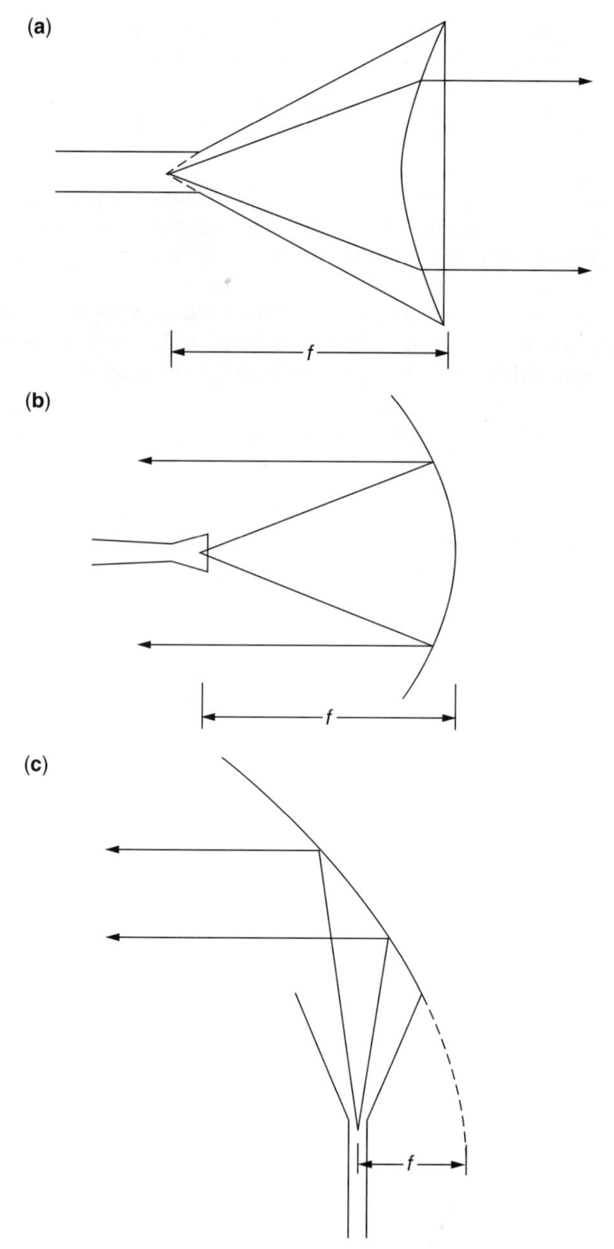

Figure 4. (**a**) Horn aperture field phase correction by a lens; (**b**) parabolic reflector fed by a horn; (**c**) horn reflector antenna (f = focal length of the lens or reflector).

are introduced by matching the lens surfaces to reduce reflections, it is seldom done. Instead, a combination of aperture width and flare length in wavelengths is chosen that provides maximum axial directivity or minimum beamwidth. This is an "optimum" horn design. To achieve higher directivity or narrower beamwidth for a given aperture width, a longer horn is required.

Sectoral horns (Fig. 1b,c) are rectangular waveguides flared in one dimension only. The incident waveguide mode becomes a radial cylindrical mode in the flared region of the horn. Since radiation pattern beamwidths are inversely proportional to aperture dimensions in wavelengths, sectoral horns have beams that are narrow in the plane containing the broad dimension. Such fan shaped beams may be useful for illuminating elongated parabolic reflectors or parabolic cylinder reflectors.

A pyramidal horn (Fig. 1d) is flared in both waveguide dimensions and so is more adaptable both as a reflector feed and on its own. The forward radiation pattern may be calculated quite accurately from Kirchhoff diffraction theory for all but small horns. The TE_{01} rectangular waveguide mode yields an aperture field uniform in one dimension (in the E plane) and cosinusoidal in the other (the H plane). A comparison of parts (a) and (b) of Fig. 3 shows that this results in a higher sidelobes in the E-plane and, for a square aperture, a narrower beam. Pyramidal horns are relatively easily constructed, and for all except small horns their axial gain can be predicted accurately. Consequently, they are used as gain standards at microwave frequencies; that is, they are used to experimentally establish the gain of other microwave antennas by comparing their response to the same illuminating field.

Most of the preceding remarks on open-ended rectangular waveguides and pyramidal horns also apply to open-ended circular waveguides and conical horns (Fig. 5a,b). For propagation of the lowest-order mode (TE_{11}) only in a circular waveguide, the interior diameter must be $0.59\lambda < a < 0.77\lambda$. This mode has a uniform aperture field in the E plane and a cosinusoidal distribution in the orthogonal H plane. This appears, modified by a quadratic phase variation introduced by the flare, in the aperture field of the horn. Consequently the E-plane radiation pattern of the horn is narrower, but with higher sidelobes than the H-plane pattern and the radiated beam is elliptical in cross-section. In addition, cross-polarized fields appear in pattern lobes outside the principal planes.

2. HORN FEEDS FOR REFLECTORS

Many refinements to horns arise from their use as efficient feeds for parabolic reflectors, particularly in satellite and space communications and radio astronomy. The phase center, where a horn's far radiation field appears to originate, must be placed at the focus of the reflector (Fig. 4b). This phase center is within the horn on the horn axis and depends on the flare angle and aperture distribution. For both rectangular and conical horns the position of the phase center is not the same in the E and H planes, or planes containing the electric and magnetic field vectors, respectively. A phase center can be calculated from the average of the positions of the E- and H-plane phase centers or determined from the position of the feed that maximizes the gain of the reflector antenna.

For efficient aperture illumination the feed horn radiation pattern should approximately match the shape of the aperture, and illuminate it essentially uniformly and with minimal spillover, or radiation missing the reflector. Pyramidal horns may seem suitable for rectangular apertures because their beams are rectangular in cross section, and conical horns may seem a natural choice for a circular aperture, but efficient aperture illumination is not obtained in either case, because their principal plane patterns differ. Both horns have high E-plane pattern

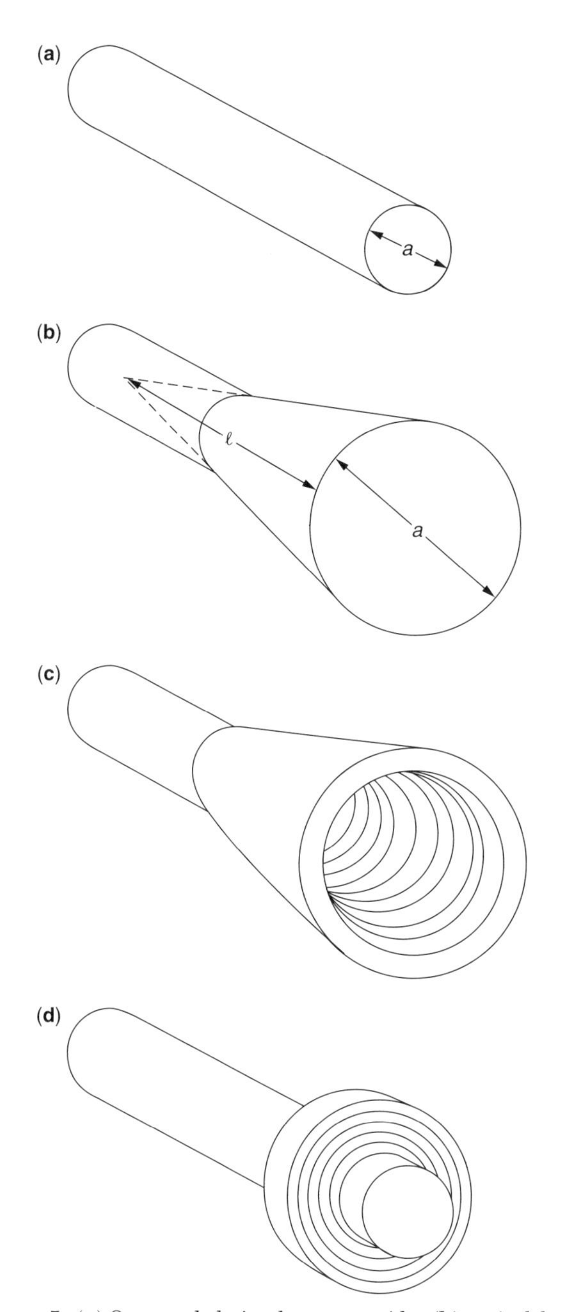

Figure 5. (**a**) Open-ended circular waveguide; (**b**) conical horn; (**c**) corrugated horn; (**d**) circular waveguide with corrugated flange.

sidelobes and low H-plane sidelobes. A dual (TE_{11}/TM_{11})-mode conical horn provides equal E- and H-plane beamwidths and equally low sidelobes, and is an efficient feed for a circular aperture over a narrow frequency band (see Love [3], p. 195; Ref. 3 also contains reprints of most earlier significant papers on horn antennas). A broadband solution achieves an axisymmetric beam with annular corrugations on the interior surfaces of a conical horn (Fig. 5c). These produce a horn aperture field distribution that is approximately cosinusoidal across the conical horn aperture in all directions and hence an axisymmetric radiation pattern with low sidelobes. Such corrugations in the E-plane interior walls only of a pyramidal horn will produce a cosinusoidal E-plane aperture distribution, and consequently similar E-plane and H-plane radiation patterns for a square horn aperture.

A feed for a small circular reflector that is more easily constructed than a corrugated conical horn but with a less axisymmetric radiation pattern, is an open-ended circular waveguide ringed by a recessed disk of approximately quarter-wavelength-deep corrugations (Fig. 5d). These corrugations suppress back radiation from the feed and so improve the aperture illumination over that of a simple open circular waveguide (Ref. 3, pp. 181, 226). Combined with dual-mode excitation, this arrangement provides a simple and efficient feed for a front-fed paraboloidal reflector.

3. RADIATION FROM APERTURES

The far-field radiation pattern of an aperture can be calculated exactly from the Fourier transform of the tangential fields in the entire aperture plane. Either electric or magnetic aperture fields may be used but for apertures in space, a combination of the two gives the best results from the usual assumption that aperture plane fields are confined to the aperture and negligible outside it. This aperture field is assumed to be the undisturbed incident field from the waveguide. For apertures with dimensions larger than several wavelengths, a further simplifying assumption usually made is that the aperture electric and magnetic fields are related as in free space.

3.1. Rectangular Apertures

With the above assumptions, at a distance much greater than the aperture dimensions, the radiated electric field intensity of a linearly polarized aperture field $E_x(x, y, 0)$ in the coordinates of Fig. 6a is

$$\overline{E}(r,\theta,\phi) = \overline{A}(r,\theta,\phi)\int_{-(b/2)}^{b/2}\int_{-(a/2)}^{a/2} E_x(x,y,0)e^{j(k_1x+k_2y)}\,dx\,dy \tag{3}$$

Here

$$\left.\begin{aligned} k_1 &= k\sin\theta\cos\phi \\ k_2 &= k\sin\theta\sin\phi \end{aligned}\right\} \tag{4}$$

and

$$\overline{A}(r,\theta,\phi) = j\frac{e^{-jkr}}{2\lambda r}(1+\cos\theta)(\hat{\theta}\cos\phi - \hat{\phi}\sin\phi) \tag{5}$$

is a vector defining the angular behaviour of the radiation polarization for an aperture in space. For an aperture in a conducting plane, it is more accurate to use

$$\overline{A}(r,\theta,\phi) = j\frac{e^{-jkr}}{\lambda r}(\hat{\theta}\cos\phi - \hat{\phi}\sin\phi\cos\theta) \tag{6}$$

which, since it is based on the aperture plane electric fields only, fully satisfies the assumption of a vanishing tangential field in the aperture plane outside the aperture. Consequently radiation fields of open-ended waveguides and small horns can be calculated accurately from (3) with (6) if they are mounted in a conducting plane.

Clearly (5) and (6) differ significantly only for large angles θ off the beam axis.

If the aperture field is separable in the aperture coordinates—that is, if in (3), $E_x(x,y,0) = E_0E_1(x)E_2(y)$ where $E_1(x)$ and $E_2(y)$ are field distributions normalized to E_0, the double integral is the product of two single integrals.

$$E(r,\theta,\phi) = \overline{A}(r,\theta,\phi)E_0F_1(k_1)F_2(k_2) \tag{7}$$

where

$$F_1(k_1) = \int_{-(a/2)}^{a/2} E_1(x)e^{jk_1y}\,dx \tag{8}$$

$$F_2(k_2) = \int_{-(b/2)}^{b/2} E_2(y)e^{jk_2y}\,dy \tag{9}$$

define the radiation field.

4. OPEN-ENDED WAVEGUIDES

4.1. Rectangular Waveguides

With the TE_{10} waveguide mode the aperture field

$$E_x(x,y,0) = E_0\cos\frac{\pi y}{a} \tag{10}$$

in (7) yields the following equations for (8) and (9):

$$F_1(k_1) = b\frac{\sin\left(\frac{k_1b}{2}\right)}{\frac{k_1b}{2}} \tag{11}$$

$$F_2(k_2) = a\left(\frac{\cos\left(\frac{k_2a}{2}\right)}{\pi^2-(k_2a)^2}\right) \tag{12}$$

This defines the radiation pattern in the forward hemisphere $-\pi/2 < \theta < \pi/2, 0 < \phi < 2\pi$. If the aperture is in space, then (5) is used for $\overline{A}(r,\theta,\phi)$, but this is not an accurate solution since the aperture dimensions are not large. Rectangular waveguides mounted in conducting planes use (6) for $\overline{A}(r,\theta,\phi)$ in (7), which then accurately provides the far field. The pattern has a single broad lobe with no sidelobes. For large apertures plots of the normalized E-plane ($\phi = 0$) and H-plane ($\phi = \pi/2$) patterns of (7) appear in Fig. 3a,b for those of a horn with no flare ($s = 0$), but without the factor $(1+\cos\theta)/2$ from (5) or $\cos\theta$ from (6).

4.2. Circular Waveguides

The dominant TE_{11} mode field in circular waveguide produces an aperture field distribution, which in the aperture coordinates ρ',ϕ' of Fig. 6b is

$$\overline{E}(\rho',\phi') = E_0\left[\hat{\rho}'\frac{J_1(k_c\rho')}{k_c\rho'}\cos\phi' + \hat{\phi}'J_1'(k_c\rho')\sin\phi'\right] \tag{13}$$

where J_1 is the Bessel function of the first kind and order, J_1' is its derivative with respect to its argument $k_c\rho'$, and $k_ca/2 = 1.841$ is its first root; E_0 is the electric field at the aperture center ($\rho' = 0$). Since (13) is not linearly polarized, its use in (3) provides only part of the total radiated far field. The total field

$$\overline{E}(r,\theta,\phi) = jkaE_0J_1(1.841)\frac{e^{-jkr}}{r}\left\{\hat{\theta}\cos\phi\frac{J_1\left(\frac{k'a}{2}\right)}{\frac{k'a}{2}} + \hat{\phi}\sin\phi\cos\theta\frac{J_1'\left(\frac{k'a}{2}\right)}{1-\left(\frac{k'a}{3.682}\right)^2}\right\} \tag{14}$$

in which $k' = k\sin\theta$.

In the E and H planes ($\phi = 0$ and $\pi/2$) the cross-polarized fields cancel and the patterns shown in Fig. 14a are similar to those of (11) and (12), respectively, but with slightly broader beams and lower sidelobes for

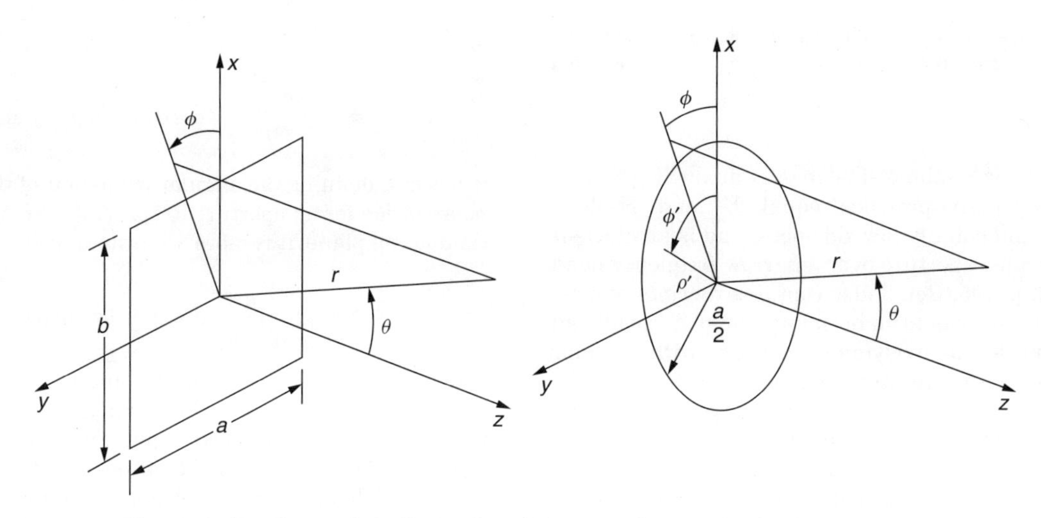

Figure 6. Coordinates for radiation from (**a**) rectangular and (**b**) circular apertures.

the same aperture dimensions. As with rectangular waveguides, open-ended circular waveguide apertures are insufficiently large for (14) to represent all the radiated fields accurately. In the principal planes ($\phi = 0, \pi/2$), it can give a reasonable approximation for the copolarized fields but fails to accurately represent the cross-polarized field patterns in $\phi = \pi/4$. This is evident from a comparison of numerical results from approximate and exact solutions (see Collin [4], p. 233).

5. PYRAMIDAL AND SECTORAL HORNS

5.1. Radiation Patterns

A pyramidal horn fed by a rectangular waveguide supporting the TE_{10} mode has an incident electric field in the aperture of Fig. 6a that is approximately the mode distribution modified by a quadratic phase variation in the two aperture dimensions:

$$E_x(x, y, 0) = E_0 \cos\left(\frac{\pi y}{a}\right) \exp\left(-jk\left(\frac{x^2}{2\ell_E} + \frac{y^2}{2\ell_H}\right)\right) \quad (15)$$

With (15), Eq. (3) becomes

$$\overline{E}(r, \theta, \phi) = \overline{A}(r, \theta, \phi) E_0 I_1(k_1) I_2(k_2) \quad (16)$$

where (5) is used for $\overline{A}(r, \theta, \phi)$ and

$$I_1(k_1) = \int_{-(b/2)}^{b/2} \exp\left(-j\left(\frac{\pi x^2}{\lambda \ell_E} - k_1 x\right)\right) dx \quad (17)$$

$$I_2(k_2) = \int_{-(a/2)}^{a/2} \cos\left(\frac{\pi y}{a}\right) \exp\left(-j\left(\frac{\pi y^2}{\lambda \ell_H} - k_2 y\right)\right) dy \quad (18)$$

The E-plane ($\phi = 0$) and H-plane ($\phi = \pi/2$) radiation patterns are, respectively

$$\frac{E_\theta(r, \theta)}{E_\theta(r, 0)} = \frac{1 + \cos\theta}{2} \frac{I_1(k \sin\theta)}{I_1(0)} \quad (19)$$

$$\frac{E_\theta(r, \theta)}{E_\theta(r, 0)} = \frac{1 + \cos\theta}{2} \frac{I_2(k \sin\theta)}{I_2(0)} \quad (20)$$

These integrals can be reduced to the Fresnel integrals

$$C(u) - jS(u) = \int_0^u e^{-j(\pi/2)t^2}\, dt \quad (21)$$

which are tabulated and for which computer subroutines are available. For example,

$$\frac{I_1(k \sin\theta)}{I_1(0)} = \frac{e^{j(\pi \ell_E/\lambda)\sin^2\theta}}{2} \frac{C(u_2) - C(u_1) - j[S(u_2) - S(u_1)]}{C(u) - jS(u)} \quad (22)$$

with

$$u = \frac{b}{\sqrt{2\lambda \ell_E}} \quad (23)$$

$$u_{\substack{2\\1}} = \pm u - \sqrt{\frac{2\ell_E}{\lambda}} \sin\theta \quad (24)$$

Figure 3a shows plots of the magnitude of (22) for various values of the E-plane flare parameter $s = b^2/8\lambda\ell_E$, while Fig. 3b shows corresponding plots of $|I_2(k \sin\theta)/I_2(0)|$ for the H-plane flare parameter $s = a^2/8\lambda\ell_H$. For no flare ($s = 0$) the patterns are those of a large open-ended rectangular waveguide supporting only the TE_{10} mode. The effect of the flare is to broaden the main beam, raise the sidelobes, and fill the pattern nulls. For larger values of s, there is enhanced pattern beam broadening and eventually a splitting of the main beam on its axis.

These curves also represent the radiation patterns of the E/H-plane sectoral horns of Fig. 1b,c. For an E-plane sectoral horn ($\ell_H \to \infty$), the E-plane pattern is given by (19) and the H-plane pattern approximately by (12). For an H-plane sectoral horn ($\ell_E \to \infty$), the E-plane pattern is given approximately by (11) and the H-plane pattern by (20).

In comparing parts (a) and (b) of Fig. 3 it is evident that E-plane beamwidths of a square aperture are narrower than H-plane beamwidths. For horns of moderate flare angle and optimum horns the E-plane half-power beamwidth is $0.89\lambda/b$ radians and the H-plane half-power beamwidth is $1.22\lambda/a$ radians. E-plane patterns have minimum sidelobes of -13.3 dB below peak power, while H-plane pattern minimum sidelobes levels are -23.1 dB.

The universal patterns of Fig. 3a,b can also be used to predict the approximate near-field radiation patterns of horns by including the quadratic phase error; which is a first-order effect of finite range r. This is done by including

$$\exp\left(-j\frac{\pi}{r\lambda}(x^2 + y^2)\right) \quad (25)$$

in (15). Then the near field principal plane patterns of a pyramidal horn are given by (17) and (18) with ℓ_E, ℓ_H replaced by

$$\ell'_H = \frac{r\ell_H}{r + \ell_H} \quad (26)$$

and

$$\ell'_E = \frac{r\ell_E}{r + \ell_E} \quad (27)$$

These near-field effects are analogous to decreasing the flare length of a horn with a fixed aperture width. The main beam broadens, nulls are filled in, and sidelobes rise.

5.2. Limitations and Extensions

Results from (16) do not apply to small horns and are limited to the forward direction ($\theta < 90°$). They are most accurate on and around the beam axis ($\theta = 0$), becoming progressively less accurate as θ increases. The simplest method for extending the analysis is by the uniform geometric theory of diffraction (see, e.g., Ref. 3, p. 66), which provides the edge-diffracted fields in the lateral and rear directions, which receive no direct illumination from the aperture. Only the edges normal to the plane of the pattern contribute significantly to the E-plane pattern, but the rear H-plane pattern requires contributions from all four aperture edges and so is difficult to calculate this way.

While the geometry of the pyramidal horn defies rigorous analysis, numerical methods have been used with some success for open waveguides and small horns. For larger horns this approach becomes computationally

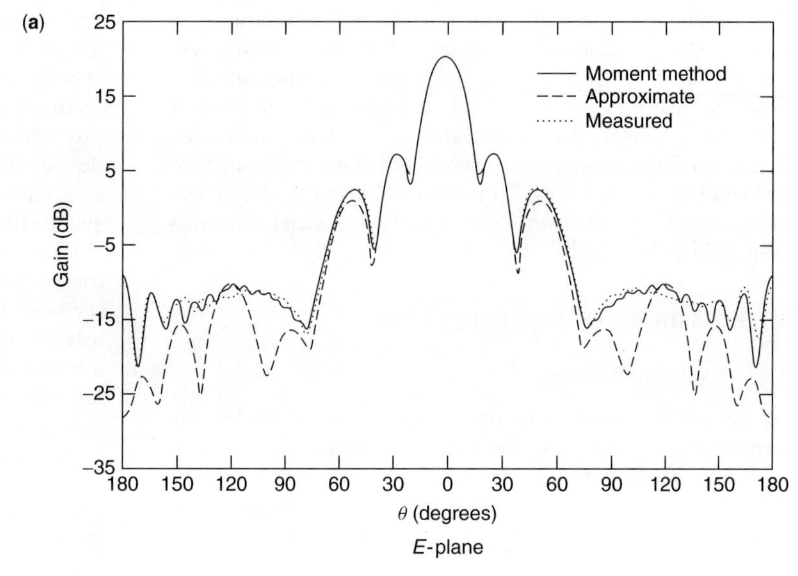

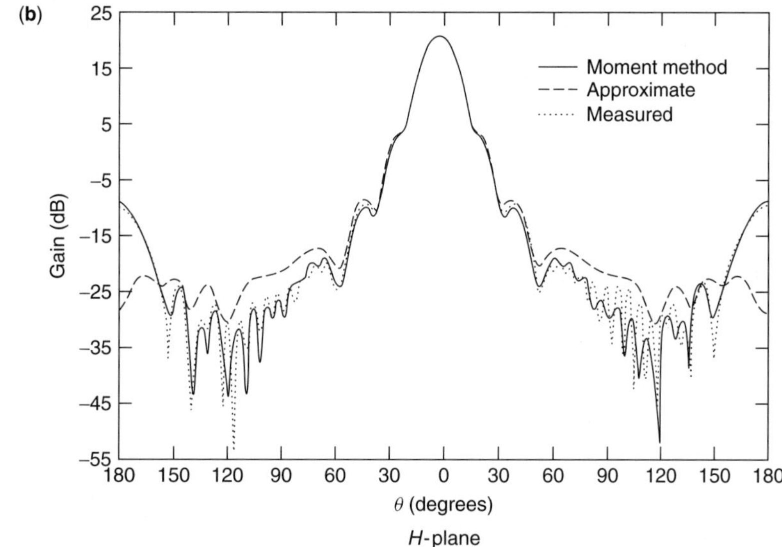

Figure 7. Calculated and measured (**a**) E-plane and (**b**) H-plane radiation patterns of a pyramidal horn of dimensions $a = 4.12\lambda$, $b = 3.06\lambda$, $\ell_E = 10.52\lambda$ $\ell_E = 9.70\lambda$. (Copyright 1993, IEEE, from Liu et al. [5].)

intensive, but some results from Liu et al. (5) are shown in Fig. 7 and compared with measurements and approximate computations. Their numerical computations and measurements by Nye and Liang (6) of the aperture fields show that higher-order modes need to be added to the dominant mode field of (15) and that the parabolic phase approximation of (1) improves as the aperture size increases.

5.3. Gain

Pyramidal horns are used as gain standards at microwave frequencies because they can be accurately constructed and their axial directive gain reliably predicted from a relatively simple formula. The ratio of axial far-field power density to the average radiated power density from (16) yields

$$G = G_0 R_E(u) R_H(v, w) \tag{28}$$

where $G_0 = 32ab/(\pi\lambda^2)$ is the gain of an in-phase uniform and cosinusoidal aperture distribution. The reduction of this gain due to the phase variation introduced by the E-plane flare of the horn is

$$R_E(u) = \frac{C^2(u) + S^2(u)}{u^2} \tag{29}$$

where the Fresnel integrals and their argument are defined by (21) and (23). Similarly the gain reduction factor due to the H-plane flare of the horn is

$$R_H(v, w) = \frac{\pi^2}{4} \frac{[C(v) - C(w)]^2 + [S(v) - S(w)]^2}{(v - w)^2} \tag{30}$$

where

$$\begin{matrix} v \\ w \end{matrix} = \pm \frac{a}{\sqrt{2\lambda\ell_H}} + \frac{1}{a}\sqrt{\frac{\lambda\ell_H}{2}} \tag{31}$$

A plot of R_E and R_H in decibels as a function of the parameter $2d^2/\lambda\ell$, where d is the appropriate aperture dimension b or a and ℓ the slant length ℓ_E or ℓ_H, respectively, is shown in Fig. 8. Calculation of the gain

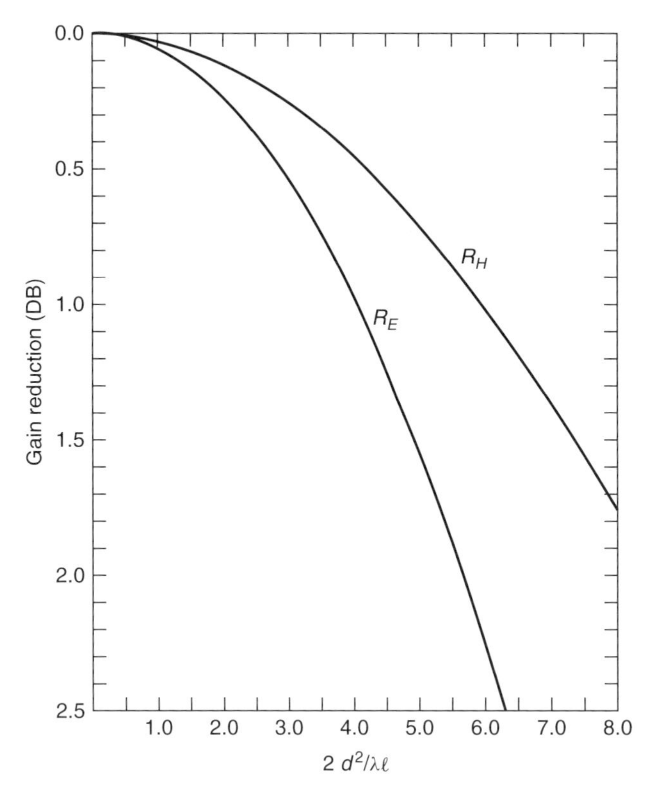

Figure 8. E- and H-plane flare and near-field gain reduction factors R_E and R_H of pyramidal and sectoral horns in decibels. (Copyright 1981, IEE, from Jull [11].)

from (28) is accurate to about ± 0.1 dB for 22 dB standard gain pyramidal horns: optimum horns with dimensions of at least 5λ. For 18-dB-gain horns, the accuracy is about ± 0.2 dB, and for 15 dB horns, ± 0.5 dB. Since optimum gain pyramidal horns have an aperture efficiency of approximately 50%, the gain is approximately

$$G = 0.5\frac{4\pi}{\lambda^2}ab \tag{32}$$

For an E-plane sectoral horn $\ell_H \to \infty$ and $R_H(v, w) \to 1$ the axial gain is then $G_E = G_0 R_E(u)$, an inaccurate formula because aperture dimension a is less than a wavelength. A result that includes the fact that aperture electric and magnetic fields are not related by free-space conditions and that interaction occurs across the narrow aperture of the horn is

$$G_E = \frac{16ab}{\lambda^2(1 + \lambda g/\lambda)} R_E(u') \exp\left[\frac{\pi a}{\lambda}\left(1 - \frac{\lambda}{\lambda_g}\right)\right] \tag{33}$$

where

$$u' = \frac{b}{\sqrt{2\lambda_g \ell_E}} \tag{34}$$

and

$$\lambda_g = \frac{\lambda}{\sqrt{1 - (\lambda/2a)^2}} \tag{35}$$

is the guide wavelength. The accuracy of (33) is comparable to that of (28) for the horns of similar b dimension.

The gain of an H-plane sectoral horn, obtained by letting $\ell_E \to \infty$ so that $R_E(u) \to 1$, is $G_H = G_0 R_H(v, w)$. It probably is reasonably accurate, but there appears to be no experimental evidence available to verify it.

The near-field gain of pyramidal and sectoral horns can be calculated from the preceding expressions by replacing ℓ_E and ℓ_H by (26) and (27), respectively.

6. CONICAL HORNS

The aperture field of a conical horn fed by a circular waveguide supporting the TE_{11} mode is approximately

$$E(\rho', \phi') \exp\left(\frac{-jk\rho'^2}{2\ell}\right) \tag{36}$$

where $\overline{E}(\rho', \phi')$ is given by (13) and ℓ is the slant length of the horn. Numerical calculation of the radiation patterns is necessary. In the example of Fig. 9 [7] with a flare angle $\alpha = 5^\circ$ and aperture width $a = 4\lambda$, the E-plane ($\phi = 0$) pattern is narrower than the H-plane ($\phi = \pi/2$) pattern as in square rectangular horns. The cross-polar ($\phi = \pi/4$) radiation pattern peak level is -18.7 dB relative to the copolar pattern peak levels, a level typical of conical horn apertures larger than about 2λ. Smaller conical horns can have more axisymmetric patterns. E- and H-plane patterns have equal beamwidths for an aperture diameter $a = 0.96\lambda$, and cross-polarized fields cancel for $a = 1.15\lambda$. This makes small conical horns efficient as reflector feeds and as array elements with high polarization purity.

Larger conical horns are similar to rectangular horns in their lack of axial pattern symmetry. Optimum gain conical horns have an aperture efficiency of about 54% and half-power beamwidths in the E and H planes of $1.05\lambda/a$ and $1.22\lambda/a$ radians, respectively, for aperture diameters of more than a few wavelengths.

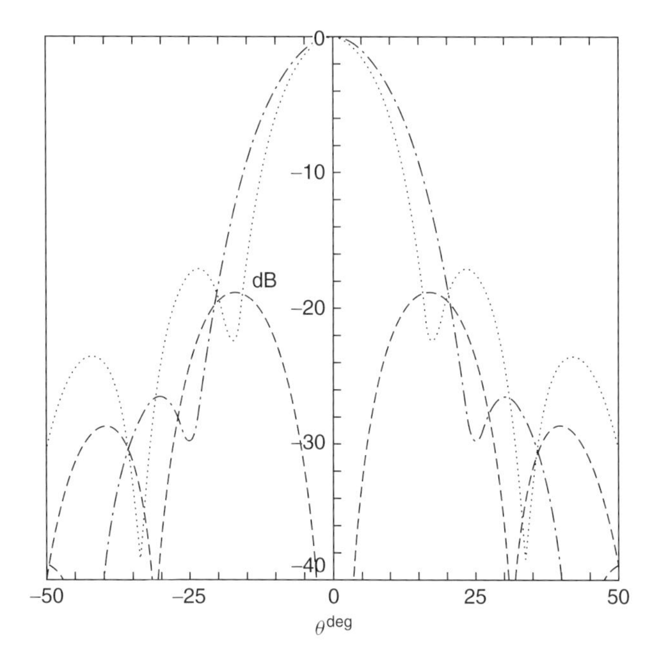

Figure 9. Copolar and crosspolar radiation patterns for a conical horn with dimensions $a = 4\lambda$, $\ell = 23\lambda$ - - - - E-plane, — · — · — H-plane — — — — crosspolarization. (Copyright 1994, IEE, from Olver et al. [7].)

7. MULTIMODE AND CORRUGATED HORNS

Lack of axisymmetric radiation patterns make rectangular and conical horns inefficient reflector feeds. Conical horns also have unacceptably high cross-polarization levels if used as reflector feeds in a system with dual polarization. Multimode and corrugated horns were developed largely to overcome these deficiencies. In a dual-mode horn in (Ref. 3, p. 195), this is done by exciting the TM_{11} mode, which propagates for circular waveguide diameters $a > 1.22\lambda$, in addition to the TE_{11} mode, which propagates for $a > 0.59\lambda$. The electric field configuration of these modes in a waveguide cross section is shown in Fig. 10a,b. Added in phase and in the right proportion, cross-polarized and aperture perimeter fields cancel, while the copolar fields around the aperture center add, yielding the aperture field configuration of Fig. 10c. These mixed mode fields are linearly polarized and taper approximately cosinusoidally radially across the aperture. This yields the essentially linearly polarized and axisymmetric radiation patterns desired.

Partial conversion of TE_{11} to TM_{11} fields can be effected by a step discontinuity in the circular waveguide feed, as in Fig. 10d, or by a circular iris or dielectric ring in the horn. The TM_{11}/TE_{11} amplitude ratio depends on the ratio of waveguide diameters, and the relative phase of the modes depends on the length of larger-diameter circular waveguide and the horn. This dependence limits the frequency bandwidth of the horn to about 5%. A multimode square pyramidal horn has similarly low sidelobe levels in its E- and H-plane radiation patterns because of an essentially cosinusoidal aperture distribution in both E and H planes [2]. This is achieved by excitation of a hybrid TE_{21}/TM_{21} mode either by an E-plane step discontinuity or by changes in the E-plane flare. With their bandwidth very limited, dual-mode horns have largely been replaced by corrugated horns in dual-polarization systems, except where a lack of space may give an advantage to a thin-walled horn.

Corrugated horns have aperture fields similar to those of Fig. 10c and consequently similar radiation patterns, but without the frequency bandwidth limitations of the multimode horn. This is achieved by introducing annular corrugations to the interior walls of a conical horn. There must be sufficient corrugations per wavelength (at least 3) that the annular electric field E_ϕ is essentially zero on the interior walls. The corrugations make the annular magnetic field H_ϕ also vanish. This requires corrugation depths such that short circuits at the bottom of the grooves appear as open circuits at the top, suppressing axial current flow on the interior walls of the horn. This groove depth is $\lambda/4$ on a plane corrugated surface or a curved surface of large radius. For a curved surface of smaller radius, such as near the throat of the horn, the slot depths need to be increased; For example, for a surface radius of 2λ, the depth required is 0.3λ. Usually slots are normal to the conical surface in wide-flare horns but are often perpendicular to the horn axis with small flares. To provide a gradual transition from the TE_{11} mode in the wave guide to a hybrid HE_{11} mode in the aperture, the depth of the first corrugation in the throat should be about 0.5λ so that the surface there resembles that of a conducting cone interior. Propagation in corrugated conical horns can be accurately calculated numerically by mode matching techniques. The aperture field is approximately

$$E_x(\rho') = AJ_0(k_c\rho') \exp\left(\frac{-jk\rho'^2}{2\ell}\right) \tag{37}$$

where $k_c a/2$ is 2.405, the first zero of the zero order Bessel function J_0; ℓ is the slant length of the horn; and A is a constant. This aperture field is similar to that of Fig. 10c, and the resulting E and H patterns are similarly equal down to about −25 dB. Some universal patterns are shown in Fig. 11. Cross-polarization fields are also about −30 dB from the axial values, but now over a bandwidth of 2–1 or more.

Broadband axisymmetric patterns with low cross-polarization make corrugated horns particularly attractive as feeds for reflectors. Low cross-polarization allows the use of dual polarization to double the capacity of the system. Another notable feature for this application is that the position of the E- and H-plane pattern phase centers coincide. Figure 12 shows the distance of the phase center from the horn apex, divided by the slant length, of small flare angle conical [8] and corrugated [9] horns for values of the phase parameter s given by (2). For a conical horn the E-plane phase center is significantly farther from the aperture than the H-plane phase center. Thus, if a conical horn is used to feed a parabolic reflector, the best location for the feed is approximately midway between the E- and H-plane phase centers. With a corrugated horn such a compromise is not required, so it is inherently more efficient.

Corrugated horns may have wide flare angles, and their aperture size for optimum gain decreases correspondingly. For example, with a semiflare angle of 20°, the optimum aperture diameter is about 8λ, whereas for a semiflare angle of 70° it is 2λ. Wide-flare corrugated horns are sometimes called "scalar horns" because of their low cross-polarization levels.

For radio astronomy telescope feeds and other space-science applications, efficient corrugated horns have been

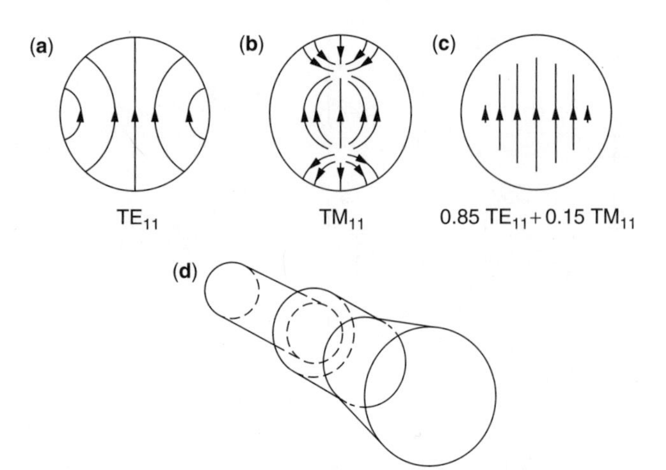

Figure 10. Excitation of axisymmetric linearly polarized aperture fields in a stepped conical horn. (Copyright 1984, McGraw-Hill, Inc. from Love [2].)

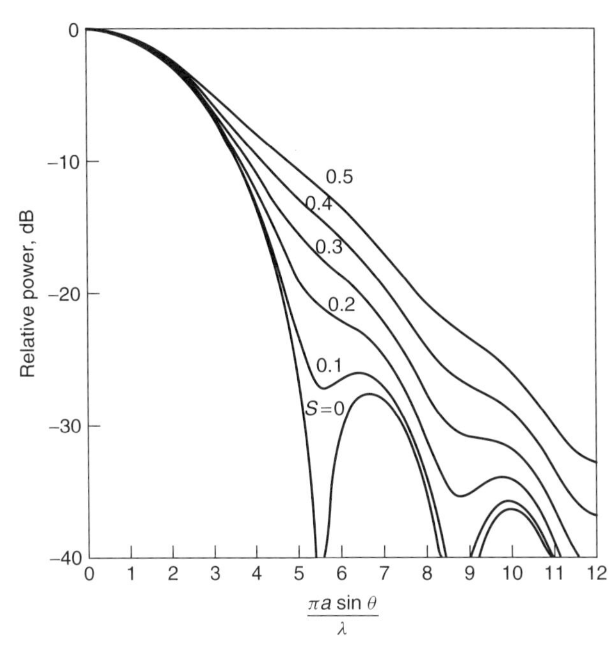

Figure 11. Universal patterns of small-flare-angle corrugated horns as a function of the parameter $s = a^2/8\lambda\ell$. (Copyright 1984, McGraw-Hill, Inc. from Love [2].)

made by electroforming techniques for frequencies up to 640 GHz. Their axisymmetric radiation patterns with very low sidelobe levels resemble Gaussian beams, which is often essential at submillimeter wavelengths.

8. PROFILE HORNS

Most corrugated horns are conical with a constant flare angle. Figure 13 shows a profile conical horn in which the flare angle varies as on a sine-squared or similar curve along its length. This arrangement provides a horn shorter than a conical corrugated horn of similar beamwidth, with a better impedance match due to the curved profile at the throat and an essentially in-phase aperture field distribution due to the profile at the aperture. Consequently the aperture efficiency is higher than that of conical corrugated horns. The phase center of the horn is near the aperture center and remains nearly fixed over a wide frequency band. Radiation patterns of a short profile horn similar to that of Fig. 13, but with hyperbolic profile curves, are shown in Fig. 14 [10]. A Gaussian profile curve has also been used. All produce patterns similar to those of a Gaussian beam, such as is radiated from the end of an optical fiber supporting the HE_{11} mode. The performance of this small horn as a feed seems close to ideal, but larger-profile horns may exhibit higher sidelobe levels due to excitation of the HE_{12} mode at the aperture.

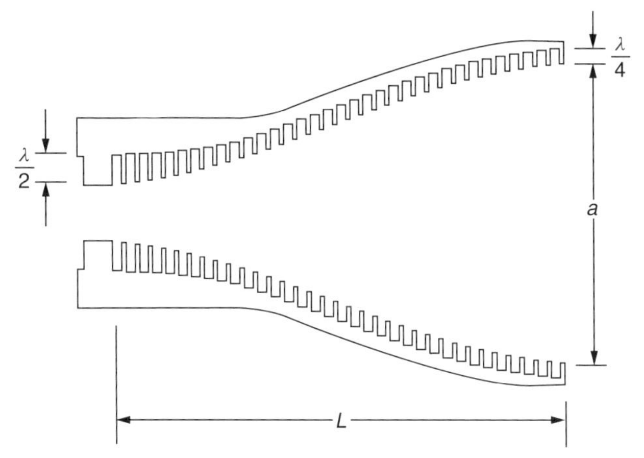

Figure 13. A profile corrugated horn. (Copyright 1994, IEE, from Olver et al. [7].)

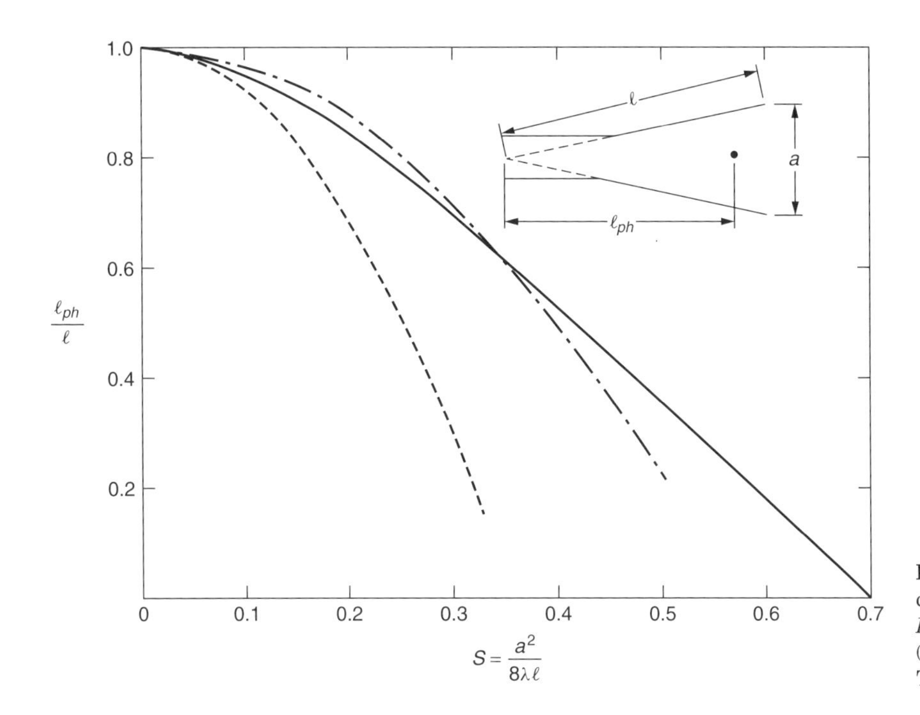

Figure 12. Normalized distance of the phase center from the apex of conical (- - - - E-plane, — · — · — H-plane and corrugated (———) horns). (Data from Milligan [8] and Thomas [9].)

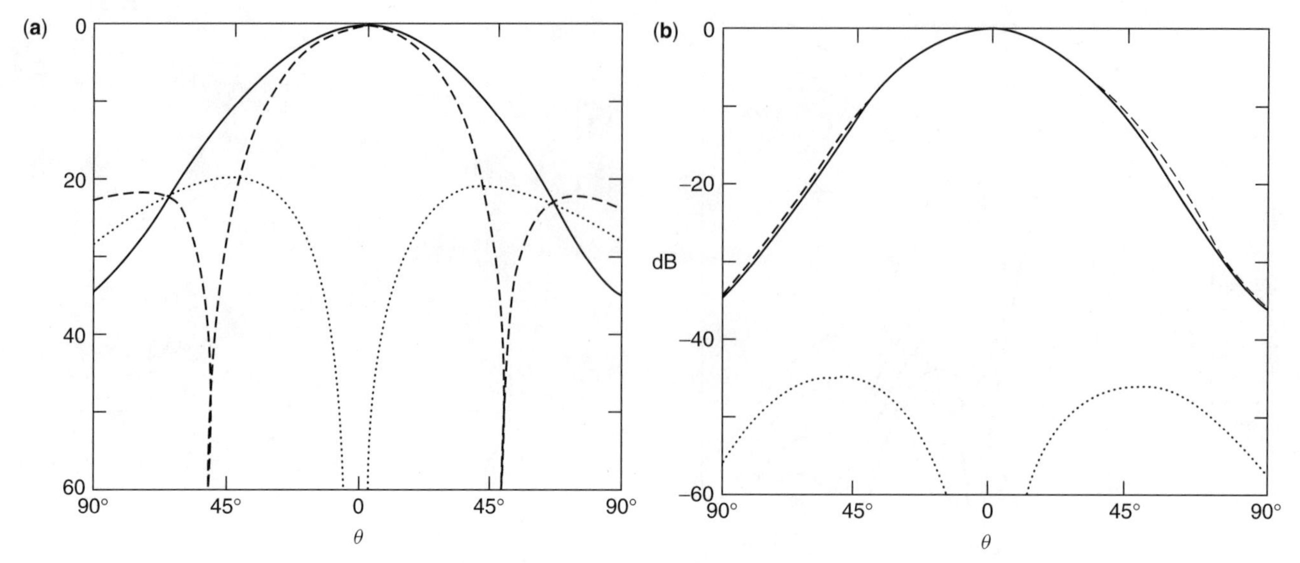

Figure 14. (**a**) Far field radiation patterns of TE_{11} mode and (**b**) radiation patterns of a profile corrugated horn of aperture $a = 15.8$ mm and length $L = 26.7$ mm at 30 GHz (- - - - - E-plane, ——— H-plane ...... crosspolarization). (Copyright 1997, IEEE, from Gonzalo et al. [10].)

9. HORN IMPEDANCE

Antennas must be well matched to their transmission lines to ensure a low level of reflected power. In microwave communications systems levels below −30 dB are commonly required.

The impedance behavior of a horn depends on the mismatch at the waveguide/horn junction and at its aperture. For an E-plane sectoral horn, reflections from these discontinuities are comparable in magnitude, and since they interfere, the total reflection coefficient oscillates with frequency and the input voltage standing-wave ratio (VSWR) may vary from 1.05 at high frequencies to 1.5 at the lowest frequency. With E-plane sectoral horns aperture reflection is much stronger than junction reflection, so their VSWRs increase almost monotonically with decreasing frequency. An inductive iris in the waveguide near the E-plane horn junction can match its discontinuity. A capacitive iris may be similarly used for an H-plane sectoral horn. Aperture reflections in these horns may be matched with dielectric covers.

Pyramidal horns of sufficient size and optimum design tend to be inherently well matched to their waveguide feeds because the E/H-plane aperture and flare discontinuities partially cancel. For example, a 22-dB-gain horn has a VSWR of about 1.04 and an 18 dB horn a VSWR of less than 1.1.

Conical horns fed by circular waveguides supporting the dominant TE_{11} mode have an impedance behavior similar to that of pyramidal horns of comparable size fed by rectangular waveguides. The waveguide/horn discontinuities of both horns may be matched by an iris placed in the waveguide near the junction. A broader bandwidth match is provided by a curved transition between the interior walls of the waveguide and the horn. Broadband reduction of aperture reflection may be similarly reduced by a curved surface of a few wavelengths' radius. Such "aperture-matched" horns also have lower sidelobe levels and less back radiation in their E-plane patterns than do conventional pyramidal and conical horns. Their H-plane flare patterns are affected little by such aperture matching because the electric field vanishes at the relevant edges.

For dual-mode and corrugated horns there are also negligible fields at the aperture edges and hence little diffraction there. Corrugated horns with initial groove depths near the throat of about a half-wavelength and which gradually decrease to a quarter-wavelength near the aperture, as in Fig. 13, are well matched at both throat and aperture. For most well-designed corrugated horns a VSWR of less than 1.25 is possible over a frequency range of about 1.5–1. Dual-mode horns using a step discontinuity as in Fig. 10d may have a VSWR of 1.2–1.4. If an iris is required for a match, the frequency bandwidth will, of course, be limited. Conical and pyramidal horns using flare angle changes to generate the higher-order modes can have VSWRs less than 1.03 and require no matching devices.

BIOGRAPHY

Edward V. Jull received his B.Sc.degree in engineering physics from Queen's University, Kingston, Ontario, Canada in 1956, a Ph.D in electrical engineering in 1960 and a D.Sc.(Eng.) in 1979, both from the University of London, United Kingdom. He was with the Division of Radio and Electrical Engineering of the National Research Council, Ottawa, Canada, from 1956 to 1957 in the microwave section and from 1961 to 1972 in the antenna engineering section. From 1963 to 1965 he was a guest researcher in the Laboratory of Electromagnetic Theory of the Technical University of Denmark in Lyngby, and the Microwave Department of the Royal Institute of Technology in Stockholm, Sweden. In 1972, he joined the Department of Electrical Engineering at the University of British Columbia in Vancouver, British Columbia, Canada. He became professor emeritus in 2000, but remains involved in teaching and research on aperture antennas and diffraction theory and is the author of a book so titled. He was president of the International Union of Radio Science (URSI) from 1990 to 1993.

BIBLIOGRAPHY

1. J. F. Ramsay, Microwave antenna and waveguide techniques before 1900, *Proc. IRE* **46**: 405–415 (1958).
2. A. W. Love, Horn antennas, in R. C. Johnson and H. Jasik, eds., *Antenna Engineering Handbook*, 2nd ed., McGraw-Hill, New York, 1984, Chap. 15.
3. A. W. Love, ed., *Electromagnetic Horn Antennas*, IEEE Press, Piscataway, NJ, 1976.
4. R. E. Collin, *Antennas and Radiowave Propagation*, McGraw-Hill, New York, 1985.
5. K. Liu, C. A. Balanis, C. R. Birtcher, and G. C. Barber, Analysis of pyramidal horn antennas using moment methods, *IEEE Trans. Antennas Propag.* **41**: 1379–1389 (1993).
6. J. F. Nye and W. Liang, Theory and measurement of the field of a pyramidal horn, *IEEE Trans. Antennas Propag.* **44**: 1488–1498 (1996).
7. A. D. Olver, P. J. B. Clarricoats, A. A. Kishk, and L. Shafai, *Microwave Horns and Feeds*, IEE Electromagnetic Waves Series, Vol. 39, IEE, London, 1994.
8. T. Milligan, *Modern Antenna Design*, McGraw-Hill, New York, 1985, Chap. 7.
9. B. MacA. Thomas, Design of corrugated horns, *IEEE Trans. Antennas Propag.* **26**: 367–372 (1978).
10. R. Gonzalo, J. Teniente, and C. del Rio, Very short and efficient feeder design from monomode waveguide, *IEEE Antennas and Propagation Soc. Int. Symp. Digest*, Montreal, 1997, pp. 468–470.
11. E. V. Jull, *Aperture Antennas and Diffraction Theory*, IEE Electromagnetic Waves Series, Vol. 10, IEE, London, 1981.

HUFFMAN CODING

En-hui Yang
Da-ke He
University of Waterloo
Waterloo, Ontario, Canada

1. INTRODUCTION

Consider a data sequence $x = x_1x_2\cdots x_n$, where each x_i is a letter from an alphabet $\mathcal{A}$. For example, the sequence x may be a text file, an image file, or a video file. To achieve efficient communication or storage, one seldom transmits or stores the raw-data sequence x directly; rather, one usually transmits or stores an efficient binary representation of x, from which the raw data x can be reconstructed. The process of converting the raw data sequence into its efficient binary representation is called *data compression*.

A simple data compression scheme assigns each letter $a \in \mathcal{A}$ a binary sequence $C(a)$ called the *codeword* of the letter a and then replaces each letter x_i in x by its codeword $C(x_i)$, yielding a binary sequence $C(x_1)C(x_2)\cdots C(x_n)$. The mapping C, which maps each letter $a \in \mathcal{A}$ into its codeword $C(a)$, is called a *code*. For example, the American Standard Code for Information Interchange (ASCII) assigns a 7-bit codeword to each letter in the alphabet of size 128 consisting of numbers, English letters, punctuation marks, and some special characters.

The ASCII code is a *fixed-length code* because all codewords have the same length. Fixed-length codes are efficient only when all letters are equally likely. In practice, however, letters appear with different frequencies in the raw-data sequence. In this case, one may want to assign short codewords to letters with high frequencies and long codewords to letters with low frequencies, thus making the whole binary sequence $C(x_1)C(x_2)\cdots C(x_n)$ shorter. In general, *variable-length codes*, in which codewords may be of different lengths, are more efficient than fixed-length codes.

For a variable-length code, the *decoding* process—the process of recovering x from $C(x_1)C(x_2)\cdots C(x_n)$—is a bit more complicated than in the case of fixed-length codes. A variable-length code (or simply code) is said to be *uniquely decodable* if one can recover x from $C(x_1)C(x_2)\cdots C(x_n)$ without ambiguity. As the uniquely decodable concept suggests, not all one-to-one mappings from letters to codewords are uniquely decodable codes [1, Table 5.1]. Also, for some uniquely decodable code, the decoding process is quite involved; one may have to look at the entire binary sequence $C(x_1)C(x_2)\cdots C(x_n)$ to even determine the first letter x_1. A uniquely decodable code with *instantaneous* decoding capability—once a codeword is received, it can be decoded immediately without reference to the future codewords—is called a *prefix code*. A simple way to check whether a code is a prefix code is to verify the *prefix-free* condition: a code C is a prefix code if and only if no codeword in C is a prefix of any other codeword. A prefix code can also be represented by a binary tree in which terminal nodes are assigned letters from $\mathcal{A}$ and the codeword of a letter is the sequence of labels read from the root to the terminal node corresponding to the letter.

Example 1. Consider a code C that maps letters in $\mathcal{A} = \{a_1, a_2, \ldots, a_5\}$ to codewords in $\mathcal{B} = \{0, 100, 101, 110, 111\}$, where $C(a_i)$ is the ith binary string in $\mathcal{B}$ in the indicated order. The code C is a prefix code. If a string 011001010110 is received, one can easily parse it into 0, 110, 0, 101, 0, 110 and decode it into $a_1a_4a_1a_3a_1a_4$. Note that the first 0 is a codeword. After it is received, it can be decoded immediately into a_1 without looking at the next digit. The binary tree corresponding to C is shown in Fig. 1.

A uniquely decodable code C satisfies the following *Kraft–McMillan inequality* [1, Chap. 5]:

$$\sum_{a\in\mathcal{A}} 2^{-l(a)} \le 1$$

where $l(a)$ is the length of $C(a)$. Conversely, it can be shown that given any set of lengths $l(a)$ satisfying the Kraft–McMillan inequality, one can find a prefix code C such that $C(a)$ has length $l(a)$ for any $a \in \mathcal{A}$.

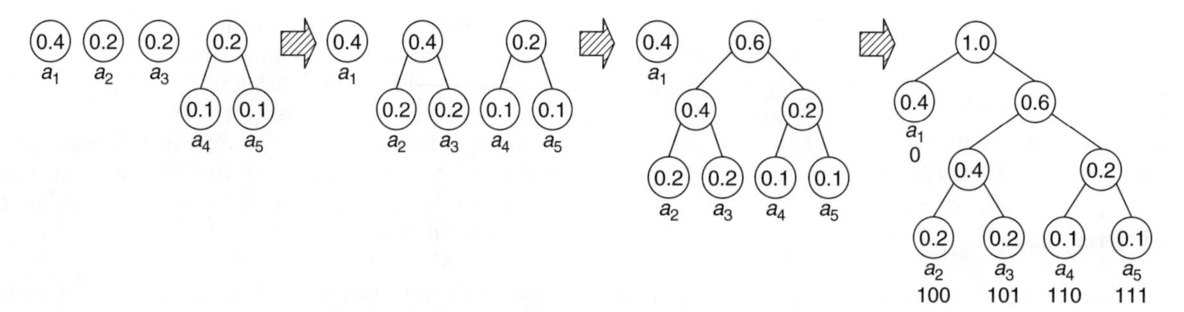

Figure 1. The Huffman code C generated by the Huffman coding algorithm.

Therefore, in view of the above result, it suffices to find an optimal prefix code C such that the total length $nr(C)$ of $C(x_1)C(x_2)\cdots C(x_n)$ is a minimum, where

$$r(C) = \sum_{a\in\mathcal{A}} p(a)l(a)$$

denotes the *average codeword length* of C in bits per letter, and $p(a)$ is the normalized frequency of a in x if x is deterministic and the probability of a if x is random.

Given a probability distribution or a set of normalized frequencies $\{p(a) : a \in \mathcal{A}\}$, an interesting problem is how to find such an optimal prefix code C. This problem is solved when the size of A is finite. Instead of performing an exhaustive search among all prefix codes, Huffman [2] proposed an elegant algorithm in 1952, which is now known as the *Huffman coding algorithm*, to generate optimal prefix codes based on a set of probabilities or frequencies. The resulting optimal codes are called *Huffman codes*.

2. HUFFMAN CODING ALGORITHM

Any elegant algorithm has its recursive procedure. There is no exception in the case of the Huffman coding algorithm. Let $\mathcal{A} = \{a_1, a_2, \ldots, a_J\}$ with $2 \le J < \infty$. For each $1 \le j \le J$, rewrite $p(a_j)$ and $l(a_j)$ as p_j and l_j, respectively. Let C be an optimal prefix code such that

$$r(C) = \sum_{j=1}^{J} p_j l_j \tag{1}$$

is minimized among all prefix codes. The following properties of the optimal code C are helpful to uncover the recursive procedure of the Huffman coding algorithm:

P1. C is a complete prefix code, that is, the binary tree corresponding to C is a complete binary tree in which every node other than the root has its sibling.

P2. If $p_i > p_j$, then $l_i \le l_j$.

P3. The two longest codewords of C, which correspond to the two least likely letters, have the same length.

P4. If in the binary tree corresponding to C, terminal nodes with the same depth are rearranged properly, then the two least likely letters are sibling terminal nodes.

Property P1 is straightforward. If a node other than the root has no sibling, then this node can be merged with its parent. As a result, all codewords of the terminal nodes in the subtree rooted at this node are shortened by 1, and the resulting prefix code will be better than C, which contradicts to the optimality of C. Property P2 is equivalent to the principle of assigning short codewords to highly likely letters and long codewords to less likely letters. Properties P3 and P4 are implied by properties P1 and P2.

Property P4 implies a recursive procedure to design an optimal prefix code C given the probabilities $p_1, p_2, \ldots, p_J$. Rewrite C as C_J. Suppose that p_{J-1} and p_J are the two least probabilities. From property P4, we know that a_{J-1} and a_J are sibling terminal nodes in the binary tree corresponding to C_J. Merge a_{J-1} and a_J with their common parent, which then becomes a terminal node corresponding to a merged letter with probability $p_{J-1} + p_J$. The new binary tree has $J-1$ terminal nodes and gives rise to a new prefix code C_{J-1} with $J-1$ codewords of lengths $l_1, l_2, \ldots, l_{J-2}, l_{J-1} - 1$. The average codeword length of C_J is related to that of C_{J-1} by

$$r(C_J) = r(C_{J-1}) + p_{J-1} + p_J$$

Since the quantity $p_{J-1} + p_J$ is independent of C_{J-1}, minimizing $r(C_J)$ is equivalent to minimizing $r(C_{J-1})$. Consequently, in order to design an optimal code C_J for J letters, we can first design an optimal code C_{J-1} for $J-1$ letters, and then extend C_{J-1} to C_J while maintaining optimality. Recursively reduce the alphabet size by merging the two least likely letters of the corresponding alphabet. We finally reduce the problem to design an optimal code C_2 for two letters, for which the solution is obvious; that is, we assign 0 to one letter and 1 to the other. Since $r(C_j)$ is minimized for each $2 \le j \le J$, we find that $C_2, C_3, \ldots, C_J$ are all optimal prefix codes with respect to their corresponding probability distributions. This is the essence of the recursive procedure of the Huffman coding algorithm.

To summarize, given a set of probabilities, the Huffman coding algorithm generates optimal prefix codes for $\mathcal{A}$ according to the following procedure:

Step 1. Start with $m = J$ trees, each of which consists of exactly one node corresponding to a letter in $\mathcal{A}$. Set the weight of each node as the probability of the corresponding letter.

Step 2. For $m \geq 2$, find the two trees $T_1^{(m)}$ and $T_2^{(m)}$ with the least weights at their roots. Combine $T_1^{(m)}$ and $T_2^{(m)}$ into a new tree so that in the new tree, the roots of $T_1^{(m)}$ and $T_2^{(m)}$ are siblings with the root of the new tree as their parent. The root of the new tree carries a weight equal to the sum of the weights of its two children. The total number of trees m is reduced by 1.

Step 3. Repeat step 2 for $m = J, J-1, \ldots, 2$ until only one tree is left. The final tree gives rise to an optimal prefix code.

The following example further illustrates the procedure.

Example 2. Consider an alphabet $\mathcal{A} = \{a_1, a_2, a_3, a_4, a_5\}$ with the probability distribution $(0.4, 0.2, 0.2, 0.1, 0.1)$. The Huffman coding algorithm generates the prefix code C in Example 1 recursively as shown in Fig. 1. By convention, the left and right branches emanating from an internal node in a binary tree are labeled 0 and 1, respectively.

It is worthwhile to point out that the Huffman codes generated by the Huffman coding procedure are not unique. Whenever there are two or more pairs of trees having the least weights at their roots in step 2, one can combine any such pair of trees into a new tree, resulting in a possibly different final tree. In Example 2, after merging a_4 and a_5 into one letter, say, a'_4, with probability 0.2, we can choose to merge a_3 and a'_4 instead of merging a_2 and a_3 as in Fig. 1. Continuing the algorithm, we can get another prefix code C' with the binary codeword set $\mathcal{B}' = \{0, 10, 110, 1110, 1111\}$ in which the jth string represents the codeword for a_j in $\mathcal{A}$, $1 \leq j \leq 5$. It is easy to verify that these both C' and C have the same average codeword length of 2.1 bits. This non-uniqueness will allow us later to develop *adaptive Huffman coding algorithms*.

3. ENTROPY AND PERFORMANCE

The entropy of a probability distribution $(p_1, p_2, \ldots, p_J)$ is defined as

$$H(p_1, p_2, \ldots, p_J) \triangleq -\sum_{j=1}^{J} p_j \log p_j$$

where log stands for the logarithm with base 2 and the entropy is measured in *bits*. The entropy represents the ultimate compression rate in bits per letter one can possibly achieve with all possible data compression schemes.

A *figure of merit* of a prefix code is its performance against the entropy. It can be shown [1] that if C is a Huffman code with respect to the distribution $(p_1, p_2, \ldots, p_J)$, then its average codeword length $r(C)$ in bits per letter is within one bit of the entropy.

Theorem 1. The average codeword length $r(C)$ of a Huffman code with respect to $(p_1, p_2, \ldots, p_J)$ satisfies

$$H(p_1, p_2, \ldots, p_J) \leq r(C) < H(p_1, p_2, \ldots, p_J) + 1 \quad (2)$$

The upper bound in (2) is uniform and applies to every distribution. For some distributions, however, the difference between $r(C)$ and $H(p_1, p_2, \ldots, p_J)$ may be well below 1. For detailed improvements on the upper and lower bounds, the reader is referred to Gallager [3], Capocelli et al. [4], and Capocelli and De Santis [5].

Another method to improve the upper bound is to design a Huffman code C' for an extended alphabet $\mathcal{A}^N$, where $\mathcal{A}^N$ consists of all length N strings from $\mathcal{A}$. In this case, one assigns a codeword to each block of N letters rather than a single letter. Accordingly, $r(C')$ is the average codeword length in bits per block and is within one bit of the block entropy. Thus, $r(C')/N$, the average codeword length in bits per letter, is within $1/N$ bits of the entropy per letter. As N increases, $r(C')/N$ can be arbitrarily close to the entropy per letter. Since the complexity of the Huffman coding algorithm grows exponentially with respect to N, this method works only when both N and the size of $\mathcal{A}$ are small.

4. HUFFMAN CODING FOR AN INFINITE ALPHABET

Since the Huffman coding algorithm constructs a binary tree using the bottom–up approach by successively merging the two least probable letters in the alphabet, it cannot be applied directly to infinite alphabets. Let $\mathcal{A} = \{0, 1, 2, \ldots\}$. Indeed, given an arbitrary distribution $(p_0, p_1, p_2, \ldots)$, the problem of constructing an optimal prefix code C with respect to $(p_0, p_1, p_2, \ldots)$ is still open at the writing of this article, even though it can be shown [6] that such an optimal prefix code exists.

However, when the distribution is a geometric probability distribution

$$p_i = (1-\theta)\theta^i, \qquad i \geq 0 \quad (3)$$

where $0 < \theta < 1$, a simple procedure does exist for construction of the optimal prefix code C. In this case, given each $i \in \mathcal{A}$, one can even compute the codeword $C(i)$ without involving any tree manipulation; this property is desirable in the case of infinite alphabets since there is no way to store an infinite binary tree. The procedure was first observed by Golomb [7] in the case of $\theta^k = \frac{1}{2}$ for some integer k, and later extended by Gallager and Voorhis [8] to general geometric distributions in (3). The corresponding optimal prefix code is now called the *Golomb code* in the case of $\theta^k = \frac{1}{2}$ for some integer k, and the *Gallager–Voorhis code* in the general case.

The procedure used to construct the Gallager–Voorhis code for a geometrical probability distribution with the parameter θ is as follows:

Step 1. Find the unique positive integer k such that

$$\theta^k(1+\theta) \leq 1 < \theta^{k-1}(1+\theta) \quad (4)$$

This unique k exists because $0 < \theta < 1$.

Step 2. If $k = 1$, let $C_k(0)$ denote the empty binary string. Otherwise, let $k = 2^n + m$, where n and m are positive integers satisfying $0 \leq m < 2^n$. For any integer $0 \leq j < 2^n - m$, let $C_k(j)$ be the binary representation of j padded with possible zeros to the left to ensure that the length of $C_k(j)$ is n. For $2^n - m \leq j < k$, let $C_k(j)$ be the binary representation of $j+2^n-m$ padded with possible zeros to the left to ensure that the length of $C_k(j)$ is $n+1$. The constructed code C_k is a prefix code for $\{0,1,\ldots,k-1\}$.

Step 3. To encode an integer $i \geq 0$, we find a pair of nonnegative integers (s,j) such that $i = sk + j$ and $0 \leq j < k$. The Gallager–Voorhis code encodes i into a codeword $G_k(i)$ consisting of s zeros followed by a single one and then by $C_k(i)$.

Example 3. Suppose that $\theta^3 = \frac{1}{2}$. In this case, $k = 3$, and

$$C_3(0) = 0, C_3(1) = 10, C_3(2) = 11$$

Table 1 illustrates Golomb (Gallager-Voorhis) codewords $G_3(i)$ for integers $i = 0, 1, 2, \ldots, 11$.

Table 1. Golomb Codewords $G_3(i)$ for Integers $i = 0, 1, 2, \ldots, 11$

Integer i	$G_3(i)$	Integer i	$G_3(i)$
0	10	6	0010
1	110	7	00110
2	111	8	00111
3	010	9	00010
4	0110	10	000110
5	0111	11	000111

5. ADAPTIVE HUFFMAN CODING

In previous sections, we have assumed that the probability distribution is available and known to both the encoder and decoder. In many practical applications, however, the distribution is unknown. Since Huffman coding needs a distribution to begin with, to encode a sequence $x = x_1x_2\cdots x_n$, one way to apply Huffman coding is to use the following *two-pass* approach:

Pass 1. Read the sequence x to collect the frequency of each letter in the sequence. Use the frequencies of all letters to estimate the probabilities of these letters, and design a Huffman code based on the estimated probability distribution. Send the estimated probability distribution or the designed Huffman code to the user;

Pass 2. Use the designed Huffman code to encode the sequence.

This two-pass coding scheme is not desirable in applications such as streaming procedures, where timely encoding of current letters is required. It would be nice to have a *one-pass* coding scheme in which we estimate the probabilities of letters on the fly on the basis of the previously encoded letters in x and adaptively choose a prefix code based on the estimated probability distribution to encode the current letter. Faller [9] and Gallager [3] independently developed a one-pass coding algorithm called the *adaptive Huffman coding algorithm*, also known as the *dynamic Huffman coding algorithm*. It was later improved by Knuth [10] and Vitter [11].

Suppose that the sequence $x = x_1x_2\cdots x_n$ is drawn from the alphabet $\mathcal{A} = \{a_1, a_2, \ldots, a_J\}$ with $2 \leq J < \infty$. Before encoding x_1, it is reasonable to assume that all letters $a \in \mathcal{A}$ are equally likely since we have no knowledge about x other than the alphabet $\mathcal{A}$. Maintain a counter $c(a_j)$ for each letter a_j, $1 \leq j \leq J$. All counters are initially set to 1. The initial probability distribution is

$$\left(\frac{c(a_1)}{\sum_{j=1}^{J} c(a_j)}, \frac{c(a_2)}{\sum_{j=1}^{J} c(a_j)}, \ldots, \frac{c(a_J)}{\sum_{j=1}^{J} c(a_j)}\right) = \left(\frac{1}{J}, \frac{1}{J}, \ldots, \frac{1}{J}\right)$$

Pick an initial Huffman code C_1 for $\mathcal{A}$ based on the initial probability distribution. To encode $x = x_1x_2\cdots x_n$, the adaptive Huffman coding algorithm works as follows:

Step 1. Use the Huffman code C_i to encode x_i.

Step 2. Increase $c(x_i)$ by 1.

Step 3. Update C_i into a new prefix code C_{i+1}, so that C_{i+1} is a Huffman code for the new probability distribution

$$\frac{c(a_1)}{\sum_{j=1}^{J} c(a_j)}, \frac{c(a_2)}{\sum_{j=1}^{J} c(a_j)}, \ldots, \frac{c(a_J)}{\sum_{j=1}^{J} c(a_j)}$$

Step 4. Repeat steps 1–3 for $i = 1, 2, \ldots, n$ until all letters in x are encoded.

It is clear that step 3 is the major step. To understand it properly, let us first discuss the sibling property of a complete binary tree.

5.1. Sibling Property

Assign a positive weight w_j to each letter a_j, $1 \leq j \leq J$. Let C be a complete prefix code for $\mathcal{A}$. Let T be the binary tree corresponding to C. Label the nodes of T by an index from $\{1, 2, \ldots, 2J-1\}$. Assign a weight to each node in T recursively; each terminal node carries the weight of the corresponding letter in $\mathcal{A}$, and each internal node carries a weight equal to the sum of the weights of its two children. A complete prefix code, C, is said to satisfy the *sibling property* with respect to weights $w_1, w_2, \ldots, w_J$ if the nodes of T can be arranged in a sequence $i_1, i_2, \ldots, i_{2J-1}$ such that (continuing the "properties" list from Section 2, above)

P5. $w(i_1) \leq w(i_2) \leq \cdots \leq w(i_{2J-1})$, where $w(i_j)$ denotes the weight of the node i_j.

P6. nodes i_{2j-1} and i_{2j} are siblings for any $1 \leq j \leq J$; and the parent of nodes i_{2j-1} and i_{2j} does not precede nodes i_{2j-1} and i_{2j} in the sequence.

The sibling property is related to the Huffman coding algorithm.

Example 4. We now revisit the Huffman code in Example 2. It is not hard to see that the Huffman code shown in Fig. 1 satisfies the sibling property with respect to weights $w_1 = 4$, $w_2 = 2$, $w_3 = 2$, $w_4 = 1$, and $w_5 = 1$.

Example 5. Let us now increase the weight w_2 in Example 4 by 1. Accordingly, the weight of each node along the path from the terminal node corresponding to a_2 to the root increases by 1, as shown in Fig. 2. It is easy to see that the prefix code in Fig. 2 is not a Huffman code for the probability distribution

$$\left(\frac{w_1}{\sum_{j=1}^{5} w_j}, \frac{w_2}{\sum_{j=1}^{5} w_j}, \ldots, \frac{w_5}{\sum_{j=1}^{5} w_j}\right) = \left(\frac{4}{11}, \frac{3}{11}, \frac{2}{11}, \frac{1}{11}, \frac{1}{11}\right)$$

Moreover, this prefix code does not satisfy the sibling property with respect to weights 4, 3, 2, 1, and 1, either.

In general, the following theorem is implied by the Huffman coding procedure.

Theorem 2. A complete prefix code C is a Huffman code for the probability distribution

$$\frac{w_1}{\sum_{j=1}^{J} w_j}, \frac{w_2}{\sum_{j=1}^{J} w_j}, \ldots, \frac{w_J}{\sum_{j=1}^{J} w_j}$$

if and only if C satisfies the sibling property with respect to weights $w_1, w_2, \ldots, w_J$.

Update C_i into C_{i+1}: Note that C_i is a Huffman code for the current probability distribution

$$\frac{c(a_1)}{\sum_{j=1}^{J} c(a_j)}, \frac{c(a_2)}{\sum_{j=1}^{J} c(a_j)}, \ldots, \frac{c(a_J)}{\sum_{j=1}^{J} c(a_j)}$$

Think of $c(a_j)$ as the weight of a_j, that is, $w_j = c(a_j)$, $1 \leq j \leq J$. Let the nodes of the binary tree corresponding to C_i be arranged in a sequence $i_1, i_2, \ldots, i_{2J-1}$ such that properties P5 and P6 are satisfied. Let $j_0, j_1, \ldots, j_l$ be a sequence such that $i_{j_0}, i_{j_1}, \ldots, i_{j_l}$ is the sequence of nodes leading from the terminal node corresponding to x_i to the root.

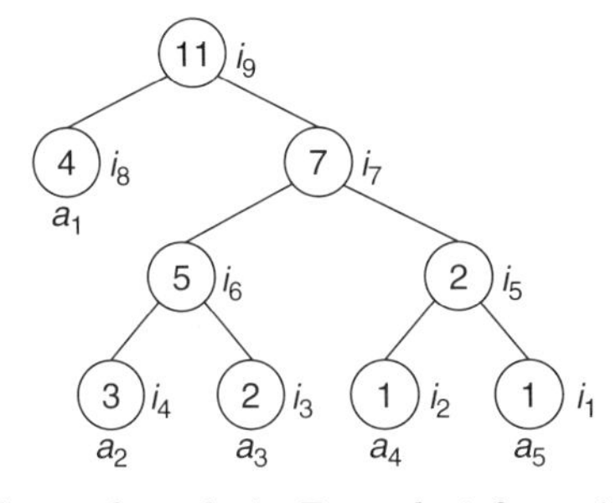

Figure 2. The prefix code in Example 4 for weights 4, 3, 2, 1, and 1.

Case 1. If for any $0 \leq k < l$

$$w(i_{j_k}) < w(i_{j_k+1}) \tag{5}$$

then after the weights of nodes i_{j_k}, $0 \leq k < l$, are updated, C_i still satisfies the sibling property. The update of the weights of nodes is due to the increment of $c(x_i)$ by 1 after x_i is encoded by C_i. In this case, C_i is also a Huffman code for the new probability distribution after $c(x_i)$ increases by 1. Hence $C_{i+1} = C_i$.

Case 2. If the inequality (5) is not valid for some k, then we can obtain C_{i+1} from C_i by exchanging some subtrees rooted at nodes of equal weight. Let j'_0 be the largest integer such that $w(i_{j'_0}) = w(i_{j_0})$. Having j'_k defined, we let j'_{k+1} be the largest integer such that $w(i_{j'_{k+1}})$ is equal to the weight of the parent of the node $i_{j'_k}$. If $j'_k = 2J - 1$, that is, if the node $i_{j'_k}$ is the root of the binary tree, the procedure terminates. Denote the maximum k by l'. It is easy to see that $l' < l$. In this case, the following must be done:

Step 1. Exchange the subtree rooted at the node i_{j_0} with the subtree rooted at the node $i_{j'_0}$.

Step 2. For $k = 0, 1, \ldots, l' - 1$, exchange (in the new binary tree resulting from the last exchange operation) the subtree rooted at the parent of the node $i_{j'_k}$ with the subtree rooted at the node $i_{j'_{k+1}}$. (The two roots of the two subtrees are not exchanged since they have the same weight.)

Step 3. Update the weight of each node along the path from node $i_{j'_0}$ to the root.

The final binary tree satisfies properties P5 and P6, and gives rise to the Huffman code C_{i+1}.

Example 6. Suppose that a_2 is the current letter to be encoded by the prefix code in Fig. 2. Denote the code by C_i. In Example 5, we know that C_i does not satisfy the sibling property with respect to weights 4, 3, 2, 1, and 1. Exchange the subtree rooted at node i_4 with the subtree rooted at node i_5, and update relevant weights accordingly. We get the two binary trees shown in Fig. 3. The binary tree on the right-hand side of Fig. 3 gives rise to a Huffman code for the probability distribution $\left(\frac{4}{11}, \frac{3}{11}, \frac{2}{11}, \frac{1}{11}, \frac{1}{11}\right)$ after the current letter a_2 is encoded by C_i.

The following example illustrates the complete process of the adaptive Huffman coding algorithm.

Example 7. Let $\mathcal{A} = \{a_1, a_2, a_3, a_4, a_5, a_6\}$. Apply the adaptive Huffman coding algorithm to encode $x = a_1a_1a_2a_3$. The initial counters are $c(a_1) = c(a_2) = c(a_3) = c(a_4) = c(a_5) = c(a_6) = 1$. Pick C_1 in Fig. 4 to be the Huffman code for $\mathcal{A}$, based on the probability distribution $\left(\frac{1}{6}, \frac{1}{6}, \frac{1}{6}, \frac{1}{6}, \frac{1}{6}, \frac{1}{6}\right)$. Encode the first letter a_1 by C_1, and then increase the counter $c(a_1)$ by 1. In this case, $l = 3, j_0 = 1, j_1 = 7, j_2 = 10, j_3 = 11, j'_0 = 6, j'_1 = 9, j'_2 = 11$, and $l' = 2$.

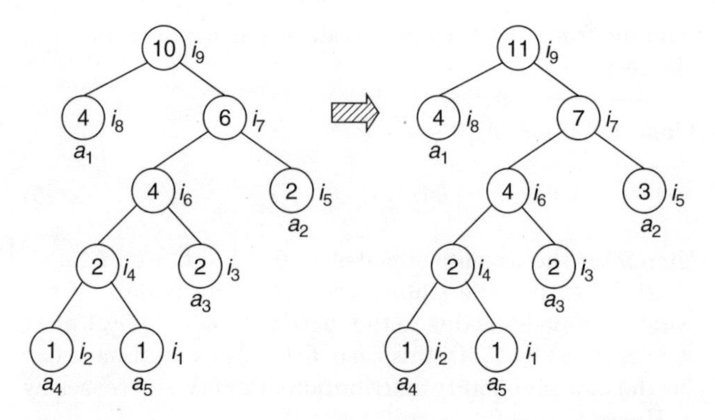

Figure 3. Exchange of subtrees and update of relevant weights.

By exchanging subtrees and updating all relevant weights, we get the Huffman code C_2 in Fig. 4. Encode the second letter a_1 by C_2, and then increase the counter $c(a_1)$ by 1. C_2 is updated into C_3 in Fig. 4 by exchanging subtrees rooted at nodes i_6 and i_8, and increasing all relevant weights by 1. Encode the third letter a_2 by C_3, and then increase the counter $c(a_2)$ by 1. C_3 is updated into C_4 in Fig. 4 by exchanging subtrees rooted at nodes i_2 and i_5, and updating relevant weights. Finally, encode the fourth letter a_3 by C_4. The codeword sequence is

$$C_1(a_1)C_2(a_1)C_3(a_2)C_4(a_3) = 00011001110$$

A very interesting fact about the adaptive Huffman coding algorithm is that as i is large enough, the adaptive prefix code C_i converges and is indeed a Huffman code for the true distribution. This is expressed in the following theorem.

Theorem 3. Apply the adaptive Huffman coding algorithm to encode a stationary ergodic source $X_1X_2\cdots X_n\cdots$ taking values from the finite alphabet $\mathcal{A}$ with a common distribution $(p_1, \ldots, p_J)$. Then with probability one, C_i converges and for sufficiently large i, C_i itself is a Huffman code for the true distribution $(p_1, \ldots, p_J)$.

6. APPLICATIONS

The years since 1977 have witnessed widespread applications of Huffman coding in data compression, in sharp contrast with the first 25 years since the appearance of Huffman's groundbreaking paper [2]. This phenomenon can be explained by the increasing affordability of computing power and the growing demand of data compression to save transmission time and/or storage space. Nowadays, Huffman coding competes with state-of-the-art coding schemes such as arithmetic coding and Lempel–Ziv coding in applications requiring data compression. Since each letter in a data sequence to be compressed must be encoded into an integer number of bits, the compression performance of Huffman coding is often worse than that of arithmetic coding or Lempel–Ziv coding. However, in practical applications, the selection of a data compression scheme is based not only on its compression performance but also on its computational speed and memory requirement. Since a fixed Huffman code can be easily implemented as a lookup table in practice, Huffman coding has the advantages of real-time computational speed and the need of only a fixed amount of memory. Thus, Huffman coding remains a popular choice in time-critical applications or applications in which computing resources such as memory or computing power are limited. Some typical applications of Huffman coding in telecommunications are described in the following paragraphs.

6.1. Facsimile Compression

One of the earliest applications of Huffman coding in telecommunications is in *facsimile compression*. In 1980, the CCITT Group 3 digital facsimile standard was put into force by the Consultative Committee on International

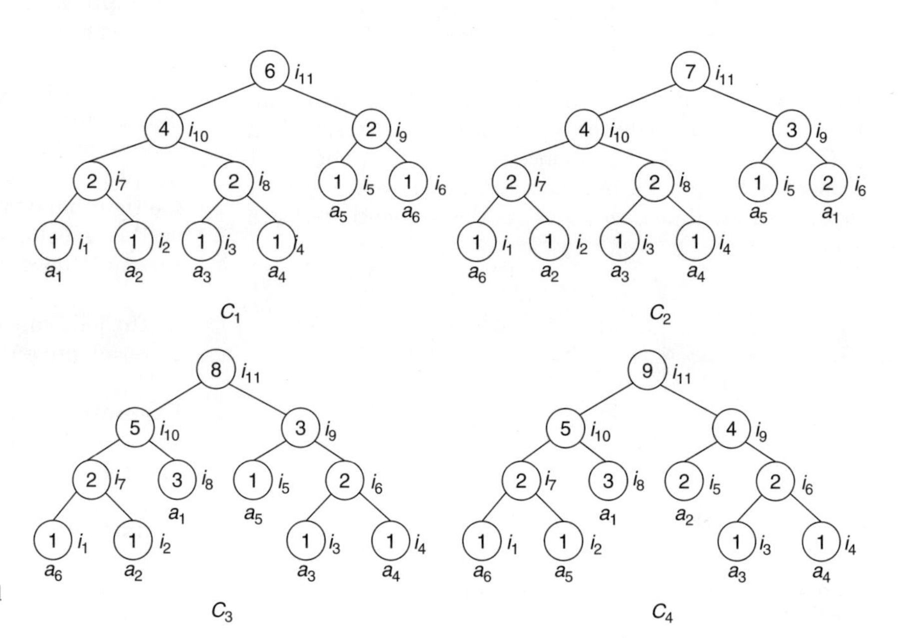

Figure 4. The Huffman codes C_1, C_2, C_3, and C_4 in Example 6.

Telephony and Telegraphy (CCITT), now the International Telecommunications Union (ITU). In the CCITT Group 3 standard, a predefined Huffman code was designed based on 11 typical fax documents recommended by the CCITT. For real-time processing of fax images, this Huffman code is implemented as a lookup table. It is estimated that a Group 3 fax system saves more than 85% of transmission time by compressing a typical business document of letter size.

6.2. Modem Data Compression

Early Modem data compression is another application of Huffman coding in telecommunications. The Microcom Networking Protocol (MNP) is a de facto standard for the modem industry. In MNP, MNP 5 is a modem data compression method that uses a much simplified variant of adaptive Huffman coding. In MNP 5, a set of 256 predefined prefix-free codewords are maintained in a table. When a letter from an alphabet of size 256 is to be encoded, MNP 5 selects a codeword according to the letter's recorded frequency. Thus, the mapping between codewords and letters are adaptively changing according to the data sequence. The compression performance of MNP 5 varies for different data. For standard text data, a modem generally can double its transmission speed by applying MNP 5. However, for some data like compressed images, MNP 5 will result in actual expansions of data, and thus slow down the modem's transmission.

6.3. Image Compression

The JPEG image compression standard developed by the Joint Photographic Experts Group uses Huffman coding as a residual coder after discrete-cosine transform (DCT), quantization, and run-length encoding [7]. Since the encoding of an image is not necessarily real-time, the JPEG standard allows two-pass Huffman coding in addition to the use of a predefined Huffman code, which is generated based on a group of typical images. The more recent JPEG-LS standard defines a new image compression algorithm that allows any image to be encoded, and then decoded, without any loss of information. In the JPEG-LS standard, variants of Golomb codes, which are called *Golomb–Rice codes* [12], are used to efficiently encode integers into easily computed codewords.

6.4. Audio/Video Compression

International standards for full-motion video compression include MPEG 1 and 2, which are both developed by the Moving Pictures Experts Group. MPEG layer 3, also known as MP3, is now a very popular audio coding standard. Similar to JPEG, MPEG 1 and 2 use Huffman coding as a residual coder after DCT, quantization, and run-length encoding. MP3 combines Huffman coding with modified DCT and quantization. A critical requirement for MP3, MPEG 1, and MPEG 2 is that the decoding of compressed audio and video data must be real-time. Therefore, in these standards Huffman codes are all predefined and implemented as lookup tables.

BIOGRAPHIES

En-hui Yang received the B.S. degree in applied mathematics from HuaQiao University, Qianzhou, China, and Ph.D. degree in mathematics from Nankai University, Tianjin, China, in 1986 and 1991, respectively. He joined the faculty of Nankai University in June 1991 and was promoted to Associate Professor in 1992. From January 1993 to May 1997, he held positions of Research Associate and Visiting Scientist at the University of Minnesota, Minneapolis–St. Paul (USA), the University of Bielefeld, Bielefeld, Germany, and the University of Southern California, Los Angeles (USA). Since June 1997, he has been with the Department of Electrical and Computer Engineering, University of Waterloo, Ontario, Canada, where he is now a Professor and leading the Leitch — University of Waterloo multimedia communications lab. He also holds the position of Canada Research Chair in information theory and multimedia compression, and is a founding partner of SlipStream Data Inc, a high-tech company specializing solutions to reduce communication bandwidth requirements and speed up Internet connection. His current research interests are multimedia compression, digital communications, information theory, Kolmogorov complexity theory, source and channel coding, digital watermarking, quantum information theory, and applied probability theory and statistics. Dr. Yang is a recipient of several research awards, including the 2000 Ontario Premier's Research Excellence Award, Canada, and the 2002 Ontario Distinguished Researcher Award, Canada.

Da-ke He received the B.S. and M.S. degrees in electrical engineering from Huazhong University of Science and Technology, China, in 1993 and 1996, respectively. He joined Apple Technology, Zhuhai, China, in 1996 as a software engineer. Since 1999, he has been a Ph.D. candidate in the Electrical and Computer Engineering Department at the University of Waterloo, Canada, where he has been working on grammar-based data compression algorithms and multimedia data compression. His areas of interest are source coding theory, multimedia data compression, and digital communications.

BIBLIOGRAPHY

1. T. M. Cover and J. A. Thomas, *Elements of Information Theory*, Wiley, New York, 1991.
2. D. A. Huffman, A method for the construction of minimum redundancy codes, *Proc. IRE* **40**: 1098–1101 (1952).
3. R. G. Gallager, Variations on a theme by Huffman, *IEEE Trans. Inform. Theory* **IT-24**: 668–674 (1978).
4. R. M. Capocelli, R. Giancarlo, and I. J. Taneja, Bounds on the redundancy of Huffman codes, *IEEE Trans. Inform. Theory* **IT-32**: 854–857 (1986).
5. R. M. Capocelli and A. De Santis, New bounds on the redundancy of Huffman codes, *IEEE Trans. Inform. Theory* **IT-37**: 1095–1104 (1991).
6. T. Linder, V. Tarokh, and K. Zeger, Existence of optimal prefix codes for infinite source alphabets, *IEEE Trans. Inform. Theory* **IT-43**: 2026–2028 (1997).

7. S. W. Golomb, Run-length encodings, *IEEE Trans. Inform. Theory* **IT-12**: 399–401 (1966).
8. R. G. Gallager and D. C. Van Voorhis, Optimal source codes for geometrically distributed integer alphabets, *IEEE Trans. Inform. Theory* **IT-21**: 228–230 (1975).
9. N. Faller, An adaptive system for data compression, *Record 7th Asilomar Conf. Circuits, Systems and Computers*, 1973, pp. 593–597.
10. D. E. Knuth, Dynamic Huffman coding, *J. Algorithms* **6**: 163–180 (1985).
11. J. S. Vitter, Design and analysis of dynamic Huffman codes, *J. Assoc. Comput. Mach.* **34**: 825–845 (1987).
12. M. Weinberger, G. Seroussi, and G. Sapiro, The LOCO-I lossless image compression algorithm: Principles and standardization into JPEG-LS, *IEEE Trans. Image Process.* **9**: 1309–1324 (2000).

I

IMAGE AND VIDEO CODING

SHIPENG LI
Microsoft Research Asia
Beijing, P. R. China

WEIPING LI
WebCast Technologies, Inc.
Sunnyvale, California

1. INTRODUCTION

Image and video coding has been a very active research area for a long time. While transmission bandwidth and storage capacity have been growing dramatically, the demand for better image and video coding technology has also been growing. The reason is the ever-increasing demand for higher quality of images and video, which requires ever-increasing quantities of data to be transmitted and/or stored.

There are many different types of image and video coding techniques available. The term coding used to solely refer to compression of image and video signals. However, in recent years it is generalized more toward representation of image and video data that provides not only compression but also other functionalities. In this article, we still focus on discussions in the traditional coding sense (i.e., compression). We briefly touch on the topic of coding with different functionalities at the end.

To better understand the details of different image and video coding techniques and standards, many of which seem to be more art than science, a good understanding of the general problem of space of image and video coding is extremely important. The basic problem is to reduce the data rate required for representing images and video as much as possible. In compressing image and video data, often some distortion is introduced so that the received image and video signals may not be exactly the same as the original. Therefore, the second objective is to have as little distortion as possible. In theoretic source coding, data rate and distortion are the only two dimensions to be considered, and the objective is to have both rate and distortion as small as possible. Image and video coding is a type of source coding that also requires other practical considerations. One of the practical concerns is the complexity of a coding technique, which is further divided into encoding complexity and decoding complexity. Therefore, in image and video coding, complexity is the third dimension to be considered, and the additional objective is to have complexity lower than a given threshold. Yet another practical concern is the delay (or latency) of a coding technique, which is the time between when an image or a video frame is available at the input of the encoder and when the reconstructed image or video frame is available at the output of the decoder, excluding the transmission time from the output of the encoder to the input of the decoder. Delay is a critical parameter for two-way communications. For such applications, delay is the fourth dimension in image and video coding, and the objective is to have the delay lower than a given threshold. Therefore, in general, the problem space has four dimensions, namely, rate, distortion, complexity, and delay. The overall problem is to minimize both rate and distortion under a constraint of complexity and possibly another constraint of delay.

This article is organized as follows. In Section 2, some basic concepts of image and video signals are presented. Its goal is to establish a good basis for the characteristics of the source in image and video coding. Section 3 reviews some basic principles and techniques of image and video coding. This is to present some components in an image or video coding system. Section 4 is devoted to the existing and emerging image and video coding standards that exemplify the usage of many basic principles and techniques. Because image and video coding is a part of a communication system, standards are extremely important for practical applications. Section 5 concludes with some thoughts on the possible future research in image and video coding.

2. BASIC CONCEPTS OF IMAGE AND VIDEO SIGNALS

Before discussing image and video coding, some basic concepts about the characteristics of image and video signals are briefly presented in this section. Images and video are multidimensional signals. A grayscale image can be considered as a function of two variables $f(x, y)$ where x and y are the horizontal and vertical coordinates, respectively, of a two-dimensional plane, and $f(x, y)$ is the intensity of brightness at the point with coordinates (x, y). In most practical cases, the coordinates x and y are sampled into discrete values so that a grayscale image is represented by a two-dimensional array of intensity values. Each of the array elements is often called a picture element or pixel. By adding a new dimension in time, an image sequence is usually represented by a time sequence of two-dimensional spatial intensity arrays (images) or, equivalently, a three-dimensional spacetime array. Figure 1 illustrates the concepts of image and image sequence. A video signal is a special type of image sequence. It is different from a film, which is another type of image sequence. In addition to such a simple description, more aspects of image and video are presented in the following subsections.

2.1. Imaging

The term *imaging* usually refers to the process of generating an image by a certain physical means. Images and video may come from a rich variety of sources, from natural photographs to all sorts of medical images, from microscopy to meteorology, not necessarily directly

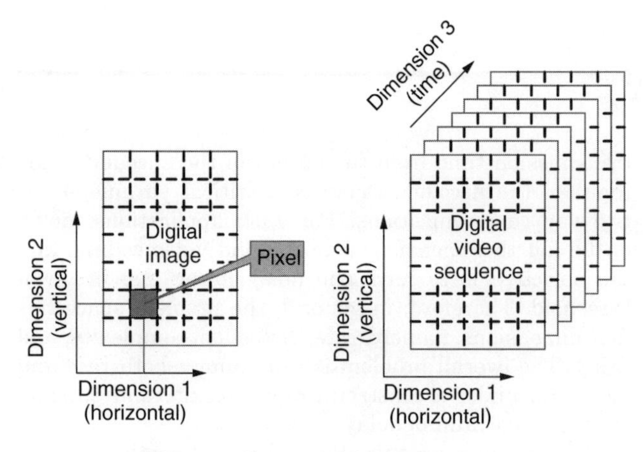

Figure 1. Image and image sequence.

perceptible by human eyes. They can be informally classified into three types according to their radiation sources: reflection sources, such as natural photographs and video; emission sources, such as MRI images; and absorption sources, such as X-ray images. Moreover, with the rapid deployment of multimedia computers, more and more artistic images and video are generated or processed synthetically with or without natural images or video as a basis.

2.2. Color Space

The visual experience of human eyes is much enriched with color information. Corresponding to the human perception system, a color image is most often represented with three primary color components: red (R), green (G), and blue (B). A number of other color coordinate systems can also be used in image processing, printing, and display systems. One particularly interesting color space is YIQ (luminance, in-phase chromatic, quadratic chromatic, also referred to as YUV, or YCbCr) commonly used in television or video systems. Luminance represents the brightness of the image and chrominance represents the color of the image. Conversion from one color space to another is usually defined by a color-conversion matrix. For example, in ITU-R Recommendation BT.709, the following color conversion is defined:

$$\begin{cases} Y = 0.7152G + 0.0722B + 0.2126R \\ Cb = -0.386G + 0.500B - 0.115R \\ Cr = -0.454G - 0.046B + 0.500R \end{cases} \quad (1)$$

2.3. Color Subsampling

Because chrominance is usually associated with slower amplitude variations than luminance and the human eyes are more sensitive to luminance than to chrominance, image and video coding algorithms can exploit this feature to increase coding efficiency by representing the chromatic components with a reduced spatial resolution or allocating fewer bits for them. Figure 2 gives some examples of different sampling schemes in the YUV color space. Three sampling schemes are commonly used in a video system. They are: 4:4:4 (Y, U, and V with the same resolutions, Fig. 2 (a)); 4:2:2 (U and V with half the horizontal resolution of Y, but the same vertical resolution, Fig. 2 (b)); and 4:2:0 (U and V with both half the horizontal and vertical resolutions of Y, Fig. 2 (c)).

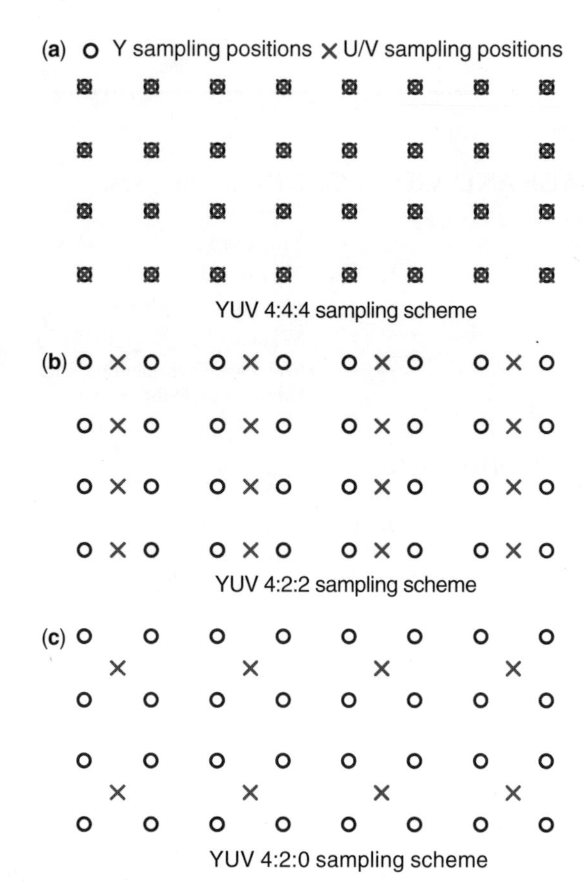

Figure 2. Examples of different sampling schemes in YUV color space.

2.4. Pixel Quantization

The image and video signals that exist abundantly in the environment are naturally analog. An analog image or video signal is a continuous function in a space/time domain and takes values that come from a continuum of possibilities. Before an analog image or video signal can be processed digitally, it must be converted to a digital format (or digitized) so that it becomes enumerable in pixel values in addition to space and time dimensions. Such an analog-to-digital (A/D) conversion process for a pixel value is called quantization. A commonly used quantization scheme when digitizing an analog image is uniform quantization, which maps the continuous-valued intensity to a finite set of nonnegative integers $\{0, \ldots, N-1\}$ through rounding operations, where N is a power of 2: $N = 2^n$. N is usually referred to as the number of gray levels and n is the number of bits allocated to each pixel. The most commonly used bit-depth is $n = 8$ for natural gray level images, whereas $n = 1$ is used for binary images, $n = 10$ is often used for studio quality video, and $n = 12$ is often used for medical images and infrared images. For color images, although different color components can be quantized jointly, most often they are quantized individually. For example, an RGB color image is frequently represented with 24 bits per pixel, with 8 bits for each color component, which is commonly called a true color image. For an image with a YUV 4:2:2 color space and 8 bits per color component, there are $4^*8 + 2^*8 + 2^*8 = 64$ bits for 4 pixels and equivalently 16 bits per pixel, which is commonly called a 16-bit color image. Similarly, an image with a YUV 4:2:0 color space and 8 bits per color component is commonly called a 12-bit color image.

2.5. Video Scanning and Frame Rate

Besides these commonalities with image signals, video signals have some special features. Unlike an image sequence obtained from film that is a time sequence of two-dimensional arrays, a video signal is actually a one-dimensional function of time. A video signal is not only sampled along the time axis, but also sampled along one space dimension (vertical). Such a sampling process is called scanning. The result is a series of time samples, or frames, each of which is composed of space samples, or scan lines. Therefore, video is a one-dimensional analog signal over time. There are two types of video scanning: progressive scanning and interlaced scanning. A progressive scan traces a complete frame, line by line from top to bottom, at a high refresh rate (>50 frames per second to avoid flickering). For example, video displayed on most computer monitors uses progressive scanning. It is well known that the frame rate of a film is 24 frames per second because the human brain interprets a film at 24 or more frames per second as "*continuous*" without a "*gap*" between any two frames. A major difference between a (scanned) video signal and a film is that all pixels in a film frame are illuminated at the same time for the same period of time, and the pixels at the upper left corner and the lower right corner of a frame of the (scanned) video signal are illuminated at different times due to the time period for scanning from the upper left corner to the lower right corner. This is why the frame rate of a video signal must be more than 50 frames per second to avoid flickering, while 24 frames per second is a sufficient rate for film.

2.6. Interlaced Scanning

A TV signal is a good example of interlaced (scan) video. Historically, interlaced video format was invented to achieve a good balance between signal bandwidth, flickering, and vertical resolution. As discussed in the previous subsection, the refresh rate for a video frame must be more than 50 times per second. The number of scan lines per frame determines the vertical resolution of a video frame. For a given refresh rate, the number of scan lines determines the bandwidth of the video signal, because, within the same time period of scanning one frame, more scan lines result in a faster change of intensity (i.e., higher bandwidth). To reduce the video signal bandwidth while maintaining the same refresh rate to avoid flickering, one must reduce the number of scan lines in one refresh period. The advantage of using interlaced scanning is that the vertical resolution is not *noticeably* reduced while the number of scan lines is reduced. This is possible because interlaced scanning refreshes every other line at each frame refresh and the full vertical resolution is covered in two refresh periods. However, the interlaced scan of more than two lines would result in noticeable reduction of the vertical resolution. The subframes formed by all the even or odd scan lines are called fields. Correspondingly, fields can be classified into even fields and odd fields, or top fields and bottom fields according to their relative vertical positions. The top and bottom fields are sent alternately to an interlaced monitor at a field rate equal to the refresh rate. Figure 3 (a) and (b) depict the progressive and interlaced video scanning processes, respectively. A video signal of either scan type can be digitized naturally by sampling horizontally along the scan line, which results in a single rectangular frame for progressive scan and two interlaced fields (in one frame) for interlaced scan. Figure 3 (c) and (d) illustrate such digitized progressive and interlaced video frames, respectively.

2.7. Uncompressed Digital Video

With the image and video signals in digital format, we can process, store, and transmit them digitally using computers and computer networks. However, the high volume nature of image and video data is prohibitive to many applications without efficient compression. For example, considering a 2-hour video with spatial resolution of 720×480 pixels per frame, YUV 4:2:2 format, 30 frames per second frame rate, and each color component quantized to 8 bits (1 byte), the total uncompressed data size of such a video sequence is $2 \times 60 \times 60 \times 30 \times 720 \times 480 \times 2 = 149$ Gbytes with a bit rate of $30 \times 720 \times 480 \times 2 \times 8 = 165$ Mbps, and it is still nowhere near the cinema quality. This is a huge amount of data for computers and networks with today's technology.

2.8. Redundancy and Irrelevancy

Fortunately, image and video data contain a great deal of redundancy and irrelevancy that can be reduced or removed. Redundancy refers to the redundant or duplicated information in an image or video signal and can generally be classified as spatial redundancy (correlation between neighboring pixel values), spectral redundancy (correlation between different color planes or spectral bands), temporal redundancy (correlation between adjacent frames in a sequence of images), statistical redundancy (nonuniform distribution of image and video pixel value), and so on. Irrelevancy refers to the part of the signal that is not noticeable by the receiver (e.g., the human visual system, or HVS). Image and video compression research aims at reducing the number of bits needed to represent an image or video signal by removing the redundancy and irrelevancy as much as possible, with or without noticeable difference to the human eye. Of course, with the rapid developments in high-capacity computers and high-speed computer networks, the tractable data size and bit rates will be increased significantly. However,

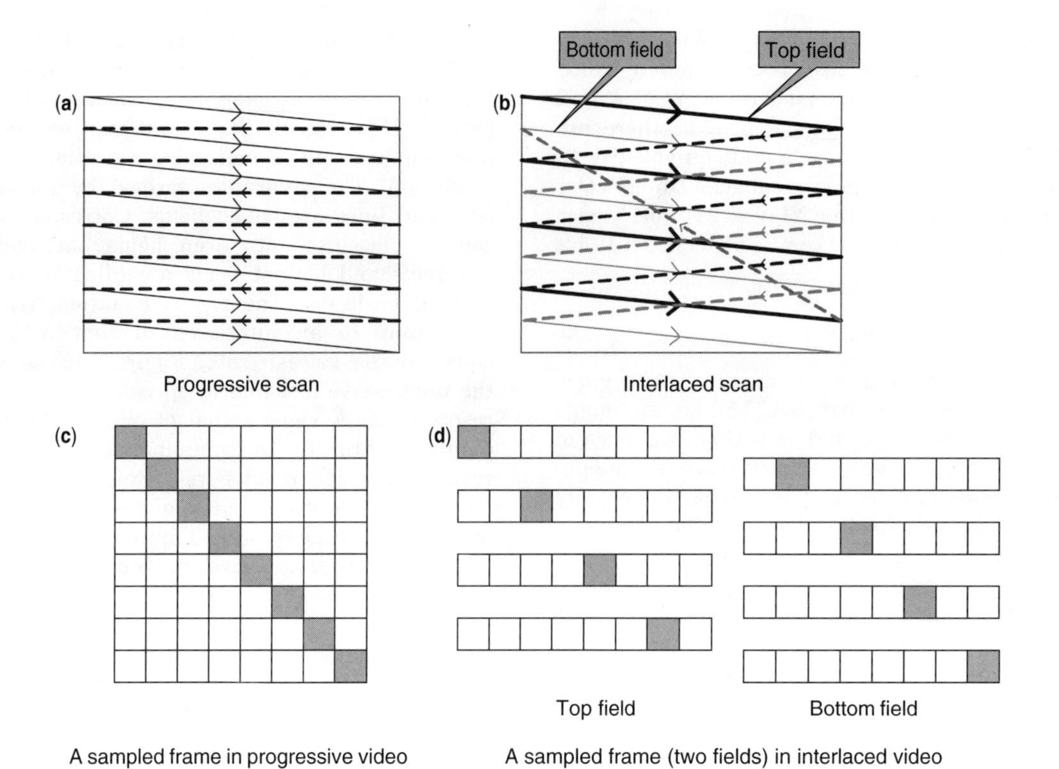

Figure 3. The comparison of progressive video and interlaced video.

the research and development of advanced algorithms for digital image and video compression continue to be necessary in the context of providing a novel and rich multimedia experience with much improved quality more efficiently, more flexibly, more robustly, and more ubiquitously.

2.9. Evaluation of Image and Video Coding Schemes

In practice, we face many choices of various image and video coding schemes. Even for the same coding scheme, we have the choice of different sets of parameters. It is important to first establish measures for quality and performance before any attempt to select the ones that best fit our needs. Some basic measurements commonly used in image and video compression are explained as follows.

The effectiveness of an image or video compression session is normally measured by the bit rate of the compressed bit stream generated, which is the average number of bits representing the compressed image or video signal, with average number of bits per pixel (bpp) for images and average number of bits per second (bps) for video. It can also be measured by the compression ratio defined as follows:

$$\begin{aligned}\text{Compression Ratio} &= \frac{\text{total size in bits of the original image or video}}{\text{total size in bits of the compressed bitstream}} \\ &= \frac{\text{bit rate of the original image or video}}{\text{bit rate of the compressed bitstream}} \quad (2)\end{aligned}$$

Ultimately, the quality of a compressed image or video bit stream should be judged subjectively by human eyes. However, it is normally very costly and time-consuming to perform a formal subjective test. Although subjective quality evaluation is still a necessary step for formal tests in various image and video coding standardization processes, many objective measurements can also be used as rough quality indicators. For example, it can be measured by the distortion of the decoded image with reference to the original image under different criteria, and a particular criterion among them is the mean squared error defined as follows:

$$D = \frac{\sum_{i,j}[(r(i,j) - o(i,j)]^2}{N} \quad (3)$$

where $r(i,j)$ and $o(i,j)$ are the reconstructed and original image intensities of pixel position (i, j), respectively, and N is the number of pixels in the image, and summation is carried out for all pixels in an image. Another commonly used measurement is PSNR (peak signal to noise ratio) defined as follows:

$$PSNR = 10 \log_{10} \frac{P^2}{D} \quad (4)$$

where D is the mean squared error calculated in Eq. (3) and P is the maximum possible pixel value of the original image; for example, if the intensities of the original images are represented by 8-bit integers, then the peak value would be $P = 2^8 - 1 = 255$.

The performance or coding efficiency of an image or video coding scheme is best illustrated in a rate-distortion or bit rate versus PSNR curve, where the encoder would encode the same image or video signal at a few bit rates and the decoded quality is measured using the above-defined criteria. This makes it a very intuitive tool for evaluating various coding schemes.

3. Basic Principles and Techniques for Image and Video Coding

Image and video coding techniques can be classified into two categories: lossless compression and lossy compression. In lossless compression, video data can be identically recovered (decompressed) both quantitatively (numerically) and qualitatively (visually) from a compressed bit stream. Lossless compression tries to represent the video data with the smallest possible number of bits without loss of *any* information. Lossless compression works by removing the *redundancy* present in video data information that, if removed, can be recreated from the remaining data. Although lossless compression preserves exactly the accuracy of image or video representation, it typically offers a relatively small compression ratio, normally a factor of 2 or 3. Moreover, the compression ratio is very dependent on the input data, and there is no guarantee that a given output bit rate can always be achieved.

By allowing a certain amount of distortion or information loss, a much higher compression ratio can be achieved. In lossy compression, once compressed, the original data cannot be identically reconstructed. The reconstructed data are similar to the original but not identical. Lossy compression attempts to achieve the best possible fidelity given an available bit-rate capacity, or to minimize the number of bits representing the image or video signal subject to some allowable loss of information. Lossy compression may take advantage of the human visual system that is insensitive to certain distortion in image and video data and enables rate control in the compressed data stream. As a special case of lossy compression, perceptually lossless or near lossless coding methods attempt to remove redundant, as well as perceptually irrelevant, information so that the original and the decoded images may be visually but not numerically identical. Unfortunately, the measure of perceptive quality of images or video is a rather complex one, especially for video. Many efforts have been devoted to derive an objective measurement by modeling the human visual system, and an effective one is yet to be found. Many practical coding systems still need close supervision by so-called compressionists.

Lossy compression provides a significant reduction in bit rate that enables a multitude of real-time applications involving processing, storing, and transmission of audio-visual information, such as digital camera, multimedia web, video conferencing, digital TV, and so on. Most image and video coding standards, such as JPEG, JPEG2000, MPEG-1, MPGE-2, MPEG-4, H.26x, and the like, are all examples of lossy compression.

Although a higher compression ratio can be achieved with lossy compression, there exist several applications that require lossless coding, such as digital medical imagery and facsimile. A couple of lossless coding standards have been developed for lossless compression, such as lossless JPEG, JPEG-LS, ITU Tele-Fax standards, and the JBIG. Furthermore, lossless compression components generally can be used in lossy compression to further reduce the redundancy of the signal being compressed.

3.1. A Generic Model for Image and Video Coding

To help better understand the basic ideas behind different image and video coding schemes. Figure 4 illustrates the components and their relationship in a typical image and video coding system. Depending on the target applications, not all these components may appear in every coding scheme. In Figure 4, the encoder takes as input an image or video sequence and generates as output a compressed bit stream. The decoder, conversely, takes as input the compressed bit stream and reconstructs an image or video sequence that may or may not be the same as the original input. The components in Fig. 4 are explained in this subsection.

An image or video coding scheme normally starts with certain preprocessing of the input image or video signal, such as denoising, color space conversion, or spatial resolution conversion to better serve the target applications or to improve the coding efficiency for the subsequent compression stages. Normally, preprocessing is a lossy process.

The preprocessed visual signal is passed to a reversible (one-to-one) transformation stage, where the visual data are transformed into a form that can be more efficiently compressed. After the transformation, the correlation (interdependency, redundancy) among transformed coefficients is reduced, the statistical distribution of the coefficients can be shaped to make subsequent entropy

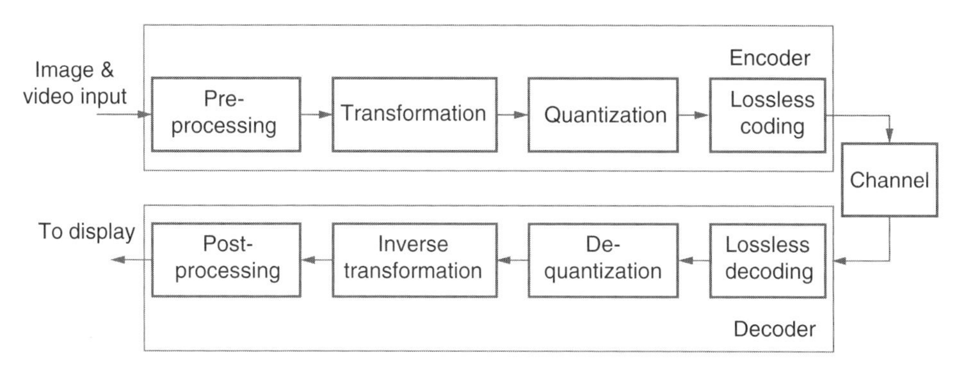

Figure 4. A generic image and video codec model.

coding more efficient, and/or most energy is packed into only a few coefficients or subband regions so that the majority of the transformed coefficients are zeros. Depending on the transform type and the arithmetic precision, even if this step is not lossless, it is close to it. Typical transformations include *differential* or *predictive* mapping (spatially or temporally); unitary transforms such as the discrete cosine transform (DCT); subband decomposition such as wavelet transform; and adaptive transforms such as adaptive DCT, wavelet packets, fractals, and so on. Modern compression systems normally use a combination of these techniques and allow different modes in the transformation stage to decompose the image or video signal adaptively. For example, there are intramodes and intermodes in MPEG video coding standards.

The transformed signal is fed into a quantization module where the coefficients are mapped into a smaller set of discrete values, thus requiring fewer bits to represent these coefficients. Because quantization causes loss of information, it is a lossy process. Quantization is actually where a significant amount of data reduction can be attained. Rate control in lossy compression can be easily achieved by adjusting the quantization step size. There are two main types of quantization: scalar quantization (SQ), where each coefficient is quantized independently, and vector quantization (VQ), where several coefficients are grouped together and quantized jointly.

In some cases, especially in predictive coding, the differential or predictive operation and the quantization may work iteratively in a feedback loop to prevent error propagation, because otherwise the original reference could not be reconstructed perfectly in the decoder after quantization.

Although not explicitly shown in Fig. 4, in practical lossy compression systems, there is always an optimization process that is closely related to rate control. The goal of this process is to minimize the bit rate given a certain quality constraint, or to maximize the decoded quality given a certain bit rate budget, by adjust encoding parameters such as transformation or prediction modes, quantization parameters, and bit allocation among different portions of the image or video signal. As a matter of fact, many compression standards specify only the decoding process and leave the encoding process, especially the optimization part, open to the implementers as long as they produce a bit stream that can be decoded by a compliant decoder. This allows the encoding algorithms to be improved over time, and yet compliant decoders will continue to be able to decode them. The compression process rarely just stops after quantization. The redundancy of the quantized coefficients can be further removed by some generic or specific lossless data compression schemes. Lossless data compression normally involves two parts, symbol formation and entropy coding.

The symbol formation part tries to convert data or coefficients into symbols that can be more efficiently encoded by entropy coding. Such a mapping can be achieved through, for example, coefficients partitioning, or run length coding (RLC). Image or video coefficients can be partitioned and grouped into data blocks based on their potential correlation. Such data blocks can be either mapped into a single symbol or formed as a context to address different probability distribution tables for one or more of the coefficients to be entropy coded. In either case, the correlation between the coefficients can be exploited to result in higher compression ratios by the entropy coder that follows. The entropy coding part tries to generate the shortest binary bit stream by assigning binary codes to the symbols according to their frequency of occurrence. Entropy coding can usually be achieved by either statistical schemes or dictionary-based schemes. In a statistical scheme, fixed-length symbols are coded using variable-length codewords (VLC), where the shorter codewords are assigned to the symbols that occur more frequently (e.g., Huffman coders and arithmetic coders). Alternatively, in a dictionary-based scheme, variable-length strings of symbols are coded using fixed-length binary codewords, where *a priori* knowledge is not required (e.g., Lempel Ziv coder).

After the lossless data compression stage, the image or video data are encoded into a bit stream that can be either directly sent to, or wrapped with some system layer information such as synchronization information or encryption, and then sent to a channel that can be either a storage device, a dedicated link, an IP network, or a processing unit. The decoder receives the bit stream from the channel with or without possible corruptions and starts to decode the received bit stream. The decoding is just the reverse process of the encoding except for the postprocessing part.

Applications of image and video coding may be classified into symmetric or asymmetric. For example, video conferencing applications are symmetric because both ends of the communication have the same requirements. In video streaming applications, where the same video content can be preencoded and stored in a server and accessed by many users, encoding may be much more complicated than decoding, because encoding is only required to be performed once but decoding is required by many users for many times. With more allowable encoding complexity, asymmetric algorithms usually lead to much better coding efficiency than the symmetric ones.

For predictive video coding, since previous decoded frames are normally needed as prediction references, the encoder would include the major parts of a decoder in it. In this sense, modules designed for decoder are also parts of the encoder, and to prevent drifting errors, the decoder must also follow the same procedures as in the decoding loop of the encoder.

There are many reasons for postprocessing. For example, if the decoded video format does not match the display, a scaling of spatial resolution or temporal frame rate, or a de-interlacing operation will be used for the conversion. If the bit-rate is so low that there are many artifacts in the decoded image or video, some de-blocking or de-ringing operations will be used to reduce these artifacts. If there are channel errors or packet losses, there would be some distortions in the decoded image or video, then an error concealment operation is required to repair the image or video to the best possible extent. If there are multiple image or video objects that need to be presented on the same display, a composition operation will be used to generate a meaningful scene to present to the user. Moreover, it has been shown that in-loop filtering of the decoded frames can improve the efficiency of predictive video coding significantly at low bit rates. Although the in-loop filtering results may or may not be output directly to the display, we regard such filtering as a kind of postprocessing as well. Postprocessing is a stage that prepares the decoded image or video data into a more favorable form for improving visual quality or subsequent coding efficiency.

As mentioned before, not all stages would appear in every image or video coder. For example, a lossless coder normally does not contain lossy stages, such as, preprocessing and quantization.

So far, we only provided a generic overview of what components are involved in an image and video compression system. In the following subsections, we present common principles and basic techniques of image and video coding in some details.

Although color components in images or video can be coded jointly as a vector based signal to achieve improved compression efficiency, practical image and video coding systems normally choose to compress them independently due to its simplicity. From now on in this article, unless especially noted, the algorithms and schemes described are for a single color component.

3.2. Entropy Coding

Entropy is a measure of disorder, or uncertainty. In information systems, the degree of unpredictability of a message can be used as a measure of the information carried by the message. In 1948, Shannon defined the information conveyed by an event $I(E)$, measured in bits, in terms of the probability of the event $P(E)$,

$$I(E) = -\log_2(P(E)). \tag{5}$$

The physical meaning of the above definition is not hard to understand. The higher the probability of an event (i.e., the more predictable event), the less information is conveyed by that event when it happens. Moreover, information conveyed by a particular sequence of independent events is the sum of the information conveyed by each event in the sequence, whereas the probability of the sequence is the product of the individual probabilities of the events in the sequence, which is exactly what the log function reflects. A discrete memoryless source (DMS) generates symbols from a known set of alphabet symbols one at a time. It is memoryless since the probability of any symbol being generated is independent of the past history. Assume a DMS U_0 with alphabet $\{a_0, a_1, \ldots, a_{K-1}\}$ and probabilities $\{P(a_0), P(a_1), \ldots, P(a_{K-1})\}$, the entropy of such an information source is defined as the average amount of information conveyed by each symbol output by the source,

$$H(U_0) = -\sum_{k=0}^{K-1} P(a_i)\log_2(P(a_i)). \tag{6}$$

Generally, a source with nonuniform distribution can be represented or compressed using a variable-length code where shorter code words are assigned to frequently occurring symbols, and vice versa. According to Shannon's noiseless source encoding theorem, the entropy $H(U_0)$ is the lower bound for the average word length of a uniquely decodable variable-length code for the symbols. Conversely, the average word length can approach $H(U_0)$ if sufficiently large blocks of symbols are encoded jointly.

3.2.1. Huffman Coding. One particular set of uniquely decodable codes are called prefix codes. In such a code, one code word cannot be the prefix of another one. In 1952 Huffman proposed an algorithm for constructing optimal variable-length prefix codes with minimum redundancy for memoryless sources. This method remains the most commonly used today, for example, in the JPEG and MPEG compression standards.

The construction of a Huffman code is as follows:

1. Pick the two symbols in the alphabet with lowest probabilities and merge them into a new combined symbol. This generates a new alphabet with one less symbol. Assign "0" and "1" to the two branches linking the two original symbols to the new combined one, respectively.
2. Calculate the probability of the combined symbol by adding up the probabilities of the two original symbols.
3. If the new alphabet contains more than one symbol, repeat steps 1 and 2 for the new alphabet. Otherwise, the last combined symbol becomes the Huffman tree root.
4. For each original symbol in the alphabet, traverse all the branches from the root and append the assigned "0" or "1" for each branch along the way to generate a Huffman codeword for the original symbol.

We now have a Huffman code for each member of the alphabet. Huffman codes are prefix codes and each of them is uniquely decodable. A Huffman code is not unique. For each symbol set, there exist several possible Huffman codes with equal efficiency. It can be shown that it is not possible to generate a code that is both uniquely decodable and more efficient than a Huffman code [16]. However, if the probability distribution somehow changes, such preconstructed codes would be less efficient and sometimes would even bring expansion. Moreover, Huffman codes are most efficient for data sources with nonuniform probability distributions. Sometimes, we may have to manipulate the data so as to achieve such a distribution.

In many cases, the data source contains a large alphabet but with only a few frequent symbols. Huffman codes constructed for such a source would require a very large code table and it would be very difficult to adapt to any probability distribution variations. An alternative method used in many image and video coding systems is to group all infrequent symbols as one composite symbol and construct a Huffman table for the reduced alphabet. The composite symbol is assigned a special escape code used to signal that it is followed by a fixed-length index of one of the infrequent symbols in the composite group. Only a very small code table is used and the statistics of the vast majority of infrequent symbols in the alphabet is shielded by the composite symbol. This greatly simplifies the Huffman code while maintaining good coding efficiency.

3.2.2. Arithmetic Coding. Huffman codes and derivatives can provide efficient coding for many sources. However, the Huffman coding schemes cannot optimally adapt to given symbol probabilities because they encode each input symbol separately with an integer number of bits. It is optimal only for a "quantized" version of the original probability distribution so the average code length is always close to but seldom reaches the entropy of the source. Moreover, no code is shorter than 1 bit, so it is not efficient for an alphabet with highly skewed probability distribution. Furthermore, there are no easy methods to make Huffman coding adapt to changing statistics.

On the other hand, an arithmetic encoder computes a code representing the entire sequence of symbols (called a string) rather than encodes each symbol separately. Coding is performed by representing the string by a subinterval through a sequence of divisions of an initial interval according to the probability of each symbol to be encoded.

Assume that a string of N symbols, $S = \{s_0, s_1, \ldots, s_t, \ldots, s_{N-1}\}$ are from an alphabet with K symbols $\{a_0, a_1, \ldots, a_i, \ldots, a_{K-1}\}$ with a probability distribution that is varied with time t as $P(a_i, t)$. Then the arithmetic encoding process of such a string can be described as follows:

1. Set the initial interval $[b, e)$ to the unit interval [0,1) and $t = 0$.
2. Divide the interval $[b, e)$ into K subintervals proportional to the probability distribution $P(a_i, t)$ for each symbol a_i at time t, that is,

$$b_i = b + (e - b)\sum_{j=0}^{i-1} P(a_j, t) \quad \text{and}$$

$$e_i = b + (e - b)\sum_{j=0}^{i} P(a_j, t).$$

3. Pick up the subinterval corresponding to symbol s_t, say $s_t = a_i$, update $[b, e)$ with $[b_i, e_i)$ and $t = t + 1$.
4. Repeat step 2 and 3 until $t = N$. Then output a binary arithmetic code that can uniquely identify the final interval selected.

Disregarding the numerical precision issue, the width of the final subinterval is equal to the probability of the string $P(S) = \prod_{t=0}^{N-1} P(s_t, t)$. It can be shown that the final subinterval of width $P(S)$ is guaranteed to contain one number that can be represented by B binary digits, with

$$-\log_2(P(S)) + 1 \leq B < -\log_2(P(S)) + 2, \tag{7}$$

which means that the subinterval can be represented by a number which needs 1 to 2 bits more than the ideal code word length. Any number within that interval can now be used as the code for the string (usually the one with the smallest number of digits is chosen). In the encoding process, there is no assumption that the probability distribution would stay the same at all time. Therefore, by nature the arithmetic encoding can well adapt to the changing statistics of the input and this is a significant advantage over Huffman coding. From an information theory point of view, arithmetic coding is better than Huffman coding; it can generate fractional bits for a symbol, and the total length of the encoded data stream is minimal. There are also many implementation issues associated with arithmetic coding, for example, limited numerical precision, a marker for the end of a string, multiplication free algorithm, and so on. For a detailed discussion of arithmetic coding, please see [17]. In practice, arithmetic and Huffman coding often offer similar average compression rates while arithmetic coding is a little better (ranging from 0 to 10% less bits). Arithmetic coding is an option for many image and video coding standards, such as JPEG, MPEG, H.26x, and so forth.

3.2.3. Lempel–Ziv Coding. Huffman coding and arithmetic coding require *a priori* knowledge of the probabilities or an accurate statistical model of the source which in some cases is difficult to obtain, especially with mixed data types. Conversely, Lempel–Ziv (LZ) coding developed by Ziv and Lempel [18] does not need an explicit model of the source statistics. It is a dictionary-based universal coding that can dynamically adapt to any sources.

In LZ coding, the code table (dictionary) of variable-length symbol strings is constructed dynamically. Fixed-length binary codewords are assigned to the variable-length input symbol strings by indexing into the code table. The basic idea is always to encode a symbol string that the encoder has encountered before as a whole. The longest symbol string the encoder has not seen so far is added as a new entry in the dictionary, and will in turn be used to encode all future occurrences of the same string. At any time, the dictionary contains all the substrings (prefixes) the encoder has already seen. With the initial code table and the indices received, the decoder can also dynamically reconstruct the same dictionary without any overhead information.

A popular implementation of LZ coding is the Lempel–Ziv–Welch (LZW) algorithm developed by Welch [20]. Let A be the source alphabet consisting K symbols $\{a_k, k = 0, \ldots, K - 1\}$. The LZW algorithm can be described as follows,

1. Initialize the first K entries of the dictionary with each symbol a_k from A and set the scan string w to empty.
2. Input the next symbol S and concatenate it with w to form a new string wS.
3. If wS has a matching entry in the dictionary, update the scan string w with wS, and go to 2. Otherwise, add wS as a new entry in the dictionary, output the index of the entry matching w update the scan string w with S and go to 2.

4. When the end of the input sequence is reached, process the scan string w from left to right, output the indices of entries in the dictionary that match with the longest possible substrings of w.

If the maximum dictionary size is M entries, the length of the codewords would be $\log_2(M)$ rounded to the next smallest integer. The larger the dictionary is, the better the compression. In practice the size of the dictionary is a trade-off between speed and compression ratio. It can be shown that LZ coding asymptotically approaches the source entropy rate for very long sequences [19]. For short sequences, however, LZ codes are not very efficient because of their adaptive nature. LZ coding is used in the UNIX compress utility and in many other modern file compression programs.

3.3. Markov Sources

Though some sources are indeed memoryless, many others, for example image and video data where the probability distribution of values for one symbol can be very dependent on one or more previous values, are sources with memory. A source with memory can be modeled as a Markov source. If a symbol from a source is dependent on N previous value(s), the source is known as an Nth-order Markov source. Natural or computer rendered images and video data are examples of Markov sources.

Conversely, joint sources generate N symbols simultaneously. A coding gain can be achieved by encoding those symbols jointly. The lower bound for the average code word length is the joint entropy,

$$H(U_1, U_2, \ldots, U_N) = -\sum_{u_1}\sum_{u_2}\cdots\sum_{u_N} P(u_1, u_2, \ldots, u_N) \times \log_2(P(u_1, u_2, \ldots, u_N)). \tag{8}$$

It generally holds that

$$H(U_1, U_2, \ldots, U_N) \leq H(U_1) + H(U_2) + \cdots + H(U_N) \tag{9}$$

with equality, if $U_1, U_2, \ldots, U_N$ are statistically independent. This states that for sources with memory, they can be best coded jointly with a smaller lower bound (the joint entropy) for the average word length than otherwise coded independently. Moreover, the word length of jointly coding the memoryless sources has the same lower bound as independently coding them. However, coding each memoryless source independently is much easier to implement than coding jointly in real applications.

For an image frame or a video sequence, since each pixel in it is correlated with neighboring pixels, to obtain the best coding efficiency, it is ideal to encode all the pixels in the whole image frame or video sequence jointly. However, practical implementation complexity prohibits us to do so. Fortunately, with the Markov model for image and video data, the problem can be much simplified.

For an Nth-order Markov source, assume the first symbol starts at time T_0 the conditional probabilities of the source symbols are,

$$\begin{aligned} P(u_T, Z_T) &= P(u_T \mid u_{T-1}, u_{T-2}, \ldots, u_{T-N}) \\ &= P(u_T \mid u_{T-1}, u_{T-2}, \ldots, u_{T-N}, u_{T-N-1}, \ldots u_{T_0}) \end{aligned} \tag{10}$$

where Z_T represents the state of the Markov source at time T The conditional entropy of such an Nth-order of Markov source is given by,

$$\begin{aligned} H(U_T, Z_T) &= H(U_T \mid U_{T-1}, U_{T-2}, \ldots, U_{T-N}) \\ &= E(-\log_2(P(u_T \mid u_{T-1}, u_{T-2}, \ldots u_{T-N}))) \\ &= -\sum_{u_T}\cdots\sum_{u_{T-N}} P(u_T, u_{T-1}, u_{T-2}, \ldots u_{T-N}) \\ &\quad \times \log_2(P(u_T \mid u_{T-1}, u_{T-2}, \ldots u_{T-N})) \\ &= \sum_{u_{T-1}}\cdots\sum_{u_{T-N}} P(u_{T-1}, u_{T-2}, \ldots u_{T-N}) \\ &\quad \times H(U_T \mid u_{T-1}, u_{T-2}, \ldots u_{T-N}). \end{aligned} \tag{11}$$

Moreover, it can be shown that,

$$H(U_T, U_{T-1}, \ldots, U_{T_0}) = \sum_{t=T_0}^{T} H(U_t, Z_t). \tag{12}$$

From the above equation, we can clearly see that for a Markov source, the complicated joint entropy of a whole symbol sequence can be simplified to the sum of the conditional entropy of each symbol in the sequence, which means that the entropy coding of such joint Markov sources can be simplified to the conditional entropy coding of each symbol in the sequence given N previous symbols. In addition, the last equation in (11) suggests a simple conditional entropy coding method: any entropy coding method for a memoryless source can be used to encode the symbol at time T as long as the probability distribution used is switched according to the contexts (or *states*) of N previous symbols.

Image and video data are normally highly correlated. The value of a pixel has dependency with a few neighboring pixels. Even for the simplest first-order Markov separable model, a pixel in a 2-D image would have dependency with at least 3 neighboring pixels and a pixel in a 3-D video sequence would have dependency with at least 7 neighboring pixels. If each pixel is represented by 8 bits, in order to most efficiently compress the pixel value with the above derived simplified context-based entropy coding method, it would require 256^3 probability tables for image coding and 256^7 probability tables for video coding. Apparently, it is impractical for the encoder or decoder to maintain such a huge number of probability distribution tables. For an efficient, independent coding of symbols, statistical dependencies should be reduced.

3.4. Predictive Coding

Predictive coding is a way to reduce correlation between data from Markov sources. It is much simpler than a conditional entropy coder described in the previous section. Instead of coding an original symbol value, the difference or error between the original value and a predicted value based on values of one or more past symbols is encoded. The decoder will perform the same prediction and use

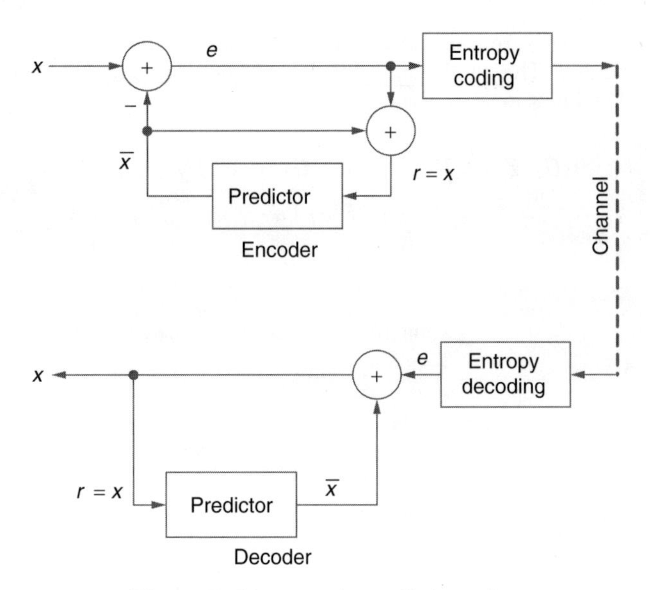

Figure 5. Diagram of a predictive coder.

the encoded error to reconstruct the original value of that symbol. Figure 5 illustrates the predictive coding process.

The linear predictor is the most used one in predictive coding. It creates a linear weighting of the last N symbols (Nth-order Markov) to predict the next symbol. That is,

$$\hat{S}_0 = \alpha_{-1}S_{-1} + \alpha_{-2}S_{-2} + \cdots + \alpha_{-N}S_{-N}, \quad (13)$$

where $\{S_{-1}, S_{-2}, \ldots, S_{-N}\}$ are the last N symbols and α_{-i} are the weights in the linear predictor. Without loss of generality, assume the input symbols have zero mean, i.e., $E\{S\} = 0$. The variance of the prediction error $e_0 = S_0 - \hat{S}_0$ is given by,

$$\sigma_{e_0}^2 = \sum_{i=0}^{N}\sum_{j=0}^{N} \alpha_{-i}\alpha_{-j}R_{ij}, \quad (14)$$

where R_{ij} is the covariance of symbols S_{-i} and S_{-j}, and $\alpha_0 = -1$.

Minimization of the prediction error variance leads to the orthogonality principle,

$$E\{e_0S_{-i}\} = 0, \text{ for all } i = 1, 2, \ldots, N. \quad (15)$$

The optimum coefficients $\{\alpha_{-1}, \alpha_{-2}, \ldots, \alpha_{-N}\}$ can be obtained by solving the above equation. Moreover, the orthogonality principle implies de-correlation of errors,

$$E\{e_0e_{-i}\} = E\{e_0\}E\{e_{-i}\} = 0, \text{ for all } i = 1, 2, \ldots, N. \quad (16)$$

For Gaussian random processes, de-correlation means statistical independence. Clearly, after optimum linear prediction, the prediction errors are uncorrelated and the much simpler and independent entropy coding can be used to efficiently encode these errors. Intuitively, after the process we have transformed a source of highly-correlated values with any possible distribution into a source of much less correlated values but with consistently high probability of being small. Such transformations that make the source more suitable for subsequent compression are very common in an image or video compression system.

For nonstationary sources, adaptive prediction can be used for more accurate prediction where the system switches between several predefined predictors according to the characteristics of the data being compressed. The choices of the predictors can be either explicitly coded with a small overhead or implicitly derived from the reconstructed values to avoid the overhead.

Predictive coding is important in a number of ways. In its lossless form, predictive coding is used in many sophisticated compression schemes to compress critical data, such as DC coefficients and motion vectors. If some degree of loss is acceptable, the technique can be used more extensively, (e.g., interframe motion compensated prediction in video coding).

3.5. Signal Models for Images and Video

In general, there is no good model to describe exactly the nature of real world images and video. A first-order Gaussian-Markov source is often used as a first order approximation due to its tractability. Though crude, this model provides many insights in understanding image and video coding principles.

A one-dimensional (1-D) Gaussian-Markov (AR(1) source) can be defined as,

$$x(n) = \alpha x(n-1) + \varepsilon(n), \text{ for all } n > n_0, \quad (17)$$

where $|\alpha| < 1$ is the regression coefficient, and $\{\varepsilon(n)\}$ is the i.i.d (independent identically distributed) zero mean normal random process with variance σ_ε^2, and $x(n_0)$ is a zero mean finite variance random variable. Such a source is known to be asymptotically stationary [1]. For a two-dimensional (2-D) image signal, the simplest source model is a two-dimensional separable correlation AR(1) model,

$$x(m,n) = \alpha_h x(m-1,n) + \alpha_v(m,n-1) - \alpha_h\alpha_v x(m-1,n-1) + \varepsilon(m,n), \quad (18)$$

where $\varepsilon(m,n)$ is an i.i.d zero mean Gaussian noise source with variance σ_N, and α_h, α_v denote the first order horizontal and vertical correlation coefficients, respectively. Its autocorrelation function is separable and can be expressed as a product of two 1-D autocorrelations. The separable correlation model of a 2-D image enables us to use 2-D separable transforms or other signal processing techniques to process the 2-D sources using separate 1-D processing in both horizontal and vertical directions, respectively. Therefore, in most cases, 1-D results can be generalized to 2-D cases in accordance with the 2-D separable correlation model.

For video signals, a 3-D separable signal model could still apply. However, the special feature of moving pictures in natural video is the relative motion of video objects in adjacent frames. A more precise signal model should incorporate the motion information into the 3-D signal model, for example, forming signal threads along the temporal direction and then applying the 3-D separable signal model. Fortunately, most modern video compression technologies have already explicitly used such motion information in the encoding process to take advantages of the temporal redundancy.

3.6. Quantization

The coding theory and techniques we discussed so far are mostly focused on lossless coding. However, the state-of-the-art lossless image and coding schemes exploiting the statistical redundancy of image and video data can only achieve an average compression factor of 2 or 3. From Shannon's rate distortion theory, we know that the coding rate of a data source can be significantly reduced by introducing some numerical distortion [4]. Fortunately, because the human visual system can tolerate some distortion under certain circumstances, such a distortion may or may not be perceptible. Lossy compression algorithms reduce both the redundant and irrelevant information to achieve higher compression ratio. Quantization is usually the only lossy operation that removes perceptual irrelevancy. It is a many-to-one mapping that reduces the number of possible signal values at the cost of introducing some numerical errors in the reconstructed signal. Quantization can be performed either on individual values (called scalar quantization) or on a group of values (a coding block, called vector quantization). The rate-distortion theory also indicates that, as the size of the coding block increases, the distortion asymptotically approaches Shannon lower bound; in other words, if a source is coded as an infinitely large block, then it is possible to find a block-coding scheme with rate $R(D)$ that can achieve distortion D where $R(D)$ is the minimum possible rate necessary to achieve an average distortion D [4]. This states that vector quantization is always better than scalar quantization. However, due to the complexity issue, many practical image and video coders still prefer to use scalar quantization. The basics on scalar and vector quantization techniques are discussed as follows.

3.6.1. Scalar Quantization. An N-point scalar quantization is a mapping from a real one-dimensional space to a finite set C of discrete points in the real space, $C = \{c_0, c_1, \ldots, c_{N-1}\}$. Normally, the mapping is to find the closest match in C for the input signal according to a certain distortion criterion, for example, mean squared error (MSE). The values of c_i are referred to as reproduction values. The output is the index of the best matched reproduction value. The transmission rate $r = \log_2 N$ is defined to indicate the number of bits per sample. The uniform quantizer with c_i distributed uniformly on the real axis is the optimal solution when quantizing a uniformly distributed source. For a random nonuniformly distributed source (such as image luminance levels) and even for a source with an unknown distribution, the Lloyd–Max quantizer design algorithm [2] provides an essential approach to the (locally) optimal scalar quantizer design. As a special case of the generalized Lloyd algorithm to be discussed for vector quantization, the Lloyd-Max quantizer tries to minimize the distortion for a given number of levels without the need for a subsequent entropy coder. Though optimal, it involves a complicated iterative training process. On the other hand, the study of entropy-constrained quantization (ECQ) shows that, if an efficient entropy coder is applied after quantization, the optimal coding gain can be always achieved by a uniform quantizer [3]. In other words, for most applications in image and video compression, the simplest possible quantizer, followed by variable-length coding, produces the best results. In order to take advantage of the subsequent entropy coder, many practical scalar quantizers have included a dead-zone, where a relatively larger partition is allocated for zero. Note that if the probability distribution of the data to be quantized is highly skewed, the inverse quantizer might achieve smaller distortion if choosing a *biased* reconstruction point towards the higher probability end rather than the usual mid-point for uniform distribution.

As shown above, the uniform quantizer combined with entropy coding is simple and well suited in image and video coding. However, new applications such as delivery of image or video over unstable or low bandwidth networks require progressive transmission and/or exact rate control of the bit stream. To equip the image or video coding with these new functionalities, progressive quantization strategies have to be adopted where a coefficient is quantized in a multi-pass fashion and the quantization error can be reduced by successive enhancement information obtained from each pass. Successive-Approximation Quantization (SAQ) of the coefficients is one such approach and is widely used in image and video coding systems [37,38,44,47]. As a special case of SAQ, bit-plane coding encodes each bit in a binary representation of the coefficient from the most significant bit (MSB) to the least significant bit (LSB) at each quantization scan.

3.6.2. Vector Quantization. In contrast to a scalar quantizer that operates upon a single, one-dimensional, variable, a vector quantizer acts on a multidimensional vector. Vector quantization is a mapping of k-dimensional Euclidean space R^k into a finite subset $\mathbf{C}$ of R^k, where $\mathbf{C} = \{\mathbf{c}_i : i = 1, 2, \ldots, N\}$ is the set of N reproduction vectors or codebook, and the element $\mathbf{c}_i$ is referred to as a codeword or a code-vector. An encoder $Q(\mathbf{x})$ takes an input vector $\mathbf{x}$ and generates the index i of the best matched vector $\mathbf{c}_i$ to $\mathbf{x}$ in the set $\mathbf{C}$ according to certain distortion criteria, for example, MSE. A decoder $Q^{-1}(i)$ regenerates the codeword $\mathbf{c}_i$ using the input index i Fig. 6 illustrates the VQ encoding and decoding processes.

Vector quantization always outperforms the scalar quantization in terms of error measurement under the same bit rate by Shannon's rate-distortion theory [3–6]. Vector quantization takes advantage of the joint probability distribution of a set of random variables while scalar quantization only uses the marginal probability distribution of a one-dimensional random variable [3]. In [7], Lookabaugh and Gray summarized the vector quantization advantages over scalar quantization in three categories: memory advantage (correlation between vector components, Fig. 7), shape advantage (probability distribution, Fig. 8) and space filling advantage (higher dimensionality, Fig. 9). Table 1 lists the high-rate approximation of coding gain brought by these VQ advantages over a scalar quantizer for different VQ dimensionalities. The results

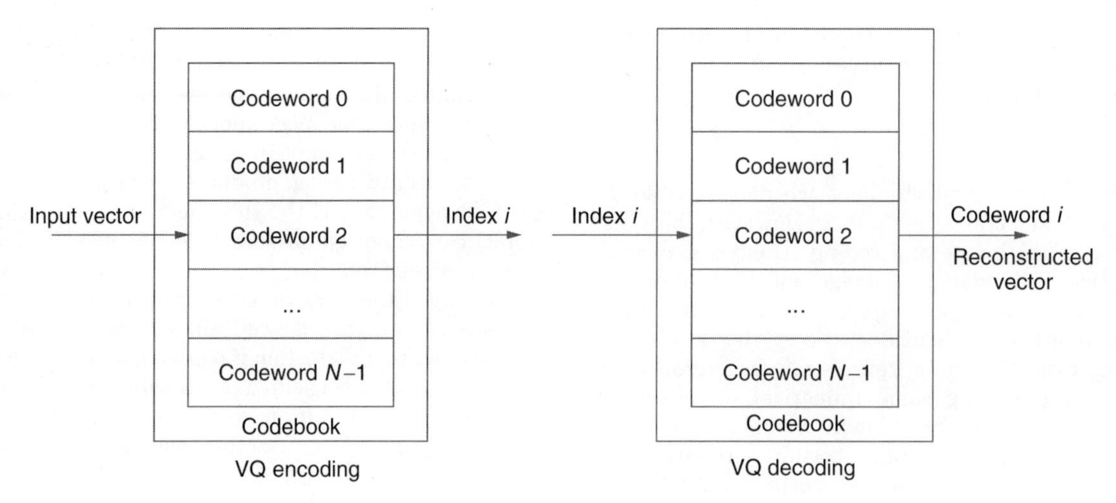

Figure 6. Vector quantization encoding and decoding processes.

are based on a first-order Gaussian-Markov source with regression coefficient $\rho = 0.95$, which is a typical value for a natural image or video source. Table 1 provides us some very useful insights on vector quantization. Firstly, as the VQ dimensionality increases, the coding gain also increases. The higher the VQ dimensionality is, the better the VQ performance is. Secondly, the increase in coding gain slows down beyond a certain VQ dimensionality. There is a delicate tradeoff between the increased complexity and the extra coding gain. The practical rule of thumb in image and video coding is that the coding gain of vector dimensionality beyond 16 may not be worth the added complexity. Thirdly, most of the VQ coding gain comes from the memory advantage. If we can completely decorrelate the samples within a VQ vector, and apply a scalar quantizer, we can still achieve most of the coding gain.

A good codebook design is crucial for the performance of a vector quantizer. The ideal codebook should minimize the average distortion for a given number of codewords. The

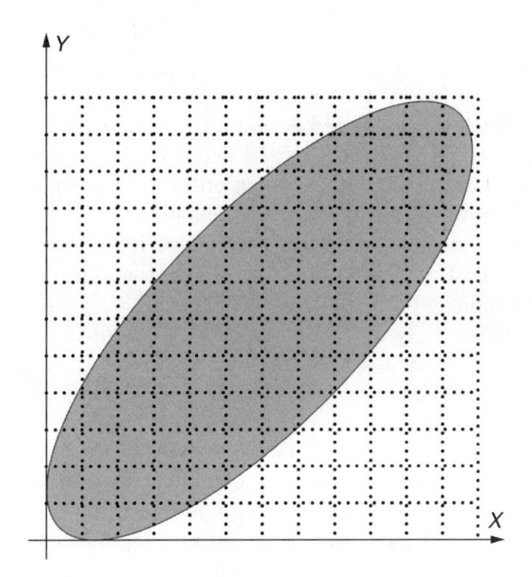

Figure 7. Memory advantage of vector quantization.

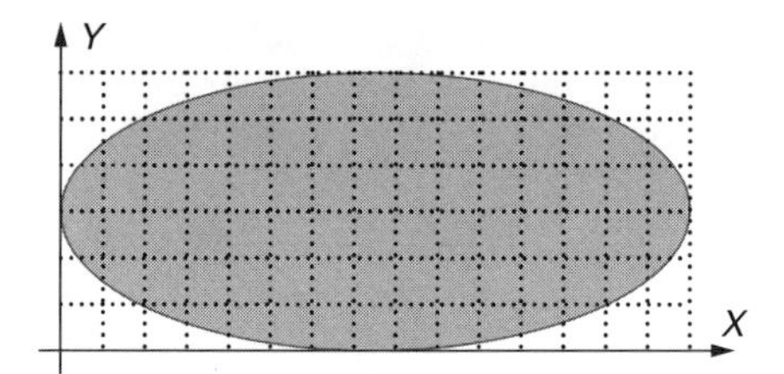

Figure 8. Shape advantage of vector quantization.

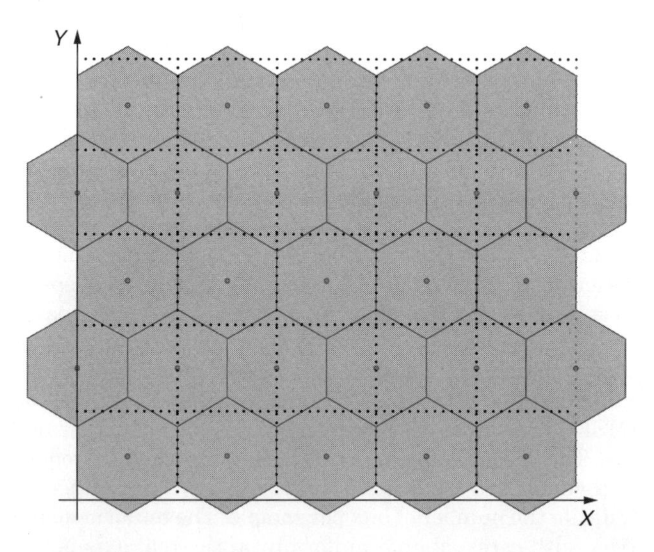

Figure 9. Space-filling advantage of vector quantization.

most commonly used algorithms for codebook generation is the LBG algorithm (Linde, Buzo, and Gray) [11], which is also referred to as Generalized Lloyd Algorithm (GLA). The LBG algorithm is an iterative process based on the two necessary conditions for an optimal codebook: the nearest neighbor condition, where the optimal partition cell for a codeword should cover any vector that has the shortest distance to it; the centroid condition, where the optimal codeword should be the centroid of the partition cell.

Table 1. The Coding Gain (dB) of Vector Quantization with Different Dimensionalities Over Scalar Quantization

VQ Dimension	Space Filling Advantage (dB)	Shape Advantage (dB)	Memory Advantage (dB)	Total Coding Gain (dB)
1	0	0	0	0
2	0.17	1.14	5.05	6.36
3	0.29	1.61	6.74	8.64
4	0.39	1.87	7.58	9.84
5	0.47	2.04	8.09	10.6
6	0.54	2.16	8.42	11.12
7	0.60	2.25	8.67	11.52
8	0.66	2.31	8.85	11.82
9	0.70	2.36	8.99	12.05
10	0.74	2.41	9.10	12.25
12	0.81	2.47	9.27	12.55
16	0.91	2.55	9.48	12.94
24	1.04	2.64	9.67	13.35
100	1.35	2.77	10.01	14.13
∞	1.53	2.81	10.11	14.45

Although vector quantizers offer unparallel quantization performance, they can be very complex in both codebook design and encoding (searching for the best match). Normally, they would be applied in very low bit rate coding case where only a small codebook size is required. There are many continuing investigations on how to reduce the complexity of codebook training and codebook searching [3,12–15,43]. Some efforts have also been put on the analogy of a uniform scalar quantizer in multidimensions — lattice VQ (LVQ) [8–10,45], where a codebook is not necessary. However, it seems that LVQ just puts off the burden of designing and searching for a large codebook to the design of a complex entropy coder.

3.7. Predictive Coding with Quantization

From the discussion on VQ, we know that the optimal (lossy) compression of an image is to take the image as a whole vector and perform vector quantization (VQ). However, the complexity of such a vector quantizer always prohibits us to do so in practice. We are constrained to very small dimensional VQ, or in the extreme case, scalar quantization. However, the performance of such a small dimensional VQ would degrade too much if there is no proper signal processing to decorrelate successive vectors. The ideal signal processing scheme should totally decorrelate the quantization unit (scalars or vectors) so that there is not much performance loss when quantizing independently and encoding with a DMS entropy coder or an entropy coder with low-order models.

Let's revisit the predictive coding discussed before but now combined with quantization. We have seen that predictive coding as a powerful lossless coding technique could decorrelate source data by prediction. The resultant error values could be regarded as a DMS with quite low entropy. However, we can save more bits if we can tolerate some small errors in the reconstructed signal. Predictive coding with linear prediction is also referred to as Differential Pulse Code Modulation (DPCM).

The difference between lossy and lossless DPCM lies in the handling of the prediction error. In order to lower the bit rate, the error in lossy DPCM is quantized prior to encoding. A block diagram for a basic DPCM encoder and decoder system is shown in Fig. 10, where e^* represents the quantized prediction error.

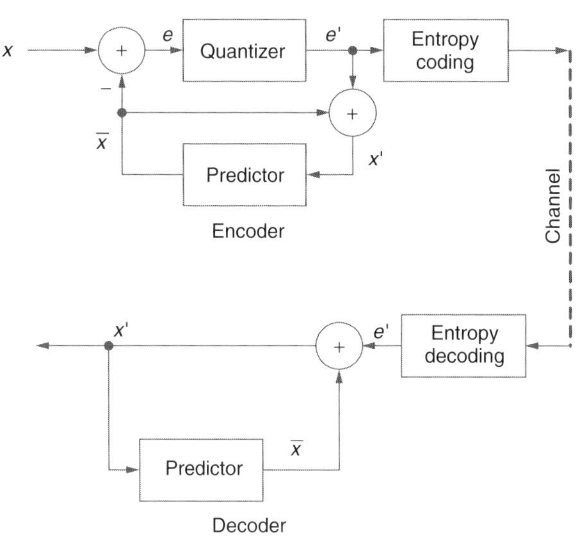

Figure 10. Diagram of a predictive coder combined with quantizer.

It is important to realize that in forming a prediction, the decoder only has access to the reconstructed values. Because the quantization of the prediction error introduces distortion, the reconstructed values typically differ from the original ones. To assure that identical predictions are formed at both the decoder and the encoder, the encoder also bases its prediction on the reconstructed values. This is accomplished by including the quantizer within the prediction loop as shown in Fig. 10. Essentially, each DPCM encoder includes the decoder within its structure. For the case of a successive-approximation quantizer, the encoder should fix a quantization level to be included in the prediction loop and the decoder should make certain the same quantization level can be transmitted to avoid mismatch errors.

The design of a DPCM system should consist of optimizing both the predictor and the quantizer jointly. However, it has been shown that under the mean-squared error optimization criterion, independent optimizations of the predictor and the quantizer discussed in previous sections are good approximations to the jointly optimal solution. Because of the reconstruction dependency of a predictive coder, any channel errors could be propagated throughout the remainder of the reconstructed values. Usually, the sum of the coefficients is made slightly less than one (called leaky prediction) to reduce the effects of channel errors.

3.8. Linear Transformations

Predictive coding offers excellent de-correlation capability for sources with linear dependence. However, it has several drawbacks. First, its IIR filtering nature decides that the correct reconstruction of future values is always dependent on the previously correctly reconstructed values. Thus, channel errors are not only propagated but also accumulated to future reconstructed values. This makes predictive coding an unstable system under channel errors. Secondly, for lossy coding, because of the iterative prediction and quantization processes, it is hard to establish a direct relation between average distortion and rate of the quantizer, which in turn makes it hard for optimal rate control. Thirdly, predictive coding is a model-based approach and is less robust to source statistics. When source statistics changes, adaptive predictive coding normally has to choose different predictors to match the source. Moreover, the prediction coding is a waveform compression technique. Since the human visual system model is best described in the frequency domain, it is difficult to apply visual masking to the prediction error.

Alternatively, transform coding techniques can be used to reduce the correlation in source data. Transform coders take an M input source samples and perform a reversible linear transform or decomposition to obtain M transform domain coefficients that are decorrelated and more energy compacted for better compression. They decorrelate coefficients to make them amendable to efficient entropy coding with low-order models, and distribute energy to only a few coefficients and thus make it easy to remove redundancy and irrelevancy. Theoretically, the asymptotic MSE performance is the same for both predictive coding and transform coding [21]. However, transform coding is more robust to channel errors and source statistics. There is an explicit relationship between distortion and data rate after transformation, and optimal bit allocation or rate control can be easily implemented. The subjective quality is better at low bit rates since transform coding is normally a frequency domain approach and the HVS model can be easily incorporated in the encoding process.

Figure 11 illustrates the advantage of transformation over a highly correlated two-dimensional vector source with scalar quantization. We can see that the transformed signal would require much fewer bits to code than

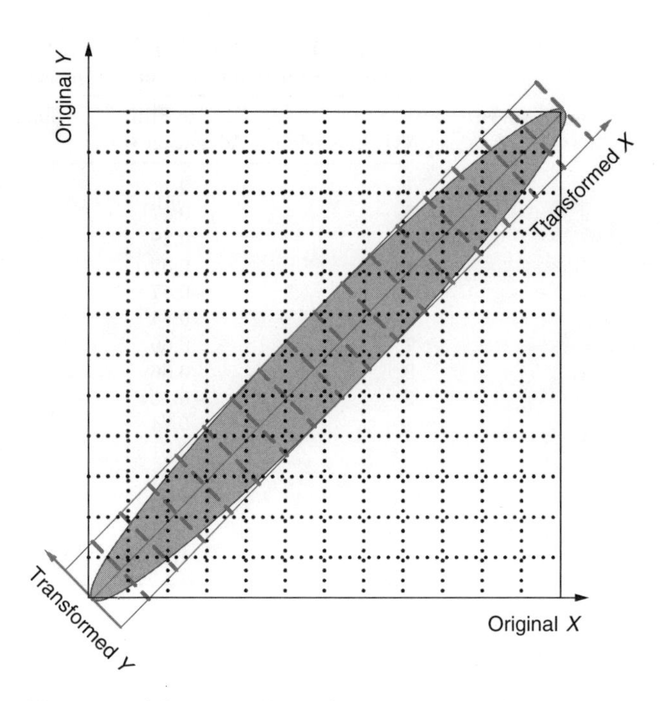

Figure 11. Advantage of transformation in signal compression.

the original signal with the same distortion although they both use a uniform scalar quantizer in each vector dimension. Furthermore, the complexity of entropy coding following the quantization step is reduced in the transformed domain since the number of codewords is reduced.

The efficiency of a transform coding system will depend on the type of linear transform and the nature of bit allocation for quantizing transform coefficients. Most practical systems are based on suboptimal approaches for transform operation as well as bit allocation. There are mainly two forms of linear transformations that are commonly used in transform coding: block transforms and subband decompositions.

3.8.1. Block Transforms. Block transform coding, also called block quantization, is a widely used technique in image and video compression. A block of data is transformed so that a large portion of its energy is packed in relatively few transform coefficients, which are then quantized independently.

In general, a 1-D transformation scheme can normally be represented by a kernel matrix $T = \{t(k,n)\}_{k,n=0,1,\ldots,N-1}$ and the transformed results $Y = \{y_0, y_1, \ldots, y_{N-1}\}^T$ can be represented as the multiplications of the transform matrix T and the signal vector $X = \{x_0, x_1, \ldots, x_{N-1}\}^T$. That is,

$$Y = T \bullet X. \tag{19}$$

The original signal can be recovered by multiplying the inverse matrix of T with the transformed signal without any distortion if we ignore the rounding errors caused by limited arithmetic precision. From a signal analysis point of view, the original signal is represented as the weighted sum of basis vectors, where the weights are just the transform coefficients.

Block transforms are normally orthonormal (unitary) transforms, which means that,

$$T \bullet T^T = T^T \bullet T = I_{N\times N}, \quad (20)$$

where $I_{N\times N}$ is the identity matrix. Orthogonality is clearly a necessary condition for basis vectors to decompose an input into uncorrelated components in an N-dimensional space. Orthonormality of basis vectors is a stronger condition that leads to the signal energy preservation property in both the signal domain and the transform domain. Moreover, the mean squared error caused by quantization in the transform domain is the same as that in the signal domain and is independent of the transform. This greatly eases the encoding process where otherwise an inverse transform is needed to find the distortion in the original domain.

Image signals are two-dimensional signals. By nature the correlation between pixels is not separable, so a non-separable transform should be applied to decorrelate the pixels. However, since a 2-D separable correlation model provides practical simplicity and sufficiently good performance, 2-D separable transforms are widely used in practical image coding systems. A 2-D separable transforms can be easily implemented with two steps of 1-D transforms for both all rows and all columns subsequently. Similarly, for 3-D video signals, 3-D separable transforms can be implemented with 1-D transforms in each direction: horizontal, vertical and temporal, separately.

In practice, the block transforms are not applied to a whole image itself. Normally, the image is divided into subimages or blocks and each block is then transformed and coded independently. The transform coding based on a small block size does not necessarily degrade too much the coding efficiency, because: (1) from the VQ theory, we learned that beyond a certain vector size, the coding gain increase tends to saturate, so in this case, larger block sizes don't bring us significant additional coding gain anyway; (2) natural images and video are generally nonstationary signals, dividing them into small blocks is particularly efficient in cases where correlations are localized to neighboring pixels, and where structural details tend to cluster; and (3) the inter block correlation can still be exploited by predictive coding techniques for certain transform domain coefficients, for example, the DC component. From a theoretical analysis and simulation results, it is shown that for natural images the block size is optimal around 8 to 16. In most image and video coding standards such as JPEG, MPEG, a value of 8 has been chosen. Recently, there is also a trend to use adaptive block transforms where transform blocks may adapt to signal local statistics. The block processing has a significant drawback, however, since it introduces a distortion termed blocking artifact, which becomes visible at high compression ratios, especially in image regions with low local variance. Lapped Orthogonal Transforms (LOT) [30] attempt to reduce the blocking artifacts by using smoothly overlapping blocks. However, the increased computational complexity of such algorithms does not seem to justify wide replacement of block transforms by LOT.

The optimal block transform is the Karhunen-Loeve Transform (KLT) that yields decorrelated transform coefficients and optimum energy concentration. The basis vectors of KLT are eigenvectors of the covariance matrix of the input signal. However, the KLT depends on the second-order signal statistics as well as the size of the block, and the basis vectors are not known analytically. Even when a transform matrix is available, it still involves quite a large amount of transformation operations because the KLT is not separable for image blocks and the transform matrix cannot be factored into sparse matrices for fast calculation. Therefore, the KLT is not appropriate for image coding applications. Fortunately, there exists a unitary transform that performs nearly as well as the KLT on natural images but without the disadvantages of KLT. This leads us to the Discrete Cosine Transforms (DCT).

The Discrete Cosine Transform kernel matrix is defined as follows,

$$t_{kn} = u(k)\cos\left(\frac{\pi(2n+1)k}{2N}\right), \quad (21)$$

where

$$u(k) = \begin{cases} \dfrac{1}{\sqrt{N}}, & \text{if } \ k = 0; \\ \sqrt{\dfrac{2}{N}}, & \text{if } \ k \neq 0. \end{cases} \quad (22)$$

The DCT has some very interesting properties. First, it is verified that the DCT is very close — in terms of energy compaction and decorrelation — to the optimal KLT for a highly correlated first-order stationary Markov sequence [22]. Secondly, its transform kernel is a real function, so only the real part of the transform domain coefficients of a natural image or video must be coded. Moreover, there exist fast algorithms for computing the DCT in one or two dimensions [25–28]. All these have made DCT a popular transform in various image and video coding schemes. For a natural image block, after the DCT, the DC coefficient is typically uniformly distributed, whereas the distribution for the other coefficients resembles a Laplacian one.

There are some other transforms that also could be used for image and video coding, such as Haar Transform, or a Walsh-Hadamard Transform. The reason to use them is not because of their performance but because of their simplicity.

3.8.2. Subband Decomposition. Another form of linear transformation that brings energy compaction and decorrelation is subband decomposition. In subband decomposition (called analysis process), the source to be compressed is passed through a bank of analysis filters (filter bank) followed by critical subsampling to generate signal subbands. Each subband represents a particular portion of the frequency spectrum of the image. At the decoder, the subband signals are decoded, upsampled and passed through a bank of synthesis filters and properly summed up to yield the reconstructed signal. This process is called the synthesis process. The fundamental concept behind subband coding is to split up the frequency band of a signal and then to code each subband using a coder and bit rate accurately matched to the statistics of the band.

Compared with block transforms, subband decomposition is normally applied to the entire signal and thus it can decorrelate a signal across a larger scale than block transforms, which translates into more potential coding gain. At high compression ratios, block transform coding suffers severe blocking artifacts at block boundaries whereas subband coding does not. The capability to encode each subband separately in accordance with its visual importance leads to visually pleasing image reconstruction. As a subset of subband decomposition, wavelet decomposition provides an intrinsic multi-resolution representation of signals that is important for many attractive image and video coding functionalities, such as adaptive coding, scalable coding, progressive transmission, optimal bit allocation and rate control, error robustness, and so forth.

It has been shown that the analysis and synthesis filters play an important role in the performance of the decomposition for compression purposes. One of the original challenges in subband coding was to design subband filters that cover well the desired frequency band but without aliasing upon the reconstruction step caused by the intermediate subsampling. The key advance was the development of quadrature mirror filters (QMF) [29]. Although aliasing is allowed in the subsampling step at the encoder, the QMF filters cancel the aliasing during the reconstruction at the receiver. These ideas continue to be generalized and extended.

A two-band filter bank is illustrated in Fig. 12. The filters $F_0(\omega)$ and $F_1(\omega)$ are the analysis lowpass and highpass filters, respectively, while $G_0(\omega)$ and $G_1(\omega)$ are the synthesis filters. In this system, the input/output relationship is given by

$$\begin{aligned} X'(\omega) &= \frac{1}{2}[F_0(\omega)G_0(w) + F_1(\omega)G_1(\omega)]X(\omega) \\ &+ \frac{1}{2}[F_0(\omega+\pi)G_0(w) \\ &+ F_1(\omega+\pi)G_1(\omega)]X(\omega+\pi), \end{aligned} \tag{23}$$

where the underlined term is where the aliasing occurs. Perfect reconstruction can be achieved by removing the aliasing distortion. One such condition could be, $G_0(\omega) = F_1(w+\pi)$ and $-G_1(\omega) = F_0(w+\pi)$. Furthermore QMF filters achieve aliasing cancellation by choosing, $F_0(\omega) = F_1(\omega+\pi) = -G_0(\omega) = G_1(\omega+\pi)$, where the highpass band is the mirror image of the lowpass band in the frequency domain.

As a special case to subband decomposition, wavelet decomposition allows nonuniform tiling of the time-frequency plane. All wavelet basis functions (baby wavelets) are derived from a single prototype (mother wavelet) by dilations (scaling) and translations (shifts). Besides the perfect reconstruction property, wavelet filters used for image and video compression face additional and often conflicting requirements, such as, compact support (short impulse response) of the analysis filters to preserve the localization of image features; compact support of the synthesis filters to prevent spreading of ringing artifacts; linear phase to avoid unpleasant waveform distortions around edges; and orthogonality to provide preservation of energy. Among them, orthogonality is mutually exclusive with linear phase in two-band FIR systems, so it is often sacrificed for linear phase. More information on wavelets and filter banks can be found in [146–148]. Wavelet decomposition of discrete sources is also often referred to as Discrete Wavelet Transforms (DWT).

Two-band systems are the basic component of most subband decomposition schemes. Recursive application of a two-band filter bank to the subbands of the previous stage yields subbands with various tree structures. Examples are uniform decomposition, octave-band (pyramid) decomposition, and adaptive or wavelet-packet decomposition. Among them, pyramid decomposition is the most widely used in image and video coding where a multi-resolution representation of image and video is generated. Based on the fact that most of the frequency of an image is concentrated in low-frequency regions, pyramid decomposition further splits the lower frequency part while keeping the high-pass part at each level untouched. Figure 13 illustrates such 2-D separable pyramid decomposition intuitively. Pyramid decomposition provides inherent spatial scalability in the encoded bit stream. On the other hand, an adaptive wavelet transform or wavelet-packet decomposition chooses the band splitting method according to the local characteristics of the signal source in each subband to achieve better compression.

Although applied to a whole image frame, the discrete wavelet transform is not as complicated as it seems. Lifting [48] is a fast algorithm that can be used to efficiently compute DWT transforms while providing some new insights on wavelet transforms.

As we mentioned before, one advantage of subband coding over block transform coding is the absence of the blocking effect. However, it introduces another major artifact — a ringing effect which occurs around high-contrast edges due to the Gibbs phenomenon of linear filters. This artifact can be reduced or even removed by an

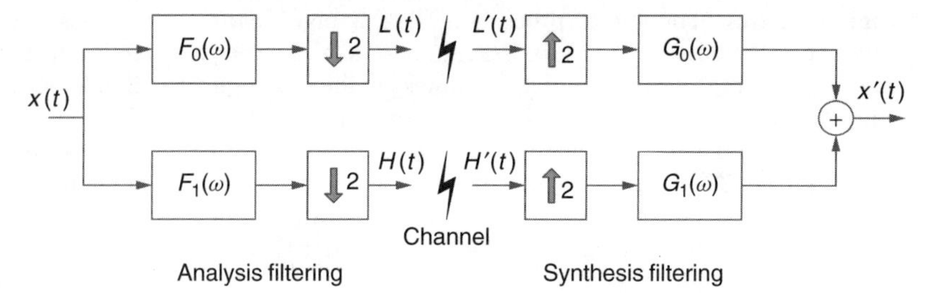

Figure 12. Two-band wavelet analysis/synthesis system. $F_0(\varpi)$ and $F_1(\varpi)$ are the lowpass and highpass analysis filters, respectively; $G_0(\varpi)$ and $G_1(\varpi)$ are the lowpass and highpass synthesis filters, respectively.

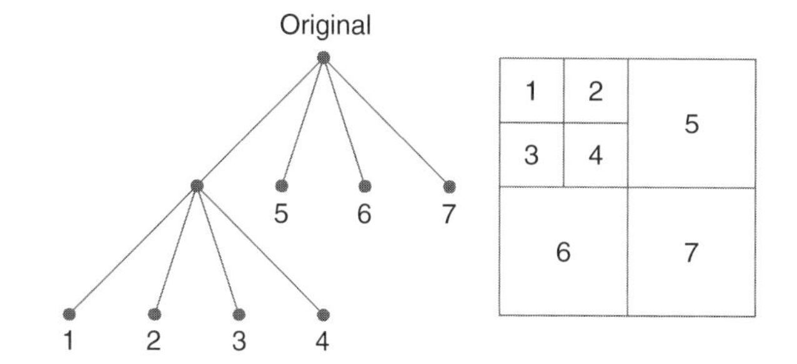

Figure 13. Illustration of a wavelet decomposition of depth two.

appropriate design of the filter bank or alleviated by some de-ringing post-filtering.

It is worth mentioning that to avoid the ringing artifacts caused by linear filtering, some nonlinear morphological filter banks are also being investigated [31–33].

It can be shown that block transforms are a special case of subband decompositions where the synthesis filters' impulse responses are the transform basis functions, the analysis filters' impulse responses are the time-reversed basis functions, and the decimation factor in each subband is the transform block length.

Note that the transformations (block transforms or subband decompositions) themselves do not compress data. The compression occurs when quantization and/or entropy coding techniques are applied to the transform coefficients. In fact, with sufficient arithmetic precision, the transformations should losslessly recover the original signal. There is also a class of transformations (reversible transforms) that can perfectly recover the original signal with limited precision. They are especially useful for lossless compression.

3.8.3. Vector Transformations. It is known from both information theory and practice that vector quantization always outperforms scalar quantization. However, when we replace the scalar quantization with vector quantization in the traditional predictive coding and transform coding schemes, the improvement is not significant. The main reason is because various transformations have a good decorrelation capability and have already decorrelated the components in the vectors to be vector quantized. As we learned from the vector quantization discussion in subsection 3.6.2 the most significant gain of VQ is from memory advantage. Is there a way to jointly optimize the transformation stage and the VQ stage so that the overall coding gain is maximized? The answer to this question is vector based signal processing [39], including vector transform coding [41,42], and vector wavelet/subband coding [40,46]. There are various kinds of vector transformation schemes, but the principle is the same, that is, (1) reduction of intervector correlation: the signal processing operations reduce correlation between the vectors as much as possible (2) preservation of intravector correlation: the signal processing operations preserve correlation between the components of each vector as much as possible. Image and video coding experiments show that vector-based transformation can achieve additional gain compared with scalar transformation followed by vector quantization with the cost of more complexity.

3.8.4. Shape-Adaptive Transformations. New applications in multimedia communications result in the need for object-based functionalities. Object-based functionalities require a prior segmentation of a scene into objects of interests. The objects are normally of arbitrary shapes. A new challenge that arises is how to efficiently compress the texture information within an arbitrarily-shaped object because the signal processing techniques we discussed so far are all based on rectangular regions. Of course, various padding schemes can be applied to expand the arbitrarily-shaped region into a rectangular one. However, such schemes are not very efficient because they need to code more coefficients than pixels within that region.

The objective of transformation over an arbitrarily-shaped region is still to decorrelate the data and achieve a high energy compaction. To be consistent with the techniques for rectangular regions, various approaches have been proposed to extend the block transforms or subband decomposition techniques to support arbitrarily-shaped regions. Among them, most notably are POCS-based block transforms (PBT) [22], shape-adaptive DCT (SA-DCT) [34,35] and shape-adaptive DWT (SA-DWT) [36,136,149,150].

The PBT, where POCS stands for projection onto convex sets, is based on two iterative steps to determine the best transform domain coefficient values. The first step transforms the pixels within an arbitrarily-shaped region through a normal rectangular transform and resets a selected set of coefficients to zeros, so that the number of nonzero coefficients equals to the number of pixels within the region. The second step performs an inverse transform of the coefficients obtained and resets the pixels within the arbitrarily-shaped region back to their original values. These two steps are iterated until the result converges to within a certain criterion. Although the convergence of the algorithm is guaranteed, this transform cannot achieve perfect reconstruction. However, as most of the energy of a natural image is concentrated in the low-frequency coefficients, the loss is minimized.

The SA-DCT is based on rectangular DCT transforms. A 2-D SA-DCT operates first vertically then horizontally. The SA-DCT begins by flushing all the pixels of the region up to the upper border of a rectangular block so that there are no holes within each column. Then an n-point DCT transform is applied to each column and where n is the number of pixels in that column. The column-transformed coefficients are normalized for the subsequent row processing. The same procedure repeats horizontally for each row to obtain the final transformed coefficients. The advantages of the SA-DCT algorithm include: computational simplicity, reversibility as long as shape information is available, effectiveness and the same number of transform coefficients as that of the pixels within the original shape. The main drawbacks of this approach lie on decorrelation of non-neighboring pixels and potential blocking artifacts.

On the other hand, the SA-DWT is an extension of the normal DWT transforms. Each level of SA-DWT is

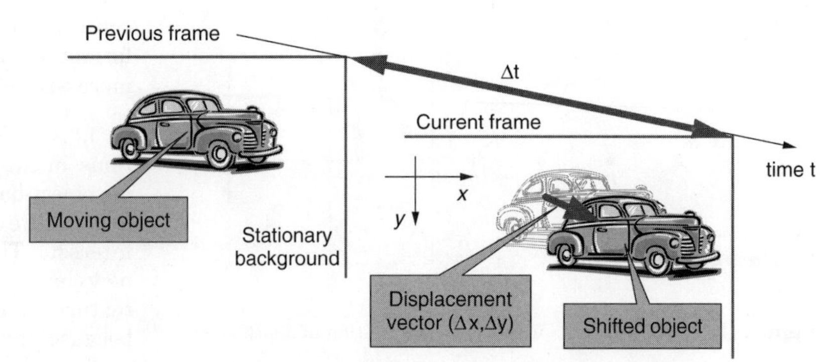

Figure 14. Illustration of motion compensation and estimation concept.

applied to each row of an image and then on each column. The key concept of SA-DWT is to keep the phase of each transformed coefficient aligned so that locality and spatial relation between neighboring pixels are well preserved. Based on the available shape information, SA-DWT starts by searching disjoined pixel segments in each row of an arbitrarily-shaped region and applies the normal DWT to each of the segments separately but the subsampling positions of each segment are aligned no matter where each segment starts. The subsampled coefficients are put in their corresponding spatial positions so that relative positions are preserved for efficient decorrelation in later steps. The same operation is then applied for each column of the row-transformed coefficients. SA-DWT also keeps the number of transform coefficients equal to the number of pixels in an original shape but with much improved decorrelation properties across an arbitrarily-shaped object texture. It has been shown that SA-DWT combined with advanced quantization and entropy coding techniques can achieve much improved efficiency over SA-DCT for still image object coding. Moreover, it will not suffer from the blocking artifacts existing in SA-DCT.

3.9. Motion Estimation and Compensation

The prediction or transformation techniques discussed so far all aim at de-correlating highly correlated *neighboring* signal samples. They are especially effective for spatial neighbors within an image or video frames. Video is a set of temporal samples that capture a moving scene across time. In a typical scene there is a great deal of similarity or correlation between neighboring images of the same sequence. However, unless the video sequence has slow motion, directly extending the above de-correlating techniques to the temporal direction would not always be very effective. The direct temporal neighboring samples that are spatially collocated are not necessarily highly correlated because of the motion in the video sequence. Rather, the pixels in neighboring frames that are more correlated are along the motion trajectory or optical flow and there usually exists a spatial displacement between them. Therefore, an efficient signal decorrelation scheme in the temporal direction should always operate on pixels along the same motion trajectory. Such a spatial displacement of visual objects (pixels, blocks or objects) is called a motion vector and the process of determining how objects move from one frame to another or finding the motion vector is called motion estimation. The operation of de-correlating samples in different frames using motion information is called motion compensation. The concept of motion estimation and compensation is illustrated in Fig. 14. Motion compensation is probably the most important factor in video coding that dominates most of the significant video coding advancements in recent years. It seems that further improvements of motion compensation are a continuing need.

The analogies to the normal prediction techniques and transformations in the temporal direction considering motion information are motion compensated predictions or motion predictions and motion compensated transformations or spatiotemporal transformations, respectively.

Motion estimation is the first step of these techniques, which is essentially a search process that normally involves heavy computation. Fortunately, there are many fast yet efficient motion estimation algorithms available today.

Ideally, each pixel in a video frame should be assigned a motion vector that refers to a pixel most correlated with it. However, it requires not only increases in the already heavy computational load for motion vector search, but also a significant amount of bits to encode the motion vector overhead. Therefore, in practice, pixels are usually grouped into a block with fixed or variable size for the motion estimation. Variable block size can adapt to the characteristics and object distribution in video frames and achieve better estimation accuracy and motion vector coding efficiency.

There are many motion estimation algorithms available, for example, block matching, where the motion vector is obtained by searching through all the possible positions according to a matching criterion; hierarchical block matching, where the estimated motion vectors are successively refined from larger size blocks to smaller size ones; gradient matching, where the spatial and temporal gradients are measured to calculate the motion vectors, and phase correlation, where the phases of the spectral components in two frames are used to derive the motion direction and speed. Among them, gradient matching and phase correlation techniques actually calculate the motion vectors of moving objects rather than estimating, extrapolating or searching for them. Block matching is the simplest and best studied one, and it is being widely used in practical video coding systems. To reduce the complexity of a brute-force full search while keeping the motion vector accuracy, many fast search schemes have been developed for block matching algorithms, for example,

3-step search (TSS) [49], Diamond search (DS) [50,51], Zonal-based search [52], and so on. Advanced search algorithms [53–55] even take advantage of the interframe motion field prediction to further speed up the motion estimation process.

Moreover, because the motion of objects in a video scene is not just limited to 2D translation, global motion estimation algorithms [56] try to capture the true object motion and further improve video coding efficiency by using an extended set of motion parameters for translation, rotation, and zooming.

3.9.1. Motion Compensated Prediction. The most widely used technique for de-correlating the samples along a temporal motion trajectory is motion compensated prediction. Motion compensated prediction forms a prediction image based on one or more previously *encoded* frames (not necessarily past frames in display order) in a video sequence and subtracts it from the frame to be encoded to generate a *residue* or *error* image to be encoded by various coding schemes. In fact, motion compensated prediction can be viewed as an adaptive predictor where the goal is to find minimum difference between the original image and the predicted image. Different techniques developed for improving motion prediction are just variations of the adaptive predictor. For example, integer-pel motion compensation is the basic form of the adaptive predictor where the choices of different predictions are signaled by the motion vectors; fractional-pel (half-pel, quarter-pel, or 1/8-pel) motion prediction is a refinement of integer-pel motion prediction where the choices of predictions are increased to provide more accurate predictions and they are signaled by the additional fractional-pel precision [57]; P-type prediction is formed by predictors using only a past frame as references; B-type prediction uses both the past frame and future frame; long-term prediction maintains a reference that is used frequently by other frames; advanced video coding schemes also use multiple frames as the input to the adaptive predictor to maximize the chance of further reducing the prediction errors with the expense of increased encoding complexity and more overhead for predictor parameters; overlapped block motion compensation (OBMC) [57] and de-blocking or loop filters in the prediction loop are ways to exploit more spatially correlated pixels at the input of the adaptive predictor without the need to transmit the overhead for more predictor parameters, while improving the visual quality of the decoded image; adaptive block size in motion prediction forces the motion predictor to adapt to the local characteristics of an image; global motion based prediction is not a linear prediction anymore and it forms more accurate prediction with complex warping operations considering the fact that the motion of video objects is not limited to 2-D translations. There are continuing research efforts on choosing better adaptive predictors, and there is always a delicate trade-off between complexity, overhead bits for predictor parameters, and prediction errors.

Ideally, if the motion compensated prediction simply removes the temporal correlation without affecting the spatial correlation between pixels, any efficient image coding schemes can be used to efficiently encode the prediction residue images with a little modification on the parts that heavily depend on the statistics of pixel values, for example, the entropy coding tables. Indeed, many existing video coding systems extend the still image coding methods to encode the residue image and achieve great results, for example, MPEG and H.26x coding standards using block-based motion compensation and block transformation for the residue coding.

Unfortunately, block-based systems have trouble coding sequences at very low-bit rates and the resultant coded images have noticeable blocking artifacts. The high-frequency components caused by the blocking artifacts in the residue image partly change the high-correlation nature of the residue. One solution to this problem is a lowpass filter in the motion compensation loop (either on the reference image or on the predicted image) to remove the high-frequency components. Another solution is to apply new signal decomposition techniques that match the statistics of the residue. Matching Pursuit [61–63] is one of these techniques that can be used to efficiently encode the prediction residue image. Instead of expanding the motion residual signal on a complete basis such as the DCT, an expansion on an overcomplete set of separable Gabor functions, which do not contain artificial block edges, is used. Meanwhile, it removes grid positioning restrictions, allowing elements of the basis set to exist at any pixel resolution location within the image. All these allow the system to avoid the artifacts most often produced by low-bit rate block-based systems.

Motion compensated prediction can work efficiently with dependency only on one immediate previous frame. This makes it very suitable for real-time video applications, such as video phone and video conferencing, where low delay is a critical prerequisite. As with any prediction coding schemes, the major drawback of predictive coding is the dependency on previously coded frames. This is especially more important for video. Features like random access and error robustness are limited in motion compensated predictive coding and transmission errors could be propagated and accumulated (called error drifting). Moreover, a predictive scalable coder must design ways to compensate the loss caused by the unavailability of part of the reference bit stream.

3.9.2. Motion Compensated Transformations. As we mentioned, direct extension of 2-D spatial signal decomposition to 3-D spatiotemporal decomposition without motion compensation [64] is not efficient especially for moderate to high motion sequences, though they offer computational simplicity and freedom from motion artifacts. Without motion compensation, the decoded video may normally present severe blurring and even ghost image artifacts.

With added complexity, motion compensated 3-D transformations align video samples along the motion trajectory to better decorrelate them for better coding efficiency. In addition, motion compensated transformations provide enhanced functionalities like scalability, easier rate-distortion optimization, easier error concealment and limited error propagation in error-prone channels, and so on. A particularly interesting example is the spatiotemporal subband/wavelet coding of video, where 3-D subband/wavelet transforms are applied to the motion aligned video data [65–68]. One of the key issues is to align the pixels in successive frames with motion information. This is still an active research topic and some good results have been reported. Another issue is related to the boundary effect that may be present in the temporal direction due to limited-length wavelet transforms. A solution is proposed in [69] to use a lifting algorithm.

One of the major drawbacks of 3-D transformations is that normally a few frames are involved and there exist considerable encoding and decoding delays. Therefore, it is not suitable for real-time communications.

3.10. Fractal Compression

Quite different from all the other schemes we have presented so far, fractal compression exploits the piecewise self-transformability (self-similarity) property existing in natural images that each segment of an image can be properly expressed as a simple transformation (rotation, scaling, translation) of another part having a higher resolution. Fractal compression is based on the iterated functions systems (IFS) theory pioneered by Barnsley [70] and Jacquin [71] and followed by numerous contributions [72]. The fundamental idea of fractal coding is to represent an image as the attractor of a contractive function system through a piecewise matching algorithm. There is no need to encode any pixel level in fractal compression, and the encoded image can be retrieved simply by iterating the IFS starting from any initial arbitrary image.

An advantage of fractal coding is that image at different levels of resolution can be computed through the IFS, without using interpolation or the duplication of pixel values. However, since it involves a search process, it requires quite intensive computation. Also, self-similarity is not self-identity, and fractal coding is always lossy.

Fractal compression is in fact related to vector quantization, but in contrast to classical vector quantization it uses a vector codebook drawn from the image itself rather than a fixed codebook. Fractal-based image compression techniques have been shown to achieve very good performance at high compression ratios (about 70–80) [73,74].

3.11. Bit Allocation and Rate Control

The ultimate goal of lossy image and video compression is to squeeze image or video data into as few bits as possible under certain quality criteria, or to get as much as possible quality under certain bit budget constraint. Besides choosing the best combinations of transformations, quantizers, and entropy coders, we have to optimally distribute the available bits across different components so as to achieve the best overall performance. This brings us to bit allocation and rate control which are indispensable for practical image and video coding systems. Rate control usually regulates the bit rate of a compression unit according to conditions not just target bit rate, but also encoder and decoder buffer models, and constant quality criteria, and so forth. On the other hand, bit allocation tries to make the quality of a picture as good as possible given a bit budget that may be assigned by the rate control.

In fact, if the target is just concerned about bit rate versus quality, rate control and bit allocation are closely related and sometimes it is hard to distinguish the difference between them. In this case, a generic rate control (bit allocation) problem can be formulated as the problem to minimize the overall distortion (quantization error),

$$D = \sum_i D_i, \tag{24}$$

under the constraint of a given total bit rate

$$R = \sum_i R_i, \tag{25}$$

by assigning to each compression unit the appropriate quantizer having a distortion D_i and a rate R_i. It has to be emphasized here that the only assumption made is that the overall distortion can be written as a sum of the individual distortions. No assumption about the nature of the distortion measure and the quantizers is made. Each compression unit can have its own distortion measure and its own admissible quantizer. Normally, the MSE is used as a measure of the distortion, although it is not a good measure for the quality of natural images or video to be evaluated by the human visual system.

There are two scenarios when dealing with bit allocation in practice: independent coding and dependent coding. Independent coding refers to the cases where the compression units (image pixels, blocks, frames, or subbands) are quantized independently. However, many popular schemes involve *dependent* coding frameworks, that is, where the R-D performance for some compression units depends on the particular choice of R-D operating points for other units. Typical examples of dependent coding are various predictive coding schemes such as DPCM and motion compensated predictive video coding. For the simple independent coding case, the optimization problem leads to a necessary *Pareto* condition,

$$\frac{\partial D_i}{\partial R_i} = \frac{\partial D_j}{\partial R_j}, \text{ for all } i \text{ and } j, \tag{26}$$

which states that when the optimal bit allocation is achieved, the slopes of the rate-distortion curves at the optimal points should be the same.

The rate-distortion (R-D) theory is a powerful tool for bit allocation. Under the R-D framework, there are two approaches to the bit allocation problem: an analytical model-based approach and an operational R-D based approach. The model-based approach assumes various input distribution and quantizer characteristics [75–78]. Under this approach, closed-form solutions can be obtained based on the assumed models using continuous optimization theory. Conversely, the operational R-D based approach [79–81] considers practical coding environments where only a finite set of quantizers is admissible. Under the operational R-D based approach, the admissible quantizers are used by the bit allocation algorithm to determine the optimal strategy to minimize the distortion under the constraint of a given bit budget. Integer programming theory is normally used in this approach to find the optimal discrete solution. Because the operational R-D approach can achieve exactly the practical optimal performance for completely arbitrary inputs and choices of discrete quantizers, it is often preferred in a practical coding system. However, it requires that: (1) the number of admissible quantizers or quantizer combinations is tractable for practical coding systems; (2) the coding system is capable of providing all the operational R-D points easily or tolerates the possible long delays and high complexity caused by calculation of these R-D points. Therefore, practically the operational approach is normally applied to independent coding cases or scalable coding schemes where a set of operational points can be easily obtained from the embedded bit streams. The model-based approach is normally applied in scenarios where delay and complexity cannot be tolerated or dependent coding cases where the combinations of admissible quantizers are out of control because of the dependency. If the input statistics are known, model-based bit allocation can be used to derive a predetermined bit allocation strategy that best fits the input source. Otherwise, it will work in a *feed-forward* fashion based on the heuristics where the parameters of the model can be updated adaptively and hopefully converge to an optimal bit allocation plan in the long run. The performance of model-based bit allocation can be also improved through multipass coding to refine the bit allocation among compression units.

For operational R-D based bit allocation, the problem normally involves finding the *convex hull* where the optimal points lie on [82]. Figure 15 shows an example of the convex hull on an R-D plot. Finding a convex hull of a set of points has been extensively investigated and many fast algorithms have been designed [83]. In order to avoid the impractical exhaustive computation of all the combinations, an algorithm was designed to find the convex hull in a limited number of computations [84]. For a single optimization point, the *BFOS* algorithm can be used to quickly allocate the available bits [85].

For a Gaussian source and MSE distortion criterion, assuming that the distortion error of each component should be within a threshold θ, the optimal bit allocation can be derived as,

$$R_i = \begin{cases} \frac{1}{2}\log_2\left(\frac{\sigma_i^2}{\theta}\right), & \text{if } \sigma_i^2 > \theta; \\ 0, & \text{if } \sigma_i^2 \le \theta; \end{cases} \tag{27}$$

and the distortion for each component is given by,

$$D_i = \begin{cases} \theta, & \text{if } \sigma_i^2 > \theta; \\ \sigma_i^2, & \text{if } \sigma_i^2 \le \theta; \end{cases} \tag{28}$$

This result shows that all the components should have the same quantization error except for those with energy σ_i^2 lower than certain threshold θ. By adjusting the value θ, various bit rates can be achieved using the above equations.

Using such a fixed bit allocation, the number of bits used for each component is fixed for all compression units of the source, so the allocation information only needs to be sent out once and all other bits can be used to encode the coefficient values. Such an optimal bit allocation is totally dependent on the statistical characteristics of the source; specifically, the variances of transform components are needed and in general, the source has to be stationary. While producing a constant-rate bit stream, coders using fixed bit allocation cannot adapt to the spatial or temporal changes of the source, and thus coding distortion may vary from one compression unit to another due to changes of the source. Conversely, adaptive bit allocation schemes can be used to deal with the random changes of a source. However, overhead bits will be introduced to indicate the difference in bit allocation schemes. Combining with the possible entropy coding afterwards, a closed-form formula is generally not available to explicitly indicate the relationship between bits and distortion. A feed-forward heuristic or multipass approach can be applied to hopefully obtain an optimal bit allocation on average. A simple example is the threshold coding that is used widely in many image and video coding standards nowadays. Threshold coding fully takes advantage of the energy packing property of transformations by encoding only those significant coefficients. In such a coding scheme, only those coefficients whose energy is higher than the threshold are quantized and encoded; all others will be treated as zero and discarded. This significantly increases the number of zeros and reduces that of significant coefficients to be encoded. Besides, the threshold could be much larger than the quantization step to take advantage of efficient zero coding. This is the basis for many *dead-zone* quantizers. The threshold itself does not need to be encoded in the bit stream; therefore, in the extreme case it could be optimized differently from coefficient to coefficient. Because threshold coding depends only on local energy, it can easily adapt to changes of the source. The drawback is that the bit rate cannot be predicted exactly, and it depends on the threshold, quantization steps, and side information about the locations of zeros. It generates variable bit rate for each coding unit. However, with an output buffer and a proper rate control scheme, the output rate can be kept within a bit budget while maintaining optimal performance on the average.

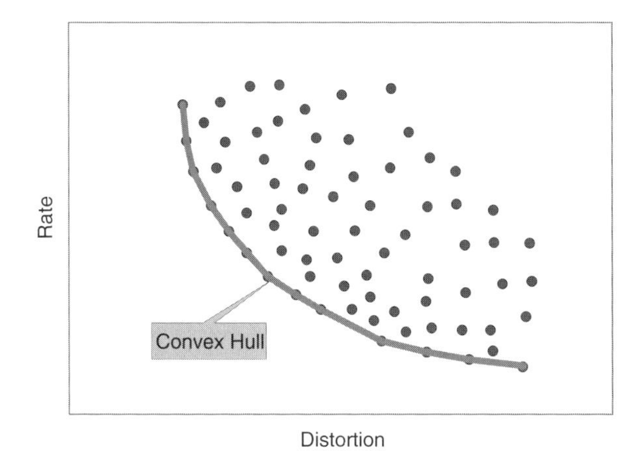

Figure 15. The convex hull of an R-D plot. The block dots represent all the admissible R-D points and the lower solid curve is the convex hull.

When delivering image or video data over time-varying channels such as wireless channels and the best effort IP networks, the bit budget that best fits the channels could not be predetermined at the encoder time. Scalable image [86] and video coding [38,87] can dynamically adjust the bit rate on the fly to adapt to the channel conditions (bandwidth, throughput, or error rate, etc.). This essentially results in a new coding scheme where the rate control is put off from the encoder time to the delivery time. How to quickly and optimally allocate the bits on the fly is an active research topic [86,88].

3.12. Symbol Formation

As an important and sometimes crucial part of final step in image and video compression systems, symbol formation essentially organizes the final coefficients that may or may not be quantized in a form that is more suitable for efficient entropy coding subsequently. Common symbol formation techniques are run-length coding (RLC), zig-zag scanning, zerotree scanning, and context formation for conditional entropy coding, and so forth.

3.12.1. Run-Length Coding (RLC). *Run-length coding (RLC)* is a very simple form of data compression. It is based on a simple principle that every segment of the data stream formed of the same data values (called *run*), that is, sequence of repeated data values, is replaced with a pair of count number (length) and value (run). This intuitive principle works best on certain data streams that contain a large number of consecutive occurrences of the same data values. For example, in the image domain, the same values or prediction differences of neighboring pixels often appear consecutively; in the transform domain, if the highly compact coefficients are sorted according to their energy distribution, after quantization, they often contain a long run of zeros.

RLC is a lossless coding scheme and is often combined with the subsequent entropy coding. It is generally believed that for a Markov source, RLC combined with entropy coding would achieve the same efficiency as we encode each data item with a conditional entropy coding scheme. However, RLC can be easily implemented and quickly executed.

3.12.2. Zigzag Scanning. As we know, block transform can concentrate the energy of the image or video data to only a few transform domain coefficients. Conversely, threshold coding employs a dead-zone in the quantizer so that many coefficients with little energy distribution can be quantized to zeros. Also, threshold coding is an adaptive quantization scheme where the locations of nonzero coefficients have to be encoded. By sorting the transform coefficients according to their energy distribution, one can get a coefficient sequence where the quantized coefficients with lower probability to be zeros are put toward the beginning of the sequence and those with higher probability are put towards the end of the sequence. For image and video data, most of the energy would be concentrated in low-frequency coefficients. Therefore, such sorting would result a zigzag scan order starting from lower frequency coefficients to higher ones and with more consecutive zeros toward the end of the sequence; thus, it makes run-length coding more efficient. Zigzag scanning takes advantage of the inherent statistics in the image and video data and arranges it in an order that is more convenient for subsequent coding. Figure 16 gives an example of such a zigzag scan order for an 8×8 DCT domain coefficients, which is commonly used in JPEG and MPEG image and video coding standards.

3.12.3. Zerotree coding. For subband/wavelet based decompositions, symbol formation is a little bit different. In the tree structure of octave-band wavelet decomposition shown in Fig. 17, each coefficient in the high-pass subbands has four coefficients corresponding to its spatial position at a higher scale. Because of this very structure of the decomposition, there should be a good way to arrange its coefficients to achieve better compression efficiency. Based on the statistics of the decomposed coefficients in each subband, it has been observed that if a wavelet coefficient at a coarse scale is insignificant with respect

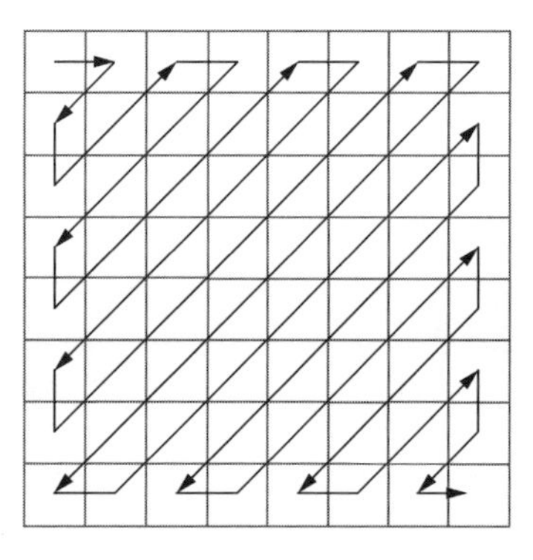

Figure 16. An exemplar zigzag scan order.

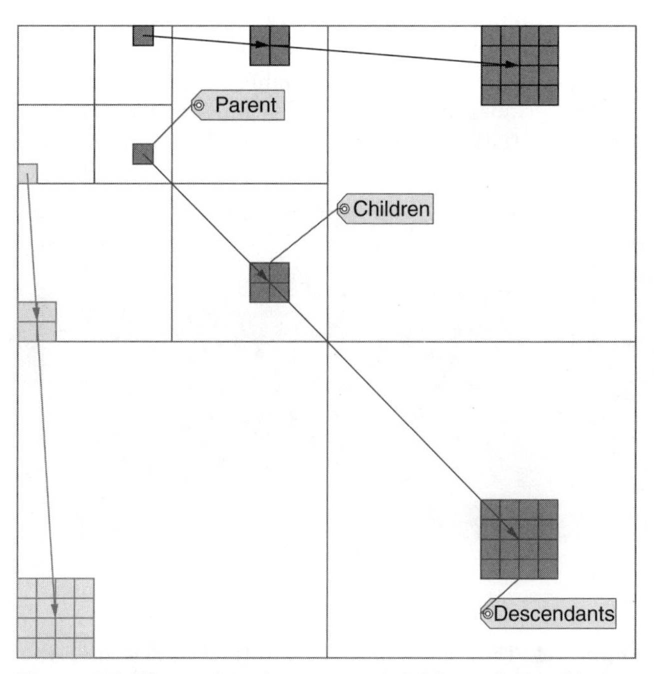

Figure 17. Illustration of parents and children relationship in a pyramid decomposition.

to a given threshold T, then all wavelet coefficients of the same orientation in the same spatial location at a finer scale are also likely to be insignificant with respect to T. EZW image compression is a wavelet based coder that was first proposed by Shapiro [89] in 1993. It uses the zerotree structure and takes advantage of the self-similarity between subbands at different scale. A zerotree [89] is formed with the root coefficient and all its children are insignificant. Figure 17 shows such a tree structure. An encoder can stop traversing the tree branch once it detects that the node is a zerotree root. Many insignificant coefficients at higher frequency subbands (finer resolutions) can be discarded. Zerotree coding can be considered as the counterpart of zigzag scan order in subband/wavelet coding. It also uses successive-approximation quantization to quantize the significant coefficients and context-based arithmetic coding to encode the generated bits. Bits are coded in order of importance, yielding a fully embedded code. The main advantage is that the encoder can terminate the encoding at any point, thereby allowing a target bit rate to be met exactly. Similarly, the decoder can also stop decoding at any point resulting in the best image at the rate of the truncated bit stream. The algorithm produces excellent results without any prestored tables or codebooks, training, or prior knowledge of the image source.

In general, ordered data are more easily compressed than un-ordered data. But when organizing data in an order that is not predefined, the order information has to be encoded in the bitstreams as overhead. Zigzag scan works well because it is a predetermined order and fits the source statistics well. On the other hand, for zerotree coding, partial order information such as a significance map has to be sent to the decoder. An efficient coding algorithm can be derived, if a good balance between overhead information for ordering and a good ordering scheme for subsequent entropy coding is achieved. Set Partitioning in Hierarchical Trees (SPIHT) algorithm [90], rearranges the transformed coefficients by partial ordering according to their magnitudes with a set partitioning sorting algorithm, transmits refinement bits in an order according to the ordered bitplane, and exploits the self-similarity of the wavelet transform across different scales. The algorithm enables progressive transmission of the coefficient values by ordering the coefficients and transmitting the most significant bits first. The ordering information also makes quantization more efficient by first allocating bits to coefficients with larger magnitudes. An efficient coding scheme of the ordering information is also included in the algorithm. The results [90] show that the SPIHT coding algorithm in most cases surpasses those schemes obtained from various zerotree algorithms [89].

3.12.4. Context Formation. It can be shown that when a data source is partitioned into different classes, the attainable entropy can be smaller than the unpartitioned one. Thus, it suggests that we can achieve more coding efficiency when a source is divided into different groups with significant different probability distributions. One way is to use a search algorithm to find which subgroup the encoded data belong to and use this information to drive a conditional entropy coding scheme. However, this would require overhead bits in explicitly coding the class of the subgroup. Fortunately, almost any data to be encoded in image and video compression are very dependent on its neighboring context. Such context information can be used to drive different entropy coders that are most suitable for the data to be coded. Context-based entropy coding is essentially a prediction coding scheme. The best part is that the prediction is *implicit* and the dependency or correlation between the data and the context does not have to be known explicitly.

As we have discussed in the entropy coding part, one challenge of context formation is to keep the contexts as small as possible while reflecting as much dependency as possible. Lower entropy can be achieved through higher order conditioning (larger contexts). However, larger context implies a larger *model cost* [92], which reflects the penalties of *context dilution* when count statistics must be spread over too many contexts, thus affecting the accuracy of the corresponding estimates, especially for adaptive context-based coding. Moreover, larger contexts means more memory is needed to store the probability tables. This observation suggests that the choice of context model should be guided by the use, whenever possible, of available prior knowledge on the data to be modeled, thus avoiding unnecessary learning costs. Often explicit prediction and context-based coding schemes can be combined to reduce the model cost even though they might be based on the same context [93].

3.13. Preprocessing and Postprocessing

In addition to the core techniques we have discussed, additional preprocessing and postprocessing stages are extensively used in image and video compression systems in order to render the input or output images in a more appropriate format for the purpose of coding or display. The possible operations include but are not limited to de-noising, format/color conversions, compression artifacts removal, error concealment, and so on.

As with any other signals captured from a natural source, image and video data normally contain noise in them. The problem becomes more severe as many low-end consumer-grade digital image and video devices are gaining in popularity. As we learned from information theory, truly random noise not only is very hard to compress but also degrades the image and video quality. It is very common to use a de-noising filter [94,95] prior to coding in order to enhance the quality of the final pictures and to remove the various noises that will affect the performance of compression algorithms.

The simplest compression techniques are based on interpolation and subsampling of the input image and video data to match with the resolution and format of the target display devices. For example, a mobile device is normally only equipped with limited display screen resolution, and a subsampled image that matches its screen size can just meet its needs while greatly reducing the bit rate. The challenge of subsampling is how to maintain the crispness of a picture without introducing aliasing, that is, how to select the lowpass filters, which is a classic image processing problem [96,97].

The human eyes are more sensitive to the difference in brightness than to the differences in color. To exploit this in image and video coding systems, the input images are often converted to YUV components (one luminance Y and two chrominance differences U and V) instead of using RGB components (red, green, blue). Furthermore, the components U and V often can be subsampled at a lower resolution.

Image format conversion is also a common postprocessing step for image and video decoding. Often the image or video signal is not encoded exactly as the same format supported by the display devices, for example, in terms of image size, frame rate, interlaced or progressive display, color space, and so on. Postprocessing must be able to convert these different formats of image video into the display format. There are active studies on this issue, especially on resizing of images [96,97], frame rate conversion [98,99], interlaced to progressive video (de-interlacing) [100–102].

It is normal to expect a certain degree of distortion of the decoded images for very low bit rate applications though a good coding scheme should introduce these distortions in areas less annoying for the users. Postprocessing can be used to further reduce these distortions. For block transform coding, solutions were proposed to reduce the blocking artifacts appearing at high-compression ratios [103–108]. Similar approaches have also been used to improve the quality of decoded signals in other coding schemes such as subband/wavelet coding, reducing different kinds of artifacts such as ringing, blurring, mosquito noise, and so on [109,110]. In addition to improving the visual quality, filtering in the prediction loop such as in motion compensation could also improve the coding efficiency [111,112].

When delivering encoded image and video bitstreams over error-prone or unreliable channels such as the best-effort Internet and wireless channels, it is quite possible to receive a partially corrupted bit stream due to packet losses or channel errors. Error concealment is a type of postprocessing technique [113–116] used to recover the corrupted areas in an image or video based on prior knowledge about the image and video characteristics, for example, spatial correlation for images or motion information for video. Other related preprocessing and postprocessing techniques are image and video object segmentation and composition for object-based video coding such as in MPEG-4 [117]; scene change detection for inserting key-frames in video coding [118]; variable frame-rate coding [119] according to the video contents and the related frame interpolation [120]; region of interests (ROI) coding of images [121]; and so forth.

Image and video coding is a very active research topic and progress on new compression technologies is being made rapidly. In a limited space, we can grasp only the fundamentals of these techniques. Hopefully, these basic principles can inspire and at least help readers to understand new image and video compression technologies.

4. IMAGE AND VIDEO CODING STANDARDS

We have introduced a number of well-known basic compression techniques commonly used in image and video coding. Practical coding systems all contain one or more of these basic algorithms. Now, we present several image and video coding standards. The importance of standards in image and video coding is due to the fact that an encoder and a decoder of a coding algorithm may be designed and manufactured by different parties. To ensure the interoperability, that is, one party's encoded bit stream can be decoded by the other party's decoder, a well-defined standard has to be established. It should be pointed out that a standard, such as MPEG-1, MPEG-2, or MPEG-4, only defines the bit stream syntax and the decoding process, leaving the encoding part open to different implementations. Therefore, strictly speaking, there is no such thing as video quality of a particular standard. The quality of a standard compliant bit stream at any given bit rate depends on the implementation of the encoder that generates the bit stream.

4.1. Image Coding Standards

4.1.1. ITU-T Group 3 and Group 4 Facsimile Compression Standards. One of the earliest lossless compression standards are the ITU-T (former CCITT) facsimile standards. The most common ones are *Group 3* [123] and *Group 4* [124] standards. Group 3 includes actually two distinct compression algorithms, known as Modified Huffman (MH) *coding* and Modified READ (Relative Element Address Designate) *coding* (MR) modes. The algorithm of Group 4 standard is commonly called Modified Modified READ (MMR) coding.

MH coding is a one-dimensional coding scheme that encodes each row of pixels in an image independently. It uses run-length coding with a static Huffman coder. It reduces the size of the Huffman code table by using a prefix markup code for runs over 63 and thus accounts for the term *modified* in the name. Each scan line ends with a unique EOL (end of line) code and it doubles as an error recovery code.

To exploit the fact that most transitions in bi-level facsimile images occur at one pixel to the right or left or directly below a transition on the line above, MR coding uses a 2-dimensional reference (Figure 18) with

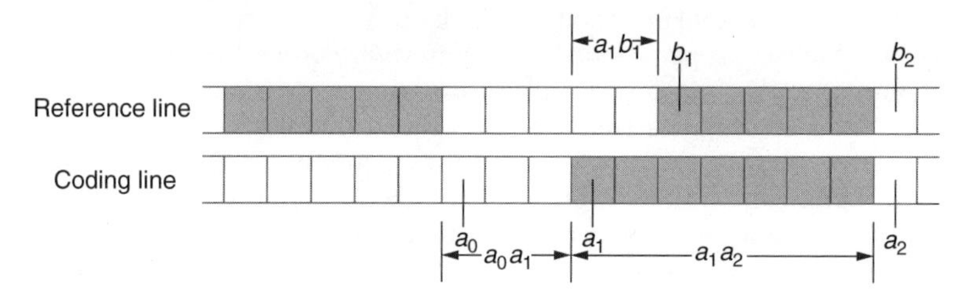

Figure 18. Reference point and lengths used during modified READ (MR) encoding.

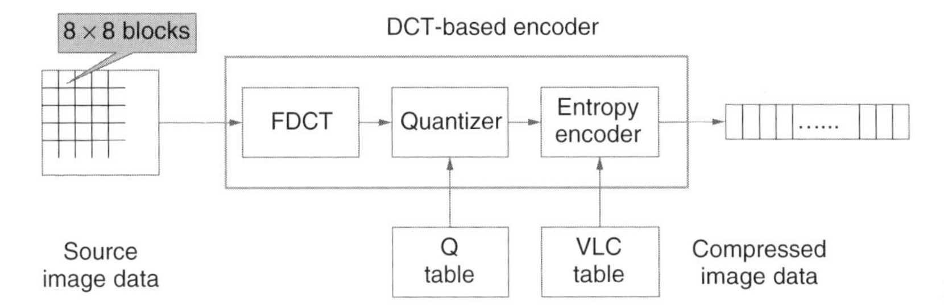

Figure 19. JPEG encoder block diagram.

run-length coding and a static Huffman coder. The name *Relative Element Address Designate* (*READ*) is based on the fact that only the relative positions of transitions are encoded. The most efficient MR coding mode is vertical mode (*interline prediction*) where only the small difference between the same transition positions in two adjacent lines is encoded. It also includes EOL codewords and periodically includes MH-coded lines (*intra-lines*) to minimize the effect of errors. The MR coding scheme, though simple, provides some very important concepts for modern image and video coding. We will see later in this section, it has much in common to the intra- and inter video coding schemes.

MMR coding in the Group 4 facsimile standard is based on MR coding in Group 3. It modifies the MR algorithm to maximize compression by removing the MR error prevention mechanisms. There is no EOL symbols and MH coding in Group 4. To enable the MR coding in Group 4, when coding the first line, a virtual white line is used as the reference. The transmission errors are corrected with lower level control procedures.

4.1.2. JBIG. Group 3 and Group 4 fax coding has proven adequate for text-based documents, but does not provide good compression or quality for documents with handwritten text or continuous tone images. As a consequence, a new set of fax standards, such as JBIG and the more recent JBIG2, has been created since the late 1980s.

JBIG stands for Joint Bi-level Image experts Group coding: The algorithm is defined in CCITT Recommendation T.82 [125,126] and it is a lossless bi-level image coding standard based on an adaptive arithmetic coding with adaptive 2D contexts.

Besides more efficient compression, JBIG provides progressive transmission of image through multi-resolution coding. As a new emerging standard for bi-level image coding, JBIG2 [127] allows both lossy and lossless bi-level image compression. It is the first international standard that provides lossy compression of bi-level images to achieve much higher compression ratio with almost no quality degradation. Besides higher compression performance, JBIG2 allows both quality-progressive coding from lower to higher (or lossless) quality, and content-progressive coding, successively adding different types of image data (e.g., first text, then halftones). The key technology in JBIG2 is pattern matching predictive coding schemes where a library of dynamically built templates is used to predict the repetitive character-based pixel blocks in a document.

4.1.3. JPEG. This is probably the best known ISO/ITU-T standard created in the late 1980s. The JPEG (Joint Photographic Experts Group) is a DCT-based standard that specifies three lossy encoding modes, namely, sequential, progressive, and hierarchical, and one lossless encoding mode. The baseline JPEG coder [122] is the sequential encoding in its simplest form. Baseline mode is the most popular one and it supports lossy coding only. Figure 19 shows the key processing steps in a baseline JPEG coder. It is based on the 8×8 block DCT, uniform scalar quantization with a perceptual weighting matrix, zigzag scanning of AC components, predictive coding of DC components, and Huffman coding (or arithmetic coding with more complexity but better performance).

The progressive and hierarchical modes of JPEG are both lossy and differ only in the way the DCT coefficients are coded or computed, when compared to the baseline mode. They allow a reconstruction of a lower quality or lower resolution version of the image, respectively, by partial decoding of the compressed bit stream. Progressive mode encodes the quantized coefficients by a mixture of spectral selection and successive approximation, while the hierarchical mode utilizes a pyramidal approach to computing the DCT coefficients in a multi-resolution way.

The lossless mode or lossless JPEG is a lossless coding scheme for continuous-tone image. It is an adaptive prediction coding scheme with a few predictors to choose from, which is similar to the DC component coding in the baseline mode but operated on the image domain pixel instead. The prediction difference is efficiently encoded with Huffman coding or arithmetic coding. It should be noted that lossless JPEG is not DCT-based as in the lossy modes. It is a pure pixel domain predictive coding.

4.1.4. JPEG-LS. Not to be confused with the lossless mode of JPEG (*Lossless JPEG*), JPEG-LS [131] is the latest and totally different ISO/ITU-T standard for lossless coding of still images and which also provides for *near-lossless* coding. The baseline system is based on the LOCO-I algorithm (LOw COmplexity LOssless COmpression for Images) [132] developed at Hewlett-Packard Laboratories.

LOCO-I combines the simplicity of Huffman coding with the compression potential of context models. The algorithm uses a nonlinear predictor with rudimentary edge detecting capability, and is based on a very simple context model, determined by quantized gradients. A small number of free statistical parameters are used to capture high-order dependencies, without the drawbacks of context dilution. The prediction residues are modeled by a *double-sided geometry distribution* with two parameters that

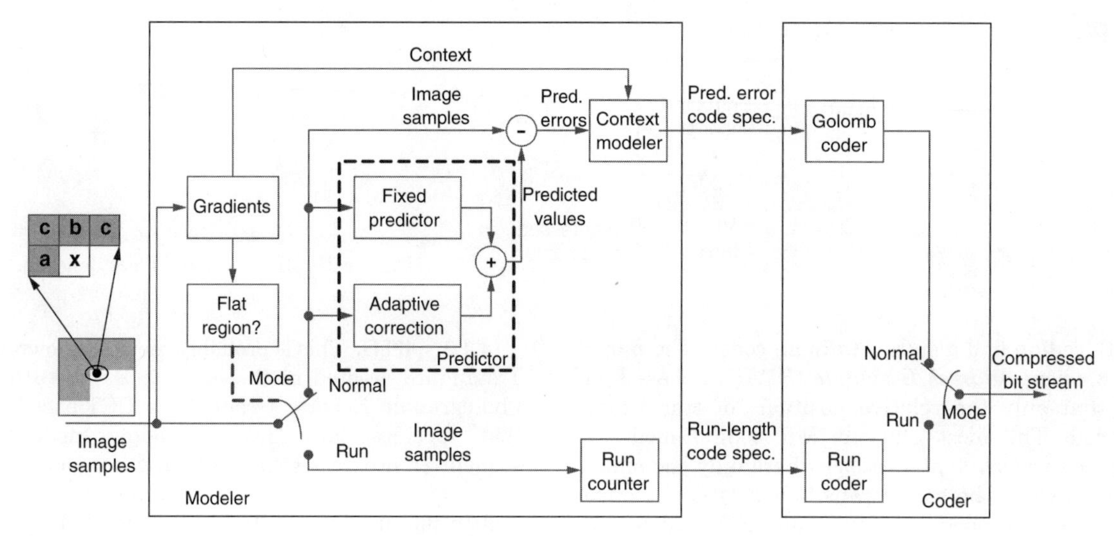

Figure 20. JPEG-LS encoder block diagram.

can be updated symbol by symbol based on the simple context, and in turn are efficiently encoded by a simple and adaptive Golomb code [133,134] that corresponds to Huffman coding for a geometric distribution. In addition, a run coding mode is defined for low-entropy flat area, where runs of identical symbols are encoded using extended Golomb coding with improved performance and adaptability. Figure 20 shows the block diagram of the JPEG-LS encoder. This algorithm was designed for low-complexity while providing high compression. However, it does not provide for scalability, error resilience, or other additional functionality.

4.1.5. Visual Texture Coding (VTC) in MPEG-4. Visual Texture Coding (VTC) [135] is the algorithm in the MPEG-4 standard (see Section 4.2.5) used to compress the texture information in photo realistic 3D models as well as still images. It is based on the discrete wavelet transform (DWT), scalar quantization, zerotree coding, and context-based arithmetic coding. Different quantization strategies are used to provide different SNR scalability: single quantization step (SQ) provides no SNR scalability; multiple quantization steps (MQ) provides discrete (coarse-grain) SNR scalability and bi-level quantization (BQ) supports fine grain SNR scalability at the bit level. In addition to the traditional tree-depth (TD) zerotree scanning similar to the EZW algorithm, band-by-band (BB) scanning is also used in MPEG-4 VTC to support resolution scalability. MPEG-4 VTC also supports coding of arbitrarily shaped objects, by the means of a shape adaptive DWT [36,150], but does not support lossless texture coding. Besides, a resolution scalable lossless shape coding algorithm [136] that matches the shape adaptive wavelet decomposition at different scales is also adopted. The scalable shape coding scheme uses fixed contexts and fixed probability tables that are suitable for the bi-level shape masks with large continuous regions of identical pixel values.

4.1.6. JPEG 2000. JPEG 2000 [86,137] is the latest emerging standard from the Joint Photographic Experts Group that is designed for different types of still images allowing different imaging models within a unified system. JPEG 2000 is intended to complement, not replace, the current JPEG standards and it has two coding modes: a *DCT-based coding* mode that uses currently baseline JPEG and a *wavelet-based coding* mode. JPEG 2000 normally refers to the wavelet-based coding mode and it is based on the discrete wavelet transform (DWT), scalar quantization, context modeling, arithmetic coding, and post-compression rate allocation techniques.

This core compression algorithm in JPEG 2000 is based on independent Embedded Block Coding with Optimized Truncation (EBCOT) of the embedded bit streams [91]. The EBCOT algorithm uses a wavelet transform to generate the subband coefficients, where the DWT can be performed with reversible filters for lossless coding, or nonreversible filters for lossy coding with higher compression efficiency. EBCOT partitions each subband into relatively small blocks of samples (called codeblocks) and encodes them independently. A multi-pass bitplane coding scheme based on context-based arithmetic coding is used to code the original or quantized coefficients in each codeblock into an embedded bit stream in the order of importance along with the rate-distortion pairs at each pass of the fractional bitplane (*truncation point*). It seems that failing to exploit the inter-subband redundancy would have a sizable adverse effect on coding efficiency. However, this is more than compensated by the finer scalability that results from the multi-pass coding.

JPEG 2000 also supports many new features, such as, compression of large images, single decompression architecture, compound documents, static and dynamic region-of-interest (ROI), multiple component images, content-based description, and protective image security.

4.2. Video Coding Standards

Most practical video coding systems including all international video coding standards are based on a hybrid coder that combines motion compensation in the temporal direction and DCT transform coding within each independent

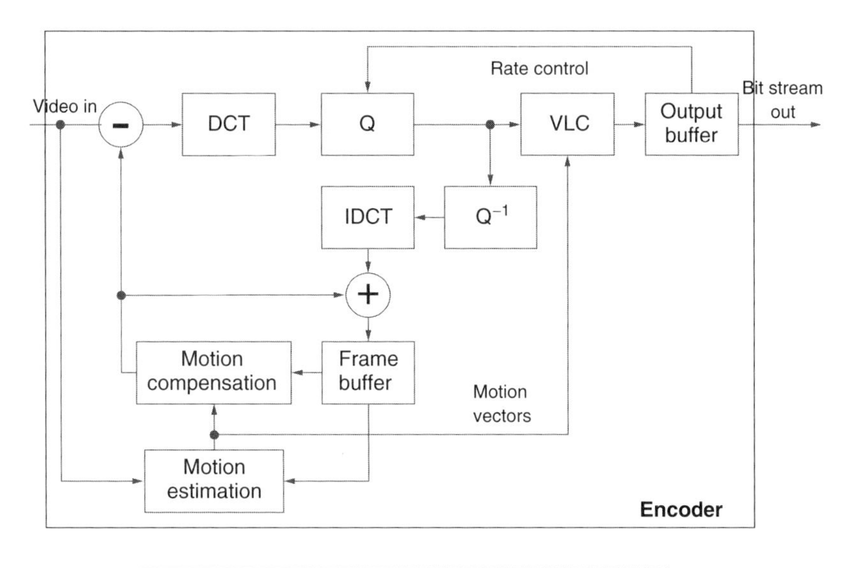

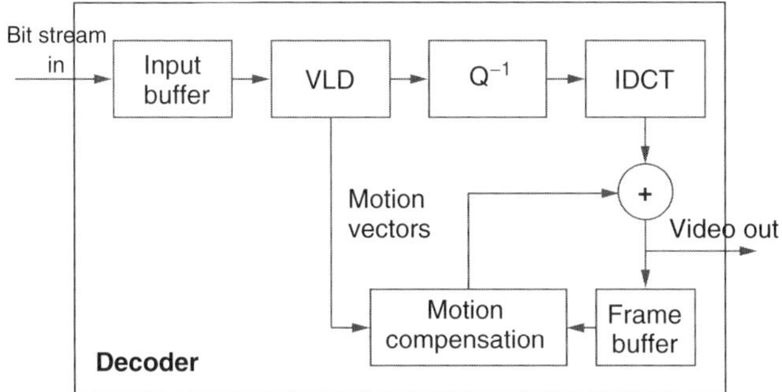

Figure 21. A generic DCT-based predictive video coder.

frame or prediction frame. A generic DCT-based predictive video coding system is illustrated in Fig. 21.

For independent frames, a DCT transform coder similar to the baseline JPEG is applied to compress the frame without using any information from other frames (referred to as intracoding). For prediction frames, a compensation module first generates a predicted image from one or more previously coded frames based on the motion vectors estimated and subtract it from the original frame to obtain the motion predicted residue image. Another DCT transform coder that fits the characteristics of the residue image is then applied to further exploit the spatial correlation to efficiently compress the residue image. In order to avoid the drifting errors caused by the mismatch between reference frames, the encoder normally embeds the same decoding structure to reconstruct exactly the same reference frames as in the decoder. Optionally, a loop filter to smooth out the blocking artifacts or generate a better reference image may be inserted in the reconstruction loop to generate a visually better image and enhance the prediction efficiency.

There have been several major initiatives in video coding that have led to a range of video standards: ITU video coding for video teleconferencing standards, H.261 for ISDN, H.262 (same as MPEG-2) for ATM/broadband, and H.263 for POTS (Plain Old telephone Service); ISO video coding standards MPEG-1 (Moving Picture Experts Group), MPEG-2 and MPEG-4; and the emerging standards such as H.264 | MPEG-4 Part 10 by JVT (Joint Video Team of ITU and ISO). These coding systems all belong to the DCT-based motion predictive coder category. In the following sections, we provide brief summaries of the video coders with emphasis on the differences of the coding algorithms.

4.2.1. H.261. The H.261 video codec [138] (1990), initially intended for ISDN teleconferencing, is the baseline video mode for most multimedia conferencing systems.

The basic coding unit in H.261 is *macroblock* that contains one 16×16 or four 8×8 luminance blocks and two 8×8 chrominance blocks. The H.261 codec encodes video frames using an 8×8 DCT. An initial frame (called an *I* or *intra* frame) is coded and transmitted as an independent frame. Subsequent frames are efficiently coded in the inter mode (*P* or *predictive* frame) using motion compensated predictive coding described above, where motion compensation is based on a 16×16 pixel block with integer pixel motion vectors and always referenced to the immediately previous frame. The DCT coefficients are quantized with a uniform quantizer and arranged in a zigzag scanning order. Run-length coding combined with variable length coding is used to compress the quantized coefficients into a video bit stream. An optional loop filter is introduced after motion compensation to improve the motion prediction efficiency.

H.261 is intended for head-and-shoulders type of scene in video conferencing applications where only small, controlled amounts of motion are present so the motion vector range can be limited. The supported video formats include both the CIF (Common Interchange Format with a resolution of 352 × 288 and YCbCr 4 : 2 : 0) and the QCIF (*quarter CIF*) format. All H.261 video is noninterlaced, using a simple progressive scanning pattern.

4.2.2. MPEG-1. The MPEG-1 standard [139] (1993) is a true multimedia standard that contains specifications for audio coding, video coding, and systems. MPEG-1 was intended for storage of multimedia content on a standard CD-ROM, with data rates of up to 1.5 Mbits/s and a storage capacity of about 600 Mbytes. Noninterlaced CIF video format (352 × 288 at 25 fps or 352 × 240 at 30 fps) is used to provide VHS-like video quality.

The video coding in MPEG-1 is very similar to the video coding of the H.261 described above with the difference that the uniform quantization is now based on perceptual weighting criteria. The temporal coding was based on both uni- and bi-directional motion-compensated prediction. A new *B* or bi-directionally predictive picture type is introduced in MPEG-1 which can be coded based on either the next and/or the previous I or P pictures. In contrast, an *I* or *intra* picture is encoded independently of all previous or future pictures and a *P* or *predictive* picture is coded based on only a previous I or P picture. MPEG-1 also allows half-pel motion vector precision to improve the prediction accuracy. There is no loop filter present in an MPEG-1 motion compensation loop.

4.2.3. MPEG-2. The MPEG-2 standard [140] (1995) was initially developed primarily for coding interlaced video at 4-9 Mbits/s for broadcast TV and high quality digital storage media (such as DVD video); it has now also been used in HDTV, cable/satellite TV, video services over broadband networks, and high quality video conferencing (same as H.262). The MPEG-2 standard includes video coding, audio coding, system format for program and transport streams, and other information related to practical implementations.

MPEG-2 video supports both interlaced and progressive video, multiple color format (4 : 2 : 0, 4 : 2 : 2 and 4 : 4 : 4), flexible picture size and frame rates, hierarchical or scalable video coding, and is backward compatible with MPEG-1. To best satisfy the needs of different applications, different profiles (subset of the entire admissible bit stream syntax) and levels (set of constraints imposed on the parameters of the bitstreams within a profile) are also defined in MPEG-2.

The most distinguishing feature of MPEG-2 from previous standards is the support for interlaced video coding. For interlaced video, it can be either encoded as a frame picture or a field picture, with adaptive frame or field DCT and frame or field motion compensation at macroblock-level. MPEG-2 also offer another new feature in providing the temporal, spatial, and SNR scalabilities. Temporal scalability is achieved with B-frame coding; spatial scalability is obtained by encoding the prediction error with a reference formed from both an upsampled low-resolution current frame and a high-resolution previous frame; SNR scalability is provided by finely quantizing the error residue from the coarsely quantized low-quality layer (there is a drifting problem).

Compared with MPEG-1, a number of improvements have been made to further improve the coding efficiency, including, a more flexible coding mode selection at the macroblock level; a nonlinear quantization table with increased accuracy for small values; an alternative zigzag scan for DCT coefficients especially for the interlaced video coding; much increased permissible motion vector range; new VLC tables for the increase bit rate range; customized perceptual quantization matrix support; and dual prime prediction for interlaced video encoded as P pictures, which mimics the B picture prediction especially for low-delay applications.

MPEG-2 also introduces some error resilience tools such as independent slice structure, data partitioning to separate data with different importance, concealment motion vectors in intra-pictures, and different scalabilities as we discussed.

MPEG-2 is probably the most widely used video coding standard so far that enables many applications, such as DVD, HDTV, and digital satellite and cable TV broadcast.

4.2.4. H.263, H.263+, and H.263++. The H.263 standard [141] (1995) was intended for use in POTS conferencing and the video codec is based on the same DCT and motion compensation techniques as used in H.261. H.263 now supports 5 picture formats including sub-QCIF (126 × 96) QCIF (176 × 144), CIF (352 × 288), 4CIF (704 × 576) and 16CIF (1408 × 1152). Several major differences exist between H.263 and H.261 including, more accurate half-pel motion compensation in H.263 but integer-pel in H.261; no loop filter in H.263 but optional in H.261 since the optional overlapped block motion compensation (OBMC) in H.263 could have the similar de-blocking filter effect; motion vector predictor in H.263 using the median value of three candidate vectors while only the preceding one being used in H.261; 3-dimensional run length coder (*last, run, level*) in H.263 versus (*run, level*) and an *eob* symbol in H.261.

As H.263 version 2, H.263+ [142] (1998) provides optional extensions to baseline H.263. These options broaden the range of useful applications and improve the compression performance. With these new options, H.263+ supports custom source format with different picture size, aspect ratio, and clock frequency. It also provides enhanced error-resilience capability with slice structured coding, prediction reference selection, motion vector coding with reversible VLC (RVLC), where RVLCs are codewords which can be decoded in a forward as well as a backward manner, and independent segment decoding. Backward-compatible supplemental enhancement information can be embedded into the video bit stream to assist operations in the receiver. In addition, temporal scalability can be provided by B picture, while SNR and spatial scalabilities are enabled by newly defined *EI* pictures (referencing to a temporally simultaneous picture) and *EP* pictures (referencing to both a temporally preceding picture and a temporally simultaneous one). Figure 22 gives an example of mixed scalabilities enabled by these new picture types.

H.263++ [143] (H.263 version 3, 2001) provides more additional options for H.263, including, *Enhanced Reference Picture Selection* to provide enhanced coding efficiency and enhanced error resilience (particularly

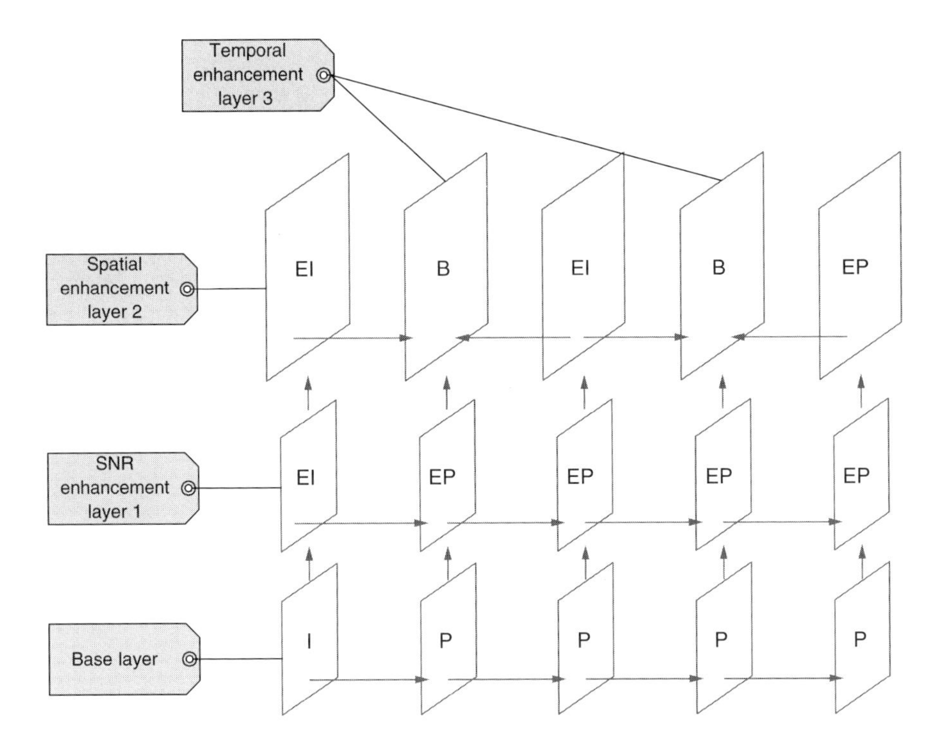

Figure 22. Example for mixed temporal, SNR and spatial scalabilities supported in H.263+.

against packet losses) through the use of multiple reference pictures; *Data Partitioned Slice* to enhance error resilience (particularly against localized corruption of bit stream) by separating header and motion vectors from DCT coefficients in the bit stream and protecting motion vectors using a reversibly decodable variable length coder (RVLC); and additional *Supplemental Enhancement Information* including support for interlaced video.

4.2.5. MPEG-4. Coding of separate audio–visual objects, both natural and synthetic, leads to the latest ISO MPEG-4 standard [144] (1998). The MPEG-4 visual standard is developed to provide users a new level of interaction with visual content. It provides technologies to view, access and manipulate objects rather than pixels, with great error robustness at a large range of bit rates. Application areas range from digital television, streaming video, to mobile multimedia, 2D/3D games, and visual communications.

The MPEG-4 visual standard consists of a set of tools supporting mainly three classes of functionalities: compression efficiency, content-based interactivity, and universal access. Among them, content-based interactivity is one of the most important novelties offered by MPEG-4. To support object-based functionality, the basic compression unit in MPEG-4 can be arbitrarily shaped video object plane (VOP) instead of always the rectangular frame. Each object can be encoded with different parameters, and at different qualities. Tools provided in the MPEG-4 standard include shape coding, motion estimation and compensation, texture coding, error resilience, sprite coding, and scalability. Conformance points, in the form of object types, profiles, and levels, provide the basis for interoperability. MPEG-4 provides support for both interlaced and progressive material. The chrominance format that is supported is 4 : 2 : 0.

For reasons of efficiency and backward compatibility, video objects are coded in a hybrid coding scheme somewhat similar to previous MPEG standards. Figure 23 outlines the basic approach of the MPEG-4 video algorithms to encode rectangular as well as arbitrarily-shaped input image sequences. Compared with traditional rectangular video coding, there is an additional shape coding module in the encoder that compresses the shape information necessary for the decoding of a video object. Shape information is also used to control the behavior of the DCT transform, motion estimation and compensation, and residue coding.

Shape coding can be performed in binary mode, where the shape of each object is described by a binary mask, or in grayscale mode, where the shape is described in a form similar to an alpha channel, allowing transparency, and reducing aliasing. The binary shape can be losslessly or lossy coded using context-based arithmetic encoding (CAE) based on the context either from the current video object (intraCAE) or from the motion predicted video object (interCAE). By using binary shape coding for coding its support region, grayscale shape information is lossy encoded using a block based motion compensated DCT similar to that of texture coding.

Texture coding is based on an 8×8 DCT, with appropriate modifications for object boundary blocks for both intramacroblocks and intermacroblocks. Low-pass extrapolation (LPE) (also known mean-repetitive padding) is used for intra boundary blocks and zero padding is used for inter boundary blocks before DCT transforms. Furthermore, shape adaptive DCT (SA-DCT) can also be used to decompose the arbitrarily shaped boundary blocks to achieve more coding efficiency. The DCT coefficients are quantized with or without perceptual quantization matrices. MPEG-4 also allows for a nonlinear quantization of DC values. To further reduce the average energy of the

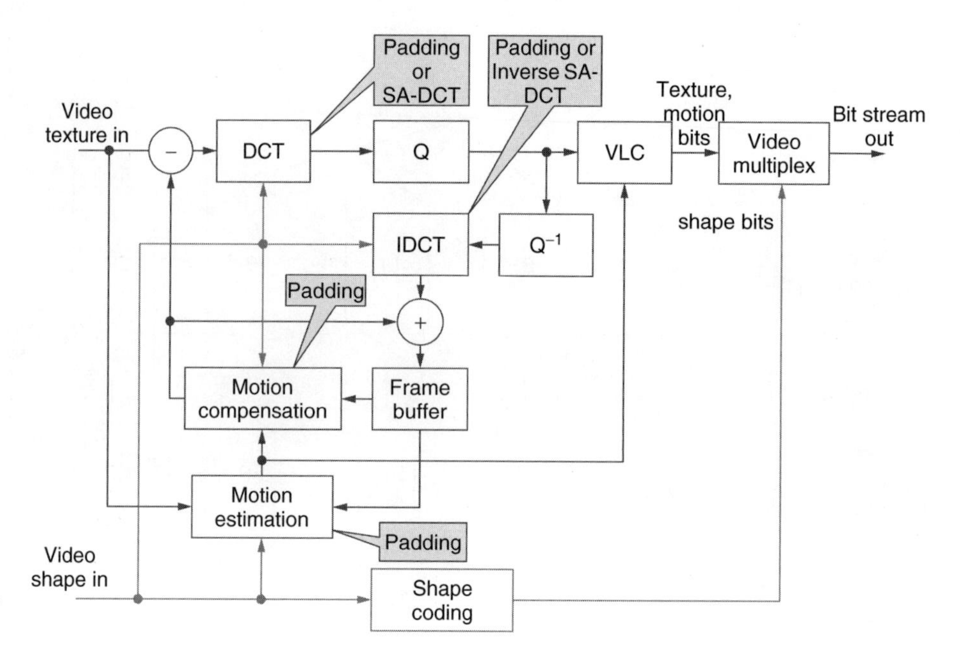

Figure 23. Block diagram of a MPEG-4 video coder.

quantized coefficients, the adaptive prediction of DC and AC coefficients from their neighboring blocks can be used to improve the coding efficiency. Three types of zigzag scans are allowed to reorder the quantized coefficients according to the DC prediction types. Two VLC tables switched by the quantization level can be used for the run length coding.

Still textures with possible arbitrary shapes can be encoded using wavelet transforms as described before with the difference that the shape-adaptive discrete wavelet transform (SA-DWT) and shape adaptive zerotree coding are used to extend the still texture coding to arbitrarily shaped image objects.

Motion compensation is block based, with appropriate modifications for object boundaries through repetitive padding and "polygon matching." Motion vectors are predictively encoded. The block size can be 16×16, or 8×8, with up to quarter pixel precision. MPEG-4 also provides modes for overlapped block motion compensation (OMBC) to reduce blocking artifacts and get better prediction quality at lower bit rates, and global motion compensation (GMC) through image warping. Moreover, static sprite can also be encoded efficiently with only 8 global motion parameters describing camera motion that represent the appropriate affine transform of the sprite.

When the video content to be coded is interlaced, additional coding efficiency can be achieved by adaptively switching between field and frame coding.

Error resilience functionality is important for universal access through error-prone environments, such as mobile communications. It offers means for resynchronization, error detection, data recovery, and error concealment. Error resilience in MPEG-4 is provided by resynchronization markers to resume decoding from errors; data partitioning to separate information with different importance; header extension codes to insert redundant important header information in the bit stream; and reversible variable length codes (RVLC) to decode portions of the corrupted bit stream in the reverse order.

MPEG-4 also includes tool for sprite coding where a sprite consists of those regions present in the scene throughout the video segment, for example, the background. Sprite-based coding is very suitable for synthetic objects as well as objects in natural scenes with rigid motion, where sprites can provide high compression efficiency by appropriate warping/cropping operations. Shape and texture information for a sprite is encoded as an intra VOP.

Scalability is provided for object, SNR, spatial and temporal resolution enhancement in MPEG-4. Object scalability is inherently supported by the object based coding scheme. SNR scalability is provided by fine granularity scalability (FGS) where the residue of the base layer data is further encoded as an out-of-loop enhancement layer. Spatial scalability is achieved by referencing both the temporally simultaneous base layer and the temporally preceding spatial enhancement layer. Temporal scalability is provided by using bi-directional coding. One of the most advantageous features of MPEG-4 temporal and spatial scalabilities is that they can be applied to both rectangular frames and arbitrarily-shaped video objects.

4.2.6. The Emerging MPEG-4 Part-10/ITU H.264 Standard. At the time of completing this manuscript, there is a very active new standardization effort in developing a video coding standard that can achieve substantially higher coding efficiency compared to what could be achieved using any of the existing video coding standards. This is the emerging MPEG-4 Part-10/ITU H.264 standard jointly developed by a Joint Video Team (JVT) [145]. In MPEG, this new standard is called the Advanced Video Coding (AVC) standard.

The new standard takes a detailed look at many fundamental issues in video coding. The underlying coding structure of the new standard is still a block-based motion compensated transform coder similar to that adopted in

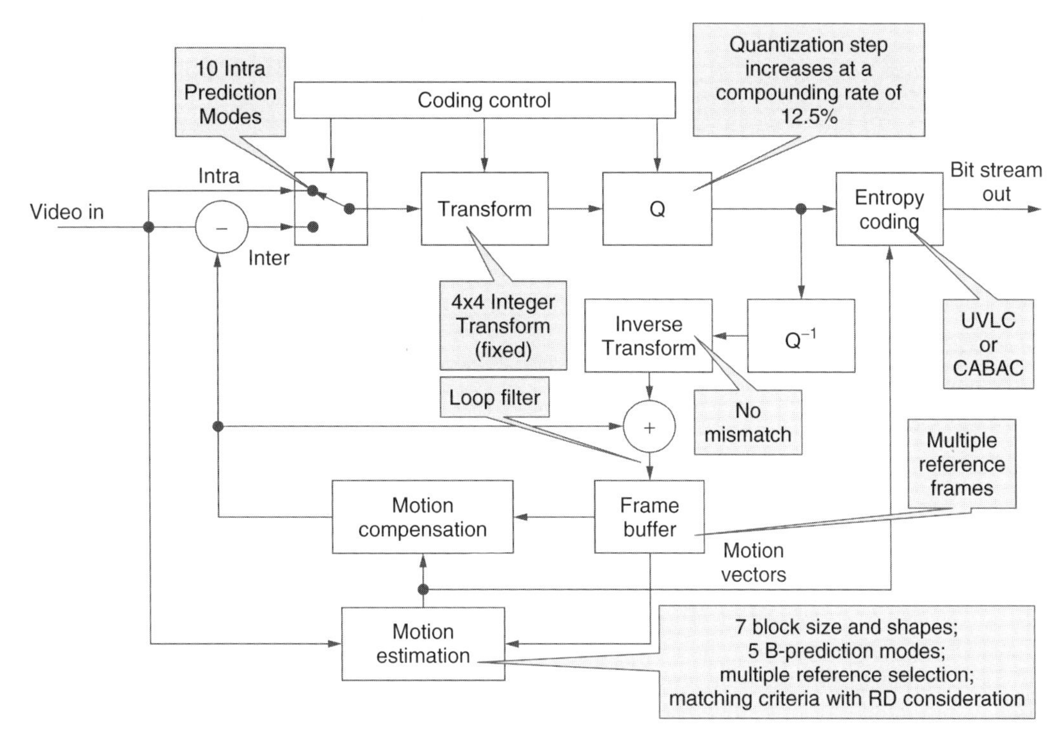

Figure 24. Block diagram of the AVC encoder.

previous standards. The block diagram of an AVC encoder is given in Fig. 24.

There are several significant advances in AVC that provides high coding efficiency. The key feature that enables such high coding efficiency is adaptivity. For intracoding, the flexible intra prediction is used to exploit spatial correlation between adjacent macroblocks and adapt to the local characteristics. AVC uses a simpler and integer-based 4×4 block transform that does not have a mismatch problem as in the 8×8 DCT due to different implementation precisions dealing with the real-valued DCT basis function. The smaller transform size also helps to reduce blocking and ringing artifacts.

Most of the coding gain achieved in AVC is obtained through advanced motion compensation. Motion compensation in AVC supports most of the key features found in earlier video standards, but its efficiency is improved through added flexibility and functionality. For motion estimation/compensation, AVC uses blocks of different sizes and shapes that provide capability to handle fine motion details, reduce high frequency components in the prediction residue and avoid large blocking artifacts. Using seven different block sizes and shapes can translate into bit savings of more than 15% as compared to using only a 16×16 block size. Higher precision sub-pel motion compensation is supported in AVC to improve the motion prediction efficiency. Using 1/4-pel spatial accuracy can yield more than 20% in bit savings as compared to using integer-pel spatial accuracy. AVC offers an option of multiple reference frame selection that can improve both the coding efficiency and error-resilience capability. Using five reference frames for prediction can yield 5-10% in bit savings as compared to using only one reference frame. The use of adaptive de-blocking filters in the prediction loop substantially reduces the blocking artifacts, with additional complexity. Because bit savings depend on video sequences and bit rates, the above numbers are only meant to give a rough idea about how much improvement one may expect from a coding tool.

In AVC, the quantization step sizes are increased at a compounding rate of approximately 12.5%, rather than increasing it by a constant increment, to provide finer quantization yet covering a greater bit rate range.

In AVC, entropy coding is performed using either variable length codes (VLC) or using context-based adaptive binary arithmetic coding (CABAC). The use of CABAC in AVC yields a consistent improvement in bit savings. AVC is still under development and new technologies continue to be adopted into the draft standard to further improve the coding efficiency. To provide a high level picture of the various video coding standards, Fig. 25 shows the approximate timeline and evolution of different standards.

5. FUTURE RESEARCH DIRECTIONS IN IMAGE AND VIDEO CODING

Further improving the coding efficiency of image and video compression algorithms continues to be the goal that researchers will be pursuing for a long time. From the success of the JPEG2000 and MPEG-4 AVC codecs, we learned that better schemes to enable rate-distortion optimization (normally generating a variable bit rate bit stream) and content-adaptivity are two areas worth further investigation. Adaptivity is the most important feature of AVC, block-sizes, intra prediction modes, motion

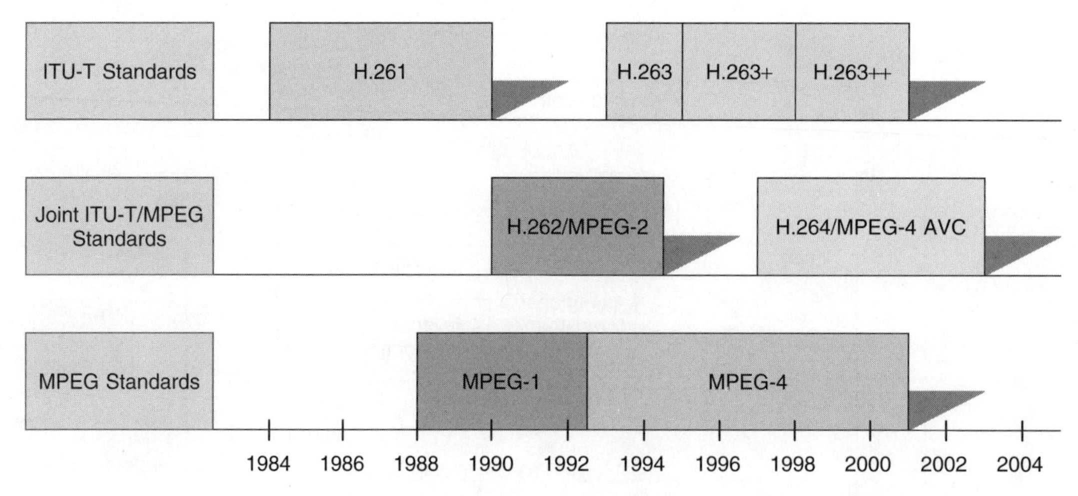

Figure 25. Progression of the ITU-T Recommendations and MPEG standards.

prediction modes, multiple reference frames, adaptive entropy coding, adaptive loop filtering, and so on, all reflect the significance of adaptivity. The main reason is that image and video signals by nature are not stationary signals, although traditional source coding theories all assume them as statistically stationary sources.

Besides coding efficiency, there are also increasing demands for more special functionalities, such as scalability, error-resilience, transcoding, object-based coding, model-based coding, and hybrid natural and synthetic coding, and so forth. These functionalities may require unconventional image and video coding techniques and may need to balance the trade-off between coding efficiency and functionalities.

Scalable coding refers to coding an image or a video sequence into a bit stream that is partially decodable to produce a reconstructed image or video sequence at possibly a lower spatial, temporal, or quantization resolution. Figure 26 shows the relationship between different scalabilities for a video bit stream. The principal concept in scalability is that an improvement in source reproduction can be achieved by sending only an incremental amount of bits over the current transmitted rate. Quantization, spatial, and temporal scalabilities are all important in applications. A key feature of scalable coding is that the encoding only needs to be done once and a single scalable bit stream will fit in all applications with different target bit rates, or spatial, or temporal resolutions. Scalable coding can bring many benefits in practical applications: potential high coding efficiency through rate-distortion optimization (e.g., JPEG-2000); easy adaptation to heterogeneous channels and user device capabilities and support for multicast; better received picture quality over dynamic changing networks through channel adaptation; robust delivery over error-prone, packet-loss-ridden through unequal error protection (UEP) and priority streaming because some enhancement bit stream can be lost or discarded without irreparable harm to the overall representation; reduced storage space required to store multiple bitstreams otherwise; no complicated transcoding needed for different bit rates and resolutions. There are two types of scalabilities, layered scalability and fine granularity scalability (FGS). Layered scalability only provides a limited number of ("quantized") layers of scalability (e.g., SNR and spatial scalabilities in MPEG-2 and temporal scalability in H.263) in which a layer is either completely useful or completely useless, while FGS provides a continuous enhancement layer that can be truncated at any bit location, and the reconstructed image or video quality is proportional to the number of decoded bits (e.g., JPEG-2000, MPEG-4 VTC, MPEG-4 FGS, 3-D wavelet video coding). Error-robust coding is necessary for error-prone channels to increase the capability of error recovery and error concealment. Error-robust features built in source coding can potentially provide better and more intelligent error resilience capability than "blind" error control channel coding. Resynchronization, data partitioning, multi-description coding are all source coding techniques aiming at error-robustness of video coding. Compared with scalable coding, error resilience coding

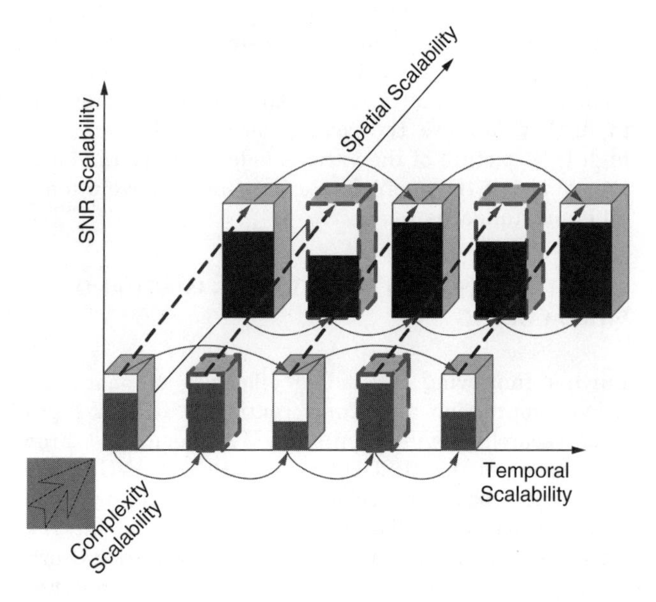

Figure 26. Relationship between different scalabilities.

may be considered as repairing damaged bitstreams while scalable coding is trying to prevent bitstreams from being damaged. For a given transmission bit rate, splitting bits between source coding and channel coding is a bit tricky. If bits are allocated to channel coding and the channel is ideal, there is a loss in performance compared to source coding alone. Similarly, if there are no bits allocated to channel coding and the channel is very noisy, there will be a loss in performance compared to using some error control coding. How to optimally allocate these bits between source coding and channel coding brings us to the problem of joint source and channel coding.

Transcoding is usually considered as converting one compressed bit stream into another compressed bit stream. There are several different reasons for using transcoding. One of them is to change the bit rate. Another reason is to convert the bit stream format from one standard to another. The challenge is to have the coding efficiency of the transcoded bit stream close to that of an encoded bit stream from the original image and video. Another challenge is to have the transcoding operation as simple as possible without a complete re-encoding process.

Object-based coding is required for object-based interactivity and manipulation. In MPEG-4 video coding standard, object-based coding is defined for many novel applications to provide content-based interactivity, content-based media search and indexing, and content scalable media delivery, and so on. Figure 27 illustrates the concept of an object-based scenario. One of the challenges in object-based coding is the preprocessing part that separates an image or video object from a normal scene. Good progresses have been reported in this direction [151–154].

For some visual objects, it is possible to establish a 2D or 3D structural model with which the movement of different parts of the object can be parameterized and transmitted. One of such successful visual models is the human face and body. In MPEG-4 visual coding standard, face and body animation is included. Using such a model-based coding, it is possible to transmit visual information at extremely low bit rates. The challenges are to establish good models for different visual objects and to accurately extract the model parameters.

Image and video coding techniques mainly deal with natural images and video. Computer graphics generate visual scenes as realistic as possible. Hybrid natural and synthetic coding is to combine these two areas. Usually, it is relatively easy for computer graphics to generate a structure of a visual object, but harder to generate realistic texture. With hybrid natural and synthetic coding, natural texture mapping can be used to make animated objects look more realistic. Latest developments in computer graphics/vision tend to use naturally captured image and video contents to render realistic environment (called image based rendering or IBR), such as lumigraph [155], light-fields [156], concentric mosaics [157], and so on. How to efficiently encode these multidimensional cross-correlated image and video contents is an active research topic [158].

In summary, image and video coding research has been a very active field with a significant impact to industry and society. It will continue to be active with many more interesting problems to be explored.

BIOGRAPHIES

Shipeng Li joined Microsoft Research Asia in May 1999. He is currently the research manager of Internet Media group. From October 1996 to May 1999, he was with Multimedia Technology Laboratory at Sarnoff Corporation in Princeton, NJ (formerly David Sarnoff Research Center, and RCA Laboratories) as a member of technical staff. He has been actively involved in research and development of digital television, MPEG, JPEG, image/video compression, next generation multimedia standards and applications,

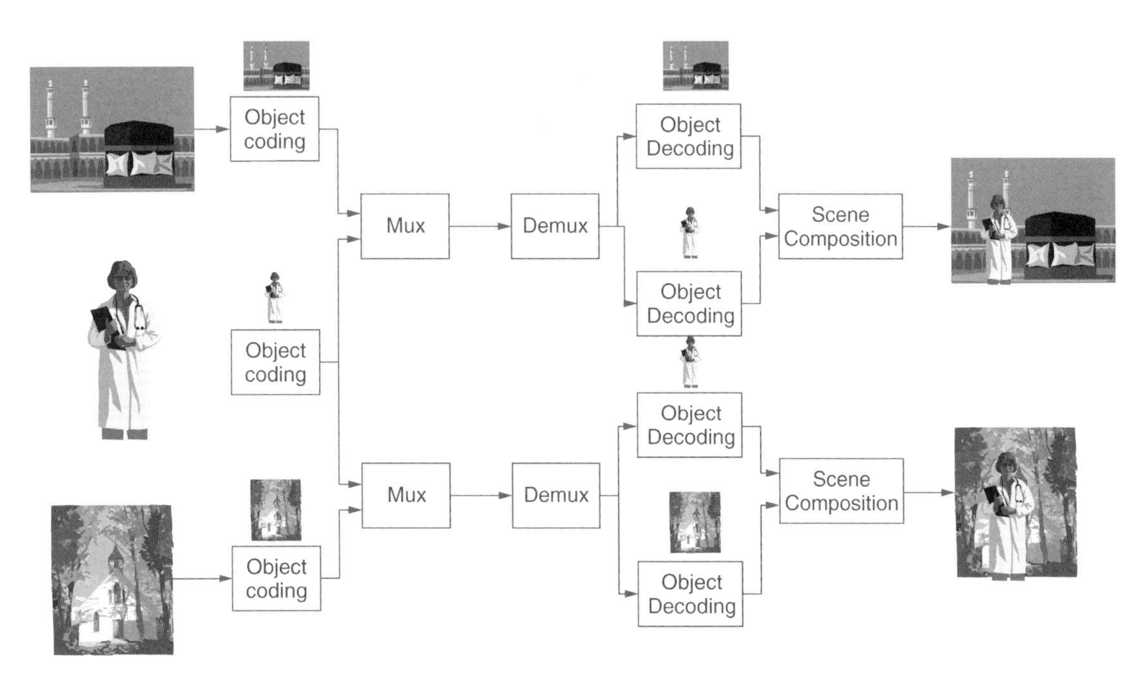

Figure 27. An object-based coding application scenario.

and consumer electronics. He made contributions in shape-adaptive wavelet transforms, scalable shape coding, and the error resilience tool in the FGS profile for the MPEG-4 standard. He has authored and coauthored more than 70 technical publications and more than 20 grants and pending U.S. patents in image/video compression and communications, digital television, multimedia, and wireless communication. He is the coauthor of a chapter in *Multimedia Systems and Standards* published by the Marcel Dekker, Inc. He is a member of Visual Signal Processing and Communications committee of IEEE Circuits and Systems Society. He received his B.S. and M.S. both in Electrical Engineering from the University of Science and Technology of China (USTC) in 1988 and 1991, respectively. He received his Ph.D. in Electrical Engineering from Lehigh University, Bethlehem, PA, in 1996. He was an assistant professor in Electrical Engineering department at University of Science and Technology of China in 1991–1992.

Weiping Li received his B.S. degree from University of Science and Technology of China (USTC) in 1982, and his M.S. and Ph.D. degrees from Stanford University in 1983 and 1988 respectively, all in electrical engineering. Since 1987, he has been a faculty member in the department of Electrical and Computer Engineering at Lehigh University. From 1998 to 1999, he was with Optivision, Inc., in Palo Alto, California, as the Director of R&D. He has been with WebCast Technologies, Inc., since 2000. Weiping Li has been elected a Fellow of IEEE for contributions to image and video coding algorithms, standards, and implementations. He served as the Editor-in-Chief of IEEE Transactions on Circuits and Systems for Video Technology from 1999 to 2001. He has served as an Editor for the Streaming Video Profile Amendment of MPEG-4 International Standard. He is a member of the Board of Directors for MPEG-4 Industry Forum. He served as one of the Guest Editors for a special issue of IEEE Proceedings on image and video compression (February 1995). Weiping Li was awarded Guest Professorships in University of Science and Technology of China and in Huazhong University of Science and Technology in 2001. He received the Spira Award for Excellence in Teaching in 1992 at Lehigh University and the Guo Mo-Ruo Prize for Outstanding Student in 1980 at University of Science and Technology of China.

BIBLIOGRAPHY

1. M. B. Priestley, *Spectral Analysis and Time Series*, Academic Press, NY, 1981.
2. S. P. Lloyd, Least squares quantization in PCM, *IEEE Trans. Inform. Theory* **28**: 127–135 (1982).
3. A. Gersho and R. M. Gray, *Vector Quantization and Signal Compression*, Kluwer Academic Publishers, 1992.
4. C. E. Shannon, Coding theorems for a discrete source with a fidelity criteria, *IRE National Convention Record, Part 4*, 142–163, 1959.
5. R. M. Gray, *Source Coding Theory*, Kluwer Academic Press, 1992.
6. T. Berger, *Rate Distortion Theory*, Prentice-Hall Inc., Englewood Cliffs, 1971.
7. T. D. Lookabaugh and R. M. Gray, High-resolution quantization theory and the vector quantizer advantage, *IEEE Trans. Inform. Theory* **IT-35**: 1023–1033 (Sept. 1989).
8. T. R. Fischer, A pyramid vector quantizer, *IEEE Trans. Inform. Theory* **32**: 568–583 (July 1986).
9. M. Barlaud et al., Pyramid lattice vector quantization for multiscale image coding, *IEEE Trans. Image Process.* **3**: 367–381 (July 1994).
10. J. H. Conway and N. Sloane, A fast encoding method for lattice codes and quantizers, *IEEE Trans. Inform. Theory* **IT-29**: 820–824 (Nov. 1983).
11. A. Buzo, Y. Linde, and R. M. Gray, An algorithm for vector quantizer design, *IEEE Trans. Commun.* **20**: 84–95 (Jan. 1980).
12. B. Ramamurthi and A. Gersho, Classified vector quantization of image, *IEEE Trans. Commun.* **34**: 1105–1115 (Nov. 1986).
13. D.-Y. Cheng, A. Gersho, B. Ramamurthi, and Y. Shoham, Fast search algorithms for vector quantization and pattern matching, *Proc. of the International Conference on ASSP*, 911.1–911.4 (March 1984).
14. J. L. Bentley, Multidimensional binary search tree used for associative searching, *Comm. ACM*, 509–526 (Sept. 1975).
15. M. Soleymani and S. Morgera, An efficient nearest neighbor search method, *IEEE Trans. Commun.* **20**: 677–679 (1987).
16. Abramson, Norman, *Information Theory and Coding*, McGraw-Hill, New York, 1963.
17. P. G. Howard and J. S. Vitter, Practical implementations of arithmetic coding, in *Image and Text Compression*, J. A. Storer, ed., Kluwer Academic Publishers, Boston, MA, 1992, pp. 85–112.
18. J. Ziv and A. Lempel, A universal algorithm for sequential data compression, *IEEE Trans. Inform. Theory* **IT-23**: 337–343 (1977).
19. R. B. Wells, *Applied Coding and Information Theory for Engineers*, Prentice-Hall, Englewood Cliffs, NJ, 1999.
20. T. A. Welch, A technique for high-performance data compression, *IEEE Trans. Comput.* 17:8–19 (1997).
21. M. Rabani and P. Jones, *Digital Image Compression Techniques*, SPIE Optical Engineering Press, Bellingham, WA, 1991.
22. O. Egger, P. Fleury, T. Ebrahimi, and M. Kunt, High-performance compression of visual information — A tutorial review — Part I: Still pictures, *Proc. IEEE* **87**(6): 976–1011 (June 1999).
23. T. Berger and J. D. Gibson, Lossy source coding, *IEEE Trans. Inform. Theory* **44**(6): 2693–2723 (1998).
24. T. Ebrahimi and M. Kunt, Visual data compression for multimedia applications, *Proc. IEEE* **86**(6): 1109–1125 (Jun. 1998).
25. M. J. Narasimha and A. M. Peterson, On the computation of the discrete cosine transform, *IEEE Trans. Commun.* **COM-26**: 934–936 (Jun. 1978).
26. W. Li, A new algorithm to compute the DCT and its inverse, *IEEE Trans. Signal Proc.* **39**(6): 1305–1313 (June 1991).
27. D. Slawecki and W. Li, DCT/IDCT processor design for high data rate image coding, *IEEE Trans. Circuits Syst. Video Technol.* **2**(2): 135–146 (1992).

28. W. B. Pennebaker and J. L. Mitchell, *JPEG Still Image Data Compression Standard*, Van Nostrand Reinhold, New York, 1993.

29. J. D. Johnson, A filter family designed for use in quadrature mirror filter bands, *Proc. Int. Conf. ASSP (ICASSP)* 291–294 (Apr. 1980).

30. H. S. Malavar, *Signal Processing with Lapped Transforms*, Artech House Norwood, MA, 1992.

31. Z. Zhou and A. N. Venetsanopoulos, Morphological methods in image coding, *Proc. Int. Conf. ASSP (ICASSP)* **3**: 481–484 (March 1992).

32. J. R. Casas, L. Torres, and M. Jare no, Efficient coding of residual images, *Proc. SPIE, Visual Communications Image Processing* **2094**: 694–705 (1993).

33. A. Toet, A morphological pyramid image decomposition, *Pattern Reconstruction Lett.* **9**(4): 255–261 (May 1989).

34. T. Sikora, Low complexity shape-adaptive DCT for coding of arbitrarily shaped image systems, *Signal Processing: Image Commun.* **7**: 381–395 (Nov. 1995).

35. T. Sikora and B. Makai, Shape-adaptive DCT for generic coding of video, *IEEE Trans. Circuits Syst. Video Technol.* **5**: 59–62 (Feb. 1995).

36. S. Li and W. Li, Shape-adaptive discrete wavelet transforms for arbitrarily shaped visual object coding, *IEEE Trans. Circuits Syst. Video Technol.* **10**(5) 725–743 (Aug. 2000).

37. J. M. Shapiro, Embedded image coding using zerotrees of wavelet coefficients, *IEEE Trans. Signal Process.* **41**: 3445–3462 (Dec. 1993).

38. W. Li, Overview of fine granularity scalability in MPEG-4 video standard, *IEEE Trans. Circuit Syst. Video Technol.* **11**(3): 301–317 (March 2001). Special issue on streaming video.

39. W. Li and Y.-Q. Zhang, Vector-based signal processing and quantization for image and video compression, *Proc. IEEE* **83**(2): 317–335 (Feb. 1995).

40. W. Li and Y.-Q. Zhang, Vector transform coding of subband-decomposed images, *IEEE Trans. Circuits Syst. Video Technol.* **4**(4): 383–391 (Aug. 1994).

41. W. Li, On vector transformation, *IEEE Trans Signal Proc.* **41**(11): 3114–3126 (Nov. 1993).

42. W. Li, Vector transform and image coding, *IEEE Trans. Circuits Syst. Video Technol.* **1**(4): 297–307 (Dec. 1991).

43. H. Q. Cao and W. Li, A fast search algorithm for vector quantization using a directed graph, *IEEE Trans. Circuits Syst. Video Technol.* **10**(4): 585–593 (June 2000).

44. F. Ling, W. Li, and H. Sun, Bitplane coding of DCT coefficients for image and video compression, *Proc. of SPIE VCIP'99* (Jan. 1999).

45. C. Wang, H. Q. Cao, W. Li, and K. K. Tzeng, Lattice labeling algorithms for vector quantization, *IEEE Trans. Circuits Syst. Video Technol.* **8**(2): 206–220 (Apr. 1998).

46. W. Li et al., A video coding algorithm using vector-based techniques, *IEEE Trans. Circuits Syst. Video Technol.* **7**(1): 146–157 (Feb. 1997).

47. F. Ling and W. Li, Dimensional adaptive arithmetic coding for image compression, *Proc. of IEEE International Symposium on Circuits and Systems* (May–June 1998).

48. I. Daubechies and W. Sweldens, Factoring wavelet transforms into lifting steps, *J. Fourier Anal. Appl.* **4**: 247–269 (1998).

49. T. Koga et al., Motion compensated interframe coding for video conferencing, *Proc. of the National Telecommunications Conference*, New Orleans, LA, pp. G5.3.1–G5.3.5 (Dec. 1981).

50. J. Y. Tham, S. Ranganath, M. Ranganath, and A. A. Kassim, A novel unrestricted center-biased diamond search algorithm for block motion estimation, *IEEE Trans. Circuits Syst. Video Technol.* **8**(4): 369–377 (Aug. 1998).

51. S. Zhu and K. K. Ma, A new Diamond search algorithm for fast block matching motion estimation, *Proc. 1st Int. Conf. Inform. Commun. Signal Process.* ICICS '97, **1**: 292–296 (Sept. 1997).

52. Z. L. He and M. L. Liou, A high performance fast search algorithm for block matching motion estimation, *IEEE Trans. Circuits Syst. Video Technol.* **7**(5): 826–828 (Oct. 1997).

53. S. Zafar, Y. Zhang, and J. S. Baras, Predictive block-matching motion estimation schemes for video compression — Part II. Inter-frame prediction of motion vectors, *IEEE Proc. SOUTHEASTCON 91* **2**: 1093–1095 (April 1991).

54. L. W. Lee, J. F. Wang, J. Y. Lee, and J. D. Shie, Dynamic search-window adjustment and interlaced search for block-matching algorithm, *IEEE Trans. Circuits Syst. Video Technol.* **3**(1): 85–87 (Feb. 1993).

55. B. Liu and A. Zaccarin, New fast algorithms for the estimation of block motion vectors, *IEEE Trans. Circuits Syst. Video Technol.* **3**(2): 148–157 (April 1993).

56. A. Smolic, T. Sikora, and J.-R. Ohm, Long-term global motion estimation and its application for sprite coding, content description and segmentation, *IEEE Trans. CSVT* **9**(8): 1227–1242 (Dec. 1999).

57. B. Girod, Motion-compensating prediction with fractional-pel accuracy, *IEEE Trans. Commun.* **41**: 604–612 (April 1993).

58. M. T. Orchard and G. J. Sullivan, Overlapped block motion compensation: an estimation-theoretic approach, *IEEE Trans. Image Proc.* **3**(5): 693–699 (Sept. 1994).

59. C. J. Hollier, J. F. Arnold, and M. C. Cavenor, The effect of a loop filter on circulating noise in interframe video coding, *IEEE Trans. Circuits Syst. Video Technol.* **4**(4): 442–446 (Aug. 1994).

60. M. Yuen and H. R. Wu, *Performance of loop filters in MC/DPCM/DCT video coding, Proceedings of ICSP'96*, pp. 1182–1186.

61. S. Mallat and Z. Zhang, Matching pursuit with time-frequency dictionaries, *IEEE Trans. Signal Proc.* **41**: 3397–3415 (Dec. 1993).

62. R. Neff and A. Zakhor, Very low bit-rate video coding based on matching pursuits, *IEEE Trans. Circuits Syst. Video Technol.* **7**(1): 158–171 (Feb. 1997).

63. O. K. Al-Shaykh et al., Video compression using matching pursuits, *IEEE Trans. Circuits Syst. Video Technol.* **9**(1): 123–143 (Feb. 1999).

64. W. E. Glenn, J. Marcinka, and R. Dhein, Simple scalable video compression using 3-D subband coding, *SMPTE J.* **106**: 140–143 (March 1996).

65. W. A. Pearlman, B.-J. Kim, and Z. Xiong, Embedded video coding with 3D SPIHT, P. N. Topiwala, ed., in *Wavelet Image and Video Compression*, Kluwer, Boston, MA, 1998.

66. S. J. Choi and J. W. Woods, Motion-compensated 3-D subband coding of video, *IEEE Trans. Image Proc.* **8**: 155–167 (Feb. 1999).

67. J. R. Ohm, Three-dimensional subband coding with motion compensation, *IEEE Trans. Image Process.* **3**(5): 559–571 (Sept. 1994).

68. J. Xu, S. Li, Z, Xiong, and Y.-Q. Zhang, 3-D embedded subband coding with optimized truncation (3-D ESCOT), *ACHA Special Issue on Wavelets* (May 2001).

69. J. Xu, S. Li, Z. Xiong, and Y.-Q. Zhang, On boundary effects in 3-D wavelet video coding, *Proc. Symposium on Optical Science and Technology*, (July 2000).

70. M. F. Barnsley, *Fractals Everywhere*, Academic Press, San Diego, CA, 1988.

71. A. E. Jacquin, Image coding based on a fractal theory of iterated contractive image transformations, *IEEE Trans. Image Process.* **1**: 18–30 (Jan. 1992).

72. Y. Fisher, A discussion of fractal image compression, in Saupe D. H. O. Peitgen, H. Jurgens, eds., *Chaos and Fractals*, Springer-Verlag, New York, 1992, pp. 903–919.

73. E. W. Jacobs, Y. Fisher, and R. D. Boss, Image compression: A study of iterated transform method, *Signal Process.* **29**: 251–263 (Dec. 1992).

74. K. Barthel, T. Voy'e, and P. Noll, Improved fractal image coding, in *Proc. Picture Coding Symp. (PCS)* **1.5**: (March 1993).

75. J. Y. Huang and O. M. Schultheiss, Block quantization of correlated Gaussian random variables, *IEEE Trans. Commun.* **11**: 289–296 (Sept. 1963).

76. N. S. Jayant and P. Noll, *Digital Coding of waveforms*, Prentice-Hall, Englewood Cliffs, NJ, 1984.

77. A. Segall, Bit allocation and encoding for vector sources, *IEEE Trans. Inform. Theory* **IT-22**: 162–169 (March 1976).

78. T. Chiang and Y.-Q. Zhang, A new rate control scheme using quadratic rate distortion model, *IEEE Trans. Circuits Syst. Video Technol.* **7**(1): 246–250 (Feb. 1997).

79. Y. Shoham and A. Gersho, Efficient bit allocation for an arbitrary set of quantizers, *IEEE Trans. ASSP* **36**: 1445–1453 (Sept. 1988).

80. K. Ramchandran, A. Ortega, and M. Vetterli, Bit allocation for dependent quantization with applications to multiresolution and MPEG video coders, *IEEE Trans. Image Process.* **3**(5): 533–545 (Sept. 1994).

81. T. Weigand, M. Lightstone, D. Mukherjee, T. G. Campbell, and S. K. Mitra, Rate-distortion optimized mode selection for very low bit-rate video coding and the emerging H.263 standard, *IEEE Trans. Circuits Syst. Video Technol.* **6**: 182–190 (April 1996).

82. H. Everett, Generalized lagrange multiplier method for solving problems of optimum allocation of resources, *Oper. Res.* **11**: 399–417 (1963).

83. G. T. Toussaint, Pattern recognition and geometrical complexity, in *Proc. 5th Int. Conf. Pattern Recognition* 1324–1347 (Dec. 1980).

84. P. H. Westerink, J. Biemond, and D. E. Boekee, An optimal bit allocation algorithm for subband coding, in *Proc. Int. Conf. Acoustics, Speech, and Signal Processing (ICASSP)* 757–760 (1988).

85. E. A. Riskin, Optimal bit allocation via the generalized BFOS algorithm, *IEEE Trans. Inform. Theory* **37**: 400–402 (March 1991).

86. M. Rabbani and R. Joshi, An overview of the JPEG 2000 still image compression standard, *Signal Process. Image Commun.* **17**(1): 3–48 (Jan. 2002). Special issue on JPEG 2000.

87. F. Wu, S. Li, and Y.-Q. Zhang, A framework for efficient progressive fine granular scalable video coding, *IEEE Trans. Circuits Syst. Video Technol.* **11**(3): 332–344 (March 2001). Special Issue for Streaming Video.

88. Q. Wang, Z. Xiong, F. Wu, and S. Li, Optimal rate allocation for progressive fine granularity scalable video coding, *IEEE Signal Process. Lett.* **9**: 33–39 (Feb. 2002).

89. J. M. Shapiro, Embedded image coding using zerotrees of wavelet coefficients, *IEEE Trans. SP* **41**(12): 3445–3462 (Dec. 1993).

90. A. Said and W. A. Pearlman, A new, fast, and efficient image codec based on set partitioning in hierarchical trees, *IEEE Trans. CSVT* **6**(3): 243–250 (June 1996).

91. D. Taubman, EBCOT (embedded block coding with optimized truncation): A complete reference, *ISO/IEC JTC1/SC29/WG1* N983, (Sept. 1998).

92. J. Rissanen, Universal coding, information, prediction, and estimation, *IEEE Trans. Inform. Theory* **IT-30**: 629–636 (July 1984).

93. M. J. Weinberger and G. Seroussi, Sequential prediction and ranking in universal context modeling and data compression, *IEEE Trans. Inform. Theory* **43**: 1697–1706 (Sept. 1997).

94. M. Mattavelli and A. Nicoulin, Pre and post processing for very low bit-rate video coding, in *Proc. Int. Workshop HDTV* (Oct. 1994).

95. M. I. Sezan, M. K. Ozkan, and S. V. Fogel, Temporally adaptive filtering of noisy image sequences using a robust motion estimation algorithm, in *Proc. IEEE Int. Conf. Acoustics, Speech, and Signal Processing (ICASSP)* **IV**: 2429–2432 (May 1991).

96. L. Chulhee, M. Eden, and M. Unser, High-quality image resizing using oblique projection operators, *IEEE Trans. Image Proc.* **7**(5): 679–692 (May 1998).

97. A. Munoz, T. Blu, and M. Unser, Least-squares image resizing using finite differences, *IEEE Trans. Image Proc.* **10**(9): 1365–1378 (Sept. 2001).

98. R. Castagno, P. Haavisto, and G. Ramponi, A method for motion adaptive frame rate up-conversion, *IEEE Trans. Circuits Systems Video Technol.* **6**(5): 436–466 (Oct. 1996).

99. Y.-K. Chen, A. Vetro, H. Sun, and S. Y. Kung, Frame-rate up-conversion using transmitted true motion vectors, in *Proc. IEEE Second Workshop on Multimedia Signal Processing*, 622–627 (Dec. 1998).

100. S.-K. Kwon, K.-S. Seo, J.-K. Kim, and Y.-G. Kim, A motion-adaptive de-interlacing method, *IEEE Trans. Consumer Electron.* **38**(3): 145–150 (Aug. 1992).

101. C. Sun, De-interlacing of video images using a shortest path technique, *IEEE Trans. Consumer Electron.* **47**(2): 225–230 (May 2001).

102. J. Schwendowius and G. R. Arce, Data-adaptive digital video format conversion algorithms, *IEEE Trans. Circuits Systs. Video Technol.* **7**(3): 511–526 (June 1997).

103. H. C. Reeve and J. S. Lim, Reduction of blocking effects in image coding, *Opt. Eng.* **23**(1): 34–37 (1984).

104. B. Ramamurthi and A. Gersho, Nonlinear space-variant postprocessing of block coded images, *IEEE Trans. Acoust. Speech Signal Process.* **34**: 1258–1268 (Oct. 1986).

105. R. L. Stevenson, Reduction of coding artifacts in transform image coding, in *Proc. IEEE Int. Conf. Acoustics, Speech, and Signal Processing (ICASSP)* **V**: 401–404 (April 1993).

106. L. Yan, A nonlinear algorithm for enhancing low bit-rate coded motion video sequence, in *Proc. IEEE Int. Conf. Acoustics, Speech, and Signal Processing (ICASSP)* **II**: 923–927 (Nov. 1994).

107. Y. Yang, N. P. Galatsanos, and A. K. Katsaggelos, Iterative projection algorithms for removing the blocking artifacts of block-DCT compressed images, in *Proc. IEEE Int. Conf. Acoustics, Speech, and Signal Processing (ICASSP)* **V**: 401–408 (April 1993).

108. B. Macq et al., Image visual quality restoration by cancellation of the unmasked noise, in *Proc. IEEE Int. Conf. Acoustics, Speech, and Signal Processing (ICASSP)* **V**: 53–56 (1994).

109. T. Chen, Elimination of subband-coding artifacts using the dithering technique, in *Proc. IEEE Int. Conf. Acoustics, Speech, and Signal Processing (ICASSP)* **II**: 874–877 (Nov. 1994).

110. W. Li, O. Egger, and M. Kunt, Efficient quantization noise reduction device for subband image coding schemes, in *Proc. IEEE Int. Conf. Acoustics, Speech, and Signal Processing (ICASSP)* **IV**: 2209–2212 (May 1995).

111. K. K. Pang and T. K. Tan, Optimum loop filter in hybrid coders, *IEEE Trans. Circuits Syst. Video Technol.* **4**(2): 158–167 (April 1994).

112. Tao Bo and M. T. Orchard, Removal of motion uncertainty and quantization noise in motion compensation, *IEEE Trans. Circuits Syst. Video Technol.* **11**(1): 80–90 (Jan. 2001).

113. S. Shirani, F. Kossentini, and R. Ward, A concealment method for video communications in an error-prone environment, *IEEE J. Select. Areas Commun.* **18**(6): 1122–1128 (June 2000).

114. S. Shirani, B. Erol, and F. Kossentini, Error concealment for MPEG-4 video communication in an error prone environment, in *Proc. IEEE International Conference on Acoustics, Speech, and Signal Processing* (2000).

115. Y. Wang and Q.-F. Zhu, Error control and concealment for video communication: a review, *Proc. IEEE* **86**(5): 974–997 (May 1998).

116. H. R. Rabiee, H. Radha, and R. L. Kashyap, Error concealment of still image and video streams with multidirectional recursive nonlinear filters, in *Proceedings of International Conference on Image Processing* (1996).

117. D. Gatica-Perez, M.-T. Sun, and C. Gu, Semantic video object extraction using four-band watershed and partition lattice operators, *IEEE Trans. Circuits Syst. Video Technol.* **11**(5): 603–618 (May 2001).

118. K. Sethi and N. V. Patel, A statistical approach to scene change detection, in *IS&T SPIE Proc.: Storage and Retrieval for Image and Video Databases III* **2420**: 329–339 (Feb. 1995).

119. J.-J. Chen and H.-M. Hang, Source model for transform video coder and its application II. Variable frame rate coding, *IEEE Trans. Circuits Syst. Video Technol.* **7**(2): 299–311 (April 1997).

120. T.-Y. Kuo, J. W. Kim, and C.-J. Kuo, Motion-compensated frame interpolation scheme for H.263 codec, in *Proc. IEEE International Symposium on Circuits and Systems (ISCAS)* (1999).

121. C. Christopoulos, J. Askelof, and M. Larsson, Efficient methods for encoding regions of interest in the upcoming JPEG2000 still image coding standard, *IEEE Signal Process. Lett.* **7**(9): 247–249 (Sept. 2000).

122. K. R. Rao and P. Yip, *Discrete Cosine Transforms—Algorithms, Advantages, Applications*, Academic Press, (1990).

123. CCITT Recommendation T.4, *Standardization of Group 3 Facsimile Apparatus for Document Transmission*, (1980).

124. CCITT Recommendation T.6, *Facsimile Coding Schemes and Coding Control Functions for Group 4 Facsimile Apparatus*, (1984).

125. ITU-T Recommendation T.82, *Information technology—Coded representation of picture and audio information-Progressive bi-level image compression*, (1993).

126. ISO/IEC-11544, *Progressive Bi-level Image Compression. International Standard*, (1993).

127. ISO/IEC-14492 FCD, *Information Technology—Coded Representation of Picture and Audio Information—Lossy/Lossless Coding of Bi-Level Images*, Final Committee Draft. ISO/IEC JTC 1/SC 29/WG 1 N 1359, (July, 1999).

128. ISO/IEC JTC 1/SC 29/WG 1, ISO/IEC FCD 15444-1: *Information technology—JPEG 2000 image coding system: Core coding system* [WG1 N 1646], (March 2000).

129. W. B. Pennebaker and J. L. Mitchell, *JPEG: Still Image Data Compression Standard*, Van Nostrand Reinhold, New York, (1992).

130. ISO/IEC, ISO/IEC 14496-2:1999: *Information technology—Coding of audio-visual objects—Part 2*: Visual, (Dec. 1999).

131. ISO/IEC, ISO/IEC 14495-1:1999: *Information technology—Lossless and near-lossless compression of continuous-tone still images: Baseline*, (Dec. 1999).

132. M. J. Weinberger, G. Seroussi, and G. Sapiro, The LOCO-I lossless image compression algorithm: Principles and standardization into JPEG-LS, *IEEE Trans. Img. Process.* **9**(8): 1309–1324 (Aug. 2000).

133. S. W. Golomb, Run-length encodings, *IEEE Trans. Inform. Theory* **IT-12**: 399–401 (July 1966).

134. R. F. Rice, *Some practical universal noiseless coding techniques*, Tech. Rep. JPL-79-22, Jet Propulsion Laboratory, Pasadena, CA, (March 1979).

135. W. Li, Y.-Q. Zhang, I. Sodagar, J. Liang, and S. Li, MPEG-4 texture coding, A. Puri and C. Chen, eds., in *Multimedia Systems, Standards, and Networks*, Marcel Dekker, Inc., New York, 2000.

136. S. Li and I. Sodagar, Generic, scalable and efficient shape coding for visual texture objects in MPEG-4, In proc. *International Symposium on Circuits and Systems 2000* (May 2000).

137. ISO/IEC FCD15444-1:2000, Information Technology—Jpeg 2000 Image Coding System, V1.0, March 2000.

138. ITU-T Recommendation H.261, Video codec for audiovisual services at p*64 kbits/sec, 1990.

139. MPEG-1 Video Group, Information Technology — Coding of Moving Pictures and Associated Audio for Digital Storage Media up to about 1.5 Mbit/s—: Part 2 — Video, ISO/IEC 11172-2, International Standard, 1993.

140. MPEG-2 Video Group, *Information Technology — Generic Coding of Moving Pictures and Associated Audio*: Part 2 — Video, ISO/IEC 13818-2, International Standard, 1995.

141. ITU-T Experts Group on Very Bit rate Visual Telephony, ITU-T Recommendation H.263: Video Coding for Low Bit rate Communication, Dec. 1995.

142. Video coding for low bit rate communication, ITU-T SG XVI, DRAFT 13, H.263+, Q15-A-60 rev. 0, 1997.

143. ITU-T Q.15/16, Draft for H.263 + + Annexes U, V, and W to Recommendation H.263, November, 2000.

144. MPEG-4 Video Group, Generic coding of audio-visual objects: Part 2 — Visual, ISO/IEC JTC1/SC29/WG11 N1902, FDIS of ISO/IEC 14496-2, Atlantic City, Nov. 1998.

145. Joint Video Team (JVT) of ISO/IEC MPEG and ITU-T VCEG T, Working Draft Number 2, Revision 8 (WD-2 rev 8), JVT-B118r8, April, 2002.

146. Y. T. Chan, *Wavelet Basics*, Kluwer Academic Publishers, Norwell, MA, 1995.

147. G. Strang and T. Nguyen, *Wavelets and Filter Banks*, Wellesley-Cambridge Press, Wellesley, MA, 1996.

148. M. Vetterli and J. Kovacevic, *Wavelets and Subband Coding*, Prentice-Hall, Englewood Cliffs, NJ, 1995.

149. S. Li and W. Li, Shape adaptive vector wavelet coding of arbitrarily shaped objects, *SPIE Proceeding: Visual Communications and Image Processing'97* San-Jose, (Feb. 1997).

150. S. Li, W. Li, H. Sun, and Z. Wu, Shape adaptive wavelet coding, *Proc. IEEE International Symposium on Circuits and Systems* ISCAS'98 5: 281–284 (May 1998).

151. C. Gu and M.-C. Lee, Semiautomatic segmentation and tracking of semantic video objects, *IEEE Trans. Circuits Syst. Video Technol.* **8**(5): (Sep. 1998).

152. J.-H. Pan, S. Li, and Y.-Q. Zhang, Automatic moving video object extraction using multiple features and multiple frames, *ISCAS 2000* (May 2000).

153. H. Zhong, W. Liu, and S. Li, A semi-automatic system for video object segmentation, *ICME 2001* (Aug. 2001).

154. N. Li, S. Li, and W. Liu, A novel framework for semi-automatic video object segmentation, *IEEE ISCAS 2002* (May 2002).

155. S. J. Gortler, R. Grzeszezuk, R. Szeliski, and M. F. Cohen, The lumigraph, *Computer Graphics* (*SIGGRAPH'96*) 43–54 (Aug. 1996).

156. M. Levoy and P. Hanrahan, Light field rendering, *Computer Graphics* (*SIGGRAPH'96*) (Aug. 1996).

157. H. Shum and L. He, Rendering with concentric mosaics, *Computer Graphics* (*SIGGRAPH'99*) 299–306 (Aug. 1999).

158. J. Li, H. Shum, and Y.-Q. Zhang, On the compression of image based rendering scene, in *Proc. International Conference on Image Processing*, **2**: 21–24 (Sept. 2000).

159. P. D. Symes, *Video Compression*, McGraw-Hill, New York, 1998.

160. J. Watkinson, *The MPEG Handbook*, Focal Press, 2001.

161. W. Effelsberg and R. Steinmetz, *Video Compression Techniques*, dpunkt-Verlag, 1998.

162. A. Bovik, ed., *Handbook of image and video processing*, Academic Press, 2000.

IMAGE COMPRESSION

Alfred Mertins
University of Wollongong
Wollongong, Australia

1. INTRODUCTION

The extensive use of digital imaging in recent years has led to an explosion of image data that need to be stored on disks or transmitted over networks. Application examples are facsimile, scanning, printing, digital photography, multimedia, Internet Websites, electronic commerce, digital libraries, medical imaging, and remote sensing. As a result of this ever-increasing volume of data, methods for the efficient compression of digital images have become more and more important. To give an example, a modern digital camera stores about 3.3 megapixels per image. With three color components and 8-bit resolution per color component, this makes an amount of approximately 10 MB (megabytes) of raw data per image. Using a 64-MB memory card, direct storage would allow only six images to be stored. With digital image compression, however, the same camera can store about 60 images or more on the memory card without noticeable differences in image quality. High compression factors without much degradation of image quality are possible because images usually contain a large amount of spatial correlation. In other words, images will typically have larger areas showing similar color, gray level, or texture, and these similarities can be exploited to obtain compression.

Image compression can generally be categorized into lossless and lossy compression. Obviously, lossless compression is most desirable as the reconstructed image is an exact copy of the original. The compression ratios that can be obtained with lossless compression, however, are fairly low, typically ranging from 3 and 5, depending on the image. Such low compression ratios are justified in applications where no loss of quality can be tolerated, as is often the case in the compression and storage of medical images. For most other applications such as internet browsing or storage for printing some loss is usually acceptable. Allowing for loss allows for much higher compression ratios.

Early image compression algorithms have mainly focused on achieving low distortion for a given (fixed) rate, or conversely, the lowest rate for a given maximum distortion. While these goals are still valid, modern multimedia applications have led to a series of further requirements such as spatial and signal-to-noise ratio (SNR) scalability and random access to parts of the image data. For example, in a typical Internet application the same image is to be accessed from various users with a wide range of devices (from a low-resolution palmtop computer to a multimedia workstation) and via channels ranging from slow cable modems or wireless connections to high-speed wired local-area networks. To optimally utilize available bandwidth and device capabilities it is desirable to transcode image data within the network to the various resolutions that best serve the respective users. On the fly transcoding, however, requires the compressed image data to be organized in such a way that different content variations can be easily extracted from a given high-resolution codestream. The new JPEG2000 standard takes a step in this direction and combines abundant functionalities with very high compression ratio and random codestream access.

The aim of this article is to give an overview of some of the most important image compression tools and techniques. We start in Section 2 by looking at the theoretical background of data compression. Section 3 then reviews the discrete cosine and wavelet transforms, the two most important transforms in image compression. Section 4 looks at embedded state-of-the-art wavelet compression methods, and Section 5 gives an overview of standards for the compression of continuous-tone images. Finally, Section 6 gives a number of conclusions and outlook.

2. ELEMENTS OF SOURCE CODING THEORY

2.1. Rate Versus Distortion

The mathematical background that describes the tradeoff between compression ratio and fidelity was established by Shannon in his work on rate–distortion (RD) theory [1]. The aim of this work was to determine the minimum bit rate required to code the output of a stochastic source at a given maximum distortion. The theoretical results obtained by Shannon have clear implications on the performance of any practical coder, and in the following we want to have a brief look at what the RD tradeoff means in practice. For this we consider the graphs in Fig. 1, which show the distortion obtained with different coders versus the rate required to store the information. Such graphs obtained for a specific image and encoding algorithm are known as operational distortion–rate curves. The distortion is typically measured as the mean-squared error (MSE)

$$D = \frac{1}{NM} \sum_{m=0}^{M-1} \sum_{n=0}^{N-1} |\hat{x}_{m,n} - x_{m,n}|^2 \tag{1}$$

where $x_{m,n}$ is the original image and $\hat{x}_{m,n}$ is the reconstructed image. However, also of interest are other distortion measures that possibly better reflect the amount of distortion as perceived by a human observer. The rate is measured as the required average number of bits per pixel (bpp). As one might expect, all coders show the same maximum distortion at rate zero (no transmission). With increasing rate the distortion decreases and different coders reach the point of zero distortion at different rates. Coder 1 is an example of a coder that combines lossy and lossless compression in one bit stream and that is optimized for lossless compression. Coder 2, on the other hand, is a lossy coder that shows good performance for low rates, but reaches the lossless stage only at extremely high rates. The lowest curve shows the minimum rate required to code the information at a given distortion, assuming optimal compression for all rates. This is the desired curve, combining lossy and lossless compression in an optimal way.

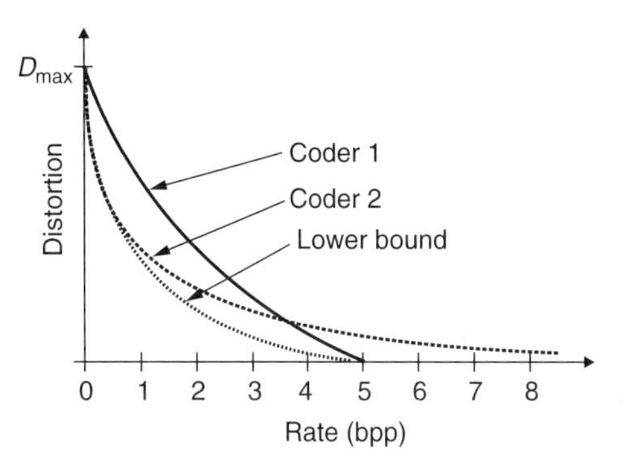

Figure 1. Distortion-rate curves for different coders.

2.2. Coding Gain Through Decorrelation

The simplest way to compress an image in a lossy manner would be to quantize all pixels separately and to provide a bit stream that represents the quantized values. This strategy is known as *pulse code modulation* (PCM). For example an 8-bit gray-scale image whose pixels $x_{m,n}$ are integers in the range from 0 to 255 could be compressed by a factor of four through neglecting the six least significant bits of the binary representations of $x_{m,n}$, resulting in an image with only four different levels of gray. Entropy coding (see Section 2.4) of the PCM values would generally allow increased compression, but such a strategy would still yield a poor tradeoff between the amount of distortion introduced and the compression ratio achieved. The reason for this is that the spatial relationships between pixels are not utilized by PCM.

Images usually contain a large amount of spatial correlation, which can be exploited to obtain compression schemes with a better RD tradeoff than that obtained with PCM. For this the data first needs to be decorrelated, and then quantization and entropy coding can take place. Figure 2 shows the basic structure of an image coder that follows such a strategy. The quantization step is to be seen as the assignment of a discrete symbol to a range of input values. In the simplest case this could be the assignment of symbol a through the operation $a = round(x/q)$, where q is the quantization step size. The inverse quantization then corresponds to the recovery of the actual numerical values that belong to the symbols (e.g., the operation $\hat{x} = q \cdot a$). The entropy coding stage subsequently endeavors to represent the generated discrete symbols with the minimum possible number of bits. This step is lossless. Errors occur only due to quantization. More details on entropy coding will be given in Section 2.4.

One of the simplest decorrelating transforms is a *prediction error filter*, which uses the knowledge of neighboring pixels to predict the value of the pixel of interest and then outputs the prediction error made.

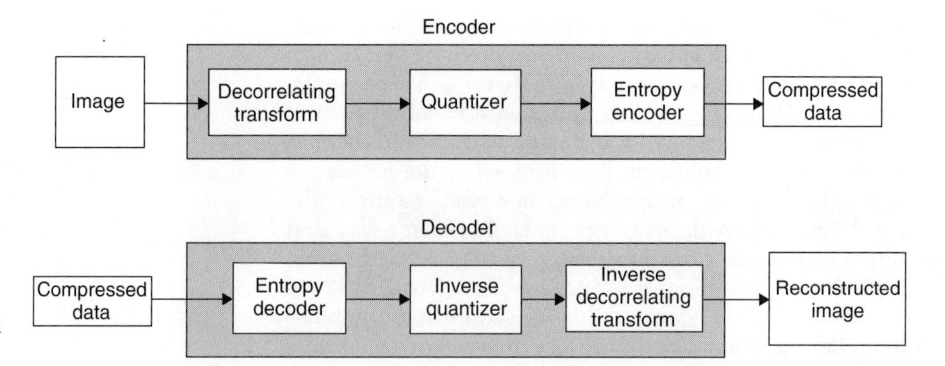

Figure 2. Typical image encoder and decoder structures.

In this case the coding paradigm changes from PCM to differential PCM, known as DPCM. To give a practical example, Fig. 12 shows the neighborhood relationship used in the JPEG-LS standard where the pixel values a, b, c, and d are used to predict an estimate $\hat{x}$ for the actual value x. Under ideal conditions (stationary source and optimal prediction), the output values $e = x - \hat{x}$ of a prediction error filter would be entirely mutually uncorrelated, but in practice there will still be some correlation left, for example due to spatially varying image properties. A coding gain of DPCM over PCM arises from the fact that the prediction error usually has much less power than the signal itself and can therefore be compressed more efficiently.

Modern state-of-the-art image compression algorithms use transforms like the discrete cosine transform (DCT) or the discrete wavelet transform (DWT) to carry out decorrelation [2–4]. These transforms are extremely efficient in decorrelating the pixels of an image and are employed in a number of compression standards [5,6]. To see how and why transform coding works, we consider the system in Fig. 3. The input is a zero-mean random vector $\boldsymbol{x} = [x_0, x_1, \ldots, x_{M-1}]^T$ with correlation matrix $\boldsymbol{R}_{xx} = E\{\boldsymbol{x}\boldsymbol{x}^T\}$, where $E\{\cdot\}$ denotes the expectation operation. The output $\boldsymbol{y} = \boldsymbol{T}\boldsymbol{x}$ of the transform is then a zero-mean random process with correlation matrix $\boldsymbol{R}_{yy} = \boldsymbol{T}\boldsymbol{R}_{xx}\boldsymbol{T}^T$. Because the random variables $y_0, y_1, \ldots, y_{M-1}$ stored in $\boldsymbol{y}$ may have different variances, they are subsequently quantized with different quantizers $Q_0, Q_1, \ldots, Q_{M-1}$. In the synthesis stage the inverse transform is applied to reconstruct an approximation $\hat{\boldsymbol{x}}$ of the original vector $\boldsymbol{x}$ based on the quantized coefficients $\hat{y}_0, \hat{y}_1, \ldots, \hat{y}_{M-1}$. The aim is to design the system in such a way that the mean-squared error $E\{\|\boldsymbol{x} - \boldsymbol{x}\|^2\}$ becomes minimal for a given total bit budget. Questions arising are (1) what is the optimal transform $\boldsymbol{T}$ given the properties of the source, (2) how should an available bit budget B be distributed to the different quantizers, and (3) what are the optimal quantization levels? Answers to these questions have been derived for both unitary and biorthogonal transforms. We will briefly sketch the derivation for the unitary case. To simplify the expressions we assume that $x_0, x_1, \ldots, x_{M-1}$ and thus also $y_0, y_1, \ldots, y_{M-1}$ are zero-mean Gaussian random variables.

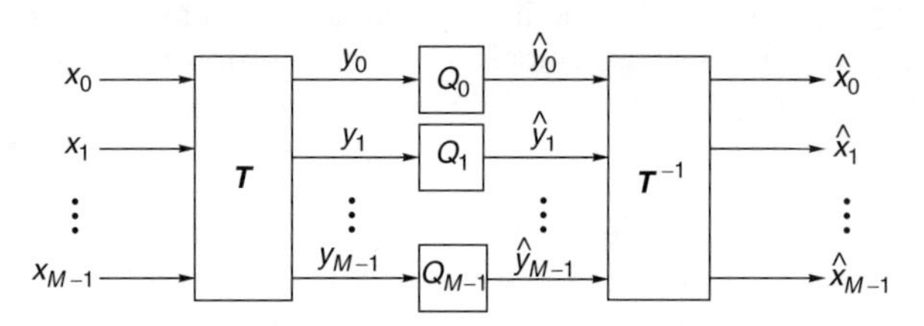

Figure 3. Transform coding.

The optimal scalar quantizers Q_k that minimize the individual error variances $\sigma_{q_k}^2 = E\{q_k^2\}$ with $q_k = y_k - \hat{y}_k$ for a given number of quantization steps are known as *Lloyd–Max quantizers* [7,8]. Important properties of these optimal quantizers are $E\{q_k\} = 0$ (zero-mean error) and $E\{q_k\hat{y}_k\} = 0$ (orthogonality between the quantized value and the error) [9].

The bit allocation can be derived under the assumption of mutually uncorrelated quantization errors and RD relationships of the form

$$\sigma_{q_k}^2 = \gamma_k \sigma_{y_k}^2 2^{-2B_k}, \quad k = 0, 1, \ldots, M-1 \tag{2}$$

for the individual quantizers. The term $\sigma_{q_k}^2$ in (2) is the variance of the quantization error produced by quantizer Q_k, $\sigma_{y_k}^2$ is the variance of y_k, and B_k is the number of bits spent for coding y_k. The values γ_k, $k = 0, 1, \ldots, M-1$ depend on the PDFs of the random variables y_k. Because of the Gaussian assumption made earlier we have equal γ_k for all k. The assumption of uncorrelated quantization errors stated above means that $E\{q_i q_k\} = 0$ for $i \neq k$, which is usually satisfied if the quantization is sufficiently fine, even if the random variables y_k are mutually correlated. Minimizing the average quantization error

$$\sigma_q^2 = \frac{1}{M}\sum_{k=0}^{M-1} \sigma_{q_k}^2 \tag{3}$$

under the constraint of a fixed average bit rate

$$\frac{1}{M}\sum_{k=0}^{M-1} B_k = B \tag{4}$$

using the Lagrange multiplier method yields the bit allocation [9]:

$$B_k = B + \frac{1}{2}\log_2 \frac{\sigma_{y_k}^2}{\left(\prod_{j=0}^{M-1} \sigma_{y_j}^2\right)^{1/M}}, \quad k = 0, 1, \ldots, M-1. \tag{5}$$

According to (5) the bits are to be allocated such that $2^{B_k} \sim \sigma_{y_k}^2$, which intuitively makes sense as more bits are assigned to random variables with a larger variance. Another interesting property of optimal bit allocation and quantizer selection can be seen when substituting B_k according to (5) into (2). One obtains

$$\sigma_{q_k}^2 = \gamma 2^{-2B} \left(\prod_{j=0}^{M-1} \sigma_{y_j}^2 \right)^{1/M}, \quad k = 0, 1, \ldots, M-1 \tag{6}$$

which means that in the optimum situation all quantizers contribute equally to the final output error.

In order to answer the question of which transform is optimal for a given coding task we consider the coding gain defined as

$$G_{\mathrm{TC}} = \frac{\sigma_{\mathrm{PCM}}^2}{\sigma_{\mathrm{TC}}^2} \tag{7}$$

where σ_{PCM}^2 is the error variance of simple PCM and σ_{TC}^2 is the error variance produced by transform coding, both at the same average bit rate B. Assuming that all random variables x_k have the same variance σ_x^2 the error of PCM amounts to $\sigma_{\mathrm{PCM}}^2 = \gamma 2^{-2B} \sigma_x^2$. Using (3) and (6) the error of transform coding can be written as $\sigma_{\mathrm{TC}}^2 = \gamma 2^{-2B} \left(\prod_{j=0}^{M-1} \sigma_{y_j}^2 \right)^{1/M}$, and the coding gain becomes

$$G_{\mathrm{TC}} = \frac{\sigma_x^2}{\left(\prod_{k=0}^{M-1} \sigma_{y_k}^2 \right)^{1/M}} \tag{8}$$

Bearing in mind that for a unitary transform

$$1/M \sum_{k=0}^{M-1} \sigma_{y_k}^2 = \sigma_x^2$$

the coding gain can be seen as the quotient of arithmetic and geometric mean of the coefficient variances. Among all unitary transforms $\boldsymbol{T}$, this quotient is maximized when $\boldsymbol{R}_{yy}$ is diagonal [10] and thus when the transform coefficients are uncorrelated. This decorrelation is accomplished by the Karhunen–Loève transform (KLT). The rows of the transform matrix $\boldsymbol{T}$ are then the transposed eigenvectors of the eigenvalue problem $\boldsymbol{R}_{xx}\boldsymbol{t}_k = \lambda_k \boldsymbol{t}_k$.

The Eq. (8) can be interpreted as follows:

1. A maximum coding gain is obtained if the coefficients y_k are mutually uncorrelated.
2. The more unequal the variances $\sigma_{y_k}^2$ are, the higher the coding gain is, because a high dissimilarity leads to a high ratio of arithmetic and geometric mean. Consequently, if the input values x_k are already mutually uncorrelated (white process), a transform cannot provide any further coding gain.
3. With increasing M one may expect an increasing coding gain that moves toward an upper limit as M goes to infinity. In fact, one can show that this is the same limit as the one obtained for DPCM with ideal prediction [9].

It is interesting to note that the expression (8) for the coding gain also holds for subband coding based on uniform, paraunitary filterbanks (i.e., filterbanks that carry out unitary transforms). The term σ_x^2 is then the variance of an ongoing stationary input process, and the values $\sigma_{y_k}^2$ are the subband variances. A more general expression for the coding gain has been derived by Katto and Yasuda [11]. Their formula also holds for biorthogonal transform and subband coding as well as other schemes such as DPCM.

2.3. Vector Quantization

Vector quantization (VQ) is a multidimensional extension of scalar quantization in that an entire vector of values is encoded as a unit. Let $\boldsymbol{x}$ be such an N-dimensional vector of values, and let $\boldsymbol{x}_i$, $i = 1, 2, \ldots, I$ be a set of N-dimensional codevectors stored in a codebook. Given $\boldsymbol{x}$ a vector quantizer finds the codevector $\boldsymbol{x}_\ell$ from the codebook that best matches $\boldsymbol{x}$ and transmits the corresponding index ℓ. Knowing ℓ the receiver reconstructs $\boldsymbol{x}$ as $\boldsymbol{x}_\ell$. An often used quality criterion to determine which vector from the codebook gives the best match for $\boldsymbol{x}$ is the Euclidean distance $d(\boldsymbol{x}, \boldsymbol{x}_i) = \|\boldsymbol{x} - \boldsymbol{x}_i\|^2$. Theoretically, if the vector length N tends to infinity, VQ becomes optimal and approaches the performance indicated by rate distortion theory. In practice, however, the cost associated with searching through a large codebook is the major obstacle. See Ref. 12 for discussions of computationally efficient forms of the VQ technique and details on codebook design.

2.4. Entropy Coding

Assigning the same code lengths to all symbols generated by a source is not optimal when the different symbols occur with different probabilities. In such a case it is better to assign short codewords to symbols that occur often and longer codewords to symbols that occur only occasionally. The latter strategy results in variable-length codes and is the basic principle of entropy coding.

A simple source model is the discrete memoryless source (DMS), which produces random variables X_i taken from an alphabet $\mathcal{A} = \{a_1, a_2, \ldots, a_L\}$. The symbols a_i may, for example, identify the various steps of a quantizer and are assumed to occur with probabilities $p(a_i)$. The entropy of the source is defined as

$$H = -\sum_{i=1}^{L} p(a_i) \log_2 p(a_i) \tag{9}$$

and describes the average information per symbol (in bits). According to this equation, the more skewed the probability distribution, the lower the entropy. For any given number of symbols L the maximum entropy is obtained if all symbols are equally likely. The entropy provides a lower bound for the average number of bits per symbol required to encode the symbols emitted by a DMS.

The most popular entropy coding methods are Huffman coding, arithmetic coding, and Lempel–Ziv coding; the first two are frequently applied in image compression. Lempel–Ziv coding, a type of universal coding, is used more often for document compression, as it does not require a priori knowledge of the statistical properties of the source.

Huffman coding uses variable-length codewords and produces a uniquely decodable code. To construct a Huffman code, the symbol probabilities must be known a priori. As an example, consider the symbols a_1, a_2, a_3, a_4 with probabilities 0.5, 0.25, 0.125, 0.125, respectively. A fixed-length code would use two bits per symbol. A possible Huffman code is given by $a_1 : 0$, $a_2 : 10$, $a_3 : 110$, $a_4 : 111$. This code requires only 1.75 bits per symbol on average, which is the same as the entropy of the source. In fact, one can show that Huffman codes are optimal and reach the lower bound stated by the entropy when the symbol probabilities are powers of $\frac{1}{2}$. In order to decode a Huffman code, the decoder must know the code table that has been used by the encoder. In practice this means that either a specified standard code table must be employed or the code table must be transmitted to the decoder as side information.

In arithmetic coding there is no one-to-one correspondence between symbols and codewords, as it assigns variable-length codewords to variable-length blocks of symbols. The codeword representing a sequence of symbols is a binary number that points to a subinterval of the interval $[0, 1)$ that is associated with the given sequence. The length of the subinterval is equal to the probability of the sequence, and each possible sequence creates a different subinterval. The advantage of arithmetic over Huffman coding is that it usually results in a shorter average code length when the symbol probabilities are not powers of $\frac{1}{2}$, and arithmetic coders can be made adaptive to learn the symbol probabilities on the fly. No side information in form of a code table is required.

3. TRANSFORMS FOR IMAGE COMPRESSION

3.1. The Discrete Cosine Transform

The discrete cosine transform (DCT) is used in most current standards for image and video compression. Examples are JPEG, MPEG-1, MPEG-2, MPEG-4, H.261, and H.263. To be precise, there are four different DCTs defined in the literature [2], and in particular it is the DCT-II that is used for image compression. Because there is no ambiguity throughout this text we will simply call it the DCT. The DCT of a two-dimensional (2D) signal $x_{m,n}$ with $m, n = 0, 1, \ldots, M-1$ is defined as

$$y_{k,\ell} = \frac{2\gamma_k\gamma_\ell}{M}\sum_{m=0}^{M-1}\sum_{n=0}^{M-1} x_{m,n}\cos\frac{k(m+\frac{1}{2})\pi}{M}\cos\frac{\ell(n+\frac{1}{2})\pi}{M}, \quad k, \ell = 0, 1, \ldots, M-1 \tag{10}$$

with

$$\gamma_k = \begin{cases} \frac{1}{\sqrt{2}} & \text{for } k = 0 \\ 1 & \text{otherwise} \end{cases}$$

The DCT is a unitary transform, so that the inverse transform (2D IDCT) uses the same basis sequences. It is given by

$$x_{m,n} = \sum_{k=0}^{M-1}\sum_{\ell=0}^{M-1} \frac{2\gamma_k\gamma_\ell}{M} y_{k,\ell}\cos\frac{k(m+\frac{1}{2})\pi}{M}\cos\frac{\ell(n+\frac{1}{2})\pi}{M}, \quad m, n = 0, 1, \ldots, M-1 \tag{11}$$

The popularity of the DCT comes from the fact that it almost reaches the coding gain obtained by the KLT for typical image data while having the advantage of fast implementation. In fact, for a 1D (one-dimensional) first-order autoregressive input process with autocorrelation sequence $r_{xx}(m) = \sigma_x^2\rho^{|m|}$ and correlation coefficient $\rho \to 1$ it has been shown that the DCT asymptotically approaches the KLT [9]. Fast implementations can be obtained through the use of FFT (Fast Fourier Transform) algorithms or through direct factorization of the DCT formula [2]. The latter approach is especially interesting for the 2D case where 2D factorizations lead to the most efficient implementations [13,14].

In image compression the DCT is usually used on nonoverlapping 8×8 blocks of the image rather than on the entire image in one step. The following aspects have lead to this choice. First, from the theory outlined in Section 2.2 and the good decorrelation properties of the DCT for smooth signals, it is clear that in order to maximize the coding gain for typical images, the blocks should be as big as possible. On the other hand, with increasing block size the likelihood of capturing a nonstationary behavior within a block increases. This however decreases the usefulness of the DCT for decorrelation. Finally, quantization errors made for DCT coefficients will spread out over the entire block after reconstruction via the IDCT. At low rates this can lead to annoying artifacts when blocks consist of a combination of flat and highly textured regions, or if there are significant edges within a block. These effects are less visible if the block size is small. Altogether, the choice of 8×8 blocks has been found to be a good compromise between exploiting neighborhood relations in smooth regions and avoiding annoying artifacts due to inhomogeneous block content.

Figure 4 shows an example of the blockwise 2D DCT of an image. The original image of Fig. 4a has a size of 144×176 pixels (QCIF format) and the blocksize for the DCT is 8×8. Figure 4b shows the blockwise transformed image, and Fig. 4c shows the transformed image after rearranging the coefficients in such a way that all coefficients with the same physical meaning (i.e., coefficients $y_{k,\ell}$ from the different blocks) are gathered in a subimage. For example the subimage in the upper left corner of Fig. 4c contains the coefficients $y_{0,0}$ of all the blocks in Fig. 4b. These coefficients are often called *DC coefficients*, because they represent the average pixel value within a block. Correspondingly, the remaining 63 coefficients of a block are called *AC coefficients*. From Fig. 4c one can see that the DC coefficients contain the most important information on the entire image. Toward the lower right corner of Fig. 4c the amount of signal energy decreases significantly, represented by the average level of gray.

(a)  (b) 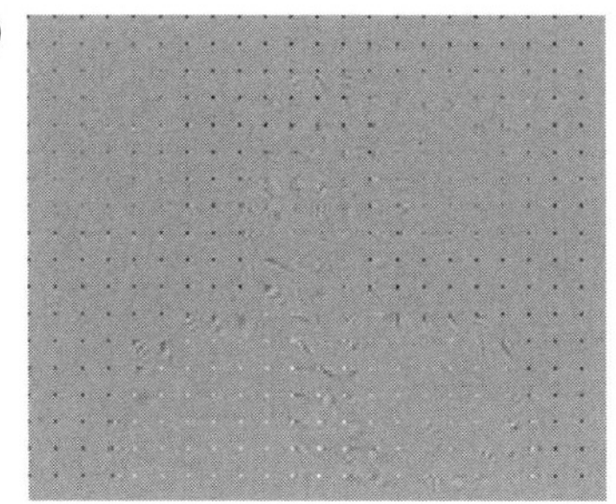 (c) 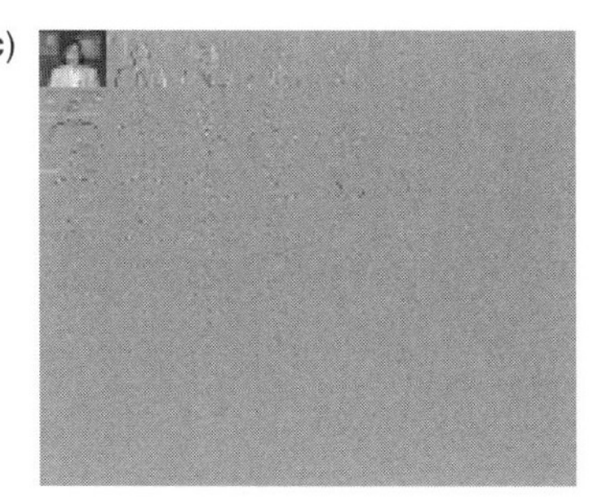

Figure 4. Example of a blockwise 2D DCT of an image: (**a**) original image (144 × 176 pixels); (**b**) transformed image (blocksize 8 × 8, 18 × 26 blocks); (**c**) transformed image after reordering of coefficients (8 × 8 subimages of size 18 × 26).

3.2. The Discrete Wavelet Transform

The discrete wavelet transform (DWT) is a tool to hierarchically decompose a signal into a multiresolution pyramid. It offers a series of advantages over the DCT. For example, contrary to the blockwise DCT the DWT has overlapping basis functions, resulting in less visible artifacts when coding at low bit rates. Moreover, the multiresolution signal representation offers functionalities such as spatial scalability in a simple and generic way. While most of the image compression standards are based on the DCT, the new still image compression standard JPEG2000 and parts of the MPEG4 multimedia standard rely on the DWT [6,15].

For discrete-time signals the DWT is essentially an octave-band signal decomposition, carried out through successive filtering operations and sampling rate changes. The basic building block of such an octave-band filterbank is the two-channel structure depicted in Fig. 5. $H_0(z)$ and $G_0(z)$ are lowpass, while $H_1(z)$ and $G_1(z)$ are highpass filters. The blocks with arrows pointing downward indicate downsampling by factor 2 (i.e., taking only every second sample), and the blocks with arrows pointing upward indicate upsampling by 2 (insertion of zeros between the samples). Downsampling serves to eliminate redundancies in the subband signals, while upsampling is used to recover the original sampling rate. Because of the filter characteristics (lowpass and highpass), most of the energy of a lowpass-type signal $x(n)$ will be stored in the subband samples $y_0(m)$. Because $y_0(m)$ occurs at half the sampling rate of $x(n)$ it appears that the filterbank structure concentrates the information in less samples, as required for efficient compression. More efficiency can be obtained by cascading two-channel filterbanks to obtain octave-band decompositions or other frequency resolutions. For the structure in Fig. 5 to allow perfect reconstruction (PR) of the input with a delay of n_0 samples [i.e., $\hat{x}(n) = x(n - n_0)$], the filters must satisfy

$$H_0(-z)G_0(z) + H_1(-z)G_1(z) = 0 \tag{12}$$

and

$$H_0(z)G_0(z) + H_1(z)G_1(z) = 2z^{-n_0} \tag{13}$$

Equation (12) guarantees that the aliasing components that occur due to the subsampling operation will be compensated at the output, while (13) finally ensures perfect transmission of the signal through the system. In addition to the PR conditions (12) and (13) the filters should satisfy $H_0(1) = G_0(1) = \sqrt{2}$ and $H_1(1) = G_1(1) = 0$, which are essential requirements to make them valid wavelet filters. Moreover, they should satisfy some regularity constraints as outlined by Daubechies [3]. It should be noted that (12) and (13) are the PR constraints for biorthogonal two-channel filterbanks. Paraunitary filter banks and the corresponding orthonormal wavelets require the stronger condition

$$|H_0(e^{j\omega})|^2 + |H_0(e^{j(\omega+\pi)})|^2 = 2 \tag{14}$$

to hold. Apart from the special case where the filter length is 2, Eq. (14) can be satisfied only by filters with nonsymmetric impulse responses [16]. Symmetric filters are very desirable because they allow for simple boundary processing (see below). Therefore paraunitary two-channel filterbanks and the corresponding orthonormal wavelets are seldom used in image compression.

For the decomposition of images, the filtering and down-sampling operations are usually carried out separately in the horizontal and vertical directions. Figure 6 shows an illustration of the principle that yields a 2D octave-band decomposition that corresponds to a DWT. In order to ensure that the analysis process results in the same number of subband samples as there are input pixels, special boundary processing steps are required. These will be explained below. An example of the decomposition of an image is depicted in Fig. 7. One can see that the DWT concentrates the essential information on the image in a few samples, resulting in a high coding gain.

When decomposing a finite-length signal (a row or column of an image) with a filterbank using linear convolution the total number of subband samples is generally higher than the number of input samples. Methods to

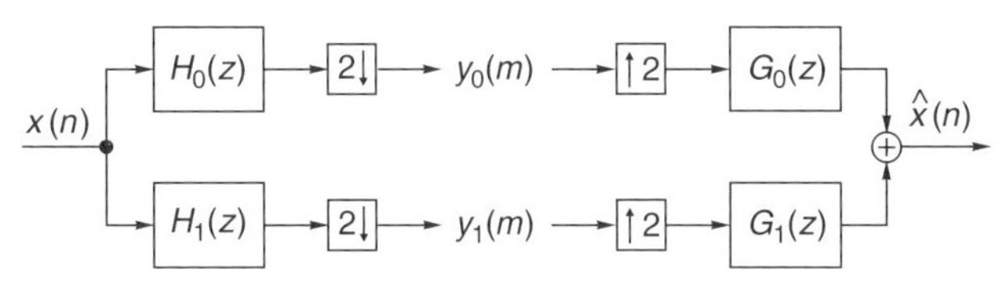

Figure 5. Two-channel filterbank.

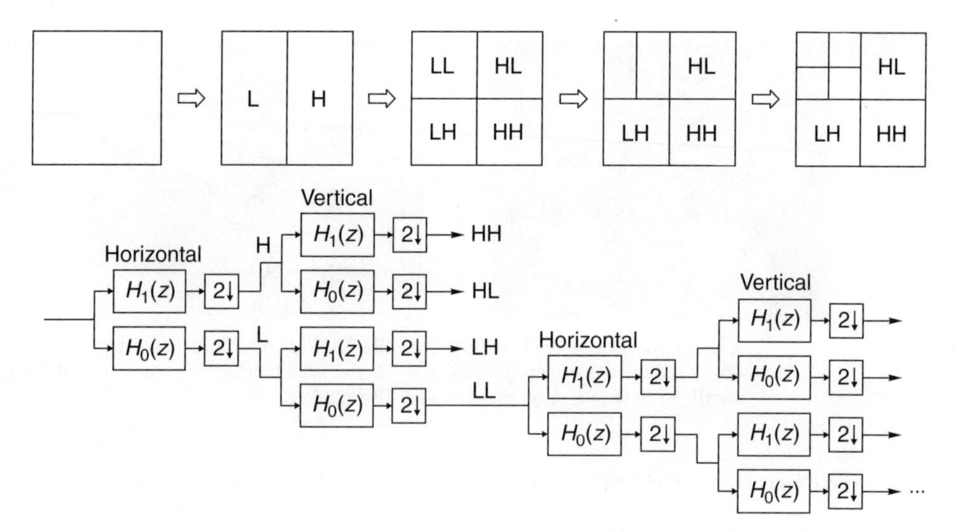

Figure 6. Separable 2D octave-band filterbank.

Figure 7. Example of a 2D octave-band decomposition.

resolve this problem are circular convolution [17], symmetric extension [18,19], and boundary filtering [20–22]. In the following we will describe the method of symmetric extension, which is the one most often used in practice. It requires the filters in the filterbank to have linear phase, which means that biorthogonal filters/wavelets have to be used. We will address the procedure for filters with odd and even length separately.

If the filter impulse responses have odd lengths, linear phase means that they obey the whole-sample symmetry (WSS) shown in Fig. 8a. For these filters the symmetric extension of the signal also has to be carried out with WSS, as depicted in Fig. 8b. A signal of length N is thus extended to a periodic signal $x_{\mathrm{WSS}}(n)$ with period $2N-2$ and symmetry within each period. When analyzing such an extended signal with the corresponding filterbank, the obtained subband signals will also be periodic and will show symmetry within each period. The type of subband symmetry depends on the filter and signal lengths. For the case where N is even, the obtained subband symmetries are depicted in Figs. 8c,d. One can easily see that only a total number of N distinct subband samples needs to be stored or transmitted, because from these N samples the periodic subband signals can be completely recovered. Feeding the periodic subband signals into the synthesis filterbank, and taking one period of the output finally yields perfect reconstruction.

Linear-phase filters with even length have the half-sample symmetry (HSS) in Fig. 9a for the lowpass and the half-sample antisymmetry (HSAS) in Fig. 9b for the highpass. In this case the HSS extension in Fig. 9c is required, resulting in an extended signal with period $2N$ and symmetry within each period. Again, the subband signals show symmetries that can be exploited to capture the entire information on a length-N input signal in a total number of N subband samples. The subband symmetries obtained for even N are depicted in Figs. 9d,e.

Note that the extension schemes outlined above can also be used for cases where N is odd, resulting in different subband symmetries. Moreover, with the introduction of two different subsampling phases, they can be used to define nonexpansive DWTs for arbitrarily shaped objects, see e.g., [23]. Such a scheme is included in the MPEG-4 standard [15].

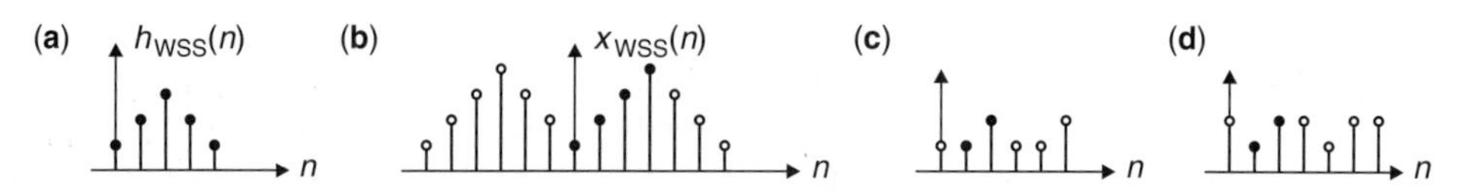

Figure 8. Symmetric extension for odd-length filters: (**a**) impulse response with whole-sample symmetry (WSS); (**b**) periodic signal extension with WSS (the original signal is marked with black dots); (**c**) HSS-WSS subband symmetry, obtained for filter lengths $L = 3 + 4k$, where k is an integer; (**d**) WSS-HSS subband symmetry, obtained for filter lengths $L = 5 + 4k$.

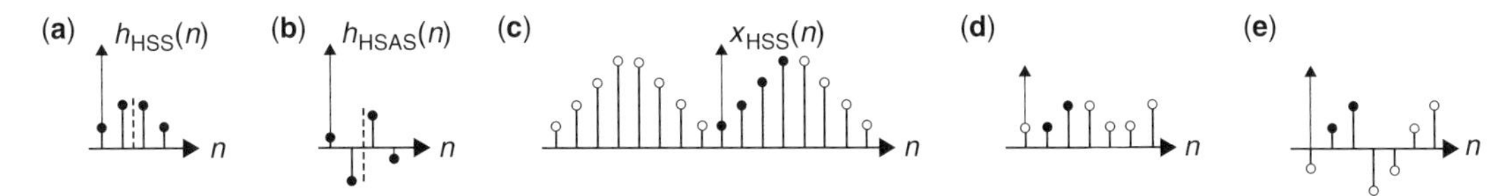

Figure 9. Symmetric extension for even-length filters: (**a**) impulse response with half-sample symmetry; (**b**) half-sample antisymmetry; (**c**) periodic signal extension with half-sample symmetry; (**d**) HSS-HSS, obtained with HSS filter; (**e**) HSAS-HSAS, obtained with HSAS filter.

4. EMBEDDED WAVELET CODING

Early wavelet image coders used to allocate bits to the various subbands according to the principles outlined in Section 2, followed by runlength and entropy coding for further compression of the quantized wavelet coefficients. Runlength coding was found particularly useful for coding of long stretches of zeros that frequently occur within the higher bands. Although such coders can perform reasonably well for a fixed bit rate, they do not offer much flexibility, and especially, they do not allow for progressive transmission of the wavelet coefficients in terms of accuracy. A new era of wavelet coders started with Shapiro's embedded zerotree wavelet (EZW) coder [24], which was the first coder that looked at simultaneous relationships between wavelet coefficients at different scales and produced an entirely embedded codestream that could be truncated at any point to achieve the best reconstruction for the number of symbols transmitted and/or received. The key idea of the EZW coder was the introduction of zerotrees, which are sets of coefficients gathered across scales that are all quantized to zero with regard to a given quantization step size and can be coded with a single symbol. All coefficients within a zerotree belong to the same image region. The formation of zerotrees and the parent–child relationships within a zerotree are shown in Fig. 10a. From looking at the wavelet transform in Fig. 7, it is clear that it is quite likely that all coefficients in such a tree may be quantized to zero in a smooth image region. The concept of EZW coding was refined by Said and Pearlman, who proposed a coding method known as *set partitioning in hierarchical trees* (SPIHT), a state-of-the-art coding method that offers high compression and fine granular SNR scalability [25]. Both the EZW and SPIHT coders follow the idea of transmitting the wavelet coefficients in a semiordered manner, bitplane by bitplane, together with the sorting information required to identify the positions of the transmitted coefficients. In the following we will take a closer look at the SPIHT coder, which is more efficient in transmitting the sorting information. In fact, the SPIHT algorithm is so efficient that additional arithmetic coding will result in only marginal improvements [25].

The SPIHT coder uses three lists to organize the sorting of coefficients and the creation of the bit stream. These are a list of insignificant pixels (LIP), a list of insignificant sets (LIS), and a list of significant pixels (LSP). During initialization the coordinates of the coefficients in the lowest band are stored in the LIP, and 3/4 of them are also stored in the LIS where they are seen as roots of insignificant sets. Figure 10b illustrates the structure of a

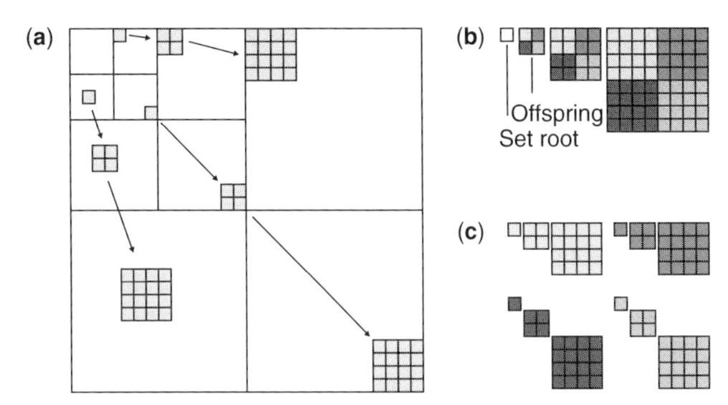

Figure 10. Formation of sets: (**a**) sets/zerotrees within the wavelet tree; (**b**) set; (**c**) set of (**b**) partitioned into four subsets.

set where each coefficient has four offspring. The LSP is left empty at this stage. A start threshold T is defined as $T = 2^n$, where $n = \lfloor \log_2(y_{\max}) \rfloor$ and $y_{\max}$ is the magnitude of the largest subband coefficient. After initialization the algorithm goes through sorting and refinement stages with respect to T. During the sorting phase each coefficient in the LIP is compared with the threshold T and the result of the comparison (a symbol being 0 or 1) is sent to the channel. If a coefficient exceeds T, its sign is transmitted and its coordinate is moved to the LSP. In a second phase of the sorting pass, each set having its root in the LIS is compared with T, and if no coefficient exceeds T, a zero is sent to the channel. If at least one coefficient within a set is larger than T, then a one is sent and the set is subdivided into the four offspring and four smaller sets. The offspring are tested for significance and their coordinates are moved to the appropriate list (LIP or LSP). The offspring coordinates are also used as roots for the four smaller sets; see Fig. 10, Parts (b) and (c) for an illustration of set partitioning. The significance test and subdivision of sets are carried out until all significant coefficients with respect to T have been isolated. At the end of the procedure with threshold T, all coordinates of significant coefficients are stored in the LSP and their signs have been transmitted along with the sorting information required to identify the positions of the transmitted coefficients. In the next stage the threshold is halved and the accuracy of the coefficients in the LSP is refined by sending the information of whether a coefficient lies in the upper or lower halves of the uncertainty interval. Then the next sorting pass with the new threshold is carried out, and so on. The procedure is repeated until a bit budget is exhausted or the threshold falls below a given limit.

The SPIHT decoder looks at the same significance tests as the encoder and receives the answers to the tests from the bit stream. This allows the decoder to

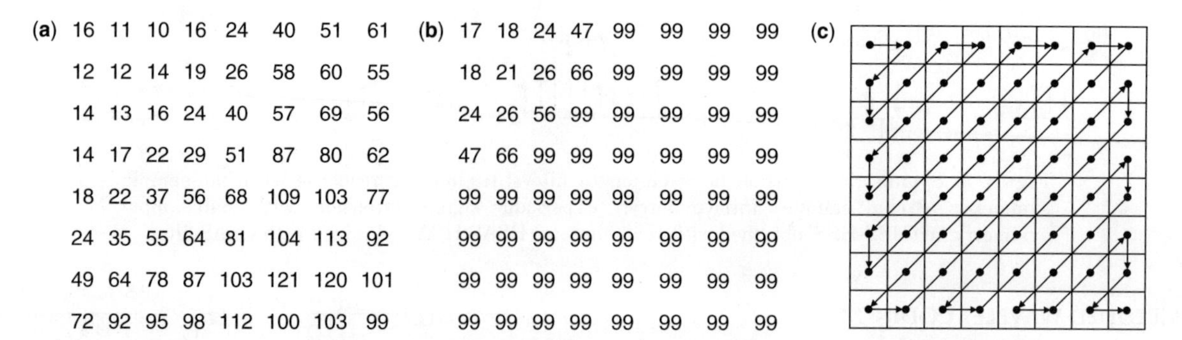

(a)

16	11	10	16	24	40	51	61
12	12	14	19	26	58	60	55
14	13	16	24	40	57	69	56
14	17	22	29	51	87	80	62
18	22	37	56	68	109	103	77
24	35	55	64	81	104	113	92
49	64	78	87	103	121	120	101
72	92	95	98	112	100	103	99

(b)

17	18	24	47	99	99	99	99
18	21	26	66	99	99	99	99
24	26	56	99	99	99	99	99
47	66	99	99	99	99	99	99
99	99	99	99	99	99	99	99
99	99	99	99	99	99	99	99
99	99	99	99	99	99	99	99
99	99	99	99	99	99	99	99

(c)

Figure 11. JPEG quantization matrices: (**a**) luminance; (**b**) chrominance; (**c**) zigzag scanning order.

reconstruct the subband coefficients bitplane by bitplane. The reconstruction levels are always in the middle of the uncertainty interval. This means that if the decoder knows for example that a coefficient is larger than or equal to 32 and smaller than 63, it reconstructs the coefficient as 48. The bit stream can be truncated at any point to meet a bit budget, and the decoder can reconstruct the best possible image for the number of bits received.

An interesting modification of the SPIHT algorithm has been proposed [26], called *virtual SPIHT*. This coder virtually extends the octave decomposition until only three set roots are required in the LIS at initialization. This formation of larger initial sets results in more efficient sorting during the first rounds where most coefficients are insignificant with respect to the threshold. Further modifications include 3-D extensions for coding of video [27].

5. COMPRESSION STANDARDS FOR CONTINUOUS-TONE IMAGES

This section presents the basic modes of the JPEG, JPEG-LS, and JPEG2000 compression standards for still images. There also exist standards for coding of bilevel images, such as JBIG and JBIG2, but these will not be discussed here.

5.1. JPEG

JPEG is an industry standard for digital image compression developed by the Joint Photographic Experts Group, which is a group of experts nominated by leading companies and national standards bodies. JPEG was approved by the principal members of ISO/IEC JTC1 as an international standard (IS 109181) in 1992 and by the CCITT as recommendation T.81, also in 1992. It includes the following modes of operation [5]: a sequential mode, a progressive mode, a hierarchical mode, and a lossless mode. In the following we will discuss the so-called baseline coder, which is a simple form of the sequential mode that encodes images block by block in scan order from left to right and top to bottom. Image data are allowed to have 8-bit or 12-bit precision. The baseline algorithm is designed for lossy compression with target bit rates in the range of 0.25–2 bits per pixel (bpp). The coder uses blockwise DCTs, followed by scalar quantization and entropy coding based on run-length and Huffman coding. The general structure is the same as in Fig. 2. The color space is not specified in the standard, but mostly the YUV space is used with each color component treated separately.

After the blockwise DCT (block size 8×8), scalar quantization is carried out with uniform quantizers. This is done by dividing the transform coefficients $y_{k,\ell}$, $k, \ell = 0, 1, \ldots, 7$ in a block by the corresponding entries $q_{k,\ell}$ in an 8×8 quantization matrix and rounding to the nearest integer: $a_{k,\ell} = \text{round}(y_{k,\ell}/q_{k,\ell})$. Later in the reconstruction stage an inverse quantization is carried out as $\hat{y}_{k,\ell} = q_{k,\ell} \cdot a_{k,\ell}$. The perceptually optimized quantization matrices in Fig. 11 have been included in the standard as a recommendation, but they are not a requirement and other quantizers may be used. To obtain more flexibility, the entire quantization matrices are often scaled such that step sizes $q'_{k,\ell} = D \cdot q_{k,\ell}$ are used instead of $q_{k,\ell}$. The factor D then gives control over the bit rate.

The quantized coefficients in each block are scanned in a zigzag manner, as shown in Fig. 11, and are then further entropy encoded. First the DC coefficient of a block is differentially encoded with respect to the DC coefficient of the previous block (DPCM), using a Huffman code. Then the remaining 63, quantized AC coefficients of a block are encoded. Because the occurrence of long stretches of zeros during the zigzag scan is quite likely, zero runs are encoded using a runlength code. If all remaining coefficients along the zigzag scan are zero, a special end-of-block (EoB) symbol is used. The actual coefficient values are Huffman-encoded.

5.2. JPEG-LS

JPEG-LS is a standard for lossless and near-lossless compression [28]. In the near-lossless mode the user can specify the maximum error ε that may occur during compression. Lossless coding means $\varepsilon = 0$. The performance of JPEG-LS for lossless coding is significantly better than the lossless mode in JPEG.

To achieve compression, context modeling, prediction, and entropy coding based on Golomb–Rice codes are used. Golomb–Rice codes are a special class of Huffman codes for geometric distributions. The context of a pixel x is determined from four reconstructed pixels a, b, c, d in the neighborhood of x, as shown in Fig. 12. Context is used to decide whether x will be encoded in the run or regular mode.

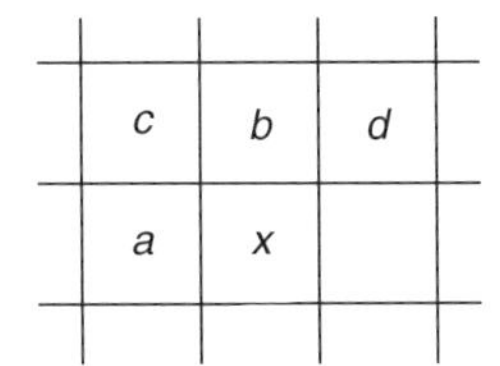

Figure 12. Context modeling in JPEG-LS.

The run mode (run-length coding) is chosen when the context determines that x is likely to be within the specified tolerance ε of the last encoded pixel. This may be the case in smooth image regions. The run mode is ended either at the end of the current line or when the reconstruction error for a pixel exceeds ε. To encode the runlength a modified Golomb–Rice code is applied.

The regular mode is used when, based on the context, it is unlikely that x lies within the tolerance ε of the previously encoded pixel. In this mode, a prediction for x is computed based on the three neighboring values a, b, and c according to

$$\hat{x} = \begin{cases} \min(a,b) & \text{if } c \geq \max(a,b) \\ \max(a,b) & \text{if } c \leq \min(a,b) \\ a+b-c & \text{otherwise} \end{cases} \tag{15}$$

This predictor adapts to local edges and is known as a median edge detection predictor. In a second step the prediction is corrected by a context-dependent term to remove systematic prediction biases. The difference between the bias-corrected prediction and the actual value is then encoded using a Golomb–Rice code.

5.3. JPEG2000

JPEG2000 is a new standard for still-image compression that provides a series of functionalities that were not addressed by the original JPEG standard [6,29]. It is meant to complement JPEG and not to replace it altogether. The main features of JPEG2000 are as follows:

- Any type of image (bilevel, continuous-tone, multi-component) with virtually no restriction on image size
- A wide range of compression factors from 200 : 1 to lossless
- Progressive transmission by accuracy (signal-to-noise ratio) and spatial resolution
- Lossless and lossy compression within one bit stream
- Random codestream access
- Robustness to bit errors
- Region of interest with improved quality

To be backward-compatible with JPEG, the new JPEG2000 standard includes a DCT mode similar to JPEG, but the core of JPEG2000 is based on the wavelet transform. JPEG2000 is being developed in several stages, and in the following we refer to the baseline JPEG2000 coder as specified in Part I of the standard [6]. Part II includes optional techniques such as trellis-coded quantization [30], which are not required for all implementations. Further parts address "Motion JPEG2000," conformance, reference software, and compound image file formats for prepress and faxlike applications. The structure of the baseline JPEG2000 coder is essentially the one in Fig. 2 with the wavelet transform of Fig. 6 as the decorrelating transform. To allow for processing of extremely large images on hardware with limited memory resources, it is possible to divide images into tiles and carry out the wavelet transform and compression for each tile independently. The wavelet filters specified are the Daubechies 9-7 filters [4] for maximum performance in the lossy mode and the 5-3 integer-coefficient filters [31] for an integrated lossy/lossless mode. Boundary processing is carried out using the extension scheme for odd-length filters discussed in Section 3.2.

The quantization stage uses simple scalar quantizers with a dead zone around zero. The step sizes for the quantizers are determined from the dynamic ranges of the coefficients in the different subbands. In the lossless mode where all subband coefficients are integers, the step size is one. The quantized coefficients within each subband are then grouped into codeblocks that are encoded separately. The compression of a codeblock is carried out bitplane by bitplane using a context-dependent arithmetic coding technique. This results in independent embedded bit streams for each codeblock. Independent compression of subbands in codeblocks is the key to random codestream access and to simple spatial resolution scalability. Only the codeblocks referring to a certain region of interest at a desired spatial resolution level of the wavelet tree need to be transmitted or decoded. In order to facilitate for SNR scalability, a layer technique is used where each layer is composed of parts of the blockwise embedded bit streams. This is illustrated in Fig. 13, which shows the formation of layers from the individual bit streams. To ensure that the final codestream is optimally embedded (layer by layer) and that a target bit rate or distortion is met, the truncation points for the individual embedded bit streams can be determined via an operational postcompression rate–distortion optimization [32]. However, although the rate allocation proposed by Taubmann [32] is used in the JPEG2000 verification model [33], other methods may also be employed. Rate allocation is an encoder issue, and the standard specifies only the decoder and the structure of the codestream.

To demonstrate the performance of JPEG2000 and provide a comparison with the older JPEG standard, Fig. 14 shows some coding examples. One can see that JPEG produces severe blocking artifacts at low rates while JPEG2000 tends to produce slightly blurry images. At

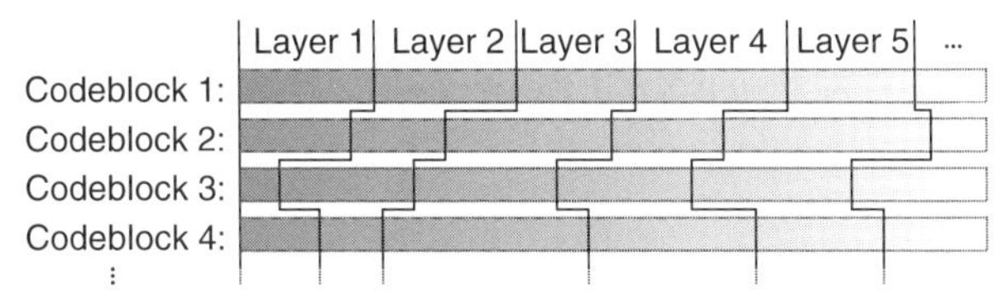

Figure 13. Formation of layered bit stream from embedded bit streams of individual codeblocks.

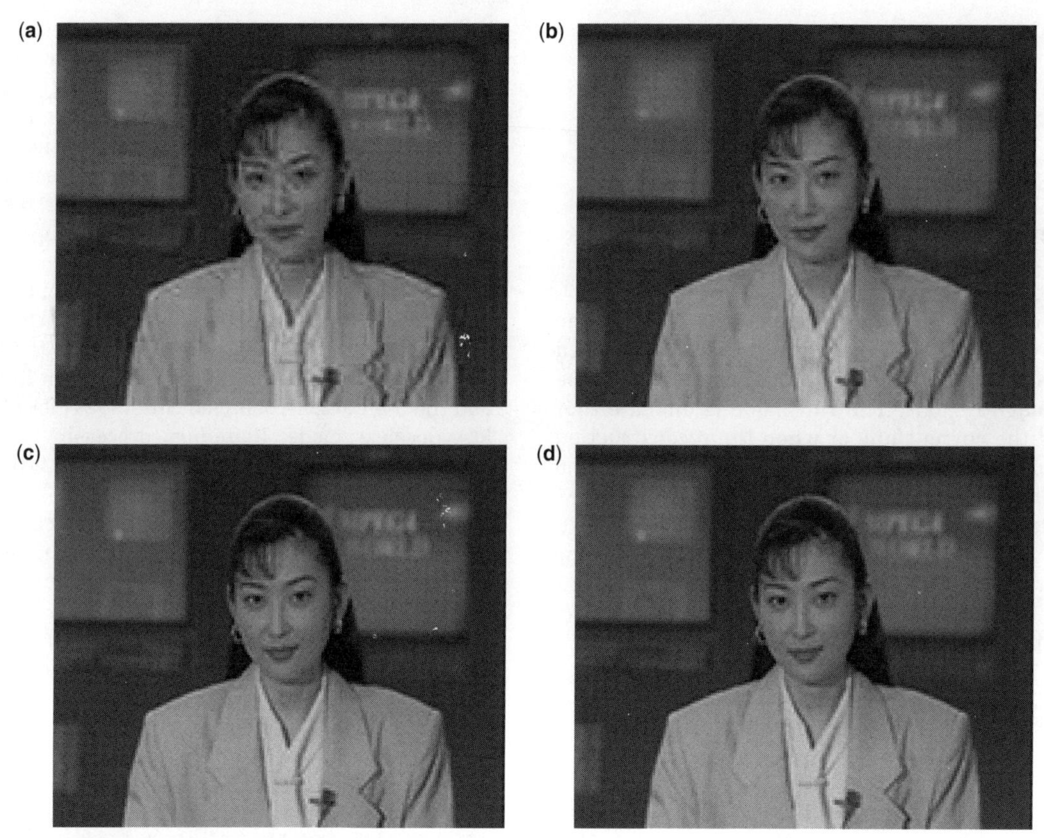

Figure 14. Coding examples (QCIF format): (**a**) JPEG at 0.5 bpp; (**b**) JPEG2000 at 0.5 bpp; (**c**) JPEG at 2 bpp; (**d**) JPEG2000 at 2 bpp.

higher rates both standards yield good-quality images with JPEG2000 still having the better signal-to-noise ratio.

6. CONCLUSIONS

This article has reviewed general concepts and standards for still-image compression. We started by looking at theoretical foundations of data compression and then discussed some of the most popular image compression techniques and standards. From todays point of view the diverse functionalities required by many multimedia applications are best provided by coders based on the wavelet transform. As demonstrated in the JPEG2000 standard, wavelet-based coders even allow the integration of lossy and lossless coding, which is a feature that is very desirable for applications such as medical imaging where highest quality is needed. The compression ratios obtained with lossless JPEG2000 are in the same order as the ones obtained with dedicated lossless methods. However, because lossless coding based on the wavelet transform is still in a very early stage, one may expect even better integrated lossy and lossless wavelet-based coders to be developed in the future.

BIOGRAPHY

Alfred Mertins received the Dipl.-Ing. degree in electrical engineering from University of Paderborn, Germany, and the Dr.-Ing. and Dr.-Ing. habil. degrees in electrical engineering from Hamburg University of Technology, Germany, in 1984, 1991, and 1994, respectively. From 1986 to 1991 he was with the Hamburg University of Technology, from 1991 to 1995 with the Microelectronics Applications Center Hamburg, Germany, from 1996 to 1997 with the University of Kiel, Germany, and from 1997 to 1998 with the University of Western Australia, Crawley. Since 1998, he has been a senior lecturer at the School of Electrical, Computer, and Telecommunications Engineering, University of Wollongong, New South Wales, Australia. His research interests include digital signal processing, wavelets and filter banks, image and video processing, and digital communications.

BIBLIOGRAPHY

1. C. E. Shannon, Coding theorems for a discrete source with a fidelity criterion, *IRE Nat. Conserv. Rec.* **4**: 142–163 (1959).

2. K. R. Rao and P. Yip, *Discrete Cosine Transform*, Academic Press, New York, 1990.

3. I. Daubechies, *Ten Lectures on Wavelets*, SIAM, 1992.

4. M. Antonini, M. Barlaud, P. Mathieu, and I. Daubechies, Image coding using wavelet transform, *IEEE Trans. Image Process.* **1**(2): 205–220 (April 1992).

5. G. K. Wallace, The JPEG still picture compression standard, *IEEE Trans. Consumer Electron.* **38**(1): 18–34 (Feb. 1992).

6. Joint Photographic Experts Group, *JPEG2000 Final Draft International Standard, Part I*, ISO/IEC JTC1/SC29/WG1 FDIS15444-1, Aug. 2000.
7. S. P. Lloyd, Least squares quantization in PCM, *Institute of Mathematical Statistics Society Meeting*, Atlantic City, NJ, Sept. 1957, pp. 189–192.
8. J. Max, Quantizing for minimum distortion, *IRE Trans. Inform. Theory* 7–12 (March 1960).
9. N. S. Jayant and P. Noll, *Digital Coding of Waveforms*, Prentice-Hall, Englewood Cliffs, NJ, 1984.
10. R. Bellman, *Introduction to Matrix Analysis*, McGraw-Hill, New York, 1960.
11. J. Katto and Y. Yasuda, Performance evaluation of subband coding and optimization of its filter coefficients, *Proc. SPIE Visual Communication and Image Processing*, Nov. 1991, pp. 95–106.
12. A. Gersho and R. M. Gray, *Vector Quantization and Signal Compression*, Kluwer, Boston, 1991.
13. P. Duhamel and C. Guillemot, Polynomial transform computation of the 2-D DCT, *Proc. IEEE Int. Conf. Acoustics, Speech, and Signal Processing*, Albuquerqe, NM, April 1990, pp. 1515–1518.
14. W.-H. Fang, N.-C. Hu, and S.-K. Shih, Recursive fast computation of the two-dimensional discrete cosine transform, *IEE Proc. Visual Image Signal Process.* **146**(1): 25–33 (Feb. 1999).
15. *MPEG-4 Video Verification Model, Version 14. Generic Coding of Moving Pictures and Associated Audio*, ISO/IEC JTC1/SC 29/WG 11, 1999.
16. P. P. Vaidyanathan, On power-complementary FIR filters, *IEEE Trans. Circuits Syst.* **32**: 1308–1310 (Dec. 1985).
17. J. Woods and S. O'Neil, Subband coding of images, *IEEE Trans. Acoust. Speech Signal Process.* **34**(5): 1278–1288 (May 1986).
18. M. J. T. Smith and S. L. Eddins, Analysis/synthesis techniques for subband coding, *IEEE Trans. Acoust. Speech Signal Process.* 1446–1456 (Aug. 1990).
19. J. N. Bradley, C. M. Brislawn, and V. Faber, Reflected boundary conditions for multirate filter banks, *Proc. Int. Symp. Time-Frequency and Time-Scale Analysis*, Canada, 1992, pp. 307–310.
20. R. L. de Queiroz, Subband processing of finite length signals without border distortions, *Proc. IEEE Int. Conf. Acoustics, Speech, and Signal Processing*, San Francisco, March 1992, Vol. IV, pp. 613–616.
21. C. Herley, Boundary filters for finite-length signals and time-varying filter banks, *IEEE Trans. Circuits Syst. II* **42**(2): 102–114 (Feb. 1995).
22. A. Mertins, Boundary filters for size-limited paraunitary filter banks with maximum coding gain and ideal DC behavior, *IEEE Trans. Circuits Syst. II* **48**(2): 183–188 (Feb. 2001).
23. A. Mertins, *Signal Analysis: Wavelets, Filter Banks, Time-Frequency Transforms and Applications*, Wiley, Chichester, UK, 1999.
24. J. M. Shapiro, Embedded image coding using zerotrees of wavelet coefficients, *IEEE Trans. Signal Process.* **41**(12): 3445–3462 (Dec. 1993).
25. A. Said and W. A. Pearlman, A new fast and efficient image codec based on set partitioning in hierarchical trees, *IEEE Trans. Circuits Syst. Video Technol.* **6**(3): 243–250 (June 1996).
26. E. Khan and M. Ghanbari, Very low bit rate video coding using virtual SPIHT, *Electron. Lett.* **37**(1): 40–42 (Jan. 2001).
27. B.-J. Kim, Z. Xiong, and W. A. Pearlman, Low bit-rate scalable video coding with 3-d set partitioning in hierarchical trees (3-D SPIHT), *IEEE Trans. Circuits Syst. Video Technol.* **10**(8): 1374–1387 (Dec. 2000).
28. Joint Photographic Experts Group, *JPEG-LS Final Committee Draft*, ISO/IEC JTC1/SC29/WG1 FCD14495-1, 1997.
29. C. Christopoulos, A. Skodras, and T. Ebrahimi, The JPEG-2000 still image coding system: An overview, *IEEE Trans. Consumer Electron.* **46**(4): 1103–1127 (Nov. 2000).
30. J. H. Kasner, M. W. Marcellin, and B. R. Hunt, Universal trellis coded quantization, *IEEE Trans. Image Process.* **8**(12): 1677–1687 (Dec. 1999).
31. D. LeGall and A. Tabatabai, Sub-band coding of digital images using short kernel filters and arithmetic coding techniques, *Proc. IEEE Int. Conf. Acoustics, Speech, and Signal Processing*, 1988, pp. 761–764.
32. D. Taubmann, High performance scalable image compression with EBCOT, *IEEE Trans. Image Process.* **9**(7): 1158–1170 (July 2000).
33. Joint Photographic Experts Group, *JPEG2000 Verification Model 8*, ISO/IEC JTC1/SC29/WG1 N1822, July 2000.

IMAGE PROCESSING

MAJA BYSTROM
Drexel University
Philadelphia, Pennsylvania

Image processing for telecommunications is a broad field that can be divided into four general categories: acquisition, analysis, compression, and reconstruction. Images can likewise be classified by their color content into binary, monochrome, or color images, or by their method of acquisition or generation, such as natural, computer-generated, radar, or ultrasonic. Following acquisition, image data may be processed for efficient storage or representation. Image analysis may be employed to extract desired information for compression or further processing. Reconstruction or enhancement is posttransmission or postacquisition processing to recover lost or degraded data or to emphasize visually important image components. More recently significant attention has focused on the related field of digital watermarking or data hiding, in which marks or identification patterns are embedded in images for security purposes.

1. IMAGE ACQUISITION AND REPRESENTATION

Images are acquired through a variety of methods such as analog or digital cameras, radar, or sonar. Regardless of the method of image generation, in order to provide for digital transmission and storage, all input analog signals must be discretized and quantized. It is well known that for perfect reconstruction, images must be sampled above the Nyquist rate, specifically, twice the greatest frequency in a band-limited signal, and infinite-duration interpolation functions must be employed. In practice, however, infinite-duration functions are infeasible and images are often represented with fewer than the optimum number of samples for conservation of storage space or transmission bandwidth. Subsampling of images reduces the number of picture elements, *pixels*, used to represent an image. However, interpolation of a subsampled image may result in visually apparent degradation, typically in the form of blockiness or blurring. Jain discusses image sampling and basic interpolation functions [1].

Following sampling, an image is represented by a two-dimensional signal denoted by $x_{i,j} = x(i,j)$; $i = 1 \cdots V$, $j = 1 \cdots H$, where V and H are the vertical and horizontal length in pixels, respectively. By convention, the upper left corner of the image is pixel $x_{1,1}$. The pixel values are continuous and must be quantized to further limit storage requirements. The most basic quantizer is the uniform quantizer in which the continuous range of values is subdivided into a finite number of equal-length intervals. All pixels with values falling within each interval are then assigned the same value. If the quantization is fine, that is, if a large number of quantization intervals is employed, then no subjective degradation will be apparent. In practice, sample values often have a nonuniform distribution such as Laplacian or Gaussian. In this case, better subjective results may be obtained with a nonuniform quantizer and quantization intervals are determined by the Lloyd–Max algorithm.

There is a significant trade-off between the storage requirement, which is a function of the number of quantization intervals, and the subjective quality of the image. The shades of gray in black-and-white (B/W) images are typically represented by 256 quantization levels. The storage requirement is then $\log_2 256 = 8$ bits per pixel. Thus, even a small image requires significant memory for storage. However, reducing the representation to 7 bits per pixel may result in loss of small objects in the image [2].

Color images can be represented in many color spaces. One of the most frequently used color spaces in image processing is the RGB color space, which indicates the proportion of the red, green, and blue components. The value of an image pixel, $x_{i,j}$, is then a vector in three dimensions. Each vector component assumes a value in the range $[0, max]$, where max is typically normalized to 2^n for n quantization levels. For full-color RGB each color is represented by 8 bits for a total of 24 bits or 2^{24} color combinations; this number of colors is significantly more than the human eye can recognize. Other, more visually intuitive, color spaces such as hue, saturation, and intensity (HSI) or hue, lightness, and saturation (HLS) can be employed as well.

A drawback of the RGB colorspace is that for natural images, there is significant correlation between the color components. Other spaces exploit this correlation and thus are more common for applications that require efficient color representation. In the YIQ space used in North American television, the YUV color space used in European television system, and the YCbCr used in image and video compression standards, there is a luminance or B/W component and two chrominance components.

To further reduce the storage space or transmission bandwidth required for each image, the chrominance components can be subsampled, typically by a factor of 2 in each direction, with little loss in the subjective quality.

2. FILTERING AND MORPHOLOGICAL OPERATORS

Sampling and interpolation are two examples of image filtering. Many other image processing applications, such as object identification, edge enhancement, or artifact removal, also rely on filtering. Given an original image, x, the filtering operation is

$$\hat{x}_{i,j} = \sum_{k,l \in \mathcal{N}} W_{k,l} x_{i+k,j+l}$$

where $\mathcal{N}$ is a set of pixels in the designated neighborhood of the (i,j)th pixel, $W_{k,l}$ is the kernel, or weighting function, and $\hat{x}$ is the filtered image. The neighborhood, $\mathcal{N}$, can be as small as one pixel or as large as the entire image. In the case of sampling, the kernel is a two-dimensional comb function; for interpolation, the kernel can be a two-dimensional rectangle or sinc function. Appropriate filter shapes for particular applications will be mentioned in the following sections.

For a class of images, the binary images in which pixels can be only black or white, morphological operators can be a valuable tool for many of the analysis and reconstruction processing operations. Morphological operators essentially perform filtering with different shape and size kernels to switch the binary value of the pixel, based on the pixel's neighbors. Through iterations of erosion, the shrinking of an object, or dilation, the expanding of an object, and other image operations, such as addition or subtraction, this class of operators can enhance lines, remove noise, and segment images. Morphological operators have been extended to grayscale images. Dougherty gives an introduction to this class of filters [3].

3. IMAGE TRANSFORMS

Often image processing applications, such as compression and reconstruction, are performed in a transform domain. The three most commonly used transforms are the discrete Fourier transform (DFT), the discrete cosine transform (DCT), and the discrete wavelet transform (DWT). The discrete Fourier transform over an $N \times N$ region of image pixels is given by

$$X_{k,l} = \sum_{m=0}^{N-1} \sum_{n=0}^{N-1} x_{m,n}\, e^{-j2\pi(mk+nl)/N}$$

while the inverse DFT is given by

$$x_{m,n} = \frac{1}{N^2} \sum_{k=0}^{N-1} \sum_{l=0}^{N-1} X_{k,l}\, e^{j2\pi(mk+nl)/N}.$$

The DCT is more commonly used in image and video compression standards than the DFT, since it has excellent energy compaction properties and has performance close to the optimal Karhunen–Loeve transform. Given a pixel block of size $N \times N$, the DCT is given by

$$X_{k,l} = \frac{2}{N}c(k)c(l)\sum_{m=0}^{N-1}\sum_{n=0}^{N-1} x_{m,n}\cos\left(\frac{k\pi(2m+1)}{2N}\right) \times \cos\left(\frac{l\pi(2n+1)}{2N}\right)$$

while the inverse transform is given by

$$x_{m,n} = \frac{2}{N}\sum_{k=0}^{N-1}\sum_{l=0}^{N-1} c(k)c(l)X_{k,l}\cos\left(\frac{k\pi(2m+1)}{2N}\right) \times \cos\left(\frac{l\pi(2n+1)}{2N}\right)$$

where $c(i) = 1/\sqrt{2}$ for $i = 0$ and $c(i) = 1$ otherwise. The DCT basis functions are shown in Fig. 1.

A final transform currently under investigation for many image processing applications and that will form the basis of the JPEG-2000 image compression standard is the DWT [4]. The DWT utilizes a combination of lowpass and highpass filters with downsampling and interpolation to separate an image into frequency bands that may be processed independently. The wavelet decomposition or analysis process is performed by filtering first along the rows and then the columns of an image and downsampling by a factor of 2. A two-level decomposition is given in Fig. 2. Note that each subband contains different frequency information. The lowest subband, in the upper left of the figure, contains the lowest frequencies, and appears as a smoothed version of the original image. Since this is a two-level transform, the lowest subband has been downsampled twice and thus is one-sixteenth the size of the original image.

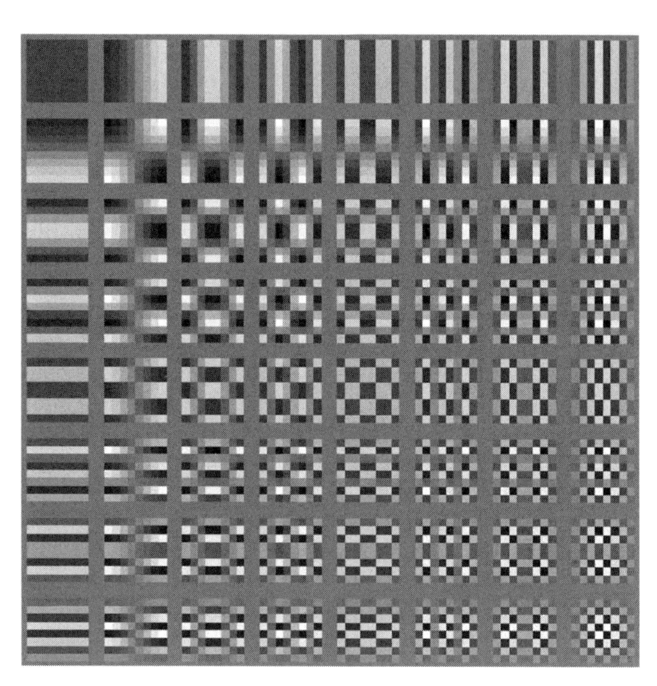

Figure 1. The DCT basis functions for an 8×8 transformation.

Synthesis, or reconstruction, is performed in the opposite manner. The synthesis filters are quadrature mirror filters designed to cancel the aliasing effects of the interpolation process. With appropriate filter design, that is, choice of wavelet basis, applications such as compression can produce perceptually better results than can DCT-based compression at the same rate, since the transform can efficiently act on large regions of the image.

4. IMAGE ANALYSIS

Image analysis is used to provide information about the image under consideration. This information could range from identification of a target in an image, segmentation of an object from a scene, or motion tracking of an object between video frames. Applications range from military target tracking, face recognition for security purposes, segmentation for compression, or object recognition for digital library indexing.

Object matching for target tracking and general object recognition is a challenging field because of the possibility of object movement, rotation, shape change, occlusion by other objects, and noise and clutter in the image. Typically for these applications, landmarks on the object are determined, or a model—based on pixel intensity, statistics, or other image features—is developed. The object is matched to known references taking into account the possible size and shape variations in the target as well as the other possible image degradation. Discussions of object recognition and target tracking are available in the literature [5–7].

Images may be segmented for applications such as efficient encoding or target location. For encoding, the foreground or more important objects are located and compressed less to maintain better quality. In target identification, targets are recognized from an often noisy and cluttered image and separated from the background. A number of segmentation methods ranging from segmentation on the basis of texture or color discrimination, motion between video frames, region growing from a starting point in a readily identifiable area, to boundary, feature, or edge identification are currently utilized. Often these methods are combined for better performance. Segmentation either is performed automatically by an algorithm or may be supervised by a user who provides input such as a starting point in a region or the number of objects to be located. An ongoing research focus is the development of unsupervised segmentation algorithms that extract perceptually important objects.

5. IMAGE COMPRESSION

Compression of an image involves reducing redundancy in the image through either lossless or lossy means. In lossy compression information is discarded through quantization or many-to-one mappings of component values. Lossy compression is typically used in applications such as broadcast images or video, since significant compression gains can be achieved at the expense of image quality. On the other hand, lossless compression is an

Figure 2. A two-level wavelet decomposition with the Haar filter.

invertible transform that preserves all image information and is employed on images in which all information is valuable, such as security, medical, or technical images.

Lossless compression typically involves some form of entropy coding such as Huffman or arithmetic coding. Image symbols are assigned codewords on the basis of their probabilities; more likely symbols are assigned shorter codewords. Often a form of prediction is employed as well. For example, pixel values can be predicted on the basis of their neighbors' values. The difference between the prediction and the actual pixel value is then entropy-coded. Naturally, these methods only work well if the symbol probability estimation and the prediction are good.

Lossy standards such as JPEG, JPEG-2000, MPEG-2, and MPEG-4 (JPEG — Joint Photographic Experts Group; MPEG — Moving Picture Experts Group) rely on a combination of either the DCT or DWT, quantization, and entropy coding in order to compress images or video. In block-based techniques, images are divided into blocks of pixels. These blocks are DCT transformed and the resulting coefficients are quantized and either Huffman or arithmetic-coded. Thus, spatial redundancy is taken advantage of and high frequencies are removed from the image. For video coding, temporal redundancy is utilized. Blocks of pixels in neighboring images tend to be similar, so for many blocks, a motion vector can be determined. This motion vector indicates the displacement of a block to a similar block in a neighboring frame. The difference between the two blocks is then transformed and coded.

The more recently developed lossy standards are wavelet-based. Because the lowest subband is a smoothed version of the original image, the higher subbands can be discarded or coarsely quantized with little effect on the reconstructed image quality. The wavelet coefficients are quantized and coded independently for each subband. The quantization of coefficients can be controlled by an algorithm such as the embedded zero-tree wavelet algorithm, which iteratively determines which coefficients are most significant and increases the precision of these coefficients. These wavelet compression techniques are flexible in that they permit coding of nonsquare regions of interest. Thus, arbitrarily shaped objects in an image may be segmented by an analysis algorithm and more important objects compressed less than others for better subjective quality. Future compression gains will likely be made through improved segmentation algorithms and region-of-interest compression.

6. IMAGE ENHANCEMENT AND RECONSTRUCTION

Because the human visual system is more sensitive to certain image components than others, the more perceptually important components of an image can be enhanced, often at the expense of the others, in order to provide a subjectively higher-quality image. Either point operations or histogram equalization can be used to modify the contrast and brightness of an image in order to enhance or emphasize objects that are washed out or hidden in a dark region. Contrast enhancement, either contrast stretching or brightness enhancement, is performed by manipulating pixel grayscale levels, that is, by nonlinear or linear mappings of grayscale values.

If the slope of a linear mapping is negative, then the pixel values are reversed and the mapping will result in a negative image. A similar process, histogram equalization, which stretches an image histogram so that it spans the color value range, will also result in the increase of image contrast.

Since the human visual system is sensitive to image edges, these high-frequency image components can be enhanced, typically by taking a local directional derivative, in order to make the image appear sharper. The Laplacian operating on the image luminance is an approximation to the second derivative of the brightness. It acts as a highpass filter, by increasing the contrast at distinct lines in images and setting pixels in homogeneous areas to zero. For edges that are step functions or plateaus in brightness, the Roberts and Sobel operators result in better performance than does the Laplacian operator. These filters approximate the first derivative and are applied oriented in multiple directions to capture diagonal edges.

Following compression, transmission over a noisy or fading channel, or capture by imperfect methods, there may be degradation in the received image. This resulting degradation may result from noise or speckling, coding artifacts, or data loss. Figure 3 illustrates three types of degradation. Error reconstruction or concealment measures must be taken to restore these degraded images.

There are various methods for noise removal; perhaps the simplest is lowpass filtering or smoothing by neighborhood averaging in the spatial domain. Common kernels employed in the filtering operation are uniform, Gaussian, and Savitsky–Golay. More complex iterative or adaptive noise and blur removal methods can be employed as well. Both stochastic and deterministic filtering and estimation have been performed with much success [8,9].

Low-bit-rate compression by standard methods often results in either block artifacts or ringing due to quantization of transform coefficients. An example of blockiness resulting from JPEG compression is shown in Fig. 3c. Techniques for compensating for this type of degradation are local adaptive filtering in the spatial or frequency domains. Molina et al. provide a survey of deblocking techniques [10].

When block-based coding schemes are employed, channel errors may affect the compressed bitstream and cause the decoder to lose synchronization and thus lose an entire block or multiple blocks. A typical example is shown in the left side of Fig. 3d. For intraimage recovery, typically some form of replacement or smoothing is employed. Blocks of pixels can be replaced by surrounding blocks, or interpolated from surrounding pixels, taking

(a)

(b)

(c)

(d)

Figure 3. Selected examples of image degradation. (**a**) original image; (**b**) image with noise; (**c**) image with block artifacts; (**d**) image with block loss.

into consideration image features such as lines. If the errors arise in coded video, a combination of temporal and spatial error concealment is often employed.

7. QUALITY EVALUATION

A measure of image quality is vital for either evaluation of the appropriateness of an image for a particular application, or for evaluation of the effects of processing on an image. These measures differ broadly depending on the method of image acquisition and the image processing application. The simplest metrics are objective measures. For radar images, measures include the clutter : noise ratio, the resolution, and metrics of additive and multiplicative noise [11]. If, following image processing, the original image, x, is available for comparison, the quality of a reconstructed image denoted by $\hat{x}$ can be measured by one of four common distortion metrics:

$$\text{SNR} = 10\log_{10}\frac{\sum_{i=1}^{V}\sum_{j=1}^{H}x_{i,j}^{2}}{\sum_{i=1}^{V}\sum_{j=1}^{H}(x_{i,j}-\hat{x}_{i,j})^{2}}$$

$$\text{PSNR} = 10\log_{10}\frac{(max_{i,j}x_{i,j})^{2}}{\sum_{i=1}^{V}\sum_{j=1}^{H}(x_{i,j}-\hat{x}_{i,j})^{2}}$$

$$\text{MSE} = \frac{1}{HV}\sum_{i=1}^{V}\sum_{j=1}^{H}(x_{i,j}-\hat{x}_{i,j})^{2}$$

$$\text{MAD} = \frac{1}{HV}\sum_{i=1}^{V}\sum_{j=1}^{H}|x_{i,j}-\hat{x}_{i,j}|$$

where SNR = signal-to-noise ratio, PSNR = peak signal-to-noise ratio, MSE = mean-squared error, and MAD = mean absolute difference.

A more intuitive measure of quality would be a subjective measure. The (U.S.) National Image Interpretability Rating Scale (NIIRS) is a 10-level scale used to rate images for their usefulness in terms of resolution and clarity [12]. This scale is typically used to evaluate images acquired from airborne or space-based systems. For compression and enhancement applications a quality measure such as the five-point ITU-R BT.500-10 scale can be employed. Subjective quality is determined by taking a large sample of evaluations; that is, a large number of viewers rate the image on a selected scale, and the result is averaged. However, care must be taken in selecting evaluators, since trained viewers tend to notice different effects than do novices.

8. DIGITAL WATERMARKING AND DATA HIDING

With the growth in image transmission, reproduction, and storage capabilities, digital image information hiding and watermarking have become necessary tools for maintaining security and proving ownership of images. The primary goal in image watermarking and data hiding is to embed into an image an identifying mark that may or may not be readily identifiable to a viewer, but is easily detectable either when compared with the original image or with knowledge of a key. Typically, the watermark is spread throughout the image with the use of a pseudonoise key. Keys may be private or public; however, public keys permit the deletion of watermarks since the user can readily identify the location of the watermark and the method of watermarking. Because images may be stored, transmitted, copied, or printed, the watermark must be robust in the face of scanning, faxing, compression/decompression, transmission over a noisy channel, and the further addition of watermarks [13].

To effectively and imperceptibly embed data within an image, image components must be modified only slightly through the use of a key known to the watermark generator. Since many image operations, such as compression or filtering, tend to remove image components that are perceptually insignificant, the watermark must be embedded into perceptually important components. The image data are divided into perceptual components, such as the frequency or color components, and the watermark is embedded into one or more components depending on the components' robustness to distortion. Watermarking or data hiding may be performed in the spatial or transform domains. In the spatial domain one common watermarking technique is to amplitude-modulate a regular pattern of blocks of pixels by a small amount another is to increment or decrement the means of blocks. In the transform domain, the DFT, DCT, and DWT techniques are commonly used and information is embedded in the transform phase or by imposing relationships between transform coefficients.

BIOGRAPHY

Maja Bystrom received a B.S. in computer science and a B.S. in communications from Rensselaer Polytechnic Institute, Troy, New York, in 1991. She joined NASA-Goddard Space Flight Center where she worked as a computer engineer until August 1992. She then returned to Rensselaer where she received M.S. degrees in electrical engineering and mathematics, and a Ph.D. in electrical engineering. In 1997 she joined Drexel University, Philadelphia, Pennsylvania, as an assistant professor. She has received NSF CAREER and Fulbright awards. Her research interests are image processing for communications, and joint source-channel coding/decoding.

BIBLIOGRAPHY

1. A. K. Jain, *Fundamentals of Digital Image Processing*, Prentice-Hall, Englewood Cliffs, NJ, 1989.
2. M. A. Sid-Ahmed, *Image Processing: Theory, Algorithms, & Architectures*, McGraw-Hill, New York, 1995.
3. E. R. Dougherty, *An Introduction to Morphological Image Processing*, SPIE Optical Engineering Press, Bellingham, WA, 1992.
4. G. Strang and T. Nguyen, *Wavelets and Filter Banks*, 2nd ed., Wellesley-Cambridge Press, Wellesley, MA, 1997.

5. R. Nitzberg, *Radar Signal Processing and Adaptive Systems*, Artech House, Boston, 1999.
6. H. Wechsler, P. J. Phillips, V. Bruce, F. F. Soulie, and T. S. Huang, eds., *Face Recognition: From Theory to Applications*, Springer-Verlag, Berlin, 1998.
7. J. Weng, T. S. Huang, and N. Ahuja, *Motion and Structure from Image Sequences*, Springer-Verlag, Berlin, 1993.
8. A. K. Katsaggelos, ed., *Digital Image Restoration*, Springer-Verlag, Berlin, 1991.
9. G. Demoment, Image reconstruction and restoration: Overview of common estimation structures and problems, *IEEE Trans. Acoust. Speech Signal Process.* **37**: 2024–2036 (Dec. 1989).
10. R. Molina, A. K. Katsaggelos, and J. Mateos, Removal of blocking artifacts using a hierarchical bayesian approach, in A. Katsaggelos and N. Galatsanos, eds., *Signal Recovery Techniques for Image and Video Compression*, Kluwer, Boston, 1998.
11. W. G. Carrara, R. S. Goodman, and R. M. Majewski, *Spotlight Synthetic Aperture Radar: Signal Processing Algorithms*, Artech House, Boston, 1995.
12. J. Pike, National image interpretability rating scales (Jan. 10, 1998), Federation of American Scientists, *http://www.fas.org/irp/imint/niirs.htm* (posted Aug. 10, 2000); accessed 8/10/2000, updated 1/16/98.
13. S. Katzenbeisser and F. A. P. Petitcolas, eds., *Information Hiding Techniques for Stegnography and Digital Watermarking*, Artech House, Boston, 2000.

FURTHER READING

Aign S., Error concealment for MPEG-2 video, in A. Katsaggelos and N. Galatsanos, eds., *Signal Recovery Techniques for Image and Video Compression*, Kluwer, Boston, 1998.

Bhaskaran V. and K. Konstantinides, *Image and Video Compression Standards: Algorithms and Architectures*, Kluwer, Boston, 1997.

Bowyer K. and N. Ahuja, eds., *Advances in Image Understanding*, IEEE Computer Society Press, Los Alamitos, CA, 1996.

Giorgianni E. and T. Madden, *Digital Color Management: Encoding Solutions*, Addison-Wesley, Reading, MA, 1998.

Home site of the JPEG and JBIG committees, *http://www.jpeg.org* (Aug. 10, 2000).

Netravali A. and B. Haskell, *Digital Pictures: Representation, Compression and Standards*, 2nd ed., Plenum Press, New York, 1995.

Lindley C. A., *Practical Image Processing in C*, Wiley, New York, 1991.

Pennebaker W. B. and J. L. Mitchell, *JPEG Still Image Data Compression Standard*, Van Nostrand Reinhold, New York, 1993.

Rihaczek A. and S. Hershkowitz, *Radar Resolution and Complex-Image Analysis*, Artech House, Boston, 1996.

Russ J. C., *The Image Processing Handbook*, 2nd ed., CRC Press, Boca Raton, FL, 1995.

Sangwine S. J. and R. E. N. Horne, eds., *The Colour Image Processing Handbook*, Chapman & Hall, London, 1998.

Sezan M. I. and A. M. Tekalp, Survey of recent developments in digital image restoration, *Opt. Eng.* **29**: (May 1990).

Vetterli M. and J. Kovacevic, *Wavelets and Subband Coding*, Prentice-Hall, Upper Saddle River, NJ, 1995.

IMAGE SAMPLING AND RECONSTRUCTION

H. J. Trussell
North Carolina State University
Raleigh, North Carolina

1. INTRODUCTION

Images are the result of a spatial distribution of radiant energy. We see, record, and create images. The most common images are two-dimensional color images seen on television. Other everyday images include photographs, magazine and newspaper pictures, computer monitors, and motion pictures. Most of these images represent realistic or abstract versions of the real world. Medical and satellite images form classes of images where there is no equivalent scene in the physical world. Computer animation produces images that exit only in the mind of the graphic artist.

In the case of continuous variables of space, time, and wavelength, an image is described by a function

$$f(x, y, \lambda, t) \tag{1}$$

where x, y are spatial coordinates (angular coordinates can also be used), λ indicates the wavelength of the radiation, and t represents time. It is noted that images are inherently two-dimensional (2D) spatial distributions. Higher-dimensional functions can be represented by a straightforward extension. Such applications include medical CT and MRI, as well as seismic surveys. For this article, we will concentrate on the spatial and wavelength variables associated with still images. The temporal coordinate will be left for another chapter.

In order to process images on computers, the images must be sampled to create digital images. This represents a transformation from the analog domain to the discrete domain. In order to view or display the processed images, the discrete image must be transformed back into the analog domain. This article concentrates entirely on the very basic steps of sampling an image in preparation for processing and reconstructing or displaying an image. This may seem to be a very limited topic but let us consider what will not be covered in this limited space.

We introduced images as distributions of radiant energy. The exact representation of this energy and its measurement is the subject of radiometry. For this article, we will ignore the physical representation of the radiant source. We will treat the image as if everyone knows what the value of $f(x, y)$ means and how to interpret the two-dimensional gray-level distributions that will be used in this chapter to demonstrate various principles.

If we include the frequency or wavelength distribution of the energy, we can discuss spectrometry. Images for most consumer and commercial uses are the color images that we see everyday. These images are transformations of continuously varying spectral, temporal and spatial distributions. In order to fully understand the effects of sampling and reconstruction of color images, more understanding of the human visual system is required than can be presented here. Satellite images are now being recorded in multispectral and hyperspectral bands. In this terminology, a hyperspectral image has more than 20 bands. We will only touch on the basics of color sampling.

All images exist in time and change with time. We're all familiar with the stroboscopic effects that we see in the movies that make car wheels and airplane propellers appear to move backward.[1] The same sampling principles can be used to explain these phenomena as will be used to explain the spatial sampling that is presented here. The description of object motion in time and its effect on images is another rich topic that will be omitted here.

Before presenting the fundamentals of image presentation, it necessary to define our notation and to review the prerequisite knowledge that is required to understand the following material. A review of rules for the display of images and functions is presented in Section 2, followed by a review of mathematical preliminaries and sampling effects in Section 3. Section 4 discusses simpling on a nonrectangular lattice. The practical case of using a finite aperture in the sampling process is presented in Section 5. Section 6 reviews color vision and describes multidimensional sampling with concentration on sampling color spectral signals. We will discuss the fundamental differences between sampling the wavelength and spatial dimensions of the multidimensional signal. Finally, Section 7 contains a mathematical description of the display of multidimensional data. This area is often neglected by many texts. The section will emphasize the requirements for displaying data in a fashion that is both accurate and effective.

2. PRELIMINARY NOTES ON DISPLAY OF IMAGES

One difference between 1D and 2D functions is the way they are displayed. One-dimensional functions are easily displayed in a graph where the scaling is obvious. The observer need examine only the numbers that label the axes to determine the scale of the graph and get a mental picture of the function. With two-dimensional scalar-valued functions, the display becomes more complicated. The accurate display of vector-valued two-dimensional functions, including color images, will be discussed after covering the necessary material on sampling and colorimetry.

Two-dimensional functions can be displayed as an isometric plot, a contour plot, or a grayscale plot. Since we are dealing with images, we will use the grayscale plot for images and the isometric plot for functions. All three types are supported by MATLAB [1]. The user should choose the right display for the information to be conveyed. For the images used in this article, we should review some basic rules for display.

Consider a monochrome image that has been digitized by some device, such as a scanner or camera. Without knowing the physical process that created the image, it is impossible to determine the best way to display the image.

[1] We used to use the example of stagecoach wheels moving backward, but, alas, there are few Western movies anymore. Time marches on.

The proper display of images requires calibration of both the input and output devices [15,16]. This is another topic that must be omitted for lack of space. For now, it is reasonable to give some general rules about the display of monochrome images:

1. For the comparison of a sequences of images, it is *imperative* that all images be displayed using the same scaling.
2. Display a step-wedge, a strip of sequential gray levels from minimum to maximum values, with the image to show how the image gray levels are mapped to brightness or density. This allows some idea of the quantitative values associated with the pixels.
3. Use a graytone mapping which allows a wide range of gray levels to be visually distinguished. In software such as MATLAB, the user can control the mapping between the continuous values of the image and the values sent to the display device. It is recommended that adjustments be made so that a user is able to distinguish all levels of a step-wedge of about 32 levels.

3. SPATIAL SAMPLING

In most cases, the multidimensional process can be represented as a straightforward extension of one-dimensional processes. Thus, it is reasonable to mention the one-dimensional operations that are prerequisite to understanding this article and will form the basis of the mutidimensional processes.

3.1. Ideal Sampling in One Dimension

Mathematically, ideal sampling is usually represented with the use of a *generalized function*, the Dirac delta function, $\delta(t)$ [2]. The function is defined as zero for $t \neq 0$ and having an area of unity. The most useful property of the delta function is that of sifting, for instance, extracting single values of a continuous function. This is defined by the integral

$$s(t_0) = \int_{-\infty}^{\infty} s(t)\delta(t - t_0)\,dt = \int_{-\infty}^{\infty} s(t_0)\delta(t - t_0)\,dt \quad (2)$$

This shows the production of a single sample. We would represent the sampled signal as a signal that is zero everywhere except at the sampling time, $s_{t_0}(t) = s(t_0)\delta(t - t_0)$. The sampled signal can be represented graphically by using the arrow, as shown in Fig. 1.

The entire sampled sequence can be represented using the *comb* function

$$comb(t) = \sum_{n=-\infty}^{\infty} \delta(t - n) \quad (3)$$

where the sampling interval is unity. The sampled signal is obtained by multiplication

$$s_d(t) = s(t)comb(t) = s(t)\sum_{n=-\infty}^{\infty} \delta(t - n) = \sum_{n=-\infty}^{\infty} s(t)\delta(t - n) \quad (4)$$

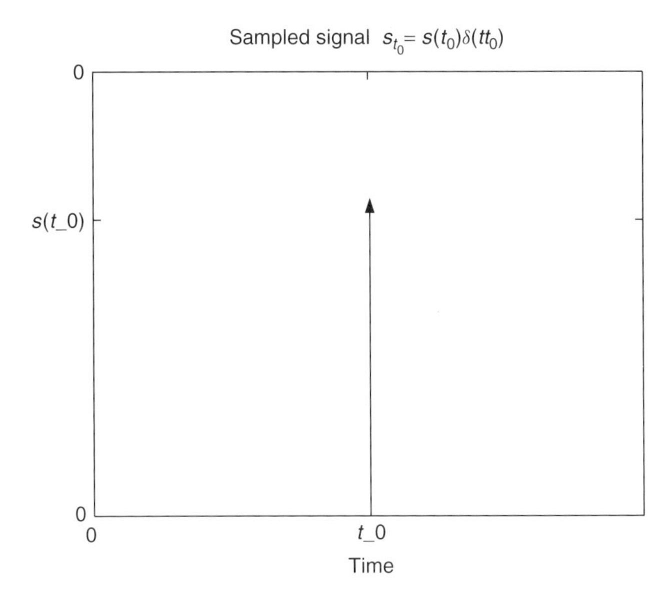

Figure 1. Sampled signal at $t = t_0$.

The sampling is represented graphically in Fig. 2. It is common to use the notation of $\{s(n)\}$ or $s(n)$ to represent the collection of samples in discrete space. The arguments n and t will serve to distinguish the discrete or continuous spaces, respectively.

The 1D effects in the frequency domain are shown in most undergraduate signals and systems texts. Briefly, we will review this graphically by considering the frequency domain representation of the signals in Fig. 2. The spectrum of the analog signal, $s(t)$ is denoted $S(\omega)$; the Fourier transform of the *comb(t)* is $2\pi comb(\omega)$.[2] The frequency-domain representation of the 1D sampling

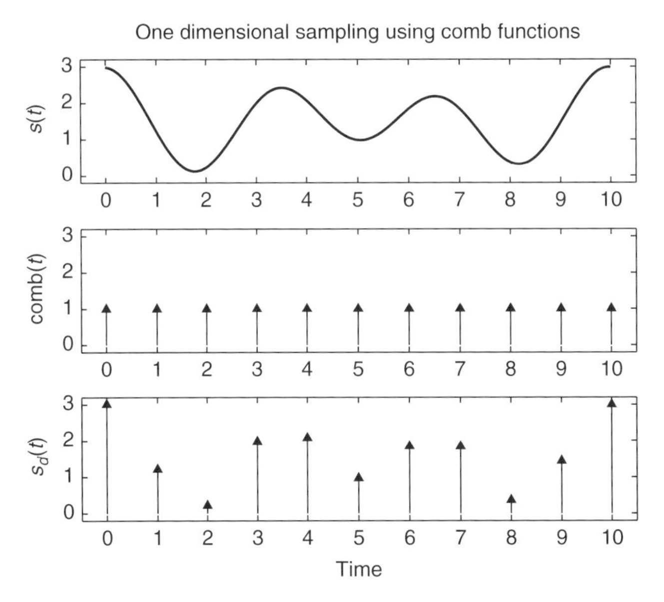

Figure 2. One-dimensional sampling.

[2] The proof of this is also available in the undergraduate signals and systems texts [e.g., 2].

process is shown in Fig. 3. The spectra in this figure correspond to the time-domain signals in Fig. 2. The most important feature of the spectrum of the sampled signals is the replication of the analog spectrum. Mathematically, if the spectrum of $s(t)$ is denoted $S(\omega)$, then the spectrum of the sampled signal, $s_d(t)$, is given by

$$S_d(\omega) = \sum_{k=-\infty}^{\infty} S(\omega - k2\pi F_s)$$

where F_s is the sampling rate. Note that reconstruction is possible only if there is no overlap of the replicated spectra. This, of course, corresponds to having a sampling rate that is greater than twice the highest frequency in the analog signal, $F_{\max}$, namely, $F_s > 2F_{\max}$.

From the frequency domain figures, it is easy to see that reconstruction of the original signal requires that the fundamental spectrum, the one centered at zero, be retained, while the replicated spectra by eliminated. In the time domain, this can be accomplished by passing the sampled signal through a lowpass filter. While ideal lowpass filters are not possible, it is possible to realize sufficiently good approximations that the reconstruction is close enough to ideal for practical applications. This is a major difference with two-dimensional image reproduction. There is no equivalent analog low pass filter that can be used with optical images. This will be addressed in a later section.

If the sampling rate is not adequate, then the original signal cannot be reconstructed from the sample values. This is seen by considering the samples of a sinusoid of frequency, F, which are given by

$$s_F(n) = \cos\frac{2\pi Fn}{F_s} + \theta\cos\left(\frac{2\pi Fn}{F_s} + \theta\right)$$

where F_s is the sampling rate and θ is the phase of the sinusoid. The samples are taken at $t_n = n/F_s$. We see that the samples are the same for all frequencies $F = F_m$ that are related to the sampling frequency by $F_m = F_0 + mF_s$. The samples of these sinusoids are all identical to those of the sinusoid of frequency F_0. We will refer to F_0 as an alias of the frequencies F_m under the sampling rate of F_s.

3.2. Ideal Sampling in Two Dimensions

The two-dimensional Dirac delta function can be defined as the separable product of one-dimensional delta functions, $\delta(x, y) = \delta(x)\delta(y)$. The extension of the comb function to two dimensions should probably be called a "brush," but we will continue to use the term comb and define it by

$$comb(x, y) = \sum_{m=-\infty}^{\infty}\sum_{n=-\infty}^{\infty} \delta(x - m, y - n)$$

The equation for 2D sampling is

$$s(m, n) = s_d(x, y) = s(x, y)comb(x, y) \tag{5}$$

where a normalized sampling interval of unity is assumed. We have the same constraints on the sampling rate in two

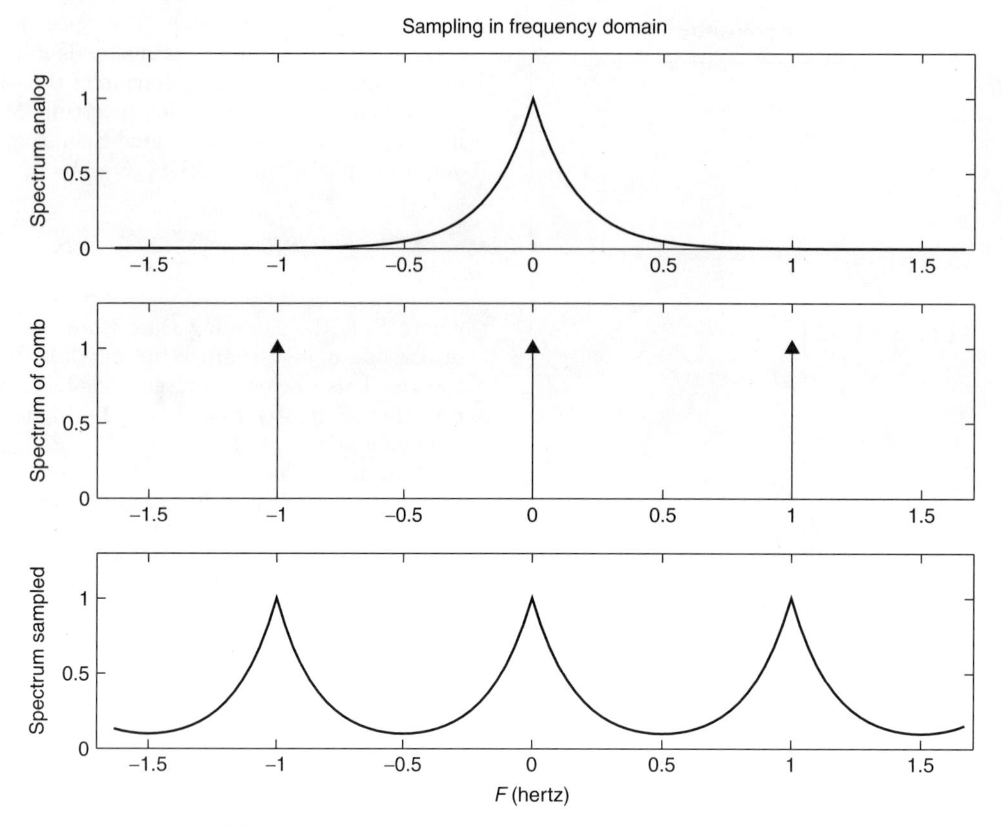

Figure 3. One-dimensional sampling: frequency domain.

dimensions as in one. Of course, the frequency is measured not in hertz, but in cycles per millimeter or inch.[3] Spatial sampling is illustrated in Figs. 4–6. Undersampling in the spatial domain signal results in spatial aliasing. This is easy to demonstrate using simple sinusoidal images. First, let us consider the mathematics.

Taking Fourier transforms of Eq. (5) yields

$$S_d(u,v) = S(u,v) * comb(u,v) = \sum_{k=-\infty}^{\infty}\sum_{l=-\infty}^{\infty} S(u-k, v-l) \tag{6}$$

where the asterisk ($*$) denotes convolution. Note that the sampled spectrum is periodic, as in the one-dimensional case.

Consider the effect of changing the sampling interval

$$s(m,n) = s(x,y) comb\left(\frac{x}{\Delta x}, \frac{y}{\Delta y}\right) \tag{7}$$

which yields

$$\begin{aligned} S_d(u,v) &= S(u,v) * comb(u,v) \\ &= \frac{1}{|\Delta x \Delta y|}\sum_{k=-\infty}^{\infty}\sum_{l=-\infty}^{\infty} S(u\Delta x - k, v\Delta y - l) \end{aligned} \tag{8}$$

Figure 7 shows the spectrum of a continuous analog image. If the image is sampled with intervals of δx in each direction, the sampled image has a spectrum that shows periodic replications at $1/\delta x$. The central portion of the periodic spectrum is shown in Fig. 8. For Fig. 8, we have used $\delta x = \frac{1}{30}$ mm.

Note that if the analog image, $s(x, y)$, is bandlimited to some 2D region, it is possible to recover the original signal from the samples by using an ideal lowpass filter. The proper sampling intervals are determined from the requirement that the region of support in the frequency domain (band limit) is contained in the rectangle defined by $|u| \le (1/2\Delta x)$ and $|v| \le (1/2\Delta\, y)$.

The effect of sampling can be demonstrated in the following examples. In these examples, aliasing will be demonstrated by subsampling a high-resolution digital image to produce a low resolution image. First let us consider a pure sinusoid. The function

$$s(x,y) = \cos\left[2\pi\left(\frac{36x}{128} + \frac{24y}{128}\right)\right]$$

where x is measured in mm, is sampled at 1 mm spacing in each direction. This produces no aliasing. The function and its spectrum are shown in Figs. 9 and 10, respectively.[4] Note that the frequency of the spectrum is in normalized

[3] Image processors use linear distance most often, but occasionally use angular measurement, which yields cycles per degree. This is done when considering the resolution of the eye or an optical system.

[4] The spectra of the sinusoids appear as crosses instead of points because of the truncation of the image to a finite region. The full explanation is beyond the scope of this article.

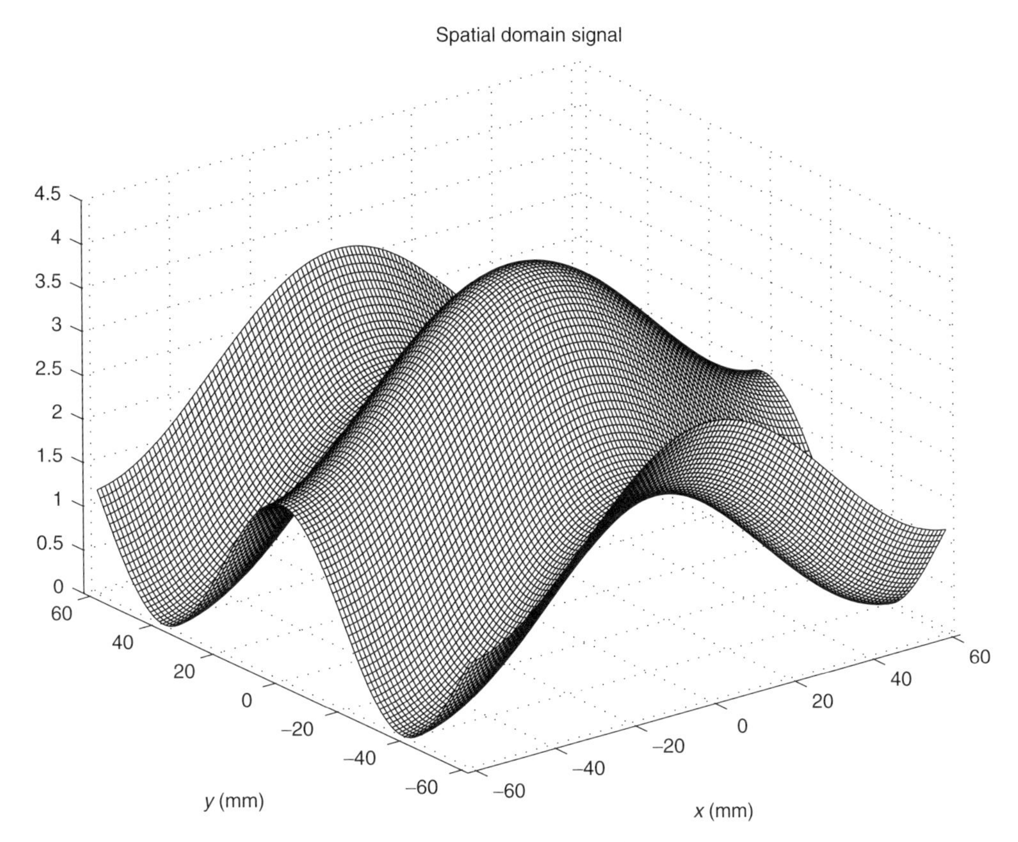

Figure 4. Two-dimensional analog signal.

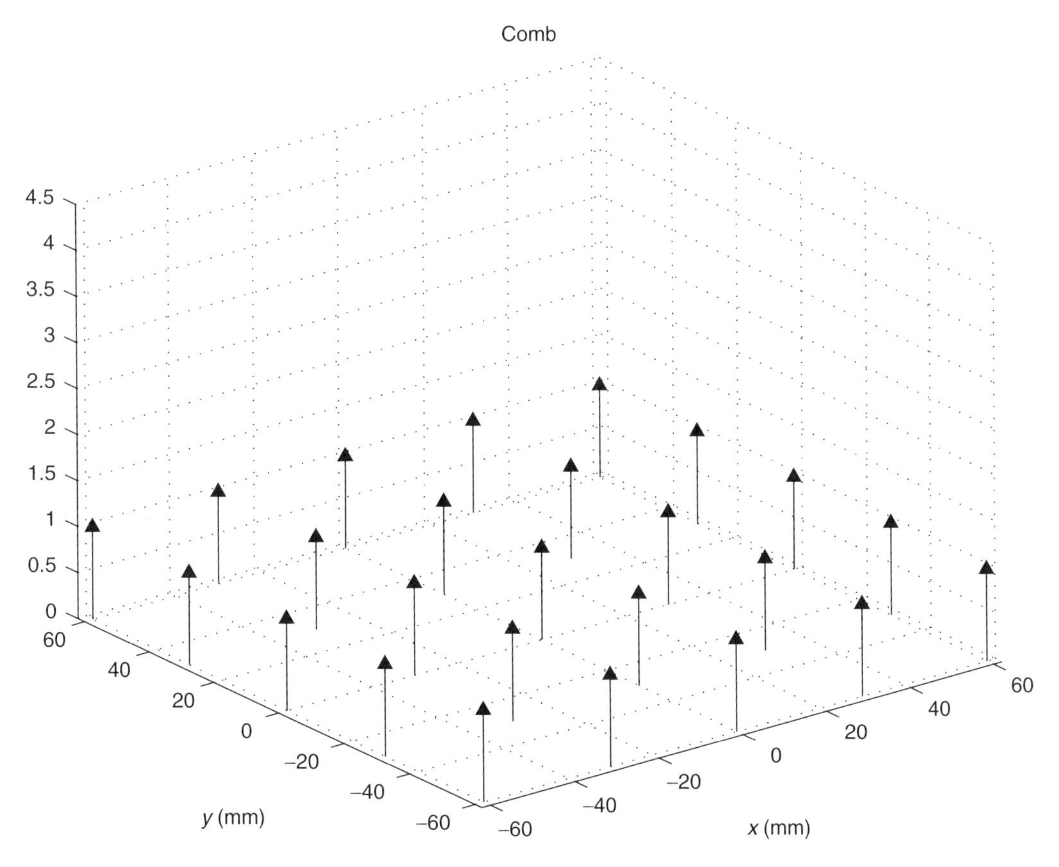

Figure 5. Two-dimensional comb.

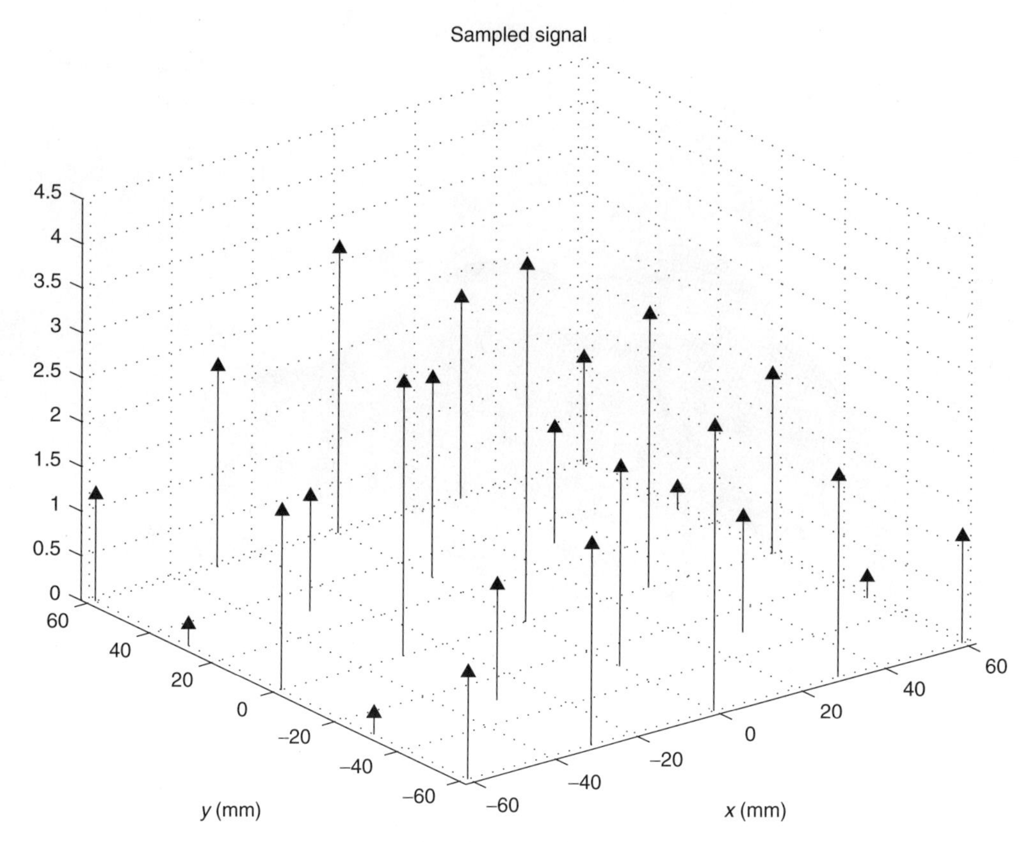

Figure 6. Two-dimensional sampled signal.

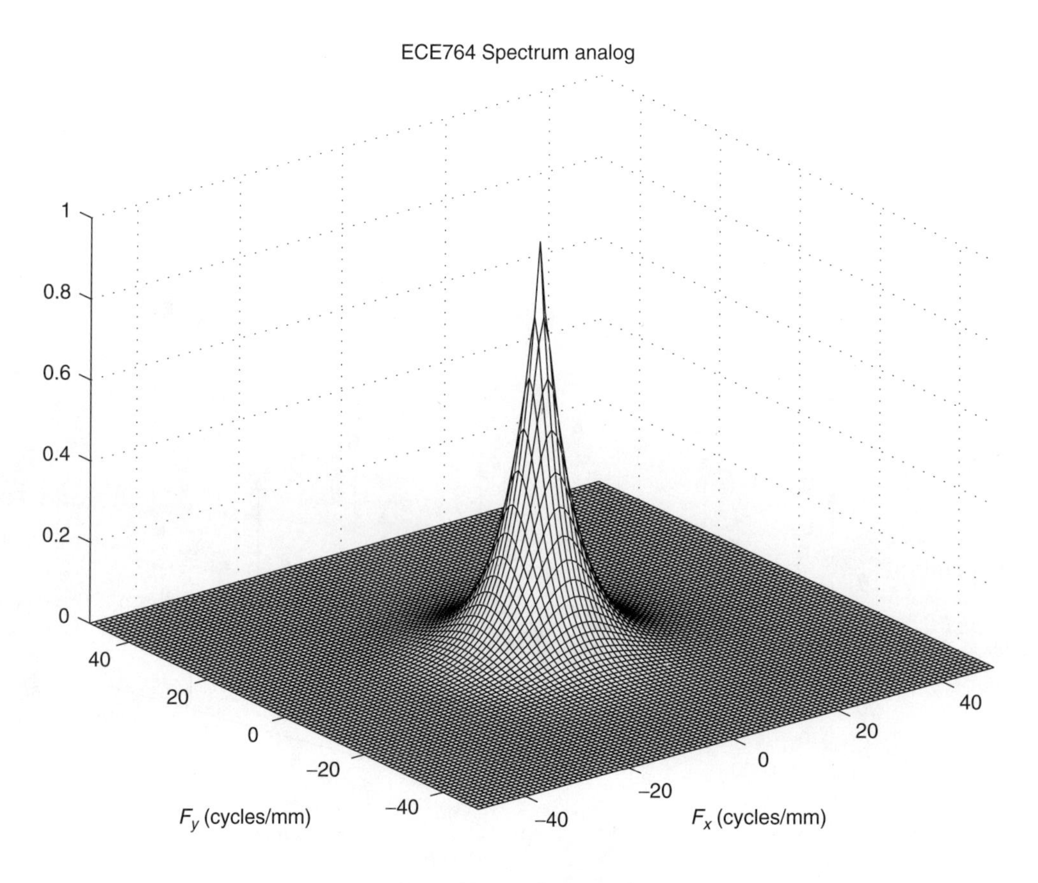

Figure 7. Analog spectrum.

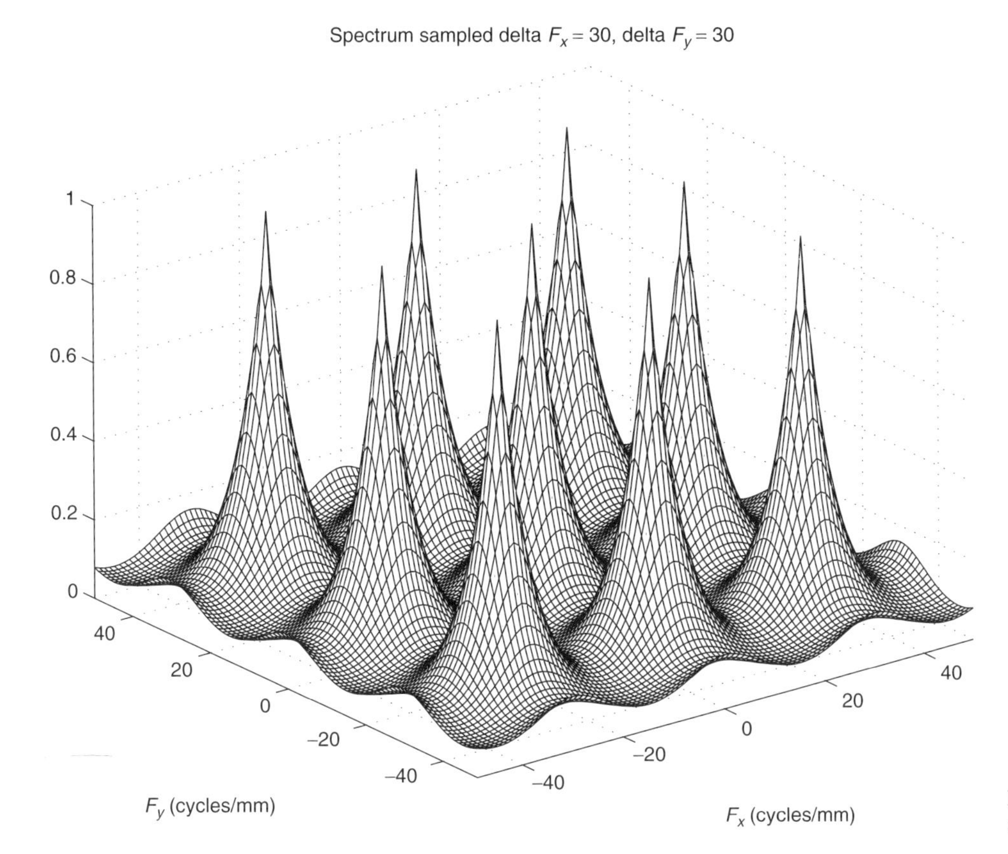

Figure 8. Digital spectrum with aliased images of analog spectrum.

digital frequency.[5] For this case, the analog frequency, F and the digital frequency, f, are the same. This yields $F_x = f_x = \frac{36}{128} = 0.28125$ and $F_y = f_4 = \frac{24}{128} = 0.1875$. The spectrum shows peaks at this 2D frequency. The image is subsampled by a factor of 4 and shown in Fig. 11. This is equivalent to a sampling of the analog signal with an interval of 4mm in each direction. The aliased 2D frequency can be found by finding k and l so that both $|F_x - kF_s| < 0.5F_s$ and $|F_y - lF_s| < 0.5F_s$ hold. For this case, $k = l = 1$ and the aliased analog 2D frequency is $(F'_x, F'_y) = (0.03125, -0.0625)$. This means that the function

$$s'(x, y) = \cos[2\pi(0.03125x - 0.0625y)]$$

will yield the same samples as $s(x, y)$ above when sampled at 4mm intervals in each direction. The spectrum of the sampled signal is shown in Fig. 12. The digital frequencies can be found by normalizing the aliased analog frequency by the sampling rate. For this case, $(f'_x, f'_y) = (0.03125/0.25, -0.0625/0.25) = (0.125, -0.25)$.

An example of sampling a pictorial image is shown in Figs. 13–16, where Fig. 13 is the original; Fig. 14 is its spectrum; Fig. 15 is a 2:1 subsampling of the original; Fig. 16 is the spectrum of the subsampled image. For this case, we can see that the lower frequencies have been preserved but the higher frequencies have been aliased.

[5] Normalized digital frequency is denoted by F and has the constraint $|F| \leq \frac{1}{2}$ [2].

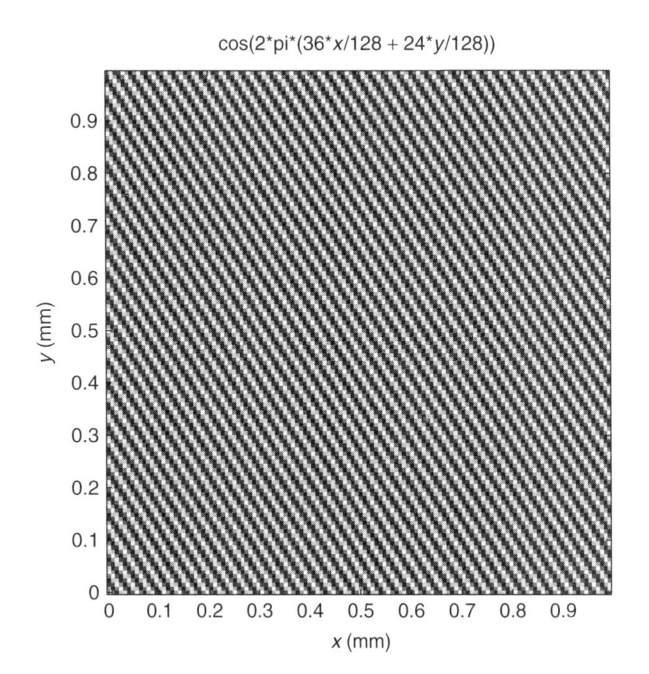

Figure 9. $\cos[2\pi(36x/128 + 24y/128)]$.

4. SAMPLING ON NONRECTANGULAR LATTICES

Because images may have oddly shaped regions of support in the frequency domain, it is often more efficient to sample with a nonrectangular lattice. A thorough discussion of this concept is found in the book by Dudgeon and Mersereau [3]. To develop this concept, it is convenient to

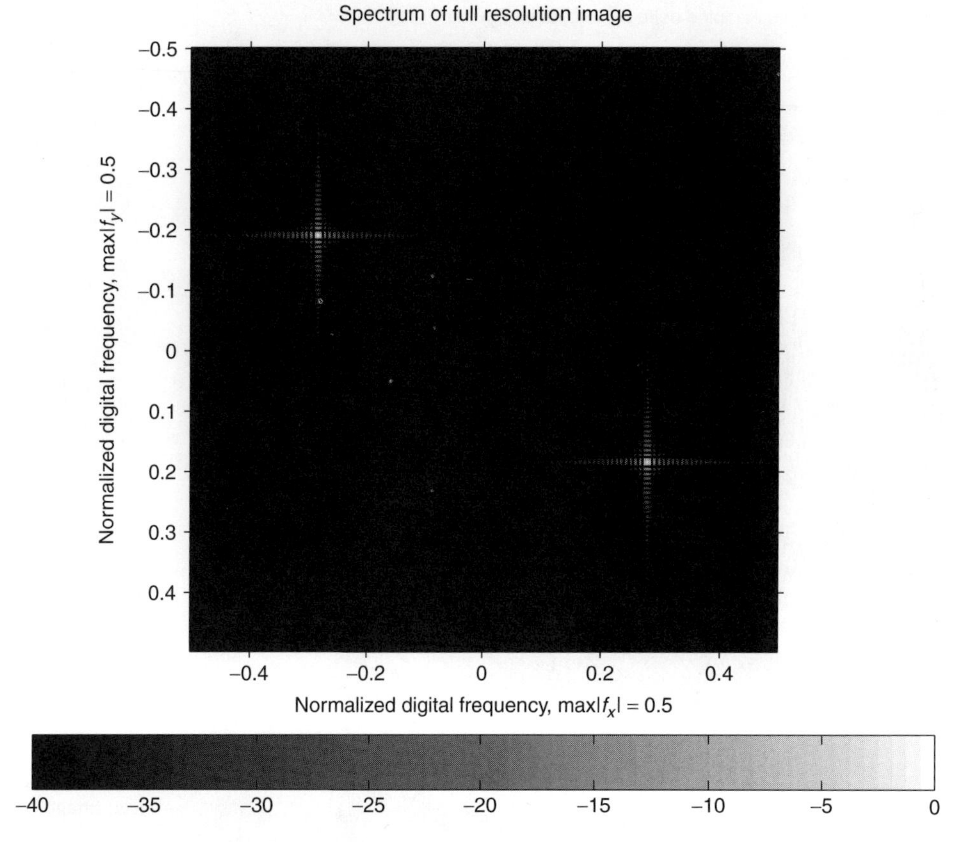

Figure 10. Spectrum of $\cos[2\pi(36x/128 + 24y/128)]$.

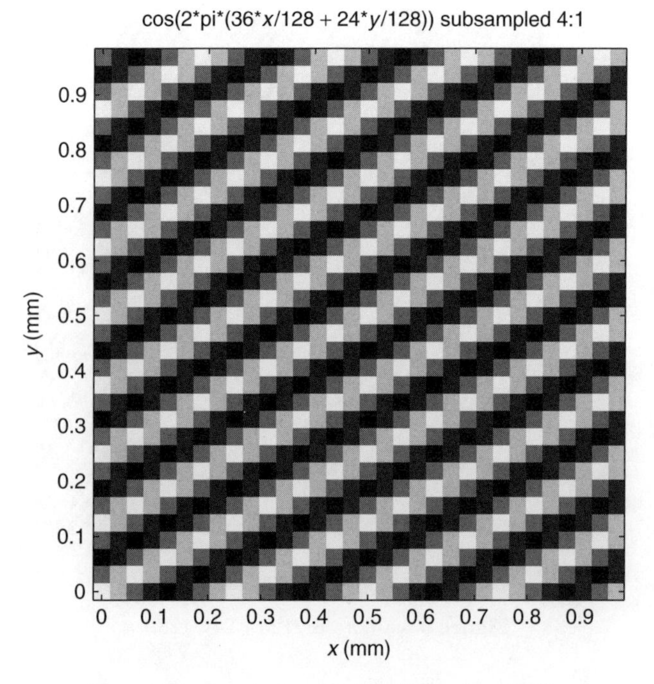

Figure 11. Sampled $\cos[2\pi(36x/128 + 24y/128)]$.

write the sampling process in vector form. Let $\mathbf{x} = [x, y]$ and the basis vectors for sampling in the space domain be given by $\mathbf{x}_1$ and $\mathbf{x}_2$. The sampling function or comb can be written

$$comb(\mathbf{r}) = \sum_{m=-\infty}^{\infty} \sum_{n=-\infty}^{\infty} \delta(\mathbf{r} - m\mathbf{x}_1 - n\mathbf{x}_2) \tag{9}$$

This yields the functional form

$$s(m, n) = s(m\mathbf{x}_1 + n\mathbf{x}_2) \tag{10}$$

writing this in matrix form

$$s(\mathbf{n}) = s(\mathbf{Xn}) \tag{11}$$

where $\mathbf{X} = [\mathbf{x}_1, \mathbf{x}_2]$ and $\mathbf{n} = [m, n]$.

The basis vectors in the frequency domain, $\mathbf{w}_1$ and $\mathbf{w}_2$, are defined by the relation

$$\mathbf{x}_k \mathbf{w}_l^T = \delta(k - l) \tag{12}$$

or using matrix notation

$$\mathbf{X}^T\mathbf{W} = \mathbf{I} \tag{13}$$

The Fourier transform in matrix notation is written

$$S(\mathbf{w}) = \int_{-\infty}^{\infty}\int_{-\infty}^{\infty} s(\mathbf{x}) \exp(-j2\pi\mathbf{W}^T\mathbf{x})\, d\mathbf{x} \tag{14}$$

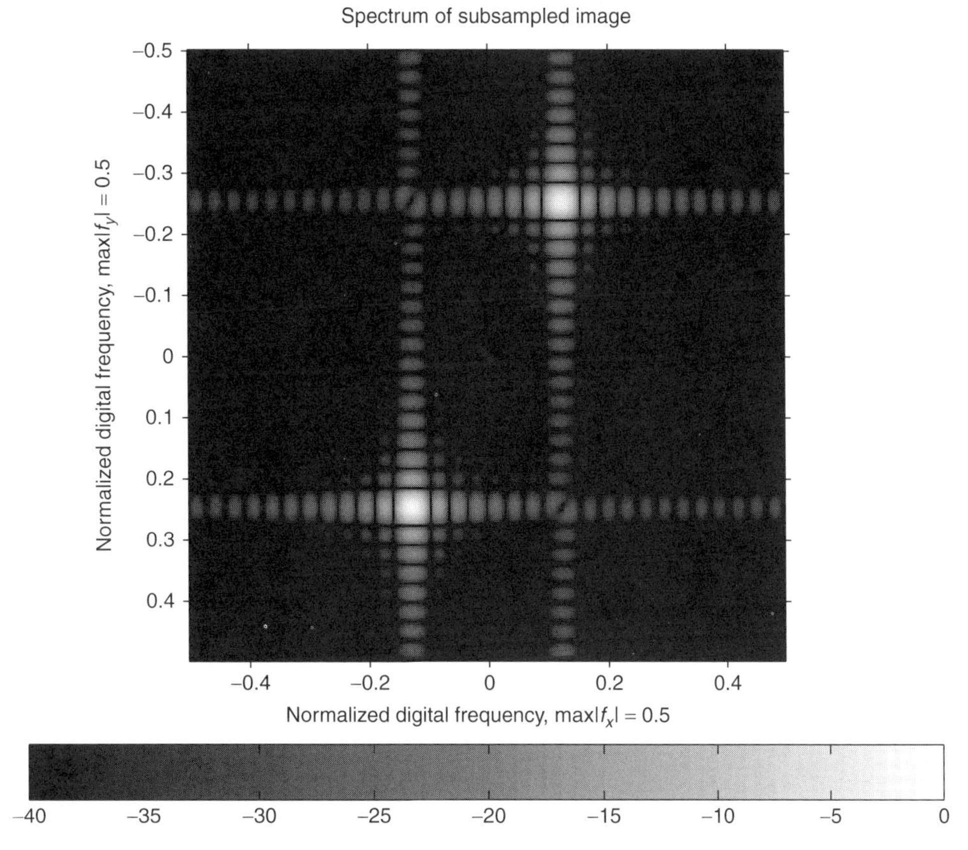

Figure 12. Spectrum of Subsampled $\cos[2\pi(36x/128 + 24y/128)]$.

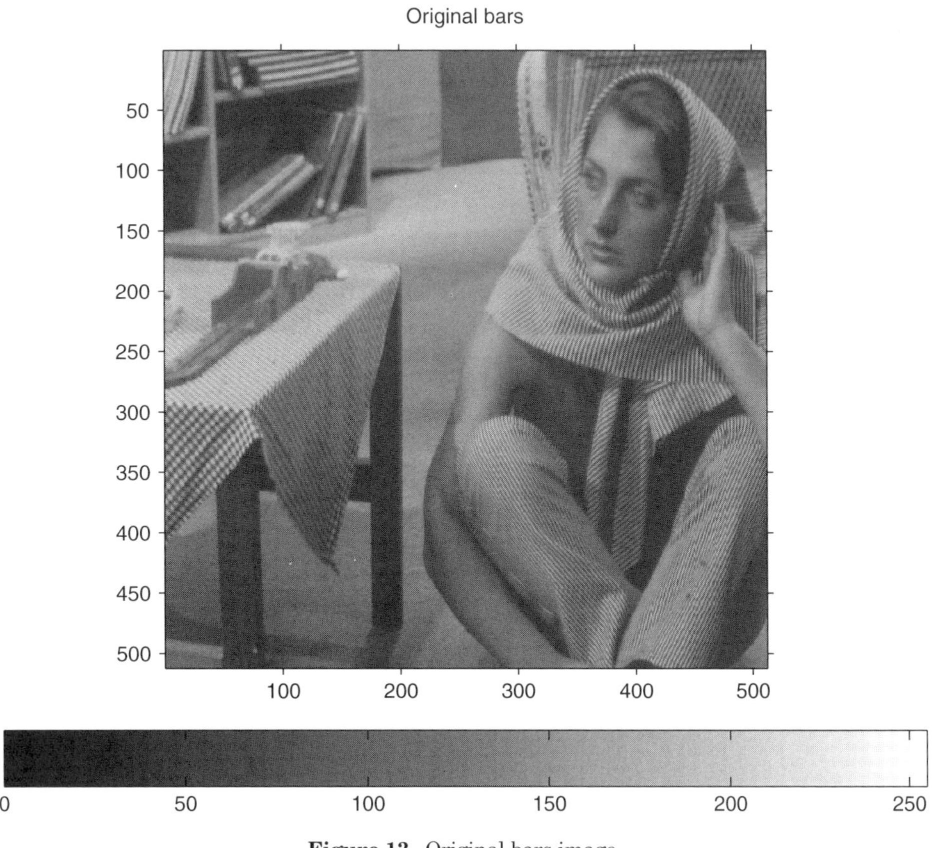

Figure 13. Original bars image.

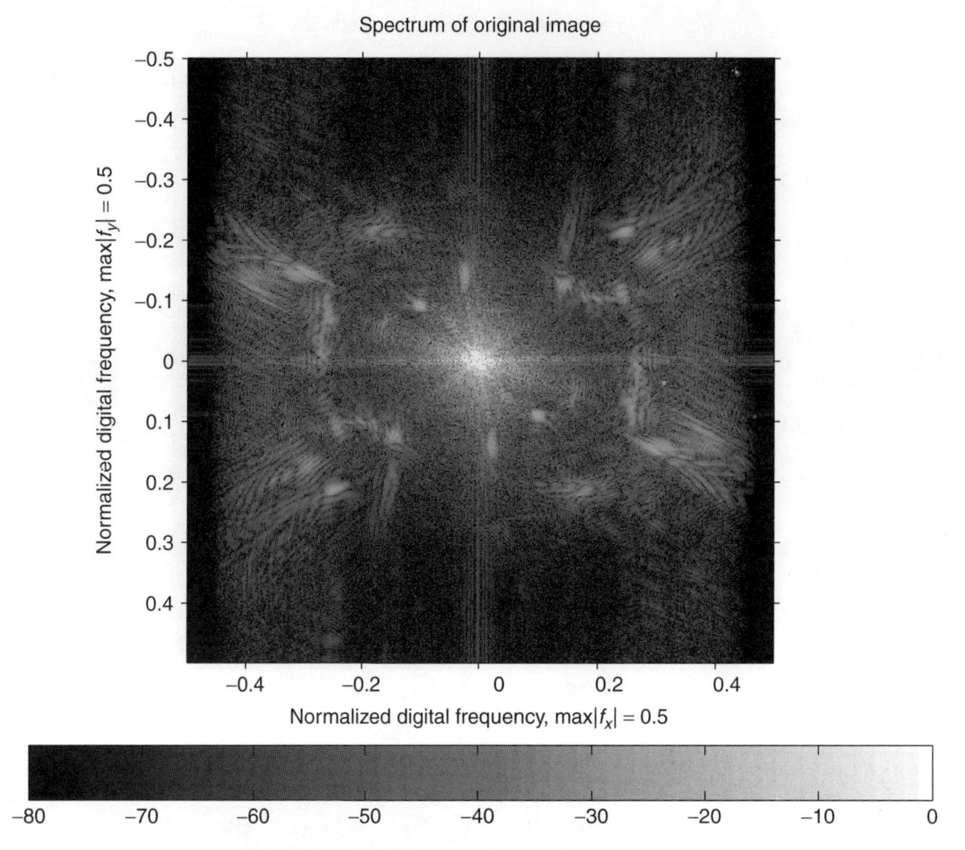

Figure 14. Spectrum of original bars image.

Figure 15. Sampled bars image 2 : 1.

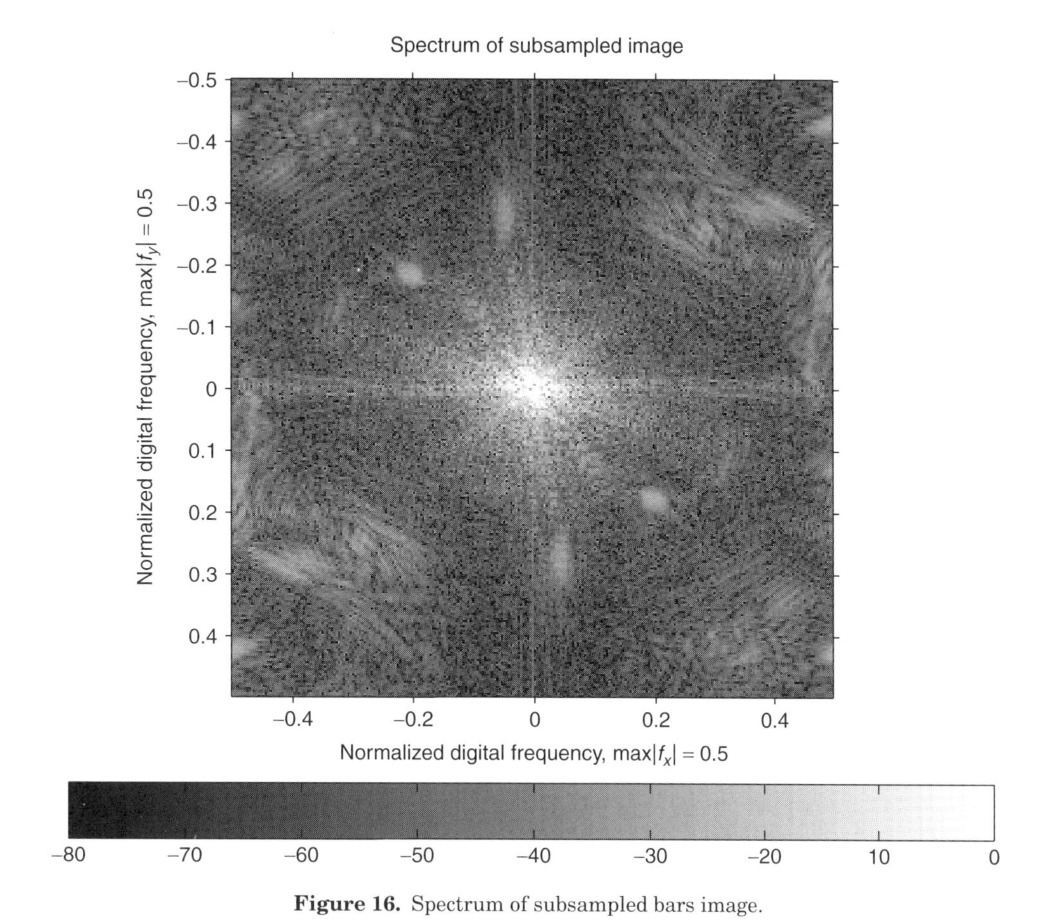

Figure 16. Spectrum of subsampled bars image.

The sampled spectrum can be written

$$\begin{aligned} S(\mathbf{w}) &= S(\mathbf{w}) * comb(\mathbf{w}) \\ &= \frac{1}{|\mathbf{X}|} \sum_{k=-\infty}^{\infty} \sum_{l=-\infty}^{\infty} S(\mathbf{w} - k\mathbf{w}_1 - l\mathbf{w}_2) \end{aligned} \tag{15}$$

(see the books by Dudgeon and Mersereau [3] and Jain [8]).

5. SAMPLING USING FINITE APERTURES

Practical imaging devices, such as videocameras, CCD (charge-coupled device) arrays, and scanners, must use a finite aperture for sampling. The *comb* function cannot be realized by actual devices. The finite aperture is required to obtain a finite amount of energy from the scene. The engineering tradeoff is one of signal-to-noise ratio (SNR) versus spatial resolution. Large apertures, receive more light, and thus, will have higher SNR's than smaller apertures, while smaller apertures permit higher spatial resolution than will larger ones. This is true for apertures larger than the order of the wavelength of light. For smaller apertures, diffraction limits the resolution.

The aperture may cause the light intensity to vary over the finite region of integration. For a single sample of a one-dimensional signal at time nT, the sample value can be obtained by

$$s(n) = \int_{(n-1)T}^{nT} s(t)a(t - NT)\,dt \tag{16}$$

where $a(t)$ represents the impulse response (or light variation) of the aperture. This is simple correlation and assumes that the same aperture is used for every sample. The mathematical representation can be written as convolution if the aperture is symmetric or, we replace the function $a(t)$ with $a(-t)$. The sampling of the signal can be represented by

$$s(n) = [s(t) * a(t)]comb\frac{t}{T} \tag{17}$$

where * represents convolution. This model is reasonably accurate for spatial sampling of most cameras and scanning systems.

The sampling model can be generalized to include the case where each sample is obtained with a different aperture. For this case, the samples which need not be equally spaced, are given by

$$s(n) = \int_{l_n}^{u_n} s(t)a_n(t)\,dt \tag{18}$$

where the limits of integration correspond to the region of support for each aperture. A common application of this representation in two dimensions is the finite area of a CCD element of an imaging chip. The aperture function $a(t)$ may also take into account the leakage of charge from one cell to another. Equation (18) is also important in representing sampling the wavelength dimension of the image signals.

The generalized signal reconstruction equation has the form

$$s(t) = \sum_{n=-\infty}^{\infty} s(n)g_n(t) \tag{19}$$

where the collection of functions, $\{g_n(t)\}$, provide the interpolation from discrete to continuous space. The exact form of $\{g_n(t)\}$ depends on the form of $\{a_n(t)\}$. For sampling using the ideal *comb* function with a sample interval of T, $g_n(t)$ is a shift of the sinc function that represents the ideal band-limited filter

$$g_n(t) = \frac{\sin(2\pi(t - nT)/T)}{2\pi(t - nT)/T} \tag{20}$$

The two-dimensional aperture in the mathematical model of sampling can be written

$$s(m, n) = [s(x, y) * a(x, y)]comb(x, y) \tag{21}$$

where $*$ represents convolution and $a(x, y)$ represents the aperture. Note that using the finite aperture model can be written

$$s(m, n) = \iint_A s(x - m, y - n)a(x, y)\, dx\, dy \tag{22}$$

where the 2D integral is taken over the region of support of the aperture denoted by A. This equation is actually a correlation. The model can be written as a convolution with a space reversed aperture, $a_r(x, y) = a(-x, -y)$:

$$s(m, n) = \iint_A s(m - x, n - y)a_r(x, y)\, dx\, dy \tag{23}$$

For a symmetric aperture, which is most often the case in optical systems, $a_r(x, y) = a(x, y)$. Commonly used apertures include circular disks, rectangles, and Gaussians. Note that these have some symmetry that permits the substitution of convolution for correlation.

The Fourier representation of the sampled image is now given by

$$\begin{aligned} S_d(u, v) &= S(u, v)A(u, v) * comb(u, v) \\ &= \sum_{k=-\infty}^{\infty} \sum_{l=-\infty}^{\infty} S_a(u - k, v - l)A(u - k, v - l) \end{aligned} \tag{24}$$

With a finite aperture, the band-limited function to be sampled is the convolution $s(x, y) * a(x, y)$. The common aperture functions are generally lowpass in character; thus, the sampled function is more nearly band-limited. The aperture is effectively a filter. While aliasing caused by undersampling is diminished, the resultant spectrum is still a distorted version of the original.

The recovery of the original image must not only eliminate the periodic replications of the product $S(u, v)A(u, v)$ but also compensate for the effect of the aperture. The straightforward approach is to filter the sampled image with a kernel of the form

$$H(u, v) = \frac{1}{A(u, v)} \tag{25}$$

The problem with this approach is that the spectrum $A(u, v)$ often has values that are very small or zero which makes the inverse filter ill-conditioned; that is, $H(u, v)$ will have very large values that will amplify noise. Since most apertures are lowpass, the small or zero values usually occur at higher frequencies. A common modification of the above correction is to include a term to make the filter well-conditioned. Such a form is given by

$$H(u, v) = \frac{H_{lp}(u, v)}{A(u, v)} \tag{26}$$

where $H_{lp}(u, v)$ is a lowpass filter.

6. COLOR SAMPLING

There is a fundamental difference of philosophy about sampling in the wavelength domain from that of sampling in the spatial domain. To understand this difference, it is necessary to describe some of the fundamentals of color vision and color measurement. A more complete description of the human color visual system can be found in the books by Wandell [13] and Wyszecki and Stiles [14].

The retina contains two types of light sensors, rods, and cones. The rods are used for monochrome vision at low light levels; the cones are used for color vision at higher light levels. There are three types of cones. Each type is maximally sensitive to a different part of the spectrum. They are often referred to as long, medium, and short wavelength regions. A common description refers to them as red, green, and blue cones, although their maximal sensitivity is in the yellow, green, and blue regions of the spectrum. The visible spectrum extends from about 400 nm (blue) to about 700 nm (red).

Grassmann formulated a set of laws for additive color mixture in 1853 [5,6,15]. In addition, Grassmann conjectured that any additive color mixture could be matched by the proper amounts of three primary stimuli. Considering what was known about the physiology of the eye at that time, these laws represent considerable insight.[6] There have been several papers which have taken a linear systems approach to describing Grassmann's laws and color spaces as defined by a standard human observer, [4,7,10,12]. For the purposes of this work, it is sufficient to note that the spectral responses of the three types of sensors are sufficiently different so as to define a three-dimensional vector space. This three-dimensional representation is the basic principle of color displays in television, motion pictures, and computer monitors.

The mathematical model for the color sensor of a camera or the human eye can be represented by

$$v_k = \int_{-\infty}^{\infty} r(\lambda)m_k(\lambda)d\lambda, \; k = 1, 2, 3 \tag{27}$$

where $r(\lambda)$ is the radiant distribution of light as a function of wavelength and $m_k(\lambda)$ is the sensitivity of the kth color

[6] The laws are not exact and there is considerable debate among color scientists today about their most accurate form.

sensor. The sensitivity functions of the eye are shown in any of the references [9–14].

Sampling of the radiant power signal associated with a color image can be viewed in at least two ways. If the goal of the sampling is to reproduce the spectral distribution, then the same criteria for sampling the usual electronic signals can be directly applied. However, the goal of color sampling is not often to reproduce the spectral distribution but to allow reproduction of the color sensation. The goal is to sample the continuous color spectrum in such a way that the color sensation of the spectrum can be reproduced by the monitor. To keep this discussion as simple as possible, we will treat the color sampling problem as a subsampling of a high-resolution discrete space, that is, the N samples are sufficient to reconstruct the original spectrum using the uniform sampling of Section 3.

Let us assume that the visual wavelength spectrum is sampled finely enough to allow the accurate use of numerical approximation of integration. A common sample spacing is 10 nanometers over the range 400–700 nm. Finer sampling is required for some illuminants with line emitters. Sampling of color signals is discussed in detail in Ref. 9. With the assumption of proper sampling, the space of all possible visible spectra lies in an N-dimensional vector space, where $N = 31$. The spectral response of each of the eye's sensors can be sampled as well, giving three linearly independent N vectors that define the visual subspace.

Under the assumption of proper sampling, the integral of Eq. (27) can be well approximated by a summation

$$v_k = \sum_{n=L}^{U} r(n\Delta\lambda) m_k(n\Delta\lambda) \tag{28}$$

where $\Delta\lambda$ represents the sampling interval and the summation limits are determined by the region of support of the sensitivity of the eye. The sensor $m_k(\cdot)$ can represent the eye as well as a photonic device.

The response of the sensors can be represented by a matrix, $\mathbf{M} = [\mathbf{m}_1, \mathbf{m}_2, \mathbf{m}_3]$, where the N vectors, $\mathbf{m}_i$, represent the response of the ith-type sensor (or cone). Any visible spectrum can be represented by an N vector, $\mathbf{r}$. The response of the sensors to the input spectrum is a 3 vector, v, obtained by

$$v = \mathbf{M}^T\mathbf{r} \tag{29}$$

Since we are interested in sampling to represent human color sensitivity, let the matrix $\mathbf{S} = [s_1, s_2, s_3]$, represent the sensitivity of the eye. The result of sensing with the eye, $\mathbf{t}$

$$\mathbf{t} = \mathbf{S}^T\mathbf{r} \tag{30}$$

is given the special name of *tristimulus* vector or values.

Two visible spectra are said to have the same color if they appear the same to the human observer. In our linear model, this means that if $\mathbf{r}_1$ and $\mathbf{r}_2$ are two N vectors representing different spectral distributions, they are equivalent colors if

$$\mathbf{S}^T\mathbf{r}_1 = \mathbf{S}^T\mathbf{r}_2 \tag{31}$$

It is clear that there may be many different spectra that appear to be the same color to the observer. Two spectra that appear the same are called *metamers*. Metamerism is one of the greatest and most fascinating problems in color science. It is basically color "aliasing" and can be described by the generalized sampling described earlier.

The N-dimensional spectral space can be decomposed into a 3D subspace known as the *human visual subspace* (HVSS) and an $N - 3$D subspace known as the *black space*. All metamers of a particular visible spectrum, $\mathbf{r}$, are given by

$$\mathbf{x} = \mathbf{P}_v\mathbf{r} + \mathbf{P}_b\mathbf{g} \tag{32}$$

where $\mathbf{P}_v = \mathbf{S}(\mathbf{S}^T\mathbf{S})^{-1}\mathbf{S}^T$ is the orthogonal projection operator to the visual space, $\mathbf{P}_b = [\mathbf{I} - \mathbf{S}(\mathbf{S}^T\mathbf{S})^{-1}\mathbf{S}^T]$ is the orthogonal projection operator to the black space, and $\mathbf{g}$ is any vector in N space.

It should be noted that humans cannot see (or detect) all possible spectra in the visual space. Since it is a vector space, there exist elements with negative values. These elements are not realizable and thus cannot be seen. All vectors in the black space have negative elements. While the vectors in the black space are not realizable and cannot be seen, they can be combined with vectors in the visible space to produce a realizable spectrum.

If sampling by an optical device, with sensor $\mathbf{M}$, is done correctly, the tristimulus values can be computed from the optical measurements, $\mathbf{v}$, that is, $\mathbf{B}$ can be chosen so that

$$\mathbf{t} = (\mathbf{S}^T\mathbf{r}) = \mathbf{B}\mathbf{M}^T\mathbf{r} = \mathbf{B}\mathbf{v} \tag{33}$$

From the vector space viewpoint, the sampling is correct if the three-dimensional vector space defined by the cone sensitivity functions is the same as the space defined by the device sensitivity functions. Using matrix terminology, the range spaces of the $\mathbf{S}$ and $\mathbf{M}$ are the same.

When we consider the sampling of reflective spectra, we note that a reflective object must be illuminated to be seen. The resulting radiant spectra is the product of the illuminant and the reflection of the object

$$\mathbf{r} = \mathbf{L}\mathbf{r}_0 \tag{34}$$

where $\mathbf{L}$ is diagonal matrix containing the sampled radiant spectrum of the illuminant and the elements of the reflectance of the object are constrained, $0 \leq \mathbf{r}_0(k) \leq 1$. The measurement of the appearance of the reflective object can be computed in the same way as the radiant object with the note that the sensor matrices now include the illuminant, where $\mathbf{LS}$ and $\mathbf{LM}$ must have the same range space.

It is noted here that most physical models of the eye include some type of nonlinearity in the sensing process. This nonlinearity is often modelled as a logarithm; in any case, it is always assumed to be monotonic within the intensity range of interest. The nonlinear function, $\mathbf{v} = V(\mathbf{c})$, transforms the 3-vector in an element independent manner:

$$[v_1, v_2, v_3]^T = [V(c_1), V(c_2), V(c_3)]^T \tag{35}$$

Since equality is required for a color match by Eq. (31), the function $V(\cdot)$ does not affect our definition of equivalent colors. Mathematically

$$V(\mathbf{S}^T\mathbf{r}_1) = V(\mathbf{S}^T\mathbf{r}_2) \tag{36}$$

is true if, and only if, $\mathbf{S}^T\mathbf{r}_1 = \mathbf{S}^T\mathbf{r}_2$. This nonlinearity does have a definite effect on the relative sensitivity in the color matching process and is one of the causes of much searching for the "uniform color space."

7. PRACTICAL RECONSTRUCTION OF IMAGES

The theory of sampling states that a band-limited signal can be reconstructed if it is sampled properly. The reconstruction requires the infinite summation of a weighted sum of sinc functions. From the practical viewpoint, it is impossible to sum an infinite number of terms and the sinc function cannot be realized with incoherent illumination. Let us consider the two problems separately.

The finite sum can be modelled as the truncation of the infinite sum

$$\hat{s}(x,y) = \sum_{m=-M}^{M}\sum_{n=-N}^{N} s(m,n)\frac{\sin[\pi(x-m)]}{\pi(x-m)}\frac{\sin[\pi(y-n)]}{\pi(y-n)} \tag{37}$$

This is equivalent to truncating the number of samples by the use of the $rect(\cdot)$ function:

$$\hat{s}(x,y) = \left[s(x,y)comb(x,y)rect\left(\frac{x}{M},\frac{y}{N}\right)\right] * \sin\mathrm{c}(x,y) \tag{38}$$

Clearly, as the number of terms approaches infinity, the estimate of the function improves. Furthermore, the $\mathrm{sinc}(x,y)$ is the optimal interpolation function for the mean-square error measure. Unfortunately, truncation by the ideal lowpass filter produces ringing at high-contrast edges caused by Gibbs phenomenon.

The inclusion of a practical reconstruction interpolation function can be modeled by replacing the $sinc(x,y)$ by the general function $h(x,y)$. The model is now given by

$$\hat{f}_a(x,y) = \left[f_a(x,y)comb(x,y)rect\left(\frac{x}{M},\frac{y}{N}\right)\right] * h(x,y) \tag{39}$$

The common forms of the interpolation function are the same as the sampling aperture when considering actual hardware, such as circular, rectangular, and Gaussian. These are used when considering output devices, such as CRT (cathode ray tube) or flat-panel monitors, photographic film, and laser printers. The optimal design of these apertures is primarily a hardware or optical problem. Software simulation usually plays a significant role in developing the hardware. Halftone devices, such as inkjet printers, offset and gravure printing, use more complex models that are a combination of linear and nonlinear processes [18]. There is another application that can be considered here that uses a wider variety of functions.

Image interpolation is often done when a sampled image is enlarged many times its original size. For example, an image may be very small, say 64×64. If the image is displayed as one screen pixel for each image pixel, the display device would show a reproduction that is too small, say, 1 in. $\times$ 1 in. The viewer could not see this well at normal viewing distances. To use the entire area of the display requires producing an image with more pixels. For this example, $8\times$ enlargement would produce a 512×512 image that would be 8×8 in. This type of interpolation reconstruction is very common when using variable sized windows on a monitor.

The simple method of pixel replication is equivalent to using a rectangular interpolating function, $h(x,y)$. This is shown in Figs. 9 and 11. The square aperture is readily apparent. The figure of the pictorial image, Fig. 13, has more pixels and uses a proportionally smaller reproduction aperture. The aperture is not apparent to the eye.[5] In the case of the sinusoidal images of Figs. 9 and 11, the purpose of the image was to demonstrate sampling effects. Thus, the obvious image of the aperture helps to make the sampling rate apparent. If we desire to camouflage the sampling and produce a smoother image for viewing, other methods are more appropriate.

Bilinear interpolation uses a separable function composed of triangle functions in each coordinate direction. This is an extension of linear interpolation in one dimension. The image of Fig. 11 is displayed using bilinear interpolation in Fig. 17. The separability of the interpolation is noticed in the rectilinear artifacts in the image.

A more computationally expensive interpolation is the cubic spline. This method is designed to produce continuous derivatives, in addition to producing a continuous function. The result of this method is shown in Fig. 18. For the smooth sinusoidal image, this method works extremely well. One can imagine that images exist where the increased smoothness of the spline interpolation would produce a result that appears more blurred than the bilinear method. There is no interpolation method that is guaranteed to be optimal for a particular image. There are reasons to use the spline method for a wide variety of images [19]. There are many interpolating functions that have been investigated for many different applications [20].

8. SUMMARY

The article has given an introduction to the basics of sampling and reconstruction of images. There are clearly several areas that the interested reader should expand on by additional reading. In-depth treatment of the frequency domain can be obtained from many of the common image processing texts, such as that by Jain, [8]. The processing of video and temporal imaging is covered well in Tekalp's text [21]. Color imaging is treated in a special issue of the *IEEE Transactions on Image Processing* [17]. A survey paper in that issue is a good starting point on the current state of the art. Sampling and reconstruction of medical images is treated in the treatise by Macovski [22].

[5] This is true even when the image is viewed without the effect of halftone reproduction, which is used here.

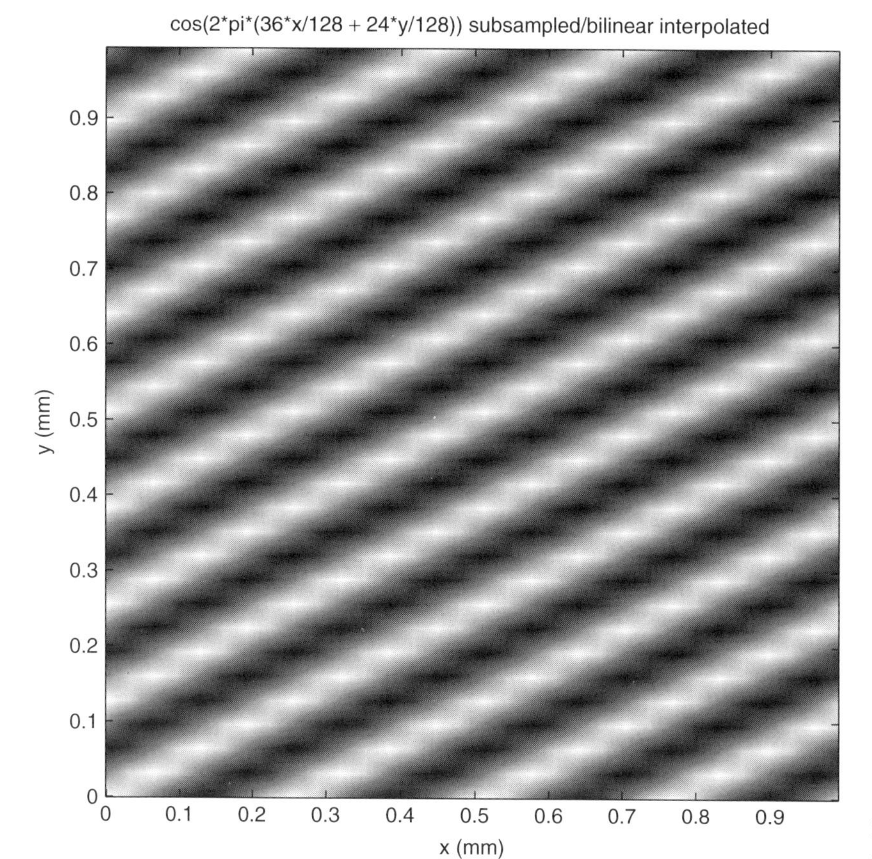

Figure 17. Linear interpolation of subsampled $\cos[2\pi(36x/128 + 24y/128)]$.

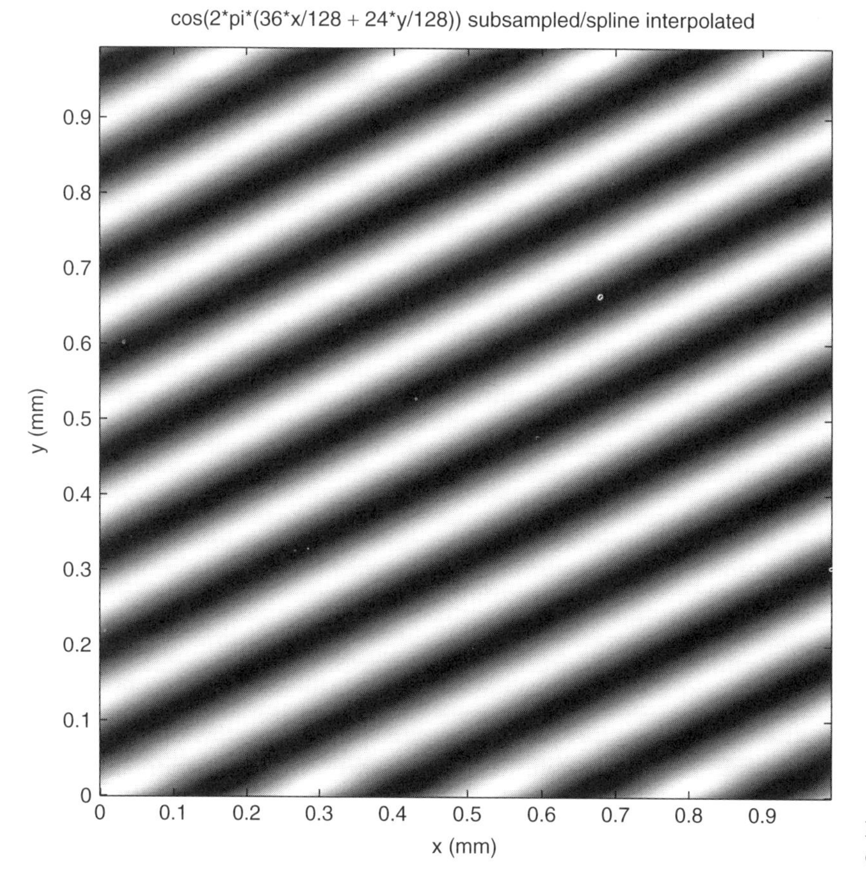

Figure 18. Spline interpolation of subsampled $\cos[2\pi(36x/128 + 24y/128)]$.

BIOGRAPHY

Joel Trussell received degrees from Georgia Tech (1967), Florida State (1968), and the University of New Mexico (1976). In 1969 he joined the Los Alamos Scientific Laboratory, Los Alamos, New Mexico, where he began working in the image and signal processing in 1971. During 1978/79, he was a Visiting Professor at Heriot-Watt University, Edinburgh, Scotland, where he worked with both the university and with industry on image processing problems. In 1980, he joined the Electrical and Computer Engineering Department at North Carolina State University, in Raleigh, where is now a Professor. During 1988/89, he was a Visiting Scientist at the Eastman Kodak Company in Rochester, New York, and in 1997/98 was a Visiting Scientist at Color Savvy Systems in Springboro, Ohio. He is a past Associate Editor for the journals *Transactions on ASSP* and *Signal Processing Letters*. He is a past Chairman of the Image and Multidimensional Digital Signal Processing Committee of the Signal Processing Society of the IEEE. He founded and edited the electronic newsletter published by this committee. He is Fellow of the IEEE and has shared the IEEE-ASSP Society Senior Paper Award (1986, with M. R. Civanlar) and the IEEE-SP Society Paper Award (1993, with P. L. Combettes).

BIBLIOGRAPHY

1. MATLAB, *High Performance Numeric Computation and Visualization Software*, The Mathworks Inc., 24 Prime Park Way, Natick, MA 01760.
2. A. V. Oppenheim and A. S. Willsky, *Signals and Systems*, 2nd ed., Prentice-Hall, Englewood Cliffs, NJ, 1997.
3. D. E. Dudgeon and R. M. Mersereau, *Multidimensional Digital Signal Processing*, Prentice-Hall, Englewood Cliffs, NJ, 1984.
4. J. B. Cohen and W. E. Kappauf, Metameric color stimuli, fundamental metamers, and Wyszecki's metameric blacks, *Am. J. Psychol.* **95**(4): 537–564 (1982).
5. H. Grassmann, Zur Therorie der Farbenmischung, *Annalen der Physik und Chemie* **89**: 69–84 (1853).
6. H. Grassmann, On the theory of compound colours, *Philos. Mag.* **7**(4): 254–264 (1854).
7. B. K. P. Horn, Exact reproduction of colored images, *Comput. Vision Graph. Image Proc.* **26**: 135–167 (1984).
8. A. K. Jain, *Fundamentals of Digital Image Processing*, Prentice-Hall, Englewood Cliffs, NJ, 1989.
9. H. J. Trussell and M. S. Kulkarni, Sampling and processing of color signals, *IEEE Trans. Image Proc.* **5**(4): 677–681 (April 1996).
10. H. J. Trussell, Application of set theoretic methods to color systems, *Color Res. Appl.* **16**(1): 31–41 (Feb. 1991).
11. P. L. Vora and H. J. Trussell, Measure of goodness of a set of colour scanning filters, *J. Opt. Soc. Am.* **10**(7): 1499–1508 (July 1993).
12. B. A Wandell, The Synthesis and Analysis of Color Images, *IEEE Trans. Patt. Anal. Mach. Intel.* **PAMI-9**(1): 2–13 (Jan. 1987).
13. B. A Wandell, *Foundations of Vision*, Sinauer Assoc. Inc, Sunderland, MA, 1995.
14. G. Wyszecki and W. S. Stiles, *Color Science: Concepts and Methods, Quantitative Data and Formulae*, 2nd ed., Wiley, New York, 1982.
15. W. B. Cowan, An inexpensive scheme for calibration of a color monitor in terms of standard CIE coordinates, *Comput. Graph.* **17**: 315–321 (July 1983).
16. M. J. Vrhel and H. J. Trussell, Color device calibration: A mathematical formulation, *IEEE Trans. Image Process.* **8**(12): 1796–1806 (Dec. 1999).
17. *IEEE Trans. Image Proc.* **6**(7): (July 1997).
18. R. Ulichney, *Digital Halftoning*, MIT Press, Cambridge, MA, 1987.
19. M. Unser, Splines — a perfect fit for signal and image processing, *IEEE Signal Process. Mag.* **16**(6): 22–38 (Nov. 1999).
20. R. N. Bracewell, *Two-Dimensional Imaging*, Prentice-Hall, Englewood Cliffs, NJ, 1995.
21. A. M. Tekalp, *Digital Video Imaging*, Prentice-Hall, Englewood Cliffs, NJ, 1995.
22. A. Macovski, *Medical Imaging Systems*, Prentice-Hall, Englewood Cliffs, NJ, 1983.

IMT-2000 3G MOBILE SYSTEMS

ANNE CERBONI
JEAN-PIERRE CHARLES
France Télécom R&D
Issy Moulineaux, France

PIERRE GHANDOUR
JOSEP SOLE I TRESSERRES
France Télécom R&D
South San Franciso, California

1. INTRODUCTION

The ability to communicate anywhere, at any time, with anyone in the world, using moving pictures, graphics, sound, and data has been a longstanding challenge to telecommunications operators. With the advent of IMT-2000 third-generation mobile systems, people on all continents will be able to take advantage of most of these capabilities. Standardization, ensuring that customers' terminals are compatible with IMT-2000 networks throughout the world, has had to accommodate regional differences. While ongoing work within different standards forums is striving to achieve a high degree of universality, IMT-2000 in fact covers a family of standards. Relevant standards issues and their implications in different regions are discussed in Section 3. The emergence of third-generation (3G) mobile systems is, of course, rooted in the development and increasingly widespread use of previous generations. The migration paths toward 3G systems are reviewed in Section 4. Enhanced data rates offered by IMT-2000 are expected to provide a broad range of mobile multimedia services on a multiple choice of terminals. These new facilities and applications are examined in Section 5. Section 6 describes IMT-2000 radio access and core network architecture. In Section 7, the authors address the most salient economic implications of migration toward 3G, with emphasis on license costs and new business models. Section 8 discusses evolution of mobile systems beyond 3G, which involves not only an all-IP core network but also optimization of scarce spectral resources through cooperation among heterogeneous networks.

2. DRIVING FORCES

Two major trends are reshaping the world of telecommunications. On one hand, mobile services have made great strides throughout the world, with penetration rates exceeding 50% in many countries. On the other hand, the explosion of Internet traffic testifies to the rapid development of the multimedia market. The prospect for convergence of these two trends, giving rise to mobile multimedia services, and the resulting need for greater spectral resources, has driven equipment suppliers, operators, standards bodies, and regulators throughout the world to develop a new generation of mobile systems. The stakes are considerable: around 2010, mobile traffic should be equal to that of fixed telephony [1]. The convergence of the mobile and Internet worlds, the strong dynamics of innovation, and anticipated cost reductions in these areas have opened new opportunities for 3G services as of 2001 in Japan and possibly in the United States, and 2002 in Europe.

3. STANDARDIZATION

3.1. IMT-2000 Frequency Spectrum

The International Telecommunications Union (ITU) initiated 3G mobile standardization with the ambition of defining a global standard replacing the broad variety of second-generation (2G) mobile systems, which implied common spectrum throughout the world. Hence, the first efforts on 3G systems truly started once the World Administrative Radio Conference (WARC) 92 had identified a total of 230 MHz for IMT-2000, as illustrated in Fig. 1.

IMT-2000 standardisation activities for the radio interface are conducted in ITU-R/WP 8F. At WARC 2000, additional frequency bands totalling 400 MHz were allocated for IMT-2000.

3.2. IMT-2000 Radio Interface

The early impetus for standardizing an IMT-2000 radio interface, specifically, the interface between the mobile terminal and the base station, can be attributed to an observation of the contrasting situation of 2G systems in Europe and in the United States. In Europe, the Global System for Mobile Communications (GSM) standard was developed and universally adopted before 2G systems were first deployed in 1991, ensuring that customers' terminals are compatible with mobile networks throughout the continent. In the United States, the lack of country or continentwide harmonization, and the relatively greater success of first-generation analog systems such as advanced mobile phone service (AMPS), led to the parallel development of three different 2G digital standards:

- Time-division multiple access (TDMA) including digital AMPS (DAMPS) and IS136
- Code-division multiple access (CDMA), known as IS95
- Global System for Mobile Communications (GSM)

Table 1 provides a brief overview of the market share of these standards in the United States.

Despite the initial goal of a single worldwide 3G air interface, standardization was strongly influenced by

Table 1. Market Share of 2G Cellular Standards in the United States

Technology	1999	2002
GSM	4%	11–15%
AMPS	55%	20%
TDMA	25%	36%
CDMA	16%	28%
Total subscribers (millions)	85	135

Source: France Télécom North America.

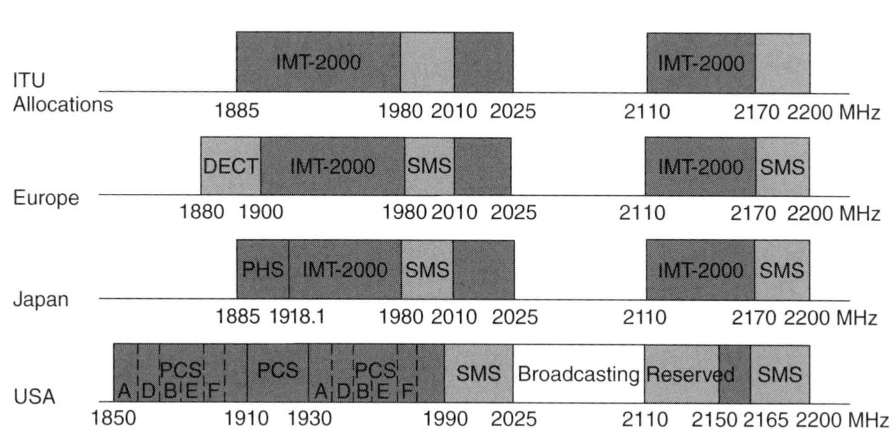

Figure 1. IMT-2000 spectrum.

mobile operators' need to leverage their 2G investment. Specifically, this meant that 3G systems must ensure backward compatibility with existing systems, providing seamless handover to and from 2G systems, in addition to sharing tower infrastructure and transmission resources. As a result, in November 1999, the International Telecommunications Union — Radiocommunications (ITU-R) could not establish a consensus on any one 3G air interface and thus adopted five different solutions:

- Universal Mobile Telecommunications System/Wideband CDMA (UMTS/WCDMA) one of the two modes supported by NTT DoCoMo, Nokia, and Ericsson, and developed by the Third-Generation Partnership Project (3GPP).
- cdma2000, an evolution of the American CDMA IS95A solution originally developed by Qualcomm and currently standardized by the Third-Generation Partnership Project 2 (3GPP2).
- UMTS/TD-CDMA: UMTS mode combining time-division (TD) and code-division multiple access, supported by Siemens. This solution also includes a specific option developed for China called TDSCDMA (the S stands for synchronous). This mode is also developed by 3GPP.
- Enhanced Data for GSM Evolution (EDGE) or UWC-136. This solution represents an evolution of both TDMA and GSM.
- Digital enhanced cordless telephony DECT developed by ETSI.

Among these solutions, cdma2000, WCDMA, and EDGE are discussed in greater detail in the following paragraphs.

3.3. Regional Standardization

3.3.1. Europe. In Europe, impetus for the development of IMT-2000 was largely provided by European Commission research programs in the late 1980s and early 1990s (RACE I and II,[1] ACTS/FRAMES[2] project). In 1991, with the deployment of the first 2G systems, the European Telecommunications Standards Institute (ETSI) created a subcommittee to develop an IMT-2000 system called *UMTS*. Efforts first focused on defining the technical requirements for the radio interface, and various solutions were presented at the December 1996 ETSI conference, including three proposals by ACTS/FRAMES. Following a vote in January 1998, a compromise was found based on two harmonized modes: WCDMA and TDCDMA. WCDMA was adopted for the frequency-domain duplex (FDD) mode, namely, one frequency per transmission direction, and TDCDMA for the time-domain duplex (TDD) mode, specifically, time-division multiplexing of the two directions on the same frequency. This combined solution offers the advantage of enabling full use of the IMT-2000 frequency bands; the FDD mode is used in priority in the paired bands and the TDD mode, in unpaired bands. This compromise was then submitted to ITU-R as the European proposal for the IMT-2000 radio interface.

Standardization in Europe was strongly influenced by lobbying within forums such as the GSM Association and the UMTS Forum, which strove to federate the stances of GSM operators and manufacturers regarding the development of 3G systems. National regulation authorities also played a fundamental role in defining the use of the spectrum identified by the WARC 92, and for the attribution of UMTS licenses.

3.3.2. Japan, Korea, and China. In Japan, most 3G developments were financed by mobile operator NTT DoCoMo. Japanese industry supported this R&D effort to develop a new standard and take the lead in this very competitive market. European manufacturers Nokia and Ericsson took part in this effort, which led to the establishment of a common solution between them and Japan, based on WCDMA. This compromise was reached just as ETSI was seeking candidates for its 3G mobile system, leading to a convergent view between Japan and Europe in favor of WCDMA for the air interface. Although other carriers like Japan Telecom followed this direction, KDDI, the second largest Japanese carrier, is strongly involved in cdma2000.

In Korea, the Telecommunications Technology Association (TTA) kept two paths open for the evolution of the country's CDMA mobile networks. Both WCDMA and cdma2000 are officially supported by the Ministry of Information and Communications. In fact, the three mobile carriers (SK Telecom, KT Freetel and LG Telecom) have already started to offer cdma2000 1x services in overlay of their 2G networks. The first two operators obtained 3G licenses to deploy WCDMA, mainly for ease of roaming. The government plans to grant another 3G license based on cdma2000.

In China, work on a 3G standard started in 1992 within the First Research Institute of the Datang Group, now the Chinese Academy of Telecommunication Technology (CATT). After studying the European GSM standard in detail, the group developed its own standard, Synchronous code-division multiple access (SCDMA), which became the basis for subsequent 3G developments. With the support of Siemens, TDSCDMA was adopted as an official ITU 3G standard. The Chinese government is devoting significant resources and energy to building a national industry around this standard, instead of relying on imported network equipment and terminals. The country which, in 2001, represents the world's second largest mobiles market, can afford to develop its own standard, touted as offering greater spectral efficiency than rivals WCDMA and cdma2000 and being more cost-effective and more easily integrated in the GSM environment. Carriers such as state-owned China Mobile and China Unicom may potentially deploy TDSCDMA.

3.3.3. United States. In the United States, tough and unrestricted competition driven by the various

[1] Research on Advanced Communication Technologies in Europe.
[2] Advanced Communication Technologies and Services/Future Radio Wideband Multiple Access System). The main partners were France Télécom, Nokia, Siemens, Ericsson, and CSEM/Pro Telecom.

manufacturers led to the creation of several standards committees. Initially, the Telecommunication Industry Association (TIA) was in charge of standardizing TDMA IS136 (TIA/TR43.3) and CDMA IS95A (TIA/TR45.5) while T1P1, the GSM Alliance, and the GSM Association were involved in GSM standardization. When 3GPP2 was created in 1999, the TIA/TR45.5 working group became a major player in cdma2000 standardization. TIA/TR45.34, on the other hand, decided that 3GPP2 was not the suitable group for TDMA standardization and joined standardization efforts within the Universal Wireless Communications Consortium (UWCC).

Establishing digital cellular coverage in the United States is costly and carriers have had to invest a substantial amount of money. It is thus not surprising to note that the primary recommendations of American proposals for IMT-2000 often correspond to evolutions of existing second-generation systems maintaining backward compatibility in order to capitalize on the initial investment. In particular, cdma2000 1x was designed for smooth evolution from the second-generation IS95A standard.

The United States is facing a spectrum shortage for 3G systems. A large part of the frequency band allocated by WARC 92 (see Fig. 1) is currently used by second-generation personal communication systems. Although the United States has allotted only 210 MHz of spectrum for mobile wireless use, compared to an average of 355 MHz per country in Europe, companies are currently pressing forward with their plans for third-generation (3G) networks, while industry efforts at obtaining more spectrum, led by the Cellular Telecommunications and Internet Association (CTIA), continue. Sprint PCS and Cingular, for example, have announced plans to squeeze more capacity out of existing networks by upgrading technology, although spectrum in the United States is already more crowded than in other markets. The United States has nearly 530,000 mobile customers per megahertz of spectrum while the United Kingdom has just more than 80,000 users, and Finland, the world's leader in wireless penetration, has only 15,000 users per megahertz.

In October 2000, a memorandum was issued to define the need for a radiofrequency spectrum for future mobile voice, high-speed data, and Internet-accessible wireless services. The memorandum directed the Secretary of Commerce to work cooperatively with the Federal Communications Commission (FCC).

Various frequency bands had been identified for possible 3G use. The FCC and the National Telecommunications and Information Administration (NTIA) undertook studies of the 2500–2690 MHz and the 1755–1850 MHz frequency bands in order to provide a full understanding of all the spectrum options available. Both bodies stated possible sharing and segmentation options, but a review of the reports showed that, for every option, there were several caveats. Therefore, 3G carriers will mainly count on existing spectrum for 3G services. Sprint PCS and Cingular are heading this way when releasing 3G services toward the end of 2001.

3.3.4. 3GPP and 3GPP2. In this international context, standardization activities led within the ITU and regional entities[3] developed with increasingly close contacts. In 1998, ETSI, the Association of Radio Industries and Businesses (ARIB, Japan), TTC, as well as the Telecommunications Technology Association (TTA) of Korea and T1P1 of the United States founded the Third Generation Partnership Project (3GPP) as a forum to develop a common WCDMA standard, assuming GSM as a basis. The following year, the American National Standards Institute (ANSI) International Committee initiated 3GPP2, geared toward developing standards for the cdma2000 standard, in continuity with the CDMA IS95A standard. Member organizations include ARIB, China Wireless Telecommunication Standards Group (CWTS), Telecommunications Industry Association (TIA) from North America, Telecommunications Technology Association (TTA) from Korea and Telecommunications Technology Committee (TTC) from Japan.

Harmonization efforts between 3GPP and 3GPP2 resulted in a number of common features in the competing technologies, and enabled work to be divided between them such that 3GPP handles direct-sequence WCDMA and 3GPP2 focuses on the multicarrier (MC) mode in cdma2000. 3GPP successfully produced a common set of standards for the WCDMA air interface, known as "Release 99." Meanwhile, 3GPP2 issued Release A of cdma2000 1x and is currently (at the time of writing) working on Release B. In parallel, 3GPP2 also released an evolution of cdma2000 1x called "cdma2000 1xEV" (1xEVOLUTION) Phase 1 [also called "HDR (High Data Rate standard) Data Only"] and is currently working on Phase 2 (Data and Voice). Section 4 provides more information on these releases. Table 2 presents an overview of the different CDMA technologies envisioned for 3G.

An Operators Harmonization Group (OHG) was also created in 1999 for promoting and facilitating convergence of 3G networks One goal is to provide seamless global roaming among the different CDMA 3G modes (cdma2000, WCDMA, and TDD modes). The OHG was involved in specifying what mode should be used for the 3G radio access network. While cdma2000 was specifically oriented toward the multicarrier mode, WCDMA was to be only direct-spread. The objective is to achieve a flexible connection between the radio transmission technologies (RTTs) and the core networks (either evolved GSM MAP (mobile application part) or evolved ANSI-41).

4. CONTEXT AND EVOLUTION PATHS

4.1. Development Context

Why are IMT-2000 systems called "3G"? First-generation mobile systems were based on analog standards including AMPS in the United States, TACS (Total Access Communication System) in the United Kingdom, CT-2

[3] ETSI for Europe, the Telecommunications Technology Committee (TTC) and the Association for Radio Industries and Businesses (ARIB) for Japan, the Telecommunication Industry Association (TIA) and American National Standards Institute (ANSI) for the United States.

Table 2. Comparison of Different CDMA Technologies

	CDMA Technology Comparisons			
CDMA technology	Peak Data Rate	Average Data Throughput	Approved Standard?	Company
cdma2000-1x Phase 1	153.6 kbps	150 kbps	Yes	—
cdma2000-1x RTT A	614.4 kbps	415 kbps	Yes	—
1X Plus Phase 1	1.38 Mbps	560 kbps	—	Motorola
WCDMA (5 MHz)	2.048 Mbps	1.126 Mbps	Yes	—
HDR	2.4 Mbps	621 kbps	—	Qualcomm
cdma2000-3x multicarrier	2.072 Mbps	1.117 Mbps	—	—
1x Plus (Phase 2)	5.184 Mbps	1.200 Mbps	—	Motorola

Source: Soundview Technology Group and Motorola (published in *Global Wireless Magazine*).

in Europe, and that of NTT (Nippon Telephone and Telegraph) in Japan. The term "second generation" refers to digital mobile systems such as TDMA and CDMA in the United States, GSM (USA, Europe, China) and Personal Digital Cellular (PDC) in Japan. These systems provide mobile telephony, short-message services (SMSs) and low rate data services relying on standards such as the Wireless Applications Protocol (WAP) and I-mode. WAP is a standard for delivering Internet content and applications to mobile telephones and other wireless devices such as personal digital assistants (PDAs). It can be used independently of the type of terminal and network. WAP content is written in wireless markup language (WML), a version of HTML designed specifically for display on wireless terminal screens. To provide WAP content, the operator must implement a server between the wireless network and the Internet, which translates the protocol and optimizes data transfer to and from the wireless device. However, slow data rates, poor connections, and a limited number of services have significantly limited the use of mobiles for data applications. I-mode, on the other hand, attracted millions of users within its first year of existence, starting in 1999, in Japan. Despite the 9.6-kbps (kilobits per second) data, this proprietary packet-data standard developed by NTT DoCoMo met with widespread success for several reasons: no dial-up, volume-based billing, services adapted to the low bit rate, and an extensive choice of content providers.

4.2. Migration Paths

The technological options taken by different cellular operators to deploy 3G networks depend on the 2G technology employed. In Europe, the starting point is GSM. In the United States, in addition to GSM, operators are focusing on two other paths, TDMA, and IS95A (CDMA). Each of these migration paths is described hereafter.

4.2.1. GSM Migration Path. The first step was high-speed circuit-switched data (HSCSD), introduced commercially in Finland in 1999. HSCSD supports data rates of up to 57.6 kbps by grouping four GSM time slots. Access to HSCSD, however, involves a new terminal for the customer. This service has not been developed extensively, and operators are putting more energy into developing packet-based data technologies such as General Packet Radio Service (GPRS or GSM phase 2+).

GPRS theoretically supports up to 115.2 kbps packet-switched mobile data alongside circuit-switched telephony. The principle is to utilize GSM time slots to carry data. The amount of data per time slot varies from about 9 to 21 kbps, depending on the coding scheme used. GPRS is well adapted to asymmetrical traffic, with a greater number of time slots dedicated to the downlink. It also enables per volume billing, which, according to the I-mode example, encourages use of the service. GPRS requires new equipment in the GSM base station subsystem (BSS) and network subsystem (NSS) in order to handle the packet data. The packet control unit (PCU) located in the BSS handles the lower levels of the radio interface: the radio-link control (RLC) and medium access control (MAC) protocols. In the NSS, there are two important elements, which are also used in 3G networks. The first is the Serving GPRS Support Node (SGSN), basically an IP (Internet Protocol) router with specific functional features. For the subscribers in a given area of the mobile network, it handles authentication and security mechanisms, mobility management, session management, billing functions, in addition to the transmission of data packets. The second element is the Gateway GPRS Support Node (GGSN), another IP router which acts as gateway for data transmission between the GPRS network and other packet data networks. Other elements in the GPRS network include a domain name server (DNS) and a legal interception gateway. The home location register (HLR) of the mobile network is modified to take into account the data capabilities of GPRS customers. Mobile terminals (MTs) must, of course, be GPRS-compatible. There are several MT classes; the most basic is a GPRS radio modem for a laptop or handheld device, the most sophisticated handles both voice and data flows simultaneously.

EDGE or Enhanced GPRS (EGPRS), one of the five standard IMT-2000 radio interfaces, represents an upgrade not only from GSM but also from TDMA. The EDGE approach is similar to that of GPRS, in that it makes use of existing time slots for packet data transmission. However, EDGE uses a more elaborate coding scheme providing up to 48 kbps per time slot, yielding an overall rate of up to 384 kbps. A potential drawback of EDGE is that, as data rates increase, range decreases. While services are not interrupted, this fact may require the operator to build out a denser network. Furthermore, as in the case of HSCSD and GPRS, a specific terminal is required. EDGE is described in greater detail in Section 6.3.

In its Release 99 (R99), UMTS, taken to be synonymous with WCDMA, offers symmetrical links at 384 kbps for customers at a speed of about 120 km/h, and 128 kbps for customers at a speed of up to 300 km/h. Technically, this technology allows a data rate of up to 2 Mbps for a quasistationary user. In order to ensure compatibility with existing GSM networks, dual band, dual-mode UMTS/GSM terminals are required for access to nationwide voice and data (GPRS or EDGE) services, at least until UMTS coverage attains a significant percentage of the country, and GSM networks are progressively phased out.

While intermediate steps HSCSD, GPRS, and EDGE are overlaid onto the GSM radio network, UMTS requires an entirely different radio access network. The network architecture is described in Section 6.2.

4.2.2. TDMA Migration Path. The first step in the evolution of TDMA is IS136+, which advances the data speed to 64 kbps through packet switching over a 200-kHz channel. Since IS136+ require a 200-kHz channel, instead of the 30-kHz bandwidth previously used by TDMA, the base stations must be upgraded with new hardware, which is expensive. The next phase is IS136 HS (high-speed), which uses EDGE technology, allowing the network to reach a theoretical data rate of 384 kbps, and also requires a hardware upgrade to the network's base stations. From there, the network requires another expensive hardware (base station) upgrade to support WCDMA. As a result, even if a carrier skipped IS136+, the TDMA migration involves two expensive hardware upgrades and therefore represents the most costly evolution path.

4.2.3. CDMA Migration Path. Given that the two stages after CDMA IS95A — CDMA IS95B and cdma2000 1x RTT (also known as IS2000 or IS95C) — operate on the same 1.25-MHz channel bandwidth as CDMA IS95A, the migration of this network architecture is the easiest to implement. Indeed, both IS95B and cdma2000 1x[4] require only a relatively cheap software upgrade.

However, those stages are completely independent; it is not necessary for a CDMA IS95A carrier to move to IS95B before moving to cdma2000. As a matter of fact, IS95B standardization work took such a long time that, when it was released, U.S. carriers chose to bypass IS95B, judging that IS95A was sufficient and that they could wait for cdma2000. Indeed, while IS95B offers a speed of up to 64 kbps, cdma2000 1x more than doubles the data speed to 144 kbps, and doubles the network's voice capacity as well. Finally, CDMA 3x RTT, which supplies a peak data rate of up to 2 Mbps, operates on a 3.75 MHz (3×1.25 MHz) frequency channel. Given the larger channel requirements, hardware upgrades of base stations are necessary for this transition.

[4] 1x means one times 1.25 MHz, the bandwidth used for CDMA IS-95.

Another potential alternative called "cdma2000 1x EV" (1x Evolution) also uses a 1.25-MHz channel. This transition includes two phases. For Phase 1, the High Data Rate (HDR) standard was released in August 2000. This technology, also named 1xEV-DO (meaning 1x Evolution Data Only), is supported by Qualcomm, Ericsson and Lucent. It is expected to increase peak data rates of up to 2.4 Mbps. Phase 2 (1xEV-DV meaning 1xEvolution Data and Voice), under standardization, will enable both voice and data channels (up to around 5 Mbps), while enhancing capacity and coverage.

Figure 2 illustrates the main alternatives for operators of 2G cellular networks.

5. IMT-2000 SERVICES AND TERMINALS

5.1. Services

A key feature of IMT-2000 is the wide range of services offered. In addition to voice, videotelephony, and videoconferencing applications, it covers asymmetrical data flows (Web browsing, video or audio streaming), as well as low-data-rate machine-to-machine communications (metering, e-wallet applications). Unlike 2G networks and fixed networks, IMT-2000 does not preassign a bit rate and service quality level to each type of service, but rather provides a framework within which communications can be characterized in terms of their requirements. In a mobile context, in particular, services should remain available in a flexible manner (at a lower rate or with less error protection, e.g., in case of degraded radio conditions) to ensure optimal use of the allocated frequency bands while guaranteeing an acceptable service level from the user's viewpoint. These considerations led to the definition of four quality-of-service (QoS) classes:

- Conversational
- Streaming
- Interactive
- Background

The main characteristics of each class are indicated in Table 3 [2], along with examples:

The characterization in Table 3 enables the mapping of applications onto the UMTS and radio access bearer services. There are no specifications set out by 3GPP for this mapping, even for a service as simple as voice. Instead, the framework for mapping (QoS, transport formats and channels, channel coding, physical channel, etc.) is described and the actual choice of forward error correction code and physical bearer service is left up to the manufacturer and/or operator; the idea is to leave as much latitude as possible for the implementation of existing and especially new services.

In the 3GPP2 Environment, the *quality of service* refers to a set of capabilities that a network may provide to a communications session. These capabilities can be specified so that particular applications (e.g., voice, video, streaming audio) fulfill human factors or other requirements with respect to fidelity and performance.

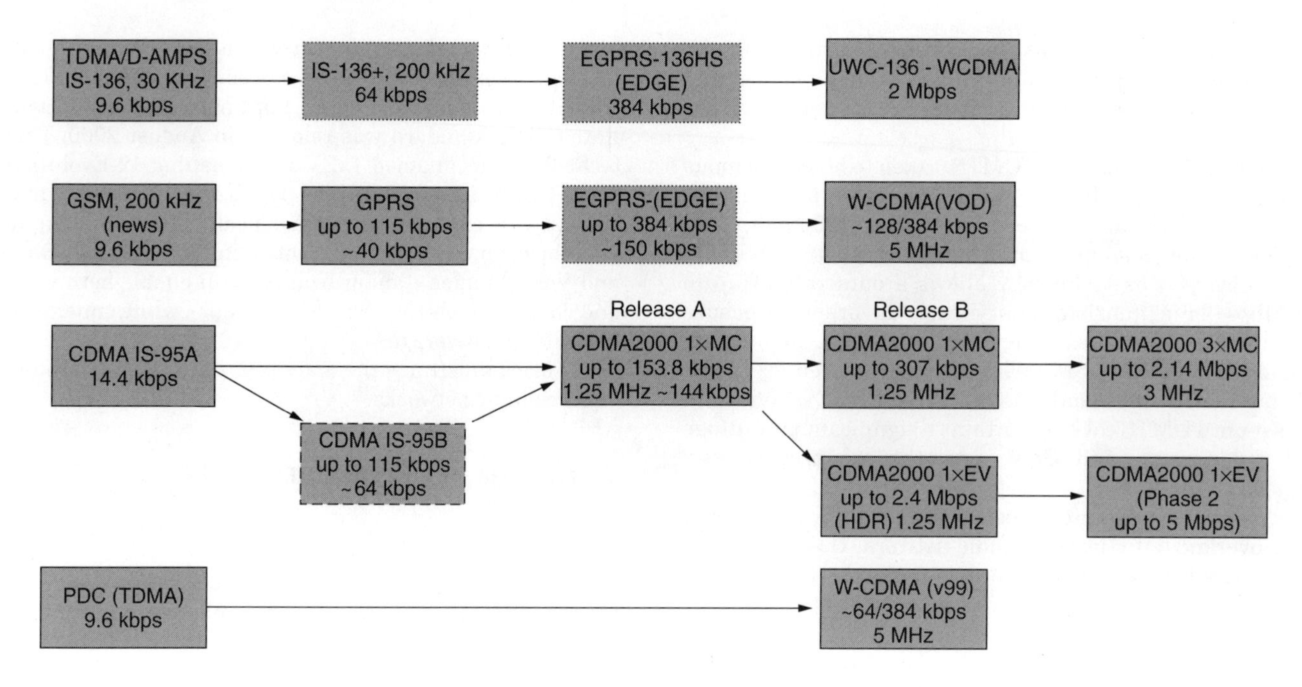

Figure 2. Possible migration paths from 2G to 3G.

Table 3. UMTS Quality-of-Service Classes

Traffic Class	Conversational Class Real Time	Streaming Class Real Time	Interactive Class Best Effort	Background Best Effort
Fundamental characteristics	Preserve time relation (variation) between information entities of the stream Conversational pattern: stringent and low delay	Preserve time relation (variation) between information entities of the stream One-way flow	Request response pattern Preserve payload content	Destination is not expecting the data within a certain time Preserve payload content
Example of the application	Voice, video telephony	Video, audio streaming	Web browsing	Background download of emails

Quality of service in a packet network consists of at least the following components:

Bandwidth—data rate

Delay—end-to-end or round-trip latency

Jitter—interpacket latency variation

Loss—rate at which packets are dropped

Additionally, this QoS may be

Unidirectional or bidirectional

Guaranteed or statistical

End-to-end or limited to a particular domain or domains

Applied to all traffic or just to a particular session or sets of sessions.

5.1.1. Open Service Architecture (OSA). Second-generation mobile systems offered fully standardized services such as voice, fax, short messages, and supplementary services (call hold, call forward, call conference, etc.). However, it was difficult for operators to propose innovative services to attract the customer. To provide greater flexibility in service creation, the second phase of GSM standardization included the introduction of "toolkits": CAMEL (*c*ustomized *a*pplications of *m*obile network *e*nhanced *l*ogic[5]), an intelligent network concept for GSM, SIM toolkit (STK), and MExE (mobile execution environment), which includes WAP. These toolkits were used in GSM to introduce prepaid services (CAMEL) and mobile internet portals (WAP). In 3G, these principles are still valid, but efforts are focused on integrating the various toolkits in a single one called Open Service Architecture (OSA), which is, in fact, an application programming interface (API) based on PARLAY, a forum developing a

[5] CAMEL is based on an intelligent network architecture that separates service logic and database from the basic switching functions, and implements the Intelligent Network Application Protocol (INAP).

common API for the different networks. This new concept is still under development in 3GPP and will be introduced in post-R99 UMTS releases.

An important feature of open service architecture is that service design and provision can be ensured by companies other than the network operator.

5.1.2. Virtual Home Environment. The virtual home environment (VHE) concept, based on CAMEL, will provide customers with the same set of services whatever the location.

When a subscriber is roaming, his or her service profile, or even the service logic registered in the home location register (HLR), is transferred to the visited network to offer the required services with the same ergonomics.

5.2. Terminals

Mobile customers will be able to use one or several mobile terminals (Fig. 3), including regular mobile phones, pocket videophones, and PDAs to manage agendas, addresses, transportation, and email, and to send and receive multiple types of information. Typical handsets are smaller and lighter than 2G handsets: 100 g and 100 cm^3 [3] versus about 130 g on average for 2G. IMT-2000 enabled laptops will provide traveling professionals, executives, and employees with direct access to their corporate intranets, offering videoconferencing, cooperative worktools, and shared programs and network resources facilitating work outside the office. Specific applications will use IMT-2000 capacity to provide data, sound, and fixed or moving images. Among the most widely cited services are location-specific information services, remote control and monitoring, remote health maintenance services, and driving assistance (navigation aids, traffic and weather information, engine maintenance information, etc.). In these instances, IMT-2000 terminals can be standard equipment in vehicles, or coupled with the application devices used (e.g., health monitors).

With large-scale 3G network deployment and mass production of 3G terminals, significant cost reductions are expected to open up a mass market for personal multimedia tools. Young people, who have been nurtured with today's mobile phones and game consoles, will undoubtedly drive the development of this market through their needs for entertainment, education, and information.

A major change in 3G terminals with respect to second-generation mobile terminals lies in the replacement of the 2G subscriber identity module (SIM) card with a more general-purpose card called a universal integrated circuit card (UICC) of the same size. The UICC contains one or more user services identity modules (USIMs) as well as other applications. Communications-related advantages include enhanced security, the ability to use the same handset for business and private use and to roam from UMTS to GSM networks as needed. The UICC can also contain payment mechanisms (micro-payment, credit card), access badge functions, and user profile information.

6. IMT-2000 NETWORK ARCHITECTURE

This section describes the network architecture of the three most prevalent IMT-2000 standards: cdma2000, W-CDMA, and EDGE.

Network architecture is divided into the radio access network (RAN) and the core network (CN). To an increasing extent, efforts are focused on developing standard interfaces between functional domains to enable interworking between network elements manufactured by different suppliers.

6.1. CDMA 2000

6.1.1. Radio Access Network. The cdma2000 radio access network architecture, illustrated in Fig. 4, is similar to that of CDMA IS95A. In cdma2000 1x is mainly a software upgrade.

The key difference between the two architectures is the Ater reference point (A3 and A7), which allows the source BSC (base station controller) to manage soft handover of communications between two base transceiver stations (BTSs) belonging to different BSCs. In Fig. 4, the source BTS is the initial BTS managing the communication. The target BTS is the BTS asked to enter in communication with the mobile. As shown, the source BTS is still managing the communication and remains the primary link with the mobile switching center (MSC) whereas, in IS95A, the MSC always manages the soft handover between two different BSCs (traffic and signaling go from the MSC to the two BSCs).

Figure 3. Examples of mobile multimedia terminals for UMTS.

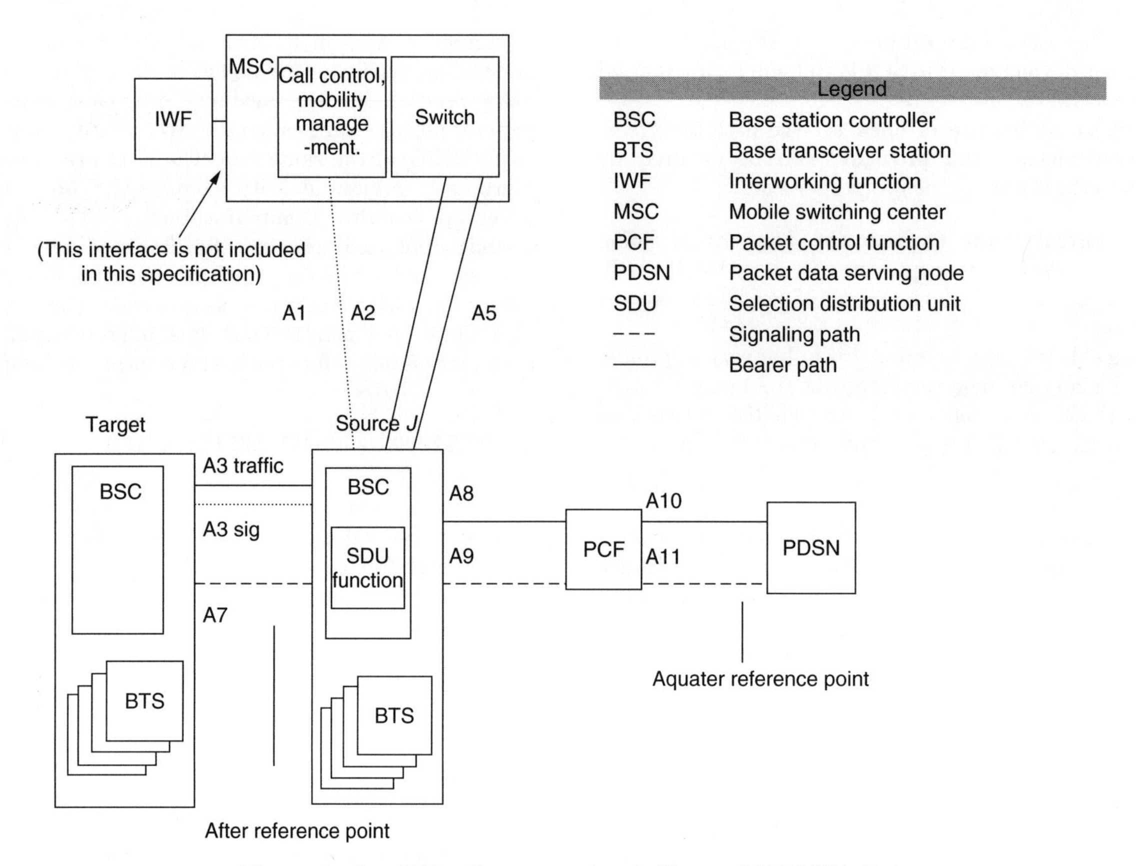

Figure 4. cdma2000 radio access network. (*Source*: 3GPP2 Website.)

Figure 4 depicts the logical architecture of the radio access network. It describes the overall system functions, including services and features required for interfacing a BTS with other BTSs, with the MSC for the circuit transmission mode, and with the packet control function (PCF) and the packet data serving node (PDSN) in the packet transmission mode.

The interfaces defined in this standard are described below.

A1 Carries signaling information between the call control (CC) and mobility management (MM) functions of the MSC and the call control component of the BTS (BSC).

A2 Carries 64/56-kbps pulse-code modulation (PCM) information (voice/data) or 64-kbps unrestricted digital information (UDI, for ISDN) between the MSC switch component and one of the following:

- The channel element component of the BTS (in the case of an analog air interface)
- The selection/distribution unit (SDU) function (in the case of a voice call over a digital air interface)

A3 Carries coded user information (voice/data) and signaling information between the SDU function and the channel element component of the BTS. The A3 interface is composed of two parts: signaling and user traffic. The signaling information is carried across a separate logical channel from the user traffic channel, and controls the allocation and use of channels for transporting user traffic.

A5 Carries a full duplex stream of bytes between the interworking function (IWF) and the SDU function.

A7 Carries signaling information between a source BTS and a target BTS.

A8 Carries user traffic between the BTS and the PCF.

A9 Carries signaling information between the BTS and the PCF.

A10 Carries user traffic between the PCF and the PDSN.

A11 Carries signaling information between the PCF and the PDSN.

A8, A9, A10, and A11 are all based on the use of Internet Protocol (IP). IP can operate across various physical layer media and link layer protocols. Conversely, A3 and A7 are based on ATM Transport; and A1, on SS7 Signaling.

In 2000, the Abis interface (between the BTS and the BSC) and Tandem Free Operation (TFO) for cdma2000 systems were also standardized.

6.1.2. Core Network. The 3GPP2 architecture is based on the same wireless intelligent network (WIN) concept as that developed by 2G mobile networks, but this structure is partially modified to introduce packet data technologies such as IP. The objective of these circuit-switched networks was to bring intelligent network (IN) capabilities, based on ANSI-41, to wireless networks in a seamless manner without making the network infrastructure obsolete. ANSI-41 was the standard backed by wireless providers because it facilitated roaming. It has capabilities for switching and connecting different systems

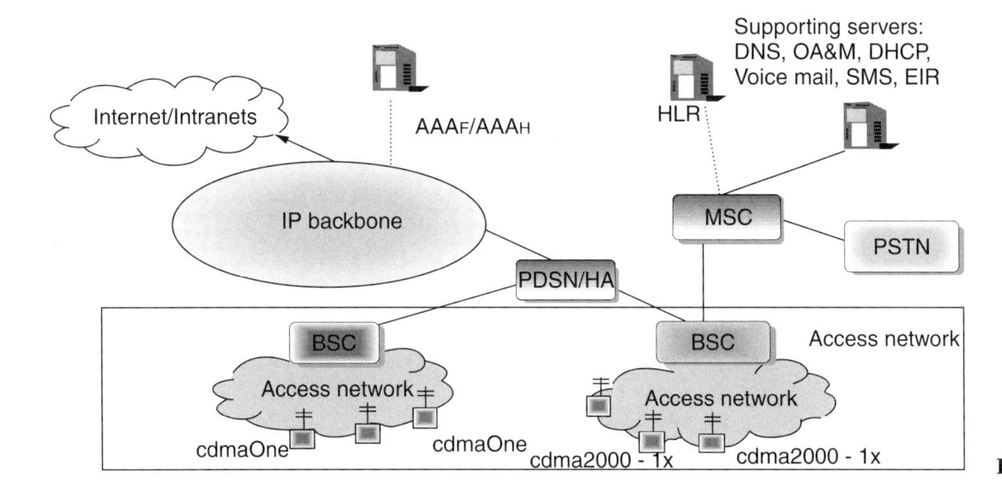

Figure 5. 3GPP2 architecture.

and the ability for performing direct connections between different elements of the WIN based on SS7 Signaling such as GSM/MAP.

Wireless data packet networking is based on the IS835 standard, whose architecture is illustrated in Fig. 5. This architecture provides access, via the cdma2000 air interface, to public networks (Internet) and private networks (Intranets) considering the required quality of service (QoS) and the pertinent accounting support. 3GPP2 strives to reuse IETF open standards whenever possible in order to keep a high level of interoperability with other cellular standards, and to increase marketability. These standards include mobile IP for interPDSN (packet data serving node) mobility management, radius for authentication, authorization and accounting, and differentiated services for the QoS.

Figure 6 introduces the general model for the 3GPP2 packet domain. This domain includes two modes, "simple IP" and "mobile IP." Only the mobile IP mode requires the use of home agent (HA) and PDSN/foreign agent (PDSN/FA) entities. It is clear that these two modes are not equivalent from a service point of view. Mobile IP supports mobility toward different PDSNs during a session. In this sense, the mobile IP solution is similar to the General Packet Radio System (GPRS — see Section 4.2). Simple IP only offers a connection to the packet domain without any real-time mobility between PDSNs. Simple IP supports mobility within a given PDSN.

It is important to note that this standard uses circuit-switched MSC resources (ANSI-41 Domain and SS7 Signaling) to handle both the voice and packet radio control resources (access registration, QoS profile based verification, paging, etc.).

This model will certainly be modified with the emergence of the all-IP network, which will support data capacities largely exceeding those offered in the

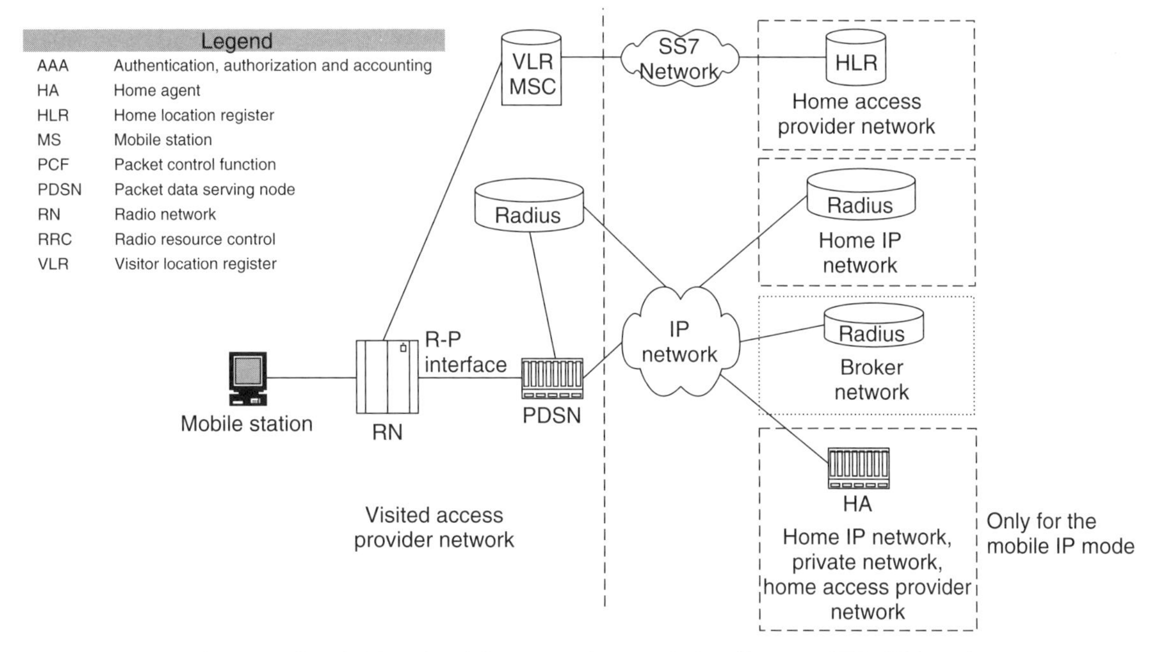

Figure 6. Simple IP and mobile IP according to 3GPP2. (*Source*: 3GPP2 Website.)

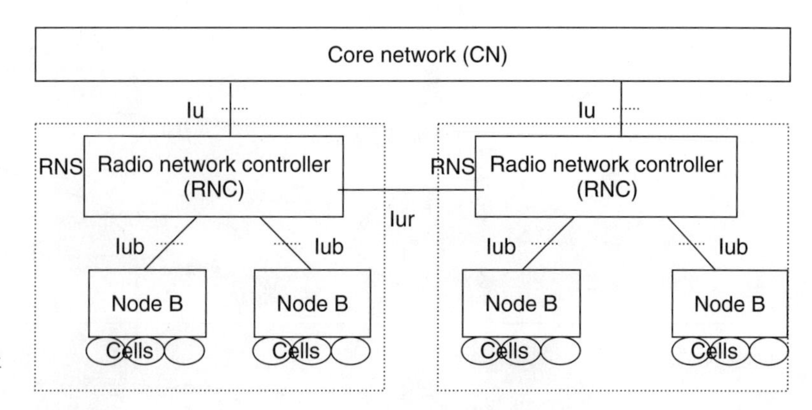

Figure 7. UMTS (W-CDMA) radio access network architecture

Table 4. Main Characteristics of TDD and FDD Modes [4]

Mode	FDD	TDD
Multiple-access method	DS-CDMA (direct-sequence CDMA)	TD/CDMA
Carrier chip rate	3.84 Mchips/s	
Channel spacing	5 MHz (nominal)	
Frame length	10 ms	
Frame structure	15 slots/frame	
Modulation	QPSK	
Spreading factors	4–512	1–16
Forward error correction (FEC) codes	Convolutional coding ($R = \frac{1}{2}$ or $\frac{1}{3}$, constraint length $K = 9$) Turbo coding for BER $<10^{-3}$ (8-state PCCC $R = \frac{1}{3}$) Service-specific coding	
Frequency bands	1920–1980 MHz — uplink (mobile to BTS) 2110–2170 MHz — downlink (BTS to mobile) Duplex separation — 190 MHz	1900–1920 MHz 2010–2025 MHz

mixed circuit/packet data scheme. This new architecture is expected to be ready by the end of 2002. The future network will enable services such as voice over IP, multimedia calls, and video streaming, based on full use of IP Transport in both the radio access and core networks. This step will mark the end of the classical circuit-switched core network.

6.2. WCDMA

6.2.1. Radio Access Network

6.2.1.1. Deployment, Duplex Mode. WCDMA deployment involves a multilayer cellular network, with macrocells (0.5–10 km in range) for large-scale coverage, microcells (50–500 m) for hotspots, and picocells (5–50 m) for indoor coverage. Handover is ensured both among WCDMA cells and between WCDMA and GSM cells, without any perceptible cut or degradation of quality.

As indicated in Section 4, the air interface adopted by ETSI in January 1998 is based on two harmonized modes: FDD/WCDMA for the paired bands and TDD/TDCDMA for the unpaired bands. UMTS is to be deployed using at least two duplex 5-MHz bands, and must ensure interworking with GSM and dual-mode FDD/TDD operation.

FDD mode is appropriate for all types of cells, including large cells, but is not well adapted to asymmetric traffic. TDD is, by definition, more flexible to support traffic asymmetry, but it requires synchronization of the base stations, and is not appropriate for large cells because of the limited guard times between time slots. Table 4 lists the main characteristics of the two modes.

The WCDMA FDD mode is based on CDMA principles with a bandwidth of 5 MHz. One major difference with the IS95 standard is that no synchronization is required among base stations, thus allowing easier deployment for operators. One of the key advantages of WCDMA is its high spectral efficiency, or capacity per unit of spectrum [typically expressed in kilobits per second per megahertz or (kbps/MHz)]. Depending on the services offered, WCDMA offers 2–3 times the capacity of GSM with the same amount of spectrum.

The TDD mode is based on a mix between TDMA and CDMA. The TDD frame has 15 time slots and each time slot supports several simultaneous CDMA communications.

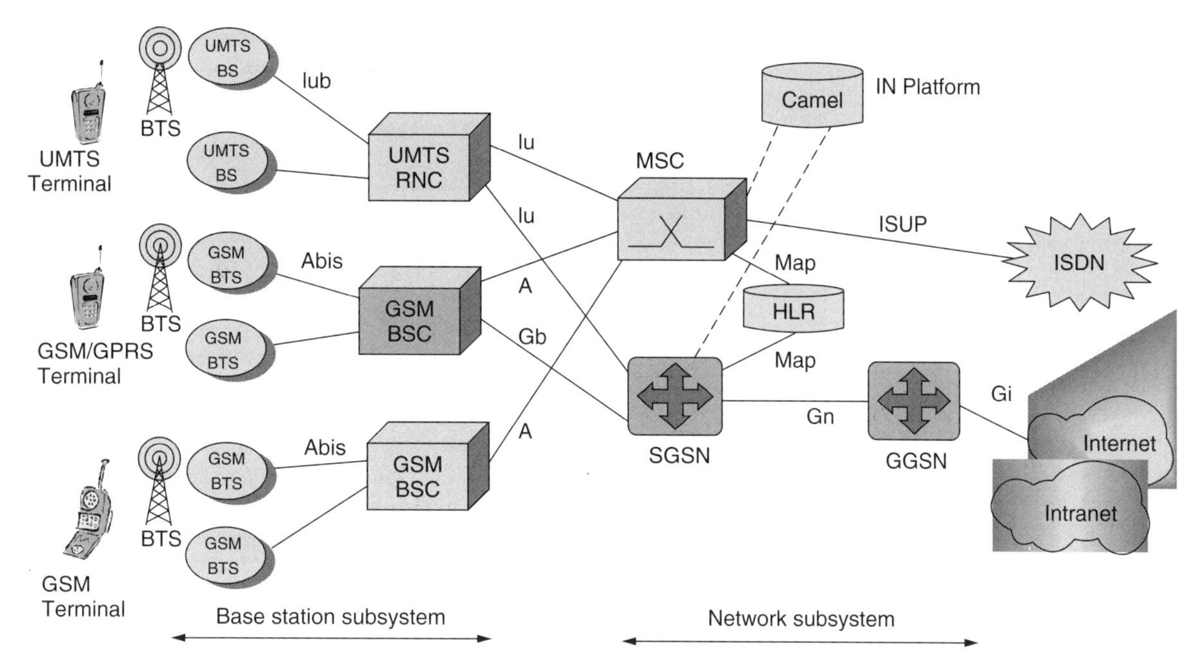

Figure 8. General architecture of the UMTS network (release 99).

6.2.1.2. Radio Access Network Architecture. Figure 8 represents the logical architecture of the UMTS radio access network. The radio network subsystem includes the radio base stations (node B) and the radio network controller (RNC).

This architecture is similar to that of the GSM radio access network. Iu represents the interface between the RNC and the core network. Iub represents the interface between the node B and the RNC. One key difference with GSM is the existence of the Iur interface between RNCs. This interface enables the management of soft handover between two node Bs belonging to two separate RNCs, independently from the core network. "Soft handover" means that the mobile terminal moving from one cell to another has links with both base stations during handover.

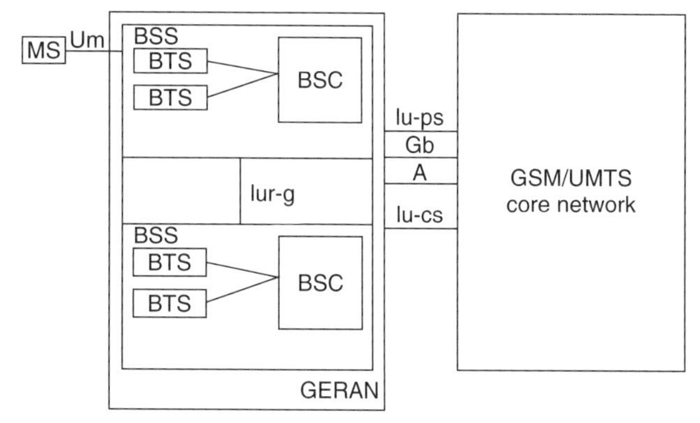

Figure 9. GSM/EDGE radio access network (GERAN) architecture.

6.2.1.3. WCDMA Radio Dimensioning. Dimensioning of the radio access network takes into account both coverage and capacity. To ensure physical coverage, the cell radius is based on link budgets calculated according to the different service types, propagation environment, indoor penetration assumptions, and loading factor. It is recalled that, in CDMA networks, each user contributes to the interference level, which causes the cell to "shrink," a phenomenon called "cell breathing." Generally, a loading factor of 50–70% is taken. This factor is based on a theoretical "100%" load at which interference tends to infinity.

On the basis of geomarketing data, the number of users per square kilometer and the average data rate per user during busy hour yields a load in terms of data rate (Mbps or kbps) or erlangs per square kilometer. The corresponding cell size is computed to handle this load. This cell area is matched against that computed from the link budgets, and when the offered load exceeds the capacity made available based on the coverage criterion, it is necessary to add carriers and/or sites.

6.2.2. WCDMA Core Network. The WCDMA network architecture is illustrated in Fig. 9. As mentioned, the radio access network is comprised of specific WCDMA base stations and RNCs. The network subsystem requires an SGSN (described in Section 4.2), which may be specific to the UMTS system or shared with the GPRS service, and a specific operations and maintenance center (OMC). Other NSS components, such as the DNS and GGSN, can also be shared with the existing GPRS system.

The UMTS Release 99 core network comprises two distinct domains: circuit-switched (CS) and packet-switched (PS), as in GSM/GPRS networks. The core network elements are the same as in these networks: MSC (mobile switching center) for CS services, and SGSN and GGSN for PS services.

In Release 99, ATM was chosen for transport in the access network. This choice makes it possible to support all the types of services offered: voice, circuit-switched data, and packet-switched data. Different ATM adaptation layers (AALs) are used: AAL2 for the voice or data on the Iu-cs circuit-switched domain interfaces, Iur and Iubis. AAL5 is used for signaling and for user data on the Iu-ps packed-switched domain interface.

6.3. EDGE

EDGE, as mentioned in Section 3.2., is a convergent solution for TDMA (IS136), and GSM.

Since July 2000, the 3GPP has been responsible for standardization of the GSM/EDGE Radio Access Network (GERAN). In order to harmonize this access technology with others developed for IMT-2000, notably WCDMA and cdma2000, the 3GPP chose to align the QoS requirements for EDGE with those of the other technologies (see Table 1). However, because the data rate varies with distance from the base station in EDGE, only interactive and background type services are expected to be offered initially, even though the EDGE standard encompasses enhanced circuit-switched data (ECSD) services.

The core network does not differ from that of a GPRS core network; however, because of the increased data rate available, capacity of the data links between base stations and BSCs, and between the BSCs and MSC and SGSN must be dimensioned appropriately. The radio access network is based on existing 2G infrastructure, as illustrated in Fig. 9.

The introduction of the Iu interfaces enables fast handover in the packet-switched domain, real-time services in the packet-switched domain, and enhanced multiplexing capabilities in both the packet- and circuit-switched domains. The A and Gb interfaces are maintained for compatibility with legacy systems.

The GERAN has been standardized for the following GSM frequency bands: 900, 1800, 1900, 400, and 700 MHz. Note that these bands do not include those allocated to IMT-2000 systems; hence this technology can be used by operators without IMT-2000 spectrum. However, it is recalled that the highest data rate possible is 384 kbps versus the theoretical maximum of 2 and 2.4 Mbps, for WCDMA and cdma2000, respectively.

7. LICENSE ASSIGNMENT AND ECONOMIC IMPLICATIONS OF IMT-2000 (4/2001)

As mentioned in Section 3.3, national regulators played a major role in shaping the development of IMT-2000 3G networks and services. In Europe, they focused on three important factors: 3G license cost and allocation mechanism, number of 3G operators per country, and rollout/coverage obligations. With an initial goal of encouraging competition to enhance service offerings and ensure reasonable prices for the end customer, regulatory agencies chose to allocate spectrum to at least one or two new entrants in each country, requiring, in most cases, incumbent GSM operators to sign roaming agreements with these new entrants, enabling them to offer their customers national coverage for voice service. The choice of assignment method — auctions or comparative hearings — led to highly divergent license costs in Europe. While auctions generated overall state revenues ranging from 85 to 630 Euros per capita (with Germany and the United Kingdom gaining the highest amounts), comparative hearings yielded revenues ranging from approximately 0 to 45 Euros per capita; a notable exception was France, where the license fee was set, shortly after the English auctions, at 335 Euros per capita. However, only two candidates submitted applications for the four licenses available; other candidates cited the high cost as the main deterrent. Rollout requirements, set primarily in conjunction with comparative hearings, generally call for initial deployment as of 2002, with coverage of the main cities and extension to 50–80% of the population within the following 6–8 years. In Japan, license costs were minimal, while in the United States, 3G spectrum auctions led to per capita costs similar to those observed in the United Kingdom and Germany.

The combined obligations of paying license fees and building out entirely new radio access networks, which represent some 80% of the initial IMT-2000 investment within a set time frame, have placed a considerable burden on both operators and equipment manufacturers. In Sweden, for example, operators have formed joint ventures to build the radio infrastructure in areas outside of the major metropolitan centers. In such cases, each operator exploits its own spectrum, but shares the cost of buildout in order to focus on service development [6]. Opinions diverge as to the economic prospects for IMT-2000. The outlook is optimistic in Japan, where customer awareness of mobile data services is high because of I-mode, and where willingness to pay for such services is among the highest in the world. Although Europe is well advanced in terms of second-generation mobiles penetration, mobile data services have gotten off to a slow start, with the disappointingly limited scope and speed of WAP. As better data rates and volume-based billing become available with GPRS, customers may more willingly adopt new data services offered by a wide panel of providers [6]. This "education" stage is considered crucial to the rapid adoption of IMT-2000.

In the United States, mobile data have generally been restricted to very low-data-rate exchanges, while fixed lines remain the preference for Internet access. Moreover, customers are used to free Internet content. This means that innovative location-based, customized services will have to be developed for operators to recoup their expenses.

In all cases, with respect to 2G, new players will be joining the value chain: content suppliers, service brokers, virtual mobile network operators, and network resource brokers. While some of these new players may simply be affiliates of today's major mobile operators, some will be entirely new entities such as retailers, banks, insurance companies, and entertainment companies. This means that the corresponding 3G revenues, whether from the end user, a third party, or advertising, will be spread more thinly than in the 2G world, leading in all likelihood to continentwide consolidation [7] of operators.

8. BEYOND 3G

With many uncertainties remaining as to the economic viability of IMT-2000 in the near term, research and development efforts are already turning toward the next phase, coined "beyond 3G." This phase, expected to emerge as of 2005, is based on seamless roaming among heterogeneous wireless environments, including 3G mobile networks and indoor wireless facilities such as radio LANs and Bluetooth networks. Another topic currently being explored is the cooperation of 3G networks and broadcast standards such as digital video or audio broadcasting (DVB, DAB).

The goal is to optimize the service offered to the end customer by taking advantage of the spectral resources available. For example, a train passenger can connect with the company intranet via the UMTS network with the available bit rate and quality inherent in this support service; then, when this passenger enters a train station or airport lounge equipped with a wireless LAN, the terminal detects the new network and vice versa, and switches over to this new broadband resource. If any background tasks are being performed, they are uninterrupted when the terminal goes from one environment to the other.

As a complement to IMT-2000, DVB-T can provide fast one-to-many services such as weather, sports scores, or stockmarket values. It can thus significantly ease the burden on IMT-2000 frequency resources which are then used for bi-directional wideband links.

9. CONCLUSION AND DISCUSSION

IMT-2000, a family of third-generation mobile standards, is on the verge of deployment. The prospect of offering a wide range of mobile multimedia services prompted numerous incumbent operators and new entrants to spend sometimes exorbitant amounts on 3G licenses. Undoubtedly, the first years will be characterized by turbulence in infrastructure rollout; technical teams will have to be trained, the manufacturers will have to ensure adequate levels of production, and — most importantly — new sites will have to be negotiated. Operators' finances will be strained as they seek to attract customers and generate adequate revenues early enough to offset the investments made. Smaller operators and new entrants with no infrastructure may find the expenditure and task too daunting and, willingly or unwillingly, merge with a larger competitor.

As networks and services reach cruising speed, operators will focus on enhancing services by finding new spectral resources in other frequency bands (e.g., GSM bands) and by encouraging cooperation among different wireless networks, both fixed and mobile. Underlying the success of this evolution are the continued efforts of standards organizations such as 3GPP and 3GPP2, the policies developed by national and international regulatory bodies that will have learned the lessons of 3G, the ability of operators and service suppliers to anticipate customers' needs and desires and, most importantly, the customers themselves.

Acknowledgments
The authors would like to thank their colleagues at France Télécom R&D as well as the partners of the ACTS 364/TERA project for their useful support and discussions.

BIBLIOGRAPHY

1. UMTS Forum, Report 5, 1998.
2. *3rd Generation Partnership Project Technical Specification* TS 23.107 v3.3.0, June 2000.
3. *Global Wireless Magazine*, Crain Communications Inc., March–April 2001, p. 18.
4. H. Holma and A. Toskala, eds., *WCDMA for UMTS — Radio Access for Third Generation Mobile Communications*, Wiley, Chichester, UK, 2000, p. 2285.
5. *Pyramid Research Perspective*, Europe, Feb. 2, 2001.
6. *Strategy Analytics SA Insight*, Dec. 8, 2000.
7. L. Godell et al., *Forrester Report: Europe's UMTS Meltdown*, Dec. 2000.

FURTHER READING

Useful Websites

Standards

ITU: *http://www.itu.int/imt/*
3GPP: *http://www.3gpp.org*
3GPP2: *http://www.3gpp2.org*
(member organizations can be accessed from these sites)

Manufacturers (Network Equipment and/or Terminals)

Nokia: *http://www.nokia.com*
Ericsson: *http://www.ericsson.com*
Motorola: *http://www.motorola.com*
Lucent Technologies: *http://www.lucent.com*
Qualcomm: *http://www.qualcomm.com*
Alcatel: *http://www.alcatel.com*
NEC: *http://www.nec.com*
Nortel Networks: *http://www.nortel.com*
Siemens: *http://www.siemens.com*
Toshiba: *http://www.toshiba.co.jp/worldwide/*
Sharp: *http://www.sharp-usa.com*
Sanyo: *http://www.sanyo.com*
Sony: *http://sony.com*
Panasonic, Matsushita Electric Industrial Co., Ltd.: *http://www.panasonic.co.jp/global/*

Kyocera Wireless Corp.: *http://www.kyocera-wireless.com*

Philips: *http://www.philips.com*

Journals

Global Wireless Magazine: *http://www.globalwireless-news.com*

Books

T. Ojanperä and R. Prasad, eds., *Wideband CDMA for Third Generation Mobile Communications*, Artech House, Boston, 1998.

ACRONYMS

AMPS	American Mobile Phone System
ANSI	American National Standards Institute
API	Application programming interface
ARIB	Association of Radio Industries and Businesses (Japan)
BSC	Base station controller
BSS	Base station subsystem
BTS	Base transceiver station
CAMEL	Customized Applications of Mobile Network Enhanced Logic
CATT	Chinese Academy of Telecommunication Technology
CDMA	Code-division multiple access
CN	Core network
CS	Circuit switched
CWTS	China Wireless Telecommunication Standards Group
DAB	Digital audio broadcasting
D-AMPS	Digital AMPS
DECT	Digital European Cordless Telephony
DHCP	Dynamic Host Configuration Protocol
DNS	Domain name server
DVB	Digital video broadcasting
DVB-T	Digital video broadcasting-terrestrial
ECSD	Enhanced circuit-swtiched data
EDGE	Enhanced Data for GSM Evolution
FCC	Federal Communications Commission (USA)
FDD	Frequency-division duplex
GERAN	GSM EDGE Radio Access Network
GGSN	Gateway GPRS Support Node
GPRS	General Packet Radio Service
GSM	Global System for Mobile communications
HA	Home agent
HDR	High Data Rate standard
HLR	Home location register
HS	High-speed
HSCSD	High-speed circuit-switched data
HTML	Hypertext markup language
IETF	Internet Engineering Task Force
IMT	International Mobile Telecommunications
IN	Intelligent network
INAP	Intelligent network application protocol
IP	Internet Protocol
ITU	International Telecommunications Union
IWF	Inter-working function
LAN	Local area network
MAC	Medium access control
MAP	Mobile application part
MC	Multicarrier
MExE	Mobile execution environment
MSC	Mobile switching center
MS	Mobile station
MT	Mobile terminal
NSS	Network subsystem
NTIA	National Telecommunications and Information Administration (USA)
OA&M	Operations, administration and maintenance
OHG	Operators Harmonization Group
OMC	Operations and maintenance center
PCF	Packet control function
PCM	Pulse code modulation
PCS	Personal communications system
PCU	Packet control unit
PDA	Personal digital assistant
PDSN	Packet data serving node
PS	Packet-switched
QOS	Quality of service
RAN	Radio access network
RLC	Radio link control
RN	Radio network
RNC	Radio network controller
RRC	Radio resource control
RTT	Radio transmission technology
SDU	Selection distribution unit
SGSN	Serving GPRS Support Node
SIM	Subscriber identity module
SMS	Satellite-based mobile service; Short message service
STK	SIM tool kit
TACS	Total Access Communication System (UK)
TDD	Time-division duplex
TDMA	Time Division Multiple Access
TD-SCDMA	Time-division synchronous CDMA
TFO	Tandem Free Operation
TIA	Telecommunications Industry Association (North America)
TTA	Telecommunications Technology Association (Korea)
TTC	Telecommunications Technology Committee (Japan)
UDI	Unrestricted digital information
UICC	Universal integrated circuit card
UMTS	Universal Mobile Telecommunication System
USIM	User services identity module
UWCC	Universal Wireless Communications Consortium
VHE	Virtual home environment
WAP	Wireless Applications Protocol
WARC	World Administrative Radio Conference
W-CDMA	Wideband code-division multiple access
WIN	Wireless intelligent network
WML	Wireless markup language

INFORMATION THEORY

Randall Berry
Northwestern University
Evanston, Illinois

1. INTRODUCTION

The key concepts for the field of information theory were introduced by Claude E. Shannon in the landmark two-part paper "A mathematical theory of communication" [1,2] published in 1948. The emphasis of Shannon's work was on understanding the fundamental limits of communication systems. Subsequently, this theory has expanded and is now widely recognized as providing the theoretical framework for modern digital communication systems. The following surveys some of the basic ideas of information theory. In addition to communication theory, information theory has proved useful in a variety of other fields, including probability, statistics, physics, linguistics, and economics. Our emphasis here is on those aspects related to communication systems, specifically, the areas closest to Shannon's original work; such topics are also referred to as Shannon theory.

The basic problems to be discussed concern a generic communication system as in Fig. 1. The object is to convey a message, generated by the information source, to the destination. Information sources are modeled probabilistically, namely, the source generates a message from a set of possible messages according to a given probability distribution. The transmitter takes this message and maps it into a signal that is then transmitted over the communication channel. The received signal may differ from the transmitted signal as a result of additive noise and other channel impairments; these are also modeled in a probabilistic framework. The receiver attempts to extract the original message from the received signal. The above framework is general enough to accommodate a wide range of systems. For example, the message could be written text, an analog speech signal or digital data; possibilities for the channel include an optical fiber, the atmosphere or a storage medium.

As shown in Fig. 1, the transmitter and receiver are often divided into two stages. At the transmitter, the source coder first represents the incoming message as a binary sequence; the channel coder maps this binary sequence into the transmitted signal. The corresponding inverse operations are performed at the receiver. This two-stage implementation has many practical advantages. For example, this allows the same channel coder/decoder to be used for several different information sources. Moreover, a key result of information theory, the *joint source channel coding theorem*, states that for a wide range of situations, the source coder and channel coder can be designed separately without any loss of optimality.

Two of the main questions addressed by this theory concern the achievable performance of the source coder/decoder and channel coder/decoder:

1. What is the minimum number of bits needed to represent messages from a given source?
2. What is the maximum rate that the information can be transmitted over the channel with arbitrarily small probability of error?

The first question addresses data compression; the second, reliable communication. In Shannon's original work, both of these questions where posed and largely answered for basic models of the source and channel. The answer to these questions is given in terms of two basic information measures, *entropy* and *mutual information*.

In the following, we first discuss the basic information measures used in this theory. We then discuss the role of these quantities in the problems of source coding and channel coding.

2. INFORMATION MEASURES

In information theory, information sources and communication channels are modeled probabilistically. The information measures used are defined in terms of these stochastic models. First, we give the definitions of these measures for discrete random variables, For example, a random variable X taking values in a finite set χ. The set χ is often called the *alphabet* and its elements are called *letters*. In Section 2.2, continuous random variables are addressed.

2.1. Discrete Models

The first information measure we consider is the *entropy* $H(X)$ of a random variable X with alphabet χ. This

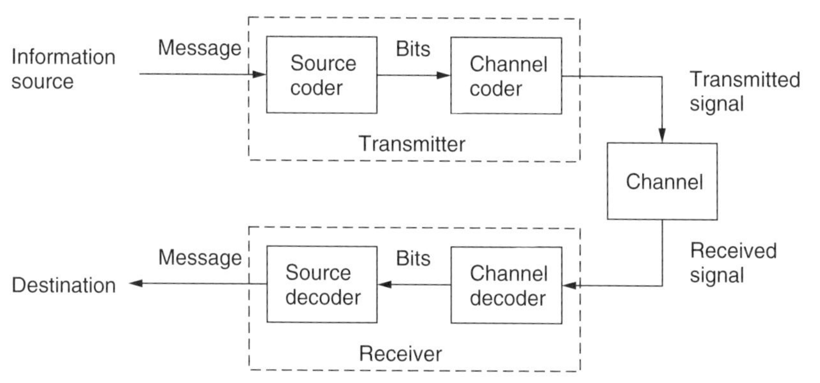

Figure 1. Model of a generic communication system.

quantity is defined by

$$H(X) = -\sum_{x \in \chi} p(x) \log p(x)$$

where $p(x) = \Pr(X = x)$ is the probability mass function of X. The base of the logarithm in this definition determines the units in which entropy is measured. Common bases are 2 and e with corresponding units called *bits* and *nats*, respectively. Entropy can be interpreted as a measure of the uncertainty in a random variable. For example, entropy satisfies

$$0 \leq H(X) \leq \log |\chi|$$

where $|\chi|$ denotes the size of the alphabet. The upper bound is achieved if and only if the random variable is uniformly distributed on this set; this corresponds to maximum uncertainty. The lower bound is achieved if and only if the random variable is deterministic, which corresponds to minimum uncertainty.

The significance of entropy arises from considering long sequences of random variables. Using entropy, the set of all sample sequences of a given length N can be divided into a set of *typical* sequences, which occur with high probability, and a set of atypical sequences, which occur with a probability that vanishes as N becomes large. Specifically, let $\{X_i\}_{i=1}^{\infty}$ be a sequence of independent, identically distributed (i.i.d.) random variables, where each X_i has entropy $H(X)$. We denote by $p(x^N)$ the probability of a sequence x^N of length N. For a given $\delta > 0$, the set of all sequences x^N whose probability satisfies

$$\left| \frac{-\log p(x^N)}{N} - H(X) \right| < \delta$$

is called the *typical set*.[1] For any $\delta > 0$, as N increases, the probability of the typical set approaches one. This result, proved by Shannon, is known as the "asymptotic equipartition property" (AEP). A consequence of the AEP is that the size of the typical set is approximately $2^{NH(X)}$; thus, entropy characterizes the rate at which the typical set grows with N. As will be discussed in the next section, this result has a natural source coding interpretation.

The entropy of two or more random variables is defined analogously; For instance, for two random variables X and Y, we obtain

$$H(X, Y) = -\sum_{x,y} p(x, y) \log p(x, y)$$

where $p(x, y)$ is the joint probability mass function. For a discrete stochastic process, $\{X_i\}_{i=1}^{\infty}$, the *entropy rate* is defined to be the asymptotic per letter entropy:

$$H_{\infty}(X) = \lim_{n \to \infty} \frac{1}{n} H(X_1, \ldots, X_n)$$

Assuming that the process is stationary will ensure that this limit exists. Using the entropy rate, the AEP can be generalized to every stationary ergodic process with a finite alphabet; this result is known as the *Shannon–McMillan theorem* [4].

The *conditional entropy* of Y given X is defined by

$$H(Y \mid X) = -\sum_{x,y} p(x, y) \log p(y \mid x)$$

where $p(y \mid x) = \Pr(Y = y \mid X = x)$. Conditional entropy represents the average uncertainty in Y when X is given. From these definitions, various relations can be derived such as

$$H(X, Y) = H(X) + H(Y \mid X)$$

The next information measure we discuss is the *mutual information* between two random variables X and Y. This is defined by

$$I(X; Y) = H(X) - H(X \mid Y)$$

or equivalently

$$I(X; Y) = \sum_{x,y} p(x, y) \log \frac{p(x, y)}{p(x)p(y)}$$

Mutual information is a measure of the reduction of uncertainty in X when Y is given. This quantity is symmetric: $I(X;Y) = I(Y;X)$. It is also nonnegative and equal to zero only when X and Y are independent.

Mutual information is a special case of another quantity called *relative entropy*.[2] Relative entropy is a measure of the difference between two probability mass functions, $p(x)$ and $q(x)$. This quantity is defined as

$$D(p\|q) = \sum_{x} p(x) \log \left(\frac{p(x)}{q(x)} \right).$$

Relative entropy is nonnegative and only equal to zero when $p = q$. The mutual information between X and Y is given by the relative entropy between the joint distribution $p(x, y)$ and the product of the marginal distributions $p(x)p(y)$.

2.2. Continuous Models

Corresponding information measures can be defined for continuous valued random variables. In this case, instead of entropy, the appropriate measure is given by the *differential entropy*.

Differential entropy is defined by

$$h(x) = -\int f(x) \log f(x)\, dx$$

where $f(x)$ denotes the probability density function of the random variable X. Differential entropy has many similar properties to entropy for a discrete random variable,

[1] These sequences are also called *weakly typical*. A stronger version of typicality can be defined using the method of types [3].

[2] Relative entropy is also referred to by a variety of other names including *Kullback–Leibler distance* and *information divergence*.

however it does not share all of the properties. One important difference is that differential entropy depends on the choice of coordinates, while entropy is invariant to coordinate transformations.

The mutual information between two continuous random variables is given by

$$I(X;Y) = h(X) - h(X \mid Y)$$
$$= \int f(x,y) \log \frac{f(x,y)}{f(x)f(y)}\, dx\, dy$$

Unlike differential entropy, mutual information is invariant to coordinate transformations, and is a natural generalization of its discrete counterpart.

3. SOURCE CODING

Source coding or *data compression* refers to the problem of efficiently representing messages from a given information source as binary sequences.[3] Here efficiency refers to the length of the resulting representation.

Information sources are modeled as random processes, depending on the situation several different models may be appropriate. One class of model consists of discrete sources; in this case, the source is modeled as a sequence of discrete random variables. When these random variables are i.i.d. (independent, identically distributed), this model is called a *discrete memoryless source* (DMS). Another class of sources are analog source models, where the output of the source is a continuous-time, continuous-valued waveform. For discrete sources, compression may be *lossless*; that is, the original source output can be exactly recovered from the encoded bit sequence. On the other hand, for analog sources, compression must be *lossy*, since an infinite number of bits are required to represent any real number. In this case, source coding consists of sampling the source and quantizing the samples. We first discuss lossless source coding of discrete sources, here the information theoretic limitations are provided by the source's entropy rate. In Section 3.5 we consider analog sources, in this case the fundamental limits are characterized by the *rate distortion function*, which is defined using mutual information.

3.1. Classification of Source Codes

For a discrete source, a source code is a mapping or encoding of sequences generated by the source into a corresponding binary strings. An example of a source code for a source with an alphabet $\{a, b, c, d\}$ is

$$a \leftrightarrow 00 \qquad b \leftrightarrow 01 \qquad c \leftrightarrow 10 \qquad d \leftrightarrow 11$$

Notice this code is lossless, because each source letter is assigned a unique binary string. This is a fixed-length code, since each codeword has the same size. Source codes may also be variable-length, as in the Morse code used for English text. The performance of a variable-length source code is given by the expected code length, averaged over the source statistics. This will clearly be smaller if shorter codewords are used for more likely symbols. Instead of encoding each source letter individually as in the above example, a source code may encode blocks of source letters at once.

For variable-length codes, it is important to be able to tell where one codeword ends and the next begins. For example, consider a code which assigns 0 to a and 00 to b; in this case the string 00 could represent aa or b. A code with the property that any sequence of concatenated codewords can be correctly decoded is called *uniquely decodable*. A special class of uniquely decodable codes are codes where no codeword is the prefix of another; these are called *prefix-free codes* or *instantaneous codes*. The class of uniquely decodable codes is larger than the class of prefix-free codes. However, it can be shown that the performance attained by any uniquely decodable code can also be achieved by a prefix-free code, that is, it is sufficient to only consider prefix-free codes. This results follows from the *Kraft inequality*, which states that a prefix-free code can be found with codewords of lengths $l_1, l_2, \ldots, l_k$ if and only if

$$\sum_{i=1}^{k} 2^{-l_i} \le 1$$

The set of codeword lengths for any uniquely decodable code must also satisfy this inequality.

3.2. Variable-Length Source Codes

In this section we discuss the performance of variable-length, uniquely decodable codes. Suppose that the code encodes a single letter from a DMS at a time. From the Kraft inequality, it can be shown that the minimum average length of any such code must be greater than or equal to the source's entropy. For a given distribution of source letters, a variable-length source code with minimal average length can be found using a simple iterative algorithm discovered by Huffman [5]. The average length of a Huffman code can be shown to be within one bit of source's entropy

$$H(X) \le L_H \le H(X) + 1$$

where $H(X)$ is the source's entropy and L_H is the average codeword length of the Huffman code. The above can be extended by considering a code that encodes blocks of N source letters at a time. A Huffman code can then be designed by treating each block as a single "supersymbol." In this case, if L_H^N is the average codeword length, then the compression ratio (the average number of encoded bits per source symbol) satisfies

$$H(X) \le \frac{L_H^N}{N} \le H(X) + \frac{1}{N}$$

Hence, as N becomes large, one can achieve a compression ratio that is arbitrarily close to the source's entropy. At times, this result is referred to as the *variable-length source coding theorem*. The converse to this theorem

[3] This easily generalizes to the case where messages are to be represented as strings in an arbitrary, finite-sized *code alphabet*; we focus on the binary case here.

states that no lossless, variable-rate code can achieve a compression ratio smaller than the sources entropy. This can be generalized to discrete sources with memory; in this case, the entropy rate of the source is used.

3.3. The AEP and Shannon's Source Coding Theorem

The AEP, discussed in Section 2.1, provides additional insight into the attainable performance of source codes. Consider a DMS with an alphabet of size K. There are K^L possible sequences of length L that the source could generate. Encoding these with a fixed-length, lossless source code requires approximately $L\log_2(K)$ bits, or $\log_2(K)$ bits per source letter. However, from the AEP, approximately $2^{LH(X)}$ of these sequences are typical. Therefore a lossy source code can be designed that assigns a fixed-length codeword to each typical sequence and simply disregards all nontypical sequences. This requires approximately $H(X)$ bits per source letter. The probability of not being able to recover a sequence is equal to the probability that the sequence is atypical, which becomes negligible as L increases. Making this more precise, it can be shown that for any $\varepsilon > 0$ a source code using no more than $H(X) + \varepsilon$ bits per source letter can be found with arbitrary small probability of decoding error. This is the direct part of *Shannon's source coding theorem*. The converse to this theorem states that any code that uses fewer than $H(X)$ bits per source letter will have a probability of decoding failure that increases to one as the block length increases.[4]

3.4. Universal Source Codes

Approaches such as Huffman codes require knowledge of the source statistics. In practice it is often desirable to have a data compression algorithm that does not need a priori knowledge of the source statistics. Among the more widely used approaches of this type are those based on two algorithms developed in 1977 and 1978 by Lempel and Ziv [6,7]. For example, the 1978 version is used in the UNIX compress utility. The 1977 algorithm is based on matching strings of uncoded data with strings of data already encoded. Pointers to the matched strings are then used to encode the data. The 1978 algorithm uses a dictionary that changes dynamically based on previous sequences that have been encoded. Asymptotically, the Lempel–Ziv algorithms can be shown to compress any stationary, ergodic source to its entropy rate [8]; such an approach is said to be *universal*.

3.5. Lossy Compression: Rate Distortion Theory

Next we consider encoding analog sources. We restrict our attention to the case where the source generates real-valued continuous-time waveforms.[5] The basic approach for compressing such a source is to first sample the waveform, which yields a sequence of continuous-valued random variables. If the source is a band-limited, stationary random process, then it can be fully recovered when sampled at twice its bandwidth (the Nyquist rate). After sampling, the random variables are then quantized so that they can be represented with a finite number of bits. Quantization inherently introduces some loss or distortion between the original signal and the reconstructed signal at the decoder. Distortion can be reduced by using a finer quantizer, but at the expense of needing more bits to represent each quantized value. The branch of information theory called rate-distortion theory investigates this tradeoff.

A scalar quantizer can be regarded as a function that assigns a quantized value or reconstruction point $\hat{x}$ to each possible sample value x. A rate R quantizer has 2^R quantization points. The loss incurred by quantization is quantified via a distortion measure, $d(x, \hat{x})$, defined for each sample value and reconstruction point. A common distortion measure is the squared error distortion given by

$$d(x, \hat{x}) - (x - \hat{x})^2$$

Given a probability distribution for the sample values and a fixed rate, the set of quantization points that minimize the expected distortion can be found using an iterative approach call the Lloyd–Max algorithm [9].

Instead of quantizing one sample at a time, performance can be improved by using a vector quantizer, that is, by quantizing a block of N samples at once. A rate R vector quantizer consists of 2^{NR} reconstruction points, where each point is an N dimensional vector. In this case, the distortion between a sequence of sample values and the corresponding reconstruction points is defined to be the average per sample distortion. A distortion D is said to be achievable at rate R if there exists a sequence of rate R quantizers with increasing block size N for which the limiting distortion is no greater than D. For each distortion D, the infimum of the rates R for which D is achievable is given by the *operational rate distortion function*, $R_{\text{op}}(D)$.

For an i.i.d. source $\{X_n\}$, the *(Shannon) rate distortion function*, $R(D)$, is defined to be

$$R(D) = \min I(X; \hat{X})$$

where the minimization is over all conditional distributions of $\hat{X}$ given X which yield a joint distribution such that the expected value of $d(X, \hat{X})$ is no greater than D. Here X is a scalar random variable with the same distribution as X_n. For a large class of sources and distortion measures, rate distortion theorems have been proved showing that $R_{\text{op}}(D) = R(D)$. Therefore, the fundamental limits of lossy compression are characterized by the information rate distortion function. As an example, consider an i.i.d. Gaussian source with variance σ^2. In this case, for squared error distortion, the rate distortion function is given by

$$R(D) = \begin{cases} \dfrac{1}{2}\log\dfrac{\sigma^2}{D}, & 0 \le D \le \sigma^2 \\ 0, & D > \sigma^2 \end{cases}$$

[4] This is sometimes referred to as the *strong converse* to the source coding theorem. The *weak converse* simply states that if fewer than $H(X)$ bits per source letter are used then the probability of error can not be zero.

[5] More generally one can consider cases where the source is modeled as a vector-valued function of time or a vector field.

In addition to i.i.d. sources, rate distortion theory has been generalized to a large variety of other sources [e.g., 10].

3.6. Distributed Source Coding

An interesting extension of the preceding models are various distributed source coding problems, that is situations where there are two or more correlated information sources being encoded at different locations. One information theoretic result for this type of situation is known as the *Slepian–Wolf theorem* [11]; this theorem implies the surprising result that the best (lossless) compression achievable by the two sources without cooperating is as good as what can be achieved with complete cooperation. For distributed lossy compression, characterizing the rate distortion region remains an open problem.

4. CHANNEL CODING

We now turn to the problem of communicating reliably over a noisy channel. Specifically given a bit stream arriving at the channel coder, the object is to reproduce this bit stream at the channel decoder with a small probability of error. In addition, it is desirable to transmit bits over the channel with as high a rate as possible. A key result of information theory is that for a large class of channels, it is possible to achieve arbitrarily small error probabilities provided that the data rate is below a certain threshold, called the *channel capacity*. As with the rate distortion function, capacity is defined in terms of mutual information. Conversely, at rates above capacity, the error probability is bounded away from zero. Results of this type are known as *channel coding theorems*.

Mathematically a channel is modeled as a set of possible input signals, a set of possible output signals and a set of conditional distributions giving the probability for each output signal conditioned on a given input signal. As with sources, several different types of channel models are studied. One class consists of discrete channels where the input and output signals are sequences from a discrete alphabet. Another class of channels consists of waveform channels where the input and output are viewed as continuous functions of time; the best-known example is the band-limited additive white Gaussian noise channel.

5. DISCRETE MEMORYLESS CHANNELS

A *discrete memoryless channel* (DMC) is a discrete channel, where the channel input is a sequence of letters $\{x_n\}$ chosen from a finite alphabet. Likewise, the channel output is also a sequence $\{y_n\}$ from another finite alphabet. Each output letter y_n depends statistically only on the corresponding input letter x_n. This dependence is specified by a set of transition probabilities $P(y \mid x)$ for each x and y. Two examples of DMC's are shown in Fig. 2. On the left is a binary symmetric channel, where both the input and output alphabet are $\{0, 1\}$. Each letter in the input sequence is reproduced exactly at the output with probability $1 - \varepsilon$ and is converted into the opposite letter with probability ε. On the right of Fig. 2 is a binary erasure channel; in this case the input alphabet is $\{0, 1\}$, but the output alphabet is $\{0, 1, E\}$. Each input letter is either received correctly, with probability $1 - \varepsilon$, or is received as an erasure E, with probability ε.

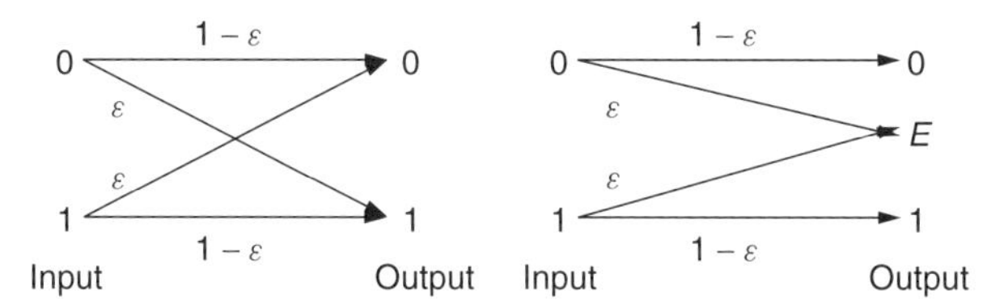

Figure 2. A binary symmetric channel (left) and a binary erasure channel (right).

Uncoded transmission over a binary symmetric channel consists of simply mapping each bit arriving at the channel coder into the corresponding input symbol. The decoder would map the received signal back into the corresponding bit. In this case, one bit is sent in each channel use and the probability of error is simply ε. To reduce the probability of error, an error-correcting code may be used. For example, suppose each bit is repeated 3 times when transmitted over the channel. In this case, if at most one bit out of three is in error, the decoder can correct the error. This reduces the error probability, but also reduces the transmission rate by a factor of 3.

The preceding example is called a *repetition code*. More generally, a (M, n) block code for a discrete channel consists of a set of M codewords, where each codeword is a sequence of n symbols in the channel input alphabet. The encoder is a function that maps each sequence of $\log_2 M$ data bits into one of these codewords and the decoder maps each possible received sequence of length n back into one of the original sequences of data bits. In the previous example $M = 2$ and $n = 3$. The *rate* of the code is

$$R = \frac{\log_2 M}{n} \quad \text{bits per channel use}$$

In a (M, n) code, the probability of error for each codeword, $e_i, i = 1, \ldots, M$, can be calculated from the channel model and the specification of the decoder. The *maximal error probability* for a code is defined to be $\max_i e_i$, and the *average error probability* of the code is defined to be $(1/M)\sum_i e_i$.

For a DMC, if the input letters are chosen with the probability distribution $p(x)$, then the joint probability mass function of the channel input and output is given by $p(x, y) = p(x)p(y \mid x)$. Using this probability mass function, the mutual information between the channel input and output can be calculated for any input distribution. The channel capacity of a DMC is defined to be the maximum mutual information

$$C = \max I(X; Y)$$

where the maximization is over all possible probability distributions on the input alphabet. In many special cases, the solution to this optimization can be found in closed form; for a general DMC, various numerical approaches may be used, such as the Arimoto–Blahut algorithm [12].

The definition of capacity is given operational significance by the channel coding theorem, proved by Shannon for DMCs. The direct part states that for any $R < C$, there exists rate R codes with arbitrarily small probability of error. This applies for either the maximal or the average probability of error. The converse to the channel coding theorem[6] states that for any rate $R > C$, the probability of error is bounded away from zero. The proof of the converse relies on Fano's inequality, which relates the probability of error to the conditional entropy of the channel output given the channel input.

Shannon's proof of the direct part of the coding theorem relied on a *random coding argument*. In this approach, the expected error probability for a set randomly chosen codes is studied. By showing that this expectation can be made small, it follows that there must be at least one code in this set with an error probability that is also small. In fact, it can be shown that most codes in this set have low error probabilities.

At rates below capacity, the probability of error for a good code can be shown to go to zero exponentially fast as the block length of the code increases. For a each rate R, the fastest rate at which the probability of error can go to zero is given by the channel's *reliability function*, $E(R)$. Various upper and lower bounds on this function have been studied [e.g., 13].

In addition to DMC, coding theorems have been proven for a variety of other discrete channels, we briefly mention several examples. One class of channels are called *finite-state channels*. In a finite state channel each input symbol is transmitted over one a set of possible DMCs. The specific channel is determined by the channel state that is modeled as a Markov chain. The capacity of a finite state channel will depend on if the transmitter or receiver has any *side information* available about the channel state [14]. A related model is a *compound channel*. In this case, one channel from a set is chosen when transmission begins; the channel then remains fixed for the duration of the transmission. *Universal coding* for compound channels has also been studied [15], which parallels the universal source coding discussed in Section 3.4. Another example is a channel with feedback, where the transmitter receives information about what is received. For a DMC it has been shown that feedback does not increase capacity [16]; however, feedback may reduce the complexity needed.

5.1. Gaussian Channels

A widely used model for communication channels is the additive white Gaussian noise (AWGN) channel shown in Fig. 3. This is a waveform channel with output $Y(t)$ given by

$$Y(t) = X(t) + Z(t)$$

where $X(t)$ is the input signal and $Z(t)$ is a white Gaussian noise process. Assuming that the input is limited to a bandwidth of W, then using the sampling theorem this

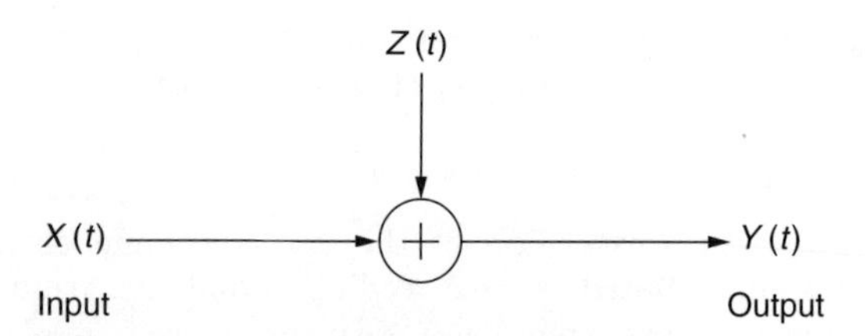

Figure 3. An additive white Gaussian noise channel.

channel can be reduced to the equivalent discrete-time channel:

$$Y_n = X_n + Z_n$$

Here, $\{X_n\}$ and $\{Y_n\}$ are sequences representing the input and output samples respectively and $\{Z_n\}$ is a sequence of i.i.d. Gaussian random variables with zero mean and variance $N_0/2$, where N_0 is the one-sided power spectral density of $Z(t)$. The samples occur at a rate of $2W$ samples per second.

The input to a channel is often required to satisfy certain constraints. A common example is an average power constraint, namely, a requirement that

$$\frac{1}{T}\int_T X^2(t)\,dt \leq P$$

or, for the discrete time channel,

$$\frac{1}{N}\sum_{n=1}^{N} X_n^2 \leq \frac{P}{2W}$$

The capacity of the AWGN channel with an average power constraint is given by the well-known formula

$$C = W\log\left(1 + \frac{P}{N_0 W}\right) \quad \text{bits per second (bps)}$$

where P/N_0W is the signal-to-noise ratio (SNR).

Many variations of the AWGN channel have also been studied, including the generalization of those discussed for DMCs above. One variation is channel with colored Gaussian noise, where the noise has power spectral density $N(f)$. In this case, achieving capacity requires that the transmission power be allocated over the frequency band. The optimal power allocation is given by

$$P(f) = \max\{\lambda - N(f), 0\}$$

where λ is chosen to satisfy the average power constraint. This allocation is often referred to as the "water-pouring" solution, because the power used at each frequency can be interpreted as the difference between $N(f)$ and the "water level" λ. Motivated in part by wireless applications, time-varying Gaussian channels and multipleinput/multipleoutput Gaussian channels have also been widely studied.

6. NETWORK INFORMATION THEORY

Network information theory refers to problems where there are more than one transmitter and/or receiver.

[6] This is also known as the "weak converse to the coding theorem." The "strong converse" states that at rates above capacity the probability of error goes to one for codes with long enough lengths.

A variety of such problems have been considered. This includes the *multiaccess channel*, where several users are communicating to a single receiver. In this case, a *capacity region* is specified indicating the set of all rate pairs (for the two-user case). Coding theorems have been found that establish the multiple access capacity region for both the discrete memoryless case [17] and for the Gaussian case [18]. Another example is the *interference channel*, where two senders each transmit to a separate receivers. The complete capacity region for this channel remains open.

7. FURTHER STUDIES

For more in-depth reading there are a variety of textbooks on information theory including Cover [19] and Gallager [13]. Shannon's original papers [1,2] are quite readable; all of Shannon's work, including [1,2], can be found in the compilation by Sloane and Wyner [20]. A collection of good survey articles can also be found in the volume edited by [21].

BIOGRAPHY

Randall Berry received the B.S. degree in Electrical Engineering from the University of Missouri-Rolla in 1993 and the M.S. and Ph.D. degrees in Electrical Engineering and Computer Science from the Massachusetts Institute of Technology in 1996 and 2000, respectively. Since 2000, he has been an Assistant Professor in the Department of Electrical and Computer Engineering at Northwestern University. In 1998 he was on the technical staff at MIT Lincoln Laboratory in the Advanced Networks Group. His primary research interests include wireless communication, data networks, and information theory.

BIBLIOGRAPHY

1. C. E. Shannon, A mathematical theory of communication (Part 1), *Bell Syst. Tech. J.* **27**: 379–423 (1948).
2. C. E. Shannon, A mathematical theory of communication (Part 2), *Bell Syst. Tech. J.* **27**: 623–656 (1948).
3. I. Csiszar and J. Korner, *Information Theory: Coding Theorems for Discrete Memoryless Systems*, Academic Press, New York, 1981.
4. B. McMillan, The basic theorems of information theory, *Ann. Math. Stat.* **24**: 196–219 (June 1953).
5. D. Huffman, A method for the construction of minimum redundancy codes, *Proc. IRE* **40**: 1098–1101 (Sept. 1952).
6. J. Ziv and A. Lempel, A universal algorithm for sequential data compression, *IEEE Trans. Inform. Theory* **24**: 337–343 (May 1977).
7. J. Ziv and A. Lempel, Compression of individual sequences via variable-rate coding, *IEEE Trans. Inform. Theory* **24**: 530–536 (Sept. 1978).
8. A. Wyner and J. Ziv, The sliding window Lempel-Ziv algorithm is asymptotically optimal, *Proc. IEEE* **82**(6): 872–877 (June 1994).
9. S. Lloyd, *Least Squares Quantization in PCM*, Bell Laboratories Technical Note, 1957.
10. T. Berger, *Rate Distortion Theory: A Mathematical Basis for Data Compression*, Prentice-Hall, Englewood Cliffs, NJ, 1971.
11. D. Slepian and J. Wolf, Noiseless coding of correlated information sources, *IEEE Trans. Inform. Theory* **19**: 471–480 (1973).
12. R. Blahut, Computation of channel capacity and rate distortion functions, *IEEE Trans. Inform. Theory* **18**: 460–473 (1972).
13. R. Gallager, *Information Theory and Reliable Communication*, Wiley, New York, 1968.
14. J. Wolfowitz, *Coding Theorems of Information Theory*, Prentice-Hall, Englewood Cliffs, NJ, 1978.
15. A. Lapidoth and P. Narayan, Reliable communication under channel uncertainty, *IEEE Trans. Inform. Theory* **44**: 2148–2175 (Oct. 1998).
16. C. Shannon, The zero error capacity of a noisy channel, *IRE Trans. Inform. Theory* **2**: 112–124 (Sept. 1956).
17. R. Ahlswede, Multi-way communication channels, *Proc. 2nd Int. Symp. Information Theory*, 1971, pp. 103–135.
18. T. Cover, Some advances in broadcast channels, *Adv. Commun. Syst.* **4**: 229–260 (1975).
19. T. Cover, *Elements of Information Theory*, Wiley, New York, 1991.
20. N. J. A. Sloane and A. D. Wyner, eds., *Claude Elwood Shannon: Collected Papers*, IEEE Press, New York, 1993.
21. S.Verdú, ed., *IEEE Trans. Inform. Theory (Special Commemorative Issue)* **44**: (Oct. 1998).

INTERFERENCE AVOIDANCE FOR WIRELESS SYSTEMS

Dimitrie C. Popescu
Christopher Rose
Rutgers WINLAB
Piscataway, New Jersey

1. INTRODUCTION

Interference from natural sources and from other users of the medium have always been factors in the design of reliable communication systems. While for wired or optical systems the amount of interference may be limited through hardware means that restrict access to the medium and/or reduce relative noise energy, the wireless medium is shared by all users. Usually, restrictions on use are legislative in nature and imposed by government regulating agencies such as the Federal Commission for Communications (FCC). Specifically, the oldest method of interference mitigation for wireless systems is spectrum licensing and the implied exclusive spectrum use. Unfortunately, licensing can indirectly staunch creative wireless applications owing to high licensing fees and the concomitant need for stable returns on large investments.

In an effort to promote creative wireless applications, the FCC has released 300 MHz of unlicensed spectrum in the 5 GHz range [1] with few restrictions other than absolute power levels — a "Wild West" environment of sorts where the only restriction is the size of the weapon! Thus, no central control is assumed. Needless to say, such a scenario seems ripe for chaos. Specifically, self-interest by individual users often results in unstable behavior. Such behavior has been seen anecdotally in distributed channelized systems such as early cordless telephones in apartment buildings where the number of collocated phones exceeded the number of channels available. Phones incessantly changed channels in an attempt to find a usable one. Thus, some means of assuring efficient use in such distributed wireless environments is needed.

Reaction to mutual interference is the heart of the shared-spectrum problem, and traditional approaches to combating wireless interference start with channel measurement and/or prediction, followed by an appropriate selection of modulation methods and signal processing algorithms for reliable transmission — possibly coupled to exclusive use contracts (licensing). Interestingly, since the time of the first radio transmissions, the methods used to deal with interference can be loosely grouped into three categories:

- Build a fence (licensing)
- Use only what you need (efficient modulation, power control)
- Grin and bear it (signal processing at the receiver)

Examples of the first item are legion, while examples of the last two items include single-sideband amplitude modulation, frequency modulation with preemphasis at the transmitter and deemphasis at the receiver, power control in cellular wireless systems [2], and code-division multiple access (CDMA) coupled to sophisticated signal processing algorithms for interference suppression/cancellation at the receiver [3].

However, Moore's law advances in microelectronics have led to the emergence of new transceiver hardware that add a new weapon to the interference mitigation arsenal. Specifically, a class of radios that can be programmed to transmit almost arbitrary waveforms and can act as almost arbitrary receiver types is emerging — the so-called software radios [4–7]. So, as opposed to traditional radios, which owing to complex transceiver hardware are difficult to modify once a modulation method has been chosen, one can now imagine programming transceivers to use more effective modulation methods. Thus, wireless systems of the near future will be able to choose modulation methods that *avoid* ambient interference as opposed to precluding it via sole-use licenses, overpowering it with increased transmission power, or mitigating it with receiver signal processing.

Interference avoidance is the term used for adaptive modulation methods where individual users — simply put — employ their signal energy in "places" where interference is weak. Such methods have been shown to optimize shared use. More precisely, iterative interference avoidance algorithms yield optimal waveforms that maximize the signal-to-interference plus noise-ratio (SINR) for all users while maximizing the sum of rates at which all users can reliably transmit information (sum capacity). In other words, interference avoidance methods, through the self-interested action of each user, lead to a socially optimum[1] equilibrium (Pareto efficient [9,10]) in various mutual interference "games."

Interference avoidance was originally introduced in the context of "chip-based" DS-CDMA systems [11] and minimum mean-square error (MMSE) receivers, but was subsequently developed in a general signal space [12–14] framework [15,16] which makes them applicable to a wide variety of communication scenarios. Related methods for transmitter and receiver adaptation have also been used in the CDMA context for asynchronous systems [17] and systems affected by multipath [18].

The relationship between codeword assignment in a CDMA system and sum capacity has been studied in several papers [19,20]; the paper by Viswanath and Anantharam [20] provides an algorithm to obtain sum capacity optimal codeword ensembles in a finite number of steps — and perhaps more importantly, also shows that the optimal linear receiver for such ensembles is a *matched filter* for each codeword. Interference avoidance algorithms also yield optimal codeword ensembles but seem conceptually simpler and suitable for distributed adaptive implementation in multiuser systems.

It is worth expanding upon this last point. Interference avoidance is envisioned as a distributed method for unlicensed bands as opposed to a centralized procedure done by an omnescient receiver. Of course, we will see that the mathematics of the algorithm also lends itself to central application, so the distinction is really only important for practical application. However, throughout we will assume distributed application which implicitly suggests that each user knows its associated channel and in addition that each user has access to the whole system covariance through a side-channel beacon. The receiver can adaptively track codeword variation in a manner reminiscent of adaptive equalization. Since communication is two-way and physical channels are reciprocal, it is not unreasonable to assume that both the user and the system can know the channel. More important is the rate at which the channel varies. We will assume that channel variation is slow relative the frame rate [21,22], or if the channel variation rate is rapid that the average channel varies slowly enough for interference avoidance to be applied [23].

2. THE EIGEN-ALGORITHM FOR INTERFERENCE AVOIDANCE

We consider the uplink of a synchronous CDMA communication system with L users having signature waveforms $\{S_\ell(t)\}_{\ell=1}^{L}$ of finite duration T, with equal received power at the base station and ideal channels. The received signal is

$$R(t) = \sum_{\ell=1}^{L} b_\ell S_\ell(t) + n(t) \tag{1}$$

[1] Maximum sum capacity or *user capacity* [8] in a single-receiver multiuser system.

where b_ℓ represents the information symbol sent by user ℓ with signature $S_\ell(t)$, and $n(t)$ is an additive Gaussian noise process. We assume that all signals are representable in an arbitrary N-dimensional signal space. Hence, each user's signature waveform $S_\ell(t)$ is equivalent to an N-dimensional vector $\mathbf{s}_\ell$ and the noise process $n(t)$ is equivalent to a noise vector $\mathbf{n}$. The equivalent received signal vector $\mathbf{r}$ at the base station is

$$\mathbf{r} = \sum_{\ell=1}^{L} b_\ell \mathbf{s}_\ell + \mathbf{n} \tag{2}$$

By defining the $N \times L$ matrix S having as columns the user codewords $\mathbf{s}_\ell$

$$\mathbf{D} = \begin{bmatrix} | & | & & | \\ \mathbf{s}_1 & \mathbf{s}_2 & \ldots & \mathbf{s}_L \\ | & | & & | \end{bmatrix} \tag{3}$$

the received signal can be rewritten in vector matrix form as

$$\mathbf{r} = \mathbf{S}\mathbf{b} + \mathbf{n} \tag{4}$$

where $\mathbf{b} = [b_1 \cdots b_L]^T$ is the vector containing the symbols sent by users.

Assuming simple matched filters at the receiver for all users and unit energy codewords $\mathbf{s}_k$, the SINR for user k is

$$\gamma_k = \frac{(\mathbf{s}_k^T \mathbf{s}_k)^2}{\sum_{j=1, j\neq k}^{L} (\mathbf{s}_k^T \mathbf{s}_j)^2 + E[(\mathbf{s}_k^T \mathbf{n})^2]} = \frac{1}{\mathbf{s}_k^T \mathbf{R}_k \mathbf{s}_k} \tag{5}$$

where $\mathbf{R}_k$ is the correlation matrix of the interference plus noise seen by user k having the expression

$$\mathbf{R}_k = \mathbf{S}\mathbf{S}^T - \mathbf{s}_k \mathbf{s}_k^T - \mathbf{W} \tag{6}$$

where $\mathbf{W} = E[\mathbf{n}\mathbf{n}^T]$ is the correlation matrix of the additive Gaussian noise.

Interference avoidance algorithms maximize the SINR through adaptation of user codewords. This is also equivalent to minimizing the inverse SINR defined as

$$\beta_k = \frac{1}{\gamma_k} = \mathbf{s}_k^T \mathbf{R}_k \mathbf{s}_k \tag{7}$$

Note that for unit power codewords, Eq. (7) represents the Rayleigh quotient for matrix $\mathbf{R}_k$, and recall from linear algebra [21, p. 348] that equation (7) is minimized by the eigenvector corresponding to the minimum eigenvalue of the given matrix $\mathbf{R}_k$. Therefore, the SINR for user k can be maximized by replacing codeword $\mathbf{s}_k$ with the minimum eigenvector of the correlation matrix $\mathbf{R}_k$. That is, user k *avoids interference* by seeking a place in the signal space where interference is least. Sequential application by all users of this greedy procedure defines the minimum eigenvector algorithm for interference avoidance, or the *eigen-algorithm* [15], formally stated below:

1. Start with a randomly chosen codeword ensemble specified by the codeword matrix $\mathbf{S}$.
2. For each user $\ell = 1 \cdots L$, replace user ℓ's codeword $\mathbf{s}_\ell$ with the minimum eigenvector of the correlation matrix $\mathbf{R}_k$ of the corresponding interference-plus-noise process.
3. Repeat step 2 until a fixed point is reached for which further modification of codewords will bring no improvement.

It has been shown [22] that in a colored noise background a variant of this algorithm, in which step 3 is augmented with a procedure to escape suboptimal fixed points, converges to the optimal fixed point where the resulting codeword ensemble "waterfills" over the background noise energy and maximizes sum capacity. If the background noise is white and the system is not overloaded (fewer users than signal space dimensions $L \leq N$), the algorithm yields a set of orthonormal codewords that corresponds to an ideal situation when users are orthogonal and therefore noninterfering. In the case of overloaded systems ($L > N$) in white noise, the resulting codeword ensembles form Welch bound equality (WBE) sets [19], which also minimize total squared correlation [15], a measure of the total interference in the system. In both underloaded and overloaded cases, the absolute minimum attainable total squared correlation is often used as a stopping criterion for the eigen-algorithm.

Finally, we note that signal space "waterfilling" and the implied maximization of sum capacity are *emergent* properties of interference avoidance algorithms. Thus, individual users do not attempt maximization of sum capacity via an individual or ensemble waterfilling scheme, but rather, they greedily maximize the SINR of their own codeword. In fact, individual waterfilling schemes over the whole signal space are impossible in this framework since each user's transmit covariance matrix $\mathbf{X}_\ell = \mathbf{s}_\ell \mathbf{s}_\ell^T$ is of rank one and cannot possibly span an N-dimensional signal space. So, emergent waterfilling and sum capacity maximization is a pleasantly surprising property of interference avoidance algorithms.

3. GENERALIZING THE EIGENALGORITHM

In order to extend application of the eigen-algorithm to more general scenarios, we consider the general multiaccess vector channel defined by [23]

$$\mathbf{r} = \sum_{\ell=1}^{L} \mathbf{H}_\ell \mathbf{x}_\ell + \mathbf{n} \tag{8}$$

where $\mathbf{x}_\ell$ of dimension N_ℓ is the input vector corresponding to user ℓ ($\ell = 1, \ldots, L$), $\mathbf{r}$ of dimension N is the received vector at the common receiver corrupted by additive noise vector $\mathbf{n}$ of the same dimension, and $\mathbf{H}_\ell$ is the $N \times N_\ell$ channel matrix corresponding to user ℓ. It is assumed that $N \geq N_\ell, \forall \ell = 1, \ldots, L$. This is a general approach to a multiuser communication system in which different users reside in different signal subspaces, with possibly different dimensions and potential overlap between them, but all of which are subspaces of the receiver signal space. We note that each user's signal space as well as the receiver signal space are of finite dimension — implied by a finite transmission frame T, finite bandwidths W_ℓ for each user ℓ, respectively, and by a finite receiver bandwidth W (which includes all W_ℓ values corresponding to all users) [24]. We also note that for memoryless channels the channel matrix $\mathbf{H}_\ell$ merely relates the bases of user ℓ's signal space and receiver signal space, but a similar model applies to channels with memory, in which case the channel matrix $\mathbf{H}_\ell$ also incorporates channel attenuation and multipath [16,18,23,25]. Figure 1 provides a graphical illustration of such a signal space configuration for two users residing in 2-dimensional subspaces with a three-dimensional receiver signal space.

In this signal space setting we assume that in a finite time interval of duration $\mathcal{T}$, each user ℓ sends a "frame" of data using a multicode CDMA approach wherein each symbol is transmitted using a distinct signature waveform that spans the frame. This scenario is depicted in Fig. 2. In other words, the sequence of information symbols $\mathbf{b}_\ell = [b_1^{(\ell)} \cdots b_{M_\ell}^{(\ell)}]^T$ is transmitted as a linear superposition of distinct, unit energy waveforms $s_m^{(\ell)}(t)$

$$x_\ell(t) = \sum_{m=1}^{M_\ell} b_m^{(\ell)} s_m^{(\ell)}(t) \tag{9}$$

as if each symbol in the frame corresponded to a distinct virtual user.

In the N_ℓ-dimensional signal space corresponding to user ℓ, each waveform can be represented as an N_ℓ-dimensional vector, thus the input vector $\mathbf{x}_\ell$ corresponding to user ℓ is equivalent to a linear superposition of unit norm codeword column vectors $\mathbf{s}_m^{(\ell)}$ scaled by the corresponding $b_m^{(\ell)}$. Therefore, each user uses an $N_\ell \times M_\ell$ codeword matrix $\mathbf{S}_\ell$

$$\mathbf{S}_\ell = \begin{bmatrix} | & | & & | \\ \mathbf{s}_1^{(\ell)} & \mathbf{s}_2^{(\ell)} & \cdots & \mathbf{s}_{M_\ell}^{(\ell)} \\ | & | & & | \end{bmatrix} \tag{10}$$

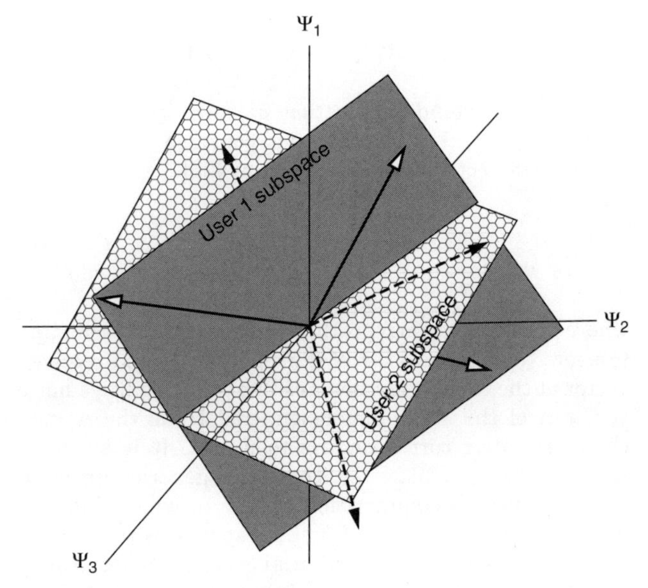

Figure 1. Three-dimensional receiver signal space with two users residing in two-dimensional subspaces. Vectors represent particular signals in user 1 (continuous line), respectively, user 2 (dashed line) signal spaces.

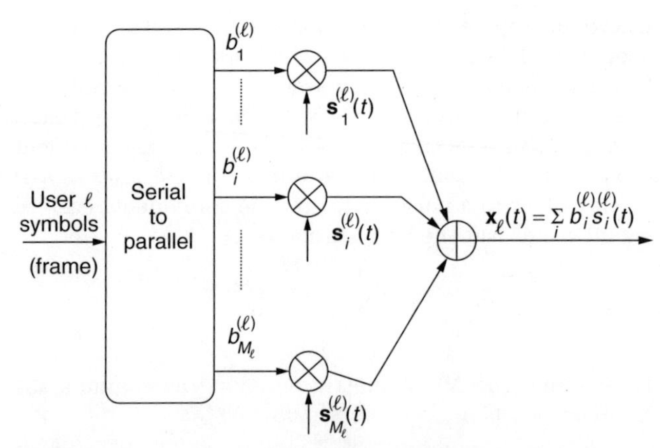

Figure 2. Multicode CDMA approach for sending frames of information. Each symbol in user ℓ's frame is assigned a distinct signature waveform, and the transmitted signal is a superposition of all signatures scaled by their corresponding information symbols.

so that

$$\mathbf{x}_\ell = \mathbf{S}_\ell \mathbf{b}_\ell \tag{11}$$

Therefore, the received signal can be rewritten as

$$\mathbf{r} = \sum_{\ell=1}^{L} \mathbf{H}_\ell \mathbf{S}_\ell \mathbf{b}_\ell + \mathbf{n} \tag{12}$$

Note that under the assumption that $M_\ell \geq N_\ell$, the $N_\ell \times N_\ell$ transmit covariance matrix of user ℓ, $\mathbf{X}_\ell = E[\mathbf{x}_\ell \mathbf{x}_\ell^T] = \mathbf{S}_\ell \mathbf{S}_\ell^T$, has full rank and spans user ℓ's signal space.

Extension of the eigen-algorithm to this general multiaccess vector channel setting is presented elsewhere in the literature [16,26]. The procedure starts by separating the interference-plus-noise seen by a given user k and rewriting the received signal in equation (12) from the perspective of user k as

$$\mathbf{r} = \mathbf{H}_k \mathbf{S}_k \mathbf{b}_k + \underbrace{\sum_{\ell=1,\ell\neq k}^{L} \mathbf{H}_\ell \mathbf{S}_\ell \mathbf{b}_\ell + \mathbf{n}}_{\mathbf{z}_k = \text{interference + noise}} = \mathbf{H}_k \mathbf{S}_k \mathbf{b}_k + \mathbf{z}_k \tag{13}$$

The covariance matrix of the interference-plus-noise seen by user k

$$\mathbf{Z}_k = E[\mathbf{z}_k \mathbf{z}_k^T] = \sum_{\ell=1,\ell\neq k}^{L} \mathbf{H}_\ell \mathbf{S}_\ell \mathbf{S}_\ell^T \mathbf{H}_\ell^T + \mathbf{W} \tag{14}$$

is then used to define a whitening transformation $\mathbf{T}_k$ of the interference-plus-noise seen by user k. The equivalent problem in which user k sees white interference-plus-noise is then projected onto the user k signal space using the singular value decomposition (SVD) [21, p. 442] of user k's transformed channel matrix. This reduces the problem to an equivalent one given by an equation identical in form to Eq. (4) and therefore allows straightforward application of the eigen-algorithm for optimization of user k's codewords.

One possible generalized eigen-algorithm is formally stated below:

1. Start with a randomly chosen codeword ensemble specified by user codeword matrices $\{\mathbf{S}_k\}_{k=1}^{L}$.
2. For each user $k = 1 \cdots L$
   a. Compute the transformation matrix $\mathbf{T}_k$ that whitens the interference-plus-noise seen by user k.
   b. Change coordinates and compute the transformed user k channel matrix $\tilde{\mathbf{H}}_k = \mathbf{T}_k\mathbf{H}_k$.
   c. Apply SVD for $\tilde{\mathbf{H}}_k$ and project the problem onto user k's signal space.
   d. Adjust user k's codewords sequentially using the greedy procedure of the basic eigen-algorithm; the codeword corresponding to symbol m of user k is replaced by the minimum eigenvector of the autocorrelation matrix of the corresponding interference-plus-noise process.
   e. Iterate the previous step until convergence (making use of escape methods [22] if the procedure stops in suboptimal points).
3. Repeat step 2 until a fixed point is reached.

We note that in steps 2d,e of the algorithm, user k waterfills its signal space by application of the basic eigen-algorithm [22] for interference avoidance by regarding all other user's signals as noise. Thus, the generalized eigen-algorithm *iteratively waterfills* each user's signal space. It has been shown [23] that an *iterative waterfilling algorithm* converges to a fixed point where the sum capacity of the vector multiaccess channel is maximized, which implies that the generalized eigen-algorithm will always yield a codeword ensemble that maximizes sum capacity.

4. THE GENERALIZED EIGEN-ALGORITHM: A VERSATILE TOOL FOR CODEWORD OPTIMIZATION

The generalized eigen-algorithm is a powerful tool that enables application of interference avoidance methods to various communication problems in which the underlying model is a multiaccess vector channel. Among these we mention codeword optimization in the uplink of a CDMA system with nonideal (dispersive) channels, multiuser systems with multiple inputs and outputs (MIMO), and asynchronous CDMA systems, for which an appropriate selection of signal space basis functions leads to particular cases of the general multiaccess vector channel model for which application of the generalized eigen-algorithm to codeword optimization becomes straightforward.

For the uplink of a CDMA system with dispersive channels considered in earlier studies [16,25], the spanning set of the signal space consists of a set of real sinusoids (sine and cosine functions) that are approximately eigenfunctions for all uplink channels corresponding to all users. Introduced in 1964 [27], channel eigenfunctions form an orthonormal spanning set with the property that their corresponding channel responses are also orthogonal, thus allowing convenient representation of channel outputs as scaled versions of the input vectors. In this case, the channel matrix of a given user is a diagonal matrix in which diagonal elements correspond to channel gain factors for the frequencies that define the signal space, and the modulation scheme turns out to be a form of multicarrier CDMA [28]. We note that even though interference avoidance is applied in this multicarrier modulation framework [25], the method is completely general and applicable to various scenarios with appropriate selection of signal space basis functions. For example using sinc functions the method is applicable to time-domain representations in which the vector channel model is obtained from Nyquist-sampled waveforms (see, e.g., Ref. 29). Using "time chips" as basis functions a vector model for DSCDMA systems with multipath is obtained [16,18].

Application of the generalized eigen-algorithm to multiuser MIMO systems is also possible [16,30]. The same multicarrier modulation framework with multiple antennas at the transmitter and receiver imply a MIMO channel matrix composed of block diagonal matrices corresponding to each transmit/receive antenna pair. We mention again that other MIMO channel models — for example, the spatiotemporal MIMO channel model [31], in which the MIMO channel matrix is composed of convolution matrices corresponding to each transmit/receive antenna pair — are perfectly valid and the generalized eigen-algorithm can be used for codeword optimization in conjunction with such models as well.

A similar signal space approach is used to apply the generalized eigen-algorithm for codeword optimization in very general asynchronous CDMA system models [16,32]. We note that for particular cases less general interference avoidance algorithms can be used [33].

5. CONCLUSION

Motivated by the emergence of software radios and the desire to foster creative uses of wireless spectrum in unlicensed bands, interference avoidance methods have been developed for wireless systems. The underlying idea of interference avoidance is for each user to greedily optimize spectrum use (SINR or capacity) through appropriate signal placement in response to interferers. Interference avoidance is applicable to a wide range of communications scenarios including dispersive channels, multiple antenna systems, and asynchronous systems.

The utility of interference avoidance methods in real wireless systems with multiple receivers — such as is found in a cellular environment — although currently under study, is still an open question in general. Specifically, theoretical results have been established for application of interference avoidance methods in a *collaborative* scenario wherein information from all receivers is pooled and used to decode all users in the system [34,35]. Since all information is available for use in decoding users, the collaborative scenario constitutes a best case of sorts. Unfortunately, real systems may not be collaborative or even cooperative.

Early experiments with geographically dispersed users assigned to different bases showed unstable behavior under direct application of the eigen-algorithm, mirroring the anecdotally reported behavior of early cordless phones (see Section 1, above). However, by allowing each user to send and adapt *multiple codewords* — a multicode approach similar to that used for dispersive channels — the algorithm became stable [36] since each user could then waterfill their signal energy over the entire signal space when necessary as opposed to choosing exactly one channel. Of course, such convergence though welcome is a bit chimeric since the implied multiple receiver system model is an instance of the interference channel [37] for which very little is known in general. Regardless, the fact that interference avoidance can attain any equilibrium [9,10] in the implied "game" of mutually interfering wireless access is interesting, and can perhaps illuminate paths toward greater understanding of efficient and peaceful coexistence in unlicensed wireless systems.

BIOGRAPHIES

Dimitrie C. Popescu received the Engineering Diploma and M.S. degrees in 1991 from the Polytechnic Institute of Bucharest, Romania, and the Ph.D. degree from Rutgers University in 2002, all in Electrical Engineering. He is currently an Assistant Professor in the Department of Electrical and Computer Engineering, the University of Texas at San Antonio. His research interests are in the general areas of communication systems, control engineering, and signal processing. In the summer of 1997 he worked at AT&T Labs in Florham Park, New Jersey, on signal processing algorithms for speech enhancement, and in the summer of 2000 he worked at Telcordia Technologies in Red Bank, New Jersey, on wideband CDMA systems. His work on interference avoidance and dispersive channels was awarded second prize in the AT&T Student Research Symposium in 1999.

Dr. Christopher Rose received the B.S. (1979), M.S. (1981), and Ph.D. (1985) degrees all from the Massachusetts Institute of Technology in Cambridge, Massachusetts. Dr. Rose joined At&T Bell Laboratories in Holmdel, New Jersey as a member of the Network Systems Research Department in 1985 and in 1990 moved to Rutgers University, where he is currently an Associate Professor of Electrical and Computer Engineering and Associate Director of the Wireless Networks Laboratory. He is Editor for the *Wireless Networks* (ACM), *Computer Networks* (Elsevier) and Transactions on Vehicular Technology (IEEG) journals and has served on many conference technical program committees. Dr. Rose was technical program Co-Chair for MobiCom'97 and Co-Chair of the WINLAB Focus'98 on the U-NII, the WINLAB Berkeley Focus'99 on Radio Networks for Everything and the Berkeley WINLAB Focus 2000 on Picoradio Networks. Dr. Rose, a past member of the ACM SIGMobile Executive Committee, is currently a member of the ACM MobiCom Steering Committee and has also served as General Chair of ACM SIGMobile MobiCom 2001 (Rome, July 2001). In December 1999 he served on an international panel to evaluate engineering teaching and research in Portugal.

His current technical interests include mobility management, short-range high-speed wireless (Infostations), and interference avoidance methods for unlicensed band networks.

BIBLIOGRAPHY

1. Federal Communications Commission, *FCC Report and Order 97-5: Amendment of the Commission's Rules to Provide for Operation of Unlicensed NII Devices in the 5 GHz Frequency Range*, ET Docket 96–102, 1997.
2. R. Yates, A framework for uplink power control in cellular radio systems, *IEEE J. Select. Areas Commun.* **13**(7): 1341–1348 (Sept. 1995).
3. S. Verdu, *Multiuser Detection*, Cambridge Univ. Press, 1998.
4. I. Seskar and N. Mandayam, A software radio architecture for linear multiuser detection, *IEEE J. Select. Areas Commun.* **17**(5): 814–823 (May 1999).
5. I. Seskar and N. Mandayam, Software defined radio architectures for interference cancellation in DS-CDMA systems, *IEEE Pers. Commun. Mag.* **6**(4): 26–34 (Aug. 1999).
6. Special issue on software radio, *IEEE Pers. Commun. Mag.* **6**(4) (Aug. 1999) K.-C. Chen, R. Prasad, and H. V. Poor, eds.
7. J. Mitola, The software radio architecture, *IEEE Commun. Mag.* **33**(5): 26–38 (May 1995).
8. P. Viswanath, V. Anantharam, and D. Tse, Optimal sequences, power control and capacity of spread spectrum systems with multiuser linear receivers, *IEEE Trans. Inform. Theory* **45**(6): 1968–1983 (Sept. 1999).
9. E. M. Valsbord and V. I. Zhukovski, *Introduction to Multi-Player Differential Games and Their Applications*, Gordon and Breach Science Publishers, 1988.
10. R. B. Meyerson, *Game Theory: Analysis of Conflict*, Harvard Univ. Press, 1991.
11. S. Ulukus, *Power Control, Multiuser Detection and Interference Avoidance in CDMA Systems*, Ph.D. thesis, Rutgers Univ., Dept. Electrical and Computer Engineering, 1998 (thesis director: Prof. R. D. Yates).
12. J. G. Proakis, *Digital Communications*, 4th ed., McGraw-Hill, New York, 2000.
13. S. Haykin, *Communication Systems*, Wiley, New York, 2001.
14. H. L. Van Trees, *Detection, Estimation, and Modulation Theory*, Part I, Wiley, New York, 1968.
15. C. Rose, S. Ulukus, and R. Yates, Wireless systems and interference avoidance, *IEEE Trans. Wireless Commun.* **1**(3): (July 2002). preprint available at http:/steph.rutgers.edu/~crose/papers/avoid17.ps.
16. D. C. Popescu, *Interference Avoidance for Wireless Systems*, Ph.D. thesis, Rutgers Univ., Dept. Electrical and Computer Engineering (2002) (thesis director: Prof. C. Rose).
17. P. B. Rapajic and B. S. Vucetic, Linear adaptive transmitter-receiver structures for asynchronous CDMA systems, *Eur. Trans. Telecommun.* **6**(1): 21–27 (Jan.–Feb. 1995).
18. G. S. Rajappan and M. L. Honig, Signature sequence adaptation for DS-CDMA with multipath, *IEEE J. Select. Areas Commun.* **20**(2): 384–395 (Feb. 2002).
19. M. Rupf and J. L. Massey, Optimum sequence multisets for synchronous code-division multiple-access channels, *IEEE Trans. Inform. Theory* **40**(4): 1226–1266 (July 1994).
20. P. Viswanath and V. Anantharam, Optimal sequences and sum capacity of synchronous CDMA systems, *IEEE Trans. Inform. Theory* **45**(6): 1984–1991 (Sept. 1999).
21. D. C. Popescu and C. Rose, Interference avoidance and multiaccess dispersive channels. In *Proc. 35th Annual*

Asilomar Conference on Signals, Systems, and Computers, Pacific Grove, CA, November 2001.

22. D. C. Popescu and C. Rose, CDMA codeword optimization for uplink dispersive channels through interference avoidance. *IEEE Transactions on Information Theory*. (Submitted 12/2002, revised 10/2001. Preprint available at *http://www.winlab.rutgers.edu/~cripop/papers*).
23. D. C. Popescu and C. Rose, Fading channels and interference avoidance. In *Proc. 39th Allerton Conference on Communication, Control, and Computing*, Monticello, IL, October 2001.
24. G. Strang, *Linear Algebra and Its Applications*, 3rd ed., Harcourt Brace Jovanovich College Publishers, 1988.
25. C. Rose, CDMA codeword optimization: Interference avoidance and convergence via class warfare, *IEEE Trans. Inform. Theory* **47**(4): 2368–2382 (Sept. 2001).
26. W. Yu, W. Rhee, S. Boyd, and J. M. Cioffi, Iterative water-Filling for Gaussian Vector multiple access channels, *2001 IEEE Int. Symp. Information Theory, ISIT'01*, Washington, DC, June 2001 (submitted for journal publication).
27. H. J. Landau and H. O. Pollack, Prolate spheroidal wave functions, Fourier analysis and uncertainty—III: The dimension of the space of essentially time- and band-limited signals, *Bell Syst. Tech. J.* **40**(1): 43–64 (Jan. 1961).
28. D. C. Popescu and C. Rose, Interference avoidance for multiaccess vector channels, *2002 IEEE Int. Symp. Information Theory, ISIT'02*, Lausanne, Switzerland, July 2002 (submitted for publication in *IEEE Trans. Inform. Theory*).
29. J. L. Holsinger, *Digital Communication over Fixed Time-Continuous Channels with Memory—with Special Application to Telephone Channels*, Technical Report 366, MIT—Lincoln Lab., 1964.
30. N. Yee and J. P. Linnartz, *Multi-Carrier CDMA in an Indoor Wireless Radio Channel*, Technical Memorandum UCB/ERL M94/6, Univ. California, Berkeley, 1994.
31. S. N. Diggavi, On achievable performance of spatial diversity fading channels, *IEEE Trans. Inform. Theory* **47**(1): 308–325 (Jan. 2001).
32. D. C. Popescu and C. Rose, Interference avoidance and multiuser MIMO systems, 2002. Preprint available at *http://www.winlab.rutgers.edu/~cripop/papers*).
33. G. G. Raleigh and J. M. Cioffi, Spatio-temporal coding for wireless communication, *IEEE Trans. Commun.* **46**(3): 357–366 (March 1998).
34. D. C. Popescu and C. Rose, Codeword optimization for asynchronous CDMA systems through interference avoidance, *Proc. 36th Conf. Information Sciences and Systems, CISS 2002*, Princeton, NJ, March 2002.
35. S. Ulukus and R. Yates, Optimum signature sequence sets for asynchronous CDMA systems, *Proc. 38th Allerton Conf. Communication, Control, and Computing*, Oct. 2000.
36. O. Popescu and C. Rose, Minimizing total squared correlation for multibase systems, *Proc. 39th Allerton Conf. Communication, Control, and Computing*, Monticello, IL, Oct. 2001.
37. O. Popescu and C. Rose, Interference avoidance and sum capacity for multibase systems, *Proc. 39th Allerton Conf. Communication, Control, and Computing*, Monticello, IL, Oct. 2001.
38. D. Tabora, *An Analysis of Covariance Estimation, Codeword Feedback, and Multiple Base Performance of Interference Avoidance*, Master's thesis, Rutgers Univ., Dept. Electrical and Computer Engineering, 2001 (thesis director: Prof. C. Rose).
39. T. M. Cover and J. A. Thomas, *Elements of Information Theory*, Wiley-Interscience, New York, 1991.

INTERFERENCE MODELING IN WIRELESS COMMUNICATIONS

XUESHI YANG
ATHINA P. PETROPULU
Drexel University
Philadelphia, Pennsylvania

1. INTRODUCTION

In wireless communication networks, signal reception is often corrupted by interference from other sources that share the same propagation medium. Knowledge of the statistics of interference is important in achieving optimum signal detection and estimation. Construction of most receivers is based on the assumption that the interference is i.i.d. (independent, identically distributed) Gaussian. The Gaussianity assumption is based on the central-limit theory. However, in situations such as underwater acoustic noise, urban and synthetic (human-made) RF noise, low-frequency atmospheric noise, and radar clutter noise, the mathematically appealing Gaussian noise model is seldom appropriate [14]. For such cases, several mathematical models have been proposed, including the class A noise model [18], the Gaussian mixture model [28], and the α-stable model [20]. All these noise models share a common feature: the tails of their probability density function decay in a power-law fashion, as opposed to the exponentially decaying tails of the Gaussian model. This implies that the corresponding time series appear bursty, or impulsive, since the probability of attaining very large values can be significant.

Existing models for non-Gaussian impulsive noise are usually divided into two groups: empirical models and statistical–physical models. The former include the hyperbolic distribution and the Gaussian mixture models [28,29]. They fit a mathematical model to the measured data, without taking into account the physical mechanism that generated the data. Although empirical models offer mathematical simplicity in modeling, their parameters are not related to any physical quantities related to the data. On the other hand, statistical–physical models are grounded on the physical noise generation process, and their parameters are linked to physical parameters. The first physical models for noise can be traced back in the works of Furutsu and Ishida [9], Middleton [18], and later in the works of Sousa [26], Nikias [20], Ilow [13], and Yang [33] and colleagues. All of these models consider the following scenario. A receiver is surrounded by interfering sources that are randomly distributed in space according to a Poisson point process. The receiver picks up the superposition of the contributions of all the interferers. By assuming that the propagation path loss is inversely proportional to the power of the distance between the receiver and the interferer, and that the pulses have a symmetrically distributed random amplitude, it has been shown that the resulting instantaneous noise is impulsive and non-Gaussian [18,20]. The power-law assumption for path loss is consistent with empirical measurement data [10,18].

The same model can be applicable for modeling interference in a spread-spectrum wireless system, where randomly located users access the transmission medium simultaneously. The user identity is determined by a signature, which is embedded in the transmitted signal. Ideally, signatures of different users are orthogonal; thus, by correlating a certain signature with the total received signal, the corresponding user signal can be recovered. However, in practice the received signals from other users are not perfectly orthogonal to the signature waveform of the user of interest, due to non-perfect orthogonal codes and channel distortion. Therefore, correlation at the receiver results in an interference term, usually referred to as *cochannel interference*. Cochannel interference is the determining factor as far as quality of service is concerned, and sets a limit to the capacity of a spread-spectrum communication system. Modeling, analysis, and mitigation of cochannel interference has been the subject of numerous studies.

The goal of this article is to provide a mathematical treatment of the interference that occurs in wireless communication systems. It is organized as follows. In Section 2, we provide a brief treatment of α-stable distributions and heavy-tail processes. We also outline some techniques for deciding whether a set of data follows an α-stable distribution, and for estimating the model parameters. In Section 3, we discuss two widely studied statistical–physical models for interference: the α-stable model and the class A noise model.

2. PROBABILISTIC BACKGROUND

2.1. α-Stable Distributions

α-Stable distributions are defined in terms of their characteristic function:

$$\Phi(\rho) = \exp\{\imath\mu\rho - \sigma^{\alpha}|\rho|^{\alpha}(1 + \imath\eta \text{sign}(\rho)\varphi(\rho,\alpha))\} \tag{1a}$$

with

$$\varphi(\rho,\alpha) = \begin{cases} \tan\dfrac{\alpha\pi}{2} & \text{if } \alpha \neq 1 \\ \dfrac{2}{\pi}\ln|\rho| & \text{if } \alpha = 1 \end{cases} \tag{1b}$$

$$\text{sign}(\rho) = \begin{cases} 1, & \text{if } \rho > 0 \\ 0, & \text{if } \rho = 0 \\ -1, & \text{if } \rho < 0 \end{cases} \tag{1c}$$

where $\alpha \in (0, 2]$: characteristic exponent — a measure of rate of decay of the distribution tails; the smaller the α the heavier the tails of the distribution

$\eta \in [-1, 1]$: symmetry index

$\sigma > 0$: scale parameter; also referred to as *dispersion*. $2\sigma^2$ equals to the variance in the Gaussian distribution case

μ : location parameter

The distribution is called symmetric α-stable ($S\alpha S$) if $\eta = 0$.

Since (1) is characterized by four parameters, we denote stable distributions by $S_{\alpha}(\sigma, \eta, \mu)$, and indicate that the random variable X is α-stable distributed by

$$X \sim S_{\alpha}(\sigma, \eta, \mu) \tag{2}$$

The probability density functions of α-stable random variables are seldom given in closed form. Some exceptions are the following:

- The Gaussian distribution, which can be expressed as $S_2(\sigma, 0, \mu)$, and whose density is

$$f(x) = \frac{1}{2\sigma\sqrt{\pi}} e^{-\frac{(x-\mu)^2}{4\sigma^2}} = N(\mu, 2\sigma^2) \tag{3}$$

- The Cauchy distribution, $S_1(\sigma, 0, \mu)$, whose density is

$$f(x) = \frac{\sigma}{\pi((x-\mu)^2 + \sigma^2)} \tag{4}$$

Two important properties of stable distributions are the stability property and the generalized central-limit theorem.

The *stability property* is defined as follows. A random variable X has a stable distribution if and only if for arbitrary constants a_1 and a_2, there exist constants a and b such that $a_1X_1 + a_2X_2$ has the same distribution as $aX + b$, where X_1 and X_2 are independent copies of X. A consequence of the stability property is the generalized central-limit theorem, according to which, the stable distribution is the only possible limit distribution of sums of i.i.d. random variables. In particular, if $X_1, X_2, \ldots$ are i.i.d. random variables, and there exists sequences of positive numbers $\{a_n\}$ and real numbers $\{b_n\}$, such that

$$S_n = \frac{X_1 + \cdots + X_n}{a_n} - b_n \tag{5}$$

converges to X in distribution, then X is stable distributed.

If the X_i are i.i.d. and have finite variance, then the limit distribution is Gaussian and the generalized central-limit theorem reduces to the ordinary central-limit theorem.

α-Stable distributions are a special class of the so called heavy-tail distributions. A random variable X is said be heavy-tail distributed if

$$\Pr(|X| \geq x) \sim L(x)x^{-\alpha}, \quad 0 < \alpha < 2 \tag{6}$$

where $L(x)$ is a slowly varying function at infinity, that is, $\lim_{x\to\infty} L(cx)/L(x) = 1$ for all $c > 0$. Thus, α-stable distributions with $\alpha < 2$ have power-law tails, as opposed to the exponential tails of the Gaussian distribution. Figure 1 illustrates the survival function $P(X > x)$ of α-stable distributions for different α ($\alpha = 0.5, 1, 1.5, 2$). We

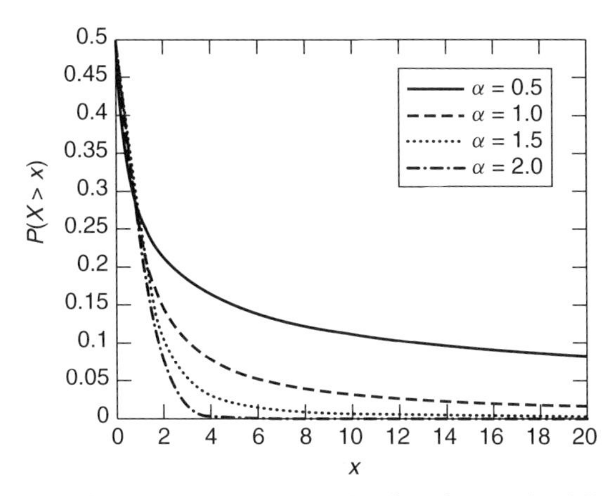

Figure 1. Survival function of α-stable distributions for different $\alpha = 0.5, 1.0, 1.5, 2.0$.

observe that the tail of the distribution decays slower as α becomes smaller.

A set of n real random variables $X_1, X_2, \ldots, X_n$ are jointly symmetric α-stable ($S\alpha S$) if and only if the linear combination $a_1X_1 + \cdots + a_nX_n$ is $S\alpha S$ for all real $a_1, \ldots, a_n$.

A random process $x(t)$ is said to be $S\alpha S$ if for any $n \geq 1$ and distinct instants $t_1, \ldots, t_n$ the random variables $x(t_1), \ldots, x(t_n)$ are jointly $S\alpha S$ with the same α.

α-Stable distributions are known for their lack of moments of order larger than α, in particular, for $\alpha < 2$ second-order statistics do not exist. In such a case, the role of covariance is played by the covariation or the codifference [24]; when $\alpha = 2$ both functions reduce to the covariance. While the covariation may not be defined for $\alpha < 1$, the codifference is defined for all $0 < \alpha \leq 2$.

The *codifference* of two jointly $S\alpha S$, $0 < \alpha \leq 2$, random variables X_1 and X_2 equals

$$R_{X_1,X_2} = \sigma_{X_1}^{\alpha} + \sigma_{X_2}^{\alpha} - \sigma_{X_1-X_2}^{\alpha} \tag{7}$$

where σ_X is the scale parameter of the $S\alpha S$ variable X. If X_1 and X_2 are independent, then their codifference equals zero. However, a zero codifference does not in general imply that the random variables are independent, except in the case $0 < \alpha < 1$.

Let $x(t)$ be a stationary $S\alpha S$ random process. The codifference, $R_{x(t+\tau),x(t)}$, provides a measure of structure of the process. For processes that are marginally only α-stable, or in general, for heavy-tail processes, the codifference is not defined. An alternative measure of structure for such processes is the so called generalized codifference [22,24]:

$$I(\rho_1, \rho_2; \tau) = -\ln E\{e^{\imath(\rho_1 x(t+\tau)+\rho_2 x(t))}\} + \ln E\{e^{\imath\rho_1 x(t+\tau)}\} + \ln E\{e^{\imath\rho_2 x(t)}\} \tag{8}$$

The generalized codifference is closely related to the codifference; if the process is $S\alpha S$, then

$$R_{x(t+\tau),x(t)} = -I(1, -1; \tau) \tag{9}$$

2.2. Estimation of Parameters of α-Stable Distributions

As was already mentioned, α-stable random variables with $\alpha < 2$ do not have finite variance. However, when dealing with a finite-length data record, the estimated variance will always be finite. A natural question is how one assesses the validity of the α-stable model based on a finite set of samples of a random process, and if the model is valid, how one determines the characteristic exponent α.

A concept that appears in all the methods that will be described next is that of *order statistics*. The order statistics of a random sequence $X_1, \ldots, X_N$ are the samples placed in ascending order, usually denoted by $X_{(1)}, \ldots, X_{(N)}$. Thus, $X_{(1)} = \min\{X_1, \ldots, X_N\}$, and $X_{(N)} = \max\{X_1, \ldots, X_N\}$.

A popular method for determining whether a given data set follows a specific distribution is the quantile–quantile (QQ) plot (see Ref. 23 for a detailed coverage of the topic). The principle of the QQ plot is based on the following observation. Assume that the random sequence $U_1, U_2, \ldots, U_N$ are uniformly distributed on [0,1]. By symmetry, the following will hold for the order statistics of U_i:

$$E(U_{(i+1)} - U_{(i)}) = \frac{1}{N+1}$$

and hence

$$EU_{(i)} = \frac{i}{N+1}.$$

Since $U_{(i)}$ should be close to its mean $i/(N+1)$, the plot of $\{(i/(N+1), U_{(i)}), 1 \leq i \leq N\}$ should be roughly linear. Now if we suspect that $X_1, X_2, \ldots, X_N$ come from a distribution G, we can plot $i/N+1$ against $G(X_{(i)})$ for $1 \leq i \leq N$. $X_{(i)}$ is the order statistics of X_i. If our suspicion is correct, the plot should be approximately linear, and should be the plot of $\{G^{-1}(i/N+1), X_{(i)}, 1 \leq i \leq N\}$. Here G^{-1} is the inverse function of the distribution function G. $G^{-1}(i/N+1)$ is the theoretical quantile, and $X_{(i)}$ is the empirical quantile, and hence the name QQ-plot.

In practice, one usually replaces the theoretical quantile, $G^{-1}(\cdot)$, with the empirical quantile of a sequence of data that are known (or simulated) to be distributed according to G. For example, to check whether a given data set follows the distribution $S_\alpha(\sigma, \eta, \mu)$, we can plot the QQ plot of the data against an ideally $S_\alpha(\sigma, \eta, \mu)$ distributed sample data set. We conclude that the data set follow the distribution of $S_\alpha(\sigma, \eta, \mu)$, if the QQ plot is close to a 45° line.

Additional tests are the probability plot [6], and the chi-square goodness-of-fit test [25]. The book by D'Agostino and Stephens [1] is an excellent reference on these tests.

Estimation of the tail index α can be obtained by several methods [20], two of which, namely the Hill plot and the QQ estimator, are briefly discussed next.

2.2.1. Hill Plot.

Let $k < N$. The Hill estimator is defined as [12]:

$$\alpha_k^{-1} = k^{-1} \sum_{j=1}^{k} \log \frac{X_{(N-k+j)}}{X_{(N-k)}}$$

Note that k is the number of upper order statistics used in the estimation.

The choice of k is a difficult issue. A useful graphical tool to choose α is to use a so called *dynamic Hill plot*, which consists in plotting α_k as a function of k. The sequence α_k^{-1} is a consistent sequence of estimators of α^{-1}, thus the Hill plot should stabilize after some k to a value roughly equal to α. In practice, finding that stable region of the Hill plot is not a straightforward task, since the estimator exhibits extreme volatility.

2.2.2. The QQ Estimator. The slope of regression $1 - \log(1 - j/(k+1))$ through the points $\log(X_{(N-k+j)}/X_{(N-k)})$ can be used as an estimator of the tail index. This estimator is referred to as the QQ estimator [15]. Computing the slope we find that the QQ estimator is given by

$$\hat{\alpha}_k^{-1} = \frac{\sum_{i=1}^{k} -\log\left(\frac{i}{k+1}\right) \times \left(n\log(X_{(N-i+1)}) - \sum_{j=1}^{k}\log(X_{(N-j+1)})\right)}{k\sum_{i=1}^{k}\left(-\log\left(\frac{i}{k+1}\right)\right)^2 - \left(\sum_{i=1}^{k} -\log\left(\frac{i}{N+1}\right)\right)^2} \quad (10)$$

The QQ estimator is consistent for i.i.d. data if $k^{-1} + k/N \to 0$. Its variance is larger than the asymptotic variance of the Hill estimator; however, the variability of the QQ estimator always seems to be less than that of the Hill estimator.

Again, the choice of k is a difficult problem. Two different plots can be constructed on the basis of the QQ estimator: (1) the dynamic QQ plot obtained from plotting $(k, \hat{\alpha}_k), l < k < N$ (similar to the Hill plot) and (2) the static QQ estimator, which is obtained by representing the data $\log(X_{(N-k+1)}/X_{(N-k)})$ as a function of $\log(1 - j/(k+1))$ together with the least-squares regression line. The slope of that line is used to compute the QQ estimator $\hat{\alpha}_{k,n}^{-1}$.

3. MODELING INTERFERENCE

3.1. A α-Stable Noise Model

Consider a packet radio network, where the users are distributed on a plane. The basic unit of time is the *slot*, which is equal to the packet transmission time T. Let us assume that all users are synchronized at the slot level.

In a typical situation, a user transmits a packet to some destination at a distance. The success of the reception depends partly on the amount of interference experienced by the receiving user. The interference at a certain network location consists of two terms: *self-interference*, or interference caused by other transmitting users; and *external interference*, such as thermal noise, coming from other systems. We are here concerned with the former type of interference.

Self-interference is shaped by the positions of the network users and the transmission characteristics of each user. Traditionally, the locations of interfering users have been assumed to be distributed on a plane according to a Poisson point process. Although this is a simplistic assumption, especially in the case of mobile users where their locations can change in a time-varying fashion, it facilitates analysis. The packets are assumed to arrive at the receiver according to a Poisson process, if continuous time is considered, or according to a Bernoulli process if discrete time is considered.

During each symbol interval T at the receiver there are a random number of transmitting users, which have emitted pulses that interfere with signal reception during the particular interval. The users are assumed to be Poisson-distributed in space with density λ:

$$P[k \text{ transmitting users in a region } \mathbf{R}] = \frac{e^{-\lambda R}(\lambda R)^k}{k!} \quad (11)$$

The signal transmitted from the ith interfering user, that is, $p_i(t)$ propagates through the transmission medium and the receiver filters, and as a result become attenuated and distorted. For simplicity, we will assume that distortion and attenuation can be separated. Let us first consider only the filtering effect. For short time intervals, the propagation channel and the receiver can be represented by a time-invariant filter with impulse response $h(t)$. As a result of filtering only, the contribution of the ith interfering source at the receiver is of the form $x_i(t) = p_i(t) * h(t)$, where the asterisk denotes convolution.

The attenuation of the propagation is determined by the transmission medium and environment. In wireless communications, the power loss increases logarithmically with the distance between the transmitter and the receiver [18]. If the distance between the transmitter and the receiver is r, the power loss function may be expressed in terms of signal amplitude loss function, $a(r)$:

$$a(r) = \frac{1}{r^{\gamma/2}} \quad (12)$$

where γ is the path loss exponent and γ is a function of the antenna height and the signal propagation environment. This may vary from slightly less than 2, for hallways within buildings, to >5, in dense urban environments and hard-partitioned office buildings [21].

Taking into account attenuation and filtering, the contribution of the ith interfering source at the receiver equals $a(r_i)x_i(t)$, where r_i, is the distance between the interferer and the receiver.

Assuming that the interferers within the region of consideration have the same isotropic radiation pattern, and that the receiver has an omnidirectional antenna, the total signal at the receiver is

$$x(t) = s(t) + \sum_{i \in \mathcal{N}} a(r_i)x_i(t) \quad (13)$$

where $s(t)$ is the signal of interest and $\mathcal{N}$ denotes the set of interferers at time t. The number of interfering users, as already discussed, is a Poisson-distributed random variable with parameter λ.

The receiver consists of a signal demodulator followed by the detector. A correlation demodulator decomposes the received signal into an N-dimensional vector. The signal is expanded into a series of orthonormal basis functions $\{f_n(s), 0 < s \leq T, n = 1, \ldots, N\}$. Let $Z_n(m)$ be the projection of $x(t)$ onto $f_n(\cdot)$ at time slot m:

$$Z_n(m) = \int_0^T x(s + (m-1)T)) f_n(s)\, ds \quad (14)$$

To compute the probability of symbol error, one needs to determine first the joint probability density function of the samples $Z_n(m)$. The samples $Z_n(m)$ may be expressed as

$$Z_n(m) = S_n(m) + \sum_{i \in \mathcal{N}} a(r_i) X_{i,n}(m) \tag{15}$$

$$= S_n(m) + Y_n(m) \tag{16}$$

where $X_{i,n}(m)$ and $S_n(m)$ are, respectively, the result of the correlations of $x_i(t)$ and $s(t)$ with the basis functions $f_n(\cdot)$, and $Y_n(m)$ represents interference.

Let us assume that the $X_{i,n}(m)$ are spatially independent [e.g., $X_{i,n}(m)$ independent $X_{j,n}(m)$ for $i \neq j$]. For simplicity, we assume tht $X_{i,k}(m)$, $X_{i,l}(m)$ are identically distributed. Therefore, we shall concentrate on one dimension only. For notational convenience, we shall drop the subscript n, thus denoting $Y_n(m)$ and $X_{i,n}(m)$ by respectively $Y(m)$ and $X_i(m)$. According to the i.i.d. assumption, we will later on denote $X_i(m)$ by $X(m)$.

For some time m, $Y(m)$ is the sum of a random number of i.i.d. random variables, which are contributions of interfering users. To compute the characteristic function of $Y(m)$, we first restrict the sum to contain interferers within a disk, D_b, centered at the receiver and has radius b. Later, we will let the disk radius $b \to \infty$. The characteristic function of interference received from D_b, to be denoted by $Y_b(m)$, is

$$\Phi_{Y_b}(\omega) = E\{e^{j\omega Y_b(m)}\}$$

$$= E\{e^{j\omega \sum_{i=0}^{N_b} a(r_i) X_i(m)}\} \tag{17}$$

where N_b is a random variable representing the number of interferers at times m in D_b. Since the interferers are Poisson-distributed in the space, given that there are k of them in a disk D_b, they are i.i.d. and uniformly distributed. Thus, the distance r_i between the ith interferer and the receiver has density function

$$f(r) = \begin{cases} \dfrac{2r}{b^2} & r \leq b \\ 0 & \text{elsewhere} \end{cases} \tag{18}$$

From (17), using the i.i.d. property of $X_i(m)$ and r_i, and also using the property of conditional expected values

$$E\{X\} = E\{E\{X \mid Y\}\}$$

we obtain

$$\Phi_{Y_b}(\omega) = E\{E\{e^{j\omega \sum_{i=0}^{N} a(r_i) X_i(m))} \mid N\}\}$$

$$= E\{[E\{e^{j\omega a(r) X(m)}\}]^N\} \tag{19}$$

where r_i and $X_i(m)$ are random variables, which, according to the i.i.d. assumption, are generically denoted by r and $X(m)$, respectively. The inner expectation in (19) equals

$$A \triangleq E\{e^{j\omega a(r) X(m)}\}$$

$$= E\{E\{e^{j\omega a(r) X(m)} \mid r\}\}$$

$$= E\{\Phi_X(a(r)\omega)\}$$

$$= \int_0^b f(r) \Phi_X(a(r)\omega)\, dr \tag{20}$$

where $\Phi_X(.)$ is the characteristic function of $X(\cdot)$.

By substituting (20) in (19), we obtain

$$\Phi_{Y_b}(\omega) = E\{A^N\}$$

$$= \sum_{k=0}^{\infty} P(N = k) A^k$$

$$= \sum_{k=0}^{\infty} \frac{(\lambda \pi b^2)^k e^{-\lambda \pi b^2}}{k!} A^k$$

$$= e^{\lambda \pi b^2 (A-1)}$$

By considering the logarithm of the characteristic function, and also using (20), we obtain

$$\Psi_{Y_b}(\omega) \triangleq \log \Phi_{Y_b}(\omega)$$

$$= \lambda \pi b^2 \left(\int_0^b f(r) \Phi_X(a(r)\omega)\, dr - 1 \right) \tag{21}$$

Taking into account (12), setting

$$\alpha = \frac{4}{\gamma} \tag{22}$$

and after a variable substitution, the equation above becomes

$$\Psi_{Y_b}(\omega) = \lambda \pi b^2 \left(\frac{\alpha \omega^\alpha}{b^2} \int_{\omega b^{-2/\alpha}}^{\infty} t^{-1-\alpha} \Phi_X(t)\, dt - 1 \right) \tag{23}$$

To obtain the first-order characteristic function of $Y(n)$, we need to evaluate this expression for $b \to \infty$. Using integration by parts, and noting that $\Phi_X(\omega b^{-2/\alpha}) - 1 \to 0$ as $b \to \infty$, we obtain

$$\Psi_Y(\omega) = \lim_{b \to \infty} \Psi_{Y_b}(\omega) = -\sigma |\omega|^\alpha, \quad \text{for } 0 < \alpha < 2 \tag{24}$$

where

$$\sigma = -\pi \lambda \int_0^\infty \frac{\partial \Phi_X(x)}{\partial x} x^{-\alpha}\, dx \tag{25}$$

Equation (25) corresponds to the log-characteristic function of a $S\alpha S$ distribution with exponent α; thus, it implies that the interference $Y(n)$, for a fixed n, is $S\alpha S$, or equivalently, marginally $S\alpha S$.

The basic propagation characteristic that led to the nice closed form result shown above is the power-law attenuation. However, the attenuation expression in (12) is valid for large values of r. As $r \to 0$, the signal amplitude appears to approach infinity. To avoid this problem, an alternative loss function was proposed [11]

$$a'(r) = \min(s, r^{-\gamma/2}) \tag{26}$$

for some $s > 0$. Under this form of attenuation, the log-characteristic function is as in (24), except that now $\sigma = \sigma(\omega)$. The difference between the two log-characteristic functions approaches zero as $\omega \to \infty$ while s is fixed. For small ω the difference tends to zeros as $s \to 0$.

The meaning of (24) is illustrated next via an example. Consider a direct-sequence spread-spectrum

(DSSS) system with large processing gain and chip synchronization. Assume that users' contributions, $X_i(m)$ values, are Gaussian, and that the attenuation law is $1/r^4$. Equation (24) suggests that the resulting interference at the receiver will be Cauchy-distributed.

The discussion above applies to both narrowband interference and wideband interference. In most communication systems, the receiver is narrowband. In those cases, we need to derive the statistics of the envelope and phase of the impulsive interference. The joint characteristic function of the in-phase and quadrature components of the interference can be found to be [20]

$$\log \Phi(\omega_1, \omega_2) = -c(\alpha)(\omega_1^2 + \omega_2^2)^{\alpha/2} \tag{27}$$

where $c(\alpha)$ is a constant that depends on α. This form of joint characteristics function is referred to as *isotropic α-stable*.

Let Y_c and Y_s denote respectively the in-phase and quadrature components of the interference. The envelope A and phase Ψ are then

$$A = \sqrt{Y_c^2 + Y_s^2}, \quad \Psi = \arctan \frac{Y_s}{Y_c} \tag{28}$$

It can be shown that the phase is uniformly distributed in $[0, 2\pi]$, and is independent of the envelope. The probability density function of the envelope cannot be obtained in closed form. However, it can be shown that

$$\lim_{x \to \infty} x^\alpha P(A > x) = \beta(\alpha, \gamma) \tag{29}$$

where β is some function that does not vary with x. According to this equation, the envelope is heavy-tailed.

To implement optimum signal detection and estimation, one needs to obtain joint statistics of the interference. In the literature, for simplicity reasons, interference is traditionally assumed to be i.i.d. However, as we will see next, such assumptions are oversimplified and it is inconsistent with practical communication systems. Temporary dependence arises when the interference sources are the cochannel users, whose activity is decided by the information type being exchanged and the user behavior.

Let us define the term *emerging interferers at symbol interval l*, which describes the interfering sources whose contribution arrive for the first time at the receiver in the beginning of the symbol interval l. It is assumed that the *emerging* interferers at some symbol interval are located according to a Poisson point process in space. In particular, at any given symbol interval, the expected number of emerging interferers in a unit area/volume is given by λ_e. Once a user initiates a transmission, it may last for a random duration of time, referred to as *session life L*. The distribution of L is assumed to be known a priori. It is not difficult to show that at any symbol interval m, the active transmitting users are Poisson-distributed in the space with density $\lambda = \lambda_e E\{L\}$.

Consider an interference source, namely, a cochannel user, whose transmission starts at some random time slot m. According to the previous analysis, it contributes to the interference observed at time m at the targeted receiver, $a(r_i)x_i(m)$ [see Eq. (15)]. Since this particular cochannel user continues to be active for some random time duration after its initiation of the session, it also contributes to the interference at time $m+l$ with probability $\overline{F}_L(l)$. Here $\overline{F}_L(\cdot)$ is the survival function of session life L. Therefore, the interference at time m and $m+l$ are correlated, in contrast to conventional i.i.d. assumption for the interference at different times.

The joint statistics of the interference can be analyzed through the joint characteristic function (JCF) of the interference at time m and n ($m < n$):

$$\begin{aligned}\Phi_{m,n}(\omega_1, \omega_2) &\stackrel{\Delta}{=} E \exp\{j\omega_1 Y(m) + j\omega_2 Y(n)\} \\ &= E \exp\left\{ j\omega_1 \sum_{i \in \mathcal{N}_n} a(r_i)X_i(m) \right. \\ &\quad \left. + j\omega_2 \sum_{i \in \mathcal{N}_n} a(r_i)X_i(m) \right\}\end{aligned} \tag{30}$$

where $\mathcal{N}_m$ and $\mathcal{N}_n$ denote the set of active interferers at m and n, respectively. It can be shown that although the discretized interference is marginally α-stable, in general, it is not jointly α-stable distributed [33].

We next consider a special case when the symbols $X_i(m)$ are symmetric binary, such as 1 or -1 with equal probability and independent from slot to slot. Then, it can be shown that [33]

$$\begin{aligned}\ln \Phi_{m,n}(\omega_1, \omega_2) &= -\sigma^\alpha \sum_{l=1}^{n-m} \overline{F}_L(l)(|\omega_1|^\alpha + |\omega_2|^\alpha) \\ &\quad - \frac{1}{2}\sigma^\alpha \sum_{l=n-m+1}^{\infty} \overline{F}_L(l)(|\omega_1 + \omega_2|^\alpha + |\omega_1 - \omega_2|^\alpha)\end{aligned} \tag{31}$$

with α as defined in (22), and

$$\sigma = \frac{1}{2}\left(\frac{\lambda\pi^{3/2}\Gamma(1-\alpha/2)}{\Gamma(1/2+\alpha/2)}\right)^{1/\alpha} \tag{32}$$

Equation (31) implies that the interference at m and n are jointly α-stable distributed [24, p. 69].

In high-speed data networks, where large variations of file sizes are exchanged, the holding time of a single transmission session exhibits high variability. It has been shown [30,34] that the holding times can be well modeled by heavy-tail-distributed random variables. As data service becomes increasingly popular in wireless communication networks, where more and more bandwidth is available for users, it should be expected that session life of users will behave similarly as in wired data networks, that is, will be heavy-tail-distributed. Indeed, preliminary verification of the latter has been presented in Ref. 16 through statistical analysis of wireless network traffic data.

By modeling the session life of the interferers as a heavy-tail-distributed random variable with tail index $1 < \alpha_L < 2$, and $X_i(m)$ as i.i.d. Bernoulli random variables taking possible values 1 and -1 with equal probability $\frac{1}{2}$, it can be shown that the resulting interference is strongly correlated. In particular, assuming a Zipf distribution for the session life,[1] the following holds [33]:

$$\lim_{\tau\to\infty} \frac{-I(1,-1;\tau)}{\tau^{-(\alpha_L-1)}} = \frac{(2-2^{\alpha-1})\sigma^{\alpha}}{\alpha_L - 1} \tag{33}$$

where $I(\cdot)$ = codifference of the resulted interference
τ = time lag between symbol intervals
α_L = tail index of the session life distribution
σ = as defined in (32)

The form of (33) implies an important phenomenon, referred to as long-range dependence (LRD) [3,22]. LRD refers to strong dependence between samples of stochastic processes. For processes with finite autocorrelation, this implies that the correlation of well separated samples is not negligible, and decays in a power-law fashion with the distance between the samples. As a result, the autocorrelation of a long-range dependent process is not summable. When the noise exhibits LRD the performance of signal detection and estimation algorithms, which are optimized for i.i.d. noise, can degrade [2,32]. For processes with infinite autocorrelation, such as marginally α-stable processes, LRD is defined in terms of a power-law decaying generalized codifference [22], which is the case in (33).

We next present some simulation results based on the model described above. A wireless network is simulated, where a receiver is subjected to interference from users that are spatially Poisson distributed over a plane with density $\lambda = 2/\pi$. The path loss is power-law with $\gamma = 4$. Once the interferers start to emit pulses, they remain active for a random time duration, which in our simulations was taken to be Zipf-distributed with $k_0 = \sigma = 1$ and $\alpha_L = 1.8$. The $X_i(m)$ values were taken to be i.i.d. Bernoulli-distributed, taking values ± 1 with equal probabilities. Then, according to (32), $\sigma = \pi$, and the instantaneous interference is Cauchy distributed with scale parameter $\sigma\mu^{1/\alpha}$, where μ is the mean of the session life. Note that $\mu = \zeta(1.8) \simeq 1.88$.

One segment of the simulated interference process is shown in Fig. 2, where it can be seen that the simulated process is highly impulsive.

We use the static Hill estimator presented in Section 2.B to estimate the tail index. Figure 3 clearly shows that the interference is heavy-tail-distributed with tail index very close to the theoretical value 1. The QQ plot of the interference against ideally Cauchy-distributed random variable with scale parameter 1.88π is also illustrated in Fig. 4, which shows that the instantaneous interference can be well modeled by Cauchy distributed random variables.

[1] A random variable X has a Zipf distribution if

$$P\{X \geq k\} = \left[1 + \left(\frac{k-k_0}{\sigma}\right)\right]^{-\alpha}, \quad k = k_0, k_0+1, k_0+2\cdots$$

where k_0 is an integer denoting the location parameter, $\sigma > 0$ denotes the scale parameter and $\alpha > 0$ is the tail index.

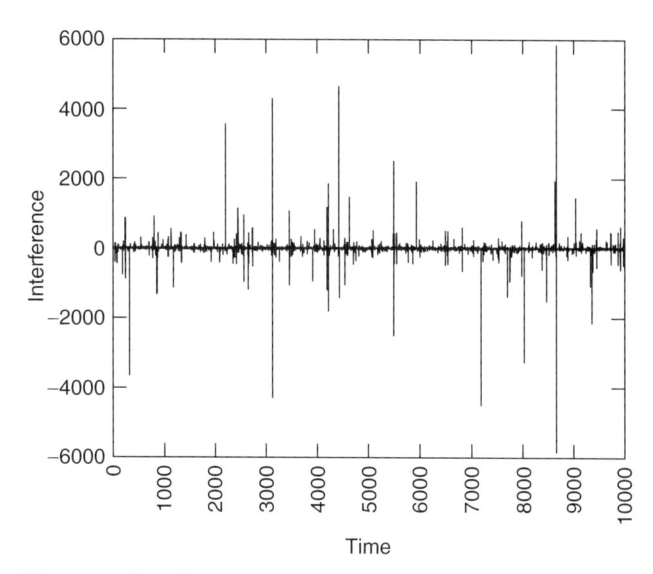

Figure 2. Interference at a receiver surrounded by Poisson distributed interferers.

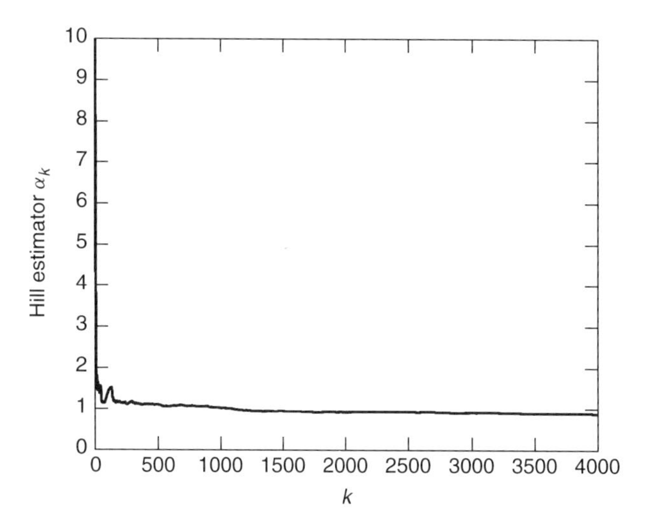

Figure 3. Dynamic Hill estimator α_k.

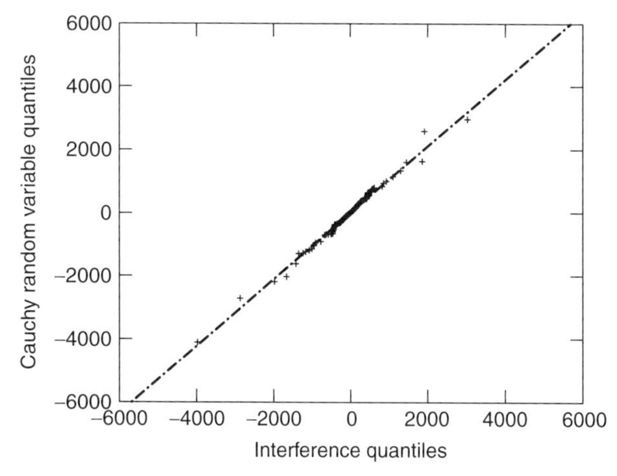

Figure 4. QQ plot of simulated interference and Cauchy random variables.

3.2. The Class A Noise Model

Another model, which has finite variance, but has been used to approximate impulsive interference, is the Middleton class A noise model [18].

The class A model assumes that a narrowband receiver is exposed to an infinite number of potential interfering sources, which emit basic waveforms with common form, scale, durations, frequencies, and so on. The parameters are randomly distributed. As assumed in the α-stable noise model, the locations of the sources are Poisson-distributed in space. The sources are statistically independent, and their emission times are Poisson-distributed in time. The propagation model employed in the class A model is power-law as defined in (12). Moreover, to take into account system noise and external interference, resultant of many independent sources none of which is exceptionally dominant with respect to the others, the class A noise includes a Gaussian component, denoted by its variance σ_G^2.

Under these assumptions, Middleton [18] derived the exceeding probability function of the envelop of the resulted noise:

$$\Pr(\varepsilon > \varepsilon_0) \cong e^{-A} \sum_{m=0}^{\infty} \frac{A^m}{m!} e^{-\varepsilon_0^2/2\sigma_m^2}, \quad 0 \leq \varepsilon_0 < \infty \tag{34}$$

with $2\sigma_m^2 = (m/A + \Gamma)/(1 + \Gamma)$. Here ε, ε_0 are normalized envelopes

$$\varepsilon \equiv \frac{E}{\sqrt{2\Omega(1+\Gamma)}} \tag{35}$$

$$\varepsilon_0 \equiv \frac{E_0}{\sqrt{2\Omega(1+\Gamma)}} \tag{36}$$

where E_0 is some preselected threshold value of the envelope E. There are three parameters involved in the class A model, namely, (A, Γ, Ω). They all have physical significance as stated next.

1. A = the *impulsive index*, which is defined as the average number of emissions multiplied by the mean duration of a typical interfering source emission. When A is large, the central-limit theory comes into play, and one approaches Gaussian statistics (for envelopes, it is Rayleigh).
2. $\Gamma \equiv \sigma_G^2/\Omega$ = the ratio of the independent Gaussian component σ_G^2 to the intensity of the non-Gaussian, impulsive component.
3. Ω = the intensity of the impulsive component.

The phase statistics of the resulting noise is uniformly distributed in $(0, 2\pi)$.

The class A noise model is a canonical model, in the sense that it is invariant of the particular noise source and associated quantifying parameters. The parameters of the class A noise model can be deduced from physical measurement, because its derivation is based on a general physical mechanism. The class A noise model fits a variety of measurements, and has been applied in various communication scenaria, where non-Gaussian noise dominates the interfering sources. However, it is often difficult to evaluate (34). A simplification has been proposed by Spaulding [27], where the infinite sum in (34) has been replaced by a finite sum. Note that the density function of class A noise can be written as

$$f(\varepsilon) = \sum_{m=0}^{\infty} a_m g(\varepsilon; 0, \sigma_k^2) \tag{37}$$

where

$$a_m = \frac{e^A A^m}{m!} \quad \text{and} \quad g(\varepsilon; \mu, \sigma_m^2) = \frac{1}{\sqrt{2\pi\sigma_m^2}} e^{-(\varepsilon-\mu)^2/2\sigma_m^2} \tag{38}$$

Essentially, (37) indicates that class A noise is a Poisson weighted sum of zero mean Gaussian functions with monotonic increasing variance σ_m^2. Since A is small, the weights in (37) for large m can be truncated without much loss in the accuracy. Spaulding [27] suggested that (37) can be well approximated (2% error) by the first two terms when A and Γ are small. The truncated class A noise now reduces to a subset of the Gauss-Gauss mixture model [14]:

$$f_{\text{mix}}(x) = ag(x; 0, \sigma_0^2) + (1 - a)g(x; 0, \sigma_1^2) \tag{39}$$

where $0 < a < 1$ and $\sigma_0^2 < \sigma_1^2$.

Efforts have been devoted to developing a multivariate class A noise model. The multivariate case arises in the scenario of communication systems with multiple antennas or antenna arrays reception, or in times where multiple time observations are available. In Ref. 7, multivariate class A models are developed based on mathematical extension of the univariate class A model. Three different extensions—independent, dependent, and uncorrelated cases—are discussed in Ref. 7. Issues of signal detection under these models are also investigated. A more physical-mechanism oriented approach is presented in Ref. 17. Under assumptions similar to those expressed in Ref. 18, the authors consider the noise presented in the intermediate-frequency (IF) stage in a communication system with two antennas. By assuming the emission times of the sources are uniformly distributed over a long time interval, Ref. 17 obtains the characteristic function of the resulting noise for cases where the antenna observations may be statistically dependent from antenna to antenna. The probability density function thus can be deduced by utilizing Fourier transform techniques.

4. CONCLUSION

In this chapter, we present two mathematical models for interference in wireless communications: the α-stable model and the class A noise model. Both models are based on the physical mechanism of the noise generation process, and lead to a non-Gaussian scenario, where noise may become impulsive, or heavy-tail-distributed. The noise process may not be i.i.d. Moreover, under heavy-tailed holding times, the noise becomes highly correlated.

Impulsiveness of noise can have severe degrading effects on system performance, particularly on most conventional systems designed for optimal or near-optimal performance against white Gaussian noise. A significant amount of research has been devoted to signal detection and estimation when the noise is non-Gaussian and/or correlated [e.g., 14,19,20].

BIOGRAPHIES

Xueshi Yang received his B.S. degree in electronic engineering from Tsinghua University, Beijing, China, in 1998, and a Ph.D. degree in electrical engineering from Drexel University, Philadelphia, Pennsylvania, in 2001. From September 1999 to April 2000 he was a visiting researcher with Laboratoire de Signaux et Systemes, CNRS-Universite de Paris-Sud, SUPELEC, Paris, France. He is currently a postdoc research associate in Electrical Engineering at Princeton University, New Jersey, and Drexel University. His research interests are in the area of non-Gaussian signal processing, self-similar processes, and wireless/wireline data networking. Dr. Yang received the George Hill Jr. Fellowship in 2001 and the Allen Rothwarf Outstanding Graduate Student Award in 2000, respectively, from Drexel University.

Athina P. Petropulu received the Diploma in electrical engineering from the National Technical University of Athens, Greece, in 1986, the M.Sc. degree in electrical and computer engineering in 1988, and her Ph.D. degree in electrical and computer Engineering in 1991, both from Northeastern University, Boston, Massachusetts.

In 1992, she joined the Department of Electrical and Computer Engineering at Drexel University, Philadelphia, Pennsylvania, where she is now a professor. During the academic year 1999–2000 she was an associate professor at LSS, CNRS-Université Paris Sud, École Supérieure d'Electrícité in France. Dr. Petropulu's research interests span the area of statistical signal processing, communications, higher-order statistics, fractional-order statistics and ultrasound imaging. She is the coauthor of the textbook entitled, *Higher-Order Spectra Analysis: A Nonlinear Signal Processing Framework*, (Englewood Cliffs, NJ: Prentice-Hall, Inc., 1993). She is the recipient of the 1995 Presidential Faculty Fellow Award.

BIBLIOGRAPHY

1. R. B. D'Agostino and M. A. Stephens, eds., *Goodness-of-fit Techniques*, Marcel Dekker, New York, 1986.
2. R. J. Barton and H. V. Poor, Signal detection in fractional Gaussian noise, *IEEE Trans. Inform. Theory* **34**: 943–959 (Sept. 1988).
3. J. Beran, *Statistics for Long-Memory Processes*, Chapman & Hall, New York, 1994.
4. K. L. Blackard, T. S. Rappaport, and C. W. Bostian, Measurements and models of radio frequency impulsive noise for indoor wireless communications, *IEEE J. Select. Areas Commun.* **11**: 991–1001 (Sept. 1993).
5. O. Cappe et al., Long-range dependence and heavy-tail modeling for teletraffic data, *IEEE Signal Process. Mag. (Special Issue on Analysis and Modeling of High-Speed Network Traffic)* (in press).
6. J. M. Chambers, *Graphical Methods for Data Analysis*, PWS Publishing, 1983.
7. P. A. Delaney, Signal detection in multivariate Class-A interference, *IEEE Trans. Commun.* **43**: 365–373 (Feb.–April 1995).
8. W. Feller, *An Introduction to Probability Theory and Its Applications*, Vol. 2, 3rd ed., Wiley, New York, 1971.
9. K. Furutsu and T. Ishida, On the theory of amplitude distribution of impulsive random noise, *J. Appl. Phys.* **32**(7): (1961).
10. A. Giordano and F. Haber, Modeling of atmosphere noise, *Radio Sci.* **7**(11): (Nov. 1972).
11. J. W. Gluck and E. Geraniotis, Throughput and packet error probability in cellular frequency-hopped spread-spectrum radio networks, *IEEE J. Select. Areas Commun.* **7**: 148–160 (Jan. 1989).
12. B. M. Hill, A simple general approach to inference about the tail of a distribution, *Ann. Stat.* **3**: 1163–1174 (1975).
13. J. Ilow, D. Hatzinakos, and A. N. Venetsanopoulos, Performance of FH SS radio networks with interference modeled as a mixture of Gaussian and Alpha-stable noise, *IEEE Trans. Commun.* **46**(4): (April 1998).
14. S. A. Kassam, *Signal Detection in Non-Gaussian Noise*, Springer-Verlag, New York, 1987.
15. M. F. Kratz and S. I. Resnick, The QQ estimator and heavy tails, *Stoch. Models* **12**(4): 699–724 (1996).
16. T. Kunz et al., WAP traffic: Description and comparison to WWW traffic, *Proc. 3rd ACM Int. Workshop on Modeling, Analysis and Simulation of Wireless and Mobile Systems*, Boston, Aug. 2000.
17. K. F. McDonald and R. S. Blum, A statistical and physical mechanisms-based interference and noise model for array observations, *IEEE Trans. Signal Process.* **48**(7): (July 2000).
18. D. Middelton, Statistical-physical models of electromagnetic interference, *IEEE Trans. Electromagn. Compat.* **EMC-19**(3): (Aug. 1977).
19. D. Middleton and A. D. Spaulding, Elements of weak signal detection in non-Gaussian noise environments, in H. V. Poor and J. B. Thomas, eds., *Advances in Statistical Signal Processing*, Vol. 2, *Signal Detection*, JAI Press, Greenwich, CT, 1993.
20. C. L. Nikias and M. Shao, *Signal Processing with Alpha-Stable Distributions and Applications*, Wiley, New York, 1995.
21. J. D. Parsons, *The Mobile Radio Propagation Channel*, Wiley, New York, 1996.
22. A. P. Petropulu, J.-C. Pesquet, X. Yang, and J. Yin, Power-law shot noise and relationship to long-memory processes, *IEEE Trans. Signal Process.* **48**(7): (July 2000).
23. J. Rice, *Mathematical Statistics and Data Analysis*, Duxbury Press, Belmont, CA, 1995.
24. G. Samorodnitsky and M. S. Taqqu, *Stable Non-Gaussian Random Processes: Stochastic Models with Infinite Variance*, Chapman & Hall, New York, 1994.
25. G. W. Snedecor and W. G. Cochran, *Statistical Methods*, Iowa State Press, 1990.
26. E. S. Sousa, Performance of a spread spectrum packet radio network link in a Poisson field of interferers, *IEEE Trans. Inform. Theory* **38**(6): (Nov. 1992).

27. A. D. Spaulding, Locally optimum and suboptimum detector performance in a non-Gaussina interference environment, *IEEE Trans. Commun.* **COM-33**(6): 509–517 (1985).

28. G. V. Trunk, Non-Rayleigh sea clutter: Properties and detection of targets, in D. C. Shleher ed., *Automatic Detection and Radar Data Processing*, Artech House, Dedham, 1980.

29. E. J. Wegman, S. C. Schwartz, and J. B. Thomas, eds., *Topics in Non-Gaussian Signal Processing*, Springer, New York, 1989.

30. W. Willinger, M. S. Taqqu, R. Sherman, and D. V. Wilson, Self-similarity through high-variability: Statistical analysis of Ethernets LAN traffic at the source level, *IEEE/ACM Trans. Network.* **5**(1): (Feb. 1997).

31. B. D. Woerner, J. H. Reed, and T. S. Rappaport, Simulation issues for future wireless modems, *IEEE Commun. Mag.* (July 1994).

32. G. W. Wornell, Wavelet-based representations for the 1/f family of fractal processes, *Proc. IEEE* **81**(10): 1428–1450 (Oct. 1993).

33. X. Yang and A. P. Petropulu, Joint statistics of impulsive noise resulted from a Poisson field of interferers, *IEEE Trans. on Signal Processing* (submitted).

34. X. Yang and A. P. Petropulu, The extended alternating fractal renewal process for modeling traffic in high-speed communication networks, *IEEE Trans. Signal Process.* **49**(7): (July 2001).

INTERFERENCE SUPPRESSION IN SPREAD-SPECTRUM COMMUNICATION SYSTEMS

Moeness G. Amin
Yimin Zhang
Villanova University
Villanova, Pennsylvania

1. INTRODUCTION

Suppression of correlated interference is an important aspect of modern broadband communication platforms. For wireless communications, in addition to the presence of benign interferers, relatively narrowband cellular systems, employing time-division multiple access (TDMA) or the Advanced Mobile Phone System (AMPS) may coexist within the same frequency band of the broadband code-division multiple access (CDMA) systems. Hostile jamming is certainly a significant issue in military communication systems. Global Positioning System (GPS) receivers potentially experience a mixture of both narrowband and wideband interference, both intentionally and unintentionally.

One of the fundamental applications of spread-spectrum (SS) communication systems is its inherent capability of interference suppression. SS systems are implicitly able to provide a certain degree of protection against intentional or unintentional interferers. However, in some cases, the interference might be much stronger than the SS signal, and the limitations on the spectrum bandwidth render the processing gain insufficient to decode the useful signal reliably. For this reason, signal processing techniques are frequently used in conjunction with the SS receiver to augment the processing gain, permitting greater interference protection without an increase in the bandwidth. Although much of the work in this area has been motivated by the applications of SS as an antijamming method in military communications, it is equally applicable in commercial communication systems where SS systems and narrowband communication systems may share the same frequency bands.

This article covers both the direct-sequence spread-spectrum (DSSS) and frequency-hopping (FH) communication systems, but the main focus is on the DSSS communication systems. For DSSS communication systems, two types of interference signals are considered, namely, narrowband interference (NBI) and nonstationary interference, such as instantaneously narrowband interference (INBI).

The early work on narrowband interference rejection techniques in DSSS communications has been reviewed comprehensively by Milstein in [1]. Milstein discusses in depth two classes of rejection schemes: (1) those based on least-mean-square (LMS) estimation techniques and (2) those based on transform domain processing structures. The improvement achieved by these techniques is subject to the constraint that the interference be relatively narrowband with respect to the DSSS signal waveform. Poor and Rusch [2] have given an overview of NBI suppression in SS with the focus on CDMA communications. They categorize CDMA interference suppression by linear techniques, nonlinear estimation techniques, and multiuser detection techniques (multiuser detection is outside the scope of this article). Laster and Reed [3] have provided a comprehensive review of interference rejection techniques in digital wireless communications, with the focus on advances not covered by the previous review articles.

Interference suppression techniques for nonstationary signals, such as INBI, have been summarized by Amin and Akansu [4]. The ideas behind NBI suppression techniques can be extended to account for the nonstationary nature of the interference. For time-domain processing, time-varying notch filters and subspace projection techniques can be used to mitigate interferers characterized by their instantaneous frequencies and instantaneous bandwidths. Interference suppression is achieved using linear and bilinear transforms, where the time–frequency domain and wavelet/Gabor domain are typically considered. Several methods are available to synthesize the nonstationary interference waveform from the time–frequency domain and subtract it from the received signal.

Interference rejection for FH is not as well developed as interference rejection for DS or for CDMA. In FH systems, the fast FH (FFH) is of most interest, and the modulation most commonly used in FH is frequency shift keying (FSK). Two types of interference waveforms can be categorized, namely, partial-band interference (PBI) and multitone interference (MTI). Typically, interference suppression techniques for FH communication systems often employ a whitening or clipping stage to reject interference, and then combined by diversity techniques.

2. SIGNAL MODEL

The received waveform consists of a binary phase shift keying (BPSK) DSSS signal $s(t)$, an interfering signal $u(t)$, and thermal noise $b(t)$. Without loss of generality, we consider the single-interferer case, and additive Gaussian white noise (AGWN) that is uncorrelated with both the DSSS and the interference signals. The input to the receiver, $x(t)$, is given by

$$x(t) = s(t) + u(t) + b(t) \tag{1}$$

The DSSS signal can be expressed as

$$s(t) = \sum_{l=-\infty}^{\infty} d(l)p(t - lT_c) \tag{2}$$

where T_c is the chip duration, $p(t)$ is the shaping waveform

$$d(l) = s(n)c(n, l) \tag{3}$$

is the chip sequence, $s(n) \in [-1, +1]$ is the nth symbol, and $c(n, l) \in [-1, +1]$ is a pseudonoise (PN) sequence used as the spreading code for the nth symbol. The PN sequence can be either periodic or aperiodic. Different types of interference signals are considered.

For discrete-time filter implementations, signals are sampled at the rate $1/T$. Typically, the sampling interval T is equal to the chip duration T_c. The input to the receiver, after sampling, becomes

$$x[n] = x(nT) \tag{4}$$

The samples of the DSSS signal, interference, and noise can be defined accordingly as $s[n]$, $u[n]$, and $b[n]$, respectively.

3. NARROWBAND INTERFERENCE SUPPRESSION

Interference suppression techniques for DSSS systems are numerous. In particular, much literature exists on the adaptive notch filtering as it relates to suppress NBI on a wideband DSSS signal. Synthesis/subtraction is another well-established technique for sinusoidal interference suppression. Other techniques include nonlinear adaptive filtering and multiuser detection techniques.

3.1. Adaptive Notch Filtering

The basic idea in employing an adaptive notch filter is to flatten the filter input spectrum. An SS signal tends to have a uniform wide spectrum and is affected little by the filtering process, whereas the NBI is characterized by spectral spikes and frequency regions of high concentrated power. The adaptive notch filter places notches at the frequency location of the NBI to bring the interference level to the level of the SS signal. At least two main approaches exist for constructing an adaptive notch filter: (1) estimation/subtraction-type filters and (2) transform-domain processing structures.

3.1.1. Estimation/Subtraction-Type Filters. If the interference is relatively narrowband compared with the bandwidth of the spread-spectrum waveform, then the technique of interference cancellation by the use of notch filters often results in a large improvement in system performance. This technique, described in many references [e.g., 5–8] uses a tapped-delay line to implement the prediction error filter (Wiener filter [9]). Since both the DS signal and the thermal noise are wideband processes, their future values cannot be readily predicted from their past values. On the other hand, the interference, which is a narrowband process, can indeed have its current and future values predicted from past values. Hence, the current value, once predicted, can be subtracted from the incoming signal, leaving an interference-free waveform comprised primarily of the DS signal and the thermal noise. A general diagram of this technique is depicted in Fig. 1. Both one-sided and two-sided transversal filters can be used for this purpose. When two-sided filters are used, the estimation of current interference value is based on both past and future values of the interference. Consider a single-sided filter as shown in Fig. 2. Define an N-dimensional vector $\mathbf{x}[n]$, denoted as

$$\mathbf{x}[n] = (x[n-1], \ldots, x[n-N])^T \tag{5}$$

where the superscript T denotes transpose of a vector or a matrix. The DSSS signal, interference, and noise vectors can be defined similarly as $\mathbf{s}[n]$, $\mathbf{u}[n]$, and $\mathbf{b}[n]$, respectively. We also define the corresponding weight vector $\mathbf{w}$ as

$$\mathbf{w} = [w_1, \ldots, w_N]^T \tag{6}$$

Hence, the output sample of the filter is

$$y[n] = x[n] - \mathbf{w}^T\mathbf{x}[n] \tag{7}$$

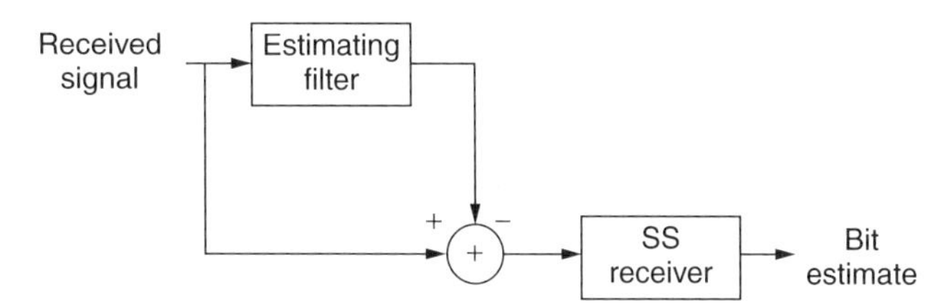

Figure 1. Estimator/subtracter-based interference suppression.

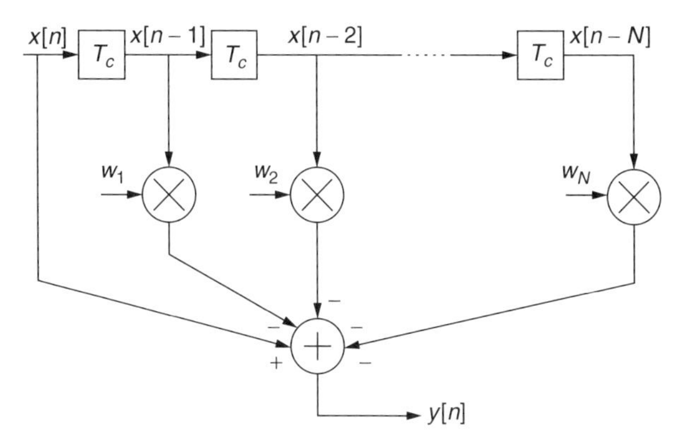

Figure 2. Single-sided transversal filter.

The mean-square value $E[y^2[n]]$, representing the output average power, is given by

$$E(y^2[n]) = E(x^2[n]) - 2\mathbf{w}^T E(x[n]\mathbf{x}[n]) + \mathbf{w}^T E(\mathbf{x}[n]\mathbf{x}^T[n])\mathbf{w}$$
$$\triangleq E(x^2[n]) - 2\mathbf{w}^T\mathbf{p} + \mathbf{w}^T\mathbf{R}\mathbf{w} \tag{8}$$

where $\mathbf{p} = E(x[n]\mathbf{x}[n])$ is the correlation vector between $x[n]$ and $\mathbf{x}[n]$, and

$$\mathbf{R} = E(\mathbf{x}[n]\mathbf{x}^T[n]) \tag{9}$$

is the covariance matrix of $\mathbf{x}[n]$. It is noted that, when the PN sequence is sufficiently long, the PN signal samples at different taps are approximately uncorrelated. On the other hand, samples of the narrowband interference at different taps has high correlations. Since the DSSS signal, interference, and thermal noise are mutually uncorrelated, it follows that

$$\begin{aligned}\mathbf{p} &= E(x[n]\mathbf{x}[n]) \\ &= E\{(s[n] + u[n] + b[n])(\mathbf{s}[n] + \mathbf{u}[n] + \mathbf{b}[n])\} \\ &= E(u[n]\mathbf{u}[n]).\end{aligned} \tag{10}$$

Minimizing the output power $E[y^2[n]]$ yields the following well-known Wiener–Hopf solution for the optimum weight vector $\mathbf{w}_{\text{opt}}$:

$$\mathbf{w}_{\text{opt}} = \mathbf{R}^{-1}\mathbf{p} \tag{11}$$

The cost of notch filtering is the introduction of some distortion into the SS signal. Such distortion results in power loss of the desired DSSS signal as well as the introduction of self-noise. Both effects become negligible when the processing gain is sufficiently large.

Note that when precise statistical knowledge of the interference cannot be assumed, adaptive filtering can be used to update the tap weights. There are a variety of adaptive algorithms and receiver structures [6,9–11]. The optimum Wiener–Hopf filter can be implemented by using direct matrix inversion (DMI) or recursive adaptation methods. For the DMI method, the covariance matrix $\mathbf{R}$ and $\mathbf{p}$ are estimated at time n by using most recent N_t data samples:

$$\hat{\mathbf{R}}[n] = \frac{1}{N_t}\sum_{l=0}^{N_t-1} \mathbf{x}[n-l]\mathbf{x}^T[n-l] \tag{12}$$

$$\hat{\mathbf{p}}[n] = \frac{1}{N_t}\sum_{l=0}^{N_t-1} x[n-l]\mathbf{x}^T[n-l] \tag{13}$$

The least-mean-square (LMS) algorithm is a simple and stable method to implement an iterative solution to the Wiener–Hopf equation without making use of any a priori statistical information about the received signal. Using the instantaneous estimates of the covariance matrix and cross-correlation vector of Eqs. (9) and (10), the LMS algorithm can be expressed as

$$\mathbf{w}[n+1] = \mathbf{w}[n] + \mu y[n]\mathbf{x}[n] \tag{14}$$

where $\mathbf{w}[n]$ is the filter weight vector of elements $w_l[n]$, $l = 1, 2, \ldots, N$, $\mathbf{x}[n]$ is the vector that includes the data within the filter, and $y[n]$ is the output, all at the nth adaptation, and μ is a parameter that determines the rate of convergence of the algorithm. It is noted that in most applications of the LMS algorithm, an external reference waveform is needed in order to correctly adjust the tap weights. However, in this particular application, the signal on the reference tap $x[n]$ serves as the external reference.

The drawback of the LMS algorithm is its slow convergence. To improve the convergence performance, techniques including self-orthogonalizing LMS, recursive least-squares (RLS), and lattice structure can be used [9,12].

3.1.2. SINR and BER Analysis. The output signal-to-interference-plus-noise ratio (SINR) and bit error rate (BER) are two important measures for communication quality and the performance enhancement using the signal processing techniques.

To derive the output SINR, we rewrite the filter output as[1]

$$y[n] = \sum_{l=0}^{N} w_l x[n-l] = \sum_{l=0}^{N} w_l(c[n-l] + u[n-l] + b[n-l]) \tag{15}$$

where $w_0 = 1$. The signal $\{y[n]\}$ is then fed to the PN correlator. Denote L as the number of chips per information bit. Then the output of the PN correlator, which is the decision variable for recovering the binary information, is expressed as

$$\begin{aligned} r &= \sum_{n=1}^{L} y[n]c[n] = \sum_{n=1}^{L} c[n] \\ &\quad \times \sum_{l=0}^{N} w_l(c[n-l] + u[n-l] + b[n-l]) \\ &= \sum_{n=1}^{L} c^2[n] + \sum_{n=1}^{L} c[n] \sum_{l=1}^{N} w_l c[n-l] \\ &\quad + \sum_{n=1}^{L} c[n] \times \sum_{l=0}^{N} w_l(u[n-l] + b[n-l]) \\ &= L + \sum_{n=1}^{L} c[n] \sum_{l=1}^{N} w_l c[n-l] + \sum_{n=1}^{L}\sum_{l=0}^{N} c[n] w_l u[n-l] \\ &\quad + \sum_{n=1}^{L}\sum_{l=0}^{N} c[n] w_l b[n-l] \end{aligned} \tag{16}$$

The first term on the right-hand side of (16) represents the desired signal component, the second term amounts to the self-noise caused by the dispersive characteristic of the filter, and the third term is the residual narrowband

[1] To keep the notation simple, we have used in (15) the same symbol w_l as in (6). The two sets of weights, however, differ in sign.

interference escaping the excision process and appearing at the output of the PN correlator. The last term in (16) is the additive noise component.

The mean value of r is

$$E(r) = L \tag{17}$$

and the variance is [6]

$$\mathrm{var}(r) \triangleq \sigma^2 = L\sum_{l=1}^{N} w_l^2 + L\sum_{n=1}^{N}\sum_{l=0}^{N} w_n w_l \rho[n-l] + L\sigma_n^2 \sum_{l=0}^{N} w_l^2 \tag{18}$$

where $\sigma_n^2 = E(b^2[n])$ is the AGWN variance and

$$\rho[n-l] = E(u[k]u[k+n-l])$$

The three terms of the right-hand side of (18) represent the mean square values caused by the self-noise, residual narrowband interference, and noise, respectively.

The output SINR is defined as the ratio of the square of the mean to the variance. Thus

$$\mathrm{SINR}_0 = \frac{E^2(r)}{\mathrm{var}(r)} = \frac{L}{\sum_{l=1}^{N} w_l^2 + \sum_{n=1}^{N}\sum_{l=0}^{N} w_n w_l \rho[n-l] + \sigma_n^2 \sum_{l=0}^{N} w_l^2} \tag{19}$$

Note that if there is no interference suppression filter, $w_l = 1$ for $l = 0$ and zero otherwise. Therefore, the corresponding output SINR is

$$\mathrm{SINR}_{\mathrm{no}} = \frac{L}{\rho[0] + \sigma_n^2} \tag{20}$$

If we assume that the self-noise, residual interference, and noise components at the output of correlator is Gaussian, then the BER can be evaluated in the same manner as the conventional BPSK corrupted only by AWGN. Under such assumption, the BER is given by

$$P_b = P(r < 0) = \int_{-\infty}^{0} \frac{1}{\sqrt{2\pi}\,\sigma} e^{-(r-L)^2/2\sigma^2}\, dr = Q(\sqrt{\mathrm{SINR}_0}) \tag{21}$$

where

$$Q(x) = \frac{1}{\sqrt{2\pi}} \int_{x}^{\infty} e^{-v^2/2}\, dv \tag{22}$$

is the Q function [12].

3.1.3. Transform-Domain Processing Structures. An alternative to time-domain excision as described in the preceding section is to transform the received signal to the frequency domain and perform the excision in that domain. Clipping and gating methods can then be applied on those transform bins contaminated by the interference.

Surface acoustic wave (SAW) device technology can be used to produce the continuous-time Fourier transform of the received waveform [13,14]. The discrete Fourier transform (DFT), with FFT implementations, is commonly applied for time-sampled signals [15]. Adaptive subband transforms generalize transform-domain processing [16,17], and can yield uncorrelated transform coefficients.

The interference-suppressed signal based on a block transform can be written as

$$\mathbf{x}_s[n] = \mathbf{BEAx}[n] \tag{23}$$

where $\mathbf{x}[n]$ is the received input vector; $\mathbf{A}$ and $\mathbf{B}$ are the forward transform matrix and inverse transform matrix, respectively; and $\mathbf{E}$ is a diagonal matrix with each diagonal element acting as a weight multiplied to the input signal at each transform bin. The weights can be controlled by different schemes. Two commonly used methods are either to set the weights binary (i.e., a weight is either one or zero) or to adjust the weights adaptively. In applying the first method, powerful NBI is detected by observing the envelope of the spectral waveform. Substantial interference suppression can be achieved by multiplying the input signal with a weight that is set to zero when the output of the envelope detector at a transform bin exceeds a predetermined level. Figure 3 illustrates the concept of transform-domain notch filtering. Adaptive algorithms, such as LMS and RLS, can be used to determine the excision weights adaptively. The application of these algorithms, however, requires a reference signal that is correlated with the DSSS signal.

When the binary weights are used, the transform-domain processing technique may suffer from the interference sidelobes. With block transforms, the energy in the narrowband interference, which initially occupies a small region of the frequency spectrum, is dispersed in a relatively large spectral region. In this case, excision of a large frequency band may become necessary to effectively remove the interference power from most transform bins. The frequency dispersion of the interference can be nearly eliminated by weighting the received signal

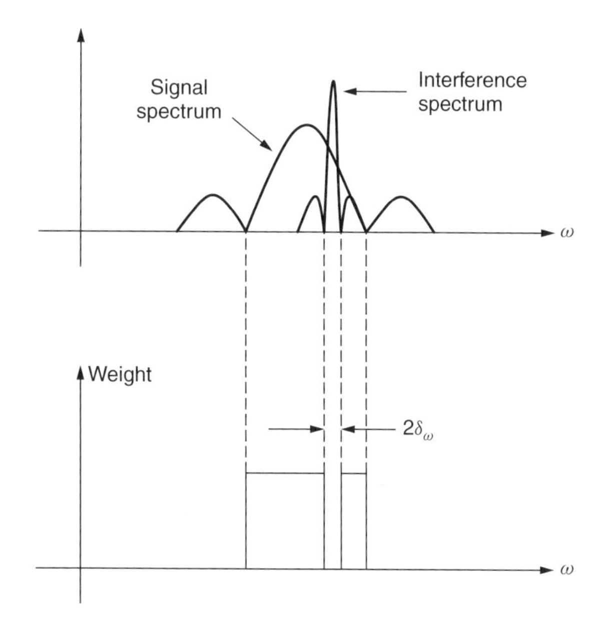

Figure 3. Transform-domain notch filtering.

in the time domain with a nonrectangular weighting function prior to evaluating the transform. In doing so, the levels of the sidelobes of the interference frequency spectrum are attenuated at the expense of broadening the mainlobe [18,14]. In this case, the conventional matched filter is no longer optimal. Using adapted demodulation accordingly can improve the receiver performance [19].

It is important to point out that for transform-domain processing, symbol detection can be performed in either the time or the transform domain. In the later case, filtering and correlation operations can be combined in one step.

The BER expression for transform-domain interference excision can be easily formulated using the Gaussian tail probability or the Q function. The residual filtered and despreaded interference is treated as an equivalent AWGN source. Typically, a uniform interference phase distribution is assumed. When transform-domain filtering is considered, the BER depends on both the excision coefficients and the error misadjustment.

3.2. Synthesis and Subtraction for Sinusoidal Interference

In this section, we view the interference signal as the one that is corrupted by the additive noise and the DSSS signal. In a typical situation, the power level of the DSSS signal is negligible relative to the power level of the interference and, in most cases, relative to the additive noise. For high interference-to-noise ratio (INR), the correlation matrix of the received signal vector consists of a limited number of large eigenvalues contributed mainly by the narrowband interference, and a large number of small and almost equal eigenvalues contributed by the DSSS signal and noise. The eigenanalysis interference canceller is designed with a weight vector orthogonal to the eigenvectors corresponding to the large eigenvalues [20]. The eigendecomposition of the correlation matrix, defined in (9), results in

$$\mathbf{R} = \mathbf{U}\mathbf{\Sigma}\mathbf{U}^H = [\mathbf{U}_r\mathbf{U}_n]\begin{bmatrix} \mathbf{\Sigma}_r & 0 \\ 0 & \mathbf{\Sigma}_n \end{bmatrix}\begin{bmatrix} \mathbf{U}_r^H \\ \mathbf{U}_n^H \end{bmatrix} \quad (24)$$

where the columns of $\mathbf{U}_r$ span the interference subspace, whereas the columns of $\mathbf{U}_n$ span the signal with noise subspace, and $\mathbf{\Sigma}_r$ and $\mathbf{\Sigma}_n$ are diagonal matrices whose elements are the eigenvalues of $\mathbf{R}$. For real sinusoidal interference, the number of dimensions of the interference subspace is twice the number of interfering tones.

The projection of the signal vector on the noise subspace results in interference suppressed data sequence

$$\hat{\mathbf{x}}[n] = \mathbf{U}_n\mathbf{U}_n^H\mathbf{x}[n] = (\mathbf{I} - \mathbf{U}_r\mathbf{U}_r^H)\mathbf{x}[n] \quad (25)$$

where $\mathbf{I}$ is the identity matrix.

The subspace projection approach can also be performed using the singular value decomposition (SVD) for the sample data matrix [21].

3.3. Nonlinear Estimation Techniques

The commonly applied predictor/subtracter technique for narrowband interference suppression previously discussed is optimum in the minimum mean-square error (MMSE) sense when trying to predict a Gaussian autoregressive process in the presence of AWGN. If the prediction is done in a non-Gaussian environment, as in the case of SS signals, linear prediction methods will no longer be optimum. In Ref. 2, depending on whether the statistics of the AR process is known or unknown, time-recursive and data-adaptive nonlinear filters with soft-decision feedback are used to estimate the SS signal. For known interference statistics, the interference suppression problem is cast in state space for use with Kalman–Bucy and approximate conditional mean (ACM) filters. A fixed-length LMS transversal filter, on the other hand, is used when there is no a priori statistical information is provided. With the same AR model, both schemes are shown to achieve similar performance, which is an improvement over the Gaussian assumed environment.

3.4. Multiuser Detection Techniques

A narrowband interference could be a digital communication signal with a data rate much lower than the spread-spectrum chip rate. This is typically the case when spread-spectrum signals are used in services overlaying on existing frequency band occupants. In this case, the narrowband interference is a strong communication signal that interferes with commercial DSSS communication systems. This type of interferer is poorly modeled as either a sinusoid or an autoregressive process. Because of the similarity of the spread spectrum signal and the digital interference, techniques from multiuser detection theory are applied to decode the SS user signal and simultaneously suppressing the interferer [22].

In order to apply methods from multiuser detection, the single narrowband interferer is treated as a collection of m spread-spectrum users, where m is a function of the relative data rates of the true SS signal and the interference. That is, m bits of the narrowband user occur for each bit of the SS user. As shown in Fig. 4, and using square waves for illustrations, each narrowband user's bit can be regarded as a signal arising from a virtual user having a signature sequence with only one nonzero entry. The virtual users are orthogonal, but correlated with the SS user.

The optimum receiver implementing the maximum-likelihood (ML) detector has a complexity that is exponential in the number of virtual users, m. To overcome such complexity, the optimal linear detector and

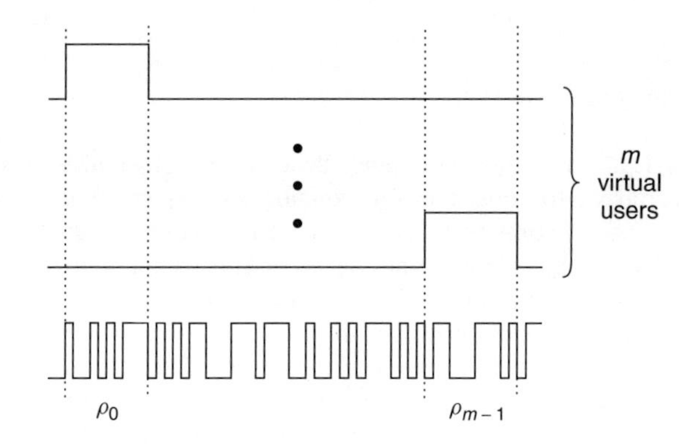

Figure 4. Virtual CDMA systems (synchronous case).

decorrelating detector are applied. While the first requires knowledge of relative energies of both the narrowband interferer and the SS user and maximizes the receiver asymptotic efficiency, the latter is independent of the receiver energies and achieves the near–far resistance of the ML detector. The asymptotic efficiency is the limit of the receiver efficiency as the AWGN goes to zero. It characterizes the detector performance when the dominant source of corruption is the narrowband interferer rather than the AWGN. The receiver efficiency, on the other hand, quantifies the SS user energy that would achieve the same probability of error in a system with the same AWGN and no other users. The input of both detectors is the output of the filter bank and consists of filters matched to the spreading codes of each active user, as depicted in Fig. 5.

The following expressions have been derived [22] for the probability of errors of four different detectors (in all four cases, it is assumed that the narrowband signal is synchronized with the SS signal):

1. *Conventional Detector* (CD), where the received signal is sent directly to a single filter matched to the spreading code. The output of the filter is then compared to a threshold to yield the spread-spectrum bit estimate. This detector is only optimum in the case of a single spread-spectrum user in AWGN. The BER is given by

$$P_{\text{cd}} = \frac{1}{2^m} \sum_{i=0}^{2^m-1} Q\left(\frac{\sqrt{w_2}(1 - \alpha \mathbf{p}^T \mathbf{q}^i)}{\sigma_n}\right) \tag{26}$$

where $\alpha = \sqrt{w_1/w_2}$, w_1 is the received energy of the narrowband interference, w_2 is the received energy of the SS user (including the process gain), $\mathbf{p}$ is the vector formed by the cross correlation between the narrowband interference waveform and the DSSS signal waveform, $\mathbf{q}$ is the narrowband interference data bits, and $\{\mathbf{q}^i\}$ is an ordering of the 2^m possible values of the vector of narrowband bits.

2. *Decorrelating Detector* (DD), where the last row of the inverse of the cross-correlation matrix of the $m+1$ users is used to multiply the output of the $m+1$ matched filters, followed by a threshold comparison for bit estimate. The BER is given by

$$P_{\text{dd}} = Q\left(\frac{\sqrt{w_2(1 - \mathbf{p}^T \mathbf{p})}}{\sigma_n}\right) \tag{27}$$

3. *Optimum Linear Detector* (OLD), where the user energies are used to maximize the asymptotic efficiency. The BER is given by

$$P_{\text{old}} = \frac{1}{2^m} \sum_{i=0}^{2^m-1} Q\left(\frac{\sqrt{w_2}(1 + \alpha \mathbf{v}^T \mathbf{p} - \alpha(\alpha \mathbf{v}^T + \mathbf{p}^T)\mathbf{q}^i)}{\sigma_n \sqrt{1 + 2\alpha \mathbf{v}^T \mathbf{p} + \alpha^2 \mathbf{v}^T \mathbf{v}}}\right) \tag{28}$$

where the ith element of vector $\mathbf{v}$ is given by

$$v_i = \begin{cases} 1 & -\rho_i > \alpha \\ -1 & \rho_i > \alpha \\ -\dfrac{\rho_i}{\alpha} & \text{otherwise.} \end{cases} \tag{29}$$

4. *Ideal Predictor / Subtracter* (IPS), which is similar to the transversal filter excision techniques described in Section 3.1. Perfect knowledge of the narrowband signal is assumed. Further, it is assume that perfect prediction to the sample interior to the narrowband bit is achieved and the only error occurs when predicting at bit transitions. The expressions have been derived [22], where one detector assumes zero bit estimate of the narrowband bit at the transition and the other detector takes this estimate to be random. For the former detector

$$P_{\text{ips}} = \frac{1}{2^m} \sum_{i=0}^{2^m-1} Q\left(\frac{\sqrt{w_2}(1 - \alpha \tilde{\mathbf{p}}^T \mathbf{q}^i)}{\sigma_n}\right) \tag{30}$$

and for the other detector

$$P_{\text{ips}} = \frac{1}{2^m} \sum_{i=0}^{2^m-1} \frac{1}{2^m} \sum_{j=0}^{2^m-1} Q\left(\frac{\sqrt{w_2}(1 - \alpha \tilde{\mathbf{p}}^T (\mathbf{q}^i - \hat{\mathbf{q}}^j))}{\sigma_n}\right) \tag{31}$$

where $\hat{\mathbf{q}}^j$ is the estimate of $\mathbf{q}^j$, and $\tilde{\mathbf{p}}$, defined only over the chip interval encompassing a narrowband bit transition, is the vector formed by the cross-correlation between the narrowband interference and the DSSS signal.

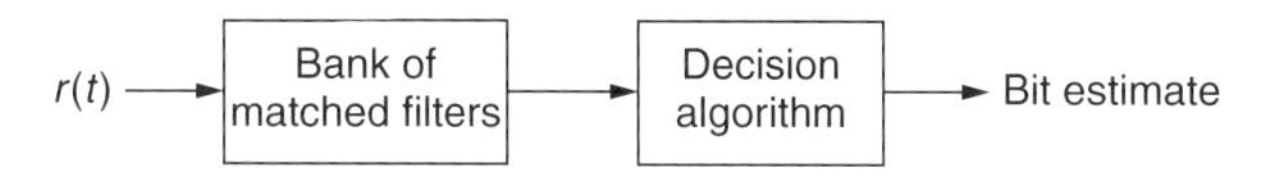

Figure 5. Multiuser detector structure.

The performance of the optimum linear and decorrelator detectors, representing the multiuser detection techniques, has been shown [22] to be similar for different interference power and bandwidth. Both techniques significantly outperform the conventional detector and the predictor/subtracter, when using a 7-tap LMS prediction filter. The improvement in BER is more pronounced for stronger and less narrowband interferers. The advantage of the decorrelator detector over the other conventional and adaptive prediction filters remains unchanged when considering asynchronous interference.

Figure 6 depicts the BER comparison between the conventional detector, decorrelating detector, optimum linear detector, and ideal predictor/subtracter with $m = 2$ and $L = 63$. This figure is in agreement with the performance figures [22] and conforms with the same observations stated above. It is clear from Fig. 6 that the matched filter performs well for weak interferers. The optimum linear detector offers slight improvement over the decorrelating detector. The ideal predictor/subtracter outperforms the decorrelating detector for moderate values of interference power. It is important to note,

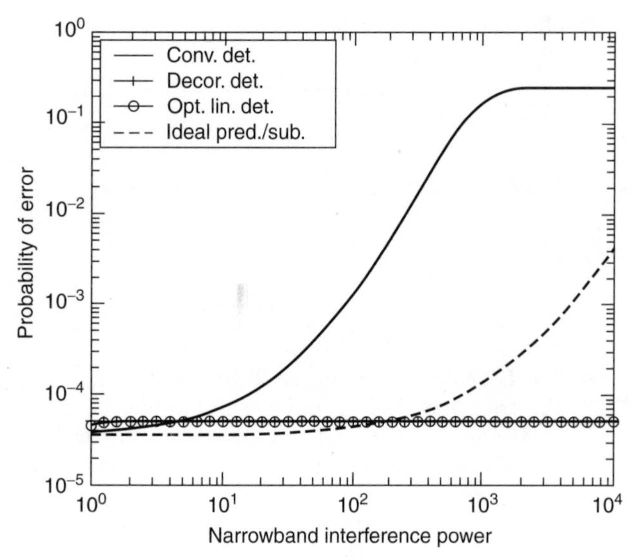

Figure 6. BER performance for different multiuser detection techniques.

however, that the actual predictor/subtracter performance will have much greater error probability [22].

3.5. Minimum Mean-Square Error Algorithm

The minimum mean-square error (MMSE) algorithm, originally proposed for suppressing multiple-access interference in CDMA multiuser detection problems, has been employed for narrowband interference mitigation [23]. Using the signal-to-interference ratio and its upper bounds as a performance measure, the MMSE has been compared with linear and nonlinear techniques in suppressing three types of interference: single-tone and multitone signals, autoregressive process, and digital communications signal with a data rate much lower than the spread-spectrum chip rate. The linear estimators include the conventional matched filter detector, the predictor/subtracter, and the interpolator/subtracter techniques. The latter is based on using a fixed number of past and future samples [23]. The nonlinear techniques include those based on prediction and interpolation. It is shown that the MMSE detector completely suppresses the digital interference, irrespective of its power, and provides performance similar to that using the nonlinear interpolator/subtracter method, when dealing with AR type of interference.

4. NONSTATIONARY INTERFERENCE SUPPRESSION

The interference excision techniques discussed in the previous sections deal with stationary or quasistationary environment. The interference frequency signature, or characteristics, is assumed fixed or slowly time-varying. None of these techniques is capable of effectively incorporating the suddenly changing or evolutionary rapidly time-varying nature of the frequency characteristics of the interference. These techniques all suffer from their lack of intelligence about interference behavior in the joint time–frequency (t-f) domain and therefore are limited in their results and their applicability. For the time-varying

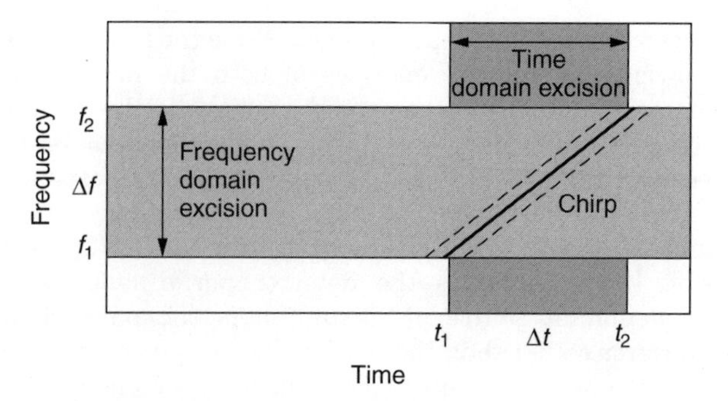

Figure 7. Excision methods for nonstationary interferers.

interference depicted in Fig. 7, frequency-domain methods remove the frequency band Δf and ignore the fact that only few frequency bins are contaminated by the interference at a given time. Dually, time domain excision techniques, through gating or clipping the interference over Δt, do not account for the cases where only few time samples are contaminated by the interference for a given frequency. Applying either method will indeed eliminate the interference but at the cost of unnecessarily reducing the desired signal energy. Adaptive excision methods might be able to track and remove the nonstationary interference, but would fail if the interference is highly nonlinear FM or linear FM, as in Fig. 7, with high sweep rates. Further, the adaptive filtering length or block transform length trades off the temporal and the spectral resolutions of the interference. Increasing the step-size parameter increases the filter output errors at convergence, and causes an unstable estimate of the interference waveform.

The preceding example clearly demonstrates that nonstationary interferers, which have model parameters that rapidly change with time, are particularly troublesome due to the inability of single-domain mitigation algorithms to adequately ameliorate their effects. In this challenging situation, and others like it, joint t-f techniques can provide significant performance gains, since the instantaneous frequency (IF), the instantaneous bandwidth, and the energy measurement, in addition to myriad other parameters, are available. The objective is then to estimate the t-f signature of the received data using t-f analysis, attenuating the received signal in those t-f regions that contain strong interference. This is depicted by the region in between the dashed lines in Fig. 7.

An FM interference in the form $u(n) = e^{j\phi(n)}$ is solely characterized by its IF, which can be estimated using a variety of IF estimators, including the time–frequency distributions (TFDs) [24,25].

The TFD of the data, $x(n)$, at time t and radian frequency ω, is given by

$$C_f(t, \omega, \phi) = \sum_{l=-\infty}^{\infty} \sum_{m=-\infty}^{\infty} \phi(m, l) x(n + m + l) \times x^*(n + m - l) e^{-j2\omega l} \quad (32)$$

where $\phi(m, l)$ is the time–frequency kernel, which is a function of the lag l and time lag m. Several requirements have been imposed on $\phi(m, l)$ to satisfy desirable distribution properties, including power localization at the IF. As shown in Eq. (32), the TFD is the Fourier transform of a time-average estimate of the autocorrelation function.

A time–frequency notch filter can be designed, in which the position of the filter notch is synchronous with the interference IF estimate. Based on the IF, two constraints should exist to construct an interference excision filter with desirable characteristics. First, an FIR filter with short impulse response must be used. Long-extent filters are likely to span segments of changing frequency contents and, as such, allow some of the interference components to escape to the filter output. Second, at any given time, the filter frequency response must be close to an ideal notch filter to be able to null the interference with minimum possible distortion of the signal. This property, however, requires filters with infinite or relatively long impulse responses.

Amin [26] has shown that a linear-phase 5-coefficient filter is effective in FM interference excision. Assuming exact IF values, the corresponding receiver SINR is given by

$$\mathrm{SINR} = \frac{L}{11/8 + 9\sigma_n^2/4} \tag{33}$$

which shows that full interference excision comes at the expense of a change in the noise variance in addition of a self-noise form, as compared with the noninterference case. The main objective of any excision process is to reduce both effects. The SINR in (33) assumes a random IF with uniform distribution over $[0, 2\pi]$. For an interference with fixed frequency ω_0, the receiver SINR becomes dependent on ω_0. The receiver performance sensitivity to the interference frequency is discussed in detail in Ref. 26.

Wang and Amin [27] considered the performance analysis of the IF-based excision system using a general class of multiple-zero FIR excision filters showing the dependence of the BER on the filter order and its group delay. The effect of inaccuracies in the interference IF on receiver performance was also considered as a function of the filter notch bandwidth. Closed-form approximations for SINR at the receiver are given for the various cases.

One of the drawbacks to the notch filter approach [26] is the infinite notch depth due to the placement of the filter zeros. The effect is a "self-noise" inflicted on the received signal by the action of the filter on the PN sequence underlying the spread information signal. This problem led to the design of an open-loop filter with adjustable notch depth based on the interference energy. The notch depth is determined by a variable embedded in the filter coefficients chosen as the solution to an optimization problem that maximizes receiver SINR. The TFD is necessary for this work, even for single component signals, because simple IF estimators do not provide energy information. Amin et al. accomplished this work [28], incorporating a "depth factor" into the analysis and redeveloping all the SINR calculations. The result was a significant improvement in SINR, especially at midrange interference-to-signal ratios (ISR's), typically around 0–20 dB.

Instead of using time-varying excision filters, Barbarossa and Scaglione [29] proposed a two-step procedure based on dechirping techniques commonly applied in radar algorithms. In the first step the time-varying interference is converted to a fixed-frequency sinusoid eliminated by time-invariant filters. The process is reversed. In the second step and the interference-free signal is multiplied by the interference t-f signature to restore the DSSS signal and noise characteristics that have been strongly impacted in the first phase.

Similar to the predictor/subtracter method discussed in Section 3, Lach et al. proposed synthesis/subtracter technique for FM interference using TFD [30]. A replica of the interference can be synthesized from the t-f domain and subtracted from the incoming signal to produce an essentially interference-free channel.

Another synthesis/subtracter method has been introduced [31] where the discrete evolutionary and the Hough transforms are used to estimate the IF. The interference amplitude is found by conventional methods such as linear filtering or singular value decomposition. This excision technique applies equally well to one or multicomponent chirp interferers with constant or time-varying amplitudes and with instantaneous frequencies not necessarily parametrically modeled.

To overcome the drawbacks of the potential amplitude and phase errors produced by the synthesis methods, Amin et al. [32] proposed a projection filter approach in which the FM interference subspace is constructed from its t-f signature. Since the signal space at the receiver is not specifically mandated, it can be rotated such that a single interferer becomes one of the basis functions. In this way, the interference subspace is one-dimensional and its orthogonal subspace is interference-free. A projection of the received signal onto the orthogonal subspace accomplishes interference excision with a minimal message degradation. The projection filtering methods compare favorably over the previous notch filtering systems.

Zhang et al. [33] proposed a method to suppress more general INBI signals. The interference subspace is constructed using t-f synthesis methods. In contrast to the work in Ref. 30, the interferer is removed by projection rather than subtraction. To estimate the interference waveform, a mask is constructed and applied such that the masked t-f region captures the interference energy, but leaves out most of the DSSS signals.

Seong and Loughlin have also extended the projection method developed by Amin et al. [32] for excising constant amplitude FM interferers from DSSS signals to the case of AM/FM interferers [34]. Theoretical performance results (correlator SNR and BER) for the AM/FM projector filter show that FM estimation errors generally cause greater performance degradation than the same level of error in estimating the AM. The lower-bound for the correlator SINR for the AM/FM projection filter for the case of both AM and FM errors is given by

$$\mathrm{SINR} = \frac{L-1}{\dfrac{1}{L} + \sigma_n^2 + A^2\left[\dfrac{1}{1+\sigma_{\Delta a}^2}(1 - e^{-\sigma_{\Delta\phi}^2}) + \sigma_{\Delta a}^2\right]} \tag{34}$$

where L is the PN sequence length, A^2 is the interference power, σ_n^2 is the variance of AWGN, and $\sigma_{\Delta a}^2$ and $\sigma_{\Delta\phi}^2$ are the variances of the estimation errors in the AM and FM, respectively.

Linear t-f signal analysis has also been shown effective to characterize a large class of nonstationary interferers. Roberts and Amin [35] proposed the use of the discrete Gabor transform (DGT) as a linear joint time–frequency representation. The DGT can attenuate a large class of nonstationary wideband interferers whose spectra are localized in the t-f domain. Compared to bilinear TFDs, the DGT does not suffer from the cross-term interference problems, and enjoys a low computational complexity. Wei et al. [36] devised a DGT-based, iterative time-varying excision filtering, in which a hypothesis testing approach was used to design a binary mask in the DGT domain. The time–frequency geometric shape of the mask is adapted to the time-varying spectrum of the interference. They show that such a statistical framework for the transform-domain mask design can be extended to any linear transform. Both the maximum-likelihood test and the local optimal test are presented to demonstrate performance versus complexity.

Application of the short-time Fourier transform (STFT) to nonstationary interference excision in DSSS communications has been considered [37,38]. In those studies [37,38], due to the inherent property of STFT to trade off temporal and spectral resolutions, several STFTs corresponding to different analysis windows were generated. Ouyang and Amin [37] used a multiple-pole data window to obtain a large class of recursive STFTs. Subsequently, they employed concentration measures to select the STFT that localizes the interference in the t-f domain. This procedure is followed by applying a binary excision mask to remove the high-power t-f region. The remainder is synthesized to yield a DSSS signal with improved signal-to-interference ratio (SIR).

Krongold et al. [38] proposed multiple overdetermined tiling techniques and utilized a collection of STFTs for the purpose of interference excision. Unlike the Ouyang–Amin procedure [37], Krongold et al. [38] removed the high-value coefficients in all generated STFTs, and used the combined results, via efficient least-square synthesis, to reconstruct an interference-reduced signal. Bultan and Akansu [39] proposed a chirplet-transform-based exciser to handle chirplike interference types in SS communications.

The block diagram in Fig. 8 depicts the various interference rejection techniques using the time–frequency methods cited above.

4.1. Example

At this point, in order to further illustrate these excision methods, the work by Amin et al. [32] will be detailed since it includes comparisons between the two most prominent techniques based on TFDs currently being studied: notch filtering and projection filtering. The signal model is, as expected, given by Eq. (1), and the major theme of the work is to annihilate interference via projection of the received signal onto an "interference-free" subspace generated from the estimated interference characteristics. This paper includes a figure, reprinted here as Fig. 9, which clearly illustrates the tradeoffs between projection and notch filtering based on the ISR. In the legend, the variable a represents the adaptation parameter for the notch filtering scheme and N represents the block size, in samples, for a 128-sample bit duration in the projection method. Thus, $N = 128$ means no block processing and $N = 2$ corresponds to 64 blocks per bit being processed for projection. Since the projection and nonadaptive notch filter techniques are assumed to completely annihilate the interference, their performance is decoupled from the interference power, and therefore correctly indicate constant SINR across the graph. The dashed line representing the notch filter with $a = 0$ really indicates no filtering at all, since the adaptation parameter controls the depth of the notch.

It is evident from Fig. 9 that without adaptation a crossover point occurs around 2 dB, where filtering with an infinitely deep notch is advantageous. Thus, when the interference power exceeds this point, presumably

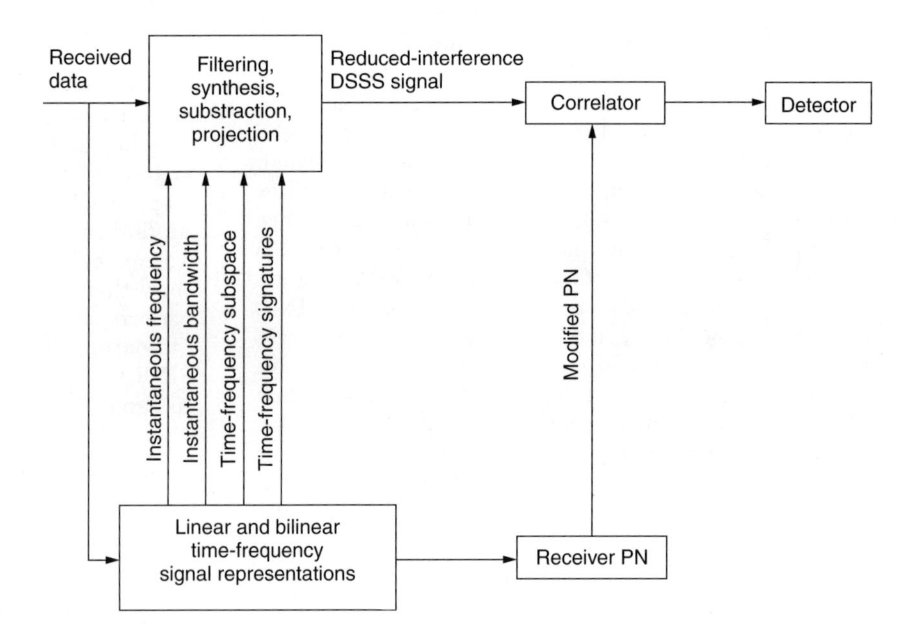

Figure 8. Interference rejection techniques.

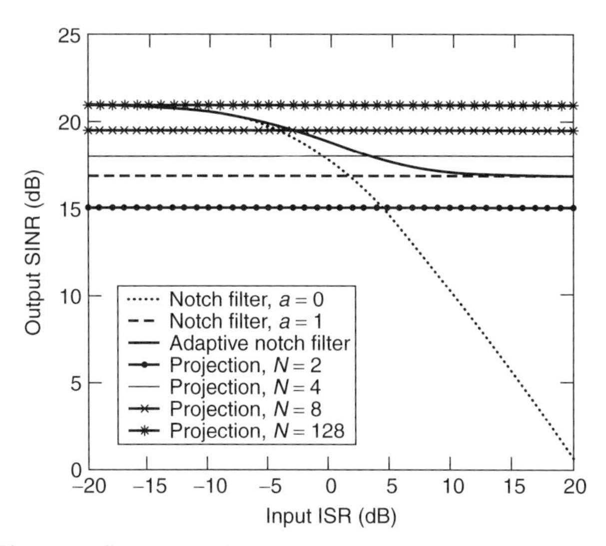

Figure 9. Comparison between projection and notch filtering excision methods.

a user would flip a switch to turn on the excision subsystem. However, with adaptation, this process happens automatically, while giving superior performance in the midrange. For the projection technique, the block size determines receiver performance conspicuously (ceteris paribus). Most important to note, however, is the superior performance of projection over all methods when the block size is equal to the bit duration, namely, no block processing. It is feasible that computational complexity may warrant a tradeoff between SINR and block size, in which case a hybrid implementation may be of benefit — one that automatically switches between adaptive notch filtering and projection depending on the desired SINR. In any case, this example illustrates the parameters involved in the design of modern excision filters for nonstationary interferers.

5. INTERFERENCE SUPPRESSION FOR FREQUENCY-HOPPING COMMUNICATIONS

Interference rejection for FH is not as well developed as interference rejection for DS or for CDMA. In FH systems, the fast FH (FFH) is of most interest, and the modulation most commonly used in FH is frequency shift keying (FSK). Two types of interference waveforms can be categorized, namely, partial-band interference (PBI) and multitone interference (MTI).

The effects of PBI and AWGN on several diversity-combining receivers in FFH/FSK SS communication systems have been investigated [40–42,44]. An alternative method using a fast Fourier transform (FFT) has been proposed [45]. An automatic gain-control (AGC) receiver using a diversity technique has also been presented [40]. In this method, each soft-decision square-law detected MARK and SPACE filter output is weighted by the inverse of the noise power in the slot prior to linear combining. This method is near-optimal (in terms of SNR) if the exact information of noise and interference power can be obtained. A similar clipped-linear combining receiver was also reported [42]. Because of the difficulty of such information, self-normalizing receivers [41] and the ratio-statistic combining technique [43] use the output values of the square-law detector in each hop to derive a weight or normalizing factor. The performance of these two methods is shown to be comparable to that of the square-law clipper receiver.

An FFH receiver that employs a prewhitening filter to reject NBI has been described [44]. For binary FSK modulations, it is shown that the FFH signal is statistically uncorrelated at lag values of $T_h/(4N_h)$, where T_h is the hop duration and $2N_h$ is the total number of frequency slots (i.e., there are N_h MARK and N_h SPACEs). Thus, as in the DS case, NBI can be predicted and suppressed independently of the desired FFH signal. Using the complex LMS algorithm to update the prewhitening filter coefficients, this technique is shown to compare favorably with the maximal-ratio combiner diversity technique. When the interferer is wide-sense stationary, the prewhitening-filter-based receiver provides performance approaching that of the AGC receiver and at least 2–3 dB superior to that of the self-normalizing receiver. However, when hostile interference is present, the adaptive prewhitening filter technique may not be able to track the interference rapidly enough. In this case, nonparametric techniques such as the self-normalizing receiver must be used to reject the jammed hops.

Reed and Agee [46] have extended and improved on the idea of whitening by using a time-dependent filter structure to estimate and remove interference, based on the interference spectral correlation properties. In this method, the detection of FHSS in the presence of spectrally correlated interference is nearly independent of the SIR. The process can be viewed as a time-dependent whitening process with suppression of signals that exhibit a particular spectral correlation. The technique is developed from the maximum-likelihood estimate of the spectral frequency of a frequency agile signal received in complex Gaussian interference with unknown spectral correlation. The resulting algorithm uses the correlation between spectrally separated interference components to reduce the interference content in each spectral bin prior to the whitening/detection operation.

An alternative approach to suppress PBI using the FFT has been proposed [45]. The major attraction of FFT-based implementation lies in the ability to achieve guaranteed accuracy and perfect reproducibility.

For suppression of MTI, basically the same processing methods applied for PBI can be employed. However, the performance analyses differ from those for PBI situations. The performance depends on the distribution of the MTI and, in turn, how many bands are contaminated by MTI. Performance analyses of FFH SS systems have been presented for linear combining diversity [47,48], for clipped diversity [49], for maximum likelihood and product-combining receivers [50,51].

BIOGRAPHIES

Dr. Moeness Amin received his B.Sc. degree in 1976 from Cairo University, Egypt, M.Sc. degree in 1980 from

University of Petroleum and Minerals, Dhahran, Saudi Arabia, and his Ph.D. degree in 1984 from University of Colorado at Boulder. All degrees are in electrical engineering. He has been on the faculty of the Department of Electrical and Computer Engineering at Villanova University, Villanova, Pennsylvania, since 1985, where is now a professor. Dr. Amin is a fellow of the IEEE and the recipient of the IEEE Third Millennium Medal. He is also a recipient of the 1997–Villanova University Outstanding Faculty Research Award as well as the recipient of the 1997–IEEE Philadelphia Section Service Award. He is a member of the Franklin Institute Committee of Science and Arts. Dr. Amin has 4 book chapters, 60 journal articles, 2 review articles, and over 150 conference publications. His research includes the areas of wireless communications, time-frequency analysis, smart antennas, anti-jamming gPS, interference cancellation in broadband communication platforms, digitized battlefield, high definition TV, target tracking and direction finding, channel equalization, signal coding and modulation, and radar systems. He has two U.S. patents and served as a consultant to Micronetics Wireless, ELCOM Technology Corporation, and VIZ Manufacturing Company.

Yimin Zhang received his M.S. and Ph.D. degrees from the University of Tsukuba, Japan, in 1985 and 1988, respectively. He joined the faculty of the Department of Radio Engineering, Southeast University, Nanjing, China, in 1988. He served as a technical manager at Communication Laboratory Japan, Kawasaki, Japan, in 1995–1997, and a visiting researcher at ATR Adaptive Communications Research Laboratories, Kyoto, Japan, in 1997–1998. Currently, he is a research fellow at the Department of Electrical and Computer Engineering, Villanova University, Villanova, Pennsylvania. His current research interests are in the areas of array signal processing, space-time adaptive processing, multiuser detection, blind signal processing, digital mobile communications, and time-frequency analysis.

BIBLIOGRAPHY

1. L. B. Milstein, Interference rejection techniques in spread spectrum communications, *Proc. IEEE* **76**(6): 657–671 (June 1988).
2. H. V. Poor and L. A. Rusch, Narrowband interference suppression in spread-spectrum CDMA, *IEEE Pers. Commun. Mag.* **1**(8): 14–27 (Aug. 1994).
3. J. D. Laster and J. H. Reed, Interference rejection in digital wireless communications, *IEEE Signal Process. Mag.* **14**(3): 37–62 (May 1997).
4. M. G. Amin and A. N. Akansu, Time-frequency for interference excision in spread-spectrum communications (section in Highlights of signal processing for communications), *IEEE Signal Process. Mag.* **15**(5): (Sept. 1998).
5. F. M. Hsu and A. A. Giordano, Digital whitening techniques for improving spread-spectrum communications performance in the presence of narrow-band jamming and interference, *IEEE Trans. Commun.* **COM-26**: 209–216 (Feb. 1978).
6. J. W. Ketchum and J. G. Proakis, Adaptive algorithms for estimating and suppressing narrow-band interference in PN spread-spectrum systems, *IEEE Trans. Commun.* **COM-30**: 913–924 (May 1982).
7. L. Li and L. B. Milstein, Rejection of narrow-band interference in PN spread-spectrum system using transversal filters, *IEEE Trans. Commun.* **COM-30**: 925–928 (May 1982).
8. R. A. Iltis and L. B. Milstein, Performance analysis of narrow-band interference rejection techniques in DS spread-spectrum systems, *IEEE Trans. Commun.* **COM-32**: 1169–1177 (Nov. 1984).
9. S. Haykin, *Adaptive Filter Theory*, 3rd ed., Prentice-Hall, Englewood Cliffs, NJ, 1996.
10. R. A. Iltis and L. B. Milstein, An approximate statistical analysis of the Widrow LMS algorithm with application to narrow-band interference rejection, *IEEE Trans. Commun.* **COM-33**: 121–130 (Feb. 1985).
11. F. Takawira and L. B. Milstein, Narrowband interference rejection in PN spread spectrum system using decision feedback filters, *Proc. MILCOM*, Oct. 1986, pp. 20.4.1–20.4.5.
12. J. G. Proakis, *Digital Communications*, 3rd ed., McGraw-Hill, New York, 1995.
13. L. B. Milstein and P. K. Das, Spread spectrum receiver using surface acoustic wave technology, *IEEE Trans. Commun.* **COM-25**: 841–847 (Aug. 1977).
14. S. Davidovici and E. G. Kanterakis, Narrowband interference rejection using real-time Fourier transform, *IEEE Trans. Commun.* **37**: 713–722 (July 1989).
15. R. C. Dipietro, An FFT based technique for suppressing narrow-band interference in PN spread spectrum communication systems, *Proc. IEEE Int. Conf. Acoustics, Speech, and Signal Processing*, 1989, pp. 1360–1363.
16. M. V. Tazebay and A. N. Akansu, Adaptive subband transforms in time-frequency excisers for DSSS communication systems, *IEEE Trans. Signal Process.* **43**: 1776–1782 (Nov. 1995).
17. M. Medley, G. J. Saulnier, and P. Das, Adaptive subband filtering of narrowband interference, in H. Szu, ed., *SPIE Proc. — Wavelet Appls. III*, Vol. 2762, April 1996.
18. J. Gevargiz, M. Rosenmann, P. Das, and L. B. Milstein, A comparison of weighted and non-weighted transform domain processing systems for narrowband interference excision, *Proc. MILCOM*, 1984, pp. 32.3.1–32.3.4.
19. S. D. Sandberg, Adapted demodulation for spread-spectrum receivers which employ transform-domain interference excision, *IEEE Trans. Commun.* **43**: 2502–2510 (Sept. 1995).
20. A. Haimovich and A. Vadhri, Rejection of narrowband interferences in PN spread spectrum systems using an eigenanalysis approach, *Proc. IEEE Signal Processing Workshop on Statistical Signal and Array Processing*, Quebec, Canada, June 1994, pp. 1002–1006.
21. B. K. Poh, T. S. Quek, C. M. S. See, and A. C. Kot, Suppression of strong narrowband interference using eigen-structure-based algorithm, *Proc. MILCOM*, July 1995, pp. 1205–1208.
22. L. A. Rusch and H. V. Poor, Multiuser detection techniques for narrow-band interference suppression in spread spectrum communications, *IEEE Trans. Commun.* **43**: 1725–1737 (Feb.–April 1995).
23. H. V. Poor and X. Wang, Code-aided interference suppression for DS/CDMA communications. I. Interference suppression capability, *IEEE Trans. Commun.* **45**: 1101–1111 (Sept. 1997).

24. B. Boashash, Estimating and interpreting the instantaneous frequency of a signal. I. Fundamentals, *Proc. IEEE* **80**: 520–538 (April 1992).
25. B. Boashash, Estimating and interpreting the instantaneous frequency of a signal. II. Algorithms and applications, *Proc. IEEE* **80**: 540–568 (April 1992).
26. M. G. Amin, Interference mitigation in spread-spectrum communication systems using time-frequency distributions, *IEEE Trans. Signal Process.* **45**(1): 90–102 (Jan. 1997).
27. C. Wang and M. G. Amin, Performance analysis of instantaneous frequency based interference excision techniques in spread spectrum communications, *IEEE Trans. Signal Process.* **46**(1): 70–83 (Jan. 1998).
28. M. G. Amin, C. Wang, and A. R. Lindsey, Optimum interference excision in spread-spectrum communications using open-loop adaptive filters, *IEEE Trans. Signal Process.* (July 1999).
29. S. Barbarossa and A. Scaglione, Adaptive time-varying cancellations of wideband interferences in spread-spectrum communications based on time-frequency distributions, *IEEE Trans. Signal Process.* **47**(4): 957–965 (April 1999).
30. S. Lach, M. G. Amin, and A. R. Lindsey, Broadband nonstationary interference excision in spread-spectrum communications using time-frequency synthesis techniques, *IEEE J. Select. Areas Commun.* **17**(4): 704–714 (April 1999).
31. H. A. Khan and L. F. Chaparro, Formulation and implementation of the non-stationary evolutionary Wiener filtering, *Signal Process.* **76**: 253–267 (1999).
32. M. G. Amin, R. S. Ramineni, and A. R. Lindsey, Interference excision in DSSS communication systems using projection techniques, *Proc. IEEE Int. Conf. Acoustics, Speech, and Signal Processing*, Istanbul, Turkey, June 2000.
33. Y. Zhang, M. G. Amin, and A. R. Lindsey, Combined synthesis and projection techniques for jammer suppression in DS/SS communications, *IEEE Int. Conf. Acoustics, Speech, and Signal Processing*, Orlando, FL, May 2002.
34. S.-C. Jang and P. J. Loughlin, AM-FM interference excision in spread spectrum communications via projection filtering, *J. Appl. Signal Process.* **2001**(4): 239–248 (Dec. 2001).
35. S. Roberts and M. Amin, Linear vs. bilinear time-frequency methods for interference mitigation in direct-sequence spread-spectrum communication systems, *Proc. Asilomar Conf. Signals, Systems, and Computers*, Pacific Grove, CA, Nov. 1995.
36. D. Wei, D. S. Harding, and A. C. Bovik, Interference rejection in direct-sequence spread-spectrum communications using the discrete Gabor transform, *Proc. IEEE Digital Signal Processing Workshop*, Bryce Canyon, UT, Aug. 1998.
37. X. Ouyang and M. G. Amin, Short-time Fourier transform receiver for nonstationary interference excision in direct sequence spread spectrum communications, *IEEE Trans. Signal Process.* **49**(4): 851–863 (April 2001).
38. B. S. Krongold, M. L. Kramer, K. Ramchandran, and D. L. Jones, Spread-spectrum interference suppression using adaptive time-frequency tilings, *Proc. IEEE Int. Conf. Acoustics, Speech, and Signal Processing*, Munich, Germany, April 1997.
39. A. Bultan and A. N. Akansu, A novel time-frequency exciser in spread-spectrum communications for chirp-like interference, *Proc. IEEE Int. Conf. Acoustics, Speech, and Signal Processing*, Seattle, WA, May 1998.
40. J. S. Lee, L. E. Miller, and Y. K. Kim, Probability of error analysis of a BFSK frequency-hopping system with diversity — Part II, *IEEE Trans. Commun.* **COM-32**: 1243–1250 (Dec. 1984).
41. L. E. Miller, J. S. Lee, and A. P. Kadrichu, Probability of error analyses of a BPSK frequency-hopping system with diversity under partial-band jamming interference — Part III: Performance of a square-law self-normalizing soft decision receiver, *IEEE Trans. Commun.* **COM-34**: 669–675 (July 1986).
42. C. M. Keller and M. B. Pursley, Clipper diversity combining for channels with partial-band interference — Part I: Clipper linear combining, *IEEE Trans. Commun.* **COM-35**: 1320–1328 (Dec. 1987).
43. C. M. Keller and M. B. Pursley, Clipper diversity combining for channels with partial-band interference — Part II: Ratio-statistic combining, *IEEE Trans. Commun.* **COM-37**: 145–151 (Feb. 1989).
44. R. A. Iltis, J. A. Ritcey, and L. B. Milstein, Interference rejection in FFH systems using least squares estimation techniques, *IEEE Trans. Commun.* **38**: 2174–2183 (Dec. 1990).
45. K. C. Teh, A. C. Kot, and K. H. Li, Partial-band jammer suppression in FFH spread-spectrum system using FFT, *IEEE Trans. Vehic. Technol.* **48**: 478–486 (March 1999).
46. J. H. Reed and B. Agee, A technique for instantaneous tracking of frequency agile signals in the presence of spectrally correlated interference, *Proc. Asilomar Conf. Signals, Systems, and Computers*, Pacific Grove, CA, Nov. 1992.
47. B. K. Livitt, FH/MFSK performance in multitone jamming, *IEEE J. Select. Areas Commun.* **SAC-3**: 627–643 (Sept. 1985).
48. R. E. Ezers, E. B. Felstead, T. A. Gulliver, and J. S. Wight, An analytical method for linear combining with application to FFH NCFSK receivers, *IEEE J. Select. Areas Commun.* **11**: 454–464 (April 1993).
49. J. J. Chang and L. S. Lee, An exact performance analysis of the clipped diversity combining receiver for FH/MFSK systems against a band multitone jammer, *IEEE Trans. Commun.* **42**: 700–710 (Feb.–April 1994).
50. K. C. Teh, A. C. Kot, and K. H. Li, Performance study of a maximum-likelihood receiver for FFH/BFSK systems with multitone jamming, *IEEE Trans. Commun.* **47**: 766–772 (May 1999).
51. K. C. Teh, A. C. Kot, and K. H. Li, Performance analysis of an FFH/BFSK product-combining receiver under multitone jamming, *IEEE Trans. Vehic. Technol.* **48**: 1946–1953 (Nov. 1999).

INTERLEAVERS FOR SERIAL AND PARALLEL CONCATENATED (TURBO) CODES

Tolga M. Duman
Arizona State University
Tempe, Arizona

1. INTRODUCTION

Interleavers are commonly used devices in digital communication systems. Basically, an interleaver is used to reorder a block or a sequence of binary digits [1].

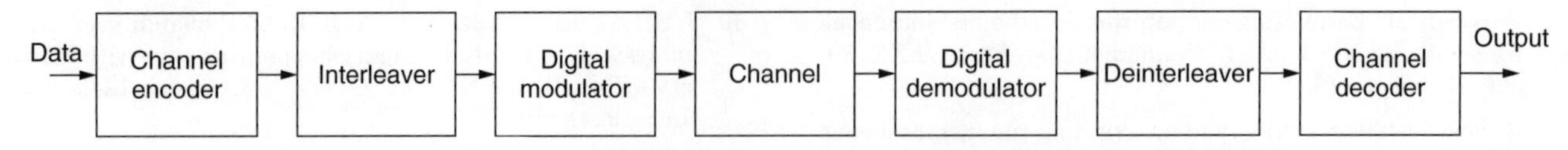

Figure 1. Usage of an interleaver in a coded digital communication system.

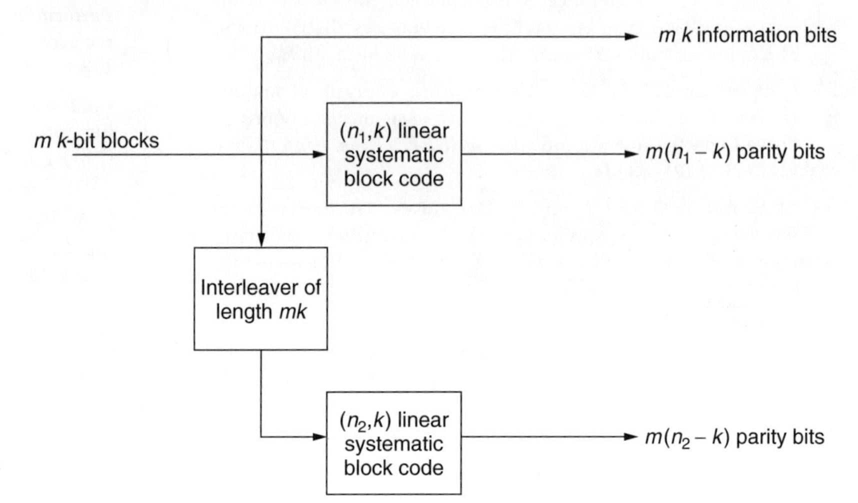

Figure 2. Parallel concatenated block code via an interleaver.

Traditionally interleavers have been employed in coded digital communications, as shown in Fig. 1, to reorder the coded bits in order to combat the burst errors that may be caused by signal fading or different types of interfering signals.

A simple way to reorder the coded bits for this purpose is to use a block interleaver where the sequence is written to a matrix rowwise, and read columnwise. For example, if the original sequence is $\{u_1, u_2, u_3, \ldots, u_{N_1N_2}\}$, the $N_1 \times N_2$ matrix

$$\begin{bmatrix} u_1 & u_2 & \cdots & u_{N_1} \\ u_{N_1+1} & u_{N_1+2} & \cdots & u_{2N_1} \\ u_{2N_1+1} & u_{2N_1+2} & \cdots & u_{3N_1} \\ \cdots & \cdots & \cdots & \cdots \\ u_{(N_2-1)N_1+1} & u_{(N_2-1)N_1+2} & \cdots & u_{N_1N_2} \end{bmatrix}$$

is constructed and read columnwise, resulting in the interleaved sequence

$$\{u_1, u_{N_1+1}, u_{2N_1+1}, \ldots, u_{(N_2-1)N_1+1}, \\ u_2, u_{N_1+2}, u_{2N_1+2}, \ldots, u_{N_1N_2}\}$$

Clearly, the order in which the data are written or read may change, resulting in different variations of the block interleaver.

A more current use of interleavers is in the construction of channel codes having very large codelengths. In his classic paper [2], Shannon showed that the codes with large blocklengths chosen randomly achieve the channel capacity. However, such codes are in general very difficult to encode or decode. As a remedy, one can construct codes with large blocklengths by concatenating two simple (short-blocklength) codes either in parallel or in series using an interleaver.

An example of a parallel concatenated code is shown in Fig. 2 [3]. Two linear systematic block codes are used as component encoders: m k-bit blocks are input to the first component encoder, which produces $m(n_1 - k)$ parity bits, and the interleaved version is input to the second encoder, which produces $m(n_2 - k)$ parity bits. The information bits together with the two sets of parity bits constitute the codeword corresponding to the original mk information sequence. The overall code is an $(m(n_1 + n_2 - k), mk)$ systematic block code. Clearly, by proper selection of the parameter m, we can obtain a large-blocklength code. Different code rates can be obtained by selecting the component codes properly, and by puncturing some of the redundant bits produced by the component codes.

Similarly, simple block codes can be concatenated in series to construct codes with large blocklengths. An example is shown in Fig. 3. In this case, m k-bit blocks are input to an outer linear systematic block code that produces m p-bit coded blocks. These coded bits are interleaved and then input to an inner encoder that generates m n-bit blocks that constitute

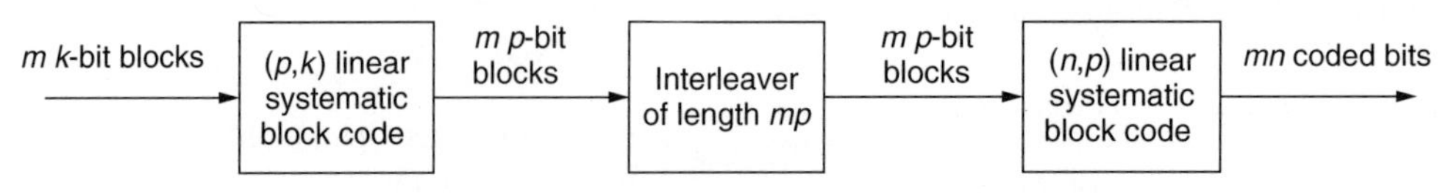

Figure 3. Serially concatenated block code via an interleaver.

the overall codeword corresponding to the original information sequence. The overall code is an (mn, mk) linear systematic block code.

As will be demonstrated later, a simple block interleaver is seldom a good choice for constructing concatenated codes because they fail to eliminate problematic error events because of their inherent structure. Therefore, we need to develop methods of generating good interleavers for code concatenation.

The article is organized as follows. In Section 2, we briefly describe Turbo codes together with the iterative decoding algorithms and explain the role of the interleaver. We summarize the results on the interleaver gain for serial and parallel concatenated convolutional codes in Section 3. We devote Section 4 to review of various promising interleaver design techniques and present several examples. We specifically focus on the use of Turbo codes over AWGN (additive white Gaussian noise) channels. Finally, we conclude in Section 5.

2. TURBO CODES

While concatenation of simple block codes with an interleaver is a viable solution for constructing large-blocklength codes to approach the Shannon limit, encoding and decoding of such codes is difficult since block codes rarely admit simple (soft-decision) maximum-likelihood decoding. Therefore, it is desirable to design concatenated codes using simple convolutional codes as the building blocks.

Parallel or serial concatenation of convolutional codes via an interleaver (i.e., Turbo codes [4,5]), coupled with a suboptimal iterative decoder, has proved to be one of the most important developments in the coding literature since the early 1990s. In particular, long "randomlike" block codes with rather simple decoding algorithms are constructed, and it is shown that their performance is very close to the Shannon limit on the AWGN channel. To be specific, at a bit error rate of 10^{-5}, performance within 1 dB of the channel capacity is common. Let us now describe these codes in more detail.

2.1. Parallel Concatenated Convolutional Codes

The idea in Turbo coding is to concatenate two recursive systematic convolutional codes in parallel via an interleaver as shown in Fig. 4. The information sequence is divided into blocks of a certain length. The input of the first encoder is the information block, and the input of the second encoder is an interleaved version of the information block. The encoded sequence (codeword) corresponding to that information sequence is then the information block itself, the first parity block, and the second parity block. The block diagram shown in Fig. 4 assumes a rate-$\frac{1}{3}$ Turbo code with $\frac{5}{7}$ (in octal notation) component convolutional codes.[1] As in the case of parallel concatenated block codes, higher rate codes can be obtained by puncturing some of the parity bits.

The main role of the interleaver in this construction is to help the code imitate a "random" code with a large blocklength.

2.2. Serially Concatenated Convolutional Codes

Convolutional codes can also be concatenated in series via an interleaver to construct powerful error-correcting

[1] The term *p/q convolutional code* is used to indicate the places of the feedforward and feedback connections in the convolutional code in octal notation.

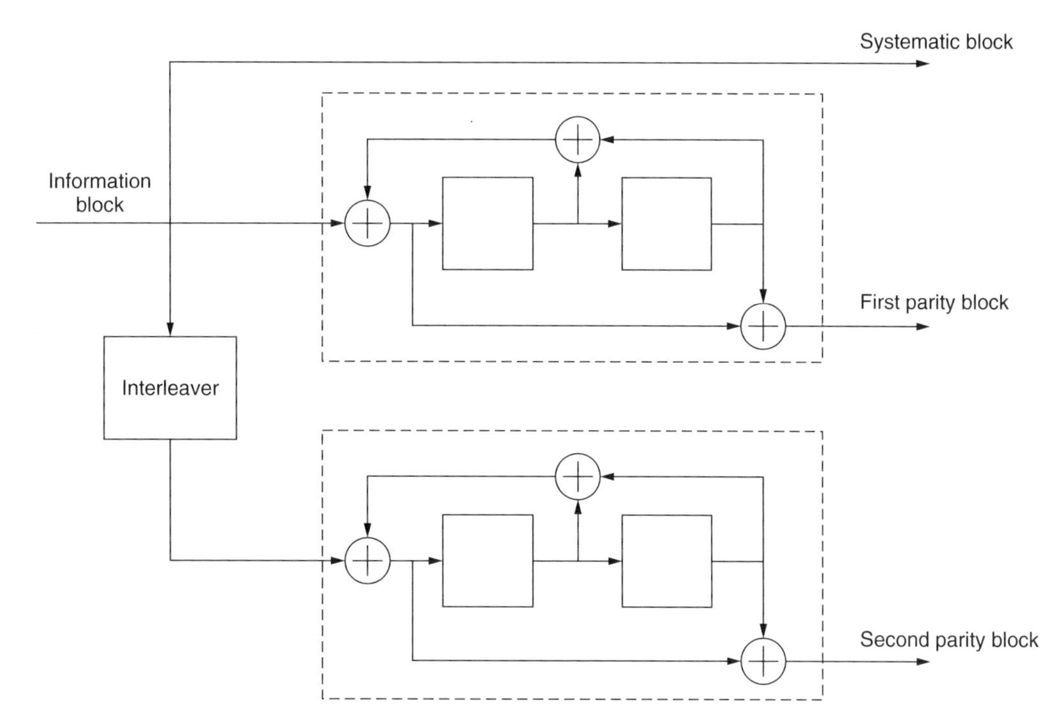

Figure 4. Rate-$\frac{1}{3}$ Turbo code with $\frac{5}{7}$ component codes.

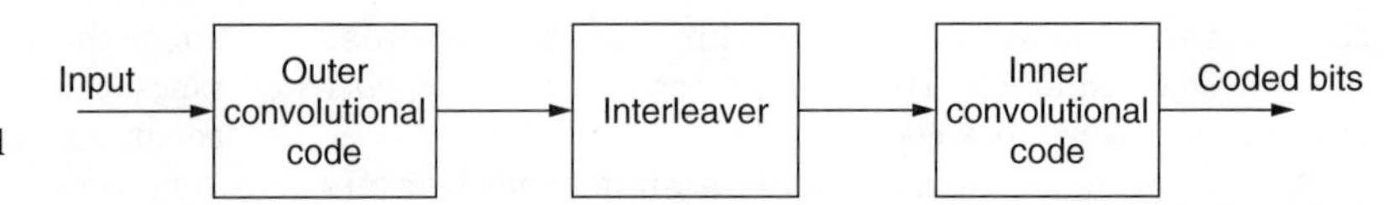

Figure 5. Block diagram for a serially concatenated convolutional code.

(Turbo) codes. In this case, the information sequence is divided into blocks, and each block is encoded using an outer convolutional code. Then, the encoded block is scrambled using an interleaver and passed to the inner convolutional code. Both inner and outer codes are usually selected to be systematic. In order to obtain a good performance, the inner code needs to be recursive, whereas the outer code may be selected as a feedback–free convolutional code [5]. The block diagram of the encoder is shown in Fig. 5.

Similarly, the interleaver is employed to generate a "randomlike" long-blocklength code.

2.3. Iterative Decoding

Assume that the parallel or serial concatenated code is used for transmission over an AWGN channel. Because of the usage of the interleaver, performing maximum-likelihood decoding is very difficult. In general, one has to consider all the possible codewords (there are 2^N possibilities, where N is the length of the information block), compute the squared Euclidean distance corresponding to each one, and declare the one with the lowest distance as the transmitted codeword. Even for short interleavers, this is a tedious task and cannot be done in practice. Fortunately, there is a suboptimal iterative decoding algorithm that achieves near-optimal performance.

Consider the case of parallel concatenation. The iterative decoding algorithm is based on two component decoders, one for each convolutional code, that can produce soft information about the transmitted bits. At each iteration step, one of the decoders takes the systematic information (directly from the observation of the systematic part) and the extrinsic loglikelihood information produced by the other decoder in the previous iteration step, and computes its new extrinsic loglikelihood information. The newly produced extrinsic information is ideally independent of the systematic information and the extrinsic information of the other decoder. The block diagram of the iterative decoding algorithm is presented in Fig. 6, where the extrinsic information of the jth component decoder about the information bit d_k is denoted by $L_{je}(d_k)$ and its interleaved version by $L'_{je}(d_k)$, $j = 1, 2$. The updated extrinsic information is fed into the other decoder for the next iteration step. The extrinsic information of both decoders is initialized to zero before the iterations start. After a number of iterations, the algorithm converges and the decision on each transmitted bit is made based on its "total" likelihood (i.e., sum of two extrinsic information terms and the systematic loglikelihood).

For serial concatenated convolutional codes, a similar iterative (suboptimal) decoder is employed. As in the case of parallel concatenation, there are two component decoders for the inner and the outer code, which can be implemented using a soft-input/soft-output decoder. However, in this case the information exchanged between the component decoders include the information about the parity bits of the outer code along with the systematic

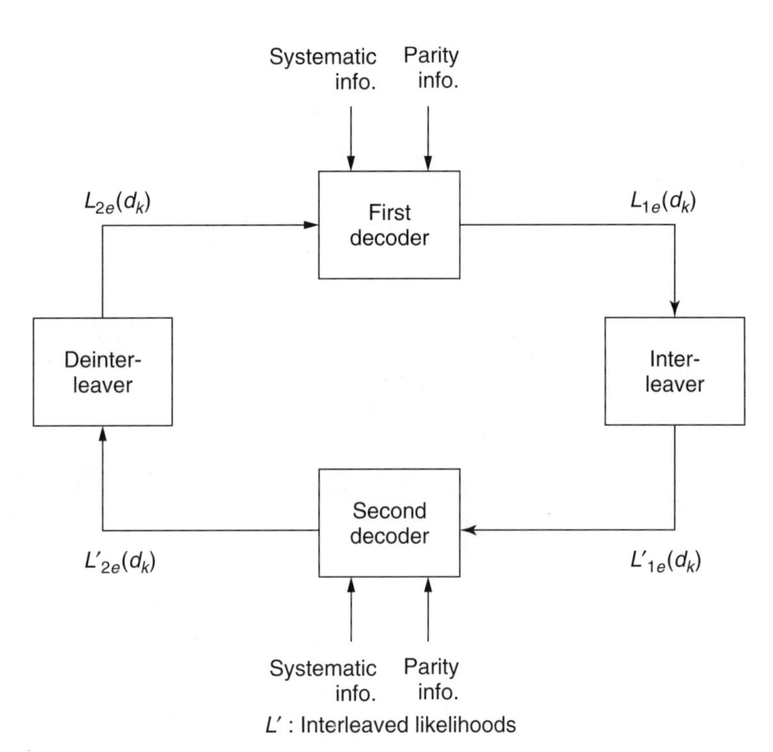

Figure 6. Block diagram of the iterative Turbo decoder for parallel concatenated convolutional codes.

bits. Therefore, the soft-in/soft-out algorithms suitable for component decoders of parallel concatenated codes should be modified accordingly.

3. UNIFORM INTERLEAVER AND INTERLEAVER GAIN

A uniform interleaver devised by Benedetto and Montorsi [6] is a probabilistic device that takes on any one of the possible $N!$ interleavers with equal probability. It is used primarily in analysis of Turbo codes, and can be employed to determine the interleaver gains provided by the parallel and serial concatenation of convolutional codes. The performance predicted by the uniform interleaver is an average over the ensemble of all possible interleavers, and therefore, an interleaver selected at random is expected to achieve this performance. Another interpretation is that there exists an interleaver whose performance is at least as good as the performance predicted by the uniform interleaver.

It is shown [7] that if two recursive convolutional codes are concatenated in parallel, the average number of lowest-weight error events for the overall code (under the assumption of uniform interleaving) is proportional to $1/N$, where N is the interleaver length. If we consider the union bound on the bit error probability, the most important terms of the bound decays with $1/N$ providing an interleaver gain. Hence, the performance of the code improves with the interleaver length (for maximum-likelihood decoding). It is also important to note that if nonrecursive component convolutional codes are employed, there is no interleaver gain. Therefore the use of the recursive codes is critical.

For serially concatenated convolutional codes [5], the outer code may be selected recursive or nonrecursive, whereas the inner code must be selected to be a recursive convolutional code to make sure that an interleaver gain is observed. Then, under the assumption of uniform interleaving, the interleaver gain is given by $N^{-\lceil d_f^o/2 \rceil}$, where d_f^o is the free distance of the outer convolutional code.

It is important to emphasize that concatenation of block codes in parallel or series does not result in any interleaver gain that increases with interleaver size [5,6]. Therefore, using convolutional codes as component codes is advantageous.

4. INTERLEAVER DESIGN FOR PARALLEL CONCATENATED CONVOLUTIONAL CODES

Mathematically, an interleaver of size N is a one-to-one mapping from the set of integers $\{1, 2, \ldots, N\}$ to the same set. Let $\pi(\cdot)$ denote this mapping; then $\pi(i) = j$ means that the ith bit of the original sequence is placed to the jth position after interleaving.

A very simple but effective way of obtaining interleavers for Turbo codes is to select them in a pseudorandom manner. To construct such interleavers, one can easily pick random integers from $\{1, 2, \ldots, N\}$ without replacement. If the ith number picked is j, then $\pi(i) = j$. Pseudorandom interleavers are shown to perform well for Turbo codes. For example, Fig. 7 shows the bit error rate versus signal-to-noise ratio (SNR) obtained by a rate-$\frac{1}{2}$ (obtained by

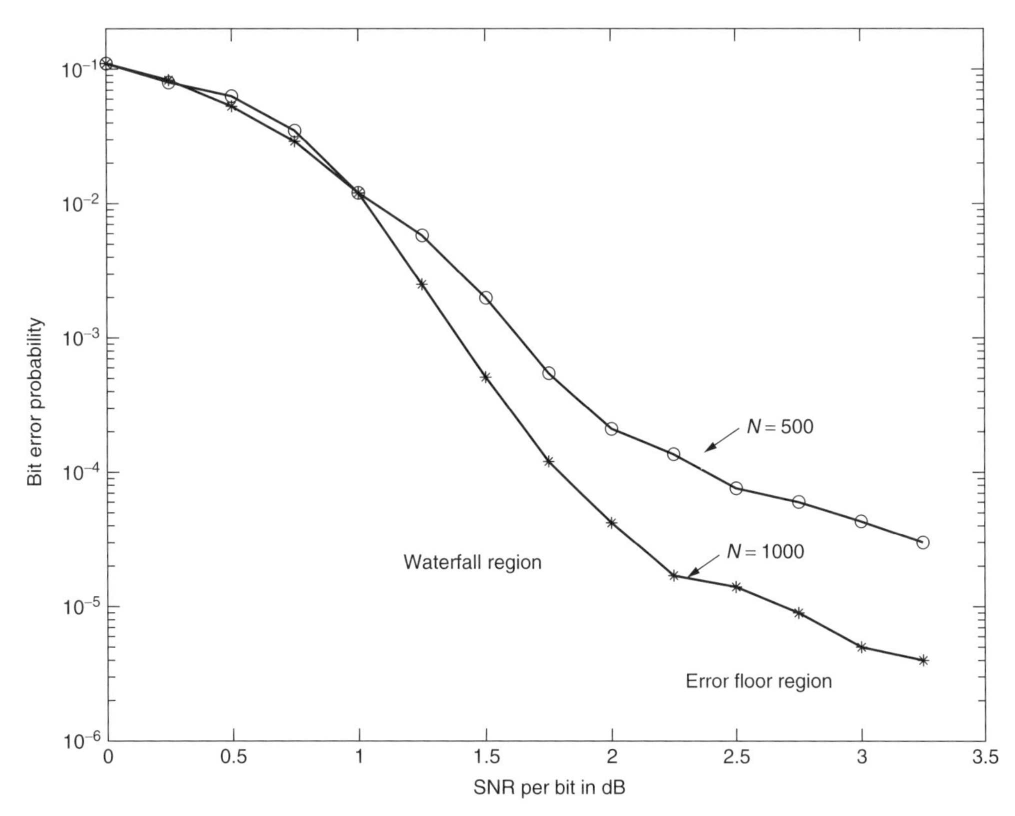

Figure 7. Probability of error for a rate-$\frac{1}{2}$ Turbo code with $\frac{5}{7}$ component codes.

puncturing half of the parity bits) parallel concatenated Turbo code with $\frac{5}{7}$ component codes. However, as we will see shortly, the Turbo code performance can be improved considerably by proper interleaver design techniques.

Before going into the details of interleaver design, it is worthwhile to mention a basic asymptotic result. Turbo codes are linear; therefore the weight distribution determines the distance spectrum of the code. Khandani has shown [8] that the weight of the systematic bitstream, and the two parity bitstreams have Gaussian distribution for large interleaver sizes, and that the correlation coefficients between these three bitstreams are nonnegative and go to zero as $N \to \infty$ for almost any random interleaver. For proper interleaver design, it is desired that these correlation coefficients be as small as possible, and asymptotically this is already satisfied by a random interleaver. Therefore, for large interleaver sizes a randomly selected interleaver is as good as it gets; that is, there are no interleavers that will perform significantly better than the average over the ensemble of all possible interleavers.

However, Khandani's result is only an asymptotic result, and for practical block sizes, it is important to design the interleavers properly. There are two basic approaches to accomplish this goal. One is to attack the weight distribution of the overall Turbo code assuming maximum-likelihood decoding, and the other is to design the interleavers by considering the suboptimal iterative decoding algorithms. We consider these two approaches separately.

4.1. Interleaver Design Based on Distance Spectrum

The recursive convolutional code used to construct the Turbo code is simply a division circuit in a Galois field GF(2). Let us denote its transfer function by $F(D)/G(D)$. Turbo codes are linear block codes; thus let us assume that the all-zero codeword is transmitted over an AWGN channel. The possible error sequences corresponding to this transmission are all the nonzero codewords of the Turbo code. Consider a codeword corresponding to a weight 1 information block. Since the component encoders are selected as recursive convolutional encoders, the parity sequences corresponding to this information sequence will not terminate until the end of the block is reached. With a good selection of the interleaver, if the single 1 occurs toward the end of the input sequence, it will occur toward the beginning of the block for the other component encoder. Therefore, the codewords with information weight 1 will have a large parity weight, hence a large total weight, provided the interleavers that map the bits toward the end of the original sequence too close to the end of the interleaved sequence are avoided.

For larger information weight sequences, the premise is that the interleaver "breaks down" the "bad" sequences. In other words, if the information block results in a lower-weight parity sequence corresponding to one of the encoders, it will have a larger weight parity sequence corresponding to the other one. Therefore, most of the codewords will have large Hamming weights, or the average distance spectrum will be "thinned" [9], and the code will perform well over an AWGN channel.

Among the larger input weight information sequences, the most problematic ones are some of the weight 2 information sequences. Clearly, if the input polynomial is divisible with $G(D)$, then the parity sequence produced by the component code terminates, resulting in a low-weight error event. For example, for the code described in Fig. 4, the feedback polynomial is $1+D+D^2$, and any input polynomial of the form $D^j(1+D^{3l})$ corresponds to a "bad" weight 2 error sequence. In general, if the degree of the feedback polynomial is m, then it divides the input polynomials of the form $D^j(1+D^{kl})$ for some k with $k \le 2^m - 1$. If the feedback polynomial is primitive, then the lowest-degree polynomial that is divisible by the feedback polynomial is $1+D^{2^m-1}$. Therefore, as a side note, we emphasize that in order to reduce the number of "bad" error events, the feedback polynomial is usually selected to be primitive.

The information weight 2 sequences are the most difficult to break down because the percentage of the self-terminating weight 2 sequences is much larger than that for the larger input weight sequences. For example, it is easy to see that with uniform interleaving the number of error events with input sequence weight 2 drops only with $1/N$, whereas the number of higher input weight error events with low overall Hamming weights reduce with $1/N^l$, where $l \ge 2$. Therefore, asymptotically, the performance of the turbo code is determined by the error events of information weight two.

Figure 7 illustrates the bit error rate versus SNR for a typical Turbo code. From the figure two regimes are apparent: the waterfall region and the error floor region. The waterfall region is due to the "thin" distance spectrum of the Turbo code, and the error floor is due to the fact that the minimum distance of the Turbo code is usually small caused by "bad" weight 2 information sequences.

Interleaver design based on the distance spectrum of the overall Turbo code is concerned mainly with lowering the error floor present by attacking the problematic lower information weight sequences, in particular sequences with information weight 2.

4.1.1. Block Interleaver. A simple block interleaver is not suitable for use in Turbo codes as there are certain "bad" information sequences that cannot be broken down. For example, consider the information sequence written rowwise

$$\begin{bmatrix} 0 & \cdots & 0 & 0 & 0 & 0 & 0 & 0 & \cdots & 0 \\ \cdot & \cdots & \cdot & \cdot & \cdot & \cdot & \cdot & \cdot & \cdots & \cdot \\ \cdot & \cdots & \cdot & \cdot & \cdot & \cdot & \cdot & \cdot & \cdots & \cdot \\ 0 & \cdots & 0 & \mathbf{1} & 0 & 0 & \mathbf{1} & 0 & \cdots & 0 \\ 0 & \cdots & 0 & 0 & 0 & 0 & 0 & 0 & \cdots & 0 \\ 0 & \cdots & 0 & 0 & 0 & 0 & 0 & 0 & \cdots & 0 \\ 0 & \cdots & 0 & \mathbf{1} & 0 & 0 & \mathbf{1} & 0 & \cdots & 0 \\ \cdot & \cdots & \cdot & \cdot & \cdot & \cdot & \cdot & \cdot & \cdots & \cdot \\ \cdot & \cdots & \cdot & \cdot & \cdot & \cdot & \cdot & \cdot & \cdots & \cdot \\ 0 & \cdots & 0 & 0 & 0 & 0 & 0 & 0 & \cdots & 0 \end{bmatrix}$$

and read columnwise. The pattern formed by four 1s shown in boldface cannot be broken down by the block interleaver since both the original sequence and the interleaved version contain two terminating weight 2 error patterns (for the $\frac{5}{7}$ convolutional code) regardless of the size of the interleaver. Since we have to consider all possible binary N-tuples, there are many information sequences containing the same problematic pattern. Clearly, for any component code, there always exists similar problematic weight 4 information sequences.

Another important information sequence that cannot be broken down is the one with a single 1 at the end of the block.

Although the block interleavers are well suited for coded communications over bursty channels (e.g., fading channels), or breaking up bursts errors due to an inner code decoded using the Viterbi algorithm in a concatenated coding scheme, they are seldom appropriate for Turbo codes.

4.1.2. Reverse Block Interleavers. The problematic information sequence that contains a single 1 at the end can be accommodated easily by a "reverse" block interleaver as proposed by Herzberg [10]. In this case, the information bits are written rowwise and read columnwise; however, the last column is read first. Clearly, this approach does not address problematic weight 4 sequences; nevertheless it is shown to perform well for very short-blocklength codes, and for moderate- and long-blocklength codes at very low probability-of-error values.

Figure 8 shows the performance of the reverse block interleavers together with the block interleavers and the pseudorandom interleavers. The Turbo code in the example is a rate-$\frac{1}{3}$ Turbo code with $\frac{7}{5}$ component codes. This example clearly justifies the superiority of reverse block interleaver for short blocklengths and for larger blocklengths at low bit error probabilities.

4.1.3. *s*-Random Interleaver and Its Variations. As we have discussed in the previous section, the main problem for Turbo codes is the "bad" weight 2 information sequences. If the two 1s resulting in the terminating parity sequence are close to each other in both the original sequence and the interleaved version, both sequences result in low parity weights, and thus the overall codeword has a low Hamming weight. In order to increase the Hamming distance and thus reduce the error floor, an s-random interleaver [11] ensures that any pair of positions that are close to each other in the original sequence are separated by more than a preselected value s (called "spread") in the interleaved version. More precisely, we want

$$\max_{i,j}\{|i-j|, |\pi(i)-\pi(j)|\} > s$$

This condition will ensure that the two 1s do not occur in close proximity in both the original and the interleaved information sequences. Thus, at least, one of the parity weights will be relatively large because of the weight that the first 1 will accumulate until the second 1 is inserted.

An s-random interleaver may be designed as follows [11]. Assume that $\pi(1), \pi(2), \ldots, \pi(n-1)$ are selected. To select $\pi(n)$, we pick a random number $j \in \{1, 2, \ldots, N\} \setminus \{\pi(1), \pi(2), \ldots, \pi(n-1)\}$. If the number selected does not violate the spread constraint, then we let $\pi(n) = j$. If it violates the spread constraint, we reject this number, and select a new one at random. We continue this process until all the integers that describe the interleaver are selected. Obviously, if the desired spread is selected too large, the algorithm may not terminate. However, it is shown by experiments that if $s < \sqrt{N/2}$, then the algorithm converges in reasonable time.

The performance of an s-random interleaver is significantly better that of the pseudorandom interleaver. In particular, the error floor is dramatically reduced since

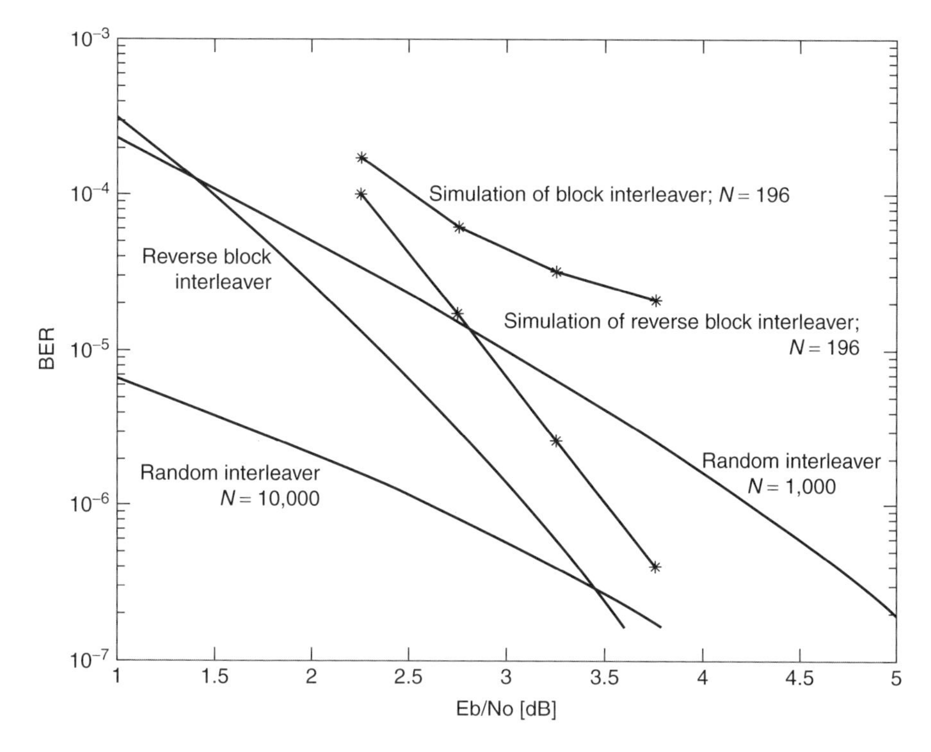

Figure 8. Performance of the reverse block interleaver, block interleaver, and pseudorandom interleaver. (From Herzberg [10], © 1998 IEEE.)

the most problematic weight 2 information sequences that cause the error floor are accommodated. We will give a specific example in the following section and compare it with other interleaver design methods.

We note that block interleavers can be used to achieve a good spread, up to $\sqrt{N}$ as opposed to $\sqrt{N/2}$, which is achieved with the s-random interleavers. However, with block interleaving, the higher-weight sequences become very problematic and the overall interleaver does not perform very well.

In addition to the weight 2 input sequences, we can consider larger input weight sequences to improve the interleaver structure. For instance, Fragouli and Wesel [12] consider the multiple error events and ensure that the interleavers that map the problematic information sequences with multiple error events to another problematic sequence are avoided. It is shown that such interleavers, when concatenated with a higher-order modulation scheme, perform well compared to the s-random interleaver. However, there is no significant improvement when they are used for binary communications over an AWGN channel.

Another approach [13] is to consider "bad" [i.e., divisible by $G(D)$] information sequences of higher weights, and make sure that the interleaver does not map such sequences to another similar sequence. Simulations show that such interleavers perform slightly better than do the s-random interleavers.

"Swap" interleavers are considered in another study [14], where the interleaver is initialized to a block interleaver, and two randomly selected positions are swapped. If the swapping results in violation of the spread constraint, the modification is rejected, and a new pair of positions are selected. After a number of iterations the interleaver is finalized. With swap interleavers, a slight performance improvement over the s-random interleaver is observed (in the order of 0.05 dB).

4.1.4. Iterative Interleaver Growth Algorithms. Daneshgaran and Mondin [15,16] have proposed and studied the idea of iterative interleaver growth algorithms in a systematic way. The main idea is to argue that we should be able to construct good interleavers of size $n+1$ from good interleavers of size n. They exploit the algebraic structure of interleavers and represent any interleaver with an equivalent "transposition vector" of size N that defines the permutation precisely. They prove what is called the "prefix symbol substitution property," which intuitively states that if a new transposition vector is obtained by appending an index to an existing transposition vector, the new vector defines a permutation of one larger size, and it is very closely related to the old one. For details, see Ref. 15.

They then define a cost function that is closely related to the bit error rate that is the ultimate performance measure. Problematic error sequences are identified using the component code structure, and elementary cost functions for each error pattern are defined. The elementary cost functions are directly related to the pairwise error probability that corresponds to the particular error pattern over an AWGN channel. The overall cost function is formed as the sum of these elementary cost functions. Finally, since for the parallel concatenation, both the actual information sequence and its interleaved version are inputs to convolutional codes, the cost function is defined for the inverse interleaver as well. The sum of the two are considered for the overall interleaver design. Extensions of these cost functions for designing interleavers for other channels are straightforward by considering the new appropriate pairwise probability of error expressions.

Now that a cost function is defined, and a method of increasing the interleaver size by one is described, the details of the interleaver growth algorithm can be presented. We start with a small-size interleaver optimized with respect to the cost function defined by an exhaustive search over all interleavers. Then, we extend this interleaver by appending the "best" prefix (with respect to the cost function defined) to the transposition vector that describes the original permutation. Since there are only a limited number of possible prefixes to consider, this step is simple to implement. We repeat this process until an interleaver of a desired size is obtained. We simply are looking for a "greedy" solution to find a good interleaver. Clearly, the process is not optimal; however, experiments show that it results in very good interleavers. We also note that the overall algorithm has only polynomial complexity.

In Fig. 9, we present the performance of the Turbo code with $\frac{5}{7}$ component codes with three different interleavers of size $N = 160$, that is, with the interleaver designed using the algorithm in Ref. 15, the reverse block interleaver and the s-random interleaver. We observe that the interleaver designed using the iterative interleaver growth algorithm performs significantly better than the others. In particular, it is $\sim$0.3 dB better than the reverse block interleaver and $\sim$0.5 dB better than the s-random interleaver at a bit error rate of 10^{-6}. Also shown in the figure is the ML (maximum-likelihood) decoding bound for a uniform interleaver (i.e., the average performance of all the possible interleavers). We observe that the interleavers designed outperform the average significantly, particularly, in the error floor region.

4.1.5. Other Deterministic Interleavers. Several other deterministic interleaver design algorithms are proposed in the literature with varying levels of success [e.g., 17–20].

4.2. Interleaver Design Based on Iterative Decoder Performance

The previous interleaver design algorithms described improve the distance spectrum of the turbo code by avoiding or reducing the number of problematic low-input-weight error sequences. They consider the maximum likelihood decoding performance and inherently assume that the performance of the suboptimal iterative decoding employed will be close to the optimal decoder. Another approach to interleaver design is to consider the performance of the iterative decoder as proposed by Hokfelt et al. [21], and design interleavers according to their suitability for iterative decoding.

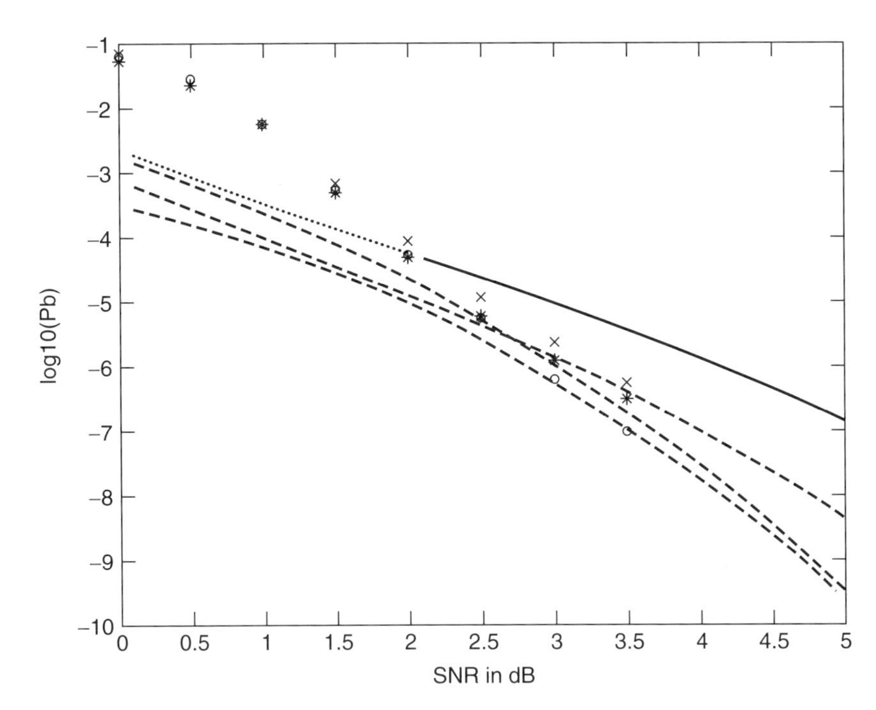

Figure 9. Performance of the Daneshgaran–Mondin (DM) interleaver (simulated points shown by "o"), the s-random (SR) interleaver (simulated points shown by "*"), and the reverse block (RB) interleaver with $N = 169$ (simulated points shown by "x"). The asymptotic BER curves are shown by dashed lines; at 5 dB, the uppermost curve is for SR, the middle one is for RB, and the lowermost one is for DM. The union bound for the uniform interleaver is shown by the solid line. (From Daneshgaran and Mondin [15], © 1999 IEEE.)

Consider the iterative decoding algorithm summarized in Section 2.3 (Fig. 6). Hokfelt et al. [21] showed that the extrinsic information produced at the output of the component decoders correlates with the systematic input. Let us denote the jth systematic information input to the lth component decoder with $x_{j,l}$, $j = 1, 2, \ldots, N$, $l = 1, 2$. The extrinsic information produced by the lth decoder about the ith bit is denoted by $L_{le,i}$.

Empirical results show that after the first decoding step (first half of the iteration), the correlation coefficient between the ith extrinsic information produced by the decoder and the jth systematic input can be approximated by [21]

$$\rho^{(1)}_{L_{e,i},x_j} = \begin{cases} ae^{-c|i-j|} & \text{if} \quad i \neq j \\ 0 & \text{if} \quad i = j \end{cases}$$

where $i, j = 1, 2, \ldots, N$, and the subscript l denoting the component decoder index is suppressed. The constants a and c depend on the particular component code selected, and can be computed using simulations.

Similarly, after the second iteration step, these correlations can be approximated by

$$\rho^{(2)}_{L_{e,i},x_j} = \frac{1}{2}ae^{-c|\pi(j)-i|} + \frac{1}{2}\sum_{\substack{m=1\\ m\neq i}}^{N} a^2 e^{-c(|\pi^{-1}(m)-j|+|i-m|)}$$

where π^{-1} denotes the inverse of the interleaver.

The nonzero correlation between the systematic input and the extrinsic information to the component decoders deteriorates the performance of the iterative decoder. With this interleaver design approach, the objective of the interleaver is to make the nearby extrinsic inputs as uncorrelated to each other as possible, or to ensure that the extrinsic information at the output of the second decoder is uniformly correlated with the systematic information. The interleaver is designed as follows [21]. Starting with $i = 1$, the ith entry of the interleaver is selected as

$$\pi(i) = \underset{j}{\text{argmin}} \sum_{m} e^{-c(|\pi(m)-j|+|i-m|)}$$

where the summation is performed over predefined interleaver elements, and the minimization is over all permissible positions to choose from.

We note that this interleaver design technique tries to minimize the number of short cycle error events that deteriorate the iterative decoder performance due to the excessive correlation of the extrinsic information. We also note that the interleaver designed in this manner competes very well with the s-random interleaver. In particular, the iterative decoding algorithm converges faster, and BER performance improves in the order of 0.1 dB.

Sadjadpour et al. [22] combined the two basic interleaver design approaches to improve the interleaver structure; that is, both the distance spectrum of the code and the correlation of the extrinsic information are considered. In addition to the iterative decoding suitability criterion of Hokfelt et al. [21], Sadjadpour et al. [22] propose another related criterion that tries to ensure that the extrinsic information is uniformly correlated with the input, and that the "power" of the correlation coefficients (sum of squares) is minimized. Then, the interleaver is designed in two steps: (1) an s-random interleaver is selected and then (2) all the low-weight error sequences that result in a lower weight than a predetermined value — typically, problematic weight 2 information sequences — are considered. On the basis of these problematic sequences, the interleaver is updated by using the iterative decoding suitability criterion. This operation is continued until all the problematic sequences are considered. Finally, step 2 is repeated with the new interleaver until there are no low-weight error patterns with a minimum distance less than the preselected value. Clearly, this procedure may not converge if the desired minimum distance of the overall code is selected to be too large.

Examples presented by Sadjadpour et al. [22] indicate that this two-step s-random interleaver design technique has a better performance than that of the s-random interleaver, typically in the error floor region.

5. INTERLEAVER DESIGN FOR SERIALLY CONCATENATED CONVOLUTIONAL CODES

Although there are many results available for the interleaver design for parallel concatenated convolutional codes, the literature on interleaver design for serially concatenated convolutional codes is very scarce. The main reason for this fact is that for serial concatenation the interleaver gain is much larger than that of parallel concatenation. To reiterate, for serial concatenation, under uniform interleaving, the probability of error decays with $N^{-\lceil d_f^o/2 \rceil}$ asymptotically, whereas for parallel concatenation this is only $1/N$. As a result, the error floor observed for parallel concatenation is much less of an issue for serial concatenation.

Clearly, there are certain interleavers that need to be avoided. For instance, the identity operation (no interleaving) does not result in any interleaver gain and is not useful. Also, we need to avoid interleavers that map the last few bits of the sequence to the end of the frame, since for input sequences that contain 1s only at the end of the block, the minimum distance of the overall concatenated code will be small. However, a pseudorandom interleaver selected will perform well; that is, the error floor will be much lower than the one observed in the parallel concatenated Turbo codes. If the interleaver is further selected to be an s-random interleaver, it should perform even better as the short error events of the outer convolutional code will be distributed across the entire sequence at the input of the inner encoder, and likely will result in a large Hamming weight at the output [23].

For parallel concatenation, weight 2 error patterns are the dominant events that warrant special attention regardless of the component convolutional codes used. However, for serial concatenation, the problematic error sequences are highly dependent on the component codes, and the interleaver design should be performed specifically for the component codes.

At the time of this writing, only two papers deal with interleaver design for serially concatenated convolutional codes [23,24]. Daneshgaran et al. [23] formulate the interleaver design algorithm as an optimization problem where the cost function to be minimized considers important error events of the outer component code, and depends on the specific interleaver and the inner component code. If one has a good interleaver of size n, then this interleaver can be grown to size $n+1$ in such a way that the cost function is minimized. This process is continued until an interleaver of desired size is obtained. The interleaver growth algorithm has polynomial complexity, and therefore is easy to implement. The authors present an example design and show that much larger minimum distances can be obtained compared to codes that employ random or s-random interleaving. Therefore, the error floor of the code designed in this manner is much smaller.

Another natural approach is to design the interleaver based on the suboptimal iterative decoding algorithm. In particular, as in the case of parallel concatenation, one can select the interleaver to make sure that the short cycles are avoided. However, at this point, this approach has not been explored.

6. CONCLUSIONS

Interleavers play a major role in the construction of both parallel and serially concatenated convolutional (Turbo) codes as they provide what is called the "interleaver gain." In this article, we have discussed the role of the interleavers in Turbo codes in detail, and reviewed some of the important results. We also have identified the design of interleavers as an important issue, and summarized several promising interleaver design algorithms developed in the literature.

BIOGRAPHY

Tolga M. Duman received the B.S. degree from Bilkent University, Turkey, in 1993, and M.S. and Ph.D. degrees from Northeastern University, Boston, in 1995 and 1998, respectively, all in electrical engineering. He joined the Electrical Engineering faculty of Arizona State University as an Assistant Professor in August 1998. Dr. Duman's current research interests are in digital communications, wireless and mobile communications, channel coding, Turbo codes, coding for recording channels, and coding for wireless communications.

Dr. Duman is the recipient of the National Science Foundation CAREER Award, IEEE Third Millennium medal, and IEEE Benelux Joint Chapter best-paper award (1999). He is a member of the IEEE Information Theory and Communication Societies.

BIBLIOGRAPHY

1. S. B. Wicker, *Error Control Systems for Digital Communication and Storage*, Prentice Hall, 1995.
2. C. E. Shannon, A mathematical theory of communication, *Bell Syst. Tech. J.* 1–10 (Jan. 1948).
3. John G. Proakis, *Digital Communications*, McGraw-Hill, New York, 2001.
4. C. Berrou, A. Glavieux, and P. Thitimajshima, Near Shannon limit error-correcting coding and decoding: Turbo-codes, *Proc. IEEE Int. Conf. Communications (ICC)*, 1993, pp. 1064–1070.
5. S. Benedetto, D. Divsalar, G. Montorsi, and F. Pollara, Serial concatenation of interleaved codes: Performance analysis, design and iterative decoding, *IEEE Trans. Inform. Theory* 909–929 (May 1998).

6. S. Benedetto and G. Montorsi, Unveiling Turbo codes: Some results on parallel concatenated coding schemes, *IEEE Trans. Inform. Theory* 409–428 (March 1996).
7. S. Benedetto and G. Montorsi, Design of parallel concatenated convolutional codes, *IEEE Trans. Commun.* 591–600 (May 1996).
8. A. K. Khandani, Optimization of the interleaver structure for Turbo codes, *Proc. Canadian Workshop on Information Theory*, June 1997, pp. 25–28.
9. L. C. Perez, J. Seghers, and D. J. Costello, Jr., A distance spectrum interpretation of Turbo codes, *IEEE Trans. Inform. Theory* 1698–1709 (Nov. 1996).
10. H. Herzberg, Multilevel Turbo coding with short interleavers, *IEEE J. Select. Areas Commun.* 303–309 (Feb. 1998).
11. S. Dolinar and D. Divsalar, *Weight Distributions for Turbo Codes Using Random and Nonrandom Permutations*, TDA Progress Report 42–122, JPL, Aug. 1995.
12. C. Fragouli and R. D. Wesel, Semi-random interleaver design criteria, *Proc. IEEE Global Communications Conf. (GLOBECOM)*, 1999, pp. 2352–2356.
13. F. Said, A. H. Aghvami, and W. G. Chambers, Improving random interleaver for Turbo codes, *Electron. Lett.* **35**(25): 2194–2195 (Dec. 1999).
14. B. G. Lee, S. J. Bae, S. G. Kang, and E. K. Joo, Design of swap interleaver for Turbo codes, *Electron. Lett.* 1939–1940 (Oct. 1999).
15. F. Daneshgaran and M. Mondin, Design of interleavers for Turbo codes: Iterative interleaver growth algorithms of polynomial complexity, *IEEE Trans. Inform. Theory* **45**(6): 1845–1859 (Sept. 1999).
16. F. Daneshgaran and M. Mondin, Optimized Turbo codes for delay constrained applications, *IEEE Trans. Inform. Theory* **48**(1): 293–305 (Jan. 2002).
17. D. Wang and H. Kobayashi, On design of interleavers with practical size for Turbo codes, *Proc. IEEE Int. Conf. Communications (ICC)*, Oct. 2000, pp. 618–622.
18. M. Z. Wang, A. Sheikh, and F. Qi, Interleaver design for short Turbo codes, *Proc. IEEE Global Communications Conf. (GLOBECOM)*, Dec. 1999, pp. 894–898.
19. A. K. Khandani, Group structure of Turbo codes with applications to the interleaver design, *Proc. IEEE Int. Symp. Information Theory (ISIT)*, Aug. 1998, p. 421.
20. S. Crozier, J. Lodge, P. Guinand, and A. Hunt, Performance of Turbo codes with relative prime and golden interleaving strategies, *Proc. 6th Int. Mobile Satellite Conf. (IMSC99)*, June 1999, pp. 268–275.
21. J. Hokfelt, O. Edfors, and T. Maseng, A Turbo code interleaver design criterion based on the performance of iterative decoding, *IEEE Commun. Lett.* 52–54 (Feb. 2001).
22. H. R. Sadjadpour, N. J. A. Sloane, M. Salehi, and G. Nebe, Interleaver design for Turbo codes, *IEEE J. Select. Areas Commun.* **19**(5): 831–837 (May 2001).
23. F. Daneshgaran, M. Laddomada, and M. Mondin, Interleaver design for serially concatenated convolutional codes: Theory and application, preprint, 2002.
24. R. Jordan, S. Host, and R. Johannesson, On interleaver design for serially concatenated convolutional codes, *Proc. IEEE Int. Symp. Information Theory (ISIT)*, 2001, p. 212.

INTERNET SECURITY

BÜLENT YENER*
Rensselaer Polytechnic University
Troy, New York

1. INTRODUCTION

As the Internet utilized as a new commercial infrastructure, meeting security requirements of diverse applications becomes imminent. Furthermore, the Web and browsers bring have brought the Internet to homes of average people, creating not only a surge in use of the Internet but also a risk to their privacy.

Internet security aims to ensure *confidentiality, authentication, integrity*, and *nonreputiation* of the "information" carried over a collection of interconnected, heterogeneous networks via messages. Confidentiality or privacy prevents unauthorized parties from accessing the message. Authentication requires that source of a message has correct and verifiable identity. Integrity protection of information ensures that unauthorized parties cannot modify the information. Nonreputiation of information requires that the sender and receiver of the information cannot deny the transmission of the message. In general, the security attacks can be grouped into several classes:

1. *Interception* attacks, which are directed at the confidentiality of information by unauthorized access. It is a passive attack where the adversary simply observes the communication channel without modifying the information. Eavesdropping on a communication channel is a typical example.
2. *Modification* attacks, which violate the integrity of information. It is an active attack in which adversary changes the content of information. Man-in-the-middle attacks are typical examples.
3. *Fabrication* attacks, in which the adversary generates and inserts malicious information to the system. This is also an active attack and it violates the authenticity of information.
4. *Interruption*, which is also an active attack that targets the availability of the system. An example is malicious jamming in a wireless network to generate intentional interference.

1.1. Cryptography

Cryptography provides the essential techniques and algorithms to keep information secure [1,2]. Confidentiality is done by encryption and decryption. Authentication is ensured with digital certificates while integrity is protected with hash functions. Nonreputiation of messages is ensured with digital signatures.

* Department of Computer Science at Rensselaer Polytechnic Institute. This article was written in part while the author was visiting Bogazici University Department of Computer Science, Istanbul, Turkey.

The roots of cryptographic research can be traced to William F. Friedman's report *Index of Coincidence and Its Applications* [3], and Edward H. Hebern's rotor machine [4] in 1918. In 1948 Claude Shannon presented his work on the communication theory of secrecy systems in the *Bell System Technical Journal* [5]. In early 1970s work by Horst Feistel from IBM Watson Laboratory led the first U.S. Data Encryption Standard (DES) [6]. In DES both parties must share the same secret key of 56 bits before communication begins. However, Michael Weiner showed that exhaustive search can be used to find any DES key [7]. More recently, the National Institute of Standards and Technology (NIST) has selected the Advanced Encryption Standard (AES), the successor to the venerable DES. AES was invented by Joan Daemen and Vincent Rijmen [8].

1.1.1. Confidentiality. Encryption is a function E that takes plaintext message M as input and produces the encrypted ciphertext C of M: $E(M) = C$. Decryption is the function for the reverse process: $D(C) = M$. Note that $D(E(M)) = M$. In modern cryptography, a *key* K is used for encryption and decryption so that $D_K(E_K(M)) = M$. Cryptographic algorithms, based on using a single key, (i.e., both encryption and decryption are done by the same key) are called *symmetric ciphers* and they have two drawbacks: (1) an arrangement must be made to ensure that two parties have the same key prior to communicate with each other and (2) the number of keys required for a complete communication mash for an n party network is $O(n^2)$. Although a trusted third party such as a *key distribution center* (KDC) can be used to circumvent these two problems, it requires that KDC must be available in real-time to initiate a communication.

Public key cryptography proposed by Whitfield Diffie and Martin Hellman in 1975 [9] is based on *asymmetric ciphers*. In such systems, the encryption key K_1 is different from the decryption key K_2 so that $D_{K_2}(E_{K_1}(M)) = M$. The encryption key K_1 is called the *public key*, and it is not secret. The second key K_2 is called the *private key*, and it is kept confidential. The best known public key crypto system, proposed by Ronald Rivest, Adi Shamir, and Leonard Adleman (RSA) [10]. Although the public key ciphers reduce the number of keys to $O(n)$, they also suffer two problems: (1) key size is much larger than in symmetric systems and (2) the encryption and decryption are much slower. These two issues become problematic in a bandwidth processing-constrained environments such as wireless networks. Thus, the main use of public key systems is limited to distribution of symmetric cipher keys.

1.1.2. Integrity. Cryptographic solutions for integrity protection are based on *one-way hash functions*. A hash function is a one-way function that is easy to compute but significantly harder to compute in reverse (e.g., a^x mod n). It takes a variable-length input and produces a fixed-length hash value (also known as "message digest"). In general it works as follows. The sender computes the hash value of the message, encrypts it using the receiver's public key, and appends it to the message. The receiver decrypts the message digest using his/her private key and then computes the hash value of the message. If the computed hash value is the same as the decrypted one, the message integrity is considered to be preserved during transmission. However, this is not an absolute guarantee since a hash collision is possible (i.e., a modified or fabricated message may have the same hash value as the original. Furthermore, the hash function is public so that the attacker can intercept a message, modify it, and compute a new hash value for the modified message. Thus, it would be a good idea to encrypt the message as well or use a message authentication code (MAC), which is a one-way function with a key.

1.1.3. Nonreputiation. In order to prove the source of a message, one-way functions called *digital signatures* can be used in conjunction with public key cryptography. To stamp a message with its digital signature, the sender encrypts the message digest with its private key. The receiver first decrypts the message using the sender's public key and then computes the message digest to compare it to the one that arrives with the message.

1.1.4. Authentication. To prevent an attacker from impersonating a legitimate party, *digital certificates* are used. At the beginning of a secure Internet session, sender transmits its digital certificate to have his/her identity to be verified. Digital certificates may be issued in a hierarchical way for distributed administration. A digital certificate may follow the ITU standard X.509 [11,12] and is issued by a *certificate authority* (CA) as a part of public key infrastructure (PKI). The certificate contains the sender's public key, the certificate serial number and validity period, and the sender's and the CA's domain names. CA must ensure integrity of the issued certificate; thus it may encrypt the hash value of it using its private key and append it to the certificate.

1.2. Cryptography and Security

Although cryptography provides the building blocks, there is much to consider for Internet security. First, the rapid growth of the Internet increases its heterogeneity and complexity. A communication channel between a pair of users may pass through different network elements running diverse protocols. Most of these protocols are designed with performance considerations and carry design and implementation holes from security point of view. For example, consider the *Anonymous File Transfer Protocol* (FTP) [13], which provides one of the most important services in the Internet for distribution of information. There are several problems with FTP and its variants. For example, the *ftpd* deamon runs with superuser privileges for password and login processing. Thus, leaving a sensitive file such as the password file in the anonymous FTP site will be a serious security gap. Another example is the *Transport Control Protocol* (TCP) [14], which provides a connection between a pair of users. In TCP each connection is identified with a 4-tuple: *<source (local) host IP address, local port number, destination (remote) host IP address ID, remote port number>*. Since the same 4-tuple can be reused, the *sequence numbers* are used to detect the lingering packets from the previous uses of the tuple. There is a potential threat here since an attacker can "guess" the initial sequence number and convince the remote host that it is communicating with a trusted host. This is known as a *sequence number attack* [15]. A remedy for this attack would be to hide the target host behind a dedicated gateway called a *firewall* to prevent direct connections [16].

Also, the network software is hierarchical and layered. At each layer a different protocol is in charge and interacts with the protocols in the adjacent layers. In the following sections we will examine layer-specific security concerns.

2. LINK-LAYER SECURITY

Link-layer security issues in *local-area networks* (LANs) and *wide-area networks* (WANs) are fundamentally different. In a WAN link-layer security requires that end point of each link is secure and equipped with encryption devices. Although institutions such as the military have been using link-layer encryption, it is not feasible in the Internet.

In a LAN hosts share the same communication medium that has (in general) a broadcast nature (i.e., transmission from one node can be received by all others on the same LAN). Thus eavesdropping is easy and, to ensure confidentiality encryption, is required. For example, in a LAN it is better to use the SSH protocol [17] instead of Telnet to avoid compromising passwords.

2.1. Access Control

Typically, a LAN connects hosts who are in the same security or administrative domain. While allowing legitimate user accessing to a LAN from outside (via a dialup modem, or a DSL, or a cable modem), it is crucial to prevent unauthorized access. Firewalls are dedicated gateways used for access control and can be grouped into three classes [16]: packet filter gateways, circuit-level gateways, and application-layer gateways. Packet filters operate by selectively dropping packets based on source address, destination address, or port number. In a firewall the security policies can be specified in a table that contains the filtering rules to which the incoming or outgoing packets are subject. For example, all outgoing mail traffic can be permitted to pass through a firewall while Telnet requests from a list of hosts can be dropped. Filtering rules must be managed carefully to prevent loopholes in a firewall. In case of multiple firewalls managed by the same security domain, it is crucial to eliminate inconsistencies between the rules.

Application- and circuit-level gateways are firewalls that can secure the usage of a particular application by screening the commands. Logically it resides in the middle of a protocol exchange and ensures that only valid commands are sent. For example, it may monitor a FTP session to ensure that only a specific file is accessed with read-only permission.

3. NETWORK-LAYER SECURITY

The Internet is composed of many independent management domains called *autonomous systems* (ASes). Internet routing algorithms within an AS and among ASes are different. Most important intra-AS routing (interior routing) and inter-AS (exterior routing) protocols are the Open Shortest Paths First (OSPF), and Border Gateway Protocol (BGP), respectively. However, the common theme in these protocols is the exchange of routing information to converge in a stable routing state. However, because of the lack of scalable authenticity check, routing information exchanged between the peers is subject to attacks. The attacker can eavesdrop, modify, and reinject the exchanged messages. Most of these attacks can be addressed by deployment of public key infrastructures (PKI) and certificates for authentication and validation of messages. For example, Kent et al. [18] discuss how to secure BGP protocol using PKI with X.509 certificates, and IPsec protocol suite. The solution proposes to use a new BGP path attribute to ensure the authenticity and integrity of BGP messages and validate the source of UPDATE messages. However, if a legitimate router is compromised, then such cryptographic mechanisms cannot be sufficient and the security problem degenerates to the Byzantine agreement problem [19] in distributed computing.

Next we discuss the IPsec protocols for securing IP-based intranets and then review the security issues in ATM networks.

3.1. IP Security:IPsec

The suite of IPsec protocols are designed to provide security for Internet Protocol version 4 (IPv4) and version 6 (IPv6) [20]. IPsec offers access control, connectionless integrity, source authentication, and confidentiality. IPsec defines two headers that are placed after the IP header and before the header of layer 4 protocols (i.e., TCP or UDP). These headers are used in two traffic security protocols: the *authentication header* (AH) and the *encapsulating security payload* (ESP). AH is recommended when confidentiality is not required, while ESP provides optional encryption. They both ensure integrity and authentication using tools such as keyed hash functions. Both AH and ESP use a simplex connection called a *security association* (SA). An SA is uniquely identified by a triple that contains a security parameter index (SPI), a destination IP address, and an identifier for the security protocol (i.e., AH or ESP). The negotiation of security association between two entities and exchange of keys can be done by using the Internet Key Exchange (IKE) protocol [22]. Conceptually, an SA is a virtual tunnel based on encapsulation. Two types of SAs are defined in the standard: *transport mode*, and *tunnel mode*. The former is a security association between two hosts, while the latter is established between network elements. Thus, in the transport mode the security protocol header comes right after the IP header and encapsulates any higher-level protocols. In the tunnel mode there is an outer header and an inner IP header. The *outer* header specifies the next hop, while the *inner* header indicates the final destination. The security protocol header in the tunnel mode is inserted after the outer IP header and before the inner one, thus protecting the inner header.

There are successful attacks on IPsec in spite of the secure ciphers used by the protocol. For example, consider the cut-and-paste attack by Bellovin [21]. In this attack, an encrypted ciphertext from a packet carrying sensitive (targeted) information is cut and pasted into the ciphertext of another packet. The objective is to trick the receiver to decrypt the modified ciphertext and reveal the information.

3.2. ATM Security

Asynchronous transfer mode (ATM) technology is based on establishing switched or permanent virtual circuits (SVCs, PVCs) to transmit fixed-size (53-byte) cells. There are no standards for ATM security, and work is in progress at the ATM Forum [23]. Next we review some of the security threats inherent in the architecture. All the cells carrying the same VPI/VCI (virtual path identifier, virtual connection identifier, respectively) are carried on the same virtual channel. Thus, eavesdropping and integrity violation attacks can be mounted to *all* the cells of a connection from a single point. In particular the cells carrying signaling information can be used to identify communicating parties. For example, capturing CONNECT or CALL PROCEEDING messages during signaling will reveal the VPI/VCIs assigned by the network to a particular connection. Flooding network with SET UP requests can be used to achieve denial-of-service attacks. Management cells can be abused to disrupt or disconnect legitimate connections. For example, by tampering with AIS/FERF cells, the attacker can cause a connection to be terminated.

3.2.1. IP over ATM. ATM networks has been deployed in high-speed backbone as the switching plane for IP traffic using IP over ATM protocols. The IP over ATM suite brings security concerns in ATM networks, many of which are similar to those in IP networks; however, their remedies are more difficult. For example, firewalls and packet filters used for access control in IP networks will require termination of ATM connection, inducing large delays and overhead. Authentication between ATMARP (ATM Address Resolution Protocol) server and hosts is a must for preventing various threads, including *spoofing, denial-of-service*, and *man-in-the-middle* attacks. For example, it is possible to send spoofed IP packets over an ATM connection, if the ATM address of an ATMARP server is known. The attacker can first establish a virtual connection to the server and then use the IP address of the victim to spoof the packets on this connection. Since the server will reply back to the attacker using the same connection the victim may not even know the attack. Similarly, the attacker can use the IP addresses of victims to register them with the ATMARP server. Since each IP address can be used only once, the victims will be denied service.

4. TRANSPORT-LAYER SECURITY

The Secure Sockets Layer (SSL) protocol [24] was developed to ensure secure communication between the Internet browsers and servers by Netscape Corporation. Protocols such as HTTP run over SSL to provide secure connections. The Transport Layer Security (TLS) protocol [25] is expected to become the standard for secure client/server applications over the Internet. TLS v.1.0 is based on SSL v.3.0 and considered to be SSL v.3.1. Both SSL and TLS provide encryption, authentication, and integrity protection over a public network. They are composed of two subprotocols (layers): the Record Protocol and the Handshake Protocol. The *Record Protocol* is at the lowest layer and resides above a reliable transport layer protocol such as TCP. It provides encryption using symmetric cryptography (e.g., DES), and message integrity check using a keyed MAC. The *Handshake Protocol* is used to agree on cryptographic algorithms, to establish a set of keys to be used by the ciphers, and to authenticate the client. The Handshake Protocol starts with a *ClientHello* message sent to a server by a client. This message contains a random number, version information, encryption algorithms that the client supports, and a session ID. The server sends back a *ServerHello* message that also contains random data, session ID, and indicates selected cipher. In addition, the server sends a *Certificate* message that contains the server's RSA public key in an X.509 [11,12] certificate. The client verifies the server's public key, generates a 48-byte random number called the *premaster key*, encrypts it using server's public key, and sends it to the server. The client also computes a *master secret* and uses the master key to derive at a symmetric *session key*. The server performs similar operations to compute a master key and a symmetric key. After the keys are installed to the record layer, the handshake is completed. Although SSL and TLS are similar there are also important differences between them. For example, in SSL each message is transmitted with a new socket while in TLS multiple messages can be transmitted over the same socket. Security of the SSL protocol is well examined and reported to be sound, although there are some easy-to-fix problems [26]. For example, unprotected data structures (e.g., server key exchange) can be exploited to perform cryptographic attacks (e.g., *ciphersuite rollback attack* [26]).

4.1. Multicast Security

Multicasting is a group communication with single or multiple sources and multiple receivers. It has considerably more challenging security concerns than does a single source–destination communication:

1. Message authentication and confidentiality in a multicast group requires efficient key management protocols. Establishing a different key between each pair of multicast members is not scalable. Thus, most of the solutions focus on a *shared* key [27–29]. However, a single-key approach is not sufficient to authenticate the identity of a sender since it is shared by all the members. Signing each message using a public key scheme is a costly solution; thus MAC-based solutions have been proposed [30].
2. The membership dynamics (i.e., joining and leaving a multicast group) requires efficient key revocation

algorithms. In particular deletion of a user from the multicast group must not reset all the keys. The solution is based on assigning multiple keys to each member and organizing the key allocation into a data structure that is easy to update [32–34]. For example, the Wallner et al. [33] key allocation scheme uses a (binary) tree structure. The group members are the leaves, and each intermediate node represents a distinct key. Each user will receive all the keys on the path to the root of the tree, and the root contains the shared group key. A group controller manages the data structure for delete and insert operations. Thus, in the key-based scheme each user gets $\log(n+1)$ keys and deletion of a user cost $2\log n - 1$ key encryptions.

5. APPLICATION-LEVEL SECURITY: KERBEROS

Kerberos is an authentication service that allows users and services to authenticate themselves to each other [31]. It is typically used when a user on a network requests a network service, and the server needs to ensure that the user is a legitimate one. For example, it enables users to log in to remote computers over the network without exposing their passwords to network packet-sniffing programs. User authentication is based on a "ticket" issued by the Kerberos key distribution center (KDC), which has two modules: authentication server (AS) and ticket-granting server (TGS). Both the user and the server are required to have keys registered with the AS. The user's key is derived from a password that is seen by only the local machine; the server key is selected randomly. The authentication between a user u and server S has the following steps:

1. The user u sends a message to the AS specifying the server S.
2. The AS produces two copies of a key called the *session key* to be used between u and S. AS encrypts one of the session keys and the identity S of the server using the user's key. Similarly, it encrypts the other session key and identity of the user with the server key. It sends both of the encrypted messages, called the "tickets" (say, m_1 and m_2, respectively) to u.
3. u can decrypt m_1 with its own key, extracting the session key and the identity of the server S. However, u cannot decrypt m_2 instead it timestamps a new message m_3 (called the *authenticator*), encrypts it with the session key and sends both m_2 and m_3 to S.
4. S decrypts m_2 with its own key to obtain the session key and the identity of user u. It then decrypts m_3 with the session key to extract the timestamp in order to authenticate the identity of the user u.

Following Step 4, all the communication between u and S will be done using the session key. However, in order to avoid performing all the steps above for each request, the TGS module in KDC issues a special ticket called the *ticket-granting ticket* (TGT). TGT behaves like a temporary password, with a lifetime of several hours only, and all other tickets are obtained using TGT.

6. WIRELESS SECURITY

Security in wireless networks is a challenging problem — the bandwidth and power limitations encourage the use of weaker cryptographic tools or keys with smaller sizes; also, the lack of point-to-point links makes it more difficult to protect the communication.

Elliptic curve crypto (ECC) systems [35] provide a remedy to some of these problems. ECC is based on discrete logarithm problem defined over the points on an elliptic curve. It is considered to be harder than the factorization problem and can provide works with much smaller key size than can other public key crypto systems [36]. Smaller key size reduces the processing overhead, and smaller digital signatures save on the bandwidth consumption.

6.1. Wireless LAN (WLAN)

WLANs use RF technology to receive and transmit data in a local-area network domain. In contrast with a wired LAN, a WLAN offers mobility and flexibility due to lack of any fixed topology. IEEE 802.11 is the most widely adapted standard for WLANs and it operates in the 2.4–2.48-GHz band.

There are several security vulnerabilities of a WLAN due to its nature: (1) any node within the transmission range of the source can eavesdrop easily, (2) unsuccessful attempts to access to a WLAN may be interpreted as a high *bit error rate* (BER) — this misinterpretation can be used to conceal an intruder's unauthorized access attack to a WLAN, and (3) the transmission medium is "shared" among the users. Thus, intentional interference (called "jamming") can be produced in a WLAN for denial of service attacks.

The *spread-spectrum* transmission technology helps countermeasure some of these problems in the WLANs. In spread spectrum, a signal is spread over the channel using two different techniques: (1) the frequency-hopping spread spectrum (FHSS), and (2) the direct-sequence spread spectrum (DSSS). An attacker must know the hopping pattern in FHSS or the codewords in DSSS to tune into the right frequency for eavesdropping. (Ironically these parameters are made public in the IEEE 802.11 standard.) Additional help comes from sophisticated network interface cards (NICs) of IEEE 802.11b devices. These cards can be equipped with a unique public and private key pair, in addition to their unique address, to prevent unauthorized access to a WLAN.

IEEE 802.11 standard provides a security capability called *wired equivalent privacy* (WEP). In WEP there is a secret 40-bit or a 128-bit key that is shared between a wireless node and an access point. Communication between a wireless station and its access point can be encrypted using the key and RSA's RC4 encryption algorithm. RC4 is a stream cipher with a variable key size and uses an *initialization vector* (IV). IV is used to produce different ciphertexts for identical plaintexts by initializing the shift registers with random bits. IV does not need to be secret but it should be unique for each transmission. However, IEEE 802.11 does not enforce the uniqueness of IV. Thus, one potential problem with the WEP is the reuse of IV, which may be exploited for cryptanalyzing and for fabricating new messages [37].

6.2. Wireless Transport Layer (WTLS)

The Wireless Transport Layer Security (WTLS) [38] protocol provides authentication, privacy and integrity for the Wireless Application Protocol (WAP) [39]. The WTLS is based on TLS v.1.0 and takes into account the characteristics of wireless world (e.g., low bandwidth, limited processing and power capacity, and connectionless datagram service). WTLS supports a rich set of cryptographic algorithms. Confidentiality is provided by using block ciphers such as DES CBC, integrity is ensured by SHA-1 [41] and MD5 [40] MAC algorithms, and the authentication is checked by RSA and Diffie–Hellman-based key exchange algorithms. WTLS does not contain any serious security problems to force an architectural change. Nevertheless there are several weak points of the protocol: (1) the computation of initialization vector (IV) is not a secret, (2) some fields in the data structures used by the protocol are not protected (one example is the sequence numbering, which enables an attacker to generate replay attacks), and (3) the key size should be at least 56 bits since 40-bit keys are not sufficient.

7. CONCLUSIONS

Heterogeneity of the Internet requires a skillful integration of the cryptographic building blocks with protocols for ensuring end-to-end security. Thus, deployment of security in the Internet cannot be confined to a particular crypto algorithm or to a particular architecture. The limitations on the processing capability or bandwidth forces sacrifices on the security (e.g., smaller key sizes, IV reuse, CRC for integrity check). Some of these problems can be addressed by efficient cryptographic tools such as ECC, and some will disappear as the technology improves. Many attacks exploit the way protocols are designed and implemented, even if these protocols may use very secure ciphers. Examples include unprotected fields in the data structures (e.g., SSL 3.0 server key exchange message) and lack of authentication in ATMARP between server and client. Finally, the security problem in the Internet degenerates to the distributed consensus problem if network elements are compromised by the adversary. For example there is no easy way to check the "correctness" of a routing exchange message if it is signed by a once-legitimate-but-compromised router.

Thus the Internet will never be absolutely secure, and creating a high cost–benefit tradeoff for the attacker, to reduce the incentive, will always remain a practical security measure.

BIOGRAPHY

Bulent Yener is an Associate Professor at the Computer Science Department at Rensselaer Polytechnic Institute. Dr. Yener received B.S. and M.S. degrees in Industrial Engineering from the Technical University of Istanbul, Turkey, and M.S. and Ph.D. degrees in Computer Science, both from Columbia University, in 1987 and 1994, respectively. He was a Member of Technical Staff at the Bell Laboratories in Murray Hill, New Jersey during 1998–2001. Before joining to the Bell Laboratories in 1998, he served as an Assistant Professor at Lehigh University and NJIT. His current research interests include quality of service in the IP networks, wireless networks, and Internet security. He has served on the Technical Program Committee of leading IEEE conferences and workshops. Dr. Yener is a member of the IEEE and serves on the editorial boards of the *Computer Networks Journal* and the *IEEE Network Magazine*. He is a member of IEEE.

BIBLIOGRAPHY

1. B. Schneier, *Applied Cryptography*, 2nd ed., Wiley, New York, 1996.
2. D. R. Stinson, *Cryptography: Theory and Practice (Discrete Mathematics and Its Applications)*, Chapman & Hall, 1995.
3. William F. Friedman, *Index of Coincidence and Its Applications in Cryptography*, Riverbank Publication 22, Riverbank Labs., 1920, reprinted by Agean Park Press, 1976.
4. E. H. Hebern, Electronic coding machine, U.S. Patent 1,510,441,30.
5. C. E. Shannon, in N. J. A. Sloane and A. D. Wyner, eds., *Collected Papers: Claude Elmwood Shannon*, IEEE Press, New York, 1993.
6. H. Feistel, Cryptography and computer privacy, *Sci. Am.* **228**(5): 15–23 (1973).
7. M. J. Weiner, Efficient DES key search, *Proc. CRYPTO'93*, 1993.
8. J. Daemen and V. Rijmen, *Rijndael Home Page* (online) *http://www.esat.kueven.ac.be/rijmen/rijndael* (March 26, 2002).
9. W. Diffie and M. E. Hellman, New directions in cryptography, *IEEE Trans. Inform. Theory* **IT-22**: 644–654 (1976).
10. R. Rivest, A. Shamir, and L. Adleman, A method for obtaining digital signatures and public-key cryptosystems, *Commun. ACM* **21**(2): 120–126 (1978).
11. ITU-T Recommendation X.509, *Information Technology — Open System Interconnection — The Directory: Authentication Framework*, 1997.
12. R. Housley, W. Ford, W. Polk, and D. Solo, *Internet X.509 Public Key Infrastructure Certificate and CRL Profile*, IETF RFC 2459, 1999.
13. J. Postel and J. Reynolds, File Transfer Protocol, *IETF RFC 959*, 1985.
14. J. Postel, *Transmission Control Protocol*, IETF RFC 791, 19.
15. S. M. Bellovin, Security Problems in the TCP/IP Protocol Suite, *ACM Comput. Commun. Rev.* **19**(2): 32–48 (1989).
16. S. M. Bellovin and W. R. Cheswick, *Firewalls and Internet Security*, Addison-Wesley, New York, 1994.
17. T. Ylonen et al., *SSH Protocol Architecture* (online) *draft-ietfsecsh-architecture-12.txt*, 2002.

18. S. Kent, C. Lynn, and K. Seo, Secure Border Gateway Protocol (S-BGP), *IEEE JSAC Network Security* **18**(34): 582–592 (2000).
19. M. Raynal, *Distributed Algorithms and Protocols*, Wiley, New York, 1988.
20. S. Kent and R. Atkinson, *Security Architecture for the Internet Protocol*. IETF RFC 2401, 1998.
21. S. Bellovin, Problem areas for the IP security protocols, *Proc. 6th USENIX Security Symp.* 1996, pp. 205–214.
22. D. Harkins and D. Carrel, *The Internet Key Exchange*, IETF RFC 2409, 1998.
23. ATM Forum. *http://www.atmforum.org.*
24. A. Frier, P. Karlton, and P. Koccher, *The SSL3.0 Protocol Version 3.0,* Netscape, 1996 (online) *http://home.netscape.com/eng/ssl3/* (March 26, 2002).
25. T. Dierks and C. Allen, *The TLS Protocol Version 1.0*, IETF RFC 2246, 1999.
26. D. Wagner and B. Schneier, Analysis of the SSL 3.0 protocol, *Proc. 2nd USENIX Workshop on Electronic Commerce*, USENIX Press, 1996, pp. 29–40 (online) *http://citeseer.nj.nec.com/article/wagner96analysi.html* (March 26, 2002).
27. H. Harney and C. Muckenhirn, *Group Key Management Protocol (GKMP) Specification*, IETF RFC 2093, 1997.
28. H. Harney and C. Muckenhirn, *Group Key Management Protocol (GKMP) Architecture*, IETF RFC 2094, 1997.
29. S. Mittra, Iolus: A framework for scalable secure multicasting, *Proc. ACM SIGCOMM'97*, 1997.
30. R. Canetti et al., Multicast security: A taxonomy and efficient constructions, *Proc. IEEE INFOCOM'99*, 1999.
31. J. Kohl and C. Neuman, *The Kerberos Network Authentication Service (V5)*, IETF RFC 1510, 1993.
32. A. Fiat and M. Naor, Broadcast encryption, *Advances in Cryptography — Crypto'92*, 1995, Vol. 8, pp. 189–200.
33. D. M. Wallner, E. J. Harder, and R. C. Agee, *Key Management for Multicast: Issues and Architectures*, IETF RFC 2627, 1999.
34. C. K. Wong and S. Lam, Digital signature for flows and multicasts, *Proc. IEEE ICNP'98'*, 1998.
35. N. Koblitz, Elliptic curve cryptosystems, *Math. Comput.* **48**: 203–209 (1987).
36. Certicom White Paper, *Current Public-Key Cryptographic Systems*, 1997 (online) *http://www.certicom.com* (March 26, 2002).
37. N. Borisov, I. Goldberg, and D. Wagner, Intercepting mobile communications: The insecurity of 802.11, *Proc. Mobile Computing and Networking*, 2001.
38. WAP Forum, *Wireless Application Protocol — Wireless Transport Layer Security Specification version 1* (online) *http://www.wapforum.org* (March 26, 2002).
39. WAP Forum, *Wireless Application Protocol* (online) *http://www.wapforum.org* (March 26, 2002).
40. R. Rivest, *The MD5 Message-Digest Algorithm*, IETF RFC 1321, 1992.
41. Federal Information Processing Standard Publication 180-1, 1995 (online) *http://www.itl.nist.gov/fibspubs/fip180-1.htm* (March 26, 2002).

INTERSYMBOL INTERFERENCE IN DIGITAL COMMUNICATION SYSTEMS

John G. Proakis
Northeastern University
Boston, Massachusetts

1. INTRODUCTION

Intersymbol interference arises in both wireline and wireless communication systems when data are transmitted at symbol rates approaching the Nyquist rate and the characteristics of the physical channels through which the data are transmitted are nonideal. Such interference can severely limit the performance that can be achieved in digital data transmission, where performance is measured in terms of the data transmission rate and the resulting probability of error in recovering the data from the received channel corrupted signal.

The degree to which one must be concerned with channel impairments generally depends on the transmission rate of data through the channel. If R is the transmission bit rate and W is the available channel bandwidth, intersymbol interference (ISI) caused by channel distortion generally arises when $R/W > 1$. Since bandwidth is usually a precious commodity in communication systems, it is desirable to utilize the channel to as near its capacity as possible. In such cases, the communication system designer must employ techniques that mitigate the effects of ISI caused by the channel.

This article provides a characterization of ISI resulting from channel distortion.

2. CHANNEL DISTORTION

Wireline channels may be characterized as linear time-invariant filters with specified frequency-response characteristics. If a channel is band-limited to W Hz, then its frequency response $C(f) = 0$ for $|f| > W$. Within the bandwidth of the channel, the frequency response $C(f)$ may be expressed as

$$C(f) = |C(f)|e^{j\theta(f)} \tag{1}$$

where $|C(f)|$ is the amplitude-response characteristic and $\theta(f)$ is the phase-response characteristic. Furthermore, the envelope delay characteristic is defined as

$$\tau(f) = -\frac{1}{2\pi}\frac{d\theta(f)}{df} \tag{2}$$

A channel is said to be *nondistorting* or *ideal* if the amplitude response $|C(f)|$ is constant for all $|f| \leq W$ and $\theta(f)$ is a linear function of frequency, that is, if $\tau(f)$ is a constant for all $|f| \leq W$. On the other hand, if $|C(f)|$ is not constant for all $|f| \leq W$, we say that the channel *distorts the transmitted signal in amplitude*, and, if $\tau(f)$ is not constant for all $|f| \leq W$, we say that the channel *distorts the signal in delay*.

As a result of the amplitude and delay distortion caused by the nonideal channel frequency-response characteristic $C(f)$, a succession of pulses transmitted through the channel at rates comparable to the bandwidth W are smeared to the point that they are no longer distinguishable as well-defined pulses at the receiving terminal. Instead, they overlap, and, thus, we have intersymbol interference. As an example of the effect of delay distortion on a transmitted pulse, Fig. 1a illustrates a band-limited pulse having zeros periodically spaced in time at points labeled $\pm T$, $\pm 2T$, and so on. If information is conveyed by the pulse amplitude, as in pulse amplitude modulation (PAM), for example, then one can transmit a sequence of pulses each of which has a peak at the periodic zeros of the other pulses. However, transmission of the pulse through a channel modeled as having a linear envelope delay characteristic $\tau(f)$ [quadratic phase $\theta(f)$] results in the received pulse shown in Fig. 1b having zero crossing that are no longer periodically spaced. Consequently, a sequence of successive pulses would be smeared into one another and the peaks of the pulses would no longer be distinguishable. Thus, the channel delay distortion results in intersymbol interference.

Another view of channel distortion is obtained by considering the impulse response of a channel with nonideal frequency response characteristics. For example, Fig. 2 illustrates the average amplitude response $|C(f)|$ and the average envelope delay $\tau(f)$ for a medium-range (180–725-mi) telephone channel of the switched telecommunications network. It is observed that the usable band of the channel extends from ~300 to ~3000 Hz. The corresponding impulse response of this average channel is shown in Fig. 3. Its duration is about 10 ms. In comparison, the transmitted symbol rates on such a channel may be of the order of 2500 pulses or symbols per second. Hence, intersymbol interference might extend over 20–30 symbols.

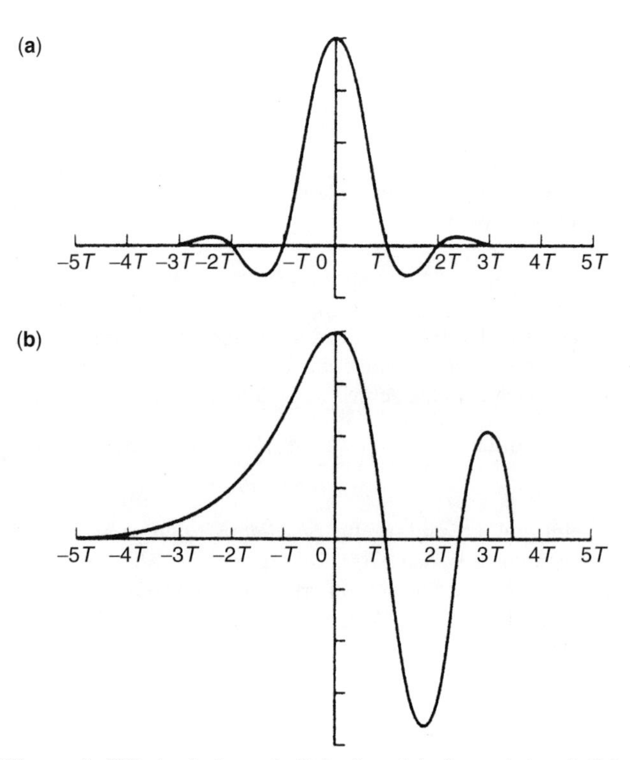

Figure 1. Effect of channel distortion (**a**) channel input (**b**) channel output.

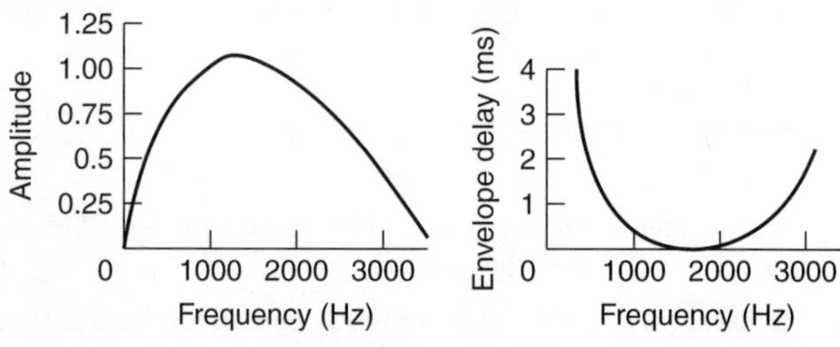

Figure 2. Average amplitude and delay characteristics of medium-range telephone channel.

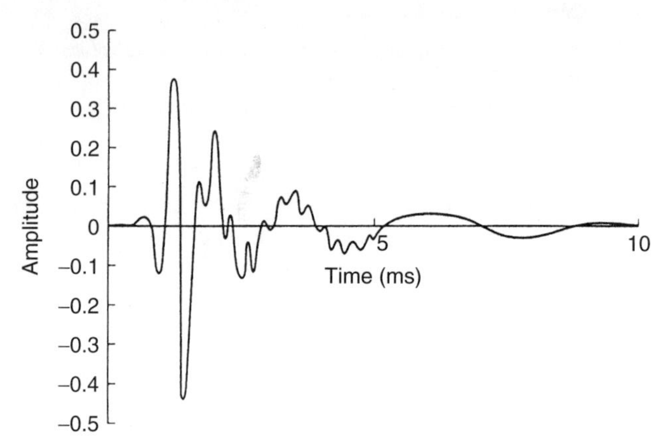

Figure 3. Impulse response of medium-range telephone channel with frequency response shown in Fig. 2.

Besides wireline channels, there are other physical channels that exhibit some form of time dispersion and, thus, introduce intersymbol interference. Radio channels such as shortwave ionospheric channels (HF), tropospheric scatter channels, and mobile cellular radio channels are examples of time-dispersive channels. In these channels, time dispersion and, hence, intersymbol interference are the result of multiple propagation paths with different path delays. The number of paths and the relative time delays among the paths vary with time, and, for this reason, these radio channels are usually called *time-variant multipath channels*. The time-variant multipath conditions give rise to a wide variety of frequency-response characteristics and the resulting phenomenon called *signal fading*.

In the following section, a mathematical model for the intersymbol interference (ISI) is developed and transmitted signal characteristics are described for avoiding ISI.

3. CHARACTERIZATION OF INTERSYMBOL INTERFERENCE

In conventional linear digital modulation such as pulse amplitude modulation (PAM), phase shift keying (PSK), and quadrature amplitude modulation (QAM), the transmitted signal is generally represented as an equivalent lowpass signal (prior to frequency conversion[1] for transmission over the bandpass channel) of the form

$$v(t) = \sum_{n=0}^{\infty} I_n g(t - nT) \tag{3}$$

where $\{I_n\}$ represents the discrete information-bearing sequence of symbols and $g(t)$ is a modulation filter pulse that, for the purposes of this discussion, is assumed to have a band-limited frequency-response characteristic $G(f)$, specifically, $G(f) = 0$ for $|f| > W$. For PAM, the information-bearing sequence $\{I_n\}$ consists of symbols taken from the alphabet $\{\pm 1, \pm 3, \ldots, +(M-1)\}$ for M-level amplitude modulation. In the case of PSK, the information-bearing sequence $\{I_n\}$ consists of symbols taken from the alphabet $\left\{e^{j\theta_m}, \theta_m = \frac{2\pi}{M}m,\ m = 0, 1, \ldots, M-1\right\}$. QAM may be considered as a combined form of digital amplitude and phase modulation, so that the sequence $\{I_n\}$ takes values of the form $\{A_m e^{j\theta_m},\ m = 0, 1, \ldots, M-1\}$.

The signal given by Eq. (3) is transmitted over a channel having a frequency response $C(f)$, also limited to $|f| < W$. Consequently, the received signal can be represented as

$$r(t) = \sum_{n=0}^{\infty} I_n h(t - nT) + z(t) \tag{4}$$

where

$$h(t) = \int_{-\infty}^{\infty} g(\tau) c(t - \tau)\, d\tau \tag{5}$$

and $z(t)$ represents the additive white Gaussian noise that originates at the front end of the receiver. The channel impulse response is denoted as $c(t)$.

Suppose that the received signal is passed first through a filter and then sampled at a rate $1/T$ samples. Since the additive noise is white Gaussian, the optimum filter at the receiver is the filter that is matched to the signal pulse $h(t)$; that is, the frequency response of the receiving filter is $H^*(f)$. The output of the receiving filter[2] may be expressed as

$$y(t) = \sum_{n=0}^{\infty} I_n x(t - nT) + v(t) \tag{6}$$

where $x(t)$ is the pulse representing the response of the receiving filter to the input pulse $h(t)$ and $v(t)$ is the response of the receiving filter to the noise $z(t)$.

[1] The frequency upconversion that is performed at the transmitter and the corresponding downconversion performed at the receiver may be considered transparent, so that transmitted and received signals are treated in terms of the equivalent lowpass characteristics.

[2] Often, the frequency-response characteristics of the channel are unknown to the receiver. In such a case, the receiver cannot implement the optimum matched filer to the signal pulse $h(t)$. Instead, the receiver may implement the filter matched to the transmitted pulse $g(t)$.

Now, if $y(t)$ is sampled at times $t = kT + \tau_0, k = 0, 1, \ldots$, we have

$$y(kT + \tau_0) = y_k = \sum_{n=0}^{\infty} I_n x(kT - nT + \tau_0) + v(kT + \tau_0) \tag{7}$$

or, equivalently

$$y_k = \sum_{n=0}^{\infty} I_n x_{k-n} + v_k, \quad k = 0, 1, \ldots \tag{8}$$

where τ_0 is the transmission delay through the channel. The sample values can be expressed as

$$y_k = x_0 \left(I_k + \frac{1}{x_0} \sum_{\substack{n=0 \\ n \neq k}}^{\infty} I_n x_{k-n} \right) + v_k, \quad k = 0, 1, \ldots \tag{9}$$

We regard x_0 as an arbitrary scale factor, which can be set equal to unity for convenience. Then

$$y_k = I_k + \sum_{\substack{n=0 \\ n \neq k}}^{\infty} I_n x_{k-n} + v_k \tag{10}$$

The term I_k represents the desired information symbol at the kth sampling instant, the term

$$\sum_{\substack{n=0 \\ n \neq k}}^{\infty} I_n x_{k-n} \tag{11}$$

represents the ISI, and v_k is the additive Gaussian noise variable at the kth sampling instant.

The amount of intersymbol interference and noise in a digital communication system can be viewed on an oscilloscope. For PAM signals, we can display the received signal $y(t)$ on the vertical input with the horizontal sweep rate set at $1/T$. The resulting oscilloscope display is called an "eye pattern" because of its resemblance to the human eye. For example, Fig. 4 illustrates the eye patterns for binary and quaternary PAM modulation. The effect of ISI is to cause the eye to close, thereby reducing the margin for additive noise to cause errors. Figure 5 graphically illustrates the effect of intersymbol interference in reducing the opening of a binary eye. Note that intersymbol interference distorts the position of the zero crossings and causes a reduction in the eye openings. Thus, it causes the system to be more sensitive to a synchronization error.

For PSK and QAM it is customary to display the "eye pattern" as a two-dimensional scatter diagram illustrating the sampled values $\{y_k\}$ that represent the decision variables at the sampling instants. Figure 6 illustrates such an eye pattern for an 8-PSK signal. In the absence of intersymbol interference and noise, the superimposed signals at the sampling instants would result in eight distinct points corresponding to the eight transmitted signal phases. Intersymbol interference and noise result in a deviation of the received samples $\{y_k\}$ from the desired 8-PSK signal. The larger the intersymbol interference

(a)

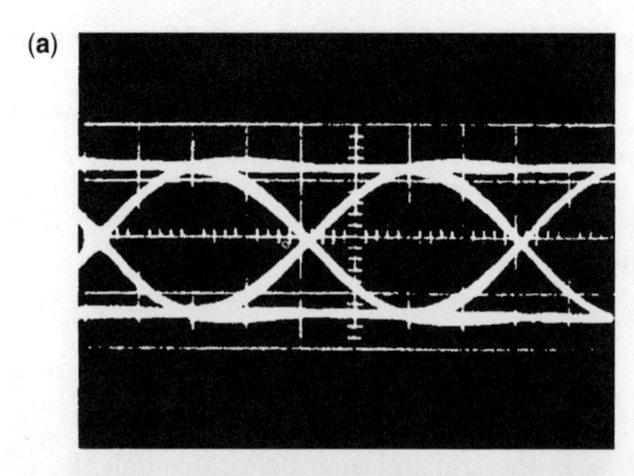

(b)

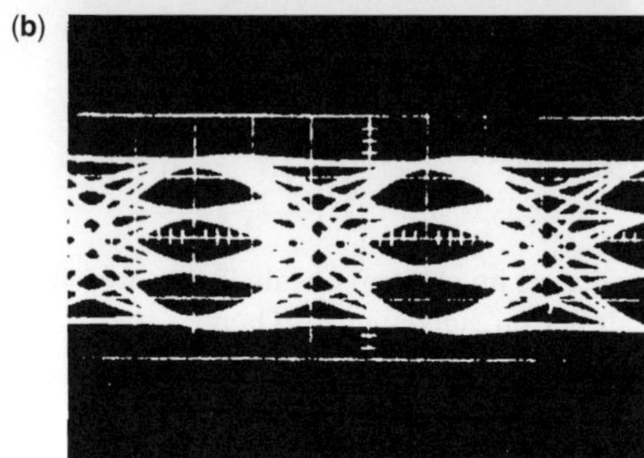

Figure 4. Examples of eye patterns for binary (**a**) and quaternary (**b**) PAM.

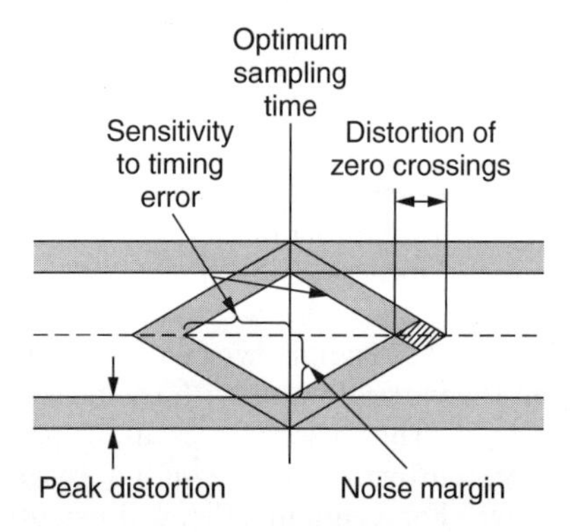

Figure 5. Effect of intersymbol interference on eye opening.

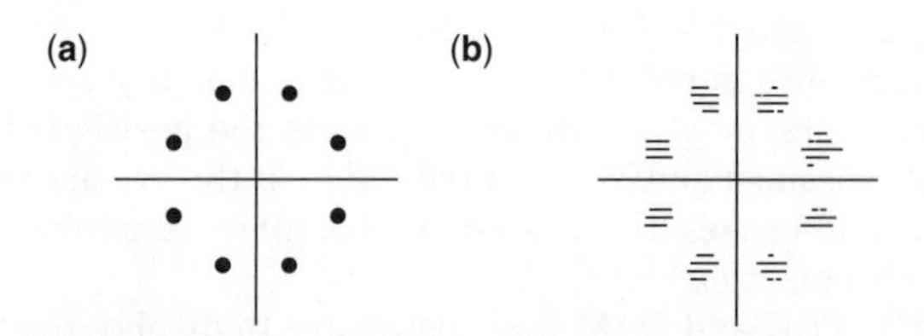

Figure 6. "Eye Patterns" for a two-dimensional signal constellation: (**a**) transmitted eight-phase signal; (**b**) received signal samples at the output of demodulator.

and noise, the larger the scattering of the received signal samples relative to the transmitted signal points.

The following section considers the design of transmitted pulses that result in no ISI in the transmission through a band-limited channel.

4. SIGNAL DESIGN FOR ZERO ISI — THE NYQUIST CRITERION

The problem treated in this section is the design of the transmitter and receiver filters in a modem so that the received signal has zero ISI, assuming the condition that the channel is an ideal channel. The subsequent section treats the problem of filter design when there is ISI due to channel distortion.

Under the assumption that the channel frequency response is ideal, i.e., $C(f) = 1$ for $|f| \leq W$, the pulse $x(t)$ has a spectral characteristic $X(f) = |G(f)|^2$, where

$$x(t) = \int_{-W}^{W} X(f)e^{j2\pi ft}\, df \tag{12}$$

We are interested in determining the spectral properties of the pulse $x(t)$ and, hence, the transmitted pulse $g(t)$, which results in no intersymbol interference. Since

$$y_k = I_k + \sum_{\substack{n=0 \\ n\neq k}}^{\infty} I_n x_{k-n} + v_k \tag{13}$$

the condition for no intersymbol interference is

$$x(t = kT) \equiv x_k = \begin{cases} 1 & (k = 0) \\ 0 & (k \neq 0) \end{cases} \tag{14}$$

Nyquist [1] formulated and solved this problem in the late 1920s. He showed that a necessary and sufficient condition for zero ISI, given by Eq. (14) is that the Fourier transform $X(f)$ of the signal pulse $x(t)$ satisfy the condition

$$\sum_{m=-\infty}^{\infty} X(f + m)T) = T \tag{15}$$

Nyquist also demonstrated that if the symbol transmission rate $1/T > 2W$, where $2W$ is called the Nyquist rate, it is impossible to design a signal $x(t)$ that has zero ISI. If the symbol rate $1/T = 2W$, the only possible signal pulse that yields zero ISI has the spectrum

$$X(f) = \begin{cases} T, & |f| \leq W \\ 0, & \text{otherwise} \end{cases} \tag{16}$$

and the time response

$$x(t) = \frac{\sin(\pi t/T)}{t\pi/T} \equiv \text{sinc}\left(\frac{\pi t}{T}\right) \tag{17}$$

This means that the smallest value of T for which transmission with zero ISI is possible is $T = 1/2W$, and for this value, $x(t)$ has to be the sinc function. The difficulty with this choice of $x(t)$ is that it is concausal and, therefore, nonrealizable. To make it realizable, usually a delayed version of it, $\text{sinc}[\pi(t - t_0)/T]$, is used and t_0 is chosen such that for $t < 0$, we have $\text{sinc}[\pi(t - t_0)T] \approx 0$. Of course, with this choice of $x(t)$, the sampling time must also be shifted to $mT + t_0$. A second difficulty with this pulseshape is that its rate of convergence to zero is slow. The tails of $x(t)$ decay as $1/t$; consequently, a small mistiming error in sampling the output of the receiver filter at the demodulator results in an infinite series of ISI components. Such a series is not absolutely summable because of the $1/t$ rate of decay of the pulse, and, hence, the sum of the resulting ISI does not converge. Consequently, the signal pulse given by Eq. (17) does not provide a practical solution to the signal design problem.

By reducing the symbol rate $1/T$ to be slower than the Nyquist rate, $1/T < 2W$, there exists numerous choices for $X(f)$ that satisfy Eq. (17). A particular pulse spectrum that has desirable spectral properties and has been widely used in practice is the raised-cosine spectrum. The raised-cosine frequency characteristic is given as

$$X_{rc}(f) = \begin{cases} T & \left(0 \le |f| \le \frac{1-\beta}{2T}\right) \\ \frac{T}{2}\left\{1+\cos\left[\frac{\pi T}{\beta}\left(|f| - \frac{1-\beta}{2T}\right)\right]\right\} & \left(\frac{1-\beta}{2T} \le |f| \le \frac{1-\beta}{2T}\right) \\ 0 & \left(|f| > \frac{1+\beta}{2T}\right) \end{cases} \tag{18}$$

where β is called the *rolloff factor* and takes values in the range $0 \le \beta \le 1$. The bandwidth occupied by the signal beyond the Nyquist frequency $1/2T$ is called the *excess bandwidth* and is usually expressed as a percentage of the Nyquist frequency. For example, when $\beta = \frac{1}{2}$, the excess bandwidth is 50% and when $\beta = 1$, the excess bandwidth is 100%. The pulse $x(t)$, having the raised-cosine spectrum, is

$$\begin{aligned} x(t) &= \frac{\sin(\pi t/T)}{\pi t/T}\,\frac{\cos(\pi\beta t/T)}{1-4\beta^2t^2/T^2} \\ &= \mathrm{sinc}(\pi t/T)\frac{\cos(\pi\beta t/T)}{1-4\beta^2t^2/T^2} \end{aligned} \tag{19}$$

Note that $x(t)$ is normalized so that $x(0) = 1$. Figure 7 illustrates the raised-cosine spectral characteristics and the corresponding pulses for $\beta = 0, \frac{1}{2}$, and 1. Note that for $\beta = 0$, the pulse reduces to $x(t) = \mathrm{sinc}(\pi t/T)$, and the symbol rate $1/T = 2W$. When $\beta = 1$, the symbol rate is $1/T = W$. In general, the tails of $x(t)$ decay as $1/t^3$ for $\beta > 0$. Consequently, a mistiming error in sampling leads

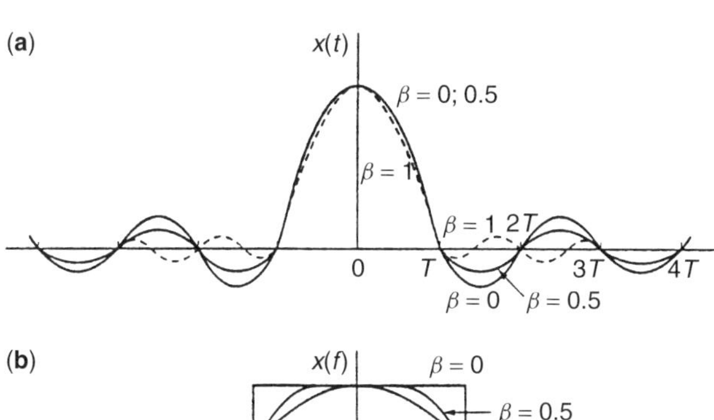

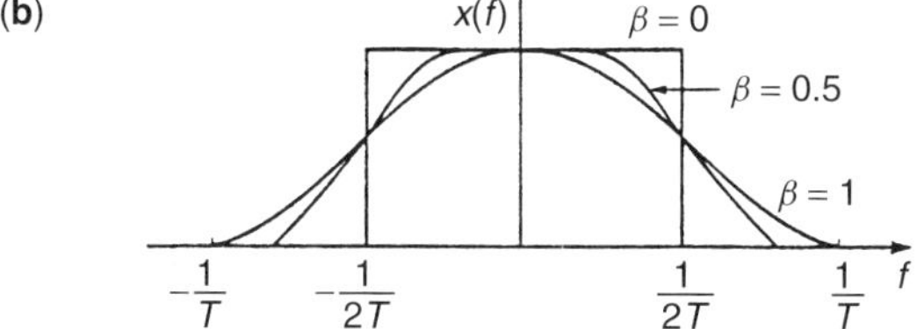

Figure 7. Signal Pulses with a Raised-Cosine Spectrum.

to a series of ISI components that converges to a finite value.

Because of the smooth characteristics of the raised-cosine spectrum, it is possible to design practical filters for the transmitter and the receiver that approximate the overall desired frequency response. In the special case where the channel is ideal, $C(f) = 1, |f| \le W$, we have

$$X_{rc}(f) = G_T(f)G_R(f) \tag{20}$$

where $G_T(f)$and $G_R(f)$ are the frequency responses of the filters at the transmitter and the receiver, respectively. In this case, if the receiver filter is matched to the transmitter filter, we have $X_{rc}(f) = G_T(f)G_R(f) = |G_T(f)|^2$. Ideally

$$G_T(f) = \sqrt{|X_{rc}(f)|}\,e^{-j2\pi f t_0} \tag{21}$$

and $G_R(f) = G_T^*(f)$, where t_0 is some nominal delay that is required to ensure physical realizability of the filter. Thus, the overall raised-cosine spectral characteristic is split evenly between the transmitting filter and the receiving filter. Note also that an additional delay is necessary to ensure the physical realizability of the receiving filter.

5. CHANNEL DISTORTION AND ADAPTIVE EQUALIZATION

Modems that are used for transmitting data either on wireline or wireless channels are designed to deal with channel distortion, which usually differs from channel to channel. For example, a modem that is designed for data transmission on the switched telephone network encounters a different channel response every time a telephone number is dialed, even if it is the same telephone number. This is due to the fact that the route (circuit) from the calling modem to the called modem will vary from one telephone call to another.

Since the channel distortion is variable and unknown a priori, a modem contains an additional component, called an *adaptive equalizer*, that follows the receiving filter $G_R(f)$, which further processes the received signal samples $\{y_k\}$ to reduce the ISI. The most commonly used equalizer is a linear, discrete-time, finite-impulse-response duration (FIR) filter with coefficients that are adjusted to reduce the ISI. With $\{y_k\}$ as its input, given by Eq. (10) and coefficients $\{b_k\}$, the equalizer output sequence is an estimate of the transmitted symbols $\{I_k\}$

$$\hat{I}_k = \sum_{m=-N}^{N} b_m y_{k-m} \tag{22}$$

where $2N + 1$ is the number of equalizer coefficients.

When a call is initiated, a sequence $\{I_k\}$ of training symbols (known to the receiver) are transmitted. The receiver compares the known training symbols with the sequence of estimates $\{\hat{I}_k\}$ from the equalizer and computes the sequence of error signals

$$\begin{aligned} e_k &= I_k - \hat{I}_k \\ &= I_k - \sum_{m=-N}^{N} b_m y_{k-m} \end{aligned} \tag{23}$$

The equalizer coefficients $\{b_k\}$ are the adjusted to minimize the mean (average) squared value of the error sequence. Thus, the equalizer adapts to the channel characteristics by adjusting its coefficients to reduce ISI.

The effect of the linear adaptive equalizer in compensating for the channel distortion can also be viewed in the frequency domain. The transmitter filter $G_T(f)$, the channel frequency response $C(f)$, the receiver filter $G_R(f)$, and the equalizer $B(f)$ are basically liner filters in cascade. Hence, their overall frequency response is $G_T(f)C(f)G_R(f)B(f)$. If the cascade $G_T(f)G_R(f)$ is designed for zero ISI as in Eq. (20), then by designing the equalizer frequency response $B(f)$ such that $B(f) = 1/C(f)$, the ISI is eliminated. In this case, the adaptive equalizer has a frequency response that is the inverse of the channel response. If the adaptive equalizer had an infinite number of coefficients (infinite-duration impulse response), its frequency response would be exactly equal to the inverse channel characteristic and, thus, the ISI would be completely eliminated. However, with a finite number of coefficients, the equalizer response can only approximately equal the inverse channel response. Consequently, some residual ISI will always exist at the output of the equalizer. For more detailed treatments of adaptive equalization, the interested reader may refer to Refs. 2–4.

6. CONCLUDING REMARKS

The treatment of ISI in this article was based on the premise that the signal transmitted through the channel was carried on a single carrier frequency. The effect of channel distortion and, hence, ISI can be significantly reduced if a channel with bandwidth W is subdivided into a large number of subchannels, say, N, where each subchannel has a bandwidth $B = W/N$. Modulated carrier signals are then transmitted in all the N subchannels. In this manner, each information symbol on each subchannel has a symbol duration of NT, where T is the symbol duration in a single carrier modulated signal. The modulation of all the carriers in the N subchannels is performed synchronously. This type of multicarrier modulation is called *orthogonal frequency-division multiplexing* (OFDM). If the bandwidth B of each subchannel is made sufficiently small, each subchannel will have (approximately) an ideal frequency-response characteristic. Hence, the ISI in each subchannel becomes insignificant. However, OFDM is particularly vulnerable to interference among the subcarriers (interchannel interference) due to frequency offsets or Doppler frequency spread effects, which are present when either the transmitter and/or the receiver terminals are mobile.

ISI in digital communications systems is treated numerous publications in the technical literature and various textbooks. The interested reader may consult Refs. 2–4 for more detailed treatments.

BIOGRAPHY

Dr. John G. Proakis received the B.S.E.E. from the University of Cincinnati in 1959, the M.S.E.E. from MIT in 1961, and the Ph.D. from Harvard University in 1967. He is an Adjunct Professor at the University of California at San Diego and a Professor Emeritus at Northeastern University. He was a faculty member at Northeastern University from 1969 through 1998 and held the following academic positions: Associate Professor of Electrical Engineering, 1969–1976; Professor of Electrical Engineering, 1976–1998; Associate Dean of the College of Engineering and Director of the Graduate School of Engineering, 1982–1984; Interim Dean of the College of Engineering, 1992–1993; Chairman of the Department of Electrical and Computer Engineering, 1984–1997. Prior to joining Northeastern University, he worked at GTE Laboratories and the MIT Lincoln Laboratory.

His professional experience and interests are in the general areas of digital communications and digital signal processing and more specifically, in adaptive filtering, adaptive communication systems and adaptive equalization techniques, communication through fading multipath channels, radar detection, signal parameter estimation, communication systems modeling and simulation, optimization techniques, and statistical analysis. He is active in research in the areas of digital communications and digital signal processing and has taught undergraduate and graduate courses in communications, circuit analysis, control systems, probability, stochastic processes, discrete systems, and digital signal processing. He is the author of the book *Digital Communications* (McGraw-Hill, New York: 1983, first edition; 1989, second edition; 1995, third edition; 2001, fourth edition), and co-author of the books *Introduction to Digital Signal Processing* (Macmillan, New York: 1988, first edition; 1992, second edition; 1996, third edition), *Digital Signal Processing Laboratory* (Prentice-Hall, Englewood Cliffs, NJ, 1991); *Advanced Digital Signal Processing* (Macmillan, New York, 1992), *Algorithms for Statistical Signal Processing* (Prentice-Hall, Englewood Cliffs, NJ, 2002), *Discrete-Time Processing of Speech Signals* (Macmillan, New York, 1992, IEEE Press, New York, 2000), *Communication Systems Engineering* (Prentice-Hall, Englewood Cliffs, NJ: 1994, first edition; 2002, second edition), *Digital Signal Processing Using MATLAB V.4* (Brooks/Cole-Thomson Learning, Boston, 1997, 2000), and *Contemporary Communication Systems Using MATLAB* (Brooks/Cole-Thomson Learning, Boston, 1998, 2000). Dr. Proakis is a Fellow of the IEEE. He holds five patents and has published over 150 papers.

BIBLIOGRAPHY

1. H. Nyquist, Certain topics in telegraph transmission theory, *AIEE Trans* **47**: 617–644 (1928).
2. J. G. Proakis, *Digital Communications*, 4th ed., McGraw-Hill, New York, 2001.
3. E. A. Lee and D. G. Messerschmitt, *Digital Communications*, 2nd ed., Kluwer, Boston, 1994.
4. R. W. Lucky, J. Salz, and E. J. Weldon, Jr., *Principles of Data Communications*, McGraw-Hall, New York, 1968.

INTRANETS AND EXTRANETS

Algirdas Pakštas
London Metropolitan University
London, England

1. INTRODUCTION

The intranet is a private network that extensively uses established Web technologies based on the Internet protocols (TCP/IP, HTTP, etc.) [1]. An intranet is accessible only by the organization's members, employees, or other users who have authorization. The Internet is a public access network. Therefore, an organization's Webpage on the Internet is its public face that helps to create its image. Such pages may be built with many graphics and special features. The intranet, however, is the organization's private face where employees get their information and then get off and go back to work. An intranet's Websites may look and act just like any other Websites, but most often their appearance is simpler and more casual. Both, the Internet and an intranet use the same types of hardware and software, but they are used for two very different purposes. An intranet is protected by the firewall surrounding it from the unauthorized access. Like the Internet itself, intranets are used to share information, but additionally the intranet is the communications platform for group work — "the 'electronic brain' employees tap into," as expressed by Steve McCormick, a consultant with Watson Wyatt Worldwide, Washington, D.C., USA. Secure intranets are now the fastest growing segment of the Internet because they are much less expensive to build and manage than private networks based on proprietary protocols. Since 1996, intranets have been embraced by corporate users of information services and have made substantial inroads in strategic vision documents and procurement practices.

An extranet is an intranet that is partially accessible to authorized outsiders, such as an organization's mobile workers or representatives of partner (sharing common goals) businesses such as suppliers, vendors, customers, and so on [2]. Thus, an extranet, or extended intranet, is a private business network of several cooperating organizations located outside the corporate firewall. Whereas an intranet resides behind a firewall and is accessible only to people who are members of the same company or organization, an extranet provides various levels of accessibility to outsiders. A user can access an extranet only if he or she has a valid username and password, and the user's identity determines which parts of the extranet can be viewed or accessed. Extranets are becoming a very popular means for business partners to exchange information. Extranets are using not only internal network resources but also the public telecommunication system. Therefore, an extranet requires security and privacy. These require firewall server management, the issuance and use of digital certificates or similar means of user authentication, encryption of messages, and the use of virtual private networks (VPN) that tunnel through the public network.

The extranets were introduced shortly after intranets. At the same time, some experts are arguing that extranets have been around since time when the first rudimentary LAN-to-LAN networks began connecting two different business entities together to form WANs and that in its basic form, an extranet is the interconnection of two previously separate LANs or WANs with origins from different business entities [3].

Table 1 summarizes the discussed features of Internet, intranet, and extranet.

1.1. Examples of Extranet Applications

The most obvious examples of extranet applications are the following:

- Exchange of large volumes of data using electronic data interchange (EDI)
- Shared product catalogs accessible only to wholesalers or those "in the trade"
- Collaboration with other companies on joint development efforts, e.g., establishing private newsgroups that cooperating companies use to share valuable experiences and ideas
- Groupware in which several companies collaborate in developing a new application program they can all use
- Project management and control for companies that are part of a common work project
- Provide or access services provided by one company to a group of other companies, such as an online banking application managed by one company on behalf of affiliated banks

Table 1. Summary of the Internet, Intranet, and Extranet Features

	Internet	Intranet	Extranet
Users	Everyone	Members of the specific firm	Group of closely related firms
Information	Fragmented	Proprietary	Shared in closely trusted held circles
Access	Public	Private	Semi-private
Security mechanism	None	Firewall, encryption	Intelligent firewall, encryption, various document security standards

- Training programs or other educational material that companies could develop and share

2. INTRANET'S HARDWARE AND SOFTWARE

The intranet's hardware consists of the client/server network. The clients are computers that are connected either to the LAN or to the WAN. The servers are high-performance computers with a large hard disk capacity and RAM.

The network operating system is the software required to run the network and share its resources (printers, files, applications, etc.) among the users. There are three basic software packages needed for a corporate intranet: Web software, firewall software, and browser software. Web software allows the server to support HTTP (Hyper Text Transfer Protocol) so it can exchange information with the clients. Firewall software provides the security needed to protect corporate information from the outside world. Browser software allows the user to read electronic documents published on the Internet or corporate intranet.

Other functions can be provided by adding software for Internet access and searching, authoring and publishing documents, collaboration and conferencing, database archive and retrieval, and document management access. A properly constructed intranet allows for the following to be done effectively [1]: centralized storage of data (real-time and archived), scalability and management of application services, centralized control over access to knowledge, decentralized management of knowledge, and universal access to the information that decision makers have deemed it appropriate to view.

These of established hardware and software technologies for intranet and extranet infrastructure makes intranets and extranets very economical in comparison with the creation and maintenance of proprietary networks. The following main approaches are used for building extranet software and inevitably are linked to the backing industrial groups or corporations:

- Crossware: Netscape, Oracle, and Sun Microsystems have formed an alliance to ensure that their extranet products can work together by standardizing on JavaScript and the Common Object Request Broker Architecture (CORBA).
- Microsoft supports the Point-to-Point Tunneling Protocol (PPTP) and is working with American Express and other companies on an Open Buying on the Internet (OBI) standard.
- The Lotus Corporation is promoting its groupware product, Notes, as well suited for extranet use.

The first approach (Crossware) is oriented to development of open application standards, while the other two are more focused on a "one company's product" type of philosophy.

Thus, intranets and extranets are rather classes of applications than *categories of networks*. Applications in the public Internet, intranet, and extranet will all run on the same type of network infrastructure, but their software and data content resources will be *administered* for different levels of accessibility and security. This can be achieved with the help of tools for network management.

2.1. Extranets and Intergroupware

Workgroups may often need special tools for *communications*, *collaboration*, and *coordination*, which are referred as *groupware* [4]. Emphasis is on computer-aided help to implement message-based human communications and information sharing as well as to support typical workgroup tasks such as scheduling and routing of the workflow tasks.

In the typical enterprise communications the inter- and intraorganizational media-based communications activities are forming two separate parallel planes [4]. The dimensions of each plane are the degree of *structure* and the degree of *mutuality* in the communications activities. *Structure* lies between informal (ad hoc) and formally structured, defined, and managed or edited processes. *Mutuality* ranges from unidirectional or sequential back-and-forth message passing, to true joint work or *collaborative transactions* in a shared space of information.

The resulting four regions characterize four different kinds of interaction, which place distinct demands on their media vehicles or tools. Separate tools have been developed in each region, usually as isolated solutions, but the need to apply them to the wide spectrum of electronic documents and in coordinated form is causing them to converge. Thus, we can describe groupware in terms of the following categories representing three of the mentioned four regions. These are [4]:

- Communications or messaging (notably E-mail)
- Collaboration or conferencing (notably forums or "bulletin board" systems that organize messages into topical "threads" of group discussion, maintained in a shared database)
- Coordination or workflow and transactions (applying predefined rules to automatically process and route messages)

Figure 1 shows the distinct forms of electronic media supported by the groupware, the kinds of interactions they support, and how they are converging [4]. Intergroupware is just groupware applied with the flexibility to support multiple interacting groups, which may be open or closed, and which may share communications selectively, as appropriate (as in an extranet). It should be noted that Lotus Notes is one of the approaches to implement intergroupware, and its success is based on the recognition of the fact that while these mentioned categories have distinct characteristics, they can only be served effectively by a unified platform that allows them to interact seamlessly.

2.2. Open Application Standards for Creating Extranets

Broad use of the Internet technology in general and development of extranets in particular is now supported by the existence of *the open application standards* that offer a range of features and functionality across all client and

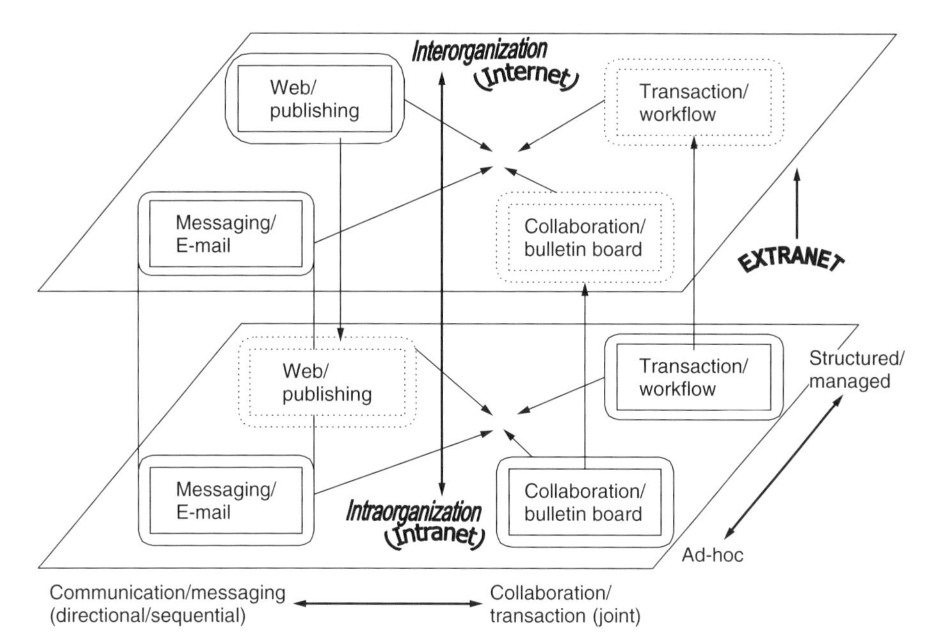

Figure 1. Relations between extranets and intergroupware. (*Source*: Teleshuttle Corporation, (4), 1997).

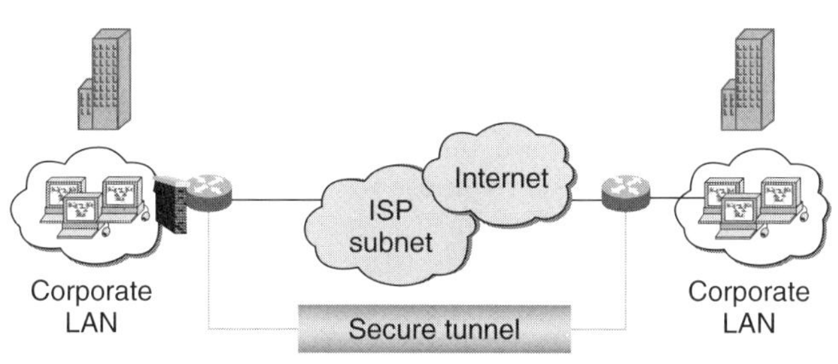

Figure 2. Site-to-site VPN. (*Source*: Core Competence, (14), 2001).

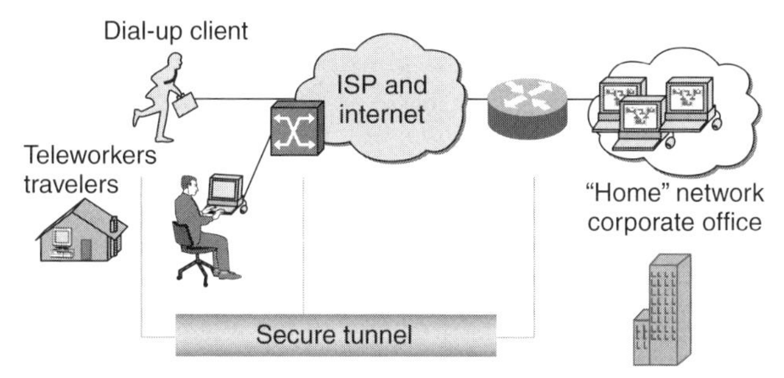

Figure 3. Remote access VPN. (*Source*: Core Competence, (14), 2001).

server platforms. Among these are the following groups of standards:

- Platform-independent content creation, publishing, and sharing of the information: HTML and HTTP
- Platform-independent software development as well as creation and deployment of distributed objects: Java, JavaScript, Common Object Request Broker Architecture (CORBA)
- Platform-independent messaging and collaboration (E-mail, discussion, and conferencing capabilities): Simple Mail Transfer Protocol (SMTP), Internet Message Access Protocol (IMAP), Multipurpose Internet Mail Extensions (MIME), Secure MIME (S/MIME), Network News Transport Protocol (NNTP), Real-Time Protocol (RTP)
- Directory and security services, network management capabilities: Lightweight Directory Access Protocol (LDAP), X.509, Simple Network Management Protocol (SNMP)

Netscape Communications has established a partnership with other companies who have agreed on a collection of standards and "best practices" for use in extranet deployment and the creation of *Crossware*. For enterprises, this offers two significant benefits:

1. An assurance of interoperability among products from multiple vendors.
2. A virtual roadmap for efficient implementation of an extranet.

Netscape's partners have committed to support the following Internet standards: LDAP, X.509 v3, S/MIME, vCards, JavaSoft, and EDI INT. Together, these standards create a comprehensive infrastructure that enables Crossware applications to interoperate across the Internet and the intranets of business partners, suppliers, and customers.

They also serve to provide a secure environment that supports much more than simple exchange of HTML pages between enterprises. In fact, the standards agreed upon by Netscape's partners represent by far the most secure and the best supported open standards technology. These standards are briefly described in the following sections.

LDAP intelligent directory services store and deliver contact information, registration data, certificates, configuration data, and server state information. These services provide support for single-user logon applications and strong authentication capabilities throughout the extranet. Key benefits include:

- Users can search for contact information across enterprises, partners, and customers using the same interface and protocols as internal corporate directories.
- A standard format for storage and exchange of X.509 digital certificates allows single-user logon applications and secure exchange of documents and information via S/MIME.
- Replication over open LDAP protocol allows secure distribution of directory data between enterprises.
- Enables extranet applications that rely on fast and flexible queries of structure information.

X.509 v3 digital certificates provide a secure container of validated and digitally signed information. They offer strong authentication between parties, content, or devices on a network including secure servers, firewalls, email, and payment systems. They are a foundation for the security in S/MIME, object signing, and Electronic Document Interchange over the Internet (EDI INT). Digital certificates can be limited to operate within an intranet or they can operate between enterprises with public certificates coissued by the company and a certification authority such as VeriSign. Certificates surpass passwords in providing strong security by authenticating identity, verifying message and content integrity, ensuring privacy, authorizing access, authorizing transactions, and supporting nonrepudiation.

Key benefits include:

- Digital certificates eliminate cumbersome login and password dialog boxes when connecting to secure resources.
- Each party can be confident of the other's identity.
- Digital certificates ensure that only the intended recipient can read messages sent.
- Sophisticated access privileges and permissions can be built in, creating precise levels of authority for Internet transactions.

S/MIME message transmission uses certificate-based authentication and encryption to transmit messages between users and applications. S/MIME enables the exchange of confidential information without concerns about inappropriate access.

vCards provide a structured format for exchanging personal contact information with other users and applications, eliminating the need to retype personal information repeatedly.

JavaSoft's signed objects allow trusted distribution and execution of software applications and applets as part of an extranet. With signed objects, tasks can be automated and access to applications and services within the extended network can be granted based on capability. A digital certificate is used with a signed object to authenticate the identity of the publisher and grant appropriate access rights to the object.

EDI INT provides a set of recommendations and guidelines that combine existing EDI standards for transmission of transaction data with the Internet protocol suite. By using S/MIME and digital signatures, EDI transactions between enterprises can be exchanged in a secure and standard fashion.

Thus, open standards provide the most flexible, efficient, and effective foundation for enterprise networking.

3. NETWORK SECURITY ISSUES

Exposing of the organization's private data and networking infrastructure to the Internet crackers definitely increases concerns of the network administrators about the security of their networks [5]. It has been very well expressed that "security should not be a reason for avoiding cyberspace, but any corporation that remains *amateurish* about security is asking for trouble" [6].

Each user who needs external access has a unique set of computing and data requirements, and the solution will not and should not be the same for all [7]. Six basic extranet components can be identified as well as some specialized and hybrid solutions [7]: (1) access to the external resources; (2) Internet protocol (IP) address filtering; (3) authentication servers; (4) application layer management; (5) proxy servers; and (6) encryption services. Each is sufficient to initiate business communications, but each carries different performance, cost, and security. Namely, security (as recent statistics shows [8]) is the main issue preventing organizations from establishing extranets.

The term *security* in general refers to techniques for ensuring that data stored in a computer cannot be read or compromised. Obvious security measures involve *data encryption* and *passwords*. Data encryption by definition is the translation of data into a form that is unintelligible without a deciphering mechanism. A password, obviously, is a secret word or phrase that gives a user access to a particular program or system. Thus, in the rest of this section we focus on the following issues: risk assessment; development of the security policy; and establishment of the authentication, authorization, and encryption.

3.1. Risk Assessment

Risk assessment procedures should answer the following typical questions:

- What are the organization's most valuable intellectual and network assets?
- Where do these assets reside?
- What is the risk if they are subjected to unauthorized access?
- How much damage could be done — can it be estimated in terms of money?
- Which protocols are involved?

3.2. Security Policy

To provide the required level of protection, an organization needs a security policy to prevent unauthorized users from accessing resources on the private network and to protect against the unauthorized export of private information. Even if an organization is not connected to the Internet, it may still want to establish an internal security policy to manage user access to portions of the network and protect sensitive or secret information. According to the FBI, 80% of all break-ins are internal.

Policy is the allocation, revocation, and management of permission as a network resource to define who gets access to what [9]. Rules and policy should be set by business managers, the chief information officer, and a security specialist — someone who understands policy writing and the impact of security decisions. Network managers can define policy for a given resource by creating an entry in access control lists, which are two-dimensional tables that map users to resources.

A firewall is an implementation of *access rules*, which are an articulation of the company's security policy. It is important to make sure that some particular firewall supports all the necessary protocols. If LANs are segmented along departmental lines, firewalls can be set up at the departmental level. However, multiple departments often share a LAN. In this case, the creation of virtual private network (VPN) for *each* person is highly advisable.

The following are recognized as basic steps for developing a security policy:

1. Assessment of the types of risks to the data will help to identify weak spots. After correction, regular assessments will help to determine the ongoing security of the environment.
2. Identification of the vulnerabilities in the system and possible responses, including operating system vulnerabilities, vulnerabilities via clients and modems, internal vulnerabilities, packet sniffing vulnerabilities, and means to test these vulnerabilities. Possible responses include encrypting data and authenticating users via passwords and biometrically.
3. Analysis of the needs of user communities:
   - Grouping data in categories and determining access needs. Access rights make the most sense on a project basis.
   - Determining the time of day, day of week, and duration of access per individual are the most typical procedures.

   Determination and assignment of the security levels can include the following, five levels:
   - Level one for top-secret data such as pre-released quarterly financials or a pharmaceutical firm's product formula database
   - Level two for highly sensitive data such as the inventory positions at a retailer
   - Level three for data covered by nondisclosure agreements such as six month product plans
   - Level four for key internal documents such as a letter from the CEO to the staff
   - Level five for public domain information

   To implement this security hierarchy, it is recommended that firewalls be placed at the personal desktop, workgroup, team, project, application, division, and enterprise level.
4. Writing the policy.
5. Development of a procedure for revisiting the policy as changes are made.
6. Writing an implementation plan.
7. Implementation of the policy.

3.3. Authentication, Authorization, Encryption

When it comes to the security aspects of teleworking and remote access, there is a tension between the goals of the participants in all aspects of information technology security. Users want access to information as quickly and as easily as possible, whereas information owners want to make sure that users can access only the information they are allowed to access. Security professionals often find it difficult to reduce this tension because of the demands of users in a rapidly changing business world. Smartcards may provide the solution to this problem [10].

Encryption requires introduction of the key management/updating procedure. Encryption can be implemented:

- *At the application*: Examples of this are Pretty Good Privacy (PGP) and Secure Multipurpose Internet Mail Extensions (S/MIME), which provide encryption for email.
- *At the client or host network layer*: The advantage of this approach is that it will provide extra protection for the hosts that will be in place even if there is no firewall or if it is compromised. The other advantage is that it allows distribution of the burden of processing the encryption among the individual hosts involved. This can be done at the client with products such as Netlock (see [11]), which provides encryption on multiple operating system platforms at the IP level. A system can be set up so that it will accept only encrypted communications with certain hosts. There are similar approaches from Netmanage and FTP Software.
- *At the firewall network layer*: The advantage to this approach is that there is centralized control of encryption, which can be set up based on IP address or port filter. It can cause a processing burden on the firewall, especially if many streams must be encrypted or decrypted. Many firewalls come with a feature called virtual private network (VPN). VPNs allows encryption to take place as data leave the firewall. It must be decrypted at a firewall on the other end before it is sent to the receiving host.
- *At the link level*: The hardware in this case is dedicated solely to the encryption process, thus offloading the burden from a firewall or router. The other advantage of this method is that the whole stream is encrypted, without even a clue as to the IP addresses of the devices communicating. This can be used only on a *point-to-point link*, as the IP header would not be intact, which would be necessary for routing.

Products such as those manufactured by Cylink [12] can encrypt data after they leave the firewall or router connected to a WAN link.

Extranet routers combine the functions of a VPN server, an encryption device, and a row address strobe [13]. The benefits of using the extranet routers include the network's ability to build secure VPNs and tunnel corporate Internet protocol traffic across public networks like the Internet. Management is much easier than it is in a multivendor, multidevice setup. Expenses are significantly lower because there is no need for leased lines.

3.4. Secure Virtual Private Networks

The goal of all VPN products is to enable deployment of logical networks, independent of physical topology [14]. That is the "virtual" part — allowing a geographically distributed group of hosts to interact and be managed as a single network, extending the user dynamics of LAN or workgroup without concern as to physical location.

Some interpret private simply as a "closed" user group — any virtual network with controlled access. This definition lets the term *VPN* fit a wide variety of carrier services, from traditional frame relay and ATM networks to emerging MPLS-based networks. Others interpret private as "secure" virtual networks that provide confidentiality, message integrity, and authentication among participating users and hosts. We focus on secure VPNs.

VPN topologies try to satisfy one of the three applications.

1. *Site-to-site connectivity*. Private workgroups can be provided with secure site-to-site connectivity, even when LANs that comprise that workgroup are physically distributed throughout a corporate network or campus. Intranet services can be offered to entire LANs or to a select set of authorized hosts on several LANs, for example, allowing accounting users to securely access a payroll server over network segments that are not secured. In site-to-site VPNs, dedicated site-to-site WAN links are replaced by shared network infrastructure — a service provider network or the public Internet.
2. *Remote access*. Low-cost, ubiquitous, and secure remote access can be offered by using the public Internet to connect teleworkers and mobile users to the corporate intranet, thus forming a Virtual Private Dial Network (VPDN). In remote access VPNs, privately operated dial access servers are again replaced by shared infrastructure — the dial access servers at any convenient ISP POP.
3. *Extranet or business-to-business services*. Site-to-site and remote access topologies can also be used to provide business partners and customers with economical access to extranet or business-to-business (B2B) services. In extranet and B2B VPNs, a shared infrastructure and the associated soft provisioning of sites and users makes it possible to respond more rapidly and cost-effectively to changing business relationships.

It is a current trend to outsource such VPN applications to a commercial service provider (e.g., regional ISP, top-tier network access provider, public carrier, or a service provider specializing in managed security services). In such outsourced VPNs, the service provider is responsible for VPN configuration (provisioning) and monitoring. Service providers may locate their VPN devices at customer premises or at the carrier's POP.

3.5. Use of Tunnels for Enabling Secure VPNs

Secure VPN applications are supported by secure, network-to-network, host-to-network, or host-to-host tunnels — virtual point-to-point connections. VPN tunnels may offer three important security services:

- Authentication to prove the identity of tunnel endpoints
- Encryption to prevent eavesdropping or copying of sensitive information transferred through the tunnel
- Integrity checks to ensure that data are not changed in transit

Tunnels can exist at several protocol layers.

Layer 2 tunnels carry point-to-point data link (PPP) connections between tunnel endpoints in remote access VPNs. In a compulsory mode, an ISP's network access server intercepts a corporate user's PPP connections and tunnels these to the corporate network. In a voluntary mode, VPN tunnels extend all the way across the public network, from dial-up client to corporate network. Two layer 2 tunneling protocols are commonly used:

- The point-to-point tunnel protocol (PPTP) provides authenticated, encrypted access from Windows desktops to Microsoft or third-party remote access servers.
- The IETF standard Layer 2 Tunneling Protocol (L2TP) also provides authenticated tunneling, in compulsory and voluntary modes. However, L2TP by itself does not provide message integrity or confidentiality. To do so, it must be combined with IPsec.

Layer 3 tunnels provide IP-based virtual connections. In this approach, normal IP packets are routed between tunnel endpoints that are separated by any intervening network topology. Tunneled packets are wrapped inside IETF-defined headers that provide message integrity and confidentiality. These IP security (IPsec) protocol extensions, together with the Internet Key Exchange (IKE), can be used with many authentication and encryption algorithms (e.g., MD5, SHA1, DES, 3DES). In site-to-site VPNs, a security gateway — an IPsec-enabled router, firewall, or appliance — tunnels IP from one LAN to another. In remote access VPNs, dial-up clients tunnel IP to security gateways, gaining access to the private network behind the gateway.

Companies with "email only" or "web only" security requirements may consider other alternatives, such as Secure Shell (SSH, or SecSH). SSH was originally developed as a secure replacement for UNIX "r" commands (rsh, rcp, and rlogin) and is often used for remote system administration. But SSH can actually forward any application protocol over an authenticated, encrypted client–server connection. For example, SSH clients can securely forward POP and SMTP to a mail server that is running SSH. SSH can be an inexpensive method of providing trusted users with secure remote access to a single application, but it does require installing SSH client software.

A far more ubiquitous alternative is Netscape's Secure Sockets Layer (SSL). Because SSL is supported by every web browser today, it can be used to secure HTTP without adding client software. SSL evolved into IETF-standard Transport Layer Security (TLS), used to "add security" application protocols like POP, SMTP, IMAP, and Telnet. Both SSL and TLS provide digital certificate authentication and confidentiality. In most cases, SSL clients (e.g., browsers) authenticate SSL servers (e.g., e-Commerce sites). This is sometimes followed by server-to-client subauthentication (e.g., user login). SSL or TLS can be a simple, inexpensive alternative for secure remote access to a single application (e.g., secure extranet "portals").

Table 2 shows a comparison of these alternatives. Combining approaches is also possible and, to satisfy some security policies, absolutely necessary. L2TP does not provide message integrity or confidentiality. Standard IPsec does not provide user-level authentication. The Windows 2000 VPN client layers L2TP on top of IPsec to satisfy both of these secure remote access requirements. On the other hand, vanilla IPsec is more appropriate for site-to-site VPNs, and SSL is often the simplest secure extranet solution.

3.6. Summary of the Weak Points and Security Hazards

Table 2 provides a summary of the weak points in the system security, identifies and shows how these problems can be addressed, and suggests technical solutions for them. Additional information for various platforms can be obtained in Refs. 15–19.

4. COST OF RUNNING WEBSITES

Evolutionary scale of Websites suggested by the Positive Support Review Inc. of Santa Monica, California [20] includes the following:

1. *Promotional*: A site focused on a particular product, service, or company. Cost: $300,000 to $400,000 per year (17–20% on hardware and software, 5–10% on marketing, and the balance on content and servicing).
2. *Knowledge-based*: A site that publishes information that is updated constantly. Cost: $1 to $1.5 million annually (20–22% on hardware and software, 20–25% on marketing, and 55–60% on content and servicing).
3. *Transaction-based*: A site that allows surfers to shop, to receive customer services, or to process orders. Cost: $3 million per year (20–24% on hardware and software, 30–35% on marketing, and 45–50% on content and servicing).

A similar classification by Zona Research Inc. of Redwood City, California (cited in Ref. 21) divides Websites into:

1. *Static presence* (*"Screaming and Yelling"*). According to Zona Research, the page cost for such sites is less than $5,000. At present, the over whelming majority of Websites belong to this category.
2. *Interactive* (*"Business Processes and Data Support"*), with page costs ranging from $5,000 to $30,000. Perhaps 15 to 20% of all current Websites are in this category.
3. *Strategic* (*"Large-Scale Commerce"*), with dynamic pages that cost more than $30,000 each to produce

Table 2. VPN Protocol Comparison

Feature	L2TP	IPsec with IKE	SSL/TLS
System-Level Authentication	Control Session: Challenge/Response	Mutual Endpoint Authentication: Preshared Secret Raw Public Keys, Digital Certificates	Server Authentication: Digital Certificates
User-Level Authentication	PPP Authentication: PAP/CHAP/EAP	Vendor Extensions: XAUTH, Hybrid, CRACK, etc.	Client Sub Authentication: Optional
Message Integrity	None (use with IPsec)	IP Header & Payload: IPsec AH or ESP, Keyed Hash, HMAC-MD5, or SHA-1	Application Payload: Keyed Hash, MD5 or SHA-1
Tunnel Policy Granularity	Network Adapter: Tunnels all packets in PPP session, bidirectional	Security Associations: Unidirectional policies defined by IP address, port, user id, system name, data sensitivity, protocol	Application Specific
Data Confidentiality	None (use with IPsec)	IP Header & Payload: IPsec ESP, DES-CBC, 3DES, other symmetric ciphers	Application Stream: RC4, RC2, DES, 3DES, Fortezza
Compression	IPPCP	IPPCP	LZS

Source: Core Competence, (14), 2001.

and maintain. Currently, less than 0.5% of all Websites are in this category.

Thus, electronic commerce sites are not toys, and before entering these *WWWaters* organization must develop a clear idea about current and strategic investments to this part of business.

4.1. ISDN Cost Model

The mobile workers should look at ISDN as an important technology for building extranet infrastructure because of its ability to support flexible access. The ISDN cost model should consider the following:

- Service fees from local telecom for each location (installation + monthly + per minute)
- Long-distance charges, if applicable
- Cost of equipment (NT1 + TA, NT1 + bridges, NT1 + routers, etc.)
- Cost of Internet Service Provider (ISP) services, if applicable

Thus, planning a budget for a month, for example, will include \$30 a month + 3 cents per minute per channel. Therefore, it is \$3.60 per hour for two B connections. If the user needs to be connected three hours per day, 20 days per month, it will cost \$16 + \$30 = \$246 for a month of service, just for the local telecom portion of your ISDN connection. Long-distance fees and ISP charges naturally would be added on the top.

4.2. Mobile Connection Cost Model

Mobile workers can consider wireless communications as another option that can be highly cost-effective, but its costs generally are *higher* than wireline communications. It should be mentioned that with packet data the modem occupies the radio channel only for the time it takes to transmit that packet. In ordinary data networks, users are usually billed for the amount of data they send. In contrast, for data communication over a wireless link, there is established a *circuit connection* and *payment is based on the duration* of the call just as with an ordinary voice call. The per-minute charges are usually the same.

Wireless modems are complex electronic devices containing interface logic and circuitry, sophisticated radios, considerable central processing power, and digital signal processing. As such, they cost more than landline modems.

4.3. Cost of Downtime

Electronic commerce is difficult in case of unreliable infrastructure. In this section, we use an example from Ref. 22 to examine the cost of downtime for some consumer-oriented business, such as an airline or hotel reservation center. If customers have a choice, they will call a competitor and place their order there.

We will use an example where customer service center has a staff of 500 people, each of whom carries a burdened cost of \$25 an hour. They make an average of 60 transactions per hour and an average of three high-priced sales per hour. Hours of operation are 24 hours a day, seven days a week, 365 days a year.

In actuality, the line managers of the site, not the IT staff, should calculate the costs of downtime. This information often is not forthcoming, however. So, this example can be presented to give a general sense of the impact that downtime has on the company's finances as well as a guideline for estimation of the cost of outages.

The cost of outages in the hypothetical network with an availability rate of 99.9% is about \$500,000 a year (see Table 3 to Table 5). If hardware and software necessary to do the job have already been bought, this estimate can be considered as a guideline for the additional budget to be spent on providing redundancy. This is separate and apart from the funds required to provide a base level of network functionality. Thus, it really is not worth rushing headlong into designing a fault-tolerant network unless all parties agree on all the implications that downtime has on the operation.

Table 3. Weak Points and Security Hazards

Weak Point/Hazard	Technical Solution
Operating system/ applications on servers	Research vulnerabilities; Monitor CERT advisories; Work with vendors; Apply appropriate patches or remove services/applications; Limit access to services on host and firewall; Limit complexity
Viruses	Include rules for importing files on disks and from the Internet in security policy; Use virus scanning on client, servers, and at Internet firewall
Modems	Restrict use; Provide secured alternatives when possible (such as a dial-out only modem pool)
Clients	Unix: Same as server issues above; Windows for Workgroups, Win95, NT: Filter TCP/UDP ports 137,138,139 at firewall; Be careful with shared services, use Microsoft's Service Pack for Win95 to fix bugs
Network snooping	Use encryption; Isolate networks with switch or router
Network attacks	Internet firewall; Internal firewall or router; Simple router filters that do not have an impact on performance
Network spoofing	Filter out at router or firewall

Table 4. Cost of Outages

Downtime Percentage	0.9990
Number of hours/year	8.76
Number of employees	500
Average burdened cost	825
Idle sale	\$109,500
Impact to production	\$131,400
Opportunity lost	\$262,500
Total downtime impact	**\$503,700**

Table 5. Impact to Production

Profit Per Transaction	0.5
Transactions per hour per employee	60
Missed transactions per hour	30,000
Total missed transactions	262,800
Impact of missed transactions	$131,400

Table 6. Opportunity Lost

Profit Per Sale	20
Sales per hour per employee	3
Missed sales per hour	1,500
Total missed sales	13,140
Impact of missed sales	$262,600

5. INTRANET AND EXTRANET TRENDS

Mindbridge.com Inc. has been analyzing and describing trends typical for intranets [1]. We suggest that these finding are also applicable to extranets.

Trend 1: Shifting the focus around the customer. Enhancing and simplifying customer interactions with the supplier or dealer has become a major focus of many industries with the proliferation of the Internet. The primary focus has been to increase the success of the customer, which in turn creates a greater sense of loyalty and increases profits.

Trend 2: Automate routine tasks. No longer is it necessary to fill out and file through request forms for daily routine activities. Now employees can complete requests for products such as office supplies when they log onto the intranet or extranet site and click on what is needed. This saves not only time but also considerable money by allowing the purchasing department to order in bulk, thus acquiring a rebate on such orders.

Trend 3: Delivering information where it is needed. No longer is location an issue. Organizations can deliver information where it is needed by an upload to the network server. Communications have become streamlined and efficient as virtual teams are now connected and able to collaborate without concern to time or distance.

Trend 4: The acceptance of the intranet or extranet by top management. Now that the intranet or extranet has had a chance to prove itself, top management and many businesses are buying in. It has been identified as a critical part of the organization, and many have moved to improve and expand their existing intranet or extranet capabilities.

Trend 5: More interesting features. Web construction kits, tools that help customers create their own homepages, and tickers that track the status of major projects are just a few of the new and interesting features being added to the intranet or extranet. This should continue as the intranet or extranet becomes an increasingly vital tool for doing business.

Trend 6: Knowledge management is growing. Knowledge management is growing, which is helping company communications between parts and developing a set of tools to help the staff find out who is doing what. These tools include a project registry, a technology and a methodology library, employee profiles, and discussion areas for consultants. Building these communities creates groups who can help each other out so as not to repeat mistakes previously made or to prevent others from reinventing the wheel.

Trend 7: New business opportunities and revenues. As time continues and organizations continue to implement and enhance their intranet or extranet technologies, capturing and sharing information will become more efficient. New opportunities can arise from this effectiveness, such as learning that the organization had internal skills that were not obvious, leading it to extend its offerings.

Trend 8: Enhancing learning environments. The intranet or extranet has entered the educational arena as it has become feasible to do so with the vast numbers of adolescents online. This will empower educators to develop their own intranet or extranet pages and update them accordingly. This will increase student–teacher interaction and develop a stronger and more productive relationship. Distance learning, particularly in regard to overseas students, will also be affected, as the boundaries of the classroom walls will continue to fade.

CONCLUSION

Currently, the extranet is conceptualized as the key technology that can enable development of the third wave of electronic commerce sites. Although technical and cost advantages are of very high importance, the real significance of the extranet is that it is the first nonproprietary technical tool that can support rapid evolution of electronic commerce.

It is already clear that the Internet impacted retail sales, the use of credit cards, and various digital cash and payment settlement schemes. However, the experts predict that the real revolution will be in systems for global procurement of goods and services at the wholesale level and that the role of extranets is crucial for this. It is also expected that on a more fundamental level the extranets will stimulate the business evolution of conventional corporations into knowledge factories.

BIOGRAPHY

Algirdas Pakštas received the M.Sc. degree in radiophysics and electronics in 1980 from Irkutsk State University, Irkutsk, Russia; Ph.D. degree in system programming from the Institute of Control Sciences in 1987; and DrTech degree from the Lithuanian Science Council in 1993, respectively. He first joined the Siberian Institute of Solar-Terrestrial Physics (SibIZMIR) in 1980. At SibIZMIR he worked on the design and development of distributed data acquisition and visualization systems. Since 1987, he has been a senior research scientist and later head of the department at the Institute of Mathematics and Informatics, Lithuanian Academy of Sciences, where he has been working on software engineering for distributed computer systems. From 1991 to 1993, he worked as research fellow at the Norwegian University of Technology (NTH),

Trondheim, and from 1994 to 1998 as professor and later full professor at the Agder University College, Grimstad, Norway. Currently, Pakstas is with London Metropolitan University, London, England. He has published 2 research monographs, edited 1 book, and authored more than 130 publications. His areas of interest are communications software, multimedia communications, as well as enterprise networking and applications.

BIBLIOGRAPHY

1. Mindbridge, (2000). Intranet WhitePapers. [Online]. Mindbridge.com Inc. *http://www.mindbridge.com/whitepaperinfo.htm* [2001, August 2].
2. R. H. Baker, *Extranets: The Complete Sourcebook*, McGraw-Hill Book Company, New York, 1997.
3. P. Q. Maier, Implementing and supporting extranets, *Information-Systems-Security* **7**(4): 52–59 (1997).
4. R. R. Reisman, (1997, March 21). Extranets and intergroupware: a convergence for the next generation in electronic media-based activity. [Online]. Teleshuttle Corporation. *http://www.teleshuttle.com/media/InterGW.htm* [2001, August 2].
5. R. Herold and S. Warigon, Extranet audit and security, *Computer Security Journal* **14**(1): 35–40 (1998).
6. J. Martin, *Cybercorp: the New Business Revolution*, Amacom Book Division, New York, 1996.
7. S. Trolan, Extranet security: what's right for the business?, *Information-Systems-Security* **7**(1): 47–56 (1998).
8. T. Lister, Ten commandments for converting your intranet into a secure extranet, *UNIX Review's: Performance Computing* **16**(8): 37–39 (1998).
9. R. Thayer, Network security: locking in to policy, *Data Communications* **27**(4): 77–8 (1998).
10. D. Birch, Smart solutions for net security, *Telecommunications* **32**(4): 53–54, 56 (1998).
11. Netlock Version 1.4.1, (1997). [Online]. Interlink Computer Sciences, Inc. *http://www.interlink.com/NetLOCK/* [1998, September 21].
12. Cylink Corporation, (1998). Global Network Security Products. [Online]. *http://www.cylink.com/products* [2001, August 2].
13. E. Roberts, Extranet routers: the security and savings are out there, *Data Communications* **27**(12): 9 (1998).
14. L. Phifer, (2001, April 12). VPNs: Virtually Anything? [Online]. Core Competence. *http://www.corecom.com/html/vpn.html* [2002, May 17].
15. S. Castano, ed., *Database Security*, Addison Wesley Publishing Co. - ACM Press 1995.
16. R. Farrow, *Unix Systems Security*, Addison-Wesley Publishing Co 1991.
17. TruSecure Media Group (2000). TruSecure publications. [Online]. *http://www.trusecure.com/html/tspub/index.shtml* [2001, August 2].
18. ALC Press, (2001) Linux boot camp. [Online]. *http://alcpress.com/* [2001, August 2].
19. S. A. Sutton, *Windows NT Security Guide*, Addison Wesley Developers Press 1995.
20. V. Junalaitis, The true cost of the Web, *PC Week* **18**: 85 (Nov. 1996).
21. E. Shein, Natural selection, *PC Week* **14**: E2 (Oct. 1996).
22. B. Walsh, (1996). Fault-Tolerant networking. [Online]. CMP Net. *http://techweb.cmp.com/nc/netdesign/faultmgmt.html* [1997, September 21].

IP TELEPHONY

Matthew Chapman Caesar
University of California at Berkeley
Berkeley, California

Dipak Ghosal
University of California at Davis
Davis, California

1. INTRODUCTION

IP telephony systems are designed to transmit packetized voice using the *Internet Protocol* (IP). IP telephony networks can offer video, high-quality audio, and directory services, in addition to services traditionally offered by the public switched telephone network (PSTN). Enterprises implement IP telephony networks to avoid the cost of supporting both a circuit switched network and a data network. Individual users save money by using IP telephony software to bypass long distance charges in the PSTN. IP telephony makes use of statistical multiplexing and voice compression to support increased efficiency over circuit-switched networks. However, many IP networks do not support quality-of-service (QoS) guarantees. Care must be taken to design these networks to minimize factors such as packet delay, jitter, and loss that decrease the quality of the received audio [29].

IP telephony services can be implemented over campus networks and wide-area networks (WANs). In a campus, a converged network with IP telephony services can take the place of a private branch exchange (PBX). These networks usually consist of one or more Internet telephony gateways (ITGs) to interface with the PSTN and a set of intelligent telephony devices offering users connectivity to the ITG. Campus networks can be interconnected over a WAN to more efficiently utilize leased lines and to bypass tariffs imposed by regulatory agencies.

The functionality of IP allows for IP telephony systems to offer a wide variety of services in addition to those offered by the PSTN. From a telephony provider's perspective, IP telephony allows integration of data, fax, voice, and video onto a single network resulting in cost savings. Converged networks eliminate the need to purchase equipment for different types of traffic, and derive increased efficiency by allowing the bandwidth to be shared among the various traffic types. From a user's perspective, the increased intelligence of the network endpoints allows for next-generation services to be implemented and quickly deployed [4,5,8]. Examples of these services include caller ID, higher quality of sound, video telephony, unified messaging, Web interfaces to call centers, virtual phone lines, and real-time billing. The increased capabilities of the endpoints allow for better user interfaces than can be supported in the PSTN. Cable providers could deploy an interactive set-top box to offer directory services and caller ID on the television screen [30]. Unlike the PSTN, new services can be deployed without making expensive, time-consuming modifications to the network. In addition, third parties may deploy services on an IP telephony service provider's network.

Several difficulties can arise in deploying IP telephony networks:

1. Delay, loss, and jitter can decrease audio quality in IP networks. Audio quality can be improved by increasing network bandwidth, implementing QoS enhancements inside the network, applying protocol enhancements such as forward error correction (FEC) to send redundant data, and using strategies to limit and load balance congestion in the network.
2. Some standards for IP telephony are complex, partially developed, or vague. Many companies are resorting to proprietary protocols that limit interoperability. Thirdly, in order to recover the cost of deployment and recover their huge sunken costs in PSTN equipment, providers need to implement competitive and economically feasible pricing strategies.
3. It is not clear how IP telephony will be regulated. IP telephony traffic may be unregulated, tariffed, or even outlawed by national governments [19].

This article gives an overview of IP telephony. The architecture and the related protocols are discussed in Section 2. Section 3 describes several common deployment scenarios. Delivering good audio quality over best effort IP networks is a challenging task, and Section 4 highlights the key issues. Section 5 discusses design issues of the various entities in the IP telephony architecture. Section 7 gives an overview of security considerations, while Section 8 lists regulatory agencies and standardization bodies guiding and accelerating the deployment of IP telephony solutions.

2. ARCHITECTURE AND PROTOCOLS

A typical IP telephony architecture is shown in Fig. 1. The key network elements include the *terminal*, the *IP telephony gateway* (ITG), *the multipoint control unit* (MCU), and the *gatekeeper*. Terminals may be a specialized device such as an IP phone, or a personal computer (PC) equipped with IP telephony software. Terminals form the endpoints of an IP telephony call. The ITG provides a bridge between the IP network and the PSTN, which uses the Signaling System 7 (SS7) protocol to set up and tear down dedicated circuits to service a call. A gatekeeper is used to support admission control functionality, while the MCU enables advanced services such as multiparty conference calls.

The two key standards for real-time voice and video conferencing are (1) *H.323*, which is developed by the International Telecommunications Union's Telecommunication Standardization Sector (ITU-T) [14]; and (2) *Session Initiation Protocol* (SIP), which is developed by the Internet Engineering Task Force (IETF) [15]. There is some overlap between these two standards and combinations of the two may be used. Many vendors are moving to support both these protocols to increase interoperability, and several application programming interfaces (APIs) have been developed to speed development of IP telephony systems [3,27,28].

2.1. Functional Entities

2.1.1. H.323. H.323 operations require interactions between several different components, including terminals, gateways, gatekeepers, and multipoint control units.

2.1.1.1. Terminals. IP telephony terminals are able to initiate and receive calls. They provide an interface for a user to transmit bidirectional voice, and optionally video or data: (1) the terminal must use a compressor/decompressor (codec) to encode and compress voice for

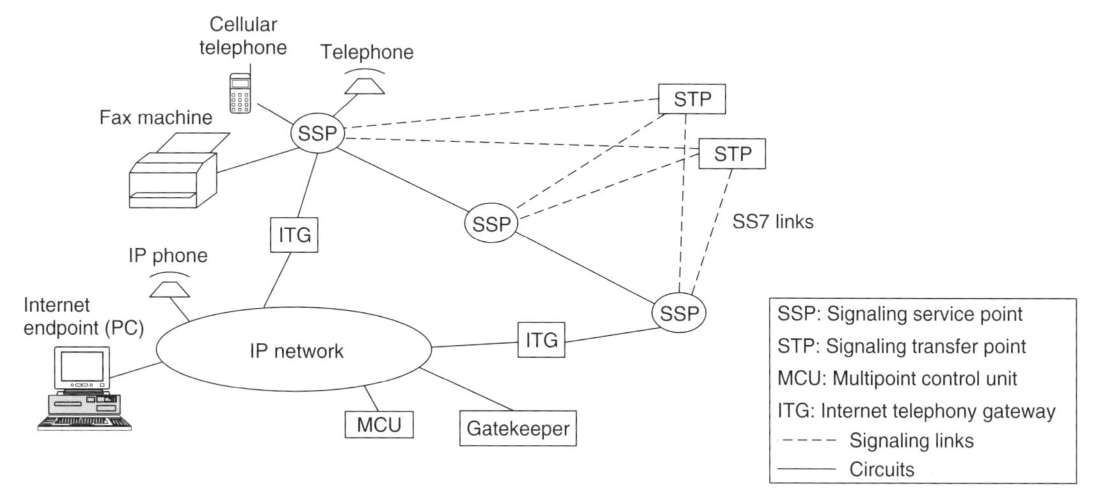

Figure 1. IP telephony architecture.

transport across the network—the receiving host must decode and decompress the voice for playback to the user; (2) terminals must perform signaling to negotiate codec type, data transfer speed, and perform call setup procedures; and (3) the terminal may provide an interface to choose the remote host and services desired. The interface may also display real-time feedback as to the connection quality and price charged.

H.323 supports several codecs for voice and video in the terminal, including G.711, G.723.1, and G.722. Codecs vary in bit rate, processing delay, frame length and size, memory and processor overhead, and resulting sound quality [21].

H.323 networks can support a wide variety of end systems. A PC may be configured to act as an IP telephony end system by installing appropriate client software. Many of these clients may be downloaded free of charge, but may require multimedia hardware such as a microphone, speakers, and possibly a videocamera. Specialized sound cards are available that allow the use of a standard telephone in place of a microphone. Performance may be improved and software cost reduced by installing a card to perform voice encoding in hardware. PC-less solutions are also available; specialized IP phones are manufactured that can make voice calls across a campus data network. Standard PSTN telephones may be equipped with an adapter to support similar functionality.

2.1.1.2. Gateway. A gateway connects IP networks to other types of networks, such as the PSTN: (1) it uses a signaling module to translate signaling information such as call setup requests from one network to another, (2) it uses a media gateway to perform translations between different data encoding types, and (3) it allows for real-time control of the data stream through the use of a media controller. The signaling module and the media controller are sometimes collectively referred to as the *signaling gateway*. Home users may purchase telephony cards to transform their PC into a gateway. More commonly, gateways are sold as specialized devices and deployed in service provider networks.

For example, a gateway might bridge the PSTN network to an H.323 network by translating PSTN Signaling System 7 (SS7) into H.323 style signaling, and encoding PSTN voice. Endpoints communicate with the gateway using H.245 and Q.931. The PSTN user dials a phone number to connect to the gateway. Address translation may be performed by requesting a name or extension from the user, and forming a connection to the appropriate IP address. An IP user may call a PSTN user by requesting the gateway to dial the appropriate phone number.

Telecommunications companies (telcos) are often hesitant to expose their SS7 network because of its critical nature. A preferred deployment method is to install the signaling gateway at the telco premises and allow it to control media gateways in data service provider networks [7,21].

2.1.1.3. Gatekeeper. A gatekeeper is responsible for managing and administrating the initiated calls in the H.323 network. It is often implemented as part of a gateway. Each gatekeeper has authority over a portion of the H.323 network called a *zone*. It performs admissions control of incoming calls originating in the zone based on H.323 guidelines and user defined policies. These policies may control admission based on bandwidth usage, network load, or other criteria. The gatekeeper is also responsible for address translation from alias addresses such as an H.323 identifier, URL (Uniform Resource Locator), or email address to an IP address. The gatekeeper maintains a table of these translations, which may be updated using the registration–admission–status (RAS) channel. The gateway also monitors the bandwidth and can restrict usage to provide enhanced QoS to other applications on the network.

Several enhancements may be made to gatekeepers to improve utility:

1. It may act as a proxy to process call control signals rather than passing them directly between endpoints.
2. Calls may be authorized and potentially rejected. Policies may be implemented that reject a call based on a user's access rights, or to terminate calls if higher priority ones come in.
3. The gatekeeper may send a billing system call details on call termination.
4. The gatekeeper may perform call routing to collect more detailed call information, or to route the call to other terminals if the called terminal is not available. The gatekeeper may perform load balancing strategies across multiple gatekeepers to avoid congestion.
5. The gatekeeper may limit the bandwidth of a call to less than what was requested to decrease network congestion.
6. The gatekeeper may keep a database of user profiles for directory services. Additional supplementary services such as call park/pickup may also be implemented with a gatekeeper.

2.1.1.4. Multipoint Control Unit (MCU). A *multipoint control unit* (MCU) is used to manage conference calls between 3 or more H.323 endpoints. It may be implemented as a standalone unit, as part of the gateway, or as part of the terminal. It performs mixing, transcoding, and redistribution operations on a set of media streams.

2.1.2. SIP. The SIP architecture [15] is composed of terminals, proxy servers, and redirect servers. Unlike H.323, terminals are the only required entities. Terminals have functionality similar to that of an H.323 terminal. Servers are used to implement directory services. Redirect servers inform the calling SIP terminal the location of the called party, and the terminal may then connect directly to that location. Proxy servers act as an intermediary for the connection, and forward the call setup request to a new location. SIP allows any codec registered with the Internet Assigned Numbers Authority (IANA) to be used, in addition to the ITU-T codecs supported in H.323.

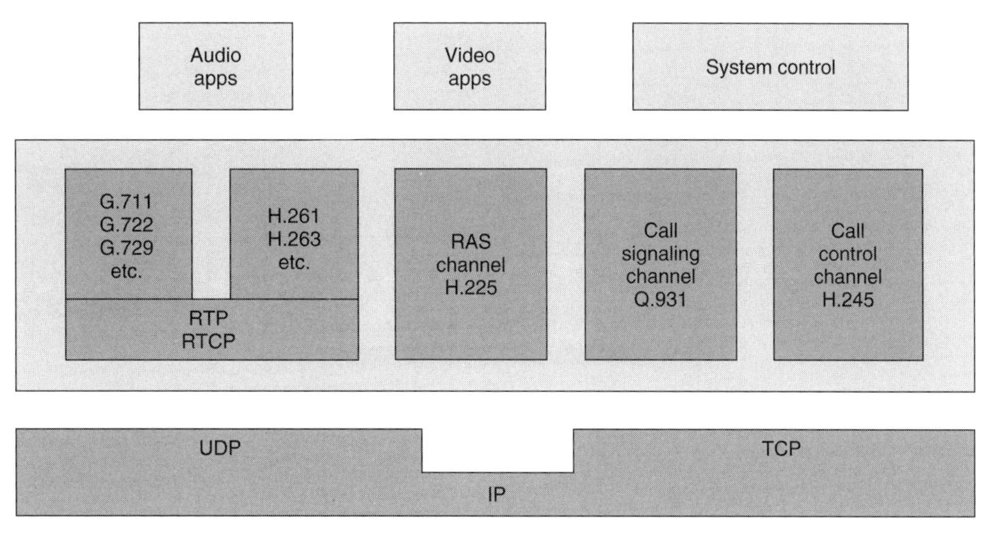

Figure 2. The H.323 protocol architecture.

2.2. Protocols

2.2.1. H.323. H.323 is an umbrella specification developed by the ITU-T which defines several protocols for IP telephony. The overall protocol architecture is shown in Fig. 2 [1]. *Q.931* is specified for call signaling, which is used to establish a connection between two terminals or between a terminal and a gateway; Q.931 implements traditional telephone functionality by exchanging signals for call setup and termination. *H.245* signaling is used for controlling media between endpoints. The control messages can be used to exchange capabilities, start and terminate media streams, and send flow control messages. *H.225* RAS is used for communication between a gatekeeper and an H.323 endpoint. RAS channels are established over the User Datagram Protocol (UDP) and are used to perform admission control, status queries, bandwidth control, and management of connections to gatekeepers. The *Real-Time Transport Protocol* (RTP) used to transport and encapsulate media streams between endpoints [13]. RTP provides payload type identification, sequence numbers, and timestamps. It is often used in conjunction with the Real-Time Transport Control Protocol (RTCP), a protocol that allows receivers to provide connection feedback information regarding the quality of the connection. *H.323* defines a set of codecs to be used to encode voice and video. These codecs vary by processor and memory utilization, bit rates, and resulting audio quality. At a minimum, an H.323 system must support the G.711 codec, RAS, Q.931, H.245, and RTP/RTCP. Finally, it specifies negotiation of codec type and supplementary services. An H.323 network must support reliable communication for control signaling and data, unreliable communication for the RAS channel, and voice and video data.

A sample H.323 call is shown in Fig. 3. Suppose that user A at terminal 1 (T1) wishes to contact user B at terminal 2 (T2). First, T1 sends an admission request (ARQ) over the RAS channel to the zone's gatekeeper asking for the IP address where user B can be reached. The gatekeeper chooses to accept the call by returning an admission confirm (ACF) with T2's

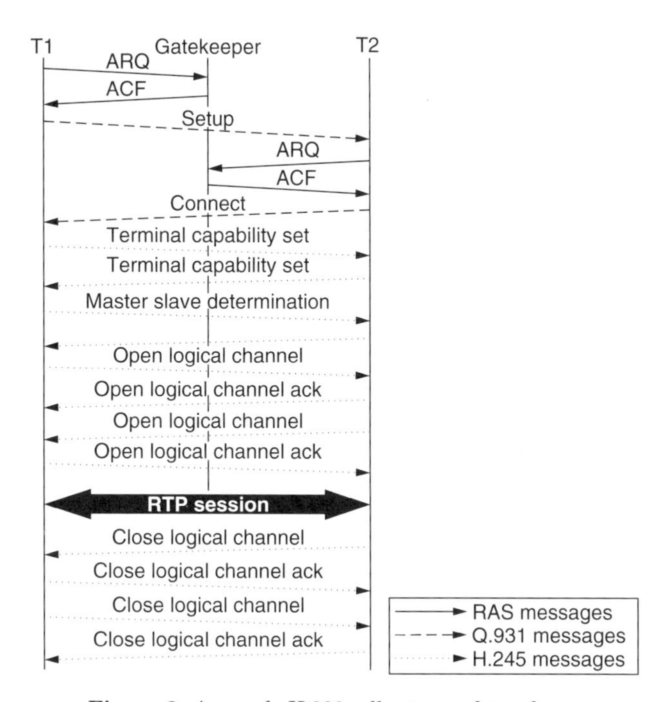

Figure 3. A sample H.323 call setup and teardown.

IP address. T1 opens a Q.931 channel to T2 over the Transmission Control Protocol (TCP) and sends a setup message. When T2 receives the message it sends an ARQ over RAS to the gatekeeper requesting permission to communicate with T2. The gatekeeper replies with an ACF and T2 returns a connect to T1 containing the TCP address it wishes to use for the H.245 channel. T1 opens an H.245 TCP channel to the specified port, and capabilities are exchanged. Either side may initiate a master/slave determination procedure or a capabilities exchange procedure. Both terminals must perform a capabilities exchange but only one master/slave determination is required. At the completion of these exchanges, logical channels may then be opened. T1 sends an *open logical channel* over the H.245 channel

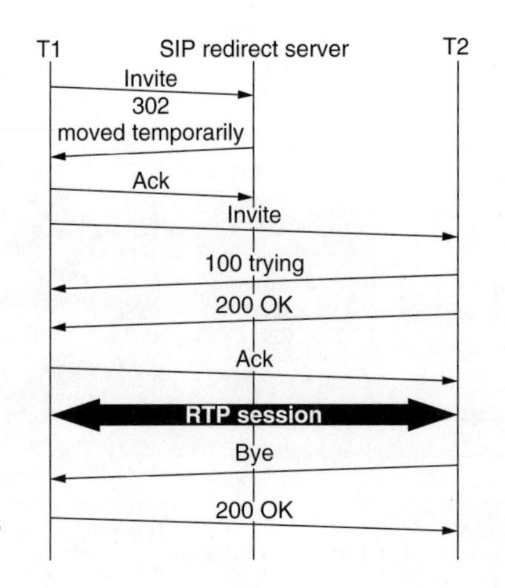

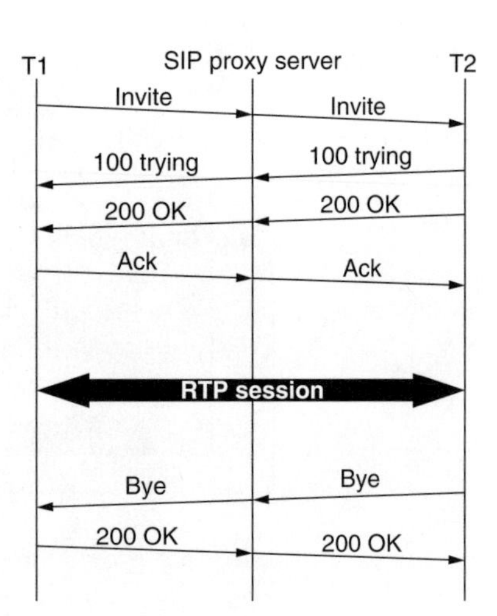

Figure 4. Sample SIP call establishment and teardown.

to request an audio channel. This message contains the UDP ports it wishes to use for the RTP audio stream and RTCP messages. T2 acknowledges this and attempts to set up a channel in the reverse direction to complete the bidirectional call, and RTP flows are established. Eventually, user B wishes to terminate the call, and causes T2 to send a *close logical channel*, which is acknowledged by T1. T1 sends a similar message to T2, and both sides send disengage requests to the gatekeeper.

2.2.2. SIP. Like H.323, SIP is a standard for signaling and control for IP telephony networks. SIP may be used in conjunction with H.323. SIP is based on HTTP, and provides call setup and termination, call control, and call data transfer. SIP is a client/server protocol, and can operate over reliable and unreliable transport protocols. SIP 2.0 contains client requests to invite a user to join a session, to acknowledge a new connection, to request the server's capability information, to register a user's location for directory services, to terminate a call, and to terminate directory searches. SIP uses the Session Description Protocol (SDP) for session announcement and invitation, and to choose the type of communication [43].

There are several differences between SIP and H.323. First, H.323 uses a binary format to exchange information. SIP is text-based, making parsers easier to build and the protocol easier to debug. Although H.323's format reduces some bandwidth and processing overhead, this overhead is minimal given that signaling and control happens infrequently during a call. SIP is considered to be more adaptable, as H.323 is required to be fully backward compatible. SIP exchanges fewer messages on connection setup, and hence can more quickly establish connections. H.323 may use only ITU-T developed codecs, whereas SIP may use any codec registered with IANA. Furthermore, SIP is considered to be more scalable, allows the servers to be stateless, provides loop detection, and has a smaller call setup delay. On the other hand, following the early publication of the standard, H.323 has been more widely supported. Furthermore, the H.323 specification provides improved fault tolerance as well as more thorough coverage of data sharing, conferencing, and video. Recent revisions of H.323 are addressing some of its shortcomings.

A sample SIP call through a redirect server is shown in Fig. 4. Suppose user A at T1 wants to contact user B at T2. T1 sends an "Invite" message to the SIP server running in redirect mode to indicate user B is being invited to participate in a session. This message usually contains a session description written in SDP including the media types T1 is willing to receive. The redirect server contacts a location server to lookup user B's current address, and returns a 302 "Moved temporarily" message containing the location of user B. T1 then sends a second invite message to T2. When T2 receives the message it sends back a 100 "trying" message and tries to alert the user. When user B picks up, T2 sends a "200 OK" message containing its capabilities. T1 responds with an acknowledgement (Ack) and RTP flows are established, completing the call. Either side may send a "Bye" message to terminate the call, which must be answered with a 200 OK.

A call handled by a SIP proxy server is also shown in Fig. 4. All messages pass through the server, and the calling terminal exchanges connection setup messages as if it is connecting directly to the callee's terminal. This scheme can cause the SIP server to become a bottleneck, limiting scalability of the IP telephony network. However, proxy servers allow for the accurate measurement of usage necessary for billing services. Furthermore, the use of a proxy server can enhance privacy by hiding each party's IP address and identification.

3. DEPLOYMENT SCENARIOS

In order to understand how an IP telephony system is constructed, it is useful to review several common scenarios for voice telephony. We will note how IP telephony can affect each of these service markets.

3.1. Data Service Provider

Data service providers can provide telephony services to increase revenue, and can provide voice with advanced supplementary services and a regulatory price advantage over the PSTN carriers. Traditionally this has been done by becoming a competitive local exchange carrier (CLEC) and installing PSTN services in rented space in a central office (CO). Today, data service providers that wish to offer voice services can implement IP telephony solutions on top of their existing data networks, eliminating the need to purchase equipment for and train personnel to run a separate PSTN network. The unreliability of IP networks requires an investment in upgrading network element software and hardware to increase availability and service quality. To succeed, data service providers must convince the public that IP telephony is not a low-quality service, reengineer their networks to give good service quality under high load, and implement reliable account control and billing systems [6].

The major players in the data service provider space include companies that deploy data lines to customer premises, and those that provide long-haul data transport services. Many cable television (CATV) providers use their hybrid fiber coaxial (HFC) networks for bidirectional data transfer to subscribers. These networks consist of an optical node that converts optical signaling from the control center transmitter to electrical signaling on coaxial cable. Instead of deploying specialized HFC telephony equipment, CATV providers are considering using Voice over IP (VoIP) to decrease cost and attract customers with a set of supplementary services [30]. Problems that need to be solved include updates to the data over cable service interface specification (DOCSIS) to support voice requirements, and forming interconnection agreements to more effectively route the IP telephony calls. Also, IP telephony support will need to be deployed through DOCSIS-compliant products at the client and inside the network. Because of the large investment in circuit switched products and the strict QoS requirements of voice traffic, many CATV providers favor initially using IP telephony for local calls. The revenue generated may then be used to implement a nationwide IP telephony network. Finally, providers will need to decide between an all-IP-based telephony solution or a hybrid IP/circuit switched solution involving a network call signaling gateway (NCSG). Other providers that deploy data connections directly to the subscriber, like digital subscriber line (DSL) and wireless data service providers, face similar issues.

Internet service providers (ISPs) are also considering implementing IP telephony solutions in their networks. ISPs provide long-haul data transport. There are several types of ISPs, including backbone ISPs and access ISPs. A typical access ISP consists of *points of presence* (POPs) to which users may connect through modem, DSL, cable, Integrated Services Digital Network (ISDN), or other types of leased lines. Access ISPs can terminate IP telephony calls at gateways placed at one or more POPs. IP telephony is expected to significantly increase the operating costs of an ISP, so it is critical that the ISP implement good pricing strategies to recover these costs. Backbone ISPs provide a long-distance network to interconnect access ISPs. Backbone ISPs can expect to generate more revenue with the introduction of IP telephony services because of the resulting increase in data traffic carrying voice flowing between POPs. There are several pure Internet telephony service providers (ITSPs) that lease capacity from existing data network to provide IP telephony service. ITSPs have recently entered the market and are facing similar issues.

3.2. Local Exchange Carrier (LEC)

Local exchange carriers (LECs) provide local telephone service. Phone lines run from the customer premises to a CO. COs are connected in a *local access and transport area* (LATA) or to an interexchange carrier (IXC) for long-distance traffic. LECs already provide voice services to a large number of customers, giving them a significant advantage over data service providers in the IP telephony market. LECs also have well-developed operations for marketing, billing, and complying with federal regulations. Implementing an IP telephony network can improve a LEC's efficiency by eliminating congestion points in and decreasing costs associated with their PSTN network.

LECs face several hurdles in implementing an IP telephony solution. LECs have a large investment in circuit switched voice services, and implementing VoIP diverts revenue away from these services. Furthermore, in the United States, LECs are required by law to subsidize certain users, and hence have higher operating costs. Also the Regional Bell Operating Companies (RBOCs) are forbidden by law from offering long distance services. This may interfere with their ability to deploy large-scale IP telephony networks. Finally, LECs must build a reliable, overprovisioned data network to handle the stringent requirements of voice traffic.

3.3. Interexchange Carrier (IXC)

IXCs are long distance carriers. They provide transport for inter-LATA calls. Like the LECs, IXCs lack an economic incentive to implement IP telephony networks, due to their large investment in PSTN equipment. IXCs must also build a large-scale management architecture for billing, provisioning, and monitoring [5] and face similar problems in developing or partnering with the provider of a data network scalable enough to handle a large number of calls.

IP telephony can provide considerable benefits to an IXC:

(1) IXCs are required by law in several countries by law to pay a minimum fixed price per minute to the LECs terminating and originating the long-distance call — costs may be reduced by using a data network to bypass the LECs on each end of the call;
(2) IP telephony allows IXCs to implement a bundled solution for nationwide data, video, and voice transport with supplementary services;
(3) cost savings may be achieved by multiplexing PSTN-originated voice calls over a nationwide data network.

3.4. IP Host to IP Host

Two users may install IP telephony software on their computers to communicate. If the call does not need to pass through the PSTN, there is no need for any sort of network intelligence, and only plain data services need to be purchased from the service provider. It is not clear how an ISP should treat calls that do not require transit on the PSTN network. Although these calls increase network load, the ISP may not be able to distinguish VoIP traffic, and hence may be unable to charge separately for IP telephony services. However, users are often willing to pay more for high-quality connections in IP telephony networks, and so ISPs may wish to charge more for increased QoS [32]. Furthermore, the IP hosts may wish for part of the call path to be routed over the PSTN, improving QoS.

4. ISSUES OF PACKET AUDIO OVER IP NETWORKS

IP networks do not provide the QoS guarantees found in the PSTN. Hence, they are inherently limited in meeting the requirements of voice transport. However, proper network and end system design can provide an infrastructure with good end-to-end voice quality.

4.1. Factors Affecting Audio Quality

Several factors affect the quality of the received audio. Delay and echo can interfere with normal conversations and make communication over the network unpleasant. Packet-switched networks suffer from packet jitter and loss, which further decrease audio quality.

4.1.1. Delay. Delay is the time for the voice signal to be encoded, sent through the network, and decoded at the receiver. One way delays greater than 150 ms cause the listener to hesitate before responding, making the mood of the conversation sound cold and unfriendly [25]. PSTN delays are typically 30 ms, whereas Internet delays tend to range from 50 to 400 ms. Delay in IP telephony networks is caused by IP network delays and signal processing delays.

There are five types of IP network delays: propagation delays, packet capture delays, serialization delays, switching delays, and queuing delays. *Propagation delay* is the time for the signal to propagate through a link, and is a function of transmission distance. *Packet capture delay* is the time required for the router to receive a complete IP packet before processing it and forwarding it towards the destination. *Serialization delay* is the time it takes to place a packet on the transmission link. *Switching delay* is the time for the router to choose the correct output port on the basis of information contained in the IP packet header. *Queuing delays* occur when packets are buffered at the router while other packets are being processed.

Signal processing delays occur in the codec algorithms implemented at the gateway or end-user systems. Codecs perform encoding and compression operations on the analog voice signal, resulting in digital signal processing (DSP) delays. These delays include the time to detect dual-tone multifrequency (DTMF) tones, detect silence, and cancel echo. DSP algorithms depend on processing an entire frame at a time, and the time to fill one of these frames before passing it to the DSP algorithm increases delay. The time for a transmitter to fill a packet with voice data is the packetization delay. Using shorter packet sizes decreases this delay, but increases network load, as more bandwidth must be expended on sending packet headers. Some coders examine the next frame to exploit any correlation with the current frame, causing lookahead delays.

4.1.2. Loss. IP network congestion can cause router buffers to overflow, resulting in packet loss. Unlike the PSTN, no end-to-end circuits are established, and IP packets from many links are queued for transmission over an outgoing link. Packets are dropped if there is no space in the queue. Packet loss interferes with the ability for the receiving host's codec to reconstruct the audio signal. Each packet contains 40–80 ms of speech, matching the duration of the phoneme, a critical phonetic unit used to convey meaning in speech. If many packets are lost, the number of lost phonemes increases, making it more difficult for the human brain to reconstruct the missing phonemes and understand the talker. Packet loss in the Internet is improving, and even long distance paths average less than 1% packet loss.

4.1.3. Jitter. Random variation in packet interarrival times at the receiver is referred to as *jitter*. Jitter occurs as a result of the variance in queueing delays at IP network routers. If this delay is too long, the packet will be considered lost by the receiver, decreasing audio quality. Packets that arrive late require complex processing in the receiver's codec. Jitter increases with both network load and distance. Jitter tends to range from near-zero values for overprovisioned local area networks (LANs) and 20 ms for intercity links up to 99 ms for international links [32].

4.1.4. Echo. Echo is the sound of the talker's voice being reflected back to the talker. The most common cause of echo in IP telephony systems is an impedance mismatch in the network, causing some of the speech energy to be reflected back to the talker. Echo may also be caused by acoustic coupling between the terminal's speaker and microphone. Echo is seldom noticed in the PSTN, because low delays allow it to return quickly to the talker's ear. The long delays of IP networks make echo more perceptible and annoying.

4.1.5. End-System Design. Several other factors can influence packet audio. Poor design of end-user equipment can accentuate background noise, decreasing intelligibility of speech. Noise can also originate from analog components or bit errors introduced by the end-user system. Finally, voice activity detectors (VADs) decrease bandwidth utilization by only sending packets when the user is speaking. Improper design of these devices can lead to inadvertently clipping parts of the audiostream.

4.2. Measurement Techniques

The quality of the received audio affects the ability of the listener to understand the talker. Hence, appropriate measurement algorithms are needed to evaluate audio quality in telephony networks. These measurement algorithms often operate by comparing an encoded waveform to an uncoded reference, and often give outputs in terms of a mean opinion score (MOS).

The ITU-T defines the *perceptual speech quality measurement* (PSQM) method to provide a relative score of the quality of a voice signal, taking into account cognitive factors and the physiology of the human ear. Output and reference waveforms must be time synchronized before they are input to the algorithm. PSQM+ was later defined as an improvement to PSQM. PSQM+ more accurately reflects audio quality in the presence of packet drops and other quality impairments. The *perceptual analysis measurement system* (PAMS) is similar to PSQM, but uses a different signal-processing model. PAMS performs time alignment to eliminate delay effects, and improves on the accuracy of PSQM+. A second alternative to PSQM is *measuring normalized blocks* (MNB). PESQ combines the cognitive modeling of PSQM with the time alignment techniques of PAMS. PESQ has been found to give, on average, the most accurate measurements of these techniques [26].

5. SYSTEM DESIGN

IP telephony service providers must implement well-designed systems to meet the strict requirements of voice traffic. Data service providers must provide networks capable of handling a large number of voice calls. Equipment vendors must implement hardware and software that is highly available and routes voice traffic with low delay. Finally, protocols and standards must be developed to support scalable, robust, and secure wide-area telephony systems.

5.1. IP Network

High availability is a prerequisite for telecommunications. The PSTN has been engineered to provide virtually uninterrupted dial tone to over a billion users worldwide. Matching this impressive achievement with a fully IP-based network will be a challenging task. If IP telephony is to be accepted as a viable alternative to the PSTN, IP network hardware and software must be engineered to provide similar reliability, scalability, and ease of use. This can be achieved through redundancy, intelligent network design, and failover to PSTN networks in cases of severe congestion or failure.

The network must provide good QoS to the voice application. Designers of campus networks should over-provision to avoid call blocking, and should mark traffic to support routing QoS enhancements [5,8]. WANs may run at near-link capacity to reap the benefits of statistical multiplexing, and should use QoS enhancements such as traffic prioritization and QoS routing to improve service [2,38]. Gateways should be placed close to terminals to decrease loss, jitter, and delay. Network topologies that duplicate voice traffic over several links can improve performance at the cost of decreased efficiency. Packet classification may also be used to provide priority service to voice traffic. QoS improvements may be made in IP networks following the Integrated Services (IntServ) or Differentiated Services (DiffServ) models, or operating over ATM. Finally, increasing link speeds decreases serialization delay, further increasing service quality.

5.2. End System

IP telephony endpoints are much more intelligent than their PSTN counterparts, allowing us to implement sophisticated algorithms to achieve improved QoS and more effective use of bandwidth.

Proper design of codecs can greatly improve service quality. Using compression decreases network utilization, but can lower audio quality. Furthermore, it increases delay, sensitivity to dropped packets, and computational overhead. Silence suppression can decrease bandwidth utilization by up to 50% by only transmitting packets when the speaker is talking. However, quality of the signal may be reduced if the transmitter does not accurately detect when speech is present. Increasing the number of frames in an IP packet improves network utilization by removing packet overhead, but increases packetization delay and sensitivity to dropped packets [21]. For example, the ITU-T's G.711 codec gives the best received audio quality under poor network conditions, while the G.723.1 codec offers much lower network utilization at the expense of audio quality. Furthermore, codec DSP algorithms may be implemented in hardware, significantly decreasing the processing latency.

Packet loss in today's IP networks is fairly common. Voice transport in IP networks usually takes place over unreliable transport mechanisms, as the overhead for reliable transport significantly increases delay. However, packet loss may be acceptable because of the ability of the human brain to reconstruct speech from lost fragments and the advanced processing capabilities at IP telephony terminals. FEC can be used to reconstruct lost packets by transmitting redundant information along with the original information [22–24]. The redundant information is used to reconstruct some of the lost original data. For example, a low-quality copy of the previous packet may be transmitted with the current packet.

Packet jitter can be alleviated by using a playback buffer at the receiver. The sender transmits RTP packets with a sequence number and a timestamp, and the receiver queues each packet to be played at a certain time in the future. This playout delay must be long enough so that the majority of packets are received by their playout times. However, too much delay will be uncomfortable to the listener. Receivers may estimate the network delay and jitter, and calculate this playout delay accordingly.

To deal with unwanted echo, echo cancelers may be placed in the PSTN, the gateway, or the end-user terminal. These devices should be placed as close as possible to the source of the echo. Long-distance networks usually implement two echo cancelers per path, one for each direction. A good echo canceler delivers good audio quality without echo, distortion, or background noise. Echo cancelers work by obtaining a replica of the echo by a linear filter and subtracting it from the original signal, then using a nonlinear processor (NLP) to eliminate the remaining echo by attenuating the signal. PSTN echo cancelers are seldom constructed to recognize the large delays from packet networks and do not attempt to perform echo cancellation in such circumstances. Echo cancellation must therefore also be implemented in the IP network or terminal.

6. RESOURCE MANAGEMENT

IP telephony networks consist of limited resources that must be allocated to a group of subscribers. These resources include IP network bandwidth, PSTN trunks, gateway functionality, and access to MCUs. IP telephony service providers must design and implement schemes to allocate resources to the users with greater demand. IntServ and DiffServ are two architectures useful for designing networks in which resources may be easily allocated. Pricing may then be used to control the way in which these resources are allocated to different users.

6.1. Resource Allocation

6.1.1. Integrated Services (IntServ). A simple way to partition resources among a group of users is to let the network decide whether a call will be admitted on the basis of factors such as user requirements and current network load. In the IntServ model, each router in the network is required to know what percentage of its link bandwidths are reserved [18]. An endpoint wishing to place a call must request enough bandwidth for the call, and the network may decide whether there are sufficient resources. If not, the call is blocked. The Resource Reservation Protocol (RSVP) is used in an IntServ framework to deliver QoS requests from a host to a router, and between intermediate routers [11]. QoS routing may be used to find the optimal path for the call to take through the IP network, as RSVP does not determine the links in which reservations are made.

Some networks, such as the Internet, are not constructed using the IntServ model. However, an administrator defined policy may be implemented at the H.323 gatekeeper or SIP server to provide admission control and resource reservation. For example, the gatekeeper may choose to block incoming calls when the total bandwidth in use by callers passing through that gateway exceeds a certain threshold.

6.1.2. Differentiated Services (DiffServ). The Differentiated Services (DiffServ) model provides different services for different classes of traffic by marking packets with the service class they belong to [17]. This class may be chosen on the basis of payload type, price charged to user, or price charged to network provider. The network allocates a set of resources for each service class, and each router provides services associated with the packet markings. Both H.323 and SIP may be used over networks architected in the IntServ or DiffServ models.

Although IntServ can provide QoS guarantees to network applications, it requires intermediate routers to store per-flow information, decreasing scalability. DiffServ provides enhanced scalability by allowing the routers to be stateless. DiffServ further improves on the IntServ model by providing more flexible service models and eliminating the need for QoS request signaling and channel setup delay. However, since DiffServ does not perform admission control, it cannot provide QoS guarantees to VoIP applications.

6.2. Pricing

It is expected that a moderate use of IP telephony will bring about an increase of 50% in an ISP's operating costs [37]. It is important that these costs be passed on to the users. The signaling architecture along with the greater intelligence of IP telephony endpoints allows service providers to formulate and implement complex pricing models that take into account both the dynamic congestion at the gateway along with the desired QoS of the call [33–35]. The price charged to the user should take into account the amount of resources consumed as well as the expected revenue lost to the provider because higher-paying users are blocked from using those resources [43].

Several pricing schemes have been proposed for IP telephony services. Although flat pricing schemes are well accepted by users and tend to dominate in most Internet services, light users subsidize heavy users, making them economically inefficient [41]. Pricing resources based on the current congestion in the network causes low paying users to back off, freeing resources for high-paying users. In congestion sensitive pricing, users may be charged per byte or per minute. In "smart market" (SM) pricing, the user puts a bid into each packet. During times of congestion, the router will drop the packet if it is below the current congestion price. Charging users a higher price for a higher-QoS connection leaves high-quality connections free for higher-paying users. In QoS-sensitive pricing, the user may be charged a higher price for closer gateways or gateways offering supplementary services. In Paris Metro Pricing [36], the network bandwidth is partitioned into several subnetworks, each priced differently. Users with low demand will tend to congregate in the lower priced partitions, increasing the service quality of more expensive partitions. The provider should choose a pricing scheme that is economically efficient and accepted by users. These pricing schemes may be implemented using the Gateway Location Protocol (GLP) to distribute price information in real time [10,12,16].

7. SECURITY ISSUES

IP telephony service providers must implement a secure system to ensure authentication, integrity, privacy, and nonrepudiation. This system must prevent eavesdropping, avoiding payment, and calls being made by nonsubscribers. Replay, spoofing, man-in-the-middle, and denial of service (DoS) attacks must also be constrained. The PSTN provides a vertical, intelligent, circuit-switched network robust against many types of attack. In the IP network, most of the intelligence is concentrated at the endpoints, making it more challenging to implement secure services. These services may be implemented in the IP telephony software, and may rely on lower-layer protocols such as IP Security (IPSec) [39] or Transport Layer Security (TLS) [40].

In the H.323 protocol architecture, security of the videostream, call setup, call control, and communications with the gatekeeper is achieved by H.235. This is achieved by having the endpoints authenticate themselves with gateways, gatekeepers, and MCUs. Encrypting data and control channels can prevent attackers from making unauthorized modifications to the information in transit, thereby providing data integrity and privacy. This can, for example, prevent attackers from using a LAN analyzer to listen to an unencrypted call on a broadcast network. Finally, nonrepudiation ensures that no endpoint can deny that it participated in a call, and is provided by the gatekeeper. Being sure of a caller's identity is critical for billing services.

SIP derives most of its security properties from HTTP [15]. In addition to these, SIP also supports public key encryption. SIP does not specify the security of user sessions, but SDP messages may be exchanged to allow the endpoints to decide on a common scheme. Data integrity and privacy can be implemented with encryption. Authentication of end users may be performed with HTTP authentication or PGP. Having callers register with a SIP server provides nonrepudiation.

8. REGULATORY AND STANDARDIZATION ISSUES

There are several standards organizations designing standards for IP telephony. The Internet Engineering Task Force (IETF) develops IP related standards for the Internet. The IETF publishes *requests for comments* (RFCs) on many areas, including IntServ, DiffServ, Gateway location, Telephony Routing over IP (TRIP), RTP, and SIP. The Asynchronous Transfer Mode (ATM) Forum deals with IP and ATM issues. Unlike IP, ATM was designed to support applications with different characteristics and service requirements, and networks may use IP over ATM networks to improve QoS. Telephony and Internet Protocol Harmonization Over Networks (TIPHON) develops specifications covering interoperability, architecture, and functionality for IP telephony networks. The Digital Audio Video Council (DAVIC) was created to address end-to-end interoperability of audiovisual information and multimedia communication. Its charter is to develop specifications for development based on Moving Picture Experts Group version 2 (MPEG-2) coding. The ITU-T is concerned with developing standards for all fields of telecommunications except radio aspects. The ITU-T began to study IP related projects in 1997, and established formal collaborations with the IETF, ATM Forum, and DAVIC.

National governments have a wide range of Internet and telephony regulatory policies. The technological advance that brought about IP telephony has moved much faster than most countries were able to change policy. Furthermore, it is not clear whether VoIP should be regulated and taxed as voice in the PSTN. Countries such as the United States, Sweden, Italy, and Russia do not restrict access to the Internet nor regulate IP telephony services. China disallows ISPs from operating international gateways or providing a nationwide infrastructure, although they may run their own switches. Canada allows ISPs to establish international gateways. However, they must pay tariffs to interface with the PSTN. Japan requires ISPs to obtain a license to establish and operate circuits and facilities. Regulatory issues are difficult to resolve and need to be reviewed in detail.

BIOGRAPHIES

Matthew Chapman Caesar completed his B.S. degree in computer science at the University of California at Davis in 1999. During 2000, he was a member of technical staff at iScale, Mountain View, California. Currently, he is enrolled in the Ph.D. program in computer science at the University of California at Berkeley. In 2001, he was awarded the National Science Foundation (NSF) Graduate Research Fellowship and the Department of Defense National Defense Science and Engineering Graduate (NDSEG) Fellowship. His research interests include Internet economics, congestion control, and real-time multimedia services.

Dipak Ghosal received his B.Tech degree in electrical engineering from Indian Institute of Technology, Kanpur, India, in 1983, his M.S. degree in computer science from Indian Institute of Science, Bangalore, India, in 1985, and his Ph.D. degree in computer science from University of Louisiana, Lafayette, in 1988. From 1988 to 1990 he was a research associate at the Institute for Advanced Computer Studies at University of Maryland (UMIACS) at College Park. From 1990 to 1996 he was a member of technical staff at Bell Communications Research (Bellcore) at Red Bank. Currently, he is an associate professor of the Computer Science Department at the University of California at Davis. His research interests are in the areas of IP telephony, wireless network, web caching, and performance evaluation of computer and communication systems.

BIBLIOGRAPHY

1. J. Kurose and K. Ross, *Computer Networking: A Top-Down Approach Featuring the Internet*, Addison-Wesley, 2000.
2. B. Li, M. Hamdi, D. Jiang, and X. Cao, QoS enabled voice support in the next generation internet: Issues, existing approaches and challenges, *IEEE Commun. Mag.* **38**(4): 54–61 (April 2000).
3. D. Bergmark and S. Keshav, Building blocks for IP telephony, *IEEE Commun. Mag.* **38**(4): 88–94 (April 2000).
4. F. Anjum et al., ChaiTime: A system for rapid creation of portable next-generation telephony services using third-party software components, *Proc. 2nd IEEE Conf. Open Architectures and Network Programming (OPENARCH)*, New York, March 1999.
5. M. Hassan, A. Nayandoro, and M. Atiquzzaman, Internet telephony: Services, technical challenges, and products, *IEEE Commun. Mag.* **38**(4): 96–103 (April 2000).
6. A. Rayes and K. Sage, Integrated management architecture for IP-based networks, *IEEE Commun. Mag.* **38**(4): 48–53 (April 2000).

7. M. Hamdi et al., Voice service interworking for PSTN and IP networks, *IEEE Commun. Mag.* (May 1999).
8. Cisco Systems, *Architecture for Voice, Video and Integrated Data*, technical white paper, *http://www:cisco:com/warp/public/cc/so/neso/vvda/iptl/avvid_wp:htm.*
9. M. Korpi and V. Kumar, Supplementary services in the H.323 IP telephony network, *IEEE Commun. Mag.* **37**(7): 118–125 (July 1999).
10. J. Rosenberg and H. Schulzrinne, Internet telephony gateway location, *IEEE INFOCOM*, San Francisco, March–April 1998.
11. R. Braden et al., *Resource reservation protocol (RSVP)—Version 1 Functional Specification*, RFC 2205, September 1997.
12. J. Rosenberg and H. Schulzrinne, *A Framework for Telephony Routing over IP*, IETF, RFC 2871, June 2000.
13. H. Schulzrinne, S. Casner, R. Frederick, and V. Jacobson, Audio-Video Transport Working Group, *RTP: A Transport Protocol for Real-Time Applications*, IETF, RFC 1889, Jan. 1996.
14. Recommendation H.323, *Visual Telephone Systems and Equipment for Local Area Networks Which Provide a Non-guaranteed Quality of Service*, International Telecommunications Union, Telecommunications Standardization Sector (ITU-T), Geneva, Switzerland, Nov. 1996.
15. M. Handley, H. Schulzrinne, E. Schooler, and J. Rosenberg, *SIP: Session Initiation Protocol*, IETF, RFC 2543, March 1999.
16. C. Agapi et al., *Internet Telephony Gateway Location Service Protocol*, IETF, Internet Draft, November 1998.
17. S. Black et al., *An Architecture for Differentiated Service*, IETF, RFC 2475, December 1998.
18. R. Braden, D. Clark, and S. Shenker, *Integrated Services in the Internet Architecture: An Overview*, IETF, RFC 1633, June 1994.
19. Organization for Economic Co-operation and Development, *OECD Communications Outlook 1999*, OECD, 1999.
20. M. Arango et al., *Media Gateway Control Protocol (MGCP)*, IETF, RFC 2705, Oct. 1999.
21. M. Perkins, C. Dvorak, B. Lerich, and J. Zebarth, Speech transmission performance planning in hybrid IP/SCN networks, *IEEE Commun. Mag.* **37**(7): 126–131 (July 1999).
22. J. C. Bolot, S. Fosse-Parisis, and D. Towsley, Adaptive FEC-based error control for Internet telephony, *Proc. IEEE INFOCOM*, New York, March 1999.
23. M. Podolsky, C. Romer, and S. McCanne, Simulation of FEC-based control for packet audio on the Internet, *Proc. IEEE INFOCOM*, San Francisco, March–April 1998.
24. J. C. Bolot and H. Crepin, Analysis and control of audio packet loss over packet-switched networks, *Proc NOSSDAV*, Durham, NC, April 1995.
25. A. Percy, *Understanding Latency in IP telephony*, technical white paper, *http://www:brooktrout.com/whitepaper/iptel_latency.htm.*
26. J. Anderson, *Methods for Measuring Perceptual Speech Quality*, technical white paper, *http://onenetworks.comms.agilent.com/downloads/PerceptSpeech2.pdf.*
27. G. Herlein, The Linux telephony kernel API, *Linux J.* **82**: (Feb. 2001).
28. Microsoft Corp., *IP telephony with TAPI 3.0 (Microsoft Corp)*, *technical white paper*, *http://www.microsoft.com/windows2000/techinfo/howitworks/communications/telephony/iptelephony.asp.*
29. Cisco, *IP telephony Solution Guide: Planning the IP telephony Network*, technical white paper, *http://www.cisco.com/warp/public/788/solution_guide/3_planni.htm.*
30. G. Cook, Jr., Taking the hybrid road to IP telephony, *Commun. Eng. Design Mag.* (Dec. 2000).
31. W. Matthews, L. Cottrell, and R. Nitzan, 1-800-CALL-HEP, presented at Computing in High Energy and Nuclear Physics 2000 (CHEP'00), Feb. 2000 (full text on website *http://www.slac.stanford.edu/pubs/slacpubs/8000/slac-pub-8384.html).*
32. J. Altmann and K. Chu, A proposal for a flexible service plan that is attractive to users and Internet service providers, *IEEE INFOCOM*, Anchorage, April 2001.
33. X. Wang and H. Schulzrinne, Pricing network resources for adaptive applications in a differentiated services network, *IEEE INFOCOM*, Anchorage, April 2001.
34. M. Falkner, M. Devetsikiotis, and I. Lambadaris, An overview of pricing concepts for broadband IP networks, *IEEE Commun. Surv.* **3**(2): (April 2000).
35. L. DaSilva, Pricing for QoS-enabled networks: A survey, *IEEE Commun. Surv.* **3**(2): (April 2000).
36. A. Odlyzko, Paris Metro Pricing for the Internet, *Proc. ACM Conf. Electronic Commerce*, Denver, Nov. 1999, pp. 140–147.
37. L. McKnight, Internet telephony: Costs, pricing, and policy, *Proc. 25th Annual Telecommunications Policy Research Conf.*, 1997.
38. A. Dubrovsky, M. Gerla, S. Lee, and D. Cavendish, Internet QoS routing with IP telephony and TCP traffic, *Proc. ICC, New Orleans, June 2000.*
39. S. Frankel, *Demystifying the Ipsec Puzzle*, Artech House, Boston, 2001.
40. T. Dierks and C. Allen, *The TLS Protocol: Version 1.0*, IETF, RFC 2246, Jan. 1999.
41. G. Huston, *ISP Survival Guide: Strategies for Running a Competitive ISP*, Wiley, New York, 1999.
42. M. Handley, and V. Jacobson, *SDP: Session Description Protocol*, RFC 2327, April 1998.
43. M. Caesar, S. Balaraman, and D. Ghosal, "A Comparative Study of Pricing strategies for IP telephony", IEEE Globecom 2000, Global Internet Symposium, San Francisco, Nov. 2000.
44. M. Caesar and D. Ghosal, "IP Telephony Annotated Bibliography", *http://www.cs.berkeley.edu/~mccaesar/research/iptel_litsurv.html.*

ITERATIVE DETECTION ALGORITHMS IN COMMUNICATIONS

KEITH M. CHUGG
University of Southern California
Los Angeles, California

1. INTRODUCTION

Iterative algorithms are those that repeatedly refine a current solution to a computational problem until an optimal or suitable solution is yielded. Iterative algorithms have a long history and are widely used in modern computing applications. The focus of this article is iterative algorithms applied to communication systems, particularly to receiver signal processing in digital communication systems. In this context, the basic problem is for the receiver to infer the digital information that was encoded at the transmitter using only a waveform observed after corruption by a noisy transmission channel. Thus, one can imagine that an iterative receiver algorithm would obtain an initial rough estimate of the transmitted block of data, and then refine this estimate. The refinement process can take into account various sources of corruption to the observed signal and various sources of structure that have been embedded into the transmitted waveform. Even within this fairly narrow context, there are a large number of distinct iterative algorithms that have been suggested in the literature.

There is one elegant and relatively simple iterative approach that has recently emerged as a powerful tool in modern communication system design. This approach, which is the focus of this article, is known by many names in the literature, including iterative detection and decoding [1], belief propagation [2–4], message-passing algorithms [5], and the "turbo principle."[1] Appreciation for the power and generality of this approach in the communications and coding literature resulted from the invention of "turbo codes" in 1993 [6,7] and the associated decoding algorithm, which is a direct application of the standard iterative algorithm addressed in this article [8]. Turbo codes are error-correction codes with large memory and strong structure that are constructed using multiple constituent codes, each with relatively low complexity, connected together by data permutation devices (interleavers). Turbo codes and similar turbo-like codes were found to perform very close to the theoretical limit, a feat that evaded coding theorist for years and was thought by many to be practically impossible.

Intrigued by the effectiveness of the iterative decoding algorithm, researchers in the field of data detection and decoding pursued several parallel directions shortly after the invention of turbo codes. One subject addressed was the general rules for the iterative algorithm, i.e., *Is there a standard view of the various iterative algorithms suggested to decode turbo-like codes?* A related issue is *In what sense and under what conditions is the algorithm optimal and, otherwise, how may it be viewed as a good approximation to optimal processing?* A third area of research was the application of the iterative decoding paradigm to other receiver processing tasks such as channel equalization, interference mitigation, and parameter tracking. By the late 1990s, all of these issues had been relatively well addressed in the literature [1]. A more recent accomplishment was the development of tools for predicting the convergence properties of the standard iterative algorithms [9,10].

Currently, there is a consensus in the literature regarding the standard iterative detection paradigm and an understanding of its practical characteristics and range of application. The transfer of this understanding in the research community to engineering practice is well under way. The use of turbo-like codes is now common is systems and standards designed after 1996. Practical implementations of other applications and a more aggressive exploitation of the potential benefits of iterative detection are currently emerging in industry. The motivation, basic ideas underlying the approach, and potential advantages are described in the following subsections.

[1] Unless emphasized otherwise, these terms are used interchangeably in this article.

1.1. Motivation for Iterative Detection

A common, generic, digital communication system block diagram is shown in Fig. 1 (a) and (b). In fact, the receiver block diagram in Fig. 1 (b) mirrors the processing performed in most practical receiver implementations. This *segregated* design paradigm allows each component of the receiver to be designed and "optimized" without much regard to the inner workings of the other blocks of the receiver. As long as each block does the job it is intended for, the overall receiver should perform the desired task: extracting the input bits.

Despite the ubiquity of the diagram in Fig. 1 (b) and the associated design paradigm, it clearly is not optimal from the standpoint of performance. More specifically, the probability of error for the data estimates is not minimized by this structure. This segregated processing is adapted for tractability — both conceptual tractability and tractability of hardware implementation. The optimal receiver for virtually any system is conceptually simple, yet typically prohibitively complex to implement. For example, consider the transmission of 1000 bits through a system of the form shown in Fig. 1 (a). These bits may undergo forward error-correction coding (FEC), interleaving, training insertion (pilots, synchronization fields, training sequences, etc.) before modulation and transmission. The channel may corrupt the modulated signal through random distortions (possibly time-varying and nonlinear), like-signal interference (co-channel, multiple access, crosstalk, etc.), and additive noise. The point is, regardless of the complexity of the transmitter and/or channel, the optimal receiver would compute 2^{1000} likelihoods and select the data sequence that most closely matches the assumed model. Intuitively, this is one of the primary advantages of digital modulation techniques — i.e., the receiver knows what to look for and there are but a finite number of possibilities. This is shown in Fig. 1 (c). Ignoring the obvious complexity problems, this requires a good model of the transmitter formatting and the channel effects. For example, the aforementioned likelihood computation may include averaging over the statistics of a fading channel model or the possible data values of like-signal interferers.

The iterative approaches described in this article are exciting because they enable receiver processing that can closely approximate the above optimal solution with feasible conceptual and implementation complexity. Specifically, data detection and parameter estimation are done using the entire global system structure. Unlike the direct approach in Fig. 1 (c), the iterative receiver in Fig. 1 (d) exploits this structure indirectly. The key

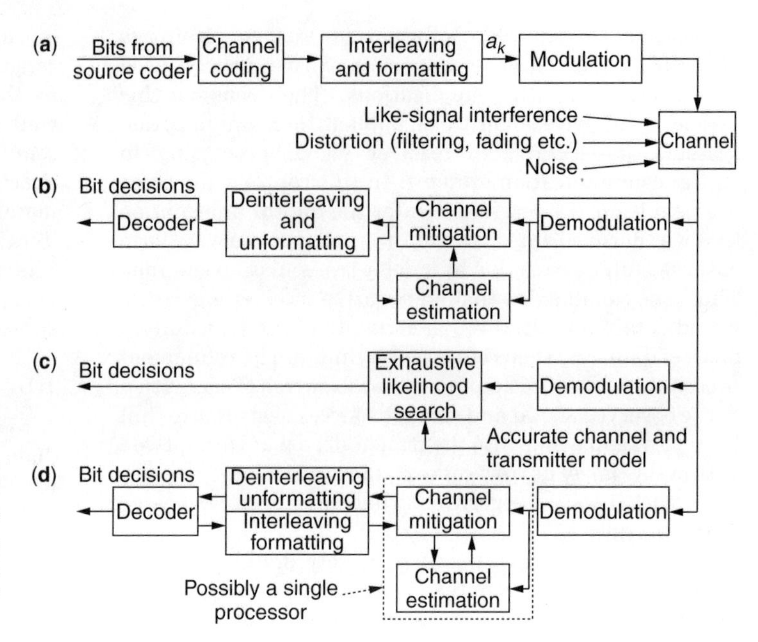

Figure 1. (**a**) A typical communication block diagram, (**b**) a traditional segmented receiver design, (**c**) the optimal receiver processing, and (**d**) a receiver based on iterative detection principles. The formatting includes insertion of framing and training information (overhead). Channel mitigation includes tasks such as equalization, diversity combining, array combining, and so on.

concept in this approach is the exchange and updating of "soft information" on digital quantities in the system (e.g., the coded modulation symbols). This concept is shown in Fig. 1 (d). The iterative detection receiver is similar to the conventional segregated design in that, for each subsystem block in the model, there is a corresponding processing block. In fact, each of these corresponding processing blocks in the receiver of Fig. 1 (c) exploits only *local* system structure (e.g., the FEC decoder does not use any explicit knowledge of the channel structure). As a consequence, the complexity of the receiver in Fig. 1 (d) is comparable to the traditional segregated design in Fig. 1 (b) [i.e., the increase in complexity usually is linear as opposed to the exponential increase in complexity associated with the optimal processing in Fig. 1 (c)].

1.2. Components of the Standard Iterative Detection Algorithm

There are several basic concepts that form the basis of the standard iterative detection paradigm. At the core of this paradigm is the exchange and update of soft information on digital variables. A digital variable is one that takes only a finite number of values, which are known to the receiver. Soft information on a digital variable is an array of numbers that gives a measure "belief" that a given conditional value is correct. For example, soft information on a binary variable $b \in \{0, 1\}$ can be represented by two numbers. If such soft information were represented as probabilities, $P[b = 0] = 0.2$ and $P[b = 1] = 0.8$ would represent the case where the current belief is that $b = 1$ is four times as likely as the zero hypothesis. Soft information can be represented in various forms and therefore is also referred to as beliefs, reliabilities, soft decisions, messages, and metrics. Soft information should be contrasted with hard decision information. For the example discussed previously, the associated hard decision information would simply be the best current guess for the value of b—i.e., $\hat{b} = 1$.

The standard iterative detection paradigm defines a way of refining soft information on the system variables, most importantly the digital data inputs to the system. The components of this approach are:

- **Modeling System Structure:** This involves specifying the structure of the system or computational problem of interest such that local dependencies are captured and the global system structure is accurately modeled. For example, in Fig. 1 (a) this is accomplished by modeling the system (global structure) as a concatenated network of subsystems (local structure). This may seem like a trivial step, but, in fact, this is possibly the most important and least well-understood aspect of the field. As will become apparent, once the model has been selected, the standard iterative detection algorithm is defined for the most part. The properties of the model selected determine in large part the complexity of the processing, the optimality or the effectiveness of a suboptimal algorithm, and many relevant implementation properties such as latency and area tradeoffs. This basic problem has provided the impetus for applying graph theory to this field. Specifically, graphical models are simply explicit index block diagrams that capture local dependencies between relevant system variables.
- **Directly and Fully Exploiting Local Structure:** For each defined local subsystem or node in the model, the associated processing performed is defined in terms of the optimal receiver for that subsystem. This leads to the notion of a *marginal soft inverse* (or, for brevity, the soft inverse) of a subsystem. The soft inverse takes in marginal soft information on each of the its digital inputs and outputs and updates this information by exploiting the local subsystem structure. The term *marginal soft information* means that the soft information is provided separately for

each variable indexed in a sequence. For example, if $\{b_0, b_1, \dots b_{1023}\}$ is a sequence of binary inputs to the system, the marginal soft information is 1024 arrays of size two as opposed to one array of size 2^{1024} (i.e., the latter is joint soft information). The subsystem soft inverse is defined based only on the local subsystem structure. Specifically, it is based on the optimal data detection algorithm under the assumption of a sequence of independent inputs and a memoryless channel corruption of the outputs. Thus, the soft inverse takes in "soft-in" information on the system inputs and "soft-in" information on the system outputs, and it produces "soft-out" information on the system inputs and "soft-out" information on the system outputs. For this reason, the soft inverse processor is sometimes referred to as a soft-in/soft-out (SISO) processor.

- **Exploit Global Structure via Marginal Soft-Information Exchange:** Since the soft inverse does not take into account the overall global system structure, this structure should be accounted for somehow if the globally optimal receiver is to be achieved or well approximated. This occurs by the exchange of soft information between soft inverse processors that correspond to subsystems having direct dependencies on the same variables. For example, if the output of one subsystem is the input to another subsystem, they will exchange soft information on these common variables. The soft information can be viewed as a method to bias subsequent soft inverse processing. Specifically, soft-in information on the system inputs plays the role of a priori probabilities on inputs and soft-in information on the system outputs plays the role of channel likelihoods.

Consider the preceding general description in the context of the example in Fig. 1. As mentioned, the concatenated block diagram defines the model. For each block in the model, there is a soft inverse processor in the iterative receiver of Fig. 1 (d). The arrows leading into each soft inverse block correspond to the soft-in information and the arrows departing represent soft-out information. The task of these processing units is to *update* the beliefs on the input and output variables of the corresponding system sub-block in Fig. 1 (a). Each sub-block processing unit will be activated several times, each time biased by a different (updated) set of beliefs.

The iterative receiver offers significant performance advantages over the segregated design. For example, suppose that a system using convolutional coding and interleaving experiences severe like-signal interference and distortion over the channel. In this case, the channel mitigation block in Fig. 1 (b) will output hard decisions on the coded/interleaved bit sequence a_k. Suppose that, given the severity of the channel, the error probability associated with these coded-bit decisions will be approximately 0.4. Deinterleaving these decisions and performing *hard-in* (Hamming distance) decoding of the convolutional code will provide a very high bit error rate (BER) (i.e., nearly 0.5).

For the receiver in Fig. 1 (d), however, the channel mitigation block produces soft-decision information on the coded/interleaved bit sequence a_k. For example, this may be thought of as two numbers $\mathrm{P}[a_k = 1]$ and $\mathrm{P}[a_k = 0]$ that represent a measure of current probability or belief that the k-th coded bit a_k takes on the value 1 or 0, respectively. Clearly, soft decisions contain more information than the corresponding hard decisions. In this example, it is possible that even though the hard decisions on a_k associated with the receiver of Fig. 1 (b) are hopelessly inaccurate, the soft-decision information contains enough information to jump-start a decoding procedure. For example, two different possible sequences of soft-decision information are shown in Table 1. Note that each of these correspond to exactly the same hard-decision information (i.e., the hard decisions obtained by *thresholding* the soft information is the same). However, the soft information in case B is much worse than that in case A. Specifically, for case A, there is a high degree of confidence for correct decisions and very low confidence for incorrect decisions. For case B, there is little confidence in any of the decisions.

A receiver of the form in Fig. 1 (d) would pass the soft information through a deinterleaver to a soft-in/soft-out decoder for the convolutional code. Thus, after activation of this decoder, one could make a decision on the uncoded bits. Alternatively, the updated beliefs on the coded bits could be interleaved and used in the role of a priori probabilities to bias another activation of the channel mitigation processing unit in Fig. 1 (d). In fact, this process could be repeated with the channel mitigation and FEC decoder exchanging and updating beliefs on the coded bits through the interleaver/deinterleaver pair. After several iterations, final decisions can be made on the uncoded bits by thresholding the corresponding beliefs generated by the code processing unit. This is what is meant by iterative detection.

Note that in this example, even though the hard-decision information on the coded bits after activating the channel mitigation processing unit is very unreliable (e.g., an error rate of 0.4), the soft information may allow

Table 1. Example of Two Sequences of Soft Information Implying the Same Hard Decisions, But Containing Very Different Soft Information. The Soft Information is Given as ($P[a_k = 0]$, $P[a_k = 1]$)

k:	0	1	2	3	4...
true data:	0	0	1	0	1
case A:	(0.99, 0.01)	(0.97, 0.03)	(0.51, 0.49)	(0.48, 0.52)	(0.03, 0.97)
case B:	(0.51, 0.49)	(0.55, 0.45)	(0.51, 0.49)	(0.48, 0.52)	(0.48, 0.52)
decisions:	0	0	0	1	1

the FEC decoder to draw some reasonable inference. For example, if the soft information is that of case A in Table 1, then the FEC decoder may update the beliefs as to overturn the unreliable (incorrect) decisions and reenforce the reliable decisions (i.e., the correct decisions in this example). Note that this updating takes into account only these marginal beliefs on the coded bits and the local code structure.

In summary, the receiver processing in Fig. 1 (d) closely approximates the performance of the optimal processing in Fig. 1 (c), with complexity roughly comparable to that of the traditional segmented design in Fig. 1 (b). It does so by performing locally optimal processing that exploits the local system structure and by updating and exchanging marginal soft information on the subsystem inputs and outputs.

1.3. Summary of Article Contents

The remainder of this article expands on the preceding material and discusses some modifications that may be required in practical applications. Specifically, in Section 2 the soft inverse operation is defined more precisely and some important special cases are presented. The standard rules for iterative detection are described in Section 3, which also contains a discussion of graphical models and a sufficient condition for optimality of the processing. A brief summary of selected applications of iterative detection is presented in Section 4.

2. SYSTEM SOFT INVERSE

Consider a system comprising a concatenated network of subsystems as illustrated in Fig. 2. To illustrate that the soft inverse is the key concept underlying iterative detection, the standard iterative detector for the system of Fig. 2 is illustrated in Fig. 3. Note that the iterative detector is specified by replacing each subsystem in the system model by the corresponding soft inverse in the iterative detector. Subsystems connected in the model correspond to soft inverse processors that directly exchange soft information messages. Given the soft inverse concept, the only remaining issues are related to scheduling, specifically in what order the soft inverses are activated and when to stop iterating. These scheduling issues are discussed in Section 3, while the soft inverse is defined more precisely in this section.

A system with digital inputs a_m and outputs x_n is shown in Fig. 4 along with the corresponding soft inverse. Two conventions for block diagrams are used. The implicit

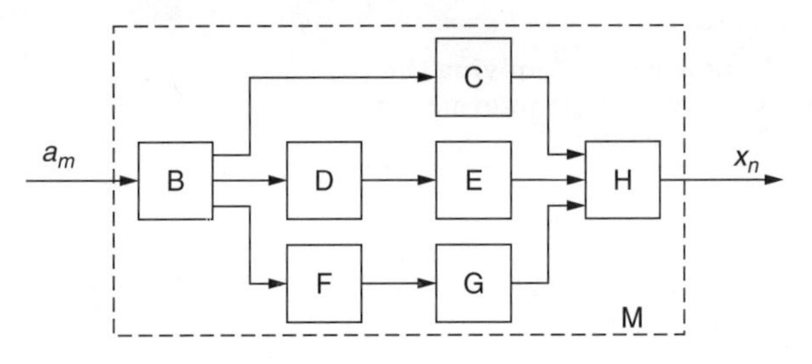

Figure 2. The block diagram of a generic concatenated network.

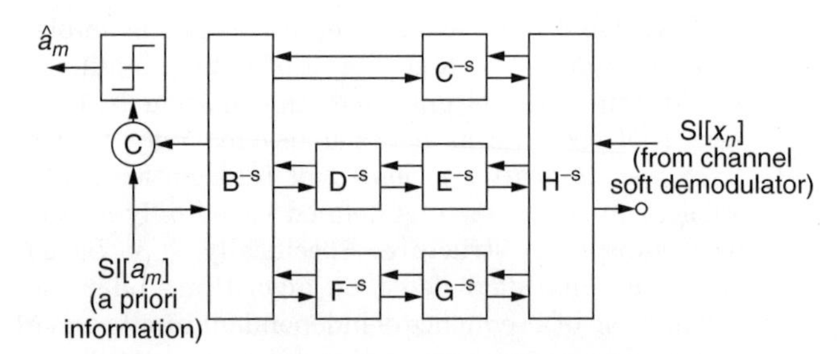

Figure 3. The iterative detector implied by the Fig. 2.

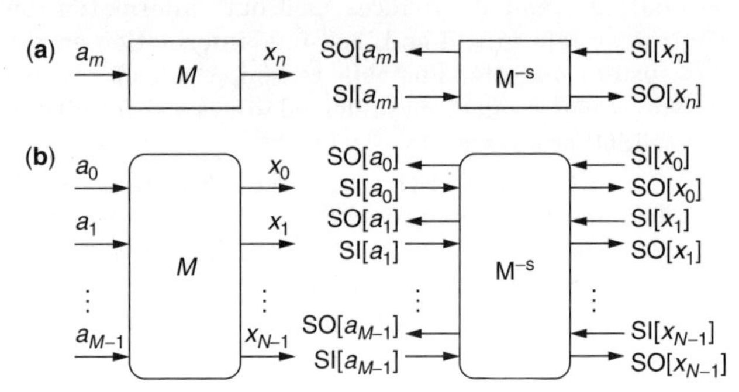

Figure 4. The marginal soft inverse of a system in the (**a**) implicit and (**b**) explicit index block diagram conventions.

(time) index convention is that most commonly used in signal processing and communications where, for example, the label a_m implies a sequence of variables. In the explicit (time) index convention, each variable in this sequence is shown explicitly. Each of these conventions has its own advantages, with the latter leading directly to the graphical modeling approaches popularized by the connection to work in computer science. The notation SI[·] and SO[·] denotes the soft-in and soft-out information, respectively. For example, if x_{10} takes on eight values, then SI[x_{10}] and SO[x_{10}] each corresponds to an array of eight numbers, much like a probability mass function.

There are many possible variations on how one represents soft information. The two most important conventions are representation in the probability domain and representation in the metric domain. In the probability domain, the soft information may be viewed as beliefs with larger numbers implying more confidence. For the special case of soft-in/soft-out information in the probability domain, the notation PI[·] and PO[·] is adopted and P[·] is used for generic probability domain soft information. Most practical software and hardware implementations are based on representation in the log-probability domain. The term *metric* is used herein to refer to the negative log of a probability domain quantity. For the special case of soft-in/soft-out information in the metric domain, the notation MI[·] and MO[·] is adopted and M[·] is used for generic metric domain soft information.

Soft inversion is a computational problem as opposed to a computational algorithm. A soft inverse computes the equivalent of a two-step process. The first step is combining of marginal soft-in information to obtain (joint) soft information on each possible system configuration. A system configuration is defined by an allowable input-output sequence pair [e.g., $(\mathbf{a}, \mathbf{x}(\mathbf{a}))$ where boldface represents the vector of variables]. In the probability domain, combining is performed by multiplication, in the same way that the joint probability is the product of the marginal probabilities for independent quantities. Thus, for each configuration one could compute

$$\begin{aligned} \mathrm{P}[\mathbf{a}, \mathbf{x}(\mathbf{a})] &= \mathrm{P}[\mathbf{x}(\mathbf{a})] \times \mathrm{P}[\mathbf{a}] \\ &= \left(\prod_{n=0}^{N-1} \mathrm{PI}[x_n(\mathbf{a})]\right) \times \left(\prod_{m=0}^{M-1} \mathrm{PI}[a_m]\right) \end{aligned} \quad (1)$$

where each element in the second equality is consistent with the configuration $(\mathbf{a}, \mathbf{x}(\mathbf{a}))$.

The second step is marginalization of this joint soft information to produce updated marginal soft information. A natural way to marginalize is to sum over all $\mathrm{P}[\mathbf{a}, \mathbf{x}(\mathbf{a})]$ consistent with a specific conditional value of a specific input or output variable. For example, to get $\mathrm{PO}[x_{10} = 3]$, one could sum $\mathrm{P}[\mathbf{a}, \mathbf{x}(\mathbf{a})]$ over all configurations with $x_{10} = 3$. The notation $z : y$ is used to denote "all z consistent with y." Thus, marginalization to obtain $\mathrm{PO}[\cdot]$ for the input and output variables is

$$\mathrm{PO}[a_m] = \left(\sum_{\mathbf{a}:a_m} \mathrm{P}[\mathbf{a}, \mathbf{x}(\mathbf{a})]\right) \div \mathrm{PI}[a_m] \quad (2a)$$

$$\mathrm{PO}[x_n] = \left(\sum_{\mathbf{a}:x_n} \mathrm{P}[\mathbf{a}, \mathbf{x}(\mathbf{a})]\right) \div \mathrm{PI}[x_n] \quad (2b)$$

Thus, for example, $\mathrm{PO}[x_{10} = 3]$ is computed by summing $\mathrm{P}[\mathbf{a}, \mathbf{x}(\mathbf{a})]$ over all configurations with $x_{10} = 3$ and then dividing by $\mathrm{PI}[x_{10} = 3]$.[2] This division converts the soft-out information to the so-called *extrinsic form*, which as will be discussed, can be viewed as a form of likelihood. The convention is to refer to processing by the marginalization and combining operators; thus, the above is *sum-product* processing.

Another reasonable marginalization operator for the probability domain is the max operator. Without altering the combining, therefore, *max-product* soft inversion is obtained by replacing the summations in Eq. (2) by max operations.

Soft inversion using sum-product or max-product marginalization and combining is based on optimal decision theory. For example, assume that $\mathrm{PI}[a_m] = p(a_m)$ is the a priori probability for the independent input variables. Further, assume that $\mathrm{PI}[x_n] = p(z_n \mid x_n)$ is the channel likelihood for system output x_n based on channel observation z_n, where the channel in memoryless. Then, using sum-product processing, $\mathrm{PO}[a_m] \times \mathrm{PI}[a_m] \equiv p(a_m \mid \mathbf{z}) = \mathrm{APP}[a_m]$ or the a posteriori probability (APP) of input a_m given the entire observation sequence $\mathbf{z}$. Thresholding this soft information therefore yields the maximum a posteriori probability (MAP) decision on a_m. So, sum-product processing under the assumption of independent inputs and a memoryless channel yields MAP symbol detection (MAP-SyD). Under the same assumptions, max-product processing yields soft information that, when thresholded, yields decisions optimal under the MAP sequence decision (MAP-SqD) criterion. In summary, while the soft inversion problem was presented as a computational problem, it is based on Bayesian decision theory for the system in isolated, ideal conditions.

As mentioned previously, for numerical stability and hardware efficiency, the soft inversion processing is almost always implemented in the log domain. Both sum-product and max-product processing can be equivalently carried out in the metric domain. For the max-product case, this is straightforward since the negative-log and max operations commute. Furthermore, in the metric domain, the product combining becomes sum combining. Thus, in the metric domain, max-product processing corresponds to *min-sum* processing of metrics that are defined as the negative-log of the probability-domain quantities defined above [e.g., $\mathrm{MI}[a_m] = -\ln(\mathrm{PI}[a_m])$]. In particular, the min-sum soft inversion problem is

$$\mathrm{M}[\mathbf{a}, \mathbf{x}(\mathbf{a})] = \mathrm{M}[\mathbf{x}(\mathbf{a})] + \mathrm{M}[\mathbf{a}] \quad (3a)$$

$$= \left(\sum_{n=0}^{N-1} \mathrm{MI}[x_n(\mathbf{a})]\right) + \left(\sum_{m=0}^{M-1} \mathrm{MI}[a_m]\right) \quad (3b)$$

$$\mathrm{MO}[a_m] = \left(\min_{\mathbf{a}:a_m} \mathrm{M}[\mathbf{a}, \mathbf{x}(\mathbf{a})]\right) - \mathrm{MI}[a_m] \quad (3c)$$

$$\mathrm{MO}[x_n] = \left(\min_{\mathbf{a}:x_n} \mathrm{M}[\mathbf{a}, \mathbf{x}(\mathbf{a})]\right) - \mathrm{MI}[x_n] \quad (3d)$$

Conversion of the sum-product to the metric domain is more complicated because the negative-log operation does not commute with the summation operator. This is handled nicely by introducing the $\min^*(\cdot)$ operation [11] as

$$\min{}^*(x, y) \triangleq -\ln\left(e^{-x} + e^{-y}\right) \quad (4a)$$

$$= \min(x, y) - \ln(1 + e^{-|x-y|}) \quad (4b)$$

$$\min{}^*(x, y, z) \triangleq -\ln\left(e^{-x} + e^{-y} + e^{-z}\right) \quad (4c)$$

$$= \min{}^*(\min{}^*(x, y), z) \quad (4d)$$

Then, the metric domain version of the sum-product soft inversion problem is the $\min^*$-sum problem obtained by replacing the min operations in (3) by $\min^*$ operations. Notice that $\min^*(x, y)$ is neither x nor y in general. Also, when $|x - y|$ is large $\min^*(x, y) \approx \min(x, y)$, which implies that the two basic marginalization approaches should yield similar results at moderate to high SNR (i.e., MAP-SqD and MAP-SyD should yield similar decisions).

The soft inversion problem as described here is actually very general and arises in many problems of interest inside and outside the area of communications. This problem is sometimes referred to as a Marginalize a Product Function (MPF) problem [5], based on the sum-product version. In the min-sum version, the problem has a very intuitive form. For example, $\mathrm{MO}[a_m = 0] + \mathrm{MI}[a_m = 0]$ is the minimum total configuration metric over all system configurations consistent with $a_m = 0$. For this reason, the

[2] Note that this division can be avoided if the term in the denominator is excluded from the combining operation.

term Minimum Sequence Metric or Minimum Sum Metric (MSM) is used to denote the minimum configuration metric consistent with a particular conditional value of a variable. The notation MSM[u] is used to denote this quantity. Thresholding MSM[a_m] yields MAP-SqD. For this reason, the min-sum version of the soft inversion problem is referred to as a shortest path problem [12].

A very useful result is that under certain conditions on the marginalization and combining operators, an algorithm to perform the soft inversion under one convention can be directly converted to another convention. Specifically, the marginalization and combining operators together with the soft information representation should form a commutative semi-ring [5,12]. In this case, any algorithm that uses only the properties of the marginalization-combining semi-ring[3] can be converted to any other marginalization-combining convention that forms a semi-ring. For example, this allows one to change a proper MAP-SyD algorithm to a MAP-SqD algorithm by simply redefining operators. This allows one to work with the most convenient form to obtain a soft inverse algorithm, then this algorithm can be converted as necessary. For example, min-sum algorithms can often be derived using pictures and intuition, while sum-product algorithms are the most straightforward to prove analytically.

Consider a toy example of a system and its min-sum soft inverse described in Table 2. This system has two inputs, $a \in \{0, 1\}$ and $b \in \{0, 1, 2, 3\}$ and one output $c \in \{0, 1\}$, with soft-in metrics as defined in the caption. Note that the best of the eight configurations is ($a = 1, b = 1, c = 1$). It follows that MSM[$a = 1$] = MSM[$b = 1$] = MSM[$c = 1$] = -5 and MO[$a = 1$] = -7, MO[$b = 1$] = -2, and MO[$c = 1$] = -1. Other values can be computed in a similar manner. For example, MSM[$b = 2$] = -2 (when $a = 0$ and $c = 1$) and MO[$b = 2$] = -4. Note that if one desires the best hard decision for a given variable, the soft-in and soft-out information should be combined and this information should be thresholded (this is the MSM information in the min-sum case and the APP information in the sum-product case). This information is sometimes referred to as *intrinsic* information, while the soft-in/soft-out information to be passed to other soft inverse modules is in extrinsic (likelihood) form.

In summary, the soft inverse of any system is defined based on the optimal receiver processing (MAP) for the system in isolation under the assumption of independent inputs and a memoryless channel. The exact form of the soft inversion problem depends on the optimality criterion (i.e., MAP-SyD or MAP-SqD) and the format used to represent the soft information (i.e., metric or probability domain). For the marginalization-combining operators discussed, the semi-ring properties hold. Thus, most soft inversion algorithms of interest can be converted by simply replacing the combining and marginalization operators.

Table 2. Toy Example of a System with Two Inputs a and b and One Output c. The Soft-In Information is MI[$a = 0$] = 0, MI[$a = 1$] = 2, MI[$b = 0$] = 0, MI[$b = 1$] = −3, MI[$b = 2$] = 2, MI[$b = 3$] = 6, MI[$c = 0$] = 0, MI[$c = 1$] = −4

a	b	c	Configuration Metric
0	0	0	$0 + 0 + 0 = 0$
0	1	0	$0 - 3 + 0 = -3$
0	2	1	$0 + 2 - 4 = -2$
0	3	0	$0 + 6 + 0 = 6$
1	0	1	$2 + 0 - 4 = -2$
1	1	1	$2 - 3 - 4 = -5$
1	2	0	$2 + 2 + 0 = 4$
1	3	0	$2 + 6 + 0 = 8$

[3] These are called semi-ring algorithms in Ref. [1].

2.1. Specific Example Sub-Subsystems

It is important to note that the soft inversion [e.g., as stated in Eq. (3)] is a computational problem rather than an algorithm. The problem statement does suggest a method of computing the soft inverse, but this brute-force approach will typically be prohibitively complex. For example, if the system in Fig. 4 has M binary inputs, it will have 2^M configurations. Computing soft information for each of these configurations and then performing the subsequent marginalization will be prohibitively complex even for moderate values of M. This brute-force method is referred to as *exhaustive combining and marginalization*. In many cases, it is possible to compute the soft inverse with dramatically less computational effort by exploiting the special structure of the system. Thus, while there is really one unique computational problem, it is useful to consider special cases of systems, find efficient algorithms for their soft inversion, and then use these as standard modules for iterative detection–based receivers.

For the implicit block diagram convention, Benedetto et al. [13] defined the marginal soft inverse for a variety of systems commonly encountered. These are shown in Fig. 5 with minor modification. Some of these are quite trivial. For example, the soft inverse of an interleaver is an interleaver/deinterleaver pair. Similarly for rate converters, no computation is required for the soft inversion.

The memoryless mapper is a mapper that maps small blocks of inputs onto outputs without memory between blocks. For example, if four bits are collected and mapped onto a 16-ary constellation, then the memoryless mapper is the proper subsystem model. The soft inverse of the memoryless mapper is computed using exhaustive combining and marginalization over the small block size (e.g., 16 configurations in the modulator example). Another example in which the memoryless mapper can be applicable is block error-correction codes. A common special case of the mapper is the broadcaster or repeater where all inputs and outputs are necessarily equal.

One of the most important subsystems is the finite state machine (FSM), which is discussed in detail in the next section. With the modules shown in Fig. 5, a large number

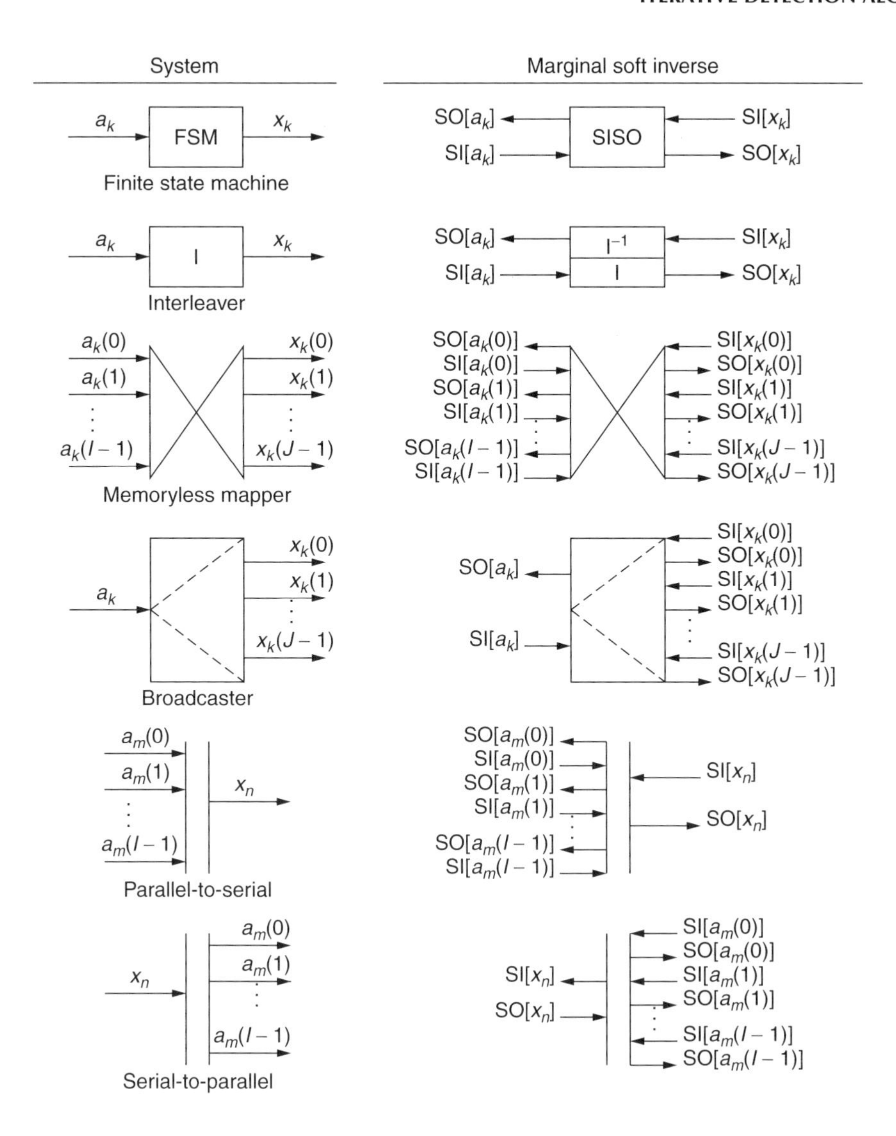

Figure 5. Several common systems and the associated marginal soft inverses in implicit index block diagrams.

of practical systems can be modeled and the corresponding standard iterative receivers defined.

2.1.1. Soft Inversions of an FSM via the Forward–Backward Algorithm. An FSM is a common model for subsystems of digital communications systems. An FSM is a system that at time k has a current state s_k that takes on a finite number of possible values. Application of the input at time k, a_k, results in the system producing output x_k and transitioning to the next state s_{k+1}. The transition at time k is denoted by t_k, which is defined by (s_k, a_k) or with some redundancy (s_k, a_k, s_{k+1}). It is common to represent an FSM by a diagram that is explicit in time and in value, the so-called trellis diagram.

Convolutional codes and related trellis-coded modulation schemes are naturally modeled as FSMs. Certain modulation formats and precoding methods such as differential encoding, line coding, and continuous phase modulation (CPM) are also routinely defined in terms of FSMs. Channel impairments are also often well modeled as FSMs. This includes, for example, the intersymbol interference (ISI) channel [14], the multiple access interference channel [15], and even channels with random fading effects [16]. So, an efficient method for soft inversion of an FSM is desirable.

Using the development of the soft inverse notion and a presumed familiarity with the Viterbi algorithm [17] or similar dynamic programming tools [12], an efficient algorithm can be derived pictorially. This is the *forward–backward algorithm* (*FBA*), which can be used to compute the MSM of the inputs and/or outputs of an FSM [18–20]. First, note that for a given transition, the FSM output x_k and input a_k are specified uniquely. Thus, each well-defined trellis transition can be assigned a transition metric of the form

$$\mathrm{M}_k[t_k] = \mathrm{MI}[a_k(t_k)] + \mathrm{MI}[x_k(t_k)]. \tag{5}$$

Second, the metric of the shortest path through a given transition t_k can be obtained if one has the metric of the shortest path to the states s_k from the left edge of the index range and the metric of the shortest path from the right edge to s_{k+1}. This concept is shown in Fig. 6, where $\mathrm{MSM}_i^j[\cdot]$ denotes the MSM using input metrics over the indices from i to j inclusive. Mathematically, the claim is that

$$\mathrm{MSM}_0^{K-1}[t_k] = \mathrm{MSM}_0^{k-1}[s_k(t_k)] + \mathrm{M}_k[t_k] + \mathrm{MSM}_{k+1}^{K-1}[s_{k+1}(t_k)] \quad (6)$$

Third, the MSM of the input a_k or output x_k can be obtained by marginalizing (minimizing in this min-sum case) over all transitions t_k consistent with those conditional values. Finally, the quantities $F_{k-1}[s_k] = \mathrm{MSM}_0^{k-1}[s_k]$ and $B_{k+1}[s_{k+1}] = \mathrm{MSM}_{k+1}^{K-1}[s_{k+1}]$ can be updated by a forward recursion and a backward recursion, respectively. The forward recursion is identical to that of the Viterbi algorithm and the backward recursion is the same, only run in reverse.

In summary, the soft inverse of an FSM can be computed via the FBA using three steps: a forward state metric recursion, a backward state metric recursion, and a completion operation that performs the marginalization over transitions to obtain the desired soft outputs. Given the transition metrics defined in (5), the steps are

$$F_k[s_{k+1}] = \min_{t_k:s_{k+1}} (F_{k-1}[s_k] + \mathrm{M}_k[t_k]) \qquad k = 0, 1, 2, \ldots, K-1 \quad (7a)$$

$$B_k[s_k] = \min_{t_k:s_k} (\mathrm{M}_k[t_k] + B_{k+1}[s_{k+1}]) \qquad k = K-1, K-2, \ldots, 0 \quad (7b)$$

$$\mathrm{MO}[a_k] = \min_{t_k:a_k} (F_{k-1}[s_k] + \mathrm{M}_k[t_k] + B_{k+1}[s_{k+1}]) - \mathrm{MI}[a_k] \quad (7c)$$

$$\mathrm{MO}[x_k] = \min_{t_k:x_k} (F_{k-1}[s_k] + \mathrm{M}_k[t_k] + B_{k+1}[s_{k+1}]) - \mathrm{MI}[x_k] \quad (7d)$$

Initialization of the forward and backward metrics is performed according to available initial edge information [1]. This is a semi-ring algorithm, so it may be directly converted to other marginalization-combining forms by simply replacing the min-sum operations appropriately. One step through a four-state trellis for the FBA is shown in Fig. 7.

In hardware implementations, it often is useful to approximate the FBA using only a portion of the soft inputs to compute a given soft output. For example,

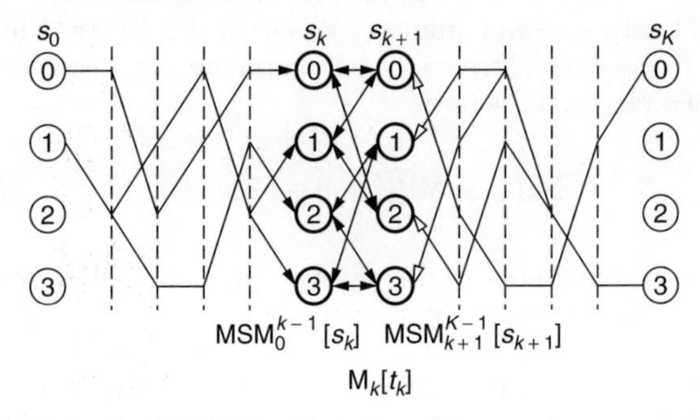

Figure 6. The MSM for a given transition may be computed by summing the transition metric and the forward and backward state metrics.

a fixed-lag algorithm uses soft-in information for times $\{0, 1, \ldots, k + D\}$ to compute the soft-out information for the input or output variables at time k. For sufficiently large lag D, no significant performance degradation will occur. A particularly attractive algorithm for hardware is the min-lag/max-lag algorithm suggested by Viterbi [21]. In this case, the lag varies with index k, but is bounded below by D and above by $2D$. This algorithm can be implemented in hardware with one forward state metric processor and two backward state metric processors in such a way that computation, memory, and speed are attractively balanced [21,22].

Finally, note that the FBA is just one algorithm for computing the soft inverse of an FSM. While the FBA solution has low complexity, it has a bottleneck in the state metric recursions (i.e., the "ACS bottleneck"). An alternative algorithm that computes the soft inverse based on a low-latency tree structure was suggested in Refs. 23 and 24.

3. STANDARD RULES FOR ITERATIVE DETECTION

The standard iterative detection technique can be summarized as follows:

- Given a system comprising a concatenated network of subsystems, construct the marginal soft inverse of each subsystem. The marginal soft inverse is found by considering the subsystem in isolation with independent inputs and a memoryless channel. Using these operators, specify an algorithm to compute the extrinsic soft outputs for the system inputs and outputs.
- Construct the block diagram of the iterative detector by replacing each subsystem by the corresponding marginal soft inverse and connecting these soft inverses accordingly. Specifically, each connection between subsystems in the system block diagram is replaced by a corresponding pair of connections in the iterative detector block diagram so that the soft-out port of each is connected to the soft-in port of the other.
- Specify an activation schedule that begins by activating the soft inverses corresponding to some subsystems providing the global outputs and ends with activation of some soft inverses corresponding to subsystems with global inputs.
- Specify a stopping criterion.
- Take the soft-inputs on global output symbols as the channel likelihoods (metrics) obtained by appropriate soft-demodulation. The soft-inputs for the global inputs are the a priori probabilities (metrics), which are typically uniform.
- At the activation of each subsystem soft inverse, take as the soft-in on the digital inputs/outputs the soft-outputs from connected subsystem soft inverses. If no soft information is available at the soft-in port, take this to be uniform soft-in information (i.e., this applies to the first activation of soft inverses that

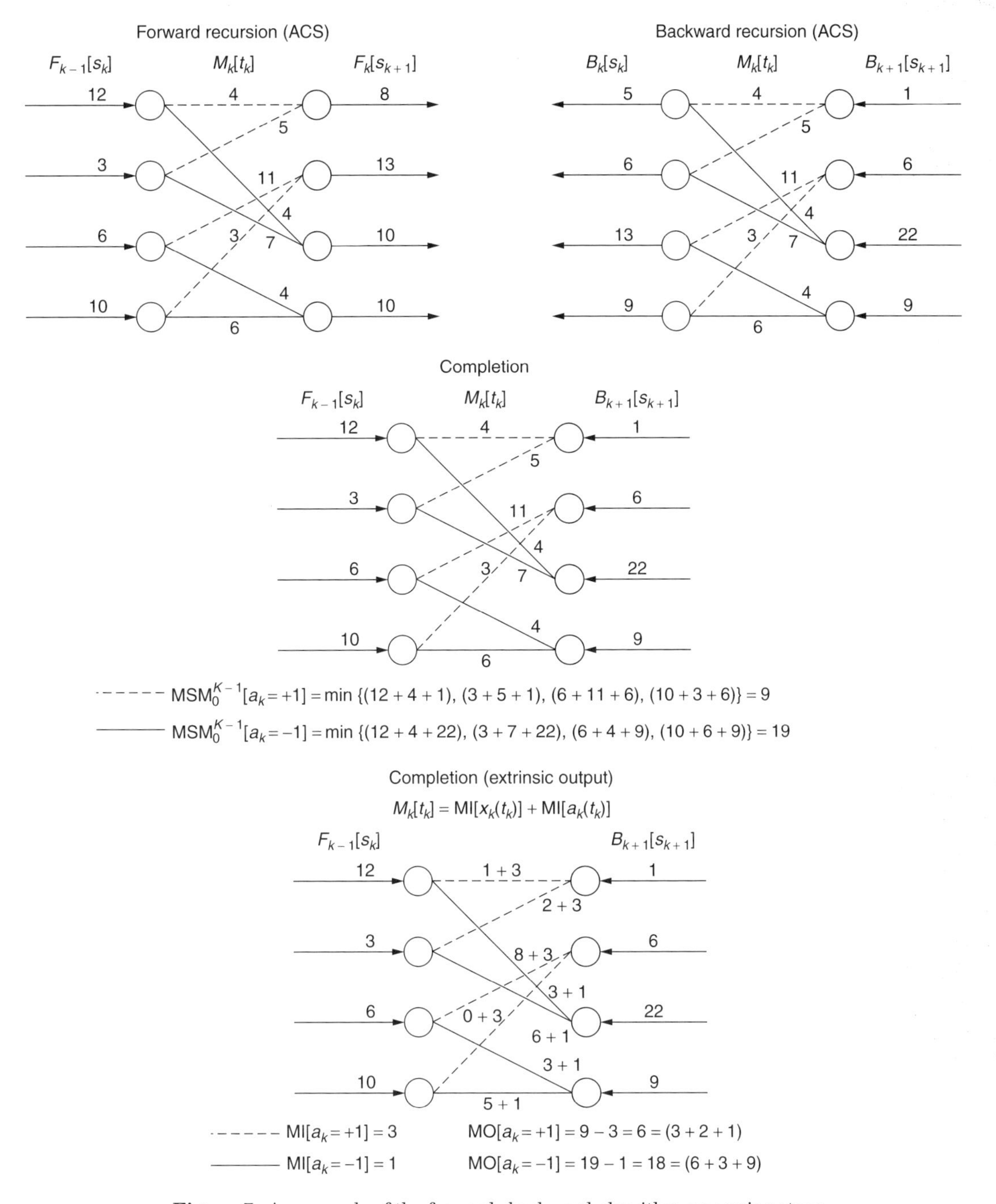

Figure 7. An example of the forward–backward algorithm processing steps.

have inputs or outputs that are internal or hidden variables).

A common stopping criterion is that a fixed number of iterations are to be performed, with this number determined by computer simulation. For example, while formal proofs of convergence for complicated iterative detectors are difficult, in most cases of practical interest, the performance improvement from iteration reaches a point of diminishing returns. Alternatively, performance may be sacrificed to reduce the number of iterations. It also is possible to define a stopping rule that results in variable number of iterations.

In most cases of practical interest, there either is a natural activation schedule or different activation schedules produce similar results. Thus, for the most part, the iterative detector is specified once the block diagram is given and the subsystem soft inverses are determined. For example, the iterative detector for the general concatenated system in Fig. 2 is shown in Fig. 3.

As a simple example of this paradigm, a simple turbo code, or parallel concatenated convolutional code (PCCC) is considered. The encoder is shown in Fig. 8 and the corresponding standard decoder is shown in Fig. 9. The blocks used in the encoder are two FSMs, a one-to-two broadcaster and a puncture/binary mapper. The

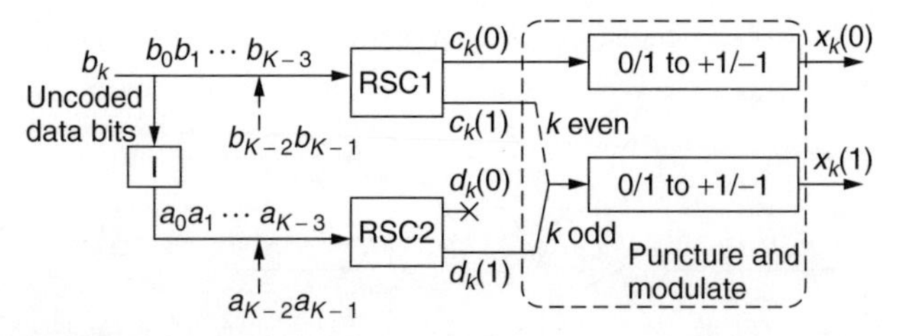

Figure 8. A rate one-half PCCC encoder with four state constituent recursive systematic convolutional codes.

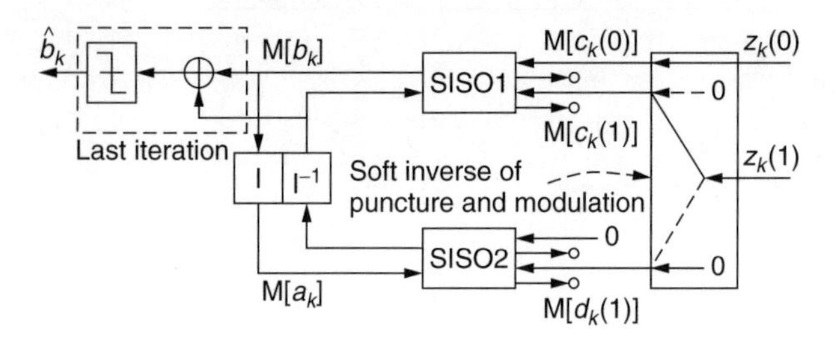

Figure 9. The min-sum iterative decoder for the PCCC in Fig. 8.

decoder exploits the special properties of a one-to-two broadcaster and efficient min-sum metric representation using normalization techniques. Even such a simple code is capable of approaching the theoretical limits within 1 dB of SNR.

3.1. Sufficient Condition for Optimality of the Standard Rules

Note that the iterative detector in Fig. 3 has the same input/output format as a soft inverse. Specifically, the detector takes in soft-in information on the global system inputs and outputs and performs an update to produce soft-out information on these same variables. Since the soft inverse is defined relative to well-established optimal receiver criteria, the natural question arises: *When does applying the standard iterative detection rules to a system model yield the global system soft inverse, and hence the optimal receiver for the global system?* Interestingly, there is a simple sufficient condition for this to occur. This is developed through the following examples.

Note that the general rules were not dependent on the implicit index convention and apply as well to explicit index diagrams. Consider, for example, the explicit index diagram for the general FSM as sown in Fig. 10 (a) in which the FSM has been decomposed or modeled by a series of small transition nodes or subsystems, each defining one transition in the trellis. Applying the standard definitions, it can be shown that the soft inverse of each of these transition nodes performs one forward state recursion step, one backward state recursion step, and a completion step for both the FSM input and output. This is illustrated in Fig. 11. As a result, running the standard iterative detection rules on the receiver shown in Fig. 10 (b), with a specific activation schedule, yields exactly the FBA-based soft inverse of the FSM. Even more remarkable, it can be shown that after some point, further activation of the soft inverse nodes does not change the soft information. In fact, the soft information will stabilize

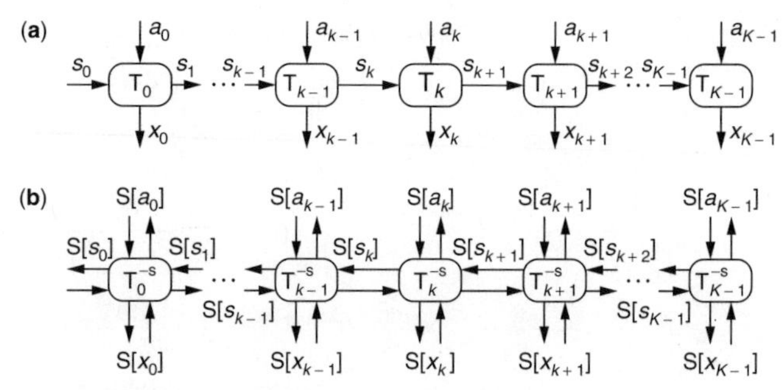

Figure 10. (**a**) An explicit index block diagram for an arbitrary FSM, and (**b**) the associated concatenated detector.

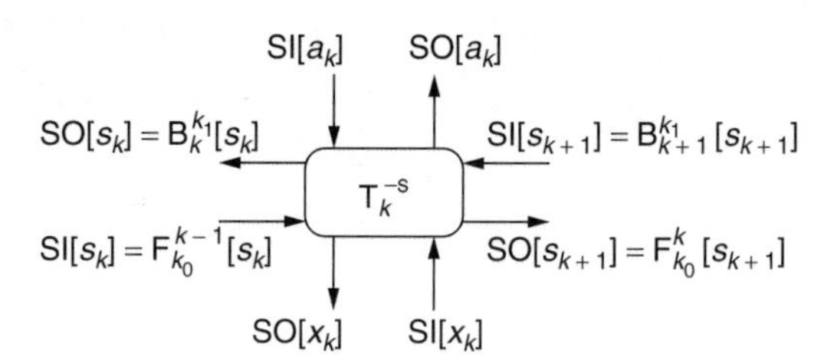

Figure 11. The soft inverse of the transition subsystem. Activation is equivalent to one update of the backward and forward state metric recursions, and completion for both the FSM input and output.

under any activation schedule as long as it satisfies some basic requirements. Intuitively, the soft-in information for each node must be passed to all other nodes before the values stabilize. So, in this example, the globally optimal solution is achieved by applying the standard rules to a concatenated system model.

This is not the case in general. For example, for the iterative decoder shown in Fig. 9, one will observe conditions where the soft information passed does not completely stabilize. Consider the explicit index diagram for the PCCC encoder of Fig. 8 as illustrated in Fig. 12. Note that this contains two FSM subgraphs of the form in Fig. 10 (a), but with connections due to the interleaver. Ignoring the directions of the arrows, one can trace a loop by following the connections in the diagram. Intuitively,

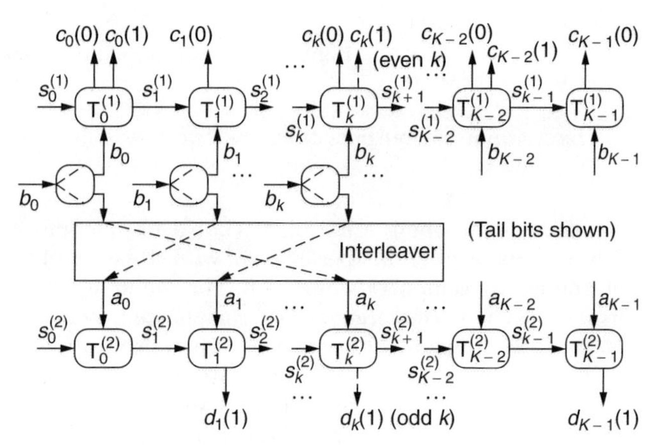

Figure 12. The explicit index block diagram of the PCCC shown in Fig. 8.

one may understand how this could compromise the optimality of the global detector. Specifically, the initial belief obtained from the channel for a particular variable may propagate through the processing and return to be "double counted" by the original processing node. This can lead to convergence to a false minimum.

In fact, it can be shown rigorously that if no cycles exist in the explicit index model (or equivalently the graphical model), then under some simple requirements on the activation schedule, the standard rules result in convergence to the globally optimal solution. This fact was known in computer science for some time [2,25]. Two remarkable facts did arise, however, from the research efforts in communications and coding. First, at a conceptual level, there really is only one algorithm. For example, the FBA and the turbo decoding algorithm are both just examples of running the standard rules on different system models [3,5,8,26]. Second, although this processing on graphical models with loops is suboptimal in general, it is highly effective in many practical scenarios. This can be motivated intuitively in a manner similar to the finite traceback approximation in the Viterbi algorithm or the fixed-lag approximation in the FBA. Specifically, if the local neighborhood of all nodes is cycle-free, then one can expect the processing to well-approximate the optimal solution.

In summary, there is one standard iterative detection algorithm that is optimal when applied to cycle-free graphs and generally suboptimal, yet very effective, when applied to graphical models with loops. One reasonable convention for terminology is to refer to the algorithm on cycle-free graphs as *message passing* and on loopy graphs as *iterative message passing*. Also note that selection of the system model is critical in determining the complexity and performance of the algorithm. For example, use of the approach on models with cycles is most effective when there exists some sparse structure in the system that can be exploited locally by proper reindexing of the variables.

3.2. Modified Rules

Despite the elegance of the single set of message-passing rules, in some applications modification of these rules is necessary or desirable. One example is when there is some uncertainty in the structure of one or more subsystems. This occurs, for example, when some channel parameters are not completely known. In this case, the input–output relation for subsystems in contact with the channel may not be completely defined. This requires consideration of the proper marginal soft inverse definition in the presence of parametric uncertainty. Typically, the theory will suggest that exhaustive combining and marginalization is required to perform soft inversion in the presence of unknown parameters. However, greedy approximations to this processing work well in practice. Such solutions typically are based on applying decision feedback or memory truncation techniques to recursive formulations of the soft inverse computation. This yields practical *adaptive SISO* modules that approximate the soft inverse. Incorporating the parameter estimation and tracking tasks into the iteration process is a powerful tool. Such *adaptive iterative detection* algorithms allow tracking of severe parameter dynamics at very low SNR. Work in this area can be found in [1,27–31].

A similar situation can arise when all parameters are known, but there is a subsystem that has a prohibitively complex soft inverse. In such cases, one can attempt to approximate the soft inverse with an algorithm of reasonable complexity. One important case is an FSM subsystem with a large number of states. One approach is to use decision feedback techniques similar to those used in reduced state sequence detectors [32]. There are a number of variations on this theme suggested for *reduced state SISO* algorithms primarily applied to ISI mitigation (equalization) [33,34]. Another approach is to use an approximate soft inverse based on a constrained receiver structure. For example, one may use a linear or decision feedback equalizer, modified to update soft information, in place of an FBA-based soft inverse [35].

Another case in which some modification of the rules can be helpful is when there are short cycles in the underlying graphical models. In such cases, it is often observed that convergence occurs quickly, but performance is poor. As is the case in many iterative algorithms, this effect can be alleviated somewhat by attempting to slow down the convergence of the algorithm. This can be accomplished in a number of ways. For example, filtering the messages over iterations so as to slow their evolution or simply scaling them to degrade the associated confidence can provide significant improvements in applications where the basic approximations break down [1].

4. APPLICATIONS AND IMPACT

Once a general view of the iterative detection algorithm is understood, applying the approach to various problems is relatively straightforward. A detailed description of these applications is beyond the scope of this article. Instead, a brief summary of the applications, the performance gains, complexity issues, and representative references is given in this section.

- **Turbo-Like Codes:** Following the invention of turbo codes, many variations on the theme were described in the literature. This includes serial concatenated convolutional codes (SCCCs) [36], low-density parity check (LDPC) codes [37,38], and product codes [39]. The most significant difference in the decoding algorithm from what has been presented herein is the use of indirect system models to perform the soft inversion. For example, it is possible to perform soft inversion using the parity check structure of a block code. This is possible because one needs only to be able to identify allowable configurations to combine and marginalize over. Decoding of LDPC codes and product codes based on high-rate block codes can be performed efficiently using this approach. Codes similar in performance and structure to LDPC codes, but with simpler encoding have also been suggested. These include the generalized repeat-accumulate codes [40,41] and parallel concatenated zig-zag codes [42], which are both based on a recursive single parity check codes and broadcasters.

Code design techniques have developed to the point where turbo-like codes are attractive alternatives to previous approaches at virtually all code rates and reasonable input block sizes (e.g., >64). For reasonable block sizes and rates, it is not unusual to achieve 3 to 8 dB of additional coding gain relative to more conventional FEC designs. Codes based on PCCCs and SCCCs are commonly adopted for standardized systems. In the coding application, code construction allows a designer to achieve *interleaver gain* as well as *iteration gain*. A system with interleaver gain will perform better as the size of the interleaver increases, while iteration gain simply means that iteration improves the performance.

- **Modulation and Coding:** Based on the knowledge of the design rules for turbo-like codes and the standard iterative detection paradigm, several common modulation and coding system designs have been demonstrated to benefit greatly from iterative processing. One example is the serial concatenation of a convolutional code, an interleaver, and a recursive inner modulator, which can be viewed as an effective, simple SCCC. This has been exploited in the case of the inner system being a differential encoder [43] and in the case of the inner modulation being CPM, which has a recursive representation [44,45]. Similarly, simple schemes such as bit interleaved coded modulation (BICM), where coded bits are interleaved and then mapped onto a nonbinary constellation, have been shown to benefit significantly from iterative decoding/demodulation [46]. The benefits in these various applications range from 1 to 6 dB of SNR improvement for typical scenarios.
- **Equalization and Decoding:** Many systems are designed with coding, interleaving, and a propagation channel that results in ISI. A number of researchers realized the applicability of the standard iterative approach to such a scenario, typically with convolutional coding and a relatively short ISI channel delay spread [47–49]. In this case, the ISI and code SISOs are both implemented using the FBA. For typical scenarios, soft-out equalization provides approximately 2.5 dB of gain in SNR over hard decision processing and iteration provides another 4 dB in SNR. If the channel is fading, then most of the iteration gain is lost when the average performance is considered. This is because the worst-case fading conditions, for which there is little iteration gain, dominate the performance [1]. In the case of severe channel dynamics, significant iteration gain will be achieved in fading if a properly designed adaptive iterative detection algorithm is used. Unless combined with a turbo-like code or equivalent modulation techniques, there is no interleaver gain in this application.
- **Interference Mitigation:** Similar to the previous application, one can use the standard iterative algorithm to perform joint decoding and like-signal interference mitigation. Specifically, considering the case where each user's data is coded and interleaved, they can be effectively separated by using an interference mitigation SISO module and a bank of code SISO modules. Utilizing the code structure is especially helpful if there is a high degree of correlation between the signals on the multiple access channel (i.e., the channel is heavily loaded). In fact, it has been demonstrated that users can be separated when the only unique feature is an interleaver pattern (see [1] and references therein).

In applications where the operating SNR is very low, there can be an advantage to using min*-sum processing over min-sum processing. This advantage typically is 0.2 to 1.0 dB for these applications, which are the turbo-like codes and similar coding-modulation constructions. For other applications, such as joint equalization and decoding or interference mitigation, there is little practical advantage to using min*-sum processing over min-sum processing.

The number of iterations required varies significantly with application as well. Generally, systems with weak local structure converge more slowly. This is the case with LDPC and similar codes that are based on single-parity-check codes. For a typical SCCC or PCCC 6 to 10 iterations will yield the majority of the gains for most practical scenarios. In many of the interference mitigation and equalization applications, most of the gains are achieved with 3 to 5 iterations.

In all cases, the complexity increase relative to a standard solution is moderate. In the case of turbo-like codes, it is not uncommon to achieve better performance with less complexity than conventional designs. In other applications, the soft inverse typically is 2 to 4 times as complex as a subsystem processor based on the segregated design paradigm. Substantially more memory is also required. While not insignificant, these increases in memory and computation requirements impact digital circuitry, which continues to experience a steady, rapid improvement in capabilities.

5. CONCLUSION

Iterating soft-decision information to improve the performance of practical digital communication systems can be motivated as an approach to approximate the optimal receiver and improve upon the performance of the segregated design. This intuitive notion can be formalized using the standard tools of communication receiver design, namely, Bayesian decision theory. This results in a standard approach that is based on the notion of the soft inverse of a system and the exchange and update of soft information. This approach, viewed generally, describes a standard algorithm for optimal or near-optimal data detection for complex systems. A sufficient condition for optimality (i.e., minimum error probability) is that the underlying graphical model describing dependencies between system variables be cycle-free. In the case where cycles exist, which is the case in many practical applications of interest, the performance of the standard approach is often extremely good. The approach has been demonstrated to improve performance substantially in a number of relevant applications. The general approach is becoming better known to researchers in the field, and the results are finding adoption in engineering practice rather quickly.

Acknowledgments

The author thanks John Proakis for his encouragement and understanding during the preparation of this article; Kluwer Academic Publishers for permission to use material from [1]; and the co-authors of [1], Achilleas Anastasopoulos and Xiaopeng Chen.

BIBLIOGRAPHY

1. K. M. Chugg, A. Anastasopoulos, and X. Chen, *Iterative Detection: Adaptivity, Complexity Reduction, and Applications*, Kluwer Academic Publishers, 2001.
2. J. Pearl, *Probabilistic Reasoning in Intelligent Systems: Networks of Plausible Inference*, Morgan Kaufmann, San Francisco, 1988.
3. R. J. McEliece, D. J. C. MacKay, and J. F. Cheng, Turbo decoding as an instance of Pearl's "belief propagation" algorithm, *IEEE J. Select. Areas Commun.* **16**: 140–152 (Feb. 1998).
4. F. Kschischang and B. Frey, Iterative decoding of compound codes by probability propagation in graphical models, *IEEE J. Select. Areas Commun.* **16**: 219–231 (Feb. 1998).
5. S. M. Aji and R. J. McEliece, The generalized distributive law, *IEEE Trans. Inform. Theory* **46**: 325–343 (March 2000).
6. C. Berrou, A. Glavieux, and P. Thitmajshima, Near shannon limit error-correcting coding and decoding: turbo-codes, in *Proc. International Conf. Communications*, (Geneva, Switzerland), pp. 1064–1070, May 1993.
7. C. Berrou and A. Glavieux, Near optimum error correcting coding and decoding: turbo-codes, *IEEE Trans. Commun.* **44**: 1261–1271 (Oct. 1996).
8. N. Wiberg, *Codes and Decoding on General Graphs*. PhD thesis, Linköping University (Sweden), 1996.
9. T. Richardson and R. Urbanke, The capacity of low-density parity-check codes under message-passing decoding, *IEEE Trans. Inform. Theory* **47**: 599–618 (Feb. 2001).
10. H. E. Gamal and J. A. R. Hammons, Analyzing the turbo decoder using the Gaussian approximation, *IEEE Trans. Inform. Theory* **47**: 671–686 (Feb. 2001).
11. P. Robertson, E. Villebrum, and P. Hoeher, A comparison of optimal and suboptimal MAP decoding algorithms operating in the log domain, in *Proc. International Conf. Communications*, (Seattle, WA), pp. 1009–1013, 1995.
12. T. H. Cormen, C. E. Leiserson, and R. L. Rivest, *Introduction to Algorithms*, Cambridge, Mass.: MIT Press, 1990.
13. S. Benedetto, G. Montorsi, D. Divsalar, and F. Pollara, Soft-input soft-output modules for the construction and distributed iterative decoding of code networks, *European Trans. Telecommun.* **9**: 155–172 (March/Apr. 1998).
14. G. D. Forney, Jr., Maximum-likelihood sequence estimation of digital sequences in the presence of intersymbol interference, *IEEE Trans. Inform. Theory* **IT-18**: 284–287 (May 1972).
15. S. Verdú, Minimum probability of error for asynchronous Gaussian multiple-access channels, *IEEE Trans. Inform. Theory* **32**: 85–96 (Jan. 1986).
16. J. Lodge and M. Moher, Maximum likelihood estimation of CPM signals transmitted over Rayleigh flat fading channels, *IEEE Trans. Commun.* **38**: 787–794 (June 1990).
17. G. D. Forney, Jr., The Viterbi algorithm, *Proc. IEEE* **61**: 268–278 (March 1973).
18. R. W. Chang and J. C. Hancock, On receiver structures for channels having memory, *IEEE Trans. Inform. Theory* **IT-12**: 463–468 (Oct. 1966).
19. P. L. McAdam, L. R. Welch, and C. L. Weber, M.A.P. bit decoding of convolutional codes, *Proc. IEEE Int. Symp. Info. Theory* (1972).
20. L. R. Bahl, J. Cocke, F. Jelinek, and J. Raviv, Optimal decoding of linear codes for minimizing symbol error rate, *IEEE Trans. Inform. Theory* **IT-20**: 284–287 (March 1974).
21. A. J. Viterbi, Justification and implementation of the MAP decoder for convolutional codes, *IEEE J. Select. Areas Commun.* **16**: 260–264 (Feb. 1998).
22. G. Masera, G. Piccinini, M. R. Roch, and M. Zamboni, VLSI architectures for turbo codes, *IEEE Trans. VLSI* **7**: (Sept. 1999).
23. P. A. Beerel and K. M. Chugg, A low latency SISO with application to broadband turbo decoding, *IEEE J. Select. Areas Commun.* **19**: 860–870 (May 2001).
24. P. Thiennviboon and K. M. Chugg, A low-latency SISO via message passing on a binary tree, in *Proc. Allerton Conf. Commun., Control, Comp.* (Oct. 2000).
25. F. V. Jensen, *An Introduction to Bayesian Networks*, Springer-Verlag, 1996.
26. F. Kschischang, B. Frey, and H.-A. Loeliger, Factor graphs and the sum-product algorithm, *IEEE Trans. Inform. Theory* **47**: 498–519 (Feb. 2001).
27. A. Anastasopoulos and K. M. Chugg, Adaptive soft-input soft-output algorithms for iterative detection with parametric uncertainty, *IEEE Trans. Commun.* **48**: 1638–1649 (Oct. 2000).
28. A. Anastasopoulos and K. M. Chugg, Adaptive iterative detection for phase tracking in turbo-coded systems, *IEEE Trans. Commun.* **49**: 2135–2144 (Dec. 2001).
29. M. C. Valenti and B. D. Woerner, Refined channel estimation for coherent detection of turbo codes over flat-fading channels, *IEE Electron. Lett.* **34**: 1033–1039 (Aug. 1998).
30. J. Garcí a-Frí as and J. Villasenor, Joint turbo decoding and estimation of hidden Markov sources, *IEEE J. Select. Areas Commun.* 1671–1679 (Sept. 2001).
31. G. Colavolpe, G. Ferrari, and R. Raheli, Noncoherent iterative (turbo) detection, *IEEE Trans. Commun.* **48**: 1488–1498 (Sept. 2000).
32. M. V. Eyuboğlu and S. U. Qureshi, Reduced-state sequence estimation with set partitioning and decision feedback, *IEEE Trans. Commun.* **COM-38**: 13–20 (Jan. 1988).
33. X. Chen and K. M. Chugg, Reduced state soft-in/soft-out algorithms for complexity reduction in iterative and non-iterative data detection, in *Proc. International Conf. Communications*, (New Orleans, LA), 2000.
34. P. Thiennviboon, G. Ferrari, and K. Chugg, Generalized trellis-based reduced-state soft-input/soft-output algorithms, in *Proc. International Conf. Communications*, (New York), pp. 1667–1671, May 2002.

35. M. Tuchler, R. Koetter, and A. Singer, Turbo equalization: principles and new results, *IEEE Trans. Commun.* **50**: 754–767 (May 2002).

36. S. Benedetto, D. Divsalar, G. Montorsi, and F. Pollara, Serial concatenation of interleaved codes: performance analysis, design, and iterative decoding, *IEEE Trans. Inform. Theory* **44**: 909–926 (May 1998).

37. R. G. Gallager, Low density parity check codes, *IEEE Trans. Inform. Theory* **8**: 21–28 (Jan. 1962).

38. D. J. C. MacKay, Good error-correcting codes based on very sparse matrices, *IEE Electron. Lett.* **33**: 457–458 (March 1997).

39. J. Hagenauer, E. Offer, and L. Papke, Iterative decoding of binary block and convolutional codes, *IEEE Trans. Inform. Theory* **42**: 429–445 (March 1996).

40. H. Jin, A. Khandekar, and R. McEliece, Irregular repeat accumulate codes, in *Turbo Code Conf.*, (Brest, France), 2000.

41. K. R. Narayanan, I. Altunbas, and R. Narayanaswami, On the design of LDPC codes for MSK, in *Proc. Globecom Conf.*, (San Antonio, TX), pp. 1011–1015, Nov. 2001.

42. L. Ping, X. Huang, and N. Phamdo, Zigzag codes and concatenated zigzag codes, *IEEE Trans. Inform. Theory* **47**: 800–807 (Feb. 2001).

43. P. Hoeher and J. Lodge, Turbo DPSK: iterative differential PSK demodulation and channel decoding, *IEEE Trans. Commun.* **47**: 837–843 (June 1999).

44. K. Narayanan and G. Stuber, Performance of trellis-coded CPM with iterative demodulation and decoding, *IEEE Trans. Commun.* **49**: 676–687 (Apr. 2001).

45. P. Moqvist and T. Aulin, Serially concatenated continuous phase modulation with iterative decoding, *IEEE Trans. Commun.* **49**: 1901–1915 (Nov. 2001).

46. X. Li and J. A. Ritcey, Trellis-coded modulation with bit interleaving and iterative decoding, *IEEE J. Select. Areas Commun.* **17**: 715–724 (Apr. 1999).

47. C. Douillard, Iterative correction of intersymbol interference: Turbo equalization, *European Trans. Telecommun.* **6**: 507–511 (Sept. 1995).

48. A. Anastasopoulos and K. M. Chugg, Iterative equalization/decoding of TCM for frequency-selective fading channels, in *Proc. Asilomar Conf. Signals, Systems, Comp.*, pp. 177–181, Nov. 1997.

49. A. Picart, P. Didier, and A. Glavieux, Turbo-detection: a new approach to combat channel frequency selectivity, in *Proc. International Conf. Communications*, (Montreal, Canada), 1997.

ITERATIVE DETECTION METHODS FOR MULTIUSER DIRECT-SEQUENCE CDMA SYSTEMS

Lars K. Rasmussen
University of South Australia
Mawson Lakes, Australia

1. INTRODUCTION

The continuing development of the Internet, wireless communication and wireless Internet is rapidly increasing the demands on enabling communication networks. The continuous development of ever-faster computers allows for the development of ever larger communication systems to meet the ever-increasing demand for capacity to support the never-ending supply of new communications services. The third-generation (3G) mobile network, the so-called IMT2000 (International Mobile Telecommunication 2000) system, is the next step to bringing wireless Internet to the general consumer [1]. The 3G cellular mobile network provides up to 2 megabits per second (Mbps) for indoor environments and 144 kilobits per second (kbps) for vehicular environments. The two dominating standards for 3G networks are UMTS (Universal Mobile Telephone System) and cdma2000, respectively [1]. They are both based on direct sequence, code-division multiple-access (DSCDMA) technology, which was considered to provide the best alternative within the standardization process. DSCDMA is a spread-spectrum transmission technology where each user in principle is assigned a unique signature waveform, creating a distinguishing feature separating multiple users and thus providing multiple access [2].

In popular terms, these principles may be likened to conversations at a cocktail party where each conversation is conducted in a unique language. Even though a particular conversation can be clearly heard by surrounding people engaged in other conversations, they do not become disturbed since they do not understand the language and thus consider it as background noise. In case two languages are closely related, two simultaneous conversations may interfere with each other. The same phenomenon occurs in CDMA if two users are assigned signature waveforms that are closely correlated. This is termed *multiple-access interference* (MAI) and is one of the performance-limiting factors when conventional single-user technologies are used in CDMA systems [3,4]. The basics of spread-spectrum and CDMA technologies are described in more details elsewhere in this book. For further information, see also Refs. 1,2,5 and 6.

Third-generation mobile cellular systems are designed for multimedia communications. The standards of person-to-person communications can be improved through better voice quality and the possibility of exchanging high-quality images and video. In addition, access to information and services on public and private networks will improve through higher data rates and variable data rate options, introducing a high level of flexibility. Through the standardization process, CDMA technologies came out as the overall winners. The UMTS system is based on so-called wideband CDMA (WCDMA) [7]. There are no conceptual differences between WCDMA and CDMA. The former merely uses a bandwidth that is significantly larger than second-generation CDMA systems, leading to additional spread-spectrum advantages such a robustness toward hostile mobile radio channels.

The flexible data rates, and especially the high data rates of 2 Mbps, represent significant challenges for equipment manufacturers. For the system load to be commercially viable, technologies providing considerable system capacity improvements are required. Three areas of technology have been identified as enabling techniques for 3G CDMA systems:

- Error control coding [8,9]
- Multiuser detection [5]
- Space–time processing [10]

In this article, we discuss the use of multiuser detection for CDMA systems in general, and the use of iterative multiuser detection strategies in particular.

1.1. Multiuser Detection

Historically, multiple-access systems have been designed to avoid the MAI problem. This is accomplished by dividing the available system resources into dedicated portions for the exclusive use by a designated communication connection. In a CDMA system, we break with these principles and allow users to utilize all resources simultaneously. Under certain system conditions, it is possible to avoid MAI. However, such conditions are in general impossible to achieve in a practical setting, and thus, a certain level of MAI is to be expected.

Initially, the MAI was considered as unavoidable interference and assumed to possess similar statistical characteristics as thermal background noise generated in electronic components. Based on such arguments, the optimal receiver structures developed for the case of thermal noise only, are also optimal in the case of MAI with similar statistical behavior [11]. The performance of such systems are determined by the signal power to noise power ratio. It is therefore tempting to increase the transmitted signal power to improve performance. However, if all active users do that, the power of the interfering MAI increases with the same amount, providing no performance gains at all. On the basis of these assumptions and techniques, it follows that the systems are interference limited. A strict upper limit on active users, corresponding to a certain signal to total noise level, decides the system capacity [5].

The problem with this approach is that each active user is making decisions in isolation, regarding the corresponding MAI as unstructured noise. As an alternative, a joint decision among all users simultaneously takes the known structure of the MAI into account, treating the MAI as additional information in the decision process. Assuming binary transmission (two possible signal waveform alternatives for transmission for each user), a decision made in isolation is based on a choice between which of the two signal waveform alternatives were transmitted, ignoring the known structure of the MAI. In contrast, a joint multiuser decision is based on evaluating a suitable cost function for all possible transmitted waveform combinations between the active users, selecting the combination that maximizes the cost function. For optimal detection, we shall attempt to maximize the probability of a certain combination of data symbols being transmitted, given the particular received signal. Assuming that all data symbols are equally likely, this optimization criterion is equivalent to maximizing the corresponding likelihood function [11]. For three active users, each using binary transmission, we need to evaluate the likelihood function for 2^3 combinations of data symbols, or in general for K active users, 2^K combinations of data symbols in order to find the combination that is maximum likely. Maximum-likelihood (ML) joint multiuser detection has been suggested by several authors in the literature [12,13], most notably by Verdú [14]. Verdú [14] suggested the use of the Viterbi search algorithm [16] for detection was as an efficient implementation of the exhaustive search.

The ML problem is known to be a so-called NP-hard problem [16] and can be solved only by an exhaustive search as described above, leading to a detection complexity that grows exponentially with the number of users. In some important cases, this complexity growth is beyond practical implementation and will remain so for the foreseeable future. To address this complexity problem, an abundance of receiver structures have been proposed [5,17]. Most of these structures are sub-optimal approximations to classic design criteria.

Among the first complexity reducing schemes, the decorrelating detector was suggested [18]. The detector decorrelates the data symbols, removing the MAI entirely through linear filtering. The filter is determined by the inverse of the channel matrix. The filter removes all MAI at the expense of noise enhancement. A slightly different approach is taken by the linear minimum mean-squared error (LMMSE) detector, which minimizes the mean squared error between detector output and the transmitted symbol [19]. This detector takes the thermal noise as well as the correlation between users into account and therefore generally performs better than the decorrelator in terms of bit error rate (BER). Both the decorrelator and the LMMSE detector require matrix inversion, which is generally also prohibitively complex for practical implementation of realistically sized systems. As an alternative, the LMMSE detector can be approximated by adaptive detectors such as described in the literature [20–22].

Since even linear detectors have complexity problems, multiuser detection for CDMA was initially considered to be prohibitively complex for practical implementation. With the massive research conducted in conjunction with IMT2000, however, this view has now changed. For practical implementation, interference cancellation schemes have been subject to most attention. These techniques rely on simple processing elements constructed around conventional receiver concepts. The main component in a conventional receiver is a filter matched to the user-specific signaling waveform. An estimate of the contribution of a specific user to the composite received signal of all users can be generated based on the corresponding matched-filter output. For a specific user, we can now subtract the contributions to the received signal made from all other users. If this estimate of the MAI experienced by that specific user is correct, we can effectively eliminate the disturbance, obtaining a clean signal with no MAI. In practise the MAI estimate is rarely completely correct, leaving some residual interference. Cancellation can then be done iteratively to improve the quality of the resulting signal for detection of a specific user [23,24]. A family of structures can be defined based on how MAI estimates are generated [23,25,26].

Let us assume that we have tentative decisions for the data symbols for all users. On the basis of these decisions we can make an estimate of the MAI experienced by each user, and subtract the corresponding estimates from the received signal. With K parallel processing units, one iteration for each user can be done in parallel. This approach is naturally denoted parallel interference cancellation (PIC) [23]. As an alternative, we can process one user at a time. From a single cancellation operation, we get an updated tentative decision for a specific user that can now be used in the process of generating an updated MAI estimate for the following user. This way, the most updated information is always used in the cancellation process. In this approach, the users are processed successively, introducing a detection delay between users. The strategy is known as successive interference cancellation (SIC) [25,26].

In the discussion above, we have assumed we have tentative decisions available. Tentative decisions are obtained based on so-called decision statistics, which are passed through a corresponding tentative decision function. The collection of all matched filter outputs represents a *sufficient statistic*, including all information required for making an optimal ML decision for all users simultaneously. The cancellation process, however, is not in general an iterative *joint* detection process. The decision statistics resulting from cancellation are thus not necessarily representing sufficient statistics. In some special cases, such as linear cancellation [27], however, they are.

Assume that each user is transmitting binary data d_k, for example represented by $d_k = +1$ or $d_k = -1$. The corresponding received decision statistic then contains contributions from the desired transmitted data symbol, noise generated in the receiver and corresponding MAI. On the basis of this received decision statistic, a tentative decision is to be made. In the literature mainly four different tentative decision functions have been suggested for interference cancellation. These are shown in Fig. 1. The first one is a linear decision where in fact the decision statistic is left untouched [28,29]. Alternatively, a hard decision can be made where it is decided whether a $+1$ or a -1 was transmitted [23,25]. This is done on the basis of the polarity of the decision statistic. These two principles can be combined to generate the linear clip function, which is piecewise linear [30]. A similar shape can be obtained on the basis of nonlinear principles as shown in Fig. 1 for a hyperbolic tangent function [31,32]. It should be noted that once the cancellation process is completed, hard decisions are applied to the resulting decision statistics in case a final decision is required. In some cases, the soft cancellation output is used directly as input to an error control decoder [33], in which case no final decisions are made in the iterative detector.

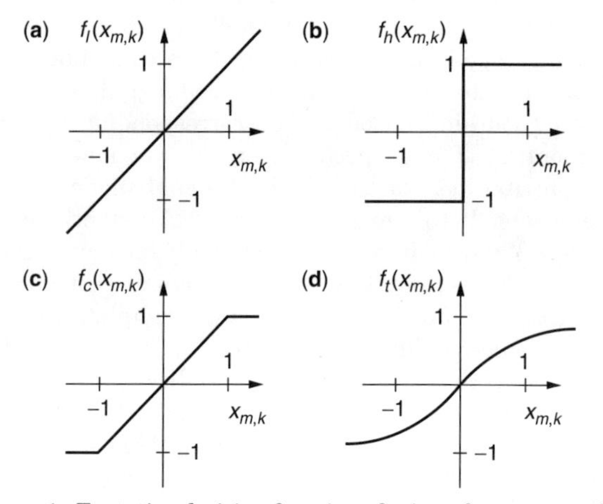

Figure 1. Tentative decision functions for interference cancellation: (**a**) linear; (**b**) hard; (**c**) clipped linear; (**d**) hyperbolic tangent.

The cancellation strategy and the tentative decision function define the character of an interference cancellation structure. The best performance in terms of bit error rate is obtained by SIC schemes as compared to PIC schemes. This advantage is, however, achieved at the expense of detection delays due to the successive processing. For the tentative decision functions, the linear clip and the hyperbolic tangent generally provide better performance than do linear and hard decisions. These issues are discussed at length in the remainder of the article.

The rest of the article is organized as follows. In Section 2, an algebraic model is derived for a simple synchronous CDMA system. The model is kept simple to more clearly illustrate the principles of iterative multiuser detection. In Section 3 the fundamental principles for interference cancellation are formalized and motivated. Corresponding modular structures are suggested, constructed around a simple interference cancellation unit. The hard tentative decision function is discussed in Section 4, while the concept of weighted cancellation is introduced in Section 5, as a powerful technique for improving convergence speed and BER performance. Linear cancellation based on the linear tentative decision function is presented in Section 6, where the connection to classic iterations for solving linear equation systems is explained. Here it is shown that the cancellation structure in fact leads to an iterative solution to the constrained ML problem. The advantages of the clipped linear tentative decision function are detailed in Section 7, and in Section 8, structures using the hyperbolic tentative decision function are discussed. Numerical examples are included in Section 9 to illustrate the characteristics of the schemes discussed, and in Section 10, concluding remarks are made.

2. SYSTEM MODEL

Let us consider a CDMA channel that is simultaneously shared by K users. Each user is assigned a signature waveform $p_k(t)$ of duration T where either $T = T_s$, or $T \gg T_s$. Here, T_s denotes the duration of one data symbol. In the first case, the signature waveform is the same for each symbol interval, while in the latter case, the signature waveform changes for each symbol interval. The former case is termed *short codes*, while the latter is termed *long codes*. For notational simplicity, the mathematical description is based on the former case. It is conceptually easy to extend the description to the long code case. A signature waveform may thus be expressed as

$$p_k(t) = \sum_{j=0}^{N-1} a_k(j)p(t - jT_c), \quad 0 \le t \le T_s \tag{1}$$

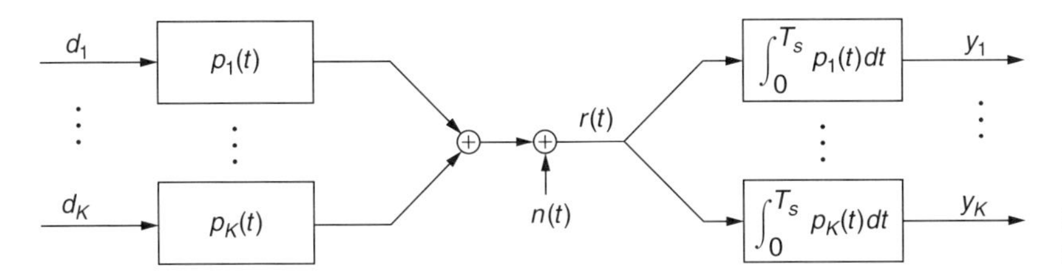

Figure 2. A simple continuous-time model for a synchronous CDMA system.

where $\{a_k(j): 0 \leq j \leq N-1\}$ is a spreading code sequence consisting of N chips that take on values $\{-1, +1\}$, and $p(t)$ is a chip pulse of duration T_c where T_c is the chip interval. Here, $p_k(t) = 0$ outside the symbol interval, $t < 0$, $t > T_s$ since a short-code system is considered. Thus, we have N chips per symbol and $T_s = NT_c$. Without loss of generality, we assume that all K signature waveforms have unit energy:

$$\int_0^{T_s} |p_k(t)|^2 dt = 1 \tag{2}$$

The information sequence of the kth user is denoted $\{d_k(l): 0 \leq l \leq L-1\}$, where the value of each information symbol may be chosen from the set $\mathcal{D}$, and L denotes the block length of the transmission. For binary transmission, we have $\mathcal{D} = \{-1, +1\}$. The corresponding equivalent lowpass transmitted waveform may be expressed as

$$s_k(t) = \sum_{l=0}^{L-1} e^{j\phi_k(l)} d_k(l) p_k(t - lT_s) \tag{3}$$

The composite signal for the K users may be expressed as follows, assuming for simplicity a single-path channel with unit magnitude:

$$s(t) = \sum_{k=1}^{K} s_k(t - \tau_k) = \sum_{k=1}^{K} \sum_{l=0}^{L-1} e^{j\phi_k(l)} d_k(l) p_k(t - lT_s - \tau_k) \tag{4}$$

where $\{\tau_k: 1 \leq k \leq K\}$ are the transmission delays, which satisfy the condition $0 \leq \tau_k \leq T_s$ for $1 \leq k \leq K$ and $\{\phi_k(l): 1 \leq k \leq K, 0 \leq l \leq L-1\}$ are the random-phase rotations, assumed constant over one symbol interval. This is the model for a multiuser signal in asynchronous mode, which is typical for an uplink scenario. In the special case of synchronous transmission, $\tau_k = 0$ for $1 \leq k \leq K$, which is a typical scenario for downlink transmission. This model can easily be extended to model transmission over multipath channels. The extension is discussion in more detail in, for example, Ref. 34.

To make the presentation in the following sections notationwise and conceptually simple, we will focus on a synchronous CDMA system without any random phase rotation, specifically, $\tau_k = 0$ and $\phi_k(l) = 0$ for $1 \leq k \leq K$, $0 \leq l \leq L-1$. A synchronous system is naturally associated with downlink transmission while multiuser detection strategies are intended primarily for uplink transmission. We still maintain a synchronous model in this presentation as it greatly simplifies notation and conception. The extensions to asynchronous systems are straightforward once the basic principles are understood.

We also assume that binary phase shift keying (BPSK) modulation formats are used: $\mathcal{D} = \{-1, +1\}$. All the presented concepts, however, generalize to more elaborate cases. For this simplified case, it is sufficient to only consider one symbol interval. The symbol interval index is therefore omitted in the following. The corresponding K user model is shown in Fig. 2, where $\phi_k = 0$ for $1 \leq k \leq K$.

The transmitted signal is assumed to be corrupted by additive white Gaussian noise (AWGN). Hence, the received signal may be expressed as

$$r(t) = s(t) + n(t) \tag{5}$$

where $n(t)$ is the noise with double-sided power spectral density $\sigma_n^2 = N_0/2$. The sampled output of a chip matched filter (CMF) for chip interval j, user k and an arbitrary symbol interval is [11]

$$r_j = \int_{jT_c}^{(j+1)T_c} r(t) p(t - jT_c) dt, \quad 0 \leq j \leq N-1 \tag{6}$$

Collecting all CMF outputs in a column vector

$$\mathbf{r} = (r_1, r_2, \ldots, r_N)^T \tag{7}$$

we have

$$\mathbf{r} = \frac{1}{\sqrt{N}} \sum_{k=1}^{K} \mathbf{a}_k d_k + \mathbf{n} \tag{8}$$

where $\mathbf{a}_k$ is a length N vector representing the code sequence $\{a_k(j): 0 \leq j \leq N-1\}$ and $\mathbf{n}$ is a length N Gaussian noise vector with autocorrelation matrix

$$\mathrm{E}\{\mathbf{n}\mathbf{n}^T\} = \sigma_n^2 \mathbf{I} \tag{9}$$

Define

$$\mathbf{s}_k = \frac{1}{\sqrt{N}} \mathbf{a}_k \tag{10}$$

On the basis of Eq. (8), discrete-time code matched filtering for user k can be conveniently described as

$$\begin{aligned} y_k = \mathbf{s}_k^T \mathbf{r} &= \sum_{i=1}^{K} \mathbf{s}_k^T \mathbf{s}_i d_i + \mathbf{s}_k^T \mathbf{n} = \sum_{i=1}^{K} \mathbf{s}_k^T \mathbf{s}_i d_i + z_k \\ &= d_k + \sum_{i \neq k} \rho_{ki} d_i + z_k = d_k + w_k + z_k \end{aligned} \tag{11}$$

where $\rho_{ki} = \mathbf{s}_k^T \mathbf{s}_i$ is the cross-correlation between the spreading codes of users k and i, z_k is the Gaussian noise experienced by user k, and w_k is the MAI experienced by user k, respectively.

Considering the algebraic structure of the discrete-time received signal y_k described in (11), we can conclude the following. Interference arises since, in general, every output, y_k, has a contribution from every input, $\{d_k: 1 \leq k \leq K\}$. Under ideal conditions where all users are orthogonal to each other, all the cross-correlations are zero and there is no interference. In practice such a scenario is virtually impossible to achieve, and thus each output has contributions from all users. When the MAI is ignored in the detection process, the performance is strongly interference limited. However, when the structure of the MAI is considered in the detector, considerable gains are possible.

The models in (8) and (11) can conveniently be extended to include all users, applying linear algebra to provide a compact description. Equation (8) can be described as

$$\mathbf{r} = \mathbf{Sd} + \mathbf{n} \tag{12}$$

where $\mathbf{S}$ is a $N \times K$ matrix containing the spreading codes (10) of all users as columns

$$\mathbf{S} = (\mathbf{s}_1, \mathbf{s}_2, \ldots, \mathbf{s}_K) \tag{13}$$

and $\mathbf{d}$ is a length K vector of user symbols

$$\mathbf{d} = (d_1, d_2, \ldots, d_K)^T \tag{14}$$

Each decision statistic is described by (11). Collecting all decision statistics in a vector

$$\mathbf{y} = (y_1, y_2, \ldots, y_K)^T \tag{15}$$

we arrive at the following model

$$\mathbf{y} = \mathbf{S}^T\mathbf{Sd} + \mathbf{S}^T\mathbf{n} = \mathbf{Rd} + \mathbf{z} \tag{16}$$

where $\mathbf{R}$ is a symmetric, positive semi-definite correlation matrix of dimension $K \times K$

$$\mathbf{R} = \begin{bmatrix} 1 & \rho_{12} & \cdots & \rho_{1K} \\ \vdots & & & \vdots \\ \rho_{K1} & \rho_{K2} & \cdots & 1 \end{bmatrix} \tag{17}$$

and $\mathbf{z}$ is a vector of length K, containing the Gaussian noise samples with autocorrelation function:

$$\mathrm{E}\{\mathbf{zz}^T\} = \sigma_n^2\mathbf{R} \tag{18}$$

An equivalent discrete-time system model can be defined on the basis of these results and is depicted in Fig. 3.

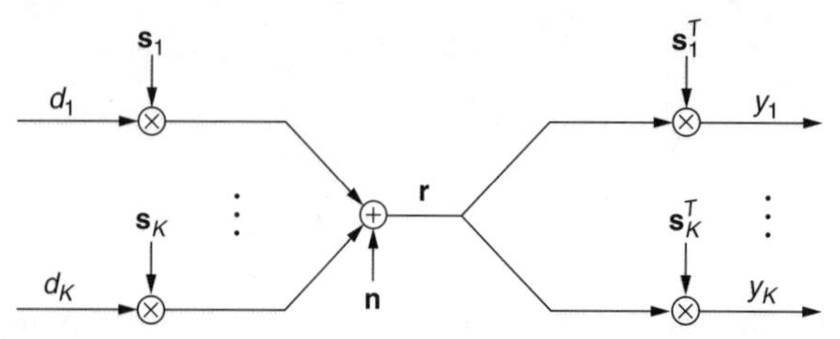

Figure 3. A simple discrete-time model for a synchronous CDMA system.

The correlation matrix can be partitioned as $\mathbf{R} = \mathbf{I} + \mathbf{M}$, where $\mathbf{I}$ is the identity matrix and $\mathbf{M}$ is the corresponding off-diagonal matrix. We can then write (16) as follows

$$\mathbf{y} = (\mathbf{I} + \mathbf{M})\mathbf{d} + \mathbf{z} = \mathbf{d} + \mathbf{Md} + \mathbf{z} \tag{19}$$

where $\mathbf{d}$ is the desired signal vector and $\mathbf{Md}$ is MAI.

3. PRINCIPLE STRUCTURE

It was argued in Section 2 that the decision statistics for detection are polluted by MAI. Let us for a moment assume that, somehow, the MAI is perfectly known at the receiver. It is then possible to eliminate the interference simply by subtracting it from the received signal:

$$x_k = y_k - \sum_{i \neq k} \rho_{ki} d_i = d_k + z_k \tag{20}$$

By subtracting the known MAI, we obtain an interference-free received signal, and thus the performance is identical to the case where only one user is present, the so-called single-user (SU) case.

Unfortunately, the MAI is not perfectly known at the receiver. To have perfect knowledge of the MAI requires perfect knowledge of the transmitted symbols, in which case there would be no information contained in the transmission. Instead of perfect knowledge, we can use an estimate of the transmitted symbols. For example, let the initial estimate for user k, $u_{1,k}$ be determined by a hard decision based on the corresponding received matched-filter output, y_k, i.e., the polarity of y_k decides whether $u_{1,k} = 1$ or $u_{1,k} = -1$

$$u_{1,k} = \mathrm{Sgn}\ (y_k) \tag{21}$$

where $\mathrm{Sgn}(\cdot)$ denotes the polarity check function, which is the same as the hard decision in Fig. 1:

$$\mathrm{Sgn}\ (x) = \begin{cases} 1 & x \geq 0 \\ -1 & x < 0 \end{cases} \tag{22}$$

An updated decision statistic, $x_{2,k}$ after one step of MAI subtraction is then

$$x_{2,k} = y_k - \sum_{i \neq k} \rho_{ki} u_{1,i} = d_k + \sum_{i \neq k} \rho_{ki}(d_i - u_{1,i}) + z_k \tag{23}$$

and

$$u_{2,k} = \mathrm{Sgn}\ (x_{2,k}) \tag{24}$$

If all the tentative decisions were correct, we have successfully eliminated all MAI and we obtain single-user (SU) performance. Each wrong decision, however, doubles the particular MAI contribution rather than eliminating it. As long as we eliminate "more" MAI than we introduce, it seems to be intuitively a good approach. The resulting decision statistic, $u_{2,k}$ can now be used to generate a new, and hopefully better, estimate of the MAI:

$$x_{3,k} = y_k - \sum_{i \neq k} \rho_{ki} u_{2,i} = d_k + \sum_{i \neq k} \rho_{ki}(d_i - u_{2,i}) + z_k. \tag{25}$$

This strategy can be continued until no further improvements are obtained. In case the updates for all the users are done simultaneously, this iterative multiuser detection scheme is also known as hard-decision, multistage parallel interference cancellation. It is called *interference cancellation* (IC) for obvious reasons as we attempt to cancel MAI. It is termed "parallel" since updated decision statistics for all the users are determined in parallel, based on the same tentative estimates of the transmitted symbols. The update process is obviously recursive or iterative, a characteristic that initially was termed "multistage detection" [23]. Finally the scheme is based on hard decisions, namely, polarity check, on the resulting decision statistics.

This is, however, not necessarily the best tentative decision strategy. In Fig. 1, four alternatives are shown: linear decision, hard decision, clipped linear decision, and hyperbolic tangent decision. We will later examine these tentative decision functions in more detail and try to establish theoretical justification.

For a general tentative decision function, the above scheme can be described by

$$x_{m+1,k} = y_k - \sum_{i \neq k} \rho_{ki} u_{m,i} \tag{26}$$

$$u_{m+1,k} = f_x(x_{m+1,k}), \tag{27}$$

or in a more compact form

$$u_{m+1,k} = f_x\left(y_k - \sum_{i \neq k} \rho_{ki} u_{m,i}\right) \tag{28}$$

with $u_{0,k} = 0$.

Cancellation can also be based on the most current estimate of the MAI. In this case, the MAI estimate is updated for each new tentative decision. As a consequence, the users are processed successively, leading to SIC in contrast to the PIC described above. An iterative SIC scheme is described by

$$x_{m+1,k} = y_k - \sum_{i=1}^{k-1} \rho_{ki} u_{m+1,i} - \sum_{i=k+1}^{K} \rho_{ki} u_{m,i} \tag{29}$$

$$u_{m+1,k} = f_x(x_{m+1,k}). \tag{30}$$

To arrive at a description convenient for implementation, we first use the fact that $y_k = \mathbf{s}_k^T \mathbf{r}$ and $\rho_{ki} = \mathbf{s}_k^T \mathbf{s}_i$:

$$x_{m+1,k} = \mathbf{s}_k^T\left(\mathbf{r} - \sum_{i=1}^{k-1} \mathbf{s}_i u_{m+1,i} - \sum_{i=k+1}^{K} \mathbf{s}_i u_{m,i}\right) \tag{31}$$

Then we add and subtract the term $u_{m,k}$:

$$x_{m+1,k} = \mathbf{s}_k^T\left(\mathbf{r} - \sum_{i=1}^{k-1} \mathbf{s}_i u_{m+1,i} - \sum_{i=k}^{K} \mathbf{s}_i u_{m,i}\right) + u_{m,k} \tag{32}$$

Finally, we define the residual error vector for user k, $\mathbf{e}_{m+1,k}$ as the term within the parentheses and get

$$x_{m+1,k} = \mathbf{s}_k^T \mathbf{e}_{m+1,k} + u_{m,k} \tag{33}$$

The residual error vector can be updated recursively:

$$\mathbf{e}_{m+1,k+1} = \mathbf{r} - \sum_{i=1}^{k} \mathbf{s}_i u_{m+1,i} - \sum_{i=k+1}^{K} \mathbf{s}_i u_{m,i} \tag{34}$$

$$= \mathbf{r} - \sum_{i=1}^{k-1} \mathbf{s}_i u_{m+1,i} - \sum_{i=k}^{K} \mathbf{s}_i u_{m,i} - \mathbf{s}_k u_{m+1,k} + \mathbf{s}_k u_{m,k} \tag{35}$$

$$= \mathbf{e}_{m+1,k} - \mathbf{s}_k(u_{m+1,k} - u_{m,k})$$

$$= \mathbf{e}_{m+1,k} - \Delta\mathbf{e}_{m+1,k} \tag{36}$$

Here, $\mathbf{e}_{m+1,K+1} = \mathbf{e}_{m+2,1}$.

The PIC structure described previously can also be described in this manner. In this case

$$\mathbf{e}_{m+1,k} = \mathbf{r} - \sum_{i=1}^{K} \mathbf{s}_i u_{m,i} \tag{37}$$

and thus the residual error signal is the same for all users. We can therefore drop the user index. Rewriting (37) for the PIC case, we get

$$\mathbf{e}_{m+1} = \mathbf{r} - \sum_{i=1}^{K} \mathbf{s}_i u_{m,i} = \mathbf{r} - \sum_{i=1}^{K} \mathbf{s}_i u_{m,i} + \sum_{i=1}^{K} \mathbf{s}_i u_{m-1,i} - \sum_{i=1}^{K} \mathbf{s}_i u_{m-1,i}$$

$$= \mathbf{e}_m - \sum_{i=1}^{K} \mathbf{s}_i(u_{m+1,k} - u_{m,k}) = \mathbf{e}_m - \sum_{i=1}^{K} \Delta\mathbf{e}_{m+1,i} \tag{38}$$

where $\Delta\mathbf{e}_{m+1,k} = \mathbf{s}_k(u_{m+1,k} - u_{m,k})$ as before. Comparing the cases for SIC and PIC, we see that for user k, the required input to make an updated tentative decision is $u_{m,k}$ and $\mathbf{e}_{m+1,k}$ while the output is conveniently $u_{m+1,k}$ and $\Delta\mathbf{e}_{m+1,k}$. We can thus define a basic interference cancellation unit (ICU) as shown in Fig. 4, where $f_x(\cdot)$ is a predetermined tentative decision function, possibly selected among the four alternatives illustrated in Fig. 1.

The SIC and the PIC structures are then obtained by different interconnection strategies of ICUs. In Fig. 5, we have an SIC structure. The residual error vector is updated according to (36), as it should be. In contrast, we have a PIC structure in Fig. 6, where the same residual error vector is input to all ICUs at the same iteration and it is updated according to (37). This modular structure is quite attractive for practical implementation and thus IC structures have received most attention

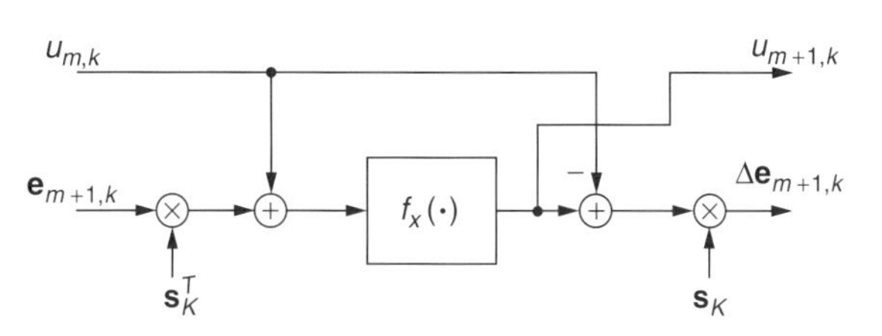

Figure 4. Basic structure of an ICU.

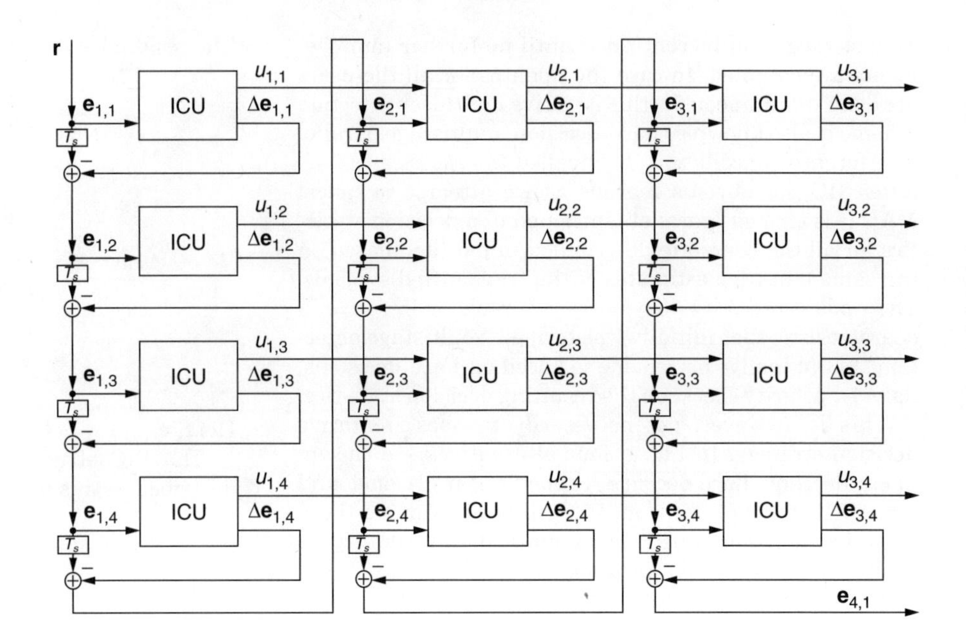

Figure 5. A modular SIC structure.

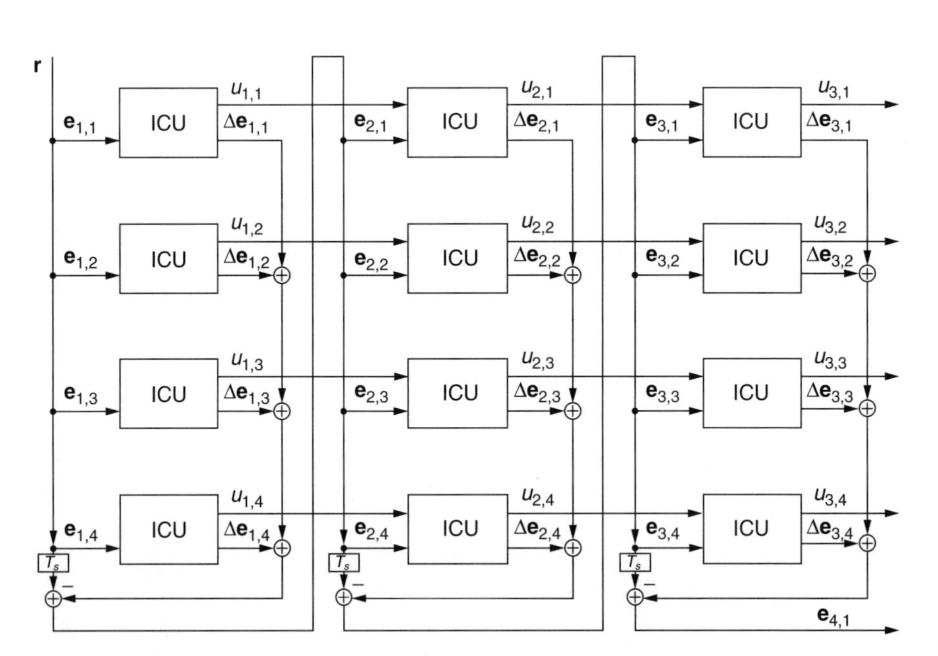

Figure 6. A modular PIC structure.

for potential commercial use. The regular structure also motivates hybrid, or group, cancellation strategies, combining successive and parallel techniques [35,36]. A groupwise cancellation structure is shown in Fig. 7.

As mentioned previously, it should be noted that once the cancellation process is completed, hard decisions are applied to the resulting decision statistics in case a final decision out of the detector is required.

As in Section 2, the decision statistics can be collected in vectors and the cancellation process can be described conveniently through matrix algebra. A PIC scheme can thus be described as

$$\mathbf{x}_{m+1} = \mathbf{y} - \mathbf{M}\mathbf{u}_m \tag{39}$$

$$\mathbf{u}_{m+1} = \mathbf{f}_x(\mathbf{x}_{m+1}) \tag{40}$$

or in a more compact form

$$\mathbf{u}_{m+1} = \mathbf{f}_x(\mathbf{y} - \mathbf{M}\mathbf{u}_m) \tag{41}$$

with $\mathbf{u}_0 = \mathbf{0}$. For a convenient algebraic description of SIC, the following partition of $\mathbf{M}$ is helpful

$$\mathbf{M} = \mathbf{L} + \mathbf{U} \tag{42}$$

where $\mathbf{L}$ is a strictly lower left triangular matrix and $\mathbf{U}$ is a strictly upper right triangular matrix, respectively. An iterative SIC scheme is then described by

$$\mathbf{x}_{m+1} = \mathbf{y} - \mathbf{L}\mathbf{u}_{m+1} - \mathbf{U}\mathbf{u}_m \tag{43}$$

$$\mathbf{u}_{m+1} = \mathbf{f}_x(\mathbf{x}_{m+1}) \tag{44}$$

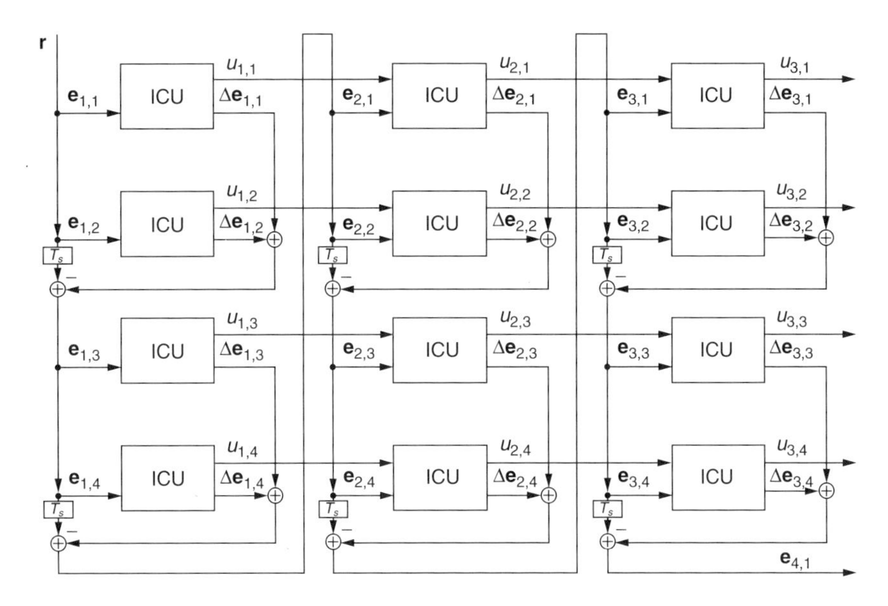

Figure 7. A modular groupwise structure.

4. HARD TENTATIVE DECISION FUNCTION

In the previous section, we defined SIC and PIC principles and established how they are related. In the continuing discussion, we focus on PIC structures. The extension to SIC schemes is usually straightforward, although more cumbersome notation is required. Known difficulties are explicitly pointed out. Also in the previous section, the principles of cancellation were presented in an intuitive manner. In the following sections, we show that certain iterative interference cancellation structures are in fact iterative realizations of known theoretically justified detector structures.

One of the first suggested IC structures was hard decision PIC [23]. This structure can be derived as an approximation to the optimal ML detector, given certain simplifying assumptions. Considering the signal model in (16), then given perfect knowledge of the correlation matrix, the decision statistics are jointly Gaussian distributed, leading to the following probability density function which is also the likelihood function:

$$p(\mathbf{y} \mid \mathbf{d}) = C \exp\left(\tfrac{1}{2}(\mathbf{y} - \mathbf{Rd})^T \mathbf{R}^{-1}(\mathbf{y} - \mathbf{Rd})\right) \quad (45)$$

Here, C is a constant which is independent of the conditional data vector $\mathbf{d}$. The corresponding log-likelihood function is [37]

$$\Lambda(\mathbf{d}) = \mathbf{d}^T\mathbf{Rd} - 2\mathbf{y}^T\mathbf{d} \quad (46)$$

and thus the optimal ML decision is the argument $\mathbf{d} \in \{-1, 1\}^K$ that minimizes the loglikelihood function:

$$\mathbf{d}_{\mathrm{ML}} = \arg \min_{\mathbf{d}\in\{-1,1\}^K} [\mathbf{d}^T\mathbf{Rd} - 2\mathbf{y}^T\mathbf{d}] \quad (47)$$

Assume now that we know all the transmitted symbols, except the symbol for user k, d_k. We want a detector that maximizes the probability of d_k given the received signal and given knowledge of all other transmitted symbols. Let us first define

$$\mathbf{d}(k) = (d_1, d_2, \ldots, d_{k-1}, d_{k+1}, \ldots, d_K)^T \quad (48)$$

We now want to maximize

$$\begin{aligned} P(d_k \mid \mathbf{y}, \mathbf{d}(k)) &= P(d_k)\frac{p(\mathbf{y}, \mathbf{d}(k) \mid d_k)}{p(\mathbf{y}, \mathbf{d}(k))} \\ &= P(d_k)\frac{p(\mathbf{y} \mid \mathbf{d}(k), d_k)P(\mathbf{d}(k) \mid d_k)}{p(\mathbf{y}, \mathbf{d}(k))} \end{aligned} \quad (49)$$

with respect to d_k. Here, $p(\cdot)$ denotes a probability density function while $P(\cdot)$ denotes a probability. The equality follows from Baye's rule [11]. Maximizing (49) is equivalent to maximizing $p(\mathbf{y} \mid \mathbf{d}(k), d_k)$, a problem that can be described by the loglikelihood function,

$$\begin{aligned} \hat{d}_k &= \arg \min_{d_k\in\{-1,1\}} [\mathbf{d}^T\mathbf{Rd} - 2\mathbf{y}^T\mathbf{d}] \\ &= \arg \min_{d_k\in\{-1,1\}} \left[d_k^2 + 2d_k \sum_{i\neq k} \rho_{ki} d_i - 2y_k d_k \right] \end{aligned} \quad (50)$$

where we have thrown away all terms independent of d_k and thus do not influence the optimization problem. We can also write this as

$$\begin{aligned} \hat{d}_k &= \arg \max_{d_k\in\{-1,1\}} \left[d_k \left(y_k - \sum_{i\neq k} \rho_{ki} d_i \right) \right] \\ &= \mathrm{Sgn}\left(y_k - \sum_{i\neq k} \rho_{ki} d_i \right) \end{aligned} \quad (51)$$

which is in fact identical in form to (23) and (24), describing hard-decision PIC. Since $\mathbf{d}(k)$ is not known, we use the most recent estimate instead and arrive exactly at (23) and (24). In conclusion, given perfect knowledge of interfering symbols, interference cancellation is an optimal structure. Using current estimates of these symbols of course leads to an approximating, suboptimal approach. Depending on the strategy of cancellation, PIC, SIC or hybrids of the two are obtained.

5. WEIGHTED CANCELLATION

Hard decision cancellation is prone to error propagation and can exhibit a significant error floor for high SNR, as it is in effect interference-limited. This is due to the coarse approximation that previous symbol estimates provide for the MAI

$$x_{m+1,k} = d_k + \sum_{i \neq k} \mathbf{s}_k^T \mathbf{s}_i (d_i - u_{m,i}) + \mathbf{s}_k^T \mathbf{n} = d_k + \hat{\mathbf{z}}_k \quad (52)$$

$$u_{m+1,k} = \mathrm{Sgn}\ (x_{m+1,k}) \quad (53)$$

where

$$\hat{z}_k = \sum_{i \neq k} \mathbf{s}_k^T \mathbf{s}_i (d_i - u_{m,i}) + \mathbf{s}_k^T \mathbf{n} \quad (54)$$

Inherent assumptions are that the cancellation error, $\sum_{i \neq k} \mathbf{s}_k^T \mathbf{s}_i (d_i - u_{m,i})$ is Gaussian and independent of the thermal noise $\mathbf{n}$. Neither of these assumptions is true. Obviously the second assumption cannot be true since each tentative decision depends on the thermal noise. This was taken into account in the improved cancellation scheme suggested by Divsalar et al. [38]. Here, it is assumed that the cancellation error and the thermal noise are correlated. Also, at iteration $(m+1)$, the detector is derived based on observing the received signal y_k as well as the previous decision statistic $x_{m,k}$. The joint likelihood function is still derived as conditioned on perfect knowledge of $\mathbf{d}(k)$, however, it now depends on the correlation between the Gaussian noise and the residual interference.

Following some manipulations, some simplifying assumptions and substituting the most current estimate $\mathbf{u}_m(k)$ in place of $\mathbf{d}(k)$, we arrive at the following revised decision statistic:

$$x_{m+1,k} = \mu_{m+1,k} \left(y_k - \sum_{i \neq k} \rho_{ki} u_{m,i} \right) + (1 - \mu_{m+1,k}) x_{m,k} \quad (55)$$

The updated decision statistic is now determined as a weighted sum of the previous decision statistic and the corresponding decision statistic determined by a traditional PIC cancellation. The weighting factor, $\mu_{m+1,k}$, is described by an involved combination of the correlation parameters between the Gaussian noise and the residual error term [38]. It may not be possible to accurately determine the weighting factor analytically, but trial-and-error selection has shown that the general structure is very powerful and provides significant performance gains over traditional hard-decision structures [38]. This technique was originally termed *partial cancellation*, and the principles are now used in most practical studies of IC techniques [39–42].

6. LINEAR TENTATIVE DECISION FUNCTION

Let us now focus on the linear tentative decision function. In this case, the corresponding iterative detectors are also linear. We therefore start by considering optimal linear detectors. Allowing the symbol estimate vector to be any real-valued vector, $\mathbf{u} \in \mathbb{R}^K$, the corresponding ML solution is easily found to be [18]

$$\mathbf{u} = \mathbf{R}^{-1} \mathbf{y} \quad (56)$$

Similarly, on the basis of the linear minimum mean-squared error criterion, the solution is [19]

$$\mathbf{u} = (\mathbf{R} + \sigma_n^2 \mathbf{I})^{-1} \mathbf{y} \quad (57)$$

Before data are delivered, the real-valued estimate must, of course, be mapped to a valid data symbol. Both optimal linear detectors rely on matrix inversion, which has a complexity of the order of $\mathcal{O}(K^3)$. The matrix inverse represents the solution to a set of linear equations [43]. There exist, however, efficient iterative techniques for solving a set of linear equations. As an example, let us focus on the implementation of the decorrelator, Eq. (56). The set of linear equations to be solved is described by

$$\mathbf{y} = \mathbf{R}\mathbf{u} = (\mathbf{I} + \mathbf{M})\mathbf{u} = (\mathbf{I} + \mathbf{L} + \mathbf{U})\mathbf{u} \quad (58)$$

where we have applied the partition of $\mathbf{R}$ described previously. A common iteration used for matrix inversion is the Jacobi iteration

$$\mathbf{u}_{m+1} = \mathbf{y} - \mathbf{M}\mathbf{u}_m \quad (59)$$

which is identical to (39) given a linear tentative decision as $\mathbf{u}_m = \mathbf{f}_x(\mathbf{x}_m) = \mathbf{x}_m$. Here, $\mathbf{f}_x(\cdot)$ denotes a vector function applying the decision function, $f_x(\cdot)$ to each element of the argument vector independently. Similarly, the well-known Gauss–Seidel (GS) iteration is described as

$$\mathbf{u}_{m+1} = \mathbf{y} - \mathbf{L}\mathbf{u}_{m+1} - \mathbf{U}\mathbf{u}_m \quad (60)$$

which, in turn, is equivalent to (43). It follows that these two classic iterations are linear PIC and linear SIC, respectively. This was first realized by Elders-Boll et al. [27]. For the linear case, IC therefore represents an iterative implementation of optimal linear detectors, given that the particular iteration converges. The GS iteration is guaranteed to converge while the Jacobi iteration is not. The linear PIC converges if the iteration matrix $\mathbf{M}^m = (\mathbf{R} - \mathbf{I})^m$ converges for increasing m. For $\mathbf{M}^m$ to converge, the maximum eigenvalue of $\mathbf{R}$ must be constrained by [44]

$$\lambda_{\max} \leq 2 \quad (61)$$

which is not true for all possible correlation matrices. To guarantee convergence for a linear PIC scheme, and in general also to increase convergence speed for linear IC, more advanced iterations can be used [44–46]. The concept of over-relaxation can be used for both PIC and

SIC structures. The Jacobi over-relaxation iteration is described by

$$\mathbf{u}_{m+1} = \mu(\mathbf{y} - \mathbf{R}\mathbf{u}_m) + \mathbf{u}_m \tag{62}$$

The corresponding iteration matrix is now $(\mu\mathbf{R} - \mathbf{I})^m$, and thus the relaxation parameter directly scales the eigenvalues of $\mathbf{R}$, allowing for tuned, guaranteed convergence. Considering the decision statistic for user k, we get

$$\begin{aligned} u_{m+1,k} &= \mu\left(y_k - \sum_{i=1}^{K} \rho_{ki} u_{m,i}\right) + u_{m,k} \\ &= \mu\left(y_k - \sum_{i \neq k} \rho_{ki} u_{m,i}\right) + (1-\mu) u_{m,k} \end{aligned} \tag{63}$$

which has the structure of (55), emphasizing that these advanced iterations correspond to weighted linear IC. The optimal weighting factor for the Jacobi overrelaxation is determined by the eigenvalue spread of the correlation matrix $\mathbf{R}$ [43]. Fastest convergence is assured when the positive and the negative mode of convergence for the iteration matrix are equal, which is obtained by

$$\mu = \frac{2}{\lambda_{\max} + \lambda_{\min}} \tag{64}$$

where $\lambda_{\max}$ and $\lambda_{\min}$ are the maximum and minimum eigenvalues of $\mathbf{R}$, respectively. An asymptotic analysis for large systems [44] has shown that this optimal weighting factor is well approximated by

$$\mu = \frac{N}{N+K} \tag{65}$$

when N and K are large. The ICU of Fig. 4 should be modified as shown in Fig. 8 to accommodate the weighted cancellation of (62).

First-order and second-order iterations have also been suggested, such as the steepest-descent iteration and the conjugant gradient iteration [45–47]. For the steepest-descent iteration, expressions for optimal weighting factors for both short and long codes have been derived Guo et al. [29,46]. For a more thorough analysis and discussion of linear IC schemes, please consult Refs. 44–46.

7. CLIPPED LINEAR TENTATIVE DECISION FUNCTION

A hard decision is effective when the decision statistic is large, in which case the corresponding decision should be

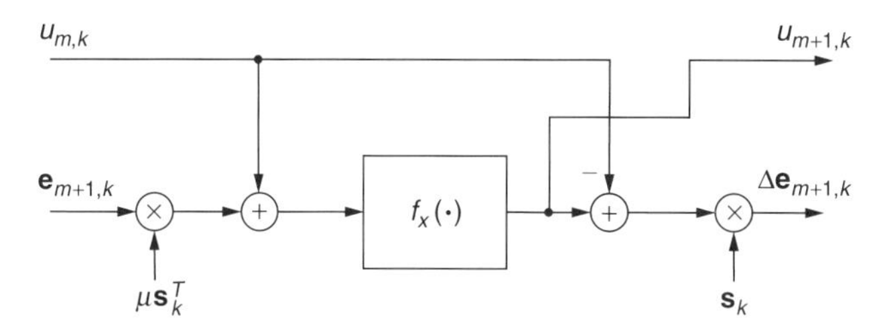

Figure 8. Basic structure of a weighted ICU.

relatively reliable. This is in contrast to a hard decision based on a decision statistic only marginally different from zero, which is bound to be relatively unreliable, potentially introducing additional interference with high probability. A linear tentative decision function is effective when the decision statistic has a magnitude less than one. For very small decision statistics, only a very small interference estimate is subtracted, limiting the potential damage of a wrong decision. As the decision statistic increases, increasingly larger interference estimates are subtracted. For decision statistics with magnitude larger than one, a linear decision will therefore invariably introduce additional interference. The transmitted symbol is limited to unit magnitude and thus we should not attempt to cancel out more than that.

The clipped linear decision function depicted in Fig. 1 seems to be an appropriate choice for combining the benefits of a hard decision and a linear decision, respectively, avoiding the inherent drawbacks. Therefore, assume now that we constrain the allowable solution for each user to $-1 \leq u_{m,k} \leq 1$ for all k and m. Enforcing these constraints for all users simultaneously describe a K-dimensional hypercube in Euclidean space. Such a constraint is denoted a box-constraint and is formally defined as

$$\mathbb{B}^K = \{\mathbf{d} \in \mathbb{R}^K : \mathbf{d} \in [-\mathbf{b}, \mathbf{b}]\} \tag{66}$$

where $\mathbf{b}$ is an all ones vector. The corresponding box-constrained optimization problem is described as

$$\mathbf{u} = \arg\min_{\mathbf{d} \in \mathbb{B}^K} [\mathbf{d}^T\mathbf{R}\mathbf{d} - 2\mathbf{y}^T\mathbf{d}] \tag{67}$$

Ahn [48] first suggested an iterative algorithm for solving the general problem of constraining the solution to any convex set. A hypercube, a so-called box, is a tight convex set. For any convex set we can define an orthogonal projection. For the box constraint, the orthogonal projection is merely the clipped linear decision function applied independently to all the users as demonstrated in Fig. 9. As shown in Refs. 30,49, and 50, the algorithm suggested by Ahn is in fact a generalization of first order iterations for solving linear equation systems:

$$\mathbf{x}_{m+1} = \mu(\mathbf{y} - \mathbf{Q}\mathbf{u}_{m+1} - (\mathbf{R} - \mathbf{Q})\mathbf{u}_m) + \mathbf{u}_m \tag{68}$$

$$\mathbf{u}_{m+1} = \mathbf{f}_x(\mathbf{x}_{m+1}) \tag{69}$$

Letting $\mathbf{Q} = \mathbf{0}$, we have a weighted PIC structure, while $\mathbf{Q} = \mathbf{L}$ leads to a weighted SIC structure. It has been shown that the same convergence conditions as for the linear case apply [48]. It follows that a traditional SIC scheme based on a clipped linear decision always converges to the solution of the box-constrained ML problem. The PIC also converges to this solution when an appropriate weighting factor is chosen. Unfortunately, no analytic results have yet been obtained for deriving optimal weighting factors. In general an SIC structure converges faster than a PIC structure. The clipped linear decision is quite attractive as it provides good performance, relatively fast convergence, and it is simple to implement in hardware. The corresponding ICU has the same structure

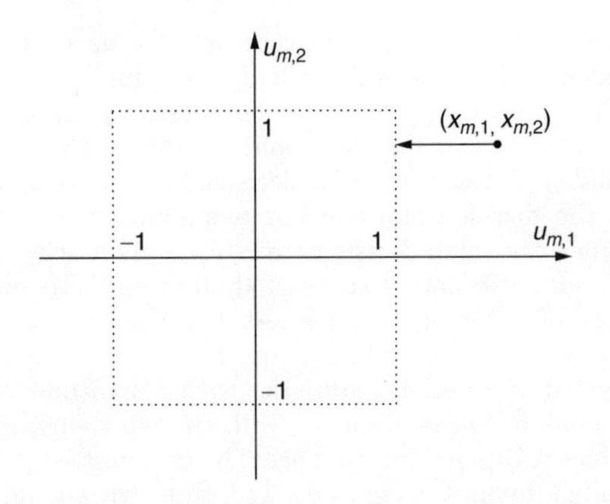

Figure 9. The orthogonal projection onto a hypercube.

as shown in Fig. 8. This approach has been suggested for practical implementation in several papers [e.g., 41].

8. HYPERBOLIC TANGENT TENTATIVE DECISION FUNCTION

The final tentative decision function we consider is based on an MMSE optimized estimate.

$$u_{m,k} = \mathrm{E}\{d_k \mid x_{m,k}\} \tag{70}$$

This was first suggested by Tarköy [51] and later developed further by other authors [31,32]. Considering the general description of PIC, we can write the decision statistic for user k and iteration m as done in (23)

$$x_{m+1,k} = d_k - \sum_{i \neq k} \rho_{ki}(d_i - u_{m,i}) + z_k = d_k - \sum_{i \neq k} \rho_{ki}\varepsilon_{m,i} + z_k \tag{71}$$

Assuming that the cancellation error is a zero-mean Gaussian random variable, independent of the thermal noise, $x_{m+1,k}$ is also a Gaussian random variable with mean d_k and variance $\sigma_{m,k}^2$:

$$x_{m+1,k} \sim N(d_k, \sigma_{m,k}^2) \tag{72}$$

$$\sigma_{m,k}^2 = \sigma_{\varepsilon,m,k}^2 + \sigma_n^2 \tag{73}$$

The expectation in (70) is obviously determined by

$$u_{m+1,k} = P(d_k = 1 \mid x_{m+1,k}) - P(d_k = -1 \mid x_{m+1,k}) \tag{74}$$

Considering one term at a time

$$P(d_k = 1 \mid x_{m+1,k}) = \frac{P(d_k)p(x_{m+1,k} \mid d_k = 1)}{p(x_{m+1,k})} \tag{75}$$

Using the fact that

$$P(d_k = 1 \mid x_{m+1,k}) + P(d_k = -1 \mid x_{m+1,k}) = 1 \tag{76}$$

we arrive at

$$P(d_k = 1 \mid x_{m+1,k}) = \frac{p(x_{m+1,k} \mid d_k = 1)}{p(x_{m+1,k} \mid d_k = 1) + p(x_{m+1,k} \mid d_k = -1)} \tag{77}$$

From (74), we then have

$$u_{m+1,k} = \frac{p(x_{m+1,k} \mid d_k = 1) - p(x_{m+1,k} \mid d_k = 1)}{p(x_{m+1,k} \mid d_k = 1) + p(x_{m+1,k} \mid d_k = -1)} \tag{78}$$

Since $x_{m+1,k}$ is assumed Gaussian with known statistics, we obtain

$$u_{m+1,k} = \tanh\left(\frac{x_{m+1,k}}{\sigma_{m,k}^2}\right) \tag{79}$$

which is a hyperbolic tangent function as shown in Fig. 1. The variance can be determined as devised by Müller and Huber [31]:

$$\sigma_{m,k}^2 = \sum_{i \neq k} \rho_{ki}^2(1 - u_{m,k}^2) + \sigma_n^2 \tag{80}$$

In this case, the corresponding ICU has the form shown in Fig. 4.

9. NUMERICAL EXAMPLES

In this section numerical examples are presented, illustrating the characteristics of the different cancellation strategies and the different tentative decision functions, respectively. The impact of weighted cancellation is demonstrated, although no attempts have been made for optimizing the weighting factors. Only general trends are illustrated, leaving the interested reader to consult the vast literature on the topic for more details regarding weight optimization.

A symbol synchronous CDMA system with processing gain $N = 32$ and the simple channel model of Eq. (16) is considered. Long codes are assumed, so a new random spreading sequence is used for each user and each symbol interval. In Fig. 10, the BER performance of PIC and weighted PIC (WPIC) is shown as a function of the number of users in the system. Here, the four tentative decision functions depicted in Fig. 1 are used, respectively. For weighted cancellation, a factor of $\mu = 0.5$ have been used for all cases. Better performance can be obtained for more carefully selected weighting factors. The PIC can be considered as the case of $\mu = 1$. With caution it is therefore possible to roughly predict performance for factors $0.5 \leq \mu \leq 1$ as the optimal weights in most cases are within this interval. The performance is captured at a bit energy to noise ratio $E_b/N_0 = 5$ dB. The iterative detectors have been restricted to five iterations which is considered reasonable for potential practical applications. In the following paragraphs we will denote the use of the tentative functions in Fig. 1 as LIN, HARD, CLIP, and TANH, respectively.

Considering first PIC, significant performance losses are observed as the system load, K/N, increases. Especially a linear tentative decision function is sensitive to the load. As K increases, it becomes more likely that the iteration matrix is diverging, leading to detector collapse with very poor performance. For the other decision functions, the performance degradation is more graceful, but still severe as the load increases. A load of 10–15% for the linear case and of 25–30% for the others can be

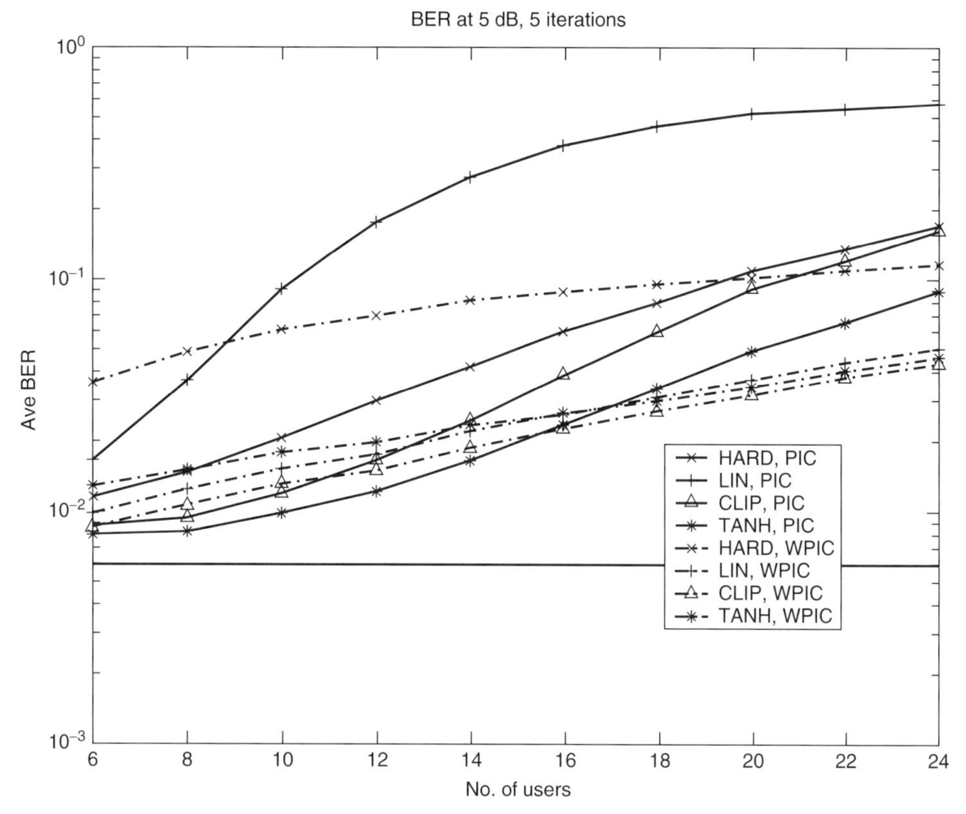

Figure 10. The BER performance for PIC and WPIC as a function of the number of users. HARD, LIN, CLIP, and TANH are used together with the cases of $\mu = 1$ and $\mu = 0.5$. The solid horizontal line represents single-user performance at 5 dB.

accommodated with reasonable losses at a processing gain of 32. The TANH shows the best performance, followed by the CLIP and HARD. For high load, CLIP and HARD provide similar performance.

Introducing a fixed weighting factor of $\mu = 0.5$ does not provide better performance for small loads. The performance degradation for WPIC as the load increases is noticeably more graceful, extending potential load with appropriately selected weights to about 50% for TANH, CLIP, and LIN. The performance of WPIC for LIN, CLIP, and TANH are quite similar to each other. This is mainly due to the limited number of iterations allowed. For a larger number of iterations, differentiating performance is obtained as illustrated in Fig. 11.

For HARD, a low weighting factor at small to moderate loads is not appropriate. For $\mu = 0.5$, the performance is considerably worse than $\mu = 1$ up to a load of 50%. This illustrates the difficulty of selecting weighting factors for HARD. Since a hard decision leads to cancellation of a "full" MAI contribution scaled by μ, regardless of the quality of the decision statistic, weighting may do more harm than good. In the first iteration decisions are based on the matched-filter output, which usually provides a BER of less than 0.5. With a weighting factor of $\mu = 0.5$, the additional MAI introduced due to wrong decisions are reduced, but at the same time only half of the MAI contributions corresponding to correct decisions are eliminated. The distinct nonlinearity of hard decisions makes the selection of weighting factors more complicated and is mainly left to a trial-and-error approach with little analytic justification. To avoid these drawbacks, the decision function should differentiate on decision statistic quality as done by the other three alternatives.

Comparing PIC ($\mu = 1$) and WPIC ($\mu = 0.5$), we can conclude that the weighting factor should decrease with load, starting at $\mu = 1$ for small loads. For appropriately varying weights, reasonable performance is to be expected at least up to a load of 50%.

It should be kept in mind that only 5 iterations are allowed in Fig. 10. In Fig. 11, the BER as a function of the number of iterations is shown for the case of $K = 24$ and $E_b/N_0 = 7$ dB. In this case, $\mu = 1$ does not provide reasonable performance for any decision function. In all cases, pingpong effects are observed where the BER oscillates with iterations [52]. For $\mu = 0.5$, HARD is still not useful as previously discussed, while LIN and TANH improve gradually up to 8 iterations, after which the performance converges to a level determined by the residual MAI. The CLIP continues to improve even beyond 15 iterations, representing the best alternative. The CLIP will, however, also converge to a level above the SU performance since this strategy provides a box-constrained ML solution and not necessarily a SU solution. At 7 dB, the SU performance is just below 10^{-3}. The benefits of a larger number of iterations at higher loads are nicely illustrated.

Successive cancellation is expected to provide better performance than PIC. This is illustrated in Fig. 12,

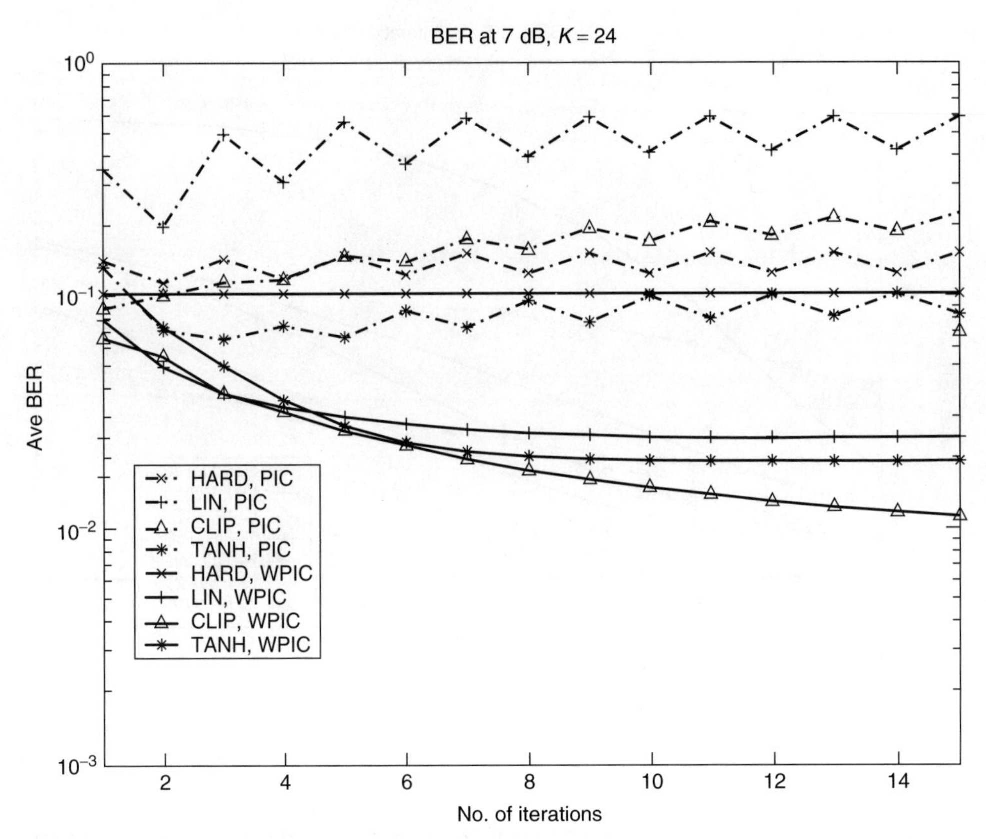

Figure 11. The BER performance for PIC and WPIC as a function of the number of iterations. HARD, LIN, CLIP, and TANH are used together with the cases of $\mu = 1$ and $\mu = 0.5$.

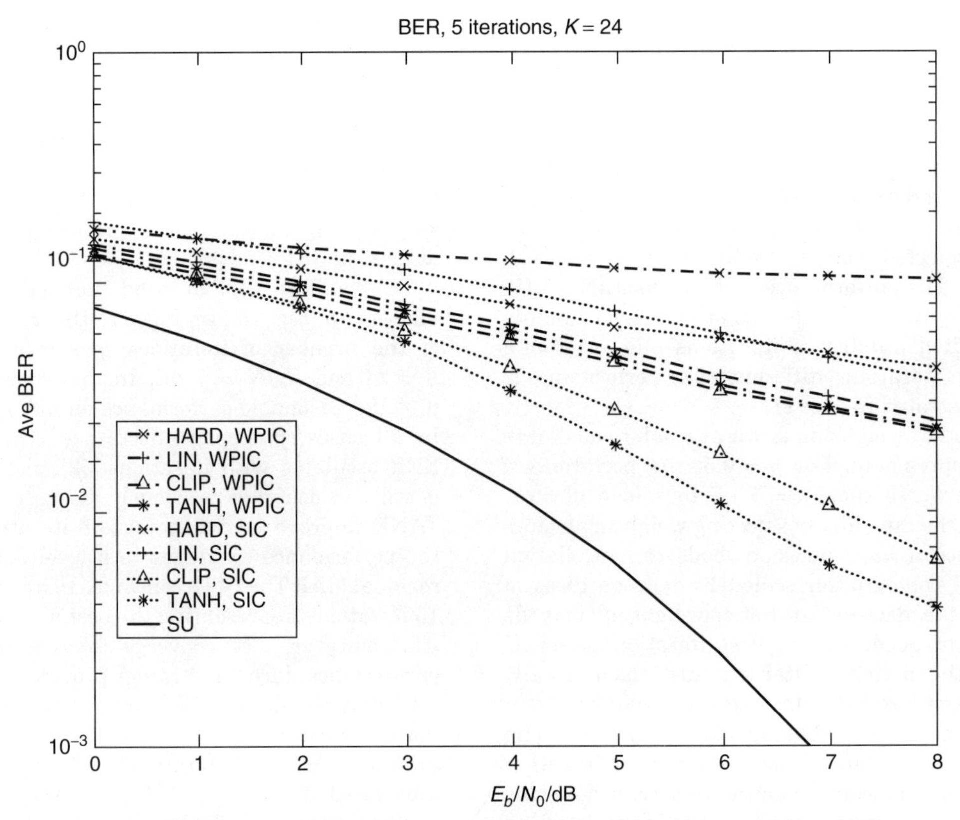

Figure 12. The BER performance for SIC and WPIC as a function of E_b/N_0, where HARD, LIN, CLIP, and TANH tentative decisions are used. For WPIC, $\mu = 0.5$.

where the BER is shown as a function of E_b/N_0. Here the performance of the SIC with the four tentative decision functions is contrasted with the performance of WPIC at a load of $K = 24$. Again, five iterations are allowed. The benefits of SIC are clear, especially for TANH and CLIP, where significant improvements in terms of E_b/N_0 are obtained.

For more detailed numerical examples and discussions on cancellation structures and tentative decision functions, the reader is referred to the open literature on the subject.

10. CONCLUDING REMARKS

In this article, we have presented the fundamental principles of iterative multiuser detection in CDMA, also known as interference cancellation strategies. The basic building block in a cancellation structure is an interference cancellation unit, taking as input the residual error signal and the tentative symbol decision from the previous iteration, giving as output an updated tentative symbol decision and an updated residual error signal. Different strategies, implementing serial and parallel cancellation structures or combinations thereof, can be constructed through different interconnections of ICUs. The ICU itself is characterized mainly by a tentative decision function. Here, we have presented four such functions, namely the linear, hard, clipped linear, and hyperbolic tangent tentative decision functions. The principles behind weighted cancellation are also presented and explained. Selected numerical examples are presented to illustrate the characteristics of each cancellation strategy and each decision function.

Acknowledgments

The author gratefully acknowledges Mr. Peng Hui Tan for providing the numerical examples and Mr. Fredrik Brännström for providing input to the presentation.

BIOGRAPHY

Lars K. Rasmussen was born on March 8, 1965 in Copenhagen, Denmark. He got his M.Eng. in 1989 from the Technical University of Denmark, and his Ph.D. degree from Georgia Institute of Technology (Atlanta, Georgia, USA) in 1993, both in electrical engineering.

From 1993 tp 1995, he was at the Mobile Communication Research Centre, University of South Australia as a Research Fellow. From 1995 to 1998 he was with the Centre for Wireless Communications at the National University of Singapore as a Senior Member of Technical Staff. He then spent 3 months at the University of Pretoria, South Africa as a Visiting Research Fellow, followed by three years at Chalmers University of Technology in Gothenburg, Sweden as an Associate Professor. He is now a professor of telecommunications at the Institute for Telecommunications Research, University of South Australia. He also maintains a part-time appointment at Chalmers.

BIBLIOGRAPHY

1. H. Holma and A. Toskala, *WCDMA for UMTS: Radio Access for Third Generation Mobile Communications*, rev. ed., Wiley, New York, 2001.
2. A. Viterbi, *CDMA, Principles of Spread Spectrum Communication*, Addison-Wesley, Reading, MA, 1995.
3. E. H. Dinan and B. Jabbari, Spreading codes for direct sequence CDMA and wideband CDMA cellular networks, *IEEE Commun. Mag.* **36**: 48–54 (Sept. 1998).
4. A. J. Viterbi, A. M. Viterbi, and E. Zehavi, Other-cell interference in cellular power-controlled CDMA, *IEEE Trans. Commun.* **42**: 1501–1504 (Feb.–April 1994).
5. S. Verdú, *Multiuser Detection*, Cambridge Univ. Press, 1998.
6. V. K. Garg, K. Smolik, and J. E. Wilkes, *Applications of CDMA in Wireless/Personal Communications*, Prentice-Hall, Englewood Cliffs, NJ, 1997.
7. F. Adachi and M. Sawahashi, Wideband wireless access based on DS-CDMA, *IEICE Trans. Commun.* **E81-B**: 1305–1316 (July 1998).
8. S. Lin and D. J. Costello, Jr., *Error Control Coding: Fundamentals and Applications*, Prentice-Hall, Englewood Cliffs, NJ, 1983.
9. C. Heegard, S. B. Wicker, and C. Heegaard, *Turbo Coding*, Kluwer, 1999.
10. P. van Rooyen, M. Lötter, and D. van Wyk, *Space-Time Processing for CDMA Mobile Communications*, Kluwer, 2000.
11. J. G. Proakis, *Digital Communications*, 3rd ed., McGraw–Hill, 1995.
12. K. Schneider, Optimum detection of code division multiplexed signals, *IEEE Trans. Aerospace Electron. Syst.* **15**: 181–185 (Jan. 1979).
13. R. Kohno, H. Imai, and M. Hatori, Cancellation techniques of co-channel interference and application of Viterbi algorithm in asynchronous spread spectrum muliple access systems, *Prod. Symp. Inform. Theory Appl.* 659–666 (Oct. 1982).
14. S. Verdú, Minimum probability of error for asynchronous Gaussian multiple-access channels, *IEEE Trans. Inform. Theory* **32**: 85–96 (Jan. 1986).
15. A. J. Viterbi, Error bounds for convolutional codes and an asymptotically optimum decoding algorithm, *IEEE Trans. Inform. Theory* **13**: 260–269 (1967).
16. S. Verdú, Computational complexity of optimum multiuser detection, *Algorithmica* **4**: 303–312 (1989).
17. S. Moshavi, Multi-user detection for DS-CDMA communications, *IEEE Pers. Commun.* **34**: 132–136 (Oct. 1996).
18. R. Lupas and S. Verdú, Linear multiuser detectors for synchronous code-division multiple-access channels, *IEEE Trans. Inform. Theory* **35**: 123–136 (Jan. 1989).
19. Z. Xie, R. T. Short, and C. K. Rushforth, A family of suboptimum detector for coherent multiuser communications, *IEEE J. Select. Areas Commun.* **8**: 683–690 (May 1990).
20. U. Madhow and M. L. Honig, MMSE interference suppression for direct-sequence spread spectrum CDMA, *IEEE Trans. Commun.* **42**: 3178–3188 (Dec. 1994).
21. T. J. Lim and Y. Ma, The Kalman filter as the optimal linear minimum mean-squared error multiuser CDMA detector, *IEEE Trans. Inform. Theory* **46**: 2561–2566 (Nov. 2000).

22. M. Honig, U. Madhow, and S. Verdú, Blind adaptive multiuser detection, *IEEE Trans. Inform. Theory* **41**: 944–960 (July 1995).

23. M. K. Varanasi and B. Aazhang, Multistage detection in asynchronous code–division multiple–access communications, *IEEE Trans. Commun.* **38**: 509–519 (April 1990).

24. K. Jamal and E. Dahlman, Multi-stage serial interference cancellation for DS-CDMA, *Proc. IEEE VTC '96*, Atlanta, April 1996, pp. 671–675.

25. P. Dent, B. Gudmundson, and M. Ewerbring, CDMA-IC: A novel code divisions multiple access scheme based on interference cancellation, *Proc. 3rd IEEE Int. Symp. PIMRC '92*, Boston, Oct. 1992, pp. 98–102.

26. P. Patel and J. Holtzman, Analysis of simple successive interference cancellation scheme in a DS/CDMA, *IEEE J. Select. Areas Commun.* **12**: 796–807 (June 1994).

27. H. Elders-Boll, H. D. Schotten, and A. Busboom, Efficient implementation of linear multiuser detectors for asynchronous CDMA systems by linear interference cancellation, *Eur. Trans. Telecommun.* **9**(4): 427–437 (Sept.–Nov. 1998).

28. L. K. Rasmussen, T. J. Lim, and A.-L. Johansson, A matrix-algebraic approach to successive interference cancellation in CDMA, *IEEE Trans. Commun.* **48**: 145–151 (Jan. 2000).

29. D. Guo, L. K. Rasmussen, S. Sun, and T. J. Lim, A matrix-algebraic approach to linear parallel interference cancellation in CDMA, *IEEE Trans. Commun.* **48**: 152–161 (Jan. 2000).

30. P. H. Tan, L. K. Rasmussen, and T. J. Lim, Constrained maximum-likelihood detection in CDMA, *IEEE Trans. Commun.* **49**: 142–153 (Jan. 2001).

31. R. R. Müller and J. B. Huber, Iterative soft-decision interference cancellation for CDMA, in Louise and Pupolin, eds., *Digital Wireless Communications*, Springer Verlag, 1998, pp. 110–115.

32. S. Gollamudi, S. Nagaraj, Y. -F. Huang, and R. M. Buehrer, Optimal multistage interference cancellation for CDMA systems using nonlinear MMSE criterion, *Proc. Asilomar Conf. Signals, Systems, Computers 98*, Oct. 1998, Vol. 5, pp. 665–669.

33. P. D. Alexander, A. J. Grant, and M. C. Reed, Iterative detection in code-division multiple-access with error control coding, *Eur. Trans. Telecommun.* **9**: (July–Aug. 1998).

34. L. K. Rasmussen, P. D. Alexander, and T. J. Lim, A linear model for CDMA signals received with multiple antennas over multipath fading channels, in F. Swarts, P. van Rooyen, I. Oppermann, and M. Lötter, eds., *CDMA Techniques for 3rd Generation Mobile Systems*, Kluwer, Sept. 1998, Chap. 2.

35. S. Sumei, L. K. Rasmussen, T. J. Lim, and H. Sugimoto, A hybrid interference canceller in CDMA, *Proc. IEEE Int. Symp. Spread Spectrum Techniques and Applications*, Sun City, South Africa, Sept. 1998, pp. 150–154.

36. S. Sumei, L. K. Rasmussen, T. J. Lim, and H. Sugimoto, A matrix-algebraic approach to linear hybrid interference canceller in CDMA, *Proc. Int. Conf. Univ. Personal Communication*, Florence, Italy, Oct. 1998, pp. 1319–1323.

37. L. K. Rasmussen, T. J. Lim, and T. M. Aulin, Breadth-first maximum-likelihood detection in multiuser CDMA, *IEEE Trans. Commun.* **45**: 1176–1178 (Oct. 1997).

38. D. Divsalar, M. Simon, and D. Raphaeli, Improved parallel interference cancellation for CDMA, *IEEE Trans. Commun.* **46**: 258–268 (Feb. 1998).

39. M. Sawahashi, Y. Miki, H. Andoh, and K. Higuchi, *Serial Canceler Using Channel Estimation by Pilot Symbols for DS-CDMA*, IEICE Technical Report RCS95-50, July 1995, Vol. 12, pp. 43–48.

40. T. Ojaperä et al., Design of a 3rd generation multirate CDMA system with multiuser detection, MUD-CDMA. *Proc. IEEE Int. Symp. Spread Spectrum Techniques and Applications (ISSSTA)*, Mainz, Germany, Sept. 1996, pp. 334–338.

41. H. Seki, T. Toda, and Y. Tanaka, Low delay multistage parallel interference canceller for asynchronous DS/CDMA systems and its performance with closed-loop TPC, *Proc. 3rd Asia-Pacific Conf. Communications*, Sydney, Australia, Dec. 1997, pp. 832–836.

42. M. Sawahashi, H. Andoh, and K. Higuchi, Interference rejection weight control for pilot symbol-assisted coherent multistage interference canceller using recursive channel estimation in DS-CDMA mobile radio, *IEICE Trans. Fund.* **E81-A**: 957–970 (May 1998).

43. O. Axelsson, *Iterative Solution Methods*, Cambridge Univ. Press, 1994.

44. A. Grant and C. Schlegel, Convergence of linear interference cancellation multiuser receivers, *IEEE Trans. Commun.* **49**: 1824–1834 (Oct. 2001).

45. R. M. Buehrer, S. P. Nicoloso, and S. Gollamudi, Linear versus nonlinear interference cancellation, *J. Commun. Networks* **1**: 118–133 (June 1999).

46. D. Guo, L. K. Rasmussen, and T. J. Lim, Linear parallel interference cancellation in random-code CDMA, *IEEE J. Select. Areas Commun.* **17**: 2074–2081 (Dec. 1999).

47. P. H. Tan and L. K. Rasmussen, Linear interference cancellation in CDMA based on iterative solution techniques for linear equation systems, *IEEE Trans. Commun.* **48**: 2099–2108 (Dec. 2000).

48. B. H. Ahn, Iterative methods for linear complementary problems with upper bounds on primary variables, *Math. Prog.* **26**(3): 295–315 (1983).

49. A. Yener, R. D. Yates, and S. Ulukus, A nonlinear programming approach to CDMA multiuser detection, *Proc. Asilomar Conf. Signals, Systems, Computers 99*, Pacific Grove, CA, Oct. 1999, pp. 1579–1583.

50. P. H. Tan, L. K. Rasmussen, and T. J. Lim, Iterative interference cancellation as maximum-likelihood detection in CDMA, in *Proc. Int. Conf. Information, Communication, Signal Processing 99*, Singapore, Dec. 1999.

51. F. Tarköy, MMSE-optimal feedback and its applications, *Proc. IEEE Int. Symp. Information Theory*, Whistler, Canada, Sept. 1995, p. 334.

52. L. K. Rasmussen and I. J. Oppermann, Ping-pong effects in linear parallel interference cancellation for CDMA, to *IEEE Trans. Wireless Commun.* (in press).

J

JPEG2000 IMAGE CODING STANDARD

B. E. Usevitch
University of Texas at El Paso
El Paso, Texas

1. INTRODUCTION

JPEG2000 is a new international standard for the coding (or compression) of still images. The standard was developed in order to address some of the shortcomings of the original JPEG standard, and to implement improved compression methods discovered since the original JPEG standard first appeared. The JPEG2000 standard offers a number of new features, with one of the most significant being the flexibility of the compressed bit stream. As a result of this flexibility, many image processing operations such as rotation, cropping, random spatial access, panning, and zooming, can be performed in JPEG2000 either directly on the compressed data, or by only decompressing a relevant subset of the compressed data. The flexible bit stream allows the compressed data to be reordered such that decompression will result in images of progressively larger size, higher quality, or more colors. All the bit streams, original and reordered, maintain a strict embedded property in which the most important bits come first in the bit stream. Consequently, truncating these embedded bit streams at any point gives an optimal compressed representation for the given bitlength. Other desirable features of JPEG2000 include

- Lossy and lossless compression using the same algorithm flow (truncated lossless bit streams give lossy image representations)
- Efficient algorithm implementation on both small memory and parallel processing devices
- Region of interest coding
- Improved error resilience.

Amazingly, all the features of JPEG2000 are realized by coding the image data only once. This contrasts sharply with previously used image compression methods where many of the image properties, such as image size or quality, are fixed at compression time. Thus, to get compressed data representing different image sizes or quality, multiple codings of the original data, and the subsequent storage of multiple compressed representations were required. Given all the advantages to JPEG2000, are there any disadvantages? One main disadvantage, which should become clear after reading this description, is that JPEG2000 is considerably more complex than its predecessor JPEG. Thus JPEG may still be the preferred method for coding images at medium compression rates where it is only slightly worse than JPEG2000 in terms of distortion performance.

This article gives a brief overview of the JPEG2000 coding algorithm, covering only Part 1 or the baseline of the standard. Section 2 first describes some basic concepts, namely, image progressions and embedded representations, that are needed to understand the standard. Section 3, the longest and most detailed section, describes the JPEG2000 coding algorithm from the perspective of encoding an image. Section 4 gives some performance results of JPEG2000 relative to the SPIHT algorithm and JPEG, and give conclusions.

2. BACKGROUND

2.1. Image Scaling

The JPEG2000 standard is capable of progressively decoding images in several different ways corresponding to different ways of scaling image data. The current standard uses four methods of image scaling: resolution, quality, position, and color. *Resolution scaling* corresponds to changing the image size, where increasingly larger resolution gives larger image sizes (see Fig. 1). Resolution

Figure 1. An example of image progression by resolution (or size).

scaling is useful in applications where an image is decoded to different display sizes, such as a palmtop display or a 21-in. monitor, and in Web serving applications where a small thumbnail image is typically downloaded prior to downloading a full-sized image.

Quality scaling, also called *signal-to-noise* (SNR) *scaling*, refers to altering the numerical precision used to represent the pixels in an image. Increasing the quality corresponds to higher fidelity images as shown in Fig. 2. Quality scaling is useful when decompressing pictures to displays having different capabilities, such as displays having only black-and-white pixels or those having 256 levels of grayscale. Position or spatial scaling refers to altering the order in which smaller portions of an image are used to progressively build up the entire image. For example, most printers print images from top to bottom in a raster scan order, corresponding to increasing the spatial scale of the image. *Spatial scaling* is useful in printers which may have to print large images with limited memory and in applications that require random access to locations of an image. *Color scaling* refers to changing the number of colors planes [such as RGB (red-green-blue)] used to represent an image and is useful when a color image is printed or displayed on a black-and-white printer or monitor.

2.2. Embedded Bit Streams

A powerful tool used in the JPEG2000 coding algorithm is that of embedded coding. A binary bit stream is said to be embedded if any truncation of the bit stream to length L results in an optimal compressed representation. By *optimal* we mean that no other compressed representation of length L, embedded or not, will have better resulting distortion than the truncated embedded bit stream. EZW [1] and SPIHT [2] are good examples of algorithms that give embedded representations that are quality scalable. Truncating the bit streams resulting from these algorithms thus gives lower SNR representations, where each representation is optimal for the truncated length. A significant advantage of embedded bit streams is that the compressed representations of an image at many different rates can be achieved by coding the image only once. The final optimal compressed representation is achieved by simply truncating the bit stream from the single coding to the desired bitlength.

Embedded bit streams can be constructed such that when sequentially decompressed they give the image progressions described above. Since the image progressions are different, it would appear that a separate coding run would be required to create the embedded bit stream corresponding to each progression. A major breakthrough achieved by the JPEG2000 coding algorithm is that the embedded bit streams corresponding to all the basic image progressions can be achieved by doing only one coding. The way JPEG2000 is able to do this is by dividing up a transformed image into a number of independent codeblocks. Each codeblock is independently compressed to form its own quality scalable bit stream called an *elementary embedded bit stream*. The set of all elementary bit streams from all codeblocks are annotated and collected together into a database called a *generalized embedded representation* [3]. The creation of this database is the initial step of the coding process. The second and final step consists of extracting, annotating, and ordering the bits from this generalized representation to give the final coded image representation. This final coded representation gives the progression desired as well as being an embedded representation, and thus optimal for any given truncation point. JPEG2000 is able to create these final coded representations by only selecting and rearranging the bits resulting from the single initial coding.

3. JPEG2000 ENCODING

3.1. Preprocessing (Tiling and Color Transform)

The first step in the JPEG2000 coding algorithm is to divide the image into rectangular regions called "tiles." These tiles are all of the same size and completely cover the image in a regular array. The tile size can be chosen to be as large as the image itself, in which case the image consists of only one tile. For purposes of coding, each

Figure 2. An example of image progression by quality (or SNR).

tile can be treated as an independent image having its own set of coding parameters. Thus tiling can be useful in coding compound documents, which are documents having separate subregions of texts, graphics, and picture data. Tiles from compound images that contain only text data can be compressed with parameters that are very different from tiles that contain only picture data. Tiling can also be used to process very large images using systems with small amounts of memory, and to give random spatial access to compressed data (although this spatial access can also be done in a single tile). The main disadvantage of tiling is that it leads to blocking artifacts and reduced compression performance similar to that found in the original JPEG standard.

After tiling, color images, or images consisting of more than one color component, can be transformed to reduce the redundancy amongst the color components. Two transforms are defined in JPEG2000 for use on standard *RGB* color data. The first is a linear transform which converts the three *RGB* components into a luminance (or black-and-white) component, and two chrominance (or color) components denoted YC_bC_r. This transform, called the *irreversible color transform* (ICT), cannot exactly recover the original data from the transformed data and therefore is used for lossy coding applications. The second transform, a nonlinear transform called the *reversible color transform* (RCT), converts RGB data into three components $Y'D_bD_r$. The RCT is an integer approximation to the ICT that maps integers to integers in a reversible manner. Because it is reversible, the RCT is the only transform used for lossless compression. The separate color components resulting from either the ICT or RCT are then coded independently. Since tiles and color components are coded independently and in the same manner, this article assumes for simplicity that the image being coded has only one tile and one color component.

The final preprocessing step is to remove any average (or DC) value in the image coefficients by adding a constant value to the image. Eliminating the average value reduces the probability of overflow and allows the JPEG2000 algorithm to take full advantage of fixed precision arithmetic hardware.

3.2. Wavelet Transform

After preprocessing the image coefficients are transformed using a standard dyadic (power of 2) discrete wavelet transform (DWT). An example three-level DWT with corresponding notation is shown in Fig. 3. Note that the number of levels of wavelet decomposition M gives rise to $M + 1$ well defined image resolutions (sizes). These resolutions are numbered from the smallest (r_0) to the full image size before transformation (r_M) as shown in Fig. 3. JPEG2000 uses a special form of the DWT called the *symmetric wavelet transform* (SWT), which is able to handle border effects and has the property that the transformed image has the same number of coefficients as the original image [4]. The wavelet transform can be implemented using either a filterbank or lifting methods, and special methods can be employed by the transform to reduce memory requirements [5].

Part 1 of the JPEG2000 standard specifies only two sets of filter coefficients to be used in the transform: the 9/7 and 5/3 filters (where the numbers correspond to filter lengths). Wavelet transforms using the 9/7 filters and finite precision arithmetic cannot exactly recover the original image from the wavelet transformed coefficients. Thus the 9/7 filters are only used in lossy coding applications. Wavelet transforms using the 5/3 filters and lifting map integers to integers such that the exact original image can be reconstructed from the wavelet-transformed coefficients. Thus the 5/3 filters are the only ones used for lossless compression. The 5/3 filters can also be used

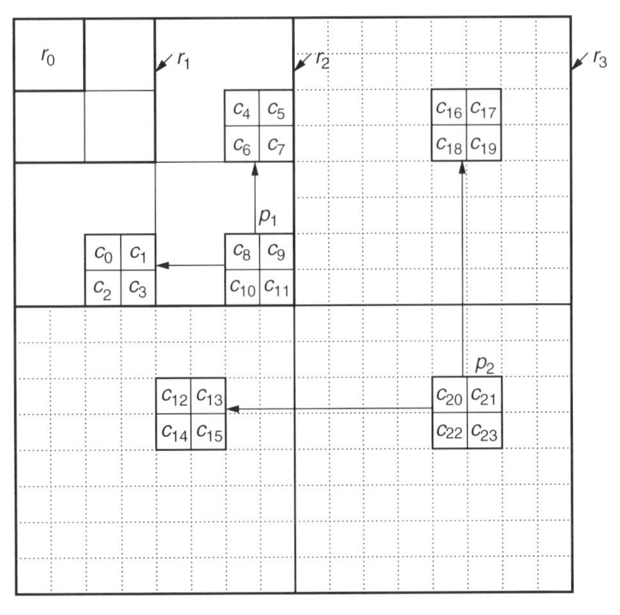

Figure 3. An example three-level wavelet decomposition and notation depicting resolutions, r_i, example codeblocks, c_i, and example precincts, p_i. Codeblocks $c_0 - c_{11}$ correspond to precinct p_1 and codeblocks $c_{12} - c_{23}$ correspond to precinct p_2.

in lossy compression but are not preferred since the 9/7 filters give better lossy compression performance.

3.3. Embedded Quantization

The wavelet coefficients are quantized using a uniform "deadzone" quantizer, where the deadzone is twice as large as the quantization step size and symmetric about zero (see Fig. 4). A deadzone quantizer is used since the deadzone improves coding performance when the quantizer is used with entropy coding. The deadzone quantizer can be represented mathematically as

$$q = \text{sign}(w)\left\lfloor \frac{|w|}{\Delta_b} \right\rfloor \tag{1}$$

where w is the unquantized wavelet coefficient in subband b, and Δ_b is the quantization step size for subband b. The quantized wavelet coefficients are represented in binary signed magnitude form. Specifically, one bit of the binary number represents the sign, zero for positive, and the remainder of the bits represent the quantized magnitude (see Fig. 4). The quantization step includes some loss of information since all wavelet coefficients in a quantization interval of size Δ_b are mapped into the same quantized value. For lossless coding applications, the wavelet coefficients are not quantized, which corresponds to setting $\Delta_b = 1$ in Eq. (1).

The quantized wavelet coefficients are coded using what is called bit plane coding [6]. In bit plane coding the most significant magnitude bits of all the coefficients are coded prior to coding the next most significant bits of all the coefficient magnitudes, and so forth. The combination of quantization and bit plane coding can be viewed as quantizing data with a set of successively finer quantizers, and this process is called embedded quantization. A set of embedded scalar quantizers is shown in Fig. 4. These quantizers have the property that finer quantizers are formed by subdividing the quantization intervals of a more coarse quantizer. The result of using embedded quantization in JPEG2000 is that the quantization using interval Δ_b includes all the quantizations having coarser quantization intervals of $2^k\Delta_b$ where $0 \le k \le k_{\max}$. Note also from Fig. 4 that the quantization interval index is formed by appending a bit to the index of the next coarser interval to which it belongs.

Embedded quantization is the method in which JPEG2000 is able to construct quality scalable compressed data. Because of embedded quantization the choice of finest quantization interval Δ_b is not critical. The interval Δ_b is typically chosen to be rather narrow (or fine), and the resulting quantization interval of the compressed image $(2^k\Delta_b)$ is determined by truncation of the embedded bit stream.

3.4. Bit Plane Encoding

Prior to coding the quantized wavelet coefficients, the wavelet subbands are divided into a regular array of relatively small rectangles called *codeblocks* (see Fig. 3). Each of these codeblocks is then coded independently to form its own elementary embedded bit stream (see Section 2.2). Coding small blocks rather than entire subbands or the whole image offers several advantages. Since the blocks are small, they can be coded in hardware having limited memory, and since the blocks are independently coded, several blocks can be coded in parallel. Independently coding blocks also gives better error resilience since the errors in one block will not propagate into other blocks (error propagation can be a significant problem in EZW and SPIHT). Having a large number of blocks makes the resulting coded bit stream more flexible. The coded blocks can be arranged such that different progressive decodings are possible. Also, by using a large number of codeblocks the rate distortion performance of the compressed image can be optimized without further coding. This is accomplished by selecting only the best bits from each compressed codeblock in a process called *postcompression rate distortion optimization* (see book by Taubman and Marcellin [3] and Section 3.6).

Each codeblock is coded to give a quality embedded bit stream. The bit stream has the property that those bits that reduce the output distortion the most come first in the bit stream. This is accomplished by bit plane encoding the wavelet coefficients starting with the most significant bit plane and ending with the least significant one. To see why the most significant bits must be coded first to get a quality embedded bit stream, consider the case of an encoder wanting to send a coefficient to a decoder. If the binary value of the coefficient is $10{,}010_2$ and only one bit could be sent, which would be the best bit? The answer from a squared error perspective is to send the most significant bit since this results in a squared reconstruction error of only $(10{,}010_2 - 10{,}000_2)^2 = (10_2)^2 = 4$ while sending the lower significant bit results in a larger squared error of $(10{,}010_2 - 10_2)^2 = (10{,}000_2)^2 = 256$.

JPEG2000 uses a quantized coefficient's significance to determine how it will be coded in a bit plane. A coefficient is defined to be significant with respect to a bit plane if the coefficient's magnitude has a nonzero bit in the current or higher bit plane. Defining the significance function $\sigma_k(q)$

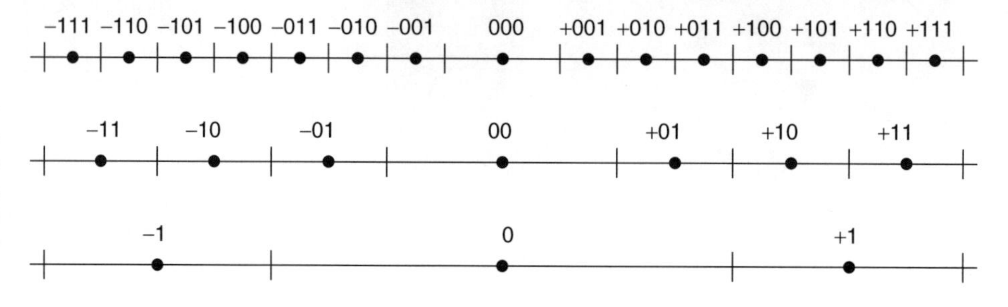

Figure 4. A set of embedded deadzone quantizers. Quantization intervals are indexed with sign magnitude binary format. Increasing quantization resolution is implemented by subdividing intervals and appending bits to interval indices.

of a quantized wavelet coefficient q at bit plane k as

$$\sigma_k(q) = \left\lfloor \frac{|q|}{2^k} \right\rfloor$$

then a wavelet coefficient is significant in bit plane k if $\sigma_k(q) > 0$. Coding of coefficients in a codeblock begins in bit plane $k_{\max}$, where $k_{\max}$ is the highest bit plane such that at least one of the wavelet coefficients in the block is significant ($k_{\max} = \lfloor \log_2(\max|q|) \rfloor$, where the maximization is over all quantized coefficients q in the codeblock).

JPEG2000 codes each bit plane using three separate passes, where each pass corresponds to what is called a *fractional bit plane*. This is different from earlier coders such as EZW and SPIHT, which used only one pass per bit plane. These previous methods argued that all bits in a bit plane reduced the output distortion by the same amount. Although this is true, further research showed that some bits in a bit plane have a higher distortion rate slope than do others [7,8]. The intuition behind this is that all bits in a bit plane (input bits) reduce the image distortion by the same amount ΔD. However, after entropy coding (discussed in the next section), some input bits require fewer output bits ΔL to encode. The result is that the distortion rate slopes $\Delta D/\Delta L$ are different for different input bits. Coding bits with higher slopes first leads to a coding advantage as illustrated in Fig. 5. The bottom curve in this figure shows the optimal distortion rate curve resulting from entropy-coded quantization and continuously varying the quantization step size. The two dark dots give the distortion rate results at the end of coding the k and $k-1$ bit planes, corresponding to quantizing with step sizes $2^k\Delta_b$ and $2^{k-1}\Delta_b$ respectively. Without fractional bit planes truncation of the bit stream results in distortion decreasing linearly with increased code length, since on the average keeping a fraction $\alpha\Delta L$ of the bits reduces the distortion by the amount $\alpha\Delta D$. By coding the higher slope bits first, truncated bit streams give distortion results that are much closer to the optimum rate distortion curve.

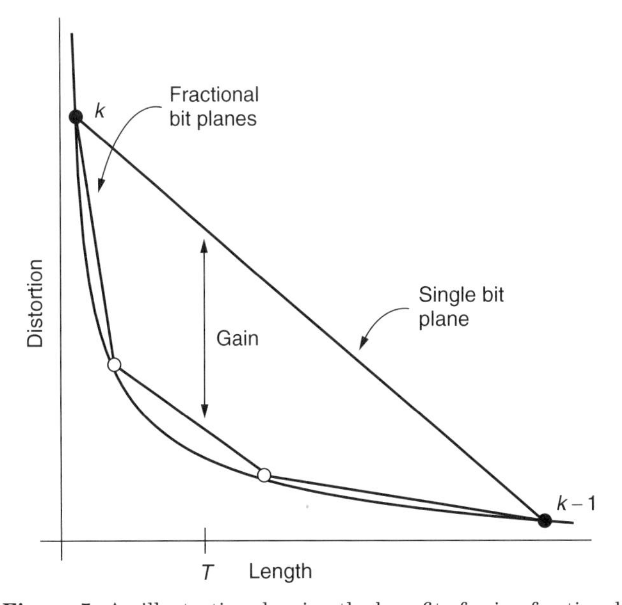

Figure 5. An illustration showing the benefit of using fractional bit plane coding when truncating to an arbitrary length T. Fractional bit planes result in reduced distortion when the code stream is not truncated at bit plane boundaries.

The 3 passes corresponding to the three fractional bit planes scan the codeblock using the scanning pattern shown in Fig. 6. The scan pattern is basically a raster scan where each scan line consists of several height four columns of coefficients. The first pass, called the *significance propagation* (SP) pass, codes all the insignificant bits that are immediate neighbors of coefficients that have been previously found significant. By "previously significant coefficients," we mean coefficients found significant in a previous bit plane or in the current bit plane earlier on in the scanning pattern of this pass. Each coefficient in this pass is coded with what is called *standard coding*, which is either a 0 to indicate that the coefficient remains insignificant, or a 1 and a sign bit to indicate that the coefficient has become significant. The second pass is called the *magnitude refinement* (MR) pass. It codes the current bit in the bit plane corresponding to coefficients found significant in a prior bit plane. That is it codes the next most significant bit in the binary representation of the these coefficients. Coefficients that became significant in the current bit plane are not coded in this pass since their values were already coded in the SP pass.

The final pass is the "cleanup" (CL) pass, which codes the significance of all coefficients in the bit plane not coded in the previous two passes. The bits from this pass are coded either using standard coding or with a special run mode designed for coding sequences of zeros. The special run mode is advantageous since at medium to high compression ratios most of the output bits from this pass will be zero [3]. Run mode is entered if a height 4 column from the scanning pattern satisfies the following: (1) the four coefficients in the column are currently insignificant (i.e., all the coefficients are to be coded in this pass) and (2) all the neighbors of the four coefficients are currently insignificant (i.e., the neighbors are either outside the codeblock and thus considered insignificant or are all to be coded in this pass and if already coded, have not become significant). Note that the conditions for run mode can be deduced at both the encoder and decoder so that no extra

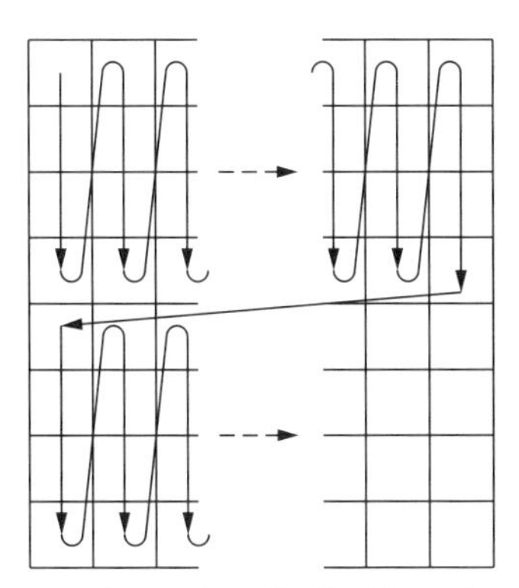

Figure 6. The scanning pattern for the three fractional bit plane passes.

side information needs to be sent. When in run mode, a 0 indicates that all four coefficients in the column remain insignificant while a 1 or "run interrupt" indicates that one or more of the coefficients have become significant. After a run interrupt 2 bits are sent to indicate the length of the run of zeros in the column (0 to 3), followed by the sign of the first significant bit. Standard coding resumes on the next coefficient.

The ordering of the passes is SP, MR, and CL for each bit plane except the first. Because there are no significant bits from previous bit planes, the first bit plane starts with the CL pass.

3.4.1. Bit Plane Coding Example. Table 1 shows the results of bit plane coding some example quantized wavelet coefficients. For clarity signs are shown as +/− instead of 0/1 and the following comments refer to the table:

1. This column satisfies the conditions for run mode, and since all the coefficients remain insignificant in this pass, they can be coded with a single 0.
2. This column satisfies the conditions for run mode but has a coefficient that becomes significant, namely, 25. The first 1 ends the run mode, and the 01 indicates a run of one insignificant coefficient. The next coefficient, 25, is known to be significant so only its sign is coded.
3. Since −11 became significant in this pass and 3 is its neighbor later in the scanning pattern, it is coded in the SP pass.
4. Since −11 was coded in the current bit plane in the SP pass, it is not coded in the MR pass.

Table 1. The First Three Passes Resulting from Bit Plane Coding the Quantized Wavelet Coefficients[a]

3	5	−1
−1	25	2
7	−2	−11
1	−6	3

	Pass						
Coef.	cl4	sp3	mr3	cl3	sp2	mr2	cl2
3	0[(1)]	0			0		
−1		0			0		
7		0			1+		
1				0	0		
5	101[(2)]	0			1+		
25	+		1			0	
−2	0	0			0		
−6	0			0	1−		
−1	0	0			0		
2	0	0			0		
−11	0	1−	[(4)]			0	
3	0	0[(3)]			0		

[a] Superscript numbers in parentheses refer to listed items in Section 3.4.1.

3.5. Entropy Coding

The sequence of ones and zeros resulting from bit plane coding is further compressed through entropy coding using a binary arithmetic encoder. Information theory results show that the minimum average number of bits required to code a sequence of independent input symbols is given by the average entropy:

$$E_{\text{av}} = -P_1 \log_2 P_1 - (1 - P_1) \log_2 (1 - P_1).$$

The input symbols are the zeros and ones from bit plane coding, and P_1 is the probability that the symbol is a 1. If $P_1 = 0.5$, the formula above shows that it requires on average 1 bit per symbol to code a sequence; thus, coding doesn't help. However, if P_1 differs or is skewed from 0.5, then it will require less than 1 bit per symbol on average to encode the binary input sequence. Arithmetic encoders are very useful since they are able to code sequences of symbols at rates approaching the average entropy. A binary arithmetic encoder requires two pieces of information in order to do its encoding: (1) the symbol (1 or 0) to be coded and (2) a probability model (the value of P_1). A great utility of the arithmetic encoder is that it can dynamically estimate P_1 by keeping a running count of the input symbols it receives. The more symbols the arithmetic encoder processes, the closer these counts will be to approximating the true probability P_1.

The specific arithmetic encoder used in JPEG2000 is called the *MQ coder*, and is the same coder used by the JBIG standard. In order to get probabilities skewed from 0.5, JPEG2000 codes bits according to their context, which is determined by the value of the eight nearest-neighbor coefficients to the coefficient being coded. A symbol probability P_1 is computed for each context, and each symbol is coded with the probability corresponding to its own context. By using contexts, the final symbol probabilities are more skewed from 0.5 than they would be without using contexts. Thus the use of contexts results in a coding gain.

In JPEG2000 contexts are formed by labeling neighbors as significant or insignificant, making a total of 2^8 contexts possible. By careful experimentation and consideration of symmetries the JPEG2000 algorithm was able to reduce this number down to only 18 contexts for arithmetic encoding: 9 for standard significance coding, 5 for sign coding, and 3 for magnitude refinement coding. A full description of the contexts and how they were derived can be found in Taubman and Marcellin's book [3]. Having a small number of contexts allows the probability models formed by counting symbols to quickly adapt to the true probability models. This is important since JPEG2000 uses small codeblocks and the probability models are reinitialized for each codeblock.

3.6. Final Embedded Bit Stream Formation

The result of bit plane coding is that each codeblock is represented by an elementary embedded bit stream. The next step is to process and arrange these elementary bit streams into a generalized embedded representation. The structure of this generalized representation is designed to make it easy to extract final coded image representations having the scaling and embedded properties discussed in Section 2. It is easy to construct resolution and scalable compressed images from the elementary bit streams. For example, resolution scalable data can be formed by concatenating elementary bit streams, starting with the lowest-resolution codeblocks and following with the higher-resolution codeblocks (extra data need to be included to indicate codeblock location and lengths). Sequentially decoding this bit stream gives resolution scaling since lower-resolution subband data appears before higher resolution subband data. Spatial scaling is implemented from this same bit stream by decoding only those blocks in the datastream associated with a particular region of interest. The only scaling not possible from this simple bit stream is quality scaling.

Quality scaling is introduced in JPEG2000 by dividing up the bits from the elementary embedded bit streams into a collection of quality layers. A quality layer is defined as the set of bits from all subbands necessary to increase the quality of the full sized image by one quality increment. These layers can be implemented by adding information to the elementary bit streams to indicate those bits belonging to each quality layer. Since each elementary bit stream is embedded, this amounts to selecting a set of increasing truncation points in each elementary embedded bit stream. These truncation points are found by selecting a set of increasing final codelengths $L_1 < L_2 < \cdots < L_N$, where L_N is the sum total of all the bits in all the elementary bit streams. The optimal set of L_1 bits from all the codeblocks that minimizes the distortion is then selected. These bits are indicated in each elementary bit stream by a truncation point and constitute the bits in the first quality layer. Next, the optimal set of L_2 bits from all the codeblocks that minimizes the distortion is selected. Since the elementary bit streams are embedded, this set consists of the L_1 bits in the first quality layer with an additional $L_2 - L_1$ bits. These additional bits are indicated in each elementary bit stream by another truncation point (one per elementary bit stream) and constitute the second quality layer. The process, called *postcompression rate distortion optimization*, is repeated for the remaining lengths L_n to form all the quality layers.

The data structure JPEG2000 uses in the generalized representation to track resolution, spatial location, and quality information is called a "packet." Each packet represents one quality increment for one resolution level at one spatial location. Since spatial locations are spread across three subbands for a particular resolution level, JPEG2000 uses precincts to refer to spatial locations. A "precinct" is defined as a set of codeblocks at one resolution level that correspond to the same spatial location as shown in Fig. 3. A full-quality layer can then be defined as one packet, from each precinct, at each resolution level.

Packets are the fundamental data structure that JPEG2000 uses to achieve highly scalable embedded bit streams since packets contain incremental resolution, quality, and spatial data. Simply rearranging the packet ordering results in final coded bit streams that are resolution, quality, or spatially scalable. The only complexity involved in forming these final coded representations is that of reordering data.

4. PERFORMANCE AND CONCLUSIONS

Table 2 shows lossy image coding results (in terms of peak SNR [6]) of JPEG2000 relative to SPIHT and the original JPEG standard. The results were generated using publicly available computer programs for JPEG [9] and SPIHT [10], and programs distributed in [3] for JPEG2000. The test images used are the monochrome, 8 bit grayscale woman and bike images shown in Figs. 1 and 2. To give an idea of visual quality versus bit rate, the bike images in Fig. 2 were coded at rates of 0.0625, 0.25, and 1 bit per pixel. The results show that the performance of SPIHT is very close to JPEG2000. However, remember that JPEG2000 is more flexible (SPIHT is not resolution or color scalable) and thus incurs some overhead due to this flexibility.

The JPEG2000 standard represents a culmination of over a decade of wavelet based compression research. It represents a fundamental shift from Fourier based techniques to wavelet based techniques that was enabled by the discovery of wavelet analysis. Research in image compression for the foreseeable future will focus on improving and extending wavelet-based techniques such as JPEG2000 until the discovery of new enabling technologies and theory.

BIOGRAPHY

Bryan E. Usevitch received the B.S. degree in electrical engineering (magna cum laude) from Brigham Young University, Provo, Utah, in 1986, and the M.S. and Ph.D. degrees from the University of Illinois at Urbana–Champaign in 1989 and 1993, respectively. From 1986 to 1987 and 1993 to 1995 he was a staff engineer at

Table 2. Peak SNR Results from Lossy Coding the Images of Figs. 1 and 2[a]

Bit Rate (bpp)	0.1	0.2	0.5	1.0
	Woman			
JPEG	—	24.88	29.68	33.55
SPIHT (lossless)	26.34	28.56	33.04	37.58
SPIHT (lossy)	26.72	28.93	33.56	38.33
JPEG2000 (lossless)	26.16	28.31	32.84	37.54
JPEG2000 (lossy)	26.76	29.02	33.62	38.48
	Bike			
JPEG	—	24.67	30.14	34.37
SPIHT (lossless)	24.67	27.64	32.64	36.98
SPIHT (lossy)	24.92	28.04	33.01	37.70
JPEG2000 (lossless)	24.98	28.00	32.95	37.34
JPEG2000 (lossy)	35.51	28.49	33.52	38.09

[a] Recall that truncating lossless encoded data gives lossy compression. JPEG values are only approximate due to the difficulty in coding at exact bit rates.

TRW designing satellite communication systems. In 1995 he joined the department of electrical engineering at the University of Texas at El Paso where he is currently an associate professor. Dr. Usevitch's research interests are in signal processing, focusing on wavelet based image compression and multicarrier modulation.

BIBLIOGRAPHY

1. J. Shapiro, Embedded image coding using zerotrees of wavelet coefficients, *IEEE Trans. Signal Process.* **41**: 3445–3462 (Dec. 1993).
2. A. Said and W. Pearlman, A new, fast and efficient image codec based on set partitioning, *IEEE Trans. Circuits Syst. Video Technol.* **6**: 243–250 (June 1996).
3. D. Taubman and M. Marcellin, *JPEG2000: Image Compression Fundamentals, Standards, and Practice*, Kluwer, Boston, 2002.
4. B. Usevitch, A tutorial on modern lossy wavelet compression: Foundations of JPEG2000, *IEEE Signal Process. Mag.* **18**: 22–35 (Sept. 2001).
5. C. Chrysafis and A. Ortega, Line based, reduced memory, wavelet image compression, *IEEE Trans. Image Process.* **9**: 378–389 (March 2000).
6. A. Jain, *Fundamentals of Digital Image Processing*, Prentice-Hall, Englewood Cliffs, NJ, 1989.
7. J. Li and S. Lei, Rate-distortion optimized embedding, *Proc. Picture Coding Symp.*, Berlin, Sept. 1997, pp. 201–206.
8. E. Ordentlich, M. Weinberger, and G. Seroussi, A low-complexity modeling technique for embedded coding of wavelet coefficients, *Proc. IEEE Data Compression Conf.*, Snowbird, UT, March 1998, pp. 408–417.
9. I. J. Group, *Home Page*, (online) (no date), *http://www.ijg.org*.
10. A. Said and W. Pearlman, *SPIHT Image Compression*, (online) (no date), *http://www.cipr.rpi.edu/research/SPIHT/spiht3.html*.

K

KASAMI SEQUENCES

Tadao Kasami
Ryuji Kohno
Hiroshima City University
Hiroshima, Japan

1. INTRODUCTION

Code-division multiple-access, (CDMA) based on spread-spectrum communication systems requires many pseudonoise sequences with sharp autocorrelation and small cross-correlation. On one hand, sharp autocorrelation results in not only reliable and quick acquisition but also tracking of sequence synchronization. On the other hand, small cross-correlation reduces interference between multiple accessing users' signals, and thus results in increasing the number of multiple accessing users in CDMA [1–5].

Kasami sequences can be generated with a binary linear feedback shift register in the same manner as maximum-length sequences abbreviated as m sequences, Gold, Gold-like, and dual-BCH sequences. Kasami sequences are classified into two sets: (1) the *small set of Kasami sequences* and (2) the *large set of Kasami sequences*, which have period $2^n - 1$, n even. For a small set of Kasami sequences, the peak correlation magnitude, the maximum absolute value of cross-correlations, is optimal and approximately half of that achieved by the Gold [6,7] and Gold-like sequences of the same period. However, its size — the number of sequences in the set — is approximately the square root of that of the Gold or Gold-like sequences. A large set of Kasami sequences contains a set of Gold or Gold-like sequences and the small set of Kasami sequences as subsets. Compared with a set of Gold or Gold-like sequences of the same period, its peak correlation magnitude is the same, and its size is approximately $2^{n/2}$ times larger.

In this article, we consider only binary sequences. Sarwate suggested extension to nonbinary or polyphase sequences in a manner similar to that for m sequences (for the term *nonbinary Kasami sequences*, see Ref. 2).

2. DEFINITIONS AND BASIC CONCEPTS

(For further details, see Refs. 1, 2, 8, and 9.) By a *sequence*, we mean a binary infinitely long sequence with a finite period. A sequence $\mathbf{u} = \ldots, u_{-2}, u_{-1}, u_0, u_1, u_2, \ldots$ is abbreviated as $\{u_j\}$. Let T denote the left-shift operator by one bit. For an integer i, T^i denotes the i-times applications of T; that is, $T^i\{u_j\} = \{u_{j+i}\}$. The period N of $\mathbf{u}$ is the least positive integer such that $u_i = u_{i+N}$ for all i. $T^i\mathbf{u}$ is called a phase shift of $\mathbf{u}$.

Define $X(0) = +1$ and $X(1) = -1$, where X represents binary shift keying modulation, that is, $X(t) = \exp\{i\pi t\} = (-1)^t$ for $t \in \{0, 1\}$, $i = \sqrt{-1}$. For a sequence $\mathbf{u}$, $wt(\mathbf{u})$ denotes the number of ones per period in $\mathbf{u}$. For sequences $\mathbf{u} = \{u_i\}$ and $\mathbf{v} = \{v_j\}$ with the same period N, the *periodic cross-correlation function* $\theta_{u,v}(\cdot)$ is defined by

$$\theta_{u,v}(\tau) \triangleq \sum_{j=0}^{N-1} X(u_j)X(v_{j+\tau}) \qquad \text{for} \qquad 0 \le \tau < N$$
$$= N - 2wt(\mathbf{u} \oplus T^\tau \mathbf{v}) \tag{1}$$

For a special case of $\mathbf{u} = \mathbf{v}$, the expression $\theta_{u,v}(\cdot)$ is called the *periodic autocorrelation function*. For a set S of sequences with period N, the *peak cross-correlation magnitude* $\theta_{\max}$ of S is defined by

$$\theta_{\max} \triangleq \max\{\theta_{u,v}(\tau) : \mathbf{u}, \mathbf{v} \in S \text{ and } 0 \le \tau < N\} \tag{2}$$

For sequences $\mathbf{u} = \{u_j\}$ and $\mathbf{v} = \{v_j\}$, $\mathbf{u} \oplus \mathbf{v}$ denotes $\{u_j \oplus v_j\}$, that is, the sequence whose jth element is $u_j \oplus v_j$, where $\oplus$ denotes addition modulo 2, that is, the exclusive OR operation. For a positive integer f, consider the sequence $\mathbf{v} = \{v_j\}$ formed by taking every fth bit of sequence $\mathbf{u} = \{u_j\}$, that is, $v_j = u_{fj}$ for every integer j. This sequence, denoted $\mathbf{u}[f]$, is said to be a decimation by f of $\mathbf{u}$.

Let $h(x) = a_0x^n + a_1x^{n-1} + \cdots + a_{n-1}x + a_n$ denote a binary polynomial of degree n where $a_0 = a_n = 1$ and other coefficients are value 0 or 1. A sequence $\mathbf{u} = \{u_j\}$ is said to be a *sequence generated by* $h(x)$ if for all integers j

$$a_0u_j \oplus a_1u_{j-1} \oplus a_2u_{j-2} \oplus \cdots \oplus a_nu_{j-n} = 0 \tag{3}$$

Subsequence $u_0, u_1, u_2, \ldots$ can be generated by an n-stage binary linear feedback shift register that has a feedback tap connected to the ith cell if $h_i = 1$ for $0 < i \le n$. If $\mathbf{u}$ is a sequence generated by $h(x)$, then its phase shift is also generated by $h(x)$. The period of a nonzero sequence $\mathbf{u}$ generated by the polynomial $h(x)$ of degree n cannot exceed $2^n - 1$. If $\mathbf{u}$ has this maximal period $N = 2^n - 1$, it is called a *maximum-length sequence* or *m sequence*, and $h(x)$ is called a *primitive* binary polynomial of degree n.

Let $gcd(i,j)$ denote the greatest common divisor of the integers i and j. Let $\mathbf{u}$ be an m sequence. For a positive integer f, if $\mathbf{u}[f]$ is not identically zero, $\mathbf{u}[f]$ has period $N/gcd(N,f)$, and is generated by the polynomial $h_f(x)$ whose roots are the fth powers of the roots of $h(x)$. The degree of $h_f(x)$ is the smallest positive integer m such that $2^m - 1$ is divisible by $N/gcd(N,f)$. For positive integers $f_1, f_2, \ldots, f_t$ such that $h_{f_1}(x), h_{f_2}(x), \ldots, h_{f_t}(x)$ are all different, the set of sequences generated by $h_{f_1}(x)h_{f_2}(x)\ldots h_{f_t}(x)$ is the set of linear (with respect to $\oplus$) sums of some phase shifts of $\mathbf{u}[f_1], \mathbf{u}[f_2], \ldots, \mathbf{u}[f_t]$.

The *linear span* of a sequence is defined as the smallest degree of polynomials that generate the sequence, and it is also called the *linear complexity*. The *imbalance* between zeros and ones per period in a sequence $\mathbf{u}$ is defined as the absolute value of the difference between the numbers of zeros and ones per period in $\mathbf{u}$, which is equal to $|N - 2wt(\mathbf{u})|$, where N is the period of $\mathbf{u}$.

3. SMALL SETS OF KASAMI SEQUENCES

See Refs. 1 and 2.

3.1. Definition

Let n be even and let $\mathbf{u} = \{u_j\}$ denote an m sequence of period $N = 2^n - 1$ generated by a primitive polynomial $h_1(x)$ of degree n. Define $s(n) \triangleq 2^{n/2} + 1$. Consider the sequence $\mathbf{w} = \mathbf{u}[s(n)]$ derived by decimating or sampling the sequence $\{u_j\}$ with every $s(n)$ bits. Then, $\mathbf{w}$ is a sequence of period $N' = N/gcd(N, s(n)) = (2^n - 1)/(2^{n/2} + 1) = 2^{n/2} - 1$ which is generated by the polynomial $h_{s(n)}(x)$ whose roots are the $s(n)$th powers of the roots of $h_1(x)$. Since $h_{s(n)}$ has degree $n/2$ and is primitive, $\mathbf{w}$ is an m sequence of period $2^{n/2} - 1$.

Now consider the nonzero sequences generated by the polynomial $h_S(x) \triangleq h_1(x)h_{s(n)}(x)$ of degree $3n/2$. As stated in Section 2, any such sequence must be one of the forms $T^i\mathbf{u}, T^j\mathbf{w}, T^i\mathbf{u} \oplus T^j\mathbf{w}, 0 \le i < 2^n - 1, 0 \le j < 2^{n/2} - 1$. Thus any sequence of period N generated by $h_S(x)$ is some phase shift of a sequence in the following set $K_S(\mathbf{u}, \mathbf{w})$ defined by

$$K_S(\mathbf{u}, \mathbf{w}) = \{\mathbf{u}, \mathbf{u} \oplus \mathbf{w}, \mathbf{u} \oplus T\mathbf{w}, \mathbf{u} \oplus T^2\mathbf{w}, \ldots, \mathbf{u} \oplus T^{2^{n/2}-2}\mathbf{w}\} \quad (4)$$

This set of sequences is called the *small set of Kasami sequences*.

3.2. Correlation Properties

It has been proved [10,11] that the periodic correlation functions $\theta_{x,y}(\tau)$ of any sequences $\mathbf{x}$ and $\mathbf{y}$ belonging to the small sets of Kasami sequences $K_S(\mathbf{u}, \mathbf{w})$ take only three values:

$$\theta_{x,y}(\tau) = -1 \quad \text{or} \quad -2^{n/2} - 1 \quad \text{or} \quad 2^{n/2} - 1 \quad (5)$$

It is obvious that the peak correlation magnitude of the periodic correlation function $\theta_{\max} = 2^{n/2} + 1 = s(n)$ for the small set of Kasami sequences is approximately one half of the values of $\theta_{\max} = 2^{(n+2)/2} + 1$ achieved by the Gold and Gold-like sequences.

The Welch bound [12] applied to a set of $M = 2^{n/2}$ sequences of period $N = 2^n - 1$ provides a lower bound of $\theta_{\max}$ for all binary sequences:

$$\theta_{\max} \ge N\left(\frac{M-1}{NM-1}\right)^{1/2} > 2^{n/2} \quad (6)$$

Since N is odd, it follows from Eq. (1) and Eq. (2) that $\theta_{\max}$ is also odd. This implies that $\theta_{\max} \ge 2^{n/2} + 1$. Comparing this Welch bound with $\theta_{\max} = 2^{n/2} + 1$ of the small set of Kasami sequences, it is noted that the small set of Kasami sequences is an optimal collection of binary sequences with respect to the bound.

Small sets of Kasami sequences contain only $M = 2^{n/2} = (N+1)^{1/2}$ sequences, while Gold and Gold-like sequences contain $N + 2$ and $N + 1$ sequences, respectively.

Regarding the linear span, since $h_S(x)$ has degree $3n/2$, the maximum linear span of sequences in the set is $3n/2$. Considering the *imbalance* between the numbers of zeros and ones per period, the maximum is $2^{n/2} + 1 : 1$.

4. LARGE SETS OF KASAMI SEQUENCES

See Refs. 1 and 2.

4.1. Definition

Let n be even. Define $t(n) \triangleq 2^{(n+2)/2} + 1$. Let $\mathbf{u}$ and $\mathbf{w}$ be defined as the nonzero sequences generated by the polynomials $h_1(x)$ of degree n and $h_{s(n)}(x)$ of degree $n/2$, respectively, as mentioned above for the small sets of Kasami sequences, and let $\mathbf{v} = \mathbf{u}[t(n)]$ be a nonzero sequence generated by $h_{t(n)}(x)$ of degree n [derived by decimating or sampling the sequence $\mathbf{u}$ with every $t(n)$ bits].

The period of $\mathbf{v}$ is given by

$$\frac{N}{gcd(N, t(n))} = \begin{cases} N/3 & \text{for } n \equiv 0 \bmod 4 \\ N & \text{for } n \equiv 2 \bmod 4 \end{cases} \quad (7)$$

Then, the set of sequences generated by $h_L(x) \triangleq h_1(x)h_{s(n)}(x)h_{t(n)}(x)$ of degree $5n/2$ has period $N = 2^n - 1$, and any sequence with the period N in the set is some phase shift of a sequence in the following set $K_L(\mathbf{u}, \mathbf{v}, \mathbf{w})$, called the *large set of Kasami sequences*. There are two cases.

Case 1. If $n \equiv 2 \bmod 4$, then

$$K_L(\mathbf{u}, \mathbf{v}, \mathbf{w}) \triangleq \{G(\mathbf{u}, \mathbf{v}), G(\mathbf{u}, \mathbf{v}) \oplus \mathbf{w}, G(\mathbf{u}, \mathbf{v}) \oplus T\mathbf{w}, \ldots, G(\mathbf{u}, \mathbf{v}) \oplus T^{2^{n/2}-2}\mathbf{w}\} \quad (8)$$

where $G(\mathbf{u}, \mathbf{v})$ is the set of Gold sequences [6,7] defined by

$$G(\mathbf{u}, \mathbf{v}) = \{\mathbf{u}, \mathbf{v}, \mathbf{u} \oplus \mathbf{v}, \mathbf{u} \oplus T\mathbf{v}, \mathbf{u} \oplus T^2\mathbf{v}, \ldots, \mathbf{u} \oplus T^{2^n-2}\mathbf{v}\} \quad (9)$$

and $G(\mathbf{u}, \mathbf{v}) \oplus T^i\mathbf{w}$ denotes the set $\{\mathbf{x} \oplus T^i\mathbf{w} : \mathbf{x} \in G(\mathbf{u}, \mathbf{v})\}$.

Regarding the size of the sequences, $K_L(\mathbf{u}, \mathbf{v}, \mathbf{w})$ contains $2^{n/2}(2^n + 1)$ sequences.

Case 2. If $n \equiv 0 \bmod 4$, then

$$\begin{aligned} K_L(\mathbf{u}, \mathbf{v}, \mathbf{w}) & \\ \triangleq \{ & H(\mathbf{u}, \mathbf{v}), H(\mathbf{u}, \mathbf{v}) \oplus \mathbf{w}, H(\mathbf{u}, \mathbf{v}) \oplus T\mathbf{w}, \\ & \ldots, H(\mathbf{u}, \mathbf{v}) \oplus T^{2^{n/2}-2}\mathbf{w} \\ & \mathbf{v}^{(0)} \oplus \mathbf{w}, \mathbf{v}^{(0)} \oplus T\mathbf{w}, \ldots, \mathbf{v}^{(0)} \oplus T^{(2^{n/2}-1)/3-1}\mathbf{w} \\ & \mathbf{v}^{(1)} \oplus \mathbf{w}, \mathbf{v}^{(1)} \oplus T\mathbf{w}, \ldots, \mathbf{v}^{(1)} \oplus T^{(2^{n/2}-1)/3-1}\mathbf{w} \\ & \mathbf{v}^{(2)} \oplus \mathbf{w}, \mathbf{v}^{(2)} \oplus T\mathbf{w}, \ldots, \mathbf{v}^{(2)} \oplus T^{(2^{n/2}-1)/3-1}\mathbf{w}\} \end{aligned} \quad (10)$$

where $H(\mathbf{u}, \mathbf{v})$ is the set of Gold-like sequences [1–2] defined by

$$
\begin{aligned}
H(\mathbf{u}, \mathbf{v}) = \{&\mathbf{u}, \mathbf{v}^{(0)} \oplus \mathbf{u}, \mathbf{v}^{(0)} \oplus T\mathbf{u}, \ldots, \mathbf{v}^{(0)} \oplus T^{(2^n-1)/3-1}\mathbf{u} \\
&\mathbf{v}^{(1)} \oplus \mathbf{u}, \mathbf{v}^{(1)} \oplus T\mathbf{u}, \ldots, \mathbf{v}^{(1)} \oplus T^{(2^n-1)/3-1}\mathbf{u} \\
&\mathbf{v}^{(2)} \oplus \mathbf{u}, \mathbf{v}^{(2)} \oplus T\mathbf{u}, \ldots, \mathbf{v}^{(2)} \oplus T^{(2^n-1)/3-1}\mathbf{u}\}
\end{aligned}
\tag{11}
$$

and $H(\mathbf{u}, \mathbf{v}) \oplus T^i\mathbf{w}$ denotes the set $\{\mathbf{x} \oplus T^i\mathbf{w} : \mathbf{x} \in H(\mathbf{u}, \mathbf{v})\}$ and $\mathbf{v}^{(i)} = (T^i\mathbf{u})[t(n)]$ is the result of decimating $T^i\mathbf{u}$ by every $t(n)$ bits.

Regarding the size of the sequences, $K_L(\mathbf{u}, \mathbf{v}, \mathbf{w})$ contains $2^{n/2}(2^n + 1) - 1$ sequences.

4.2. Correlation Properties

In either case 1 or 2, the correlation functions for any sequences $\mathbf{x}, \mathbf{y} \in K_L(\mathbf{u}, \mathbf{v}, \mathbf{w})$ take only the following five values [11]; for $0 \le \tau < N = 2^n - 1$, we obtain

$$
\theta_{x,y}(\tau) = -1 \quad \text{or} \quad -2^{(n+2)/2} - 1 \quad \text{or} \quad -2^{n/2} - 1 \quad \text{or } 2^{(n+2)/2} - 1 \quad \text{or} \quad 2^{n/2} - 1 \tag{12}
$$

Thus, although the large set of Kasami sequences involves the small set of Kasami sequences and a set of Gold or Gold-like sequences as subsets, the correlation bound equals that of Gold or Gold-like sequences, that is, $\theta_{\max} = 2^{(n+2)/2} + 1$.

The maximum linear complexity of the large set of Kasami sequences is $5n/2$. The range of imbalance between the numbers of zeros and ones per period is $2^{(n+2)/2} + 1 : 1$.

Table 1 shows several measures of Kasami sequences compared with those of Gold and Gold-like sequences [2].

5. RELATION TO BINARY CYCLIC CODES

Let $h(x)$ be a binary polynomial of degree k, and let S_0 denote the set of sequences generated by $h(x)$. The greatest period N of sequences in S_0 is the smallest positive integer such that $x^N - 1$ is divisible by $h(x)$. For a sequence $\mathbf{u} = \{u_j\}$ in S_0, define $\mathbf{u}_c \triangleq (u_0, u_1, u_2, \ldots, u_{N-1})$, and let C denote $\{\mathbf{u}_c : \mathbf{u} \in S_0\}$. There is a one-to-one correspondence between S_0 and C; C is a binary cyclic code of length N whose parity-check polynomial is $h(x)$. Sequence $\mathbf{u}_c$ is called a *codeword* of C. Corresponding to the left-shift operator T, the cyclic left-shift operator T_c is defined by $T_c(u_0, u_1, u_2, \ldots, u_{N-1}) \triangleq (u_1, u_2, \ldots, u_{N-1}, u_0)$. For $\mathbf{u_c} \in C$, $T^i\mathbf{u_c} \in C$ and $T^i\mathbf{u}_c$ is called a *cyclic shift* of $\mathbf{u}_c$. The period of $\mathbf{u}$ is the least positive integer N' such that $T_c^{N'}\mathbf{u}_c = \mathbf{u}_c$, which is the same as the period of $\mathbf{u}$. The code C can be partitioned into blocks in such a way that codewords $\mathbf{u}_c$ and $\mathbf{v}_c$ belong to a block if and only if they are cyclic shifts of the other. The period of a codeword in a block is equal to the size of the block.

Let S be a minimal subset of S_0 such that any nonzero sequence with period N in S_0 is some phase shift of a sequence in S. Define $C_s \triangleq \{\mathbf{u}_c : \mathbf{u} \in S\}$. Then, C_s consists of codewords chosen as a unique representative from each block of size N. Thus the size of S is equal to the number of blocks of size N.

For a codeword $\mathbf{u}_c$, let $wt(\mathbf{u}_c)$ denote the weight of $\mathbf{u}_c$, that is, the number of ones in $\mathbf{u}_c$. The set $W \triangleq \{wt(\mathbf{u}_c) : \mathbf{u}_c \in C\}$ is called the *weight profile* of C. From Eq.(1), we obtain

$$
\theta_{u,v}(\tau) = N - 2wt(\mathbf{u}_c \oplus T_c^{\tau}\mathbf{v}_c). \tag{13}
$$

Since a cyclic shift of a codeword has the same weight as the codeword, the set of those values on which the correlation functions for the sequences in S take can be readily found if the weight profile of C_s is known. The profile can be easily derived from weight enumerators that give the number of codewords with any weight for C and its certain subcodes. For a class R_2 of subcodes of the second order (punctured) Reed–Muller codes, weight enumerators have been derived [10,11,13]. The class R_2 contains the codes corresponding to Gold sequences generated by $h_1(x)h_f(x)$ whose f is of form $2^e + 1$, Gold-like, dual-BCH, and Kasami sequences.

As a historical remark, Sarwate and Pursley [1,14] chose two subclasses of R_2 from a point of view of sequence design for communications applications. They translated the results on weight spectra into results on correlation spectra, and named the small and large sets of Kasami sequences. They later discovered that some of the results were already known to Massey and Uhram [15].

Table 1. Parameters of Gold, Gold-like and Kasami Sequences (2)

Sequence Set		Order n	Period N	Size M	Linear Span	Peak Cross Correlation $\theta_{\max}$	Decimation Sampler f
Gold		1 (mod 2)	$2^n - 1$	$2^n + 1$	2^n	$2^{(n+1)/2} + 1$	$2^{(n+1)/2} + 1$
		2 (mod 4)				$2^{(n+2)/2} + 1$	$2^{(n+2)/2} + 1$
Gold-like		0 (mod 4)		2^n			
Kasami	small	0 (mod 2)		$2^{n/2}$	$3n/2$	$2^{n/2} + 1$	$2^{n/2} + 1$
	large	2 (mod 4)		$2^{n/2}(2^n + 1)$	$5n/2$	$2^{(n+2)/2} + 1$	$2^{(n+2)/2} + 1$,
		0 (mod 4)		$2^{n/2}(2^n + 1) - 1$	$5n/2$		$2^{n/2} + 1$

BIOGRAPHIES

Tadao Kasami received the B.E., M.E., and D.E. degrees in communication engineering from Osaka University, Osaka, Japan, in 1958, 1960, and 1963, respectively. He joined the faculty of Osaka University in 1963, and he was a professor of engineering science from 1966 to 1994. From 1992 to 1998, he was a professor at the Graduate School of Information Science of Nara Institute of Science and Technology, Nara, Japan. He is an emeritus professor of Osaka University and Nara Institute of Science and Technology. Since 1998, he has been a professor of information science at Hiroshima City University, Hiroshima, Japan. His research and teaching interests have been in coding theory and algorithms. He is a life fellow of IEEE and a recipient of the 1999 Claude E. Shannon Award from the IEEE Information Theory Socity; a fellow of the Institute of Electronics, Information and Communication Engineers in Japan; and a recipient of the 1987 Achievement Award and 2001 Distinguished Services Award from the Institute.

Ryuji Kohno received his Ph.D. degree in electrical engineering from the University of Tokyo in 1984. Since 1998, he has been a professor in the Division of Physics, Electrical and Computer Engineering, Graduate School of Engineering, Yokohama National University. Dr. Kohno was elected a member of the Board of Governors of the IEEE IT Society in 2000. He was an associate editor of the *IEEE Transactions on Information Theory* from 1995 to 1998 and an editor of the *IEICE* (Institute of Electronics, Information, Communications Engineers) *Transactions on Communications* from 1990 to 1993. He was chairman of the IEICE Professional Group on Spread Spectrum Technology from 1995 to 1998. From 1998 to 2000, he was chairman of the IEICE Technical Group on Intelligent Transport System (ITS), and currently he is chairman of the IEICE Technical Group on Software Radio. Dr. Kohno also is an associate editor of both the *IEEE Transactions on Communications* and the *IEEE Transactions of Intelligent Transport Systems* (ITS).

BIBLIOGRAPHY

1. D. V. Sarwate and M. B. Pursley, Cross-correlation properties of pseudorandom and related sequences, *Proc. IEEE* **68**(5): 593–619 (1980).
2. P. Fan and M. Darnell, *Sequence Design for Communication Applications*, Wiley, New York, 1996.
3. E. H. Dinan and B. Jabbari, Spreading codes for direct sequence CDMA and wideband CDMA cellular networks, *IEEE Commun. Mag.* **9**: 48–54 (1998).
4. M. K. Simon, J. K. Omura, R. A. Scholtz, and B. K. Levitt, *Spread Spectrum Communications*, Vol. 1, Computer Science Press, Rockville, MD, 1985.
5. R. C. Dixon, *Spread Spectrum Systems with Commercial Applications*, 3rd ed., Wiley, New York, 1994.
6. R. Gold, Optimal binary sequences for spread spectrum multiplexing, *IEEE Trans. Inform. Theory* **IT-13**: 619–621 (1967).
7. R. Gold, Maximal recursive sequences with 3 values recursive cross-correlation functions, *IEEE Trans. Inform. Theory* **IT-14**: 154–156 (1968).
8. S. W. Golomb, *Shift Register Sequences*, Holden-Day, San Francisco, 1967.
9. W. W. Peterson and E. J. Weldon, Jr., *Error-Correcting Codes*, 2nd ed., MIT Press, Cambridge, MA, 1972.
10. T. Kasami, *Weight Distribution Formula for Some Class of Cyclic Codes*, Coordinated Sci. Lab., Univ. Illinois, Urbana, Tech. Rep. R-285, 1966.
11. T. Kasami, *Weight Distribution of Bose–Chauduri–Hocquenghem Codes*, Coordinated Sci. Lab., Univ. Illinois, Urbana, Tech. Rep. R-317, 1966 (also in *Combinatorial Mathematics and Its Applications*, Univ. North Carolina Press, Chapel Hill, NC, 1969; reprinted in E. R. Berlekamp, ed., *Key Papers in the Development of Coding Theory*, IEEE Press, New York, 1974.
12. L. R. Welch, Lower bounds on the maximum cross-correlation of signals, *IEEE Trans. Inform. Theory* **IT-20**: 397–399 (1974).
13. E. R. Berlekamp, *Algebraic Coding Theory*, McGraw-Hill, New York, 1968.
14. D. V. Sarwate and M. B. Pursley, Applications of coding theory to spread-spectrum multiple-access satellite communications, *Proc. 1976 IEEE Canadian Commun. Power Conf.*, 1976, pp. 72–75.
15. J. L. Massey and J. J. Uhran, *Final Report for Multipath Study*, Contract NAS5-10786, Univ. Notre Dame, IN, 1969.